AREA ARRAY PACKAGING HANDBOOK

McGraw-Hill Packaging and Electronics Books

Coombs • PRINTED CIRCUITS HANDBOOK 5/e
Coombs • ELECTRONIC INSTRUMENT HANDBOOK 3/e
Harman • WIRE BONDING
Harper • ELECTRONICS PACKAGING AND INTERCONNECTION HANDBOOK 3/e
Harper • HIGH PERFORMANCE PRINTED CIRCUIT BOARDS
Hwang • MODERN SOLDER TECHNOLOGY
Jawitz • PRINTED CIRCUIT BOARD MATERIALS HANDBOOK
Lau • LOW-COST FLIP CHIP TECHNOLOGIES
Lau • CHIP SCALE PACKAGE
Lau • ELECTRONIC PACKAGING
Lau • BALL GRID ARRAY TECHNOLOGIES
Lau & Lee • MICROVIAS
Manko • SOLDERS AND SOLDERING 4/e
Marks • PRINTED CIRCUIT ASSEMBLY DESIGN
Martin • ELECTRONIC FAILURE ANALYSIS HANDBOOK
Smith • THIN FILM DEPOSITION
Tummala • FUNDAMENTALS OF MICROSYSTEMS PACKAGING
Van Zant • MICROCHIP FABRICATION 4/e

AREA ARRAY PACKAGING HANDBOOK

Ken Gilleo

McGraw-Hill

New York Chicago San Francisco Lisbon London
Madrid Mexico City Milan New Delhi San Juan
Seoul Singapore Sydney Toronto

Library of Congress Cataloging-in-Publication Data
Gilleo, Ken.
Area array packaging handbook : manufacturing and assembly / Ken Gilleo.
p. cm.
ISBN 0-07-137493-0
1. Ball grid array technology. 2. Microelectronic packaging. I. Title.

TK7870.15 .G54 2001
621.381'046—dc21 2001041051

McGraw-Hill
A Division of The ***McGraw-Hill*** *Companies*

1 2 3 4 5 6 7 8 9 0 AGM/AGM 0 5 4 3 2 1

ISBN 0-07-137493-0

The sponsoring editor for this book was Steve Chapman, the editing supervisors were Penny Linskey and Steven Melvin, and the production supervisor was Sherri Souffrance. This book was set in the HB1 design in Times Roman by Paul Scozzari of McGraw-Hill Professional's Hightstown, N.J. composition unit.

Printed and bound by Quebecor/Martinsburg.

This book is printed on recycled, acid-free paper containing a minimum of 50% recycled, deinked paper.

CONTENTS

Section 2 Materials

Chapter 21. Die Attach and Rework 21.1

Chapter 26. Ovens in Electronics 26.1

Chapter 27. Process Development, Control, and Organization 27.1

Section 4 Economics and Productivity

Section 5 Future

Foreword

Jan Vardaman
TechSearch International

INTRODUCTION

Just as the semiconductor industry has seen a dramatic shift from through-hole to surface-mount packages, another major change is underway—the shift from leads to balls and from wires to bumps.

Ball grid array (BGA) packages are increasingly found in products including personal computers, portable communications devices, workstations/servers, mid-range and high-end computers, network and telecommunications systems, and even automotive applications.

With the demand for smaller, lighter, thinner portable products has come the need for semiconductor packaging and assembly technology to enable the production of the new era of consumer products. Out of this need, the chip scale package (CSP) was born.

Flip chip's advantage over wire bond interconnection includes higher density mounting, improved electrical performance, and improved reliability, as well as improved manufacturability through gang bonding and the self-aligning nature of solder bumps.

Demands from the consumer segment will continue to drive packaging developments—especially for small form factor packages. Mobile phones, in particular, demand smaller, thinner, lighter packages. Many companies have adopted stacked packages and flip chip is one of the interconnect options for the near future. The rate of adoption and expanded use of ball grid arrays (BGAs), chip scale packages (CSPs), and flip chip interconnect continues to amaze even the skeptics.

BGAS

The industry has made dramatic progress from the early days of the first BGA introductions. The first plastic ball grid array (PBGA) packages were in Motorola's pagers and radios, soon followed by products from Compaq Computer. These packages typically had less than 200 balls.

In 2001 more than 1.6 billion BGA packages of all types with pin counts over 2000 will ship. Ceramic ball grid array (CBGA) and ceramic column grid array (CCGA) packages remain common for exceptionally high pin count ASICs and processors. Shipping in high volume are 1,657-ball CCGAs for workstation/servers and network system products. These packages have a 42.5 × 42.5 mm body size and a 1.0 mm column grid array pitch. Soon to be introduced are CCGAs with more than 2000 solder columns.

Tape ball grid array (TBGA) packages—mainly with wire bond interconnects—continue to ship for many computer, telecommunications, and network systems.

The largest volume shipments are the plastic BGA (PBGA) packages, and personal computers remain the largest volume application. Intel has long moved its chip set designs from plastic quad flat packages (PQFPs) to PBGAs—many pin counts are in excess of 500 I/Os. Motorola has introduced some members of its PowerPC family in PBGAs. While some of Intel's CPUs are still shipping in its organic land grid array (OLGA) package with solder balls, the majority of new Pentium shipments are supplied in plastic pin grid arrays (PPGAs) as a result of demand for socketable processors.

Perhaps the PC's greatest future rival is the game machine, and BGAs are no exception for the high pin count devices in these systems. Sony's popular PlayStation 2 contains an Emotion Engine processor packaged in a 540-ball PBGA with a 42.5 × 42.5 mm body size and a 1.27 mm solder ball pitch. The graphic chip is also packaged in a PBGA.

A variety of PBGA pin counts can be found in workstation/servers, telecommunications, and network system products. High pin count designs from ASIC makers continue to ship in BGAs. Sun Microsystems has some of the highest pin count PBGA packages, one of which has 1848 balls with flip chip inside.

CSPs

Since the package first surfaced more than five years ago, CSPs have undergone the most remarkable proliferation of any recently introduced package type. Today there are more than 100 varieties of CSPs in existence with various configurations in shipping numerous applications.

CSPs offer many of the advantages of bare die—tested die, size, weight, and low profile, but at the same time provide a packaged solution. Low pin count devices are increasingly being packaged in CSPs with larger volumes shipping every year.

Flex-based CSPs shipping in volume include Fujitsu's FBGA, NEC's D2BGA, Sharp's F.BGA, Tessera's μBGA, and Texas Instruments MicroStar™ bump chip carrier (BCC)—Tessera's patented μBGA continues to be used for flash, although laminate substrate alternatives have increased in popularity due to cost advantages and the ability to handle die shrinks. Rigid substrate CSPs include Matsushita's ceramic-based packages and Motorola's molded array package—adopted by almost every contract assembly operation. High-volume lead frame CSPs include Fujitsu's bump chip carrier (BCC), Fujitsu's SON, Matsushita's QFN, and Amkor's *Micro* leadframe (MLF) package. These packages do not have solder balls and are targeted for low I/O (<85), inexpensive devices.

Stacked packages—typically with one flash memory and one SRAM—have increased in volume dramatically in the last year. These packages use either type of laminate substrates. Shipments of CSPs with more than two stacked devices, as well as stacked CSP packages, will soon move into volume production.

One of the first products to make use of the small CSP was the camcorder. In Sony's first application, 20 of the 40 ICs were CSPs. Sony's DCR-PC7, introduced in the fall of 1996, was one of the first products that made use of CSPs. With a body size of just 59 × 129 × 118 mm, the camcorder contained 40 ICs—20 of them packaged in CSPs and all with 0.5 mm pitch. Three Micro Star™ BGAs from Texas Instruments (TI) Japan used TAB tape as the substrate and wire bonding as the chip-to-substrate interconnect. One of NEC's D2BGA packages used TAB tape as the substrate, with a bumpless TAB inner lead bonding process used to interconnect the chip to the substrate. The remainder of the small area array packages, developed by Sony, consisted of a laminate substrate and flip chip as the interconnect method.

Numerous companies followed Sony's introduction—including JVC and Matsushita—with numerous product introductions. These packages include flex-based and rigid substrate packages—all with less than 1.0 mm pitch. Third- and fourth-generation digital cameras and camcorders will continue to ship with CSPs—making extreme miniaturization possible.

The unprecedented growth in demand for CSPs is being driven by the mobile phone sector. Mobile phone production is now projected to increase beyond the wildest expectations of many manufacturers—in some cases outpacing the supply of key silicon devices such as flash memory. CSPs remain the packaging choice for mobile phones for form factors reasons. This small package with I/Os of less than 1.0 mm, used in combination with microvia substrate technology, has enabled manufacturers to shrink mobile phone size and weight—the smallest phone in Japan weights about 58 g. While several companies have considered flip chip mounting to achieve the same goals, CSPs continue to be the top packaging choice.

FLIP CHIP

Flip chip use can be categorized as either flip chip on board (FCOB) or flip chip in package (FCIP). FCOB, or a flip chip device mounted directly on a motherboard, includes such applications as automotive electronics, disk drives, driver ICs for flat panel displays, watches, pagers, cellular phones, and contactless cards or RF ID tags. FCIP is the mounting of a flip chip in packages such as ceramic ball grid arrays (CBGA), column grid arrays (CCGA), and land grid arrays (CLGA), and plastic ball grid arrays (PBGA). Inside the package, flip chip is increasingly used as an interconnect method for both performance and form factor reasons. FCIP can be found in high-end mainframes and supercomputers, workstations and servers, mid-range systems, personal computers, network systems, and a variety of consumer products.

Flip chip is increasingly the interconnect method of choice for high-performance ASICs. More than half the BGAs used by Sun Microsystems this year will be flip chip inside, and in five years 90 percent of Sun's BGAs will be flip chip. Flip chip is also finding its way into the interconnect realm previously dominated by wire bond. The potential expansion of flip chip technology into the midrange pin counts represents a shift in the adoption of the technology as well as a maturing of the industry. A recent example is the introduction of LSI Logic's flip chip package targeted at applications with between 300 and 1150 leads that require higher electrical and thermal performance than that offered by wire bonded packages.

Flip chip interconnect is found in high volume in the personal computer market. For several years, microprocessor speeds have required the use of flip chip interconnect to achieve the performance specifications designed into silicon. FCIPs include products such as AMD's processor family, Intel's Pentium products, Motorola's PowerPCs, and the new Transmetta microprocessor. In the near future chip sets are also expected to use flip chip as the interconnect method inside the package.

Flip chip will continue to be found in a variety of emerging products. Wireless applications such as Bluetooth modules are already shipping with flip chip interconnect.

Another new development is the proliferation of bumps for small die (less than 2×2 mm), sometimes considered wafer level packages. Both of these developments represent not only improvements in flip chip infrastructure but also cost reductions in the technology.

WAFER LEVEL PACKAGES: THE FUTURE STARTS TODAY

Wafer level packages (WLPs) are packaged and tested at the wafer level, before the dicing operation. Some bumped die may be considered wafer level packages if they are not mounted inside a package and are not underfilled on the product board. For example, most microprocessors and ASICs that use flip chip interconnect inside the package are not wafer level packages. Dallas Semiconductor ships many low lead count auto ID chips and battery controllers as bumped die that do not require underfill when mounted on the product boards. These devices are called wafer level packages.

Devices packaged at the wafer level, with or without bumps, continue to grow. A variety of wafer level packages are shipping in volume—EEPROMs, voltage regulators, op-amp devices, sensors, flash memory, and integrated passives.

Demands from the consumer and portable communication segments will continue to drive packaging developments—especially in wafer level CSPs and flip chip. While in the early stages today, wafer level CSPs represent the wave of the future.

S·E·C·T·I·O·N · 1

PACKAGING CONCEPTS AND DESIGNS

CHAPTER 1
INTRODUCTION TO ELECTRONIC PACKAGING

Ken Gilleo
Cookson Electronics

1.0 DEFINITION OF THE PACKAGE

The ideal electronic package is an economical and manufacturable electromechanical platform for one or more electronic devices that affords protection, facilitates handling, and provides the geometric translations and compatible interface required for connection to the next system level by practical assembly processes.

1.1 INTRODUCTION AND REQUIREMENTS

This book covers much of the field of packaging, including the new minimal category, but with a focus on area array designs. This chapter will describe the functions of the electronic package, protective requirements, and issues as well as solutions for general packaging. This chapter emphasizes area array packages because this is the central theme of this handbook.

The electronic package is designed to perform an assortment of tasks. Some functions can be at variance with one another, requiring balance and compromise for optimization. The package must first house and protect the delicate integrated electronics devices from the external environment. This generally is accomplished by enclosing the chip in a protective cocoon of electrically insulative materials. Alternatively, the package can be a metal vacuum-sealed container with appropriate internal insulation or a similar structure made of gas-impervious ceramic. The package also may need to dissipate substantial heat generated by the electronics. Heat management generally is not an issue for the hermetic metal package or the ceramic type. However, the common plastic package has good thermal insulation characteristics that prevent efficient heat transfer to the outside. Plastics have low thermal conductivity, so enhancing thermal performance of plastic packages takes ingenuity in both design and materials. The general electronics industry also seeks to reduce size and weight while the devices to be packaged become faster and have more connections. There is also strong pressure to accomplish this while lowering the cost. Thus the paradoxical phrase, "faster, smaller, cheaper." The packaging industry therefore is tasked with achieving goals that appear mutually exclusive but is asked to do so more quickly. Surprisingly, the industry has succeeded to a large extent, but solutions have required major innovations that are defined accurately as the "packaging revolutions" (Fig. 1.1).

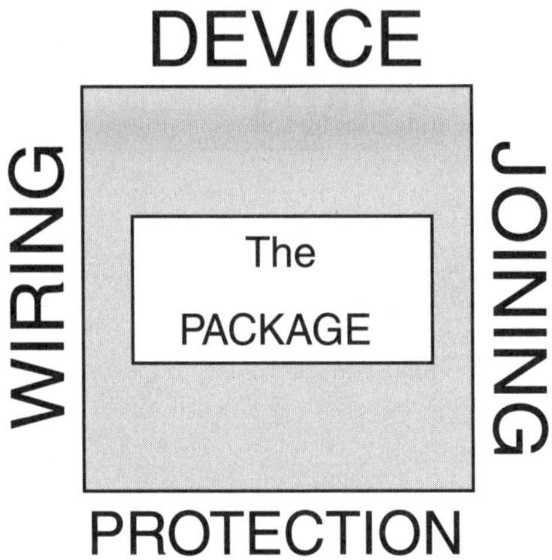

FIGURE 1.1 Packaging diagram.

1.2 PACKAGING EVOLUTIONS AND REVOLUTIONS

1.2.1 SMT

Most people are familiar with the standard molded-plastic packages seen in computers and their plug-in cards. Many of these packages are feed-through types where the metal leads fit through holes in the circuit board. The most common design in the feed-through class typically is called a *dual-in-line package* (DIP) because of the two rows of leads on each side of the plastic encasement. In the early 1980s, a more efficient packaging design strategy, called *surface-mount technology* (SMT), was introduced to solve a host of problems. The SMT process makes the connections simply by soldering the package leads to the surface of the board. The surface-mount device (SMD) can be much smaller, accommodate many more connections, is easier to assemble with automated equipment, and does not require holes through the circuit board. Absence of feed-through holes allows components to be assembled on both sides of the board. The DIP-type packages not only use up more surface area but also consume circuit density because holes must be fabricated through all layers of the board that otherwise could accommodate inner circuitry. The clear benefits of SMT have allowed this style package to capture more than half of designs. SMT has advanced over the years to reduce size and accommodate more leads. It should be noted that feed-through packages are still used, and this will continue. In fact, SMT and feed-through packaging can be complementary, and there are assembly processes for connecting both types in one operation. Figure 1.2 compares the feed-through style with SMT.

1.2.2 Perimeter Paralysis

In the early 1990s, it once again was time for a major change in packaging style. The relentless advancements in the semiconductor industry were putting an ever-increasing demand on even the densest SMD-style packages. The number of connections, or inputs/outputs (I/OS), moved continuously, while more compact electronic products demanded smaller packages. Since the packaging leads, or terminations, were on the perimeter, this meant that more leads were being placed in a smaller space. In other words, there was an ever-increasing "perimeter paralysis." Some of the packaging fabricators, most notably in Japan, responded by making smaller leads and placing them closer together for finer pitch. This approach soon ran out of steam as leads became flimsy, solder bridging defects increased, and assembly lines struggled with alignment problems. Ultrafine-pitch perimeter packages were the wrong solution.

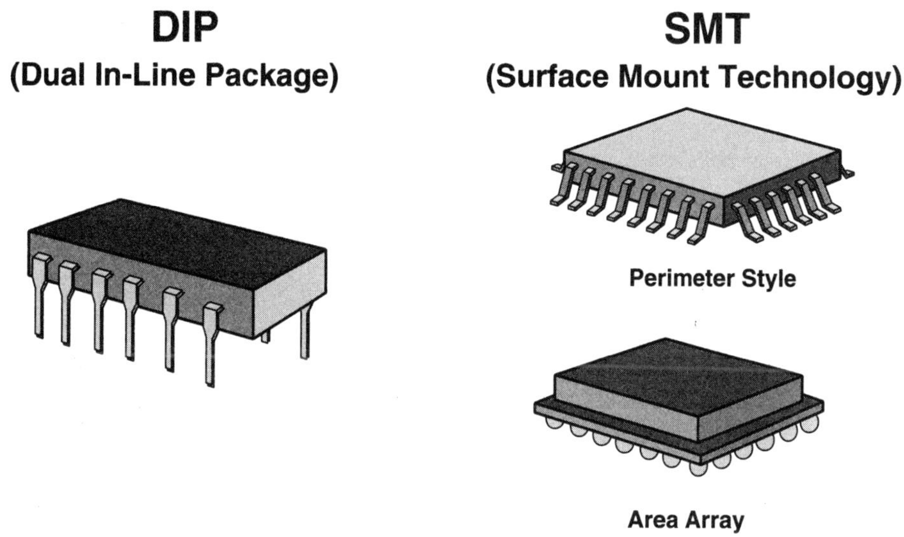

FIGURE 1.2 DIP versus SMT.

Vertically integrated electronics companies in the United States, led by IBM and Motorola, recognized that the time was right for the next packaging revolution. The simple answer for packing more connections into a shrinking housing was to use the wasted area inside the perimeter. Area array packages simply exploit the area under the package for connection sites. This was not an overlooked solution waiting to be conceived. IBM had used area array packaging since the 1960s. The issues were that a move to area array for ordinary packaging would be a hard sell and a daunting implementation task. IBM had developed ceramic area packages, with some using columns of high-melting solder, called *column grid array* (CGA). Motorola introduced a lower-cost *ball grid array* (BGA) in which solder spheres were used to make the connection to the board. Motorola's announcement of the Ompack BGA caused a major stir within the packaging and assembly industries. Many of us view Ompack as the official launch for the area array packaging revolution that is still vigorously underway. The plastic Ompack, or *over*molded *pac*kage, generally is referred to as a *PBGA,* with the *P* standing for plastic. The PBGA is the most common type of area array electronic package. This book will deal with design, manufacturing, testing, and assembly of PBGAs and other area array packages as well as the materials.

There are many subdivisions of the BGA, but all are surface mountable, and most are area array. The next discussion will include the minimal micropackage beginning with the flip chip. While packaging is necessary, many view it as a "necessary evil" that should be minimized. Admittedly, the package tends to add cost and increase the size. Many have sought to reduce it to a minimum; hence the minimal package. I will avoid the clumsy terms *minimalist package* and *minimalistic package* to stay with the more appropriate and more minimal term *minimal package.*

1.2.3 Direct Connections

I will start with the simplest and most minimal package, the flip chip that was pioneered by IBM and referred to as C4 or C^4, for *controlled collapse chip connection.* Actually, it is necessary to take just one more step back to find the beginning of area array, SMT, and BGA—all in one package. In the early 1960s, while the industry was still working on transistor packaging, IBM came up with *solid logic technology* (SLT). This package, shown in Fig. 1.3, appears to be the very first minimal package: the first BGA, the first SMD, and the original flip chip. A more generic and inclusive term for this type of package is *direct chip attach* (DCA). Some people may argue that the SLT is not a chip, or integrated circuit (IC), but rather a lowly transistor. However, a transistor requires most of the same packaging attributes as an IC, and this design approach certainly could be used on an IC. The SLT has the essential constituents for the minimal package: protection and the electromechanical interface to the circuit board. The connection balls are made of copper and are soldered to the chip pads with a high-melting alloy. The nonfusible copper provides testability, an assured standoff height, and the potential for reworking the device. While most of today's BGA spheres are made of low-melting solder, copper spheres have been reintroduced recently by Alpha-Fry Technologies, the same company that made the original microspheres for IBM. Thus the newest area array packages may be reaching back almost four decades for "new" solutions (Fig. 1.3).

The now popular DCA, or flip chip, is the only system that directly attaches an IC to the printed wiring board (PWB) or to a chip carrier. The generic and descriptive term is *direct chip attach* because the important feature is the direct connection, not flipping the chip. Other packaging, such as *tape automated bonding* (TAB), can have a flipped chip; hence the confusion. Unfortunately, the more appealing name, *flip chip,* has become dominant and also will be used in this book. The DCA is most commonly attached to the substrate through small metal bumps that are formed on the chip pads by methods that will be described in detail in Chap. 5, "Flip Chip Technology." The most popular attachment method is to simply make the DCA bumps out of solder and reflow them to the substrate pads. The assembled component needs protection and mechanical enhancement, especially when bonded to a substrate with higher thermomechanical expansion. A polymer composite called *underfill* is used in most cases. The DCA assembly, once underfilled, has reasonably good reliability. The in situ "package" created with typical underfills cannot be removed in the practical sense. This means that the underfilled chip probably cannot be reworked, although special reworkable underfills recently have come to market. Some people would argue that a package must be removable and should be reworkable in the true and practical sense of the term. So is the DCA flip chip a package? Let's define the terms of packaging requirements and try to separate absolute essentials from desirables.

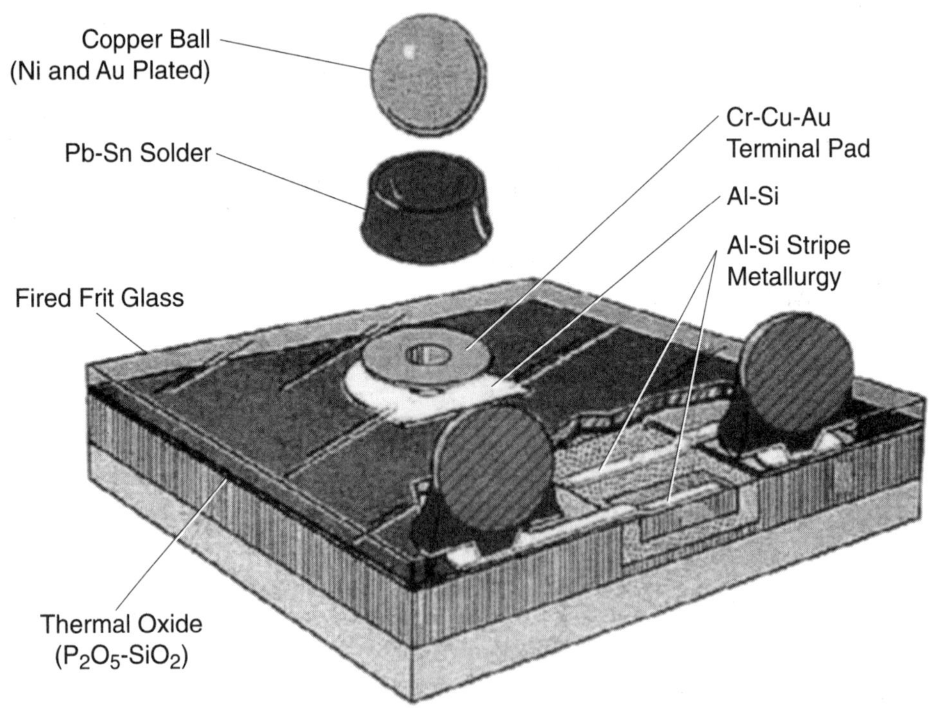

FIGURE 1.3 IBM SLT package. (*Courtesy of IBM, P. Totta.*)

1.3 USEFUL VERSUS ESSENTIAL FEATURES

The essential features of the package listed in Table 1.1 are required to make the semiconductor chip compatible with the printed circuit board (PCB and to give it reliability. The IC and PCB, or printed wiring board (PWB), used to avoid the other acronym associated with a pollutant) come from different worlds with different materials. The package bridges the inherent incompatibility between the microworld of inorganic ICs and the more macroscopic environment of organic PWBs. Not only does the most common package design expand and fan out the IC's microgeometry to fit the circuit board, it also allows the very thin aluminum (and now copper) chip pad to be connected to the relatively thick copper circuit pads. The list of useful features has become the main focus of modern packaging and will be discussed briefly here.

Table 1.1 is a list of all the both essential and useful attributes of the electronic package. In some applications, a useful attribute can be essential. Thermal management may not even be useful for a low-power memory chip but a critical requirement for a powerful central processing unit (CPU).

1.3.1 Geometric Translation

The IC device is made using processing that can produce features sized down to nearly 0.10 μm or 1×10^{-7} m, or about 1/10,000 in. Microelectronics is soon to become nanoelectronics as the feature size falls below 0.1 μm (100 nm). While the prefix *micro-* means one-millionth (of a meter), the prefix *nano-* means one-billionth (1×10^{-9} m). The printed circuit industry generally considers fine line to be about 3 to 4 mils (75 to 100 μm). There is nearly a 1000-fold feature size disparity between the two industries and their products. We can expect this ratio to remain somewhat constant or even to increase. The translator, the technology that spans the breach, is the package. The package has had to add a greater level of translation capability each time that the semiconductor industry has surged ahead with a smaller IC that required more connections or I/Os. The increasingly common strategy is to take the chip's perimeter pad connections and reroute them into an efficient area array. Although very high density interconnect (HDI) packages can accommodate chip pad dimensions, the HDI substrate is more expensive and only available from a limited number of suppliers. The economics can be better with a fan-out package, especially when many low-density components are used with one or a few high-density ones. Adding to the cost of packaging can save on the printed circuit costs to provide a net gain, especially if yield improves.

1.3.2 Material Compatibility: IC to PWB

There are two important material differences between the IC and PWB that make them incompatible. First, the IC substrate is inorganic silicon, with silicon compounds used as insulators. The most common PWB is made from epoxy resin with glass reinforcement. Silicon has a coefficient

TABLE 1.1 Essential and Useful Attributes of Electronic Packaging

Attribute	Absolutely essential	Very useful (not required)
Geometric translation	Usually	
Compatibility: IC to PWB	X	
Environmental protection	X	
Mechanical protection	Yes, most cases	
Handling ease		X
Enable standardization		X
Facilitate auto assembly		X
Removability	X	
Reworkability	?	X
Performance enhancement		X
Thermal management		In many cases

of thermal expansion (CTE) of just under 3 ppm/°C, whereas the CTE of the PWB can range from about 15 to more than 25 ppm/°C. This nearly 1:10 expansion difference can create considerable stress during temperature changes for a directly coupled chip. The package must intercede to alleviate or somehow compensate for this potentially destructive situation. The next incompatibility results from the differences in conductors. ICs typically use thin vacuum-deposited aluminum for conductor layers that results in a connectivity challenge. Even though some segments of the industry are changing to copper, its extreme thinness, a fraction of a micron, makes a direct connection difficult without preprocessing. First, these metals oxidize, and the oxidation must be removed for soldering, making the chip pad even thinner. The soldering process dissolves some of the substrate so that the thin copper or aluminum is dissolved, or leached away. In the case of aluminum, solderability is difficult at best even if the aluminum is thick. Something must be added to the pads. The package must make the chip pads compatible by adding additional conductive material, and adding more metal does this. However, the metal can be a coating, bumps, or wires, and this determines the basic connection class (covered later). Conductive adhesives cannot be used directly on IC pads because the native oxides are an electrical barrier.

1.3.3 Environmental and Mechanical Protection

All ICs are potentially sensitive to the external environment. Moisture, oxygen, and contaminants are usually harmful. Static electricity, even at low voltage, can blow apart sensitive microscopic elements within an IC. Moreover, mechanical handling can destroy many types of electrical chip connections such as wire bonds. Nearly all chips must be protected by encapsulation or by sealing within a preformed metal or ceramic hermetic housing. Even direct-attach and "chip-on-board" assemblies need protection, even though the "package" may seem to be absent. For example, a DCA assembly typically will use underfill within the gap between the chip and PWB to control strain. The underfill also serves as a protective encapsulant that prevents moisture and contaminants from reaching the active side of the chip. The underfill adds the extra strength needed for conductive adhesives and also reduces the potential for electromigration. Applying encapsulant after assembly can be called *postpackaging* or *in situ packaging.*

1.3.4 Handling Ease

The vast majority of the approximately 1 trillion chips produced each year are packaged using standard form factors that are handled easily by modern electronic component assembly lines. The common package is hard, strong, and easy to grip. The simple design allows easy recognition by visual inspection and by machine vision. The lead layout and package shape permit easy orientation. An SOT-23 transistor package, for example, has two leads on one side and a single lead on the opposite side. Not only can a pick-and-place robot easily confirm the orientation, but simple mechanical feeders also can orient this nonsymmetric package so that all are presented correctly. Packaging and machine-building associations have worked together to maximize compatibility between package style and machine, and this leads to the next subsection.

1.3.5 Standardization

Packaging standards bodies, such as the well-known Joint Electron Device Engineering Council (JEDEC), review proposals for new package designs and either provide suggestions for improvement or accept them into a standard. JEDEC is now the Solid State Technology Association (SSTA), a semiconductor engineering standardization body of the Electronic Industries Alliance (EIA), a trade association that represents all areas of the electronics industry. The approved standard is then published and distributed. Members and others agree to produce packages within the standard.

1.3.6 Facilitate Auto Assembly

The packaging and equipment makers work closely together to ensure compatibility. Several packaging standards organizations approve packages after sufficient time to comment. Assembly machine builders are represented by the Surface Mount Equipment Manufacturers Association (SMEMA) in the United States, which helps ensure that equipment meets standards that allow different brands to work together.

1.3.7 Removability/Reworkability

While most people agree that a package should be removable, not everyone insists that it be reworkable. *Removability* implies that the circuit board or substrate can be reused and that a new package can be assembled. Many people feel that *reworkability* also means that the package can be removed and then be placed and reconnected to substrate. This requires that the terminal leads be disconnected and then reconnected without damage or using a complex process. A removal process requiring chemicals and mechanical cleaning probably is not acceptable. The heating or other assembly processes should not degrade any of the structures during the removal and reattachment processes. Standards may require that the remove/add process can be repeated several times. The underfilled flip chip cannot be removed easily, and some people do not consider it to be a true package. Reworkable underfills use troublesome processes and even require mechanical cleaning steps and may not be reworkable in the accepted sense of the term.

1.3.8 Performance Enhancement

The standard package typically will degrade electrical performance, although only slightly for most applications. The connector length increase that results from the wire bonds, lead frame, or various other packaging connectors adds resistance, inductance, and time delay. The added resistance is insignificant for all but high-power applications. Inductance can add signal delay and distortion, but it has been a problem mostly at very high frequencies. Special packages are used for very high radiofrequency (rf) applications. However, as digital signal clock rates increase, the impedance characteristics (the combination of resistance, capacitance, and inductance) of the package will become more important. Packages that are designed to handle these higher frequencies can be said to enhance performance. Enhanced performance goes beyond simply protecting the chip. Some characteristic of the total system should be improved over the results in a more standard package design. A package substrate with a low dielectric constant *k* would enable better signal transmission at higher frequencies and would fall into the enhancement class. Underfill is considered packaging material with performance-enhancing qualities. The material can boost the thermal fatigue resistance of the assembly by a factor of 10 or more. Thermal transfer enhancement is such a critical area that it is a separate topic covered in the next subsection.

1.3.9 Thermal Management

All packages generate some heat because no electronic processes are 100 percent efficient. Their inefficiencies produce by-product heat. Some ICs can produce considerable heat, around 100 W for powerful new CPUs. The faster clock rates, greater number of transistors per chip, and diminishing size of the chip all come together to create a potential thermal catastrophe. The chip produces heat faster than it can be dissipated, causing the temperature to rise. The hot chip becomes even more inefficient and produces a higher output of heat. This thermal runaway condition finally reaches a level where the chip breaks down and is destroyed. A package must be designed to transfer this heat to the outside environment, either passively or with external coolers such as a fan.

The two thermal pathways are from the backside of the chip or through the metal chip pads on the front. A die, prior to wire bonding, is attached to a lead frame or chip carrier using die attachment adhesive that generally is filled with metal or mineral filler to increase thermal conductivity, as shown in Fig. 1.4. Some heat is removed through the connection pads, but this is minor when thin gold or aluminum wires are used. Tape automated bonding (TAB), covered in the next section, provides a better thermal path because more metal conductor with shorter paths is used. The best thermal conduit for the chip pads is with flip chip structures. The connection bump is very short and relatively thick. The flip chip package is one of the better thermal dissipation formats because heat also can be removed from the back of the chip that faces upward. Some flip chip packages, especially flex ball grid arrays (FBGAs), have a built-in metal heat spreader bonded to the flip chip

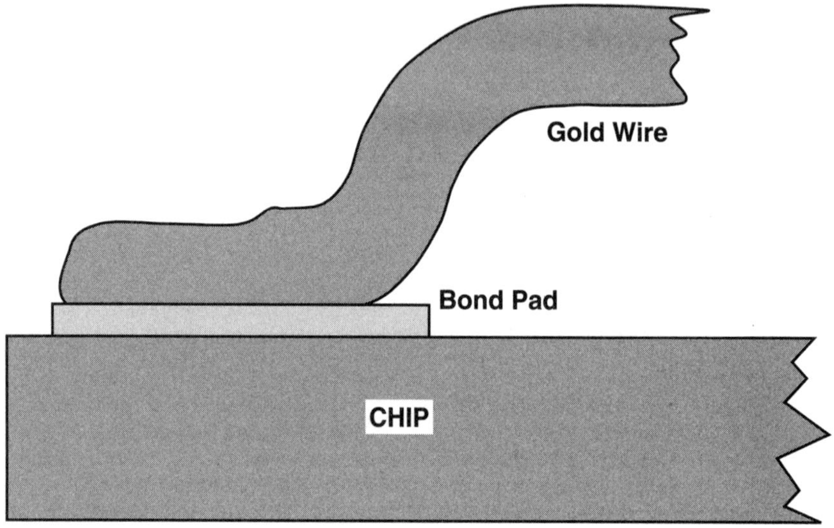

FIGURE 1.4 Thermal compression (T/C) bond with die attached.

that can be mated with a large heat sink for maximum heat management. New underfills will have increased thermal conductivity, thus adding an additional thermal path for flip chips.

1.4 EARLY PACKAGE DEVELOPMENT

Now that modern packaging principles have been discussed, let's review earlier packaging development. The need for packaging began with the first electronic device of any sophistication, and that, arguably, was the vacuum tube. The filaments, plates, and electrodes were protected in a glass vacuum tube that provided the mechanical and environmental protection while enabling easy assembly with a plug. The vacuum tube, which dates back to the twentieth century, provided all the essential features in Table 1.1 and many of the useful ones. Figure 1.5 shows an early Fleming triode in its very functional package. Vacuum tubes are still popular today for audio equipment and high-power transmitters. One can argue that the socketed vacuum tube is the first pin-grid array.

Modern packaging was launched with the solid-state electronics revolution, the first in electronics. The invention of the transistor in 1947 represented a major breakthrough that was not fully appreciated for many years. Transistor manufacturers quickly designed and implemented packages, and many are still in use today. Most were metal cylinders with three protruding wires, such as the still-popular transistor outline (TO) package. The eventual integration of the transistor to produce the IC would launch much more complex packages a decade later. However, probably IBM designed the most advanced and innovative early packages for the transistor in the early 1960s. IBM's solid logic technology (SLT) concept, described earlier, was far-reaching. This package has many of the features found in most modern micropackages. This direct-attach predecessor of the C4 flip chip is probably the first area array package ever used. Unlike most BGAs and flip chips, this package used copper balls instead of solder. Theoretically, the SLT package could be attached to substrate with eutectic solder and reworked because the balls are attached to the die with high-melting solder. Perhaps the SLT will be reborn.

Although the copper balls were attached with some level of automation, a mass bumping method was sought. Vacuum deposition was chosen because of the high degree of precision and excellent control of thickness. Depositing one metal and then another could form metal alloys. Since the metal

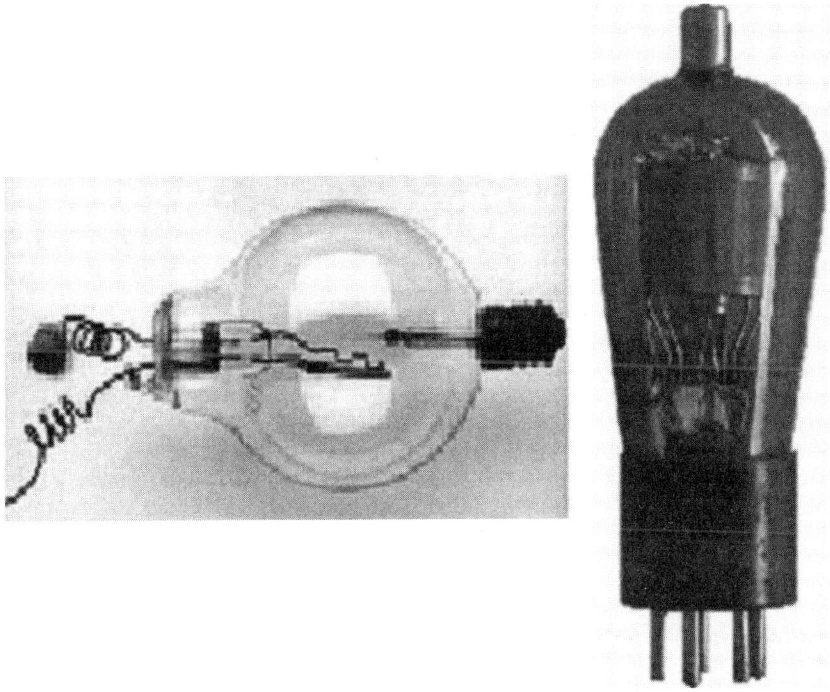

FIGURE 1.5 Fleming triode in a glass hermetic package (Prismark) and a glass vacuum tube.

columns were reflowed to form spheres, a homogeneous solder mixture resulted. These bumps allowed the chips to be soldered directly to substrate using controlled collapse chip connection (C4). C4 and flip chips are covered in detail in Chap. 5, "Flip Chip Technology."

Wire bonding (WB) was being developed independent of several DCA semiconductor producers. While IBM sought a very dense and extremely reliable main-frame computer package system, others needed efficiency, versatility, and low cost. The most important attribute of WB was the ability to "program" the wiring locations and thus eliminate tooling. WB also could be used on chips as received. WB allows the die size and pad configuration to be changed within reasonable limits without a package or lead-frame change. This feature is still extremely valuable today and accounts for the continuing popularity of the method.

Tape automated bonding (TAB) was developed shortly after WB and was sought as a means of simplifying the WB process. The idea was to fabricate a pattern of metal wires that could be bonded to the chip in a single gang-bond step. The original concept used metal with no dielectric carrier. A metal frame around the perimeter held the pattern of cantilevered wires. This frame was cut away after bonding to the chip. Now the "spider package" could be compression bonded or soldered to the circuit board. This bare-die concept evolved into an overmolded package. While several single-metal-layer packages have come along, various problems occurred, and the spiders became lead frames that were connected to the chip by WB. Not only did the wire bond allow different chips to be connected to one frame pattern, the wire served as a strain relief.

In the mid-1960s, true TAB, or flex-based, circuits were developed using conductors on a thin dielectric film. While the earliest idea used heat-sensitive DuPont Mylar (polyester), the products quickly moved to high-temperature Kapton (polyimide). Today's TAB or tape carrier packages or flex-based packages use polyimide. However, low-cost, low-lead-count polyester is used for calculators. The outer leads are connected using conductive adhesive film to avoid soldering.

1.5 ELECTRICAL CONNECTIONS

1.5.1 First Level: Chip to Carrier

The chip-to-carrier connection, generally referred to as *first level,* is a basic designed differentiation for area array packaging. There really are only two fundamental methods of connection to chip pads but several subcategories. The chip pad is connected either directly by means of a miniature bump or by a fine wire. Even TAB can be viewed not as a wire connection, but a pattern of wires that is chemically fabricated and held in a pattern. However, most authorities recognize three first-level connections: (1) DCA, (2) WB, and (3) TAB, or tape (flex-based). All three of these chip-level connections can be used for either full-sized BGAs or chip-scale packages(CSPs). The geometry of the package wiring pattern determines if the small CSP footprint is feasible.

1.5.2 DCA (Flip Chip)

DCA provides the highest electrical performance or speed. The method also allows the greatest number of chip connections and chips, and 7800 bump chips are being implemented. A disadvantage is that DCA requires precision and a high level of sophistication. We can view flip chip (FC) as a stand-alone package that is assembled *o*nto a circuit *b*oard (FCOB) or as a component *i*n *p*ackage (FCIP). Some people have argued that FCOB is not a full package because it cannot be removed once underfilled. A case is also made that the underfilling process is not part of the normal SMT assembly process and a great burden for the assembler. This may be somewhat valid, although new underfills are simplifying the process. The DCA high-density footprint also can challenge board-making capability. These issues have helped boost the chip-scale package.

However, there is another answer to these concerns: the FCIP. The FCIP is removable, reworkable, does not need to be underfilled by the assembler, and can alleviate some of the board density requirements. The resulting package typically is area array and can be designed easily as a small footprint fitting the CSP definition. The trend for all CPUs and many other flip chip components is toward FCIP.

1.5.3 Wire Bonded (WB)

While joints used in flip chip (DCA) provide a simple direct path, the alternative WB method from the late 1950s is much more popular. The amazingly simple but extremely versatile WB connection process has been the standard of the packaging industry for many decades. The method involves forming a reliable metallurgical bond between a very fine wire and a chip pad, looping the wire to the substrate, and forming the second bond. The versatility results from being able to cut the wire to a programmed length and to bond onto the desired location of the substrate. Very thin gold wire is used commonly, while aluminum is employed for simpler devices requiring thicker wire, such as power chips, but using an ultrasonic (U/S) process instead of thermal compression (T/C). The popular T/C process first forms a ball at the end of the wire using a hot flame such as hydrogen. The ball is pressed against the bond pad with heat and pressure, forming a metallurgical joint. The process is shown in Fig. 1.6, and it is used in the majority of area array and other packages for first-level assembly. The WB process can be used for just about every type of package, including micro-BGAs.

1.5.4 Tape Automated Bonding (TAB)

TAB is a flex-based package that has been around for many decades under many names. The most recent revival of this package class was by Intel, which chose to call it a *tape carrier package* (TCP), a more appropriate name. TAB, or TCP, is actually a specialized flexible circuit made with thin dielectric film with a "window" for bonding. The most common flex packaging material is high-temperature polyimide, which is able to handle processing temperatures well above 300°C. Flexible circuitry offers special benefits for packaging, including thinness, lightness, ultrafine pitch, and strain-relieving compliancy. A pattern of "wires" is formed from the copper that can be connected to the chip pads. An opening, called a *window,* is formed in the flex dielectric so that the metal conductors become overhanging cantilevered beams. Now the TAB can be placed over the chip and the beams aligned over the chip metal bonding pads. These metal beams, or fabricated wires, are then bonded using thermocompression bonding. Wire bonding and TAB inner lead bonding (ILB) have similarities. However, it is possible to gang bond the TAB leads, making this step

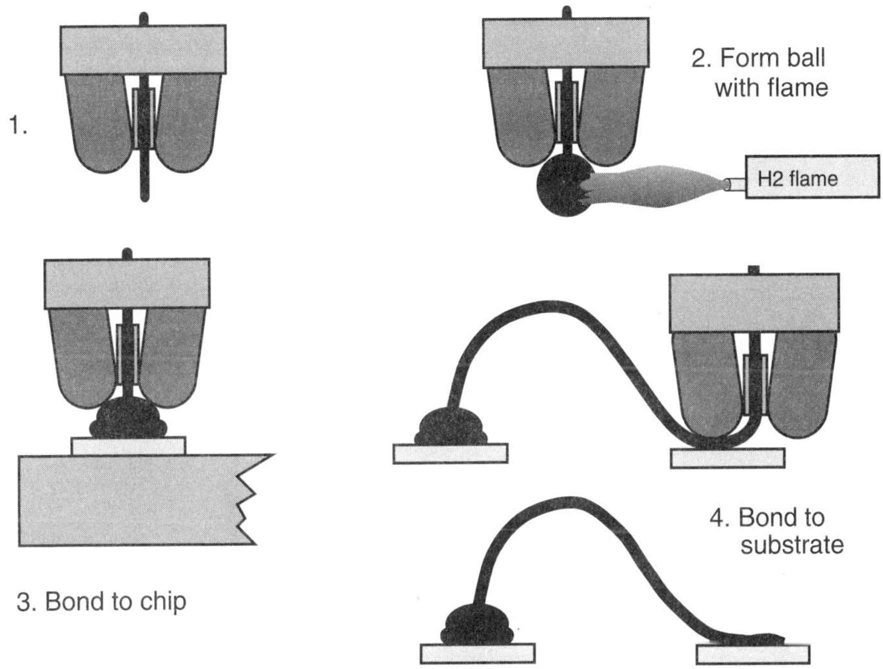

FIGURE 1.6 Thermal compression (T/C) bonding process.

very productive. The TAB inner leads typically are gold plated. The chip is often gold plated and may even have gold bumps to aid the connection process. The ILB process is efficient and reliable. The resulting package can be singulated from a reel of "tape" or kept in this form; hence the original tape name. Figure 1.7 shows TAB. The package generally is assembled to the circuit board by soldering the outer leads with a hot bar. The outer lead bonding (OLB) also may use anisotropic conductive adhesive (ACA) film. While the ILB process is a reasonably good one, the OLB step has to be the notoriously troublesome step. Later I will show how some of the newer area array packages have preserved ILB but eliminated the noisome OLB.

Density Ramifications. The DCA connection scheme provides the highest density because it is the only one that can use the full area of the chip. TAB can be problematic because it is impractical to connect to an area array chip. In principle, WB could be used to connect beyond just the perimeter of the chip, but the resulting "rats nest" would not be efficient, and signal crosstalk would be likely. However, WB or TAB over active die is considered potentially destructive. The mechanical force could fracture transistors and other devices or reduce their lifetime. The DCA bump can be produced by gentle nonmechanical processes that will not injure the delicate IC structure. The DCA bonding process is also gentle, making area array chip connection quite practical only by this method. The DCA will remain the highest-density package.

Connection Performance. All three methods have good records of reliability, provided that design and process rules are followed and the right protection is used. Protection materials and processes are covered later in this chapter.

1.5.5 Second Level: To PWB

There are two basic categories of connections for second-level, or board-level, assembly of the area array package: mechanical connections and those joined with conductive material. The most common

FIGURE 1.7 TAB (tape automated bonding).

mechanical method uses pins, and the package is called a *pin-grid array* (PGA). Other mechanical attachments have been proposed, and some have even reached prototype stage, but only the PGA has become important to date. For example, a small amount of work has been done on boards that provide plug-ins for BGAs and flip chips with hard bumps such as copper.

Connections designed for joining are either spherically shaped or columnar. The joining material can be the solder obtained by melting the interface or externally applied metal. Conductive adhesives also can be used to mate metal bumps or balls as well as columns. Other chapters will cover many aspects of BGAs.

1.6 PACKAGING MATERIALS

One can view the package as a miniature container that houses one or more semiconductor devices and provides at least the minimum requirements that were listed in Table 1.1. The container can be made of anything that will satisfy the requirements. The protective enclosure can be fabricated before assembly or after. Metal, ceramic, and plastics (polymers) are the three common categories and are now discussed in more detail.

1.6.1 Metal

Metal has nearly all the desirable properties for a packaging material. Metal is a barrier to all gases, including water vapor, one of the most common corrosivity-promoting agents. Metal is also a good heat-transfer material and can be processed into any desirable shape by common methods supported by an immense infrastructure. However, metal is also a good electrical conductor that potentially can short out connections to the semiconductor devices. The last problem is solved by using insulators, such as ceramic or PWBs. The metal package is designed to be evacuated and sealed so gases that could cause problems are reduced to an acceptable minimum. This type of package is referred to as a *hermetic* or *sealed hermetic package.* Special molecular scavengers, called *getters,* can remove contaminants throughout the lifetime of the package. A typical design is more or less that of a small box that is open at the top and finally sealed with a lid. The electronics is sealed within, and electrical leads exit by means of insulators that maintain gas-barrier properties. Alternately, an insulative base made of ceramic can be used to solve the electrical isolation issue. Figure 1.8 shows metal hermetic packages.

1.6.2 Ceramic

Ceramic material has long been used because of excellent barrier properties, good thermal conductivity, and good electrical isolation. In many ways, ceramic is the ideal packaging material. However, ceramics are more expensive, and their processing is not as versatile as that of plastics, which can be molded, fluid dispensed, and even stenciled. Ceramics also are processed at high temperatures that can exceed 1000°C. Yet ceramics enjoy high-volume use especially for larger chips despite the costs. Many CPUs use ceramic chip carriers in pin, column, and ball grid formats. Some of the newer packages use a flip chip attached to a ceramic surface in an FCIP. The low CTE, extremely good flatness, low moisture absorption, and high thermal conductivity make the ceramic carriers very suitable for the fast, high I/O, and hot running computer chips. Other versions have a cavity where the chip is

FIGURE 1.8 Metal hermetic packages.

placed and then wire bonded to conductor pads on the ceramic. This type of package can be sealed with polymer encapsulant. Figure 1.9 shows ceramic packages with an attached chip that is ready for sealing with a lid or filling with encapsulant. One must keep in mind that plastic is a good, but not a perfect barrier, so it does not provide a hermetic environment.

1.6.3 Plastic

Most packages are made of plastic because of the significant cost and manufacturing benefits, even though polymers do not provide a hermetic barrier. Polymers are extremely versatile, and there is essentially an endless number of packages that can be made with highly engineered and tailored properties. A key attribute of polymers is that they can be handled easily and shaped efficiently. Thermoplastics can be melted and formed into the desired shape and hardened again by cooling. However, most electronic polymers are thermosets that are hardened to permanent, nonmelting shapes. The thermosets can begin as liquid or low-melting solid precursors that are shaped easily and precisely by any one of numerous industrial processes. Plastic area array packages use several different plastic encapsulation processes that are fast and efficient, hence their popularity. The common starting point is an area array chip carrier that is encapsulated with a thermoset polymer.

Rigid Carriers. Rigid carriers are the most common types of plastic package platform. The carrier is basically a miniature PWB that provides the chip mounting platforms and the wiring interconnect structure. FR4 has been used, but higher-performance organic laminates such as bismaleimide-triazine (BT) have become common. BT and other advanced laminates have more thermomechanical stability and heat resistance. The chip carrier is heated four or more times during its life: die attach, chip interconnect, solder ball attachment, and assembly to board. The carriers typically are thin two- or four-layer circuits that are produced in multicircuit (or carrier) panels for manufacturing efficiency. The panels often are cut into strips that make a convenient form factor for chip-attach, ball-attach, and encapsulation machines. Figure 1.10 shows a rigid organic chip carrier just prior to encapsulation.

Flexible Carriers. Flexible circuitry has been used as a chip carrier since the 1960s, but flex-based packages are increasing in popularity today. Several commercial designs are based on thin copper/polyimide circuits. They may be divided into two categories based on the conductor geometry and the resulting package size compared with the chip. Flex with a fan-out conductor configuration produces a BGA configuration in which the package is larger than the 1.2 times value arbitrarily used to classify CSPs. A fan-in routing, in which the leads track inward under the chip, can generate less than 1.2 times chip area to achieve CSP designation (Fig. 1.11).

FIGURE 1.9 Ceramic area array packages.

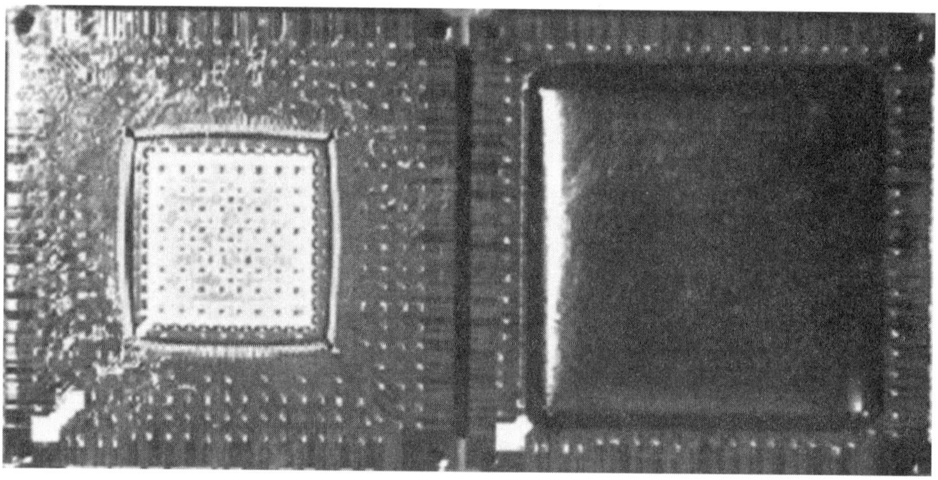

FIGURE 1.10 Rigid BGA carrier (before and after encapsulation).

All three types of first-level chip connections are used with flex, DCA, WB, and TAB. The TAB-like connection is the most unusual and does not have an equivalent in rigid systems. IBM was the first to use TAB inner-lead bonding while removing the outer-lead connection. The leads from the chip were routed into an area array pattern and connected to metal bumps by means of vias through the thin polyimide film. This innovation kept the best of TAB and eliminated the troublesome outer leads. What's more, the package was transformed into SMT. IBM named the product *tape ball grid array* (TBGA). The μBGA from Tessera also uses a TAB-like cantilevered beam array for the chip connection. Gold leads are bonded to ordinary chip pads as in TAB ILB, but the outer leads are gone. The conductors fan in underneath the chip and terminate as bumps under the chip to produce a very small area array package. Since the conductors fan in, the package is much smaller than the equivalent TAB style. The flex-CSP can be surface mounted instead of using a specialized TAB bonder. Both the TBGA and micro-CSP are successful.

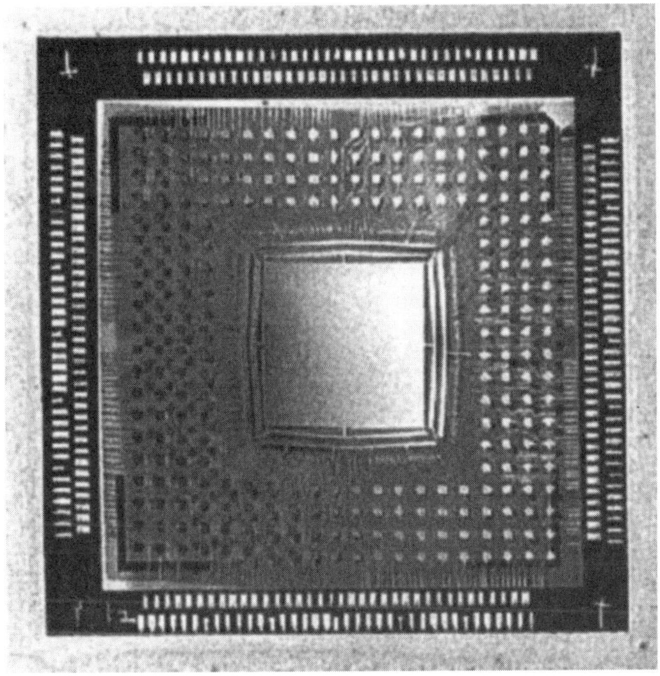

FIGURE 1.11 Flex BGA.

1.7 PACKAGE STYLES

1.7.1 Lead Frame

As suggested earlier, metal lead frames can only be used for perimeter-type packages and will only be mentioned in this book for reference purposes.

1.7.2 Chip Carriers/Platforms

Chip carriers are necessary for area array packages, but they also can be designed as perimeter-only packages. Once again, perimeter types will only be included within this handbook for reference purposes. The chip carrier serves the mechanical function of providing a platform for the chip. The chip is strongly bonded to the carrier with die-attach adhesive or other means. The chip must then be connected electrically to the carrier, which routes the lines to the correct positions for bonding to the PWB.

Perimeter. Carrier-type packages, even with metal ball connections, can be perimeter packages. The larger carrier-type designs are almost exclusively area array packages. However, small CSP designs are often perimeter style. There are two reasons. The simplified connection schemes may not allow area configurations. Some simple CSPs bring the connections to the sides of the package so that no through-the-carrier conductors are needed. The second reason is that low-count packages, such as those for single memory chips, do not always need to use area. This also makes inspection easier because the connections can be examined visually.

Area Array. Area array packaging can be as simple as a double row of connector balls around the perimeter of a chip carrier. Many of the early BGA packages used a full area deployment with continuous rows of balls or columns. The platform area where the chip was bonded adhesively was constrained from normal movement by the low-CTE silicon chip. The organic board behaved like it had a much lower CTE, more like that of the chip. Solder balls on the BGA substrate therefore were placed under greater strain when the BGA was assembled to a PWB. The organic PWB and chip carrier would be expected to expand at similar rates, except for the area under the chip, where natural movement is restricted. This localized differential rate of expansion caused sufficient under-chip stress to produce selected solder ball failure. Subsequent BGA designs relieved the under-chip area, thus avoiding full arrays.

Chip Location. The chip can be placed onto the topside of the carrier in a *die-up configuration* with electrical signals routed to the underside. The other design is *die-down,* sometimes called *cavity-down,* where the die or chip resides in a recessed cavity on the same side as the bumps. The cavity-down design precludes bumps in the center, but this is not an issue, as explained earlier under "Area Array."

Multiple Chips. Multiple chips on single carriers are becoming increasingly common. The *multichip package* (MCP) has achieved what the multichip module (MCM) failed to do. In fact, the MCP is the MCM. Many of the current CPUs have several capacitors on their carrier. Memory packages have been designed with four flip chips. The use of multiple chips is a natural evolution of the advanced area array package, and we can expect continued development here.

1.8 PROTECTION

1.8.1 Hermetic

The hermetic package is a gas-tight container that does not require an encapsulating material. The hermetic package is not a common class for area array packaging and may be viewed as a specialty type for very high reliability or unusual systems.

1.8.2 Molding Compounds

Molding compounds have represented the preferred means of package protection for several decades. The most common materials are epoxy molding compounds (EMCs). EMCs consist of solid epoxy resins, solid hardeners, fillers, and additives. Generally, they are provided in a cylinder shape referred to as a *puck* or *hockey puck.* However, they come in a myriad of shapes designed for different molding machines and run sizes. A transfer-molding machine heats the solid slug of EMC to melting and then injects it into a mold loaded with packages. Heating causes the prepolymer mix to polymerize into a hard, durable, nonmelting mass that permanently takes on the shape of the mold. Area array packages can be molded in singulated form, in strips, or in sheets. A mold can hold hundreds of packages to produce many thousands of parts per hour. More recently, strips and sheets of carrier substrate have been molded for high throughput and mold standardization. The entire carrier sheet containing the attached chips can be flood or area molded and later sawed into separate packages. Not only does this increase output, but standard molds also can be used that preclude having to build complex molds for each different package.

1.8.3 Liquid Encapsulants

Liquid encapsulants typically are made with epoxies and hardeners having a composition similar to EMCs. The polymer precursors are liquids instead of solids like the EMCs, but they can be hardened into polymers with similar properties. Liquid encapsulants typically have less filler to allow flow and do not quite measure up to the EMC properties. Liquid encapsulants are applied with needle dispensing equipment that is slower than molding machines but more versatile. Needle dispensers are programmed to apply the required pattern and therefore do not need tooling. This means that the liquids are ideal for both short runs and quick time-to-market applications. Die-up designs require that a dam first be placed around the chip to retain liquid encapsulant. The dam can be made by needle dispensing thick encapsulant in a pattern. The low-viscosity fill-type encapsulant is then dispensed over the chip. Both the dam and encapsulant are hardened in the same bake process. A die- or cavity-down package is inverted, and the bottom die cavity is filled with a needle dispenser. Recently, high-output dispensers have been offered with up to four heads.

1.8.4 Underfill

Underfill fits within a subcategory of encapsulants but has an additional role to play beyond barrier protection. The underfill acts as a mechanical coupling adhesive that anchors the chip to the carrier and prevents differential movement that could weaken and eventually destroy the solder or conductive adhesive joints of the flip chip. Underfills have become more complex, and at least three types are available with a fourth, wafer-level, film under development. Application methods include needle dispensing and even transfer molding. These are covered in Chap. 5, "Flip Chip Technology."

1.8.5 Chip Passivation Only

Early flip chips had glass or other inorganic passivation over the active side of the die. Bumps were designed to fill in and close any gap so that the flip chip was effectively sealed off from the environment. Passivation is still used today. A passivated chip may not need underfill if there is a reasonably good thermomechanical match between the chip and the substrate, as is the case with ceramic carriers. However, the common practice is to still use underfill for added cycle life and better environmental protection, especially for larger chips.

1.9 PACKAGE ASSEMBLY

Area array packages are surface mountable, although pin-grid array types could be considered through-hole in some cases. However, the more common method of board connection for PGAs is by means of a socket on the PWB. The PGA that plugs into a surface package should not be considered a feed-through type. The PGA, although one of the earlier area array packages, continues to remain popular because its pluggable feature allows easy field servicing as well as convenient upgrading, especially for computers.

1.9.1 Issues with Area Array

The biggest issue with BGAs and other nonpluggable area array packages has been inspection after assembly. The assembly industry had become so accustomed to visual inspection that a package with hidden joints was rejected by many manufacturers. While the issue has subsided to some degree, many people still view the lack of optical inspection as negative. Simplified real-time x-ray equipment has come into use that allows careful inspection, but one can argue that the high cost compared with a magnifier makes this only a partial answer.

1.9.2 Soldering

Solder joint formation is the most common method of assembly for area array packages. Most BGAs have balls or spheres made of eutectic solder so that solder paste need not be applied. Paste can be used, but a more common practice is to only apply flux.

1.9.3 Conductive Adhesives

Conductive adhesives could be used hypothetically for the large BGA packages, and work was done in this area. The concept was to eliminate the solder balls and apply patterned anisotropic conductive adhesive (PACA) film to the underside of the package. The assembly would then be bonded with heat and probably pressure. A major problem was package warp that was considerably worse several years ago. While solder balls can accommodate significant noncoplanarity, adhesive film is much less tolerant. Furthermore, a symmetric BGA creates strong alignment forces that can correct a misoriented package, and this is not the case with adhesives. However, smaller packages, especially flip chips, did not present a coplanarity problem. Both isotropic (ICAs) and anisotropic conductive adhesives (ACAs) are used commercially for flip chip assembly. When used with a nickel or gold bump, the adhesive provides a system that is ideal for memory chips because there are no radioactive emissions. Memory chips and some of the high-density CPUs do not tolerate α-particle emissions that emanate from natural lead in standard solders that contain trace amounts of radioactive nuclides.

1.10 PACKAGE RELIABILITY

Area array packages have equal or better reliability than the equivalent perimeter-style packages. Earlier problems with solder ball fatigue under the die have been corrected primarily by leaving the under-chip area open. The partial array design also relieves density and makes routing easier for the PWB manufacturer There are still some assembly concerns for plastic BGAs because moisture can enter the package through the organic carrier. Absorbed moisture can cause explosive evolution of water vapor during solder reflow heating that fractures the molding or causes delamination between encapsulant and carrier. Either storing BGAs in a dry box or predrying before assembly can eliminate this phenomenon, known as *popcorning.* While these are effective solutions, future materials and designs may eliminate these added procedures. However, the higher processing temperatures expected for lead-free solders could worsen the problem. Ceramic BGAs and CGAs do not have this problem. Also, the higher standoff of the columns used in CGAs nearly eliminates any fatigue in thermocycling conditions.

Flip chips on ceramic may have the highest record of reliability of any system. IBM has assembled millions of flip chip modules over the past three decades with essentially no package failures. However, second-generation flip chips on organic substrate encounter higher stress forces, and failure can occur sooner. The stress-relieving (diverting) solution is underfill. Underfill allows the high-modulus, low-expansion chip to restrain the carrier's typically high expansion. However, during the evolution of underfill and the related processes, problems were encountered. Some underfills lacked the high adhesion required and delaminated from the chip surface, allowing bumps to fracture during thermocycling. Voids in the underfill also were found to cause premature failure depending on their number and location. Newer predispensed underfills that have inherently higher CTEs also may result in lower reliability. However, the majority of underfill-related failures occurred in the laboratory and during development. In general, second-generation flip chips can pass 1000 thermocycles without failure, and the industry considers this acceptable. At the same time, flip chips on ceramic have been tested at high as 60,000 cycles without failure. Flip chips and underfill are covered extensively in Chap. 5, "Flip Chip Technology."

1.11 FUTURE EXPECTATIONS

Area array packaging is expected to continue to gain market share as ICs move to higher lead counts and product designs require further miniaturization. While the target moves slowly upward, better performance and economics are found with area array packaging with lead counts of over 400 for standard fan-out BGA packaging. However, area array chip-scale packages are being used for very low counts for memory chips where small size and high speed are becoming increasingly important. MCPs also will be a growth area.

Future BGAs will continue to shift to DCA for the first-level connection as lead counts grow and the flip chip infrastructure expands. This type of package, FCIP (flip chip in package), is already the preferred style for computer chips because wire bonding and TAB become impractical at the high ($>$1000) lead counts. Furthermore, CPUs in flip chip format can be designed with redundant power and ground bumps so that power does not need to be routed within the chip with lower-current-capacity submicron conductor traces. Additional bumps can be added for heat piping. The bump can be an excellent path to conduct heat away from the IC into the carrier and finally to the PWB. Additionally, heat can be removed from the back of the chip by applying a heat sink. Faster, more powerful CPUs have an increasing heat burden, and FCIP is one of the best designs for dealing with heat. This also suggests that FCIP for computer chips will continue to leave the back of the flip chip exposed for heat removal.

We will continue to see much progress in the CSP area as more cost-effective products are introduced. While the popular CSPs, such as the Tessera μBGA, are very reliable, many manufacturers seek lower-cost solutions using existing equipment. AMKOR and other packaging foundries, along with large vertically integrated electronics companies, continue to evolve lower-cost CSPs. We can expect a sorting out of the several dozens of styles in the future. Simpler, lower-cost CSPs will become available that perhaps better fit the established processes and equipment already available.

CHAPTER 2
ELECTRONICS INDUSTRY OVERVIEW

Prismark Partners LLC
Cold Spring Harbor, NY

2.0 THE ELECTRONICS INDUSTRY TODAY

Gross earnings in the electronics industry surpassed $1 trillion worldwide in 2000 and are expected to grow at a compound annual rate of 6.4 percent over the next 4 years. It is the largest and fastest growing industry in the world, yet it is interesting to note that it comprised only 3.5 percent of the world gross domestic product (GDP) output for 1998. However, it is growing twice as fast as the global economy. It is a very dynamic industry, propelled by constant technological change.

Most consumers of electronic products think of the electronics industry in terms of electronics systems, such as a personal computer, a cellular phone, or a digital camera, which are ultimately the products bought by end users. These systems, however, are assembled from components, and component manufacture is enabled by engineered materials, making the industry a layered pyramid of interdependent levels. Components represent a third of the value of the total industry, whereas specific engineered electronic materials account for less than 5 percent (Figs. 2.1 and 2.2).

Materials used in the construction of electronic components and systems also comprise structural metals, ABS case moldings, and other materials, which are not electronics industry–specific materials. Although small in percentage terms, electronic materials have great significance in the electronics industry and will play an increasingly important role in sustaining the advances in system-level performance.

The business structure of the industry is evolving rapidly. Unlike a decade ago, when companies such as IBM manufactured most critical components in-house, systems manufacturers increasingly rely on component and materials suppliers for enabling technology advances to deliver better systems performance at lower cost. Despite this growing significance, materials and component manufacturers typically remain hidden behind the systems cover, with the one notable exception of the Intel Inside branding success. Given the emerging importance of the "under-hood" engine of the electronics industry, it is useful to look more closely at the component and materials hierarchy.

2.1 THE ANATOMY OF A SYSTEM

The wonder of many of today's electronic systems would not have been possible without advances in electronic components and materials. Innovations in system architecture drive component and mate-

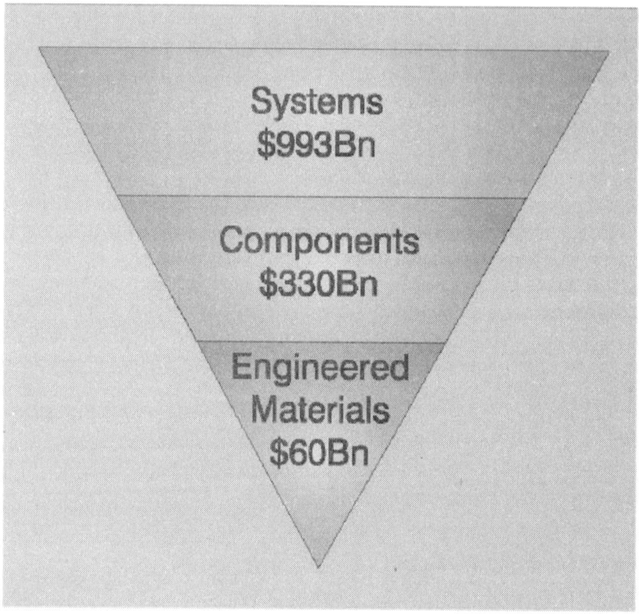

FIGURE 2.1 Electronics industry hierarchy. (*Prismark.*)

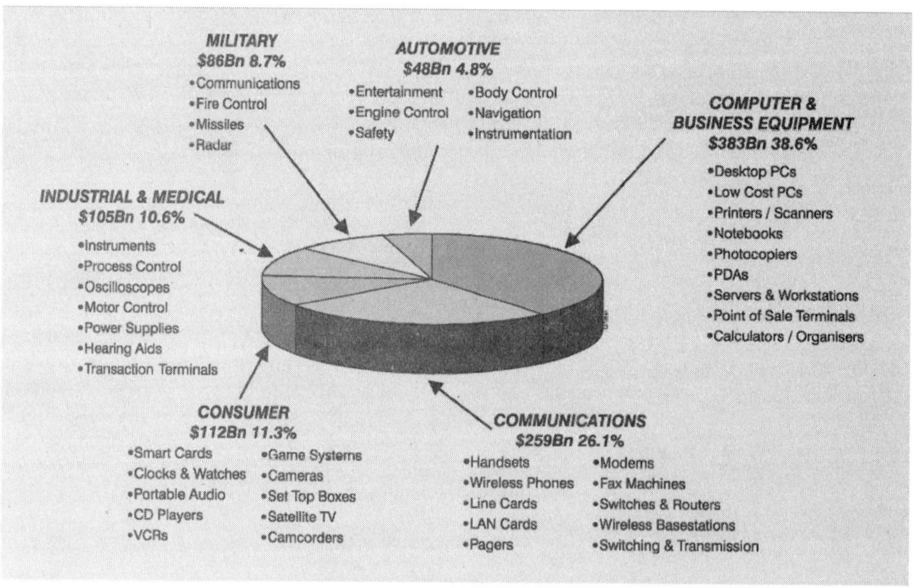

FIGURE 2.2 Electronics: a $1 trillion industry. (*Prismark.*)

rials development, and the physical realization of systems is enabled by these advances. Electronic components can be categorized roughly into active and passive components. *Active* components consume power in delivering functionality within a system, whereas *passive* components provide connection, mechanical support, filtering, noise reduction, and other functions that are critical to the performance of active devices. Passive components can be broken down further according to the function they perform: printed wiring boards (PWBs), switches and relays, connectors, resistors and capacitors, and passive microwave components, which include filters and crystal oscillators.

Fundamental to the function of electronic systems, semiconductors are the single most important segment among electronic components. On-chip integration, higher switching speeds, and greater functional density are the most important technology drivers of active components (Fig. 2.3).

With feature sizes below 0.1 μm on the horizon, the semiconductor industry is moving toward the adoption of copper conductors and low-*Dk* dielectric. Copper allows the signal to travel faster at lower power through the chip, and the lower capacitance of the dielectric makes it possible for the signals to travel faster through the interconnect layer. Microprocessors and memory are undergoing the fastest technological change, while tremendous growth is expected in the active radiofrequency (rf) arena, driven by explosive growth in wireless communications. Integrated passive networks, high-density printed circuit substrates, and high-performance connectors for optical systems are only a few examples of other actively evolving domains within the component industry. Engineered electronic materials are critical to achieving the cost, productivity, and reliability required for component and system manufacture. Prismark has segmented the electronic materials market by end use, as shown in Fig. 2.4.

The combination of materials used in the fabrication and packaging of integrated circuits (ICs) is by far the largest category of electronic materials. These materials and their delivery in packaged or bare-die form to the system assembler are critical to the evolution of semiconductor devices. Board assembly materials are consumed in the assembly of components to PWBs and their subsequent protection. Box assembly materials include electromagnetic interference (EMI) shielding and thermal interface materials. Pending environmental regulations on the proper disposal of electronic systems and components and the prospect of lead elimination altogether could force a substantial change in

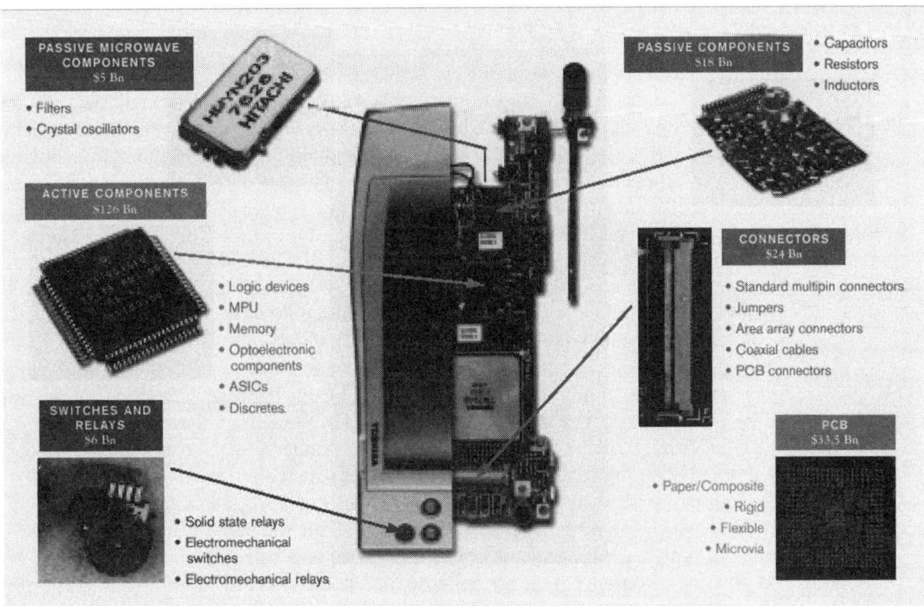

FIGURE 2.3 Component content of a typical electronic system. (*Prismark.*)

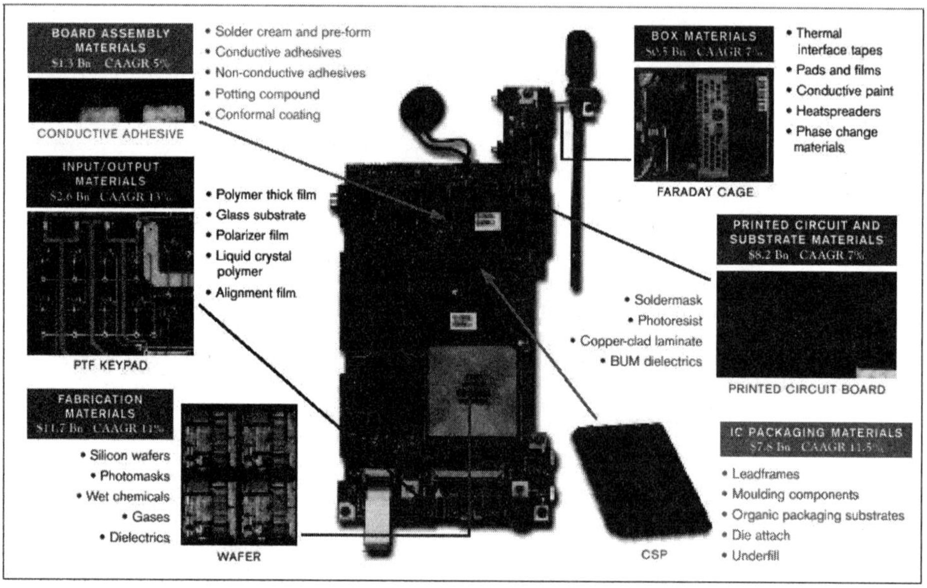

FIGURE 2.4 Electronic materials market breakdown. (*Prismark.*)

choice of materials at all levels of the industry. The continued evolution of components and materials is clearly a necessity for the future vitality of the electronics industry.

2.2 THE ELECTRONICS INDUSTRY IN 2020

Chronologically, the electronics industry is in early middle age. It will not be old in the year 2020, but it will be different. It will be bigger, and the products will have different functions, will be easier to use, and will be cheaper. Some companies will be clearly dominant, and the geographic distribution of production and consumption will have changed.

Today, the electronics industry chides the automotive industry that if cars were like computers, they would cost a few dollars and get hundreds of miles per gallon. The automotive industry responds that if cars were like computers, every mile of highway would have a frustrated driver who had crashed and was trying to reboot his or her vehicle. This friendly rivalry between the two largest manufacturing industries in the world is only about 80 years out of date. The automotive industry of 1920 was very much like the electronics industry today. Back then the "horseless carriage" was not very reliable at all. The cry, "Get a horse!" was a common admonition from the horse and buggy driver to the motorist tinkering with his Model T beside the road. The prices of automobiles were dropping precipitously. The average price of a car in the United States dropped from $1750 in 1908 to $828 in 1928 (Ford's lowest-priced vehicle was under $300 in early 1929), prices that largely parallel and predate the personal computer price experience by some 80 years (Fig. 2.5).

In looking forward to the electronics industry of 2020, we may start by looking back at the automotive industry in the early part of this century. The parallels are uncanny, and the subsequent development of the automotive industry as it matured will help to provide some level of understanding as to how the electronics industry is likely to develop over the next 20 years. The similarities between the automotive industry and the electronics industry are shown in Table 2.1.

We have used American examples in this comparison because this is where the major developments took place. Europe was an early leader in automotive technology and had an automotive

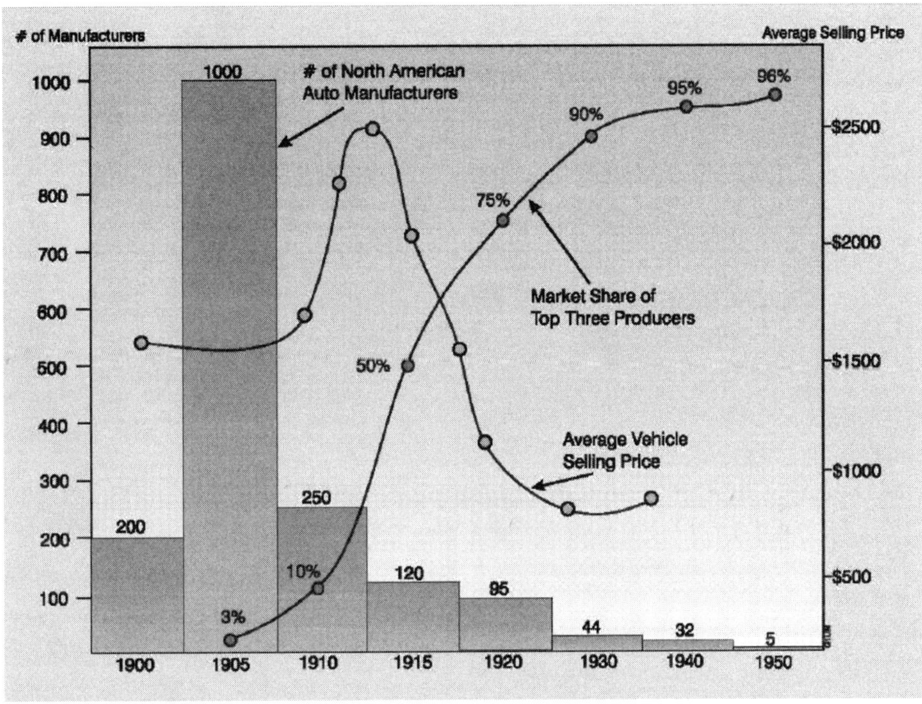

FIGURE 2.5 Automotive manufacturing in the United States. (*Prismark.*)

"industry" producing hand-crafted "horseless carriages" in the 1880s and early 1890s, almost 10 years before production started in the United States (1895 generally is recognized as the start of automotive production in the United States). However, mass production of a low-cost vehicle for "the great multitude" was pioneered in the United States. The vision, scale, and organization that enabled the American industry to drop prices to one-third and increase production 20-fold in 10 years were uniquely American. Similarly, the development of the computer and the related semiconductor design and manufacturing tools owe much to America, notwithstanding the powerful contributions made by Japan in consumer electronics and Europe in telecommunications. History, of course, does not repeat itself exactly. Intelligently applied, it merely provides a guide and reference for what might occur in the future. The early 1900s were interrupted by two major recessions (1907, 1920), one depression (1929), and one world war (1914–1918). Historians of the automotive industry regard these events as external, as events that may have hastened the conclusion of trends that were clearly already in place rather than actually precipitating them. During World War I, truck production for the war effort in Europe by U.S. companies equaled automotive production for private use in the United States. The need for truck production did not cause mass-production processes to be developed; rather, it took advantage of what Henry Ford had already developed. The dominance of the "Big Three" automakers in the 1930s after the crash of 1929 was similarly accelerated by the crash of 1929 but not caused by it. The number of major manufacturers had been in decline from several hundreds, if not thousands, early in the century as the manufacturing barrier to entry for low-cost vehicle production became higher. Mergers among the smaller vehicle manufacturers had become commonplace in the 1920s. Before the crash of 1929, the "Big Three" had a 75 percent share of the North American passenger car market. At the outbreak of World War II, the "Big Three's" market share had risen to 90 percent, with Packard, Hudson, Nash, Studebaker, and Willis mainly serving the balance. No one knows when or what major disruptive external events will be visited on the elec-

TABLE 2.1 Characteristics of the Automotive and Electronics Industries

Characteristic	Automotive example	Electronics example
Technology-based industries: Key enabling technologies predate useful product by 25 years.	Lenoir in France first patented the internal combustion engine in 1860, but it was not until 1885 (25 years later) that Daimler Benz produced the first successful gasoline-powered vehicle.	The transistor was patented in 1947, but the first commercially successful microprocessors were not produced until 1971 (24 years later).
Bold, visionary leaders: A few dedicated, visionary giants shaped each industry.	Ransom Olds, William Durant, Walter Chrysler, Alfred Sloan, and Henry Ford were all bold, visionary businessmen with an entrepreneurial talent for marshaling and directing technologies and people to achieve their goals.	Dave Packard, Andy Grove, Chairman Morita, and Al Shugart were all bold, visionary businessmen with an entrepreneurial talent for marshaling and directing technologies and people to achieve their goals.
Brutally competitive environments: Manufacturing expertise was key to creating and delivering the industry's growth.	A machinist and watchmaker from Michigan made his first car in 1896, but it was not until 1913 that the Highland Park Plant was mass producing cars. By 1920, one in every two cars produced in the United States was a Model T Ford.	A manufacturer of memory chips (which it later abandoned) was founded in 1968 and went on to become the world's most successful semiconductor company. Manufacturing skill is still key to Intel's 80 percent microprocessor market share.
Open communications: Both industries centers of growth were determined by the exchange of ideas between creative people rather than geographic attributes.	Detroit's tool and die shops and key companies such as Olds became spawning grounds for innovation. The SAE's predecessor gave free or low, royalty licenses to encourage the industry's development.	Stanford University adopted a low-threshold licensing policy for its intellectual property, helping to build industrial opportunities for its graduates in Silicon Valley. Key companies such as Fairchild became spawning grounds for technical and managerial talent in the Bay Area.
Unfettered operational freedom: During the first 35 years both industries progressed with hardly any government or labor interference.	During the New Deal, the free-wheeling automotive industry was facing the threat of tariffs, minimum wages, and child labor laws from the "yuppies" in government who "knew everything and believed nothing."	Is Microsoft's day in court the start of greater government involvement in the electronics industry? Are recycling and environmental laws going to constrain the growth of electronics?

tronics industry in the next century. We have come through the Asian crisis; will there be a crash in 2002 or a major war in the new century? We cannot tell, but we are certain that the path of the automotive industry, as it matured, provides some guideposts to understanding the electronics industry of tomorrow. There are, however, some fundamental differences. They are *physical mass, communications,* and more sophisticated *business relationships.*

By *mass,* we mean physical mass. Automobiles are bulky, heavy pieces of equipment. In the early days of the automotive industry, they often were driven from the factory to their destination or built in assembly plants close to their markets. This practice continues with the transplants today. By contrast, electronics equipment has a high value-to-weight ratio. It can be produced in one region and shipped readily to another for consumption. There are a lot of passengers in wide-bodied aircraft girdling the world every day. Below their cabin is an equal volume of cargo space that is used to ship many of the electronics industry's components from efficient manufacturing regions to the regions where they will be consumed or assembled.

Communications have been enabled by the electronics industry and will be a key technology shaping the business of electronics in the future. In the early days of the automotive industry, the Americans were largely ignorant of the technology developments that were going on in Europe. Communication was limited to the store of knowledge that individual immigrants could bring from Europe and the sharing of self-taught knowledge that was communicated among Detroit's engineers and machinists.

Business relationships were not very sophisticated. The Ford Motor Company built its own glass-making plant and developed a continuous process for the production of plate glass. This was done out of the belief that vertical integration was more efficient, as well as the paranoia of the man who controlled the company. We see vertical integration equivalents in Japanese companies and the start of deintegration in the United States and Europe (e.g., Rockwell/Conexant, Siemens/Infineon), as well as captive printed circuit and assembly shops going merchant. Belatedly, we have seen the automotive companies also deintegrating (e.g., General Motors/Delphi), and the fundamental difference today is the belief that vertical integration is not beneficial. We also have become more comfortable and sophisticated in our ability to handle arm's-length business dealings with companies that may be suppliers and competitors at one and the same time.

The early inventions and technology developments occurred in the 1860s to 1890s for the automotive industry versus the 1940s to 1970s for the electronics industry. The commercial beginnings of both industries can be pinpointed to 1895 and 1947, respectively. The real volume growth and major price reduction started to occur around 1913 with Ford's development of mass production for a standard vehicle and the availability of Intel's Pentium processors for the mass market in 1993. Based on these correlations, it is possible to believe that the electronics industry is where the automotive industry was in 1920. It can be argued that industry moves faster today and that any correlations between developments in the automotive industry and the electronics industry should have a compressed time frame when referencing a historical precedent. The timeline correlation between the early days of the automotive industry is not perfect, but it could be used to argue against the concept that "things move faster today." Most of the key developments in both industries occurred on virgin ground and were driven by people. People have finite lifespans and predictable periods in which they can engineer and accept change. We are firmly of the belief that human nature remains unchanged, particularly in making and accepting change in virgin territory. Consequently, we are comfortable with a year-by-year correlation between the automotive industry in the early part of the twentieth century and the electronics industry at the end of it. At the turn of the millennium in the electronics industry we are now in the year 1920 in the automotive industry. What is going to happen next? More subtly, what is the electronics industry going to look like over the next 20 years given its uncanny similarity to the early automobile industry and its three key differences (product mass, global electronic communications, and greater sophistication in business relationships)?

2.2.1 The Structure of the Electronics Industry in 2020

Twenty years ago, electronic equipment production was fairly evenly distributed between the different market sectors (military, industrial, medical, computer, etc.). The rapid growth of personal computers in the 1980s and 1990s and their ancillary equipment (printers, modems, etc.) has raised the computing and communications content of the industry from 36 to 65 percent today. Thus two-thirds of the electronics industry, in hardware terms, is devoted to enabling us to access, route, and store information or, in more simple terms, to communicate globally on a personal and business level.

Despite spending around $700 billion a year on hardware for this purpose, we are still not very good at it. Our cellphone batteries run down, our laptops can only just access the Internet wirelessly, our high-speed digital subscriber lines (DSLs) are expensive, and our banks and brokerage houses are only now updating our net worth every time we make a transaction. Only in the most forward-looking companies are customer relations and supply-line management starting to come online (e-business). All the pieces are there and the technologies are known, but they have yet to come together in an immediately cost-effective, ubiquitous, reliable whole.

When they do (over the next 10 years), access to and routing and storage of knowledge will be available to anyone with "a full-time job," to paraphrase Henry Ford. At that time, this segment of the electronics industry will account for 80 percent of the industry (up from 65 percent today) and will be equivalent to the car and light-truck segment of the automotive industry. Of course, there will still be medical electronics, military electronics, and industrial electronics, but these sectors will continue to decline in relative importance and will be analogous in automotive terms to ambulances, tanks, and trucks.

How will we progress from our present 1920 condition? Back then, the product definition was clear, the elements were available, and the technologies were well understood, but the deliverable was still open to the admonition, "Get a horse!" We need an electronic product that does not need a "mechanic" to operate and maintain it, that does not consume more "water than gasoline," and that can be driven from one end of the country to the other on reliable highways with service stations.

In short, we need global communications that will access, route, and store our combined knowledge at the tap of a key. Our system must start every time and operate with a minimum of maintenance. We are quite a long way away from achieving this product goal. Several companies have pieces of the pie: Microsoft (operating systems), AT&T (routing), Cisco (Internet hardware), Intel (processing), Sun (servers), Seagate (data storage), Nokia (wireless access), and Toshiba (portable access), to give some examples.

The combination and integration of these elements into a "branded" product that provides what the global customer needs is the principal structural change that will occur in the electronics industry over the next 20 years. Today, 36 companies account for 76 percent of the design of communications and computing hardware. We may expect this to fall to 20 companies over the next 20 years, not exactly the "Big Three" in the U.S. automotive industry but substantially fewer than exist today.

In this process, the companies that manage their relationships with their customers, partners, suppliers, and acquisitions most effectively will be best able to meet market need through branded global e-business-based virtual companies. Whose brand will you be using in 2020—Cisco's, Lucent's, Nokia's, NEC's?

These companies will not be manufacturing things anymore—in fact, the quicker they get out of manufacturing, the better—they will be designing, maintaining, and marketing global knowledge systems. However, someone has got to manufacture electronic products in 2020, so who will it be? It will be a handful of semiconductor companies and contract assemblers.

Will the Lucent brand-name communications system ordered by the Bavarian school district be drop shipped directly by Celestica's eastern European factory? Of course, it is almost happening today.

Supporting this whole chain will be the materials companies or, more correctly, the process suppliers and guarantors. The electronics industry has grown up by adapting materials from other industries—woven-glass fire curtains became printed circuit reinforcements, and no one in his or her right mind would have developed polyimide as a dielectric for flexible circuits. What will happen over the next 20 years is that a few major materials companies will emerge. They will have as hallmarks the following characteristics:

- Delivery of process capability with guaranteed process parameters in the much more tightly controlled manufacturing environments that will exist at the time
- Sufficient size to afford 10 percent or greater R&D-to-sales ratios across the full range of fabrication materials from semiconductor to assembly

- An ability to customize and support products throughout the world in much the same way that paint makers support the automobile industry today

Thus the electronics industry of 2020 will evolve (at least as Prismark sees it) into a four-level structure:

Level 0. This is the branded supplier of global networks. These companies design and market systems but make nothing.

Level 1. This consists of the capital-intensive semiconductor fabricators, the system-on-a-chip foundries. The manufacture of semiconductors is a hugely capital-intensive business, and the barriers to entry are increasing. Perhaps only 15 or 20 semiconductor companies will remain. Their focus will be on efficient production of customized products. Already the systems houses such as Rockwell and Siemens are recognizing the unique capital needs of semiconductor manufacture as they spin off Conexant and Infineon. Is it inconceivable that Taiwan Semiconductor Manufacturing Company could acquire Infineon and IBM's semiconductor operations to become the largest system-on-a-chip supplier in the next 10 to 20 years? Far from inconceivable, it is probable.

Level 2. This consists of the contract assemblers who embrace semiconductor assembly as well as board assembly as these disciplines merge, and the heart of the system will still need assembly. This will be contract assembly. Today, contract semiconductor assemblers account for only 30 percent of the device assembly market; 70 percent is still captive. The same is true of contract board assembly. Fast-growing companies such as Solectron and Celestica still have huge growth potential as they take over the assembly functions that are still largely captive. What will emerge over the next 20 years are combo companies that do both device-level assembly and board-level assembly as these disciplines merge. These global operations with their massive component purchasing power and ability to start up and bring on standardized assembly operations with ease, anywhere in the world, are the electronics industry's tier 1 suppliers. There still will be a greater number of tier 2 suppliers for such items as boards, connectors, etc. (Fig. 2.6).

Level 3. This consists of the major materials companies who supply and guarantee process. If this scenario plays out over the next 20 years, we will look back on today's electronics industry as a hodgepodge of many businesses with ill-defined missions. It is clearly important to recognize which of the four groups you are in to decide if you can achieve critical mass through acquisition or if you should be acquired. The financial reengineering and restructuring of the electronics industry have only just begun.

In the past, Prismark has demonstrated a very clear linkage between people's ability to consume electronics and the production of electronic equipment. As GDP rises, so does people's ability to derive the productivity, safety, or entertainment benefits of electronics. The value of electronics to people will continue to grow over the next 20 years in much the same way that the convenience of the automobile increased between 1920 and 1940. More important, as with the automobile, many more people will be able to afford the benefits.

Electronic equipment increasingly will be assembled in the region of use—much like automobiles. The exception still will be small mass components that will be produced in areas of expertise and shipped to assembly sites much as car engines may be shipped from a manufacturing site in Japan to an assembly site in Thailand.

Based on the assumptions

- That electronics equipment is only just starting to deliver its benefits
- That electronic equipment will be assembled in regions of use with some component exceptions
- That population and GDP growth will follow established trends

it is possible to envision an electronics industry of $3 trillion in 2020 (three times the size of the industry today) that is globally distributed as shown in Fig. 2.7. The inventions have been made, the manufacturing skills are in place, and the restructuring has only just begun.

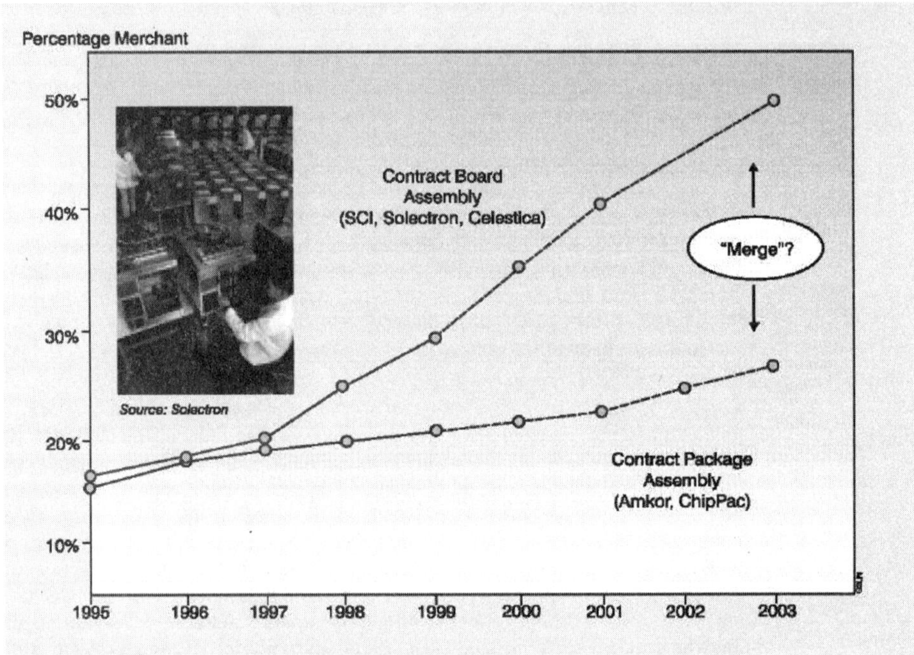

FIGURE 2.6 The trend to contract assembly (high volume: captive; lower volume: merchant). (*Prismark.*)

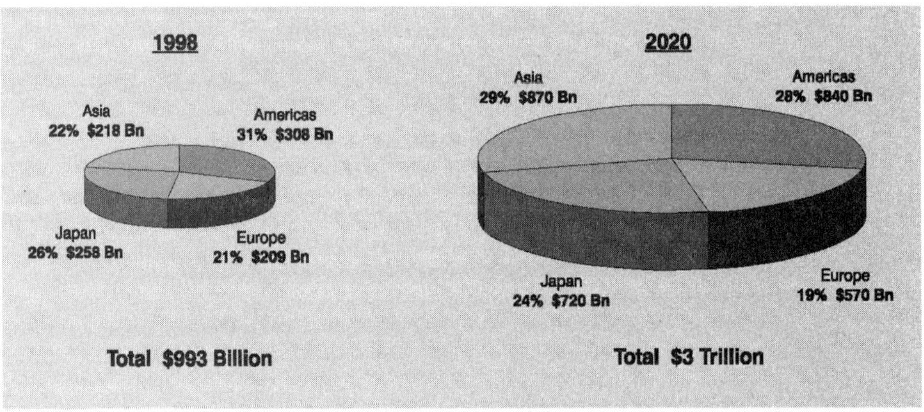

FIGURE 2.7 Total world electronic system production. (*Prismark.*)

CHAPTER 3
TRENDS/DRIVERS IN THE ELECTRONICS MANUFACTURING INDUSTRY

R. Lasky and Y. Morvan
Cookson Electronics

3.0 INTRODUCTION

As the electronics industry expands and matures, it is essential for those who wish to prosper in the future to understand the fundamental trends that are shaping this most vital industry. It is important to start the discussion of these trends by recognizing that products drive technology, not vice versa. Hence the general trends in products will drive technology and manufacturing needs. The National Electronics Manufacturing Initiative (NEMI) determined that the desire for smaller, lighter, faster, and cheaper products (hereafter referred to as "the fundamental four" or TFF) is the product driving force that will shape electronics manufacturing. Two of the hallmark products that presage these trends are cellular phones and laptop computers. It is little wonder, then, that NEMI chose an advanced personal data assistant (PDA) with a modem as a "product emulator" for its roadmapping and strategic planning exercises.

The electronics assembly capability to effectively support these changes is only now emerging. The current state of the assembly and printed wiring board (PWB) industry can be fairly compared with that of automobile assembly before Henry Ford. As in Ford's day, the continued growth and maturation of this industry will result in tremendous benefits for those who comprehend and take advantage of these opportunities. For those who stay with the old way of doing things, failure will be their lot.

The changes that we see occurring in response to the fundamental four product attributes are the following:

1. *Parallel and data-driven assembly processes.* Up to today, engineers often integrate an assembly line with disparate equipment, often using equipment cost as the main purchasing criterion. The advent of costing software such as SPACE has shown that equipment represents only 1 to 2 percent of the cost of the product produced. SPACE also has shown that increasing throughput with the right equipment can have a dramatic impact on profits. A 4-second reduction in cycle time often can increase profits by more than 25 percent. As this throughput issue is addressed, it will come to be recognized that assembly equipment that is more parallel, data-driven, and flexible is a fundamental part of the solution. Equipment cost will be a secondary consideration; cycle time will be the first concern.

2. *Software and line integration.*
   a. *Software replaces steel/modular, small size.* In the late 1970s, the fighter plane became a computer peripheral. A technology analyst noted this point because at that time, more than 50 percent of the cost of the plane was in computers and software. Assembly equipment is experiencing a similar transformation. The content of assembly machines will shift to software and vision systems replacing massive gantries or turrets and powerful motors.
   b. *Complete software line integration arrives.* Soon, individual assembly equipment must be capable of complete software line integration. This compatibility will be required for the total data-driven processes described earlier. Without this capability, the line will not be capable of the fastest throughputs and changeovers and hence will not be competitive, no matter how low the equipment cost.
3. *Slow secondary processes become niche (TAB, DCA?).* Direct chip attach (DCA) is not a new process; it was invented by IBM in the 1960s and has been used ever since. In 1995, about 450 million DCA dies were assembled, representing only 1 percent of all packages. Currently, DCA may never grow as strongly as many forecast, the reason being that slow, secondary processes are required by the assembler to essentially manufacture the first-level package for the DCA on the PWB. Without material and process breakthroughs to improve the speed of the DCA process, it will never become common. These same issues will drive the package types that will prevail among QFP, BGA, CSP, DCA, and TAB.
4. *Through-hole assembly maintains strong presence through 2005.* Through-hole assembly still accounts for nearly 35 percent of the volume in electronics. The main reason for this staying power is that so many applications do not require the high performance of surface-mount technology (SMT), and through-hole assembly can provide a low-cost solution in these cases. It is also favored in emerging economies because it is an easier technology to implement. Through-hole assembly will continue to generate considerable revenue well beyond 2000, but we must be aware that some of this business will change. An example of such change is through-hole components being assembled in single-center reflow operations. These changes should be viewed as an opportunity for new products and applications.
5. *Nonstandard assembly (optical, odd form) arrives.* Nonstandard and "odd form" assembly is an emerging field. It has been highlighted by NEMI as one of the critical enablers for future high-volume products such as PDAs. Nonstandard assembly in all manufacturing processes represents at least a $100 billion dollar per year opportunity for future nonstandard assembly machines. This opportunity will not continue to be neglected as vision, robotics, and machine intelligence become lower in cost and proficient in handling assembly tasks currently only performed by hand. Optical assembly, although smaller in dollar value at about $10 billion, may be addressed first because it may require less skilled robots. There also exist natural centering mechanisms, discovered in the DCA process, that may facilitate this new field.
6. *Throughput, yield, and cost guarantees. Total solutions.* As pitch becomes finer and input-output (I/O) counts increase, the difficulty in maintaining high yield increases exponentially. This demand and the requirement for maintained and increased throughput as pitch narrows will increase the value of process knowledge. Knowledge will be sold individually or tied into materials and equipment. Manufacturers eventually will want process guarantees from their suppliers.
7. *Knowledge.* It will become more and more clear that knowledge in electronics has a value as processes and productivity become more dependent on knowledge.
8. *PCB/flex line widths from 6 to 3 mils as standard.* The ultrafine-pitch (UFP) technologies that will emerge with chip-scale package (CSP), DCA, etc., will require substrates with line widths and spaces to 3 and even 2 mils. The standard line widths of 5 to 6 mils today will approach and even surpass 3 mils by 2000.
9. *Lead-free issues: Pb/Sn still dominates.* Environmental issues are important, but the difficulties in finding good substitutes for Pb/Sn and the increased realization that electronics only contributes a small percentage of environmental lead will ease efforts for lead-free solders. Hence electroconductive adhesives (ECAs) and lead-free solders will only enjoy niche (albeit important) markets. Were lead to be banned as waste, recycling would challenge lead-free processes as a solution.

10. *Passives continue to be needed, imbedded emerge.* Passive components will not be "integrated away." They will continue to grow at more than 20 percent a year. It is most cost-effective for LSI integrated chip (IC) manufacturers to design a few basic microprocessors or application specific integrated circuits (ASICs) and "tune" the circuit with passives. I/O count, clock frequency, lower-power circuits, and analog applications all drive the need for more passives. The trends in all these "drivers" are to increase the number of passives required. Currently, some applications, such as PCMCIA cards, do not have enough room for all the passives. Buried passives as a solution are being investigated by a DARPA-sponsored consortium. From a PWB, flex, and ceramic materials perspective, this topic offers tremendous challenges and opportunities.

We have reviewed 12 trends that are driving the future of our industry. We believe that these trends will make the future the most interesting, exciting, profitable, and potentially risky period in the history of civilization. We have purposely avoided referring to an eventual maturing of electronics because we believe that it will reinvent and redefine itself throughout the future.

3.1 PARALLEL AND DATA-DRIVEN ASSEMBLY PROCESSES

As the electronics industry matures, the fraction of the value of semiconductors in electronic equipment continues to grow. In the early 1990s, approximately 10 percent of the value was in the semiconductors, whereas today the figure has risen to over 20 percent. In portable devices such as PDAs and camcorders, the semiconductor value has skyrocketed to 50 percent. There has never been an industry so successful as the semiconductor industry in adding value to its products. In 1970, 20 MB of RAM cost $20 million, about as much as a commercial jet. Today, 20 MB of RAM costs less than a coach ticket from Boston to Atlanta.

It is difficult, then, to look to the semiconductor industry to lower the cost of the final product. This situation has increased the pressure on the packaging and assembly industry to control cost. For the first time, the cost per I/O in many packages has dropped below $0.01 per lead. The assembler also shares this cost pressure. Hence the most critical parameter for low-cost assembly is throughput. With quality as a given, throughput is what makes money.[1] It is becoming understood in the industry that the two critical parameters to ensure optimal throughput are that processes be parallel and data-driven.

3.1.1 Importance of Throughput

Although the intent of this section is to discuss parallel and data-driven processes, profit is so strongly tied to throughput that it should be discussed first as background for a thorough understanding of these two critical parameters. To understand the effect of throughput on profit, we will analyze the assembly of a typical unit—a low-price CD-ROM player. To start, consider the important elements of cost[2] in Table 3.1.

The cost parameters can be calculated with a recently developed cost-estimating spreadsheet. Table 3.2 lists some of the cost output from this scenario.

When costs are analyzed at the level of the assembly line (i.e., costs only calculated with labor for the assembly-line workers), the nonmaterial assembly cost per I/O, often called the *conversion cost,*[3] is considerably less than the industry standard at the factory level. This situation exists because there are considerable numbers of workers who support the product but do not work on the assembly line. These include designers, administration, purchasing, hand assembly, shipping, etc. In the simulation in Table 3.1, adding the 26 "overhead" or indirect workers provides an industry competitive conversion cost of $0.0144 per I/O. Strictly speaking, many of these 26 people are in manufacturing and would not be considered overhead in most analyses, but they do not work on the assembly line for this analysis.

To see the effect of throughput on profit, assume that the selling price is $110, and calculate the gross profit as a function of cycle time. For the baseline case (25 seconds = 0.416 minutes), the gross profit per year is approximately $4.5 million. The cost spreadsheet is able to plot the gross profit ver-

TABLE 3.1 Assembly Cost Elements

Item	Cost
Assembly equipment	$1.7 M
Floor space + utilities	$40,000/yr
Components	$75/unit
Stencil, squeegee	$600, $200, life = $1,000,000
Solder paste, consumables	$0.7/unit
PWB	$15/unit
7 workers	$28/ h (includes benefits)
Selling price	$110
Shifts	12 10-h shifts per week
Cycle time	25 seconds
Downtime, setup, maintenance	8%, 10 h/wk–12.5 h/wk
Workers supported	26 @ $42/h each
I/O	950
Yield	97% first pass, reworkable

TABLE 3.2 Cost Analysis

Element of cost	Cost per board ($)	Percent of total cost per board
Labor (direct and indirect)	12.44	12.0
Machine depreciation/repair	0.75	0.72
Floor space and utilities	0.06	0.06
PWB and components	90	87.0
Rework	0.35	0.30
Consumables (paste, etc.)	0.18	0.17
TOTAL	**103.78**	**100**

sus cycle time, illustrating the stunning effect of cycle time on gross profit (Fig. 3.1). Reducing the cycle time from 25 to 20 seconds (0.33 minute) increases the gross profit to almost $8 million per year! The reasons for the increase are twofold: First, more product is produced, and second, unit costs are reduced. These two factors result in an exponential increase in gross profit as cycle time is reduced.

This analysis also suggests the path to financial success in subcontract assembly. We have been in many discussions during which someone has said, "Aren't you glad your business is not subcontract assembly? The gross margins are so small." It is true that the gross margins are only 3 to 10 percent, but this point misses the essence of financial success in this business. Although the margins are low, the throughput can be so high as to produce an almost unequaled return on assets. Returning to the example, assets were about $2 million (assuming perhaps some asset valuation for the building, e.g., $300,000 per $4.5 million) If net profit is 50 percent of gross, net profit is then $2.25 million. Hence return on assets (ROA) is 112.5 percent! This ROA is clearly world class. From the previous discussion it is now clear that from an ROA perspective, ROA grows strongly with throughput.

3.1.2 Parallel and Data-Driven Processes

Parallel processes are processes that accomplish several process steps at the same time. Examples might be stencil printing (many pads printed simultaneously), "walking beam" pick-and-place machines (placing several components at the same time), and arguably, reflow ovens (reflowing all components at once). The throughput benefit of parallel processes is obvious and universal. *Universal* means that a wide variety of process lines benefit from parallel processes. Hence, for a dedicated line, the process steps being parallel usually will enhance productivity in the same product day in and day out. Even if your line produces a wide variety of products with frequent changeovers, parallel processes are still attractive.

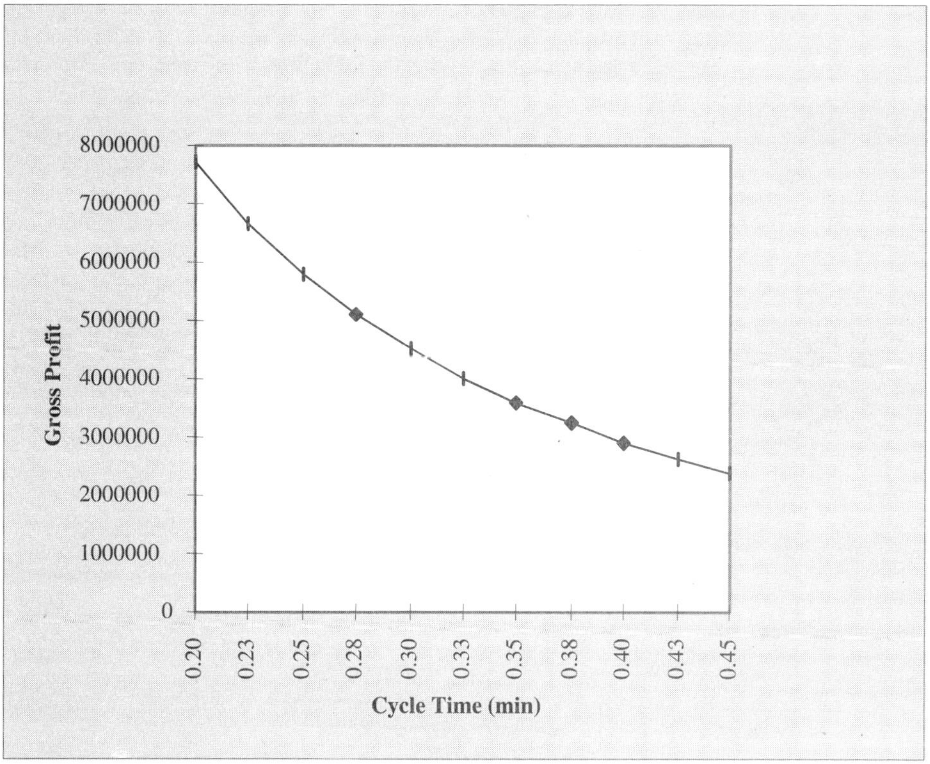

FIGURE 3.1 Gross profit versus cycle time.

The opposite of parallel processes is *serial processes*. An example of a serial process is a single-beam, gantry pick-and-place machine. From a throughput perspective, the disadvantage of serial processes is self-evident: It is difficult for something to be fast when it is done one step at a time.

Data-driven processes only require the change of a software program to assemble a different product. Examples of data-driven processes are changing the software program on an adhesive dispenser or pick-and-place machine. The advantage to throughput is less time lost in product changeovers.

Hard-tooled processes are the opposite of data-driven processes. An example of a hard-tooled process is stencil printing. The benefit of data-driven processes depends strongly on the number of product changeovers required. If there are few product changeovers, data-driven processes are not as critical to throughput. If frequent changeovers are the rule, data-driven processes can greatly improve throughput.

3.1.3 Developments in Parallel and Data-Driven Processes

Recently, there have been many efforts to develop and improve assembly processes so that they are both parallel and data-driven (P&DD). Unfortunately, few process steps have both these characteristics. Simultaneously, much R&D currently is being pursued in this direction. The following is a review of a few current developments that do fall into the P&DD realm in solder-paste printing, dispensing, pick-and-place, and reflow ovens. The most dramatic changes in parallel assembly are anticipated in pick-and-place machines

Stencil printing of solder paste is usually parallel and hard-tooled (one stencil per product design), but some current machines[4] will handle up to six separate stencils. This feature allows product changeovers to be more rapid, and thus this type of stencil printer is parallel and "near" data-driven. Metal jetting technology (MJT) is relatively new for materials deposition. MJT deposits drops of molten metal using one of two techniques: (1) drop on demand or (2) continuous stream.[5,6] It is serial (one drop at a time) but continuous stream jetting that can deposit almost 1 g/s (10,000 drops per second with 8-mil-diameter drops). In continuous-mode printing, MJT is a fast serial (near-parallel) and data-driven process. Drop-on-demand technology is slower but simpler and more mature. It is anticipated that the first manufacturing-capable MJT will be shipped this year in the drop-on-demand format, with continuous mode to follow.

Dispensing of adhesives typically is a serial data-driven process because it is usually performed with one dispensing head. However, product changes can be accomplished with the change of a software file. Dual-head dispensers are currently on the market,[7] making this process now near parallel and data-driven. The dual heads allow the dispensing of multiple adhesive dot sizes, enhancing overall throughput. This trend is expected to continue with multiple-bead dispensers in the not too distant future.

The most dramatic changes in parallel assembly are anticipated in pick-and-place machines and are data-driven because product changeovers only require changing a computer program. However, more effort will be spent on producing pick-and-place machines that are parallel. Currently, there are a few such offerings, most notably the "walking beam" and multiple-head chip shooters.[8] In the drive to produce parallel pick-and-place machines, it will become obvious that machines with 100-kg overhead gantries to place parts that weigh less than a grain of rice do not constitute optimized machine design.

Reflow ovens are clearly a parallel process, in that they melt all the solder on the board at the same time. There are also significant efforts to make reflow ovens more data-driven. The biggest change in reflow ovens, however, will be the combination of convective heat and microwave. The heat will be used to melt the solder, and the microwave will be used to cure epoxies in flip chip or other applications. This combination will allow the reflow and cure processes to be performed simultaneously. This approach would accelerate the acceptance of flip chip dramatically and would make this process closer to a standard SMT process.

The industry is only now beginning to realize that both parallel and data-driven processes are needed to achieve the productivity goals of the future. Eventually, we should expect to see process lines with all processes being parallel or data-driven.

3.2 SOFTWARE AND LINE INTEGRATION

In the 1970s, the jet fighter became a computer peripheral. A technology historian made this interesting observation. His reasoning was that at that point in time, greater than 50 percent of the value of the fighter was in computers and software. This trend certainly has continued in the aircraft industry. Many planes today cannot be flown without computers.

We expect that this trend will be more and more evident in electronics assembly equipment also. In the near future, tremendous increases in productivity and efficiency will occur by combining the increased use of computers, vision systems, and software on the individual assembly machines and the integration and control of the entire assembly line with "system" computers.

3.2.1 Software and Machine Vision Replace Steel

Software and machine vision improvements will affect most of the equipment used in PWB assembly. However, pick-and-place equipment likely will be affected the most, so we will use this machine as the basis on which to elaborate.

The pick-and-place process, as currently practiced in a typical fine-pitch placement machine, is not optimized from either a machine or process perspective. To understand these perspectives, we will look at this process in some detail.

Step 1. As the pick-and-place process begins, the machine first loads the PWB.

Step 2. The machine then finds several global fiducials on the PWB so that it can have the necessary references to properly place the components at the correct location. The machine knows the locations for the components because the PWB layout had been entered previously into the controlling computer's memory.

Step 3. The machine then picks the component from a feeder *and aligns it with respect to the placement head* in preparation for placement.

Step 4. As the machine places the components, it measures the location to place them by referencing from the location of the fiducials. This step requires a precise (and expensive) scale. *In the time domain, this step can be performed in parallel to step 5, but it is delineated separately because of its importance in relating to the complexity of machine design.*

Step 5. The component is then moved into position. The PWB has dimensional stability to about 0.001 to 1.0 in, so for finer-pitched components [say, 0.4-mm quad pack flats (QFPs)], an additional fiducial near the placement location (called a *local fiducial*) may be required to achieve proper placement. The machine has to reference to this fiducial before placing the QFP, thus taking additional time.

Step 6. The machine places the component.

These steps are shown pictorially in Fig. 3.2.

This approach has several fundamental disadvantages. Referencing to the PWB global fiducials is the most obvious. The lack of PWB dimensional stability is a critical shortcoming, making such referencing nonoptimal. This lack of dimensional stability can be such that it is not possible to place a component with ultrafine-pitch leads accurately enough. In such cases, a fiducial near the component placement area is required. The placement machine must then find this local fiducial and take the time to perform an additional positional referencing operation. Hence, as ultrafine-pitch packages become more and more common, more and more local fiducials will be required to place them, slowing the placement process down more and more. The advent of area array packages, with their coarser lead spacings (such as BGA and CSP), will only temporarily relieve this concern because in time, even these packages will evolve to finer pitches. The additional time to reference to these local fiducials will be unacceptable in an era of increased productivity demands.

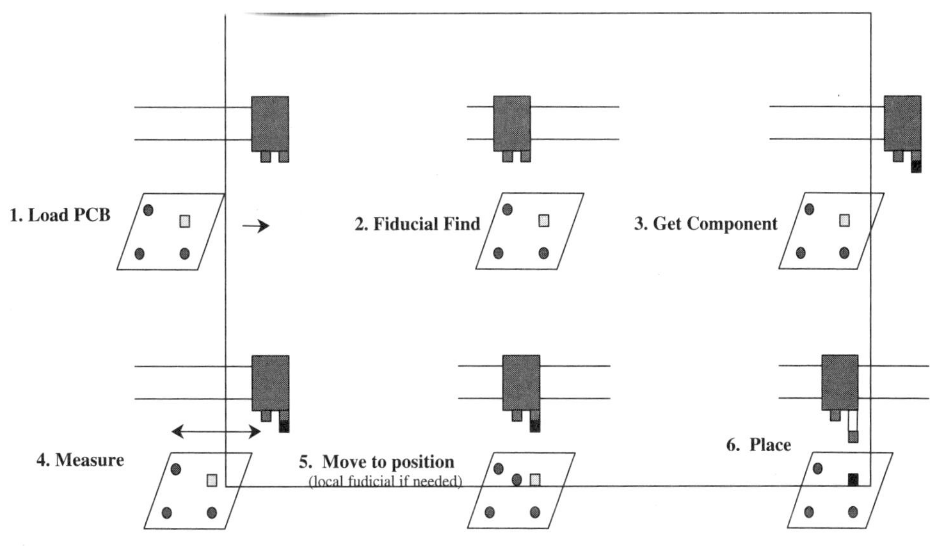

FIGURE 3.2 The steps in pick and place.

In the placement process, an expensive scale is used to measure placement positions relative to global fiducials. To obtain accurate measurements, the entire machine frame must be very stiff. This requirement may result in an overhead gantry weighing upward of 100 kg (over 200 lb) in this type of pick-and-place machine. This massive overhead gantry then requires powerful (20-hp or so) and expensive motors to drive it. These designs can result in machines costing over $300,000 to place 5000 or so components per hour. It is interesting to contemplate that when placing a 0402 passive capacitor, the gantry weighs 100 million times more than the passive (100 kg versus 1 mg). In no other technology are such ratios tolerated to deliver such a small payload.[9]

The National Center for Manufacturing Sciences (NCMS) chaired an effort called Light, Flexible, Mechanical Assembly (LFMA) in the early 1990s. The purpose of this work was to develop assembly equipment that could perform the complex assembly tasks currently only performed by humans (screwing a lid on a jar and assembling the covers on an electronic product such as a personal computer are examples). This effort spawned a small company, CAMotion,[10] that is working on this complex problem. Optimized pick-and-place machines are a simpler subset of this challenge, and this topic also is being addressed by CAMotion. This new machine-optimized approach will now be discussed. A schematic is shown in Fig. 3.3.

As with current machines, step 1 involves loading the PWB into the machine. In step 2, the machine picks the component from a feeder *and aligns it with respect to the placement head* in preparation for placement. In step 3, the machine places the component without any fiducial referencing. The machine knows approximately where the placement site is from mechanical information entered into the machine with the computer-aided design (CAD) data. It aligns precisely to the placement site by imaging the pads, *referencing to them,* and then placing the part.

This approach not only reduces the steps to three but also eliminates the need for precise scales and a stiff, exacting frame. The resulting machine would have the equivalent of a 10-lb gantry and a 1-hp motor.

Several technology enhancements are needed before the look-while-place approach is realized. The first is the availability of low-cost, light, and compact vision systems. The vision system will need to be in the 1 × 3 × 2 in range and weigh less than 1 lb. The second need is the development of machine learning algorithms that can "learn" what the typical placement sites look like, say, a QFP or BGA site. The last technology need is design improvements in the control of flexible structures. Reducing the mass and stiffness of the placement machine will enable a tremendous cost and speed performance advantage. These advantages have a price, however. The lighter, less stiff frame likely will be subject to greater vibration problems during placement. These vibrations can cause a lack of placement accuracy. The science of the control of flexible structures is maturing to a point where machines can be designed so that such vibrations are readily not excited. In addition, learning algorithms can be designed that not only recognize the placement position but also learn how to move quickly to the site while exciting minimal vibrations. We believe that these types of technologies will have a profound and perhaps disruptive effect on the assembly industry in the not too distant future.

These three technology challenges are currently being pursued, and it appears that the current approach, as described in the next subsection, has promise. However, even if this approach proves unsuccessful, another will be developed to perform look-while-place action—the prize is too valuable to ignore.

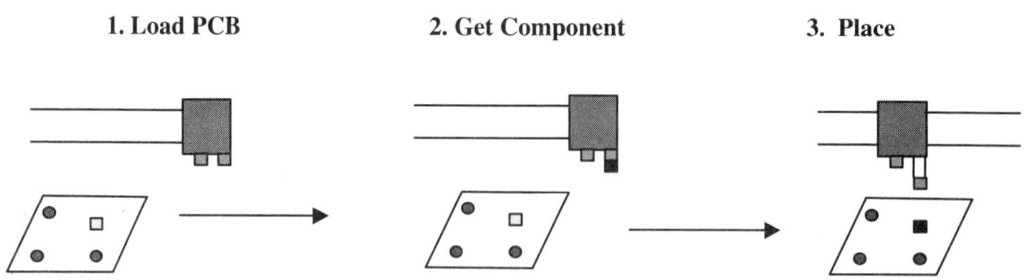

FIGURE 3.3 Look-while-placing approach.

3.2.2 Line Integration

Integration of the assembly line with software and computers is becoming common and will become ubiquitous in the near future. Productivity will be the main "driver" of this trend. This topic is broad and goes well beyond our scope here, but we will mention some of the current line-integration products that operate under the most common architecture, GEM-SECS (Generic Equipment Module–Semiconductor Equipment Communications System). This architecture was developed for semiconductor fabrication equipment and was adopted for assembly equipment. GEM-compliant architecture allows one to configure different assembly machines with different operating systems into one computer database system. Hence your dispenser, printer, and oven may run Windows 95 and your placement machine may run DOS, but they can still communicate if they are GEM-compliant.

Recent products such as GEMHost and GEMComm[11] are examples that work within this mantle. They improve integration and allow better scheduling, line balancing, statistical process control (SPC) data collection, and cost control. At this time, most of these products only allow line communication for data collection but do not enable complete line control from one computer. In the future, such total line control and even control at the factory level will become common.

3.3 SLOW SECONDARY PROCESSES BECOME NICHE

As we discussed the importance of parallel and data-driven processes in PWB assembly earlier, we emphasized the importance of productivity to profit. Thus it should not be a surprise that any process that slows down the cycle time of the assembly line erodes profits exponentially. Tape automated bonding (TAB) was a classic example of this phenomenon. Through the late 1980s and early 1990s, TAB was the "package of the future." TAB's low profile and very fine lead spacings (down to 6-mil centers) seemed to give it numerous advantages. The manufacturing of TAB packaging was performed with many pieces of equipment that were "off the shelf" because they also were used for the manufacture of 35-mm photographic film. Hence TAB packaging would appear to have many advantages; It is inexpensive to manufacture and has the finest lead spaces and lowest profile. Thus TAB was poised to take the world by storm, but it didn't. Understanding why will be very instructive in predicting which future component technologies will prevail.

An understanding of TAB's downfall comes through an understanding of the assembly process. Let us assume that a TAB package is a candidate to replace a BGA package in the example that we analyzed previously. Please refer to Table 3.1 again for a list of pertinent information.

Let us assume that the TAB substitution saves $2 on the bill for materials. What a terrific start! We again use the cost-estimating software SPACE[12] to perform the cost analysis. At first glance, it appears that we have reduced the cost by almost 2 percent. The bad news is that TAB packaging requires two new TAB machines that cost $350,000 each, and the cycle time must be slowed down to 30 seconds. These changes result in the cost and profit figures shown in Table 3.3. The results from the previous analysis (BGA) are listed for comparison.

By using the TAB process, the unit cost actually goes up by more than 1 percent, but the most stunning change is that the gross profit goes down by more than 30 percent from $4.14 million to $2.66 million. The reason is that the number of boards produced dropped from 661,914 to 551,595. Again, the profit is hit by the "double whammy" of reducing productivity—fewer units at a slightly higher unit cost. SPACE indicates that if we add yet another TAB machine and it can bring the cycle time back to 25 seconds, the profits are then restored with a slight increase. Unfortunately, the addition of three machines and the additional space and workers needed may present other challenges unrelated to cost.

This analysis indicates a strong fundamental principle. *Lasky's rule: Any process change that slows the process down usually will result in lower profits. Corollary: Slow secondary processes almost always should be avoided.*

Hence flip chip or DCA also should be a productivity concern because it has several added process steps. Typically, flux must be dispensed onto the substrate at the chip site. The die then must

TABLE 3.3 Comparison of TAB and BGA costs.

Element of cost	BGA build ($)	TAB build ($)
Labor (direct and indirect)	12.44	15.39
Machine depreciation/repair	0.75	1.23
Floor space, electric, etc.	0.06	0.078
Materials (components, PWB)	90	88
Rework	0.35	0.32
Consumables (solder paste, etc.)	0.18	0.18
Total cost per board	103.75	105.18
Number of boards produced	661,914	551,595
Gross profit	$4.14 million	$2.66 million
Delta gross profit		−$1.48 million

be placed with a pick-and-place machine that has a special head. After reflow, the die then must have underfill dispensed under it, and the underfill must be cured in a curing oven. Hence additional equipment in the form of a modified pick-and-place machine, a dispenser, and a cure oven is required. Evidence of this process slowdown exists in that it has been stated recently that the DCA process requires 38.5 percent more time in cellphone assembly. Thus it is likely that DCA will only become a mainstream technology when these processes become more like the standard SMT process. Work in this regard is currently underway.[13]

3.4 THROUGHPUT, YIELD, AND COST GUARANTEES: TOTAL SOLUTIONS

We have discussed the need for continuously improving throughput previously, so we will not emphasize it again. An additional challenge to throughput, however, will be the trend that good yields will be harder and harder to achieve. The main difficulty in maintaining good yields is that yield loss grows exponentially as pitch is reduced. To understand this situation, let us assume that yield loss per lead varies as in Eq. (3.1).

$$YL = a \exp\left(-\frac{b}{p}\right) \tag{3.1}$$

where YL is yield loss per lead, a is a constant, p is the pitch in mils, and b is also a constant. This type of relationship might be obtained from process data, where the process capability is constant and the only variable is component pitch. The constants a and b would be determined from the process data. Let's say that a given pager assembly process capability results in 1 ppm yield loss per lead in a process when the lead spacings are 65 mils and 20 ppm when the spacings are 20 mils. Usually, Eq. (3.1) and a little algebra reveal that $a = 0.265$ ppm and $b = 86.5$ mils. Applying these constants to Eq. (3.1) results in a process capability as shown in Fig. 3.4. To be fair, this analysis is highly idealized. The lead spacings in any product are not of only one value, and process capability is not as easy to model as we have shown. However, the general trends that are shown here (i.e., that yield loss per lead increases strongly as pitch decreases) are inarguable.

Compounding the preceding concern, as pitch decreases, the number of leads per area in a given package increases. It can be shown that this relationship will increase roughly as

$$P = \frac{c}{p^2} \tag{3.2}$$

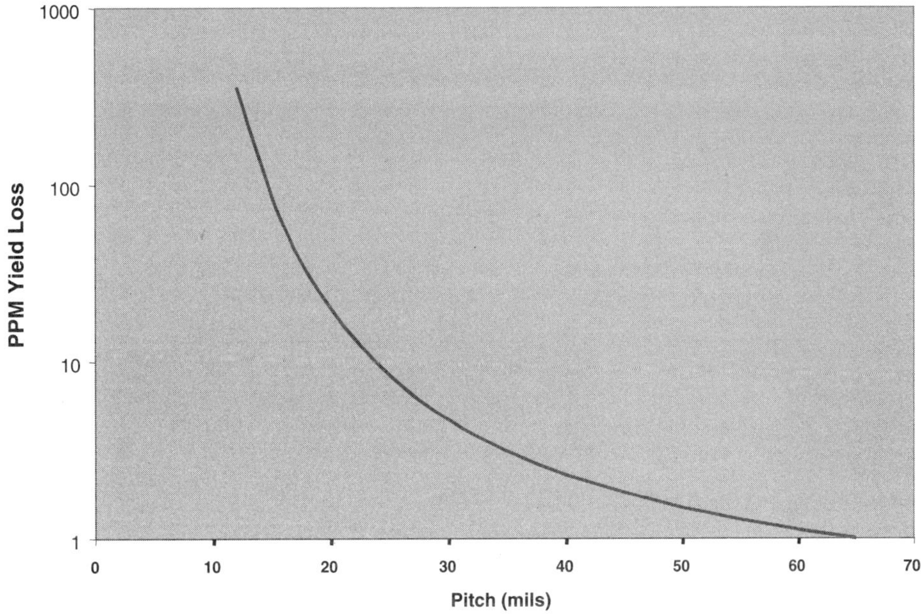

FIGURE 3.4 Yield loss (ppm) versus load spacing.

where c is a constant, and P is the number of PWB pads (i.e., leads) in the package. Thus, for an old pager technology where p might equal 25 mils and P is 150, c then equals 93,750 (i.e., 150×25^2). If the technology is reduced to 16 mils, Eq. (3.2) suggests that P then equals 366, assuming the same size package. Again, this calculation is a "ballpark figure" used to show a trend, since the packages have a mix of pitches.

Typically, a manufacturer will be concerned with end-of-the-line yield loss, i.e., how much product does the line lose during the entire process. End-of-the-line yield loss is the loss per lead times the number of leads:

$$YLEOL = a \exp\left(\frac{-b}{p}\right) \times \frac{c}{p^2} \tag{3.3}$$

In our fictitious pager example we have determined all the constants and can now plot Eq. (3.3). This plot is shown in Fig. 3.5. This figure indicates that in our pager example the end-of-line yield loss might be only 0.1 percent when the lead spacings are 25 mils, but they skyrocket to 2 percent as the lead spacings are reduced to 16 mils. Figure 3.5 suggests that if the lead spacings are reduced much below 12 mils, the yield loss will be catastrophic.

Although this analysis is fictitious, the thrust of the argument is not. As lead spacings are reduced, yield loss per lead increases, and there are more opportunities (i.e., more leads) to be at risk. Hence the combined result is that as pitch decreases, end-of-line yield loss increases exponentially if process capability is not improved.

The preceding analysis suggests strongly that process improvements will be necessary just to hold yield loss stable as pitch decreases. Area array technologies such as BGA may relax finer pitch requirements temporarily in some applications. However, with the continuing trend to produce smaller, lighter, faster, and cheaper products, even BGA eventually will have finer and finer pitch requirements. As one improves one's process capability, one must not forget Lasky's rule. Do *not* slow the process down.

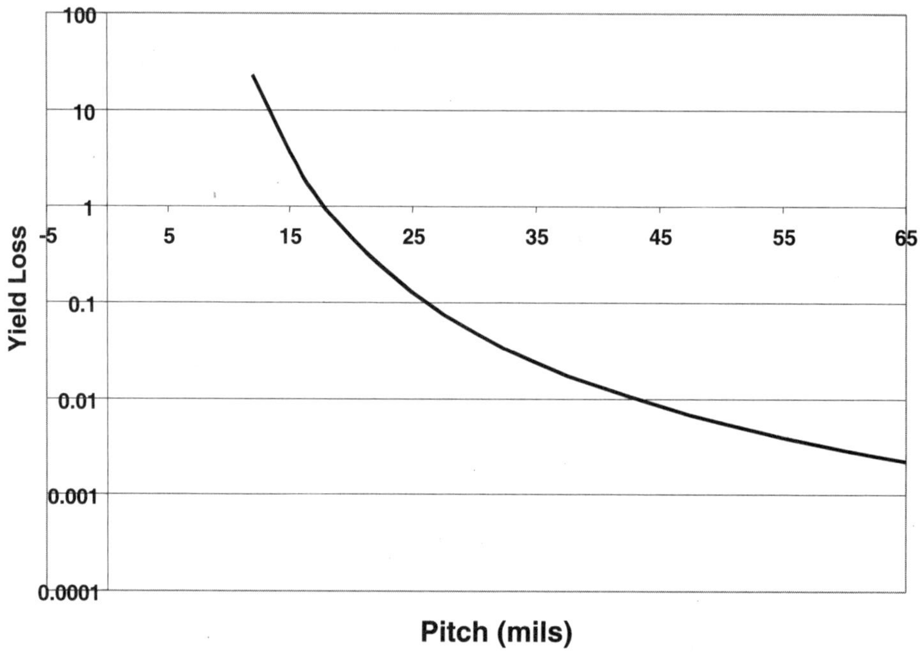

FIGURE 3.5 Product yield loss.

The twin and somewhat mutually exclusive challenges of greater productivity and improved yields can only be met by a thorough understanding of all individual (i.e., printing, placement, etc.) and systems (i.e., the effect of solder-paste parameters on the entire process, etc.) aspects that affect productivity and yield. It will be more and more difficult for assemblers to meet the twin challenges of improved productivity and yield unless they work very closely with their material and assembly equipment suppliers. This need will encourage the offering of integrated solutions (i.e., tailor-made combinations of materials and process equipment and process knowledge to fit the assembler's needs). Such offerings are now emerging.[14] Ultimately, it may be that customers demand process guarantees. Such guarantees undoubtedly will require an unprecedented closeness between assemblers and their vendors.

In conclusion, both productivity and high yields will continue to be required to compete in the electronics assembly marketplace of the future. Companies that survive and prosper will master assembly system techniques that enable greater productivity and higher yields. It will not be acceptable to trade one or the other off. Such needs will drive the offering of integrated solutions in the future and will elevate the value of process knowledge in its own right.

3.5 THROUGH-HOLE ASSEMBLY REMAINS AND ODD-FORM ASSEMBLY EMERGES

3.5.1 Through-Hole Assembly

It is interesting to reflect on the changes that have occurred in electronic assembly in the past generation. It started with armies of people on an assembly line placing single resistors, capacitors, inductors, and transistors. Today's machines can place greater than 75,000 components per hour with accuracy approaching 50 μm (2 mils). Early efforts to automate assembly were pioneered by such companies as Universal Instruments and IBM. The resulting through-hole assembly equipment concepts developed in these early years are still used today, accompanied by many innovations. The rapid acceptance of surface-mount technology (SMT) can give the false impression that through-hole assembly is "going the way of the buffalo." Figure 3.6 shows the number of though-hole and SMT components projected to be assembled. It should be noted that as recently as the early 1990s, more through-hole than SMT components were assembled. Even in the year 2000, about 25 percent of all components were through-hole.

The reasons that through-hole technology is only fading slowly are straightforward: cost and simplicity. Cost is the most obvious strong point for through-hole technology. We can all get excited about SMT and flip chip and BUM (built-up multilayer) PWBs, and they are clearly needed for laptop personal computers (PCs) and "handycams," but we must remember that a large portion of electronics is simple. An example might be a garage door opener. This small portable unit is inexpensive and relatively low tech. It can be manufactured with through-hole technology in a low-labor-rate region of the world such as China or Mexico. There is little reason to change to SMT for products such as this, especially since it would be more expensive.

Electronics has been the industry of choice for third-world countries with strategic economic growth plans. The reason is simple: Electronics historically has grown four times as fast as economies in general [i.e., 8 percent composite annual growth rate (CAGR) versus 2 percent for economies as a whole]. When a country does choose to enter electronics, the simplicity of through-hole technology makes it a natural choice. We know of examples where a small through-hole assembly line was set up in a house in a third-world country. One only needs axial and radial inserters and a wave solder machine and one is ready for business. In one such "facility," the kids played beneath the solder pot! In contrast, SMT typically requires some experienced technicians and operators and is a considerably more demanding process.

3.5.2 Through-Hole Equipment

The keen-minded process engineer who embraces today's modern arsenal of manufacturing equipment is usually geared for the assembly of advanced technology products. To many professionals of this audience, the following may not at first resemble legitimate reasons to use through-hole technology. However, we all must remember that at least in part, ingredients for productivity

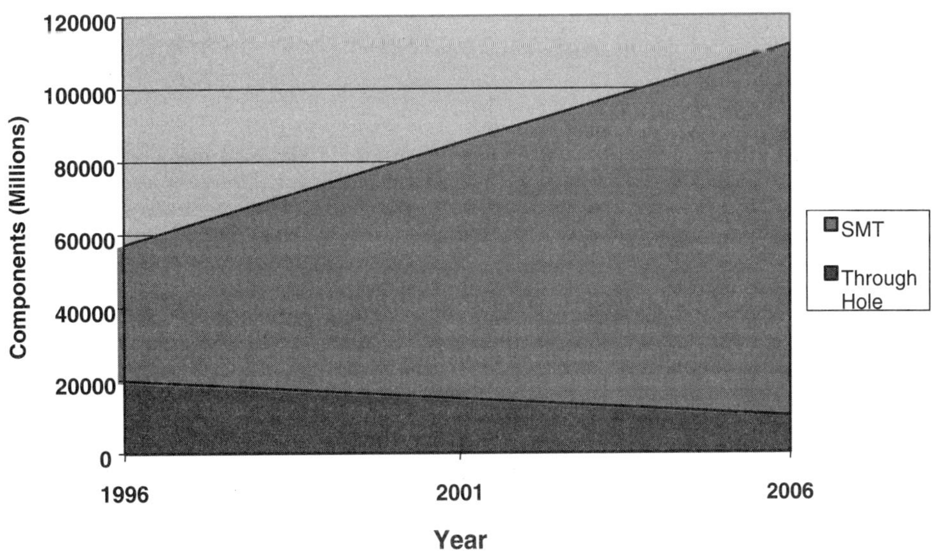

FIGURE 3.6 Through-hole and SMT component consumption.

include the manufacturing environment, the quantity of units needed to be produced, and end-product requirements.

As compared with SMT, through-hole component technology assembly is more forgiving to harsh manufacturing environments, more adaptive to batch-style manufacturing, and more conducive to manual assembly; the PWBs and components generally are cheaper; and the technology is less process-intensive. On the other hand, one reason this technology faces demise is the lack of component availability as SMT takes an ever-increasing hold. This is ironic because a decade ago, SMT was slow to get started for the same reason (lack of available components in its own domain). Yet SMT assemblies still must yield to the older through-hole technology. Most PWBs today employ both technologies. Many devices such as connectors, transformers, inductors, large electrolytic capacitors, and high-power components are through-hole in nature.

This is not intended to imply that through-hole technology assembly techniques and equipment have sat still. The complexion of automated through-hole assembly has changed dramatically. For instance, at one time it was necessary to presequence components on tape prior to the insertion operation. More modern equipment allows component sequencing, electronic verification, and insertion all on the same machine. The equipment also has become more flexible and can accommodate a much wider breadth of component specifications such as lead spacing, lead thickness, and body configuration. In most cases, the newer equipment is able to tolerate component leads that are more deformed. Historically, these machines also were stand-alone in nature and required an operator to manually mount and remove PWBs from workboard holders as a function of the process. This did not fit the trend of more modern approaches in that only batch manufacturing was possible. Even though some assembly houses still use these techniques, most modern assembly equipment facilitates "in line" manufacturing with pass-through board-handling systems. This approach reduces operator dependency and allows the PWB assembly to pass between process cells with conveyor systems instead of humans. In most instances, the dedicated workboard holder has been eliminated.

A major supplier in Japan offers a high-speed VCD axial-lead component-insertion machine with an insertion speed of 18,000 pieces per hour. This is twice that of the company's previous model. In addition, with two separate component feeder carriages, components can be exchanged and replenished during operation. A major supplier of equipment in the United States offers a radial insertion line rated at over 11,000 components per hour. This equipment handles an even wider variety of components than its predecessor, which before were inserted at a rate of about 4000 parts per hour.

3.5.3 Nonstandard Automated Assembly

The impressive state of assembly equipment and robotics and the recent chess victory of Deep Blue can give the wrong impression of the abilities of robotics and artificial intelligence (AI). Assembly equipment with its associated robotics, vision systems, and computers can accurately place thousand of components per hour on PWBs and even verify passive values. However, the most complex robot with AI and vision cannot perform the simplest task for a 3-month-old baby—to recognize her mother's face from across the room. Many of us were happy for IBM when Deep Blue became world chess champion, but we will be really impressed when computers can understand poetry and "read" nonverbal communications between people. This introduction sets the stage for the challenge of performing odd-form and final assembly with machines. Assembly robots are outstanding at rapidly assembling standard components that are simply precisely placed on a PWB. However, the assembly of multiple odd-shaped components is a great challenge due to difficulties in grasping and correctly orienting the component. Even a greater challenge for machines is final assembly and placing the finished product in a box. It is difficult for machines to master the spatial recognition and mechanical movement that a person easily performs in the final assembly of a VCR or other typical electronic product.

Presently, SMT is at somewhat of a mature state, *mature* in the sense if excluding the normal invention of related advanced processes, components, and process materials of the technology. One can say that SMT has become more "customary versus revolutionary." However, we cannot ignore the influence that other component technologies has on the SMT process, such as leaded through-hole components. The impact that through-hole and odd-form devices have on the overall SMT process can be quite significant. Many companies, consultants, and trade organizations have addressed the solder attachment of these components using conventional SMT process materials and equipment. The primary process prescribed is the use of solder paste. The paste supplies the solder volume and flux required to form the solder joint. The through-hole device is then concurrently reflow soldered with the SMT components. This process has been referred to under various names and acronyms such as *pin-in-paste, reflow of through-hole, solder paste on through-hole technology* (SPOTT), *alterative assembly reflow technology* (AART), and a host of others. Unfortunately, various issues are usually encountered: "Can the device withstand the solder paste thermal reflow profile?" "How do we get enough paste on the PWB?" "How and at what stage of the process do we mount the device on the PWB?" In many instances, a workable process can be obtained. However, because of the inherent uniqueness of each assembly, the manufacturing process should be addressed on an individual basis.

Economic opportunities for nonstandard assembly. The economic opportunities in nonstandard assembly are overwhelming and somewhat difficult to quantify. However, we will make an attempt to do so. It has been estimated by Yann Morvan of Cookson Electronics that there are approximately 25,000 SMT assembly lines in the world. Combining this number with the information in Fig. 3.1 would suggest that approximately 11,000 through-hole assembly lines exist. Hence the total number of assembly lines is in the 35,000 range. Assuming a minimum of 10 people (quite low most likely) per line performing nonstandard and final assembly and packaging, these tasks represent a 350,000-person industry. Assuming that one $200,000 machine could replace four people (the machine works faster and for three shifts), there is a $17.5 billion market for these types of assembly machines.

This analysis does not even consider the enormous market for these types of machines in other industries. The opportunities in optical assembly alone add another several billion dollars to the preceding figure. In all industries where assembly is done, the market is clearly over $100 billion for such equipment if it existed. This prize undoubtedly has encouraged such efforts as CAMotion (Atlanta, Georgia). CAMotion is a small startup company with ties to Georgia Tech that is working on the technical issues relating to nonstandard assembly challenges. The company hopes to have prototype machines in a year or so.

3.5.4 Conclusion

Through-hole assembly still will be significant in the foreseeable future. The fact that it is a rugged, low-cost process will ensure its life for low- to medium-technology products. Nonstandard and final assembly and even packaging products in shipping boxes are a field with almost limitless potential that is only now being tapped by automation. We expect this field to continue to see innovations in automation as assembly robots master the challenges of these assembly processes.

3.6 KNOWLEDGE

Predictions *circa* the 1980s:

1. The theoretically fastest speed for a modem is about 14.4 kB/s; no modem will ever be faster over a standard telephone line.
2. The theoretical limits of photolithography will never allow less than 0.3 micron -μm line widths in semiconductors.
3. It will never be possible to develop a low-cost PWB technology with pads over vias, less than 4-mil lines, and more than 50 pads per square centimeter that is fabricated in a sequential process.

The reality *circa* 1998:

1. Standard modems now reach 56.6 kB/s, while asymmetric digital subscriber line (ADSL) is a technology that is capable of 1.5 MB/s over standard phone lines! It is expected to be a consumer reality soon.[15]

2. Scientists and engineers are currently working on photolithography tools to manufacture ICs with 0.13 -μm line widths.
3. See Sec. 3.7.

The breaking of theoretical barriers occurs for very practical reasons. In the first example, the installed telephone infrastructure is worth hundreds of billions of dollars. If a technology can be developed that will use the installed hardware base and achieve high data rates, the financial reward will be tremendous. Hence individuals have a very strong incentive to try to beat the theoretical limits. The second example is similar. The next step beyond photolithography is x-ray lithography. This technology requires a synchrotron electron accelerator to generate the necessary x-ray intensities. The synchrotron is expensive ($100s of millions) and dangerous (exposure to its x-rays for a few seconds would be deadly); hence there is a tremendous incentive to break the theoretical barriers here too.

3.6.1 Knowledge: A Most Valuable Asset

All the breaking of barriers described earlier occurs by the application and development of new knowledge. As assembly and related processes become more and more challenging, the value of process knowledge also increases exponentially. This is especially true as it relates to yields and throughput, as we saw earlier. Yet it is our belief that the commodity in shortest supply in our industry today is practical yet advanced process knowledge. We make this statement after numerous tours and audits of assembly lines throughout the industry. In many cases the process engineers, technicians, and workers are assembling UFP, BGA, CSP, TAB, or other new or challenging packages and struggling with the process. Or perhaps they are implementing a new process such as the pin-in-paste process (also known as *reflow of through-hole,* which essentially strives to eliminate the wave solder and/or manual soldering operations even though through-hole parts are still on the board) and do not know where to begin.

All these trends make process knowledge more valuable than ever. Yet at this time the assembly industry has very tight profit margins, and companies often cannot afford to perform the necessary process R&D and collateral in-house training of their employees. Numerous consortia, alliances, and university centers[16–19] have risen to this challenge. These efforts have been documented in detail elsewhere.[20] All these worthy efforts, to some degree, lack one critical component, however, and that is "hands on" process education. We know of one large consortium that investigated a certain process. When the consortium finished, one of the companies hired the consortium leader to implement that very process in its factory. The irony of this situation seems to have escaped both parties. Why didn't the consortium participant implement the process on its own? Was it because of the inability to grasp the knowledge or concept? Or was it lack of internal resources to implement the new process?

We strongly believe that hands-on process training is the most important type of education and will be most sought after in the future to meet the needs of the industry. Hence, if your company needs you or your team to learn the pin-in-paste process, you should go to a hands-on workshop on the subject taught by some industry leader such as Jim McLenaghan.[21] In addition to Jim's effort, MicroTech has been a champion of hands-on state-of-the-art process workshops.[22] There is even currently a hands-on consortium being developed with regard to electronics assembly productivity.[23]

Probably the most important message in this chapter is the value of hands-on education as the most effective way to stay ahead of the competition. Another valuable way to obtain knowledge is by performing literature searches on technologies before investing your valuable R&D time. ITM will perform literature searches on PWB assembly topics for those who wish this service.[24]

And now to answer reality number 3 as listed earlier, an example of the value of knowledge follows.

3.7 PWB/FLEX LINE WIDTHS FROM 6 TO 3 MILS AS STANDARD, MICROVIAS, AND BUM PWBS PROLIFERATE

Through the 1980s, expensive high-tech products such as mainframe computers usually drove advances in packaging technology. A classic example of this is IBM's thermal conduction module

(TCM) and supporting PWB. This technology, released in 1981, is still used in mainframes today. The combination of high I/O density and heat dissipation was the envy of the industry when it was released, and with its enhancements, it is still impressive today. The economic commitment that IBM made to develop this technology was in the billions of dollars.

Today, a significant change is occurring on the packaging development horizon. The consumer product development response to the mantra of faster, cheaper, lighter, and smaller is the main driver of packaging technology. Products such as cellular phones, pagers, camcorders, Personal Computer Memory Card Industry Association (PCMCIA) cards, and laptop and palmtop PCs are requiring new technologies to satisfy market demands for smaller, lighter, faster, and cheaper models. Technologies such as chip-scale packages (CSPs), ultrafine-pitch quad flat packs (UFPQFPs), tape automated bonding (TAB), ball grid arrays (BGAs), and 0603 (60 × 30 mils), 0402, and even 0201 passive resistors and capacitors are all responses to these consumer product demands for smallness. Substrates (typically PWBs) are needed to support the wiring required by these very I/O-dense packages. The standard 5- to 6-mil line, large (>8 mils)- via PWBS cannot wire these types of packages without an excessive number of board layers and will never have enough space for the needed mounting pads for the components. These high-layer boards also would be prohibitively expensive and add extra weight and package thickness.

In addition, these product requirement changes have occurred very rapidly. Figure 3.7 is a graph of connecting pads per square centimeter versus time. Note that the historical trend in a state-of-the-art (SOA) product has had quite a gentle increasing slope. Both National Electronics Manufacturing Initiative (NEMI) and the Semiconductor Industry Association (SIA) were prescient in their predictions that the actual need for pads would skyrocket when related to this historic trend. The first product to show this high pad count was the Sony handycam in 1997. In response to these needs, fine-line PWB, microvia, and built-up multilayer (BUM) PCB technologies have been developed. These new technologies will be reviewed briefly in the following subsection.

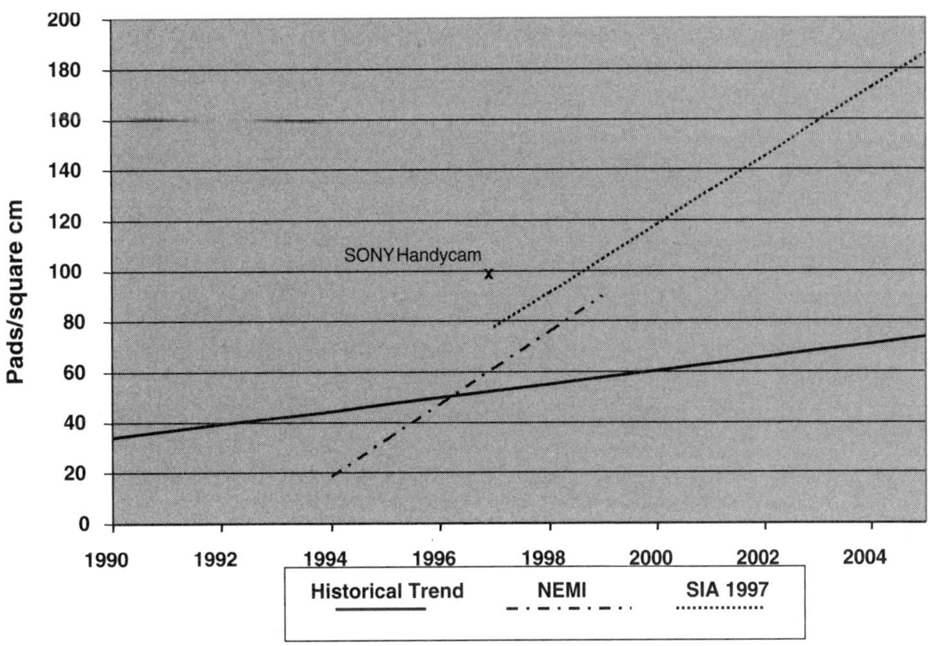

FIGURE 3.7 PWB pad density versus time.

3.7.1 Multilayer versus Built-Up

Multilayer technology has been with us for years. Much of the early work was pioneered by large OEMs such as IBM and Lucent. The driver for the development of this technology was the need for enough wiring and interconnection to support high-performance components and systems. Don Seraphim was one of the pioneers in developing the methodologies for assessing wireability needs for high-performance OEM products.[25] The basic technology is quite straightforward. A schematic of a multilayer PWB is shown in Fig. 3.8. The basic PWB process is as follows:

1. Copper foil is bonded to prepreg.
2. The copper is photolithographically processed to develop the circuit lines needed for that layer.
3. The different layers are laminated and drilled at appropriate levels to form interlayer vias. The lamination is a sequential process (i.e., the layers are laminated one at a time versus one grand lamination at the end).
4. Photolithographic and plating processes are required to create the electrical connection for the vias.
5. The plated through holes are not created until all layers are laminated. The plated through-hole (PTH) process requires drilling through the entire board and plating the hole.
6. The PTH plating operation usually requires another photolithographic process.

This entire process is sequential in that the board is built up one layer at a time. The sequential nature of this process can result in considerable end-of-the-line yield loss. This is true because in sequential processes such as this, net yield is the product of yields at each process step. Hence, if there are 10 steps in a process and the yield at each step is 98 percent, the net yield is 81.7 percent (0.98^{10}).

Conventional multilayer technology has the additional disadvantage that the vias and PTHs take up too much "real estate" that cannot be used to mount components. Although the schematic in Fig. 3.8 is not precisely to scale, it gives the correct sense of the lack of room for attachment pads and channels for wires because so much space is taken up by the PTHs and vias. This situation results in a maximum of about 20 pads/cm^2 on the top or bottom surface of the multilayer board. As noted earlier, currently, consumer products exist that require up to 100 pads/cm^2, and almost all SOA cellular phones, pagers, camcorders, PCMCIA cards, and laptop and palmtop PCs have more than 20 pads/cm^2. Hence multilayer boards, although they could

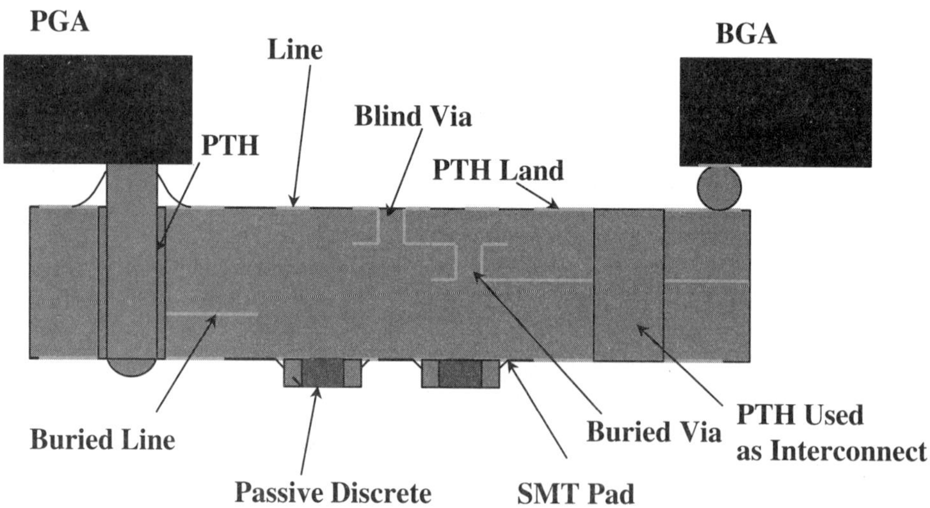

FIGURE 3.8 Standard multilayer interconnection.

have the wireability needed (i.e., length of wire), lack the number of pads to mount components on the surfaces.

In answer to this need, built-up multilayer (BUM) boards were developed. A near-to-scale assembly schematic of a BUM board is given in Fig. 3.9. This technology can achieve greater than 100 pads/cm^2. There are numerous competing processes in building BUMs; a generic process is as follows:

1. An individual prepreg layer is drilled with a laser to form the via holes.
2. Via holes are filled with conductive paste.
3. Copper foil is laminated onto the prepreg.
4. Photo processing forms the circuit lines.
5. The individual layers are tested.
6. The individual layers that form the BUM are laminated together.
7. Photo processing forms the lines and pads on the outer layers.

As noted earlier, the process is not sequential in that the individual layers are not laminated together until they are all circuitized. Since they are tested before lamination, the only yield loss is at the final lamination step. This yield loss typically is quite small. The main functional advantage of BUM over standard multilayer PWBs is that the space that the via takes up can be used by other vias and pads. This single advantage will make BUM boards the technology of the future for substrates. Prismark has contributed several excellent articles on all aspects of BUM technology in its publications for clients.

There are numerous types of BUMs today, and it would be difficult to predict which ones will prevail. In general, though, the BUM technology that achieves lowest cost will be the one to dominate. One thing is for sure — BUM technology increasingly will be the future in substrates. It cannot lose with the combination of a low-cost PWB fabrication process, high fabrication yield, and the greatest pad density and wireability.

There also will be a need for 3-mil line widths at a low cost for some technologies. Currently, line widths below 5 to 6 mils usually exact a cost premium. This need is probably not quite as great as the need for BUM technology, but with flip chip and CSP proliferating, the need for 3-mil lines cannot be too far away.

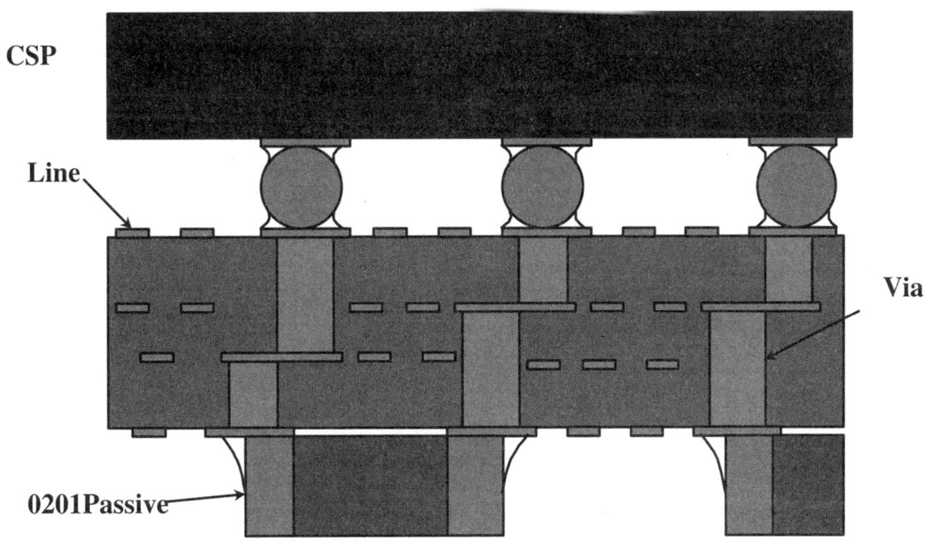

FIGURE 3.9 BUM PWB.

3.8 LEAD-FREE IMPACT

The art of soldering predates written history. Soldering was performed initially with silver, gold, and tin and typically was used as a lower-temperature attaching technique for precious metal castings. By 500 B.C., the Romans had determined that tin and lead produced a very low temperature eutectic solder that could be used to join lead for plumbing and bronze artwork. Hence some of the early cases of lead poisoning were Romans who drank the water from aqueducts. For those who are interested, Alpha Metals has an excellent article on this history on its Web site.[26]

Tin-lead solder has lasted all these years because tin and lead are inexpensive and abundant, melt at a low temperature, "wet" copper very well, and are relatively well understood. The primary negative aspect of tin-lead solder is the toxicity of the lead. There is much concern and confusion today with regard to this topic. Hence we will review some of the issues. Let's start with a true/false test.

True or False

1. Electronics is a major contributor of environmental lead.
2. The level of lead in the blood of U.S. residents has increased since 1945.
3. There are numerous cost-effective alternatives to tin-lead solder.

Answers

1. False. Battery production consumes about 4 million tons of lead, whereas electronics production consumes only 18,500 tons.[27] A significant minority of people in electronics know the qualitative nature of this fact. However, few people know that electronics is not even number two in lead consumption because the manufacture of bullets consumes 200,000 tons of lead. Hence electronics consumes less than 0.5 percent of the world's lead. These numbers do not even include all the lead used in weights to balance tires.
2. False. The level of lead in the blood of U.S. residents has experienced a stunning drop since 1945, going from close to 30 μg/dl in 1945 to less than 5 μg/dl in 1990[28] (Fig. 3.10). The likely reason for this stunning drop is the gradual elimination of lead pipes and lead paint in the environment. The additional steep drop since 1975 also may relate to the use of lead-free gasoline.
3. False. There are several lead-free alternatives such as Sn96.5/Ag3.5 for solder paste and Sn99.3/Cu0.7 for wave soldering. However, the tin-silver solder-paste alloy is twice as expensive, and both of them have melting points that are about 40°C higher than eutectic tin-lead. These and other critical issues in finding lead-free solder alternatives for electronics assembly have been reviewed thoroughly in an excellent article by Bastecki.[29] Unfortunately, in the 2500 years since the discovery of tin-lead solder, we have not been able to discover a metal system that has all its desirable features at a low cost.

Why, then, is there so much concern over the electronics contribution of lead to the environment? For one thing, the numbers in question 1 are slightly misleading. Much of the lead in batteries is recycled, but if only 5 percent of batteries are not, the amount of battery lead is still 10 times that of electronics. In addition, the discarding of old electronics devices is a most visible process. We all discard many electronic items per year. Still, all in all, it is difficult to conclude that the concern for the electronics contribution of lead does not have an emotional content to it. However, as people are inclined to say today, "perception is reality."

Recycling is an alternative to outright banning of lead in electronics. Recycling is already being practiced in some European countries. It typically works by charging an additional fee when the product is purchased; the fee is used to pay for the recycling. One of the issues in recycling is that a by-product of the process can be some toxic gases when the PWB is incinerated.

With all this information as background, we will go out on a limb. We do not think that lead will be banned in electronics in any country. If it were, the difficulty that that country would then have competing in the world electronics industry would be so great that its electronics industry would suffer to the

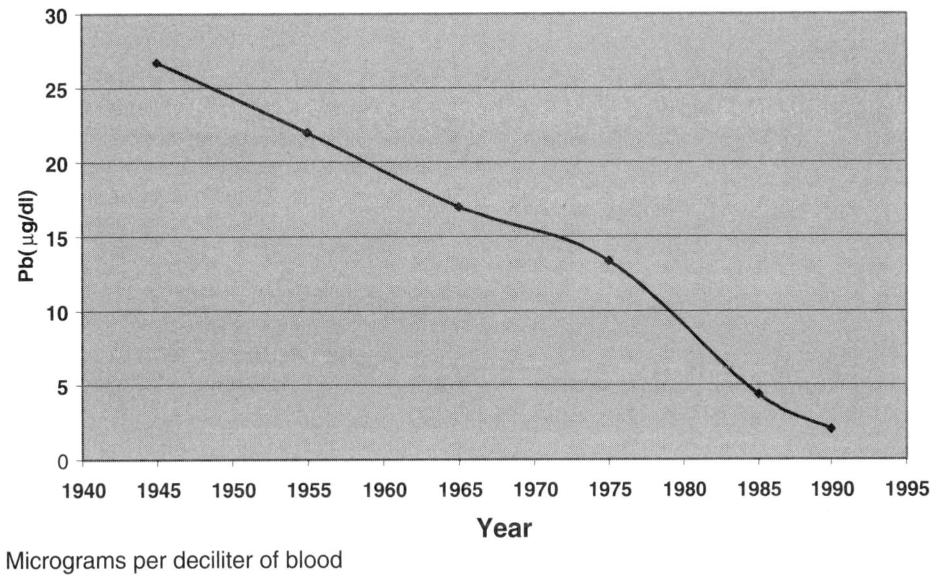

FIGURE 3.10 Average U.S. lead levels in blood.

extent that the legislation would be rescinded. Hence the only way banning could work is if it were done worldwide. We believe that this action is highly unlikely because of the information just stated. We do believe, however, that recycling will become common and even ubiquitous as a response to the just concern for lead in the environment. Recycling will be the answer to the call for banning lead. This does not mean that lead-free solders will not exist and prosper in important niche markets. One possible market is for companies that want to emphasize the environmentally friendly nature of their products. Nortel recently funded the development of a lead-free assembly process for cellular phones for just this purpose.[30] It will be interesting to see how the market receives such products. It also should be pointed out that there are already some significant applications of lead-free conductive adhesives. These adhesives are used in chip-on-glass and other applications where polyester flexible circuits are used. In chip-on-glass applications, typically a die is electrically connected to glass substrates such as in LCD displays. The polyester flex circuit applications are low-cost replacements for polyimide flex circuits. These circuits cannot be processed at reflow temperatures; hence conductive adhesives are used.

3.9 PASSIVES: THE GROWTH CONTINUES

Passives are defined as devices that do not provide power gain. They are most commonly resistors and capacitors but also include inductors and diodes. Since resistors and capacitors are the "lion's share" of passives, the remainder of this discussion will focus on them. Thus, when the term *passives* is used in the remainder of this chapter, it will mean only resistors and capacitors.

The use of passives continues at an astounding rate. A billion hours ago the stone age was the future, a billion seconds ago the iron age was maturing, a billion passives ago was this morning! Currently, more than 800 billion are used per year. This amount is more than 100 for every person on the planet, and if they were place end to end, they would circle the globe more than 20 times! In the future, some people have assumed that passives will be "integrated" away into the IC. The exact opposite is happening—passive growth continues unabated.

3.9.1 Why Is Passive Use Growing?

Passives serve several vital functions in electronic products. In layperson's terms, they are used to tune the circuit. A brief explanation of the most prominent functions performed by passives follows:

1. *Decoupling capacitance.* The constant "switching" that occurs in digital circuits creates noise that can interfere with the power supplied to the numerous circuits on the PWB. If the power supply voltage has too much noise from switching, it is possible for a digital "one" to be read as a "zero," or vice versa. This event is usually quite unlikely, but with millions of operations per second in modern microprocessors, it can occur if the power-supply voltage is not protected. To minimize this effect, capacitors are placed throughout the PWB and charged to the line voltage. The capacitors respond by either adding or subtracting current to the power grid in response to any perturbations to it. This effect is called *decoupling* (decoupling the effects of the switching from the power grid).
2. *Line termination.* One of the requirements of an electrical transmission line is that it be properly terminated. Without proper termination, electrical signals can reflect and cause false signals.
3. *Pull-up resistors.* Many ICs have numerous and different applications. These different applications require different electrical connections. Occasionally, some terminations on the IC will not be used in a specific electrical product. In most cases, the I/O termination in question needs to be set at the power grid voltage with no current flowing. This operation is performed with a large-resistance-value pull-up resistor.
4. *Filters.* Passives are often used to form electrical filters.

It is logical to ask if passives must be used to solve these design challenges. Apparently some of the passive function, especially related to electrical filters, could be integrated into the IC. However, this type of design would limit the IC's flexibility for many uses. Hence there is typically one IC design for many uses, and the circuit is "tuned" for the specific applications. However, it must be understood that even in the best of IC designs, passives that are separate from the IC are needed.

3.9.2 What Are the Trends in Passives?

Current trends in electronics are driving the increased use of passives in all the categories mentioned in the last subsection except for electrical filters. These trends are the need for faster clock speeds, low operating voltages, higher I/O counts on active devices, and increased digital-analog function combinations. Electrical filters can require fewer passives as voltages decrease, but more decoupling capacitors are needed as voltages are reduced. The reduction in filter requirements is usually quite minimal and is overshadowed by the soon to be mentioned growth in other passive uses. Figure 3.11 indicates the stunning increase in PC microprocessor speed over the last 15 years. This same chart shows the typical operating voltages of the microprocessors. These two trends independently require more passives. Together with the increase in active component I/O, they demand exponential passive growth with time. Hence we can expect the growth of passives to continue for the foreseeable future.

Figure 3.12 indicates that the percentage of passives in all SMT components in a PWB has grown from only 25 percent in 1984 to over 90 percent today. This change is arguably the most stunning change in all of electronics. This dramatic change not only requires that more passives are placed, but it inhibits miniaturization, since the passives require more and more area on the PWB. The industry has responded to this change in several ways. First, high-speed chip shooters have been developed. Currently, some chip shooters approach 100,000 placements per hour. A faster chip shooter is a logical first step in the challenge of exponential passive growth. The passives industry also has responded to the situation by producing continuously smaller passives, as shown in Fig. 3.13, and developing integrated passives. These integrated components have numerous capacitors or resistors in one package. This approach also reduces the assembly cost in that only one passive component needs to be placed to satisfy the function of numerous individual passives. Since assembly costs often are quoted as cost per placement, this approach can have a seemingly dramatic effect on cost control. On the negative side, however, these integrated passives are typically of a custom design;

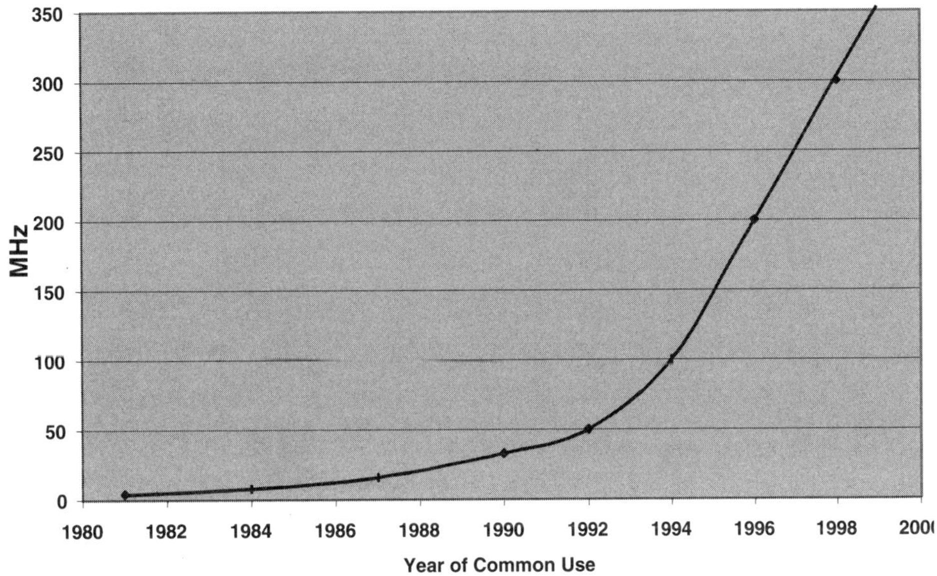

FIGURE 3.11 PC microprocessor speed increases.

hence their unit cost is high. Currently, only about 250 million integrated passives per year are used versus almost 900 billion discrete passives.

In some cases, the number of passives required is so large that there is not enough room on the PWB to place them all. To address this need, considerable work is being performed to develop the technologies required to "bury" the passives in the PWB or flex circuit material. Hadco has been a leader in buried passives in PWBs, while Sheldahl has been a champion of buried passives in polyimide flexible circuits. The buried approach is not particularly flexible from a design perspective and is still quite expensive.

3.9.3 What Is the Future for Passives?

Integrated and buried passives will hold important but niche positions in the entire passive market because of their lack of flexibility vis-á-vis discrete passives. As prices come down, integrated passives will likely grow more than buried passives because their inflexibility is less. We see difficulties in discrete passives becoming common below the 0201 (20 × 10 mils or 0.5 × 0.25 mm) size because their small size creates assembly handling difficulties. Even today, handling of 0402 passives is a challenge.

Thus we see discrete passives continuing to grow, but the 0201 and 01005 sizes being slow to emerge unless placement breakthroughs occur. Integrated passives will grow rapidly from their much smaller base, but they will never seriously threaten discretes. Buried passives will continue to be very much a niche but important solution in the future. They will be used primarily when "real estate" concerns demand their use. We believe that other approaches such as jetting of resistive material are still a long way off.

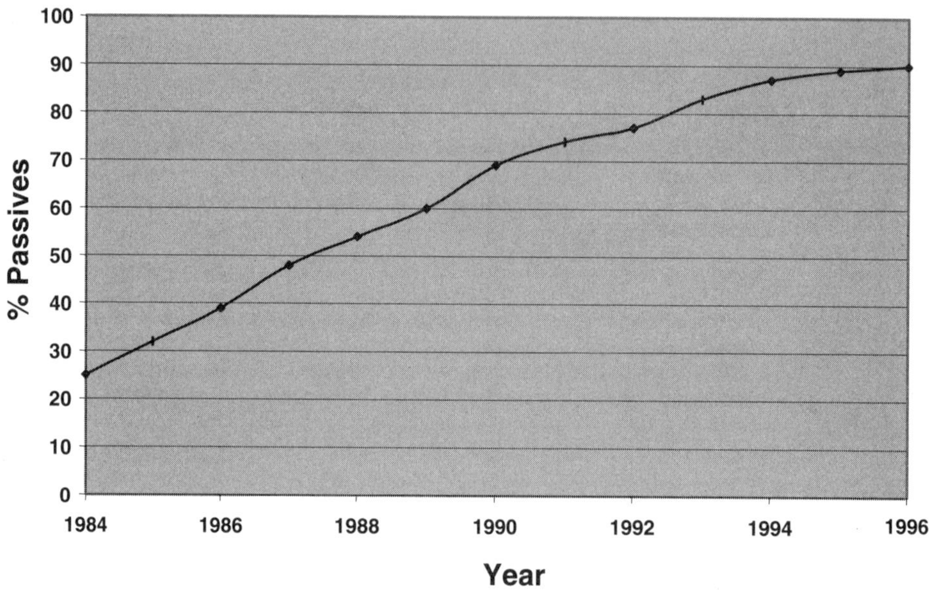

FIGURE 3.12 Passive-SMT ratio in packages.

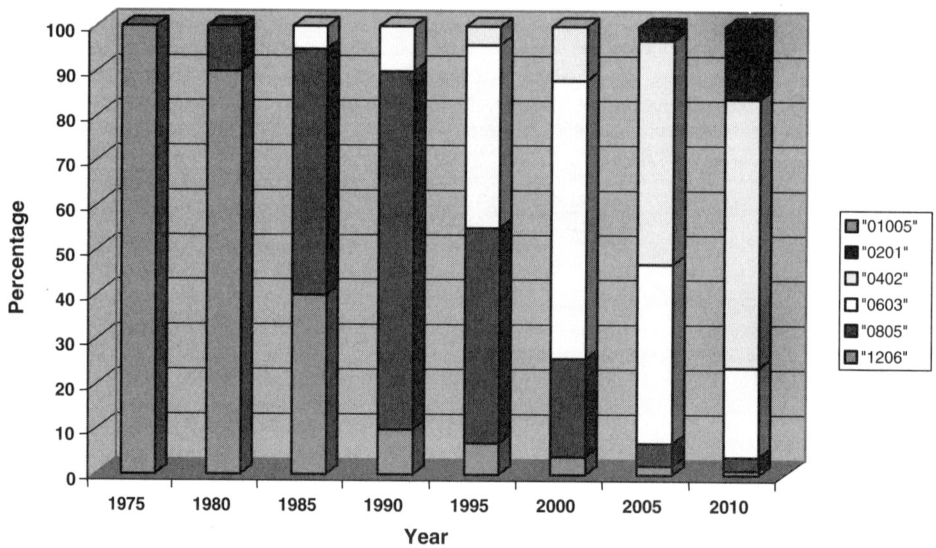

FIGURE 3.13 Percentage of passives by size.

3.10 NOTES AND REFERENCES

1. Bagley, Tom, MPM Corp., private communication, 1996.
2. Lasky, R. C., and Baldwin, D., "Emerging Packaging Technologies," SMTA Conference, Raleigh, NC, November 1996, pp. 1–4.
3. *National Electronics Manufacturing Roadmaps,* NEMI, Herndon, VA, pp. 41–51, 1996.
4. MPM UP3000 Literature, "Solder Jet Printing," MPM Corp., Franklin, MA, 1998.
5. Snyder, M., and Lasky, R., Materials Research Society Meeting, San Francisco, April 1995.
6. Godlin, R., *SMT,* January 1997, pp. 66–68.
7. Camelot Gemini Literature, Camelot Corp., Haverhill, MA, 1999.
8. Phillips FCM Literature, Philips Corp., Eindhoven, The Netherlands, 1999.
9. Another light-hearted view of the situation is that if all the energy in the 20-hp motor were used to move only the passive, it would move it at about 300 times the speed of sound (i.e., it could cross the United States in 8 minutes).
10. CAMotion, Inc., Manufacturing Research Center, Room 436, Atlanta, GA, 30332-0405, Steve Dickerson, President.
11. Speedline MPM, 16 Forge Park, Franklin, MA 02038.
12. SPACE was developed by MPM Corp. A derivative product, CoastCoach, is marketed by ITM, Durham, NH (*www.itm-smt.com*).
13. Professor Dan Baldwin, Georgia Institute of Technology, Atlanta, GA; Steve Corbett, PolyFlex, Cranston, RI.
14. Cookson Electronics, Foxboro, MA (508-541-5800).
15. *PC Magazine,* Feb. 10, 1998, p. 28.
16. George Westby, UFP and BGA/DCA Consortia, Universal Instruments, Binghamton, NY (607-779-5258).
17. Rao Tammula, PRC, Georgia Tech, 813 Ferst Drive, Atlanta, GA 30332-0560 (404-894-9097).
18. NEMI, 2214 Rock Hill Rd., Suite 110, Herndon, VA 20170-4214 (703-834-2082); e-mail: *jmcelroy@nemi.org.*
19. A registry of university alliances can be purchased from HTA, 26 Howe St., Medway, MA 02053.
20. Lasky, Ron, "Consortia, Alliances, Centers and Other Such Phenomena: A Perspective," *Circuits Assembly,* March 1998.
21. Pin-in-Paste Workshop, Jim McLenaghan, Cookson Electronics, Providence, RI 02903 (401-521-1000).
22. Martin Barton, MicroTech, 1416 Glastonbury Drive, Plano, TX 75075 (972-424-8805).
23. Dan Baldwin, Georgia Tech (404-894-4135); e-mail: *daniel.baldwin@me.gatech.edu.*
24. ITM (603-868-1754); e-mail: *theitmtcam@aol.com.*
25. Seraphim, Lasky, and Li, *Principle of Electronic Packaging,* McGraw-Hill, New York, 1989, pp. 59–67.
26. *http://www.alpha-polymers.com.*
27. Prismark Partners, 130 Main Street, Cold Spring Harbor, NY 11724 (631-367-9187).
28. *Wall Street Journal,* September 16, 1993.
29. Bastecki, Chris, *SMT,* "Lead Free Solders," May 1997, pp. 52–58.
30. *Los Angeles Times,* December 1, 1997.

CHAPTER 4
AREA ARRAY PACKAGING

Ken Gilleo
Cookson Electronics

4.0 INTRODUCTION

Area array packaging is the oldest and the newest high-density package style. The early IBM solid logic technology (SLT) package developed in 1961 appears to have been the very first area array design. SLT was used for transistors, which were the only commercialized solid-state devices at that time. The array was the simplest possible because only three inputs-outputs (I/Os) were required for the common transistor. Soldering three copper microspheres to the transistor pads created the micropackage. The package was passivated with glass, but once it was assembled by solder reflow, an underfill was applied for environmental protection. The SLT was used in the famous System 360 mainframe computer launched in 1964. Figure 4.1 shows the SLT. Note that this is not only a ball grid array (BGA), it is also a chip-scale package (CSP) and the first surface-mount technology (SMT) product.

Modern area array packages include BGAs, CSPs, and flip chips (FCs). Each has two common characteristics that are of great significance. Some or all of the area on the bottom of the package is used for connecting to the circuit board, unlike the common perimeter packages, which waste all this "real estate" and thus crowd the pitch along the perimeter. The second feature is that virtually all use SMT. One may argue that the pin-grid array (PGA) is a feed-through device and thus is not surface mounted. This may be true in some cases, but most PGAs are plugged into a surface-mount device (SMD) package, and none of the pins protrude through the board. SMT devices do not pass through the circuit board, as do the older feed-through components. Certainly all the other styles with microspheres, solder balls, studs, columns, and any similar shape are SMDs. This is an important feature because it allows area array packages to fit into modern mainstream assembly processes. SMT has been the most common electronic component assembly process for many years, and it continues to gain more share because of high density and ease of automation. Fitting the SMT manufacturing infrastructure is important to growth and helps account for the popularity of area array packaging. Now let's examine the area array package concept more closely.

4.1 BASIC ELEMENTS OF A PACKAGE

4.1.1 Device

The area array package can be divided into simple elements. First, there is always at least one electronic device. There may be more than one integrated circuit (IC), and multichip packaging (MCP) continues to gain strength. The device is typically a solid-state IC, but it also can be passive such as

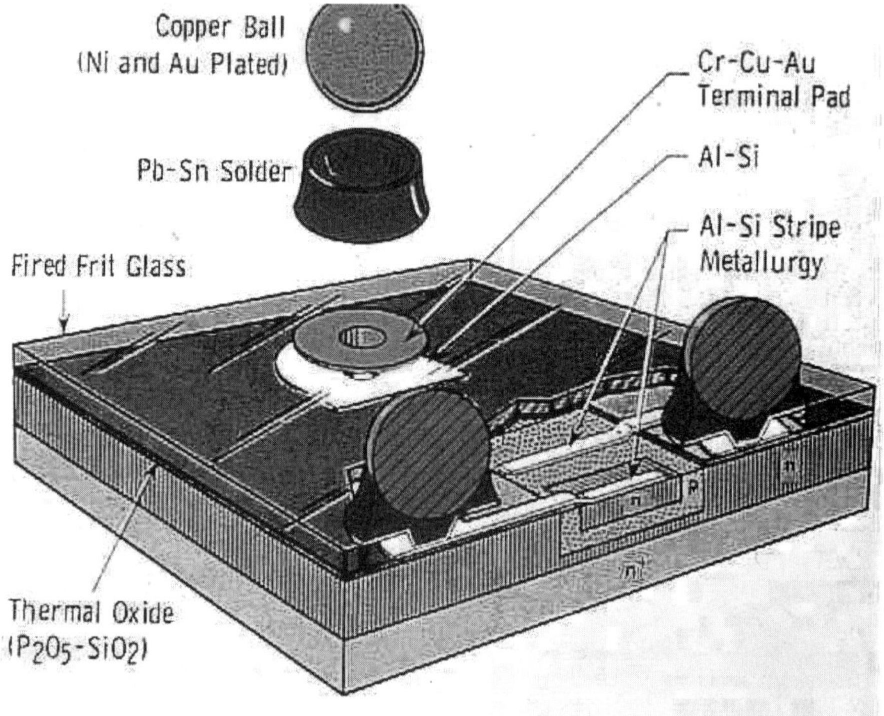

FIGURE 4.1 Solid logic technology (SLT). (*IBM Corp.*)

a resistor or capacitor array. In fact, tiny CSPs are now available to provide arrays of resistors, capacitors, diodes, light-emitting diodes (LEDs), and other optoelectronic (OE) systems. Computer chip packages often incorporate passive chips and sometimes memory.

4.1.2 Wiring or Routing

The next part of the system is some form of wiring structure that creates the pathway between the device (or devices) and the bottom of the package that ultimately will connect to the printed wiring board (PWB). There is almost always a geometric translation. The device dimensions are very small, and the wiring may route or fan out to a larger pitch that better matches the *practical* densities that can be produced using printed circuit technology. When the wiring pattern from the device spreads out and away, the design may be called a *fan-out pattern.* Conversely, the wiring may route conductors inward and under the device, and this is called a *fan-in pattern.* The fan-in pattern typically is used for very small packages and allows CSPs to be made. The common definition of a CSP is a packaged IC with a footprint not greater than 1.2 times the area of the die. An alternate but less common definition is a packaged chip where the total system is not greater than 1.5 times the volume of the die. These values are derived somewhat arbitrarily and have prompted terms such as *near-chip scale, near-chip size,* etc.

The wiring structure for area array can have several subelements. The package itself will have a pattern of electrical conductors typically produced by circuit board processes such as etching. Higher-density wiring is also being produced by additive and semiadditive methods where the conductors are produced by plating up. Other parts of the wiring structure may include wire bonds or whatever connection means is used to mate the package wiring to the chip. This is called the *first-level connection* and will be covered later. There is also the ball, bump, stud, or column connection, and while this is part of the electrical pathway, it is not included as the wiring structure. The package wiring may be made of copper, gold, or a combination. It may be possible theoretically to use polymer thick film (PTF), but the limitations of this conductor medium have excluded its use so far.

A wiring layer also can be applied directly to the chip and sometimes is referred to as *rerouting.* Several companies offer commercial rerouting services, and these are usually combined with bumping of the chip pads to allow direct assembly. The resulting "package" may be called an FC or a CSP, and this adds to the confusion in distinguishing between these two competing designs. The process usually involves depositing a thin layer of dielectric over the bottom of the die, placing a pattern of conductors on the layer, and adding bumps. Most are wafer-level processes that produce all the packages simultaneously. The last process step is separation, called *singulation,* and is a cutting or wafer dicing method that typically uses a diamond saw.

4.1.3 Packaging Enclosure

The package must have some form of enclosure to protect the device, wiring structure, and chip-level interconnects. The enclosure can have other important uses, such as facilitating handling and board-level assembly, heat management, a surface for nomenclature, and other uses. The most common enclosure is made using polymer encapsulants, especially epoxies. Both liquids and solids are used, and separate chapters are devoted to materials, such as Chap. 11 covering polymers, Chap. 22 covering liquid encapsulant dispensing, and Chap. 23 covering package molding. Some of the CSPs use special materials such as elastomers and chip coats. Although plastic packages are most common, ceramic and metal housings are also used, and these are covered elsewhere. See Chap. 9 for a detailed discussion of ceramic area array packages.

4.1.4 Board-Level Joining System

The materials and structures for connecting area array packages to circuit boards are where the most substantial differences will be found. While conventional perimeter packages use nonmelting metal leads, these are eliminated with area array. There are only two basic structures that can be used to populate underneath the package or die, and they are spheres (and bumps) and columns. Most joining structures are formed from metal spheres that are commonly referred to as balls to generate the common BGA term. Columns are also used because the greater standoff length provides special benefits in terms of joint fatigue resistance. Some FCs can have a rather flat or even mushroom-shaped bump that is produced by electroless plating without a resist to restrict the shape. However, a solder sphere or semisphere is produced on top of the structure by reflowing solder paste. Alternatively, conductive adhesive paste or film can make the connection.

4.2 OVERVIEW OF TYPES OF AREA ARRAY PACKAGES

Our definition in this book of area array packaging requires the existence of some area connections or the ability to produce them. In other words, the package technology must allow a board-level connection structure, such as bumps, to be produced within the substrate area, not just along the outer edge. Some area array layouts may only populate the outer row so that the package appears to be only a perimeter type. However, this can still meet the area array package classification if a second row of connectors can be produced. In all cases, the area array connection structure emanates from the bottom of the package, not from the sides like many perimeter designs.

The BGA is the most common type of area array package, and it offers a balance of features with reasonably good economics. The features include low profile, light weight, excellent reliability, and a quick time to market with many designs. Common ball pitches range from 1.0 to 1.50 mm. The package can accommodate over 1000 I/Os. Figure 4.2 shows a typical structure, but it should be noted that many other designs and enhancements are possible. CSPs and FCs can have much closer pitches, and the number continues to shrink.

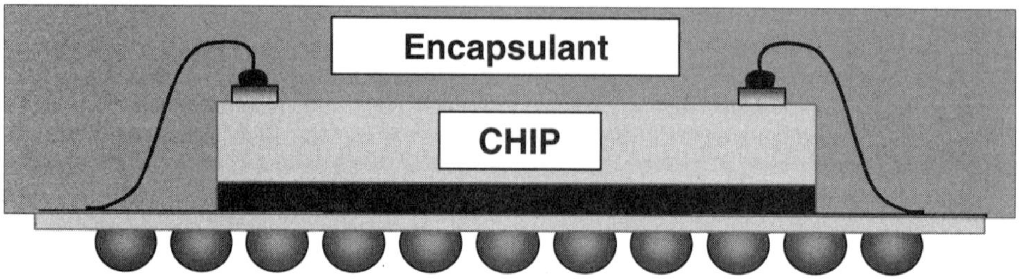

FIGURE 4.2 BGA cross section.

Multichip BGA packages became popular beginning in the late 1990s. They are the successor to the multichip modules (MCMs), and many people will say that the highly touted MCM was simply waiting for the area array package to be implemented. This may be correct, and the term *multichip module BGA* (MCM-BGA) is also used. The leading packaging foundries offer over a dozen multichip BGA styles. This package provides all the same advantages as the single-chip BGA but in addition provides simpler board layout and a reduced number of components to be assembled and often yields higher performance because of the closer proximity of the chips in package. The same Joint Electron Device Engineering Council (JEDEC) standard footprint outlines can be used that are designated for single-chip BGAs. Chips can be placed adjacent to each other in the same plane or stacked in 3D fashion (see Chap. 5).

Several companies offer thermally enhanced BGAs. A metal heat spreader or other thermally conductive material interface for coupling to a heat sink or cooler generally is incorporated. The heat spreader is nearly always placed in contact with the die for optimal heat transfer. Designs typically place the metal heat spreader on top of the package so that air can be blown across or a cooling device can be attached. The package carrier can be rigid organic, flex (tape), metal, or ceramic. However, the thin, flexible circuit–based packages can provide special thermal features. Heat also can be removed through the bottom of the package using a thermal block. Bumps, spheres, and columns also can serve as heat pipes to the PWB, and extra ones can be added even if they are not used for signal, power, or ground connections.

Tape- or flex-based BGAs are another important package, especially where enhanced thermal performance is desired. These packages are referred to as TBGAs, FBGAs, and fleX-BGAs (registered trademark of Amkor). The thin flex substrate offers very little thermal resistance. While all these BGAs are made with carriers that are tiny flexible circuits, the tape terminology has come about because of the similarity with tape automated bonding (TAB). TABs, also called *tape carrier packages* (TCPs), are really flexible circuits but without any balls or microspheres for attachment to the wiring board, and in fact, they are not even area array packages. The board-level bonding is done by soldering the perimeter leads with a heated bar. However, flex area array package designs retain the TAB inner lead bonding (ILB) format but replace the troublesome outer lead bonding (OLB) structure with area array bumps, balls, or spheres. The thinness of the flex dielectric makes it rather easy to fabricate vias through the bottom of the carrier that can accommodate attachment of solder spheres or some other interconnect structures. While TAB packages were limited particularly because they required special bonders, TBGAs have gained popularity because they are SMT components.

Several CSP designs based on flexible carriers are also available. The best known is the Tessera μBGA that is licensed to a number of companies now in high-volume production. The design is somewhat unique because the circuit pattern is fanned in to produce a nearly die-sized package. The connection to the chip uses an S-shaped lead (viewed as a cross-sectional) that is usually made from gold to avoid bond failure due to thermomechanical fatigue. These packages offer all the advantages mentioned earlier but have even a smaller footprint and lower profile. Electrical performance is considered to be modestly better than that of the larger BGAs because of shorter signal path lengths. This package is covered in Chap. 6, "Compliant IC Packages."

The flip chip BGA (FC-BGA), also called a *flip-chip-in-package* (FCIP), is an increasingly important package. It has the lowest profile and excellent electrical performance. A flip chip, or direct chip attach (DCA), rather than wire bonding is used to connect the chip to the carrier. The FC-BGA is especially useful with high lead counts. At some point between 600 and 800 I/Os, the flip chip economics become better than those of wire bonding. However, the exact number depends on many factors, and it is also a moving target. Continuous improvements in wire-bonding equipment move the crossover point to a higher number each year. However, as the die is reduced in size by "die shrink," a semiconductor densification process, and the number of I/Os goes up, a point is reached where flip chip is the right way to go or maybe the only solution. In fact, several central processing units (CPUs), or computer chips, can only be assembled by flip chip techniques. There are just too many pads to be placed along the die's perimeter even when staggered double rows are used. We should keep in mind that the flip chip is the only established chip-level connection that offers area array. Wire bonding and TAB are only perimeter methods.

The last area array package is actually the first, and it is the *flip chip on-board* (FCOB). Rather than building an FCIP, the bare chip is assembled directly to the wiring board. There are advantages but significant limitations for FCOB. First, the wiring board must have a very high density pad layout. There is no intermediate chip carrier to fan out the pitch. The next problem is that the assembled flip chip nearly always must be underfilled. Since underfilling is not part of the standard SMT assembly process, a second line generally is set up. The underfill line adds capital cost, material expense, and extra time. And the underfilled chip may not be reworkable, although new reworkable underfills are starting to be offered. Many, if not most, assemblers thus prefer the FCIP because the packaging foundry does the underfilling and the package can be reworked like any other BGA. However, when the minimum size, lowest profile, and highest electrical performance are critical, then FCOB is the right package. We should be aware that FCOB is used extensively in the mainframe computer area where it first started. Other FCOB applications include automotive controllers and cellular phones. Even though CSPs are supposed to eliminate underfill, this has not always been the case for cellular phones because CSPs can be damaged when the phone is dropped. Since many CSPs are underfilled to make them drop-proof, the flip chip becomes attractive again because it is smaller than any CSP. Also, products that have rerouting on the chip may be called CSPs, but they are not really more than a rerouted flip chip. Progress in underfilling eventually will minimize the cost impact. See these chapters for a discussion on new underfill methods: Chap. 7, "Flip Chip Technology"; Chap. 17, "Next Generation Flip Chip Materials and Processes"; and Chap. 18, "Flip Chip Assembly and Underfilling".

4.3 ADVANTAGES OF AREA ARRAY

4.3.1 Density

The area array package greatly increases density, and this is probably the most important driving force that has made it a popular package today. The total number of leads T_p or I/Os, that can be placed along the perimeter of a package is derived from $T = 4N - 4$, where there are N leads or I/Os in a row because there are four sides to the package, and we need to subtract 4 so that we do not count the corner I/Os more than once. However, the number of connections for a fully populated area is $T_a = N$. The total number goes up dramatically as N increases. Table 4.1 shows how the total number of I/Os increases geometrically with the number of balls per row compared with perimeter-only. Figure 4.3 compares this graphically. Figure 4.4 compares perimeter layout with area array. Note how dramatically the area array layout exceeds perimeter as the value of N rises to 10 and higher.

4.3.2 Thermal Management

Although many standard packages have enhanced heat-management capabilities, area array is better suited to this task. In fact, there are families of area array packages designed specifically for high

TABLE 4.1 Perimeter versus Area Array

N I/Os per row	T_p, total perimeter leads	T_a, total area bumps
2	4	4
4	12	16
6	20	36
8	28	64
10	36	100
12	44	144
14	52	196
16	60	256
18	68	324
20	76	400
30	116	900
40	156	1600
50	196	2500

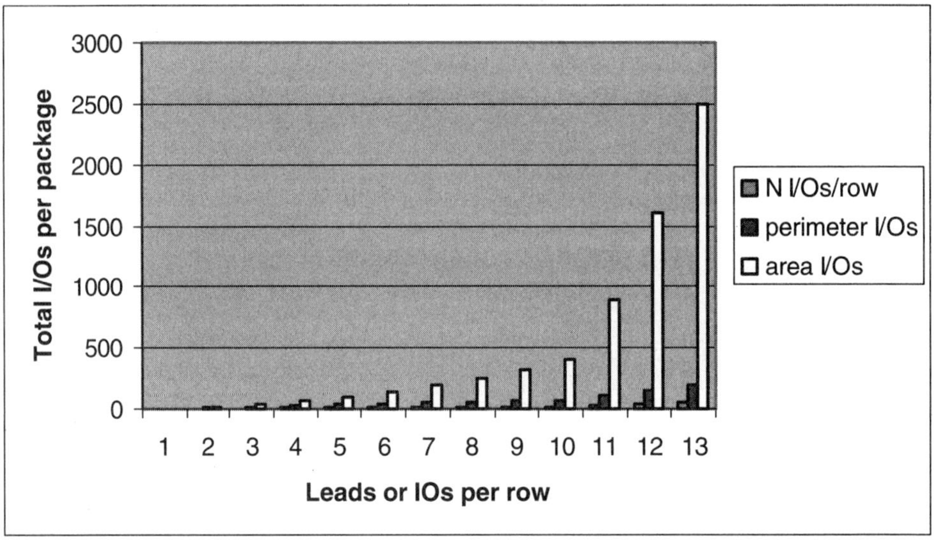

FIGURE 4.3 Perimeter versus area array (same pitch density).

heat dissipation. The flex-based BGA (also TBGA) has a metal heat spreader bonded to the thin flex carrier. Recall that heat transfer is inversely proportional to material thickness. This means that thin-flex, which can be as thin as 1 mil, is 10 times better at heat transfer than a 10-mil-thick rigid carrier with the same thermal conductivity. Since thermally enhanced flex polyimide laminate is available, the factor can be increased further. Ceramic, often an excellent thermal conductor, and highly conductive metal are also used for thermal packages with even higher efficiency, but at a cost. Figure 4.5 shows a thermally enhanced BGA.

4.3.3 Multiple Chips

Multichip BGAs, PGAs, and even CSPs have become available over the past several years, and their use continues to increase. While standard perimeter packages also can accommodate more

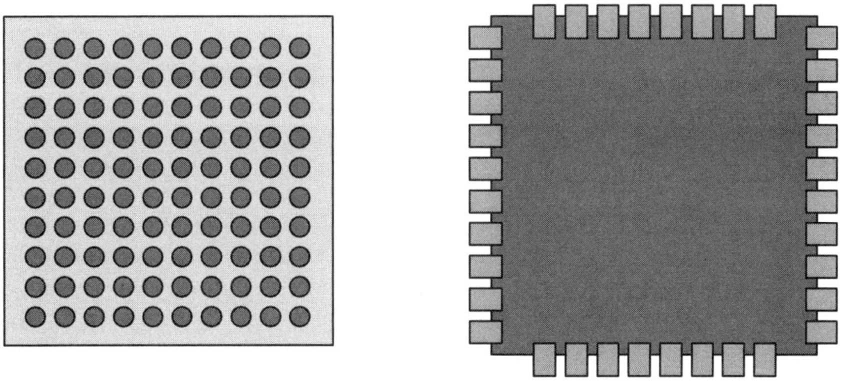

FIGURE 4.4 Perimeter versus area array layout.

than one chip, their use is more difficult. First, the use of multiple chips in a single package increases the number of package I/Os. The area array construction can accommodate many more I/Os than perimeter so that the package does not run out of capacity very quickly. Also, the ability to use FCIP with area array packaging allows closer packing than is practical with wire bonding. And finally, the ability to remove heat efficiently from the area array package allows higher density with less concern about overheating. Many people believe that that area array is the ideal form factor for multichip packages (MCPs), and the increasing popularity of this product segment supports this opinion.

4.3.4 Built-In Solder Source

Most area array packages are designed with the solder built into the package, thus eliminating the need to apply paste to the wiring board. Not only is the solder supplied with the package, the volume is much more precise than can be deposited onto the board by stenciling. However, flux generally is required, but this application process is much easier and less critical than for solder paste. Flux can applied to the bumps by an automatic coating process in which the pick-and-place machine dips the BGA into a reservoir of flux that is part of the equipment. The flux dipping process ensures that the flux is only where it is needed, on the bumps. Alternately, flux can be applied to the board, but the registration precision is low compared with solder's requirements. FCs and CSPs with eutectic solder also have integral solder, and this is even more beneficial with the typical fine pitch and small dimensions of these micropackages.

4.3.5 Self-Centering

When solder is melted to a liquid, the very high surface tension exerts considerable mechanical force. Surface tension attempts to reduce the surface area to a minimum. This is why solder forms spheres. When the solder ball of a BGA becomes molten, it wets the wiring board metal pad and then exerts force to minimize the joint area. A skewed or offset package usually will be forced into alignment. However, since area array packages have bumps on all four sides and often over most of their area, the alignment forces are stronger and more symmetric than on older perimeter-style SMDs. Work has shown that BGAs that are only half on the pad will orient. The centering works well, if not better, with CSPs and FCs because of their relatively lower weight.

4.3.6 High Assembly Yield

High assembly yields are due to some of the items listed previously. The precise amount of solder provided by the ball or bump typically precludes bridging or starved joints. The high level of self-

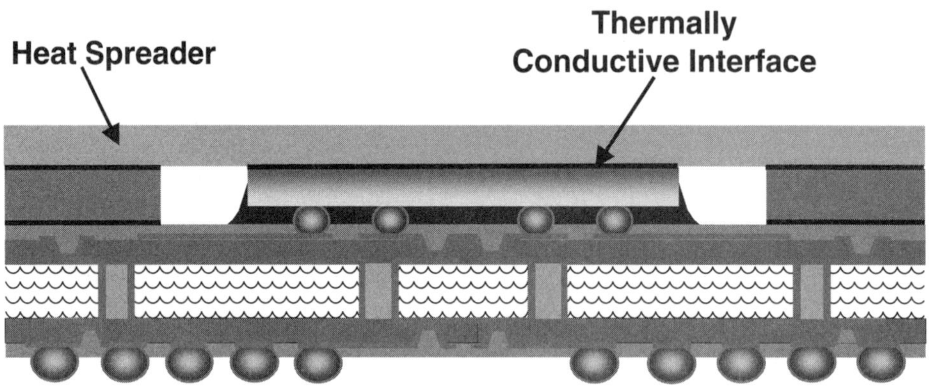

FIGURE 4.5 Thermal BGA.

orientation also is a big plus for yield. While one should attempt to place BGAs, CSPs, and FCs as accurately as possible, the self-alignment boost only adds to the high yield, where defects are often in the few parts per million range.

4.4 ISSUES WITH AREA ARRAY

4.4.1 Inspection

Many assemblers were frustrated in the early days of BGA assembly development because they could not see the solder joints under the package. We have relied on direct visual inspection for so long that it seems essential for quality, but this is not true. Difficulty of inspection still can be a problem today, although sideways-looking cameras can see further in but not perfectly. When accurate inspection is essential, real-time x-ray has become a preferred solution, albeit at a cost. Today, BGA joints can be inspected and analyzed quickly with digital x-rays with as much or more accuracy than vision systems for non-area array packages. Magnification, especially for CSPs and FCs, has been especially valuable for "seeing" small joints. Some people prefer acoustical imaging for FCs because even the underfill can be inspected. Underfill voids and even filler separation are well revealed.

4.4.2 Voids

Voids within solder balls (spheres) have been observed in BGAs before and after assembly by researchers and commercial assemblers. The source of the problem is not the solder sphere because the various manufacturing processes make it virtually impossible for voids to survive. All the solder sphere manufacturing processes go through a stage where the sphere is molten and the high surface energy of the metal quickly expels any gas or vapor. The common source of voids is volatiles from the flux either during the BGA balling process or during assembly to the PWB. Excessive flux has been the most common cause, but the wrong reflow profile contributes. Proper process and material control eliminate any voiding problems.

4.4.3 Cost

BGAs generally cost more than the equivalent perimeter SMD package. The BGA is generally more complex, and several conductor layers are common compared with a single metal lead frame for non-area array packages. While cost will continue to drop, the BGA may not appear to achieve parity. We need to keep in mind that there may not even be an alternate to the BGA, especially at higher I/O counts. The perimeter package becomes impossible to build at some point when I/Os increase and allowable footprint decreases. The BGA also may have enhancements such as heat management that may not be present in another style of package. Economics will continue to favor the use of perimeter-style packages when I/Os are low enough and area is not too restrictive. However, when certain density and performance criteria are important, then the BGA becomes the only choice. Make sure that you do not compare a BGA or CSP with a "phantom" package.

The FC is somewhat of a cost paradox. Since it is only a bare die with bumps, cost should be lower than for a standard package. Automotive electronics manufacturers such as Delphi-Delco claim that FCs are used to save money. Others complain about the high cost. The difference of opinion can relate to underfill and the added steps required. Underfill process equipment has been said to take up as much floor space as the assembly equipment. Underfilling remains an added step that imparts significant cost for many applications. However, continuous progress with underfill materials and processes will reduce and perhaps remove this cost adder. These chapters cover aspects of underfilling: Chap. 7, "Flip Chip Technology"; Chap. 17, "Next Generation Flip Chip Materials and Processing"; and Chap. 18, "Flip Chip Assembly and Underfilling."

4.4.4 Planarity

Area array packages must be reasonably planar for the solder balls to make contact with the wiring board and form joints. Once the balls collapse during solder reflow, they can accommodate significant noncoplanarity, however. There is no real problem for most ceramic and metal packages because these materials are very planar and stiff enough to resist warping stresses. The *flex-based BGA,* also called a *tape BGA,* typically is very planar because it is commonly bonded to a backer or stiffener such as metal. The *plastic BGA* (PBGA) can have planarity problems, however. Organic substrate, especially when thin, will warp out of plane if stressed. Warping stresses can be produced when encapsulant polymerizes. Nearly all polymers shrink slightly when they cure or polymerize. The shrinkage tends to warp the BGA organic carrier edges upward. This can happen with both liquid encapsulants and solid epoxy molding compounds (EMCs). The PBGA also may warp due to a thermomechanical mismatch between any of the components, but especially when the cured encapsulant and the carrier have different coefficients of thermal expansion (CTE values).

The early PBGAs had a significant warping problem due to excessive shrinkage of the liquid encapsulants that were used several years ago, as well as problems with substrate that was not yet optimized for this application. Today's encapsulants are considerably better because of improved polymers and higher filler levels. The higher the filler, the lower is the percentage of polymer that can shrink. Low-stress EMCs have come into use over the years and reduce warping even further. Substrate also has improved in terms of advanced polymers and better glass-reinforcement technology. However, a large PBGA made of thin substrate still can have warping problems.

4.4.5 Moisture Absorption

Moisture absorption by a package has at least two detrimental effects. Absorption by the dielectric can reduce electrical isolation and thus degrade the signal. A more serious problem can be the explosive evolution of water vapor in the form of steam on rapid heating. A PBGA can absorb enough moisture to delaminate or even crack the encapsulant during reflow soldering. The explosive evolution of steam by the package is known by the term *popcorning.* While the industry continues to reduce moisture absorption of BGA materials, the fix is to prebake the package just prior to solder reflow unless it has been kept dry in storage. Moisture in either the organic chip substrate, the die-attach adhesive, or the encapsulant can be a problem. Although some moisture absorption in the common epoxy encapsulants has been reduced, a different polymer with intrinsically lower moisture absorption may be the answer. At this point, moisture sensitivity is still considered one of the problem areas for PBGAs, including flex-based types. The polyimides used for flex can absorb up to nearly 3 percent moisture, making them one of the most sensitive package and circuit materials. In some cases, explosive evolution of water trapped in the polyimide film can blow metal conductors off. Predrying, while an annoyance, can be used for all packages. Flex circuitry is routinely predried before components are assembled.

4.4.6 Rework

BGAs are more complex to rework because the lead frame is the solder ball. While reworking procedures are well established and equipment is available, the process is not as simple as for perimeter packages with nonmelting connectors (see Chap. 19, "BGA Rework").

4.5 FIRST-LEVEL INTERCONNECT (CHIP TO CARRIER)

4.5.1 Wire Bonding

Wire bonding remains the most popular first-level connection process for BGAs and many CSPs. The process allows the wire-bonding machine to be programmed so that one BGA substrate design may accommodate many different chips of different sizes. Wire bonding continues to be an efficient process for all but the highest I/O types of chips or those dies with an area array pattern. Wire bonding over an active area (inside from the die's perimeter) is considered potentially destructive and not recommended even if it were practical. Continual rate improvements for die bonders allow dies with 600 I/Os to be bonded economically. The exact crossover point for moving to another first-level process, such as flip chip, continues to be a moving target as bonder rates increase. However, the point for moving away from wire bonding is probably between 600 and 800 I/Os. Wire bonding also increases the signal path length and therefore can limit the maximum external clock rate (frequency) of the chip. About 90 percent of all packages are still wire bonded, down from almost 100 percent a few years ago. The slow move away from wire bonding will continue, but it will remain a valuable process far into the foreseeable future.

4.5.2 (TAB) Tape Automated Bonding

TAB inner leads consist of very small cantilevered metal beams suspended over an opening (window) in the dielectric film of the chip carrier. The leads are patterned to correspond to the bond pads of the chip to be connected. The tiny leads typically are made of copper with a gold finish over a nickel-plated barrier. Some CSPs, like the Tessera μBGA, use pure-gold conductors. The connection is made by accurately positioning these inner leads over the chip pads and applying heat and pressure with a TAB inner lead bonder. The chip pads may be gold plated or even gold bumped to facilitate the bonding process. The pure-gold leads of the μBGA can be bonded to common aluminum pads, but TAB processing generally requires a gold-to-gold junction formation. One can view TAB bonding as a form of wire bonding where the "wires" are built into the carrier. The interconnect bonds can be formed one at a time or as an entire row (gang bonding). In some cases, all bonds are made simultaneously, making TAB bonding much faster than wire bonding. The TAB carrier is actually a flex circuit, but it often is made with thinner material and higher-precision circuitry processes. Standard TAB has a conductor pattern that fans outward and terminates as a wire pattern that can be soldered to the wiring board for second-level connections. However, this format is not area array, and the common TAB design must be modified. This will be covered under second-level, or package-to-board, connections later. Figure 4.6 shows the TAB inner lead and a "ball" outer lead that is commonly seen in many flex-based area array packages.

4.5.3 Flip Chip (FC)

FCIP generally is considered the first-level ultimate connection because the highest density can be achieved and the path length is shortest so that optimal electrical characteristics are achieved. The FC has bumps over the bond pads that are used for the connection. These bumps may be made of solder so that the bump itself is the source or of a nonfusible metal or alloy. All the FC bonds are made simultaneously, making this the fastest first-level connection method. FC is also the only area array first-level process, making it the densest format. The total number of I/Os possible for a single chip probably can exceed 10,000. IBM has built actual products with over 7000 bumps. Both organic and

FIGURE 4.6 TAB with BGA format.

ceramic carriers are used with FC. CPUs, better known as *computer chips,* are most commonly assembled as FCIP, and I/Os range from 1000 to 3000, although this number continues to climb. This technology is well covered in these chapters: Chap. 7, "Flip Chip Technology," and Chap. 18, "Flip Chip Assembly and Underfilling".

4.6 PROTECTION METHODS

4.6.1 Epoxy Transfer Molding

Epoxy molding compound (EMC) is the most common encapsulant used to protect packages, and this is true of BGAs as well as other styles. Individual packages, strips, or even full arrays of unsingulated carriers can be molded. See Chap. 23 for details on molding.

4.6.2 Liquid Encapsulation

Liquid epoxies and other polymers can be applied to encapsulate an area array package using a needle dispenser. Automated dispensers are used widely that can handle the entire process, which may require several steps. A common method called *dam and fill* involves needle dispensing a retaining bead, or "dam," of viscous encapsulant around the package perimeter. Fluid "fill" encapsulant is next deposited within the boundary to cover the chip and interconnect structure. The assembly is then heated to harden both materials. Liquid encapsulants are used more commonly for prototyping, for shorter production runs, and in situations where fast time to market and minimum production tooling are important criteria. Liquids are used almost exclusively for cavity-type packages, where the die is located in a recess in the bottom of the carrier, because molding is not practical. Chapter 22 covers liquid encapsulation in detail.

4.6.3 Lid Seal

Some metal and ceramic packages may be sealed with a cover or lid that is bonded over the opening once the chip is attached and connected. Metal lids can be soldered, welded, or bonded adhesively, although the latter method does not give an airtight hermetic package. Ceramic packages and lids also can be soldered if the appropriate metalization is applied. They also can be bonded adhesively. In all cases, an adhesive lid seal does not provide an airtight hermetic enclosure because air slowly leaks through the polymer. Continuing improvement in polymer seal materials may provide near-hermetic or quasi-hermetic packaging that is good enough for most situations that need a low-gas environment.

4.6.4 None?

Some people view the FCIP when not covered with encapsulant as unpackaged. However, the active side of the die faces downward, and the underfill serves as the protective encapsulant that seals the interconnect structure. Very few, if any, area array packages are unprotected. Chips are not only sensitive to the environment, the sensitivity also has been increasing as denser, more sophisticated products have been introduced. A few CSPs may be protected by inorganic "encapsulant" instead of polymers. The early IBM STL is an example.

4.7 SECOND-LEVEL INTERCONNECT (PACKAGE TO PWB)

4.7.1 Pins: Pin-Grid Array (PGA)

The PGA uses a mechanical connection method by incorporating an array of stiff metal pins into the package substrate that can plug into a socket on the PWB. The pins are usually a base metal with gold plating to ensure a good, unoxidized system for mating. The PBA body can be organic, ceramic, or metal. Ceramic has become popular because of strength, low thermal expansion, high elasticity modulus, good planarity, high thermal conductivity, exceptional heat resistance, and excellent electrical insulation. Since the second-level assembly process is mechanical, easy field replacement is possible, making this a good package for chips that need to be upgraded, such as CPUs. This type of package remains popular, and FCIP is common for CPUs. The higher cost, however, makes ceramic a target for organic substrate.

4.7.2 Eutectic Solder Spheres

Most BGAs, CSPs, and FCs use eutectic solder for second-level connections. Eutectic solders, by definition, have a melting point (liquidus) and freezing point (solidus) that are equal (see Chap. 14, "Solder Pastes"). Exceptions are high-melting-point alloys that require the addition of solder to make the connection to the board. Eutectic solder typically is supplied as the BGA ball or FC bump, and no additional joining material needs to be added.

4.7.3 Nonfusing Metal Spheres, Bumps, or Columns

Nonfusible means that the metal or alloy does not melt under typical solder reflow conditions. Nickel, gold, copper, high-lead solders, bumps, balls, and studs are all used. Solder or conductive adhesives are used to form the joint between the package metal and the wiring board. Advantages for nonfusible systems include an easier test interface and a predictable stand-off height, or gap, which can be valuable for FCs. The gap for FCs has a great influence on joint fatigue, and underfill cannot flow under too small a space.

4.7.4 Conductive Adhesives

Conductive adhesive bumps are used for FCs but are not common on other area array packages. Work to apply them to BGAs did not show any real advantages over solder balls.

Isotropic Balls/Bumps. Isotropic conductive adhesives (ICAs) are typified by silver-filled epoxies similar to those used for die attachment. The adhesives can be applied to the chip as a paste by stencil printing, preferably at wafer level, and then hardened thermally. The polymer-bumped chips are then bonded to PWBs by adding more paste than is usually printed onto the PWB. The bumped FC is placed into the paste and then thermally cured. Underfill is required. Some versions involve applying a nonconductive film of underfill to the wafer or PWB. The adhesive bumps are forced through the film by heating and applying force.

Anisotropic Films. Films or pastes can be applied to FCs to provide an interconnect using anisotropic conductive adhesives (ACAs). The film or paste contains a dispersion of tiny conductive particles, usually spheres, that will create a path between the chip pad and wiring board when force is applied. The conductor level is too low for paths in the *xy* plane to be produced, and it acts like an insulator. Only particles trapped between opposing chip and circuit pads form electrical paths. No underfill is needed because the adhesive layer serves this purpose. See Chap. 16 for topics relating to adhesive assembly.

4.8 THE PLASTIC BALL GRID ARRAY (PBGA)

4.8.1 Die-Up

The simplest area array package has a die-up (cavity-up) design (Fig. 4.7). The chip is bonded to the top of the carrier substrate opposite to the ball or stud connection side. The chip can be wire bonded or attached directly in an FC format. The chip and interconnect are encapsulated by liquid or EMC in the case of wire bonding. An FC may be underfilled but also overcoated with encapsulant. There are molding processes that underfill and overcoat simultaneously.

4.8.2 Die-Down

The die-down (cavity-down) design cannot have a full array of bumps because space must be reserved for the die cavity. The package must be deep enough that a die cavity can accommodate one or more chips as well as the interconnect structure such as wire bonds. The package can be organic, ceramic, or metal. Once the chip is bonded and connected, the package must be sealed with a lid or filled with liquid encapsulant to protect the die. Figure 4.8 shows a typical die-down format.

4.9 METAL PACKAGES

Metal area array packages are designed with insulation to isolate conductors. The chip is placed and bonded within the cavity and then connected electrically.

4.9.1 Hermetic

A truly hermetic seal requires that a metal lid be sealed metallurgically, with solder or by welding. Chapter 12 covers hermetic packaging.

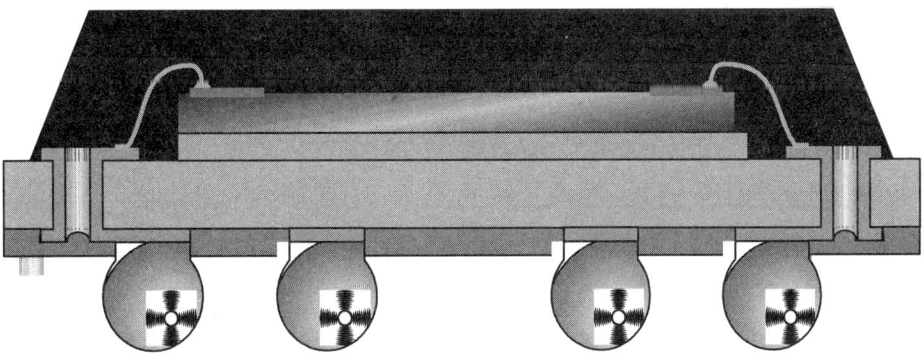

FIGURE 4.7 Die-up BGA.

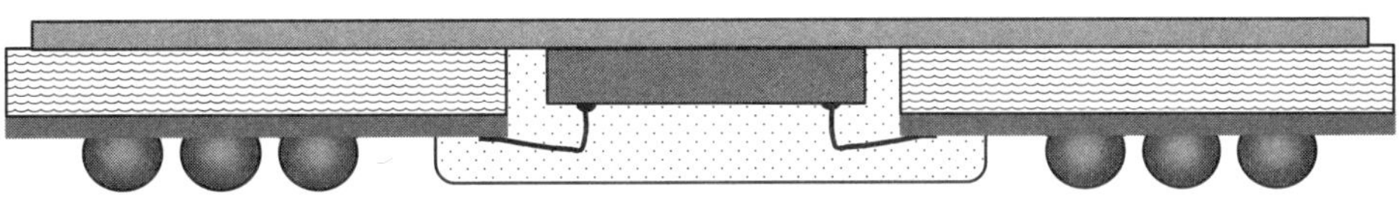

FIGURE 4.8 Die-down BGA.

4.9.2 Encapsulated

Filling with polymer encapsulant can protect the die and connection structure within the metal cavity. This is a simpler and lower-cost process, but gas molecules will diffuse slowly through the plastic. This is not a problem for most dies.

4.10 CERAMIC

Ceramic BGAs and PGAs remain popular for high-lead-count chips such as computer chips (see Chap. 10). The high degree of planarity, excellent thermal stability, and good thermal conductivity override the higher cost factor. Ceramic modules can be used as a carrier for FCs, sealed with a lid to form hermetic package, or used with underfill. The most popular format for computer chips is a carrier for an FC. The board-level connection can be mechanical pins, solder balls, or solder columns.

4.11 CSP PRODUCTS

CSPs come in all shapes and configurations. The simplest use an organic carrier platform with solder balls on the bottom. The chip is bonded to the top of the carrier and connected with short-path and low-loop wire bonding. The only difference between a BGA and a CSP is really one of relative dimensions of die versus package footprints.

4.12 FLIP CHIP: IS IT A TRUE PACKAGE?

There is no doubt that the FCIP is the genuine article, a real package. What about FCOB? The consensus is that an underfilled FC is not reworkable, at least in the common meaning of the term, and therefore is not a package. Most, but not all, packaging authorities suggest that a package should be removable. The ideal package is *reworkable,* meaning that it can be used again on a new wiring board or be repositioned if the fault was with the assembly. This is especially important for computer chips, which can have much more value than a PWB.

4.13 RELIABILITY

The area array package is one of the most reliable designs. The FC on ceramic substrate used in essentially all the IBM mainframe computers probably holds the record for reliability. The ability to distribute signal and power more evenly and to remove heat more efficiently gives area array a reliability advantage. Early BGAs had some reliability problems as the technology was just emerging. Improved designs, optimized materials, and a better understanding of the system have eliminated these initial reliability issues.

4.14 FUTURE EXPECTATIONS

The various area array designs can be expected to gain more market share as the desire for higher performance and larger-I/O-count products increases. While many people had hoped for a consolidation of packaging types, the trend will be for a continued proliferation of types. We can expect more CSP and BGA packages to be developed as the need for unique packaging solutions increases, especially since new microelectromechanical systems (MEMs) and advanced photonics products add special requirements that may be best handled with area array designs. See Chap. 31 for predictions.

CHAPTER 5
STACKED/3D PACKAGES

Lee Smith
Amkor Technology, Inc.

5.0 INTRODUCTION

Industry experts believe that the fourth wave of semiconductor package innovation is being driven by challenging new electronic applications. The term *application-specific packaging* is being used to characterize the new microelectronic innovations associated with this era of electronic system convergence. This fourth wave is enabling further silicon integration at the system level by allowing a mix of semiconductor technologies to be integrated at the package level to address specific subsystem density requirements. This chapter will look at both the mobile phone market as the new "killer" application driving the magnitude of this fourth wave and the stacked or three-dimensional (3D) package as one of the most promising new application-specific packaging innovations available. Figure 5.1 provides a definition of application-specific packaging and describes the scope of requirements being addressed through multichip integration at the package level.

5.1 THE FOURTH WAVE OF PACKAGING INNOVATION

Figure 5.2 is Prismark Partners' look at the past and projected waves of packaging innovation over a 40-year horizon.[1] The first wave of packaging was through-hole components driven by the large motherboard assemblies associated with the telephone and computer industries that dominated the late 1970s through early 1980s. When the wave of new low-cost consumer electronics began to flow in the middle to late 1980s, innovations in packaging and assembly were needed to provide the automation required to deliver the high volumes and low costs needed to satisfy strong consumer appliance demand. Thus the surface-mount component was developed and saw rapid adoption in both consumer and emerging personal computer (PC) applications. As packaging pin counts and package sizes continued to increase to support the device advances associated with very large scale integration (VLSI), the perimeter leaded surface-mount technology (SMT) component began to be a limiting factor for automatic assembly and miniaturization due to the sensitive nature of the fanned-out fine-pitch surface-mount leads. Thus a new innovation that distributed the contacts across the entire area of the component was born in the 1990s to support the increasing input-output (I/O) demands of the PC wave. This third wave was called *area array* and was characterized by the plastic ball grid array (PBGA). It provided significant size reduction and SMT throughput increases due to the benefits of the self-aligning solder balls evenly distributed under the package. Now that the mobile phone handset has become the "killer" application of the new millenium, a fourth wave of packaging innovation is required to allow handsets to continue to miniaturize while increasing in

Definition:

- A package developed or customized to meet the specific requirements of an electronic subsystem

- Scope of requirements:
  - Reliability specific to that application (1st & 2nd Level)
    - 1st or component level testing defines the reliability of the package over the range of assembly temperatures and use environments expected over the life of the product
    - 2nd or board level testing defines the reliability of the package solder joint to mother board interface over the products life
  - Performance specific to the application and devices or functions integrated within the Package
    - Form Factor, Electrical, Thermal, Mechanical, Optical
  - Designed for optimization of the total cost required for the application
    - Standards, Infrastructure, Time to Market

FIGURE 5.1 Application-specific packaging.

functional performance. In addition, the growing density gap between the semiconductor industry (which has remained on track with Moore's law for doubling every 18 months) and the slower-moving printed wiring board (PWB) industry has required that the integrated chip (IC) package provide the critical space transformer—between the submicron density of the IC industry and the much courser submillimeter density of the PWB world. One of the most effective ways for the IC packaging industry to keep pace or scale with Moore's law without outpacing the PWB electronic platform's ability to handle more silicon integration is through 3D packaging—volumetric solutions. The mobile phone market was a strong early adopter of chip-scale packages (CSPS) and also has been a leader in seeing the benefits 3D packaging provides for increasing memory capacities in shrinking handsets.

To help put Fig. 5.2 in perspective, in 1999, over 68 billion ICs were produced worldwide, 94 percent of these housed within a package. Prismark projects over 113 billion ICs in 2004; assuming a conservative estimate of 15 percent unit growth through the year 2020, then the 3D packaging share of 11 percent can total to well over 70 billion units that employ stacking technologies. Thus, even if this forecast is off by a few percentage points, the market for 3D packaging is expected to be huge going forward.

To achieve continuous size, weight, and cost reductions, wireless handset manufacturers and their semiconductor suppliers have been developing new semiconductor devices and packaging technologies that enable higher levels of functional integration at the smallest, most cost-effective level, which is through silicon (Si) integration. The higher levels of Si integration for baseband functions have developed the quickest due the system-level integration nature of application-specific ICs (ASICs) and the common CMOS processing for baseband functions. The memory and radiofrequency (rf) functional integration for cellphones is currently being accomplished through application-specific packaging advancements defined as *system-in-a-package* (SiP). The emergence of SiP

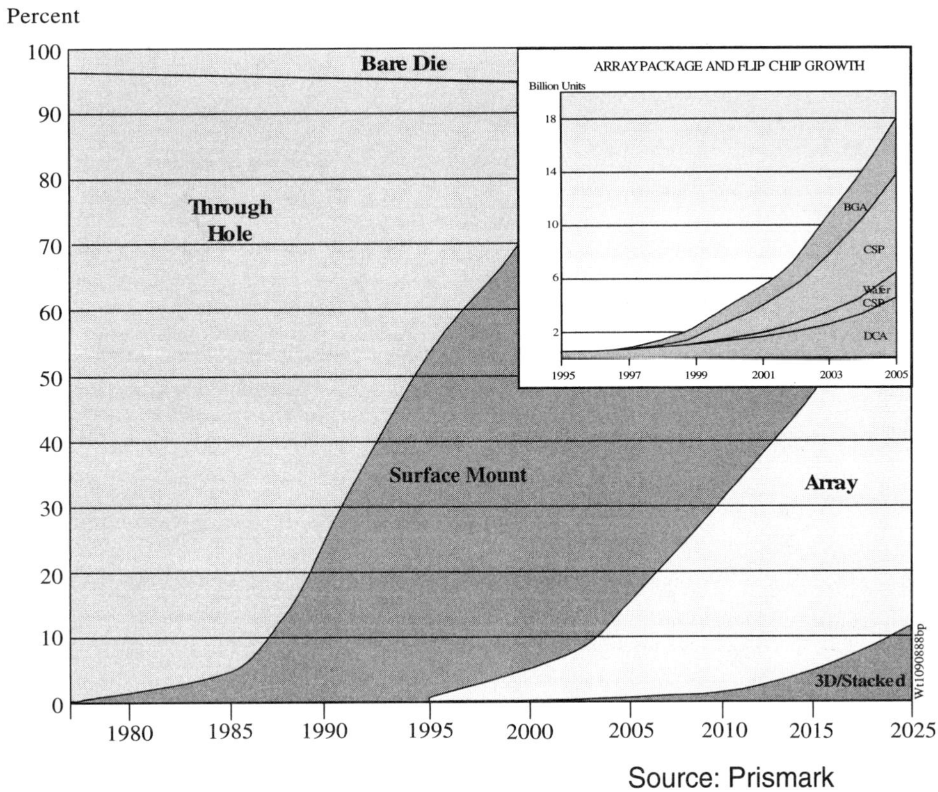

FIGURE 5.2 Waves of electronic packaging.

is due to the complications of integrating different IC technologies or densities in a single chip or system-on-a-chip (SoC) solution. This chapter will take a look at advancements in thin die stacking delivered through 3D packaging and more specifically stacked chip-scale packaging (S-CSP) technology, which is a major segment of the SiP movement and application-specific packaging trend. To avoid the confusion associated with the terms *application-specific packaging, system-in-a-package,* and *3D packaging,* consider the following guidelines:

- Application-specific packaging defines the *trends* that are driving a wide range of package advancements to meet the requirements of a specific segment or segments of device and application cost/performance requirements.
- System-in-a-package (SiP) defines the technical approach of integrating mixed semiconductor technologies at the package level where multiple ICs and their associated discrete and passive components are integrated to form a subsystem functional block that is fully characterized and tailored to meet that application's performance requirements.
- 3D packaging is a subsegment of the SiP movement where die or package stacking is employed to provide higher levels of silicon efficiency with respect to the package area by using the *z* dimension for volumetric density solutions.

3D packaging allows further system integration through stacking of diverse device technologies, including Flash, static random access memory (SRAM), and dynamic random access memory

(DRAM) memory blocks, as well as baseband, mixed signal, analog, and logic functions. Stacked chip package integration will be critical to allow the pace of handset size, weight, and cost reductions to continue, as well as new handset form factors to be developed.

5.2 REVIEW OF CHIP-SCALE PACKAGING (CSP) AND STACKED CSP (S-CSP) ADVANCES

The pace of CSP technology development is accelerating in the semiconductor industry, driven by broad adoption of CSPs in wireless handsets and handheld electronics. The following milestones are cornerstones of the foundation for a strong global infrastructure supporting CSP and S-CSP technologies. In August 1996, Sharp Corporation began production of wire-bonded, flex circuit, tape-based CSPs for initial adoption in cellphones. In March 1998, Sharp and Amkor Technology, Inc., formed a technology agreement to promote standardization and industry development for tape-based CSPs. In April 1998, Sharp introduced the S-CSP; by July 1999, Hitachi, Mitsubishi, and Intel teamed with Sharp to support S-CSP standards (industry specifications) for Flash and SRAM die combinations. In August 1999, Sharp announced the world's first triple-chip S-CSP development (triple-chip S-CSP production plans followed a year later). In September 1999, Sharp and Amkor signed a cross-licensing codevelopment agreement to promote standardization and industry development for S-CSP laminate- and tape-based technologies. Amkor is licensed to use Sharp's tape-based S-CSP technology, and Sharp is licensed to use Amkor's laminate-based ChipArray technology for S-CSP and CSP applications.

This type of S-CSP agreement and relationship between industry leaders such as Sharp and Amkor should benefit the semiconductor and electronic equipment industries by teaming to drive S-CSP cost reductions and ensure reliable supply and infrastructure development.

5.3 HANDSET FUNCTIONAL SYSTEM INTEGRATION

The continuous size, weight, and cost reduction trends for handsets would not be possible without the tremendous capabilities of the industry supply base to respond to new challenges and the insatiable levels of wireless consumer demand. The mobile phone broke the price elasticity supply-demand model and created a new model of the greatly discounted or free system carried by the wireless communications provider through the service contract. This delivery model contributed to the tremendous rates of wireless adoption and phone replacement rates as better handsets and services emerged each year.

Figure 5.3 provides a multiaxis graph of the reduction trends that have helped drive tremendous worldwide handset adoption. This graph provides a macro view of the rates of change over the past 5 to 10 years.[2]

1. 15 percent annual reduction in weight.
2. 14 percent annual reduction in PWB area.
3. 25 percent annual reduction in cost/price.
4. 60 percent annual increase in unit volume.

However, to understand the roots (enablers) of this ongoing level of change, an understanding of the functional blocks and their associated levels of Si integration and high-density interconnection (HDI) is required. Figure 5.4 summarizes the functional block diagram of a typical high-volume mobile phone, single-mode Global System for Mobil Communications (GSM) handset. Up through 1998, high levels of Si integration in the baseband subsystem have been key in enabling handset advancements. The three ASICs in the baseband subsystem of Fig. 5.4 illustrate this strong baseband integration.[3]

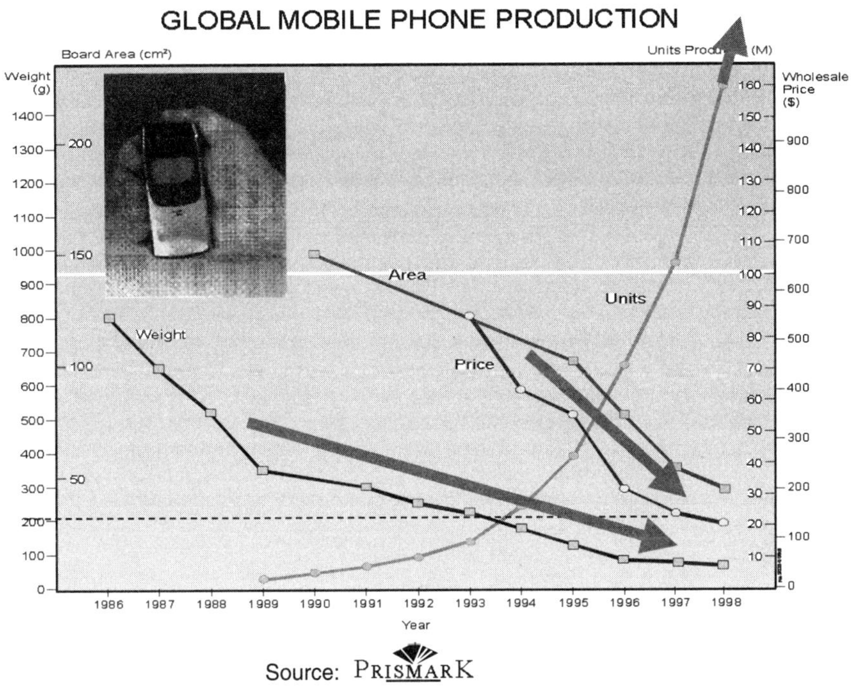

FIGURE 5.3 Global mobile phone production trends.

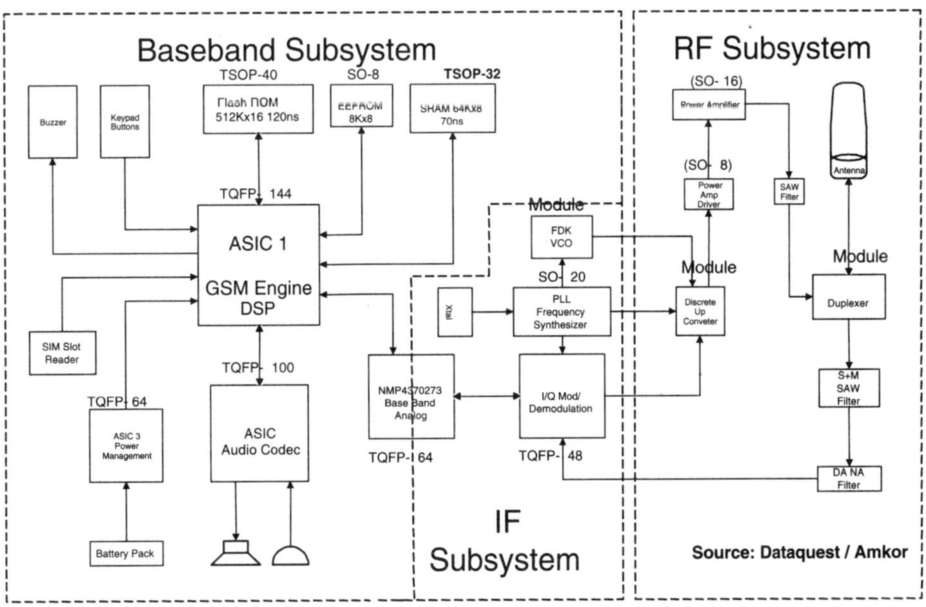

FIGURE 5.4 A wireless handset is a complex system in a few cubic inches.

The majority of 1999 and 2000 model handsets include multimode and multiband functions for broader regional service connectivity and are planning or testing features that offer generation 2.5 through 3G (third generation) voice plus data communications.

As a result, multimode (e.g., AMPS + TDMA or GSM + CDMA), multiband (e.g., 900 and 1800 MHz) handsets are becoming standard, with I-mode (Japan) and Wireless Application Protocol (WAP, Europe) Web browsing capabilities gaining strong adoption. To achieve size and cost-effective integration of multiple radios within a single handset, multichip package integration (system-in-a-package) is the current industry answer. Radiofrequency (rf) integration is being accomplished in package platforms that support the associated high-frequency materials and passive integration requirements in combination with GaAs- or high-frequency Si-based semiconductors. (Multichip rf packaging could employ stacked chip technologies in the future when the rf shielding capability is made more cost and size effective, but the rf subsystem is outside the scope of this chapter.)

To handle the additional multimode code and memory storage requirements, along with the data storage needed for Web browsing, memory capacities are increasing rapidly. To achieve higher levels of memory integration without increasing PWB area or handset size, weight, and cost, stacked chip package integration is the key enabling technology gaining worldwide adoption. Despite the strong promotion, system-on-a-chip (SoC) silicon integration approaches that require large memory blocks (or require memory technologies that differ in volatile versus nonvolatile architecture and feature size density) have proven too complex and costly for high-volume applications such as mobile phones.

5.4 CSP TO S-CSP ADOPTION IN HANDSETS

The mainstream 1998 handsets (as represented in Fig. 5.4) used traditional perimeter leaded IC packaging technologies based on copper lead frames. Small-form-factor "personal handy phones" available from Japanese suppliers in 1998 and mainstream handsets for 1999 adopted CSP technologies to reduce size and weight. The initial devices and CSPs adopters were

1. Tessera's widely licensed μBGA (TAB bonding) for Flash and SRAM packaging.
2. Wirebond on flexible printed circuits (polyimide tape) or rigid laminate-based CSPs/FBGAs (fine-pitch plastic ball grid arrays) for digital signal processors (DSPs) and SRAM and electronically erasable programmed read only memory (EEPROM) devices. Leading tape-based CSP packages include Texas Instruments' MicroStar BGA and Amkor's *fle*XBGA. Motorola's MAP (mold array package) and Amkor's ChipArray were leaders in the wirebond on rigid substrate package families, with each employing gang mold and saw singulate process technologies for improved die to package edge design rules and reduced tooling/nonrecurring engineering (NRE) costs for each new package size.

These first two classes of CSPs became widespread in late 1999 and early 2000 handsets. Before moving forward to the S-CSP, it is important to help clarify the blurring definition of a CSP. Early on, a CSP was defined by Joint Electron Device Engineering Council (JEDEC) as a package only 20 percent larger than the die it contained. However, this definition did not really lend itself to standardization and infrastructure development. From the original equipment manufacturer (OEM) and surface mount technology (SMT) perspectives they did not want to see a motherboard change every time a CSP die shrunk, thereby requiring a new footprint or package outline. The tray and test socket suppliers had to wait for clear customer demand for each new CSP, creating a longer time-to-market gap for new CSPs than the rapidly moving wireless industry could allow. Also, every new minimalist package being developed wanted to be characterized as a CSP (due to the growing interest and market demand) while at the same time solving the disadvantages of true CSPs (by allowing a wider range of die sizes and shrinks to be compatible, without the need for a package change or costly requalification). In addition, the CSP definition had no application to the emerging multichip packages being adopted in handset and portable applications. Thus the distinction or lines between

BGAs and CSPs really began to blur, making it hard for customers and market forecasters alike to characterize and segment for evaluation. Therefore, a more practical definition of the CSP was developed for area array packages. The new distinction was based on the associated solder ball pitch, with CSPs characterized as array packages with ball pitches of 0.8 mm and below and BGAs as ball pitches of 1.0 mm and above. This distinction has worked pretty well from the OEM down through the supply chain because now the distinction is more a factor of density rather than die-to-package size. A fine-pitch density segmentation also has application to the type of SMT assembly and PWB fabrication required to support the finer-pitch devices.

As soon as this new definition of CSP achieved broad acceptance, a new 3D minimalist packaging approach emerged in order to continue meeting handset cost and form-factor reduction requirements for their increasing memory capacities. Now we see S-CSP technologies being adopted by nearly all major handset manufacturers for the memory blocks and baseband section. Figure 5.5 illustrates the space savings associated with integrating Flash plus SRAM into a S-CSP versus two separate FBGAs. For a 16-Mb Flash device and a 2 Mb SRAM (common in the initial transition from CSP to S-CSP for 2000 mainstream handsets), the associated PWB area savings is 60 mm^2.[4] Figure 5.6 provides a size and weight comparison for S-CSP versus the alternative Flash and SRAM single-chip and multichip packaging approaches.[5] Two separate thin small outline packages (TSOPs) provide the baseline (100 percent) for size and weight reduction comparisons. As the first column indicates, 16 Mb of Flash and 2 Mb of SRAM integrated in a S-CSP are about one-third the area and weight of two separate TSOPs common in 1998 model handsets. In other words, S-CSP provides a factor of 3 increase in packaging density. In addition to the preceding benefits for Flash + SRAM in an S-CSP, in some cases the EEPROM (data storage) block or function is being integrated within the base Flash chip or its architecture, allowing read while write (RWW) functionality.

These Flash device and packaging advancements allow handsets to integrate higher memory capacity in a single package versus three separate components, providing greater silicon efficiency through 3D packaging density and the associated cost reduction.

5.5 S-CSP GROWTH:STANDARDS AND INFRASTRUCTURE

The electronics and semiconductor markets saw a slowing in their respective torrid growth rates in the fourth quarter of 2000. Subscriber growth rates for wireless communications also have slowed in the largest markets, the United States and Europe, due to economic softening and high rates of wireless adoption (market penetration or saturation). Coupled with a decline in replacement or upgrade

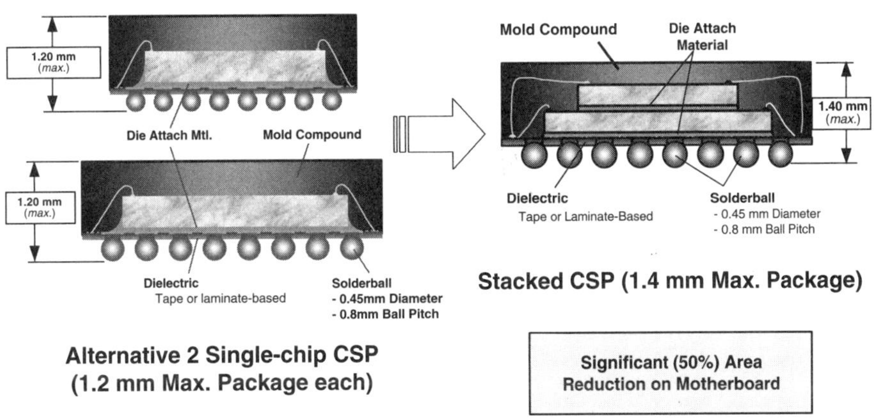

FIGURE 5.5 Flash plus SRAM integrated into S-CSP.

	Stacked CSP	TSOP x2	CSP x2	Stacked TSOP	MCM/Module
Structure (*Memory*)			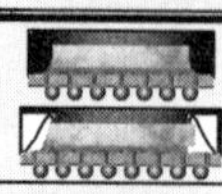		
16Mb Flash (x8)	8x10mm 80mm² 0.17g	10x14mm 140mm² 0.30g	8x10mm 80mm² 0.13g	10x14mm 140mm² 0.32g	11x14mm 154mm² 0.34g
2Mb SRAM (x8)		8x13.4mm 107.2mm² 0.25g	6x8mm 60mm² 0.08g		
Mounting Area (mm²)	80	247.2	140	140	154
	32%	100%	57%	57%	62%
Weight (g)	0.17	0.55	0.21	0.32	0.34
	30%	100%	38%	58%	62%
Package Height (mm)	1.4max. (1.25 typ.)	1.2max. (1.10typ.)	1.2max. (1.05 typ.)	1.2max. (1.10 typ.)	1.6max. (1.40 typ.)

Highest Silicon Efficiency

FIGURE 5.6 S-CSP versus alternatives.

phones and services, this has offset the tremendous rise in I-mode adoption within Japan to downgrade worldwide cellphone consumption forecasts for 2001. Industry estimates vary, but on average, forecasters and handset suppliers report that around 420 million handsets were sold in 2000. Not all these sales ended up in the hands of consumers, with service providers and distribution channels accumulating excess inventory, further causing demand adjustments throughout the supply chain. Since Flash memory has been in tight supply through 2000, inventory adjustments have had less impact on the growing demand for S-CSP.

TechSearch International estimates that 100 million S-CSP products were consumed during 2000 almost exclusively from cellphone applications. Figure 5.7 provides TechSearch's recent forecast for cellphone growth, S-CSP consumption, and the rate of penetration into cellphones.[6] Based on the slow start in 2001, this forecast is seen as optimistic for cellphones over the next 4 years. However, the forecast is conservative for 500 million S-CSPs in 2004 due to the broad range of new 3D packaging applications from personal digital assistants (PDAs), wireless base station memory blocks, Flash memory cards, and Internet switch/router memory blocks being planned for next generation systems.

This enormous 400+ million unit consumption established wireless handsets as the overwhelming highest volume electronic system and establishes cellphones as the "killer" application for the new millennium. Further, the handset becomes the dominant communications platform and subscriber base for further integration of information management applications such as Internet access (Web browsing), which is at the center for the delivery of these *new data-centric wireless services,* as can be seen in Fig. 5.8*a,* along with the innovative new handset form factors being tested or developed in Japan, as can be seen in Fig. 5.8*b.*[7,8] The following innovative new handset form factors, features, functions, and services clearly require innovative new packaging solutions, including 3D packaging, to allow added functionality without increasing handset size and weight.

- Voice mail and text messaging
- Faxes
- E-mail

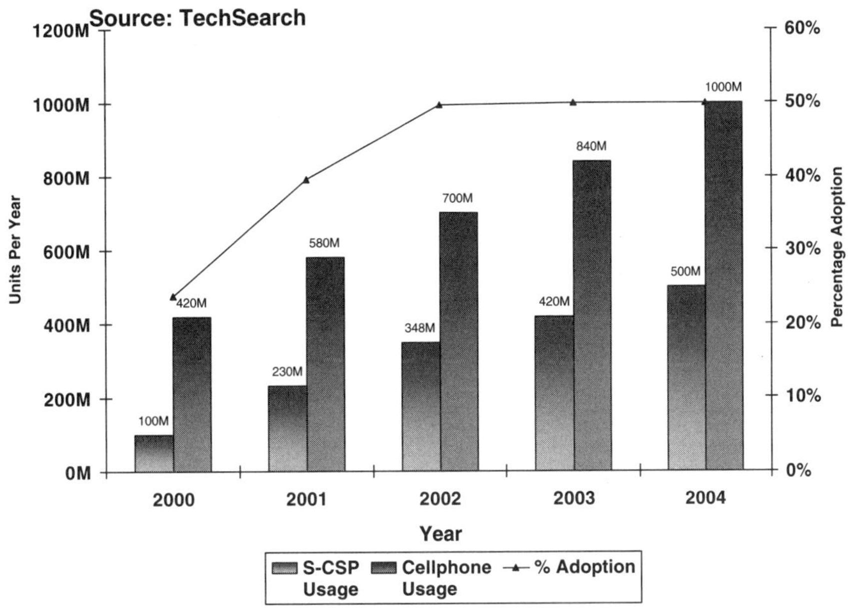

FIGURE 5.7 S-CSP market forecast.

- Internet browsing
- Maps
- Airline and other online reservations
- Telephone directories
- Over the air software (program) updates
- Stock quotes and transactions
- News clipping services
- GPS (global positioning system)

These data-centric applications will drive additional code and data-storage capacity requirements. Further, the availability of diverse services will cause application-specific segmentation of the handset platforms that must cost-effectively support their base voice communication services along with this range of additional new services. Some of the industry leaders are indicating there may be as many as five different handset segments for different levels of budget, service, features, and functions. This surely will drive the need for a robust industry infrastructure that meets not only today's need for high-volume availability of the latest semiconductor, packaging, microvia, and thin high-density substrate technologies but also tomorrow's needs for even higher levels of integration and density. Figure 5.9 provides an outlook for component and assembly densities expected for three major handset segments by 2004.

The following are some of the current items essential to ensure that the industry infrastructure is in place to support the expanding requirements of these additional cellphone segments along with their shrinking form-factor and time-to-market demands:

1. Second-source availability of Flash memory components for cellphones, since leading Flash suppliers continue to have different strategies, protocols, and pin-outs for their leading-edge designs. The handset makers drive their Flash suppliers for innovative products that solve their storage prob-

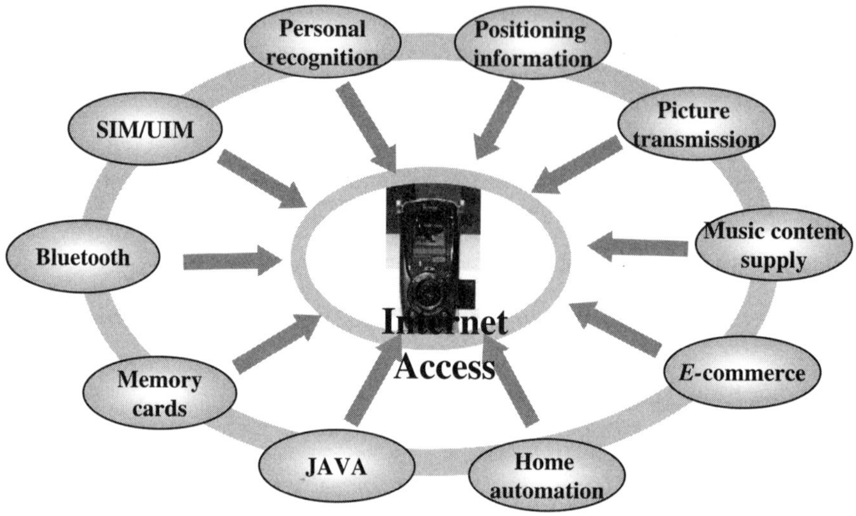

(a)

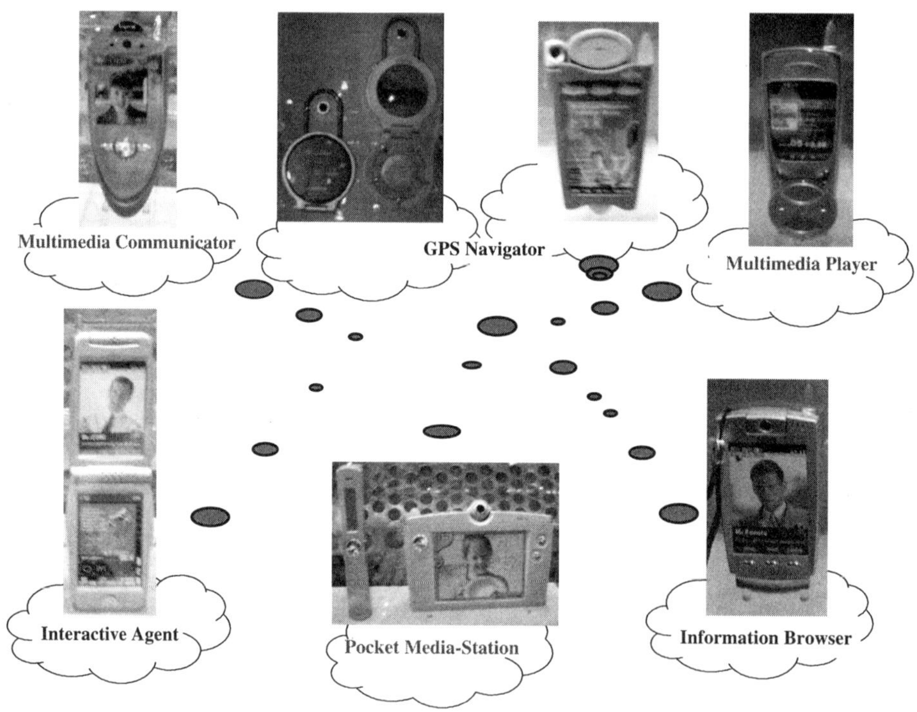

(b)

FIGURE 5.8 (*a*) New services drive proliferation of handset features and form factors. (*b*) New form factors and wireless service platforms.

- Standard Phone

Board Top Surface Area:	35 sq cm, mostly single-sided assembly
Component Pitch:	Mostly 0.8 mm CSP
Average Density:	20 pads/sq cm
Peak "Per Sq In" Density:	60 pads/sq cm
Peak "Per Comp" Density:	110 pads/sq cm

- Watch Phone

Board Top Surface Area:	12 sq cm, double-sided assembly
Component Pitch:	Mostly 0.5 mm CSP
Average Density:	40 pads/sq cm
Peak "Per Sq In" Density:	80 pads/sq cm
Peak "Per Comp" Density:	310 pads/sq cm

- PDA Phone

Board Top Surface Area:	55 sq cm, mostly single-sided assembly
Component Pitch:	0.5 mm CSP for low I/O, 0.8 mm CSP for high I/O
Average Density:	35 pads/sq cm
Peak "Per Sq In" Density:	80 pads/sq cm
Peak "Per Comp" Density:	220 pads/sq cm

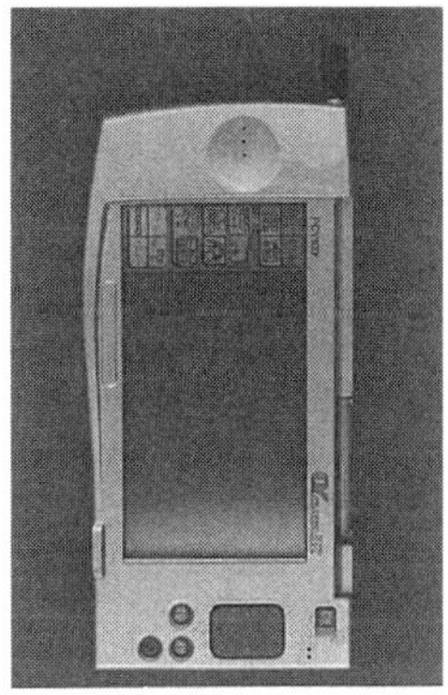

Source: Prismark Partners

FIGURE 5.9 Mobile phones in 2004.

lems as quickly as possible, thereby leaving standards in the dust.[9] Adopting a universal specification for an internal component may not be as critical as it is for end products such as VCRs, but a uniform Flash plus SRAM S-CSP specification surely would serve to facilitate expanding the industry volume readiness and infrastructure cost reductions. The same will hold true for the emerging integration of ASIC plus memory S-CSP packages. Here, most digital signal processing (DSP) and baseband processor applications can look to existing FBGA standards to adopt package outlines and pin-outs that are supported by open tool package, tray, and test socket offerings. This wide range of square FBGA packages can accommodate the logic plus memory stack size requirements without the need to invest in a new set of standards or infrastructure that was required for stacking of Flash plus SRAM in the initial S-CSP offerings. The packaging suppliers will need to develop the wafer thinning, thin die attach stacking, and low loop wire bonding and design rules to accommodate two dies in a package originally designed for one.

2. Widespread availability of thin, high-density PWBs using microvia fabrication and thin-core technologies is essential for both the needed buildup multilayer handset product boards and the thin two- to four-layer interposer substrates in laminate based S-CSPs. Substrates using 100-μm and thinner BT cores with 50-μm trace and space, 130-μm bond fingers, laser via in-pad, and tight soldermask registration control are critical to meet the advancing Si integration, density, I/O, and thin package height requirements emerging for new high-density memory stacks and ASIC plus memory stacks.

Note: Prismark forecasts the demand for microvia and HDI substrates used in product boards and packaging substrates to increase from $2.1 billion in 1998 to $7.6 billion in 2002, with 67 percent of this total consumed by the packaging substrate applications.[10] Substrate suppliers should plan aggressively to increase their capacity and capabilities to capture share in this high-growth market.

3. Widespread availability of cost-effective high-density single- and two-metal polyimide flex or TAB circuits for tape-based S-CSPs. The TAB and flex circuit industries need to invest aggressively in new one- to two-metal capabilities and capacity or will risk losing share to the emerging rigid thin-core suppliers.

4. A value-chain appreciation for the packaging and interconnection industries. Total cost-of-ownership valuations are required to support the higher levels of R&D and capital equipment required to meet the current wave of advanced 3D and packaging integration requirements. Thus 3D packages such as S-CSP, measured on a cost per pin basis, fail to account for the total cost of ownership benefits S-CSP delivers to the OEM.

5. Greater rate of attraction for students to pursue engineering and technical degrees in electronics packaging and interconnection disciplines.

These issues must be resolved in a cost-effective and timely manner to support the huge demand for S-CSP solutions required to support the current and projected handset, PDA, and memory block applications. Forecasts usually vary (especially for emerging technologies such as S-CSP), but current estimates require that the industry plan on capability and capacity to support up to 300 million S-CSP units for 2001. Equipment, test, material, and service suppliers cannot afford to wait until the demand/forecast outlook is clearer before finalizing their S-CSP strategies. Thus Sharp, Amkor, and others have developed flexible, robust S-CSP technology platforms and roadmaps to support a diverse range of new die combinations and end-application requirements. Figure 5.10 illustrates technology capabilities or building blocks Amkor has developed to provide a foundation to support the development and industrialization of 3D, SiP, and application-specific packaging platforms.

5.6 THREE-CHIP INTEGRATION IN S-CSP PLATFORMS

To understand the benefits that triple-chip stacking can provide to handsets requires a quick review of Figs. 5.3 and 5.4. Figure 5.3 shows a flattening in the rate of weight reduction below the 100-g threshold broken in 1996. As can be deduced from Fig. 5.4, further weight, size, and cost reductions for next-generation handsets can be accomplished through stacking the Flash plus SRAM combination on top of the DSP function in ASIC 1. This three-chip stacked package and wire-bond structure is represented in Fig. 5.11*a* and *b*.

For three-chip memory stacks (where die sizes and bond pads have been optimized for stacking), by using a thin, flex tape–based interposer and thin backgrind wafer technology, the triple-chip S-CSP can still meet the 1.4-mm maximum thickness goal associated with advanced handsets.

The adoption of triple-chip S-CSP is being pursued by Japanese handset makers to allow them to cost-effectively crack the 60-cc and 60-g benchmark as represented in Fig. 5.12. In addition, three-

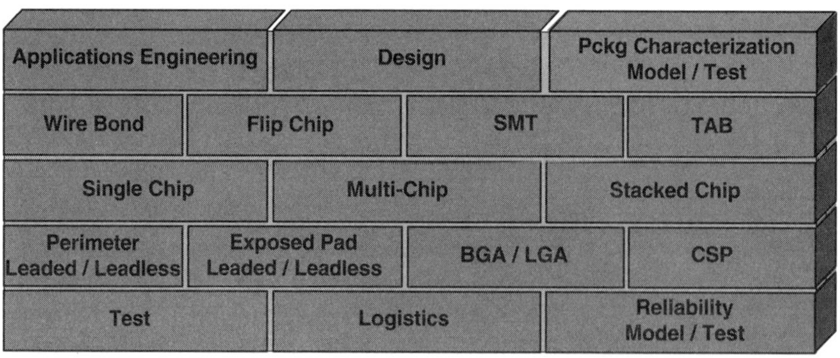

FIGURE 5.10 Capabilities required for an SiP foundation.

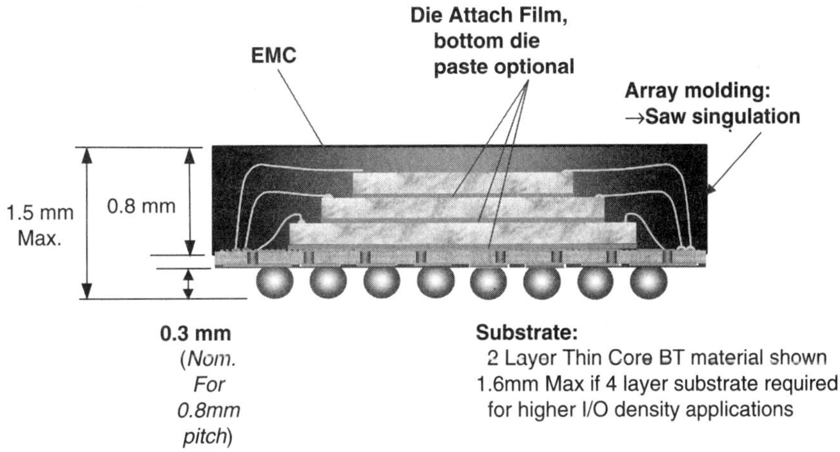

(a)

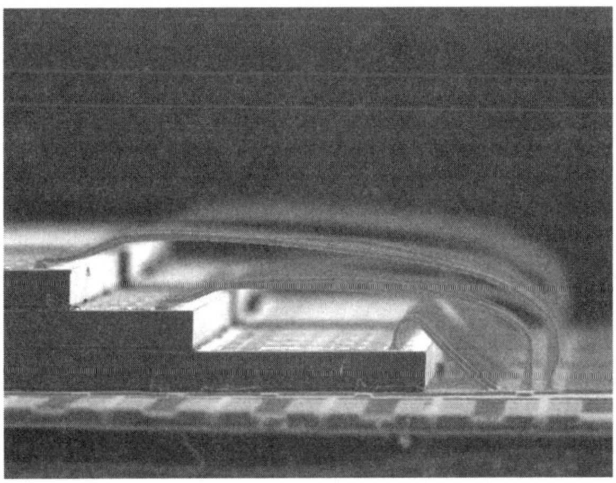

(b)

FIGURE 5.11 (*a*) Die-stacked CSP. (*b*) Die-stack wire-bond photo.

chip stacking will help enable high-end/data-centric cellphones such as the I-Mode and 3G handsets to be delivered with more features and functions in the small form factors consumers demand.

Another multichip 3D package that is gaining strong adoption for three-chip integration requirements is the so-called stacked MCM (S-MCM) depicted in Fig. 5.13. Here, die sizes, I/O, and wire-bond complexity do not allow integration in a three-die pyramid stack, so one of the dies is mounted to the side of the stack to accommodate thermal enhancement or fine-pitch wire-bonding requirements. Table 5.1 provides a comparison of the benefits and design flexibility for three-chip pyramid versus the 2 + 1 configuration associated with the stacked MCM. Figure 5.14 illustrates the various die stack and placement configurations feasible with stacked multichip packaging.

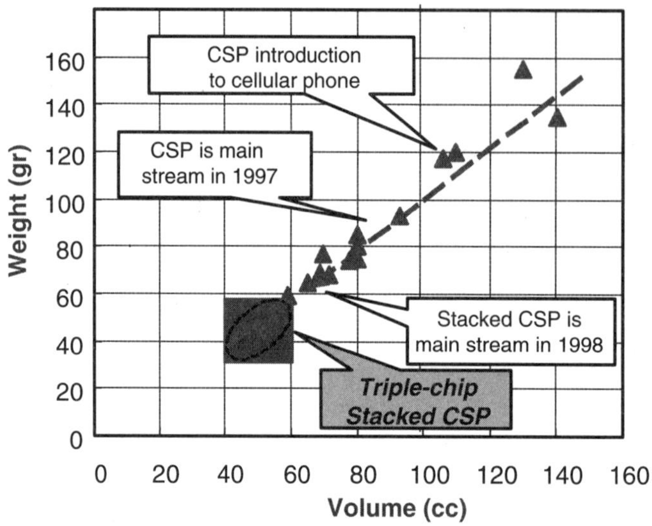

FIGURE 5.12 Triple-chip S-CSP contribution to compact, lightweight cellular phones.

5.7 FLIP CHIP (FC) AND S-CSP TECHNOLOGY ROADMAPS

A great deal of recent attention, particularly for analog and memory devices, has been placed on FC direct chip attach (DCA) or wafer-level packaging technologies for improved Si efficiency. However, as Fig. 5.15 illustrates, the best level wafer-scale or bare-chip solutions can achieve is 100 percent silicon efficiency, defined as the silicon area being equal to its associated package area. When you add the extra test and processing costs and underfill bleed buffer and rework limitations associated with current FC and wafer-level packages, they tend to be much less effective and harder to adopt than the 80 percent efficiency associated with leading single-chip CSPs. Further, the use of triple-chip stacking in emerging S-CSP technologies allows a level of 200 percent Si efficiency to be approached without having to tackle the more complex second-level interconnect and solder-joint reliability challenges associated with high-I/O 0.5-mm and finer ball pitches.

When this Si efficiency factor is looked at in combination with the widening gap between IC fabrication and PWB density capabilities, it becomes evident that 3D package integration is critical in ensuring that system suppliers can continue to leverage the unmatched rate of technology doubling represented by Moore's law. The International Technology Roadmap for Semiconductors (formerly SIA roadmap) identifies packaging as a critical enabling technology and calls for further investment and innovation in IC packaging so that packaging can scale with Si. One of the key functions of an IC package is to provide a space/density transformer between the submicron scale of the wafer-fabrication technologies and the submillimeter scale of the PWB technologies, but from a 2D perspective these technologies are not even on the same scale. One of the key enabling technologies to allow packaging to scale with Si without outpacing the PWB and printed wiring assembly capabilities to handle it is through use of the z dimension through stacked 3D packages that provide a volumetric density solution.

Figure 5.16 depicts the range of 3D packaging technologies being developed or possible within Amkor's diverse laminate-based package family to deliver higher levels of silicon efficiency to a wide range of applications. In addition to this 3D applications roadmap, Amkor has a range of stacked-die configurations available in standard and advanced lead-frame packages.

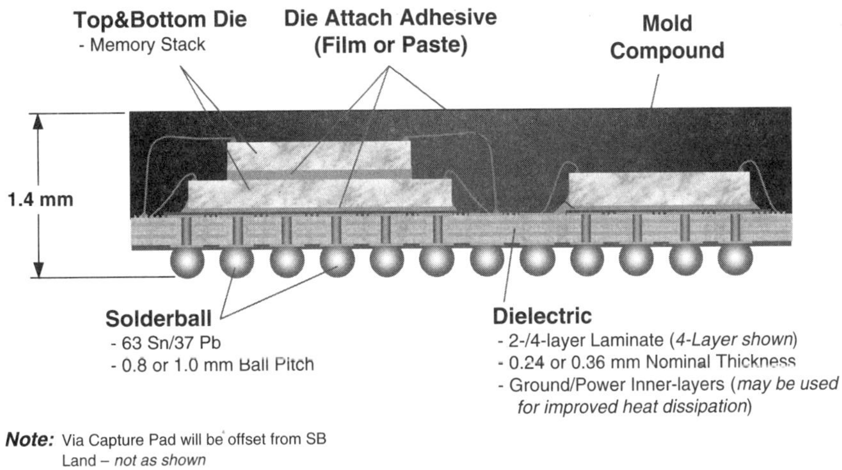

(a)

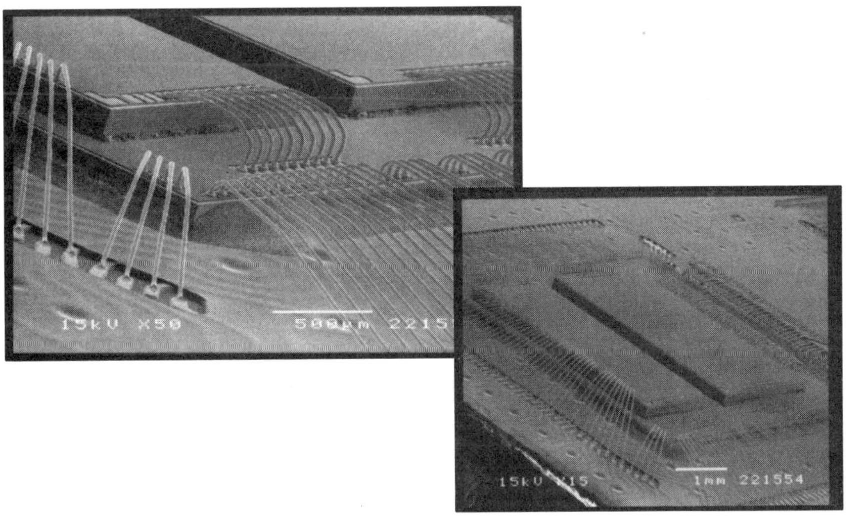

(b)

FIGURE 5.13 (*a*) S-MCM or 2 + 1 structure. (*b*) S-MCM (2 on 1) + 1 wire-bond photo.

5.8 FLIP CHIP (FC) AND WIRE-BOND STACK-DIE INTEGRATION

The current S-CSP solutions being adopted in handsets (and represented in the preceding figures) use wire-bond interconnect technologies. However, Amkor's stacked packaging roadmap uses combinations of wire-bond and flip chip interconnection to accomplish a range of stacked-chip configurations for various die size and performance requirements. The flip-stack, face-on-face structure is being developed to support rf applications through low-inductance/capacitance FC interconnections between rf and ASIC devices. This configuration can enable integration of GPS (global positioning system) receive/transmit radios in a small, low-cost form factor that can be integrated into cellphones

TABLE 5.1 S-MCM (2 + 1) versus Three-Chip Pyramid

Pros 2 + 1	Cons 2 + 1
Shared material set with two-die S-CSP, better substrate availability	Larger body size may be required, increasing board mount area and weight
Die shrink flexibility	
Use of 7-mil and thicker die	
Flexibility for different bond layouts, ease of routing, and die-to-die bonding	
Better thermal management (>1 W)	
Supports higher density requirements, two to four metal layers; easier to integrate passives	
1.4- to 1.2-mm height compatible	
Package, test socket, and tray infrastructure available	

and PDAs. This package structure also can be used to support Bluetooth radio components for a wide range of short-range wireless networking applications. The FC plus wire-bond device stack can be an effective solution for ASICs and graphics processors that are looking to integrate higher memory capacities than embedded economics can support. See Fig. 5.17 for cross sections of FC plus wire-bond die stacking. These interconnect advancements and the S-CSP infrastructure being established to support initial applications in Flash plus SRAM through ASIC plus memory for cellphones can be applied to a wider range of device combinations, package platforms, and end-market applications. Die stacking along with emerging thin package stacking such as Amkor's *et*CSP and Toshiba's paper-thin packaging will allow 3D packaging to deliver continuous improvements in optimal cost and performance for the new millennium.

5.9 CONCLUSION

Sharp Corporation and Amkor Technology have aligned S-CSP technologies to ensure that current Flash plus SRAM high-volume requirements are addressed from a global perspective. Further, the two industry leaders have reviewed and aligned their packaging roadmaps to ensure that the size, weight, and cost-reduction benefits of stacked-chip packaging technologies are extended to new device combinations and applications. Stacked-chip interconnection and BGA and CSP technology platforms are merging to create a forceful fourth wave of packaging, characterized as 3D package integration, that is critical in helping sustain Moore's law. The huge volumes and high rates of 3D packaging adoption from the mobile phone market are driving further enhancements and cost reductions in stacked-packaging technologies. Figure 5.16 illustrates both die and package stacking advancements that are being developed and adopted by a growing range of portable or memory-intensive applications. 3D packaging is becoming a critical strategy for many OEMs and integrated device manufacturers (IDMs) as a silicon and system-in-a-package integration tool as well as a cost-effective approach for stacking versus embedding large memory blocks.

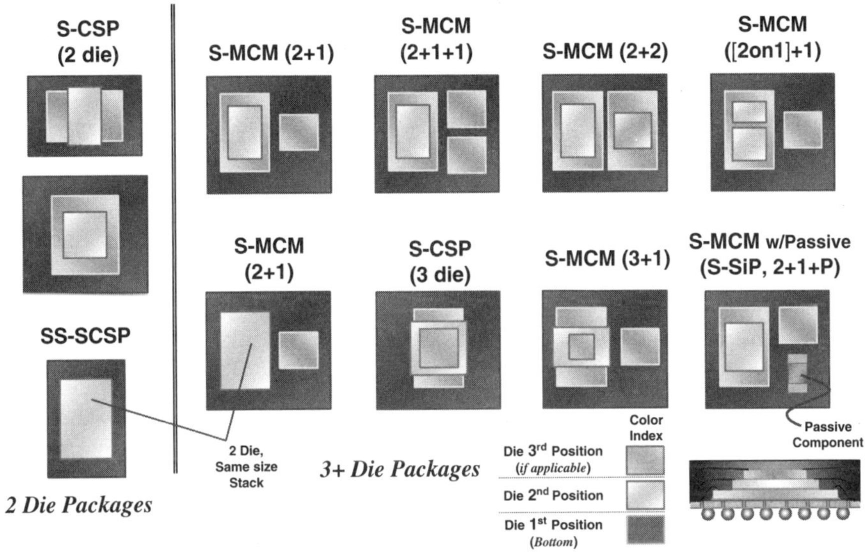

FIGURE 5.14 Stacked-package–die-position designations.

Source: Sharp

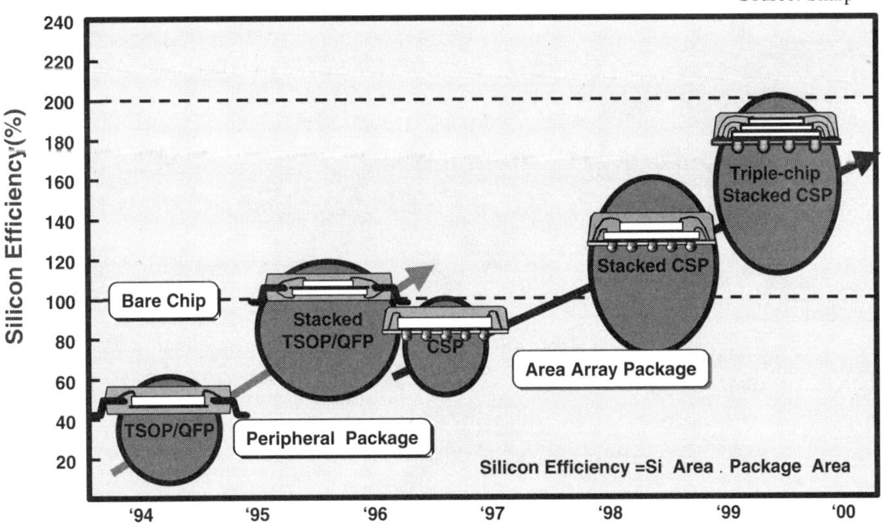

FIGURE 5.15 Silicon efficiency.

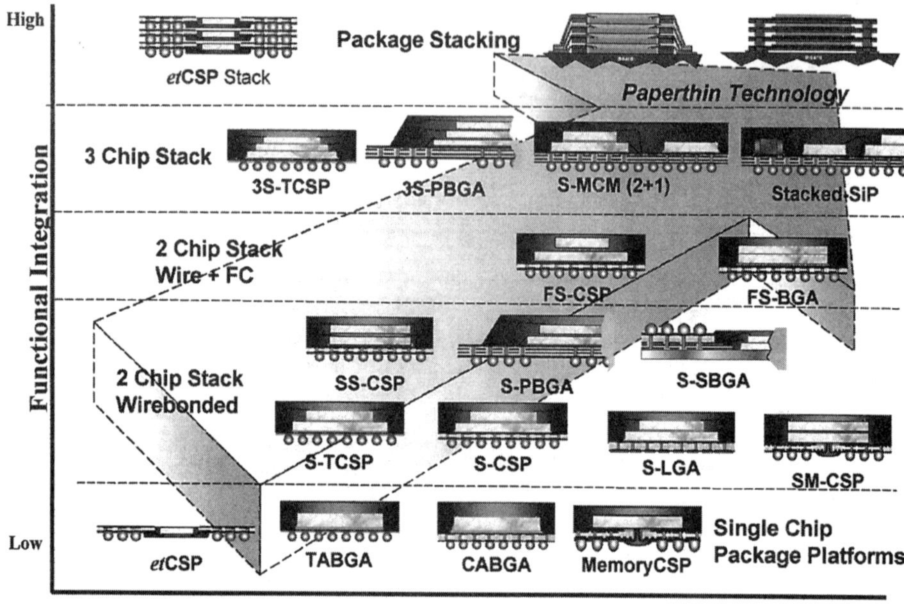

FIGURE 5.16 3D packaging application roadmap.

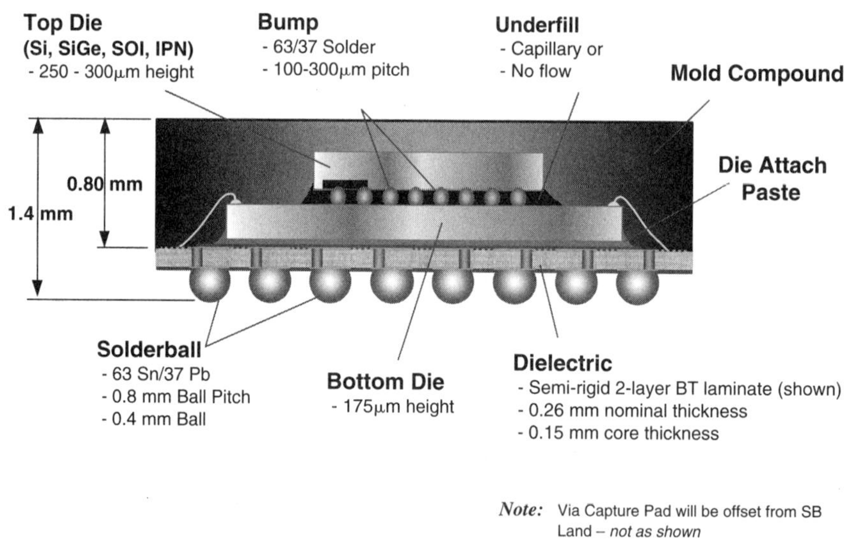

FIGURE 5.17 Flip-stacked CSP (fsCSP).

5.10 REFERENCES

1. B. Swiggett, Amkor Technology's Third Annual Customer Symposium, Santa Clara Marriott, Santa Clara, CA, November 1999.
2. B. Prior, "Trends in Electronics Packaging," SMTA Silicon Valley Chapter Meeting, August 1999.
3. J. Grenier, "The State of the Semiconductor Industry," IPC AMRC Conference, December 1998.
4. J. Woodyard, L. Smith, and K. Ramakrishna, "Stacked-CSP Advances in Handheld Electronics," *European Semiconductor Magazine,* Back-end Supplement, November 1999.
5. M. Kada, "Stacked CSP: A Solution for System LSI," *Chip Scale International,* 1999, pp. B-1–B-7.
6. TechSearch International, Inc., BGA/CSP Development Update Service, 2000.
7. E. Cuellar, "MultiMediaCard (MMC): World's Smallest Removable Storage Device," in *Proceedings of the Wireless Symposium Portable by Design Conference,* Boston, MA, Penton Media, Inc., September 1998.
8. H. Ueda, Amkor Technology Japan, Amkor internal report on mobile phone and associated packaging trends driven by I-Mode handset/service adoption in Japan.
9. B. Bonner, 1999 Flash Memory Applications Report, Dataquest Market Trends Report, Gartner Group, April 3, 2000, p. 28.
10. Sections of this chapter are taken or expanded from L. Smith and M. Kada, "Advancements in Stacked Chip Scale Packaging (S-CSP) Provides System-in-a-Package Functionality for Wireless and Handheld Applications," SMTA Pan Pacific Microelectronics Symposium, January 2000.

CHAPTER 6
COMPLIANT IC PACKAGING

Joseph Fjelstad and Gary Yasumura
Pacific Consultants, LLC.

Young Gon Kim
Tessera, Inc.

6.0 INTRODUCTION

Packaging silicon integrated circuit (IC) chips at or near the size of the chip presents very important electromechanical challenges to the electronics system assembler. The mismatch in coefficient of thermal expansion (CTE) creates significant stress on the interconnections made between the packaged device and the substrate to which it is mounted. Historically, making the leads of the packaged device flexible or compliant has served this need.[1]

Making the IC package leads compliant serves to compensate for this mismatch, making the package capable of meeting the demands of the full range of environments that the electronic assembly may encounter in use. If the package lacks a stress-management mechanism, the package assembler/silicon provider must limit the warranty of the package to only certain environmental conditions.[2]

This chapter will review a packaging concept that provides just such compliance as is necessary, obviating the need for an expensive underfill process to protect the solder joints of the chip-scale package (CSP). The packaging concept that is the subject of this chapter is one developed by Tessera, Inc. It has become one of the most recognizable and popular formats for CSPs in the world. The μBGA (microBGA) CSP is presently being used extensively in Flash memory applications, having been adopted by Intel and others for this purpose.[3] It is also the reference package for Rambus RDRAM, which is expected to dominate in next generation computer memory applications due to its fast operating speeds, which are projected to 1.6 GHz.

The fundamental structural elements of the Tessera μBGA CSP are shown in Fig. 6.1. Some licensees of the technology (e.g., Amkor, Hitachi, Sony, and others) have made modifications to the manufacturing process, but the fundamental elements remain the same.[4] There are a number of special attributes to the μBGA CSP. These attributes help to improve the performance of electronic devices packaged in μBGA format, but the provision of a strain-reduction layer of material between the flex tape and the silicon die is perhaps the most important.[5] This patented[6] technique affords a CTE mismatch strain buffer between the die and the substrate to which the device is to be mounted. It will be reviewed in more detail later in this chapter.

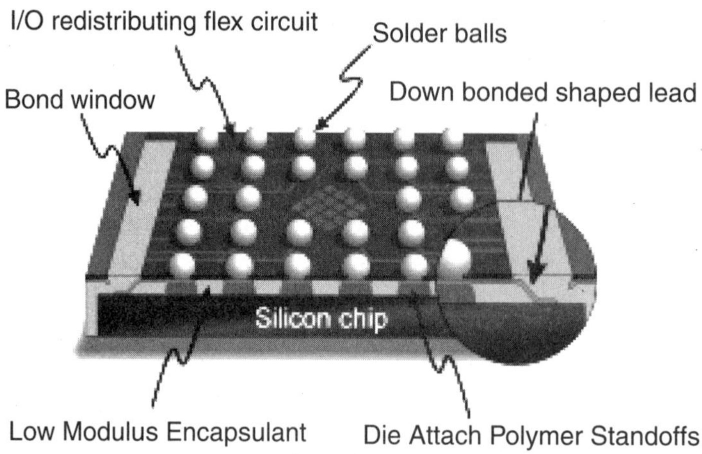

FIGURE 6.1 Construction elements of the basic μBGA CSP.

6.1 PACKAGE TECHNOLOGY OBJECTIVES

The μBGA CSP was developed in response to two different conditions that were in effect in the early 1990s. One was the lack of a readily available good die, a problem that has been a continuing concern for the multichip module (MCM) industry. The other was the fact that bumped dies for flip chip applications were not readily available. A related concern was the fact that bare dies were not standardized. This meant that each die had an individual footprint, making the prospect of standardized component libraries for printed wiring board (PWB) design impossible. These conditions combined to drive the concept of packaging ICs at chip size.

The μBGA package was developed originally for a unique MCM in development at Tessera. While electrical and thermal performance were key concerns,[7] compliance was considered to be very important as well. In fact, the μBGA CSP was originally called the *Tessera-compliant chip* (TCC). Compliance was expected to provide some relief to the electrical test process by making the device contact vertically deformable, thus facilitating reliable contact with test sockets. Testing would ensure that the dies were good at the time of assembly. In short, the package was designed to bring together the benefits of flip chip with those of standard surface-mount technology (SMT) packages (see Fig. 6.2).

6.2 I/O PLACEMENT

Array placement of input-output (I/O) terminals is the most economical use of space and is the method most favored for CSPs. Area array packaging concepts are intrinsically powerful. They allow I/Os to be positioned at a relatively course pitch, and yet they consume much less area than peripherally leaded packaging alternatives with finer lead pitches (see Fig. 6.3). Still, very small dies with low I/Os may be better served by placing I/Os around the edge of the die.

6.3 COMPLIANT CSP CONSTRUCTION

Tessera's compliant chip packaging technology went through a number of evolutionary changes from the original concept to reach its present state of development. Following is a review of the innovation path to the present product.[3]

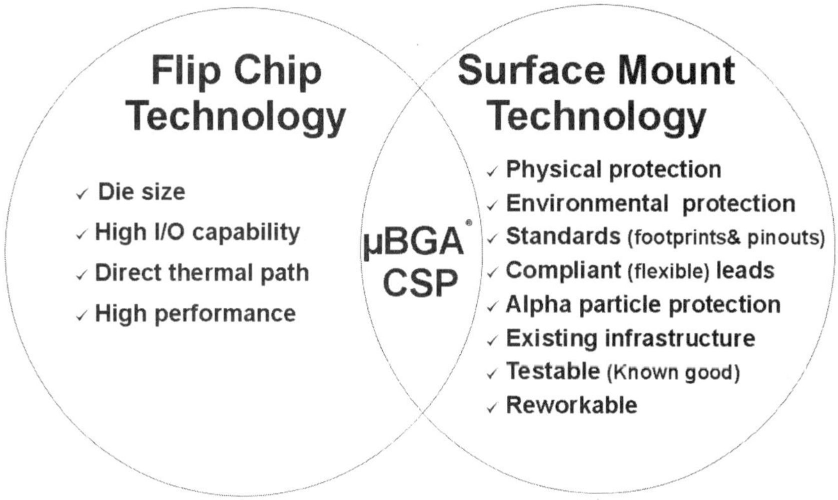

FIGURE 6.2 A simple graphic of the conceptual μBGA.

FIGURE 6.3 μBGA CSP compared with the TSOP and SOP packages. Note that the lead pitch on the μBGA is greater than that on the TSOP. (*Courtesy of Intel.*)

6.3.1 Materials of Construction

Very few materials are required to manufacture the μBGA CSP. However, each of these materials plays a significant and vital role. Primary elements include

- *Flexible base-film material.* A nonreinforced flexible base film is used as the foundation in the construction of the μBGA. The primary reason for using flexible films is that they have a lower Young's modulus than the reinforced and rigid alternatives. The low modulus contributes to strain reduction on the interconnections made at the next level (normally a solder ball).
- *Circuit redistribution wiring.* The metal circuit traces provide for the redistribution of chip I/Os from the bond pad locations to the locations required for component mounting and interconnec-

tion at the next level. The base metal is normally copper, but gold and aluminum leads also have been employed in such constructions.

- *Buffer or compliant layer.* A critical element of any compliant CSP construction is the compliant layer, also called a *buffer layer.* This is the layer that absorbs most of the strain caused by CTE mismatch during thermal excursions related to either operation of the electronic product itself or the environment in which it is operated. The result is that the interconnection to the next level (such as a solder ball) is protected. This layer theoretically can be made to vary in thickness directly with its elastic modulus, but because of system complexity, it is normally formulated to provide strain relief at a thickness of approximately 75 to 150 μm. The compliant layer also serves to protect the metal interconnection leads or flexible links. A very important requirement of the buffer layer is that it be free of alpha-particle emitters. This is so because if the buffer layer is to be used directly on the active circuit side of the die, it is a necessary condition that it be free of alpha-particle emitters. (This has been a concern in memory for many years, but as feature sizes are reduced, it becomes even more critical.) Silicone elastomers have proven highly effective in this role; moreover, because moisture moves rapidly through the material, the package can be made moisture-insensitive.
- *Flexible link.* The flexible link serves to make reliable electrical and mechanical interconnection between the chip and the flexible CSP interposer. In the original concept of the μBGA CSP, this link was to have been a gold bond wire. However, a follow-on concept that made the flexible link an integral element of the redistribution wring obviated the need and made the package higher performing, smaller, and lower cost. The lead must be strain relieved by the compliant encapsulant/buffer layer as well to protect it from early failure due to local work hardening.
- *Next-level contacts.* A wide range of options are available for contacting the μBGA CSP; planar lands, electroformed bumps, and solder balls have all been examined. Bumps or balls can facilitate testing and allow for easier interconnection to the next-level interconnection structure (e.g., PWB).

Obviously, many variations can be envisioned as to the specifics of the materials used in such constructions (Fig. 6.4).

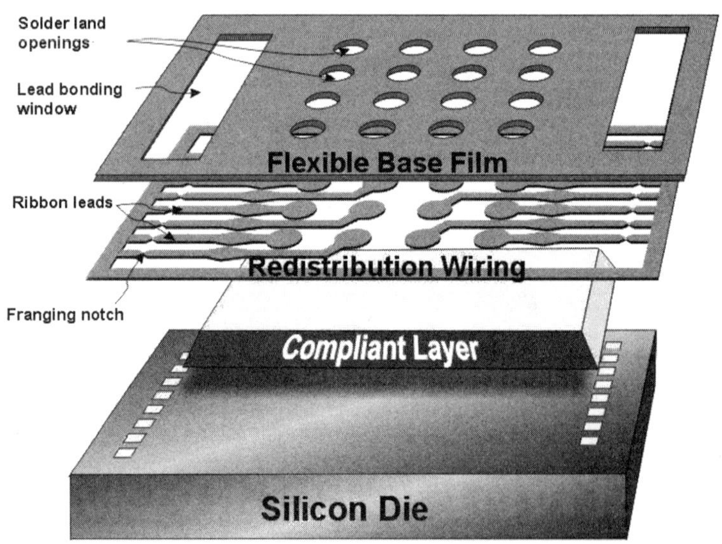

FIGURE 6.4 Elements of construction of the compliant μBGA CSP.

6.3.2 Bond Lead Design

Lead design is of great importance to successful design of the ultimate reliability of the package.[8] The bond lead must be able to move in all three dimensions of space. Lead design also must accommodate the needs of manufacturing in order to facilitate lead bonding and shaping, which have a direct impact on design. Lead design is also heavily influenced by the other elements of construction, such as whether the lead is placed on top or bottom of the flexible film. Finally, the lead design is also affected by the thickness and physical properties of the encapsulant. The key elements of lead design in three dimensions are provided in Fig. 6.5.

6.3.3 Principles of Operation

The physical principles of operation of the μBGA-compliant CSP product group are relatively simple. A compliant layer is provided between the die and the flex substrate to which the die package is mounted. The device has been engineered to protect the most vulnerable element of a mounted IC package—the solder joint. The provision of a compliant layer within the package in combination with the flexible link or bond lead serves this purpose. Figure 6.6 illustrates the concept.

This solves the problem of CTE mismatch between silicon chip and PWB by using a compliant encapsulant between die face and interposer. This takes the stress off the solder balls, which normally fail first on most CSPs. Illustrated in Fig. 6.6 is the die face-down implementation of the compliant-layer concept.

6.3.4 Other Compliant μBGA Configurations

The die face-down construction of the compliant μBGA CSP is widely known and in use around the world, but it was not the original design. Wire-bonded compliant packages preceded it in conception. Following are brief descriptions of earlier Tessera-compliant chip constructions (Fig. 6.7):

- *Die face-up.* Dice face-up is by far the most common method of presentation of the die both to the package for assembly and to the substrate on which the assembled IC package is mounted. It was

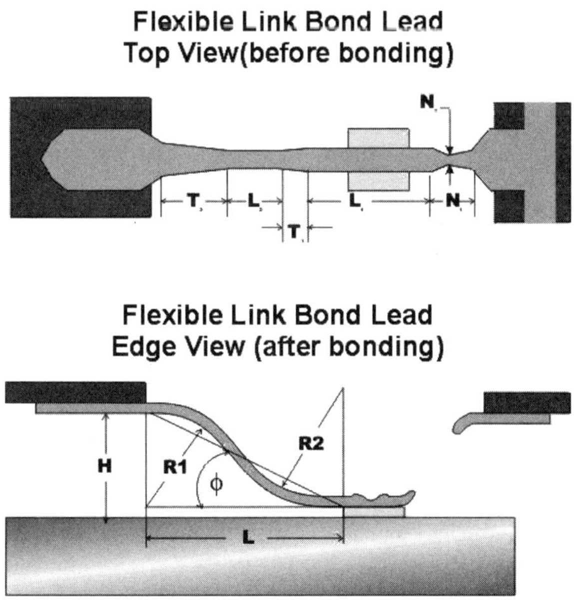

FIGURE 6.5 Design elements of the flexible link/bonding lead for the μBGA CSP.

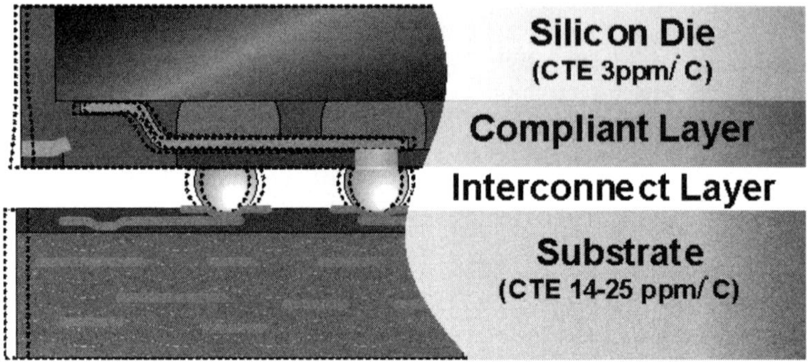

FIGURE 6.6 Structure of the μBGA CSP.

this concept that was first considered for the μBGA CSP. Two variations were examined. One was a flat, flexible film (trade named FμBGA) that used a much thicker than normal die-attach material of a modulus sufficiently low enough to reduce the stain on the solder ball beneath. Most flex circuit–based CSPs such as TI's MicroStar BGA originally used a traditional thin die-attach layer and reported lower reliability. However, TI reported in the *Proceedings of the 1999 Pan Pacific Electronics Conference* that better results are seen when a compliant layer is added.

- *Wrap-around.* This concept was examined in order to avoid having to make small holes in the flex tape. The die was mounted on the side opposite the circuits of the flex circuit, again using a thick, low-modulus die attach. The flex was then wrapped around a support ring, and the I/O bond pads of the die were wire bonded to the flex circuit. Other investigators, including Fujitsu, have reported exploring this concept since it was first examined at Tessera. However, the structure is difficult to assemble and is not a likely candidate for mass production.
- *Die face-down.* Another wire-bonded variant of the compliant μBGA CSP was described in an early patent. The original structure was the inspiration for today's version of the μBGA package. Some contract manufacturers have adopted the method as a means of performing wafer-level assembly as described in early Tessera patents. Toshiba is using the method for discrete chips to package memory. The structure resembles the standard μBGA package except that a wire bond is used in place of the integral lead. This design used wire bonding to make a connection from the interposer to the chip. The patented face-down wire-bonded structure is being used by some contact IC assemblers at present.

6.3.5 I/O Configurations

There are several different possible configurations of the compliant μBGA package. Following is a review of the most commonly used constructions. The package types are illustrated in Fig. 6.8.

- *Fan-in.* All I/Os are placed within the die area and interconnected by redistributed wiring to die bond pads located at the perimeter of the die.
- *Center bond pad.* Like fan-in, all I/Os are placed within the die area and interconnected by redistributed wiring, but die bond pads are located in the center of the die, a placement favored for memory devices. This construction easily facilitates die shrink when required.
- *Fan-in/fan-out.* Fan-in/fan-out constructions are desirable under certain conditions, such as when a peripheral lead die is shrunk to a point that makes it impossible to hold the solder ball I/O footprint on the face of the die or when a larger I/O pitch is desired.
- *Fan-out.* Fan-out-only constructions of compliant packages are used when the die is very small and

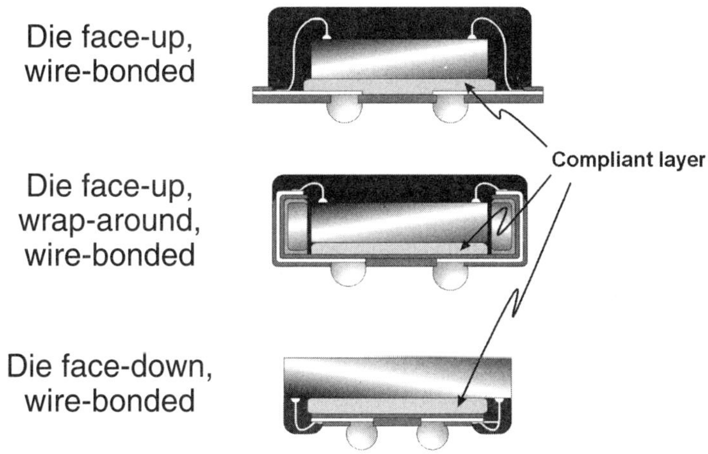

FIGURE 6.7 Early concepts of the compliant CSP.

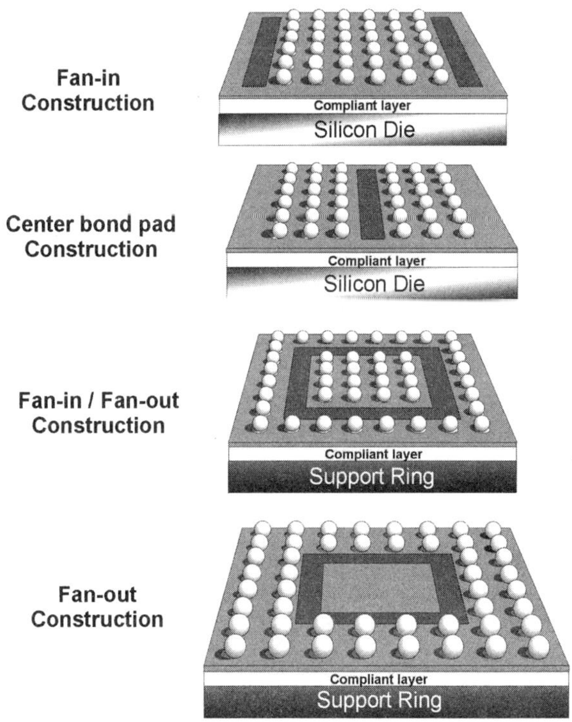

FIGURE 6.8 The compliant layer.

yet has a significant number of I/Os for its size. Fan-out design also can be applied to the replacement of rigid ball and array (BGA) packages when better mechanical compliance is desired.

The layer is a key characteristic of the μBGA package and is incorporated into all the various construction formats possible, from chip size to larger than the chip, to meet the needs of the industry infrastructure.

6.3.6 Multiple Metal Layers

The vast majority of CSP applications at present only require one metal layer for I/O redistribution, but with expected increases in complexity, two, three, or more metal layers may be required to meet next generation performance demands. Some products such as SRAM and next generation fast DRAM such as RDRAM are looking forward to such need. An example of a two-metal-layer fast SRAM device is shown in Fig. 6.9.

6.3.7 Adapting to Die Shrink

Die shrink is standard practice in semiconductor manufacture. It represents a challenge to all CSPs. CSP size obviously is determined by the die size. Thus, if an area array format is chosen for I/O presentation to the next-level assembly, die shrink translates to package outer dimension shrink as well. If the I/Os remain within the perimeter of the die at the chosen pitch, no difficulty is expected. However, if die shrink results in I/Os being located outside the perimeter of the die, a decision must be made to either shrink the I/O pitch or make the package larger than otherwise would be necessary.

Thus, for CSPs, it is necessary to establish a design strategy for the package outer dimension and to determine whether the old package outer dimension will be used or a new reduced package outer dimension will be established. Die shrink is especially challenging to die face-down constructions that use wire-bonding technologies. This is so because wire-bond pad placement often interferes with next-level I/O contact land location. Anticipating all future die shrinks and using them for establishing I/O placement solves the problem, but the problem is then transferred to the next-level PWB design, manufacture, and assembly, which can add significantly to the manufacturing cost of the assembly. Despite this problem, clever design layout and process modification can nearly subdue die shrink challenges. It is possible to stave off some of the die shrink problem by adopting a fine-pitch ball matrix from the beginning. Figure 6.10 illustrates methods for addressing die shrink for face-down constructions.

FIGURE 6.9 A larger than die size fan-in/fan-out μBGA chip package.

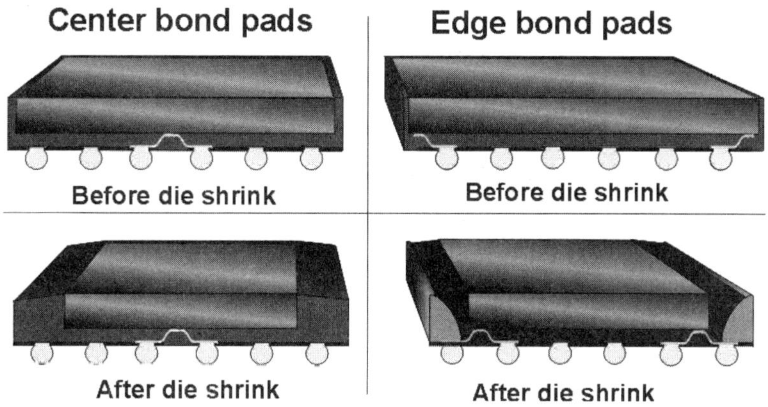

FIGURE 6.10 μBGA CSP die shrink strategy varies with location of the bond pads.

6.4 MANUFACTURING PROCESSES

A number of different manufacturing processes have been developed over the years to build the μBGA CSP. Each new process resulted in significant improvement in terms of both yield and cost reduction. While the prime objectives for the technology have remained unchanged (i.e., highest performance in the most reliable package possible), the latest process technology also excels in the area of cost reduction. The manufacturing process for the μBGA is capable of producing IC packages at less than $0.0049 per lead according to cost modeling data. As a point of reference, most IC packages are presently priced at approximately $0.01 per lead.

Even though earlier manufacturing methods did not meet the cost target, system developers found that their overall cost-performance needs nevertheless could be met by the μBGA CSP in many cases. This is so because the package could be placed at rates of up to 20,000 units per hour in assembly at very high yield, and it did not require an expensive underfill process after it was mounted on the PWB to protect the delicate solder joints.

6.4.1 Original Process

Once settled on the idea of an integral lead frame with bond ribbon ultimately being superior to wire bonding, an attempt to find a suitable process was undertaken. The fundamental notion was to have tape automated bonding (TAB)-like lead frame that fanned the fine-pitch I/O of the die bond pads inward to a relatively coarse area array of contacts rather than outward to a fine-pitch peripherally leaded device. Many challenges were in store. For example, one of the long-standing problems with TAB was the fact that the delicate cantilevered inner leads could be bent or damaged easily. If these same leads were to be cantilevered around the outer edges of the device, the problem likely would be worse. Because of this concern, it was decided that the leads should remain attached to a carrier film until the moment of bonding. However, having the lead break loose for bonding reliably was a challenge. Originally, the leads were designed to be tenuously attached to the polymer carrier film and would be pulled away from the film during the lead bonding process. However, this method was not as consistent as was desired. The concept of a frangible lead based on a notch thus was developed.

Most of the world's tape suppliers shied away from the challenges presented by the tape design. As a result, it fell to the company to develop processes to manufacture the flex circuits, as well as for assembly. Early flexible circuit tape for such material was made at the company using electroformed gold leads, laser-drilled holes and windows, and electroformed nickel-gold bumps. Gold was chosen first because of its malleability and resistance to work hardening. These were deemed important attributes for structural reliability. Lead design matured over time, and a taper was added to the lead to create a "beam of uniform strength." This feature prevented lead kinking and made lead shaping during bonding (an important design element) easier. The structure of the original flexible circuit tape μBGA package is shown in Fig. 6.11.

The μBGA CSP construction has changed over time to improve manufacturing efficiency while maintaining reliability. Also illustrated in Fig. 6.11 are the major process revisions that have been employed in the manufacture of the μBGA CSP.

6.4.2 The Zinger Assembly Process

Once TAB tape suppliers decided to produce μBGA tape in small volume, the company was able to demonstrate a more cost-effective manufacturing method. The result was a process trade named Zinger. The new process relied, to the maximum extent possible, on the existing package assembly infrastructure. Off-the-shelf copper-based TAB tape, standard stenciling equipment, pick-and-place machines, wire bonders, dispense equipment, etc., all were used to create the manufacturing line. While the first process was not fully optimized, it did accomplish its objectives and enabled the first mass production of product (Flash memory chips) by technology licensees.

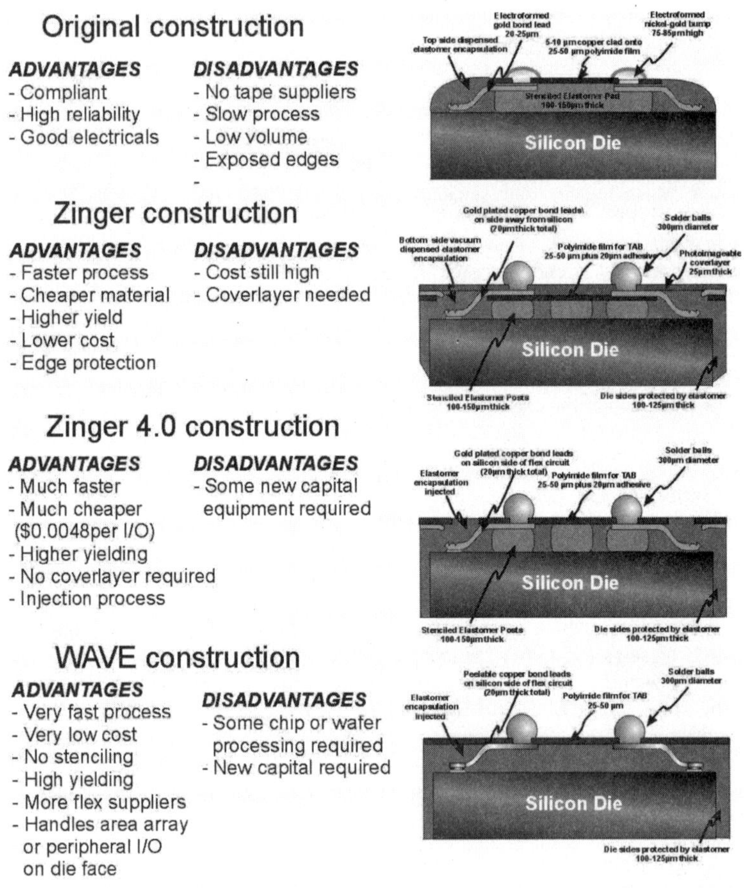

FIGURE 6.11 Construction changes over time.

- *Zinger 1.4 process.* The first process released for volume manufacture was the Zinger 1.4. This process used existing equipment with only minor modifications.
- *Zinger 3.0 process.* The second process released for μBGA CSP manufacture was the first to have equipment specifically designed for it. Innovations included in the process were the use of a novel vacuum dispenser of encapsulant.
- *Zinger 4.0 process.* The latest of the Zinger process family has further innovations and new, specifically developed manufacturing equipment such as is required for injection. This process brings the cost of the μBGA package down to $0.0048 per I/O according to cost models, with further reductions possible. The process is illustrated in Fig. 6.12.

6.4.3 The WAVE Process

The wide area vertical expansion (WAVE) process was developed originally to serve the needs of designers who sought to lay out chip I/Os in an array fashion over the face of the die. The process

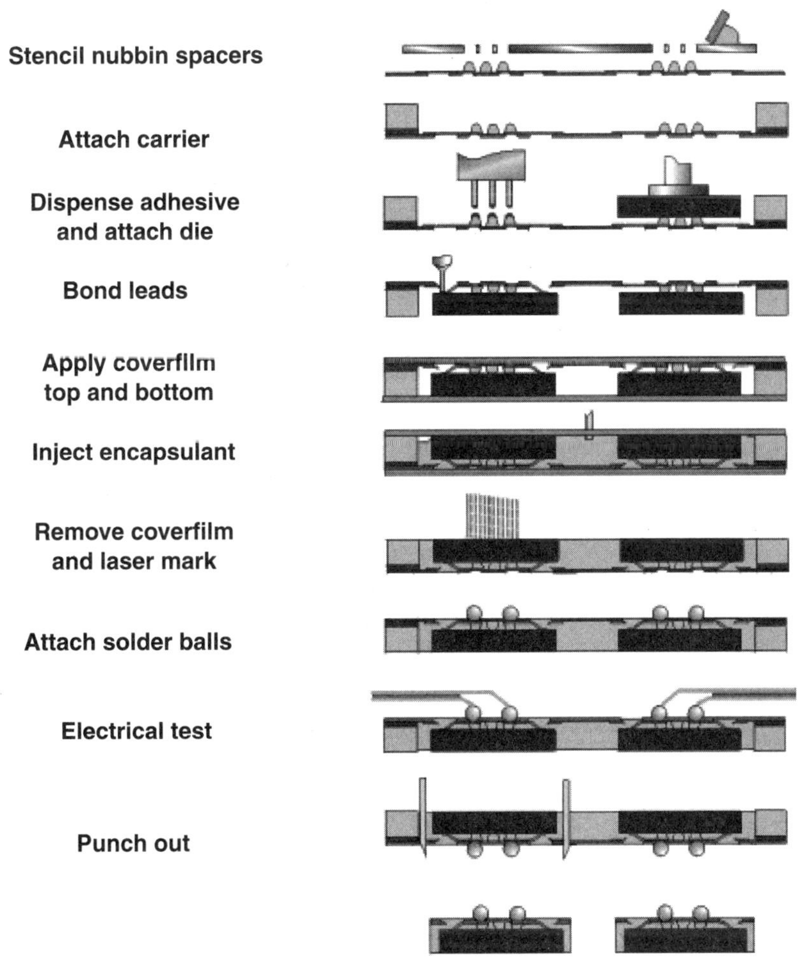

FIGURE 6.12 The Zinger 4.0 process.

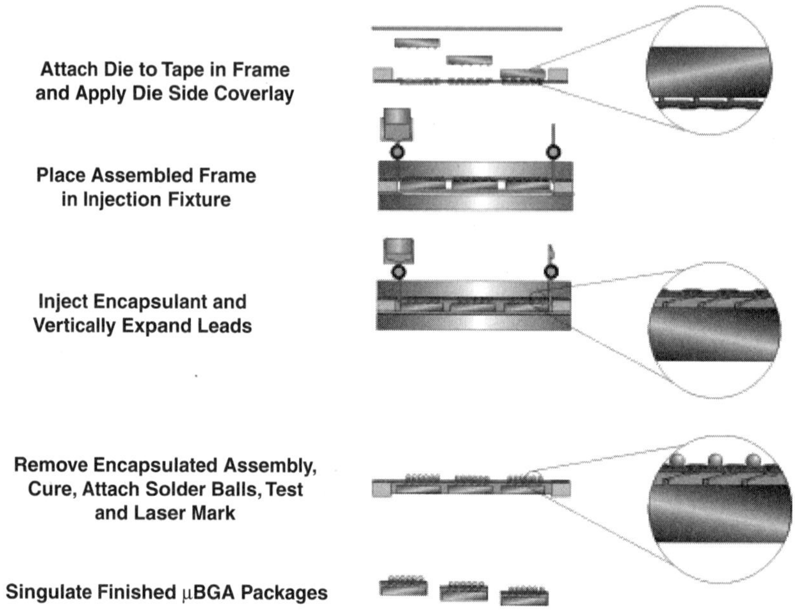

FIGURE 6.13 The WAVE process.

also was viewed as a likely solution to the underfill dilemma faced by flip chip technology users. It would in effect make flip chip devices into compliant packages. The process also was extended to include packaging directly on the wafer. The chip-at-time process is illustrated in Fig. 6.13.

The method greatly improves manufacturing efficiency by allowing for the injection of encapsulant. The method also greatly reduces the cost of chip packaging by allowing all interconnections to be made in a single step. The lifting of the leads is accomplished by injecting the encapsulant and curing it under pressure.[10] The process is suitable for both wafer-level and discrete IC packaging.

6.5 ELECTRICAL PERFORMANCE OF THE μBGA CSP

While the flexible circuit used in the construction of a μBGA CSP is capable of carrying high-speed signals from an IC chip through solder balls surface mounted to a PWB, a circuit designer can optimize the flex circuit's configuration and materials to ensure high-frequency signal integrity.

6.5.1 Software-Modeled *RCL* Parasitics

Modeling of the μBGA CSP has been performed by a number of different companies since introduction of the package to determine the common electrical parasitics of resistance, capacitance, and inductance. Pacific Consultants of Mountain View, California, used Ansoft's Q3D parameter extractor software to derive modeled *RCL* parasitics from the μBGA TV46 package. Four traces in the package were modeled. Solder balls at the beginning of each signal trace were assumed to be 0.37 mm in diameter. It also was assumed that these were surface mounted to an FR4 epoxy-glass board. The signal traces were 0.02 mm thick by 0.04 mm wide. The four bonded leads of the signal traces were bent downward and bonded to pads on the silicon die in accord with the manufacturing process. A length of 0.32 mm was used for each bond lead. While the standard TV46 does not have a ground plane, a future generation of the device is expected to incorporate one for performance reasons. Thus a ground plane situated above the polyimide layer was assumed for this model. The compliant elastomer layer fills the space between the signal traces and the silicon die. The four traces were modeled as "normal" conductors with a limited conductivity of 5.8×10^7 S/m and a dc analysis was carried out. The calculations include flux inside the conductors (Fig. 6.14 and Tables 6.1 through 6.3).[5]

Table 6.1[1] gives the thickness and dielecric constant values for TV46 while Table 6.2 indicates lengths of traces tested. Measured results are given in Tables 6.3, 6.4, and 6.5 that show resistance, capacitance, and inductance, respectively. Table 6.6 gives these values as predicated by Ansoft Field Solver.

The software's field solver automatically refines the mesh by optimizing the placement of new elements into the existing mesh. To gain the required accuracy for this model's inductance and resistance matrices, the solver created over 90,000 tetrahedral elements. The values are displayed in Tables 6.4 through 6.6.

The software-generated value of 3.1 nH results from calculating partial inductances composed of signal traces only. It is possible to use a "Reduce Matrix" command during postprocessing to include the ground planes in the loop inductances.

In a follow-on analysis, the conductors were modeled as "perfect" conductors, whose conductivity is infinite. The skin depth was an order of magnitude smaller than the cross-sectional dimensions of the signal traces. Perfect conductors are assumed in Ansoft because they provide a reasonably accurate solution. The measured values are displayed in Tables 6.7, 6.8, and 6.9 that give resistance, capacitance, and inductance for the ac case with ground. Predicated results are shown in Table 6.10 for comparison.

The ac case changes the *RCL* parasitics values. The resistance increases because current must travel in a smaller cross-sectional area and more heating occurs. Capacitance changes very little, as expected, and inductance drops by one third. Figure 6.15 visually displays these relationships.

6.6 THERMAL PERFORMANCE

With many CSPs, the backside of the IC chip is covered by plastic. However, with the μBGA CSP, the backside of the die can be accessed directly by a heat sink or thermal spreader. As a result, the μBGA package can provide excellent thermal performance. This is rapidly becoming a significant

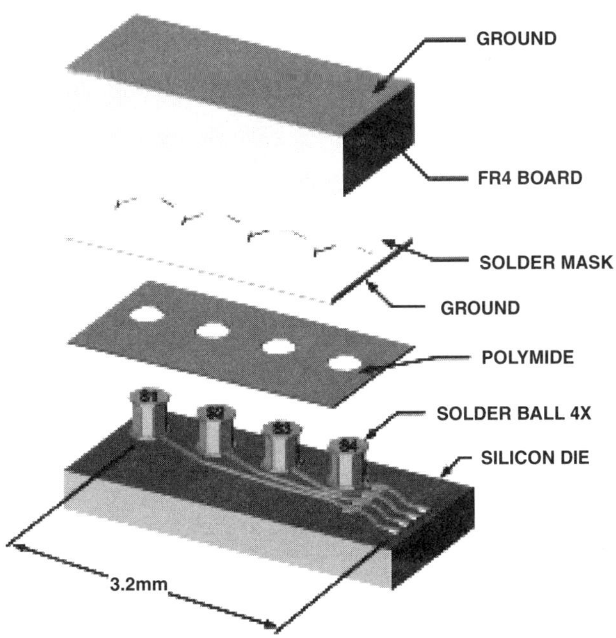

FIGURE 6.14 Test vehicle model.

TABLE 6.1 Layer Geometry

Layer	Thickness, mm	ε_r (relative dielectric constant)
Ground	0.02	NA
FR4	0.74	4.7
Solder mask	0.05	3.7
Ground	0.02	NA
Polyimide	0.02	3.3
Elastomer (not shown)	0.125	3.0
Silicon die	0.42	4.0

TABLE 6.2 Trace Lengths

Trace designation	Trace length, mm
S1	3.2
S2	2.5
S3	1.7
S4	1.0

TABLE 6.3 Resistance Matrix (ohms), Grounded Model, dc Case

	S1 trace	S2 trace	S3 trace	S4 trace
S1 trace	0.59	0	0	0
S2 trace	0	0.498	0	0
S3 trace	0	0	0.325	0
S4 trace	0	0	0	0.125

TABLE 6.4 Capacitance Matrix (picofarads), Grounded Model

	S1 trace	S2 trace	S3 trace	S4 trace
S1 trace	0.502	0.0389	0.00496	0.00176
S2 trace	0.0389	0.412	0.0328	0.00430
S3 trace	0.00496	0.0328e	0.308	0.0291
S4 trace	0.00176	0.0430	0.0291	0.185

TABLE 6.5 Inductance Matrix (henries), Grounded Model, dc Case

	S1 trace	S2 trace	S3 trace	S4 trace
S1 trace	3.10	1.41	0.728	02.77
S2 trace	1.41	2.55	0.921	03.20
S3 trace	0.728	0.921	1.63	0.379
S4 trace	0.277	0.320	0.379	0.723

TABLE 6.6 Ansoft Field Solver for dc Case for S1 Trace (3.2 mm length)

Ansoft field solver results	
S1 resistance	0.059 Ω
S1 self-capacitance	0.502 pF
S1 self-inductance	3.100 nH

TABLE 6.7 Resistance Matrix (ohms), Grounded Model, ac Case

	S1 trace	S2 trace	S3 trace	S4 trace
S1 trace	0.113	0.0658	0.00223	0.00247
S2 trace	0.0658	0.720	0.0234	0.000781
S3 trace	0.00223	0.0234	0.511	0.00955
S4 trace	0.00247	0.000781	0.00955	0.145

TABLE 6.8 Capacitance Matrix (picofarads), Grounded Model

	S1 trace	S2 trace	S3 trace	S4 trace
S1 trace	0.513	0.0396	0.00496	0.00174
S2 trace	0.0396	0.416	0.0331	0.00432
S3 trace	0.00496	0.0331	0.309	0.0293
S4 trace	0.00174	0.004.2	0.0293	0.188

TABLE 6.9 Inductance Matrix (nanohenries), Grounded Model, ac Case

	S1 trace	S2 trace	S3 trace	S4 trace
S1 trace	2.4	1.13	0.665	0.266
S2 trace	1.13	1.87	0.706	0.275
S3 trace	0.665	0.706	1.15	0.307
S4 trace	0.226	0.227	0.307	0.663

TABLE 6.10 Field Solver for ac Case for S1 Trace (3.2 mm length)

	Ansoft field solver results
S1 resistance	0.113 Ω
S1 self-capacitance	0.513 pF
S1 self-inductance	2.400 nH

concern as package wattage continues to climb due to increased operating frequencies, increases in integration, and reduction in feature sizes, causing them to become more resistive.

When applying the thermal spreader, a thin thermal pad is recommended. However, the pad should be capable of conforming well and should occupy all the space between the die and heat sink in a void-free manner.

θ_{jc} is thermal resistance of the system (package) and is given in °C rise per watt input. Low values are desirable since a minimum rise in temperature results in maximum component lifetime.

θ_{jc} can be as low as 0.8 and 0.2°C/W using a boron nitride thermal pad and zinc oxide thermal paste, respectively, as interfaces (Table 6.11). This configuration provides much better thermal performance than can be had with a cavity-down BGA.

The short thermal path has other benefits in that it facilitates both assembly and rework. Figure 6.16 illustrates this aspect of the package when compared with a plastic BGA. This illustrates how the package can facilitate both assembly and rework by requiring a lesser amount of energy to accomplish solder-joint reflow.

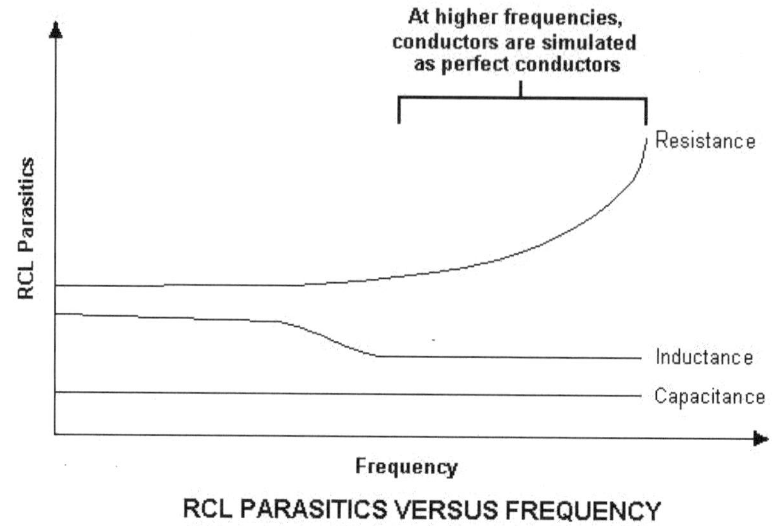

FIGURE 6.15 *RCL* parasitics versus frequency.

TABLE 6.11 Thermal Conductivity Measures of Various Heat-Sink-Enabling Materials

Material	Thermal conductivity (W/m·°C)	Pad thickness (mm)	θ_{ja} (°C/W)	θ_{jc} (°C/W)
Silicone rubber	0.4	0.20	5.8	3.8
Thermally conductive elastomer	2.2	0.20	4.2	2.2
BN-doped elastomer	6.0	0.15	2.8	0.8
Thermal grease	0.8	0.05	2.2	0.2

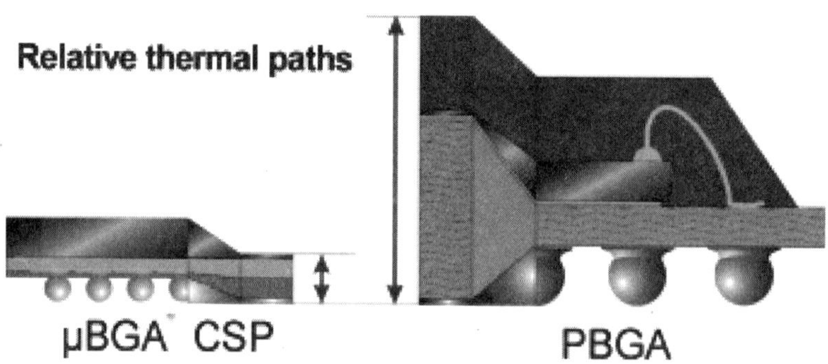

FIGURE 6.16 Relative thermal paths of the µBGA versus a plastic BGA.

6.7 RELIABILITY AND FAILURE ANALYSIS

The μBGA CSP has been proven to be a highly reliable IC package. It is considered to be among the most reliable packages ever produced. The μBGA CSP has demonstrated its reliability in numerous reliability studies by both OEMs and consortia.

Qualification testing to determine reliability normally is performed both on the package by itself to determine package-level reliability (PLR) and again with the device mounted on a substrate such as FR4 to determine board-level reliability (BLR). Qualification testing for reliability has long been an accepted practice to provide the end user with a high level of certainty that the product will not fail prematurely because of faulty package design or assembly processes.

Mil-STD-883 provides a detailed prescription for testing all types of electronic devices and has long served as the electronics industry's general qualification standard. While presently there is an effort underway to redefine qualification testing to a more benign regimen (e.g., 0 to 100°C for on-board reliability), the μBGA CSP was designed to withstand the rigors of the most stringent testing without paying a price penalty.

6.7.1 Moisture Sensitivity

Moisture can outgas explosively during reflow soldering, resulting in a cracked package. Because of this, moisture sensitivity of chip packages is an item of considerable concern to the electronics assembler. Silicon dominates the CSP, with encapsulants playing a much smaller role, and thus limits the total amount of moisture that can be absorbed. However, CSPs still can be affected adversely by moisture. The μBGA CSP, depending on the type of encapsulant used, is capable of meeting the requirements of JEDEC level 1. Level 1 is considered the most desirable designation because it implies moisture insensitivity. The JEDEC requirements are outlined in Table 6.12.

6.7.2 Conditions for Reliability Testing

The μBGA CSP has passed the full battery of tests outlined in Table 6.13. These conditions, while extreme, have been the benchmark for electronic packages for many years. The purpose of such testing is to stimulate failure so that device reliability can be determined and understood and reliability predictions made.

Warp and twist and bend testing are recent additions to the test regimen and reflect a real concern over the trend in industry to use thinner boards. Here, compliance becomes a true necessity because the silicon chip or its solder joints are at a high risk of breaking during box assembly or due to mechanical shock. Underfill provides some protection, but a compliant layer is an advantage (Fig. 6.17).

This type of stress can have a profound influence on the reliability of a CSP. Compliance in all three dimensions is required to compensate for all the various thermal and mechanical effects that can damage the package or its leads.

6.7.3 Reliability Test Results

Tessera and others have performed extensive reliability tests on the package. As with most CSPs, package-level reliability is excellent. However, the concern of most OEMs is the performance of the package on the board. Figure 6.18 shows recent reliability results. *Note:* t_0 correction was applied to the data from test condition *C* (−65 to 150°C) that resulted in significant improvement in the goodness of fit (r^2) from 0.931 to 0.989. The Weibull ordinate scale and the characteristic life (*Eta*) are plotted in the t_0 domain. The actual characteristic life (*Eta*) of data from test condition *C* (−65 to 150°C) is 2857. Data for the chart are included in Table 6.14.

6.7.4 Failure Mechanisms

While the μBGA CSP is an extremely robust package, it is prone nevertheless to certain types of wearout failure. The package works as designed, and no solder-joint failures have ever been cited as a point of wearout failure. Figure 6.19 provides a finite-element model showing the efficacy of the strain relief provided by the compliant layer. (*Note:* Some brittle failures of the solder joints were

TABLE 6.12 JEDEC Moisture Sensitivity Requirements

JEDEC level	Description	Out-of-bag floor life allowed before assembly
1	Not moisture sensitive	Unlimited at 30°C and 85% relative humidity
2	Limited moisture sensitivity	1 year at 30°C and 60% relative humidity
3	Moisture sensitive	1 week at 30°C and 60% relative humidity
4	Highly moisture sensitive	72 h at 30°C and 60% relative humidity
5	Extremely moisture sensitive	48 h at 30°C and 60% relative humidity
6	Bake before use	6 h at 30°C and 60% relative humidity

TABLE 6.13 Sample of Reliability Requirements Testing for CSPs

Test required	Conditions	Duration
High-temperature storage	150°C	1000 h
Precondition thermal cycling, condition *C*	−65–150°C	1000 cycles
Condition *B* cycling (only if condition *C* fails)	−55–125°C	1000 cycles
Thermal shock, condition *C*	−65–150°C	1000 cycles
Steam pressure test	121°C at 2 atm	168 h
Precondition temperature and humidity biased	85°C and 85%	1000 h
HAST	130°C and 85%	80 h
Mechanical shock testing	1500-*g* peak	—
Bend/bow (bidirectional)	1.0-mm deflection	2.5 million cycles
Twist (0.25-mm-thick PWB)	5° and 9° twist	100,000 cycles

Source Intel.

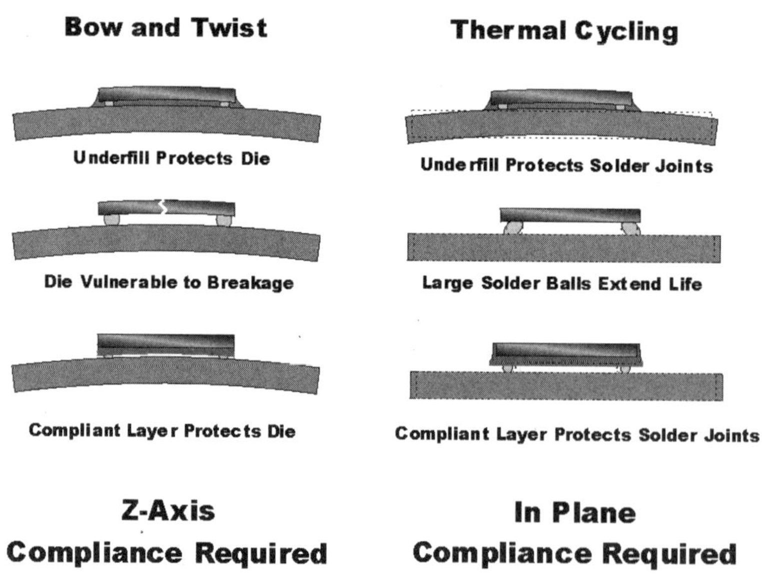

FIGURE 6.17 Warp and twist.

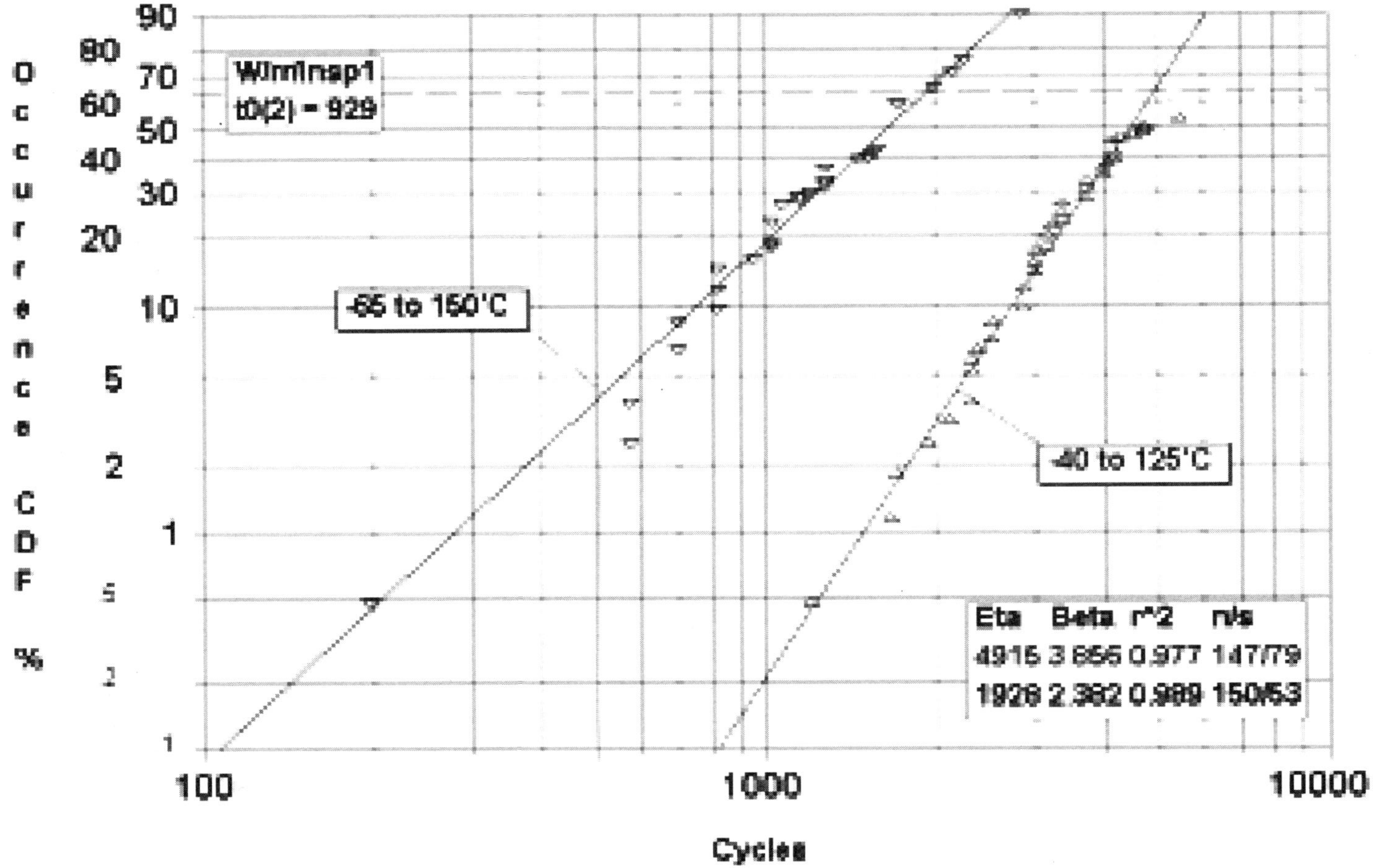

FIGURE 6.18 Example of a Weibull analysis of a μBGA CSP. Test conditions: −40 to 125°C on board.

TABLE 6.14 Data for Fig. 6.18

Run	Sample size	Fails/test point
1	30	0/2000, 2/2500, 5/3000, 11/3200
2	28	0/2000, 1/2500, 6/3000, 9/3500, 11/4000, 14/4100
3	29	0/1500, 1/2000, 3/3000, 9/3500, 12/4000, 13/4500, 14/4900
4	30	0/1000, 1/1500, 3/2500, 4/3000, 5/3500, 10/4000, 15/4500, 16/4700
5	30	0/1500, 2/2000, 3/2500, 5/3000, 7/4000, 11/4500, 12/5000, 13/5400
	147	
		Test conditions: −65 to 150°C
1	30	0/1000, 4/1500, 11/2000, 17/2400
2	30	0/1000, 2/1500, 9/2000, 17/2200
3	30	0/1500, 14/2000, 17/2400
4	30	0/2000, 3/2500, 7/3000, 14/3500, 24/3700
5	30	0/1500, 6/2000, 7/2500, 22/3000
	150	

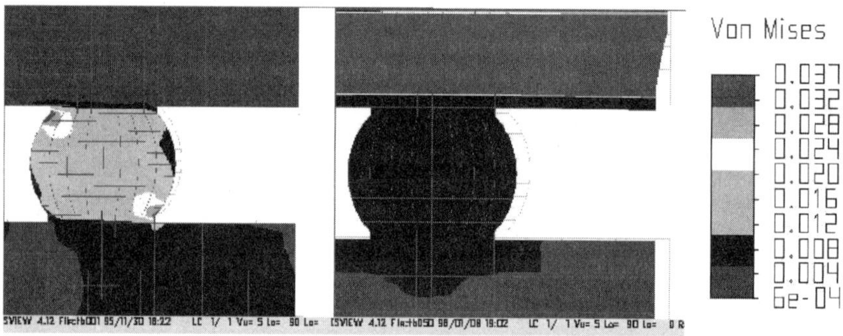

FIGURE 6.19 Solder joints on area array packages are the most vulnerable elements of the package in board-level reliability. It is desirable for the package structure to protect the solder joints by having the package absorb the strain of CTE mismatch rather than the solder balls, as shown in the right and left finite-element models, respectively. (*Courtesy Pacific Consultants LLC, Mountain View, CA.*)

reported on early versions, where gold thickness in the area of the solder joints far exceeded recommended limits.) The point of failure thus turns out to be the bond lead, which was designed to work in concert with the elastomer to protect the solder joint.

Lead failure is a classic wearout failure based on local work hardening. While the package and lead are designed to allow the elastomer to distribute the strain over the length of the lead, failure most often occurs at the heel of the lead near the bond pad. This location is the one most often predicted by finite-element modeling. However midspan breaks in the lead also have been discovered in a later test vehicle package (the TV-74). Figure 6.20 shows scanning electron micrographs of cracks in the package leads found in failure analysis. The result of this discovery was a full-scale review of the package concept and its mechanisms of failure.

Every failure that occurs in a μBGA package is driven by both the material used in its construction and the geometry of its features. There is thus a need to understand the primary factor that leads to a specific type of failure. Figure 6.21 illustrates the relationship between failure modes and the primary factors of material and geometry based on failure analysis observations.

The CTE mismatch between die and flex-circuit tape imposes stress on the leads during the thermal cycling. Depending on the geometry of the leads, the peak stress (or strain) will be on the heel, midspan, or shoulder. Therefore, lead heel break and midspan break can be classified as geometry-

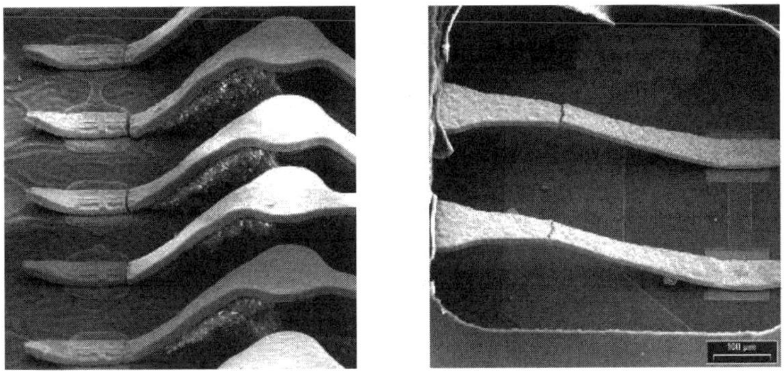

FIGURE 6.20 Lead heel break in TV-46 (*left*) and midspan break in TV-74 (*right*).

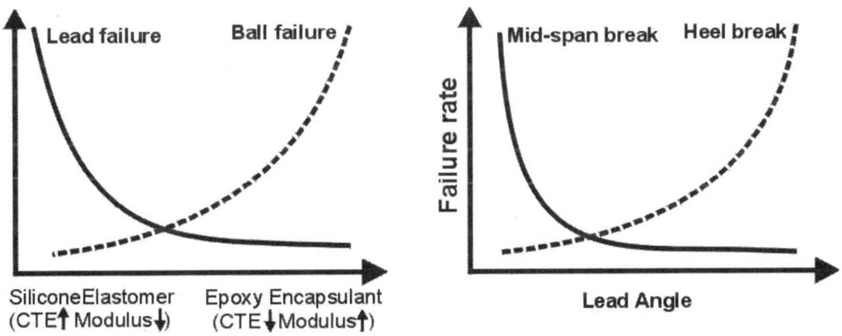

FIGURE 6.21 Failure mechanisms for the μBGA package can be changed by altering the variables of materials and design.

driven failure modes, whereas heel breaks are mainly due to fatigue from thermal cycling. The stress concentration at the heel can be minimized by reducing the lead angle, as illustrated in Fig. 6.21 (*right*). However, if the lead angle is reduced significantly, midspan break will become the major failure mode.

The failure mechanism of midspan break is not the same as that of heel break. With the reduced lead angle, the shape of the lead is not representative of the S shape typically sought. The actual lead shape more closely resembles a straight line. This condition offers no extra length to handle the CTE mismatch between die and tape. The result is a midspan break, which can be explained by the combination of tensile stress during the cooling half of a temperature cycle and lead buckling during the heating portion of a thermal cycle. The act of increasing the lead angle provides the potential for having more lead slack, thus avoiding the excessive tensile stress encountered during the cooling step. The higher lead angle also changes the lead deformation mode from buckling to bending under the heating step, resulting in reduced stress concentration at the midspan.

6.8 SUMMARY

The μBGA CSP is a highly reliable, low-cost, high-performance IC package. Moreover, it is capable of providing excellent electrical and thermal performance. Advances in manufacturing technology and refinements in package design are resulting in steady improvements in terms of both cost and performance. The assembler community is mastering the manufacturing technologies, and cost is expected to continue to decline. Next generation processes such as WAVE are in full-scale development and are expected to deliver on the promise of flip chip technology by providing a highly manufacturable device that provides standards and ease of assembly at highly productive rates.

Circuit card assemblers have reported that the μBGA package can be placed at rates of 20,000 units per hour with very few defects. Moreover, the field reliability data are extremely good because the compliance of the package protects the solder joints from failure. As cost targets are met and system designer and user demand continue to grow, the μBGA CSP is expected to play an increasingly important role in the future of electronic packaging over the full spectrum of products and pin counts.

6.9 REFERENCES

1. Fjelstad, J., "The Evolution of Area Array Packaging from BGA to CSP," in *Proceedings of SEMICON Europa*, Geneva, Switzerland, April 1998.
2. Kim, Y. G., "Low Cost and High Reliability Wafer Level CSP Technology," in *Proceedings of the IEEE Electronic Components and Technology Conference*, San Diego, CA, 1999.
3. Fjelstad, J., "Strategies for Creating Compliant IC Packages at Near Chip Size," in *Proceedings of ASME, INTERPAK,* held in Kauai, HI, 1999.
4. Johnson, H., and Graham, M., *High Speed Digital Design: A Handbook of Black Magic,* Prentice-Hall, Upper Saddle River, NJ, 1993, pp. 7–8.
5. Bogatin, E., "Electrical Principles of Packaging and Interconnects for High-Speed Digital Systems, Course 3: Creating Circuit Models with Approximations," course notes, 1994, pp. 2–21.
6. Johnson, H., and Graham, M., *High Speed Digital Design: A Handbook of Black Magic,* Prentice-Hall, Upper Saddle River, NJ, 1993, pp. 257–258.
7. Bogatin, E., "Electrical Principles of Packaging and Interconnects for High-Speed Digital Systems, Course 3: Creating Circuit Models with Approximations," course notes, 1994, pp. 4–9.
8. "Design Guidelines for Electronic Packaging Utilizing High-Speed Techniques," IPC-D-317A, Institute for Interconnecting and Packaging Electronic Circuits, Northbrook, IL, Sec. 5.7 to 5.8.3, 1995.
9. Lau, J., Wong, C. P., Prince, J., and Nakayama, W., *Electronic Packaging: Design, Materials, Process, and Reliability,* McGraw-Hill, New York, 1998, pp. 77–80.
10. Loo, M., and Gilleo, K., "Area Array Chip Carrier: SMT Package for Known Good Die," in *Proceedings of ISHM,* 1993.

ACKNOWLEDGMENTS

We wish to acknowledge the assistance of our many colleagues in the preparation of this material. Special recognition is due to Michael Perry of Pacific Consultants LLC for his contributions in the electrical and finite-element modeling. Also, recognition is due Max Nguyen and Anthony Faraci of Tessera for their work on and contributions to the reliability data and electrical modeling, respectively.

CHAPTER 7
FLIP CHIP TECHNOLOGY

Ken Gilleo
Cookson Electronics

7.0 INTRODUCTION

7.0.1 Basics

Flip chip (FC) is also called *direct chip attach* (DCA), perhaps a more appropriate term. The chip is flipped over so that the attachment side with the connection pads faces downward, giving the process its attractive name. However, it is really the method of connection that distinguishes this technology from all others. The interconnection is a joint without any wires. All other methods, including most chip-scale packages (CSPs), use wires between the chip pads and the substrate or package. There are other connection schemes, such as tape automated bonding (TAB), that connect to a circuit board with the chip face down, or "flipped," but the TAB package incorporates "wires." Thus, while TAB can have a flipped orientation, it can never have a DCA connection.

The concept of a direct chip connection to substrate adds significant benefits while also creating challenges for both assembly and the substrate. FC can be viewed as a nearly "packageless" interconnect system. The elimination of the package both adds and subtracts attributes. The package nearly always adds size and weight. Some near-chip-sized CSPs significantly increase thickness. The package adds protection and ease of handling and aids in standardization. We can see that an understanding of the package, with its penalties and benefits, is a requisite for evaluating the flip chip and other systems.

The Package. The package may add over a dozen features, but only a few are critical. We can think of the package as a structure consisting of a semiconductor device, a first-level interconnect system, a wiring structure, a second-level interconnection platform, and an enclosure that protects the system and provides the mechanical platform for the sublevel. The two critical attributes provided by most packages are environmental protection and ***geometric/mechanical translation.*** In other words, the package converts the extremely dense connection dimensions of the integrated circuit (IC) pads to the larger geometries typically used for the printed wiring board (PWB). The other important translation is mechanical, and this introduces compatibility between the ***inorganic*** world of the IC and the ***organic*** domain of most PWBs. The inorganic IC characteristics include very high modulus and low thermal expansion. The PWB is not nearly as stiff or dimensionally stable, and the coefficient of thermal expansion (CTE) is almost an order of magnitude higher than that of the chip. The package ***bridges*** together these two vastly different worlds. Elimination of the package is not a trivial matter and can require somewhat specialized circuit boards with the addition of added steps and new materials such as underfill. Figure 7.1 summarizes packaging attributes.

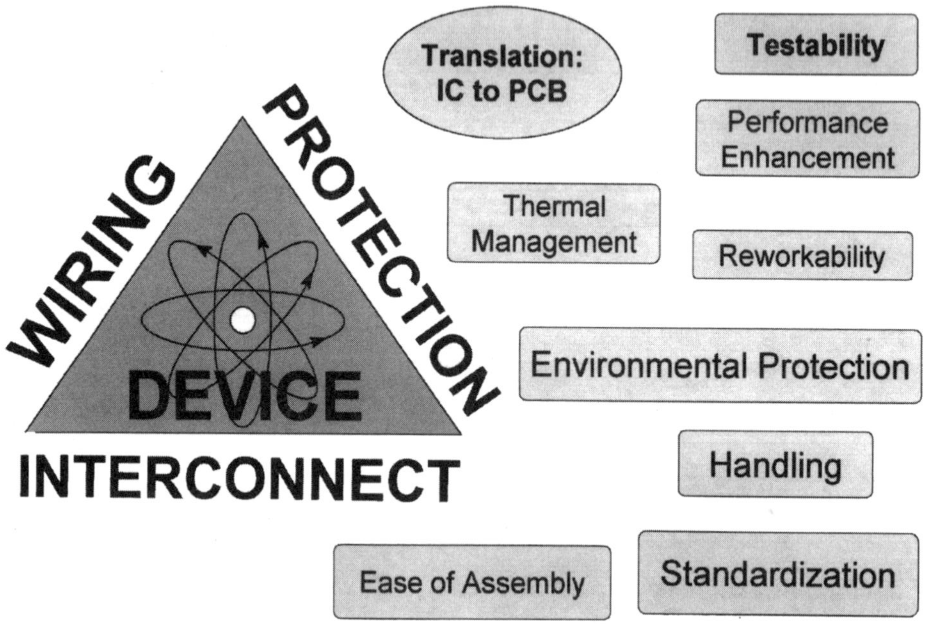

FIGURE 7.1 The package.

7.0.2 History

DCA was developed in the 1960s during the same time period that wire bonding was being established. Although a few early semiconductor manufacturers, such as Fairchild, explored the DCA concept, it was IBM who reduced it to a workable technology and a major commercial success. This was to be expected because IBM was *the* computer company, and mainframe computers were constantly pushing the envelope of performance. The FC benefits of higher performance, high I/O capability, and extreme density were just what the powerful mainframes needed. The FC also allowed heat to be removed efficiently, a feature important for computers but not for much else at that time.

The first DCA development did not even involve the IC, which was still evolving into a real product in the 1960s. Initial packaging work was done on transistors, which were being used in computers ever since IBM had made the pivotal decision to abandon vacuum tube logic for the new, but risky solid-state technology. IBM would need to move away from feed-through assembly if the room-sized mainframes were to ever get down to a manageable form factor. Therefore, much effort was directed toward directly bonding transistors to the substrate with a compact packing arrangement.

The first DCA product was the solid logic technology (SLT), a three-bump ball grid array (BGA) transistor. Earlier DCA work had involved direct soldering to planar pads on the transistor, but height, or standoff control, was a serious problem. The proposed solution was to use a nonfusible metal ball or sphere that would ensure good height control. Copper was the metal of choice because of its long use in electronics and excellent solderability. Copper microspheres were obtained from Alpha Metals, one of the specialty solder suppliers of the period. Attachment to the transistor was straightforward. The spheres were soldered to special finishes, now called *under-bump metallization* (UBM), using tin-lead alloy. High-melting alloy eventually was chosen to allow board assembly without sphere detachment from the device. Copper balls initially were supplied in bare copper, but later they were solder coated. The SLT is shown in Fig. 7.2.

The SLT is a very intriguing package—and it is a true package if we compare the attributes with those in Table 7.1. The transistor is passivated, so no additional protection is needed. The interconnect structure is made of nonfusible metal that could allow package removal, a requirement for the

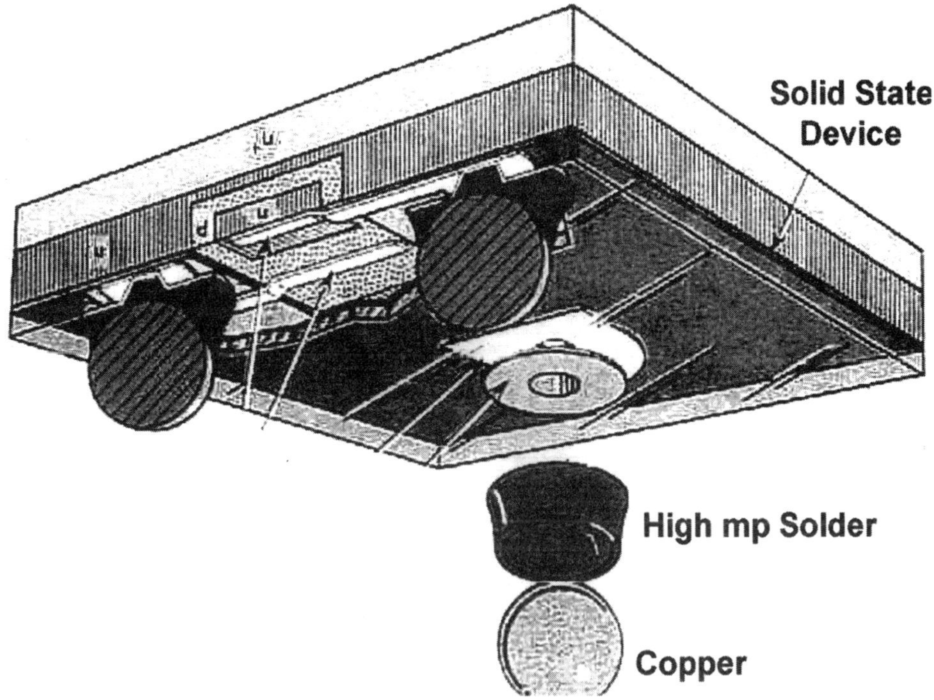

FIGURE 7.2 IBM SLT, the first FC?

true definition of packaging. Ostensibly, the SLT can be reworked, although IBM did not do this. However, look at its size. The SLT is chip size. The product certainly fits the dimensional requirements for a CSP. The SLT is probably the first commercial CSP, beating the AT&T-Allentown gold lead-frame product by less than a year. The AT&T package, with its cantilevered gold leads, was difficult to assemble and not easily reworkable and had limited success (Fig. 7.3). The SLT, the first practical FC, also adds a nice touch of irony if we look back on the CSP versus FC "battles" of the 1990s. The argument about which is better, CSP or FC, could not have been an issue 30 years ago because the FC was a CSP. The more recent FC versus CSP arguments may become moot again if FC becomes a CSP—*again.* And this is precisely what can be expected to happen, but more on the subject in later sections.

The final step in early FC development involved eliminating the discrete copper ball attachment and replacing it with a mass ball or bump formation method. Although a number of processes could be used, and many were tried, vacuum metallization won out. High-melting solder alloy could be deposited through a mask in an efficient wafer-level process. Although vacuum metallization equipment was rather expensive, the high throughput gave a good return on investment. Also, the solid-state electronics industry was much more comfortable with dry vacuum processes than with wet chemical methods such as plating that are only now becoming popular. The elimination of the nonfusible standoff would seem to bring back the height-control problem, but there was a simple solution. Height control was achieved by controlling the collapse of the bump. A high-lead alloy bump could be soldered to a circuit substrate using lower-melting solder while maintaining the height of the initial bump. Alternatively, the wet-out area on the wiring board could be designed precisely so that the amount of bump collapse could be dialed in. Not surprising, the system was called *controlled collapse chip connection* (C^4, and later, C4). The C4 process and its several variations are still used today.[1,2]

TABLE 7.1 Package Attributes

1.0 Platform/Chip Carrier
1.1 Mechanical support for electronic device(s)
1.2 Electrical ground as required
2.0 Interconnect
2.1 Means of electrical connection to device(s)
2.2 Signal paths with required transmission characteristics
2.3 Means of electrical and mechanical connection to PWB
2.4 Means of nonelectrical path when required, e.g., optical
2.5 Interface to PWB that allows removal and rework
2.6 Routing between device and PWB as required
2.7 Means of eliminating or counteracting thermomechanical stress
3.0 Protection
3.1 Mechnical, electrical, and environmental protection of device
3.2 Specific interior atmosphere as required, e.g., hermetic
4.0 Physical Structure
4.1 Robust structure sufficiently durable for assembly
4.2 Mechanical shape for handling and standardization
4.3 Surface for markings and nomenclature
4.4 Low-flammability materials and construction as required
4.5 Materials that meet environmental and work environment regulations
5.0 Thermal Management
5.1 Thermal path for heat removal as required
5.2 Integrated heat sink or spreader as required
6.0 Testability
6.1 Interconnect structure that permits electrical testing as required

7.1 UNDER-BUMP METALLIZATION (UBM)

The first step in preparing a chip for most of the bumping processes is to lay down a proper metallic base over the chip's bond pads. Most of the bumping processes are not compatible with the type and thickness of metal used in IC fabrication, which typically is very thin vacuum-deposited aluminum. Even newer copper metallization requires UBM. The UBM can be applied by standard vacuum coating techniques used in the semiconductor industry or by chemical plating. Vacuum deposition methods have been the most common, beginning with the very first developed by IBM over 30 years ago. More recently, maskless electroless plating has gained popularity because of simplicity that results in cost reduction and lower capital equipment requirements. A suitable UBM provides a barrier layer and serves as a tie coat, or "primer," for the bump that will follow. There is really no standard UBM, and many companies have their own proprietary composition of multiple metals based on many years of data accumulation. However, the electroless nickel UBM has a reasonable chance of emerging as a standard, especially among the new players in the bumping arena.

7.1.1 Problems with Aluminum

Most ICs use aluminum (Al) as the conductor layer to interconnect in-chip devices, although copper is starting to displace aluminum. Aluminum is a reasonably good conductor and is compatible with silicon materials. However, aluminum quickly forms a robust, chemically stable, and electrically insulative oxide that creates a connection challenge. Wire bonding with gold or aluminum wire produces a good, reliable connection because the aluminum oxide is removed in the bonding process. Aluminum is not soldered easily, especially because the layers used for ICs are very thin and easily damaged.

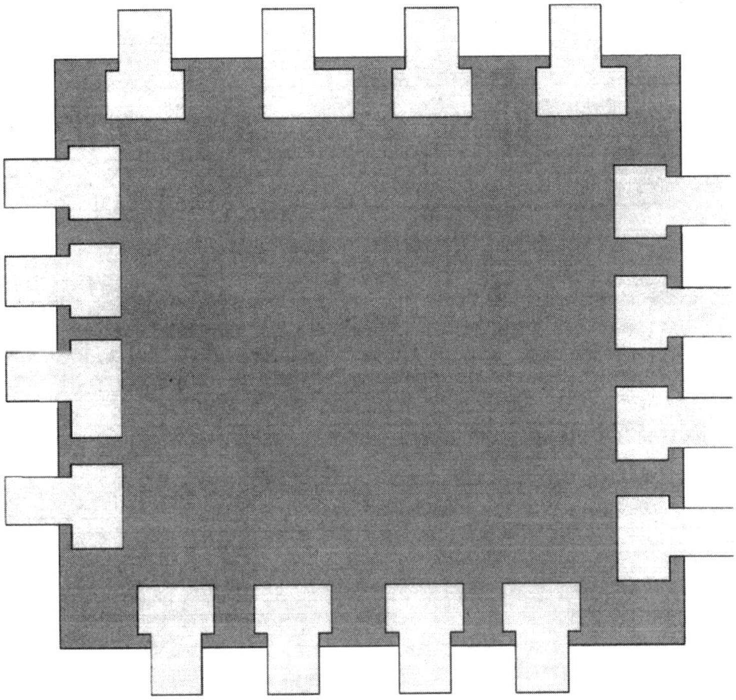

FIGURE 7.3 AT&T lead-frame CSP.

Only a few bumping or bump-forming processes can be used directly with aluminum pads. The most common is gold stud bumping,[3] a modification of gold wire bonding, although more recent work has been done with copper.[4] Virtually all other bumping methods require that a different metal finish be applied over the aluminum before creating the bump. Even conductive adhesives, also called *polymer bumping,* cannot be used directly on aluminum because the oxide insulates the conductive particles from the pure metal. Considerable research on adhesives, including those with penetrating particles, has not led to success with the bare metal. Although aluminum oxide can be removed by chemical or mechanical means, the oxide will "grow" back even with cured adhesive in place. The most common solution has been to apply a different metal or several metals to the aluminum pad prior to bumping. Since the metal finish ends up under the bump, it is referred to as *under-bump metallization* (UBM). Many different metals are used, but all serve as "primer" and sometimes as a barrier against metal diffusion.

7.1.2 Types of UBM

We can categorize UBM by either the type of metal or the method of application, but the latter is more meaningful. The early processes relied on vacuum deposition. The vacuum process is similar to the method used to put down aluminum conductor traces and pads. Metal is vaporized in a high vacuum and allowed to condense on the wafer through a patterning mask. Heating to a high temperature can vaporize the source metal, or bombardment with a stream of energetic accelerated ions or atoms can be used. The thermal evaporation used to vaporize the metal is often referred to as *metal vapor deposition* or *chemical vapor deposition* (CVD). Heat can be generated by applying an electric current to source containers called *boats* in a process called *resistive,* or *resistance, evaporation,* or *firing.* Since the heat flux and the evaporation rate fluctuate with this method, a controlled heat source, such as an electron beam, is more desirable for semiconductor work. Generally, only aluminum is vapor deposited by the IC industry because it is easily "boiled." Other methods are used to deposit copper, tungsten, and the more difficult metals.

Sputtering is a more manageable means of vacuum depositing. Virtually any metal can be sputtered, making the process ideal for UBM. The sputtering process uses high-energy gas ions to strike a metal target and dislodge the source, or target, material by transferring the high energy. Sputtering is a slower process than evaporation, but it can be controlled with extreme precision, virtually atom-by-atom. Even high-boiling metals, such as tungsten and chromium, can be sputtered because it is not a thermal energy process but rather an electrical/magnetic-powered atomic collision process. The basic principle is to energize a heavy inert gas, such as argon, and form plasma. The gaseous ions are then accelerated in electrical and magnetic fields when the common magnetron process is used. High-energy argon ions and neutralized atoms strike the metal target with enough energy to dislodge metal atoms and clusters. The metal particles can be directed at the substrate in a beam that produces a metal coating. Multiple metal targets can be used in a single chamber so that several different metal layers can be deposited in a single run. UBM may use a layer structure of two or more different metals to provide compatibility and good adhesion. Some of the common systems are aluminum-titanium-copper (Al/Ti/Cu), aluminum-nickel-copper (Al/Ni/Cu), and aluminum-palladium-copper (Al/Pd/Cu).

Metal also can be applied to aluminum pads by chemical plating. Three basic types of plating are recognized: (1) electrolytic, (2) electroless, and (3) immersion (a subcategory of electroless). All are being used for UBM. *Electrolytic plating,* also called *galvanic,* requires application of a dc current to the wafer and to a plating electrode. Positive metal ions in the plating bath are attracted to the cathode, in this case the wafer. The metal ion picks an electron or several, depending on its charge, to return to the metal state again. However, in order to connect the wafer die pads to an electrical source, a plating bus must be added. This is commonly done by vapor depositing an aluminum film on top of the wafer. A plating mask is first applied over the bus to define where plating will occur. Once the UBM is formed, the mask and bus are stripped away. The need to apply and remove a bus and mask adds steps and cost.

Electroless plating is a much simpler process that does not need current, but it is more limited.[5,6] Only a few electroless plating systems are compatible with aluminum. In fact, no commercial electroless plating systems are even compatible with aluminum. Fortunately, an immersion plating method can be used to produce a platable surface. Electroless immersion zinc plating can be applied directly to aluminum. Then electroless nickel can be applied over the zinc. *Immersion plating* is a type of electroless plating where a double ion-exchange reaction occurs between the plating bath and the metal being plated over.[7] In this case, zinc ions in solution are exchanged with aluminum atoms on the surface. The aluminum dissolves, and the zinc in solution deposits at the location vacated by the aluminum. Once the zinc plating bath has displaced the entire aluminum surface, the reaction stops. There is no more aluminum to drive the reaction, and thus it is said to be *self-limiting.* While immersion plating is self-limiting, nickel and most other electroless plating processes are *autocatalytic,* with the plating reaction continuing until the plating bath is exhausted.

These nonelectrolytic plating methods do not require a mask or bus, and this is an important benefit. One limitation is that only a few metals will plate electrolessly. Even though immersion and electroless plating generally do not require a mask, there is a thickness limit that is a direct result of the maskless method. A plating mask restricts plating in the plane of the wafer, the x and y dimensions, by providing sidewalls that restrict metal growth. Plating can only build up in the vertical or z direction. Plating buildup can continue up to the top surface of the mask. Without a mask, plating occurs in the x, y, and z directions. The deposit thus increases in width as it grows in height. Electroless plating therefore produces a mushroom shape as the reaction progresses. Continued plating will cause the metal at each pad site to grow outward and finally short out at adjacent pads. Since the UBM needs to be only a few hundred angstroms thick, this is not a problem.

The most popular newer UBM is electroless nickel with a zinc, or zincate, layer first applied to activate the nickel plating chemistry. In many ways, this is an ideal process because only chemical solutions are required. However, chemical composition control is very critical. Failure to control the zinc plating step, for example, will cause low adhesion. The nickel plating also might be deposited with high stress if the composition is not correct throughout the process. The stress will lead to adhesion loss. However, if the two plating steps are controlled correctly, an excellent nickel UBM will be produced. Since nickel will oxidize slowly, a gold flash, actually immersion gold plating, is commonly applied. The immersion gold is only about one atomic layer thick, and this is too insignificant to add cost or create soldering problems later. Nickel, with or without gold, is compatible with solder and many conductive adhesives. While some adhesives form stable junctions with a nickel surface, all perform well with a gold surface, and this is the preferred finish for polymer bumps.

7.1.3 The IC Transition to Copper

Several developments in the past few years have solved the numerous problems associated with using copper metal in place of aluminum as the IC interconnect metal. Copper is more conductive (about three times) than aluminum and allows higher frequencies to be used with smaller line widths. Many fabricators are converting to copper not just for speed but also for cost reduction. Thinner conductors allow closer spacing and smaller chips. The switch to copper allows many more dies per wafer, and this is where the savings come from.

Since copper is much more compatible with bump materials than aluminum, the transition to copper is expected to boost flip chip technology possibly by eliminating the UBM step. It remains to be seen what the final finish will be for copper ICs. The industry will continue to use wire bonding as the main interconnect method, so the final pad must be compatible with gold wire bonding. IBM has indicated that its copper chips will have aluminum as the final pad layer to accommodate wire bonding, and this may become standard practice. However, aluminum can be removed easily without affecting the copper underneath. Aluminum is an amphoteric metal that can be dissolved in both acid and base. Dilute caustic (sodium hydroxide) quickly removes aluminum. Many other reagents also can be used. Thus, even if the new copper chips come with an aluminum finish, a simple aqueous washing step will unveil the desired copper layer. In fact, the aluminum over copper would serve to protect the copper from oxidation. Gold over nickel also could be used on the copper pads similar to PWB common finishes. This too would be a very good surface for most bumps.

Conductive adhesives would receive a real boost in the switch to copper because none are compatible with aluminum. Some of the conductive adhesives form reasonably stable junctions with bare copper, especially those using an oxide-penetration mechanism. Even for those adhesives which are not suitable for bare copper, simple UBM methods could be used. Silver and other finishes can be applied to copper by electroless, maskless plating. The advent of copper-based chip interconnection metallurgy undoubtedly will simplify FC fabrication in the near future. It probably will be possible to directly bond the copper pads with conductive adhesives. This simple processing ability would have a great impact on cost and infrastructure issues by eliminating UBM and maybe the bumping step. Assemblers could run the entire FC preconditioning and bonding process.

7.2 BUMPING MATERIALS

There is some overlap between UBM, just covered, and bumps. Some UBM processes can be extended to produce complete bumps. UBM can be thought of as the very thin metal primer, or tie coat, on which the bump is deposited. The UBM will always have more conductive material added to increase thickness beyond the die passivation layer. Some UBM processes, such as electroless nickel, can be continued until a thickness is built up by many orders of magnitude to produce a bump that rises above the die passivation layer.

In theory, any conductor can be used for bumps, provided that it has sufficient strength and integrity to hold the chip to the substrate. Bump materials can be divided into fusible and nonfusible classes. The difference is quite important. Many fusible types, especially those of lower-melting alloys, provide the joining material for assembly. While it is desirable to have the joining material built into the bump, this can have a downside such as more difficult height control. A bump made totally of the same fusible alloy can totally collapse by spreading out on the connection surface if the substrate pads do not have a means to restrict the wetting area.

Nonfusible bumps, such as the copper spheres used in the SLT, solve the height-control problem. The penalty is obvious: Joining material must be added. This requires an added step and just one more place where things can go wrong if good control is not maintained. However, the nonfusible bump can add features that offset the added step cost. Proper design can permit rework. Higher standoff can improve reliability. The higher melting point of the bump can be ideal for flip-chip-in-package (FCIP), which will experience reflow temperature at least one more time. Thus each type has its advantages, and the end use often will determine which way to go.

7.2.1 Fusible Bumps

Fusible bumps are now the most widely used bump type because they can supply the joining material and are easy to make. However, some fusible bumps have a high enough melting temperature to allow use as nonfusible bumps. Thus my definition will get a little fuzzy unless the final assembly is considered.

High-Lead Solder. The original C4 process used high-melting tin-lead (Sn/Pb) solder such as 3/97. This is still a fusible alloy, although a reflow temperature as high as 350°C may be needed. This is a practical temperature for ceramic substrate, however. IBM and other mainframe computer makers have used ceramic substrate and continue to do so because of its unique attributes. One important reason is the high thermal conductivity. Some ceramic circuit materials are better thermal conductors than metals while serving as excellent dielectrics. The massive number of chips in a mainframe and the high board and system density suggest that ceramic materials will continue to be used here and that the high-melting-alloy bumps will see service into the foreseeable future.

Some of the ceramic FC BGAs also use high-lead bumps. This allows the package to be soldered to a board with eutectic solder without disturbing the FC-to-chip-carrier joints. The ceramic substrates, while more costly than organic substrates, give more planar surface and low moisture absorption. Low moisture absorption means that packages do not have to be stored in a dry atmosphere or predried prior to solder reflow. Moisture absorption, as seen in most organic BGAs, causes explosive evolution of steam and instant delamination, called *popcorning.* It also should be mentioned that the high-lead bumps are less prone to thermomechanical fatigue. All told, high-lead bumps serve a function, and their important niche uses will continue.

Stratified Bumps. One clever idea is to build a high-lead bump that is rich in tin near the top. This is also called the tin-cap process. Both IBM and Motorola have been active here. This kind of structure is said to provide the high standoff and favorable fatigue resistance of high-lead bumps while creating eutectic solder for the joint.

Eutectic Sn/Pb. Eutectic solder is the de facto standard of electronic assembly, so it is no wonder that this same material has become essentially the standard solder bump. First, a eutectic bump can be used as the source of joining material for assembly on organic PWBs. The industry has more or less geared itself to processing around the conditions required by eutectic solder. Next, eutectic tin-lead solder is readily available in a variety of forms, including fine-particle paste that can be used for bumping. Since the eutectic temperature is the natural equilibrium point of an alloy, it is the ideal region for processing. However, eutectic solder joints are prone to fatigue and will fracture eventually if subjected to continual stress.

Polymer Bumps, Thermoplastic. Polymers can be either fusible or nonfusible. Thermoplastics behave like solders in that they melt on heating and solidify on cooling in a process that can be repeated many times over. Thermosets, on the other hand, do not melt once they are set, or polymerized. Limited work has been done on thermoplastic-based conductive adhesives as bumping materials. The adhesive bump can be formed by applying an isotropic conductive adhesive (ICA) to the UBM or to a nonfusible metal bump. ICAs represent one class of conductive adhesives that are highly filled with metal and conduct equally in all planes. The other class of adhesive is the anisotropic conductive adhesive (ACA) with unidirectional electrical conductivity. The ACA is considered to be a joining material, not a bump-forming system, and will be covered in a later section.

Thermoplastics are polymerized solids, but they can be made into conducti ing them in polar solvents and adding conductive particles. The most common silver because of its stable conductive oxide surface, ability to be formed into t gy, and a reasonable balance of cost and performance. Silver filler is typically a flakes and spherical particles ranging in size from 5 μm down to even submic the ICAs is about 80 percent by weight, but excluding solvent. Once applied evaporated to form a dry bump. The bump can be attached to substrate by heating to the softening or melting point. Figure 7.4 shows bump construction.

7.2.2 Nonfusible Bumps

Nonfusible bumps make up only part of the interconnect between IC and substrate. Joining material must be added to complete the connection. Therefore, the nonfusible bump produces a more complex joint made up of a different composition at each interface and possibly intermediate regions of both compositions including intermetallic compounds. However, the nonfusible bump can provide several benefits over fusible materials. Depending on hardness and other properties, the nonfusible bump may be tested and burned in. A nickel or copper bump can be probed, socketed, and burned in much more easily than a soft solder bump. Virtually all nonfusible bumps offer controlled standoff so that a minimum height can be maintained. This can be a very significant feature because too low a standoff will increase strain dramatically. A low clearance between chip and substrate can cause underfill voids or even make filling impossible.

Gold. Gold was used to bump chips even before it became popular for flip chips. The tape automated bonding (TAB) process uses a gold-to-aluminum connection metallurgy and requires a bump somewhere in the process. Gold bumps can be created on the IC pads or on the TAB chip carrier.[8] Methods developed to form gold bumps on aluminum IC pads for TAB also can be used for flip chips. Pure gold is used typically and can be applied directly to aluminum pads without UBM. The gold bumps are highly compatible with all conductive adhesives. However, soldering requires specialty alloys.[9] Eutectic tin-lead solder will leach or dissolve gold and form metal compounds called intermetallics that greatly reduce joint reliability because of high brittleness.

Although a few companies are promoting gold bumps and solder attachment with special alloys, most use conductive adhesives as the joining material. Gold is soft and malleable, allowing bumps

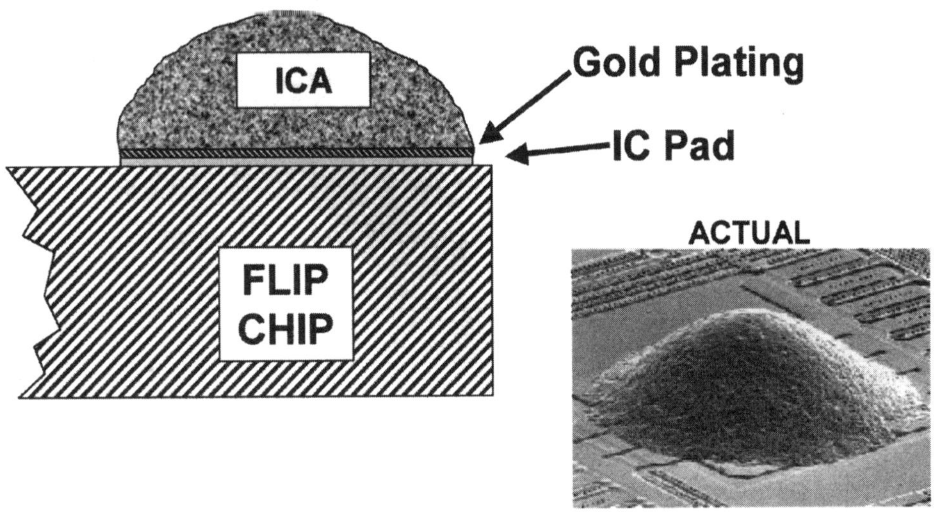

Courtesy of Polymer Flip Chip Inc.

FIGURE 7.4 Polymer bumps.

to be shaped easily. The type of joining material determines the shape requirement. ICAs work well with conical shapes, whereas flat bumps are more suitable for ACAs. I should note that unlike lead-based solders, which emit alpha particles that cause problems with many ICs, gold contains no radioactive isotopes. The use of nonemitting conductive adhesives therefore provides a system with no radioactive emissions, making it ideal for memory and other types of chips that are sensitive. Shaping will be covered under bumping processes below.

Nickel. Nickel is a very strong, hard, and high-melting metal that is used extensively in electronics. Nickel is soldered readily and also is compatible with some conductive adhesives. While nickel can be used as a UBM, it also can be formed into bumps. In most cases, the nickel serves the dual purpose of UBM and bump. The nickel plating process is simply continued until the thin UBM coating builds up to a bump. Once formed, the nickel bump usually is overplated with a thin coating of gold to prevent oxidation.

The nickel bump can be coated with solder to form a multiple-composition system. Although it can be argued that the nickel is then serving as a UBM, if there is substantial height, the nickel should be classified as a bump. The purpose of producing a bump instead of just UBM is to create a minimum standoff. At least one of the bumping processes produces shallow bumps because the bumps widen as they grow in height. The bump height may be less than 25 μm (1 mil), with the fusible portion making up most of the height.

Polymer, Thermoset. Polymer bumps usually are made with thermoset ICA materials that are stenciled or screen printed onto the UBM. A gold-finish UBM typically is required, although nickel can be used if the adhesive can form a stable junction. However, since a simple and inexpensive process is used to apply a flash of gold, bare nickel is seldom encountered. Silver-filled epoxies are by far the most common bumping material. Their conductivity is nearly the same as that of solder, and their mechanical strength is reasonable. Conductive adhesive joints still fatigue when thermomechanically stressed, although some give modestly better performance than solder bumps. Once cured, the conductive adhesive bump requires a joining material for assembly and this simple, more conductive adhesive paste. However, Fujitsu claims to make bumps with thermoset adhesive that are hardened but not fully cured (B stage). These bumps can be attached by applying heat and pressure. The B-staged thermoset bonds and then fully polymerizes.

7.3 BUMPING PROCESSES

This area of technology is still evolving, and by the time you read this, a few additional methods may have been announced. All bumps are made of metal or composites containing mostly metal. The bump may be deposited directly or as a material that is converted to a bump by a subsequent process such as solder reflow. The methods may be roughly grouped into discrete one-at-a-time singular and mass-deposition parallel processes. While mass deposition is most popular, discrete methods such as stud bumping are also enjoying success. The mass-deposition methods can be subdivided into physical material transfer (such as printing), chemical (such as plating), and vacuum deposition, although the last also can be classified as a mechanical-physical process.

7.3.1 Vacuum Deposition

Let's start with vacuum deposition, also called *vacuum plating,* because this was the most successful early process and still remains an important one today. The basics of vacuum deposition were covered earlier. A vacuum-deposited bump is always formed on the appropriate UBM. Virtually all metals can be vacuum deposited, as well as alloys, although it is more common to apply pure metals in sequence and then form the alloy by melting in a later step. Lower-boiling metals such as tin and lead can be evaporated easily in a vacuum, and this is the primary vacuum method for tin-lead alloys. The deposition time, and thus the amount, of each constituent can be varied to produce the desired alloy. The same vacuum chamber therefore can deposit 3/97 and 60/40 tin-lead just by changing the parameters.

A mask is used to define the deposition areas. A metal contact mask can be positioned over the wafer with openings that correspond to the IC pads. Masks are made from high-temperature, precisely imaged materials including titanium, tungsten, and molybdenum ("moly mask"). These masks are used repeatedly and cleaned as required. Metal deposits can be removed by melting or by chemically dissolving the bump material in reagents that do not affect the mask.

Integral masks, those applied directly to the wafer, also can be used. A photoimageable polymer is employed, although laser machining could be used with nonphotosensitive polymer films. The mask material is applied by spin coating and then imaged and developed in place. Once the bumps are formed, the mask is stripped away and cannot be used again. The integral mask can provide better registration and resolution because photolithography is used. Turnaround is quicker because a phototool is used. The process is also called *lift-off.*

Applying one metal and then the other(s) can produce alloys. The process can be alternated to provide more homogeneity, although this is not necessary because the bump is usually reflowed as part of bump manufacture and often reflowed again during assembly if it is a fusible type. A stratified bump with a low-melting tip can be produced by first applying the high-melting composition and then the lower-melting one. One process constructs a high-lead bump and ends by applying a tin cap that becomes a eutectic tin-lead composition during reflow. This approach provides an assured standoff height while allowing the bumps to be soldered at lower temperatures without adding solder pastes.

7.3.2 Plating

Chemical plating is a common industrial process whereby metal salts in solution are converted back to metal on the surface that is to be coated. The process is divided into *electrolytic,* where electric current is the controlling and driving force, and *electroless,* where chemical reactions supply the energy.

Electrolytic Plating. Most metals form at least one water-soluble salt, although the pH may need to be adjusted for optimal results. A metal salt consists of a metal cation with a positive electrical charge of 1, 2, 3, or higher corresponding to its chemical valence, the number of easily lost electrons. One or more anions balance the positive metal cation. The anion is a species of opposite charge that results in a salt that is electrically neutral. When a direct current (dc) is passed through the aqueous salt solution, positive metal cations are attracted to the negatively charged cathode. Acquiring the appropriate number of electrons to convert back to the metallic state neutralizes the cation. Since metal is being depleted from the solution, or plating bath, an electrode of the same metal typically is used for the positive anode, where metal ions lose electrons to form replenishing cations. Solder alloy can be plated, or the individual metals can be stratified and melted into solder.[10] However, it is more efficient to codeposit metals and form the alloy in situ.

Electroless Plating. Metal plating also can occur without the use of an electric current. A reduction-oxidation, or redox, reaction must occur for "automatic" electroless plating to result. Electroless plating can be further divided into autocatalytic, where any thickness is possible, and immersion, where plating stops when all the base metal is covered up. Both types of plating are used in electronics and in bumping.

Electroless nickel plating, covered earlier, became very popular in the mid-1990s as both a UBM and a bumping method. Nickel systems will plate onto a number of surfaces, but aluminum is not one of them. Using zinc immersion plating solves the problem of plating onto the still-common aluminum pads. Zinc salts will react with aluminum to convert that metal to a salt while simultaneously being reduced to metallic zinc. This is an immersion plating reaction where the zinc cation is reduced to metal, and the aluminum is oxidized to a salt, hence the term *redox.* Once the zinc has been deposited by immersing the wafer in the zincate solution, we are ready for nickel plating. It should be noted that the zinc plating, or zincate treatment, is a critical process, with cleaning and multiple treatments often required for optimal results. In a sense, the zinc layer can be thought of as the UBM for the nickel bump.

Electroless nickel plating can now be deposited over the zinc. The nickel will continue to plate out until the bath is depleted because nickel is a catalyst for its own plating reaction (autocatalytic). Time and temperature determine the plating thickness, so precise control is possible. As we saw earlier, a UBM deposit only requires that sufficient nickel be applied to serve as a base for the bump that typically is solder. A nickel bump is many orders of magnitude thicker because it must extend well beyond the die passivation layer that is several microns thick.

The electroless nickel process can be used to form bumps by increasing the plating time until the height is approximately 2 to 40 μm. However, unless a plating mask is applied, and this is not usually done, the nickel will plate upward and outward, forming a mushroom-shaped bump, as shown in Fig. 7.5. Continued plating eventually will cause the individual bumps to touch and short out. Pitch and the spacing requirement between bumps therefore limit the useful plating thickness or bump height. Although this is a restriction for high-density FCs, many classes of FCs can use the process at substantial cost savings. The flat bumps, especially with gold flash plating, are ideal for use with anisotropic conductive adhesives but also can be attached with solder paste. A common practice involves applying a very thin layer of gold over the nickel as the final step. This prevents oxidation, and the small amount of gold does not cause any problems with solder embrittlement. The gold plating process is immersion gold that stops when all the nickel is covered up.

7.3.3 Printing/Stenciling

Solder-paste printing on wafers is a very efficient and low-cost method of forming bumps. Although screen printing can be used for low-density chips, stencil printing is superior and used exclusively.[11] The advances in stencil making over the past several years have enabled the solder-paste bumping process, but stencil feature size capability is still the limiting factor for the process. Stencils made by the highly precise electroforming method (uses electroplating to form the stencil) are considered optimum, although laser-cut stencils also have been used. Specialized solder pastes give best results.[12] A solder-coated process using metal salts that are thermally decomposed to form solder also has been reported.[13]

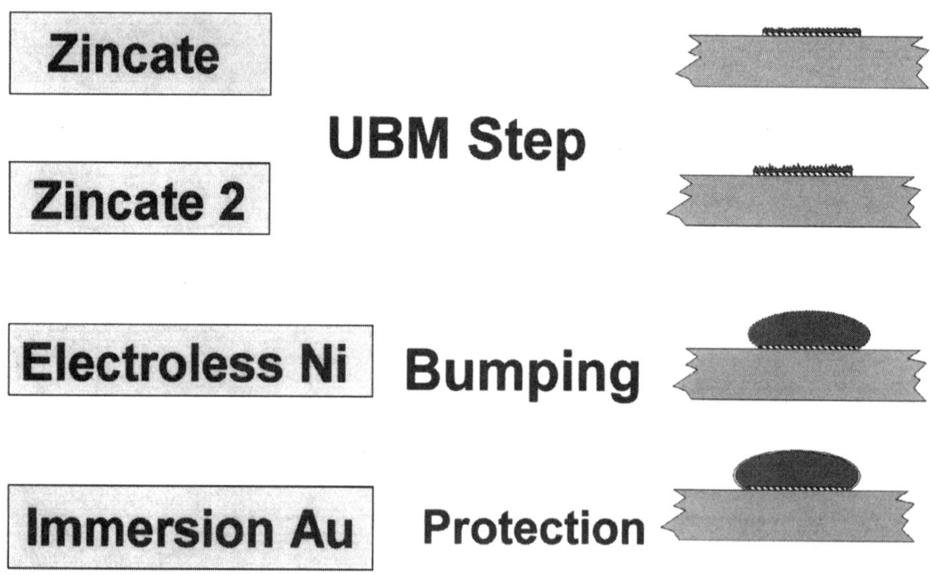

FIGURE 7.5 Electroless nickel bump process.

The precision stencil is aligned to the wafer pads, and fine solder paste, type 4 or 5, is applied after the appropriate UBM has been laid down. The paste is then reflowed to attach the solder to the UBM and form the semispherical bumps. Several groups have defined the process parameters and reported successful commercialization of the process. Solder-paste stenciling probably has become the most popular FC bumping method for all but the high-density chips.

There is one other modification of the solder-paste stencil process worth mentioning—integral mask. The stencil is created on the wafer using photoimagable resist. Only the pad areas are opened. Solder paste is applied to the openings using a doctor blade. The paste is reflowed, and the mask is stripped away. The unmasked bumps may be reflowed a second time to produce the desired semispherical shape. This method allows higher density bumping and is thought to be used by Flip Chip Technologies, Inc. Patents may be pending or have been issued by now, and anyone contemplating using integral mask deposition should check the patent literature.

7.3.4 Metal Fluid Jetting

A technique has been developed in which a microdrop of liquid metal is propelled to locations on a substrate with extreme precision.[14] The hot metal can adhere to UBM surfaces to create bumps. Reflow will provide a stronger bond and round out the bump. Equipment is now available that can be used to bump wafers at speeds of 1000 or more drops per second while still maintaining high precision. One style of equipment from MPM/Speedline moves substrate while a piezo "gun" fires the solder spheres. This *shoot-on-the-fly method* provides a good balance of productivity and precision and bump sizes down to at least 75 μm. However, this is a serial process and probably not optimum for very high I/O dies. Figure 7.6 shows "jetted" bumps.

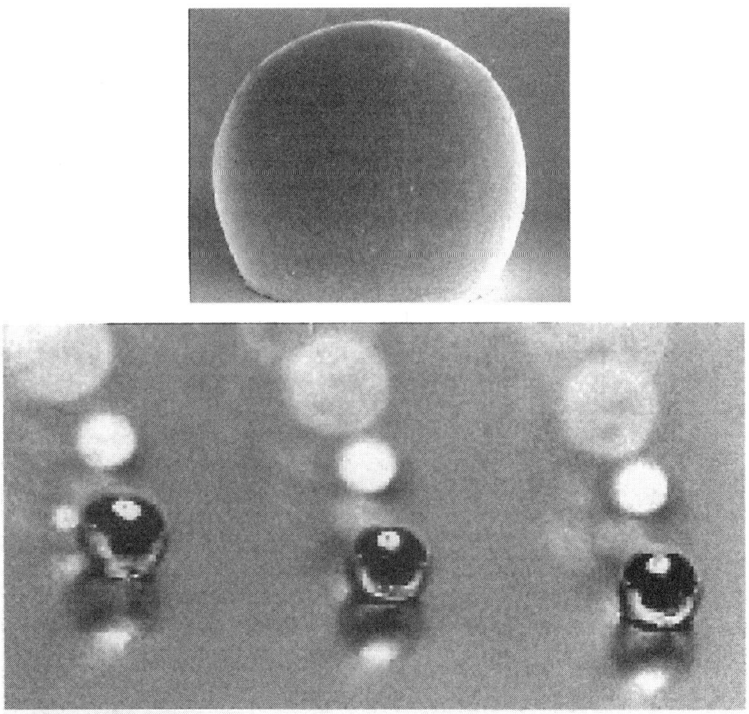

Courtesy of NIST

FIGURE 7.6 Solder jetted bumps.

7.3.5 Mechanical

There is only one important mechanical bumping method, and it is gold-stud bumping, a rediscovered process that dates back to the 1980s. The process is a modification of gold-ball bonding and has been around for a dozen or more years but is still covered by a patent.[15] Earlier use was strictly for prototyping, but several companies are using the method in production because of a number of unique features.[16] The gold-ball bond is formed, and the wire is broken close to the ball instead of making the loop. There are several important attributes to the stud-bump method:

1. No UBM is required.
2. The process is completely mechanical; no chemicals are used.
3. The process is practical for single dies as well as wafers.
4. It is lead-free, with no environmental issue or alpha-particle emission concerns.

This simple bumping method can be used without substantial investment or permits from regulatory agencies. However, there are a few significant limitations that reduce widespread use. First, a joining material must be added, and common Sn/Pb eutectic solder cannot be used because gold intermetallic compounds will form. Indium and other special alloys work, but at a higher cost and with certain performance deficiencies. Conductive adhesives have become the most common material for use with stud bumps, and products, such as memory chips, have gone with this approach. Methods have been worked out to increase stud-bump height and to control shape. A double bump can be used to increase height. A secondary shaping operation can be applied to form a conical bump that is preferred for use with adhesive pastes. Figure 7.7 shows a conical gold-bump profile. The joint is formed using isotropic conductive adhesive (ICA).

7.4 JOINING MATERIALS AND AGENTS

Ancillary joining agents such as flux may be required to form FC joints when the joining material is provided by a bump made of solder. However, several types of nonfusible and high-melting bumps are used, including high-lead solders that require joining materials such as eutectic solder alloys or conductive adhesives. This section will describe the most common materials in use today for joining FCs to substrate.

7.4.1 Flux

The presence of oxide on bumps or on the wiring board pads interferes with soldering and the formation of good metallurgical junctions. The simplest remedy is to convert the metal oxides back into pure, easily wet metal. Flux is a chemical compound that contains a reducing agent to convert metal oxides to metals just prior to solder joint formation. While flux can be incorporated into solder paste, this section will deal with stand-alone fluxes. The most common flux agent is organic acid, and nearly all fluxes for electronics use this chemistry. Once the flux agent does its job, it must be either removed or deactivated to prevent corrosion later on. The presence of acid in contact with metal can cause a host of problems later, especially in high-humidity environments. Flux can have other functions, such as holding the FC in place as the assembly moves into the reflow oven.

The FC presents special challenges to flux chemists, however, and many good materials for standard SMT cannot be used. Flux residue cannot be removed easily from under the FC because the gap can be 25 μm or less. "No clean" fluxes have become popular for FCs, and many advancements have occurred. However, if flux residue is left under the FC, it should be compatible with underfill. Many "no clean" fluxes were found to be very incompatible with underfills. Since the underfill must form a very strong bond to both the wiring board and the bottom of the chip, any flux that reduces the bond strength or performance of the underfill would degrade reliability.

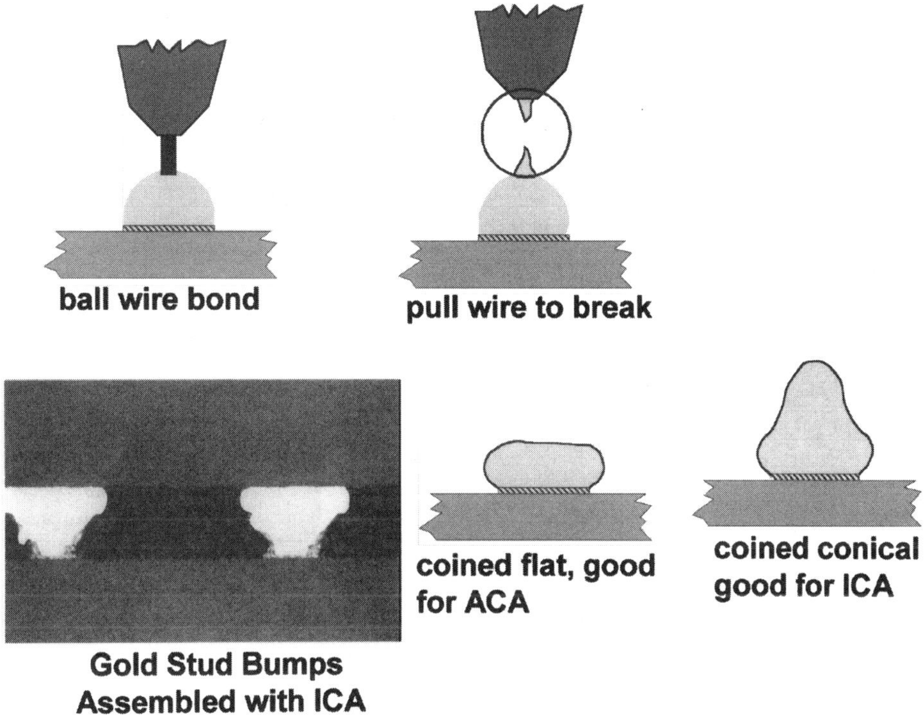

FIGURE 7.7 Gold stud bumps.

In the early 1990s, Alpha Metals hit on the idea of using underfill ingredients to make flux. Since nearly all the popular underfills are based on epoxy resins, fluxes with similar chemistries were investigated and developed. Active flux agents were selected that could polymerize with the epoxy after reducing the oxides. The most useful class of flux/hardeners is multifunctional organic carboxylic acids. A series of FC epoxy-based fluxes have been commercialized, and the concept has received broad patent coverage. The flux residue becomes a hard, strongly adhering epoxy thermoset after exposure to solder reflow temperatures. Underfill will react chemically with the flux surface so that no delamination will occur. The epoxy fluxes can be made in a solventless form over a wide range of viscosities with different tackiness values. These are the preferred fluxes for FCs using eutectic solder, although several more standard fluxes have been found to work as well.

7.4.2 Solder Paste

Solder paste is used for nonfusible bumps such as high-lead, high-melt-point solders and plated metals such as nickel. Although alloys such as 3/97 tin-lead can be melted and soldered, the reflow temperature of up to 350°C is too high for anything but ceramic substrate. A more common practice is to apply solder paste to the wiring board, place the FC, and reflow the solder paste in a process similar to SMT. Solder pastes for FC assembly should be of a finer particle class such as type 4 or 5, especially for fine-pitch work. The paste also must have enough tack to hold the FC in place during transport to the oven. The paste ingredients should be compatible with underfill. Epoxy-based flux technology has been applied to solder pastes.

7.4.3 Conductive Adhesives

Conductive adhesives offer several benefits for FC assembly. Both anisotropic (materials with unidirectional conductivity) and isotropic (equal conductivity in all directions) conductive adhesives can be used. In fact, even nonconductive adhesives have had success when used as "mechanical springs" to force bumps against pads. The limitations of adhesives when used for standard SMT, such as lower strength, are mitigated when they are used for FCs, making them an important joining material.

Anisotropic Conductive Adhesives (ACAs). First glance suggests that there is magic in an adhesive that only conducts in one direction, the *z*, or vertical, plane. A closer look shows that the *z*-axis adhesives are rather ordinary. The materials consist of an insulative binder that can bond to the chip and substrate. The conductors are small particles, preferably spherical, that are dispersed in the binder at a low enough loading so that interparticle contact cannot occur. In one sense, an ACA is an ordinary conductive adhesive without enough conductive filler. Electrical connections occur when the conductive particles are trapped between opposing bumps and circuit pads, as shown in Fig. 7.8.

One can see from Fig. 7.8 that ACAs produce a pressure-contact interconnect instead of a metallurgical junction like solder. This means that pressure must be maintained for reliability, and this is the key to good performance. ACAs only work well with highly planar systems or a substrate that is comformable, such as flexible circuits.[17] Fortunately, FCs are very planar, allowing ACAs to be used. Since ACAs are capable of connecting high-density structures, they are a good fit for FCs and have achieved success, especially for chip-on-glass (COG) used with flat-panel displays.

Modern ACAs use randomly dispersed conductive microspheres. The list of particles that have been used is extensive and is found in Table 7.2. Early ACAs used solid conductors, such as nickel, silver, or gold particles, but coated nonconductors are more common today. Metal may be plated onto glass, plastic, or other materials using electroless methods well known to the industry. A very popular system is gold plate over nickel plated onto elastomeric microspheres. The plastic spheres are cleaned, activated, and then plated with electroless nickel and a final flash of gold by immersion

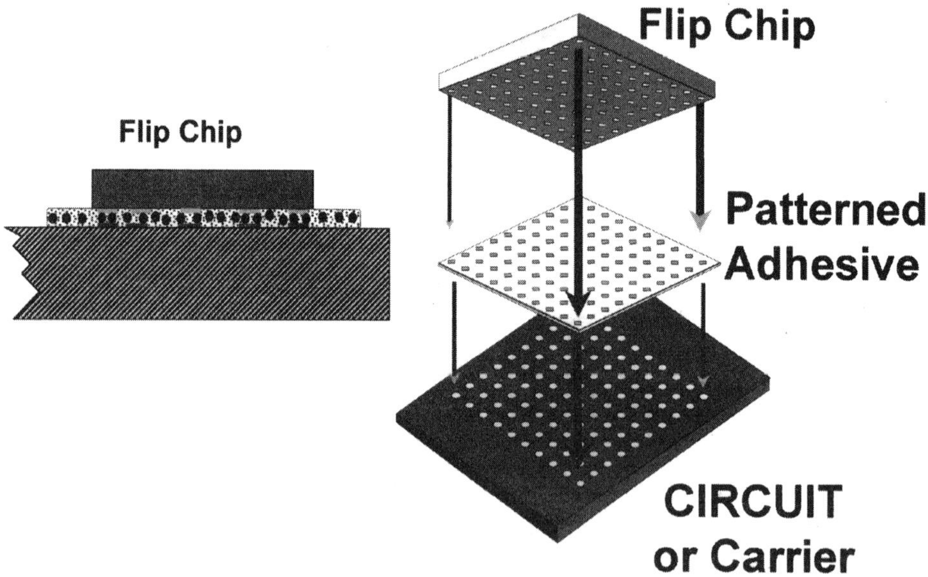

FIGURE 7.8 ACA junction.

plating to prevent oxidation. A coating of pure gold over plastic gives better performance, but electroless gold plating is more difficult and expensive.

The conductive particles are dispersed in a polymer adhesive at only a few weight percents. The adhesive can be a thermoplastic, a thermoset, or a blend of both. The thermoplastic ACAs have extraordinary shelf life at room temperature but are limited to an operating temperature of 100°C or slightly higher. Thermosets, with or without added thermoplastic, are more common and can pass aging and cycling up to 125°C. Their limitations are storage life, higher bonding temperature/pressure, and longer assembly times. Products are commonly available as films, including B-staged thermosets, but pastes are also sold. Finally, it should be noted that the ACA binder also serves as underfill, thus eliminating this bothersome step.

Isotropic Conductive Adhesives (ICAs). ICAs are exemplified by the commonly available silver-filled epoxies used for die-attach and solderless SMT assembly. These materials are highly loaded with silver filler so that the intimate contact produces good conductivity in all directions. Silver is used as the conductor because of its optimal balance of performance versus cost. Silver oxide, unlike the oxides of copper and the common base metals, is conductive. Although nonoxidizing metals such as gold can be used, silver provides equal performance at a cost savings. Silver is quite ductile and can be milled into the right flake morphology to produce high conductivity and the required rheology. Flake, because it can form an overlapping matrix, must be used for good performance. Powder alone would result in lower conductivity and a general instability. Silver flake also provides thixotropy, which allows the dispensed adhesive to stay in place. Figure 7.9 shows the model for ICAs. Special adhesives have been developed for FCs using finer silver fillers and additional milling to break up any agglomerate that would cause an open in fine-pitch assembly. The junction resistance will be higher than solder for most materials but not enough to be a concern. However, thermal conductivity is only about one-tenth that of solder, and this could be a problem for high-thermal-output devices.

Adhesives, Nonconductive. First, nonconductive adhesive is not used as part of the bump-to-board junction. The adhesive serves only as a mechanical link, pulling the chip to the substrate. A pressure contact must be created between the bump and circuit pad. Gold bumps and pads generally are used because a gold-to-gold pressure contact is extremely reliable. The adhesive should provide tensional forces, acting like a spring. This has been accomplished by using polymers that shrink on cure. While thermosets have been used, ultraviolet-cured systems have found success in chip-on-glass (COG) applications because the "light" can be transmitted through the flat-panel display from the opposite side.

Some data suggest that the tension of the adhesive relaxes after several years, causing opens. The proposed solution is to create compliant bumps from elastomers that are vacuum coated with metal,[18] but this concept has not gained commercial significance. Gold contacts tend to fuse when under pressure for extended periods, so there may not really be a problem.

TABLE 7.2 Adhesive Fillers

Conductor	Comment
Silver	Used early; can cause electromigration
Nickel	Oxidizes; may be unstable
Gold	Expensive; seldom used
Gold-over-nickel	Cost-effective, hard, nonelastic
Silver-on-glass	Low cost, hard, nonelastic
Gold/nickel-on-plastic	De facto standard; good performance
Gold-on-plastic	Appears to be the ultimate in performance
Solder	Has been used in multilayer circuit applications
Carbon fiber	Low cost, higher resistance

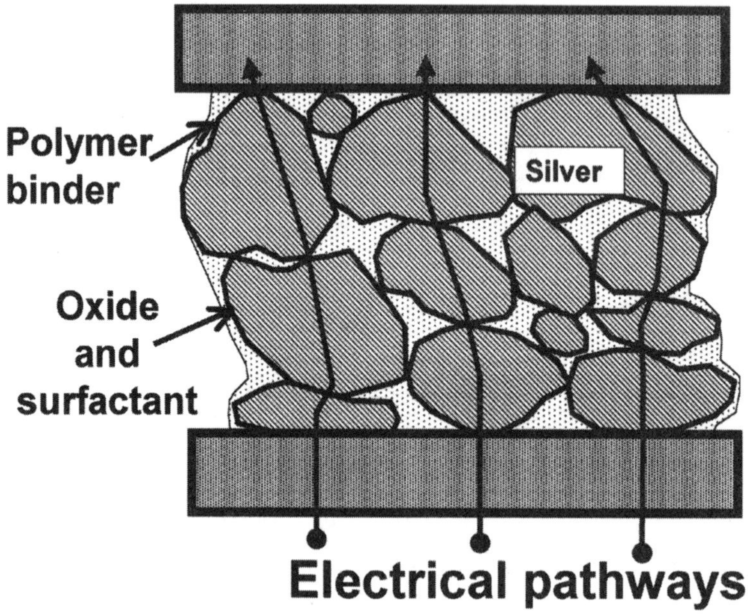

FIGURE 7.9 ICA model.

7.5 THE ASSEMBLY PROCESS

FC assembly processes are more numerous than for packaged SMT because of the various types of joining materials, the use of self-contained joining systems, and the need for underfill that can be applied at different stages in the process.[19]

7.5.1 Solder Reflow

Most FCs are joined to the PWB or package substrate by the solder reflow process. A number of solders are used, ranging from tin-lead eutectic to higher-melting allows made with a substantial lead content. Newer lead-free alloys are also available. The source of solder is typically the bump, but solder paste is also used, especially for high-melting and nonfusible bumps. Flux is applied when the solder bump is the source of solder, and its two main purposes are to hold the chip in place and to remove any oxides on the bump or substrate pads. Solder paste contains its own flux. The solder reflow process involves heating the assembly with the chip in place at a predetermined rate called a *profile* that is nonlinear but generally has a time/temperature profile resembling a steep bell curve. The first heating stage activates the flux to promote oxide reduction. The temperature is then increased, peaking at 20 to 40°C above the reflow point of the solder, allowing the bump to melt and form a metallurgical joint with the substrate metallization. The temperature is then reduced, permitting the solder to harden. The entire heating and cooling process is called *solder reflow.*

SMT Process with Eutectic Bumps. The eutectic solder reflow process is essentially identical to SMT assembly and closely resembles BGA attachment. The FC is an SMT device, the very first ever, and the modern FC is often a BGA. Eutectic solder bumps become the source of the joining material, and no additional metal needs to be added. Additional solder is sometimes provided as solder-coated pads or even solder-bumped substrate. The simplest approach is to use the bump as the joining material, however. Flux is usually required and can be added to the FC bumps or to the substrate. Many pick-and-place machines can be purchased with flux applicators, such as a rotating drum. This device provides a precise thickness of flux that can be applied to the bumps on an FC or other BGA simply by dipping the package into the material. Setting a doctor blade, positioned over the drum, to a specific gap controls flux thickness. The rotating drum therefore becomes a flux reservoir with just the right thickness of flux. Some rotating drums have channels or troughs of a specific depth that provide the reservoir. Several different depths can be used on the same drum to provide more than one thickness of flux. This feature is useful when different types of FCs, CSPs, or BGAs are being assembled onto the same PWB. The total amount of flux per bump is controlled primarily by the thickness of the reservoir, but withdrawal rate has a secondary affect. The drum rotation is synchronized with placement so that the drum is stopped when the chip is dipped. One disadvantage of the flux drum method is that the total assembly rate may be cut by up to 50 percent because of the extra step. A major advantage is that flux is controlled precisely and deposited only where needed, at the joint.

Flux also can be applied directly to the PWB by needle deposition, screen printing, or spraying. Many assembly machines can be fitted with needle dispensers for applying flux. Machines with needle dispensers only experience a minimal delay. Flux that is preapplied by a printer, high-speed automatic needle dispenser, or other means gives maximum throughput. Printing is a parallel process that can improve throughput significantly when PWBs are used with multiple bond sites, as with BGA manufacturing. Integrated lines have throughputs of over 10,000 units per hour.

Once flux is applied and the chip placed, the assembly moves through a reflow oven. A flux with chip-holding tackiness is preferred so that vibrations do not misalign the chips before solder reflow. Profiles are somewhat similar to those used for solder reflow, but fine-tuning will optimize results. The manufacturing goal is to place and reflow both FCs and SMDs on the same line concurrently.

Preapplied Flux/Underfill Assembly. Now that FC assembly is fairly well understood and reliable products are in the field, major efforts are underway to increase productivity. The flux and underfill processes are clearly the bottlenecks. Work has been underway for several years to combine these processes with some success. Preapplied flux/underfill is commercially available and can be used on standard FC and some SMT lines. The liquid flux/underfill is applied to the substrate before placing the chip. A stand-alone dispenser or screen printer can apply the materials, or the operation can be incorporated into the pick-and-place unit. Once again, note that dispensing by the pick-and-place equipment will drop productivity.

A needle dispenser can be used to apply a drop of flux/underfill onto the FC placement site. Work by Georgia Institute of Technology[20] and others has shown that the drop height should be high enough so that the FC forces liquid from the center to the perimeter of the chip. This helps displace air, a major concern for this process. Most success has been found with smaller dies, less than 10 × 10 mm, because of the increasing problem of displacing air with larger ones. Also, the preapplied underfills typically do not contain filler because it interferes with the soldering process. This means that the CTE of this class of underfill can be well above an optimal value, and reliability may not be adequate for large chips.

Once the chip has been placed, the assembly moves to a reflow oven. The chemical timing must be nearly perfect: Viscosity drops, flux activates, solder melts, joints form, and the polymer hardens. The rheology curve must be just right for good solder joint formation. Premature hardening of the polymer will inhibit good joint formation or even prevent attachment. The desire is for all the polymerization to occur during reflow, but a postcure can be required. In fact, some systems are designed to be reworkable until after a lengthy postcure, but this can be counterproductive. In some cases, the reflow profile may need to be extended to complete the cure. The preapplied materials and processes are still in a relatively early stage, and improvements can be expected.

Solder Paste Processes. Nonfusible and high-melting-alloy bump FCs generally are assembled by applying eutectic solder paste to the carrier or PWB. Both screen printing and stenciling have been used, but stenciling gives better resolution and control. Solders made with finer powder will give better results, especially for fine pitch. Vendors are providing paste specifically designed for FCs. Any paste that will be placed in production should be tested for compatibility with underfill if used. Most of the FC solder pastes, made with epoxy-based flux, are compatible with all underfills. Fine-solder dots also have been achieved by needle dispensing, but this process should be considered only for special circumstances.[21] Low-alpha-particle solder pastes are now becoming available because alpha emissions can interfere with IC operation.

7.5.2 Thermomechanical Attachment

Thermal compression bonding, long used for TAB, has been applied to small FCs on flexible substrate.[22] Gold bumps must be used with gold pads on the circuit. The chip is forced against the PWB under heating conditions. This is really a throwback to TAB, but without a "window." In fact, one of the earliest "no window" tape chip carriers was called *flip chip strip*. It may be noted that early TAB and some more modern versions did not have windows. The resulting gap between chip and substrate may be only 10 to 15 μm, and very fine filler underfill is needed, although some success has been achieved with unfilled polymers.

7.5.3 Adhesive Bonding

All types of electronic adhesives can be used here. Generally, adhesives provide the advantages of low-temperature bonding and wider compatibility. This can be crucial for certain optical FCs and temperature-limited substrates such as polyester film. Additionally, adhesives are compatible with thin-film circuits and indium–tin oxide on glass that cannot be soldered. All types of adhesives are also compatible with gold bumps that can only be soldered with very special alloys.

ICA Assembly. These paste materials are best applied by stencil printing, but screen printing, needle dispensing, and dipping are alternatives. The stencil should be no more than 100 mm thick. Laser-cut stencils will work, but electroformed stencils are considered best. The adhesive is printed onto the wiring board, and the chip is placed. The adhesive usually will be tacky enough to hold the chip in place as the assembly moves into an oven. A reflow oven can be used, but the temperature can be set much lower, typically 130 to 150°C. Many adhesives can be cured at reflow solder temperatures, and this would be an advantage where other components are being soldered. Nearly all types of bumps can be assembled with ICA, including eutectic Sn/Pb, but not all adhesives will form a highly stable junction. Virtually all are reliable with gold-stud bumps[23] and gold-plated nickel.

Once cured, the adhesive joint is fragile, more so than solder, and underfill is required. Standard underfills work well, and since no flux is used, compatibility is not an issue. The underfill will shrink slightly during curing, and this will tend to compress the adhesive joint and increase conductivity. The underfill also will greatly reduce any chance of silver migration. The underfill increases the thermomechanical reliability of the adhesive joints and greatly reduces the two most significant limitations of these solderless materials, mechanical strength and silver migration. Unlike all but the specially purified lead-based solders, silver ICAs do not emit alpha particles that are especially detrimental to memory and high-density CPUs. Adhesives are therefore a good choice for many FC applications.

ACA Assembly. ACAs have been used with FCs for a dozen or more years. In many ways, this is an ideal system. The ACAs are capable of attaching fine-pitch devices without concern for shorting, making them well suited for FCs. The ACAs require a very flat surface and do not tolerate noncoplanarity. Fortunately, FCs are quite planar, and most bumping gives narrow-range bump height tolerances. The ACA film can be preapplied to the chip or substrate by "tack bonding" with a short duration of heat and low pressure. The FC can then be placed onto the bond site. Both heat and pressure are needed, and the total force can be substantial. Dwell time can be a minute or longer, and some systems need a postcure. This means that ACA FC assembly is not really an SMT process, and assembly rates are slow. However, no underfill is needed, and this helps offset the time penalty. The ACA assembly can give reasonably good performance, especially on glass. Flat-panel displays use the ACA process to assemble FC drivers, and the concept is called ***chip-on-glass*** (COG).

Nonconductive Adhesive Assembly. The process, first reported by Matsushita, involves applying ultraviolet-cured underfill to the bond site, pressing an FC against a glass flat-panel display circuit, and curing the adhesive by exposure through the glass from the opposite side. The adhesive is designed to shrink and create a holding force that translates to pressure at the bump-pad interface. The process, although limited, has worked well for COG applications.

7.5.4 Testing and Rework

Several testing and rework methods are available. IBM developed the first one many years ago as a method for producing known good dies (KGDs). The chip is soldered to a reusable ceramic test substrate that has undersized bond pads. The chip can then be tested and burned in. The chip can be removed by breaking the purposely weak bonds. The bumps are then reflowed to regain their original spherical shape. However, most FC assemblies also can be tested and reworked provided that there is no underfilling. A bad component can be removed by applying localized heat to melt the solder joints and then pulling off the chip. Excess solder can be removed from the bond site with a heated copper block. Once tested "good," the assembly is underfilled. A few of the newer underfills are said to be reworkable, but this involves tedious steps and mechanical cleaning of the site. Test and rework therefore should be done before underfilling.

Pretesting. Pretesting of FCs before assembly probably cannot be considered by the assembler. The section, "Known Good Die Challenge," below describes available methods, however.

In-Circuit Testing. FCs can be assembled by soldering or adhesive bonding and then tested electrically prior to underfill. At this stage, rework is possible. Early FC technology used ceramic substrate with no underfill, making rework relatively easy. The modern FC on organic substrate requires underfill that generally is not reworkable, although considerable work is underway to address reworkability. The best production strategy when FC removal will be required is to assemble the FC by solder reflow, test, and rework as necessary and then underfill. Those who want to test after underfilling will need to wait for advances in underfill that will be described in the next section.

What about adhesives? Can they be reworked? Most ACAs are not reworkable in the practical sense. The thermoplastic class can be reworked, but this requires applying heat to debond and removing adhesive that can stay on the chip, substrate, or both. ICAs require underfill and are even less reworkable than solder. The best answer to a reworkable FC appears to be the FCIP (flip-chip-in-package). The FC is connected to a BGA substrate that can be soldered to the PWB and removed using BGA rework techniques. FC BGAs are becoming more prevalent and are available in CSP format and larger sizes.

7.6 ENCAPSULATION/UNDERFILL

Silicon, the common substrate for semiconductors, has a very low CTE. All other substrates, including ceramic, have higher CTEs. The popular organic PWBs have CTE values that are about an order of magnitude higher than that of FCs. The large thermal mismatch produces significant stress during thermal cycling. Heating causes the board to expand to a greater extent than the chip, whereas cooling produces the opposite result. The net effect is that the joints are strained, and fracture failures occur. Although it is possible to produce laminates with lower expansion rates, most are matched to the CTE value of their copper conductors, 17 to 18 ppm/°C. No practical joint material can survive, not solders and not conductive adhesives. The result is that low-cost FCs on organic boards cannot pass most reliability criteria. Fortunately, there is a simple solution called *underfill.* Figure 7.10 shows the thermal-mismatch problem.

Underfill can be viewed as a laminating adhesive that mechanically couples the low-expansion chip to the high-expansion board. PWB movement is constrained by the very high modulus chip. Since the board area under the chip moves approximately in step with the FC, stress is reduced, and the joint is preserved. The chip, joints, and PWB are locked in step, and the thermomechanical stress is moved away from the critical assembly zone into the laminate that can tolerate the forces. The underfill must have very special properties to accomplish this feat, including the correct modulus and CTE. A study of Fig. 7.11 will show that the CTE should match the properties of the joint, not the

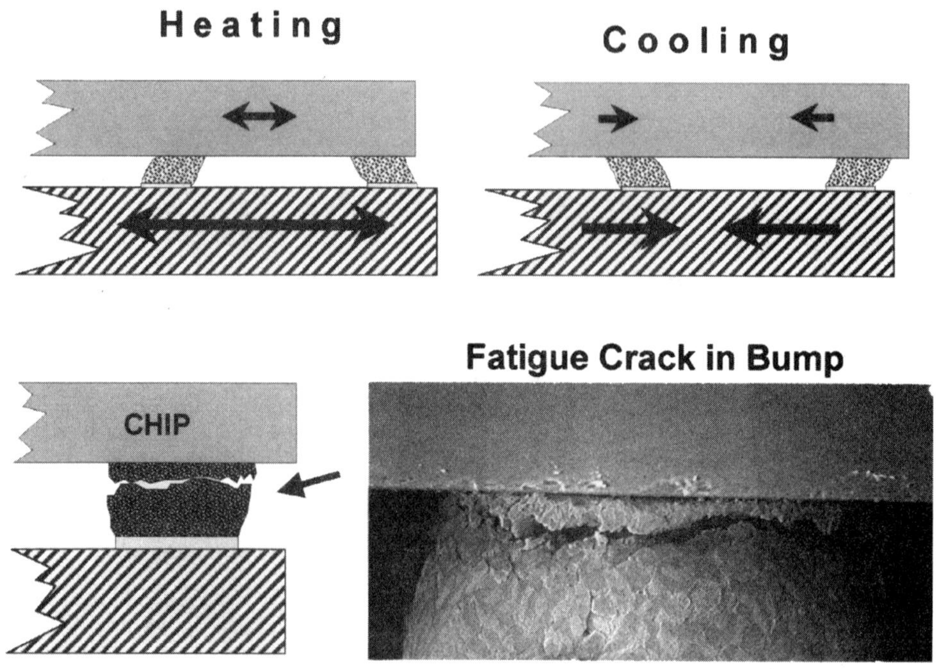

FIGURE 7.10 Thermomechanical mismatch model.

chip. If the underfill has a very high CTE, heating will elongate the joint, whereas too low a value will cause compression. Either way, fatigue will be induced in the vertical axis by the underfill, even though lateral movement will be controlled. The next subsection will describe all possible types of underfill, including those not yet invented.

7.6.1 Preapplied Flux/Underfills

This process was discussed earlier. It will suffice to mention here that the preapplied flux/underfill system eliminates underfill as a separate step. However, if a postcure is required, then only part of the underfilling process is eliminated.

Liquid on Substrate. There are several possibilities for a preapplied underfill, but the simplest is a liquid that can be deposited onto the PWB just prior to chip placement. The material can be applied by a conventional stand-alone needle dispensing machine or by a dispenser that is part of the FC placement equipment. Screen or stencil printing also can be considered if strips of substrate or large arrays are being assembled. The preapplied-on-substrate materials have been referred to as *no flows*, a handy but inaccurate term.

The process involves depositing a specific amount of flux/underfill, placing the FC carefully into position, and running the assembly through a reflow oven. There is a tendency to trap air as the chip is pushed into the liquid, and considerable work has gone into the process, especially at the Georgia Institute of Technology.[24] The best method for reducing trapped air is to deposit a glob of material onto the PWB bond site and force the chip down so that material is displaced from the center outward. Good results have been achieved with smaller dies (<10 mm). Even if better techniques will allow larger dies to be placed with few voids, there may still be a maximum die size limit related to the cured properties. The liquid preapplied products have a high CTE (60 to 80 ppm/°C) because

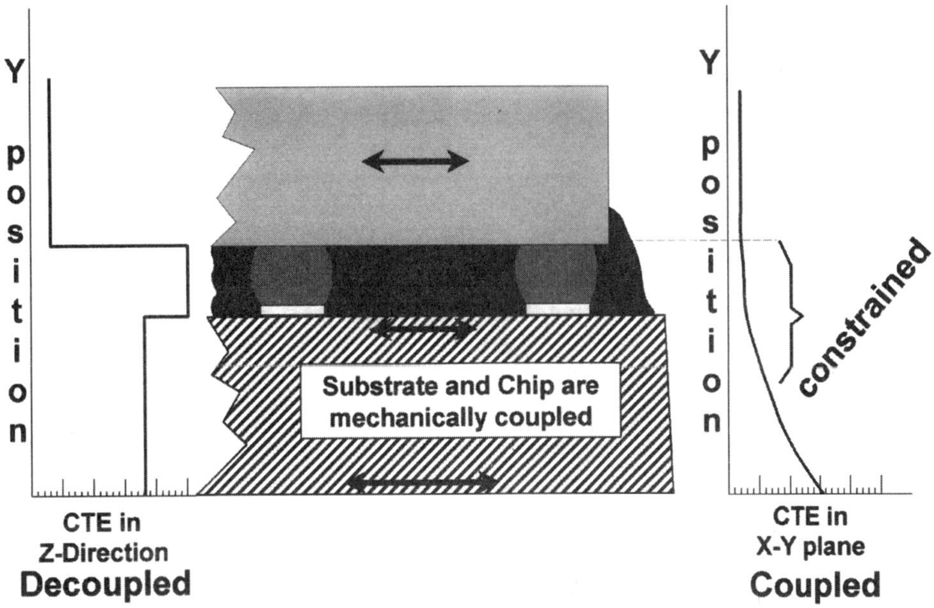

FIGURE 7.11 Underfill mechanism.

expansion-reducing filler should not be used, which would interfere with solder joint formation. Larger dies produce larger stresses, and the unfilled flux/underfill may experience problems. Nevertheless, preapplied liquid underfills enable in-line FC processing with a major boost in productivity.

Solid on Substrate. You may be surprised to find that underfills can start off as solids. Solid materials can be applied to the substrate as a film or as a liquid that is hardened before chip assembly. Solid flux/underfill on the substrate could allow wiring boards or chip carriers to be handled and even shipped in ready-to-bond form with all the materials required for assembly. The flux/underfill would need to go through a liquid stage during assembly, of course, and then return to a solid state again. The final solid state could be higher melting or even nonmelting (thermoset) depending on the polymer system. For example, low-melting, B-staged thermoset film could be used. The material would melt during solder reflow but then polymerize to a high-performance cured underfill. A thermoplastic could even be used, and this would result in a reworkable underfill. However, relatively high temperature and pressure on the FC probably would be required. A product and the necessary bonding equipment have been made available by Nitto-Denko.[25] The idea of solid underfill is analogous to ACAs, except that there is no conductor in the polymer film. While the concept of solid flux/underfill on substrate may have value, the process does not fit the infrastructure because a special bonder is required, and this may be a significant limitation.

Solid on Chip/Wafer. One also can apply solid flux/underfill to a chip or wafer.[26–28] The ramifications are profound. The flux/underfill application is moved to the semiconductor industry and away from the assembler. The resulting FC becomes a ready-to-bond component. The underfilling process appears to go away and becomes transparent to the assembler. The FC is now a CSP. Most important, the FC with integral flux/underfill fits the SMT process; FC = SMT! And if this is not enough, solid flux/underfill on chip can be made reworkable.

Work has been in progress since the mid-1990s to develop wafer-level flux/underfill.[29] Several segments of the industry, as well as consortia, are involved, and the U.S. government has contributed financial support. Some of the players are LSI Logic, Motorola, Loctite, and National Starch and Chemical.[30] Cookson Electronics, with corporate partners, including Alpha Metals and Speedline, is funding a consortium with several universities, including Georgia Tech and Binghamton University, as prime contractors. Semitech is also sponsoring work. The substantial effort by the electronics industry almost certainly will produce results. Success can shift micropackaging preference toward FCs. Others have investigated multilayer combinations of materials on-chip and on-board, although the increased complexity would seem to unnecessarily limit these systems.[31]

Several strategies are being explored for wafer-level FCs, but they can be divided into single material and multiple. The simplest multilayer material consists of separate underfill and flux layers. While the method of application is more involved, material development is straightforward. The underfill can be a filled thermoplastic (reworkable) or a thermoset. No flux activity is required, and filler can be used. The flux layer can be applied over the solid underfill layer or just on the bumps. Coatable solid flux is already available. The flux layer contains no filler because it would interfere with solder joint formation. The flux must melt, bond, deactivate, and convert to a higher-melting solid.

The single-component material must have flux properties initially but then convert to an underfill. This can be done within the realm of epoxy-based flux because flux agents such as carboxylic acids also can serve as hardeners. Work continues in this area, and a product may even be commercial by the time this book is in print. It may even be possible to produce a linear thermoplastic epoxy that will be reworkable.

7.6.2 Postapplied Materials

The term *postapplied* is used to describe the common capillary flow underfills that are dispensed under the FC after it has been assembled to substrate. This is the more common approach, and it can be viewed as the standard process for now.

Capillary Underfill. The common capillary flow underfill is a postapplied product. The FC is first assembled by soldering, conductive adhesive bonding, or other means. Then the underfill is allowed to flow under the chip. The underfill is "pulled" along by capillary action or intramolecular attraction between the liquid and the solid surfaces. Flow requires a set of criteria that include an "advancing" wetting angle that results when the surface tension of the underfill is less than that of the solid surfaces. The viscosity must be low enough to permit flow within a reasonable time, although heating can be used to reduce viscosity and increase flow rate. Filler, added to reduce the CTE, must be small enough so that particles do not block flow. A good approximation is that maximum particle size should not exceed one-third the gap size. State-of-the art materials offer fast flow and fast cure.[32–35]

The capillary underfill must be applied using a pattern that does not trap air. A general rule is to apply underfill from only one side of the die for larger (>10 mm) FCs and then dispense more material on the other three sides after the underfill has flowed to these sides and displaced the air. Smaller dies can use an L-shaped dispensing pattern. Material is dispensed starting at one corner of the chip, and the dispensing needle is moved the full length and then around the corners, as shown in Fig. 7.12. Once the underfill has flowed completely under the chip, adding the missing L can complete the fillet. It is not always necessary to add a fillet, but reliability is maximized. A faster process only requires that a dot of underfill be added to the far-side corner of the chip after flow out, as shown in Fig. 7.13. Reliability is slightly lower, but productivity is higher. The L plus dot approach works well for smaller chips used in cellular phones and pagers. The dispensing equipment industry has reported extensively on the underfilling process.[36–38]

Once underfill is cured, rework is not possible when using any of the commercial capillary flow products. However, work continues in the capillary underfill area to produce reworkable materials.[39]

Encapsulant (Over Chip). Both liquid encapsulant and solid molding compounds may be used to protect FCs used in packages. Underfill must be applied first so that there is no airspace underneath when the encapsulant is applied. Many FC BGAs have been produced with no overcoat or encapsu-

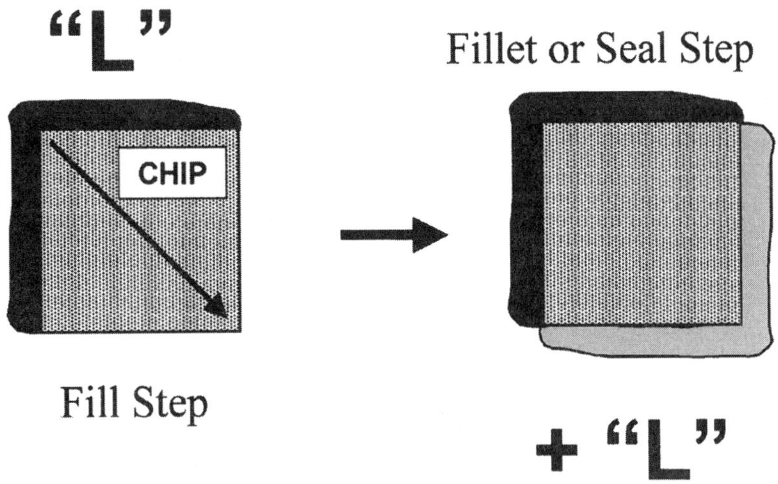

FIGURE 7.12 Double-L dispense pattern.

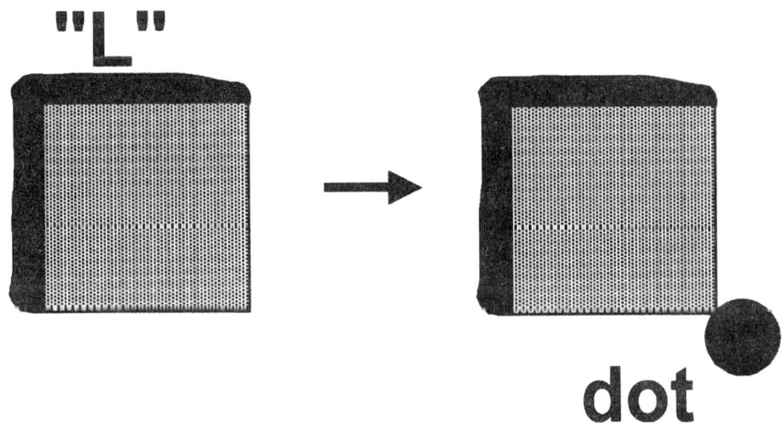

FIGURE 7.13 L and dot pattern.

lant because the underfill protects the sensitive underside of the die. Also, the backside of the die can be used to radiate heat by applying a heat sink. Once encapsulant covers the die, the heat-sink option is eliminated. Work is also underway to transfer mold encapsulant over the chip while simultaneously underfilling.

7.7 SUBSTRATES FOR FCs

The first FC substrates were ceramic, and they are still preferred from the viewpoint of performance and ease of assembly. Flatness and low CTE are two preferred attributes for FC substrates that are inherent in most ceramics. During the 1990s, organic substrates became popular for many FCs, especially the smaller, lower-lead-count variety. However, the move to organic substrate introduced the problem of significant thermomechanical mismatch because of the relatively high CTE of organic materials. However, underfilling adequately reduces joint stress, and reliability can be regained. Underfilling, while allowing popular, low-cost substrates to be used, increases processing time and adds cost. Nonetheless, underfill allows virtually all substrates to be used. The preferred organic substrates include standard FR4 circuits, higher grades of epoxy such as bismaleimide triazine (BT), and flexible circuit substrates. Large, high-performance chips, such as CPUs, continue to favor ceramic materials.

7.7.1 Ceramic

IBM first used ceramic circuits and chip carriers for FCs. Ceramic offers two important advantages for FCs. First and most important, the CTE is low enough that there is only a small thermomechanical mismatch between chip and substrate. This means that underfill is not required for thermomechanical enhancement, at least for smaller dies. The second feature of ceramic is high-temperature stability as well as superior thermal conductivity. Ceramic can handle 300°C or higher, a valuable feature for a chip carrier. Some materials conduct heat better than aluminum and therefore are ideal for heat dissipation.

Although ceramic has become less popular in recent years because of higher cost, it has been used to build FC BGAs. AMD, for example, has used ceramic BGAs for the K6 and K7 microprocessors. The high-temperature performance of ceramic allows higher-melting solders to be used for chip attach so that melting will not occur when the package is soldered to the PWB. Ceramic substrate is also much less prone to warping than organic laminate and is a good material for FC BGAs. The factor that will continue to reduce usage is cost, especially for high density. The PWB industry continues to advance methods, such as microvias, that improve the cost/density picture, and this will likely increase the cost differential.

7.7.2 Organic, Rigid

Organic laminate has become the preferred substrate for both FCOB (flip-chip-on-board) and FCIP (flip-chip-in-package). In fact, second-generation FC is really centered on organic substrate to lower cost. The advent of the BGA in the early 1990s, especially the plastic BGA (PBGA), pushed the demand for better laminates. New materials were introduced, and more are under evaluation aimed at providing improved mechanical stability and higher heat ranges. BT resin has become one of the most popular advanced laminates and is now used widely for BGAs. The same high-performance BGA substrates also provide FCs with desirable properties. Smaller chips used on cellular phones, pagers, and other consumer products do not necessarily require higher performance than offered by FR4. It is the FCIP products that need better substrates, and these have been developed for wire-bonded BGAs.

7.7.3 Organic, Flexible, High-Temperature

Flexible circuit materials have long been used for chip carriers. TAB, called *tape carrier package* (TCP) by Intel, has been used to handle high-density chips with very good results. The polyimide class of substrate, such as Dupont's Kapton, is noted for extraordinary high-temperature performance. Some of these materials have a zero strength factor at 800°C, a region usually confined to ceramics. The flex materials are also thin and compliant, making them well suited for use as chip carriers and FC assembly substrates. A number of companies, including IBM, 3M, and Amkor, have developed products that can be classed as *flex-based packages.* Both wire-bonded and FC assembles have been made. Thermal compression bonding to flex is also possible, and Motorola has developed technology here.

Flexible circuitry is an enabling technology for many industries such as disk drives. The flex allows the read/write head to move over the disk.[40,41] Faster and higher-density drives have been made possible by using lighter components placed closer to the head. Today, the FC is the optimal solution for high-density magnetoresistive disk drive technology. The FC is assembled to the polyimide drive circuit using reflow soldering. Stud bumps of various materials also have been used for flex.[42] Underfill must be used even though the thin, compliant flex produces lower thermomechanical stress than hard board. Although it is possible to make flex substrate with a low CTE approaching that of the chip, a value of around 18 ppm/°C is used in order to match the copper conductors. This means that underfill is required.

Disk drive makers, most notably Seagate, have been able to apply underfill without a fillet step. Underfill is applied from one side, and the assembly moves to the curing oven. Underfill with lower filler and hence lower viscosity can be used, and this produces self-filleting. The high-volume, low-cost requirements of the disk drive industry suggest that this segment of electronics will continue to lead productivity and cost reduction. Expect to see the implementation of preapplied flux/underfill here.

7.7.4 Organic, Flexible, Temperature-Limited

There is a large segment of the flexible circuit industry that uses very low cost polyester substrate such as Dupont's Mylar. The polyesters have a temperature-processing limit of about 150°C, so one probably can rule out solder assembly even though low-temperature alloys exist. This is the area where conductive adhesives excel. Both aniso- and isotropic types can be used. Two basic types of conductors are used with polyester film, and they are traditional subtractive copper and additive polymer thick film (PTF). Bare copper can be used with adhesives but not without junction stability concerns. A much better interface is gold plate over nickel plating on the copper. All adhesives form a stable electrical junction with gold-plated conductors.

PTF silver ink makes a very good interface with conductive adhesives, as would be expected because of the similarities. PTF ink has a silver-rich surface that mates well with the silver matrix of ICAs. Millions of such junctions have been made without any issues using SMDs. ACAs also form good connections with PTF ink because the conductive spheres tend to embed into the ink surface. The oxide that forms on silver is of no real consequence because it is electrically conductive. Radiofrequency identification (RFID) products are now being made with PTF inks and FCs bonded with adhesives, and they will be covered later under "Applications."

7.8 FEATURES AND BENEFITS

FCs offer a wide variety of benefits that often seem to be at different ends of the property spectrum. One company may use FCs to reduce cost, another for high performance, and a third to reduce size. Let's examine the features and see what benefits can be derived.

7.8.1 Geometric Considerations

The FC is used for two primary reasons: maximum performance and minimum geometry. The FC represents the simplest, smallest, and lightest minimal package that is possible. No other systems can be smaller, have a lower profile, or be lighter. This is so because the FC adds only the minimum essentials to the bare chip. The FC can be viewed as a chip with a simple connection interface for direct attachment to substrate. The bumps can have very low height so that the resulting assembly has a height only slightly greater than the thickness of the die. The footprint is exactly equal to the dimensions of the die, although some allowance may be needed for the underfill fillet that surrounds the chip. In the future, however, emerging wafer-level underfills may reduce the fillet or even eliminate it. Thus, when the very smallest size, lowest profile, or minimum weight is essential, FC is the best choice.

Footprint. There is nothing smaller than the bare die, and an FC structure adds nothing at all to size. No CSPs can be smaller than the FC. One could argue that underfill increases the footprint because of the fillet and a "keep out" area that must be reserved for excessive flow out. However, even this minor addition of area is about to change with wafer-level underfill. The wafer-level flux/underfill solids that will be an integral part of the chip are not expected to flow out to any significant degree. Applications that require minimum area can take advantage of the FC. The FC can be used for very high I/O chips; not so with most CSP technology.

Profile. Even the FC connection has the lowest profile possible. Joints as small as 10 to 15 mm are being used. Even the more typical 50-mm high joint does not add significantly to the height. Many of the CSPs, while keeping area close to chip size, can more than double the height of the final assembly. FCs have been used in handheld products where reducing thickness was more important than area. The StarTac cellular phone was said to use FC technology for height, not footprint reduction. A second, "stacked" wiring board could then be added, thereby doubling the useful board area. Smart cards and RFID products are also benefiting from the minimal profile. FCs should be considered when profile reduction is a key consideration.

Weight. Once again, nothing can be lighter than the basic die. The interconnecting structure is as minimal as possible, consisting only of the short bump that forms part of the entire bump. The bumps are often so small, less than 50 mm in diameter, that their weight is insignificant even when solder is used. However, lower-density conductive adhesive reduces weight even more, and underfill has a low density of around 3 g/cc depending on filler loading. FCs should be considered when weight reduction is an important factor.

7.8.2 Performance

Speed. Many packages limit the operational rate or clock speed of a device. Added wire length adds signal delay time. Many of the CSPs have a die-sized foot print, but their wiring structure adds length and signal delay compared with an FC. It should be mentioned that an IC designed for FC packaging will optimize the on-chip wiring so that rerouting is not required. The move to copper IC conductors also will reduce internal resistance, making in-chip routing more effective. Chips designed for perimeter pads and then rerouted externally can have a very inefficient layout that just about doubles the total wire length. Short lengths reduce performance-robbing inductance. No package has lower inductance than the simple FC bump.

I/O Density. FC is the clear winner for high density. FCs are in production with more than 2000 bumps. IBM has built working prototypes with 10,000. Advanced bumping technologies, such as solder plating,[43] allow extremely fine bump pitch. It is safe to say that the FC will never be the limit to density; it will be the substrate.

Process-Related Economics. FC has been designed as a wafer-scale process. Most bumping methods deal with the wafer and therefore enjoy considerable economics of scale. The various parallel or mass bumping processes therefore should achieve the same scaling economies as the IC. The move to 300-mm wafers will reduce bumping costs.

Underfilling is still a board-level serial process that reduces the overall economics and productivity of the FC. However, this is about to change as wafer-level flux/underfill developments reach fruition. In the future, all the materials needed for FC assembly will be applied at wafer level. Instead of applying underfill to each assembled FC, the entire wafer will be coated. Thousands of FCs will have underfill applied in minutes, not the many hours now required for the capillary flow systems. In the meantime, preapplied flux/underfill is helping to reduce the processing time and boost the productivity. However, the wafer-level integral flux/underfill FC of the future likely will provide the best economics of any micropackage.

7.9 LIMITATIONS AND ISSUES

First, let's realize that traditional packages will continue far into the foreseeable future. SMDs and even feed-through devices have a continuing role to play, and the infrastructure for their manufacturing and assembly is well established and entrenched. Therefore, let's look at FC limitations and issues compared with other micropackages.

7.9.1 Known Good Die Challenge

Can FC deliver known good product? The answer is yes and no. Wafer-level testing continues to advance, but most people feel that it is an approximation at best. Recent advances in spring probes, notably by Form Factor, may be changing this, but for now, we cannot totally rely on wafer-level testing. Several schemes have been applied to FC test and burn-in over the years that can ensure good chips, but at a cost. IBM has assembled FCs for test and burn-in on ceramic substrate using a restricted pad technique. The pad on the test module is smaller than would be used normally for the specific chip. The assembly, which can contain many chips, is run through a test/burn-in sequence. The FCs are then removed by mechanically fracturing the minimal joints.

NCMC developed substrate and a process for test and burn-in that does not require mechanical fracturing.[44] A thin-film ceramic test circuit is produced that will accommodate FCs and allow attachment by solder reflow. The assemblies are burned in. Next, the chips are removed by melting the solder joints and removing the FCs. The special pads allow complete removal with all the joint solder remaining with the FC because the special test circuit metallization produces a dewetting effect. The test substrate can be used many times before the pads need to be renewed. This process, as well as electroplating bumping services, has been commercialized by Unitive, Inc., Research Triangle Park, NC.

Various probing technologies are also available, including IBM's dendritic contact pads and Form Factor's microsprings. One can still argue that these methods are not as useful as on-site testing and burn-in afforded by CSPs, however. Yet the intensifying effort to resolve these kinds of issues is likely to produce results in the near future.

7.9.2 High-Density-Circuit Requirements

The packaging revolution of the 1990s was the result of continuing advances in the semiconductor industry as well as the consumer demand for higher performance and real portability at an attractive price. This can be summed up in the marching theme, "smaller, faster, cheaper." The perimeter package had run out of steam for the CPU, and the transition to area array was more than a nice idea—it was an absolute necessity. These major changes in packaging put significant pressure on the printed wiring industry. New materials and processes were needed both for chip carrier substrate and PWBs that would accommodate the area array and micropackages in general. Tried and true drilling is being replaced by laser machining to produce microvias. Effective use of microvia technology requires changing the old stack-up lamination method of making multilayers. Various builtup circuit conductor methods have moved into production, but only a few are able to produce the required high-density products, and they are mostly in Japan. The PWB industry eventually will add the new infrastructure required for this general area called *high-density interconnect* (HDI).

7.9.3 Assembly Difficulty

Smaller package features typically raise the bar of difficulty. However, area array products such as FCs are much better at self-alignment. FCs purposely misaligned by 50 percent are able to center during reflow. Work by Siemens indicates that a 30 percent misalignment will not reduce production yield. Many types of FCs also solve the predicament of applying solder paste to small areas. No paste is required because the bump contains the joining material. In all, the difficulty added by the FC's small size appears to be offset by the features just mentioned.

7.9.4 Ramifications of Die Shrink

The semiconductor industry achieves the objectives of smaller, faster, cheaper primarily by die shrink. The die is scaled down to a smaller size, undergoing a "shrink," to boost the speed through shorter connection paths while reducing cost because more dies can be placed on each wafer. It is possible to lay out the bump pads on an FC favoring the center area to accommodate die shrink without altering pad placement, and this is presently being done. Progressive die shrinks, often accompanied by minor IC architecture changes, inevitably require a new bump pad configuration that will not fit the original PWB pad layout. The designer is then faced with only a few choices for FCOB: Modify the PWB or reroute the FC—on-chip. The die shrink problem is reduced if FCIP is used because the chip carrier substrate can be rerouted and keep the same BGA footprint. The same kind of rerouting is also required for all of today's CSPs. Some bumping providers also offer rerouting on the FC.

7.10 PERFORMANCE AND RELIABILITY

FC holds the enviable title of logging the highest reliability of any system. IBM has established a long record of near-perfect reliability for FCs on ceramic. Insiders talk of thermocycling reaching the 60,000 mark before the test was terminated. The incredible performance on ceramic should not be a surprise, however. First, the ceramic circuits were designed to approximate the properties of the FCs with the CTE being only slightly higher. Smaller dies were used in these earlier mainframe modules, further reducing any stress. More fatigue-resistant high-lead solder was used. These MCM modules were cooled with highly efficient dynamic liquid heat exchangers. Many of the early modules are still running today.

Second-generation FC, the move to organic substrate, is not as reliable. It also should be noted that ceramic is still used today for both FCIP and FCOB. There are several factors that reduce the time to failure for modern lower-cost systems. The most obvious is the greater thermomechanical mismatch between chip and substrate. The substrate is also more prone to warping, and this can produce high strain forces. Solder joints are commonly made with eutectic tin-lead solder that work hardens and fatigues faster. The use of underfill to counteract the high mismatch is a double-edged sword. While correctly applied underfill can extend life well past 1000 cycles, voids and other defects can cause failures within the pass criteria. Small, undetected voids may reduce thermal cycle life by up to half, producing marginal reliability. Larger voids are detected more easily and are less likely to occur. Overall, FC performance is quite acceptable and steadily improving. Wafer-level flux/underfill may solve the problem of voids altogether and allow more consistent materials to be employed in the future.

7.11 APPLICATIONS

FC is somewhat remarkable in that it is used for the highest-performance, most expensive products, such as supercomputers, and the lowest-cost systems, such as smart cards. FC simplicity appeals to high-volume, low-cost products like the smart card and the RFID tag because of low cost and ease of assembly, where chips can be pulled directly from bumped wafers. The simplicity of having only a bump array structure with no external packaging provides the next desirable attribute of minimum size. The simple short-length bump and the resulting direct attachment structure also deliver the high electrical performance that is so appealing to makers of high-speed, high-end products. One added benefit of the tiny bump is that the highest density can be achieved so necessary for modern computer chips and other high-I/O products. The result is that the FC is used virtually everywhere, from lowest to highest range. All the major electronic business segments, including automotive, computers, and communications, have embraced FC. This is why many people believe that the FC is the final destination for future packaging.

7.11.1 Computers and Peripherals

Virtually all computer CPUs now require and use FCs. We have gone beyond the point where anything else can accommodate the large number of I/Os and high clock rates. Early conversions included the PowerPC from IBM/Motorola and AMD's K-6. In 1998, Intel's Pentium II went to FC. All are FCIP because this format allows test/burn-in and rework. The early FCIPs were ceramic, and some still are, but the Pentium moved to organic, and this will be the trend. The Pentium II FC has 2100 bumps, although some are redundant. This produces better heat transfer and a symmetry that allows better assembly and underfilling.

Computer components and peripherals, including disk drives, displays, and printers, have been quickly adopting FCs. Seagate went from SMT to COB (chip-on-board) to FC during the 1990s to increase performance and reduce cost. The read/write amplifier head is placed directly on the flexible circuit using solder reflow followed by underfilling. Even though flex is compliant and produces smaller thermomechanical forces than PWBs, reliability requires underfilling. However, the underfill can be applied without a secondary and time-consuming filleting step. The assembly goes directly into a curing oven once underfill is applied to one side. Even more productivity will be achieved when preapplied flux/underfill is adopted in the near future. Figure 7.14 shows an FC on flex disk drive assembly.

7.11.2 Automotive

The automotive industry followed right behind mainframe computers to become a veteran area for FCs starting in 1968. Delco (now Delphi) pioneered this area and developed novel bumping and assembly techniques. As with early computer applications, the substrate was ceramic and still is, although the industry is just now moving to organic for such items as ABS. Figure 7.15 shows an engine controller with FC on ceramic. Automotive requirements are probably the most difficult within the industry.[45]

7.11.3 Consumer Products

There are now so many consumer products using FCs that it is a task just to keep up. The digital watch is certainly worthy of note and remains the highest-volume application. Figure 7.16 shows an

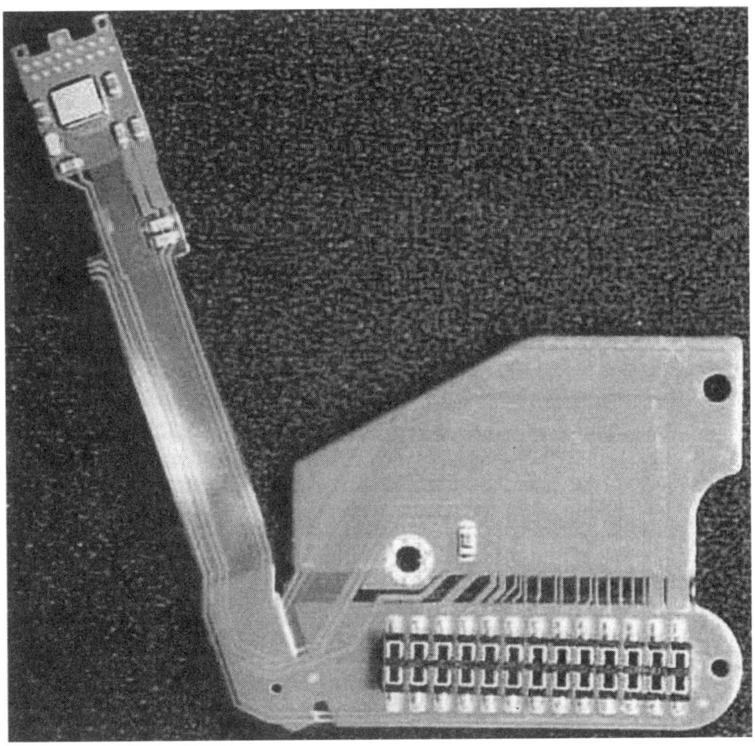

FIGURE 7.14 FC disk drive assembly.

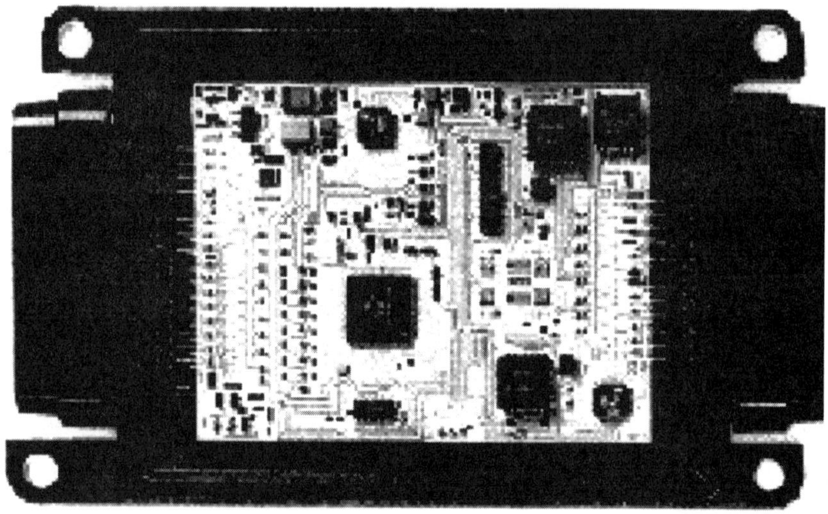

FIGURE 7.15 Engine controller.

FC watch product. Perhaps the most famous FC example is the earlier Sony HandiCam digital camera shown in Fig. 7.17. Many of the FCs are CSPs because they are FCIPs.

7.11.4 Communications

Communications continues to be one of the most exciting, fast-paced areas of electronics development. Personal communication has boosted this market to one of the most important within our industry. At the same time, convergence and incredible miniaturization require adoption of the latest packaging technology. Motorola pioneered FC in this area beginning with the Pen Pager of the mid-1990s, shown in Fig. 7.18. The StarTac followed close behind and was launched as the world's smallest and lightest (99 g with battery) in the mid-1990s. Since then, many other phones have moved to FC, not just for size and weight reduction, but also for performance. Figure 7.19 shows the famous StarTac.

7.11.5 Smart Cards/RFIDs

We have seen that the FC can accommodate thousands of I/Os and let clocks run past 1 GHz. What about the opposite end—low cost, low I/O? The same question was asked about TAB many years ago when it was the premium high-I/O package. While the Pentium portables were using TAB in the late 1990s to get performance within a low-profile package, low-cost Japanese calculators used TAB to reduce cost. FC provides the same paradox for the low-end market—the highest performance delivers lowest cost.

The emerging area of RFID tags is enabled by FC. The RFID is essentially an antenna and an IC. The antenna picks up a burst of electromagnetic energy to power the chip and a signal transmitted by the reader. The IC processes information and then returns an "answer" by transmitting an rf signal through the antenna back to the reader. The simple circuit of a loop antenna and chip is built cost-effectively using DCA. The antenna is built on low-cost flex film, and chip connections are made with low-temperature conductive adhesives.[46,47] Figure 7.20 shows an RFID device.

7.11.6 Other FC Products

Although the military is often cautious and slow to adopt newer technologies, FC work has been done under U.S. government sponsorship.[48] A consortium of military and aerospace companies has

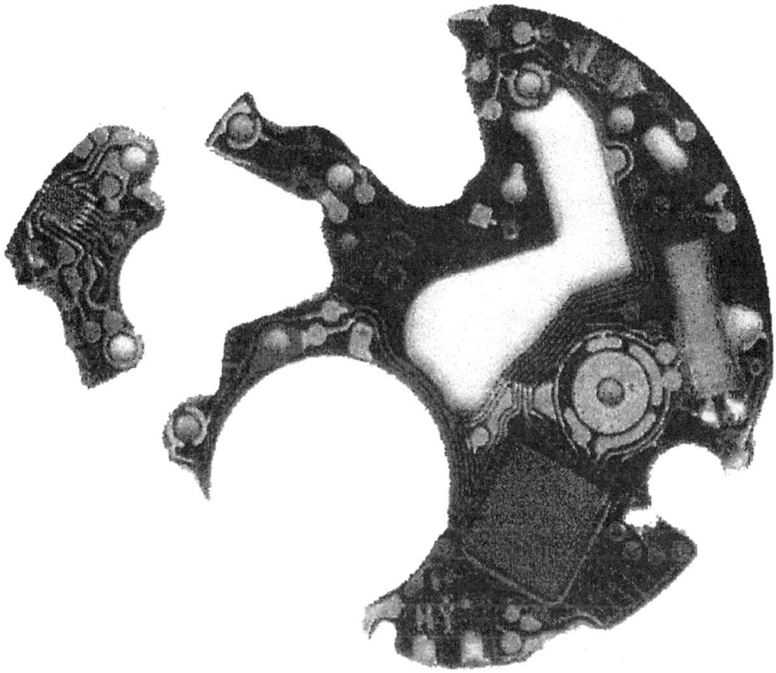

FIGURE 7.16 Watch.

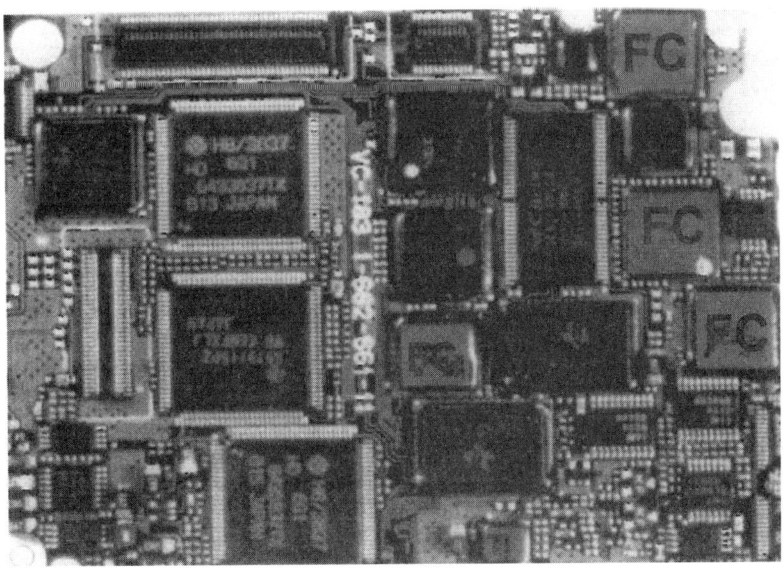

FIGURE 7.17 Sony HandiCam.

FIGURE 7.18 Motorola Pen Pager.

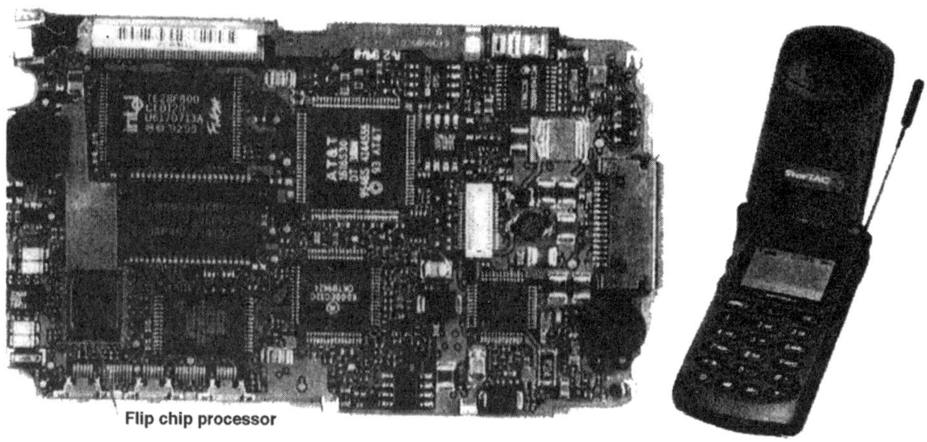

FIGURE 7.19 StarTac. (*Motorola.*)

done extensive development and evaluation of numerous FC technologies. Testing of FC on a missile demonstration module made of polyimide led to the conclusion that this area of technology can provide weight and size reduction in a durable and robust system.

The medical products field is also embracing FC technology, but perhaps at a more cautious pace common for that industry. Implants represent an ideal area where size reduction is so critical and the stable environment of the body eliminates thermomechanical stress. One fascinating product using FC is the DNA analyzer, a disposable detector that consists of a chip on flex.[49] A photograph of the detector module is shown in Fig. 7.21.

7.12 SUMMARY AND CONCLUSIONS

Many of us have long referred to FC as the ultimate package and the final destination[26] for micropackaging. Nothing can be smaller, run faster, or hit the low-cost potential promised by FC,

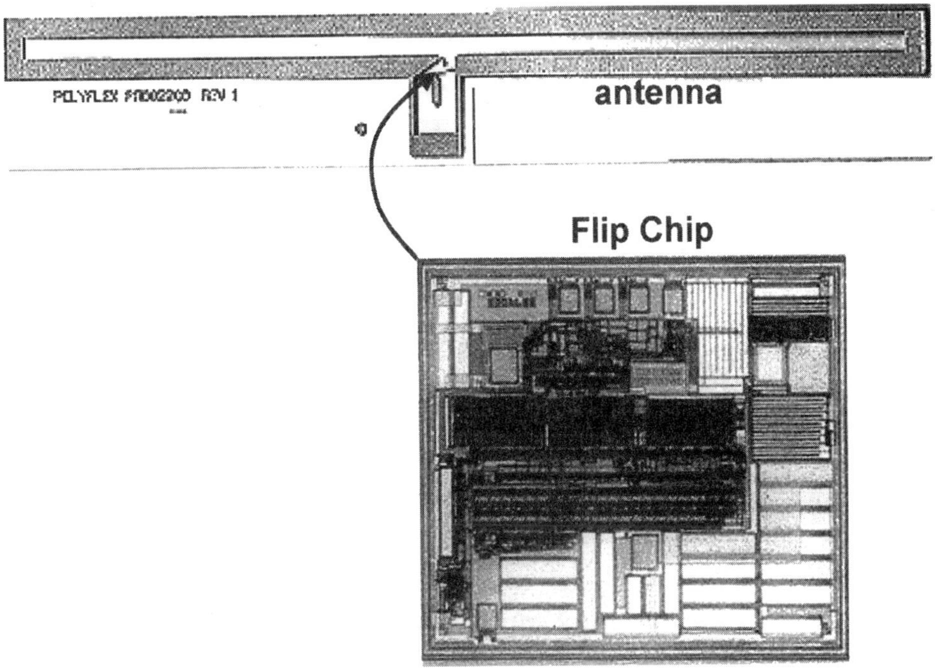

FIGURE 7.20 RFID with FC.

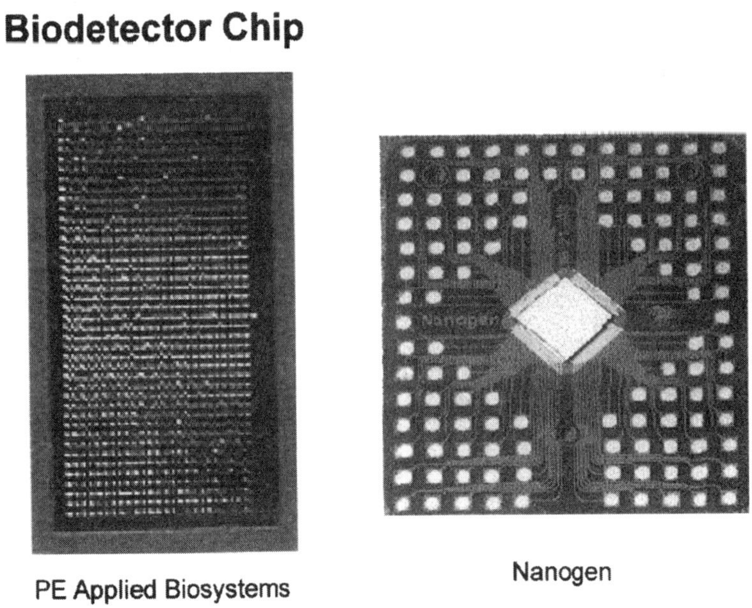

FIGURE 7.21 DNA detector, SmartFlex.

and it is certainly on the roadmap of the semiconductor industry.[50–52] And third-generation wafer-level FC will be the final enabling technology to allow FC to achieve maximum performance, manufacturability, and productivity. Just think of a chip-sized package, only 1 or 2 mils thicker than the chip, that can accommodate 10,000 I/Os and run at more than 1 GHz. Now think of the simple surface-mount assembly process, where all the necessary ingredients are right there with the chip. Solid solder, flux, and underfill are an integral part of the IC. And if this were not enough, the finished assembly would be reworkable. Should all this happen as planned, the integral flux/underfill FC would be the most important micropackage—truly the ultimate.

Although major strides are being made at a rapid pace in the FC arena, many issues impede this technology. Perhaps most important is a lack of centralized infrastructure. Major companies such as IBM, Motorola, and Amkor may have built their own infrastructure, but it is these islands of technology, often proprietary and closely guarded, that inhibit mainstream progress. Although advances have been made in bumping, much more is needed in this area, as well as in design and test. Maximum benefits from FC must start at the chip level, with area array pad layout resulting. On-chip routing of conductors to the perimeter followed by rerouting back to the center is counterproductive to say the least. The full FC infrastructure, starting with design, must be put in place to achieve the maximum payoff. This will happen, but the timetable is still unclear. What is certain is that FC will become the ultimate micropackage.

7.13 REFERENCES

1. Hayden, T., and Partridge, J., "Practical Flip Chip Integration into Standard FR-4 Surface Mount Processes: Assembly, Repair and Manufacturing Issues," in *Proceedings of the 1st International Symposium on Flip Chip Technology*, San Jose, CA, Feb. 15–18, 1994, pp. 1–7.
2. Lowe, H., "No Clean Flip Chip Attach Process," in *Proceedings of the 1st International Symposium on Flip Chip Technology,* San Jose, CA, Feb. 15–18, 1994, pp. 17–24.
3. Ono, M., T. Shiraishi, Y. Bessho, et al., "Area-Array Interconnect Using Stud-Bump-Bonding," in *Proceedings of the International Symposium on Microelectronics,* sponsored by IMAPS, San Diego, CA, Nov. 1–4, 1998, pp. 893–898.
4. Zuma, S., and Kaga, Y., "A Stud Bumping Technology Using Copper Wire," IMAPS 2nd International Advanced Workshop on Low Cost, Braselton, GA, March 13–15, 1998.
5. Oppert, T., Zakel, E., and Teutsch, T., "A Roadmap to Low Cost Flip Chip Technology and Chip Size Packaging Using Electroless Nickel Gold Bumping," IMAPS 2nd International Advanced Workshop on Low Cost, Braselton, GA, March 13–15, 1998.
6. Anhock, S., Ostmann, A., Oppermann, H., Aschenbrenner, R., and Reichl, H., "Reliability of Electroless Nickel for High Temperature Applications," in *Proceedings of the Advanced Packaging Materials,* sponsored by IMAPS, Braselton, GA, March 14–17, 1999, pp. 256–261.
7. Ostmann, A., Kloeser, J., Zakel, E., and Reichel, H., in *Proceedings of the 1996 International Flip Chip, Ball Grid Array, TAB and Advanced Packaging Symposium (ITAB 96),* Sunnyvale, CA, Feb. 14–16, 1996, pp. 152–157.
8. Simon, J., Zakel, E., and Reichl, H., "Electroless Deposition of Bumps for TAB Technology," in *Proceedings of the 40th ECTC,* Las Vegas, NV, 1990, p. 44.
9. Yoneda, Y., Kuramochi, T., Sohara, T., et al., "A Novel Flip Chip Bonding Technology Using Au Stud Bump and Lead-Free Solder," in *Proceedings of the 4th Annual Pan Pacific Microelectronics Symposium,* Kauai, HI, Feb 2–5, 1999, pp. 147–152.
10. Wolf, J., Chimiel, G., Simon, J., and Reichel, H., "Solderbumping—A Comparison of Different Technologies," in *Proceedings of the 1st International Symposium on Flip Chip Technology,* San Jose, CA, Feb. 15–18, 1994, pp. 105–111.
11. Huang, Y.-W., Collier, P., Teo, K., et al., "Wafer Bumping by Stencil Printing," in *Proceedings of the Pan Pacific Microelectronics Symposium,* Kauai, HI, Feb. 10–13, 1998, pp. 455–460.
12. Nguty, T., Riedlin, N., and Erere, N., "Solder Paste Characterization for Flip Chip Applications," in *Proceedings of the Technical Program of Nepcon West,* Anaheim, CA, Feb. 22–27, 1999, pp. 589–601.
13. Kaga, Y., and Zhang, J., "Low Cost Wafer Bumping with Maskless Process," in *Technical Proceedings of Surface Mount International,* San Jose, CA, Aug. 23–27, 1998, pp. 265–269.
14. Godin, R., "Flip Chip Bumping with High Speed Metal Jet Technology," in *Proceedings of the Technical Program of Nepcon West,* Anaheim, CA, Feb. 22–27, 1999, pp. 1489–1495.

15. Neugebauer, C. A., and J. A. Loughran, "Method of Fabricating Gold Bumps on ICs and Power Chips," U.S. Patent 4,750,666, June 14, 1988.
16. Eldring, J., Prokoph, G., Bauer, A., et al., "Flip Chip Attachment Using Mechanical Bumps," in *Proceedings of the 1st International Symposium on Flip Chip Technology,* San Jose, CA, Feb. 15–18, 1994, pp. 74–81.
17. Lyons, A., Hall, E., Wong, Y.-H., and Adams, G., "Flip Chip on Flex with Anisotropic Conductive Adhesives," in *Proceedings of Flexcon,* Sunnyvale, CA, Oct. 12–14, 1994, p. 17.
18. Keswick, K., et al., "Compliant Bumps for Adhesive Flip Chip Assembly," in *Proceedings of the 1st International Symposium on Flip Chip Technology,* San Jose, CA, Feb. 15–18, 1994, pp. 66–73.
19. Lasky, R., Morvan, Y., and Baldwin, D., "What Is Needed to Establish Flip Chip as a Standard SMT Process," in *Proceedings of Technical Program of Nepcon West,* Anaheim, CA, Feb. 22–27, 1999, pp. 1025–1030.
20. Johnson, C., and Baldwin, D., "Pre-Applied Underfills for Low Cost Flip Chip Processing," in *Proceedings of Advanced Packaging Materials,* Braselton, GA, March 14–17, 1999, pp. 73–76.
21. Ludwig, R., and Lee, N.-C., "Achieving Ultra-Fine Dot Solder Paste Dispensing," in *Proceedings of Technical Program of Nepcon West,* Anaheim, CA, Feb. 22–27, 1999, pp. 576–588.
22. Nave, J., Zakel, E., and Reichl, H., "TC-Bonding for FC-Technology on Ceramic, Silicon and Organic Substrates," in *Proceedings of the 1996 International Flip Chip, Ball Grid Array, TAB and Advanced Packaging Symposium (ITAB 96),* Sunnyvale, CA, Feb. 14–16, 1996, pp. 90–98.
23. Higashi, K., Ontani, H., Yagi, Y., et al., "Development of Conductive Stud Bump Technology for the Best Shape Design and Control," in *Proceedings of the Pan Pacific Microelectronics Symposium,* Hawaii, Feb. 10–13, 1998, pp. 449–454.
24. Baldwin, D., "Advances in Low Cost Flip Chip Process Technology," IMAPS 2nd International Advanced Workshop on Low Cost, Braselton, GA, March 13–15, 1998.
25. Ito, S., and Akizuka, S., "A New Flip Chip Packaging by Non-Conductive Resin Sheet and the Stress Management," IMAPS 2nd International Advanced Workshop on Low Cost, Braselton, GA, March 13–15, 1998.
26. Gilleo, K., "Will Flip Chip Become the Ultimate CSP?" IMAPS 2nd International Advanced Workshop on Low Cost, Braselton, GA, March 13–15, 1998.
27. Gilleo, K., "Transforming Flip Chip Into CSP With Reworkable Wafer-Level Underfill," in *Proceedings of the 4th Annual Pan Pacific Microelectronics Symposium,* Kauai, HI, Feb 2–5, 1999, pp. 159–165.
28. Gilleo, K., "New Generation Underfills Power the 2nd Flip Chip Revolution," in *Proceedings of the Pan Pacific Microelectronics Symposium,* Hawaii, Feb. 10–13, 1998, pp. 147–154.
29. Gilleo, K., and Blumel, D., "The Ultimate Flip Chip—Integrated Flux/Underfill," in *Proceedings of the Technical Program of Nepcon West,* Anaheim, CA, Feb. 22–27, 1999, pp. 152–160.
30. Crane, L., Torres-Filho, A., Yaeger, E., and Heuel, M., "Development of Reworkable Underfills, Materials, Reliability and Processing," in *Proceedings of the Technical Program of Nepcon West,* Anaheim, CA, Feb. 22–27, 1999, pp. 144–151.
31. Capote, A., "A Novel Flip-Scale Package Using Pre-Applied Multilayer Flip Chip Under-Encapsulation," in *Proceedings of the 4th Annual Pan Pacific Microelectronics Symposium,* Kauai, HI, Feb 2–5, 1999, pp. 153–158.
32. Gilleo, K., "Flip Chip 1, 2, 3: Bump, Bond and Fill," *Circuits Assembly,* June 1996, pp. 32–34.
33. Gilleo, K., "Flip or Flop?" *Circuits Assembly,* Feb. 1997, pp. 40–48.
34. Anderson, B., Bacher, B., and Gomez, M., "Development Methodology for a High-Performance, Snap-Cure Flip-Chip Underfill," in *Proceedings of the Technical Program of Nepcon West,* Anaheim, CA, Feb. 22–27, 1999, pp. 135–143.
35. Gilleo, K., and Blumel, D., "The Great Underfill Race," in *Proceedings of the International Symposium on Microelectronics,* San Diego, CA, Nov. 1–4, 1998, pp. 704–706.
36. Duck, A., "Flip Chip Underfill," IMAPS 2nd International Advanced Workshop on Low Cost, Braselton, GA, March 13–15, 1998.
37. Norris, M., "The Dispensing Process in Advanced Electronic Component Manufacturing of Ball Grid Arrays, Flip Chip and Chip Scale Packages," in *Proceedings of the Pan Pacific Microelectronics Symposium,* Hawaii, Feb. 10–13, 1998, pp. 179–185.
38. Wyllie, G., and Miquel, B., "Technical Advancements in Underfill Dispensing," in *Proceedings of the Technical Program of Nepcon West,* Anaheim, CA, Feb. 22–27, 1999, pp. 152–157.
39. Nguyen, L., Fine, P., Cobb, C., Tong, Q., Ma, B., and Savoco, A., "Reworkable Flip Chip Underfill—Materials and Processes," in *Proceedings of the International Symposium on Microelectronics,* IMAPS, San Diego, CA, Nov. 1–4, 1998, pp. 707–719.
40. Cowburn, A., et al., "Flexible Circuit Solutions for Advanced MR Head Disk Drives," in *Proceedings of Flexcon,* Sunnyvale, CA, Oct. 12–14, 1994, pp. 1–6.

41. Cowburn A., et al., "Flex Circuit Requirements for Flip Chip Attach," in *Proceedings of the 1996 International Flip Chip, Ball Grid Array, TAB and Advanced Packaging Symposium (ITAB 96),* Sunnyvale, CA, Feb. 14–16, 1996, pp. 75–79.
42. Masumoto, T., et al., "Stud Bumped Flip Chip on Flex (FOF) in Hard Drive Applications," in *Proceedings of the 1996 International Flip Chip, Ball Grid Array, TAB and Advanced Packaging Symposium (ITAB 96),* Sunnyvale, CA, Feb. 14–16, 1996, pp. 59–63.
43. Jang, S.-Y., et al., "Study on the Under Bump Metallurgy of Electroplated Eutectic Pb/Sn Solder Bumps on Organic Substrate," in *Proceedings of the 4th Annual Pan Pacific Microelectronics Symposium,* Kauai, HI, Feb 2–5, 1999, pp. 173–179.
44. Koopman, N., et al., "Solder Flip Chip Developments at MCNC," in *Proceedings of the 1996 International Flip Chip, Ball Grid Array, TAB and Advanced Packaging Symposium (ITAB 96),* Sunnyvale, CA, Feb. 14–16, 1996, pp. 64–74.
45. Rosson, J., Clawson, R., and Ihms, D., "Flip Chip Underfill Characterization methods: Developing a Test Methodology for Success in the Harsh Automotive Environment," in *Technical Proceedings of Surface Mount International,* San Jose, CA, August 23–27, 1998, pp. 295–302.
46. Price, D., and Corbett, S., "High Volume Flip Chip on Flex with Conductive Adhesives," in *Proceedings of the Technical Program of Nepcon West,* Anaheim, CA, Feb. 22–27, 1999, pp. 1502–1508.
47. Gilleo, K., Boyes, B., Corbett, S., Larson, G., and Price, D., "High Volume, Low Cost Flip Chip Assembly on Polyester Flex," *Circuit World,* 25 (2), Feb. 1999, pp. 11–17.
48. Roybal, T., "Developing Aerospace/Military Flip Chip Technologies," IMAPS 2nd International Advanced Workshop on Low Cost, Braselton, GA, March 13–15, 1998.
49. LeClair, T., Harper, A., Graham, S., and Ackley, D., "Flip Chip Interconnect of DNA Chip Devices," in *Proceedings of the International Symposium on Microelectronics,* San Diego, CA, Nov. 1–4, 1998, pp. 732–736.
50. Surface Mount Council Report, "Status of the Technology Industry Activities and Action Plan," San Jose, CA, Aug. 23–27, 1998.
51. "The National Technology Roadmap for Semiconductor Technology Needs," Semiconductor Industry Association (SIA), San Jose, CA, 1997.
52. Werner, R., Frear, D., DeRosa, D., and Sorongon, E., "Flip Chip Packaging," in *Proceedings of the Advanced Packaging Materials,* IMAPS, Braselton, GA, March 14–17, 1999, pp. 246–251.

CHAPTER 8
OPTIONS IN HIGH-DENSITY PART CLEANING

Steve Randolph and John Sanders
Speedline ACCEL

8.0 INTRODUCTION

While high-density packages such as flip chip (FC) and ball grid array (BGA) devices deliver substantial input-output (I/O) resources, their close dimensions introduce substantial cleaning challenges. These devices may have interconnect pitch dimensions of 250 μm and standoff distances in the range of 50 to 100 μm. These packages must be absolutely free of contamination in order to meet performance and dependability standards. However, contaminated interconnect areas are only accessible from the periphery, making it very difficult to reach and clean flux residues from the very center of a package.

As with other soldered packages, high-temperature reflow of high-density devices can create charred and caramelized flux residues that are difficult to remove. In addition, any residual solvent, water, or other manufacturing by-products left after BGA or FC cleaning will lead to ionic contamination and corrosion.

Process residues also interfere with subsequent underfilling, resulting in voids that can promote moisture collection, overheating, and part failure. A cleaning system must remove all flux residue, solder balls, ionic contamination, oils, and foreign matter, as well as chemical cleaning agent residues.

Manufacturers have three primary alternatives for postreflow cleaning of flux residues from large packages with fine-pitch interconnects and minimal clearances. These are in-line spray, ultrasonic bath, and centrifugal cleaning. Table 8.1 summarizes the primary characteristics of each of these cleaning methods.

8.1 IN-LINE CLEANING

In-line spray cleaners are extended, conveyorized systems that direct cleaning solvents against parts to be cleaned using vertically oriented spray nozzles. Devices being cleaned pass through wash, prerinse, rinse, and drying zones on a continuous conveyor. For difficult cleaning applications, nozzle count and spray pressure may be increased, and extra in-line cleaning stages may be added to achieve the desired results.

As additional stages are added to increase cleaning dwell time, the system footprint is extended. In-line spray systems are good for removing flux residues and oils on exposed surfaces in a continuous production setting but generally are not effective in penetrating the close tolerance and hidden gaps of FC assemblies.

TABLE 8.1 FC and BGA Postreflow Cleaning Options

Cleaning method	Advantages	Disadvantages
In-line spray	A continuous, integrated manufacturing process	Minimal cleaning penetration around die, large equipment footprint
Ultrasonic	Relatively effective cleaning	May damage delicate parts, long-term reliability concerns, typically off-line batch
Centrifugal	Very effective cleaning process in a small floor space footprint, short cycle times, complete cleaning, total drying	Batch, off-line process

8.2 ULTRASONIC CLEANING

Batch-style ultrasonic cleaners place parts to be cleaned in a solvent immersion bath, where ultrasonic energy aids in flux removal by cavitation of the liquid, which varies according to fluid density and the frequency and amplitude of the excitation (Fig. 8.1). This process is superior to in-line cleaning for cleaning complex parts. However, in the case of FC components, ultrasonic energy may not be able to fully penetrate the gap and reach hidden contaminants. In addition, this process is relatively ineffective at removing solubilized contaminants from the confines of high-density packages. Mechanical resonance induced during ultrasonic cleaning can cause microfracture of delicate parts, thereby degrading long-term reliability.

8.3 CENTRIFUGAL CLEANING

Centrifugal cleaning, another batch process, offers an important performance advantage over other FC and BGA cleaning methods: the ability to direct the solvent laterally around and between die terminations under pressure. A centrifugal cleaning system holds devices in fixtures so that the component edges are impinged directly by solvent. A robot arm assembly containing fixtured parts is lowered into a cleaning solvent bath in a sealed process chamber and rotated alternately in clockwise and counterclockwise directions. This causes the solvent to be forced through the package in a direction parallel to the plane of the substrate, driving it under and between shadowed components. Contaminants are solubilized, suspended in solution, and driven off (Figs. 8.2 and 8.3).

8.3.1 The Centrifugal Cleaning Sequence

This cleaning process combines immersion in a semiaqueous or traditional solvent with agitation and solvent flow resulting from the centrifugal force of rotation. Because the chamber is sealed and nitrogen inerted, the solvent temperature can be raised safely to a level near the flash point, and this reduces surface tension and lowers viscosity values for increased solvent action.

As immersed circuit assemblies are rotated, the spaces beneath package features and the force vectors acting on them are in the same plane. The mass of the cleaning compound together with Coriolis forces produced by rotation propel the solution into the tightest spaces to extricate entrapped flux.

Periodic reversal of rotation ensures that liquid flow is established in every direction so that flux is rapidly solubilized and flushed away. Since both the fluid and packages are in constant motion, solvent action is continuous. In contrast, solvent action in a spray cleaning system is essentially limited to the portion of a device being impinged by the solvent spray (Figs. 8.4 and 8.5).

After the centrifugal solvent washing step, cleaning solution is drained from the sealed chamber, and a distilled water or virgin solvent rinse spray is introduced as the spin rate of the fixture sus-

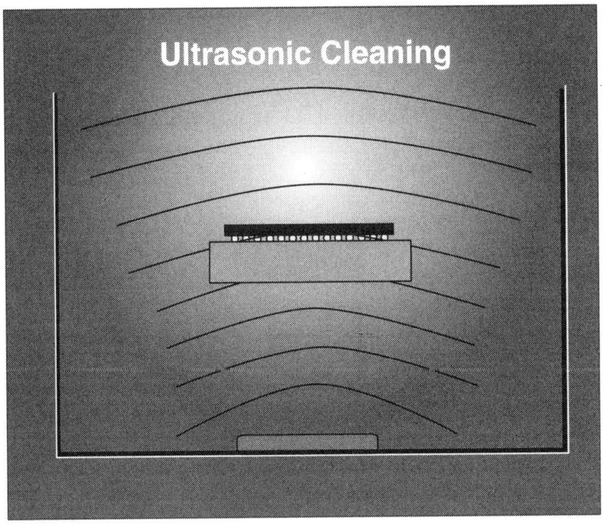

FIGURE 8.1 Ultrasonic cleaning. The energy created by an ultrasonic cleaning system displaces surface contaminants effectively but may not penetrate the hidden recesses that characterize devices that have high interconnect density.

FIGURE 8.2 The Speedline ACCEL MicroCel centrifugal cleaning system. The unit uses fixtures for BGA and FC devices that are mounted on a robot arm and rotated in the cleaning chamber.

pended from the robot arm is increased. The rinse spray orientation is lateral so that rinse water also impinges on the edge of packages and flows between interconnects. Rotation of assemblies on the robot arm further enhances penetration of the rinse agent.

The spray rinse step is carried out in an empty chamber so that cleaned parts are not recontaminated with residue from the rinse water. When rinsing is complete, parts are dried by increased centrifugal rotation, accompanied by high-volume flow of heated air or nitrogen.

Products cleaned in a centrifugal cleaner generally are held in place and secured to the rotating robot arm head using universal adjustable fixtures, standard fixtures, or custom fixtures that are specific to a product. Manufacturers generally provide hardware to handle devices, including small circuit modules, wafers, singulated packages, Auer boats, magazines, cassettes, and JEDEC trays.

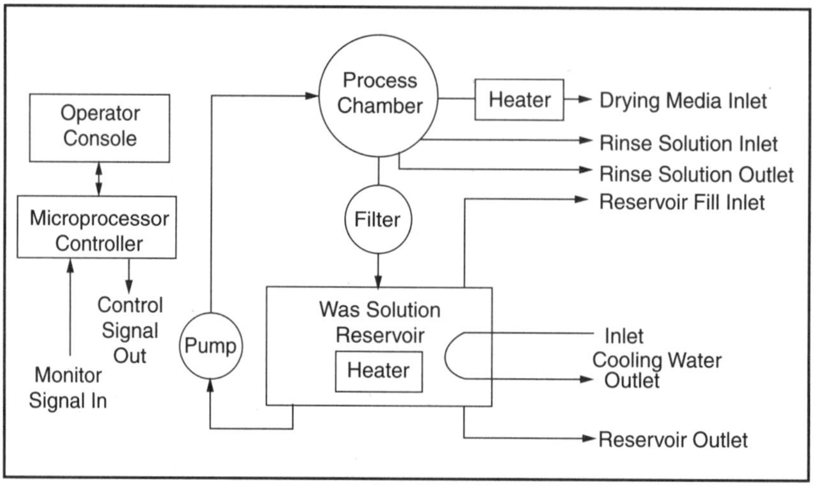

FIGURE 8.3 Machine schematic.

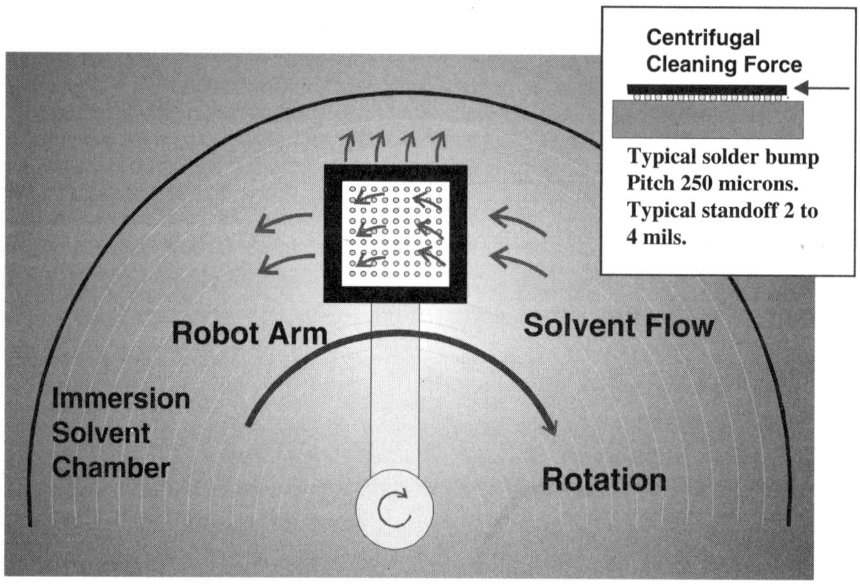

FIGURE 8.4 Movement of fixtured assemblies through the solvent bath creates positive flow of solvent between interconnections. Full coverage is ensured by alternating the direction of rotation during a cleaning cycle. Multiple components are mounted on the centrifugal system robot arm.

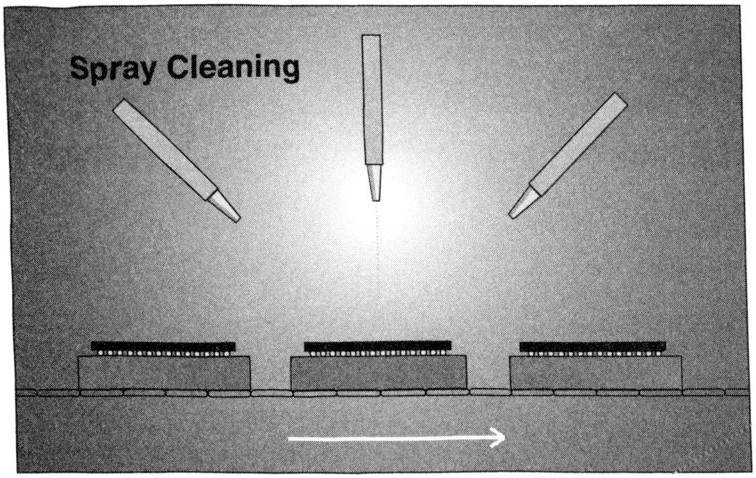

FIGURE 8.5 A spray cleaning system. This is effective for conventional circuit assemblies but may not be capable of removing flux residues and contaminants from FC or BGA assemblies.

Some centrifugal cleaning systems incorporate automatic, closed-loop, and integrated wastewater treatment. When using a solvent that will phase separate from water, such systems automatically extract it from the rinse water and return the solvent to the wash reservoir for reuse. Used rinse water may be processed through a multistage purification process that includes microbial control, microfiltration, carbon adsorption, and mixed-bed deionization to restore the water to its original level of purity. No drain or external water treatment is required for such systems, and the entire cleaning and fluid conditioning process takes place in the compact cleaning system housing.

A centrifugal cleaning system may require process recipe adjustment for optimal performance depending on part complexity, the solvent in use, and the nature of contaminants to be removed. Adjustable process variables include centrifugal rotation speed; the dwell times of wash, rinse, and dry cycles; and solvent and rinse agent temperature.

While centrifugal cleaning is a batch process, this cleaning method generally can keep pace with a conventional manufacturing line. For example, the ACCEL MicroCel centrifugal cleaning system can process hundreds of devices per cleaning/rinsing/drying cycle, which consumes approximately 7 to 15 minutes. This centrifugal system is compatible with semiaqueous, aqueous, alcohol, and traditional cleaning solvents.

A longer cycle time in a centrifugal cleaning system requires no additional floor space, unlike conveyorized systems, where space limitations may impose constraints on system size and thus the level of cleanliness that can be achieved. A closed centrifugal cleaning process provides good odor containment because cleaning, rinsing, and drying all take place in a sealed process chamber. With centrifugal cleaning, there is no aerosol phase, as with a spray system, no solvent dragout, and no objectionable odors because the solvent is never exposed to the atmosphere.

The primary solvent factors to be considered for cleaning close-tolerance assemblies are solvency for the flux/contaminants in question, solvent surface tension, vapor pressure, flash point, safety/toxicity, and cost. Because of the sealed nature of centrifugal cleaning, concerns related to solvent flash point, safety/toxicity, odor, and waste disposal are minimized.

8.3.2 Solvent Performance

An evaluation of centrifugal cleaning by Texas Instruments illustrated the cleaning performance that could be achieved with centrifugal energy. This component manufacturer tested flux removal efficiency using square glass coupons sandwiched to glass plates with clearances of 2, 5, and 10 mils.

The test contaminant was alpha 321 flux, dyed for visual inspection. This material was injected into the spaces between plate and coupon and then heated to a temperature that simulated reflow levels typical of FC manufacturing conditions.

This test sequence confirmed that average levels of ionic contamination on cleaned FC parts could be reduced to 0.0 μg per square and visible flux residue totally eliminated on inspection at 30× magnification. A hydrocarbon cleaning compound was the solvent used in the centrifugal cleaning system test, with processing taking place 24 hours or more after reflow. Results were compared with spray-cleaned parts that were processed within 1 hour of reflow and thus had reduced potential for consolidation of contaminants.

Spray cleaning was performed using an in-line system with Dupont HCFC 120 solvent. Results from this process are shown on the left in Fig. 8.6. Flux removal is apparent around the edges of all coupons, with results proportional to the clearance between glass surfaces.

The in-line-spray test plate (left side of figure) illustrates the amount of flux remaining beneath closely spaced components following this traditional cleaning method, which relies on capillary force for solvent penetration. (These tests made no attempt to simulate flow impediments such as the fine-pitch leads and area array connections that could be expected on devices of this size.)

Comparable glass plates were prepared for testing in an ACCEL MicroCel centrifugal cleaning system using alpha EC-7R semiaqueous solvent. The samples on the right in Fig. 8.6 show the amount of flux remaining following this cleaning method.

These examples graphically illustrate the improved defluxing efficiency that can be accomplished using solvent penetration propelled by centrifugal force. Additionally, the centrifugal extraction of entrapped rinse water promotes rapid drying and contributes significantly to overall throughput in aqueous and semiaqueous processes.

8.3.3 Flux/Solvent Performance

There has been an evolution in cleaning solvents for electronics applications over the past several decades. Hydrocarbon solvents were the first materials used for cleaning, and while such fluids were effective, concerns about flammability and workplace safety eventually led to adoption of nonflammable chlorinated solvents. These materials, also effective solvents, were identified eventually as potential carcinogens and replaced with chlorofluorocarbons (CFCs), which were welcomed because of their solvency, nonflammability, low vapor pressure, and favorable toxicity properties.

CFCs in various forms began to predominate in many industrial applications, and by the late 1980s, consumption of materials such as CFC-113 and 1,1,1,-trichloroethane reached nearly 1 billion pounds per year in the United States alone. However, scientists eventually discovered that the

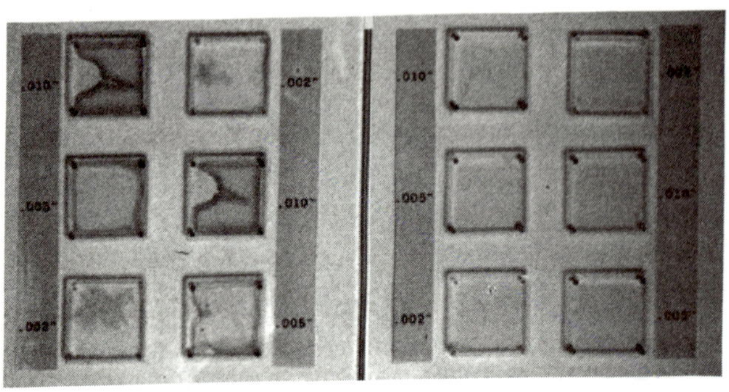

FIGURE 8.6 Glass test plates cleaned with in-line (*left*) versus centrifugal (*right*) cleaning methods.

favorable properties of CFCs came at the expense of stratospheric ozone depletion. In the mid-1990s, worldwide environmental restrictions were placed on these ozone-depleting compounds.

Today, the challenge in developing non-ozone-depleting solvents is to achieve a balance in solvent effectiveness, environmental properties, nonflammability, and safety/toxicity properties. There are a dozen or more flux formulations that can be considered for FC and BGA assemblies today, and each presents its own cleaning challenges. In the United States, the primary solvent categories used in electronics are terpenes (various citrus-derived semiaqueous solvents), long-chain alcohols, and petroleum-hydrocarbon derivatives.

Another change affecting electronics manufacturing today is the ban on use of lead-based solders in electronic manufacturing. The new lead-free solders require higher reflow temperatures, and this has created a need for modified paste/flux formulations, which, in turn, call for special solvent properties.

Manufacturers may find it useful to test the performance of the various solvent options with specific assemblies. Factors to be considered are cleaning performance with selected solder/flux materials, solvent consumption, foaming potential, environmental impact, odor, toxicity, and water treatment/disposal issues.

It may be appropriate for certain low-cost or short-term electronics applications to eliminate cleaning of high-density devices. However, the cost of effective cleaning is an incidental amount in terms of the overall manufacturing investment, and the removal of flux residues and manufacturing contaminants yields substantial benefits in package quality, reliability, and in many applications, electrical parametric performance.

8.4 SUMMARY

Centrifugal cleaning produces results that meet or exceed those of other cleaning technologies for high-density devices such as FC and BGA packages. For example, typical cleaning results measured with the ACCEL centrifugal cleaning system include ionic contamination of 0.0 $\mu g/in^2$, surface insulation resistance greater than 10^{14} Ω/in^2, and no visible residue at 30× magnification.

Centrifugal cleaning technology lends itself to a range of applications and is compatible with virtually all cleaning agents. Thus a system obtained for one application and solvent can be converted readily to another application, prolonging the useful life and productivity of the equipment. Since all processing is confined to a closed chamber, liquid storage and consumption are minimized, along with evaporative and drag-out losses.

FCs, BGAs, and other large, fine-pitch packages present some of the most difficult cleaning challenges in electronics manufacturing. However, with an appropriate cleaning solvent and a properly designed and programmed centrifugal cleaning system, it is possible to deliver contamination-free devices ready for subsequent assembly steps. The centrifugal cleaning method has been shown to support product quality, providing ease of operation, low ownership cost, and environmental compliance.

CHAPTER 9
MEMS PACKAGING AND ASSEMBLY CHALLENGES

Ken Gilleo
Cookson Electronics

9.0 INTRODUCTION

Microelectromechanical system (MEMS) may just be the ultimate enabling technology for the micro-level integration long sought for advanced products such as system-on-chip (SOC). MEMS devices operate using many forms of energy and sources of data. These include mechanical kinetic forces, light waves, sound waves, radio waves, and conventional electrical energy. This technology has been applied for complex molecular detection, a variety of analyzers, motion sensors, fluid pumping, radio signal manipulation, a host of medical applications, and photonics. When photonics or optics is added, the term *micro-optical-electromechanical systems* (MOEMS) can be used, although some prefer to use *optical MEMS.* The most extraordinary feature of MEMS is that all the diverse functions can be crafted into a single chip using modified semiconductor fabrication techniques. MEMS therefore fits the giant semiconductor infrastructure, but even older, lower-density fabrication facilities can be used. While integrated-circuit (IC) logic devices have given us the *brain,* MEMS adds the *eyes, nose, ears,* and *new senses* beyond the limits of human range. MEMS has output in the form of information that can be electronic or photonic. However, more important, MEMS also produces control functions. This adds the *muscle, hands,* and *fingers,* allowing MEMS devices to physically move internal elements and nearby objects. The *merging of motion, sensing, computation, and control* appears to represent the highest level of technology yet achieved.[1]

While packaging the versatile MEMS devices is a major challenge, old and new solutions are being deployed. Most devices are housed in costly hermetic packages, especially optical MEMS systems. Perhaps we will be able to move to a more traditional type of packaging in the future. Some less complex optical products have already been packaged in a ball grid array (BGA) format. For example, the optoBGA (iC-Haus GmbH, Bodenheim, Germany) is described as a family of optical sensors made as a micro-BGA having a glass cover.[2] The glass is placed onto the chip, which is wire bonded to a BGA substrate. The assembly is then encapsulated so that no material covers the glass, as shown in Fig. 9.1.

Many MEMS chips are hermetically sealed to exclude oxygen and moisture that can cause wear and friction problems. Most optical devices are also hermetically sealed because oxygen can degrade lasers and water can fog optics. They use atmosphere control agents such as getters (molecule-specific scavengers). Wafer-level (0-level) hermetic packaging, now under development, also may be an answer, bringing reduced cost, enabling high-volume manufacturing, and delivering the high level of protection needed. Some popular MEMS and MOEMS packaging strategies will be discussed.

However, we also face the challenge of MEMS assembly. While many packages are hand assembled, will we standardize on surface-mount technology (SMT)? Moving to SMT assembly appears

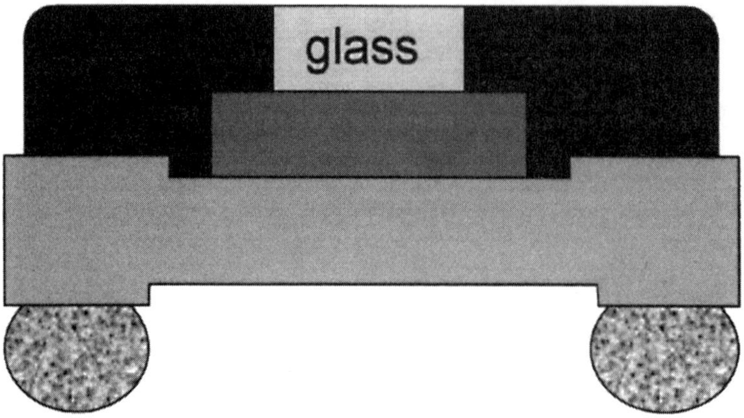

FIGURE 9.1 OptoBGA. (*iC-Haus GmbH.*)

to be a given for the high-volume markets, but there will be new rules for MEMS primarily because of the need to sense the external environment. MEMS motion-sensing products, such as automotive accelerometers for air bags, must be oriented correctly because they are highly directional. How important is package skew and tilt, and can the solder fillet even change MEMS sensitivity? We will first examine MEMS packaging and then focus on assembly issues. There may be more questions than answers at this early and evolving stage of MEMS.

MEMS may appear to be a paradox because products, mostly sensors, have been around for two decades, while forms of this technology are still emerging and in developmental form. Government and university laboratories are still working on basic problems. Sandia National Laboratories, the University of Colorado, and other research organizations have built truly remarkable devices. One example of a fascinating device is a chip-sized spectrophotometer that can detect and measure atmospheric gases from miles away. Microscopic pumps and valves also have been built that may be used for medical applications including drug delivery. Devices are so small that they can be injected into the body. The day may come when microrobots, or "nanobots," travel through the body to clear arteries and make repairs, borrowing a scene from the classic 1960s science fiction movie *Fantastic Journey.* Yet MEMS has been quietly commercialized as motion sensors for airbags and ink jet cartridges for printers. However, optical MEMS is now powering digital projectors using sophisticated micromirror arrays. These optical control modules are now being tested in cinema projectors where over 1 million miniature mirrors "paint" the movie in vivid color. Figure 9.2 shows scanning electron micrographs of the extraordinary range of MEMS devices from Sandia National Laboratory.

9.1 BACKGROUND

Electronic chips use one primary source of energy—electricity. ICs deal primarily with electronic signals. There is no mechanical movement because the flow of electrons is invisible and only manifests itself in electrical output or the emission of light in some of the electro-optical (EO) devices such as light-emitting diodes (LEDs). MEMS can represent a very high level of integration because many dissimilar kinds of functions can be combined and operated using all kinds of signals, not just electrical signals. This unification of functions includes motion, light, sound, electromagnetic radiation, and high-level analysis of external input. Computation, analysis, and central control of these I/O functions results in a fully integrated system of incredible versatility. MEMS is the convergence point of many diverse technologies. Mechanics adds gears; optics introduces mirrors, filters, and guides; the electri-

FIGURE 9.2 MEMS from Sandia.

cal field delivers motors; chemistry adds molecular detection; and general electronics adds radio computation and power to this new microworld. And we can expect biotechnology to enter the MEMS arena soon. The key to MEMS is that the technology blends and integrates complete functional systems. It synergistically merges subsystems into a fully integrated, self-contained microcosm that once only existed in the macro world. How remarkable that these once-isolated technologies can converge into the microscopic world of silicon using wafer-level mass processes.

However, there are many major challenges in nearly every area. Specialists in this field insist that MEMS fabrication, packaging, and assembly are the greatest challenges our industry has ever faced. Not only are the advanced MEMS devices small and highly complex, many must communicate with the outside world by direct-path nonelectronic signals. The burden placed on packaging and assembly is to provide electromechanical connections that do not interfere with or alter sensing and control function. Later, fiberoptic connections will be added. This challenge is new because assembly has dealt mostly with electrical signals. The additional I/O of motion and photonic and molecular "signals" adds a new twist.

Some MEMS devices send and receive light beams; others detect specific molecules, including biological analysis such as DNA. The merging of motion, sensing, and computation most certainly represents a new level in technology that is still embryonic. Surprisingly, this new technology has been around for nearly three decades but at a simple level such as motion sensing. More recently, however, optical MEMS, also called MOEMS, has become a popular area. The Internet is the new and powerful driving force. There are solid and durable opportunities in communications hardware. The modern Internet is really a fiberglass highway, a system of photonics light pipes. Light control is the master key to efficient photonics—the technology of communication using photons—and optical MEMS is poised to be the winner here. Just during the year 2000, the Internet giants spent billions of dollars to acquire MEMS companies. Thus, as light wave technology moves from the present long haul, or backbone, closer to the end user, we can expect the volume for MOEMS assemblies to increase and move out to the contract assemblers. The future wave may be light, not solder!

9.2 MEMS FABRICATION

Techniques have been derived mostly from IC fabrication with appropriate modifications to produce movable parts. Standard semiconductor bulk fabrication processes create the traditional electronic circuit portion. Subtractive etching produces the movable mechanical elements. Silicon dioxide can be formed as the sacrificial material that is removed in the final step to free up the movable components. The silicon or silicon nitride later becomes the mechanical or optical elements in a final "release" step. Subtractive methods are used to remove the elemental silicon, such as wet chemical etching or reactive ion milling. The sacrificial holding structure is sometimes etched away at the packaging foundry so that delicate parts are protected while the wafer is transported. Although some devices are very fragile initially, MEMS parts are more robust after packaging. However, many MEMS devices, even when packaged, need careful handling by the assembler compared with conventional electronic packages. However, the very low mass per unit volume makes the microstructures fairly robust. In addition, silicon is a very strong, high-modulus material that is considered ideal for these nanodevices. Table 9.1 lists some of the fabrication processes used for MEMS.

Table 9.2 lists some of the materials, masks, and etchants that have been used for MEMS fabrication. Note that silicon dioxide (SiO_2) can be used as both a mask and the sacrificial material. The high selectivity of the etchants makes it possible to produce the complex three-dimensional (3D) structures that are shown in Fig. 9.2.

9.2.1 MEMS Actuation

There are several ways of powering MEMS devices, but most are based on the input of electricity that also may be converted to thermal energy, hydrostatic pressure, electrostatic attraction, or magnetic energy. Interconversions are almost limitless. Lucent, for example, has developed optical devices that are powered by electricity generated from light fed into the system though an optical fiber. This can allow a photonic switch to be powered by energy sent down the fiber instead of requiring electrical lines. Since electric current is so readily available and easily transformed, electricity is the de facto standard "fuel" for MEMS. Several of the MEMS motors use electrostatic forces

TABLE 9.1 MEMS Fabrication Processes

Photolithography
Wafer bonding
Chemical etching
Plasma etching, reactive ion etching (RIE)
Ion beam etching
Ion milling
Plating
Vacuum metallizing
Electrical discharge machining (EDM)
Laser micromachining
Package assembly
Place getters
Thin film coating
Hermetic sealing
Molding

TABLE 9.2 Materials, Masks, and Etchants Used in MEMS Fabrication

Sacrificial layer	Etchant	Mask
Pure silicon	Amine base/pyrocatechol	SiO_2, Si_3N_4, metal
Si	KOH, alcohol	SiO_2, Si_3N_4
Si	Hydrofluoric, nitric, acetic acids	SiO_2
SiO_2	HF	Si, polyimide
Polysilicon	HF	Si_3N_4
Aluminum	HF	Polyimide

because they are substantial at the micro level. Very small objects have a relatively high surface area compared with mass. Static electricity forces are proportional to surface area and behave like movable capacitor plates. Very thin conductor plates and combs have very little mass. The electrostatic engines are capable of high speed and torque. These engines typically use linear or rotating combs. Conductive films and even magnetic ones also can be applied by vacuum techniques to produce MEMS engines. These systems are fairly efficient and do not generate much waste heat.

Thermal energy is very popular and is used in the high-volume MEMS ink jet pumps found in many disposable printer cartridges. The micro scale permits rapid conversion and transfer of thermal energy. Agilent has even applied the thermal concept to optical switches, where heat-activated "bubbles" alter the light beam. Dimorphic materials can be constructed that bend when heat from electricity is introduced. The bimaterial strips are popular because motion can be derived without "moving contact" surfaces that encounter friction and can have wear problems. Pneumatic and hydraulic engines also have been built that often are powered by heat derived from electricity. While all present packaging challenges, the "electric motors" are likely the easiest to deal with.

Once mechanical energy has been produced, it can be translated into useful motion. Just about every conceivable mechanical translator has been built. Many are just adaptations of the common mechanisms from our macro world such as gears and sliders. Others are somewhat unique to the microscale environment and would not be practical at a larger scale. Micromirrors that use thermal dimorphic beams are an example. Before moving into the two principal topics, packaging and assembly, let's briefly examine software specifically for this field.

9.3 SOFTWARE

One issue has been the lack of dedicated MEMS software for simulation and analysis. This gap is being filled rapidly. MEMS requires new software to bring order and product design efficiency to the industry. IntelliSense, a fast-rising MEMS star in Wilmington, MA, claims to be the only MEMS manufacturer with a full complement of software. The company's IntelliSuite CAD for MEMS includes a fabrication process database, materials libraries, 3D structure graphics, and a performance simulator/analyzer. The software is said to simulate and optimize MEMS to reduce cost, the number of prototypes, and time to market. However, new software is appearing almost monthly as the industry recognizes the importance of this field. Microcosm Technology, with well-known Cadance, has released MEMCAD. Apollo Photonics has just released an optical MEMS program simulator and analyzer called Photonics Suite. Figure 9.3 shows a micromirror simulation.

9.4 MEMS PACKAGING

Most packages are hermetically sealed to keep out particles and gases that would interfere with mechanical movement or fog and degrade optical systems.[1] The package complexity that can

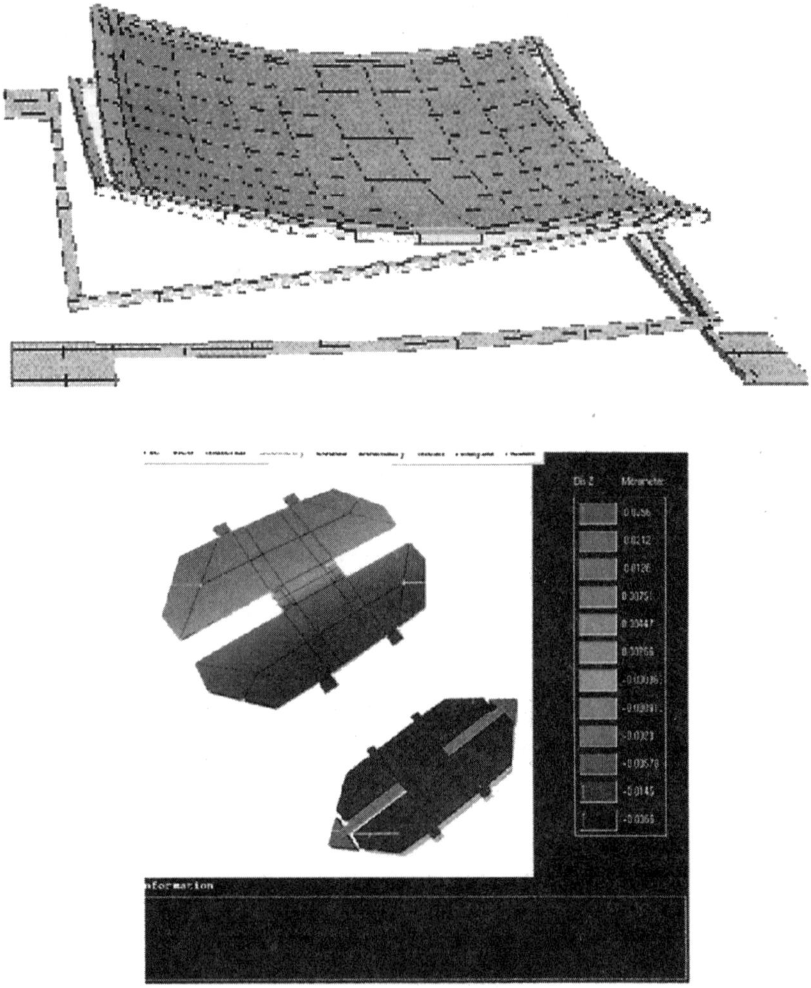

FIGURE 9.3 CAD software display of micromirror. (*From IntelliSense.*)

add assembly restrictions is highly dependent on the type of MEMS product, its functions, and the need for external access. Most MEMS devices move, but their mode and the purpose of motion determine the packaging requirements. Table 9.3 indicates some of the basic modes of motion, and several modes are shown in Fig. 9.4. Table 9.4 gives applications, I/O, and possible package types. Products with motion input and electrical output would appear to be the easiest to assemble.

The commercial MEMS industry has dealt primarily with type 1 motion devices, such as accelerometers. The industry is scaling up for two- and three-axis gyroscopes that will be used for high-volume input devices for computers, games, and all kinds of handheld products. These products are easier to handle because the package can be sealed from the outside environment. The contract assembler is likely to encounter a MEMS motion package as one of the first such packages. Packages will be both ceramic and plastic, but both will be in SMT format. Let's look at some of the old and new MEMS packaging.

TABLE 9.3 Basic Modes of Motion

1. Deformation of isolated parts: bending or twisting, no friction (accelerometers)
2. Rotating parts: friction
3. Sliding parts: friction and stiction
4. Impact: wear
5. Moving optical elements
6. Combinations of the above

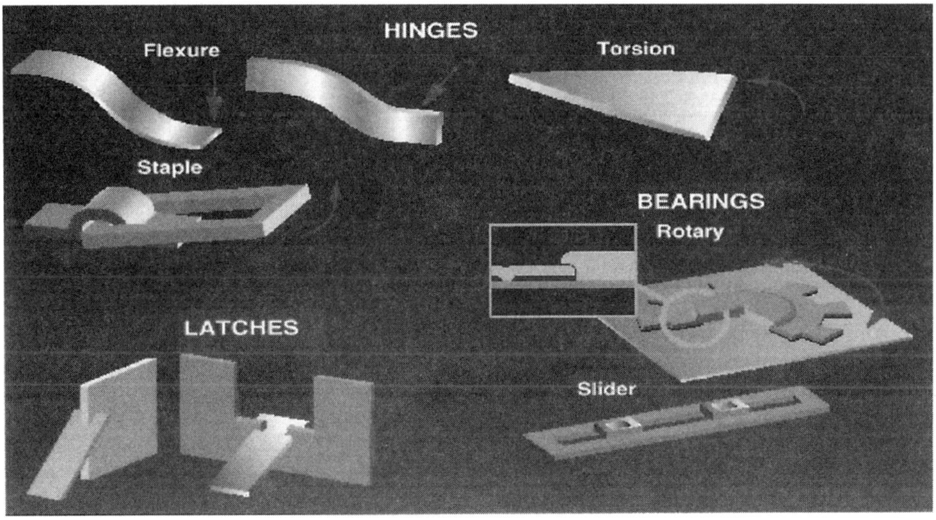

Tutorial:
MICROMACHINING IN OPTICS
Gregory A. Magel
Texas Instruments
Paper MOO1
OSA Annual Meeting
October 21, 1996

FIGURE 9.4 Modes of motion. (*From TI Tutorial.*)

9.4.1 MEMS-Specific Package Designs

Most MEMS packages started off as standard sealed hermetic containers such as a Cerdip (ceramic dual in-line) package because a ceramic or metal hermetic package offers a high level of protection but at a cost penalty. High-volume accelerometer manufacturers such as Analog Devices have converted some of these packages to SMTs called *Cerpacks.* Figure 9.5 shows the accelerometer chip. We will look at this package later to gain insight into the assembling of MEMS devices.

Now let's examine some newer packaging concepts that eventually could become standard packages that assemblers will deal with.[3] Although the MEMS package can be very application-specific, some general design concepts are emerging. The MEMS devices have a universal requirement that the "motion zone" cannot be obstructed. A protective cap over the action area can permit the component to be overmolded while still allowing unrestricted movement within the device. Once the cap is in place, standard transfer molding can be used. In some cases, liquid encapsulants may produce less stress. Liquids may be applied by automated needle dispensing equipment commonly used for some BGAs.

TABLE 9.4 Applications, I/O, and Packaging Types

MEMS device	Packaging	Input	Output
Accelerometer	EMC (epoxy mold. com'd.)	Motion	Electrical
2D and 3D motion detection	EMC	Motion	Electrical
Ink jet	Selective	Electrical	Motion
Digital mirror array	Hermetic/window	Electrical and light	Controlled light
Microspectrophotometer	Hermetic/window	Light	Electrical
DNA analyzer	Liquid access	Biological samples	Electrical
Disk drive heads	None	Magnetic	Electrical
Optical switches	Hermetic/window	Electrical and light	Electrical and light
Capacitors, tunable	EMC	Electrical/rf	rf
rf Tuners	EMC	Electrical/rf	rf
Gyroscopes	EMC	Electrical/motion	Electrical
Diagnostics (in vitro)	Open	Electrical, liquid	Electrical
Infrared imagers	Hermetic/window	IR, electrical	Electrical
Microrelay	EMC	Electrical	Electrical
Pressure sensors	Sealed with thin wall	Mechanical force	Electrical

FIGURE 9.5 Accelerometer package. (*Analog Devices.*)

Another possibility is to use flip chip (FC). Since the active surface of the device with moving parts is placed downward toward the substrate, a natural protective zone can be formed. The standoff distance, or *chip gap,* can be controlled accurately by the chip bump height. High-melting alloys or even nonfusible bumps can be used to ensure a specific minimum gap between the chip and board. The next step is to selectively dispense underfill. Normally, underfill is applied to completely fill the chip gap, but this would interfere with MEMS movement or sensing. A more viscous encapsulant, resembling damming compound, could be applied to all four edges of the chip. This sealing encapsulant, a fillet without the underfill, would then be thermally hardened. Now the package can be fully encapsulated by conventional transfer molding or by needle dispensing a liquid encapsulant. A chip access port could be added to the packaging substrate if the device must communicate with the outside atmosphere such as with a gas analyzer. A filter or semipermeable membrane possibly could be used to limit entry to only the intended molecules. Figure 9.6 shows the FC designs.[3]

A more elegant concept is to apply a microcap at the device level or, more preferably, at wafer level and then proceed to standard packaging. Several groups, including Amkor, IMEC (Belgium), and Georgia Institute of Technology, are working on or have already developed cap-on-chip, sometimes called *0-level packaging.* The cap essentially produces a silicon microhermetic package well suited for accelerometers, gyroscopes, or any other motion detectors. The present caps are made of silicon to achieve a perfect match of materials. However, the thin, flexible silicon cap may require a premolding step to prevent cave-in due to high molding pressures. Perhaps low-expansion metal caps can eliminate this problem. The alternative may be liquid encapsulation. Figure 9.7 shows the cap-on-chip, or 0-level, concept. The final package, if successfully implemented, will be similar outwardly to the common SMT plastic packages, although it is too early to know exactly what the form factors will be.

The chip scale packaging (CSP) industry also has begun to address MEMS and optical MEMS (or MOEMS). ShellCase, Inc., has developed the ShellMEMS. Windows for optical devices can be produced easily by using a transparent lid for the cap. Some package configurations use a BGA footprint. Figure 9.8 shows some of the area array designs for MEMS.

9.4.2 Packaging Atmosphere Control

The MEMS package also can consume (absorb) and emit (outgas) materials, but the atmosphere needs to be controlled within certain ranges. The atmosphere maintenance approach must be simple but effective, and a class of molecular scavengers, called *getters,* could be the best answer. Getters are agents true to their name and "get" things—primarily contaminants. Today's getters for electronic packages are very selective. These molecular scavengers are available as products that absorb gases, liquids, and solids. While the present list of commercial getters only includes moisture, particle, and hydrogen getters, additional getters can be designed as the need arises. Moisture getters use compounds that absorb and tightly bind water molecules. Zeolites and certain metal oxides are very effective. A special polymer matrix can be used as the "breathable" binder. Particle getters are made with permanently sticky polymers. This is not simply a "fly paper" material. The getter must be easy to use, maintain performance over extended time, and not produce contamination. Hydrogen getters, used to prevent poisoning of GaAs devices, are more complex. We may see the use of GaAs devices in the optical MEMS area later. Getters simply could be "printed" on the inside of the chip cap.

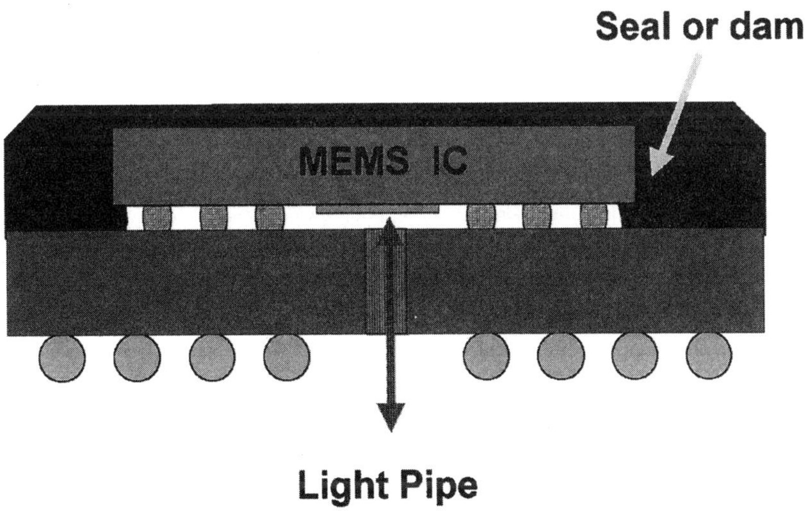

FIGURE 9.6 Flip chip MEMS packages.

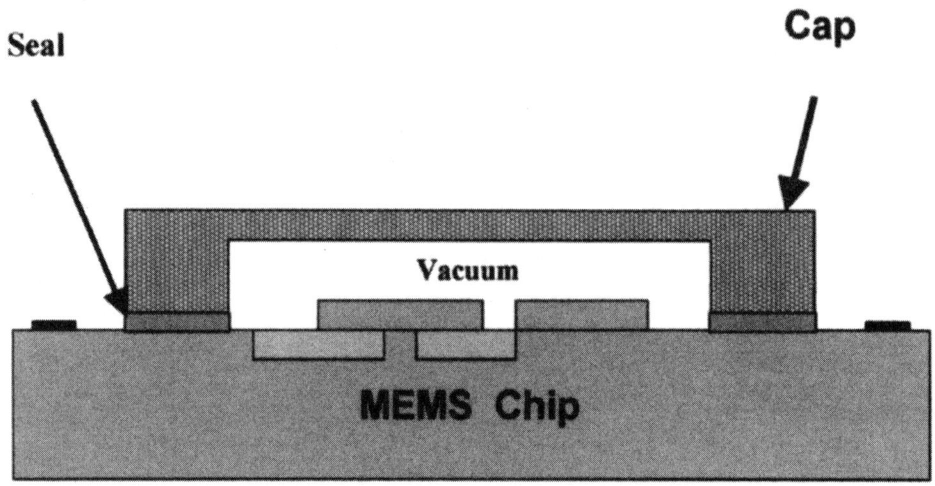

FIGURE 9.7 Cap-on-chip hermetic package.

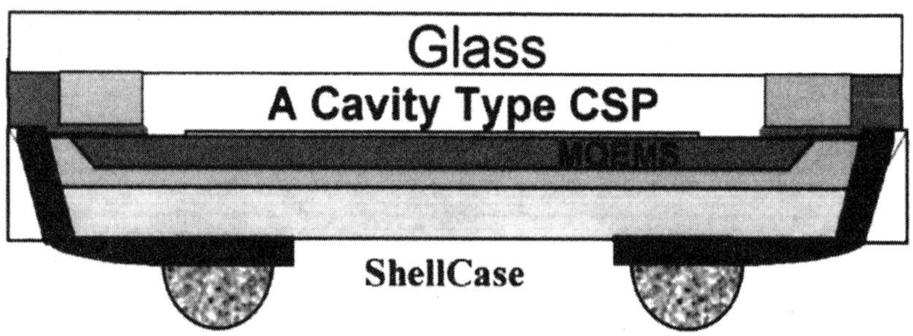

FIGURE 9.8 ShellMEMS package. (*ShellCase.*)

9.4.3 Getters

Getters have been around since the earliest phase of the electronics revolution. The early electronic vacuum tubes experienced unacceptably short lifetimes because of the oxidation of filaments and electrodes. Back in the early 1900s, it was not practical to produce a high and stable vacuum. But the life-limiting culprit was oxygen. The simple solution was to add an oxygen getter to the package. The getter was simply a metal or compound with a strong affinity for oxygen. The vacuum tube, even with a small leak, remained at a low oxygen level provided that some of the getter remained active. Now back to the twenty-first century.

Today's getter materials for modern electronic packages, while very selective, can be combined to perform multiple functions. For example, a popular product absorbs moisture and also binds microparticles. This type of getter is used for optical MEMS products because water degrades optics and particles would jamb moving parts. Moisture getters use high-capacity desiccants that absorb and tightly bind water molecules. Zeolites and certain metal oxides are very effective. A special polymer matrix can be used as a "breathable" binder, such as Staystik 415 and 482 (StayDry SD1000 and SD800, respectively). Particle getters are made with permanently sticky polymers. Creation of such polymers is a nontrivial task because outgassing must be kept low. Sticky polymer surfaces typically are produced by adding solvent or creating low-weight polymers, but this approach would produce volatiles that would be released inside the package. However, sophisticated polymer chemistry can produce high-weight, stable, but sticky surfaces. And finally, the getter must be easy to use and remain stable for years under variable conditions.

Hydrogen getters, used to prevent poisoning of gallium arsenide (GaAs) devices, are more complex. Although hydrogen-adsorbing palladium (Pd) metal can be applied to lids, a better solution is to use a polymer-bound system. This is so because the plated palladium compound has a tendency to embrittle and flake over time, leaving the possibility of unwanted particulates forming in the package. The hydrogen getter H2-3000, for example, uses palladium oxide (PdO) in combination with zeolite dispersed within a stable polymer matrix. The PdO converts H_2 gas to water that is then consumed by the zeolite, making the reaction irreversible. Incidentally, H_2 sources include the electroplating on the metal package, die-attach adhesives, and radiofrequency (rf) absorbers. Released hydrogen degrades the wire-bond interfaces of GaAs devices, reducing device performance. High-speed MOEMS optical devices could be made with gallium arsenide (GaAs) or indium phosphide (IP) that can be degraded by hydrogen. Table 9.5 summarizes the limited information available on getters.

Can getters control humidity within a specific range? This is certainly feasible, although the term *getter* may not be entirely accurate here. Specific compounds or mixtures are known that can maintain moisture within specific limits inside a sealed container. Equilibrium levels are soon reached with atmospheric water to regulate humidity. The appropriate selection of such desiccants and polymer carriers could produce atmospheric control agents for specific humidity ranges.

9.4.4 Surface Control: Friction and Stiction

As you can see, MEMS technology has its own vocabulary. *Stiction* is a form of friction in which a much higher force is required to get things moving, and the effect can be incapacitating. Once the gear, wheel, slider, or arm is unstuck, we are back to ordinary friction. Stiction goes unnoticed in the macro world, where it is a minor factor. The attractive forces in the micro world are relatively higher, and stiction is inconsequential. A high initial force may be required to start a mechanism moving if stiction cannot be overcome, and this requires bigger motors. However, microdevices have a very high surface area to mass ratio. Since friction depends on area, microdevices have stationary friction (stiction) values that can be a million times higher than friction during motion.

However, stiction can be a greater problem for even simple bending-mode devices such as accelerometers. What happens if the motion-sensing beams in the accelerometer make contact? They stick! They stay stuck! Figure 9.9 shows the stiction effect. The capacitor sensor is now a short circuit, and the device is inoperable. We cannot just pry the beams apart because the package is already sealed. But what made the beams come together? Dropping the parts or just moving them too quickly can cause the "floating" beam to move because it is designed to move with a change in motion. Thus the assembler can unknowingly "break" the part just by handling. Worse yet, the only evidence for stiction immobilization is electrical testing. While designers try to avoid MEMS devices with moving parts in contact, this is not always possible. Even motion designs with torsion beams often have stops that can stick on repeated contact.

A device that is permanently inactivated when the assembler moves it quickly is unfriendly indeed. Thus the accelerometer stiction problem is important to solve for assemblers. This will reduce handling restrictions and improve yield. There is still some debate as to what kind of atmosphere is best for reducing stiction because it can involve mechanical interlocking, atomic forces, and even chemical reactions such as hydrogen bonding. Many workers advocate the highest vacuum pos-

TABLE 9.5 Getter Types

Getter	Form	Status
Moisture (H_2O)	Film, paste	Commercial
Particle	Film, paste	Commercial
Hydrogen (H_2)	Film, paste, thin film	Commercial
Oxygen (O_2)	Film, paste, thin-film?	Feasible
Humidity range control	Film, paste, fluid?	Feasible

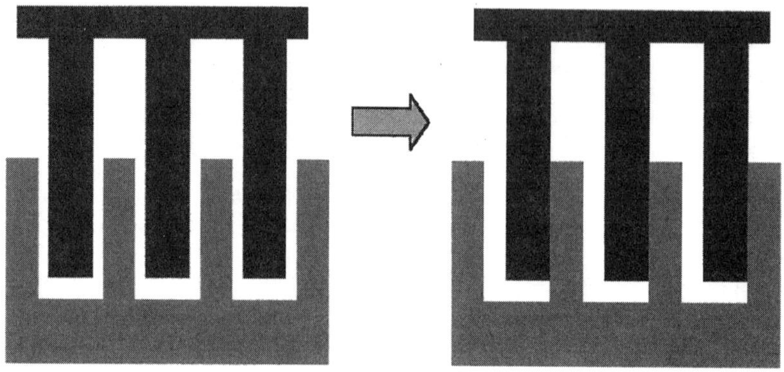

FIGURE 9.9 Stiction of an accelerometer.

sible, whereas others suggest that a small amount of moisture can reduce wear and stiction.[4] However, there are some surface chemistry complexities involved with silicon and its oxides that appear to make surface coating a logical solution. Organic coatings with low surface energy and perhaps hydrophobic properties should help.[5] The coating would need to be very thin, and this is the realm for vacuum coating technology.

9.4.5 Antifriction Coatings

Many have heard of parylene (poly-*para*-xylylene) as a special polymer film applied to substrates with a vacuum coater. The parylenes are noted for high chemical inertness, strong barrier properties, and perfect conformity to the surface. The surface tension approaches that of Teflon. A fluorinated Teflon-like version (Nova HT) has been developed recently for semiconductors that can handle over 500°C. Nova HT could be ideal for antistiction. Hopefully, work in the near future will determine if this material is a good solution to stiction and wear to deliver assembler-friendly MEMS devices.

Specialty Coating Systems, a manufacturer of chemical vapor deposition (CVD) materials and equipment, suggested that a very thin organic parylene film could solve the stiction and wear problems. Parylene, a thermoplastic polymer, has been around for years, and many assemblers are familiar with this organic CVD coating. Parylene is a high-temperature polymer film applied to substrates in a vacuum chamber by means of gas-phase polymerization that provides unusual electrical and environmental performance. This class of polymer has been used on a variety of applications, especially those involving the protection of electronic devices and circuitry. CVD is used to form an insulating thermoplastic coating with a high degree of chemical inertness, absence of pinholes, and perfect conformity to the topography of the surface. Coefficients of friction range from 0.25 to 0.33, so the lubricity is close to that of Teflon.

A fluorinated parylene, Nova HT, has been developed recently for semiconductors. Nova HT shares the unique properties of the other parylenes but offers properties that should be ideal for antistiction. The film is deposited in a molecule-by-molecule polymer process, with none of the cure-related stress that can occur with liquid polymers. There is no liquid phase, no hydraulic forces, and the coating conforms to substrate features rather than pooling or bridging in the manner of conventional liquid coatings. Free molecular dispersion of the monomer results in the development of an overlying film on all exposed surfaces, with equal thickness on inside and outside corners, on flat surfaces, and in crevices. Parylene can effectively penetrate inside surfaces through small openings.

Nova HT has a crystalline melting point above 500°C, which is at least 250°C higher than the recommended continuous exposure level for the conventional parylenes. Nova HT can be used for applications that require exposure to lasers and high-intensity lamps such as MOEMS devices with its improved ultraviolet (UV) resistance. It is particularly resistant to yellowing and physical degradation under such conditions. This advanced coating has all the useful properties of traditional parylenes, including resistance to solvents, moisture, gases, and other contaminants; high dielectric strength in very thin layers; and favorable physical and electrical properties. Its superior properties are due in part to the integration of fluorine into the parylene lattice, which results in improved polymer stability and a low dielectric constant. Hopefully, work in the near future will determine if this material is a good solution to stiction and wear.

9.5 MEMS CONSTRUCTION AND ASSEMBLY

Let's assume that the myriad of fabrication and packaging problems will be solved. The next challenge is MEMS assembly.[2] What kind special requirements and restrictions will be faced in assembly? The accelerometer and motion detectors in general are probably the best place to learn about MEMS assembly.

9.5.1 Accelerometers: Air Bag Electronics

The accelerometer is one of the earliest and most important MEMS devices. The chip typically uses a cantilevered silicon arm or a comb that bends slightly when there is a change in motion, as shown earlier. Movement is detected and translated into an electrical response. Analog Devices, a leader in this field, uses a design where a movable beam and a stationary beam form a capacitor. Motion changes the spacing and thus the capacitance to translate deceleration into an electrical signal. The system must accurately sense deceleration and send the signal that deploys the airbag. The deployment signal must only trigger under crash conditions. "Almost right" is not good enough in a life-and-death situation, as has been shown in some unfortunate incidents. The packaged product can appear very ordinary because traditional methods can be used here. This is one of the few MEMS products that can be sealed completely because motion detection does not require an opening to the outside. Still, the packaging cannot interfere with mechanical movement and must have low stress. Package stress, if present, must be predictable so that allowances can be made. The critical factor for both the packaging and assembly is *stress*. Anything that increases package stress will change sensitivity.

The accelerometer (decelerometer) must sense change in motion but not in all directions. The sensor must detect the rate of change primarily in the forward direction of the vehicle. We do not want the air bag activated if the vehicle is rear-ended or hit from the side or bounced by a pothole. Some cars are adding side air bags, but their separate sensors detect sideways motion. Anything that interferes with sensing the direction or alters the ability to detect absolute deceleration will be a problem. Let's look at how circuit assembly can affect things. Figure 9.10 shows the accelerometer diagram from Analog Devices.

9.5.2 Ink Jet

The popular ink jet chip built into most ink cartridges represents a different level of packaging challenge. The ink jet uses a MEMS chip that rapidly propels droplets when an electrical impulse is received. The chip consists of microscopic jet nozzles that discharge droplets using piezoelectric or thermomechanical pumps inside the chip. The numerous micronozzles must be kept free of packaging, but the chip-to-substrate interconnect must be protected and made robust for handling by end users. This can be accomplished by *selective packaging*. Hewlett Packard (HP) uses tape automated bonding (TAB) for the package. This flex-based package incorporates cantilevered metal beam leads that are thermocompression bonded to the IC pads. Instead of the common outer lead bond (OLB) termination, however, HP incorporates a pressure-contact connection that mates with the printer when the user installs the cartridge. This unusual and nonhermetic MEMS package therefore serves additional purposes as a 3D circuit and connector.

The IC connection zone is selectively encapsulated using robotic needle dispensing to apply liquid prepolymer sealant. Today's needle dispensing equipment allows the selective encapsulation process to be accomplished accurately, efficiently, and automatically. The deposited encapsulant is hardened thermally or with UV energy. Figure 9.11 shows the integrated flex package–connector

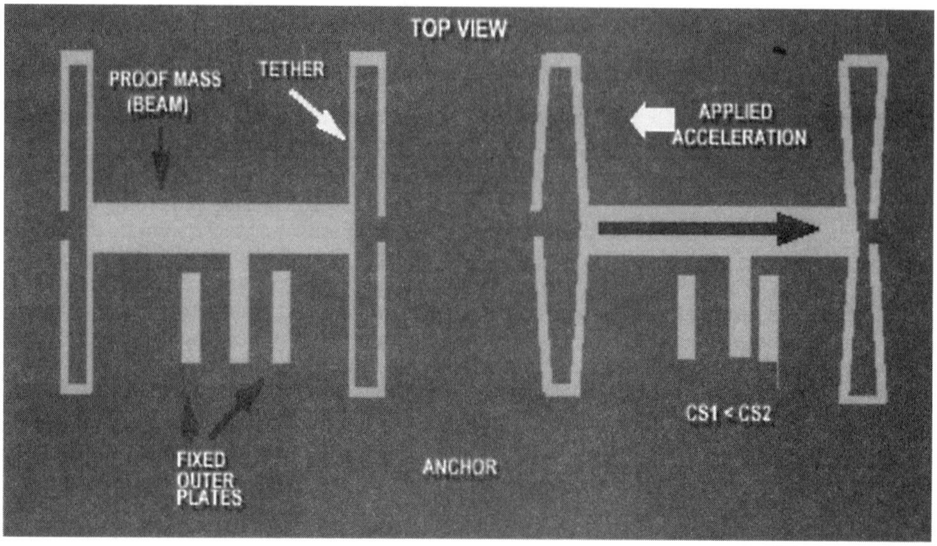

FIGURE 9.10 Accelerometer diagram. (*Analog Devices.*)

circuit, while Fig. 9.12 shows an automatic needle dispenser of the type used to selectively encapsulate a MEMS ink jet device.

9.5.3 Unique Board Assembly Issues

Analog Devices has used a Cerpack hermetic package successfully for a number of years, although considerable development was required to build a durable and reliable product. Even with a good package, assembly must be well controlled, with new rules applied. Package orientation must be correct. Skew will reduce sensitivity because the MEMS device will not point true. Hypothetically, if the package were to be skewed by 45 degrees, this would reduce the forward direction sensing by 50 percent. While it may be impossible to have this degree of skew, there is a potential for some. Component tilt is also undesirable because it also would change the effective direction of motion and reduce sensitivity. Even the solder fillet could be critical. A thicker fillet could reduce compliancy, and the sensor would receive a stronger signal because less deformation would transmit more force. Other types of MEMS products may not be as critical. On the other hand, optical systems can have more complexities for the assembler.

These are some of the more obvious assembly concerns, but there are many more that involve a change in sensing. We asked if the solder fillet was critical, and the answer is "Yes," but for mechanical not electrical reasons. Even lead compliancy plays a role in determining what the sensor "feels" in a high-impact crash. Stiffer leads will allow less deformation, and this results in a stronger signal. The assembly challenge appears to be complex.

Even the laminate characteristics must not change, but because of yet another effect. We mentioned that stress on the device, while permitted within limits, must be predictable. While Cerdip packages are popular for MEMS, there is a changeover to plastics that can increase stress and sensitivity. The manufacturer may calibrate the device during packaging as part of the test, but if the assembly process later causes unexpected changes, there will be problems. For example, board warpage and twist could be devastating. Even a change in the coefficient of thermal expansion (CTE) of the laminate would need to be taken into account. The more we delve into assembly, the more questions arise. We do not even know "what we don't know" at this point. The guiding principle comes from Amkor's Steve Anderson: "View the device, package, and assembly as an interactive system."

FIGURE 9.11 Ink jet cartridge.

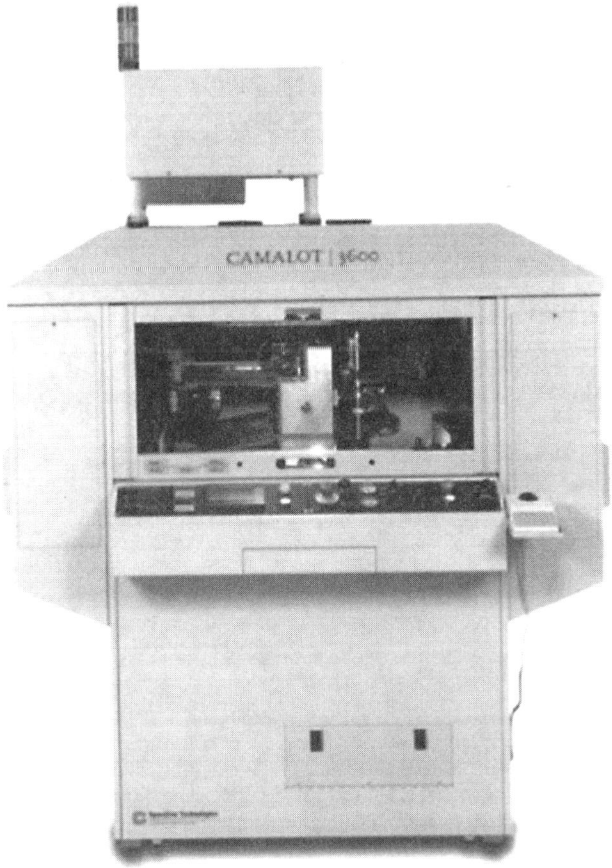

FIGURE 9.12 Needle dispensing machine. (*Speedline.*)

9.5.4 Recent MEMS Products

Camcorders with motion stabilizers use MEMS motion sensors, and Amkor is already packaging them. Again, the packaging challenge is minor for such products that can be sealed, but package-induced stress must be kept at a minimum or at least factored in. The strategy is to cap the package with silicon followed by transfer molding. Amkor points out that this area of packaging is very application-specific. Motion detectors are also being used in games.

IBM's new flash card–sized MicroDrive uses MEMS to create the world's smallest and highest-density removable hard drive (1 GB). MEMS brings the high level of precision required. The card can be used in the PC slot or in cameras, etc. Note that this is a real hard drive, not memory chips. MicroDrive is shown in Fig. 9.13.

9.6 OPTICAL MEMS: MOEMS

Optical MEMS products add one more level of complexity. The chip should be hermetically sealed, but a light path is another obvious requirement. The solution is somewhat apparent, but implementation can be a Herculean task. A light-transmissive lid or "porthole" is designed into the package. Several materials can be used. The micromirror module from Texas Instruments is one of the best examples of the packaging of complex electrophotonic products. Figure 9.14 shows the Digital Micromirror Device (DMD).[1]

The Digital Micromirror Device is very likely the most sophisticated MEMS product that has yet been commercialized and a preview of what lies ahead. The MEMS chip incorporates light beam–directing mirrors that move independently and almost instantaneously during operation. A pixel is turned "on" by pointing a mirror at a projection lens, while turning "off" involves pointing the mirror away. Let's look more closely at the optical package. Figure 9.15 shows a section of the micromirrors with some pointed "on." Large arrays are being used for digital projectors right now, but there are other applications such as digital video and optical switches.

FIGURE 9.13 IBM MicroDrive.

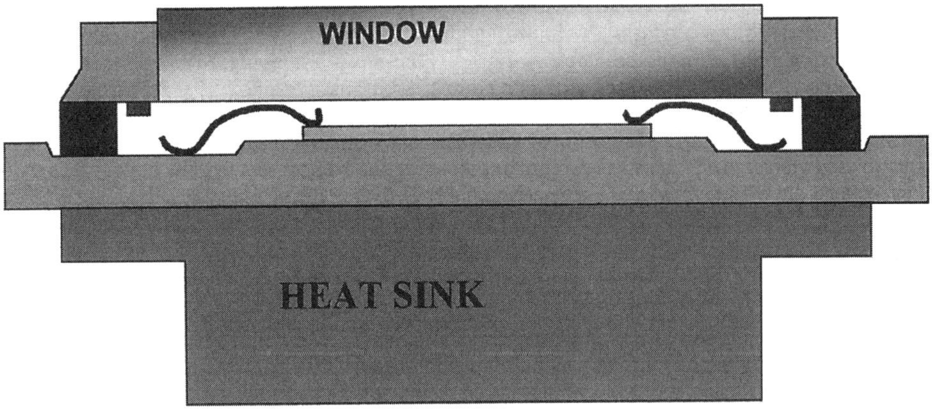

FIGURE 9.14 TI's DMD and package diagram.

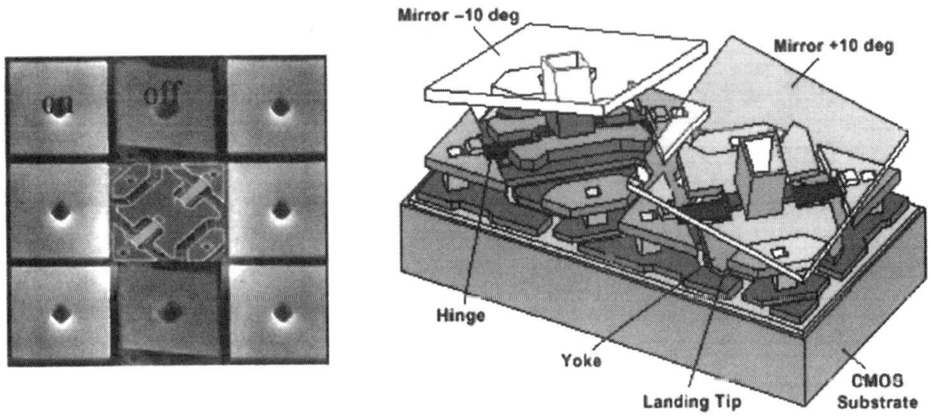

FIGURE 9.15 TI's micromirror array close up.

Most optical systems use a high-vacuum hermetically sealed package. Although many researchers recommend a high vacuum, others suggest a specific range for gases. Movable optical systems require that the package atmosphere not only start off clear but also that nothing within will later degrade lens clarity. One added problem is that materials can outgas to generate atmospheres that cause damage later, even though the initial package had a very high vacuum. Microscopic particles also can dislodge or even form during operation. How can we deal with gases and contaminants that are produced long after the package is sealed and the product is in the field? The best solution appears to be getters, which were discussed earlier.

The Internet giants, such as Cisco Systems, Nortel Networks, Lucent, Alactel, and others, have bet billions of dollars on MOEMS. Right now, the network routers and switches run optical-electronic-optical (OEO). This means that an Internet light signal must be converted to electronic format, switched, and then reconverted back to light. This adds cost and slows down the switch. The industry is intent on developing full optical switching, where photons are routed without the double conversion. One way is to use micromirrors to catch and direct the light wave. While this is easier said than done, it almost certainly will happen. While movable mirrors are the most obvious solution, there are other technologies within MOEMS that ultimately may win a piece of the light show action. So stay tuned to see who will catch the light wave. Lucent is already shipping MOEMS switches with two-axis "point anywhere" mirrors. The use of 3D mirrors greatly reduces the total number of mirrors required compared with simple on/off types. Figure 9.16 shows the two-axis optical switch mirrors.

9.7 WHY AREA ARRAY?

Will MEMS really move to area array packaging? After all, most of the packages are classic hermetic types that typically are not designed as area array or even surface mount. But it certainly is possible to design and build area array hermetic packages. Ceramic ball grid array (CBGA) and pin-grid array (PGA) are good examples of area array packages that can be hermetic and are suited for MEMS and MOEMS. Also, the micropackaging industry has already addressed the MEMS field. ShellCase has built ShellMEMS, as described earlier. Much work is also going on by packaging foundries and universities in the 0-level category, as discussed earlier.

Area array packaging, especially BGAs, bring many benefits to MEMS and could obviate several of the assembly concerns. Orientation accuracy, for example, should improve due to the highly symmetrical self-alignment. Joints would be much more controlled because BGA solder comes from the precise solder spheres. And MEMS can be expected to shrink in size while growing in complexity so that area array will be needed for the high I/O count. There is every reason to expect that area array packaging will become important for higher-level MEMS even if this does not hold for the simple sensors.

9.8 SUMMARY AND CONCLUSIONS

MEMS will be a hallmark technology for the twenty-first century. The capacity to sense, analyze, compute, and control, all within a single chip, will provide new and wonderful products during the coming decade. While package challenges are substantial, progress is accelerating. The need to control and regulate the package atmosphere will be critical. Stiction and wear problems may be solved

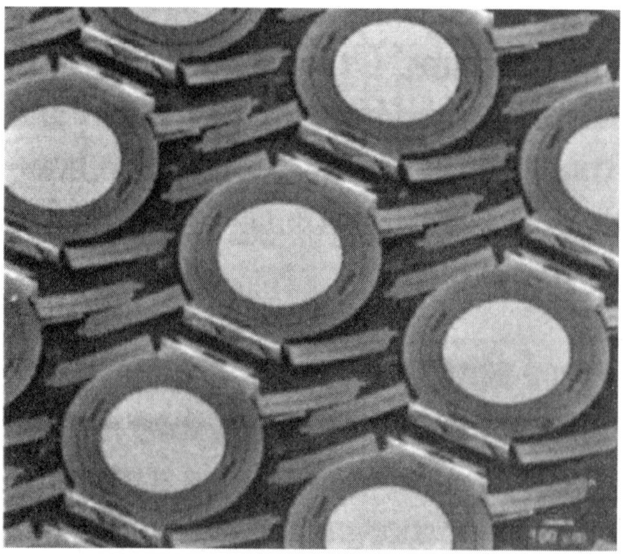

FIGURE 9.16 MOEMS switch for the Internet. (*Lucent.*)

in the future with new vacuum-applied polymers. Getters and emerging control agents appear to offer a practical and cost-effective solution today and for future generations of MEMS products. Assembly will have some different rules, but the rules will be less restrictive if package sensitivity issues can be solved. Expect to see MEMS as a high-volume component in the future, one that will usher in the new era—*beyond electronics!*

9.9 REFERENCES

1. Gilleo, K., "MEMS Packaging Solutions," *Electronic Packaging and Production,* June 2000, pp. 49, 50, 52, 53, 55, 56.
2. Quasdorf, J., et al., "BGA Goes Opto! Introducing optoBGA Packaging Technology," System Integration in Micro Electronics Exhibition and Conference, Nuremberg, Germany, June 27–29, 2000.
3. Gilleo, K., "MEMS PCB Assembly Challenge, *Circuits Assembly,* March 2000, pp. 62, 64, 66, 68, 70.
4. Miller, W. M., "MEMS Reliability and Testing," IMAPS Packaging of MEMS Microsystems Workshop, Chicago, October 23–24, 1999.
5. Martin, J., and Zhao, Y., "Micromachined Devices Packaged to Reduce Stiction," U.S. Patent No. 5,694,740, December 9, 1997.

CHAPTER 10
CERAMIC BALL AND COLUMN GRID ARRAY OVERVIEW

Marie S. Cole
IBM Microelectronics

10.0 INTRODUCTION

During the 1990s, ceramic ball grid array (CBGA) and ceramic column grid array (CCGA) packages moved from development laboratories[1] to volume factory production.[2] The steady growth in the use of these packages typically has been driven by either performance factors, process compatibility, or a combination of the two. These high-interconnection-density, surface-mount-compatible packages offer many advantages over traditional pinned or peripheral-leaded packages. CBGA and CCGA packages currently can be found in a wide range of applications spanning telecommunications and personal computers (PCs) to supercomputers. This chapter gives an overview of the package structures, range of offerings, established infrastructure, performance attributes, and package interconnection processes.

Increasing complexity of semiconductor devices has driven the requirements of the microelectronic packaging technologies beyond that of simple space transformers. Increases in the number of interconnections, interconnection density, operating speed, driver switching, and power dissipation have forced the development of more sophisticated package technologies. This trend is particularly evident in the case of logic and microprocessor functions (also memory, to some extent), where package electrical and thermal performance have become critical.

The ability and extendibility of multilayer ceramic (MLC) chip carriers to meet these electrical and thermal performance requirements and compatibility with area array flip chip dies have contributed to the rapid rise of CBGA and CCGA packages. CBGA packages are competitive for some performance-driven applications with as few as 250 inputs-outputs (I/Os) and the package of choice for flip chips (FCs) with about 350 I/Os and beyond. CCGA packages provide one of the few options available in volume production for I/O requirements above 1000 and on a 1.27-mm pitch or less. CBGA and CCGA packages are ideal for satisfying high-interconnection-density requirements and a high-lead-count capability while being compatible with standard printed wiring board (PWB) ground rules and materials, surface-mount technology (SMT) card assembly, and well-established infrastructures.

10.1 CERAMIC BALL GRID ARRAY (CBGA)

CBGA packages are composed of four basic elements: a chip carrier or substrate, attached die, die or package encapsulation, and a solder ball interconnection structure. Each of these elements is described below. Typically, FC dies are used in CBGA packages. The key feature of the CBGA interconnection structure is the use of a 90Pb/10Sn solder ball (MP > 300°C) that does not melt during package or card assembly when joined with eutectic 63Sn/37Pb solder (Figs. 10.1 and 10.2). One of the larger body sizes is the 32.5-mm CBGA package with 625 solder balls on 1.27-mm pitch (Fig. 10.3).

10.1.1 Chip Carrier

Menu of Packages. The CBGA chip carrier typically consists of a standard MLC alumina substrate. The ceramic material poses no restrictions for package or card assembly processing because it is compatible with high temperatures and process chemicals used during package fabrication and assembly. Typically, the body sizes of CBGA packages range from 18.5 to 32.5 mm, corresponding to ball counts ranging from 196 to 625 on a 1.27-mm pitch and from 292 (depopulated) to 937 on 1.0-mm pitch. Depending on die complexity and package performance requirements, chip carriers typically range between 4 and 20 ceramic layers, corresponding to a fired substrate thickness of 0.8 and 3.0 mm, respectively.

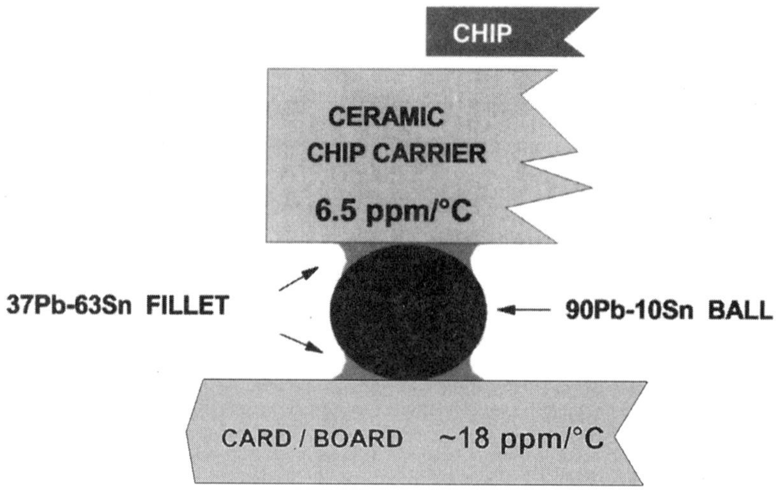

FIGURE 10.1 Schematic vertical cross section of a CBGA package joined to a PCB. (After T. Caulfield et al., Ref. [1], 1993.)

FIGURE 10.2 Metalloprahic vertical cross section of dual-metal solder joints attaching a CBGA package to PCB.

FIGURE 10.3 Photograph of a 32.5 mm CBGA package with 625 interconnections. (*Courtesy of IBM.*)

Ceramic Process. Individual ceramic layers or green sheets are punched with via patterns and screened with refractory metal–conducting paste to form lines and fill the vias. Depending on chip layout complexity, two or more layers are used to redistribute signals from chip connections through the package to the balls located on the bottom of the package. Power and ground planes are added as required to support voltage levels on the die and provide shielding between signal layers.

The personalization of individual ceramic layers is dictated by the die interconnection layout and package interconnection requirements. Typically, alumina chip carrier personalization consists of 0.08- to 0.10-mm-wide lines on a 0.15- to 0.20-mm pitch with less than 0.10-mm-diameter vias on a 0.23-mm pitch that provide the vertical communication path to attached dies and between ceramic layers.

The individual ceramic layers are laminated together and sintered at high temperatures to form a monolithic chip carrier. The last process step is overplating the die-attach pads and joining pads for solder ball attachment with nickel and gold. A generalized process flow for ceramic chip carrier fabrication starts with the casting of the raw material and ends with plating (Fig. 10.4). Personalization layers of the CBGA chip carrier typically can be categorized as signal redistribution or voltage planes (Fig. 10.5). A thorough description of the ceramic chip carrier and its processing can be found elsewhere in the literature.[3]

Materials Properties. Standard 92% alumina ceramic has a coefficient of thermal expansion (CTE) (6.5 ppm/°C) that is fairly well matched to silicon (3 ppm/°C). Solder interconnections between die and chip carrier, therefore, exhibit generally acceptable thermal cycle fatigue life, particularly with the use of die underfill materials that enhance fatigue life by an order of magnitude. The alumina ceramic CTE, however, is not so well matched to epoxy-glass PWB materials, such as FR4 (18 to 20 ppm/°C). This large CTE mismatch poses a reliability challenge for chip carrier to PWB interconnections. To achieve acceptable fatigue life, solder-joint structures joining ceramic chip carriers to PWBs must overcome significant strain. Fatigue life in area array solder joints is improved by optimizing geometry and materials factors such as joint height and alloy selection, respectively.

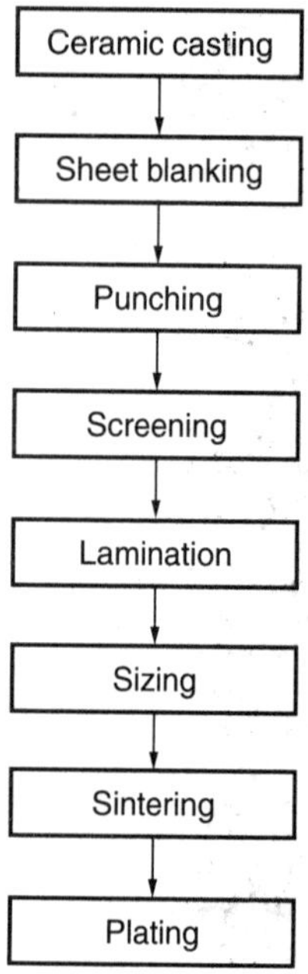

FIGURE 10.4 General process flow for ceramic chip-carrier fabrication.

CBGA and CCGA chip carriers are fabricated with either dark ceramic or white alumina ceramic. Newer ceramic materials also have been investigated recently, such as those with a higher CTE[4] or lower dielectric constant.[5]

10.1.2 Die Interconnection and Encapsulation

FC Dies. The FC proves to be the most advantageous die interconnection method for CBGA packages in terms of both performance and interconnection density. The MLC structure of chip carriers discussed earlier supports wiring-dense interconnection arrays by redistributing large numbers of signal connections and providing multiple power and ground connections. To achieve an FC attachment, a set of vias intercepts the surface MLC chip carriers and terminates at plated capture pads (termed *microsockets*) that match the array of solder bumps on an FC die. During placement, the die is faced downward, and the solder bumps are oriented to mate with corresponding chip carrier pads, thus the term *flip chip* (FC). Distortion and camber control of ceramic chip carriers across a die site are key factors in achieving high yields and reliable joints. During FC attachment reflow, 97Pb/3Sn die solder bumps are fully melted. Until recently, rosin-based flux and solvent cleaning had been required because of the high reflow temperature of high-lead solders. Advances in flux chemistries and joining processes, however, have allowed the implementation of no-clean fluxes for dies with a small number of solder bumps.

Die Underfill. CBGA FC applications are most often nonhermetic, with an epoxy encapsulant filling the gap between the die and ceramic chip carrier. The epoxy underfill completely surrounds all FC solder joints and strongly bonds to both the device and ceramic chip carrier. Underfill materials enhance thermal cycle solder-joint fatigue life by 10 times or more while providing environmental protection. In the case of capless or lidless configurations, a heat sink can be attached directly to the back side of the FC after card assembly with thermally conductive adhesive or by mechanical means.

Encapsulation Options. An encapsulation option for FC CBGA packages is to place an aluminum-alloy lid over the die, sealing it to the chip carrier with an adhesive, and filling the gap between die and lid with a thermal compound dispensed in the lid. The purpose of the thermal compound is to reduce internal thermal resistance. Adhesively joined lids will not pass a fine-leak test (i.e., are nonhermetic) but must pass a gross-leak test in order to preserve the thermal characteristics of the thermal compound. A heat sink may be attached after card assembly to provide additional cooling capability.

A variation on the nonhermetic lid structure is a lid attached directly to the backside of an FC, known as *direct lid attach* (DLA).[6] A flat lid is joined directly to the backside of an FC using a conductive adhesive material that serves to secure the lid to the die while providing a thermal path.

The choice of FC encapsulation options (Fig. 10.6) available for CBGA packages provides flexibility for a wide variety of applications. The lidless or DLA options allow CBGA packages to comply with J-STD-020 level 1 moisture specification requirements.[7,8] Nonhermetic lidded CBGA packages with FC dies typically meet the requirements of the Joint Electron Device Engineering Council (JEDEC) level 2.[9]

10.1.3 Package Interconnection Structure

The standard CBGA package interconnection structure is formed in two steps. High-melting-point solder balls are joined to chip carrier terminal pads with Sn/Pb eutectic solder. CBGA components are then placed on and reflowed to epoxy-glass PWBs via standard SMT.[10–12] The 90Pb/10Sn solder ball (MP > 300°C) does not reflow during package or card assembly (see Figs. 10.1 and 10.2). This

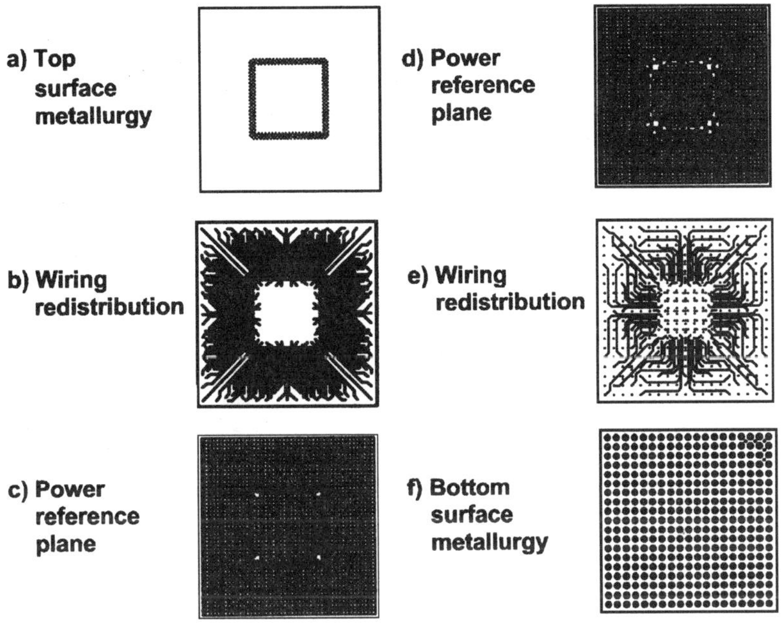

FIGURE 10.5 Various ceramic personalization layers for a CBGA chip carrier. g) Schematic vertical cross section depicting a laminated and sintered stack of ceramic layers into a monolithic chip carrier with an attached flip-chip die and package I/O solder balls.

hierarchical solder structure creates a consistent and reproducible gap or standoff between the ceramic package and the card. There is an inverse relationship between standoff height and strain generated during normal machine on/off cycles due to the thermal mismatch between the ceramic carrier and PWB. The thermal fatigue reliability of CBGA assemblies depends largely on the ability to accommodate and manage thermal-mechanical strains. Standoffs also provide clearance for cleaning after card assembly and help accommodate strains associated with card torquing and flexing during handling and box assembly. Use of meltable solder joints characteristic of plastic ball grid array (PBGA) packages that partially collapse during card assembly reflow would significantly decrease the fatigue life of CBGA packages joined to standard epoxy-glass PWBs. The increased joint height (i.e., standoff) of CBGA packages afforded by its high-melt structure aids in countering the effects of CTE mismatch between ceramic chip carriers (6.5 ppm/°C) and PWBs (18 to 21 ppm/°C) that would otherwise decrease fatigue life by two to three times.[13,14]

Typical CBGA packages use 0.89-mm-diameter balls on a 1.27-mm pitch. Increasing interconnection density by decreasing ball pitch requires smaller-diameter balls. For example, a 1.0-mm pitch requires the ball diameter to decrease to 0.8 mm. It is desirable to use the largest ball practical to obtain the highest standoff. Ball diameters must not be so large, however, that they create solder shorting concerns during package or card assembly. Metallized joining pads on chip carriers have a critical diameter. For example, 1.27-mm-pitch packages provide optimal manufacturability and reliability with approximately a 0.86-mm chip carrier pad diameter, whereas 1.0-mm-pitch packages require corresponding chip carrier pad diameters of 0.8 mm. The process for attaching solder balls to chip carriers is discussed in detail later in this chapter.

10.2 CERAMIC COLUMN GRID ARRAY (CCGA)

The CCGA package is a technology extension of the CBGA concept.[1,15,16] The solder ball array is replaced by an array of solder columns for improved thermal fatigue performance, required in the case of large ceramic body sizes. Solder columns (90Pb/10Sn) are joined with palladium-doped

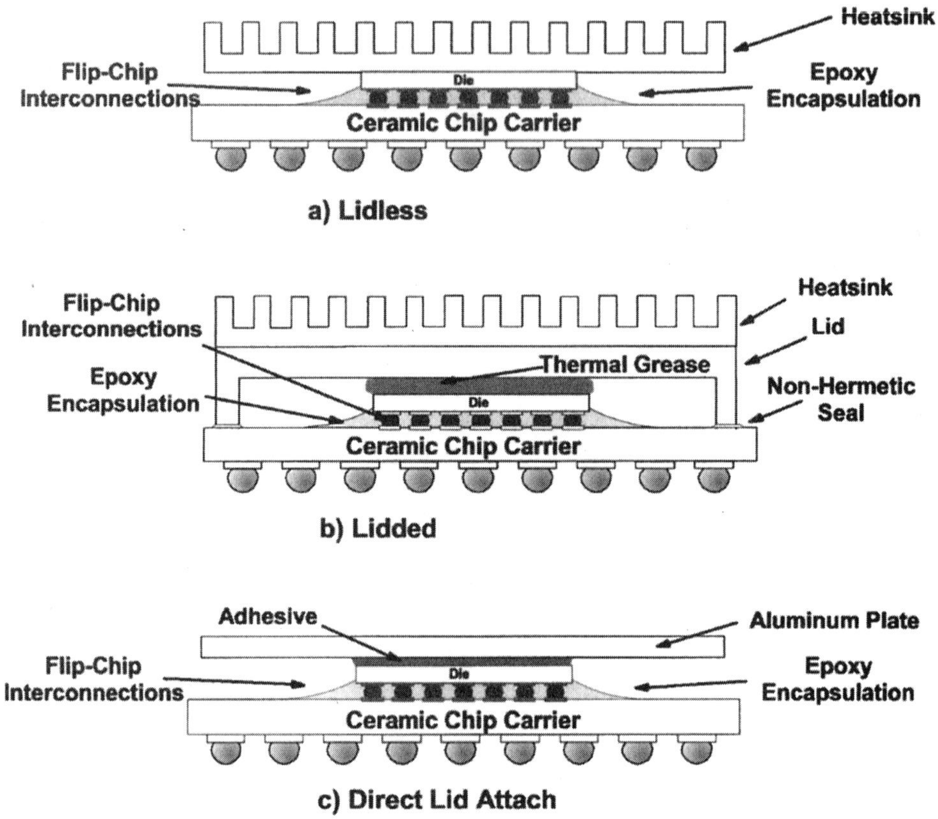

FIGURE 10.6 Schematic diagrams of various flip-chip CBGA packages. a) Lidless with a heat sink directly attached to the die after card assembly. b) Lidded with thermal compound to provide a thermal path between the die and lid-heatsink assembly. c) Lid directly attached to the die with an adhesive. (After J Zitz and F.L. Pompeo, Ref. [6], 1999.)

eutectic solder (Fig. 10.7). Similar to CBGA packages, the four basic elements of CCGA packages are the ceramic chip carrier, joined dies, encapsulation, and solder columns.

10.2.1 Chip Carrier

CCGA chip carriers are composed of the same multilayer alumina ceramic described earlier for CBGA packages. Because the primary purpose of a solder column is to extend solder-joint thermal fatigue life, CCGA body sizes typically start at 32.5 mm and currently extend to 45 mm, corresponding to 625 and 1224 I/Os a on 1.27-mm pitch, respectively. CBGA components in this range of body sizes would not be effective in dissipating strain at solder joints a large distance from the neutral point (DNP). The 42.5-mm CCGA package with 1088 I/Os (1.27-mm pitch) (Fig. 10.8), extendible to 1657 I/Os (1.0-mm pitch), illustrates the high I/O capacity of these packages. In addition, development is underway to further extend the CCGA package to a 52.5-mm body size providing more than 2500 I/Os. Since larger body sizes enable higher I/O counts, CCGA chip carriers are often more complex in terms of design (7 to 40 layers, 1.4 to 5.75 mm thick) than CBGA chip carriers.

10.2.2 Die Interconnection and Encapsulation

Similar to CBGA, the CCGA assembly processes of FC underfill and lid encapsulation are completed prior to column attachment. Encapsulation options for CCGA packages remain the same as for CBGA packages. Lids and/or heat sinks are more common with CCGA packages because the higher-function dies used with these packages dissipate more power.

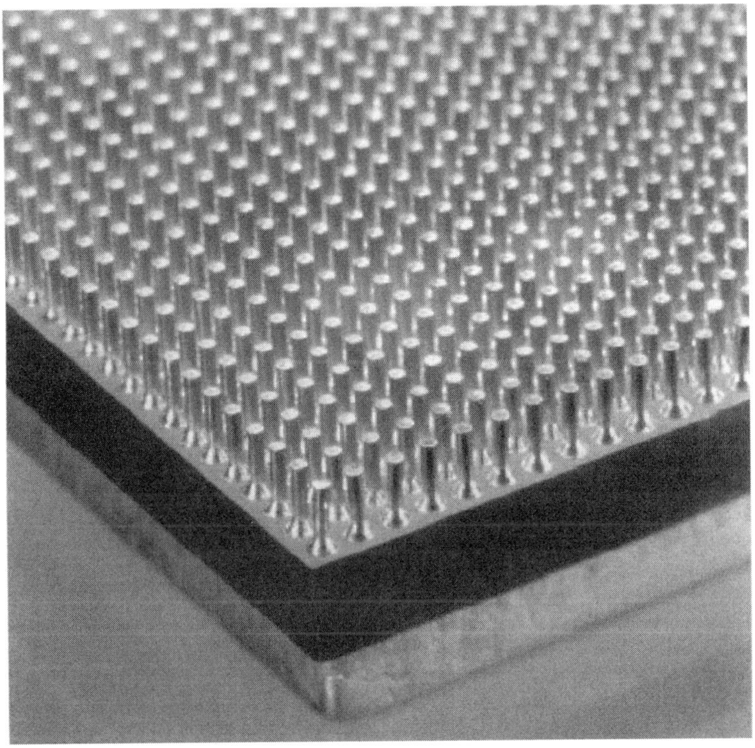

FIGURE 10.7 Legend to come.

10.2.3 Package Interconnection Structure

CCGA packages replace the high-melting-point solder ball of CBGA packages with a solder column of the same 90Pb/10Sn alloy. Solder columns, approximately 0.5 mm in diameter and either 2.2 or 1.27 mm tall, are joined to ceramic chip carriers with lead-doped Sn/Pb eutectic solder fillets in a manner similar to solder balls in the case of CBGA packages.

Because the column structure is less stable and more difficult to rework during card assembly than the ball structure, the lead is added to increase the robustness of the column interconnection in a process known as a *column last attach solder process* (CLASP).[16,17] The presence of lead in the eutectic solder creates a solder fillet with additional intermetallics that melt at a temperature higher than that of eutectic tin-lead (Fig. 10.9). These higher-melting intermetallics add structural support to the solder column joint during card assembly and rework processes.

After attachment, the columns are all uniformly shaved to the desired length on removal from the joining fixtures. With this exception, the CCGA process flow is identical to that of CBGA.

10.2.4 Effect of Pitch Reduction

For 1.0-mm-pitch packages, the only change in column structure and geometry developed for 1.27-mm-pitch CCGA packages is a reduction in the chip carrier pad diameter from 0.86 to 0.80 mm. Further reductions in column pitch, however, also would require an optimized geometry change to smaller-diameter columns because the 0.5-mm-diameter column used for 1.27- and 1.0-mm-pitch products would not easily be extendible to tighter pitches with chip carrier pad diameters smaller than 0.80 mm. It would be difficult to join columns to such small-diameter pads with consistency in a manufacturing environment because the columns are required to be captured within the perimeter of the pads.

FIGURE 10.8 Photograph of an IBM P2SC CCGA Package. (*Courtesy of IBM.*)

10.2.5 Thermal Cycle Fatigue Life

Tall, slender solder columns of CCGA packages are better able to accommodate strain arising from the CTE mismatch between a ceramic chip carrier and epoxy-glass PWB because the increased joint height increases the ease with which the solder joints can bend under shear conditions. Since joint strain is inversely proportional to joint height, the increased compliance of CCGA solder columns results in a significant enhancement in thermal cycle fatigue life over solder-ball joints. Thus, in addition to providing high I/O counts by virtue of the larger body sizes, solder-column packages also allow use in more severe application conditions where system temperature ranges or on/off cycles are more extreme.

10.3 PACKAGE STANDARDS AND OFFERINGS

10.3.1 CBGA JEDEC Standards

Two JEDEC outline standards, specified in the metric system, have been established for CBGA packages: MO-156 and MO-157, describing square and rectangular bodies, respectively. The features specified in the JEDEC outlines are package body size and tolerances, ball array size, total number of balls, ball diameter, coplanarity of the ball array, and true position (location tolerance) of the ball array. The overall package height is not specified because this dimension varies with chip carrier complexity and choice of encapsulation. The package size tolerance is ±0.2 mm, and the ball array coplanarity requirement is 0.15 mm. The location tolerance of the ball array is specified in terms of the geometric tolerance specification "true position."[18] The ball array true-position specification requires the center of each solder ball to be within a 0.30-mm-diameter circle (0.25 mm for 1.0 mm-pitch) whose center is defined by the design location relative to data defined by the package edges (Fig. 10.10). The ball array also has an offset tolerance, requiring the center of each ball to be

FIGURE 10.9 CLASP solder column joint. (*After S. Ray et al., Ref.* [17], 1999.)

within a 0.15-mm-diameter circle (0.10 mm for 1.0-mm pitch) whose center is defined by a best fit of the design ball array locations (see Fig. 10.10). The JEDEC specifications for ball coplanarity and true position have been evaluated to confirm that CBGA packages meeting these requirements can be assembled successfully.[19] Although not a JEDEC requirement, the A01 corner ball is often eliminated to easily identify the A01 corner.

The most common CBGA body sizes and their corresponding array sizes and I/O counts are given in Table 10.1. Body sizes range from 11 to 33 mm for square packages and 18.5 × 21 mm to 25 × 32.5 mm for rectangular packages. The available I/O range extends from 64 to 676 for CBGA packages with a 1.27-mm pitch. The JEDEC outline includes options for 1.5-, 1.27-, and 1.0-mm pitches, with the 1.27-mm pitch by far the most common. The I/O count for the largest CBGA package (33 mm) increases to 961 with a 31 × 31 array on a 1.0-mm pitch. Although the outline actually allows up to a 32 × 32 array for a 1.0-mm pitch on this 33-mm body size, processing and handling introduce concerns if balls are located too close to the edge of the ceramic carrier.

In addition, JEDEC standard MO-163, which defines the package outlines for PBGA memory components, is used for CBGA static random access memory (SRAM) packages. The SRAM package uses a 14-× 22-mm body size with either a 7 × 17 array (119 solder balls) or a 9 × 17 array (153 solder balls).

10.3.2 CCGA JEDEC Standards

Similarly, two JEDEC outline standards have been established for CCGA packages: MO-158 and MO-159, which describe square and rectangular bodies, respectively. The most common CCGA body sizes and their corresponding array sizes and I/O counts are given in Table 10.2. Body sizes range from 25 to 45 mm for square packages and from 25 × 32.5 to 32.5 × 42.5 mm for rectangular packages. The available I/O range extends from 361 to 1225 for a 1.27-mm pitch. While the 45-mm body size is included, with a few exceptions, the 42.5-mm square package is typically the largest CCGA in use. A 42.5-mm package (33 × 33 array) provides 1089 I/Os on 1.27-mm pitch, extendable to 1764 I/Os (42 × 42 array) if the pitch is reduced to 1.0 mm. To avoid processing and han-

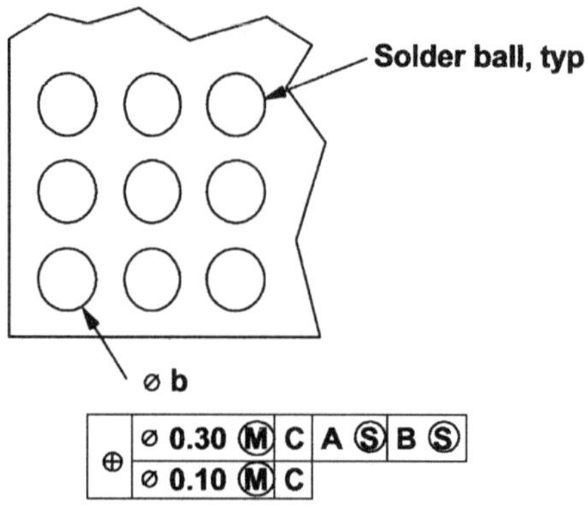

FIGURE 10.10 Ball true position as defined by JEDEC specification MO-156 and MO-157.

dling concerns, it is recommended that the 1.0-mm-pitch array on the 42.5-mm package be limited to 1657 I/Os (41 × 41 array with six columns removed in each corner).

10.4 TEST AND BURN-IN

Burn-in prior to ball or column attachment is recommended to prevent damage to nonmelting solder balls. This approach also minimizes value-added processes before burn-in. Many types of burn-in sockets are available to contact the package in this so-called land grid array (LGA) format. Final module test also can be performed prior to ball attachment if critical-speed sorts are not required. Otherwise, final module test can be performed after ball attachment using sockets with various contact schemes that minimize indenting or flattening of the solder balls. In some cases, burn-in may be performed after ball attachment, but it is critical that neither operation cause solder-ball or solder-column damage in violation of the JEDEC 0.15-mm coplanarity requirement or engineering specification.[20]

10.5 COMPARING CBGA AND CCGA

CBGA packages benefit from their low profile and the extremely robust nature of the ball array interconnection. For these reasons, CBGA packages tend to be chosen over CCGA packages. CBGA packages are limited, however, in terms of thermal fatigue reliability of the ball array interconnection. CBGA packages are reliable in body sizes up to and including 32.5 mm under typical desktop computer use conditions, considering combined effects of package temperature and number of system on/off cycles. CCGA packages are used beyond the 32.5-mm or smaller body sizes when CBGA packages fail to meet application requirements where either the use temperature range or the number of on/off cycles is excessive. CCGA packages may have either 1.27- or 2.2-mm-high columns to accommodate increased CTE mismatch–generated strains. Increases in interconnection height increase the thermal fatigue reliability of area array solder-joint interconnections.

TABLE 10.1 Array Size and I/O Count for Common CBGA Packages

	1.27-mm pitch		1.0-mm pitch*	
Body size (mm)	Array size	I/O count	Array size	I/O count
14.0 × 22.0	7 × 17	119	N/A	N/A
14.0 × 22.0	9 × 17	153	N/A	N/A
18.5 × 18.5	14 × 14	196	N/A	N/A
21.0 × 21.0	16 × 16	256	N/A	N/A
21.0 × 25.0	16 × 19	304	N/A	N/A
25.0 × 25.0	19 × 19	361	24 × 24	552
25.0 × 32.5	19 × 25	475	24 × 31	720
32.5 × 32.5	25 × 25	625	31 × 31	937

*Six balls are removed from the array in each corner.

While CCGA packages offer significant improvement in thermal fatigue reliability and are less sensitive to card assembly solder-volume requirements,[21] the columns create some difficulties. Handling concerns are increased because of column fragility. Columns also increase the package profile (height), an undesirable feature for products such as handheld electronics. For these reasons, CBGA remains the package of choice when reliability requirements permit.

10.6 COMPARING CBGA AND CCGA PACKAGES TO OTHER OPTIONS

10.6.1 Advantages

Interconnection Density: Full Array. An interconnection density advantage offered by CBGA or CCGA over most BGA packages is the ability to provide a full array of interconnections (i.e., balls, columns). Many BGA packages do not allow solder balls beneath the die area at the center of a package. Not only do CBGA and CCGA packages support the use of a full array, the multilayer structure allows a high degree of flexibility in assigning signal, power, or ground I/Os. These assignments can be made based on tradeoffs between electrical performance requirements and card wirability.

Thermal Performance. CBGA packages exhibit excellent thermal performance because of the inherently good thermal conductivity of ceramic chip carriers. The heat dissipation from BGA packages is divided among three paths: (1) through the chip carrier and solder balls into the PWB, (2) through the chip carrier into the cap and heatsink assembly, and (3) most directly through a path provided by a thermal compound between the die and heat sink into the airstream (Fig. 10.11*a*). The proportional heat flow among these paths depends on several factors, among them the heat-sinking capability of the card design (component density, number of metallized planes, and their proximity to the card surface). The heat sink design also affects the proportional heat flow. As the heat sink height increases, the percentage of heat flow into the card through the solder balls decreases (Fig. 10.11*b*).

In addition to the thermal paths through CBGA and CCGA chip carriers and their solder connections, FC packages can dissipate heat from the backside of the die through a thermally conductive compound to an aluminum lid or directly attached heat sink. In a typical desktop system, the CBGA package thermal resistance (Fig. 10.12) allows a 32.5-mm FC package to dissipate approximately 6 W without a heat sink, more than 12 W with a low-profile heat sink, and greater than 20 W with a large heat sink. Wire-bond CBGA packages can benefit thermally from a cavity-down configuration with or without a heat sink attached.

Electrical Performance. BGA packages, particularly in an FC format, provide short die-to-card signal paths, reducing electrical parasitics that affect signal time of flight and quality adversely. The importance of package electrical performance increases as the number of package interconnections increases.

Cofired molybdenum or tungsten personalization can provide strip-line signal conditions. Multiple power and ground planes provide impedance control for signal lines while reducing inductance along power paths.

TABLE 10.2 Array Size and I/O Count for Common CCGA Packages

	1.27-mm pitch		1.0-mm pitch*	
Body size (mm)	Array size	I/O count	Array size	I/O count
32.5 × 32.5	25 × 25	625	31 × 31	937
32.5 × 42.5	25 × 33	825	31 × 41	1247
42.5 × 42.5	33 × 33	1089	41 × 41	1657

Six columns are removed from the array in each corner

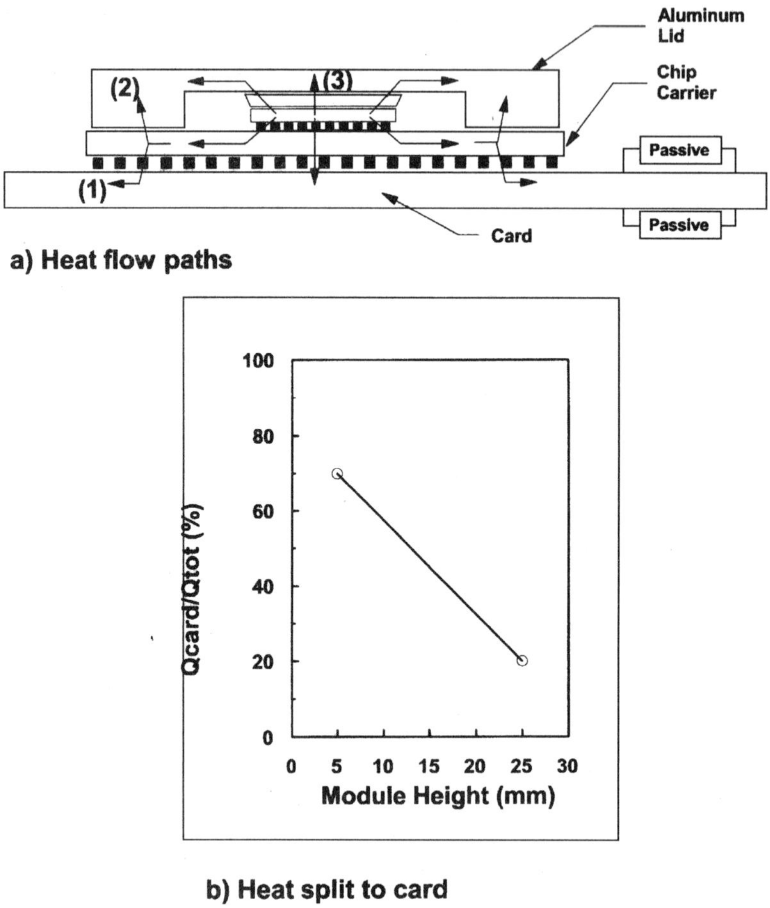

FIGURE 10.11 a) CBGA package heat flow paths. (b) Percentage of heat flow into card depending on heatsink height. (*Courtesy of J. Zitz, IBM Microelectronics.*)

CBGA packages are ideally suited to FC interconnections that reduce inductance by 1 to 2 nH over wire bonds. The multilayer structure also allows for wiring beneath the die, giving a full area array. These features provide CBGA packages with their high signal-to-power connection ratio, low signal-line noise, and high simultaneous switching capability. Modeling of CBGA package design ground rules indicates that clock frequencies into the gigahertz range can be supported. Ground rule improvements or new ceramic materials will allow frequencies higher than even 1 GHz to be achieved.

10.6.2 Disadvantages

The excellent I/O density and thermal and electrical performance attributes of CBGA and CCGA packages are all advantages in using these packages. Several disadvantages also must be considered. In the lower I/O or performance arenas, ceramic packages typically are not cost-competitive. At 256 I/Os or when voltage planes are required, the cost becomes competitive. While CBGA packages can be small (18 or 21 mm) and thin (chip carrier $<$ 1 mm), the higher-performance CBGA and CCGA packages tend to be larger (32.5 or 42.5 mm) and thicker (chip carrier 2.5 to 4.0 mm). These large packages pose some challenges for the card assembly process in component handling and require reflow profile optimization. Although a full array capability provides an electrical advantage, it poses a wiring challenge in designing the PWB. And while the CTE of alumina chip

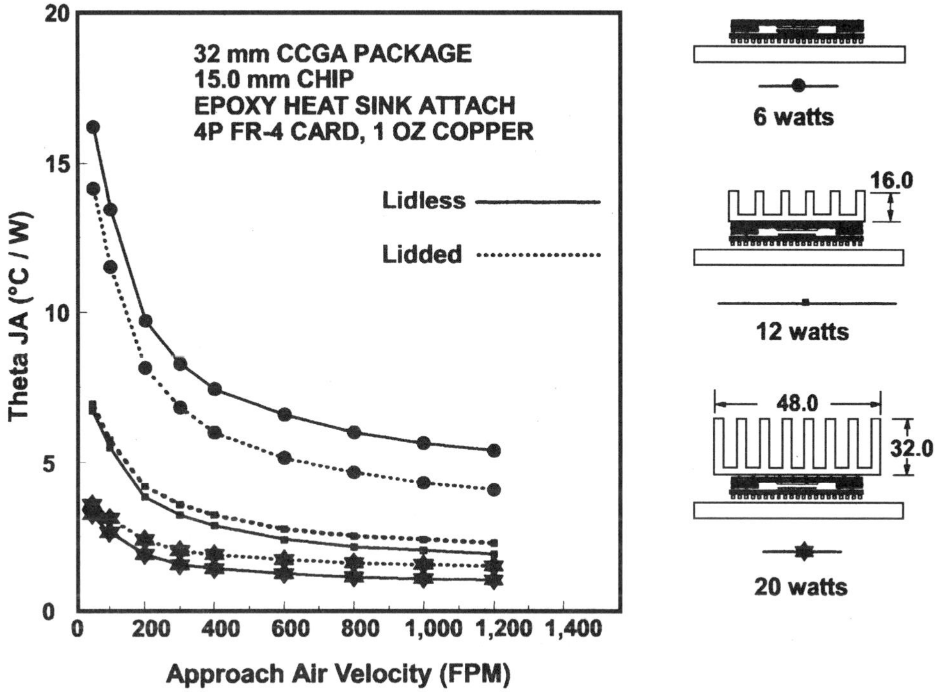

FIGURE 10.12 Thermal resistance for a 32.5 mm flip chip CBGA lidded and lidless packages for several heatsink and air flow conditions. (*After T. Caulfield et al., Ref.* [2], 1995. *Reprinted with permission of the McGraw-Hill Companies.*)

carriers supports highly reliable silicon die to package interconnections, it does limit the reliability of package to epoxy-glass PWB interconnections.

10.7 PWB REQUIREMENTS

As with other BGA packages, a dogbone card-attach configuration is required. There are key attributes and dimensions for designing dogbone card pads on 1.0- and 1.27-mm pitches for CBGA and CCGA packages (Fig. 10.13). Direct joining of BGA packages to card vias is not recommended. If solder runs into the via, depleting the card side fillet, it can result in a defective solder joint. There are techniques for maintaining a filled via to achieve direct joining, but none is currently practiced in production.

Design guidelines are found in several references.[22,23] The key elements of a dogbone card-pad design consist of a joining pad, plated through-hole (PTH) or via, wiring trace, and solder mask, as discussed below.

10.7.1 PWB Joining Pads

Joining pads are plated-copper surface pads to which BGA packages are attached by solder reflow. Solder paste is screened on these pads that are typically round but may be square or diamond-shaped.[24] The copper pads are formed as part of the personalization step during PWB fabrication with an etch angle causing the pad diameter at the bottom to be greater than the top by 0.025 mm (see Fig. 10.13, inset). While PWB fabricators tend to specify card-pad diameters at the pad bottom, PWB assemblers often refer to pad-top diameters. For reliable CBGA and CCGA solder joints on a 1.27-mm pitch, a minimum card-pad diameter of 0.68 mm (measured at the joining surface at the

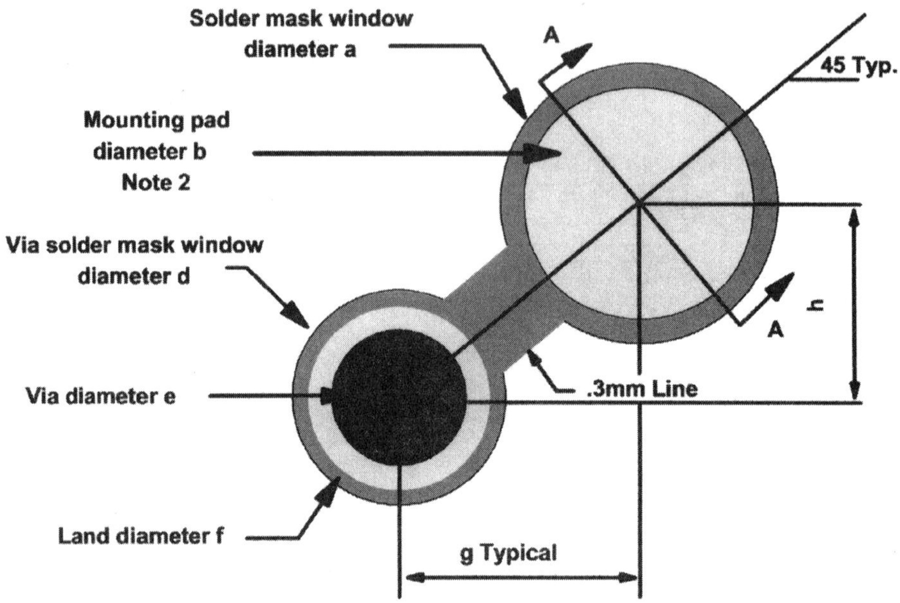

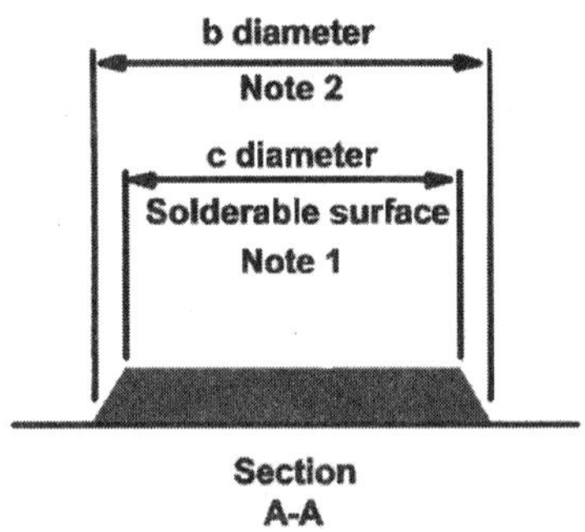

Feature	1.27mm Pitch	1.0mm Pitch
a	0.851 mm	0.80 mm
b	0.749 mm	0.69 mm
c	0.720 mm	0.67 mm
d	0.483 mm	0.38 mm
e	0.305 mm	0.20 mm
f	0.560 mm	0.46 mm
g	0.635 mm	0.50 mm
h	0.635 mm	0.50 mm

Note 1: Functional Surface
Note 2: Nominal Diameter at copper / FR4 interface with typica etch angle

FIGURE 10.13 Key attributes and dimensions of the card dogbone pad (non-solder mask defined, NSMD). (*After J. Stack, Ref.* [22], 1997.)

pad top) is required. The nominal card pad-top diameter is 0.72 mm, based on a typical tolerance range for card-pad diameters of ±0.04 mm. The maximum card-pad diameter is defined by the PWB fabricator's ground rule for copper-to-copper spacing. The closest allowable spacing typically is the distance between a joining pad and PTH land on a neighboring dogbone. Card pads larger than 0.72 mm, when allowed by supplier ground rules and capabilities, are desirable to further increase the reliability of CBGA and CCGA card side solder joints.

10.7.2 Plated Through-Hole (PTH)

Communication among inner layers of a card is made through vertically oriented via structures, called *plated through-holes* (PTHs), each connected to a surface pad by a copper wiring trace.

Typical PTH dimensions, to accommodate BGA/CGA components with a 1.27-mm I/O pitch, consist of 0.36-mm drilled holes plated to a 0.30-mm finished diameter. A copper pad with a diameter of 0.56 mm encircles the PTH at the card top and bottom surfaces. To maintain the largest viable card-pad diameter on a 1.0-mm pitch (0.69 mm nominally, measured at the bottom of the pad), the PTH size must be reduced to a 0.254-mm drilled hole and plated to a 0.20-mm finished diameter. In the case of very thick cards (>3.56 mm) for 1.27-mm-pitch arrays or even average-thickness cards (2.54 mm) for 1.0-mm-pitch arrays, the aspect ratio of PTH depth to diameter becomes a concern for PTH thermal fatigue life.[25] The quality of PTHs becomes critical in their ability to survive thermal exposures encountered during manufacture and assembly.

10.7.3 Wiring Trace/Solder Mask

Wiring traces connecting the joining pads to mating PTHs are typically 0.3 mm wide. The solder mask, to prevent solder flowing along the trace between the copper joining pad and PTH, must cover a distance of at least 0.08 to 0.10 mm between them to be effective. Care must be taken during card assembly rework not to damage the solder mask or underlying copper trace.

A solder mask is also used around joining pads. A solder mask opening larger than the copper joining-pad diameter (referred to as *non-solder-mask-defined,* NSMD; see Fig. 10.13) is preferred to an opening that is smaller than the copper surface pad (referred to as *solder-mask-defined,* SMD). The solder-joint appearance varies slightly between NSMD and SMD card pads (Fig. 10.14) because of this difference in solder mask configuration. A solder mask also may be used to "tent" or cover PTH openings to prevent fill during wave solder operations. When attempting to tent PTH vias, care must be taken to ensure that they are completely covered. Partial solder mask coverage can result in plugged vias if the solder mask material infiltrates the via, leading to reliability exposures of weak-

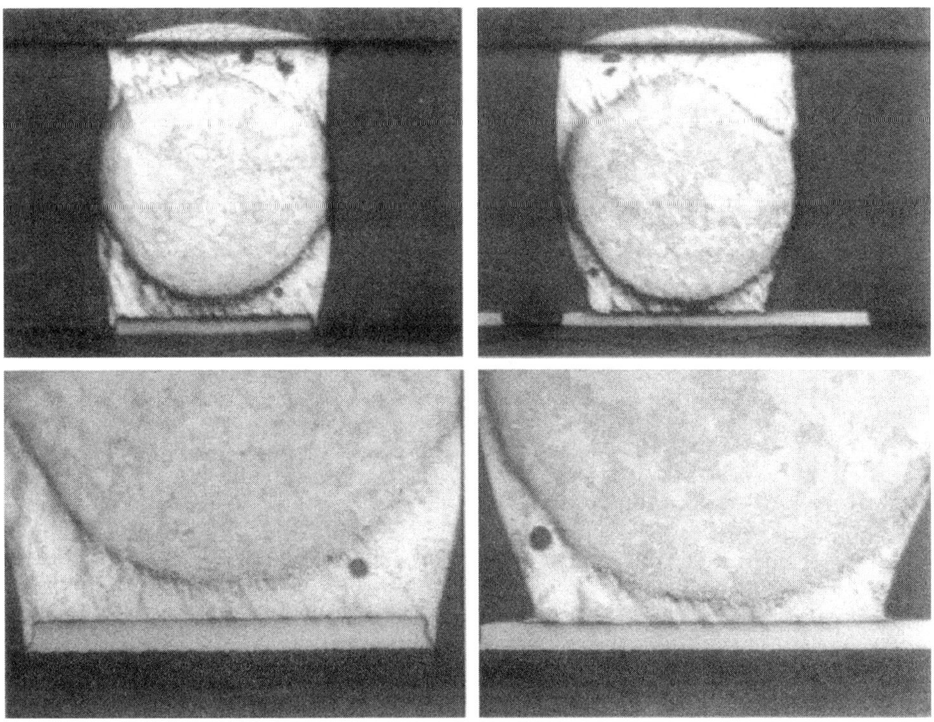

FIGURE 10.14 CBGA mounted on a non-solder mask defined (NSMD) pad (left) and a solder mask defined (SMD) pad (right). (*After R. Master et al., Ref.* [26], 1997. *Copyright 1997 IEEE.*)

ened barrels from trapped chemical residues. Likewise, solder mask openings around PTHs must be large enough to prevent unintended plugging when the vias are designed to be open.

10.7.4 Card Surface Finishes

CBGA and CCGA packages are used with a variety of card surface finishes but most commonly with an organic solder protective (OSP) finish such as Entek. Other finishes include hot-air solder leveled (HASL) or nickel-gold (NiAu) over-plate layers. Any surface finish that is solder wettable and yields relatively flat pads is acceptable for reflow attaching of CBGA or CCGA components. Surface finish quality is important because it affects the wettability and thus reliability of an assembly.[27] The HASL process adds thermal exposures during raw card processing and, as noted earlier, is cause for concern when the card thickness to PTH diameter ratio is very high (>10). For example, PTH diameters of 0.30 mm in 3.56-mm-thick PWBs or PTH diameters of 0.20 mm in 2.54-mm-thick PWBs are considered challenging.

10.7.5 Ground Rules

Lines and Spaces. Card ground rules for 1.27-mm-pitch CBGA and CCGA packages are typically 0.127-mm (5 mil) lines and 0.127- (5 mil) or 0.152-mm (6 mil) spaces, allowing two wiring lines per channel on internal PWB layers. For higher-lead-count packages, such as 625 and above, three wiring lines per channel are achieved by choosing 0.076-mm lines (3 mil) and 0.10-mm (4 mil) spaces, which may be desirable to reduce the card layer count. For 1.0-mm-pitch packages, 0.076-mm (3 mil) lines and 0.102-mm (4 mil) spaces are necessary to achieve the required wiring density.

Layup/Flatness. The full array area format of CBGA and CCGA packages allows signal count and placement to be a tradeoff between card wirability and electrical performance, so card cross sections can range from four layers with two signal and two power planes (2S2P) to over 20 layers, with 12S10P as one example. The actual card cross section for any application depends on several factors, including number of signals, location of signals, number of voltage levels, card ground rules, component-to-component wiring requirements, and wiring efficiency. To reduce the chances of card warpage, a symmetric cross section is recommended. Overly restrictive card flatness requirements are not necessary to join CBGA and CCGA packages successfully. The industry standard range of 0.7 to 1.0 percent warpage (7 to 10 mils of warpage per inch of card length) is adequate as an incoming requirement because this specification is not intended to apply locally across a BGA site. The flatness across individual CBGA or CCGA card sites typically ranges from 0.025 to 0.10 mm.

Material/Layout. Any standard epoxy-glass PWB material may be used, where the CTE typically ranges from 16 to 22 ppm/°C, and where card thickness normally ranges from 1.37 to 3.56 mm. Card cross section and thickness, in addition to material factors, affect card CTE and stiffness, thus influencing the reliability of CBGA or CCGA interconnections. The layout format of CBGA packages (single- or double-sided, packages back to back, component spacing, etc.) also can affect interconnection reliability.

Among the PWB layout considerations for ease of manufacture during card assembly are component distribution and body-to-body spacing of CBGA and CCGA packages. Avoiding regions with clusters of high-mass-packages or relatively unpopulated areas eases the degree of tuning necessary to achieve an acceptable thermal profile for reflow. A minimum clear zone of 5.0 mm between package bodies is required for rework tooling clearance. Additional spacing up to 10.0 mm between package bodies may be necessary for large clusters of high-thermal-mass CCGA packages (e.g., 42.5 mm) to prevent secondary reflow of neighboring solder joints when reworking the CCGA packages.

10.8 CARD ASSEMBLY PROCESS REQUIREMENTS

The card assembly process and equipment for CBGA and CCGA packages reflect standard SMT requirements. Key card assembly steps include solder-paste screening, placement, solder reflow, cleaning, and inspection. Rework processes for the removal and replacement of CBGA and CCGA packages also have been established. Given that incoming packages that meet the solder-ball or solder-column coplanarity requirement of 0.15 mm and cards that meet the flatness requirements described earlier, establishing a high-yield card assembly process is generally straightforward. Many captive and contract assembly manufacturers have demonstrated high first-pass yields attaching CBGA components at card assembly compared with quad flat pack (QFP) components (Fig. 10.15). Process-control monitors or inspections at several steps ensure low defect levels (i.e., high yields) and high reliability. In particular, solder-paste screening must be monitored to ensure proper volume and quality of print, key variables in the process. The reflow profile also must be monitored to ensure a proper temperature profile at both center and outer solder joints. Card assembly process details for CBGA and CCGA packages can be found in industry literature.[21,28–30]

10.9 PACKAGE INTERCONNECTION RELIABILITY

10.9.1 Geometric Factors

CBGA. The geometry of attachment pads at both the ceramic chip carrier and PWB is the result of optimization studies that include experimentation and finite-element modeling.[31] The solder-ball and fillet compositions[14] and dimensions are also the result of extensive study.

Standoff Height. The solder-ball diameter is driven by a need to achieve a maximum standoff height to accommodate strains that result from the CTE mismatch between a ceramic chip carrier and epoxy-glass PWB. A high standoff height corresponds to a large ball diameter that is limited in size to allow consistent package and card assembly processing. A 0.89-mm ball diameter has been determined to be optimal for 1.27-mm-pitch components, whereas a ball diameter of 0.8 mm is optimal for 1.0-mm-pitch components.

Pad Diameter/Solder Volume. The chip carrier pad diameter needs to be sufficiently large to capture several electrical connections, or vias, within a ceramic carrier. On the other hand, the card-

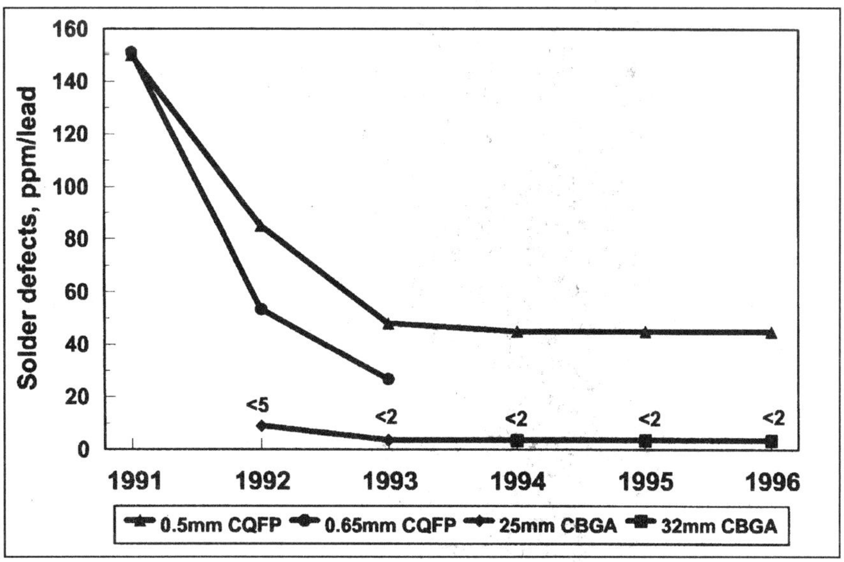

FIGURE 10.15 CBGA card assembly defect levels as compared to QFP. (*After T. Caulfield et al., Ref* [2], 1995. *Reprinted with permission of The Mcgraw-Hill Companies*).

pad diameter must be sufficiently small to meet PWB ground rules consistent with the PTH-dogbone structure and solder-mask requirements. Finite-element modeling (FEM)[31,32] and reliability testing[28] indicate a near-optimal structure with chip carrier and card pads at 0.86 and 0.72 mm, respectively, for a 1.27-mm pitch, with sufficient eutectic solder at both fillets. The recommended card-side solder-paste volume is 7000 cubic mils nominal, 4800 cubic mils minimum, for 1.27-mm pitch. The dependence of solder-joint reliability on both card-pad diameter and solder-paste volume is described by Phelan and Wang.[28] For card assembly of 1.0-mm-pitch CBGA packages, the recommended solder-paste volume is 2500 to 4600 cubic mils.[33]

CCGA. The card-pad diameter and solder volume again play a role in the interconnection reliability for both 1.27- and 1.0-mm-pitch interconnections,[21,34,35] with a recommended solder-paste volume of 3000 to 7600 and 2000 to 5000 cubic mils, respectively.

10.9.2 Metallurgical Factors

CBGA Joints. Metallurgical factors influence the reliability of CBGA solder joints. The choices of eutectic Sn/Pb solder fillets and 90Pb/10Sn solder balls are the result of metallurgical studies conducted to determine the optimal fatigue life within SMT processing constraints that require Sn/Pb-compatible alloys.[14] While silver and tin can be added to significantly increase the stiffness (i.e., creep resistance) of Pb/Sn eutectic solder fillets, it was determined that the fatigue life of dual-metal BGA solder joints is relatively insensitive to fillet stiffness. The choice of solder-ball alloy, however, has a very significant effect on joint fatigue life. The thermal fatigue life of joints with 97Pb/3Sn solder balls is about three times lower than that of joints with 90Pb/10Sn solder balls. The ductile, low-strength 97Pb/3Sn balls experience excessive deformation during thermal cycling, causing stress localization and distortion and resulting in early failure. CBGA joints consisting of 90Pb/10Sn balls and eutectic Sn/Pb fillets therefore provide an excellent balance: highly manufacturable, SMT compatible, and near-optimal fatigue life capability for alloys within the Pb/Sn system.

Single-Metal Solder Joints. Compared with an all-eutectic solder ball, a dual-metal system extends fatigue life by two to three times,[13,14] primarily due to the increased height of the noncollapsing dual-metal structure (37 versus 21 mils). The choice of single-metal alloy also influences the fatigue life.

CCGA Joints. Based on the aforementioned solder-ball alloy studies, 90Pb/10Sn solder is also used for solder columns. CCGA process studies optimized reliability,[21] including 1.0-mm-pitch components[34,35] and robustness of the solder-column interconnection to survive test and assembly.[36]

10.9.3 Additional Design Factors

More recent studies have investigated the influence of additional design factors, such as ceramic chip carrier thickness, that can have a significant effect on CBGA solder-joint reliability.[37] Based on modeling and experimental data, a 1.0-mm-thick chip carrier can have more than twice the fatigue life of a 3.0-mm-thick carrier, which is stiffer and less compliant. Factors such as the use of a lid, card thickness, and component spacing on the card, which also affect compliance, determine whether the full effect of chip carrier thickness is realized. Another aspect limiting thermal fatigue life is so-called green cycles arising from hardware or software power management adjustments to the operating temperatures and cycles, which was explored by Atwood et al.[38]

10.9.4 Mechanical Robustness

The effect of shock and vibration using MIL-810E–based conditions on both CBGA and CCGA solder-joint reliability has been studied to establish weight limits for adhesively attached heat sinks.[39] Weight limits for 32.5-mm, 625-I/O packages of 100 g for CBGA and 55 g for CCGA may be scaled based on the number of I/Os. Heat sinks attached with spring clips inserted into the PWB that apply a continuous compressive load to CBGA packages are common. In one study, a 44-g heat sink applying compressive loads of 1.8 to 5 kgf in conjunction with applied impact shocks of up to 20 and 75 g resulted in no degradation in fatigue life for a 25-mm CBGA package.[40]

10.10 SOLDER-BALL ATTACHMENT PROCESSES

10.10.1 Process Overview

The 90Pb/10Sn solder balls are joined to nickel- and gold-plated pads with 63Sn/37Pb eutectic solder. There are several process options both for eutectic solder application and for fixturing the solder balls. One approach to an automated ball attachment process flow shown in Fig. 10.16 is described in this section.[41]

Line automation requires integration of the various process sectors, in addition to prudent equipment and processing choices at each step. Belt conveyors link load stations, placement equipment, and reflow furnaces. Software controls the process flow at and between tools. High-speed vision-alignment systems can be used to place FC dies.[42]

In automating a CBGA ball attachment process to expand factory capacity, increase throughput, and decrease unit cost, the challenge is to maintain package quality. Key process and equipment choices must be made. The selection of either standard solder-paste printing or solder preform pretinning is one example.

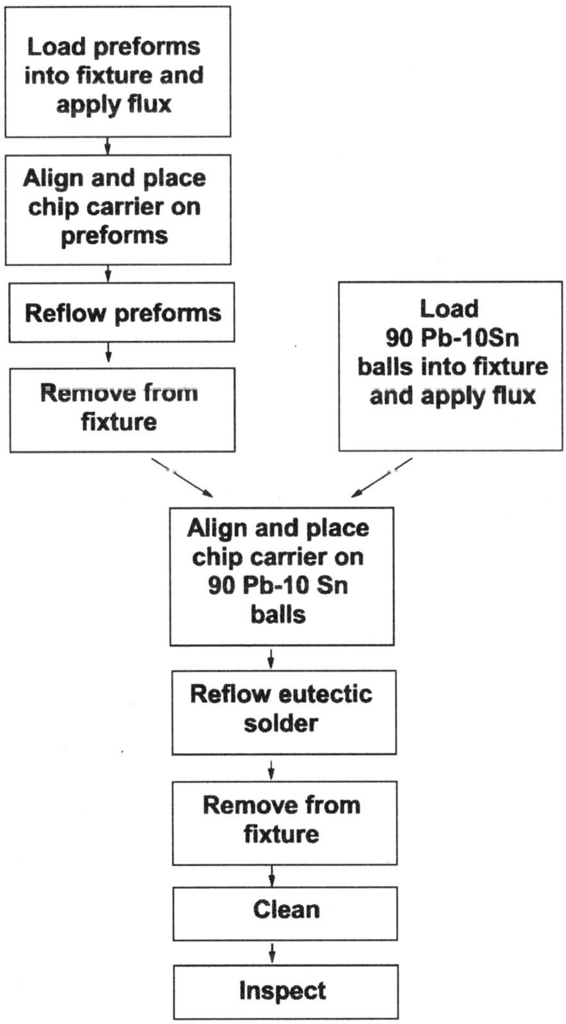

FIGURE 10.16 CBGA automated perform-process flow.

Screened Solder Paste. Concerns arise with paste screening, whether on ceramic chip carriers or on arrays of solder balls. If the choice is to screen on the ceramic chip carrier, dimensional concerns mentioned in the preceding section also apply here. To achieve alignment between chip carrier and balls, it is necessary to automate "flipping" the screened chip carriers. If the choice is to screen solder paste on a fixtured array of solder balls, then solder volume control continues to be a concern. To automate the process, screening parameters must be tightened and automated paste inspection implemented.

Solder Preform Process. A preform-based process changes the way Sn/Pb eutectic solder is supplied and alters the process flow of product through the ball attachment sector (see Fig. 10.16). Once optimized (i.e., adequately defining the sphere diameter), the preform process provides an extremely repeatable volume. Preforms are handled easily by automated equipment, allowing high production capacity.

Preform Loading. Preforms, which are small spheres of Sn/Pb eutectic solder, are loaded into universal (usable with a range of component body sizes) fixtures (Fig. 10.17), similar to the boats described previously. Large pallets are used with a preform fill pattern defined by form-factor stencils rather than dedicated fixtures. Following a controlled flux application, a standard SMT-type placement tool (with sufficient accuracy for fine dimensions) visually aligns the chip carrier to the array of preforms, placing fluxed chip carriers onto preform arrays. The assembly is placed in a reflow furnace to melt the eutectic solder preforms, pretinning the ceramic chip carrier's metallized pads.

Solder-Ball Loading. The 90Pb/10Sn solder balls are loaded into a universal pallet, fluxed, and await joining to pretinned chip carriers (see Fig. 10.17). Placement equipment, described earlier, is used to align pretinned chip carriers to a pallet loaded with solder balls. The flux must be sufficiently tacky to maintain the pretinned, slightly bump-shaped chip carrier pads centered on the solder balls.

Reflow Attachment. After aligning the solder balls to the mating chip carrier pads (prepared either with paste or preforms), the pallet assembly is placed in a reflow furnace as before, melting only the eutectic solder. Chip carriers are extracted from pallets and automatically loaded into cleaning fixtures. After cleaning, the CBGA packages are visually inspected, measured for dimensional specifications on a sampling basis, and then packed for shipment in either JEDEC trays or tape and reel format.

10.10.2 Solder-Ball Rework

Solder-ball arrays are reworkable if ball-related problems are discovered during fabrication or after removal from a PWB.[20] The ball rework process requires that all balls be removed through a squeegee reflow process (Fig. 10.18), leaving a thin solder layer on the chip carrier pads.[43] The operation is performed under a blanket of synthetic oil to prevent solder oxidation of the pads. After solder-ball and eutectic solder debris are removed from the chip carrier pads, a new array of solder balls is attached following the same process flow described earlier. Attempting to replace a single solder ball or use a chip carrier as removed from a card would risk using a ball array that does not meet coplanarity or true-position specifications.

10.11 SOLDER-COLUMN ATTACHMENT PROCESSES

CCGA interconnections are attached to ceramic chip carriers in a similar fashion to CBGA. The solder to form columns is supplied in the form of wire segments loaded into graphite boats rather than the solder ball.

10.11.1 CLASP Column Attachment Process for Automation

Automation of the CLASP column process is similar to that described for the CBGA process. The desire to maintain the CBGA process flow but increase the stability and robustness of the column to chip carrier solder joint during card assembly rework has led to the development of the column last attach solder process (CLASP).[16,17] The CLASP column attachment process replaces the standard eutectic tin-lead solder with a doped eutectic tin-lead solder. This doped eutectic solder creates a solder fillet with additional intermetallics that melt at a temperature higher than that of eutectic tin-lead

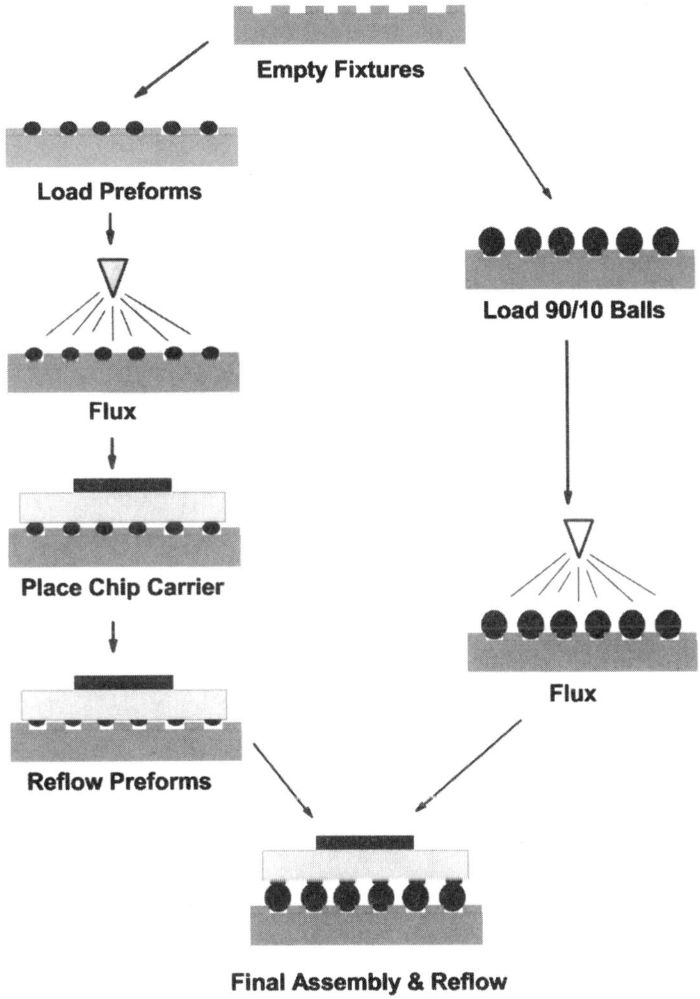

FIGURE 10.17 CBGA automated preform process schematic. (After J. Dankelman and L. Tousignant, Ref. [41], 1996.)

(see Fig. 10.9). These higher-melting intermetallics add structural support to the solder-column joint during card assembly and rework processes.

The CLASP process can be implemented in a highly automated manufacturing line. Solder-paste screening using an automated tool results in accurately located solder-paste deposits through vision alignment and high pad-to-pad solder volume uniformity, with 100 percent in-line paste inspection. The columns are also optimally centered on the chip carrier pads by the automated placement tool using vision and a best-fit algorithm. There is a significant reduction in handling of the in-process packages. An automated extraction and shave process with 100 percent automated inspection for column position after shaving also improves the quality of a finished solder-column array.

10.11.2 Column Rework

CLASP solder columns of CCGA packages can be reworked in a fashion similar to that described earlier for solder balls of CBGA packages.[20] The same rework process is used during package fabrication and after card removal.

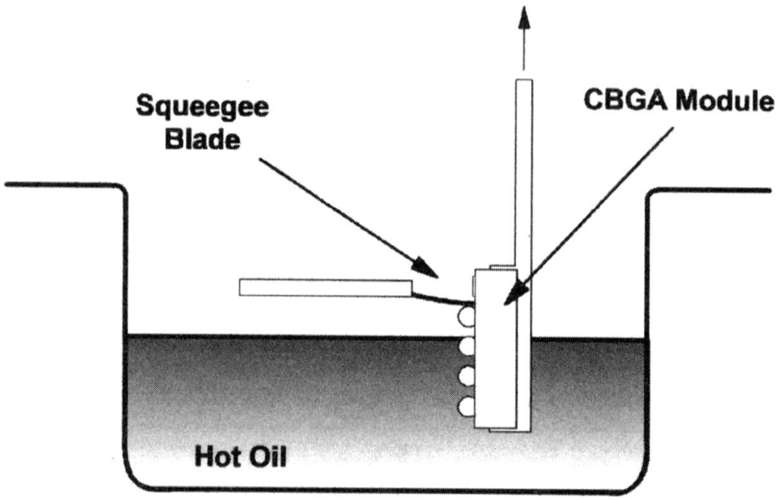

FIGURE 10.18 CBGA ball removal using a squeegee in a hot oil bath. (*After A. Downing et al., Ref* [43], 1997

10.12 SHIPPING CONTAINERS

10.12.1 CBGA Shipping Containers

Finished CBGA packages may be packed in molded JEDEC outline stacking trays compatible with automated card assembly placement equipment. The tray pockets are designed for a particular package size. Foam may be used to line the bottom of a tray, protecting the solder balls from damage. JEDEC trays have a diagonal cut at the A01 corner for ease of identification in loading tray stacks into card assembly placement equipment. CBGA packages are placed in JEDEC trays with their A01 corners in alignment with the tray A01 corner. Tape and reel packaging for high-volume applications is also an option for CBGA packages.

10.12.2 CCGA Shipping Containers

Protecting columns during package shipment and card processing is more challenging than for ball arrays. Care must be taken in loading and removing CCGA packages from trays. Ideally, a tray pocket suspends the columns above the pocket floor in an open space. If corner columns are removed or if an array size is reduced relative to the body size, the chip carrier can then rest on corner areas or edge ledges that are part of the tray design. Typically, I/O utilization required for the larger body sizes does not allow this luxury. For 1.27-mm grid packages, however, shipping tray pockets with a matrix of holes can be used if corner columns are necessary. An open pocket is required for 1.0-mm-pitch CCGA packages because mold-tolerance issues cause interference problems with a hole array. The best compromise allowing an effective tray design with the largest number of columns requires removing six columns in each corner. Corner areas rest on tray pocket ledges, which only modestly reduces the usable array size. Following these recommendations, 42.5-mm packages with 1.0-mm-pitch columns, limited to a 41 $\times$ 41 array with six columns removed in each corner, provide a total of 1657 columns.

10.13 FUTURE USE OF CBGA AND CCGA PACKAGES

Advances in CBGA and CCGA package technology are continuing to make these packages increasingly more attractive for a variety of applications. The trend toward increased package complexity will continue even at the low end. CBGA packages are as well suited to house commodities such as PC microprocessors and high-end products owing to the need for increased interconnection counts and density for products at all levels.

CCGA packages on 1.0-mm pitch and less will become more common to help fulfill future I/O requirements, although additional industry capability and cost parity for higher-density card technologies are necessary for their widespread use. High interconnection requirements for high-end servers and networking systems will be the key drivers.

An increased number of applications will require double-sided joining of CBGA and CCGA packages, including shared vias (at least for memory applications). These formats can result in lower solder-joint fatigue life and should have application-specific reliability testing and analysis.

10.14 SUMMARY

CBGA and CCGA packages are now standard in the microelectronics industry for cost-performance applications. The use of these packages in midrange servers, high-end workstations, and increasingly for network systems is driven by the need for high-I/O logic in high-I/O-density packages. CBGA and CCGA packages have demonstrated the ability to meet the electrical and thermal performance requirements of these applications while providing a highly manufacturable and reliable FC package. Growth in the use of these packages is expected to continue as new systems drive to even higher I/O and performance requirements. While other FC packages emerge in the marketplace, these ceramic packages will continue to meet the needs of advanced systems.

ACKNOWLEDGMENTS

This chapter would not have been written without the technical guidance and editorial encouragement of Karl Puttlitz. His support is especially appreciated. In addition, many IBM colleagues over the years have made significant contributions to the understanding and documentation of numerous topics relating to CBGA and CCGA packaging, including J. Acocella, D. Banks, R. Behun, J. Benenati, P. Brofman, A. Caron, T. Caulfield, J. Corbin, J. Currie, J. Dankelman, S. Dwyer, L. Goldmann, H. Harrington, L. Heck, C. Heim, K. Hoebener, T. Holmes, E. Ingalls, M. Interrante, P. Isaacs, R. Jackson, J. Jaspal, L. Jimarez, J. Jozwiak, E. Kastberg, S. Konecke, G. Martin, R. Master, C. Milkovich, G. Phelan, H. Quinones, J. Ross, S. Ray, C. Reynolds, W. Sablinski, J. Stack, K. Stalter, L. Tousignant, T. Wiggins, and W. Wildey. Special thanks to Tom Rednour for assistance with the figures.

10.15 REFERENCES

1. Caulfield, T., Benenati, J. A., and Acocella, J., "Surface Mount Array Interconnections for High I/O MCM-C to Card Assembly," in *Proceedings of the 1993 International Conference and Exhibition on Multichip Modules,* Denver, CO, April 1993, pp. 320–325.
2. Caulfield, T., Cole, M. S., Cappo, F., Zitz, J., and Benenati, J., "An Overview of Ceramic Ball and Column Grid Array Packaging," in J. H. Lau (ed.), *Ball Grid Array Technology,* McGraw-Hill, New York, 1995, Chap. 5.
3. Tummala, R. R., Garrou, P., Gupta, T., Kuramoto, N., Niwa, K., Shimada, Y., and Terasaa, M., "Ceramic Packaging," in R. R. Tummala, E. J. Rymaszewski, and A. G. Klopfenstein (eds.), *Microelectronic Packaging Handbook,* Chapman & Hall, New York, 1997, Chap. 9.
4. Yamaguchi, K., Higashi, M., Hamada, N., Yonekura, H., and Kunimatsu, Y., "Improvement of Solder Joint Reliability between Multilayer Ceramic Package and Printed Wiring Board by New Ceramic Material," in *Proceedings of the 47th Electronic Components and Technology Conference,* San Jose, CA, May 1997, pp. 1277–1282.
5. Fasano, B. V., Indyk, R., O'Connor, E., Plachy, A. L., and Reddy, S. N. S., "Glass Ceramic Substrates for Flip Chip Packages," *Future EMS International,* 1999, pp. 24–26.

6. Zitz, J., and Pompeo, F. L., "Direct Lid Attach Packaging for Ceramic Chip Carrier Applications," *IBM Microelectronics MicroNews,* 5 (2), pp. 24–26, 1999.
7. "J-STD-020, Moisture/Reflow Sensitivity Classification for Plastic Integrated Circuit Surface Mount Devices," EIA / JEDEC JC-14.1, Electronic Industries Association, Englewood, CO, October 1996.
8. "EIA / JEDEC Standard Preconditioning of Nonhermetic Surface Mount Devices Prior to Reliability Testing," JESD22-A113-B, Electronic Industries Association, Englewood, CO, March 1999.
9. Coffin, J., "Moisture Sensitivity of CBGA and CCGA Packages," in *Proceedings of International Electronics Packaging Society Conference,* Austin, TX, September 1996, pp. 94–101.
10. Behun, J. R., Cole, M. S., Hoebener, K. G., Call, A. J., Cappo, F. F., Milliken, J. C., and Klingel, B. T., "Improved Interconnection Structure and Test Method," U.S. Patent No. 5,060,844, 1991.
11. Behun, J. R., Cole, M. S., Hoebener, K. G., Call, A. J., Cappo, F. F., Milliken, J. C., and Klingel, B. T., "Interconnection Structure and Test Method," U.S. Patent No. 5147,084, 1992.
12. Cole, M. S., and Caulfield, T., "Ceramic Ball Grid Array Packaging," *Advancing Microelectronics,* ISHM, Jan.-Feb. 1995, pp. 18–19, 52–56.
13. Banks, D. R., Burnette, T. E., Gerke, R. D., Mammo, E., and Mattay, S., "Reliability Comparison of Two Metallurgies for Ceramic Ball Grid Array," in *Proceedings of the 1994 International Conference on Multichip Modules,* Denver, CO, April 1994, pp. 529–534.
14. Puttlitz, K. J., Caulfield, T., and Cole, M. S., "Effect of Material Properties on the Fatigue Life of Dual Solder (DS) Ceramic Ball Grid Array (CBGA) Solder Joints," in *Proceedings of the 45th Electronic Components and Technology Conference,* Las Vegas, NV, May 1995, pp. 1005–1010.
15. Banks, D. R., Heim, C. G., Lewis, R. H., Caron, A., and Cole, M. S., "Second-Level Assembly of Column Grid Array Packages," in *Proceedings of SMI,* San Jose, CA, August 1993, pp. 92–98.
16. Farooq, S., Interrante, M., Ray, S., and Sablinski, W., "Interconnect Structure and Process for Component Assembly and Rework," SMTA Surface Mount Technology Assoc., Docket No. 9-97-180, filed 1/98.
17. Ray, S. K., et al., "CLASP Ceramic Column Grid Array Technology for Flip Chip Carriers," in *Proceedings of the 2nd Annual Packaging Symposium at Semicon West,* San Jose, CA, July 1999, pp. F1–F7.
18. *Dimensioning and Tolerancing,* Y14.5M, ASME, New York, 1999.
19. Cole, M. S., and Hoebener, K. G., "Planarity and Centrality Requirements for Ceramic Ball Grid Array Packaging," in *Proceedings of Surface Mount International,* San Jose, CA, August 1995, pp. 273–278.
20. *CBGA and CCGA Module Limitation/Inspection Specification,* IBM Microelectronics, Hopewell Jct., NY, November 2000.
21. Phelan, G., Welch, M., Wang, S., and Cole, M., "Card Assembly and Reliability of 44-mm Ceramic Solder Column Array Modules," in *Proceedings of Nepcon West,* Anaheim, CA, February 1995, pp. 1048–1058.
22. Stack, J., *BGA/CGA Raw Card and Assembly Design Guide Covering CBGA, CCGA, TBGA, and PBGA,* IBM Document No. EN01-2711-05, March 1997.
23. Johnston, P., "Printed Circuit Board Design Guidelines for Ball Grid Array Packages," in *Proceedings of Surface Mount International,* San Jose, CA, August 1995, pp. 255–260.
24. Yee, S., and Ladhar, H., "The Influence of Pad Geometry on Ceramic Ball Grid Array Solder Joint Reliability," in *IEEE/CPMT International Manufacturing Technology Symposium,* Austin, TX, October 1996, pp. 267–273.
25. Knadle, K., and Ferrill, M. G., "Failure of Thick Board Plated Through Vias with Multiple Assembly Cycles: The Hidden BGA Reliability Threat," in *Proceedings of Surface Mount International,* San Jose, CA, September 1997, pp. 109–115.
26. Master, R. N., Dolbear, T. P., Cole, M. S., and Martin, G. B., "Ceramic Ball Grid Array for AMD K6 Microprocessor Application," in *Proceedings of the 48th Electronic Components and Technology Conference,* sponsored by SMTA Surface Mount Technology Assoc., Seattle, WA, May 1998, pp. 702–706.
27. Bradley, E., and Banerji, K., "Effect of PWB Finish on the Reliability and Wettability of Ball Grid Array Packages," *IEEE Transactions on Components, Packaging, and Manufacturing Technology, Part B,* 19 (2), May 1996, pp. 320–330.
28. Phelan, G., and Wang, S., "Solder Ball Connection Reliability Model and Critical Parameter Optimization," in *Proceedings of the 43rd Electronic Components and Technology Conference,* Orlando, FL, June 1993, pp. 858–862.
29. Heck, L., Lewis, R., and Phelan, G., "Surface Mount Assembly and Rework for Ball Grid Array Packages," in *Proceedings of New and Critical Technologies for Surface Mount,* Raleigh, NC, October 1994, pp. 11–14.
30. Heck, L., "Card Assembly Rework for Ceramic Column Grid Array Packages," in *Proceedings of Nepcon West,* Anaheim, CA, February 1995, pp. 1971–1975.
31. Corbin, J. S., "Finite Element Analysis for Solder Ball Connect (SBC) Structural Design Optimization," *IBM Journal of Research and Development,* 37 (5), September 1993, pp. 585–596.
32. Guo, Y., and Corbin, J. S., "Reliability of Ceramic Ball Grid Array Assembly," in J. H. Lau (ed.), *Ball Grid Array Technology,* McGraw-Hill, New York, 1995, Chap. 8.

33. Cole, M., Duchesne, R., Interrante, M., and Jimarez, L., "Design and Process Optimization for 1.0 mm Pitch CBGA," in *Proceedings of SMTA International,* San Jose, CA, September 1999.
34. Cole, M., and Berger, K., Brofman, P., "Design and Process Optimization for 1.0 mm Pitch CCGA," in *Proceedings of SMI,* San Jose, CA, August 1998, pp. 7–13.
35. Ingalls, E. M., Cole, M. S., Jozwiak, J., Milkovich, C., and Stack, J., "Improvement in Reliability with CCGA Column Density Increase to 1 mm Pitch," in *Proceedings of the 48th Electronic IEEE Components and Technology Conference,* Seattle, WA, May 1998, pp. 1298–1304.
36. Ingalls, E. M., and Jozwiak, J., "Robustness of Deformed Ceramic Column Grid Array Columns," in *Proceedings of IMAPS,* Philadelphia, PA, October 1997, pp. 623–627.
37. Martin, G. B., Cole, M. S., Brofman, P. J., and Goldmann, L. S., "The Effect of Substrate Thickness on CBGA Fatigue Life," in *Proceedings of Surface Mount International,* San Jose, CA, September 1997, pp. 172–177.
38. Atwood, E., Collins, E., and Quinones, H., "Area Array Interconnect Reliability Prediction Using Stochastically Defined Conditions," in *International Symposium on Microelectronics,* Minneapolis, MN, October 1996, pp. 538–543.
39. Cole, M. S., Kastberg, E. J., and Martin, G. B., "Shock and Vibration Limits for CBGA and CCGA," in *Proceedings of Surface Mount International,* San Jose, CA, September 1996, pp. 89–94.
40. Dolbear, T. P., Master, R. N., Cole, M. S., and Martin, G. B., "Effect of Mechanical Shock and Vibration on the Second-Level Temperature Cycling Reliability of Ceramic Ball Grid Arrays with a Continuous Compressive Load Applied," in *Proceedings of IMAPS,* Philadelphia, PA, October 1997, pp. 385–389.
41. Dankelman, J., and Tousignant, L., "Automation at 1st Level Assembly of Ceramic Ball Grid Array Packages," in *Proceedings of the 2nd International Assembly and Packaging Foundry Conference,* Sunnyvale, CA, 1996.
42. DiSciullo, J., and Glidden, W., "High Volume Production of Ceramic BGA Packages," *Electronic Packaging and Production,* 37 (12), September 1997, pp. 91–94.
43. Downing, A. J., Foster, D. C., and Puttlitz, K. J., "Apparatus and Method for Removing Meltable Material from a Substrate," U.S. Patent No. 5,620,132, 1997.

S • E • C • T • I • O • N • 2

MATERIALS

CHAPTER 11
POLYMER PACKAGING MATERIALS: ADHESIVES, ENCAPSULANTS, AND UNDERFILLS

Bill Bates,* Tim Chen,[1] Bruce Cotterman,[2] Dave Garrett,* K. Gilleo,[3]

Rita Mohanty,[4] John Perry, and Mike Preveti

**Cookson Semiconductor Packaging Materials; [1]Intel Corp., [2]Ambrusan Associates; [3]Cookson Electronics; [4]Arc-Less, Inc.*

11.0 INTRODUCTION

Organic polymers are a very old class of joining and assembly materials, but they are also new and modern. The paradox has to do with the source of materials. Natural polymers were discovered and came into use thousands of years ago. Many civilizations used the polymers from nature to craft works of art, make tools, and forge weapons of war. The earliest Native Americans used amber (prehistoric tree resin) to help strengthen the connection between stone spearheads and the wooden shaft. Polymer adhesives have been concocted since prehistoric times from fish, animals, tar, and agricultural materials. With a history going back at least 12,000 years, polymers became essential materials for civilization.

Modern polymers are synthesized in factories around the globe and serve just about every industry, and electronics is well served by this material class. The exciting world of electronic polymers, those magic macromolecules that make modern electronics possible, has moved into high gear as we move into the new millennium and the second century of electronics. More developments have occurred in the last 10 years than in the previous 10,000. New products, innovative electronics, and the general expansion of electronics and allied fields are creating the need for new materials and processes. The industry has responded, and the field of electronic polymers is at its highest level and is undergoing the fastest advancement ever seen. Area array packaging not only has improved electronic polymers but also has resulted in the introduction of new classes of products.

11.1 DESCRIPTION OF ELECTRONIC POLYMERS

Electronic polymers, as the name suggests, are those polymer-based materials which play a key role in electronics as insulators, attachment adhesives, encapsulants, protectants, and even electrical interconnects. They represent a wide range of basic ingredients used for circuit board substrates, solder masks, wire insulation, package molding compounds, die-attach adhesives, encapsulants, and conductive pastes for component assembly. The theme of consumer electronics has been smaller, faster, cheaper for many years. While smaller, faster may be possible without polymers, the third, all-important cheaper characteristic absolutely requires polymer-based materials because electronic infrastructures have become extremely dependent on this technology and its processing methods.

New materials are being offered in every area, and novel products are constantly being introduced, with announcements every month. Electronic polymers represent the most active and dynamic area for the plastics industry. New design challenges and performance requirements, as well as more insistent demands for protecting the environment, open up the greatest opportunities yet for electronic polymers. Polymers help miniaturize, increase performance, and drastically reduce cost when suitably employed. Perhaps the greatest attribute of electronic polymers is their ease of use and the resulting simplicity of manufacturing that accrues. Today, polymer-based electronic materials are reducing capital equipment, processing steps, errors, time to market, and total cost.

A continuing trend is the replacement of many ceramic materials with lower-cost polymers (plastics). The packaging industry has embraced this strategy with great success. The limitations of polymer-based products, real or imagined, are being swiftly resolved with design and material improvements and common sense, as was seen with ball grid arrays (BGAs). One important path is the conversion of ceramic circuits and carriers for direct chip attach to lower-cost organics. However, most polymers have much larger thermal expansion rates than silicon, and the industry must resolve the thermal mismatch issues to move forward with continued cost reductions. Fortunately, the magic elixir for solving the problem created by polymer-based substrate lies in the area of polymer chemistry itself. We will examine how polymers are now being used to literally and figuratively *bridge the gap* between silicon and polymer substrate in a very affordable way. Polymers have helped make area array packaging the great success that it is today.

11.2 POLYMER SCIENCE BASICS

Polymers are the most interesting and versatile substances in the world of material science and in the biological realm. They provide an incredibly wide range of physical and chemical properties. They can be as simple as polyolefin, made up of just carbon and hydrogen, or as complex as DNA, the blueprint and code of life. Engineering polymers can be viscous liquids, like silicones, or steel-like solids with high-temperature performance. Although there are natural polymers, such as rubber, most of the thousands of commercial products are synthesized. Our ability to design polymers with specific properties is what makes them so useful. Polymers are now the most critical ingredients for electronic circuitry and packaging.

Nearly all life forms rely on polymers to achieve the wide range of properties required to exist on earth. The human body incorporates an incredible range of biopolymers to achieve the myriad functions it performs. Skin, muscle, and even the neural network are made up of polymers that can be highly elastic or provide steel-like strength. It is no coincidence that the chemist, who mimics but builds on nature, has come to focus on polymers. Now let's briefly examine polymers at the molecular level to appreciate and understand their unusual but easily engineered properties.

Polymers are long-chain molecules typically formed by interlinked carbon atoms. These chainlike structures impart the strength, elasticity, and general resilience so common to many polymers. Virtually limitless ranges of properties are possible as molecular weight, structure, and chemical groups are varied. Polymers are everywhere, and life is impossible without them. The human body, for example, is made up of many different types of biopolymers that provide just the right properties for each specific function. Today, chemists have learned to create thousands of different polymers that are used in every industry. Blending different polymer ingredients together also can vary properties.

11.2.1 Chains and Links

Polymer macromolecules can be produced from a wide variety of both liquid and solid starting materials. It is the chainlike structure that yields strength and high-temperature stability. Polymer chain length and the level of links between chains determine the type of polymer and its major properties. Unlinked linear chains are found in the meltable and easily shaped thermoplastics that can be injection molded and thermoformed. The addition of three-dimensional interchain cross-links "sets" the shape and provides "structural memory." The cross-linked polymers represent the broad class called *thermosets* that are so widely used in electronics. FR4, an epoxy-polymer composite, is a well-known electronic thermoset, as are most encapsulants. The non-cross-linked class, called *thermoplastics,* can be remelted, making them ideal for reworkable materials. The characteristics of both these main divisions of polymers are covered in greater detail in Sec. 11.2.4.

11.2.2 Terms and Properties

Before moving on, terms that are commonly used to describe the intrinsic properties of polymers in all industries are now provided. Additives, especially fillers, modify some of these properties. Therefore, we must look at the finished or formulated product and use the inherent polymer properties as a guideline. Modulus, for example, increases with filler. The modulus value of a polymer will only increase as filler is added, and a low-modulus encapsulant requires a polymer with an inherently low modulus. Section 11.4.1 covers specific formulated property terminology and definitions related to electronic encapsulants.

Cross-Link Density. This is a measure of the relative number of interchain chemical bonds, or cross-links, in a polymer structure. The higher the cross-link density, the more rigid is the structure. Temperature performance usually increases with cross-link density, as does moisture resistance. On the down side, a highly cross-linked polymer can be brittle and inflexible. A very lightly cross-linked polymer, however, can be tough and elastomeric. Rubber is a thermoset with a low cross-link density that still provides the molecular memory to hold the structural shape. Zero cross-link density is found in the thermoplastic class that is often flexible and generally meltable. Polymers with various levels of cross-linking are used in our industry, and it is this ability to achieve a variety of properties that allows the continued expansion of electronic polymers.

Modulus. This term reflects the stiffness and is related to the chain length and the cross-link density as well as the types of groups within the polymer. A high-modulus polymer is very stiff and resists deflection. Strength is usually high. Molding compounds and encapsulants can have a high modulus of more than 1 million GPa ($N \cdot m^2$), whereas elastomeric connector materials can have a modulus as low as a few thousand gigapascals. As modulus is lowered, tensile strength also tends to drop.

Tensile Strength. As this term implies, tensile strength is a basic measure of intrinsic material strength and should not be confused with adhesive strength, a measure of the bond at an interface. High-strength materials, such as epoxies, have tensile strengths well in excess of 10,000 psi (pounds per square inch).

Glass Transition Temperature (T_g). This is the thermal point at which most polymers shift from their glassy, more rigid state to a rubber, less brittle phase. T_g is one of the more important parameters of polymers and often defines the temperature performance limits. This is so because a polymer typically undergoes a dimensional change above its T_g, and strength typically drops. There is nearly always a volume expansion and also an increased rate of expansion above the T_g that can be especially detrimental for encapsulants.

Coefficient of Thermal Expansion (CTE). The CTE is used to define the rate of a material's dimensional change per degree change in temperature. Nearly all polymers have a positive CTE, which means that they expand on heating. Several units are used, such as inch (Δ)/inch (length)/°F or mm (Δ)/m (length)/°C, but change in parts per million/°C has come into common use. CTE values of polymers range from over 500 ppm/°C for some ultraviolet (UV)-cured elastomers to near zero for liquid-crystal polymers (LCPs). However, epoxies come in at 60 to 100 ppm/°C, which is more than an order of magnitude higher than the CTE for silicon at 2 to 3 ppm/°C. This CTE disparity is the basis for the thermal-mismatch problems experienced in some assembly situations.

Ionic Content. This factor indicates the level of free ions, such as chloride (Cl^-), present in a material. High ion levels can accelerate corrosion of circuitry and chip metallization. Electronic-grade polymers, especially encapsulants and underfills that contact dies, are made with very low ion content, typically less than 5 ppm (parts per million, by weight).

Volume resistivity is the basic measure of resistance to electrical flow in a material and is expressed in ohm-centimeters. Insulators have a value of greater than 1×10^6 Ω·cm, but encapsulants can have values up to 10^{14} Ω·cm.

Thermal Conductivity. This value is important for die-attach adhesives, and values of at least 3 W/m·K are required. Encapsulants typically have low thermal conductivity, and some level of design ingenuity is needed to overcome the limitation.

11.2.3 Packaging Polymers

The major class of polymers for packaging is the encapsulant. *Encapsulants* are dielectric materials that are designed for maximum protection. Their purpose is to act as barriers for the semiconductor device and the interconnect structure. A perfect encapsulant would block everything from entering but let all the heat generated by devices escape. However, polymers are imperfect barriers that allow gases to enter and a relatively small amount of thermal energy to escape. Yet, by careful compounding and clever packaging designs, packaging polymers get the job done and are used widely. Another important product is underfill, used to increase the performance of flip chips (FCs). However, underfill may be classified as a subdivision of encapsulants.

11.2.4 Thermoplastics versus Thermosets

Polymers are readily classified into two major categories: thermoplastics and thermosets. Thermoplastics can be softened and melted by heating, but return to the original solid state when cooled. The process is repeatable and does not normally alter the polymer properties. Heating just adds the thermal energy (molecular motion) that allows the long polymer chains to move freely past one another and take on new shapes. Cooling reduces molecular motion to a level where chains no longer move past one another. Melting and hardening constitute an easy-to-control mechanical cycle. We can think of thermoplastics as "organic solders." Figure 11.1 shows the fundamental difference in structure between thermoplastics and thermosets.

One important feature of thermosets is that low-viscosity liquids can be the starting point. This means that polymer precursors can be poured, printed, needle-dispensed, and coated prior to polymerization. The end user makes the polymer by applying heat or "light" or by adding a catalyst. This feature provides incredible versatility and allows very cost-efficient manufacturing schemes to be used. Once cured (polymerized), the high-performance thermosets are nonmelting structures that can survive high-temperature environments. The degree of cross-linking is one of the factors that determine maximum temperature performance. Although epoxies have been used as printed wiring board (PWB) materials for decades, increased performance is being achieved by increasing the cross-link density to produce the so-called high-T_g products.

Thermosets do not offer the reversible phase change found in thermoplastics, but a meltable encapsulant is undesirable in most cases. Thermoset polymers form links, or chemical bonds, *between adjacent chains* during the polymerization. The result is a three-dimensional (3D) network that is much more rigid than the linear thermoplastic structure. The interlinked chains are not free to move when heat is applied, and the thermoset, as the name implies, is "set" into a permanent shape after polymerization. One valuable property of thermosets, however, is that they can start off as liquid prepolymers without using solvent. The initial liquid property has made them valuable for use in capillary flow underfills that must quickly fill the narrow gap between the assembled FC and substrate.

Although thermoplastic adhesives are used widely in all industries, they have only gained limited acceptance in electronics during this decade. Advanced thermoplastics now can provide the high

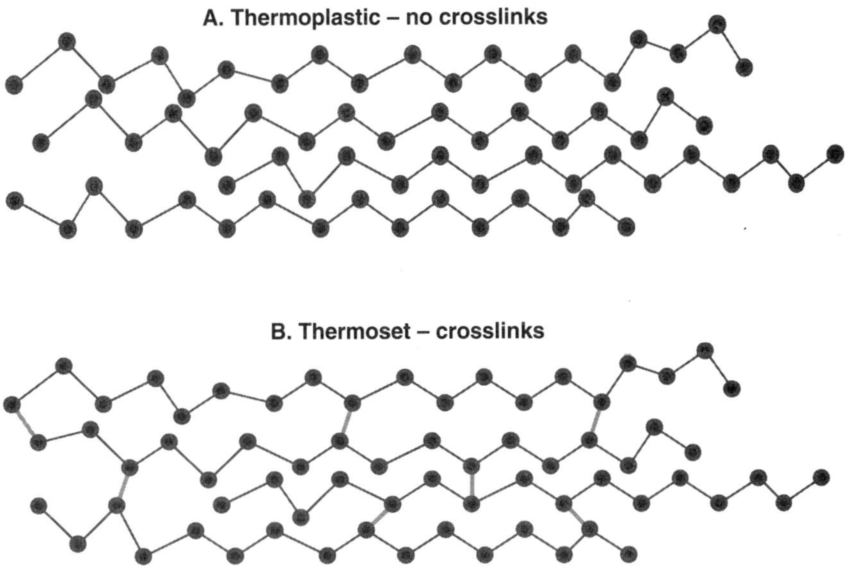

FIGURE 11.1 Thermoplastic versus thermoset polymers: (*a*) thermoplastics—no cross-links; (*b*) thermosets—cross-linked.

melting points and impressive dimensional stability that earlier polymers could not deliver. State-of-the-art thermoplastics now perform in the same temperature ranges as the thermosets. Low moisture absorption and CTEs are now possible, making thermoplastics useful for applications such as package overmolding, encapsulation, and die attach. The *remeltability* also sets thermoplastics apart and provides special advantages. Let's first see how thermoplastics have been adapted for die attach. A few thermoplastics have come into use as injection-molded encapsulants.

Polymers are by far the most popular class of adhesive throughout the world. Thermoplastic adhesives are especially valuable because they can be used in a dry form. They are already fully polymerized as received. The bonding process simply involves softening or melting the polymer while in contact with the adherents and then allowing the joined construction to cool. The structure can be disassembled easily or repositioned by reheating while applying force. Thermoplastics are the most convenient, safe, and reliable adhesives available to the electronics industry. Their use has been limited by the requirement of bonding while concurrently applying heat and pressure, however. Most of the bonders now in use cannot provide the heat requirement. Modification or purchase of a different type of bonder is necessary, and this creates an obstacle for many assemblers.

Thermally and electrically conductive thermoplastic adhesives can be used in dry-film or paste form, but with certain limits for the latter. The adhesive system can be filled with silver particles when electrical and thermal conductivity are needed. Non-electrically conductive but thermally transmissive fillers, such as aluminum nitride (AlN) and others, are also being used. Both the film form and pastes are *prepolymerized*. All the chemistry has been completed by the manufacturer, unlike thermosets, where the assembler must become a chemist of the moment.

Thermoplastics provide very fast processing, more accurately controlled properties, and virtually unlimited shelf life without refrigeration. The paste and film forms each have their own advantages and limitations. Pastes can be printed or deposited in virtually any pattern. This reduces inventory because "one container size fits all dies" and allows efficient mass-application processes to be used. However, the additional steps of deposition and drying are required, although small dies can be applied to "wet" paste. The dry-film form of adhesive is easy to use, but the film must be cut or punched into the desired shape by the supplier (preform) or by the user. However, the dry-film process is simple, clean, easy to control, and simple to automate.

11.2.5 Epoxies

Epoxies represent the most important class of electronic polymers and are the dominant materials for packaging. Nearly all encapsulants and underfills are based on epoxies. These products consist of resins, hardeners, and a catalyst/accelerator. A large number of solid and liquid epoxy resins have been offered commercially for many decades, giving epoxies a very long "experience curve." The resins are reacted with hardeners that become part of the polymer backbone. Hardeners are reactive monomers that combine with the epoxy resin molecules to form repeating polymer units; the hardener becomes an intimate part of the polymer chain. A stoichiometric (reactive equivalent) amount of hardener is needed. Many hardeners are available, ranging from very low viscosity liquids to moderately high melting solids. The reaction rate and the corresponding cure time are partly controlled by the choice of resins and hardeners. However, many resin-hardener systems react too slowly to be of practical value. This is where the catalyst or accelerator comes in. Strictly speaking, a catalyst is not consumed in the polymerization reaction. The most popular additives for boosting epoxy reactions are accelerators that act as catalysts but react with the monomers. The activators promote the chemical polymerization reaction but are consumed to become part of the polymer structure. However, a relatively small percentage of accelerator is required, so they have only a minor effect on final properties. Often, the difference between a 5-minute "snap" cure underfill and one requiring 30 minutes is the accelerator, which may make up much less than 1 percent of the prepolymer mix.

Epoxy systems continue to enjoy wide popularity because of balanced properties, a large commercial infrastructure, an extensive experience base, ease of modification, and an impressive safety record. More than 95 percent of encapsulants are based on epoxies, even though the technology is more than half a century old. Improvements continue to be made, and a more recent trend is to add copolymers, such as cyanate esters, that improve properties without giving up those hard-won attributes gained over the years. However, there are many other polymers that can be used as general electronic polymers.

11.2.6 Other High-Performance Polymers

While epoxies are the de facto standard for encapsulants, underfills, and other electronic polymers, silicones, polyimides, acrylates, polyxylylenes, and urethanes are also offered. Silicones and urethanes are the most popular after epoxies.

11.2.7 Fillers

Fillers are added to prepolymer systems to achieve properties that cannot be derived easily otherwise. Polymers tend to have higher thermomechanical expansion rates than inorganic materials such as silicon. While a few low-expansion polymers are known, a much easier path to low-expansion encapsulants is to add inorganic fillers such as silicon dioxide (SiO_2). Fillers are so important that they actually determine the basic product class. Conductive adhesives and encapsulants can be made from the same resin mix simply by using metal filler for conductivity or silica for the dielectric encapsulant. Table 11.1 lists properties that are derived from or altered by fillers.

11.3 PACKAGING ADHESIVE

11.3.1 Die-Attach Adhesive

Die-attach adhesives are polymer bonding agents that produce electromechanical junctions to create the mechanical bonds between the silicon or other semiconductor integrated circuit (IC) dies and the substrates. In the past, one die-attach material has been sufficient for a wide range of device types and package styles, but current trends toward larger dies, thinner packages, multichip modules (MCMs), and in-line processing have forced adhesive manufacturers to develop new polymer-based die-attach adhesives to meet the needs of the market. Both fundamental types of polymers are in use as the binder for die-attach adhesives—thermoplastic and thermoset. For a detailed description, refer to Chap. 23, Molding for Area Array Packages.

TABLE 11.1 Polymer Properties Derived from or Altered by Fillers

Rheology
Strength
Toughness
Modulus
Electrical conductivity
Thermal conductivity
Dielectric constant
Color
Cost
Density
Durability
Temperature capability
Many others

Thermoset Die-Attach Adhesive. Epoxies have been the dominant thermoset polymers used in die-attach adhesives for more that 40 years. During the 1950s, a number of silver-filled epoxies were developed for die attach and soon became the material of choice. One example of such an epoxy-based die-attach adhesive is Ablebond 8390A. Traditional thermoset die-attach adhesive comes in the form of paste and requires 1 to 2 hours to cure. During the cure process, the thermoset forms covalent chemical bonds between resulting adjacent large three-dimensionally cross-linked molecules that are polymerized in place. These materials are noted for good adhesion, low shrinkage, and their ability to retain shape. They are naturally rigid, inflexible polymers with a high modulus. Their high modulus appears to be creating serious problems in new packages, especially those with large dies.[1] Epoxies, although wonderful materials, do not appear to be the right choice for many new packages.

A relatively new form of thermoset die-attach adhesive is gaining popularity because of its low modulus and ability to absorb moisture. This is a silver-filled cyanate ester–based adhesive. Although thermosetting conductive adhesives have dominated the die-attach area, they are not acceptable for MCP substrates that may require removal of devices during repair and rework, especially for large devices such as application specific integrated circuits (ASICs) and microprocessors.

Thermoplastic Die-Attach Adhesive. Thermoplastic adhesives that do not cure but soften to a workable consistency at a certain temperature are high-molecular-weight polymers consisting of long linear chains without intermolecular linkages. Figure 11.1 shows the distinction between thermoplastics and thermosets.

In the solid state these long linear chains are held by hydrogen bonding and chain entanglement. When heated, the polymer chains are able to freely slide past one another with minimal applied force. When cooled, the chains are no longer able to move and freeze in place as a solid. Unlike thermoset, this melting-freezing process is completely reversible, since no chemical reaction takes place.

Thermally and electrically conductive thermoplastic adhesives can be used in dry-film or paste form to attach dies to any substrate. The adhesive system can be filled with silver particles when electrical and thermal conductivity are needed. Both the film form and pastes are prepolymerized—the manufacturer has done all the chemistry in a well-controlled environment. This equates to fast processing, more tightly controlled properties, and a long shelf life without refrigeration. One example of such a die-attach adhesive is the Staystik products from Cookson SPM.

While the performance of thermoplastic die-attach adhesives is generally comparable with that of the workhorse thermoset epoxies, one unique feature sets them apart. They can be tailored for specific solubility. Special solvent-removable thermoplastic die-attach adhesives are helping solve the known good die (KGD) challenge. A die can be bonded onto a temporary circuit and then tested and burned in. The adhesive bond is then "unglued" with safe solvent, such as alcohol.[2] For further information, refer to Chap. 23.

The electronics industry historically has relied on lead-based eutectic metallurgy for the electrical and mechanical attachment of silicon IC devices to ceramic substrates. In recent years, cheaper organic substrates have become more popular, which in turn demands an advanced polymer adhesive as a replacement for traditional eutectic metallurgy. A number of polymer-based adhesives are used in the packaging industry today. Packaging adhesives can be grouped into two broad categories: (1) mechanical attachment and (2) electrical interconnect. The first group of products is used to hold components in place on the wiring board while they are being soldered. The second category of packaging adhesive, electrical interconnect, is made up of polymer binder filled with metal particles (predominately silver) that provide the electrical pathways between component leads and circuit pads. Silver-filled epoxy, long used in the hybrid circuit industry for assembly, is finding increasing use as a solder alternative for PWBs. Low-temperature processing, no cleaning requirement, and no hazardous lead are some of the features that are moving these adhesives into mainstream surface-mount technology (SMT).

11.3.2 Conductive Adhesive for Interconnect

Electrically conductive adhesives have been used for over 20 years in the electronics industry. Conductive adhesives represent an intrinsically clean, simple, and logical solution for all kinds of electrical interconnect challenges. Adhesives not only provide a lead-free, "no clean" alternative to solder, these highly compatible materials offer viable answers to problems where solder is totally inadequate. Traditionally, their primary use has been for semiconductor die attachment and hybrid surface-mount applications. Although they have been evaluated in the past as a replacement for solder paste, only in recent years has the concept gained popularity and momentum.

A number of polymer-based conductive adhesives are employed in the electronics industry today. The most common materials are the die-attach adhesives, and these materials are called *isotropic conductors* because electrical conductivity is equal in all directions. Another important class of bonding agents with unidirectional conductivity is called *anisotropic*. These anisotropic bonding agents are experiencing significant growth because they are well suited for very fine pitch bonding and solve basic interconnect problems associated with the widely used flat-panel displays.[3] A detailed description of conductive adhesives is given in Chap. 16.

Isotropic Conductive Adhesives. Isotropic conductive adhesives produce approximately equal electrical conductivity in all directions. They are typified by the silver-filled epoxies originally used for die attach but now modified for component assembly. Epoxies have been the workhorse polymers of electronics because of their ease of use, the availability of hundreds of resin-hardener combinations, balanced properties, and generally superior bonding properties. Because die-attach adhesives usually have good electrical conductivity, they were the obvious starting point for component-assembly conductive adhesives. In recent years, a new generation of isotropic conductive adhesives has appeared on the market that has the potential of replacing some lead-based interconnects. One example of such adhesives is PolySolder from Poly-Flex Circuits. PolySolder is a silver-filled epoxy specially designed for flex substrate. Since all present types of commercial conductive adhesives require conductive particles as the electrical path, morphology and metallurgy are critical. The presence of oxide on a conductive particle increases the junction resistance because most metal oxides are good electrical insulators. One exception is silver. Silver oxide is quite conductive, which makes this metal the preferred conductor particle.

A new generation of PolySolder, designed for rigid substrate, is also available from Cookson SPM. This product has the same junction-stability mechanism as the product for flexible circuits. The unique feature of this product is its conductive particles that can break and penetrate metal oxides. A special shrinkage feature of the polymer binder system creates sufficient force so that conductive particles form connections to surfaces that behave like metallurgical junctions. Figure 11.2 shows the proposed mechanism for pseudometallurgical junction formation.

Anisotropic Conductive Adhesives. Anisotropic conductive adhesives represent the second major division of polymer bonding agents. The anisotropic class of adhesives provides unidirectional conductivity in the vertical, or *z*, axis. This directional conductivity is achieved by using a relatively low

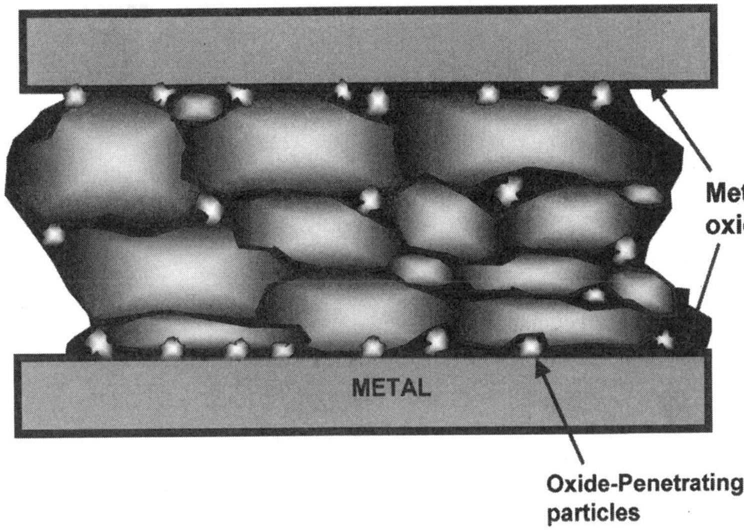

FIGURE 11.2 Proposed mechanism for stable junction formation.

volume loading of spherical conductive filler. The low-volume loading is insufficient for interparticle contact and prevents conductivity in the *xy* plane of the adhesive. The *z*-axis adhesive, in film or paste form, is interposed between the surfaces to be connected. Application of heat and pressure to this stackup causes conductive particles to be trapped between opposing conductor surfaces on the two adherents. Once electrical continuity is produced, the dielectric polymer binder is hardened by a thermally initiated chemical reaction (thermosets) or by cooling (thermoplastics). The hardened dielectric polymer holds the assembly together and helps maintain the pressure contact between conductors and particles.[4] Figure 11.3 shows a cross section.

11.4 ENCAPSULANTS

Encapsulants are nonconductive, easily shaped materials that are used to protect, increase performance of, and often provide a mechanical structure for electronic devices, packages, and subassemblies. These polymer-based electronic materials have been used for decades. Today's increased activity and the introduction of new materials are the result of strong growth in BGAs, micro-BGAs, MCPs, smart cards, and other bare-die products. The most popular encapsulants are molding compounds, but liquids, such as "glob tops" and the newer casting epoxies, also have become popular for BGAs and other area array packages.

11.4.1 Basic Properties of Electronic Encapsulants

The most important function of an encapsulant is to protect delicate electronic devices from the environment in which they are used. Moreover, the encapsulant must be able to be applied without damaging the device and must not itself adversely affect device electrical performance or long-term reliability. The ability of the encapsulant to perform these functions is largely dependent on the following basic properties.

Electrical Properties

Volume Resistivity, American Society for Testing and Materials (ASTM) D257. *Volume resistivity* is the electrical resistance in ohm-centimeters between opposite faces of a 1-cm cube of a given material:

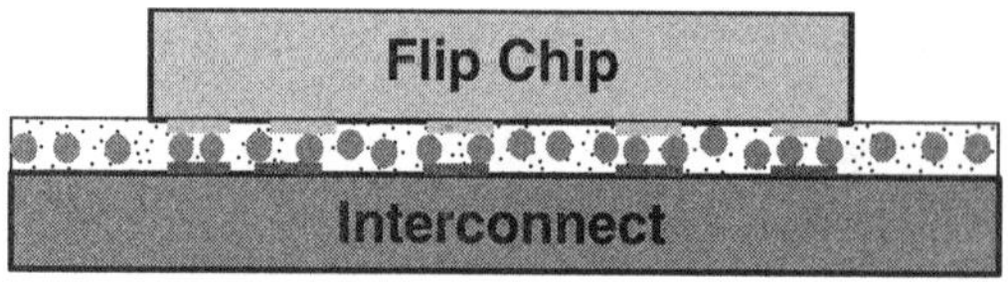

FIGURE 11.3 Anisotropic conductive adhesive.

$$V_r = \frac{A}{t\,(R_v)}$$

where V_r = volume resistivity, $\Omega \cdot$ cm
A = area
t = thickness
R_v = resistance, Ω

A high-volume resistivity is necessary to provide electrical insulation of microelectronic encapsulants. Volume resistivity decreases with increasing temperature, and for polymeric materials, this increase is more pronounced as the polymer goes through its T_g. It is therefore common practice to report volume resistivity as measured at a temperature near the T_g of the encapsulant as well as at room temperature.

Dielectric Constant and Dissipation Factor, ASTM D150. *Dielectric constant,* also called *permittivity,* is the ratio of the capacitance of a given material between two electrodes to the capacitance of an equal volume of air between those electrodes. The definition of *dissipation factor* is the ratio of the conductance of a capacitor with the material as the dielectric to its susceptance. Basically, dissipation factor is a measure of the electrical loss from the insulating material when voltage is applied. Dissipation factor is often reported as a percentage of the corresponding dielectric constant. The product of dielectric constant and dissipation is called the *power factor.* Very low dielectric constant and dissipation factor values generally are required for microelectronic encapsulants because higher values equate to higher capacitance and energy loss, which slows the transmission of electronic signals. Both dielectric constant and dissipation factor increase with temperature, ionic contamination, and water. Voiding or air pockets in an encapsulant will, on the other hand, result in a lower dielectric constant.

Dielectric Strength, ASTM D149. *Dielectric strength* in volts per mil is the highest applied voltage that a material can withstand before breakdown or burn-through occurs. Low dielectric strength in power devices can result in electrical breakdown or shorts between wire interconnects and leads. High dielectric strength is therefore very important for materials used to encapsulate devices operating at high voltages. Low dielectric strength can be the result of voids or conductive contaminants.

Arc Resistance, ASTM D495. *Arc resistance* is the time in seconds a material can withstand a high-voltage, low-current electric arc across two tungsten electrodes on the surface of that material. Failure occurs when the applied arc ceases to arc over the material and instead burns a path through the surface.

Physical Properties

Flexural Strength and Flexural Modulus, ASTM D790. *Flexural strength* is a three-point bending test run on molded test bars typically $1/8 \times 1/2 \times 5$ in. *Flexural modulus* is an indication of the elasticity of the material being tested. The higher the modulus, the less compliant is the material. The flexural strength test is often run at 215 to 260°C to simulate conditions in solder reflow. Flexural strength and modulus are important factors in controlling stress-related cracking and delamination. Many stress models have been published, with most including thermal expansion and flexural modulus as the main factors. A useful model for comparing stress of different encapsulants on a silicon die is

$$S = (E_{\alpha_1} - E_{si}) \times M$$

where S = stress index
E_{α_1} = $alpha_1$ expansion of the mold compound
E_{si} = expansion of the silicon chip
M = flexural modulus of the mold compound

Ideally, an encapsulant should have high flexural strength with low flexural modulus. High flexural strength is needed to withstand the mechanical stress of bending under pressure. Low modulus adds flexibility, allowing an encapsulant to yield without breaking.

High-Temperature Hardness, ASTM D2240. Hot hardness is a very important property of thermosetting molding compounds. This property, which is measured on molded parts immediately at the end of a molding cycle, gives a good indication of the degree of in-mold cure. Insufficient hot hardness or rigidity of molding compound runners, culls, and encapsulated devices as they are ejected from molds may result in sticking and/or physical damage to the parts being encapsulated. A Shore D durometer is used to quantify the hot hardness of cured molding compound. This handheld instrument measures the force at which a pointed indentor can just penetrate the surface of a material. Hot-hardness measurements are taken on a molded part, runner, or cull at the end of a molding cycle immediately after the press opens. In a typical transfer molding operation, a minimum Shore D hot hardness for acceptable demolding is 60, and the preferred range is 75 to 95.

Specific Gravity, ASTM D792. *Specific gravity* is the ratio of the weight of a given volume of a material in air to the weight of an equal volume of water at the same temperature. Specific gravity is important in cost-per-piece calculations because encapsulants are sold by weight but molded by volume. This means that more parts can be produced with a lower-specific-gravity encapsulant than one with an equal weight but higher specific gravity.

Linear Shrinkage, ASTM D6289. *Linear shrinkage* is the difference in length (centimeter per centimeter) of a molded part at room temperature to the length of the mold cavity at the same temperature. The molded part is allowed to equilibrate to room temperature for 1 hour before measurements are taken. Mold shrinkage is an important factor in designing the correct draft angle in mold runners and cavities for easy part ejection.

Thermal-Mechanical Properties *Coefficient of Thermal Expansion (CTE) and T_g, Semiconductor & Equipment and Materials, Inc. (SEMI) STD G13.* *Coefficient of thermal expansion* can be defined as the change in unit length of a material per temperature degree change of that material. Typically, CTE is expressed as ppm/°C. Mismatches in thermal expansion between epoxy molding compound (EMC), the lead frame, silicon die, and other components of a microelectronic device can lead to cracking of the mold compound or die, metal movement, delamination, and ball bond fractures. Table 11.2 lists the expansion coefficients of different components of a plastic molded semiconductor device.

Cured thermosetting encapsulants do not melt but when heated go through a reversible change from a hard glassy state to a rubbery state. This temperature-induced transition (T_g) results in changes in physical and electrical properties such as flexural strength and modulus, volume resistivity, and thermal expansion. Fig 11.4, a plot of CTE versus temperature, clearly shows the transition. The CTE of the section of the expansion curve before the transition is referred to as *$alpha_1$*, and the CTE of the upper portion of the curve is *$alpha_2$*. Glass transition temperature is determined from the intersection of tangent lines drawn along the $alpha_1$ and $alpha_2$ portions of the expansion curve.

Thermal expansion mainly depends on filler content and type, whereas T_g is governed by the resin, hardener, and accelerator system. Through manipulation of filler type and amount, the CTE of

TABLE 11.2 Coefficients of Thermal Expansion of Various Materials

Component	CTE, ppm/°C
Epoxy mold compound below T_g	10–20
Epoxy mold compound above T_g	40–75
Copper	16.6
Silicon chip	2.3–2.8
Aluminum	23
Gold	14.2
Palladium	11.8

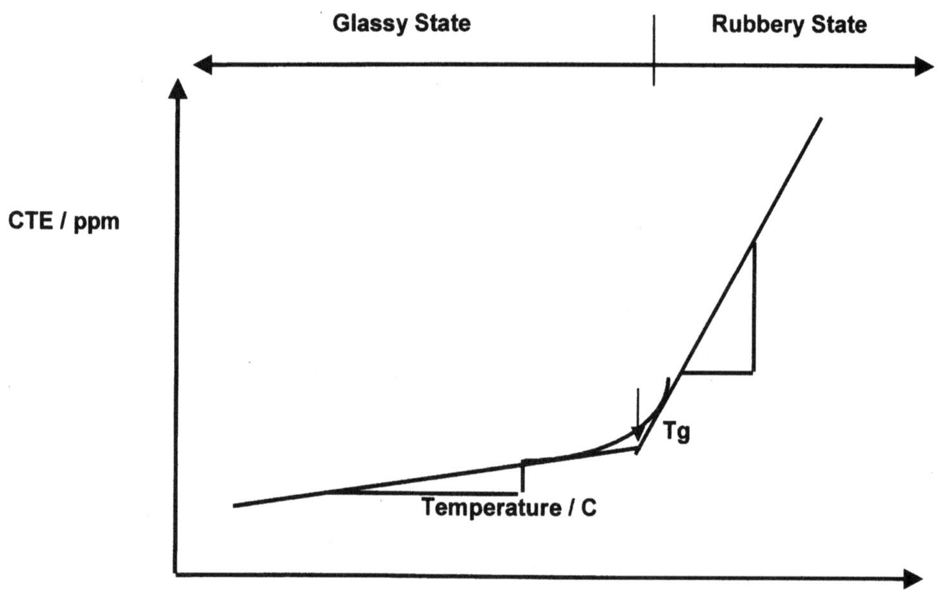

FIGURE 11.4 CTE versus temperature.

an encapsulant can be tailored to closely match that of major components of the devices being encapsulated. A high T_g is generally desirable because it extends the alpha$_1$ range and therefore reduces stress due to CTE mismatch at higher temperatures.

Thermal Conductivity, ASTM D792. *Thermal conductivity* is defined as the quantity of heat transferred through a unit of material per unit of time when exposed to a unit temperature gradient. Testing typically is done using a guarded-hot-plate apparatus, and conductivity values are expressed as watts per meter per degree kelvin or calories per degree centigrade per centimeter-second. Thermal conductivity of encapsulants is important for effective dissipation of heat generated by devices operating under high electric current. The type, amount, shape, and particle size distribution of fillers are the primary factors affecting thermal conductivity of electronic polymers.

Chemical Properties

Trace Ionic Impurities, SEMI STD G29. Trace ionic impurities such as Cl^-, Br^-, Na^+, and K^+ in polymeric packaging materials are known to contribute to electronic device failures such as corrosion, leakage current, dendrite growth, and intermetallic formation. These trace ionics can be present in any of the components of an encapsulant. Serious corrosion of aluminum circuitry and bond pads results when three factors are present: free halogen or alkali ions, water, and an electric current.

Anode:

$$Al(OH)_3 + Cl^- \rightarrow Al(OH)_2Cl + OH^-$$
$$Al + 4Cl^- \rightarrow Al\,Cl_4^- + 3e^-$$
$$AlCl4^- + 3H_2O \rightarrow Al(OH)_3 + 3H + 4Cl^-$$

Cathode:

$$Al + 3\,(OH) \rightarrow Al(OH)_3 + 3e^-$$
$$2Al + 3H_2O \rightarrow 2Al(OH)_3 + 6H^+$$
$$Na^+ + e^- \rightarrow Na$$
$$Na + H_2O \rightarrow Na^+ + OH^- + H^+$$

SEMI Standard G29 is a method commonly used in the semiconductor industry to determine the type and amount of water-extractable ionic impurities in packaging materials. In this method, a 10% solution of the encapsulation compound in water is placed in a Parr bomb and conditioned at 121°C for 48 hours. The resulting water extract is run through an ion chromatograph to determine both cation and anion impurity content.

Raw materials manufacturers strive to minimize the amount of ionic impurities in their products, but trace amounts of potentially harmful contaminates remain. Inorganic ion exchange scavengers are often added to encapsulants to effectively immobilize these trace ionic species.

Radioactive Contaminants. Trace amounts of thorium and uranium can be found in all types of electronic packaging materials. In 1978, Intel engineers May and Woods[5] reported that alpha-particle emissions from these trace contaminants were the cause of "soft errors" in memory devices. A *soft error* is the result of an alpha particle penetrating a memory well, which temporarily changes the register from a 1 to a 0.

The level of alpha-particle emission in counts per hour per square centimeter can be determined directly with a gas-proportional alpha counter, or it can be estimated from the thorium and uranium content as determined by inductively coupled plasma mass spectrophotometry (ICP-MS) or other analytical techniques.[6] A drawback to using gas-proportional alpha counters is that getting accurate results often requires a large sample size of 1000 cm and a long counting time of 168 hours or more. An assay for thorium and uranium by ICP-MS can be run on a few grams of prepared sample in minutes, and the sensitivity is about 0.01 ppb.

Specially formulated low-alpha-flux encapsulants are available that contain no more than 2 ppb of combined thorium and uranium and have an alpha flux rate of less than 0.001 counts/h/cm^2.

Water Absorption, ASTM D570 and JEDEC STD 020. Absorbed moisture is very detrimental to electronic polymer properties and performance in use. Water reduces mechanical properties such as T_g, flexural strength, and tensile strength. It also degrades electrical properties such as dielectric constant, dissipation factor, and volume resistivity. Absorbed moisture can increase ionic mobility and catalyze corrosion of metallic components of encapsulated electronic devices. In addition, absorbed water is the cause of "popcorn" cracking of devices during solder reflow.

ASTM D570 is an accelerated test used to determine the moisture absorption of polymeric materials. For electronic encapsulants, molded disks or test bars are immersed in water boiling water for 24 or 48 hours. The percentage difference in weight between the parts before immersion and the same parts after immersion is the total water absorption under those conditions.

An alternate method that is becoming a standard in the semiconductor industry is JEDEC J-STD 020A. This method is used to determine estimated time before the polymer absorbs enough moisture to become susceptible to moisture-induced cracking (popcorning) during solder reflow. Instead of the boiling water of the ASTM test, the exposure is to specific humidity and temperature conditions. For example, JEDEC level 1 specifies 168 hours of exposure at 85°C and 85 percent relative humidity.

Viscosity and Flow Properties. Melt viscosity control is a key to successful defect-free encapsulation of electronic devices. A viscosity that is too high can cause incomplete fill, metal movement, and wire sweep during mold fill. Too low of a viscosity also can lead to problems, such as air entrapment due to jetting into the cavities, excessive resin bleed, and blocked mold vents.

Automatic Capillary Rheology, ASTM D3835. Capillary rheology is a common method used to quantify melt viscosity of both thermoset and thermoplastic polymers. This test consists of forcing molten polymer through a capillary die and calculating viscosity from flow rate and extrusion time using the following formula:

$$\text{Viscosity, Pa} \cdot \text{s} = \frac{\text{Fr}^4\text{t}}{8R^2LV}$$

where F = force by ram, N
r = radius of the capillary, mm
t = extrusion time
R = radius of barrel
L = length of the capillary, mm
V = volume extruded

Viscosity control of electronic encapsulants is accomplished by precise control of the molecular weight of each of the resinous ingredients in the formulation and through control of filler particle size, shape, and size distribution. For thermosets, gel time is also an important factor because viscosity changes rapidly as cross-linking occurs. In practice, the *molding window* for a thermosetting epoxy is equal to the time at minimum viscosity.

Spiral Flow and Ram Follower Gel Time, ASTM D3123 and SEMI. The spiral flow test is a method used to quantify the flow characteristics of thermosetting epoxy molding compounds. It is a measure of the combined characteristics of gelation time and viscosity under specific conditions of temperature and pressure. The helix mold used for this test is shown in Fig. 11.5 and described in detail in ASTM D3132. To run this test, the mold is preheated to 175 ± 2°C in a transfer molding press. A predetermined amount of granular molding compound is poured into the transfer cylinder, and then the transfer plunger under 1000 ± 25 psi pressure is activated, pushing the molten epoxy molding compound into the mold and through the spiral. After allowing 90 to 120 seconds for the compound to cure, the mold is opened, and flow in centimeters or inches is read from marks on the mold or the molded spiral. A typical spiral flow of an encapsulation-grade epoxy molding compound is 25 to 35 in.

Ram follower gel time (RFGT) is another material property that can be determined using ASTM D3123. A transducer is used to track the movement of the transfer plunger during the flow test. Plunger displacement versus time data are fed into a computer. The time in seconds from when the plunger first touches the mold compound until it gels and stops moving is the ram follower gel time. Most epoxy encapsulants will have an RFGT in the range of 12 to 20 seconds.

Spiral flow and RFGT are important to both the producer and the molder of epoxy molding compounds. These two properties give insight into cure kinetics, viscosity, mold release, and batch-to-batch consistency. They can be used by the molder as guidelines in establishing molding parameters such as mold temperature, preheat temperature, transfer pressure, and transfer speed.

Flame Resistance Properties, Underwriters Laboratories (UL) 94 and Oxygen Index. Many of the base polymers used in packaging materials for electronics are inherently flammable and therefore require flame-retardant additives to meet industry flame-resistance standards. Historically, the flame retardants of choice have been halogenated compounds combined with antimony oxide. In recent years, however, there has been a strong effort to switch to more environmentally safe flame-retardant ingredients such as inorganic hydrates, borates, and phosphorus compounds.

UL 94 is the accepted universal standard for flame resistance of electronic packaging material. This is a test in which a molded bar of a material is subjected to a controlled laboratory burner flame for two applications of 10 seconds each, and time to self-extinguish is measured. The material is then classified as V0, V1, HB, etc., at a set minimum thickness based on time to self-extinguish.

Oxygen index (ASTM D2863), primarily used in Europe, is a flame-resistance test in which a "stick" of the material to be tested is ignited in a chamber filled with a controlled mixture of oxygen

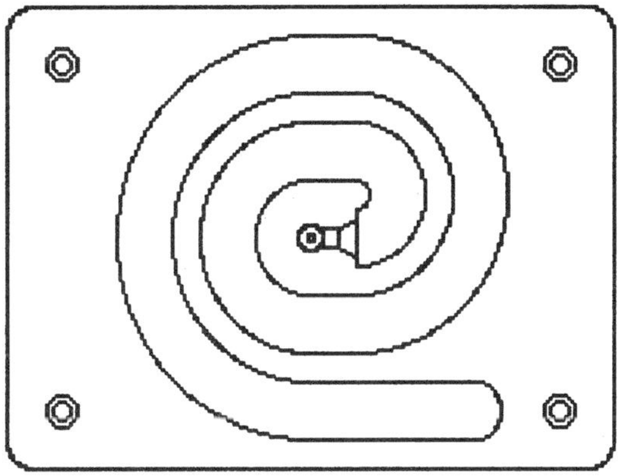

FIGURE 11.5 Spiral mold.

and nitrogen. The ratio of oxygen and nitrogen in the chamber is adjusted to the point that it just supports combustion of the test material. The percentage of oxygen in the mixture that just supports burning is the *limiting oxygen index* of the material being tested. The higher the oxygen index number, the more resistant the material is to burning.

Adhesion Properties. Adhesive strength is one of the most important properties of electronic encapsulants. It is necessary for the encapsulant to adhere strongly to all components of a device, including the lead frame, silicon die, gold wires, aluminum bond pads, etc., to secure the device mechanically and to prevent moisture ingress up the leads. Good adhesion to the underside of the die pad is essential for preventing delamination and the development of free space where moisture can accumulate and cause "popcorn" cracking during solder reflow. With the advent of area array packages such as BGAs, encapsulants face new challenges of adhering to organic substrate material such as bismaleimide triazine (BT) resin and organic solder mask.

One of the most commonly used methods of evaluating the adhesive strength of an encapsulant is the button shear test. This method is useful for testing adhesion of the encapsulant to all types of metals (Cu, Ag, Au, Pd, Ni, etc.) found in semiconductor devices, as well as ceramic and organic substrates. In this test, a small 1-cm^2 button of the encapsulant is molded onto a strip of the material being evaluated. The strip with the molded-on button is secured in a universal mechanical tester, and a shear force is applied. The minimum shear force required to break the bond is reported as the button shear in kilograms per square centimeter (Fig. 11.6).

A dilemma facing both the formulator and the molder of encapsulants is that while the encapsulant must have good adhesion to all parts of the device being encapsulated, it also must release easily from metal molds. One of the options available to formulators to overcome this problem is the use of adhesion promoters such as silanes and titanates that couple the resinous constituent of an encapsulant to targeted component(s) of the part to be encapsulated.

11.4.2 Epoxy Molding Compounds

Liquid Encapsulants. Chip-on-board products, including most of the BGAs, are protected with encapsulants made of polymer composites. Epoxies are the most popular resin system because of their balanced properties and well-established infrastructure. Fillers, such as silica, are added to reduce the CTE and thereby reduce thermomechanical stress. There are four major categories of liquid encapsulants: damming compounds, cavity fill (well fill), glob tops, and free flow/printable. Although they can have similar cured properties, their rheologic characteristics are very different, as are the dispensing processes.

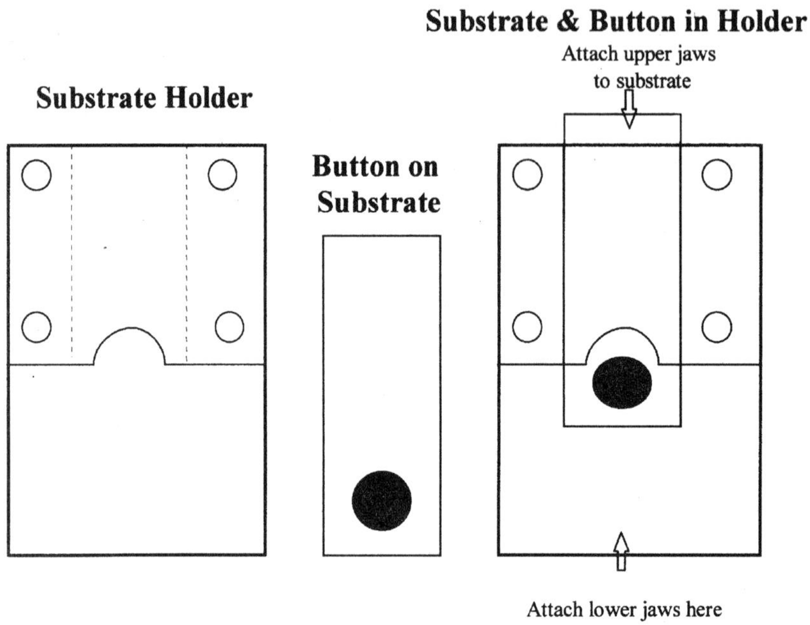

FIGURE 11.6 Button shear test.

Damming Compounds. Damming compound, as the name implies, is used to form a dam around the area to be encapsulated. In some assembly designs, the flow of the encapsulant must be restricted, especially if material will intrude into other component areas. The challenge of restricting the flow of encapsulant can be solved in several ways, but today's most common approach is to first apply a dam. The damming material must be easily dispensable but highly thixotropic with a fast rheologic hysteresis curve, which means that the fluid quickly thins on dispensing but thickens after dispensing. The damming compound, or flow containment vehicle, is first dispensed in the desired pattern. Typically, a square or rectangular dam is needle-dispensed around the perimeter of the bare die. The dam must hold up against the encapsulant in an uncured state.

Fill Materials. The fill materials have a very different rheologic characteristic than the damming materials. These are more flowable encapsulants and are dispensed within the dam area until they flow to its boundaries. Heating now hardens the dam and encapsulant so that both materials cure together. Figure 11.7 shows the principle of the damming compound. The dam and fill materials are used with BGAs and other products where the location of the encapsulant must be precise. Both materials are hardened simultaneously for efficiency. Success with the dam and fill process requires the right dispensing equipment, and the process is covered in more detail in Chap. 22, Liquid Encapsulation Equipment and Processes.

Glob Tops. One of the earliest encapsulants, but a product enjoying a new wave of growth, is the glob top, a bare-die overcoat. The old glob tops only needed to protect simple chips in products such as digital watches and game cartridges where requirements were loose and sometimes nonexistent. Today, the demands are much more stringent, and many more characteristics must be measured. Low stress, low expansion, and low ion levels are just some of the requirements. Even the rheology, not just the viscosity at one or two points, must be held to tolerance because processing characteristics are just as critical as cured properties. End-use applications include smart cards, PCMCIA cards, portable communication items, and many consumer electronics desktop and portable products (Fig. 11.8).

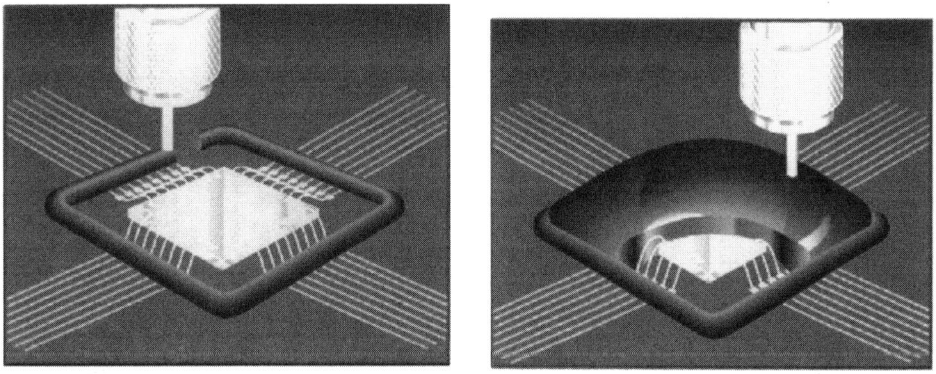

Courtesy of Loctite

FIGURE 11.7 Principle of damming compounds.

Printable Encapsulants. Although the use of damming compound is a practical method for restricting encapsulant flow, it is a time-consuming step that is often the process bottleneck. Recently, a new form of encapsulant has been developed that can be stenciled and hardened for both small products such as smart cards and larger ones including BGAs. The goal is to stencil and cure the encapsulant, thus avoiding the damming step. This approach increases throughput and substantially improves thickness control, and the finished product can have the appearance of a molded package. The key property of a printable encapsulant is its rheology, not just the viscosity at one or two points (Fig. 11.9).

11.4.3 Thermoplastic Injection Molding Materials

Plaskon TPE liquid-crystal polymers (LCPs) are a family of compounds developed for encapsulation of active, passive, and discrete components. They offer a combination of high-performance properties and processability unmatched by any other thermoplastic or epoxy molding compound. Plaskon TPE LCP resins are thermoplastic and have extraordinary flow, making it possible to use conventional injection-molding equipment in semiconductor device encapsulation.

Plaskon TPE LCPs are ultra-high-performance thermoplastic resins that combine the engineering properties of epoxy resins and ceramics with the design freedom and superior processing economies of injection-moldable plastics. The advantages of these products include high reliability, unlimited shelf life without refrigeration, a low CTE, level 2 antipopcorning on 144-lead thin quad flat pack (TQFPs), self-intumescent flame resistance (UL 94 V0 at 1/16 in) without additives, no postmold cure, and outstanding resistance to most chemicals and radiation. In addition, because Plaskon TPE LCP is a thermoplastic, culls can be recycled back into the process stream up to 30 percent weight or more.

Applications for Plaskon TPE LCP resins include SOICs, PDIPs, TQFPs, and TOs to name a few. Plaskon TPE LCP resins are suitable for active, discrete, and passive devices.

11.4.4 Thermoplastic versus Thermoset Molding Compounds

New thermoplastic materials such as Plaskon TPE have been introduced into the arena of encapsulation of electronic devices. Historically, epoxy-based resins and other thermosets have served this technology. There is a growing drive to move to new encapsulation materials that can offer benefits that the traditional thermoset compounds cannot. Two significant attributes of this drive are environmental and economic.

A version of Plaskon TPE is based on LCPs. These polymers typically are capable of meeting UL 94 V0 requirements without the addition of any flame-retardant materials. The LCPs in general are

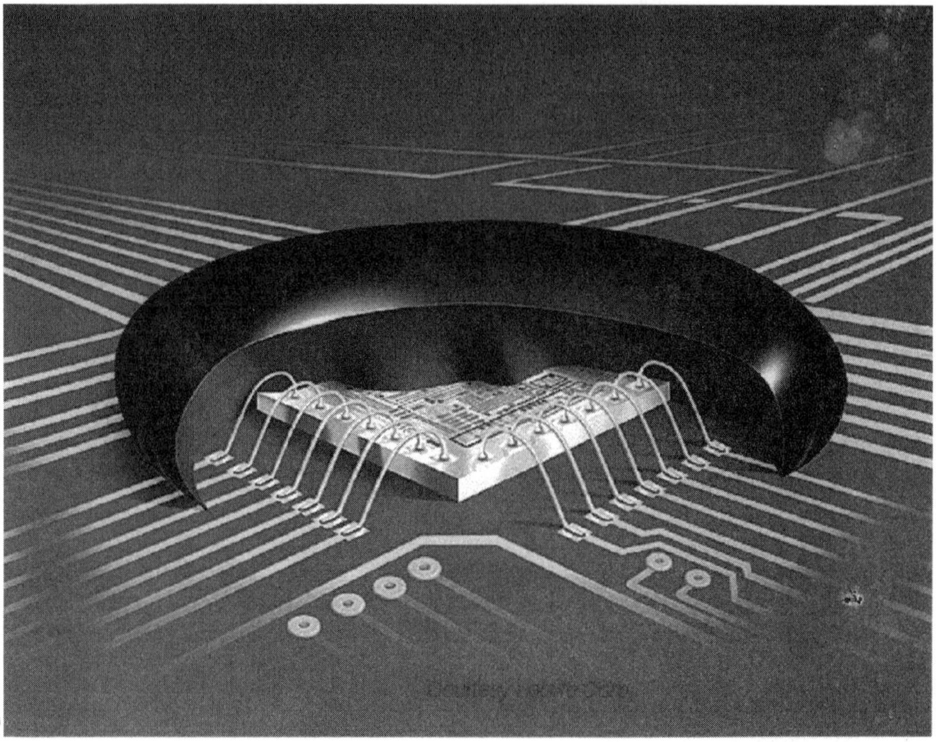

Courtesy of Loctite

FIGURE 11.8 Encapsulated chip-on-board.

relatively inert. They are wholly aromatic copolyesters that are extraordinarily stable and do not readily decompose below their melt temperatures. Usually there are no special requirements for the disposal of these materials in landfills. And because Plaskon TPE is a thermoplastic, it is recyclable. Sprues and runners (these are called *culls* in thermoset technology) are reusable for many generations. With these performance properties, Plaskon TPE using LCP resins is a good selection for environmentally conscious applications.

Shelf Life. For some of the same reasons that Plaskon TPE is a good choice environmentally, it is also a good choice economically when compared with thermosets. Unlike thermosets, the thermoplastic Plaskon TPE does not have a shelf life. Manufacturers of encapsulated electronic components are all too familiar with what happens if a shipment of thermoset is not kept refrigerated properly. The same is true even if refrigeration is maintained but the material is not consumed within the material's rated shelf life. With Plaskon TPE, there is no need for refrigeration or worry about inventory aging.

Recycling. The ability to recycle is a tremendous advantage for environmentally conscious applications, but it also can distinctly improve the cost of manufacturing an encapsulated device. The most obvious improvement is that if you can reuse all waste from the molding process, that much less material must be purchased to make parts. Secondary is reduced cost in waste disposal, which is increasing every year due to demands on limited landfill space.

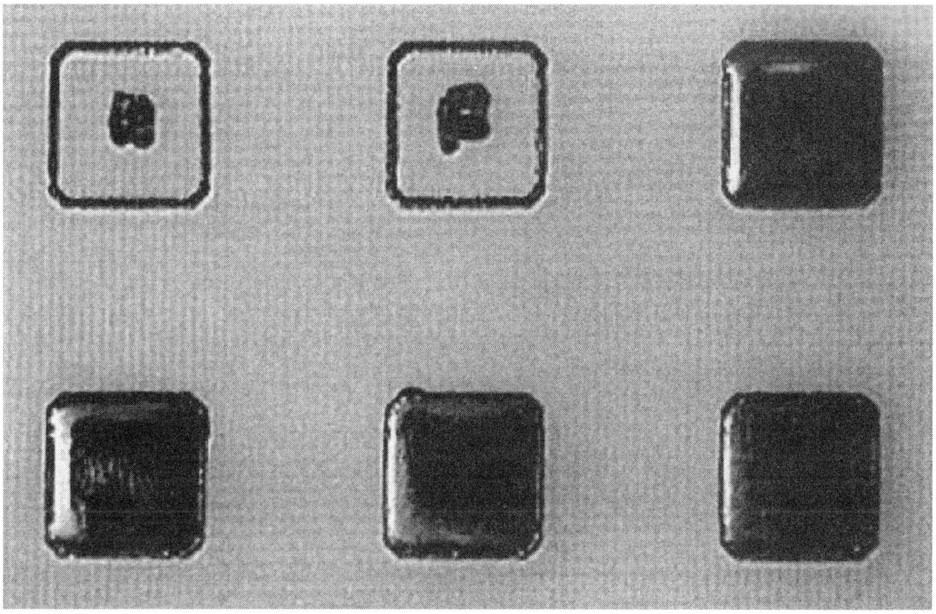

FIGURE 11.9 Stencil and cure process.

Processing. In traditional thermoset encapsulation, it is normal to use a large transfer molding machine with a mold that has a cavity count numbering in the hundreds. A machine operator will spend his or her shift at this machine loading and unloading the transfer molds, wiping down the molds between cycles, and blowing flash off the molds. A cycle on a transfer molding machine can last from 1 to 3 minutes.

In injection molding of thermoplastics for encapsulation, the molds and machines are much smaller, typically four to eight cavities, and use a reel-to-reel system or an automated cartridge feeder for cut-strip lead frames. Although cavity count is considerably less, cycle time is considerably shorter. Cycle times of 10 seconds or less are routine with Plaskon TPE. Because of the automation of the reel-to-reel system and the rapid cycle times, throughput can be significantly better than with traditional transfer molding.

It may be possible to install two injection-molding machines in the same floor space as one transfer molder, and a single operator could be expected to operate four or more machines.

Injection Molding of Thermoplastics. The processing characteristics of LCP polymers are quite different from those of thermosets and, in some sense, even other thermoplastics. LCPs are made up of rigid rod molecules, and these rods aggregate into bundles. To achieve the low-viscosity behavior of LCPs, it is necessary to have a process that develops a great deal of shear. Shear is the single largest contributor to reducing the size of the rod bundles and achieving the low viscosities required for encapsulation. Most transfer molding machines usually do not have adequate ram speed to develop this shear. In addition, transfer molds usually have large flow channels and gates that also will preclude the development of adequate shear.

Beyond the need for shear, transfer molding machines do not have a screw that melts the polymer. Without a traditional reciprocating screw found in injection-molding machines, it is almost impossible to develop a homogeneous melt. What is most often seen when an LCP-based compound is put into a shot pot of a transfer molding machine is burned polymer near the metal wall and unmelted polymer near the center.

11.5 UNDERFILL

Conventionally, an FC is attached to a carrier or circuit by solder reflow. During this process, a small gap is created ranging from 15 to 100 μm in size. Due to thermomechanical mismatch between the silicon chip (CTE = 2.3 ppm/°C) and the most cost-effective organic board (CTE = 15 to 25 ppm/°C), the FC joints are placed under severe mechanical strain during thermal cycling.[7] The CTE mismatch results in FC assembly failure during thermal cycling as the joints fatigue and fracture. Fortunately, the materials suppliers have provided a practical and cost-effective solution to this problem in the form of underfill.

Underfill is a liquid polymer–based composite that is designed to flow under the FC in a predictable manner. The underfill is then hardened by thermal polymerization so that it "locks" the chip-substrate structure together and moves stress away from the joints. The result is a substantial increase in joint reliability. Once the FC is underfilled properly, the assembly normally will pass well over 1000 thermal cycles (−55 to 125°C).

Epoxies are the dominating polymer used for underfills because of their balanced properties, long-established safety record, wide availability, high versatility, and general characteristics that are desired in this product area. Today's underfills can be divided into three distinctly different systems: capillary, predispense, and solid film. The incumbent capillary flow underfill manufacturers attempt to win by offering snap-flow/snap-cure materials and various new features.

11.5.1 Preapplied Materials

The preapplied materials are highly specialized and are covered in separate chapters. It will suffice to say they also must serve as a flux, and this can require unique chemistry not found in more traditional underfills.[8]

Solid on Substrate. Solid films have been used as underfill materials, but this approach has questionable value. Film must be produced and cast onto release liner. Then the film must be bonded to a substrate. Finally, the FC must be bonded using special equipment that applies heat and pressure.[9]

Wafer-Level Solid Underfills. Underfill can be applied to an FC as film or paste that is later hardened. There is a precedent for this type of process within the die-attach adhesive field. Thermoplastic die-attach adhesives, such as Staystik, are available as both films and pastes. The liquid form can be applied to the backside of a wafer and then hardened by solvent evaporation. Spin coating, stenciling, and screen printing have all been used successfully. The wafer is diced after the adhesive is hardened. Although die-attach adhesive is not an underfill, the properties are somewhat similar, and the same wafer-level coating methods are applicable.

The most reasonable location for solid underfill is on the bottom, or bump side, of the chip. FCs with integral underfill take on the characteristics of chip-scale packages (CSPs). The solid "flipped underfill" also should have flux characteristics. "No flow" underfills have demonstrated that fluxing agents can be incorporated. There is one more important property to add to the wafer-level solid film underfill. The product should be reworkable and remain so after assembly and any postprocessing steps. A readily reworkable underfill transforms FCs into a true package because this is really a requisite to qualify as an electronic package. The addition of a solid *reworkable flux/underfill* to the FC moves it to the CSP domain. There are very significant ramifications for such a package. Also note that wafer-level flux/underfill moves the underfill from the assembler to the semiconductor realm.

There are at least three approaches for constructing a *ready-to-assemble* FC "package." A single material strategy can be used, but flux and underfill properties must be achieved in one material. However, the goal of achieving reworkability in a single flux/underfill is more difficult but achievable.

A two-layer system is also feasible because of the solid nature of the materials. Underfill and flux can be kept separate, making the chemistry easier but application more complex. While the term *layers* is used, flux may be localized on bumps and not necessarily formed as a stratum on the underfill.

Solid underfill also can be applied to the wafer before it is bumped. Lasing or other imaging means could be used to open chip pad areas prior to bumping. Flux would be applied after bumping, and we refer to this approach as *mask, flux, and underfill.*

11.5.2 Postapplied Materials

The postapplied class is the original and simplest type of underfill. The more common name is *capillary flow* because material is "pulled" along by surface energy or capillary action. Surface tension is the "engine" that drives capillary flow, and the phenomenon is well understood within the realm of surface chemistry and fluid dynamics. Capillary flow processing uses the basic wetting principles of surface chemistry. Intramolecular attraction must exceed intermolecular forces. The underfill resin molecules (continuous phase) must be more strongly attracted to the FC and substrate surfaces than to one another so that an advancing contact angle is achieved. *Advancing contact angle* means that the underfill wets the surface and advances forward—the liquid molecules are being attracted to the surface substrate molecules or atoms. This is accomplished by ensuring that the surface tension σ of the underfill is lower than the surface energy of the solid surfaces to be wet. Wetting agents generally produce the desired low surface tension. The free energy of wetting is the source that pulls the underfill through the gap, whereas viscosity acts to resist flow.

Epoxies typically are used as A-B polymer-type systems in which resin and hardener combine in approximately equal stoichiometric proportion to produce polymer chains made up of both constituents. The most common hardeners for epoxy-based encapsulants and underfills are anhydrides. Figure 11.10 shows the polymerization reaction, and Fig. 11.11 is a graphic representation of a thermoset structure.

Capillary Underfill Materials. The underfill composition primarily consists of epoxy resins and hardeners filled with very pure and precisely shaped silica. There are critical additives, usually proprietary, and they contribute to viscosity, flow, and wetting.

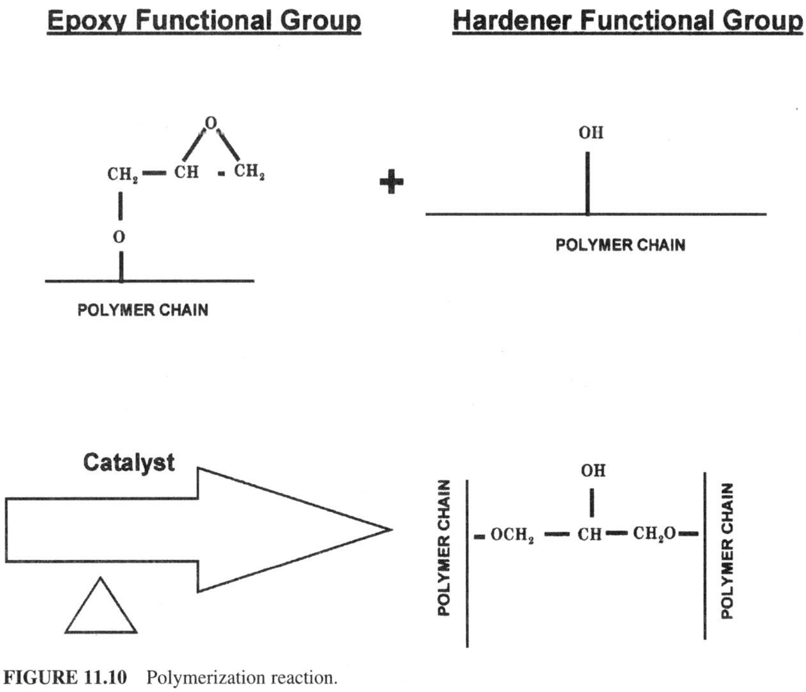

FIGURE 11.10 Polymerization reaction.

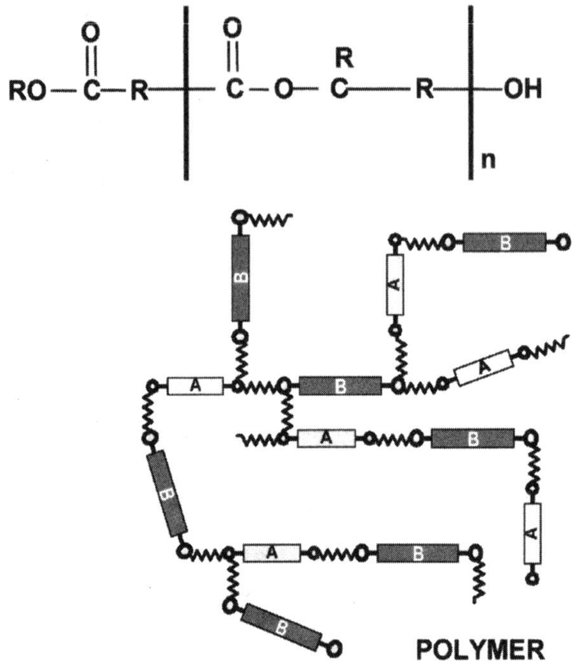

FIGURE 11.11 Thermoset structure.

The silica is the crucial component that controls the CTE. It is an inorganic powder that accounts for the highest cost of the underfill. With a CTE approaching that of silicon, the filler is loaded into the epoxy to decrease the CTE to 20 to 25 ppm/°C. The epoxy without filler has a CTE of 60 to 80 ppm/°C, so with the addition of silica, the overall underfill CTE is reduced to a level to match the CTE of the solder interconnect. The cured underfill locks the chip, bump, and wiring board surface into a low-strain structure the can be run through 1000 temperature cycles without failure.

Important parameters of the filler itself are percentage loading, particle size, particle size distribution, and morphology. These all affect CTE, viscosity, flow, and the ability to flow under the die. The silica is usually added at 62 to 70 percent by weight to achieve CTE ranges of 20 to 25 ppm/°C, matching the solder interconnect. Higher levels of filler will increase viscosity and decrease flow of the underfill. And vice versa, lower levels will decrease viscosity and increase flow.

As the interconnect changes to gold or conductive epoxies, higher-CTE materials, the expansion can be more forgiving, where the CTE range of the underfill can increase up to 40 to 50 ppm/°C—noting that the CTE of the underfill must match that of the interconnect. As the interconnect expansion increases, the underfill CTE moves up, allowing the silica level to decrease for a faster-flow underfill. For manufacturing throughput, fastest flow under the die is desirable.

FC dies are becoming larger with higher-density interconnects, where the gap between the die and substrate is forced to decrease. Typical gaps for small dies are in the 75- to 80-μm range. The filler size of the underfill used in these designs has 40-μm particle sizes. Gaps are now down to 10 to 15 μm. The underfill for these FC assemblies must be redesigned with a smaller filler particle size. As the filler size reduces, viscosity increases, slowing down flow rate. Lowering the percentage loading compensates this—CTE will increase slightly, but the underfill maintains its flow speed.

Liquid Encapsulant. Liquid encapsulants, once typified by the older glob top materials, are mixtures of low-molecular-weight monomers, hardeners, fillers, and additives that are converted into high-performance thermoset polymers by the application of heat. Although properties are similar to transfer molding compounds, their liquid nature permits a wide variety of dispensing methods to be used. The liquid encapsulants used to protect bare dies have long been applied by needle dispensing, an excellent lower-volume method. One increasingly popular process involves dispensing a highly viscous containment dam around the bare-die area and filling the inner volume with a flowable encapsulant. This process of needle dispensing a thixotropic damming compound and then applying a second, more fluid encapsulant is slow but very effective. Large panel arrays, with multiple bare-die sites, take considerable time for encapsulation because needle dispense deals with a *one-site-at-a-time serial process.* There are other application processes that allow quick, *multiple-site* deposition. One that is well suited for thick but precise deposits is *transil printing,* a screen process and stencil method. Screen and stencil printing have been used successfully for the application of electronic materials for many years.[10]

Solid Encapsulant. Solid epoxy resins and hardeners can be blended and compounded to produce solid but reactive molding compounds. The epoxy-based products are called *epoxy molding compounds* (EMCs). EMCs are made by mixing the solid prepolymers with fillers and additives and forming them into useful shapes called *preforms.* The process may require many steps, with a final preform step that involves applying heat and pressure to the mix in order to form stable shapes that often look like hockey pucks. Although the resins and hardeners are solids, chemical reactions still occur, and shelf life generally will have specific limits. Like liquid encapsulants, lifetime can be greatly extended by storing in a freezer. The solid preforms are heated, melted, injected into a mold, and then polymerized by exposing to the correct time/temperature profile. Molding is covered in Chap. 23, Molding for Area Array Packages.

11.6 AREA ARRAY MOLDING MATERIALS (BGAs)

During the past several years, a process called *flood molding,* or *area molding,* has been perfected. A newer term is *integrated molding.* Rather than fill individual cavities with EMCs, the entire array of devices is encapsulated so that the molding machine produces a sheet of packages that must be singulated later. The polymerized molded-up array is sawn into individual packages using equipment similar to that used to saw silicon wafers. The integrated molding process substantially reduces tooling cost and boosts productivity even though a sawing step is added. The process uses EMCs similar to those used in the traditional cavity molding method.

11.6.1 BGA Package Materials

EMCs used for area array packages have the highest level of requirements of any encapsulants. Large BGAs, especially on thin substrates, will easily warp if forces are generated by the encapsulant. All commercial epoxy systems shrink during curing because the resulting polymer occupies less volume than the starting materials. The most suitable EMC and liquid encapsulants produce the minimum amount of warpage. Resins and hardeners can be selected that undergo the minimum level of shrinkage. Higher levels of filler also have a positive effect because there are fewer polymers to shrink. Some resin mixes can be made to go through a gelling stage in which shrinkage occurs while the partially polymerized material is too soft to cause warpage. The oven profile typically is set with two temperature stages. The first one is lower, and it produces gelling. The second stage is higher, and it allows the polymer to fully cross-link. While some shrinkage will occur at the second stage, the absolute value is lower because significant polymerization has already occurred at the gelling stage.

11.6.2 CSP Materials

CSP materials can be product-specific. For example, the Tessera μCSP uses silicone rubber that was specially made for the application. Other CSPs also have required custom materials, and it is not really practical to cover this topic in a general way.

11.7 SUMMARY AND CONCLUSION

The advent of the plastic package many decades ago has led to a strong materials infrastructure made up of worldwide resin producers and countless formulators. Both large and small companies supply electronic polymers to the industry, and the high level of competition has resulted in continual product improvements as well as many new materials. While the most common encapsulants are solids exemplified by epoxy molding compounds, liquid products have become popular during the past decade as area array packaging has gained success. Today's materials serve the needs of virtually all the area array packages with mostly standard products, although custom products have been designed and implemented.

11.8 REFERENCES

1. Gilleo, K., Cinque, T., Corbett, S., Corey, M., and Muculich, R., "Thermoplastic Die Attach Adhesive for Today's Packaging Challenges," *Electronic Packaging & Production,* February 1994, pp. 109–112.
2. Gilleo, K., Cinque, T., Corbett, S., and Lee, C., "Thermoplastic Die Attach Adhesives: The Attachment Solution for MultiChip Modules," *IEPS,* pp. 232–242, 1993.
3. Gilleo, K., "Conductive Adhesives for Flat Panel Displays," ISHM Flat Panel Display Conference, Ismoralda, FL, November 2, 1994.
4. Gilleo, K., "SMT Assembly Using Conductive Adhesives," Workshop at EMPF Government Laboratory, Indianapolis, IN, 1995.
5. May, T. C., and Woods, M. H., "A New Physical Mechanism for Soft Errors in Dynamic Memories," in *Proceedings of the 16th International Reliability Physics Symposium,* Las Vegas, NV, April 1978, pp. 33–40.
6. Meieran, E. S., Emgel, P. R., and May, T. C., "Measurement of Alpha-Particle Radioactivity in IC Device Packages," *IEEE Transactions,* 1979, pp. 13–21.
7. Gilleo, K., "Flip or Flop," *Circuits Assembly,* February 1997, pp. 40–42, 44.
8. Preveti, M., "No Flow Underfill Reliability is Here—Finally!" in *Technical Proceedings of APEX*, Long Beach, CA, March 12–16, 2000, pp. P-MT1/1-1-1-4.
9. Ito, S., et al., "A Novel Flip Chip Technology Using Non-Conductive Resin Sheet," in *Proceedings of the 48th ECTC,* Seattle, WA, May 1998, pp. 1047–1051.
10. Godin, R., Corbett, S., Stevens, J., Gilleo, K., and Johnson, A. U.S. Patent No. 6,067,709, May 2000.

CHAPTER 12
HERMETIC PACKAGING SYSTEMS: ADHESIVE AND GETTER

Rita Mohanty and Jacob Lee
Vectron, Inc.

12.0 INTRODUCTION

Microelectronic circuits and components that are exposed to all types of environments require hermetic packaging. Besides providing mechanical protection and thermal dissipation, a hermetic package prevents the intrusion of atmospheric contaminants such as moisture, ionic contaminants, airborne particles, and unwanted gases. The fundamental construction of microelectronic circuits consists of bare semiconductor dies, epoxy-mounted components, and interconnects such as wire bonds, flip chip (FC) metal bumps, and metal traces on the substrate board (Fig. 12.1). Semiconductor dies are very susceptible to contaminants. Moisture and ionic contaminants are the known causes of wire-bond failure and leakage or inversion effect on semiconductor devices. In addition, moisture also can cause corrosion on aluminum-to-gold wire-bond interfaces and on nichrome thin-film resistor networks. Dendritic growth, as a form of silver migration on the metal trace, is also well documented as a failure mode caused by contaminants. To minimize the number of contaminants, packages with the fully populated microcircuits are carefully prebaked and sealed with a control atmosphere such as dry nitrogen or vacuum. A general guideline for hermetic packages is to pass the leak-rate requirement of 1×10^{-8} cm^3/s. A hermetic package can be either a metal, ceramic, or glass package.

12.1 HERMETIC PACKAGING

12.1.1 Methods and Technology

There are many choices of technologies in hermetic packaging. Glass-to-metal seals and ceramic-to-metal seals are the common choices to produce metal-type hermetic packages. Typical metal enclosures are formed by stamping, deep drawing, brazing, photochemical etching, and machining. Package sealing methods include projection weld, seam seal, cold weld, laser weld, and E-beam seal. For ceramic-type hermetic packages, the header can be either a single-level substrate or a multilayer high-temperature cofired ceramic (HTCC) structure. In HTCC technology, ceramic is processed in green tape form. It is a batch process, and individual units are arranged in array format. Layers are built by sequential operations of hole punching, metal filling and screening, cutting, stacking, and laminating. The entire assembly is cofired or sintered at temperatures as high as 1600°C. Electrode-nickel plating and brazing operations are performed before singulating and final gold plating operations. In order

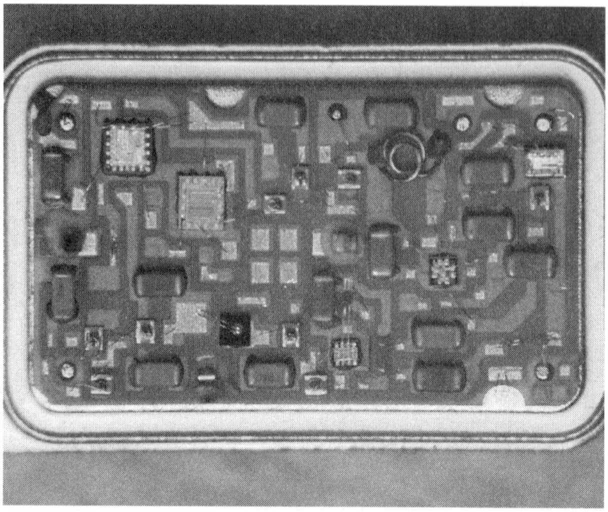

FIGURE 12.1 Hybrid microcircuit on a metal DIL package.

to bond a metal lid to a ceramic package, a brazed Kovar ring or a metallized seal area with attached solder preform is required. Seam seal, laser weld, and E-beam seal are the common choices of sealing methods.

12.1.2 Hermetic Package Types

Table 12.1 lists several common types of hermetic packages. Plastic packages are excluded because of their nonhermetic nature. Therefore, only metal and ceramic packages are accounted for. Besides categorizing packages by construction type, packages also can be divided into insertion mount and surface mount types, as shown in Table 12.1. Insertion mount packages also can be referred to as plug-in packages. The mounting pins of insertion-type packages typically are perpendicular to the base of the package. In most cases, insertion-type packages are inserted through the plated through-holes on the printed wiring boards (PWBs). Alternatively, surface mount packages are mounted directly on the surface of the mounting substrates.

Metal Package Types

Dual In-Line (DIL) Package. Standard metal DIL packages, as shown in Fig. 12.2, are a plug-in type of platform package with two row of pins centered at 0.100-in spacing. Pins typically are round in shape and have a 0.018-in outside diameter (OD). Pins are glass sealed perpendicularly to the metal platform (Fig. 12.3). DIL packages are easily solder dipped and mounted to PWBs. Due to their large 0.100-in pitch count, DIL packages are only suitable for low input-output (I/O) devices.

Cavity Plug-In Package. Cavity plug-in packages are very similar to surface mount flatpacks. The major difference is the pin arrangement. As in DIL packages, pins are sealed to the bottom of the package perpendicularly with a typical spacing of 0.100 in (Fig. 12.4). The overall construction of the metal cavity package can be divided into two types. Solid-body construction is the most common construction method. Basically, the package eyelet is stamped, deep drawn, trimmed, and hole punched in a progressive fashion. It is a one-piece construction without a bottom brazed joint. The limitations of the solid-body construction are the wall height and the inside radii sharpness of the package. In order to increase the wall height and achieve sharper inside radii, sidewall-brazed con-

TABLE 12.1 Common Types of Hermetic Packages

	Metal type	Ceramic type
Insertion mount	Dual in-line (DIL) package Cavity plug-in package Transistor outline (TO) package	Sidewall-brazed DIL package Ceramic pin grid array (CPGA) package
Surface mount	Flatpack Quad flatpack (QFP)	Small outline (SO) package Ceramic chip carrier Ceramic quad flatpack (CQFP) Ceramic ball grid array (CBGA)

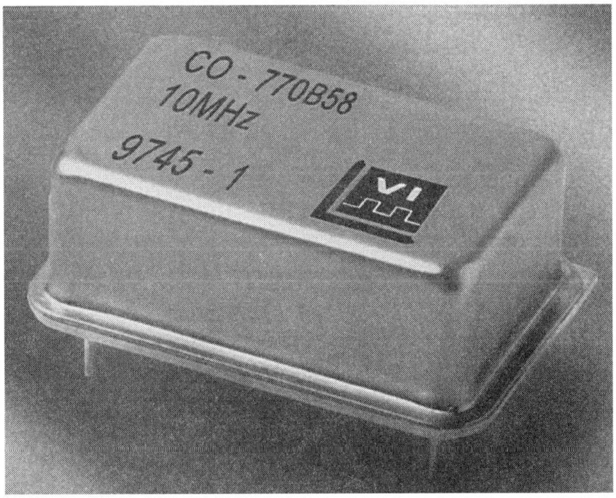

FIGURE 12.2 Metal DIL package.

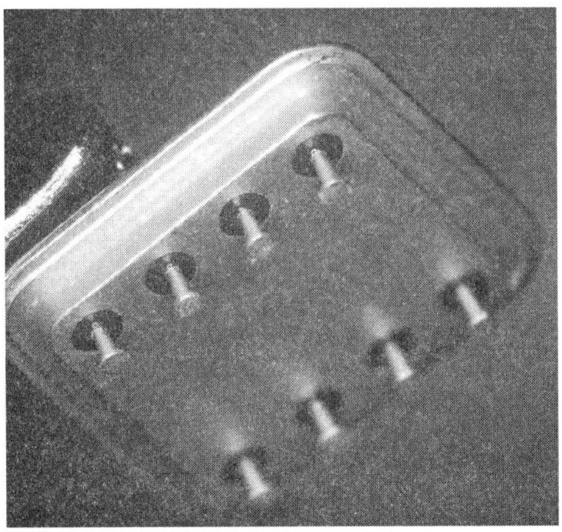

FIGURE 12.3 Glass-sealed pin of a DIL package.

FIGURE 12.4 Cavity plug-in package.

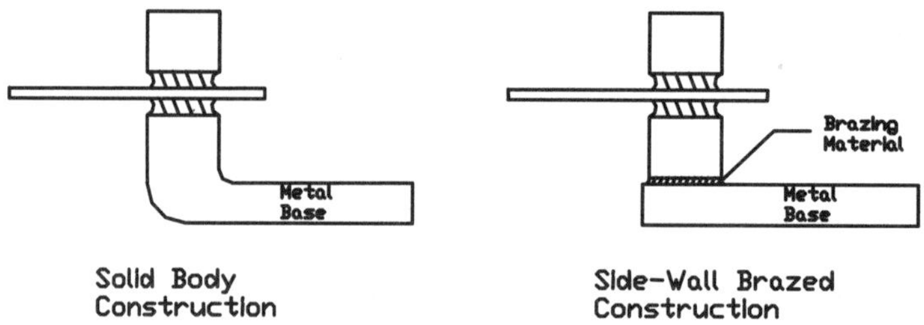

FIGURE 12.5 Cavity package construction types.

struction or two-piece construction is required. The technique is to braze a metal ring with a flat metal bottom base. Figure 12.5 illustrates both construction types.

Transistor Outline (TO) Package. TO packages were designed originally for housing single-transistor dies. With a few exceptions, most TO packages, such as TO 46, have a round header with insulated leads (Fig. 12.6). Besides mounting transistor dies, the latest applications include mounting precision quartz crystals and optical laser diodes. In the latter case, a filter glass cover is used instead of a solid metal cover.

Flatpack/Quad Flatpack (QFP). Both flatpack and quad flatpack packages consist of a cavity-type surface mount package (Fig. 12.7). Like the cavity plug-in package, the housing can be either a one-piece drawn cavity or a two-piece sidewall brazed cavity. The insulated leads extend horizontally from the sides of the cavity housing. In quad flatpacks, leads extend in all four directions from the package. The leads are formed and trimmed to a surface-mountable shape such as gullwing prior to mounting. Among the other metal packages, flatpack, especially the quad flatpack, can accommodate high-I/O devices. Typical pitch size is 0.050 in, but a 0.040-in pitch size can be accomplished easily.

Ceramic Package Types

Sidewall-Brazed Ceramic DIL Package. Instead of using a glass-to-metal seal like its metal package counterpart, leads in ceramic DIL packages are eutectically brazed to the side of the package. Although the external lead pitch remains the same, the sidewall-brazed ceramic DIL package has a smaller overall package size. This package is used in both single-chip and multichip module (MCM) solutions (Fig. 12.8).

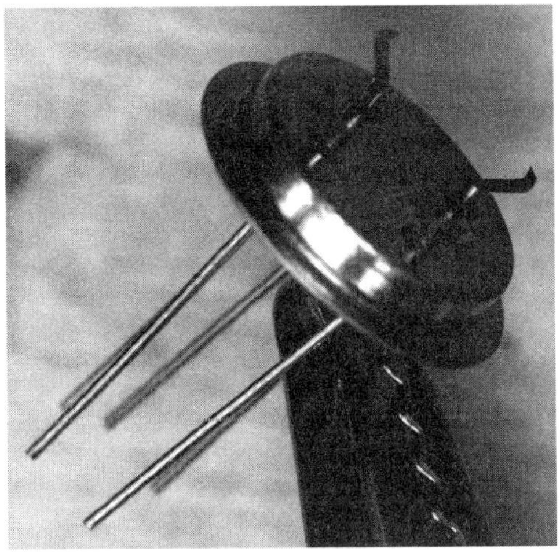

FIGURE 12.6 TO package.

FIGURE 12.7 Metal-flatpack.

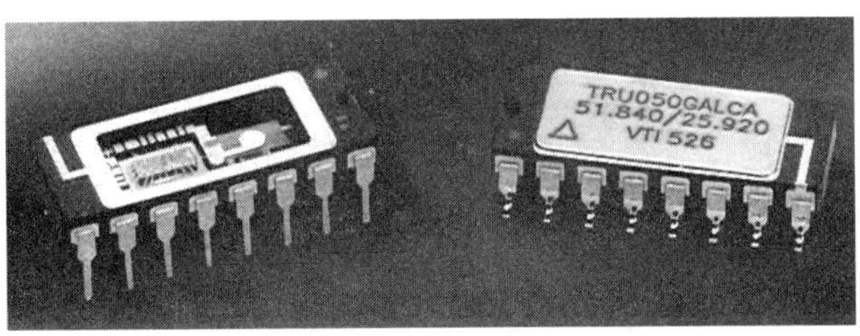

FIGURE 12.8 Sidewall-brazed ceramic DIL package.

Ceramic Pin Grid Array (CPGA) Package. CPGA packages consist of leads distributed on the metallized ceramic surface in an array format. CPGA is an insertion mount type of package. It can be either through-hole attached to the PWB directly or attached through a test socket. CPGA packages have both cavity-up and cavity-down designs. The leads are eutectically brazed to the metallized ceramic surface. The typical pin array standard is a 0.100-in distribution. The pin array also can be staggered to provide better spacing without decreasing the I/O density (Fig. 12.9).

Small Outline (SO) Package. The small outline (SO) package is a surface mount type of package. The typical pitch size of the ceramic SO package is 0.050 in. The external leads can be either a gull-wing or J-lead configuration, as shown in Fig. 12.10. SOIC is the designation for packaging semiconductor integrated circuits (ICs) in the SO-type package. Likewise, SOT is the designation for SO packages for transistor mounting.

Ceramic Chip Carrier. A ceramic chip carrier can be either a leaded type or a leadless type. Leadless chip carriers (LCCs) are soldered to the mounting substrate such as a PWB directly from their semicircular metallized contacts on the sidewalls. Alternatively, leaded chip carriers have compliant metal leads such as J-lead attached to the chip carrier as its I/O leads. Ceramic chip carriers are mainly produced by a high-temperature cofiring ceramic (HTCC) process at above 1600°C. Most JEDEC type LCCs have a pitch size of 0.050 in. (Fig. 12.11).

Ceramic Quad Flatpack (CQFP). Ceramic quad flatpacks also are constructed with an HTCC process. The lead frame is attached to either the top or the bottom of the ceramic package. External

FIGURE 12.9 Ceramic pin grid array package.

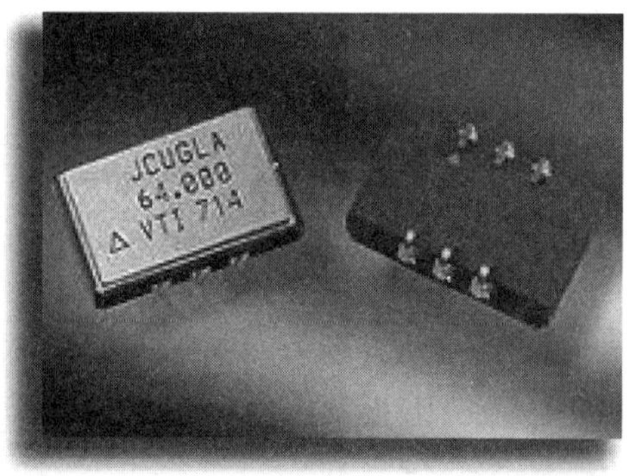

FIGURE 12.10 SO package with J-lead configuration.

FIGURE 12.11 Leadless chip carrier.

FIGURE 12.12 Ceramic quad flatpack.

lead pitch ranges from 0.020 to 0.050 in. The CQFP can accommodate high-density circuitry. Compared with its metal counterpart, it is lighter in weight and lower in profile (Fig. 12.12).

Ceramic Ball Grid Array (CBGA). A CBGA is a surface mount counterpart of the CPGA. Instead of using brazed metal pins, solder balls are used to increase the compliance of the solder joints. The array pitch can be as high as 0.050 in (Fig. 12.13).

12.1.3 Hermetic Seals

Glass-to-Metal Seals. Glass-to-metal seals are vacuum-tight assemblies of glasses with metals used to feed electrical conductors through the walls of hermetically sealed packages. Cold-rolled steel (CRS), ASTM F15 alloy (Kovar), and Alloy 52 are the three common metals that can seal directly to the glass without additional brazing compound. In general, there are two types of glass-to-metal seals. The first type is a compression seal, and the second type is a match seal.

For the compression-seal technique, CRS is the preferred metal for the enclosure. Then Alloy 52 is commonly selected as the I/O lead material, and 9013 or an equivalent type of glass is used as the insulator. The compressive strength of glass is generally 10 to 20 times as high as its tensile strength, which ranges from 20 to 80 MPa. This fact is used in compression glass-to-metal seals by applying uniform compressive stress to the glass element, which prevents the occurrence of tensile stress in the seal, even when exposed to relatively severe mechanical and thermal loads.[1] In compression seals, all materials are selected purposely for the coefficient of thermal expansion (CTE) mismatch. Because of the differential in expansion rates, the glass-to-metal seal assembly will become vacuum-tight after the cool-off period. After the glass is sealed, the metal elements shrink to form a compression ring around the glass element. In order to prevent edge cracking, a recessed design of the glass in relation to the metal ring is used (Fig. 12.14). The advantage of compression seals is that the enclosure and leads can be preplated prior to the glass-to-metal seal operation for better corrosion resistance.

For the match-seal technique, ASTM F15 alloy (Kovar) for the enclosure and leads and 7052 or borosilicate-type glass are the common choices. The CTE of all materials used in the match seal must

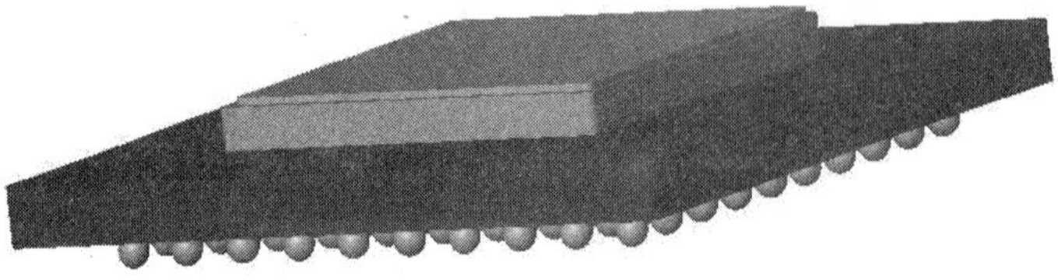

FIGURE 12.13 Ceramic ball grid array.

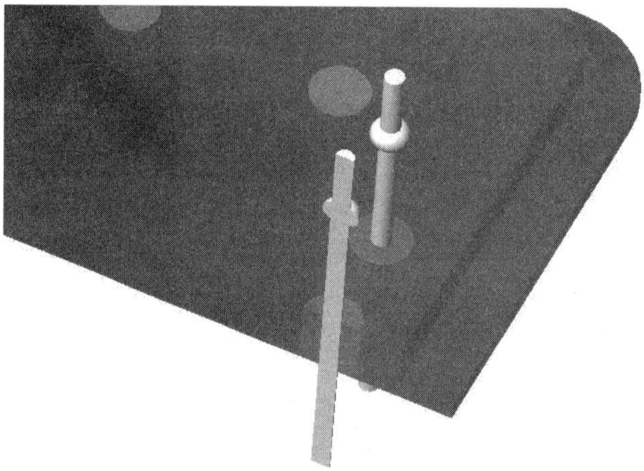

FIGURE 12.14 Glass-to-metal seal cross-sectional view.

be closely matched at any given temperature between the ambient temperature and the transformation temperature of the glass. In this way, mechanical stresses in the seal never exceed the stress limit of the glass while the glass-to-metal seal cools from the setting temperature of the glass (viscosity ranging from 10^{13} to $10^{14.5}$ dPas) to ambient temperature.[1] The conventional match-seal process consists of three consecutive steps: decarburization, oxidation, and sealing. Each employs a high-temperature furnace with a control atmosphere. The typical atmosphere is moist hydrogen or dissociated ammonia for decarburization, nitrogen containing O_2 or H_2/H_2O for oxidation, and nitrogen with H_2 plus H_2O for sealing.[2] ASTM F15 alloy (Kovar) enclosures and leads have to be oxidized in the furnace before assembly. The hermatic seal basically is provided by the oxide seal between the 7052 glass and the preoxidized ASTM F15 alloy (Kovar) enclosures and leads (Tables 12.2 and 12.3).

Ceramic-to-Metal Seals. Ceramic-to-metal seals typically involve a brazing process. Brazed ceramic-to-metal lead seals are a viable choice for high-current applications. In high-power environments, a large-diameter and high-conductivity lead is required to carry high current and dissipate heat. Traditional Kovar leads in glass-to-metal seal technology may not be sufficient for such applications. In addition, a brazed ceramic-to-metal seal provides higher mechanical strength than a glass-to-metal seal. In order to braze metal components such as leads to the ceramic, the ceramic must be metallized. Ceramics such as alumina and aluminum nitride are metallized with refractory metals such as tungsten and molybdenum for high-temperature sintering above 1400°C. To improve its wettability by brazing alloys, the ceramic metallization is electrode plated with a layer of nickel. After proper fixturing, the metal components are brazed to the ceramic metallized area with the brazing fillers in a hydrogen furnace environment. Depending on the brazing filler selected, the firing temperature can be range from 700 to 1100°C. Copper-silver eutectic is used commonly as the filler, and

TABLE 12.2 Sealing Metal Properties

	CR Steel (A-29)	Kovar (F-15)	Ceramic type
Thermal conductivity (W/in-°C)	1.48	0.44	0.39
CTE (ppm/°C)	11.7	5.1	9.8
Electrical resistivity (μΩ/cM)	17	49 4	3
Tensile strength (MPa)	448	534	534

TABLE 12.3 Sealing Glass Properties

	Corning 7052	Corning 7070	Corning 9010
Mass density	1.48	2.13	2.66
CTE (ppm/°C)	4.60	3.20	8.90
Strain temp. (°C)	435	455	408
Dielectric constant @1 MHz	4.9	4.1	6.3

its brazing temperature is 779°C. Depending on the ceramic-to-metal seal design, certain restrictions apply to selection of the lead-pin metal. The CTE values of the pin and ceramic must be well matched. Copper-cored Kovar and Alloy 42 are the common choices for the pins. Even with a favorable CTE match, the outside diameter of the pins is somewhat limited to 0.040 in.[3] The configuration of the ceramic-to-metal leads can be either a single feedthrough or a multipin feedthrough in a common ceramic strip (Fig. 12.15).

12.1.4 Package Sealing Technologies

Projection Welding. Projection welding is one of the widely used techniques of resistance welding. Like other resistance welding methods, projection welding requires physical contact between current-carrying electrodes and the parts to be welded. With the aid of pressure, welding occurs by the passage of low voltage and high current through the high-resistance projection ring of the package from the top electrode to the bottom electrode. The full welding cycle includes the approach, weld, and postweld periods for initial contact establishment, projection collapsing, and forging after weld solidification. Typical electronic packages such as metal DIL packages are sealed by the projection welding technique. In order to provide high-resistance point contact, either the header or the lid must possess a continuous weld projection ring along the welding flange (Fig. 12.16).

Power supplies for projection welders basically can be divided into ac and dc capacitor discharge types. Normally, ac is used where a relatively slow heat input is required. The dc capacitor discharge welder is used for rapid heat input. CRS and Kovar are the common material choices for the header. Typical surface finishes of the header include nickel plating with a gold overlay. The condition and design of the weld projection ring are critical for a reliable weld, as shown in Fig. 12.17. Nicks, scratches, and organic contaminations on the projection ring will create localized hot spots during welding. As a result, excessive weld spatters and microcracks will occur. Projection welding equipment is relatively inexpensive and easy to operate (Fig. 12.18). Since it is a "one shot" welding operation, it can accommodate high production throughput. The disadvantages are the limitation of package design and the high rate of wear of the electrodes.

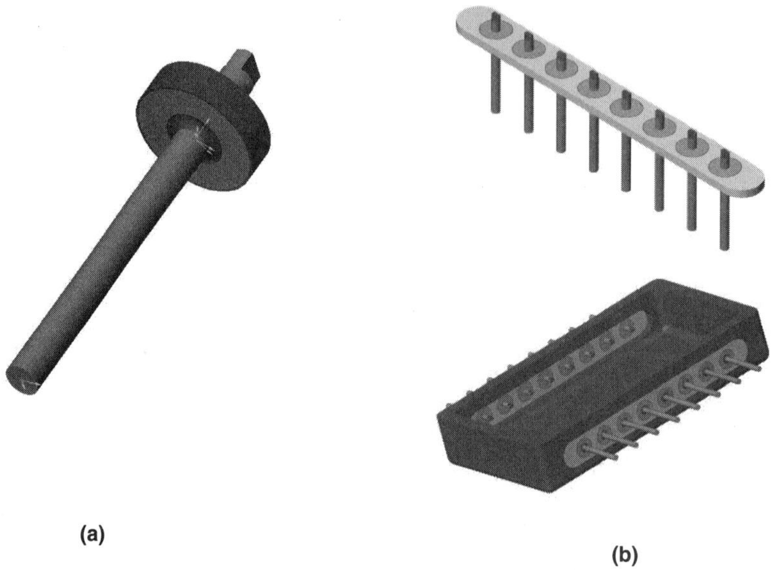

FIGURE 12.15 (*a*) Single feedthrough. (*b*) Multipin feedthrough module.

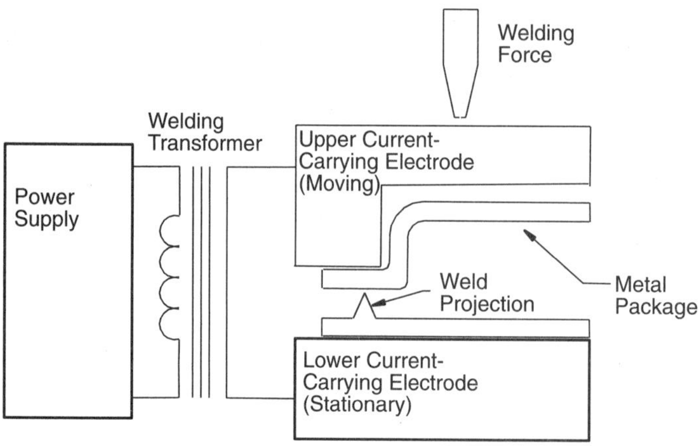

FIGURE 12.16 Projection welding.

Parallel Seam Sealing. Parallel seam sealing is a reliable method to seal a microcircuit hermetic package with a controlled internal atmosphere. Within the optimized process setting, parallel seam sealing equipment can perform weld sealing as well as braze sealing. Weld sealing involves welding two metal surfaces directly. Weld sealing is a fusing process to join two similar metal surfaces together by melting the interface. On cooling, the molten interface will become a solid metal structure again. Braze sealing involves a third metallic material as a brazing compound. In general, the eutectic temperature of the third metal is much lower than the melting temperature of the package materials. By reaching the eutectic temperature, the brazing compound will wet and seal the other two material surfaces together. Compared with weld sealing, braze sealing requires much lower sealing temperatures.

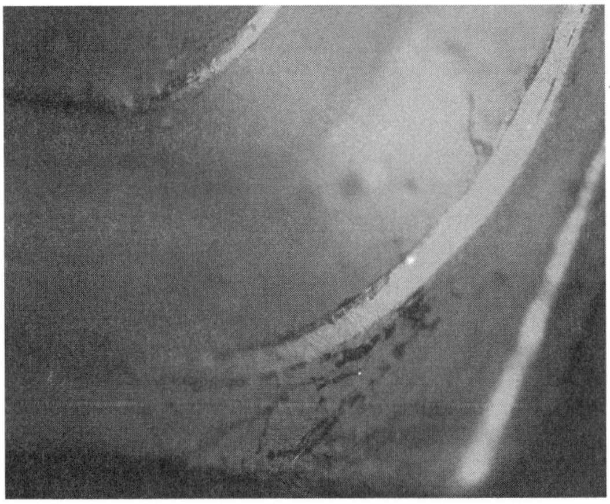

FIGURE 12.17 Projection ring.

FIGURE 12.18 Projection welding machine.

The mechanism of parallel seam sealing involves applying current through a set of conical rolling electrodes. The power supply produces a series of energy pulses that are conducted from one electrode across the package lid to the other electrode. By precisely overlapping the weld spots, a continuous weld joint is formed (Fig. 12.19). Extremely localized heat is generated right at the point of contact between the electrodes and the package lid when the electrode pair rolls along the edges of the metal lid. Localized heat is generated by the I^2R losses at the high-resistance contact surfaces between the electrodes and the package lid. The travel speed of the electrodes, duration and intensity of the pulses, and time between pulses are process variables that can be adjusted to optimize formation of the weld overlap (Fig. 12.20). Minimizing welding heat is extremely important because overheating will cause microcracking due to metal grain growth. In addition, prolonging welding time also will heat up the package very rapidly and will cause thermal damage and degradation of the microcircuit inside the package. Therefore, the material must flow in a shorter time than the thermal time constant of the package.

The relationship of the welding parameters is defined by the energy equation of the parallel seam sealing process[4]:

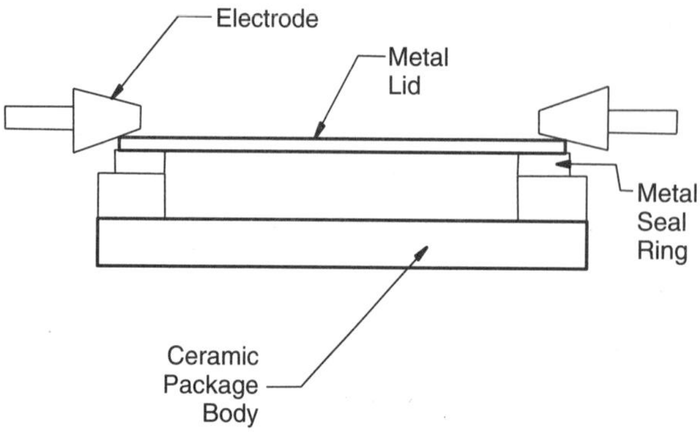

FIGURE 12.19 Parallel seam seal.

FIGURE 12.20 Seam seal weld joint.

$$\begin{aligned}\text{Energy (W}\cdot\text{s)} &= (\text{avg. power}) \times t \\ &= (V^2_{\text{RMS}}/R_c) \times t \\ &= (V^2_{\text{RMS}}/R_c) \times (PW/PRT) \times t\end{aligned}$$

where V = voltage across electrodes
R_c = contact resistance between electrode and edge of lid
PW = pulse width, in milliseconds
PRT = pulse repetition time, in milliseconds—defined as the time interval between the leading edges of successive pulses

Parallel seam sealing techniques can be applied to seal metal and ceramic packages, providing that the lid is of metal construction. Kovar is a commonly used material for seam-sealable packages because of its high resistance. The typical metal package suitable for parallel seam sealing is flat-pack package. Ceramic packages such as LCC and SOJ types also can be seam sealed by providing a metal top surface such as a Kovar seal ring. Matching package lids are mainly divided into two types: step lid and flat lid. A step lid can be fabricated by photochemical etching or machine milling processes. The typical step thickness is 3 to 5 mils. A flat lid is produced mostly by stamping processes. Depending on the package size, flat lids can range from 5 to 15 mils thick. Although thicker lids can withstand higher pressures, the packages require much more power to seal. The lid design rela-

FIGURE 12.21 Seam sealing equipment.

tive to the package is very critical. Improper lid design can result in weld arcing and reduce the efficiency of thermal energy transfer.

The advanced parallel seam sealing equipment shown in Fig. 12.21 incorporates different sealing modes such as parallel mode, rotary mode, and matrix mode. Parallel mode is the basic mode in which the electrodes seal along two parallel edges and rotate 90 degrees to complete sealing the remaining two edges. The weld overlap is located in all four corners of the package. For rotary mode, the package and lid are rotated 185 degrees relative to the stationary electrodes to provide a seal with a 5-degree overlap area. Matrix mode involves sealing the packages in an array format with the basic parallel seal mode.

Rework is relatively simple for a parallel seam sealing package. One industrial practice involves fixturing the welded package upside down and using a precision milling tool to remove the edge of the lid. The entire operation can be performed inside a vacuum enclosure to ensure removal of the delidding metallic particulates from the package cavity. After reworking, the package can be reattached with a replacement lid.

Cold Welding. Cold welding is a true solid-phase weld between two ductile metals at ambient temperature. The weld joint is consolidated by pressure. Under the high pressure introduced through the indentation ring on the welding die, a plastic flow of material takes place on the mating surfaces (Fig. 12.22). The deformation at the weld joint is in the range of 30 to 80 percent (Fig. 12.23). The end result is a hermetically sealed enclosure without contamination from weld spatters, dust, or vapor. More important, cold-welded enclosures can achieve a high level of vacuum tightness. Radiflow tests have been performed on cold-welded enclosures without detecting leaks at the instrument threshold of 10^{-11} atm·cc/s.[5] Surface preparation, deformation, oxide solubility, crystal structure, and pressure are key variables in the cold-welding process. Common package materials for cold-welded enclosures include pure copper or cladded materials such as copper-cladded Kovar and copper-cladded nickel. In most cases, the package is nickel plated to prevent oxidation on the copper surface. To prevent cracking and failure of the glass-to-metal seal, the weld decoupling effect should be considered carefully in the cold-weld package design phase.

Induction Heat Sealing. Induction heating techniques can be applied in sealing hermetic packages such as all-glass packages for quartz crystals (Fig. 12.24). The induction heating process is a noncontact technique that provides localized heating via electromagnetic waves. A custom-designed coil needs to be located strategically to provide the right amount of heat at the exact location for the exact

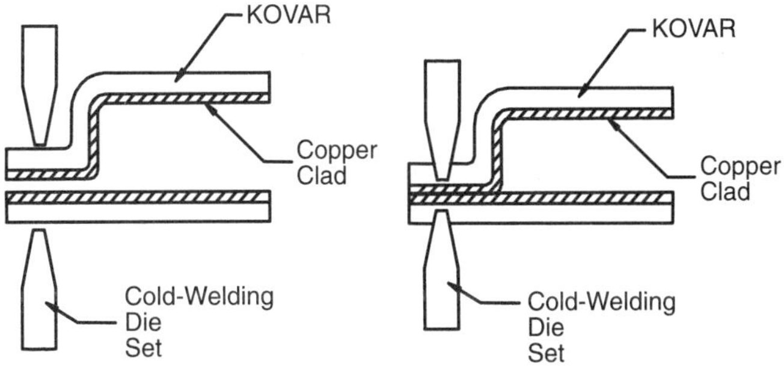

FIGURE 12.22 Cold welding.

FIGURE 12.23 Cold weld flange deformation.

FIGURE 12.24 All-glass crystal hermetic package.

amount of time. The packages can be sealed inside an enclosed chamber such as an evaculated bell jar with a vacuum or inert gas backfill. A glass-coated Kovar ring is used as a susceptor between the glass lid and the glass header. A steatite nest held in a plated-type induction coil fixtures the glass header. The glass lid is held down by a spring-loaded copper block that provides pressure to ensure intimate contact during the plastic flow phase of the glass after localized heating. Figure 12.25 illustrates the process. The induction heat sealing technique is highly repeatable.

Laser Welding. Laser welding is mostly performed by a solid-state neodymium-YAG (yttrium-aluminum-garnet) laser. In neodymium-YAG, neodymium is an impurity that takes the place of some yttrium atoms in the YAG ($Y_3Al_5O_{12}$) crystal. Nd-YAG laser rods are relatively small compared with the ruby rods. However, most of the energy can be removed from the rod in a Q-switched pulse, and the energy can be replenished in the millisecond duration of a flashlamp pulse. As a result, a repetitively pulsed Nd-YAG rod can generate high average powers, as well as high peak power in Q-switched pulses.[6] In order to perform reliable and highly repeatable hermetic package welding, the welding laser must have sufficient energy for heat transfer, relatively long pulse length, selection of a laser wavelength for optimal material absorption, and control of the pulse shape. Similar to parallel seam welding, laser welding produces a series of overlapping weld spots. Laser welding can join a variety of dissimilar metals due to its high energy and weld penetration. The heat generated is highly localized, and therefore, the package will not experience excessive heat rise. Since laser welding is a noncontact method, the gap between the lid and the header is highly critical to provide a reliable seal. An excessive gap will cause voids and potential leaks because the weld materials do not fill the gap properly. A gap of 7 to 8 mils will provide a uniform weld with a laser penetration depth of 40 to 50 mils.[7] In addition to welding dissimilar metals, laser welding is very effective in welding multiple-cavity packages.

E-Beam Sealing. Besides laser welding, E-beam sealing offers an alternative high-energy-density source for fusion welding. In principle, the electron (E) beam is generated by heating a tungsten filament. The E-beam is accelerated electrostatically by a voltage between a cathode and an anode. A

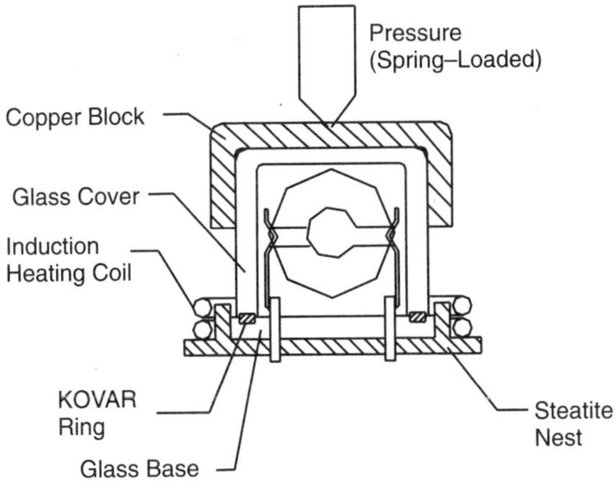

FIGURE 12.25 Induction heat sealing.

magnetic focus coil guides the E-beam to the surface of the workpiece. Positional control is provided by a separate beam deflection coil. To avoid the electron diffraction effect, the entire welding operation must be conducted inside a vacuum chamber. Beam power, beam traverse rate, and beam deflection pattern are the major control parameters. Beam spinning is an additional parameter in which a circular deflection pattern is superimposed on the E-beam while traversing the weld joint.[8] E-beam welding has advantages in welding dissimilar metal and irregularly shaped enclosures. It is very accurate and has high weld purity resulting from the under-vacuum operation. The main disadvantage is the high capital equipment expense.

Solder Lid Sealing. Gold-silicon, gold-tin, and gold-germanium solder preforms are usually required for solder lid sealing of packages. Solder preforms can either preattach to the metal lid or can be inserted between the lid and the package. Proper fixtures are required for accurate alignment and lid-holding pressure. Solder lid sealing can be conducted with several different types of equipment, as shown in Table 12.4.

Glass Lid Sealing. Glass lid sealing requires a ceramic frit lid or cap, as shown in Fig. 12.26. The typical sealing temperature is in the range of 280 to 430°C. Frit seal glass can be divided into two different types: lead-based and lead-free. The low-melting-point sealing glasses are printed at the sealing portion of the ceramic and then baked.[9] Mechanical fixtures are required to provide adequate holding pressure for ensuring a hermetic seal. Due to CTE mismatch between the ceramic and the glass, this technique is used mostly on relatively small ceramic packages. In addition, degradation of the gold-aluminum wire-bond interface also should be considered because of the high temperature of the process.

12.1.5 Hermeticity Testing

Leak-Rate Detection Criteria. The seal-test methods of MIL-STD-202 and MIL-STD-883 are the most common specifications to determine the hermeticity requirements of sealed packages. MIL-STD-883 requires the hermetically sealed packages to pass both a gross leak test and a fine leak test. The testing order is defined to be the fine leak test followed by the gross leak test. As described in test condition A, helium tracer gas is used, and the passing condition is based on either a fixed condition schedule or a flexible method.

According to the fixed condition, the sealed package should be tested using the appropriate bomb condition specified in Table 12.5 based on the internal cavity volume of the package under test.

TABLE 12.4 Solder Lid Sealing Equipment

Equipment	Maximum process temperature	Atmospheric control	Manufacturing mode
Convection oven	350°C	Inert gases, vacuum	Batch
Conveyor furnace	1200°C	Inert gases	Continuous
E-beam sealer	N/A	Inert gases	Batch
Seam sealer	N/A	Inert gases	Continuous (single)
Resistive graphite heating equipment	1000°C	Inert gases, vacuum	Batch

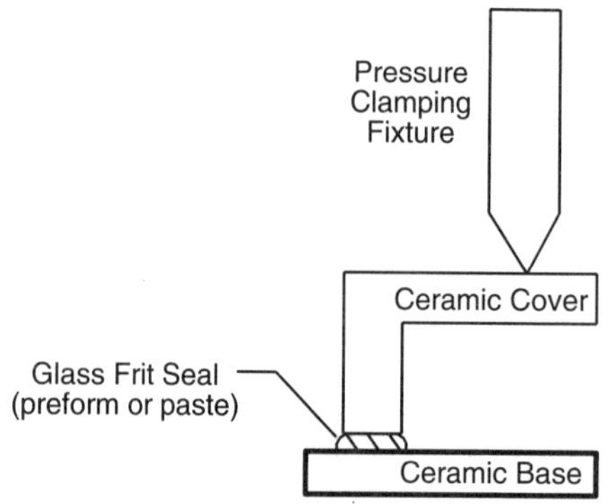

FIGURE 12.26 Glass lid seal.

TABLE 12.5 Fixed Conditions for Test Condition A1 (MIL-STD-883 Method 1014.9)

	Bomb condition				
Volume of package (V), in cm^3	Psi, ±2	Minimum exposure time, hours (t_1)	Maximum dwell time, hours (t_2)	R_1, reject limit, atm · cc/s He
<0.05	75	2	1	5×10^{-8}
≥0.05–<0.5	75	4	1	5×10^{-8}
≥0.5–<1.0	45	2	1	1×10^{-7}
≥1.0–<10.0	45	5	1	5×10^{-8}
≥10.0–<20.0	45	10	1	5×10^{-8}

The flexible method is defined in test condition A2 of MIL-STD-883 Method 1014.9. The measured leak-rate (R_1) limit is calculated by the following equation:

$$R_1 = LP_E/P_o \times (M_A/M)^{1/2} \times \{ 1 - e^- [L_{t1}/VP_o \times (M_A/M)^{1/2}] \} \times e^- [L_{t1}/VP_o \times (M_A/M)^{1/2}]$$

where R_1 = the measured leak rate of tracer gas (He) through the leak, in atm · cc/s He
L = the equivalent standard leak rate, in atm· cc/s air
P_E = the pressure of exposure, in atmospheres absolute
P_o = the atmospheric pressure, in atmospheres absolute
M_A = the molecular weight of air, in grams
M = the molecular weight of tracer gas (He) , in grams
t_1 = the time of exposure to P_E, in seconds
t_2 = the dwell time between release of pressure and leak detection, in seconds
V = the internal volume of the device package cavity, in cubic centimeters

Types of Tests

Gross Leak Test. In order to detect gross leaks, packages are immersed in a container filled with fluorocarbon fluid such as FC-40 or FC-43. The fluid is heated and maintained at 125°C. The test packages should be held at a minimum of 2 in below the surface of the fluid with the lid side downward. As the test packages are heated, the expanded internal gas will leak into the fluid through the openings on the packages. Leakers are detected visually by a pattern of a single bubble or a stream of bubbles under a magnifier. Gross leak test equipment is shown in Fig. 12.27.

Fine Leak Test (Helium Bomb). In fine leak testing, a mass spectrometer is used to detect fine leaks. Helium typically is used as the tracer gas. The procedure starts with pressurizing the test packages with a bomb schedule such as MIL-STD-883 Method 1014. After helium gas bombing, test packages are placed inside a vacuum chamber that is coupled with the mass spectrometer. If a leak exists, the helium tracer gas will pass through the test package into the vacuum chamber and finally will be detected by the mass spectrometer. Test packages should be rejected if the measured leak-rate (R_1) limit is exceeded. Figure 12.28 illustrates a commercial fine leak detector.

Radioisotope Leak Test. Similar to the helium leak test, the radioisotope leak test is used to detect package leak rates below 10^{-5} atm·cc/s. A tracer gas consisting of a mixture of dry nitrogen and krypton 85 is used to pressurize the test packages. The concentration of krypton gas in dry nitrogen should be no less than 100 μCi/atm·cc, according to MIL-STD-202F. The test packages are pressurized with the krypton 85 mixture for a few minutes to allow the mixture to enter leaky packages. The test packages are then removed and transferred to a counting station for identifying the rejects, which contain a measurable quantity of krypton 85 gas. The advantage of the radioisotope method is the high test throughput rate. The main disadvantages are the licensing and safety requirements in handling such equipment.

Laser Shearography. Laser shearography is a nondestructive leak testing method that measures the real-time deformation of the lid of a hermetic package under different pressure settings. The test packages are placed inside a variable-pressure chamber. Once pressure is applied, the lids of the hermetic packages will deform.

FIGURE 12.27 Gross leak tester.

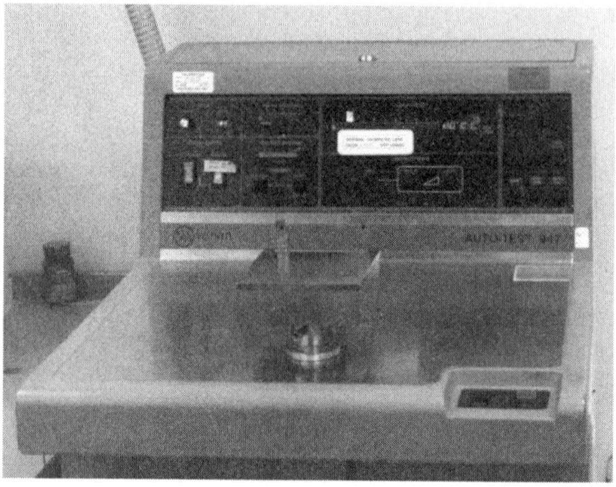

FIGURE 12.28 Fine leak detector.

Permanent deformation will occur with a hermetically sealed package. For defective packages, the deformation will recover in a time-dependent fashion. Shearography uses the inteference of monochromatic laser light to detect finite deformation in the submicron range. Digital images are recorded before and after the pressure is applied to the test packages for identification of defective packages. Laser shearography can detect leaks in the range from gross to fine.

12.2 PERFORMANCE

Modern packaging used for electronic devices, such as integrated microcircuits (silicon chips or dies), hybrid microcircuits, and surface mounted devices used as transistors, diodes, resistors, capacitors, transducers, and such, is designed both to protect the sensitive components and circuits mechanically and environmentally and to provide a functional interface to macroscopic applications, such as PWBs. Such packages usually are constructed from metals or ceramics and hermetically sealed in a moisture-free atmosphere to minimize the possibility of corrosion during use. Hermetic seals may be formed by soldering, welding, brazing, and sealing through glass.

There has been continuing concern regarding the presence of water vapor and other harmful gases such as carbon dioxide, sulfur dioxide, ammonia, and hydrogen in hermetically sealed semiconductor devices. This concern is based on both theoretically possible failure modes and actual observations of failure caused by corrosion due the presence of moisture or other corrosive gases.

12.2.1 Degradation in Devices

Degradation in hermetic packages occurs for various reasons. Moisture, particulates, and harmful gases such as hydrogen are some of the primary causes. Research shows that intrusion of moisture into microelectronic products is a major problem in the manufacture, performance, and reliability of electronic devices. Hermetic packages used in military, space, medical, and other applications requiring high reliability have an upper limit of 5000 ppm of water vapor content at the time of fabrication.[10] The package leak rate is limited to 10^{-8} atm·cc/s maximum to prevent leakage of significant amounts of moist ambient air into the packaging during the device's useful lifetime. Despite extreme precautions, it is difficult to manufacture a hermetic package for microelectronic devices with low water vapor content and to maintain the seal during the package's useful lifetime. There are various channels by which water vapor finds its way inside of the enclosure:

1. The various seals in a package usually are not perfect, and it has a small but sufficient leak to let ambient air, containing moisture, inside the enclosure.
2. Many epoxies used to bond dies and substrates outgas moisture into the packaging with time.
3. The packaging material itself outgases a certain amount of moisture. Prebaking prior to sealing may not liberate all the absorbed moisture.
4. The sealing atmosphere may be contaminated with moisture.
5. The testing process itself for leak rate may introduce moisture inside the package if not done with adequate care.

12.2.2 Effect of H_2 on Devices

Recently, small, lightweight, high-performance microwave amplifiers and other circuits have been made possible by the use of GaAs or other compound semiconductor devices. Reliability investigations, which started in the early 1980s, concluded that hermetically sealed semiconductor devices based on GaAs with palladium and platinum metal layers are significantly degraded by hydrogen. The effect of the desorbed hydrogen inside the hermetically sealed package has been well documented. Hydride formation, oxide reduction, and an excessive level of moisture that eventually lead to device failure are being attributed to the desorbed hydrogen source. Failure analysis of failed devices showed that the titanium resistors had undergone physical change. The thin-film titanium resistor gradually degraded via hydride formation. Degradation of GaAs MMICs caused by the desorbed hydrogen also was reported. Hydrogen gas in quantities of less than 0.5 percent can significantly degrade devices at an elevated temperature of 125°C for 168 h in sealed Kovar packages.[11] Residual gas analysis correlated the quantity of hydrogen with the gain reduction in the MMICs. Finally, oxide reduction due to the desorbed hydrogen can contribute to the moisture level increase inside sealed packages.

12.2.3 Sources of H_2 in Electronic Devices

The primary source of hydrogen has been found to be ferrous packaging materials such as Kovar and CRS,[b] electroplating, epoxy adhesives, and radiofrequency (rf) absorbers. Most metal housings contain structural imperfections, grain boundaries, precipitate interfaces, dislocation cores, etc., that can trap hydrogen. This trapped hydrogen can be desorbed during heating, such as during burn-in. Gold and nickel electroplating used in hermetic packages is permeable to hydrogen diffusion and also may be a source of hydrogen.[12] In addition, rf absorbers have been shown to release molecular hydrogen that shortens device life when converted to atomic hydrogen that poisons the semiconductor material. Epoxy adhesives used in assemblies have been known to outgas hydrogen, but not to the same extent as Kovar and other packaging materials.[13]

12.2.4 Mechanism of Degradation

A primary mechanism for device degradation is believed to be the transformation of molecular hydrogen into atomic hydrogen by platinum or palladium gate metallization that serves as the conversion catalyst. The highly reactive hydrogen radical ($H^{\cdot}$) diffuses into the semiconductor and the silicon dopant (donors). This results in reductions in current and gain of the device. A secondary mechanism is a reduction in Schottky gate barrier height due to modification of the gate-semiconductor interfacial layer.[14] Below is a schematic representation of this reaction:

$$H_2 \text{ (molecular hydrogen)} + \text{Pd (or Pt)} \rightarrow 2\text{H (atomic hydrogen)}$$

$$H\cdot + \text{GaAs}$$

Reactions occur that degrade semiconductor performance

12.3 RELIABILITY

Hermetically sealed semiconductor packages used in military, space, medical, and other applications have very low tolerance for water vapor, foreign particles, and other gaseous contamination such as ammonia, sulfur dioxide, carbon dioxide, and hydrogen. As was mentioned earlier, high-reliability hermetic packages have an upper requirement limit of 5000 ppm of water vapor content at the time of fabrication. Research shows that hydrogen concentrations as low as 0.5 percent at elevated temperatures can cause significant performance reduction in GaAs semiconductor devices. It is difficult to produce a hermetic package with low water vapor content and to maintain that level during the package's useful life. The various ways that water vapor and harmful gases (hydrogen) can enter the enclosure are described in Sec. 12.2.

Proposed solutions to the hydrogen and moisture problem include thermal treatment of the metal container, the use of low-hydrogen-absorption packages, substitution of noncatalytic gate metals, the use of nonhermetic packages, the use of low-outgasing adhesive, and placement of getter materials (moisture, hydrogen) inside the package. The most practical solution appears to be the in situ chemical removal of harmful agents, since sources are difficult to eliminate. The focus of this discussion will be low-outgasing adhesive and getters.

12.3.1 Use of Low-Outgasing Adhesive

Traditionally, gold-silicon eutectic metallurgy has been the primary means of mechanical and electrical attachment of silicone IC devices to ceramic substrates in hermetic packages. This process requires high temperatures (above 400°C) and long process times (over 90 min). Due to recent advancements in high-temperature polymer technology, new die-attach materials such as thermosets and thermoplastics are now available for hermetic packaging. Thermoset (typically epoxy), as the name implies, requires heat to set or cure. This process not only outgases during the curing process but also outgases harmful gases such as hydrogen during the lifetime of the package.

Staystik. On the other hand, thermoplastic adhesives, such as Staystik from Cookson SPM, are fully cured hot-melt adhesives that take on new shapes as they are heated. Since there is no curing involved, little or no outgasing is observed with this adhesive. For hermetic packaging, this is an attractive alternative to traditional metal or epoxy bonding.

12.3.2 Use of Getters

Getters are agents that counteract harmful contaminants inside a sealed package; this includes solids, liquids, gases, and combinations. With the recent advancement in MEMS and optical MEMS, controlling the atmosphere inside the package is critical. MEMS/optical MEMS devices typically are constructed of a silicon IC in a hermetic package. Moisture appears to be the culprit in package degradation. A simple but effective solution is use of a moisture getter.

Types of Getters

Particle Getters. Particle getters are tacky materials that, when cured, hold the particles in place.

Moisture Getters. As the name implies, these materials "get" or remove moisture from a sealed package. Typically, moisture getters consist of desiccant dispersed in a polymer matrix.

Gas Getters. The most commonly used gas getter is the oxygen getter. It has been in used since the earliest days of electronic vacuum tubes. Recent advancements in compound semiconductor devices, such as GaAs with a palladium and platinum metal layer, have brought the issue of degradation due to hydrogen into focus. Hydrogen getters have proven to be an effective solution to this problem.

12.4 PARTICLE/MOISTURE GETTERS

12.4.1 Moisture Getters

SD800 and SD1000. SD800 and SD1000, manufactured by Cookson SPM, are designed for use in a variety of high-performance electronic applications. Both these materials are nontacky and function as *moisture* getters, absorbing excess moisture released from assembled materials and adhesives during environmental conditioning. Both these materials are supplied as a thick paste that can be screen printed or dispensed.

SD800 is designed for low-temperature application (up to 375°C), whereas SD1000 is designed for high-temperature application (up to 400°C). These products also ingest other corrosive gases present in the environment, such as ammonia, sulfur dioxide, and hydrogen sulfide. When fully cured, the moisture-gettering capacity of SD1000 is more than 300,000 ppm/in^2/mil of coating in a 1 × 1 in package of 1.5 cm^3 free internal volume at 150°C and 5 times more at room temperature. Table 12.6 shows some of the key features of these getters.

12.4.2 Particle and Moisture Getters

GA2000-2. GA2000-2 is a two-part dual-function system that becomes tacky when cured. It functions both as a *particle and moisture* getter for applications requiring PIND testing and increased operating life in hostile environments. GA2000-2 meets or exceeds the limits for a getter material as stipulated in MIL-STD-883D Method 5011.3. Similar to the SD series, GA2000-2 is also printable and dispensable.

12.5 HYDROGEN GETTERS

As the name implies, these getters are specially designed to remove hydrogen gas from a sealed device. Several hydrogen getter materials were developed and include flexible films, pastes, coatings, and sputtered film. Among the choices available, flexible film appears to provide maximum protection and optimal processing, whereas sputter coating appears to be the second choice.

12.5.1 Methods for Hydrogen Capture

Sputtering of Palladium. The most common application of this method is a sandwichlike structure of titanium and palladium using sputtering or evaporation. In a hermetic package, the getter is usually applied to the lid used to seal the package. Titanium has been reported to absorb up to

TABLE 12.6 Key Features of SD Series and GA2000-2 Getters

	GA2000-2			SD series	
Typical properties	Part A	Part B	Mixed at 10:1	SD800	SD1000
Appearance	Blue	Clear	Blue	Blue	Blue
Viscosity (kcps)	(Premix)	(Premix)	50–60 at 5 rpm	144–216	75–150
Storage temperature	25°C	25°C	(Use once mixed)	25°C	25°C
Thermal stability at 250°C	N/A	N/A	<0.3% loss	<0.3% loss	<0.3% loss
Getter cure or dry temperature	N/A	N/A	2 h at 150°C	30 min at 150°C	30 min at 150°C
Getter activation temperature	N/A	N/A	30 min at 225°C	15 min at 170°C	15 min at 225°C
Shelf life (months)	12	12	24 h at 25°C	12	12

67 atomic percent hydrogen.[15] The palladium layer has two functions: (1) it prevents oxidation of titanium, and (2) it catalytically converts hydrogen, allowing it to diffuse into and be absorbed by the titanium. Typically, gettering capacity of sputtered palladium is much lower than that of palladium oxide.

Palladium Oxide Getters. In 1997, Cookson SPM acquired a hydrogen getter system from Allied Signal (H_2-3000). This product consists of an active oxide and desiccant in a flexible silicon polymer matrix. The active palladium oxide reacts with hydrogen, converting it to H_2O, and the desiccant absorbs the moisture to permanently remove the harmful gases. The reaction is irreversible and therefore drives hydrogen to a minimum. There are no hazardous or problematic side effects because the gases are converted to inert compounds that are retained in the getter system.

12.5.2 Getter Mechanisms

Getter mechanisms can be expressed schematically as shown in Fig. 12.29. As the figure shows, palladium oxide reacts with the hydrogen, reducing it to pure metal and water.

12.5.3 Manufacture of Getters. All getters (moisture, particle, and hydrogen) are produced using high-purity electronics or spacegrade polymer binders.

H_2-3000. Hydrogen getter is manufactured in a clean-room environment to minimize contamination. Table 12.7 lists the features of H_2-3000, Table 12.8 lists general data, and Table 12.9 lists available forms.

12.5.4 Application of Getters

Attach Process The hydrogen getter film can be attached to the lead or package before sealing. One of the standard recommended procedures for attachment is as follows:

TABLE 12.7 H_2-3000 Features

Hydrogen capacity: >45 cm^3/g
Moisture capacity: >5% by weight
Irreversible hydrogen reaction
Polymer matrix
Flexible
Film form
Highly permeable to both moisture and hydrogen

TABLE 12.8 H_2-3000 General Data

Temperature range: −55 to 150°C
Maintain H_2 levels <1 ppm
Maintain dew point <−100°F
CTE: 327 μm/°C
Density: 1.35 g/cc
Electrical properties (at 10 MHz)
Volume resistivity: $1.76 \times 10^{12}\ \Omega \cdot cm$
Loss tangent: 4.14×10^{-6}
Dielectric constant: 3.76
Dielectric breakdown: 1350 V

TABLE 12.9 H_2-3000 Forms

Sheets: Up to 8 × 10 in
Custom preform
Waffle packed
Bulk packed
Laser cut
Die cut
Standard thickness: 0.008, 0.010, 0.020 in*

*Other thicknesses are possible.

$$PdO + H_2 \longrightarrow Pd + H_2O \text{ (removed by desiccant)}$$

The by-product Pd (palladium can absorb additional H_2 (hydrogen)

FIGURE 12.29 Schematic of getter mechanism.

- Thoroughly clean the surface to which the film will be bonded.
- Apply a thin layer of adhesive, such as Dow Corning DC1204 primer, to both the H_2-3000 and the surface of the substrate, and allow to dry in a room with 50 percent or greater relative humidity for 4 h.
- Apply Dow Corning RTV 3145 (gray) to the surface of the H_2-3000 or the surface of the substrate.
- Press getter, with adhesive, to the surface of the substrate. Apply squeeze-out pressure (clean up excess, if needed).
- Cure in humidity cabinet for 16 h (85 percent relative humidity at 76°F).

Activation. Place the bonded assembly into a vacuum oven at 150°C for a minimum of 16 h. (Recommended vacuum is 500 mmHg or less.) Seal package *immediately* after activation step. Assemble in an inert atmosphere (90 percent of gettering capacity is gone after 45-min exposure to normal humidity at 25°C). Preassembled material can be preapplied or attached to the device by curing for 2 h at 150°C. Store in an inert atmosphere until ready to assemble (Fig. 12.30).

12.6 CONCLUSION

Harmful agents such as particles, moisture, and hydrogen gas typically found in hermetically sealed packages from numerous sources can degrade electronic devices over time. One successful solution is the installation of getters within the enclosure. Getters have the ability to dramatically increase device life by immobilizing solid particles and ingesting moisture and hydrogen gas. Palladium oxide, the preferred getter for hydrogen, irreversibly converts the hydrogen to water that is then consumed by a desiccant.

12.7 REFERENCES

1. "Schott Electronic Packaging: Glass-to-Metal Seals," Schott Glaswerke, Landshut, Germany.
2. Tamhandar, S.S., Bandyadhyay, N., and Kirschner, M.J., "Glass-to-Metal Sealing by a Single Atmosphere, Single-Firing Process," *International Society of Hybrid Manufacturers,* 11 (2), 1988, p. 31.
3. Anderson, C., "Improved Technology for Construction of Hermetic Power Hybrid Packages," *Hybride Circuit Technology,* June 1991, pp. 21-23.

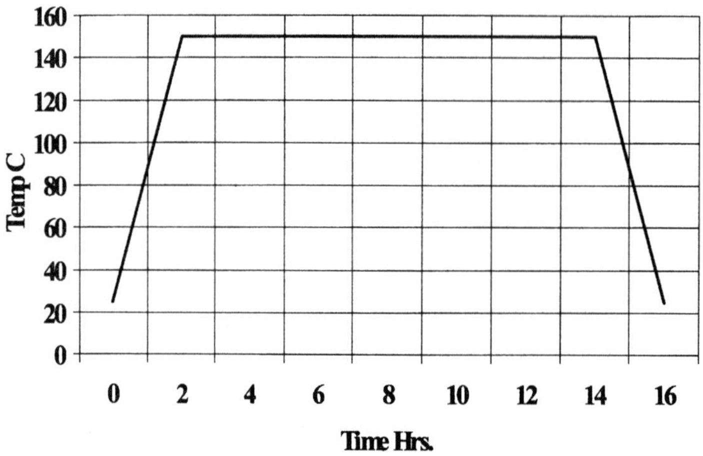

FIGURE 12.30 H_2-3000 activation profile.

4. Parallel Seam Sealing: Low Thermal Stress, High Reliability Hermetic Sealing," *Solid State Equipment Corporation Bulletin.*
5. Associate General Laboratories, Inc., Product Catalog, Franklin Park, IL.
6. Hecht, J., *Understanding Lasers,* J. Howard W. Sams & Company, 1988.
7. Mendicino, P., "Low Cost Hermetic Sealing of Microwave Modules,"
8. Zimmerman, D. D.,*The Fundamentals of Microjoining Processes,* ISHM Technical Monograph Series 6983-003. *ISHM*, Silverspring, MD, 1983.
9. Kyocera, "Semiconductor Components," brochure,
10. Shores, A., and Miculich, R., "An Effective Moisture Getter Coating for Hermetic Packages," in *ISHM 1992 Proceedings,* Pasadena, CA.
11. Camp, W., Lasater, R., and Hume, G., "Hydrogen Effects on GaAs Microwave Semiconductors," California Institute of Technology and Jet Propulsion Laboratory Report Number SMC97-0701.
12. Schuessler, P., and Gonya, S., "Hydrogen Desorption from Base and Processed Packaging Alloy," NIST Conference, Washington DC, April1993.
13. Saito, Y., Griese, R., Kono, R., and Fang J. "Hydrogen Degradation of GaAs MMICs and Hydrogen Evolution in Hermetic Package," *Microwave and Millimeterwave Monolitic Circuits Symposium Digest,* 1995, p. 119.
14. "Hydrogen Effects on Reliability of GaAs MMICs," 11th Annual GaAs IC Symposium, *Technical Digest,* 1989, pp. 203-206.
15. Hansen, M., *Constitution of Binary Alloys,* McGraw-Hill, New York, 1958.

CHAPTER 13

AREA ARRAY SOLDER SPHERES, PASTES, AND FLUXES

Gerard Minogue
Alpha-Fry Technology

13.0 INTRODUCTION

Ball grid array (BGA) technology is currently perhaps the most innovative practical means of addressing issues of package and board interconnect problems as pitches decrease, input-output (I/O) densities increase, and signal speeds increase. BGA technology is an evolution of pin grid array (PGA) technology, substituting solder spheres for brazed-on pins made of Alloy 42 or Kovar. As a rule, the use of solder spheres is less expensive than a pin in socket connection. BGA technology is completely amenable to surface mount technology (SMT) assembly and equipment as well.

13.1 SOLDER METALLURGY BASICS

13.1.1 Solder Alloys for Spheres

The selection of solder alloy is driven primarily by the assembly process and the materials to be joined. The most common material is the 63Sn/37Pb tin-lead eutectic alloy. As a close alternative, some manufacturers prefer to use 62Sn/36Pb/2Ag due to fatigue resistance and the dispersion and solution hardening assumed to be imparted by the addition of silver. The materials properties of the most common electronic solder alloys are summarized in Table 13.1.

63Sn/37Pb, 95.5Sn/3.5Ag, and 62Sn/36Pb/2Ag. These alloys differ metallurgically from the other BGA sphere materials in that they are eutectic alloys. The *eutectic* is a combination of the two or three constituent metals that remain liquid at a lower temperature than any other percentage combination of those metals. The liquid-alloy solution will freeze solid from the liquid solution at a distinct temperature, just as if one were dealing with a single molten metal. This point is referred to as the *liquidus-solidus transition.* Eutectic alloys are unique from other alloys in that their solidus value is the same as their liquidus value, and this value in turn is lower than the melting points of any of the constituent metals comprising the eutectic alloy. The solid solution following cooling will consist of two phases in the case of 63Sn/37Pb or three phases in the case of 62Sn/36Pb/2Ag. The eutectic compositions have lower viscosity and a higher degree of fluidity than other similar but noneutectic compositions comprised of the same metals. Noneutectic alloys have a gap between the solidus and the liquidus points that is referred to as the *pasty range.*

TABLE 13.1 Properties of Some Common Solder Alloys

Solidus	Liquidus	Alloy
118	125	50In/50Sn
118	118 (eutectic)	52In/48Sn
118	131	52Sn/48In
118	145	58Sn/42In
141	237	90In/10Ag
143	143 (eutectic)	97In/3Ag
157	157	100In
179	179 (eutectic)	62Sn/36Pb/2Ag
183	183 (eutectic)	63Sn/37Pb
188	188 (eutectic)	77.2Sn/2.8Ag/20In
219	219 (eutectic)	96.2 Sn/0.5Sb/0.8Cu/2.5Ag
217	222	95.75Sn/3.5Ag/0.75Cu
217	222	95.5Sn/4Ag/0.5Cu
221	295	90Sn/10Ag
221	240	95Sn/5Ag
221	221 (eutectic)	96.5Sn/3.5Ag
221	221	96Sn/4Ag
221	226	97.5Sn/2.5Ag
227	349	95.5Sn/0.5Ag/4Cu
227	300	97Sn/3Cu
227	227	99.3Sn/0.7Cu
227	232	99Sn/1Cu
232	232	65Sn/25Ag/10Sb
232	240	95Sn/5Sb
232	238	97Sn/3Sb
232	232	100Sn
234	245	91.5Sn/8.5Sb
280	280	20Sn/80Au
275	303	10Sn/90Pb

High-Lead 10Sn/90Pb. This is probably the most common high-lead material. Its high liquidus onset temperature (303°C) limits its direct application in direct reflow primarily to ceramic systems (CBGAs). High-lead solders are soft, and care must be taken not to mechanically dent or damage the spheres if this will pose concerns later during the manufacturing process. The high-lead spheres also have a greater propensity to darken during handling than the higher-tin-content spheres. This alloy is metallurgically compatible with 63Sn/37Pb and is commonly employed jointly with the tin-lead eutectic in joining applications in which the 63/37 accomplishes the joint and the 10/90 serves as a controlled dimension mechanical standoff. 10Sn/90Pb is also metallurgically compatible with 63Sn/36Pb/2Ag in those instances where the addition of silver is desired for improved fatigue resistance. The high-lead alloy is also compatible with no-lead solders, although in practical usage this fact is generally only of academic interest.

Lead-Free Sn/Ag and Sn/Ag/Cu Alloys. The prime driver away from the incorporation of lead in solders is governed by the desire to remove electronic assemblies as a contributor to lead in the environment. There are persuasive arguments for and against the use of lead in solder, but it is a reality that at present the elimination of lead from electronic solders has been legislated for the year 2008.

Lead-free solders are less desirable as measured by many of the metrics valued by electronic assemblers. In general, 63Sn/37Pb has superior plastic and creep behaviors than any of the popular no-lead solders (Table 13.2). The differential in mechanical behaviors between tin-lead eutectic and no-lead solders is maintained throughout the full range of temperatures encountered in electronic assembly. Mechanical strength of all common no-lead alloys is superior to that of 63Sn/37Pb (Table 13.3).

One possible risk that must be understood as inherent in the selection of some no-lead solders is that their greater creep resistance and lower creep rate combined with higher strength and stiffness relative to 63Sn/37Pb carries with it the risk of transferring mechanical strain to the chip or package via the BGA sphere, which may predispose the assembly to failure if it exceeds the ultimate strength of the pad metallizations.

13.2 BGA SOLDER SPHERE TECHNOLOGY AND METALLURGY

Many of the metallurgical principles governing conventional bulk solder and solder paste applications also apply to BGA spheres.

13.2.1 General Characteristics of Solders

All solders function as solvents for the metal surfaces to be joined. The solder permits the creation of a liquid-phase intermetallic at a temperature far below the normal melting temperature of the metals to be joined. It is this intermetallic, serving as the intermediary between the solder and the joined metals, that actually accomplishes the solder bond. Solders will only react with and wet pure metal surfaces. For most metals, this means that prior to solder wetting and bonding, a flux must act on the surfaces to be bonded to remove metal oxides on the surface. In practice, this is done by placing a flux under or alongside the BGA sphere. The flux activity is increased as the joint is heated to melt the BGA sphere so that by the time the sphere has reached the liquidus point, both the sphere surfaces and the pad surfaces are free of oxide and primed for solder wetting.

Behavior of Molten Solder Metals. Metallurgically and chemically, either leaded or lead-free solders may be defined as fusible alloys that manifest a melting point significantly below the melting point of the metal surfaces that are to be joined. Two types of melting behaviors are seen in solders. A true eutectic solder alloy accomplishes a sharp transformation from solidus to liquidus at a given temperature that is significantly lower than the melting points of any of the metals constituting the

TABLE 13.2 Mechanical Behavior of Sn63/Pb37 and Selected No-Pb Solders

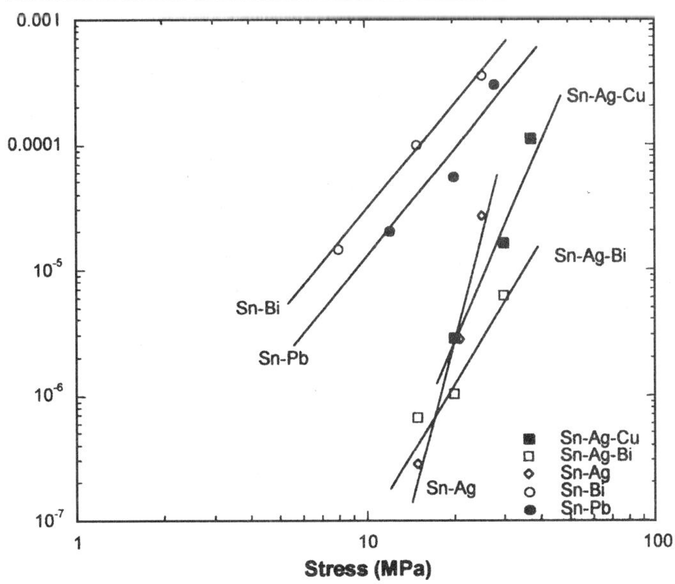

TABLE 13.3 Tensile Strength Values for 63Sn/37Pb and Common Lead-Free Alloys

Alloy	Tensile strength (ksi)
63Sn/37Pb	3.7
96.5Sn/3.5Ag	6.1
Sn/3Ag/0.5Cu	7.8
Sn/3.5Ag/0.5Bi	10.2
Sn/Ag/Sb/Cu	5.8
Sn/0.9Zn	9.6

solder alloy. There are a larger number of noneutectic solder alloys that have two critical temperatures: a solidus transition temperature and a liquidus transition temperature. Below the solidus temperature, the alloy is completely solid. Above the liquidus temperature, the alloy is fully in the liquid state. The temperature range between these two points is referred to as the *pasty region.* Within this range, the solder alloy is in a viscous, semifluid state. Noneutectic alloys frequently consist of ternary (three element) or quaternary (four element) alloys that are specifically engineered to make a solidus and liquidus transition at a specific temperature required by a particular manufacturing process. One drawback to solder alloys with a wide pasty region is that significant scavenging of other metals into the solder may occur during the time required to make the full transition to the liquid state. Alternate metallic phases may be created in the solder by the presence of the scavenged metals that may make the solder gritty and less prone to wet properly, may suddenly and dramatically raise the liquidus point of the alloy, or may make the resulting joint mechanically weak.

Solder creates the metallurgical bond at the device interconnection by dissolving into the solder a thin layer of the metal on the pad or device (typically copper or tin) at temperatures very much lower than the melting point of the copper alone. Soldering is accomplished by initiation of a dissolution of the device or pad surface metal, with the liquid solder serving as the solvent. An intermetallic layer is formed between the solder mass and the substrate metal. The intermetallic is wetted by both the solder and the substrate metals. The resulting solder joint constitutes a true chemical bond, not merely a physical adhesion of two surfaces.

The intermetallic layer properties in the created joint differ chemically and physically from both the solder and the substrate metal.

The key intermetallic system for both tin-lead alloys and the tin-containing lead-free solders is the tin-copper system. A number of intermetallic stoichiometries between tin and copper can exist depending on the soldering conditions employed. These intermetallics can have very different properties with regard to solderability and the mechanical integrity of the solder joint. In general, Cu_6Sn_5 is very solderable, whereas low surface energy tends to make Cu_3Sn bearing surfaces poorly wettable solder joints. Exposure of pad surfaces containing either Cu or Cu_6Sn_5 to heat and humidity can over a period of months to years induce the formation of Cu_3Sn with concomitant lack of wettability and degradation of the solder joint.

Surface Oxide. The primary objection to the presence of metal oxides on the surface of the BGA spheres and the mounting pad is that the surface energy mismatch is so great as to inhibit or even prevent wetting on reflow. DiGiacomo demonstrated that both electrical and thermal resistance can increase with the log of oxide thickness in solder. The formation of tin oxide on BGA spheres and joints, particularly as SnO_2, is kinetically favored over lead in tin-lead systems. The oxide growth rate on sphere surfaces and on sphere joints is greatest in the first 20 min to 4 h after sphere fabrication or joint reflow. The oxide growth proceeds more slowly over a period of weeks and approaches an asymptotic level after approximately 3 months of storage (Table 13.4).

Initial oxide thickness in BGA spheres immediately following manufacture is typically between 0.5 to 1.5 nm, rising to as much as 5 nm in extreme cases. The stoichiometry of oxide growth in tin-lead systems initially favors tin oxide due to kinetic advantages; however, the lead oxide component makes more and more of a contribution as time goes on and the sphere ages.

TABLE 13.4 Mechanical Behavior of Sn63/Pb37 Surface Oxide Growth

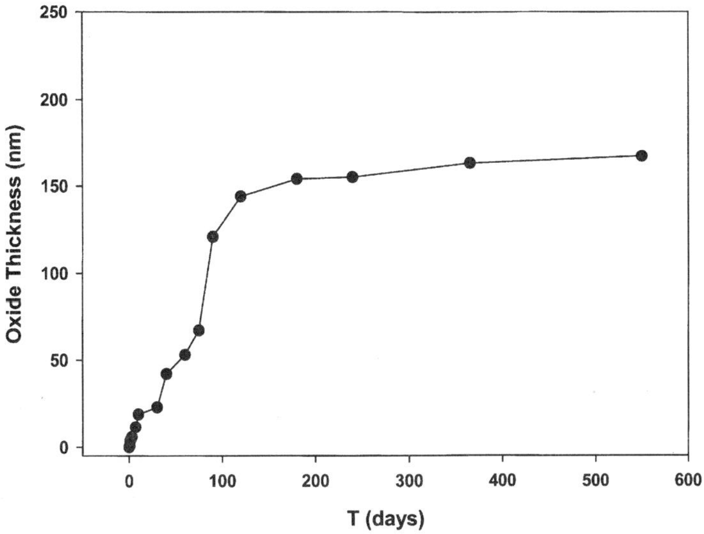

In general, a low burden of sphere oxide is desirable, but low oxide does not ensure superior wetting or a superior joint. Mechanical pull strength tests did not reveal a strong correlation across the board between bond strength and sphere oxide levels. High relative humidity by itself increased sphere oxide burden only slightly over a period of 2 weeks, providing the spheres were maintained in a relatively quiescent state in or out of their storage jars. High relative humidity combined with periodic mechanical agitation demonstrated a potent synergistic effect on both sphere surface oxidation and sphere surface darkening when compared with agitation in a low relative humidity environment or storage in a high relative humidity environment without agitation.

The oxidation of tin is kinetically preferred over most of the other species in both tin-lead and common no-lead, typically Sn/Ag/Cu (SAC), alloys. Table 13.5 presents the comparative Gibbs free energies of formation for the common solder sphere alloy constituents.

Surface analysis work of both solder joint and solder sphere surfaces conducted using x-ray photoelectron spectroscopy and Auger electron spectroscopy suggest that in the case of solder spheres, the tin phase is preferentially oxidized at the surface, and the oxidation of other metal species only occurs after the immediately available amounts of tin on the surface are significantly depleted. This appears to be true whether one is observing one of the alloys of tin and lead or one of the tin-based non-lead solders increasingly coming into use. Surface oxides of Sn/Ag and Sn/Ag/Cu (SAC) alloy BGA spheres are notably more resistant to reduction than Sn/Pb binary systems. This necessitates a judicious choice of BGA flux when reflowing no-lead spheres.

Grain Structure. Sphere grain structure is a corollary to the time-temperature environment seen by the spheres before and during processing. There are two times in which grain structure is relevant to the processing of BGA spheres: during manufacture and during recrystallization taking place after reflow. The grain size seen in the joint after recrystallization is partly determined by the grain structure of the BGA sphere prior to reflow and partly shaped by the time-temperature environment of the reflow cycle. Grain structure affects the physical appearance of the sphere prior to reflow, mostly as the degree of specular reflectivity (brightness). Typical grain size in tin-lead eutectic spheres is on the order of 10 to 20 μm with a homogeneous distribution of tin and lead phases. Grain size in other tin-lead binaries as well as 62Sn/36Pb/2Ag is comparable (Fig. 13.1).

TABLE 13.5 Thermodynamics of Common Solder Constituents

Oxidation reaction	ΔG°
$Sn + O_2 \rightarrow SnO_2$	−515.8
$2Sn + O_2 \rightarrow 2SnO$	−515.0
$2Pb + O_2 \rightarrow 2PbO$	−378.9
$2Cu + O_2 \rightarrow Cu_2O$	−292.9
$2Cu + O_2 \rightarrow 2CuO$	−254.6
$2Ag + O_2 \rightarrow Ag_2O$	−11.2

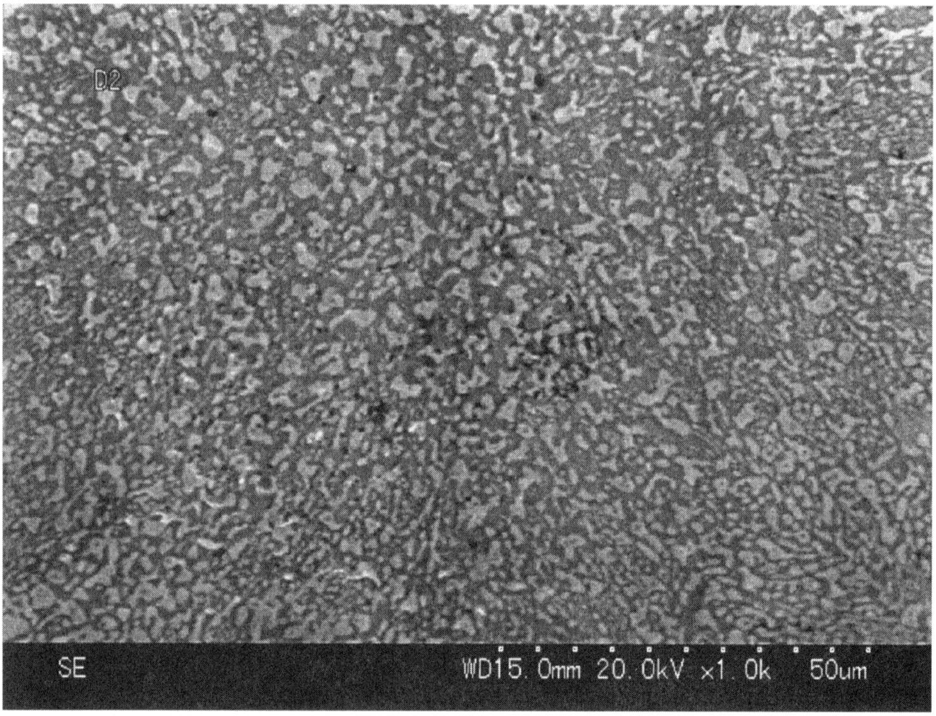

FIGURE 13.1 Scanning electron micrograph of 63Sn/37Pb BGA sphere surface.

Grain size distribution in spheres ideally is homogeneous. Spheres produced via reflow of preforms or wire may have less of a grain-size difference between the surface and the interior than jetted spheres when examined on cross section. This is due to the more rapid surface cooling regime experienced by the jetted spheres compared with spheres made by preform reflow. Tin-lead spheres held at elevated temperature approximating reflow temperature for extended periods and then recooled will show evidence of recrystallization usually seen as lead-phase coarsening (Fig. 13.2).

Often, 62Sn/36Pb/2Ag alloy spheres will show simultaneous tin-phase coarsening combined with precipitation of the silver phase into coarse dendrites if the sphere is not cooled rapidly enough on initial manufacture or on recrystallization following reflow (Fig. 13.3).

The main practical difference between spheres of a given alloy composition with different nominal grain sizes and grain-size distributions is brightness and aesthetic appearance. Extensive reflow testing of BGA spheres with tin-lead eutectic, 62Sn/36Pb/2Ag, and tin-silver-copper compositions has shown that so long as the stoichiometry of the feedstock alloy is correct, the brightness of the surface does not appreciably affect the reflow liquidus temperature or the nature of the final joint.

13.3 SOLDER SPHERE MANUFACTURING METHODS

13.3.1 Stamping and Reflow of Solder Preforms

The stamping and reflow of solder performs are the most common manufacturing methods used for volume production of BGA solder spheres. The process typically begins with either the batch or continuous casting of solder slabs or ingots. Following certification of composition and homogeneity, the slabs are passed through sets of progressive rollers that incrementally reduce the slab cross section to the typical target thickness of between 0.010 and 0.020 in. Process control of target thickness and thickness consistency (run-out) is key, and among present manufacturers, the solder sheet output thickness is monitored continuously via laser micrometer gauging that feeds back to the adjustable rollers in real time to maintain solder sheet thickness within specification. The rolled solder sheet is coiled onto take-up reels for die punching of solder reflow preforms.

The coiled solder sheet, which is between 2 and 12 in wide, is next passed into the die stamping operation to create the final sphere perform. Sets of circular punch dies arranged in arrays of between 20 and 200 individual dies punch out the preforms from the solder sheet at a feed rate of approxi-

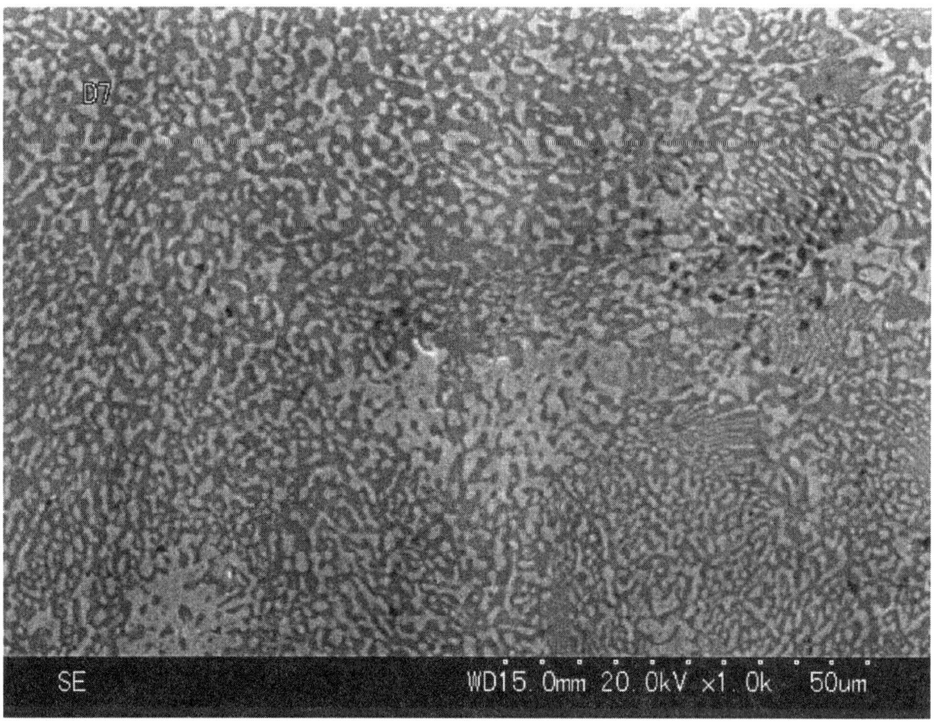

FIGURE 13.2 Scanning electron micrograph of 63Sn/37Pb BGA sphere surface showing grain coarsening after reflow.

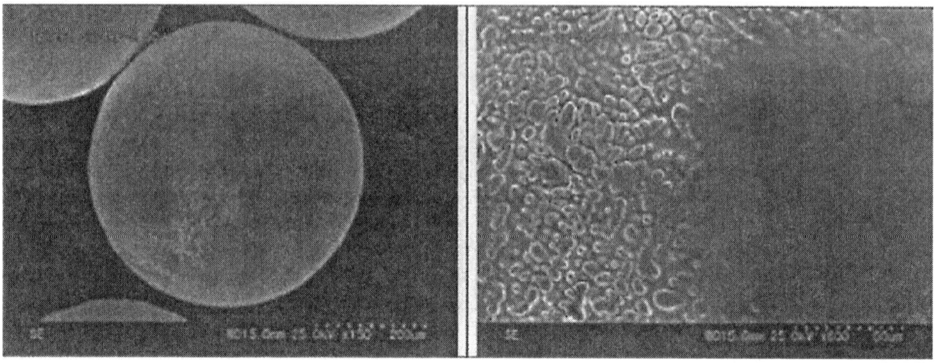

FIGURE 13.3 Scanning electron micrograph of 62Sn/36Pb/2Ag BGA sphere surface showing dendritic coarsening on cooling.

mately 12 in/s. Following die punching, the preforms are washed and sent on to the sphere-forming operation. The balance of the solder sheet left over following punching (the web) is recycled back into the slab casting operation.

The small solder preforms are disks that must be transformed into spheres with a high degree of part-to-part consistency. This is done by dropping the spheres into a liquid thermal transfer medium that is chemically and physically stable at the liquidus temperature of the solder and which is essentially nonreactive toward the solder. Commonly used reflow liquids are perfluoroether thermal transfer medium (3-M Fluorinert and Montedison Galden fluids), high-melting-point aliphatic long-chain organic molecules, high-temperature stable nonionic surfactants, and high-temperature silicone oils. The heat-transfer liquid must be resistant to oxidation at the high temperatures (220 to 340°C) employed and ideally should have a low solubility to oxygen at elevated temperatures so as not to oxidize the solder during the spheridization process.

Once the preforms are dropped into the thermal transfer liquid and reach their liquidus point, the molten solder will begin to spheridize spontaneously under the effects of surface tension at the droplet surface and the thermodynamic driver to minimize the free surface area in contact with the heating liquid. This brings up another key property of the thermally conductive liquid, namely, that it should not wet the liquid-solder surface or should wet it to the minimum extent possible in order to promote rapid spheridization. Once the liquid solder has spheridized, the temperature of the thermal liquid must be reduced below the liquidus point of the solder (or below the liquidus-solidus pasty region in the case of noneutectic solders). Typically this is accomplished by allowing the newly formed sphere to free fall under gravity into a region where the liquid temperature is below the solder solidus. At this point, the spheres are harvested, and the residual thermal liquid is washed away using a variety of conventional industrial cleaning methods appropriate to the chemistry of the liquid employed.

13.3.2 Solder Jetting

An alternative to the production of spheres via remelting of stamped preforms is to produce the spheres directly by the controlled breakup of a liquid-solder stream under tightly controlled conditions. The basic physics associated with this process is not new. The essentials for describing the fluid breakup behavior of liquids with viscosities approximating that of water was described by Lord Rayleigh in a paper presented to the Royal Academy in 1878. In it he characterized the relationship between a column of water ejected from an orifice or aperture and the droplets that are formed as the column breaks up under the effects of the fluid forces on the column and the surface tension on the free surface of the liquid. He discovered that for water (and other fluids with approximately the same viscosity), the relationship between the size of the droplets and the opening from which the liquid column originated is $D_d = 1.89D_a$, where D_d, represents the mean diameter of the droplets produced after fluid column breakup, and D_a represents the diameter of the aperture from which the fluid column originated. Both the stream and the droplets are subjected to a variety of hydrodynamic and aerodynamic forces (the precise discussion of which is outside the scope of this book) that cause a droplet-to-droplet variation as the stream breaks up.

At the extreme, where a transition is made from laminar to turbulent flow in the fluid column, fluid ligature formation and partial fluid atomization will take place. However, even under less vigorous conditions, where laminar flow is more or less maintained, spontaneous Rayleigh fluid breakup will result in a statistical variation in the droplet sizes produced. It is possible, however, to impart a periodic mechanical perturbation onto a fluid column and override the variations in Rayleigh breakup to produce monomodal droplets that individually spheridize under the driving force of surface tension and surface area minimization to produce essentially identical spheres.

In actual practice, solder is melted inside a sealed chamber that has been purged of air by an inert gas such as nitrogen or argon to prevent oxidation of the solder melt. The gas is also used to pressurize the container to force the molten solder out through an orifice in the wall of the container. Before the solder leaves the melt chamber, a periodic impulse is conducted to the liquid stream by any number of electromechanical or piezoelectric impulse conduction devices. The final size of the spheres is controlled by the combination of orifice size, velocity of the fluid stream ejected from the orifice (which, in turn, is a function of the combination of gas and hydrostatic pressure), and the frequency of the periodic mechanical impulse imparted to the ejected solder stream. The breakup of individual droplets from the continuous stream is driven primarily by the combination of periodic impulse value imparted to the liquid and the partial pressure of oxygen in the environment into which the spheres are jetted. For most of the commonly used electronic solders, the surface tension of the solder increases inversely with the partial pressure of oxygen in the jetting environment. The spheres cool through a combination of conduction and convection in the inert gas environment until they are solid and then can be collected for inspection, sorting, and packing.

13.3.3 Other Methods of Sphere Production

One alternative method for producing spheres that has been employed by some manufacturers is to substitute lengths of cut wire for rolled and stamped preforms as feedstock for sphere production. Solder feedstock is die drawn into wire and fed into a guillotine cutter that chops the continuous solder wire into segments. The solder wire segments are then fed into a bath of thermally conductive liquid in a process identical to that used with solder preforms. One advantage to the use of cut wire is that the forming technologies for wire generally are simpler and less expensive than those for stamped preforms. The main drawback to the use of cut wire for spheres is the difficulty of consistently achieving clean, controlled cuts on very small wire segments produced at high speed in sustained production runs lasting many hours.

13.4 SIZE, SHAPE, AND VOLUME

13.4.1 Quality Criteria

Three key criteria for sphere quality are diameter, sphericity, and surface condition. Typical industry specifications for diameter are expressed as a tolerance range from the nominal specification value. The tolerance range may be anywhere from ±0.0003 to 0.002 in depending on the nominal sphere specification size and the sphere application.

Standard Sizes. Standard sizes are dictated primarily by the demands of the various BGA interconnect technologies and secondarily by individual manufacturer's preferences and requirements for proprietary technologies. Common standard BGA sphere sizes are 0.010, 0.012, 0.014, 0.016, 0.018, 0.020, 0.025, 0.030, and 0.035 in. Standards are less evolved for sphere sizes under 0.010, but 0.004, 0.005, 0.006, and 0.008 have all been produced in quantity for dedicated micro-type BGA sphere applications. Spheres smaller than 0.012 in are more difficult to manufacture to a high tolerance and therefore command premium pricing.

13.5 ALLOYS: ATTRIBUTES AND APPLICATIONS

13.5.1 Eutectic Tin-Lead

Eutectic 63Sn/37Pb solder is presently the most common interconnect solder in use. Eutectic 63Sn/37Pb solder is distinguished by its ideal reflow temperature of 183°C, its reliable mechanical performance, its excellent wetting onto a variety of pad substrate materials with only minimal flux activation required, its ease of availability, and its relatively lower cost.

13.5.2 High-Lead

High-lead alloys are used in BGA sphere applications primarily as controlled dimension standoffs in CBGA (ceramic BGA) assemblies. The most common high-lead alloy employed in BGA is 10Sn/90Pb, which has an onset of liquidus of 303°C, far beyond that of 63Sn/37Pb (183°C) with which it is commonly paired.

13.5.3 Lead-Free

The move toward lead-free BGA alloys is driven primarily by environmental concerns to eliminate the introduction of elemental lead into the environment from electronic assemblies. The consensus choice for lead-free solder alloys appears at this time to be one or more of the tin-silver-copper ternary alloys in which the copper level varies from 0.5 to 1.0 percent. An excellent lead-free alternative to the ternary alloys is the 96.5Sn/3.5Ag eutectic.

13.5.4 Specialty Alloys

With the exception of the lead-free alloys coming into increasing use for environmental reasons, there are few if any other specialty solder alloys being employed for BGA sphere fabrication.

13.6 PACKAGING, HANDLING, AND STORAGE OF SOLDER SPHERES

13.6.1 Static Electricity Issues

Static electricity is an issue that definitely must be factored into BGA sphere handling once the spheres are removed from the container. Electrostatic dispersive (ESD) packaging is recommended. The propensity of BGA spheres to charge is derived from the capacitance effect stemming from the growth of surface oxide on the spheres. The initial oxide growth is quite rapid, occurring essentially simultaneously with the production of the spheres from the molten state. Once the spheres have solidified and cooled, the oxide growth stabilizes and tends to become self-limited after several weeks.

Spheres stored in glass or conventional plastic jars will charge tribolelectrically on agitation of the container. The net static charge in the spheres can range from tens to hundreds of volts. The natural surface oxide layer present on solder spheres following manufacture will make the spheres behave both as resistors and capacitors, the resistance being proportional to the stoichiometry and thickness of the oxide and the capacitance deriving from the dielectric properties of the oxide and the mirror charge induced in the spheres as they are agitated. Agitation of the spheres against one another and against the walls of a container during shipping not only will result in the darkening of the spheres from fret corrosion but also will increase the oxide thickness over time, making the spheres increasingly predisposed to static charging during handling.

The use of ESD containers is recommended for the storage and transport of solder spheres. However, the use of ESD containers does not always guarantee that all the spheres in a container will be static-free; since the oxide coat also has a characteristic breakdown voltage, the spheres in the center of the jar may retain a residual static charge even though the spheres in direct contact with the ESD material usually will discharge immediately.

The use of vibratory devices on feed funnels or hoppers for BGA sphere pick-and-place hardware should be avoided, if possible, or at least minimized. If a vibratory funnel or hopper is used, electrically grounded electropolished stainless steel is preferable to plastic. If plastic is used, a grounded dangler wire or strap should be positioned to discharge the spheres. Alternatively, an ionization gun-type device can be used to continuously discharge the spheres as they are fed into the pick-and-place hardware.

13.6.2 Denting

Denting of spheres during handling is an ever-present risk, particularly for high-lead alloys, which are softer than either tin-lead eutectic or most no-lead alloys. Denting occurs commonly during sorting and handling operations conducted by the manufacturer; however, spheres dented during manufacturing are by design removed prior to shipment to the customer. The major risk for sphere denting occurs during mechanical handling by pick-and-place hardware. Vibratory feed apparatus should be adjusted so as to instill the minimum amount of mechanical energy into the spheres being fed. Sphere denting in itself will not necessarily result in an improperly formed BGA joint, but it can result in jams or misfeeds of the BGA sphere placement hardware. Dents that occur after manufacturing should be differentiated from surface artifacts that are normally a part of sphere cooling and solidification, particularly in the case of silver-containing alloys or no-lead alloys. In these cases, the specular reflectivity commonly associated with tin-lead alloys typically is not seen due to the differing cooling regimes and mechanics of alloy segregation on the sphere surface.

13.6.3 Oxidation

In the absence of extensive posthandling, the majority of sphere oxidation occurs in the first seconds to minutes after manufacture, at least in the case of tin-lead and other common alloy systems used in electronic assembly. The kinetics of oxidation for the tin-lead systems favor the initial formation of tin oxide.

13.6.4 Discoloration

The major discoloration phenomenon seen in BGA spheres, excepting the possibilities of alloy contamination or extrinsic coatings applied following the manufacture of the sphere, is sphere darkening due to fret corrosion. The degree of darkening or discoloration will vary in direct response to the degree of surface oxidation induced by agitation. Discoloration varies from slight dulling of surface specularity to near black in tin-containing systems. High-lead solders such as 10Sn/90Pb will develop a purplish discoloration on agitation that will turn black rapidly. In general, high-tin solders are more prone to rapid surface discoloration and darkening than are high-tin solders. High-tin lead-free solder systems tend to darken in a manner analogous to tin-lead eutectic solder. Solder spheres that are kept in a dry atmosphere (50 percent or less relative humidity) and are not mechanically agitated will keep their initial surface brightness for 6 months or more based on tests conducted at Alpha Metals laboratories. Unless special dulling flux compositions are used, the joints created by reflow of BGA solder spheres are as bright as those formed by solder paste reflow.

Atmosphere. The standard processing atmosphere for both tin-lead and no-lead solder sphere reflow is nitrogen gas, with the ambient oxygen level ideally at 1000 ppm or lower. The driving force for reduction of oxygen during the reflow profile is the trend toward milder and milder activator systems in BGA fluxes, particularly the water-washable and low-residue no-clean classes of fluxes.

Storage atmosphere for BGA spheres prior to placement and reflow ideally should be in as cool and as dry a location as possible. If the spheres are not to be used for some period of time following receipt, the container should be left sealed. If opened for incoming inspection, it should not be left open. High relative humidity (>60 percent) exposure of solder spheres that are stationary in a storage container ordinarily will not cause extensive surface oxidation and solderability problems within a period of 2 days to 2 weeks, a typical residence period for a high-volume manufacturing environment. High relative humidity combined with high temperatures (>30°C) and mechanical agitation or vibration of the spheres can cause rapid acceleration of both surface oxidation and surface darkening. The agitation exacerbates the oxidation process by repeatedly cracking and fretting away surface oxide, exposing fresh metal for further oxidation. The fretted oxide removed from the spheres is occasionally visible as a fine black powder in the container holding the sphere product.

Temperature. Experimental evidence suggests that temperature excursions below those encountered in the assembly process itself do not appreciably affect BGA spheres unless combined with high humidity and mechanical agitation. In this instance, significant levels of sphere surface oxidation and darkening can be expected if the spheres are also stored above 30°C.

Storage Limits. Tin-lead spheres left intact inside sealed shipping jars may not retain solderability indefinitely, but they will definitely remain intact for periods of at least 4 to 6 weeks if not disturbed and if not exposed directly to high temperatures (>30°C) and high relative humidity (>50 percent). Spheres that have been opened for use may begin to experience solderability loss within days at high temperature and high relative humidity, but this does not usually present a problem in most manufacturing environments where sphere consumption is much greater than one jar every few days. If assembly operations are to be interrupted for more than several hours, solder spheres should be returned to their packaging and preferably stored in cool, dry conditions to maintain maximum solderability.

Assembly Machine Damage (Balling). The risk of assembly machine or pick-and-place damage comes from two possible sources: the mechanical energy imparted to the spheres by vibratory generators intended to keep the spheres feeding smoothly without bridging and the mechanical action of the vacuum pickup chucks or positioning plate depending on the assembly hardware manufacturer chosen. Spheres produced via conventional manufacturing technologies involving oil reflow of cut preforms on occasion may demonstrate less of a tendency to bind or gall in assembly hardware than do jetted spheres, the tradeoff being that jetted spheres may have more of a chemically pristine surface than oil reflowed spheres by virtue of the manufacturing technology employed.

13.7 FLUXES

13.7.1 For Sphere Attachment

Sphere attachment fluxes are similar in chemistry and composition to fluxes used for corresponding solder powder alloys in solder pastes. The chief distinction between BGA fluxes and fluxes for other soldering applications lies in the degree of tack engineered into the flux, which not only must activate the sphere and pad surfaces but also must act temporarily as a glue to hold the sphere in place between the time it is placed and the time it is reflowed. Sphere attachment fluxes are almost entirely either WS (water soluble) or NC (no clean) compositions. Extensive testing performed at Alpha Metals has shown the WS BGA fluxes and the OMNIX family of fluxes to yield exceptional performance in BGA sphere reflow applications for both 63Sn/37Pb and tin-silver and tin-silver-copper no-lead alloy formulations.

CHAPTER 14
MODERN SOLDER AND SOLDER PASTE

Jennie S. Hwang
H-Technologies Group, Inc.

14.0 INTRODUCTION

Within the scope of electronics and microelectronics applications, solder has been serving successfully as a primary interconnecting material for all three different levels of interconnections: die (semiconductor) level, carrier (package) level, and main circuit board level. In addition to serving as interconnecting material, solder also has performed as the surface coating for components and boards.

This chapter is an overview of solder, particularly solder paste for both lead-containing and lead-free solder, with the focus on the properties and characteristics that are important to the applications in area array packaging and assembly.

One general definition of solder is a fusible alloy with a liquidus temperature below 400°C (750°F). The successful alloys for solder bumps at the die level (particularly the flip chip) consist of high-temperature compositions (for lead-bearing alloys, high-lead compositions such as 5Sn/95Pb or 10Sn/90Pb; for lead-free alloys, some proprietary alloy may apply) as well as eutectic or near-eutectic alloys such as 60Sn/40Pb, 62Sn/36Pb/2Ag, and 63Sn/37Pb. The solder bump on the underside of the carrier chip-scale package–ballgrid array (CSP/BGA) substrate, for example, also can either be high-temperature or near-eutectic tin-lead or lead-free equivalents. Because of the temperature tolerance level of conventional board materials such as FR-4, board-level solder for attaching components and integrated circuit (IC) packages is limited to eutectic, near-eutectic tin-lead, or tin-lead-silver solders or the lead-free equivalents.

Solder can be made in various physical forms, including bars, ingots, wire, powder, preform in designated shapes and dimensions, balls, and paste. In addition to the elemental compositions and physical form, the performance of the solder material is determined by the specific land pattern such as for BGAs that solder will be applied on, and the soldering process as discussed in Sec. 14.3. Soldering process is in turn distinguished by the type of chemistry or cleaning methodology. Three main chemistries, rosin mild activation (RMA) chemistry with solvent cleaning, water-soluble chemistry with water cleaning, and no-clean chemistry without cleaning, are considered the state-of-the-art technologies.

14.1 SOLDER MATERIAL

The intrinsic materials properties can be grouped into three categories: physical, metallurgical, and mechanical.[1]

14.1.1 Physical Properties

Although other properties can be contributing factors to the overall performance of solder, five physical properties—phase-transition temperature, electrical conductivity, thermal conductivity, coeffcient of thermal expansion, and surface tension—are discussed in this section because of their special importance to today's and future electronic packaging and assembly.[1]

Metallurgical phase-transition temperatures have practical implications. In application, the liquidus temperature can be considered to be equivalent to the melting temperature and the solidus to the softening temperature. At a given composition, the temperature range between liquidus and solidus is referred to as the *plastic range* or *pasty range.*

It is apparent that the solder alloy selected as the interconnecting material must be compatible with the service temperature in the worst-case condition. It is advisable that the alloy have a liquidus temperature at least two times higher than the upper limit of the expected service temperature. As the service temperature gets closer to the liquidus temperature, solder generally becomes mechanically and metallurgically "weaker."

The electrical conductivity of solder interconnections contributes to the performance of transmission of electrical signals. Solder, a metal, can be viewed as an assembly of positive ions immersed in a cloud of electrons and metallic crystals held together by the electrostatic attraction between the negatively charged electron cloud and the positively charged ions. By definition, electrical conductivity is the result of movement of electrically charged electrons or ions from one location to another under an electrical field. Electron conductivity is predominant in metals, whereas ionic conductivity is responsible for oxides and nonmetallic metals. For solders where electrical conductivity is contributed primarily by electrons, the resistivity (the reciprocal of electrical conductivity) increases with increasing temperature due to the reduction of electron mobility that is directly proportional to the mean free path of electron motion as temperature increases. The electrical resistivity of solders also can be affected by the amount of plastic deformation; resistivity increases with increasing amounts of plastic deformation.

The thermal conductivity of metals normally correlates well with electrical conductivity due to the fact that the electrons that carry both thermal and electrical energy are primarily responsible for thermal conductivity as well as electrical conductivity. For insulators, however, the phonon activity predominates. The thermal conductivity of solders decreases with increasing temperature.

Coefficient of thermal expansion (CTE) issues have been under observation and an area of effort by the industry since the inception of surface mount technology (SMT) in the printed circuit industry. This is due to the large difference in CTEs of materials that are interconnected. A typical assembly consists of FR-4 board, solder, BGA, CSP, and other leadless and leaded components. Their respective CTEs are FR-4, 16.0×10^{-6} μm/°C; 63Sn/37Pb, 23.0×10^{-6} μm/°C; Cu leads, 16.5×10^{-6} μm/°C; and Al_2O_3 leadless components, 6.4×10^{-6} μm/°C. Under temperature fluctuation and power on/off, this mismatch in CTE increases the stress and strain imposed on the solder joint, which consequently may shorten its service life and lead to premature failure. Two major material properties dictate the magnitude of CTE, namely, crystal structure and melting point. When materials have similar lattice structures, the thermal expansion of the materials varies inversely with their melting point.

The surface tension of molten solder is a key parameter related to the wetting phenomenon, thus also to solderability. The relative strength of attraction forces acting between molecules of the surface is weaker than that of molecular forces in the interior due to the broken bonds at the surface. Thus the free surface of a material has higher energy than the interior. Surface tension is a direct measure of the intermolecular forces acting at the surface. A simple but important concept is that the wetting/spreading occurs when the free energy of the newly formed system after wetting is lower than that before wetting. In other words, in order for molten solder to wet the substrate, the substrate surface must have higher surface energy than the molten solder. In view of this requirement, the lower the surface energy of the molten metal or the higher the surface energy of the metal substrate, the more favorable is the wetting. It should be noted that fluxing is intended to maximize the surface energy of the metal substrate, not to lower its surface energy, as occasionally misrepresented in the literature. In conjunction with the proper metallurgical reaction, this is how the flux/fluxing plays a role in wetting.

14.1.2 Metallurgical Properties

Under the environment and conditions to which solder interconnections are exposed during their service life on the circuitry, the metallurgical phenomena commonly occurring in solder materials include plastic deformation, strain hardening, recovery, recrystallization, solution hardening, precipitation hardening or softening, and superplastic deformation. When solder is exposed to an applied force, be it the result of mechanical or thermal stress, solder deforms irreversibly. This irreversible deformation is called *plastic deformation.* Plastic deformation is usually initiated through shearing on a number of parallel planes of its crystal structure. The plastic deformation may proceed globally or locally within the solder joint, depending on the stress level, strain rate, temperature, and material characteristics. Continued or cyclic plastic deformation eventually leads to solder joint fracture.

As a result of plastic deformation, solder may be hardened (strain hardening), as often observed in the stress versus strain relationship. A counteracting phenomenon to strain hardening is the recovery process, which is a softening process. The solder tends to release the stored strain energy. The recovery process is driven by thermodynamics. This energy-releasing process that starts at a rapid rate and proceeds at a slow rate is called *recovery.* During the recovery state, the physical properties that are sensitive to joint defects tend to be restored to their original value; however, this does not impart any detectable change in microstructure.

The recrystallization process is another phenomenon often observed in a tin-lead solder joint during its service life. It usually occurs at relatively high temperatures and involves a larger amount of energy release from the strained materials than the recovery process. During recrystallization, in addition to the energy release, a new set of essentially strain-free crystal structures is formed, which obviously involves both a nucleation and a growth process. The effect of solid-solution alloying results is an increase in yield stress. A typical example of solution hardening is that tin-lead compositions are strengthened by addition of antimony. Solution hardening can occur at an even larger extent in the well-designed lead-free solder alloys. Another strengthening effect can come from a structure with well-distributed fine precipitates (precipitation hardening).

In general, for a system with liquid to wet the solid substrate, the spreading occurs only if the surface energy of the substrate to be wetted is higher than that of the liquid to be spread. As the molten solder solidifies during cooling to form solder joints, the cooling process, such as the cooling rate, has a direct bearing on the resulting solder joint as to its microstructure and voids development.

14.1.3 Mechanical Properties

Three fundamental mechanical properties of solders include stress versus stress behavior, creep resistance, and fatigue resistance. Although stress can be applied by tension, compression, or shear force, most alloys are weaker in shear than in tension or compression. Shear strength is important because most solder joints are subjected to shear stress during service.

Creep is a global plastic deformation that results when both temperature and stress (load) are kept constant. This time-dependent deformation can occur at any temperature above absolute zero. However, creep phenomena only then become significant at "active" temperatures.

Fatigue is the failure of alloys under alternating stresses. The stress that an alloy can tolerate under cyclic loading is much less than that under static loading. Therefore, the *yield strength,* a measure of the static stress that solders will resist without permanent deformation, often does not correlate with fatigue resistance. The fatigue crack usually starts as several small cracks, which grow under repeated applications of stress, resulting in a reduction of the load-carrying cross section of the solder joint.

Solder in electronic packaging and assembly applications normally undergoes low cycle fatigue (a fatigue life of less than 10,000 cycles) and is subjected to high stresses. Thermomechanical fatigue is another test mode used to characterize the behavior of solder. It subjects the material to cyclic temperature extremes, i.e., a thermal fatigue test mode. Either method has its features and merits, yet both impose strain cycling on solders.[2]

In addition to the intrinsic bulk material strength, the strength of solder joints is often affected by joint configuration, metallurgical reactions, interfacial wettability, interfacial effect, and the characteristics of other materials incorporated in the assembly.

Alloys of tin-silver, tin-antimony, and 5Sn/85Pb/10Sb are found to impart high creep resistance. This is primarily attributed to solution hardening, as substantiated by their high strength and low elongation. When load is applied, the deformation is hindered by means of either interaction of solute atoms with dislocations or interaction with the formation and movement of vacancies, resulting in the impediment of the dislocation movement.

Alloys 10Sn/90Pb and 5Sn/95Pb, however, are benefited by the high melting point of their microstructural continuous phase, resulting in the more sluggish steady-state creep. This is attributed to lower self-diffusion, although the alloys are ductile and have moderate strengths. Alloy 62Sn/36Pb/2Ag has the highest creep resistance. Its mechanism, whether through the impediment of grain-boundary sliding due to silver segregation or the result of high activation energy for the dislocation movement, is not substantiated.

Bismuth alloys, 42Sn/58Bi and 43Sn/43Pb/14Bi, although having high tensile strength, are found to be prone to creep. This may be due primarily to their low melting temperature and the predominance of the diffusion-controlled process. The low melting point of their microstructural continuous phase is considered a main factor.

As the testing temperature or the applied load changes, a change in the creep behavior of the alloys may result. With continued development of lead-free solders, new ternary, quaternary, and pentanary systems will proliferate.[3] Sections 14.13 through 14.17 summarize the development and selection of lead-free alloys.

14.1.4 Solder Alloy Selection: General Criteria

Generally, the alloy selection is based on the following criteria:

- Alloy melting range in relation to service temperature
- Mechanical properties of the alloy in relation to service conditions
- Metallurgic compatibility, consideration of leaching phenomenon, the potential formation of intermetallic compounds
- Rate of intermetallic formation in relation to service temperature
- Other service compatibility considerations, such as silver migration
- Wettability on specified substrate
- Eutectic versus noneutectic compositions
- Ambient environment stability

14.2 SOLDER PASTE

Solder in paste form is of particular importance to the industry due to its unique virtues. Solder paste, with its deformable viscoelastic form, can be applied in a selected shape and size and can be adapted readily to automation. Its tacky characteristic provides the capability of holding parts in position without additional adhesives before the permanent metallurgic bond is formed. The metallic nature of solder paste offers relatively high electrical and thermal conductivity. The combined features in adoption to automation, tackiness, and high conductivity make solder paste the most viable material for SMT assembly manufacturing. It provides electrical, thermal, and mechanical interconnections for electronic packages and assemblies.

Paste technology is an interplay of several scientific disciplines, as depicted in Fig. 14.1. From the concept of paste technology, many commercial product lines are derived. These include thick-film materials, polymer thick-film products, conductive adhesives, electromagnetic interference (EMI) shielding materials, brazing pastes, and other products that are composed of metallic or oxide particles uniformly distributed and imbedded in the organic-polymeric matrix. Each of these product lines has its unique performance requirements, processing parameters in its making process, and end-use function. However, one thing in common is paste technology. The dedicated discussion can be found in the literature.[4]

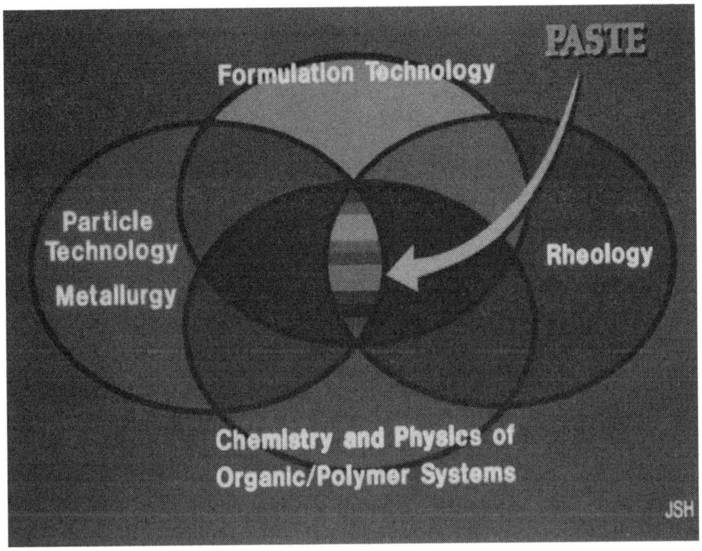

FIGURE 14.1 Solder paste technology—an interplay of multiple technologies.

14.2.1 Solder Powder

Alloy powders can be produced by one of the common techniques: chemical reduction, electrolytic deposition, mechanical processing of solid particulates, and atomization of liquid alloys.[4,5] Alloy powders made from chemical reduction under high temperature generally are spongy and porous. The fine particles of noble metal powders frequently are precipitated by reduction of the salts in aqueous solution with proper pH. The precipitate slurry is then filtered, washed, and dried under highly controlled conditions. A mechanical method generally is used to produce flakelike particles. The metals possessing high malleability, such as gold (Au), silver (Ag), copper (Cu), and aluminum (Al), are most suitable for making flakes.

The electrolytic deposition process is characterized by dendrite particles, and it produces high-purity powders. The resulting particle sizes are affected by the type, strength, and addition rate of the reducing agent and by other reaction conditions. The characteristics of the particles are also affected by current density, electrolytes, additives, and temperature. The principle of atomization is to disintegrate the molten metal under high pressure through an orifice into water or into a gaseous or vacuum chamber. The powders produced by this method have relatively high apparent density, good flow rate, and smooth surface, and are spherical in shape, as shown in Figs. 14.2 and 14.3. Powders to be used in solder paste are mostly produced by atomization because of its desirable inherent morphology and the shape of the resulting particles. Hence the discussion that follows is concerned with the atomization technique only.

An inert gas atomization system with options of a bottom-pouring system and a tilting-crucible system normally consists of a control cabinet, vacuum induction furnace, tundish, argon (or other inert gas) supply line, ring nozzle, atomization tower, cyclone, and powder collection container. The alloy is melted under inert gas at atmospheric pressure to avoid the evaporation of component ingredients. A high melt rate can be achieved. The molten material is then charged into the atomization tower. The melt is disintegrated into powder at atmospheric pressure by an energy-rich stream of inert gas. The process, conducted in a closed system, is able to produce high-quality powder.

In addition to inert gas and nitrogen atomization, centrifugal and rotating electrode processes have been studied extensively. The atomization mechanisms and the mean particle diameter are related to the operating parameters (diameter of electrode D, melting rate Q, and angular velocity w of the rotating electrode) and to the material parameters (surface tension at melting point g, dynamic

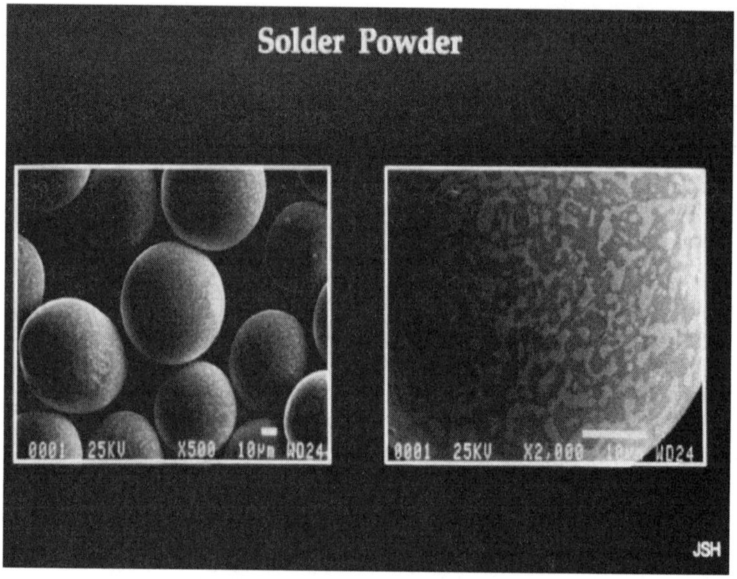

FIGURE 14.2 SEM of 63Sn/37Pb solder powder (×500).

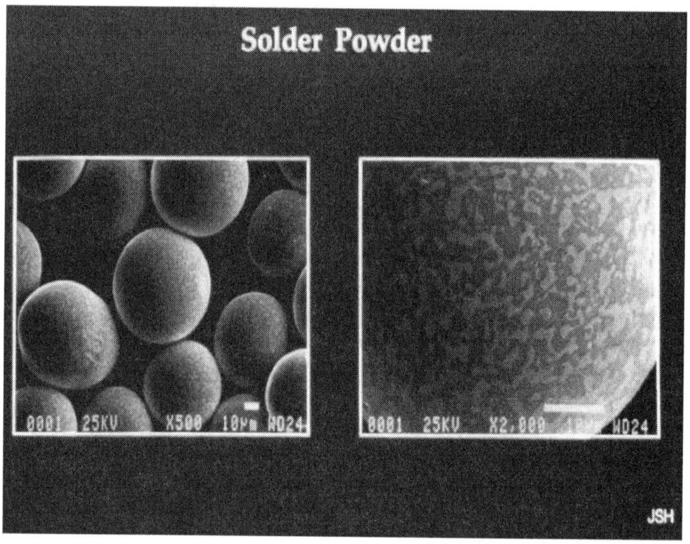

FIGURE 14.3 SEM of 63Sn/37Pb solder powder (×2000).

viscosity h, and density at melting point r of the atomized liquid). The relationships among these parameters are interdependent.

It has been found that the mean volume-surface diameter d is proportional to the surface tension of the atomized liquid and the melting rate but inversely proportional to the angular velocity of the rotating electrode, the diameter of the electrode, and the density of the atomized liquid.

14.2.2 Interplay of Multiple Technologies

In addition to metallurgy and particle technology, the physical, chemical, thermal, and rheologic properties of a vehicle-flux system are equally important to the performance and characteristics of the resulting paste. Furthermore, what ingredients are to be composed (i.e., formulation) and how the formula is put together (i.e., processing) are crucial to the properties and parameters of vehicle-flux systems. With respect to performance, solder paste is categorized into four areas: applicability, solderability, residue characteristics, and joint integrity.

Applicability refers to the ability of a paste to be adapted to a specific paste-applying technique such as dot dispensing or stencil printing. *Solderability* is intended to be used in a broad sense—the ability of a paste to wet the surfaces to be joined with complete coalescence of solder powder particles and to achieve a reliable metallurgic bond. Further discussion on solderability is covered in the literature.[6] *Residue characteristics* cover the physical and chemical properties of the resulting chemical mixture after soldering (e.g., corrosivity, activity, tackiness, hardness, compatibility with cleaning process). Solder *joint integrity* is the ultimate performance of the solder joint after the soldering process in terms of mechanical properties, resistance to adverse environments, and compatibility with service conditions.

A working paste under real-world conditions is quite intricate in nature. Its complexity and variability are further augmented by the dependency of performance parameters on the variables of paste handling and the soldering process. With this in mind, it is apparent that an understanding of the fundamental technologies involved is a necessity. Through the understanding of metallurgy, the solder alloy is selected with the consideration of solderability and joint integrity. Through particle technology, the size distribution, shape, and morphology of the alloy powder are considered to formulate the desired paste applicability and solderability. Through chemistry, chemical properties (e.g., reactivity with solder powder and surfaces to be joined and reactivity in relation to temperature) that affect solderability, residue characteristics, and joint integrity can be better understood. In addition, the functional groups and structure of chemicals in relation to a specific performance characteristic can be correlated and anticipated in principle. The physical properties, including surface and interfacial phenomena of individual ingredients and of a system as a whole, have significant effect on the paste performance in solderability and residue properties.

The rheology, not only as a result of a designed composition of a flux-vehicle system and solder powder but also as a result of paste processing, controls the paste applicability, solderability, and even joint integrity. Thermal properties, such as stability versus temperature and reactivity versus temperature, contribute to the residue characteristics and solderability.

It is worth noting that the paste is considered mostly as being kinetically stable rather than thermodynamically stable. This is in contrast to a true "solution" or to other multicomponent systems such as microemulsions. Therefore, formulation and processing are crucial to the consistency and properties of the paste and, in most cases, even to its rheology and shelf stability.

A viable solder paste should be constituted by using fundamental technologies in selecting starting raw materials, in anticipating the interactions among these raw materials, and in understanding interrelations between starting materials and end-use performance. Solder paste should be handled and used with an understanding of the characteristics, techniques, and technologies involved to ensure its optimal performance.[4]

14.2.3 Chemical and Physical Properties

Chemical and physical properties of the flux-vehicle system are an essential part of solder paste performance. The basic physical properties to be considered include

- Melting point
- Boiling point
- Softening point
- Glass transition temperature
- Vapor pressure
- Surface tension
- Viscosity
- Miscibility and dispersability

All these properties are more or less controlled by intermolecular forces, although not to great extent in the formation of chemical bonds. For example, the tendency of a molecule to volatilize from its liquid is a function of a total transitional energy, which in turn depends on temperature. The boiling point depends on the relation of such transitional energy and the cohesive energy as a result of intermolecular interactions. For a polymer, molecular weight plays an important role in the occurrence of decomposition or volatization as temperature rises; it is also important to the tackiness, viscosity, and rheology of the resulting system. The viscosity of a polymer in solution is a function of molecular weight and its distribution in addition to other characteristics of the polymer and solvent being used. The viscosity is related to molecular weight by

$$\eta = KM^a$$

where K and a are empirical constants, and M is the molecular weight.[4] For chemicals in solid form, the melting point, in addition to cohesive energy, is influenced by the orderliness of molecules or expressed as entropy. Although melting point and boiling point in relation to chemical structure are complicated, the higher boiling point in general is associated with high melting point, and symmetric molecules melt at higher temperature due to the low entropy of fusion, when other factors are the same. The melting point of polymers generally increases with increasing molecular weight. The softening point also increases with molecular weight and increased crystallinity.

In solder paste, the vapor pressure of every ingredient affects more than one aspect of paste performance, including the selection of reflow method and void development in the solder joint. The boiling point of the liquid phase as a whole or of an individual liquid ingredient can alter the flux activity and the residue characteristics. It also may affect compatibility with a specific reflow method. The melting or softening point of the solid phase as a whole or of individual solid ingredients has a direct impact on the flux activity.

14.2.4 Rheologic Flow Property

The flow property that is expressed practically as cold slump or hot slump is a major contributor to a common manufacturing defect—pad bridging. In addition to paste deposition performance during the dynamic printing or dispensing operation, the slump or not-to-slump behavior is an important paste property. Ideally, the paste is desired to retain its original shape after deposition before formation of the final solder joint. This requires the paste to have zero cold slump under ambient temperature between the printing/dispensing step and reflow as well as zero slump when temperature rises during reflow.

It should be noted that spreading in the paste state and in the molten state is dominated by different driving forces. In the paste state, the interparticle forces dominate, and in the molten solder state, the relative interfacial tension (force per unit area) between flux-vehicle liquid, substrate, and molten solder governs the spreading. Therefore, the objective is to design a flux-vehicle system capable of intimately wetting the surface of solder particles while providing adequate surface tension in the liquid phase so that the solder particles are imbedded in the matrix of the flux-vehicle system with high cohesive force.

To illustrate the effect, the difference in the spreading factor between system A and system B using Antonow's approximation is

$$S_A - S_B = [\gamma_{s1} - (g_{sm} + \gamma_{m1}\cos\theta)\,]_A - [\gamma_{s1} - (\gamma_{sm} + \gamma_{m1}\cos\theta]_B$$
$$\cong (\gamma_{1B} - \gamma_{1A})\,(1 - \cos\theta)$$

Thus, when

$$\gamma_{1B} > \gamma_{1A}$$
$$\therefore S_A > S_B$$

and when

$$\gamma_{LB} < \gamma_{LA}$$

$$\therefore S_A < S_B$$

The spreading factor in a system with relatively higher surface tension is smaller than in one with lower surface tension.

Since most solder pads, as on printed wiring boards (PWBs) or on hybrid circuits, are individual pads surrounded by an unwettable surface (solder mask or bare alumina surface), it is perceived that the molten solder should have retracted back to its own pad after reflow due to its high surface tension. This perception does not always hold, as in the cases where the solder volume exceeds the tolerance of the pad area or a component exerts sufficient weight on solder. Again, the balance of the forces is the criterion. Furthermore, when we examine the spread phenomenon that causes pad bridging or binding between two solder joints, the primary factor is a result of hot slump in the paste state before the solder melts. *Hot slump* is defined as the shape change when the paste sinks in height and expands in area in response to temperature rise before the solder melts. Since the paste deposit between adjacent pads is in physical contact due to hot slump, it is likely to cause solder bridging between pads. The likelihood of this depends on the volume of solder applied in relation to pad area and the spacing between pads, as well as on the weight of the component exerted on the paste deposit per unit area. Thus, as the circuitry becomes more intricate and shrinks in size, it is more prone to bridging problems.

Solder paste slump also contributes to both types of solder balling: large solder balls that are only associated with small capacitors and resistors and the general solder balling phenomena as discussed in Secs. 14.3 and 14.7.

14.2.5 Solder Paste: Formulation

Four states are involved in the formulation of a solder paste:

- *Design stage.* Utilizing fundamental technologies in chemistry, physics, and materials science for the targeted performance
- *Formulation stage.* Making up a formula through a systematic and meticulous process
- *Fine-tune stage.* Working for the best balance and/or the specific performance
- *Scale-up stage.* Setting up a process to produce the formula consistently and reproducibly

The first step in formulating a paste product is to define the performance objective to be achieved. With a clear objective in mind, a paste can be designed to meet the performance parameters by using the fundamental technologies, by understanding how the selected raw materials contribute to the characteristics of a paste, and by anticipating any synergistic or antagonistic interactions among raw materials. A product involves many performance parameters, and some of them may be tradeoffs. For example, high metal content is beneficial to solder joints in volume and void and in residue yet may make the paste more prone to crusting and difficult to apply. High-viscosity paste may improve flow control against temperature yet may cause the paste to be difficult to apply. Using highly active fluxing chemicals may improve solderability in some cases yet may leave more corrosive residue. In such a case, improving the solderability by selecting proper ingredients without the use of highly active fluxing chemicals is a skill. Increasing flux content does not always improve solderability in terms of wetting and/or solder balling elimination.

The ability to achieve an ultimate balance among the performance parameters and to prioritize them for a specific group of applications is the key. Once a prototype formula capable of providing all functions is constituted, the next step is to fine-tune the formula to coincide with specifications or any specific values designated.

With the accomplishment of a specific composition formula, it is equally important to develop a reproducible process for making a paste with consistent performance. It is not an exaggeration but an indication of the importance of the role of the process to state that an identical composition can produce different products when the process is allowed to vary. The required skill is to be able to use the technologies to make the best balance of all desired performance parameters and to meet the performance requirements in the real world, as well as to develop a controlled process in making end-use products.

To make the after-reflow residue water cleanable under prevailing manufacturing environments, the key to formulating a successful water-soluble paste is to have an intact chemical matrix that is water cleanable, even if the composition does not necessarily comprise ingredients that are all water soluble.

The unique demand of no-clean paste is to minimize the amount of after-reflow residue. Two logical approaches are (1) to minimize the total solid contents in the flux-vehicle system and (2) to maximize the metal load in the paste. Solid content of the flux-vehicle system is, to a certain extent, related to fluxing efficiency, which commonly reflects the number of solder balls that occurred during reflow. As noted before, fluxing efficiency is not solely related to the amount of flux. The rest of the chemical system that serves as a protective barrier from ambient air during reflow also plays an important role. The acceptability of metal load in the flux-vehicle system depends on

- The physical and chemical nature of the flux-vehicle system
- Metal particle size and distribution
- The surface condition of the metal particles
- The process of paste making

Formulating a good solder paste not only requires knowledge and experience but also demands a high level of patience and good instinct.

14.2.6 Solder Paste: Performance Parameters

Performance parameters can be grouped in three states: paste state, soldering state, and postsoldering state. Under each state, there are a number of important parameters.

In the paste state:

- Paste stability and shelf life
- Dispensability through a fine needle
- Stencil printability
- Tack time
- Exposure life
- Quality and consistency

In the soldering state:

- Compatibility with substrates
- Solder balling phenomenon
- Wicking phenomenon
- Dewetting phenomenon
- Wettability
- Flow property before molten
- Residue corrosivity
- Residue cleanability

In the postsoldering state:

- Solder joint appearance

- Solder joint voids
- Solder strength
- Solder joint microstructure
- Solder joint integrity versus mechanical fatigue/creep
- Solder joint integrity versus thermal fatigue/creep

14.3 REFLOW SOLDERING

During soldering, a series of reactions and interactions occur in sequence or in parallel. These can be chemical or physical in nature in conjunction with heat transfer. The mechanism behind fluxing is often viewed as the reduction of metal oxides. Yet, in many situations, chemical erosion and dissolution of oxides and other foreign elements act as the primary fluxing mechanisms. Using a more complex fluxing process in solder paste as an example, the primary steps are represented by the flowchart in Fig. 14.4.

Several events occur during this stage, as shown in steps (II) and (IV) of the flowchart. These include temperature set to fit the specific flux activation temperature of the chemical system of the paste and the time at heat to fit the constitutional makeup of the paste. Improper preheating often causes various problems, such as the spattering problem, which manifests itself as discrete solder balls and component damage. Too high a temperature or too long a time at the elevated temperature results in insufficient fluxing and/or overdecomposition of organic acid, causing solder balling or hard-to-clean residue (if the no-clean route is adopted). The third stage is to spike quickly to the peak reflow temperature at a rate normally practiced of 1.0 to 4.0°C/s. The purpose of temperature spiking is to minimize the exposure time of the organic system to high temperature, thus avoiding charring or overheating. Another important characteristic is the dwell time at the

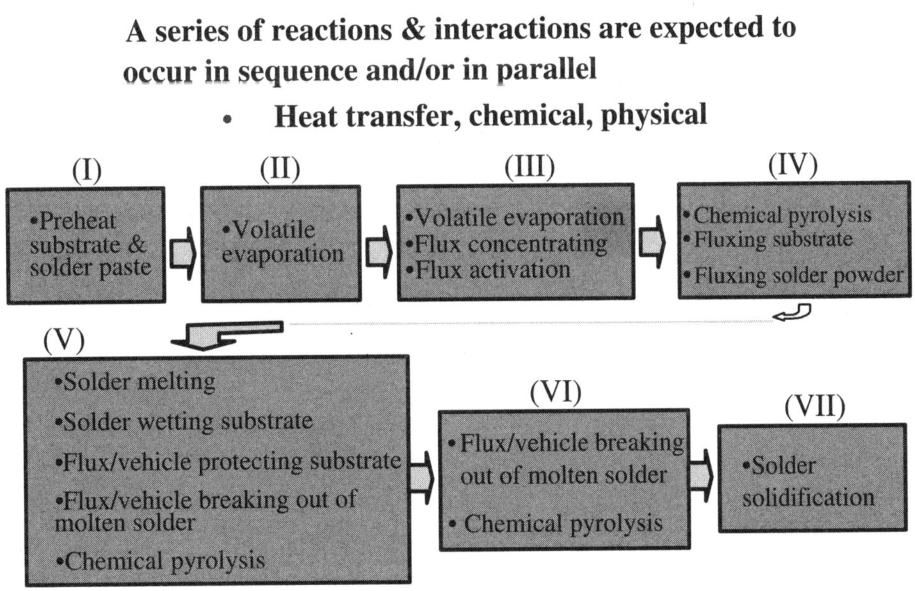

FIGURE 14.4 Key steps in the oven reflow process.

peak temperature. The rule of thumb in setting the peak temperature is 20 to 50°C above the liquidus or melting temperature; e.g., for the eutectic tin-lead composition, the range of peak temperatures is 203 to 233°C.

The wetting ability is related directly to the dwell time at the specific temperature in the proper temperature range and to the specific temperature being set. Other conditions being equal, the longer the dwell time, the more wetting is expected but only to a certain extent; the same trend applies at higher temperatures. However, as the peak temperature increases or the dwell time is prolonged, the extent of the formation of intermetallic compounds also increases. An excessive amount of intermetallics can be detrimental to long-term solder joint integrity. Peak temperature and dwell time should be set to reach a balance between good wetting and to expel any nonsolder (organics) ingredients from the molten solder before it solidifies, thus minimizing void formation.

With the prevalence of oven reflow, a few more words about oven heating profile and operating parameters are pertinent. It should be stressed that reflow is a dynamic heating process in that the condition of the workpiece is constantly changing as it travels through the furnace in a relatively short reflow time. The momentary temperature that the workpiece experiences determines the reflow condition and therefore the reflow results.

It is ultimately important to establish a correlation between setting the temperature of a given oven, the measured temperature of the workpiece at each specified belt speed, and the soldering performance. The resulting correlation between soldering performance and temperature setting or profile provides a "workable range" for the assembly.

Under mass reflow operation, both heating and cooling steps are important to the end result. It is generally understood that the heating and cooling rate of reflow or soldering process essentially contributes to the compositional fluctuation of the solder joint. This is particularly true when there are significant levels of metallurgic reactions occurring between the tin-lead solder and substrate metals. In the meantime, the cooling rate is expected to be responsible for the evolution of the microstructure.

14.3.1 Process Parameters

The key process parameters that affect the production yield as well as the integrity of solder joints include

- Preheating temperature
- Preheating time
- Peak temperature
- Dwell time at peak temperature
- Cooling rate

For a given system, cooling rate is directly associated with the resulting microstructure, which in turn affects the mechanical behavior of solder joints.[7] It was found that the microstructural variation and corresponding failure mechanisms of solder joints that were made under various reflow temperature profiles are complex. Nonetheless, some correlation between the cooling rate and the basic properties can be obtained.

14.3.2 Reflow Temperature Profile

Reflow temperature profile representing the relationship of temperature and time during the reflow process depends not only on the parameter settings but also on the capability and flexibility of the equipment. Specifically, the instantaneous temperature conditions that a workpiece experiences are determined by

- Temperature settings to all zone controllers
- Ambient temperature
- Mass per board

- Total mass in the heating chamber (load)
- Efficiency of heat supply and heat transfer

14.3.3 Effects of Reflow Profile

The reflow profile used for SMT manufacturing has a direct bearing on manufacturing yield, solder joint integrity, and the reliability of the assembly. Specific areas that are affected by reflow profile follow. Each area may be affected, to a different degree, by one or more of the three heating stages:

- Temperature distribution across the assembly
- Plastic IC package cracking
- Solder balling
- Solder beading
- Wetting ability
- Residue cleanability
- Residue appearance and characteristics
- Solder joint voids
- Metallurgic reaction between solder and substrate surface
- Microstructure of solder joints
- Board warpage
- Residual stress level of the assembly

Uniformity of Temperature Distribution. In a normal reflow environment, temperature differential across the assembly is inevitable. This is due to the large disparity in mass and the characteristics of the components coupled with the relatively short total reflow time (the entire cycle lasts only several minutes). A large temperature differential causes uneven soldering, resulting in localized cold joints or overheated joints. These problematic joints may contribute to manufacturing defects or jeopardize the long-term integrity of the solder joints under service conditions if they are not detected as manufacturing defects and corrected.

For a given oven, the rate of natural warm-up (°C/s) and the intended preheating temperature and time are the main factors that control temperature uniformity across the assembly. A slower heating rate in the warm-up state is desired to reach a more uniform board temperature distribution.

Plastic IC Package Cracking. Along with factors such as die size, the moisture sensitivity of the molding compound, and its thickness, reflow profile plays an important role in causing or preventing plastic IC package cracks.[8] When the IC package (e.g., BGA, QFP, SOIC) absorbs a certain level of moisture during storage, handling, or transit (without proper dry pack), the absorbed moisture may cause package cracking during reflow. Setting a proper reflow profile can mitigate the cracking problem; the heating rate from ambient temperature to 140 to 150°C is most critical.

Solder Balling. Elevated temperatures or excessive time at those temperatures during the warm-up and preheating stages can result in inadequate fluxing activity or insufficient protection of solder spheres in the paste, causing solder balling. In addition to the quality of solder paste, the presence of solder balls essentially may be related to the compatibility between the paste and the reflow profile. On the other hand, inadequate preheating or heating too fast may cause spattering, evidenced by random solder balls. The two heating stages preceding the spike/reflow zone are primarily responsible for this phenomenon.

Solder Beading. *Solder beading* refers to the occurrence of large-sized solder balls (usually larger than 0.005 in in diameter) that are always associated with small and low-clearance passive components (capacitors and resistors). This problem will occur even when the paste otherwise may perform perfectly, i.e., free of solder balls at all other locations (components) on the board and with good wetting. The trouble with solder beading is that it may occur in most or all board assemblies, rendering the first-time yield to nearly zero. The current remedy on the production floor is to remove the beads manually.

The formation of solder beads near or under capacitors and resistors is largely attributed to paste flow into the underside of the component body between two terminations aided by capillary effect. As this portion of paste melts during reflow, it becomes isolated away from the main solder on the wettable solder pads, forming large, discrete solder beads. Besides other factors, reflow profile is one of the areas that contributes to this phenomenon. The practice of adopting a slower preheating rate and a lower reflow peak temperature can reduce solder beading. However, if the reflow profile is at its optimum and the problem still persists, a new paste with a strengthened chemistry is the solution. In addition, the pad design and the volume of the paste deposit, which is, in turn, determined by the stencil design, are other factors contributing to this beading phenomenon.

Wettability. The temperature setting and time spent in both preheating and spike/reflow affect wettability. However, each stage works by a separate mechanism. In the preheating stage, the range of temperature and the time spent in this range directly affect the activity of flux. Wettability, in turn, is affected through the fluxing action. However, in the spike/reflow zone, wetting on the "cleaned" surface is influenced by the peak temperature because of the intrinsic wetting ability of molten solder alloy. This ability increases on a wettable substrate with higher temperature. With all other conditions being equal, a longer dwell time can, to a limited degree, further enhance wetting. Modification of the spike/reflow zone sometimes may solve a minor wetting problem.

Cleanability. For solder paste designed to be cleaned, particularly water cleaned, excessive heat may make it difficult for the residue to be removed, rendering a normal cleaning process ineffective. In this case, all stages of the reflow profile can be contributors.

Residue Appearance and Characteristics. The importance of the compatibility of the solder paste's chemical composition with the reflow profile can be readily demonstrated using a no-clean soldering process. For instance, if the paste was reflowed with a temperature profile below the heat requirement, a higher amount of residue than expected will remain. In addition, the characteristics of that residue may range from tacky to ionically active.

Solder Joint Voids. Incomplete outgasing (gases entrapped in the solder joint) is the main cause of voiding. In addition to design factors, the compatibility between the reflow profile and the chemical makeup of the solder paste is important. There also should be sufficient dwell time in the molten state (above 183°C for 63Sn/37Pb) to ensure that the gases have enough time to separate and escape from the molten solder. With respect to other factors contributing to voiding in solder joints, readers are referred to the literature.[9]

14.3.4 Optimal Reflow Profile

The heat transfer from the surrounding hot air to the various components on the board, such as leaded packages, array packages, and discretes, differs during the process where a thermal equilibrium hardly exists. This disparity can be compensated for by setting a reflow profile either with a higher heat supply rate and higher temperature or with a slower heating rate and lower temperature.[10,11] On most manufacturing lines, unfortunately, a reflow profile with a higher heating rate and higher temperature have been used, as shown in Fig. 14.5. This disparity in the heat transfer may be heightened as large or heavy array packages are incorporated. Although increasing the temperature has accommodated most reflow results, the approach will not work well with heat-sensitive components or with PWBs that contain increasingly versatile components, particularly BGAs, CSPs, and flip chips (FCs).

The initial warm-up state plays a far more influential role in the quality and reliability of assembled boards than was first thought. An initial heating rate of less than 1°C/s in conjunction with a heating rate for the rest of the profile of not more than 3°C/s is considered most beneficial and is recommended. Under SMT environments, the small degree of reduction in heating rate would not be a bottleneck for production throughput. By using the slower rate in the warm-up and preheating stages prior to reaching 183°C, the peak temperature can be maintained in the range of 210 to 215°C, in contrast to 215 to 230°C. The total dwell time above the liquidus temperature (183°C) falls in the range of 30 to 90 s.

Reflow profiles based on slower heating rates and cooler temperatures, as one example, as shown in Fig. 14.6, will be more in line with today's complex assemblies, minimizing in-process heat exposure as well as residual stress.

14.4 INERT AND REDUCING ATMOSPHERE SOLDERING

At the soldering temperature, the atmosphere surrounding the workpiece protects or interacts with the surface of substrates, the solder alloys, and the chemical ingredients in the flux-vehicle system. These interactions determine the chemical and physical phenomena in terms of volatilization, thermal decomposition, and surface-interfacial tension. A controlled atmosphere is expected to deliver a more consistent soldering process.[12–16]

In addition to consistency, the inert or reactive atmospheres potentially may offer further merits, including

- Improved temperature uniformity
- Solderability enhancement
- Solderability uniformity
- Minimal solder balling

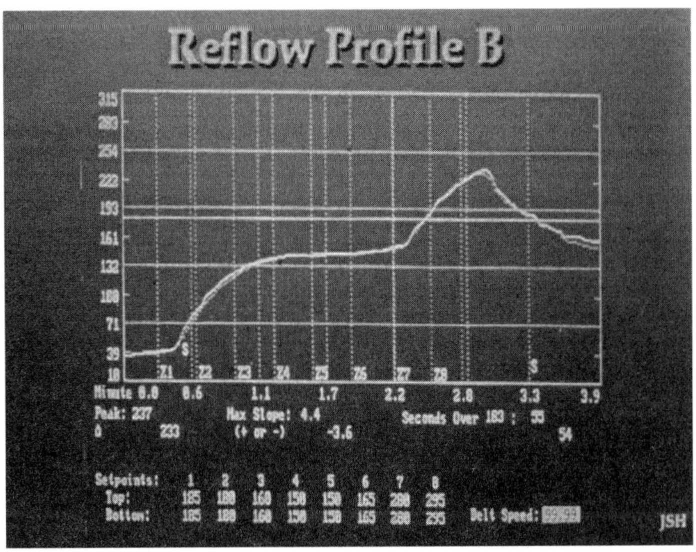

FIGURE 14.5 Reflow profile—with faster initial heating rate.

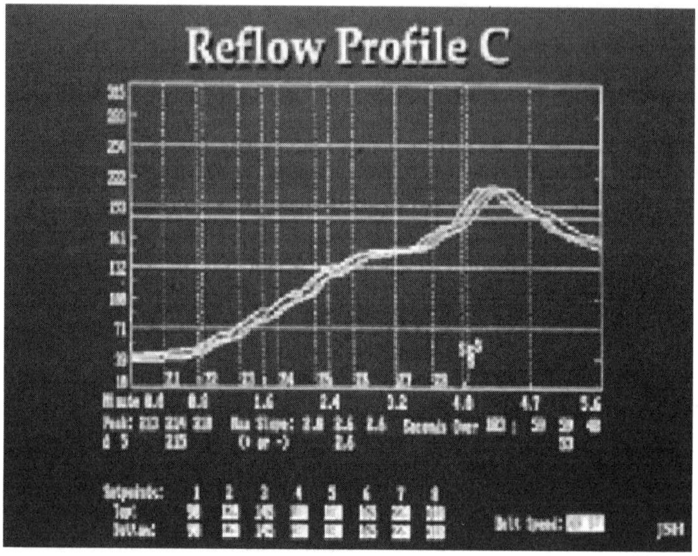

FIGURE 14.6 Reflow profile—with slower initial heating rate.

- Irregular residue charring prevention
- Reduced residue
- Polymer-based board discoloration prevention
- Wider process window
- Overall quality and yield improvement

The inert and reactive atmospheres are expected to facilitate conventional fluxing efficiency during soldering. It should be noted, however, that performance results rely greatly on the specific atmospheric composition and its compatibility with the solder material, substrate, and chemicals incorporated in the system, which also must be compatible with the soldering temperature profile.[1] Figure 14.7 shows that solderability under an N_2 atmosphere is significantly improved because solder balls that are formed under ambient air are eliminated.

14.4.1 Process Parameters

The additional process parameters for inert and reducing atmosphere soldering include

- Gas flow rate
- Humidity and water vapor pressure
- Oxygen level
- Belt speed
- Oven temperature settings

Gas Flow Rate. The gas flow rate required to achieve a specific level of oxygen in the dynamic state of the reflow oven is largely controlled by the type of oven: categorically closed system, semiclosed system, or open system. For a given oven, the required flow rate increases when the allowable oxygen level is lowered. At a given flow rate, when the air tightness in oven construction is reduced, the achievable oxygen level will be higher.

As expected, for a given reflow system, the oxygen level is inversely related to gas flow rate. The gas flow rate also affects the temperature distribution and temperature uniformity of the assembly. The temperature gradient is reduced when the gas flow rate is incresed. However, the downside of using a high flow rate goes to the higher gas consumption, thus increasing cost. The cost impact may be mitigated when the design of the oven is capable of internal gas recirculation in an efficient fashion.

Humidity and Water Vapor Pressure. Water vapor pressure inside the soldering oven can be contributed from

- The composition and purity of the atmosphere
- The reaction product of the flux-vehicle chemical system with metal substrates
- The moisture released from the assembly, including components and board
- The ambient humidity

Because water vapor is essentially oxidizing to metal substrates that are to be joined by soldering, its partial pressure in the oven affects the overall function of the atmosphere. The partial pressure of water vapor in an atmosphere gas is conveniently expressed as the dew point, the temperature at which condensation of water vapor in air takes place. The dew point can be measured by a hygrometer or dewpointer by means of a fog chamber, chilled mirror aluminum oxide technique. The purity of incoming gas in terms of moisture normally is monitored by measuring the dew point.

Belt Speed. For an evenly spaced loading on the belt, the belt speed not only determines the throughput but also affects other operating parameters that can alter the soldering results. For example, the parameters that are affected by the change in belt speed include

- *Peak temperature.* At fixed temperature settings, increasing belt speed results in a decrease in peak temperature.

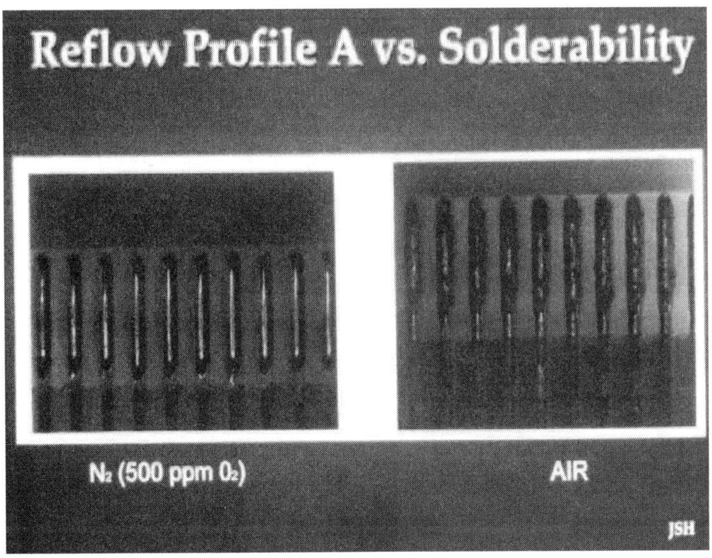

FIGURE 14.7 Comparison of solderability (solder balling) in air and nitrogen environments.

- *Atmosphere composition.* While other conditions are equal, the change in belt speed may alter the oxygen level (including moisture content).

Temperature Settings. The operating temperature or temperature profile is an integral part of the soldering process. It affects the physical activity and chemical reaction of the organic system in solder paste or flux. The operating temperature, particularly peak temperature, changes the wetting ability of molten solder on the metal substrate; wetting ability generally increases with increasing temperature. Chemical reactions and thermal decompositions respond to the rising temperature and the temperature profile.

Oxygen Level. Various studies have focused on application of the no-clean process and on the determination of the maximum oxygen level allowed for using a nitrogen-based no-clean soldering process in solder paste reflow and in wave soldering.[17] Each study was performed with a specific solder paste and flux or with a selected series of pastes and fluxes. Tests were conducted with specific equipment and under a designated process. In view of the continued introduction of new equipment and the diversity of processes coupled with the versatility of solder paste and flux compositions, the test results are expected to represent the specific system (paste, oven, process, assembly) and at best to provide a guideline reference point. For example, a solder paste from vendor *I* to be used with process *A* may require a maximum of 50 ppm oxygen level in order to obtain good solderability, grossly solder-ball-free, and acceptable after-soldering residue. To achieve the similar results, the same paste to be used with process *B* may only need a maximum of 300 ppm oxygen. The same could be true for a different paste used in the same process.

The precise oxygen level requirement for no-clean soldering depends on the characteristics of the system. The general principle and trends in the relationship between the performance feature and the allowable maximum oxygen level can be derived. Section 14.4.2 illustrates the range of the optimal oxygen level.

14.4.2 Optimal O_2 Level

In general, with higher than 2000 ppm O_2, the effect of nitrogen hardly may be detected.[17] Below 20 ppm O_2, the process will become difficult to control and, needless to say, too costly. For a given oven and process, the O_2 level required is essentially controlled by the chemistry and makeup of the solder paste. For example, a solder paste from supplier *A* may require a maximum level of 800 ppm O_2 to obtain the desirable results (good wetting, no solder balls, etc.). To achieve similar results, solder paste from supplier *B* may need a maximum of 200 ppm O_2. In practice, O_2 levels in the range of 20 to 2000 ppm should be able to accommodate most applications.

Soldering under nitrogen poses two additional demands: more stringent process control and higher operating costs. However, its potential effects on solderability, heat transfer, PWB materials, and process window may bring benefits in mounting large-area and heavy BGAs, as well as in connecting small and delicate CSPs onto complex PWBs.

14.4.3 Temperature Measurement

A low-mass, direct, and firm contact without the need for extraneous attaching material is one way to meet the criteria for achieving accurate temperature measurement.[18]

14.5 PRINTING

In addition to selection of the solder paste and printer and the settings of printing parameters, stencil thickness versus aperture design, stencil aperture versus land pattern, and stencil selection are major factors contributing to the printing results of solder paste.

TABLE 14.1 Guidelines in Designing Stencil versus Aperture Size

Component lead pitch		Aperture width		Maximum stencil thickness	
inch	mm	inch	mm	inch	mm
0.005	1.26	0.023	0.58	0.014	0.35
0.025	0.63	0.012	0.30	0.0075	0.19
0.020	0.50	0.010	0.25	0.0063	0.16
0.015	0.38	0.007	0.18	0.0043	0.11
0.008	0.20	0.004	0.10	0.0025	0.06

14.5.1 Stencil Thickness versus Aperture Design

When printing solder paste, the relative dimensions of stencil thickness and stencil aperture are chosen to achieve a balance between printing resolution and the proper amount of solder deposit in order to avoid starved solder joints or pad bridging. For a selected stencil thickness, too small a stencil aperture width leads to open joints or starved joints. Too large an aperture width causes pad bridging. Table 14.1 provides guidelines for designing stencil thickness in relation to aperture.

14.5.2 Stencil Aperture Design versus Land Pattern

To make solder joints with a one-pass printing process, the stencil thickness must be selected for transferring a sufficient amount of paste onto the non-fine-pitch solder pads while avoiding an excessive amount of paste deposited onto the fine-pitch pads. There are several options to achieve the deposition of a proper amount of solder paste on the land pattern to accommodate a mix of sizes of solder pads. These are as follows:

1. *Step-down stencil.* This is commonly achieved by chemically etching the non-fine-pitch pattern area from one side of the stencil while etching the step-down area for the fine-pitch pattern on the other side durng a double-sided etch process. Alternatively, the step-down area is etched in one foil, and the non-fine-pitch pattern is etched in the other foil. Then the two foils are registered and glued together.

   The practical step gradient is 0.002 in (0.05 mm); some common combinations are

   0.008 in (0.20 mm) for non-fine-pitch

   0.006 in (0.10 mm) for fine-pitch, or

   0.006 in (0.15 mm) for non-fine-pitch

   0.004 in (0.15 mm) for fine-pitch

2. *Uniform reduction on four sides of apertures.* The dimensions of the fine-pitch aperture on the stencil are reduced by 10 to 30 percent in relation to those of the land pattern. This reduces the amount of paste deposition on the fine-pitch land pattern and also provides some room for printing misregistration and paste slump, if any.

3. *Staggered print.* The opening of the stencil is only one-half the length of the solder pad and is arranged in an alternate manner as shown in Fig. 14.8. For tin-lead coated solder pads, when the paste starts to melt during reflow, the molten solder is expected to flow to the other half of the pad, providing complete coverage. With a bare copper or nickel surface, the molten solder may not flow out to cover the area on which the paste is not printed.

4. *Length or width reduction.* The dimensions of the stencil opening are reduced along the length or along the width by 10 to 30 percent in relation to that of solder pads, achieving a reduction in the amount of paste deposited.

5. *Other shapes.* The stencil openings are made with selected shapes such as a triangle or a teardrop in order to achieve reduced solder paste deposition on the fine-pitch pattern.

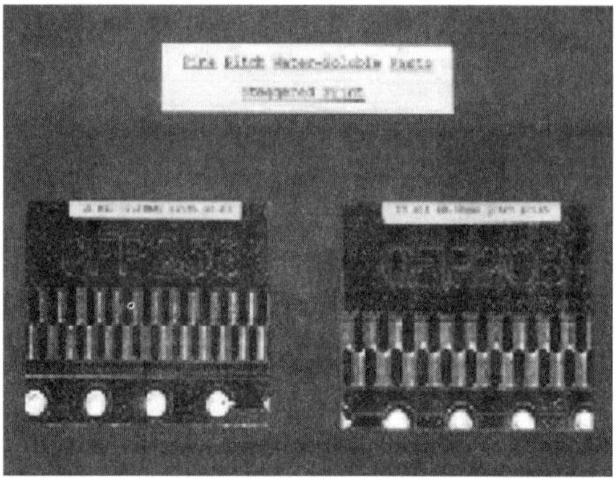

FIGURE 14.8 Staggered print.

6. *Compromised stencil thickness.* Instead of using the specific thickness that is considered the most suitable for a specific land pattern, select a thickness that is practical to both fine-pitch and non-fine-pitch patterns.

14.5.3 Stencil Selection

The performance of stencils is driven primarily by the foil metal and the process being used to create the printing pattern. Currently, five types of stencil materials are available commercially: brass, stainless steel, molybdenum, Alloy 42, and electroformed nickel. The processes making the stencils may involve chemical etching, laser cutting, electropolishing, electroplating, and electroforming. Each type of foil or fabricating process possesses inherent merits and limitations. The key performance of a stencil is assessed by the straight vertical wall, wall smoothness, and dimensional precision. In addition, durability, chemical resistance, fine opening capability, and cost are also important factors.

Table 14.2 compares various stencil materials, and Table 14.3 summarizes the relative performance characteristics of stencil-making techniques.[19]

14.6 DESIGN AND USE OF SOLDER PASTE FOR SYSTEM RELIABILITY

With the versatility of components and the vast variation in their solderability, it is tempting to formulate a flux chemistry with high activity. It is also convenient to incorporate the halide-containing organic ingredients to enhance the activity without adversely affecting the test results in the content of ionic species. This is so because organic halides can be very effective fluxing agents in very low dosages, thus relieving the level of elaboration (skills) in formulation technology. In addition, in various chemical makeups, low-dosage halides may be able to pass the "standard" tests. It is a well-established fact that among the chemical families, mobile halides are the most reactive species toward metals that make up the circuitry. Thus their use should be discouraged.

Finer powder obviously facilitates the fine-pitch deposition via printing or dispensing. Undesirably, the paste with finer powder results in higher demand in the content, as well as in the activity of flux, and is more often prone to solder balling during reflow. Finer powder is also associ-

TABLE 14.2 Comparison of Various Stencil Materials

Performance measure	Material				
	Brass	Stainless steel	Molybdenum	Alloy 42	Ni (electroforming)
Mechanical strength	Unfavorable	Favorable	Favorable	Favorable	Favorable
Chemical resistance	Unfavorable	Favorable	Unfavorable	Favorable	Favorable
Etchability	Favorable	Less favorable	Favorable	Favorable	N/A
Sheet stock availability	Favorable	Favorable	Unfavorable	Favorable	N/A
Cost	Favorable	Less favorable	Unfavorable	Less favorable	N/A
Fine pitch	Favorable	May need electro-polishing	Favorable	May need electro-polishing	Most favorable
Unique feature	Lowest cost	Durable	Self-lubricating smooth wall	Durable	Finest opening

TABLE 14.3 Comparison of Stencil-Making Techniques

Technique	Characteristics	Superior capabilities or features
Chemical etching	Most established process; sensitivity of fine-pitch capability to process and control; sensitivity of aperture size and vertical wall control	Versatile, economic
Laser cut	Grainy wall surface; sequential cut, not concurrent formation of openings; higher cost; difficulty in making step stencil	Fine-pitch capability; no photoresist needed
Electroforming	Additive process via electrodeposition; concern about fine foil strength; difficulty in making step stencil; suitable for stencil of less than 0.004 in	Gasket effect, minimizing bleeding; no need for electropolishing
Electropolishing	Complementary step to produce smooth wall surface	Smooth wall surface
Ni plating on aperture wall (polished or unpolished)	Reducing aperture opening; smooth surface	Finer opening

ated with higher cost. Thus it is always advantageous to use the coarsest powder that is allowable by a flux-vehicle system for achieving printability and dispensability so that reduced cost and proper flux activity can be obtained.

As large or heavier array packages are incorporated into assemblies, the disparity in heat transfer is heightened. In such cases, increasing the temperature has indeed accommodated most reflow results. However, from a reliability standpoint, the approach may not be sound when considering the advent of heat-sensitive components, more complex PWB designs, and increasingly versatile components contained in an assembly.[20]

Reflow profiles based on slower heating and cooling temperatures will be more in sync with today's complex assemblies, minimizing in-process heat-induced damage, as well as the level of residual stress that may cause problems, such as plastic BGA package cracking, board warpage, and board delamination. These three areas affect not only the production floor first-pass defect rate and yield but also the long-term performance of the assembly.

The principles of design and use of solder paste—mild flux, coarse powder, and low-temperature reflow profile—work in tandem toward achieving the highest system reliability.

14.7 SOLDERING-RELATED ISSUES

14.7.1 Intermetallics versus Solder-Joint Formation

Intermetallic compounds often have been observed at or near the solder-substrate interface as well as in the interior of solder joints. Metallurgically, an intermetallic compound is one type of intermediate phase that is a solid solution with intermediate ranges of composition. Intermetallic compounds form when two metal elements have a limited mutual solubility. These compounds possess a new composition of a certain stoichiometric ratio of the two elements.

The new compositions have a different crystal structure from those of their elemental components. The properties of the resulting intermetallic compounds also differ from those of the component metals in that they exhibit reduced ductility, density, and conductivity. Tin or tin-lead solder is metallurgically active with most metals that are commonly used in electronics packaging and assembly. Various intermetallic compositions have been identified under the equilibrium condition between tin and substrate metals, such as Au, Ag, Cu, Pd, Ni, and Pt. Indium-based solders also interact with these substrate metals, often forming intermetallics. One should note that thermodynamically stable compounds may not always be present and that some intermetallics that do not appear in the equilibrium phase diagram have been identified in soldered systems. Relating to electronics packaging and assembly, intermetallic compounds may come from one or more of the following processes and sources:

- Intermetallics are formed at the solder-substrate interface during soldering.
- Intermetallics are carried over from the component and board coating processes.
- Intermetallics are present in the interior of the solder joint as the inherent metallurgical phases of a given solder composition, such as 95Sn/5Sb and 96Sn/4Ag solder.
- Intermetallics are developed during service life either along the interface and/or in the interior of the solder joint.

When solder comes in contact with a common metal substrate for a sufficient amount of time at a high enough temperature, intermetallic compounds may form. Below a solder's liquidus temperature, formation is primarily a solid-state diffusion process and thus depends highly on temperature and time. Figure 14.9 illustrates the rate of formation of Sn-Cu intermetallics, depending on temperature, and Fig. 14.10 depicts the relationship between the rate of formation and the tin content at a given temperature. While solder is in a molten state, the solubility of the element from substrate into molten solder accelerates the rate of intermetallic formation.

External factors such as the temperature of exposure and the time at the elevated temperature also affect the rate of intermetallic compound formation. Thus solder reflow conditions such as peak temperature and total dwell time at the elevated temperature influence the rate and extent of intermetallic growth. Also, while in storage or service, the exposure of the assembly is a factor for intermetallic growth in systems.

The thickness of growth between eutectic tin-lead and copper is proportional to the square root of time, coinciding with the diffusion-controlled kinetics. As temperature rises, the rate of formation increases, with the higher tin promoting the process.

The composition of intermetallics at the interface may differ from those of the solder-joint interior. Furthermore, the surface condition of the substrate affects the kinetics of intermetallic develop-

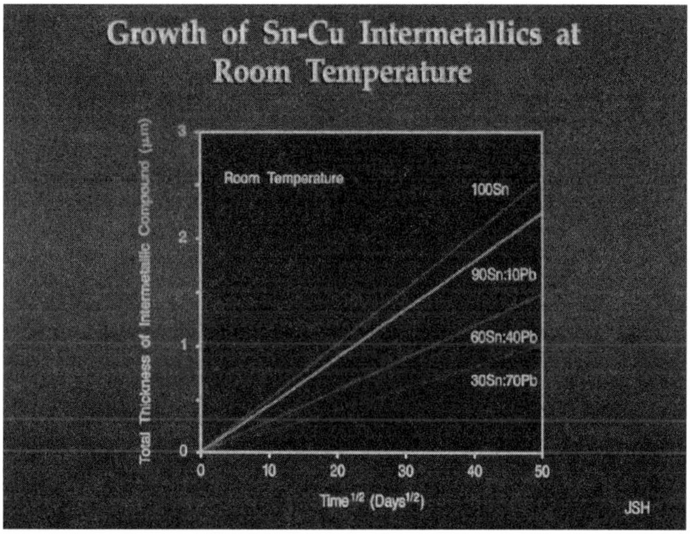

FIGURE 14.9 Rate of intermetallic formation between tin-lead solder and copper versus temperature.

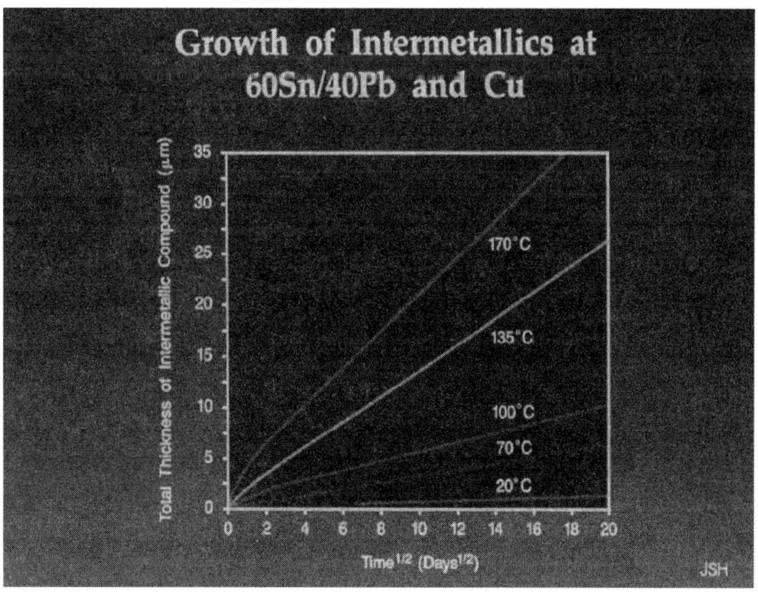

FIGURE 14.10 Rate of intermetallic formation between tin-lead solder and copper versus tin content.

ment. For example, an oxidized surface may show a delayed development of the intermetallic phases, making a thinner layer as compared with a clean surface at a given amount of time. Unlike high-tin solder, which tends to form intermetallic compounds with small crystal structures, high-lead solder forms high, needlelike crystals.

In brief, the extent of intermetallic formation, the composition of the compounds, and their morphology depend on intrinsic factors. These factors include the following:

- The metallurgical reactivity of a solder with a substrate
- Soldering (reflow) peak temperature
- Dwell time at peak temperature
- The surface condition of a substrate—clean versus oxidized
- The postsoldering storage and service conditions

Intermetallics at the interface can be beneficial or detrimental. Wetting on the substrate followed by the formation of a thin layer of intermetallics is the prevalent mechanism in making permanent solder bonds. However, adverse effects may occur if the intermetallic layer becomes too thick. Generally acceptable thicknesses fall in the range of 1 to 5 μm.

The morphology, size, and distribution of intermetallics in solder determine their beneficial or detrimental effects on solder-joint integrity. In proper properties, the intermetallics in the interior of the solder joint (away form the interface) act as a strengthening phase. In contrast, large and needle-shaped compounds generally weaken the mechanical properties of a solder joint.

The formation of excessive intermetallic compounds has proven to be a frequent source of solder-joint failure. Cracks often are initiated around the interfacial area under stressful conditions when an unacceptable amount of intermetallics develops along the solder-substrate interface.

The adverse effect of intermetallic compounds on solder-joint integrity is believed to be attributed to the brittle nature and thermal expansion properties of such compounds, which may differ from the interior solder. The difference in thermal expansion contributes to a solder's internal stress development. In addition, excessive amounts of intermetallic compounds impair the solderability of some systems, depleting one element of the contact surface. For instance, tin depletion from tin-lead coating on copper leads causes the exposure of Cu_3Sn to oxidation, resulting in inconsistent and/or poor solderability of component leads. In this case, the interfacial area is composed of gradients with Cu_3Sn phase next to the copper substrate followed by Cu_6Sn_5 phase and lead-rich phase away from the interface line. Also, excessive intermetallics render a dull, rough look to solder joints.

14.7.2 Gold-Plated Substrates versus Solder-Joint Formation

Using gold (Au) as a surface coating to resist the oxidation of underlying metals in semiconductor packages and electronics assemblies is a routine practice.[21] Common applications include gold plating on PWBs, gold-containing thick-film circuitry on hybrids, soft gold (24-karat) wire bonding, and hard gold (cobalt or nickel gold) for edge fingers as connectors. However, many workers in the industry are concerned or uncertain about the full role gold plays in solder.

When a gold-coated substrate is in direct contact with a tin-containing solder, the gold combines with the tin of the solder at a rapid rate due to the metallurgic affinity between tin and gold, forming gold-tin intermetallics. Gold-tin intermetallics can affect a solder's physical and mechanical properties and alter a solder joint's appearance and microstructure.

A gold concentration below 10 percent by weight in tin-lead solder slightly increases that solder's initial tensile strength. However, beyond 3 percent, a solder's shear strength slowly drops. Normally, its hardness increases with the addition of gold. This effect is enhanced as the gold content exceeds 7 percent. A solder's ductility is slowly reduced with gold concentrations below 7 percent by weight and then drops rapidly as the gold content exceeds 7 percent by weight.

Gold can affect a solder's ability to wet and spread. Although a 2 percent gold concentration has no effect on 63Sn/37Pb, concentrations above 2 percent reduce the solder's spreadability and fluidity. For copper plated with gold, a pure-gold coating has shown better wetting and spread than alloy gold when soldering with 63Sn/37Pb under identical conditions.

The dissolution rate of gold in solder depends on temperature, time, and solder composition. Foreign elements, such as gold, indium, and zinc, in tin-lead solder retard this dissolution. During a reflow process with a long heating time, the quick dissolution of gold in molten solder causes that solder to wet directly onto the base metal and not the gold coating.

Although one might expect gold's inert nature to provide a base metal with full protection, tests on the aging of gold-electroplated copper indicate that the solderability as measured by wetting time degrades with aging at a temperature of 170°C. Solderability degrades due to the following causes: (1) the diffusion of atmospheric contaminants through the porous gold film results in the oxidation of the base material, (2) the diffusion of base metal reaches the surface through the coating, and (3) the diffusion rate is associated with the gold coating grain size, with smaller grain sizes favoring diffusion.

Gold dissolved in solder alters that solder's microstructure. As the gold content reaches 1 percent by weight, the characteristic needle-shaped phase found in eutectic solder becomes readily detectable in microstructure. The amount of hard phase increases with elevated gold concentrations. At room temperature, the composition of these intermetallics is a mixture of $AuSn_4$ and tin.

The incorporation of gold may or may not change a solder's physical properties. At concentrations below 10 percent by weight, gold does not significantly affect a solder's electrical or thermal conductivity.

Gold can lower a solder's solidus temperature and increase its liquidus temperature, thereby widening the paste range or creating a pasty range for eutectic solder. This affects a solder's application performance, particularly for solder interconnections. Lowering a solder's softening temperature changes its mechanical response to rising temperatures. A eutectic solder is required for applications demanding high solder fluidity, while assemblies with a wide gap to fill find solder with a wide pasty range preferable.

Overall, gold has the most pronounced effect on solder joints in the following areas:

- Fluidity
- Wettability and spread
- Mechanical properties
- Phase transition temperature
- Microstructure
- Appearance

An overly thick gold coating results in a higher gold concentration in solder and an increase in material cost. If the coating is too thin, the surface protection effectiveness may suffer considering the state-of-the-art process techniques. One also should take into account that the surface condition of the gold, particularly its porosity, is equally important to surface protection. An optimal gold application balances surface intactness, concentration in solder after dissolution, and cost.

When a solder's gold content is excessive, the following mechanical and/or metallurgical phenomena may occur:

- Premature solder-joint fracture due to embrittlement
- Void creation
- Microstructure coarsening

The upper limit of gold concentration is assessed to be approximately 3 percent by weight. Above 3 percent, deleterious effects could occur in one or more of the aforementioned areas. The 3 percent limit cited here is only a guideline. As a rule, one should verify the effect of gold concentration in solder for its performance in a specific electronics package and assembly under a given set of conditions. This is particularly important when the solder joint is small, such as for CSPs and FCs.

To ensure that gold concentrations do not exceed acceptable levels, industry standards call for gold removal immediately prior to soldering. The general guidelines for gold removal are as follows:

- A double tinning process of dynamic solder wave must be used for proper gold removal.

- A gold removal procedure is unnecessary for through-hole components intended for dip or wave soldering attachment, provided that the gold on the leads is less than 0.0025 mm.
- For surface-mounted parts, ideally, gold should be removed from at least 95 percent of the surface to be soldered.

14.7.3 Solder-Joint Voids

Voiding is one of the adverse phenomena in solder-joint integrity and reliability. It is generally expected that a low volume of small, well-dispersed voids has little effect on solder-joint integrity; however, high-volume or large-sized voids would degrade the joint with respect to its electrical, thermal, or mechanical properties.

For solder joints made from solder paste, the flow characteristics and the thermal and physical properties of the vehicle-flux system as well as the metal load are important factors. To minimize voiding, the processing parameters and the joint design should be optimized. These include the dosage of paste deposit, deposit thickness, joint configuration, reflow time, cooling rate, and wettability. Different solder pastes contribute to a different level of voids in the solder joint. However, the same paste could generate different voiding in size and concentration if used under different conditions. A quality joint therefore is influenced equally by the solder-joint assembling process, as outlined earlier, and by a compatible quality paste.

14.7.4 Solder Balling and Beading

When using solder paste, solder balling in the reflow process is a common phenomenon, as discussed in Sec. 14.3.3. It has been a continuous effort in soldering process control, in component and board quality, and in solder paste design to minimize the occurrence of solder balling.

The solder balling phenomenon can be defined as the situation that occurs when small spherical particles with various diameters are formed away from the main solder pool during reflow and do not coalesce with the solder pool after solidification. Versatile manufacturing environments have revealed two distinct types of solder balling in terms of physical characteristics:

1. Solder balling around any components and over the board
2. Large-sized solder balls associated with small and low-clearance passive components (e.g., 0603, 1206), being mostly larger than 0.005 in (0.13 mm)

The type 1 solder balls normally can be removed during the cleaning process; type 2 solder beads, however, are difficult to remove using a normal cleaning process. With implementation of the no-clean process, it is obviously desirable to avoid the occurrence of both types of solder balling. With the use of array packages (BGAs, CSPs), solder balling also becomes more troublesome. In the presence of solder balls, the assembly may encounter the risk of electrical short when any solder balls become loose and mobile during service. Excessive solder balling also may deprive solder from making good solder-joint fillets. In general, type 1 solder balls can be formed for different reasons. The following are the likely sources to be considered:

- Solder paste with inefficient fluxing with respect to solder powder or substrate or reflow profile, resulting in discrete particles that do not coalesce, due either to paste design or to subsequent paste degradation.
- Incompatible heating with respect to paste prior to solder melt (preheating or predry), which degrades the flux activity.
- Paste spattering due to too fast heating, forming discrete solder particles or aggregates outside the main solder pool.
- Solder paste contaminated with moisture or other high-energy chemicals that promote spattering.
- Solder paste containing extrafine solder particles that are carried away from the main solder by the organic portion (flux/vehicle) during heating, resulting in small solder balls.

- Interaction between solder paste and the solder mask.

The appearance of solder balls and their distribution often reveal the cause. Solder balls as a result of spattering are usually irregular and relatively large in size (larger than 20 μm is not uncommon) and are scattered over a large area of the board; solder balls caused by fine powder in the paste often form a halo around the solder; ineffective or insufficient fluxing results in small solder balls scattered around the joint; and solder mask–related solder balls leave discoloration marks on the board.

Solder paste spattering during reflow can be caused by

- Incompatibility between paste and reflow profile, such as too-fast heating or a high volatile content in the paste.
- Hygroscopicity of the paste. When the open time during assembly exceeds the capability of the paste or the paste is exposed to a temperature and/or humidity beyond its tolerance level, the moisture absorbed by the paste can cause spattering.

To minimize solder balling during board assembly, several issues need to be addressed:

- Selection of a solder paste that is able to deliver performance under the specific production conditions
- Understanding of the characteristics of the solder paste selected
- Setup of the reflow process that best fits the solder paste selected
- Assurance of consistency and quality of the solder substrate, including boards and components
- Control of ambient conditions (temperature and humidity)
- Control of open time that the paste can accommodate
- Assurance of solder mask compatibility with the solder paste
- Assurance of complete cure of the solder mask

For large solder beads associated with small passive components, their formation is largely attributed to the paste slump and flow under the component body between two terminations via capillary effect. The slump and flow dynamics also can be affected by the reflow temperature profile, the volume of paste, and component placement. In order to reduce the occurrence of these large solder beads, the following parameters are recommended for consideration:

- *Solder paste rheology*—minimizing paste slump
- *Amount of solder paste deposited*—avoiding excess paste
- *Component placement*—avoiding paste spreading during placement
- *Reflow profile*—reducing preheating temperature exposure

14.8 MICROSTRUCTURE

When dealing with physical objects in the linear dimension larger than 10^2 μm, we work on structure engineering. If we desire to view an object in the scale of 10^{-10} to 10^2 μm, we study material science and nuclear physics. As the scale shrinks to less than 10^{-10} μm, the object becomes intangible and immeasurable. Microstructure essentially falls in the range of 10^{-10} to 10^2 μm. Hence the understanding of solder joints within this range of dimensions is generally considered adequate in relating material properties to end-use applications.

Solders are normally polycrystallines that consist of an aggregate of many small crystals or grains. Most solder compositions contain multiple metallurgic phases that are physically distinct and formed and distributed according to given thermodynamic and kinetic conditions. For example, 63Sn/37Pb typically is composed of lead-rich and tin-rich phases in solid state below eutectic temperature. The finer structure with features smaller than grains and phases is called the *submicrostructure. Macrostructure* is coarser than microstructure and is discernible to the human eye.

The parameters affecting the formation of a microstructure during the joint-making process include heating and cooling. For an assembly prone to the formation of intermetallic compounds at the interface or in the intrinsic solder composition, prolonged heating may produce excessive intermetallic compounds at the interface or in the solder joint. When the solder is liquid, intermetallic compounds at the interface may continue to grow and migrate toward the solder-joint interior. In extreme cases, intermetallics may emerge onto the free surface of the solder, causing a change in solder-joint appearance. As to the cooling effect, the faster its rate, the finer the microstructure becomes. When the cooling rate is slow enough and approaches equilibrium, the microstructure of the eutectic composition normally consists of characteristic lamellar colonies. As the cooling rate increases, the degree of lamellar structure degeneration increases, and colonies eventually disappear. Although it is generally accepted that a faster cooling rate creates a finer grain structure in bulk solder, this rule is often complicated by the interfacial boundary and metallurgic reaction at the interface of solder joints.

The nature of the substrate and its metallurgic affinity to solder composition can affect solder-joint microstructure development. It would not be a surprise to see the microstructure of a 63Sn/37Pb joint interfacing with a nickel-plated substrate differ from that of a copper-plated substrate. Figure 14.11 shows a scanning electron micrograph (SEM) of the microstructure of a 62Sn/37Pb solder joint on copper substrate.

During service life, the integrity of joints made with sound fillet design and good wetting at the interfaces is affected by compatibility between the solder alloy and the substrate metal and subsequent in-circuit and external conditions such as heat dissipation, mechanical load, and environmental temperature fluctuation. Heat, load, time, and extensive metallurgic interaction between the solder and the substrate metal cause changes in microstructure. Failed solder joints have revealed significant degradation in microstructure otherwise hidden by its as-solidified counterpart. In most cases where the failure is a result of fatigue (fatigue-creep) phenomenon, grain (phase) coarsening has been observed to be a precursor of solder cracks, as shown in Fig. 14.12.

If we assess the mechanical properties of a solder joint by using commonly established techniques, then shear strength, creep, isothermal low-cycle fatigue, and thermomechanical creep are the top four parameters. For a eutectic solder composition, the shear strength of the solder joint is improved by a very slow cooling rate, which results in the formation of a near-equilibrium lamellar eutectic structure. On the other hand, strength is also enhanced by using a very fast cooling rate as a

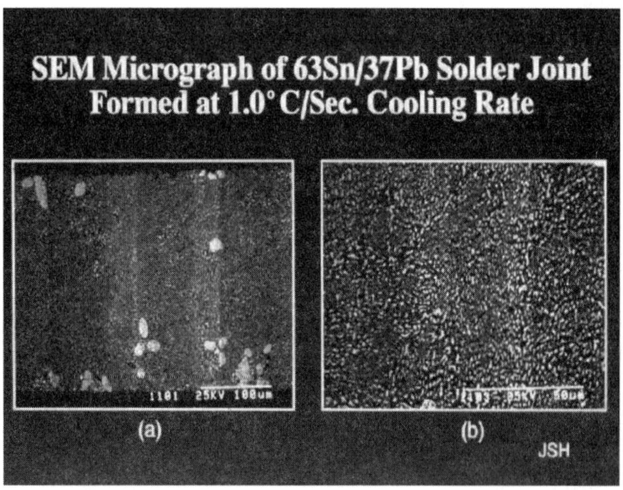

FIGURE 14.11 SEM of microstructure of as-reflowed 63Sn/37Pb solder joint.

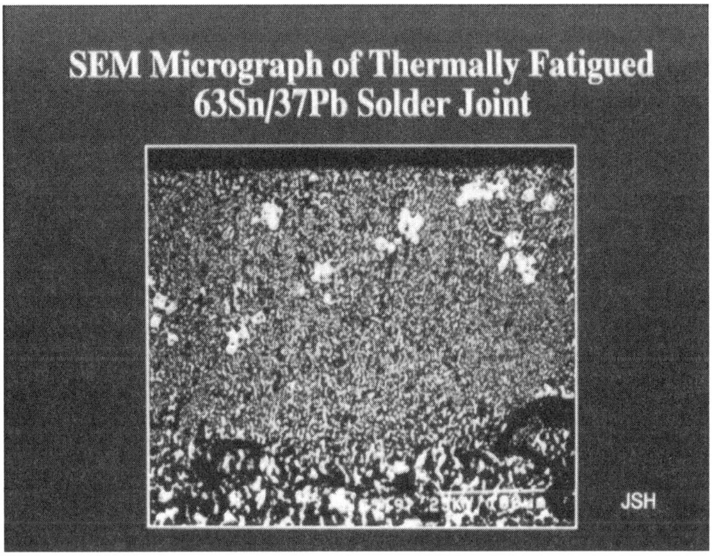

FIGURE 14.12 SEM of microstructure of fatigued 63Sn/37Pb solder joint.

result of grain-size refining. For plastic deformation under creep mode, creep resistance depends on the operating mechanism. When the lattice or vacancy diffusion process is the predominating step, creep resistance is often lower with a finer microstructure. This is due to an increased vacancy concentration resulting from a faster cooling rate. Under an isothermal fatigue environment, the relation between microstructure and fatigue resistance is not always straightforward. Nonetheless, microstructure homogeneity is more important to low-cycle fatigue resistance. With thermal cycling, a high fatigue resistance is often associated with decreased grain size.

To examine microstructural features, magnification of 100× to 5000× is needed. The characterization can use optical (light) or electron microscopy (desirably, both). For light microscopy, the solder specimen must be prepared carefully through metallagraphic techniques involving successive grinding and polishing. The technique uses ascending levels of abrasive particle fineness bonded on papers or used as a slurry on a cloth-covered wheel. The size of the abrasive particles can range from 23 μm to submicrometer size. Then the specimen goes through an etching process. In comparison, SEM requires little sample preparation when the sectioning (cutting) of the specimen is performed properly. Images from either secondary or backscattered electron signals can be obtained readily. Either provides informative characteristics with distinctive features. By combining information from both images, the microstructure and morphology of a solder joint can be better understood.

For solder joints, the two most information-revealing parameters are elemental composition and microstructure. For a given solder composition, the microstructure in the form of a quality microgram provides "sights" and "insights" into the state of solder-joint integrity.

14.9 SOLDER-JOINT INTEGRITY

Solder-joint integrity can be affected by the intrinsic nature of the solder alloy, the substrates in relation to the solder alloy, the joint design or structure, the joint-making process, and the external environment to which the solder joint is exposed. Therefore, to ensure the integrity of a solder joint, a step-by-step evaluation of the following items is warranted:

- Suitability of the solder alloy for mechanical properties
- Suitability of the solder alloy for substrate compatibility
- Adequacy of the solder wetting on substrates
- Design of joint configuration in shape, thickness, and fillet area
- Optimal reflow method and reflow process in temperature, heating time, and cooling rate
- Conditions of storage in relation to the aging effect on the solder joint
- Conditions of actual service in terms of upper temperature, lower temperature, temperature cycling, vibration, and other mechanical stresses
- Performance requirements under the conditions of actual service
- Design of viable accelerated testing conditions that correlate with actual service conditions

14.10 RELIABILITY OF BGA SOLDER INTERCONNECTIONS

By virtue of array packages' attributes, BGA interconnections on the main cards generally consist of relatively high numbers of solder joints per device in comparison with small outline integrated circuit (SOICs), plastic leaded chip carriers (PLCCs), or quad flatpacks (QFPs).[22] The higher number constitutes a higher probability of defect occurrence. This, coupled with less accessibility for inspection, rework, and repair, makes the consistency of forming array interconnections and their quality and integrity critically important.

The main concern for the reliability of array interconnections stems from two areas. First, array solder interconnections are less compliant than conventional peripheral-leaded interconnections. The decreased compliance may contribute to reduced performance under a fatigue environment due to the cyclic thermal stress and strain imposed on the system by temperature fluctuations and in-circuit power on/offs. The surface-mount array interconnection is also relatively new, and its applications are still in the infant stage for board-level assembly. Statistically substantiated data are lacking in terms of field performance. A common failure mode of BGAs on PWB interconnections is shown in Fig. 14.13.

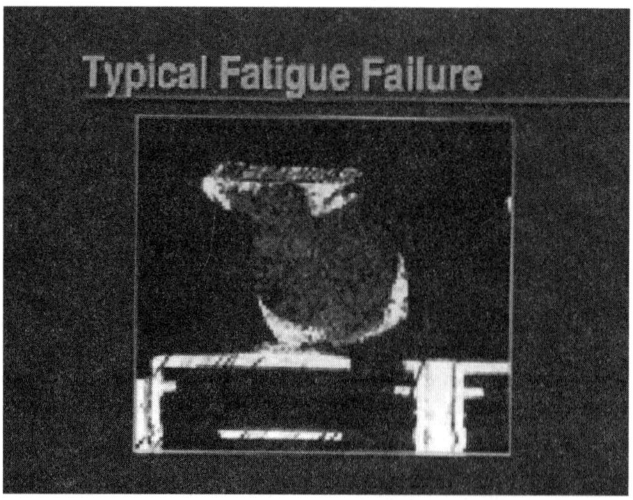

FIGURE 14.13 Thermal fatigued BGA solder joint.

Factors affecting long-term reliability of array solder joints include features of the component, board material, solder composition, solder-joint configuration and volume, underfilling, and manufacturing process.

14.10.1 Components

The temperature profile to which each solder joint is exposed is a major contributor to the distribution of thermal stress and strain among array solder joints. In addition to the external temperature change, the temperature profile depends on the functional characteristics of the die, the ratio of die to package size, the thermal property of the carrier substrate, and power dissipation. A study of IC packages ranging from 81 to 421 pins under power cycling demonstrated that outermost solder joints reached 84.3°C, whereas center joints were 98.9°C (1.2°C below the junction temperature) for an 81-pin package.

However, the 421-pin package, having a lower die-to-package size ratio, experienced a large temperature differential between the outer joints (56.2°C) and the center joints (98.5°C).[20] It was found that a 165-pin device that had the largest die size in the components under study had the earliest failure, and its cycle to failure was lower than that of the 225-pin device.[22]

It also was found that the solder joints directly underneath the perimeters of the die failed first under temperature cycling. This indicates that solder-joint fatigue life depends more on die size than on package size and that the relative location of solder joints to the edge of the die plays an important role in the fatigue performance of solder joints.

14.10.2 Board Materials

Two characteristics of board materials that are most influential to the long-term performance of solder interconnections are planarity and CTE. Poor board planarity adds to the coplanarity problem of the BGA package, contributing to the occurrence of solder-joint distortion, which in turn may lead to early failure of the solder joint under cyclic stresses. The CTE of conventional board material (FR-4) is approximately 15×10^{-6}/°C, whereas the ceramic carrier substrate of CBGA has a nominal CTE of 6 to 7×10^{-6}/°C. The CTE of solder material falls in the range of 21 to 30 $\times 10^{-6}$/°C, depending on the alloy. The differential in CTE between the board and carrier substrate results in an additional force of cyclic plastic deformation in solder joints under temperature-imposed conditions. A closely matched CTE between the board and carrier substrate reduces thermally induced stresses.

14.10.3 Solder Composition

The solder composition of the BGA carrier solder bumps affects the mechanical behavior of the solder interconnections. In general, solder that is "stronger" in fatigue and creep resistance is expected to deliver a better service life. The thickness (height) of solder joints between the BGA and board is much larger than that of a fine-pitch QFP. The actual BGA solder joint height depends on the diameter of the bumps and the dimensions of solder pads; for example, the 0.022-in (0.55-mm) BGA solder height compares with a 0.003-in (0.08-mm) height for the QFP. Since the solder height for BGAs is larger, the effect of the intrinsic properties of the solder material on BGAs is expected to be more pronounced than on QFPs.

14.10.4 Solder-Joint Configuration and Volume

The shape or configuration of solder joints can change the stress distribution and consequently affects failure mode development. Solder-joint volume contributes to the kinetics of solder-joint crack propagation. In addition, uniformity and consistency in volume and configuration among array solder joints within a package are important.

14.10.5 Other Material: Underfill

Several studies demonstrate that an epoxy that fills the air gap between the solder and the underside of the component is beneficial to the fatigue life of solder joints. For plastic BGAs, the fatigue life of solder joints for over-molded pad array carriers (OMPACs) under temperature cycles of −40 to 125°C improved nearly twofold with epoxy underfill. The underfill around the chip and the solder of the slightly larger than IC carrier (SLICC) assembly, which is a combination of FC and ball array technology, was used to enhance solder-joint reliability.[20] The eventual solder joint failure under thermal shock of −55 to 125°C was attributed to the separation of the underfill from the die surface. This loss of adhesion was further related to foreign contamination not thoroughly removed during the cleaning procedure. A similar enhancement in solder-joint performance by means of underfill also was observed for the assembly of ceramic BGAs.

14.10.6 Manufacturing Process

In addition to the factors that contribute to the long-term performance of solder joints during their service life, the ability to make high-quality solder joints at the point of production is equally important. Although an existing installed surface-mount operation can be used directly to mount BGAs on a mother board, the setup of process parameters (particularly reflow temperature profile), control of the process, and proper ambient conditions are key to making quality solder joints. High humidity and high temperature are generally detrimental to surface-mount manufacturing.

Material behavior—in relation to temperature change, component effect, and design—is a significant factor in the reliability of interconnections. Understanding each of these areas in conjunction with establishing a quality manufacturing process is the means to full utilization of the merits of BGA packages.

14.11 CHALLENGES IN MODELING SOLDER-JOINT LIFE PREDICTION

It is well recognized that solder-joint reliability relies not only on intrinsic material properties but also on design, the component type, the process that makes solder connections, and the long-term service conditions. As electronic IC packages and components continue to change at a rapid pace, it is highly desirable to have a model able to predict the service life and reliability of solder joints under a specific set of conditions. However, to derive such a model is an ever-daunting task. This is primarily due to the complex nature of solder materials in conjunction with the "active" service conditions. Solder materials impart more complex behavior in response to temperature, stress, and time than high-temperature materials such as steel. Much is to be understood. The challenges are further complicated by the high level of versatility in circuit boards with various materials and designs.

For a given solder composition and design, the main physical factors affecting solder material performance are temperature, ambient environment, strain range, strain rate, loading waveform, intrinsic microscopic structure, and surface condition of the solder joints. Furthermore, the solder joint is expected to behave differently from bulk solder materials. Hence some established mechanical and thermal behavior of solders may need to be modified for the solder-joint design. This is presumably due to the following causes:

- The presence of a high ratio of substrate surface to solder volume, resulting in a large number of heterogeneous nucleation sites during solidification
- A concentration gradient of elemental or metallurgic composition when the solder joint is formed

Either one of the preceding conditions may lead to a structure that is not homogeneous. As solder-joint thickness decreases, the interfacial effect is more pronounced. Accordingly, the properties of solder joints may be altered, and the failure mechanism may be incongruent with that derived for bulk solder. It is generally accepted that under cyclic strain conditions, the creep-fatigue process essentially accounts for solder-joint degradation, assuming that the interfacial problems, such as those caused by excessive intermetallics or poor wetting, are not the determining factor for failure. Consequently, most studies have been carried out under the creep-fatigue testing mode. The goals in studying the creep-fatigue process are

- To understand material behavior under cyclic strains that are inevitable encounters during solder-joint service life in electronic assembly
- To develop or improve the resistance to degradation under cyclic strains by taking a systems approach
- To predict the fatigue life of solder joints so that performance reliability at a given set of service conditions can be designed and ensured

Numerous fatigue-life prediction methods have been proposed, including the frequency-modified Coffin-Manson (C-M) method, strain-range partitioning, fracture mechanics, and finite-element analysis (FEA). The methodologies are largely borrowed from the established fatigue and creep phenomena of steels as a result of extensive studies coupled with the field data obtained over a longer period of time. The frequency-modified C-M method and the fracture mechanics–based methods are not capable of handling complex loading waveforms, although fracture mechanics can monitor the effect of interfacial crack initiation and propagation on life in a comprehensive manner, and the frequency-modified C-M method takes frequency effect into consideration. Strain-range partitioning is able to deal with the strains in any waveform, yet separating the total inelastic strain range per cycle into creep strain and plastic strain is not easy. FEA also lacks the capability of including complex waveforms. Increased efforts at tailoring the basic lifeprediction models established for steels are burgeoning in the electronics industry. Although the result of efforts may have generated the models that predict solder-joint life in a comparative sense, a true working model has yet to be found.

Service conditions under which solder joints must perform in electronics packages and assemblies often involve random multiaxial stresses, and they expose solder joints to creep range in addition to cyclic strains. At this time, sufficient and integrated data on solder-joint behavior under such conditions and corresponding damage evolution are lacking. Consequently, some important areas and conditions are grossly ignored in the modeling scheme. Listed below are the areas that either have not been included or have not been covered adequately. In turn, they are considered to contribute to the limitations of the single model to wide applications.

1. Effect of initial microstructure
2. Effect of grain size
3. Effect of microstructure that is not homogeneous
4. Change in microstructure versus external conditions
5. Multiaxial creep-fatigue
6. Identification of presence or absence of crack-free materials at the starting point
7. Size of existing cracks, if present
8. Effect of interfacial metallurgic interaction
9. Joint thickness versus interfacial effect (thinner solder joint imposes increased interfacial effect and decreased conventional fatigue-creep phenomena)
10. Damage mechanism—transgranular or intergranular
11. Potential damage mechanism shift (from transgranular or intergranular)
12. Presence or absence of grain boundary cavitation
13. Effect of fillet geometry
14. Effect of free surface condition

15. Correlation of accelerated testing conditions and the actual service condition
16. Testing condition versus damage mechanism
17. Service condition to include possible variation in chip power dissipation over time in addition to ambient temperature change and the number of on/off power cycles
18. Effect of variation in coplanarity among solder joints

Including the above-listed areas in modeling is not only overwhelmingly time-consuming but also extremely difficult. It is a challenge indeed. However, the inclusion of all the preceding areas in devising a model is necessary to achieve a model that ultimately predicts the service life of solder joints for a specific application.

14.12 CREEP AND FATIGUE INTERACTION

Serving as interconnections in electronics packaging and assembly, solder materials usually entail a simultaneous exposure to more than one hostile environment, such as temperature and stress. Most solder materials, even under ambient temperature (298°K), reach homologous temperatures (T/T_m^*) well beyond 0.5. Under these service conditions, both creep and fatigue processes may exist and operate interactively. These situations would be equivalent to creep under cyclic loading or fatigue at high temperature.

Whether a wearout phenomenon should be viewed as creep-aggravated fatigue or fatigue-accelerated creep depends on several factors. Generally, when the cyclic stress (or strain) amplitude is small compared with the mean stress (or strain) or the applied frequency is low and/or the temperature is high, the phenomenon can be treated as creep perturbed by fatigue. In contrast, when the cyclic stress amplitude is large or the applied frequency is high and/or the temperature is low, the degradation phenomenon should be considered fatigue accelerated by creep behavior.

Solder material of electronic interconnections may undergo changes through one of the two interactive behaviors involving both creep and fatigue. The readily measurable material properties for obtaining maximum creep resistance often differ from those for obtaining maximum fatigue resistance. The development of improved materials should target enhancing both creep and fatigue resistance.

14.13 LEAD-FREE SOLDER VERSUS LEAD-BEARING SOLDER

The driving forces for the lead-free solders are primarily two: performance demands and environmental/health concerns.[3,23–78] It is reasonably well substantiated that the common thermal fatigue failure for solder interconnections is linked with the lead-rich phase. This lead-rich phase cannot be effectively strengthened by tin solute atoms due to limited solubility and tin precipitation. At room temperature, the limited solubility of lead in tin matrix renders it incapable of improving the plastic deformation slip. Under temperature cycling (thermomechanical fatigue) conditions, this lead-rich phase tends to coarsen and eventually leads to the solder-joint crack. It is therefore expected that the absence of the lead phase of a properly designed lead-free tin-based solder may impart improved mechanical behavior, resulting in strengthened solders. The new alloys selected and listed below manifest the improved performance over 63Sn/37Pb.

On the international landscape, some companies have implemented lead-free solders in commercial products. Some manufacturers have their own effort in developing or selecting a suitable lead-free alloy composition. Many companies plan to convert to lead-free technology by 2001. The industry in Japan has started producing lead-free products, including the Panasonic minidisk players. Reportedly, Toshiba has developed lead-free solders suitable for high-density cellular phones.

Nortel Networks has produced lead-free Meridian phones at four different sites in North America and Europe.

14.14 FUNDAMENTAL TECHNOLOGY FOR LEAD-FREE

Any viable lead-free solders in lieu of tin-lead eutectic or near-eutectic compositions could not escape from being a tin-based system (i.e., a minimum of 60 wt% of tin). This was concluded based on both fundamental materials science and practical perspectives. Fundamentals include metallurgic bonding capability on commonly used substrates, dynamic wetting ability under practical reflow conditions, and metallurgic "interactions" or alloying phenomena between elements. Practical factors cover the availability of natural resources, manufacturability, toxicity, and cost.

In selecting any element, its alloying ability with tin and its property in melting-point depression while alloying with tin are the two crucial material characteristics for selecting the constituent elements and their specific dosages in designing lead-free solders. Based on metallurgy, elements such as In, Bi, Mg, Ag, Cu, Al, Ga, and Zn are the candidates that can lower the melting temperature of tin to create the tin-based alloys that possess the required properties for electronics packaging and assembly. Table 14.4 lists the speculated melting-point depressions with tin at the selected temperature ranges for the candidate elements.[23]

14.14.1 Strengthening Approaches

Under the high-temperature conditions (even above room temperature) to which solder joints normally are exposed, the mobility of atoms increases, and so do the dislocations. Other crystallinic defects such as vacancies also increase. Additional slip systems are introduced, and metallurgic stability is affected unfavorably. In addition, environmental effects (oxidation, corrosion) also become more pronounced.

Approaches that potentially can hinder these material phenomena are expected to enhance the performance of solders, which in turn will meet the level of performance required for new and future applications. These include (1) microscopic incorporation of nonalloying dopant, (2) microstructural strengthening, (3) alloy strengthening, and (4) macroscopic blending of selected fillers. These approaches involve both process and material factors. For example, solid solutioning, where solute atoms normally reduce the stacking fault energy and favorably control the diffusion behavior, is one of the well-adopted strengthening mechanisms. In any of the approaches, the resulting objective of the alloy design is to achieve the proper parameters for the following properties:

- Phase-transition temperatures (liquidus and solidus temperature) to be as close to lead-bearing counterparts as practical

TABLE 14.4 Estimated Melting-Point Depression of Tin by the Selected Elements at the Specified Temperature Ranges

	Melting-point depression		
Element	160 to 183°C	183 to 199°C	200 to 230°C
In	2.3	2.1	1.8
Bi	1.7	1.7	1.7
Mg	—	—	16.0
Ag	—	—	3.1
Cu	—	—	7.1
Al	—	—	7.4
Ga	2.6	2.5	2.4
Zn	—	3.8	3.8

- Suitable physical properties, specifically electrical and thermal conductivity and CTE
- Compatible metallurgic properties with the interfacial substrates of components and boards
- Adequate mechanical properties, including shear strength, creep resistance, isothermal fatigue resistance, thermomechanical fatigue resistance, and microstructural stability
- Intrinsic wetting ability
- Environmental shelf stability
- Relatively low (or no) toxicity
- Acceptable cost

14.14.2 Alloy Design

For a tin matrix, candidates that can serve as viable alloying elements are quite small in number, practically limited to Ag, Bi, Cu, In, and Sb. However, doping elements may extend to a larger group of elements and compounds, such as Ga and Se. Metallurgic interactions (reactions) and microstructure evolution in relation to rises in temperature are the critical scientific basis for developing new lead-free solders.

Binary phase diagrams provide the general information about the conditions and extent of metallurgic interactions, although complete phase diagrams beyond the binary system are scarce. Nonetheless, binary phase diagrams offer a useful starting point.

After a decade of research,[3] we found that the actual test results of the designed multiple-element alloy compositions came very close to the anticipated features in properties and performance between a candidate element and tin matrix. To illustrate the point, as examples, Se and Te were found to readily embrittle the tin-based alloys. Antimony in an improper amount quickly jeopardizes the alloy's wetting ability. The distribution of indium atoms in the tin host lattice is sensitively reflected in the fatigue performance. The level of bismuth second-phase precipitation is closedly associated with the mechanical properties of the bismuth-bearing alloys. The formation of intermediate phases and intermetallic compounds between tin and copper, silver, or antimony remarkably affect the strength and fatigue life of the alloy, which in turn depends on the concentration of each element as well as on the relative concentration among the elements.

Since the general performance is as predictable as stated, a high-performance alloy composition, however, demands a stunningly intricate balance of the elemental constituents. In each compositional system, the useful products are often a specific composition or a narrow range of compositions at best.[67–69]

New solder alloys must possess characteristics that are compatible with practical manufacturing techniques and end-use environments. The basic material properties such as liquidus/solidus temperature, electrical/thermal conductivity, intrinsic wetting ability on surfaces that are commonly used, mechanical properties, and environmental shelf stability must be gauged. Under the current framework, conductivity and shelf stability are not as sensitive to the makeup of a specific system as intrinsic wetting ability, mechanical performance, and phase-transition temperatures. An ability to optimize these properties through indepth application of materials science and metallurgic phenomena is the key.

14.15 PWB SURFACE FINISHES

For making sound interconnections, the characteristics and properties of the PWB surface finish are as important as the component leads and termination. Hot-air solder-leveled Sn/Pb (HASL) has been used successfully as the surface finish for surface-mount and mixed PWBs. As the need for a flat surface with uniform thickness becomes increasingly important to forming consistent and reliable fine-pitch solder joints, the HASL process often falls short. Alternatives to HASL include immersion Sn, electroplated Sn/Pb (reflowed or non-reflowed), electroplated Au/Ni, electroless Au/electroless Ni, immersion Au/electroless Ni, immersion Pd, immersion Pb/electroless Ni, electroplated Sn/Ni alloy, and organic coating. When selecting an alternative surface finish for PWB assembly, the key parameters of solderability, ambient stability, high-temperature stability, suitability of the use as contact/switch surface, solder-joint integrity, wire bondability of those assemblies that involve wire bonding, and cost are to be considered.

Ideally, a PWB surface finish fulfills all four functions:

- Solderability protection
- Contact/switch
- Wire bonding
- Solder-joint interface

However, practically, some surface finish systems are designed primarily for solderability protection.

For solderability retention and protection, the HASL process has been successfully used as the PWB surface finish for surface-mount and mixed PWB assemblies. As the industry continues to evolve, the following forces primarily drive the development of HASL alternatives:

- Increased demands for flat and uniform solder pads
- Increased demands for consistent thickness of surface finish
- Obtaining the same metal system and process for contact/switch
- Less thermal stress process for temperature-vulnerable PWBs, such as PCMCIA
- Eliminating lead

14.15.1 Solderability Factors

Several factors affect the solderability of PWB solder pads:

- *Pad surface composition:* copper, tin-lead-coated copper, antioxidant-coated copper, gold-nickel-copper, palladium-nickel-copper
- *Surface conditions:* oxide and sulfide content, organic contaminants, intermetallics, other contaminants
- *Thickness of coating:* determination of a proper thickness in relation to storage conditions
- *Storage condition:* time, temperature, humidity

Although the required coating thickness may vary, a proper thickness is one that is compatible with the time and condition of storage to avoid excessive formation of intermetallics and exposure of intermetallics to the ambient environment. Generally, the lower the temperature and the humidity, the less degradation of solderability there is with time.

14.15.2 Basic Processes

Three basic techniques to deposit metallic surface finish are electroplating, electroless plating, and immersion. Inherently, electroplating using electric current is able to economically deposit coatings up to 0.000400 in thick. The exact thickness depends on metal and process parameters. Electroless plating, requiring the presence of a proper reducing agent in the plating bath, converts metal salts into metal and deposits them on the substrate. The immersion plating process, in the absence of both electric current and the reducing agent in the bath, deposits a new metal surface by replacing the base metal. In this process, plating stops when the surface of base metal is completely covered; thus only a limited coating thickness can be obtained through the immersion process. For both electroless and immersion processes, the intricate chemistry and control of the kinetics are vital to the plating results. Further, the process parameters and chemistry, including pH and chemical ingredients, must be compatible with the solder mask and PWB materials.

To summarize the respective characteristics of these three processes:

Electrolytic:
- Requires electric current
- Economic
- Wide operating window
- Can serve as etch resist

Electroless:
- Requires abundant reducing agent in plating bath
- Reducing agent is metal-specific, e.g., Ni: hypophosphite, dimethylamine borane; Cu: formaldehyde
- Chemistry is critical
- Higher cost
- Uniform coating

Immersion:
- No electric current used
- Does not require reducing agent
- Limited thickness
- Chemistry is critical
- Temperature is important

14.15.3 Metallic Systems

Available metallic surface finishes on copper traces include Sn, Sn/Pb alloy, Sn/Ni alloy, Sn/Bi, Sn/Sb, Au/Ni, Au/Pd, Pd/Ni, and Pd. The systems containing noble or seminoble metals, such as Au/Ni, Au/Pd, Pd/Ni, and Pd/Cu, are capable of delivering the coating surface with uniform thickness. Systems imparting a pure and clean surface also provide wire-bondable substrate. In addition, wire bonding generally requires thicker coatings, namely, more than 0.000020 in. A unique feature of an Au/Ni system is its stability toward elevated temperature exposure during the assembly process as well as during its subsequent service life. When in contact with molten solder of Sn/Pb, Sn/Ag, or Sn/Bi, surfaces coated with Sn and Sn/Pb are normally associated with better spreading and a lower wetting angle than others. Of the metallic systems, those containing a nickel interlayer are expected to possess more stable solder-joint interfaces, and in these systems, solder is expected to wet on nickel during reflow. For a phosphorus-containing plating bath, a balanced concentration of phosphors in electroless nickel plating is needed. When the phosphorus content is too high, wetability suffers, and when it is too low, thermal stress resistance and adhesion strength are sacrificed. In addition, noble metals in solder need to be noted in order to prevent any adverse effects in solder joining integrity.

Another characteristic that is important to solderability is the porosity on the surface. Thinner coatings are more prone to porosity and therefore porosity-related problems, although the surface density and texture can be controlled, in part, by the chemistry and kinetics.

Below is a listing of a number of specific systems, their characteristics, and the functions fulfilled.

Electroless Pd/Cu
- 4 to 20 μin of Pd
- Solderability protection
- Contact/switch

Electroless Pd/Ni/Cu
- The Pd thickness is a function of dwell time; deposition rate depends on temperature

- 4 to 20 μin of Pd, 150 to 200 μin of Ni
- Processing temperature at 60 to 70 °C
- Solderability protection
- Contact/switch
- Wire bonding[80,81]

Immersion Pd/Cu
- Dissolve Cu to deposit Pd
- 2.5 to 3.0 μin of Pd
- Controlled base metal plating bath
- A dense and low-porosity coating achieved
- Can reach maximum deposit thickness in 2 min at 50°C
- Porosity related to the plating time and strength of the solution
- Surface may discolor after multiple passes in reflow over, but solderability retained
- Solderability protection

Immersion Au/electroless Pd/Cu
- Less than 0.1 μin of Au, 10 to 25 μin of Pd
- Solderability protection
- Contact/switch
- Wire bonding

Immersion Au/electroless Pd/electroless Ni/Cu
- Less than a 0.1μin of Au, 10 to 25 μin of Pd
- 125 to 150 μmin of Ni
- Solderability protection
- Contact/switch
- Wire bonding
- Stable solder-joint interface

Immersion Au/electroless Ni/Cu
- 3 to 5 μin of Au, 50 to 200 μin of Ni
- Vulnerable to Au porosity
- Solderability protection
- Contact/switch
- Wire bonding
- Stable solder-joint interface

Electrolytic Au/electrolytic Ni
- Harsh plating condition to solder mask
- Difficult to get adequate Ni plating in small vias in thick board
- Avoid "black pad" problem for some BGA assemblies
- Solderability protection
- Contact/switch
- Wire bonding
- Stable solder-joint interface

Immersion Au/electrolytic Ni
- Electrolytic Ni has good solderability (versus electroless)
- Solderability protection
- Contact/switch
- Wire bonding
- Stable solder-joint interface

Electrolytic Au/electroless Ni
- Characteristics of electrolytic Au (10 to 100 μin)

Immersion Bi[82]
- 4 to 6 μin of Bi
- Flat surface

Immersion Sn[83–85]
- Sn is deposited on Cu, while Cu is transferred into solution
- 2.5 to 35 μin of Sn
- Incorporating an organometallic complex
- Plating bath temperature around 65°C
- Plating solution with stannous sulfate or chloride
- Sn/Cu intermetallics
- Flat surface
- Not wire bondable
- Not a good switch material
- Cost competitive

Immersion Ag[86,87]
- Ag is deposited on Cu, while Cu is transferred into solution
- About 5 μin of Ag
- Incorporating an oxidation inhibitor layer
- Plating solution with silver nitrate
- Flat surface
- May be wire bondable
- Not a good switch material at thin deposition (3 to 9 μin)

HASL Sn/Cu eutectic[88]
- Operating temperature is higher than Sn/Pb, as shown in Table 14.5
- Potential effects of higher temperature on PWB are to be considered

14.15.4 Organic Systems

Benzotriazole has been well recognized as an effective copper antitarnish and antioxidation agent for decades. Its effectiveness, attributable to the formation of benzotriazole complex, is largely limited to ambient temperature.

As temperature rises, the protective function disintegrates. Azole derivatives, such as imidazole (mp 90°C, bp 257°C) and benzimidazole (mp 17°C, bp 360°C) have been used to increase the stability under elevated temperature. SMT assembly of mixed boards involves three stages of temperature excursion—reflow, adhesive curing, and wave soldering. The reflow step, however, is considered to be potentially most harmful to the intactness of the organic coating because it is the step with the highest temperature and longest exposure time. In addition to its vulnerability to high temperature, this family of chemicals has an appreciable solubility in alcohol and water.

Although the performance of organic coatings varies with the formula and process, the general behavior of organic coatings falls in the following regimen:

- Need compatible flux (generally more active flux).
- May need more active flux in wave soldering for mixed boards.
- Thicker coating is more resistant to oxidation and temperature but also may demand more active flux.
- Organic coating needs to be preformed as the last step of PWB fabrication.
- At temperatures higher than 70°C, coating may degrade. However, the degradation may or may not reflect on solderability.
- May be sensitive to PWB presoaking process (e.g., 1250, 1 to 24 h).
- For no-clean chemistry, may require N_2 atmosphere or higher solids content in no-clean paste.

TABLE 14.5 Operating Parameters of HASL Sn/Cu versus HASL Sn/Pb[88]

	Sn/Pb, °C	Sn/Cu, °C
Bath temperature	250	280
Air knife temperature	250	280
Oil temperature	230	255
Air heat exchanger	250	300
PWB preheat	150	200

TABLE 14.6 Relative Performance of PWB Surface Finishes

HASL	Au/Ni	Pd/Ni	Pd/Cu	Organic
Pros: Most solderable	Uniform thickness Wire bondable Most stable to temperature	Uniform thickness Wire bondable	Uniform thickness Wire bondable	Uniform thickness Low cost Easy inspection
Cons: Nonuniform thickness Potential IMC problem Unsuitable for COB PCB exposed to high temperature	Higher cost	Higher cost	Higher cost (thicker coating)	Unsuitable for COB Flux and reflow process sensitive High-temperature degradation Cu reaches upper limit in solder bath Required as a last board fabrication step

- Steam aging test is not applicable
- Not suitable for chip-on-board (COB), where wire bonding is required.

Nonetheless, when fluxing activity and the process parameters are compatible, an organic coating can be a viable surface finish for PWBs. An additional bonus effect is that the bare copper appearance of the organic coated surface enhances the ease of visual inspection of peripheral solder fillets.[1,89]

14.15.5 Comparison of PWB Surface Finish Systems

When selecting an alternative surface finish for PWB assembly, the key parameters in terms of solderability, ambient stability, high-temperature stability, suitability for use as contact/switch surface, solder-joint integrity, and wire bondability for those assemblies which involve wire bonding, as well as cost, are to be considered.

Table 14.6 summarizes the relative performance of PWB surface finish systems. Regardless of other deficiencies, however, HASL provides the most solderable surface. When comparing metallic systems with HASL, the HASL process subjects PWBs to high temperature (above 200°C), producing inevitable thermal stress in the PWB. HASL is also not suitable for wire bonding. To make a choice in replacing HASL, many variables are to be assessed. Understanding the fundamentals behind each variable in conjunction with setting proper priorities among the variables for a specific application is the way to reach the best balanced solution.

Wetting spread is compared among various PWB surface finishes for three solder alloys—under convection air reflow. As shown in Table 14.7, less spread is associated with the lead-free alloys in comparison with 62Sn/36Pb/2Ag, and the extent of wetting spread was essentially similar except that Sn/Ag/Cu on immersion Pd is the lowest among the systems tested. Equivalent tests were conducted under vapor-phase soldering, which provided a somewhat protective atmosphere during soldering. Table 14.8 shows that the wetting spread improved over air reflow, and still the Sn/Ag/Cu alloy on immersion Pd showed the lowest spread. Furthermore, Ni/Au showed the most consistent wetting spread for all three alloys.

14.16 SELECTION MENU: LEAD-FREE SOLDER JOINT

From the simplest alloy that is a binary system to incrementally complex systems containing more than two elements, lead-free materials have been thoroughly explored, designed, and studied.[3,67–69] Based on more than a decade of study and progressive research and development work, application of the basic materials fundamentals coupled with understanding of practical and process parameters lead to an array of designs. Six systems with the corresponding compositions stand out on their performance merits. Their strengths and comparison with the established alloy compositions are summarized below. These six systems are

Sn/Ag/Bi

Sn/Ag/Cu

Sn/Ag/Cu/Bi

Sn/Ag/Bi/In

Sn/Ag/Cu/In

Sn/Cu/In/Ga

TABLE 14.7 Wetting Spread* of Three Alloys on Various PWB Surface Finishes under Convection Air Reflow[90]

PWB surface finish	62Sn/36Pb/2Ag	Sn/3.8Ag/0.7Cu	Sn/3.3Ag/3.0Bi/1.1Cu
OSP	4.5	4.2	4.0
Immersion Ag	4.7	4.55	4.6
Immersion Pd	4.4	3.9	4.4
Ni/Au	5.0	4.4	4.7

*On relative scale of 1 to 5, with 5 being complete spread.

TABLE 14.8 Wetting Spread* of Three Alloys on Various PWB Surface Finishes under Vapor-Phase Reflow[90]

PWB surface finish	62Sn/36Pb/2Ag	Sn/3.8Ag/0.7Cu	Sn/3.3Ag/3.0Bi/1.1Cu
OSP	5.0	4.3	4.5
Immersion Ag	4.7	4.8	5.0
Immersion Pd	4.7	3.9	4.7
Ni/Au	5.0	5.0	5.0

*On relative scale of 1 to 5, with 5 being complete spread.

A discussion of each of these systems is omitted in this text. Readers can obtain the detailed illustrations and data in the literature.[3] Some compositions are covered under the patents.[59–69] The selected compositions from each of the systems also are compared with the pertinent known lead-free alloys as well as with 63Sn/37Pb. Figures 14.14 through 14.23 summarize the relative performance of these selected compositions with the established solder alloys. An overall comparison will then be provided among these six systems, leading to the ranking by melting temperature (Table 14.9) and fatigue life (Table 14.10), and to the final slate of selections.

Figure 14.24 is an exemplary manufacturing process using a reflow solder of Sn/Ag/Bi/In and a wave solder of Sn/Cu composition to successfully produce the Panasonic Panasert image acquisition card. According to Dr. Ken Suetsugu and Tom Baggio of Panasonic-Matsushita, the process enjoyed a relatively low melting point of 210°C while offering workability, quality, and reliability in the finished product similar to 63Sn/37Pb.

Figures 14.25 through 14.27 show the Panasonic product evolution from player type (MJ30, MJ70) to player/recorder (MR100) products, and Figs. 14.28 through 14.30 show the corresponding main circuit cards. All were produced in lead-free Sn/Ag/Bi/Cu alloy compositions.[91]

The alloy also has impressively higher strength than any of the binary alloys—63Sn/37Pb, 96.5Sn/3.5Ag, or 99.3Sn/0.7Cu.

14.17 LEAD-FREE RECOMMENDATIONS

- An optimal composition should be determined by the performance level of a specific application. Tables 14.6 and 14.10 provide the relative performance of the selected alloys that show the most promise.
- A slate of composition as listed below can be considered:

  Sn/3.0–3.5Ag/3.0–3.5Bi/0.5–0.7Cu

  Sn/3.3–3.5Ag/1.0–3.0Bi/1.7–4.0In

  Sn/3.0–3.5Ag/0.5–0.5Cu/4.0–8.0In

  Sn/0.5–0.7Cu/5.0–6.0In/0.4–0.6Ga

  Sn/3.0–3.5Ag/0.5–1.5Cu

  Sn/3.0–3.5Ag/1.0–4.8Bi

  99.3Sn/0.7Cu

  96.5Sn/3.5Ag

- Melting temperature (liquidus temperature) is an important selection criterion.
- A proper reflow profile is able to compensate to some extent for the higher melting temperature (higher than 183°C) associated with lead-free alloys.
- For surface-mount PWB assembly, the melting temperature of solder alloys below 215°C provides the necessary process window.
- Alloy flow property, and thus wetting behavior of lead-free solder, differs from that of lead-bearing alloys.

Overall, the lack of universality and standardization and the hope for a quick conversion appear to be the top three hurdles in accepting or implementing lead-free products. Search for a universally

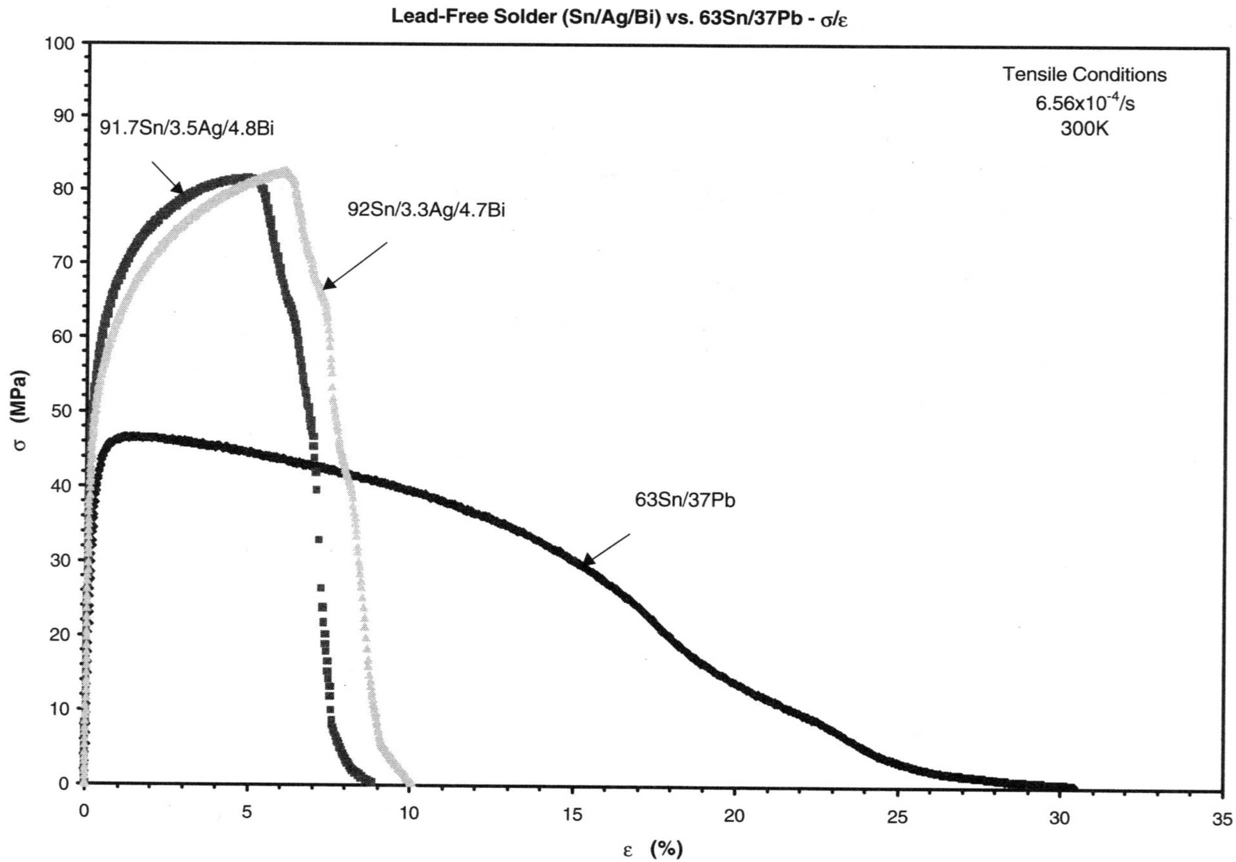

FIGURE 14.14 Tensile stress σ versus strain ε at 300 K and 6.56×10^{-4}/s for tin-silver-bismuth alloys and 63Sn/37Pb.

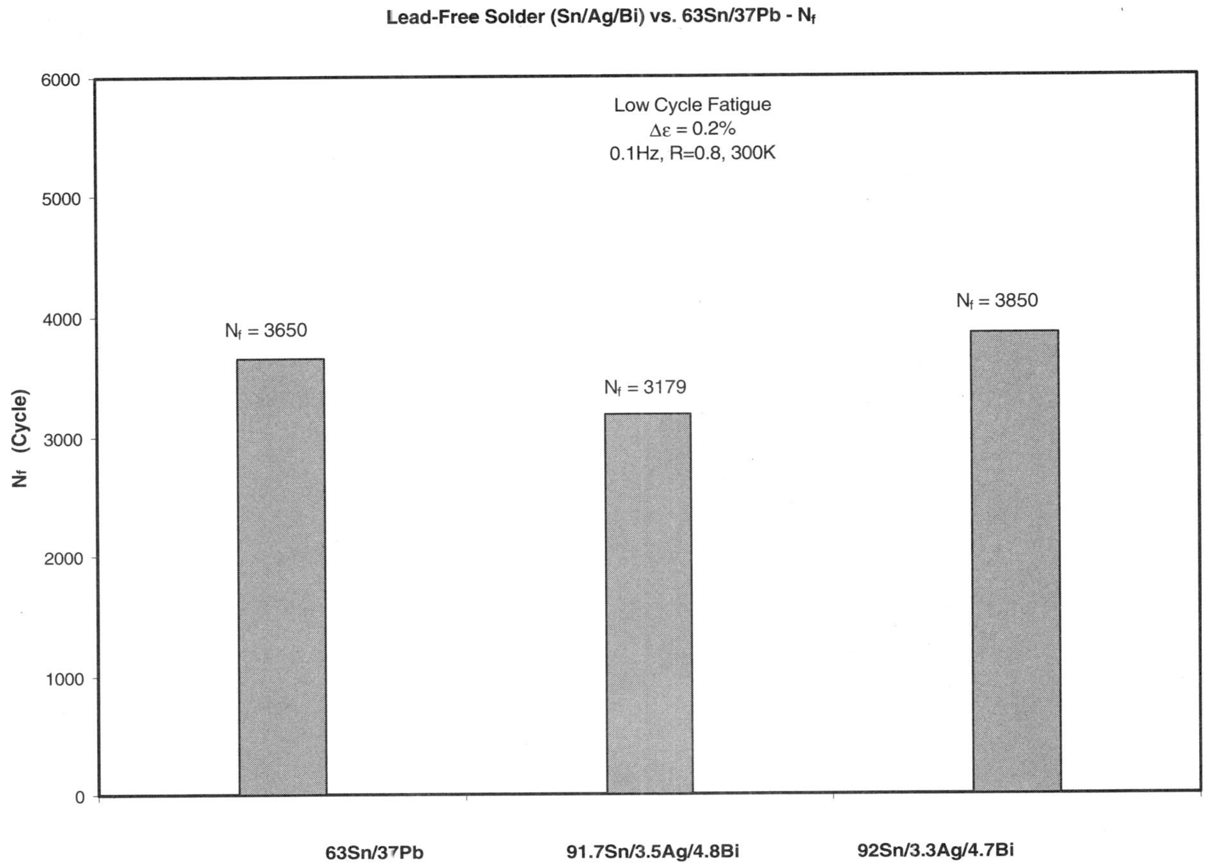

FIGURE 14.15 Comparison of fatigue life of tin-silver-bismuth alloys with 63Sn/37Pb.

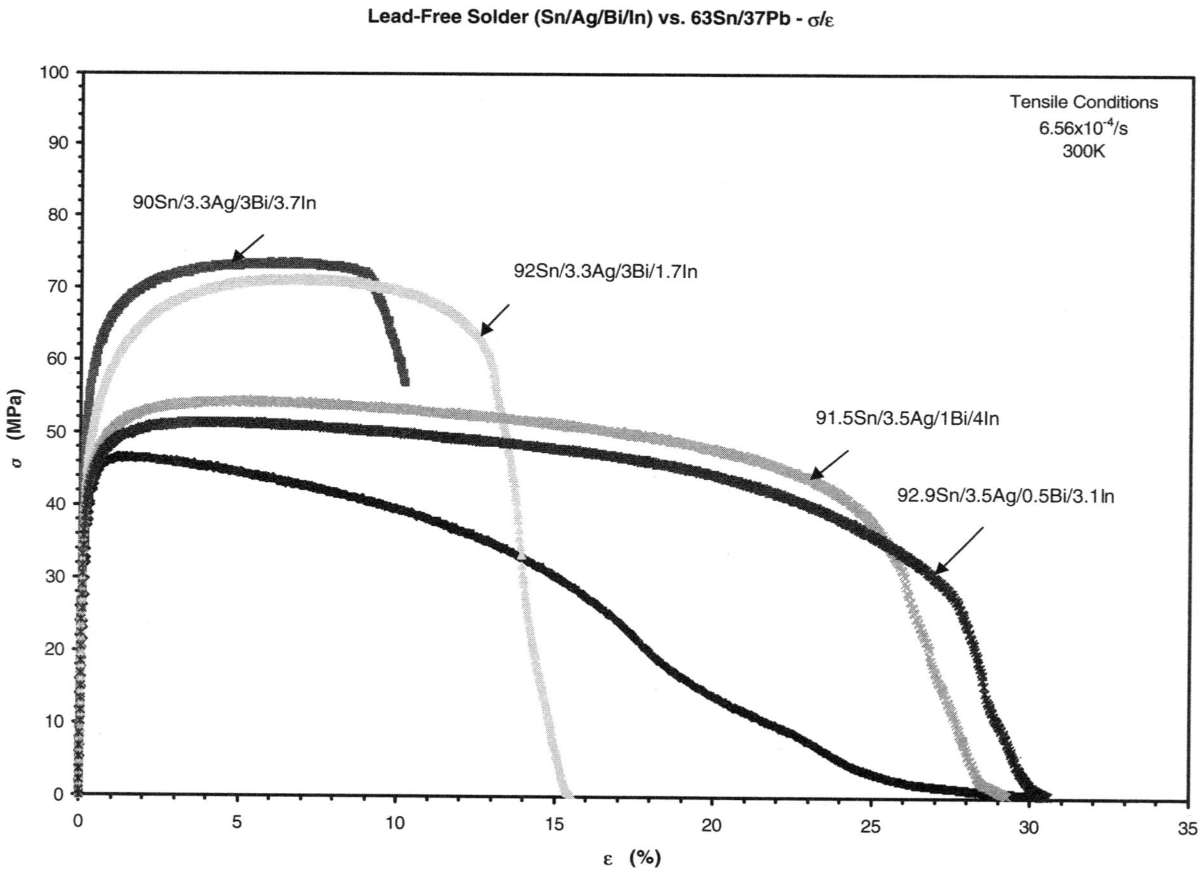

FIGURE 14.16 Tensile stress σ versus strain ε at 300 K and 6.56×10^{-4}/s for tin-silver-bismuth alloys and 63Sn/37Pb.

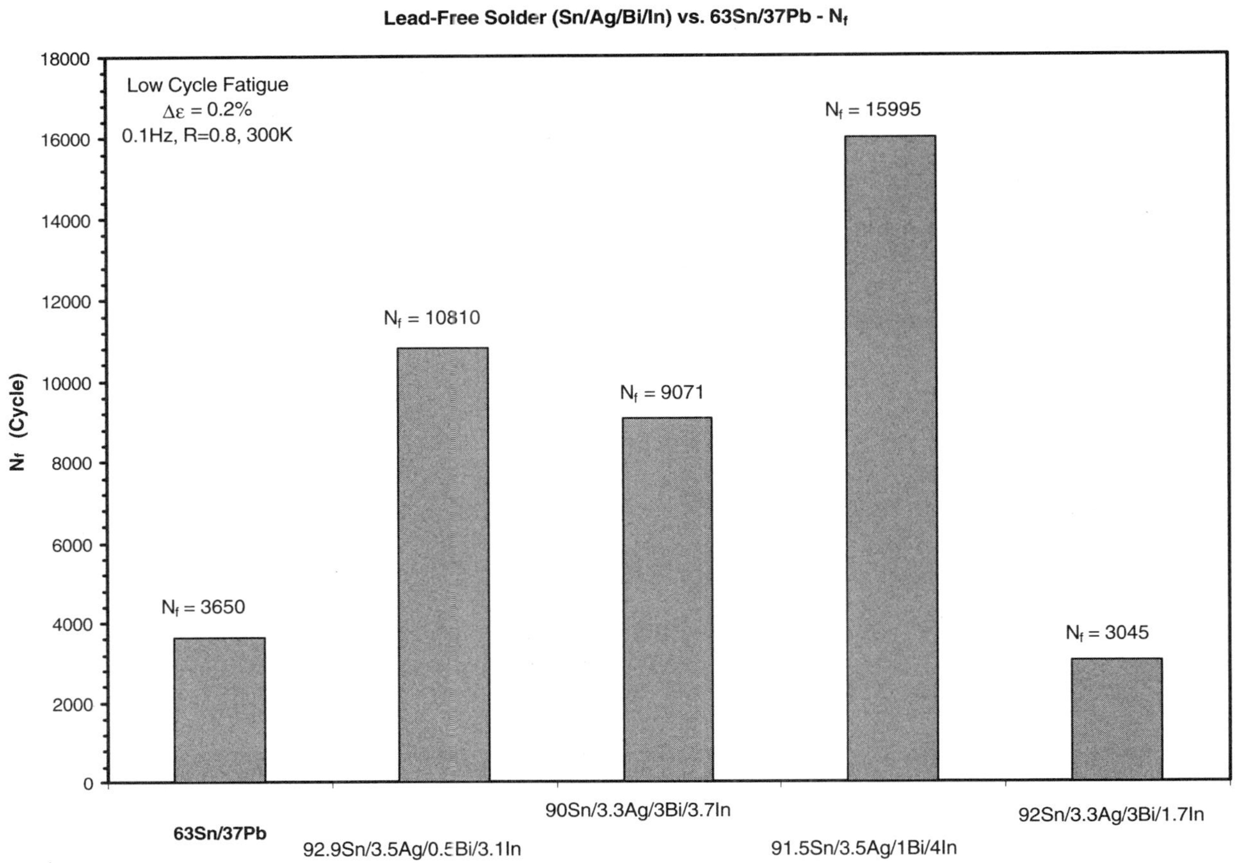

FIGURE 14.17 Comparison of fatigue life of tin-silver-bismuth-indium alloys with 63Sn/37Pb.

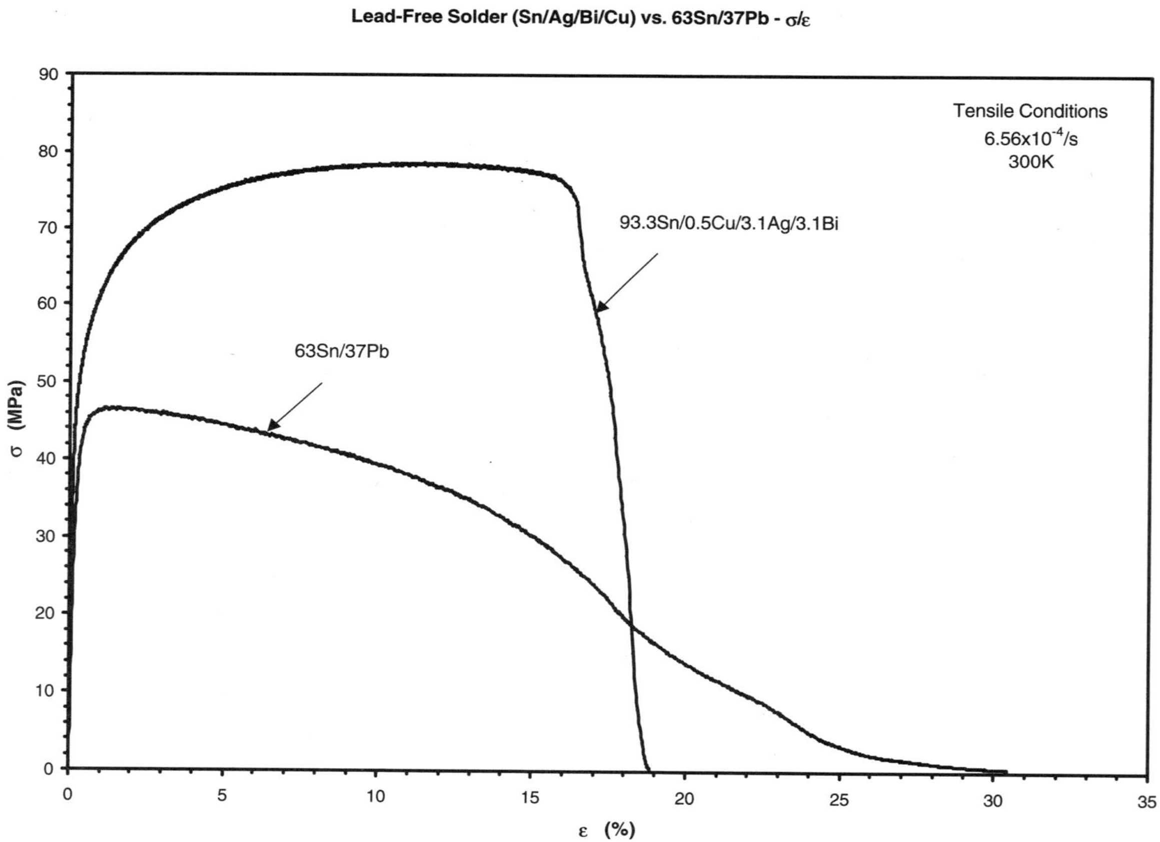

FIGURE 14.18 Tensile stress σ versus strain ε at 300 K and 6.56×10^{-4}/s for tin-silver-copper-bismuth alloys and 63Sn/37Pb.

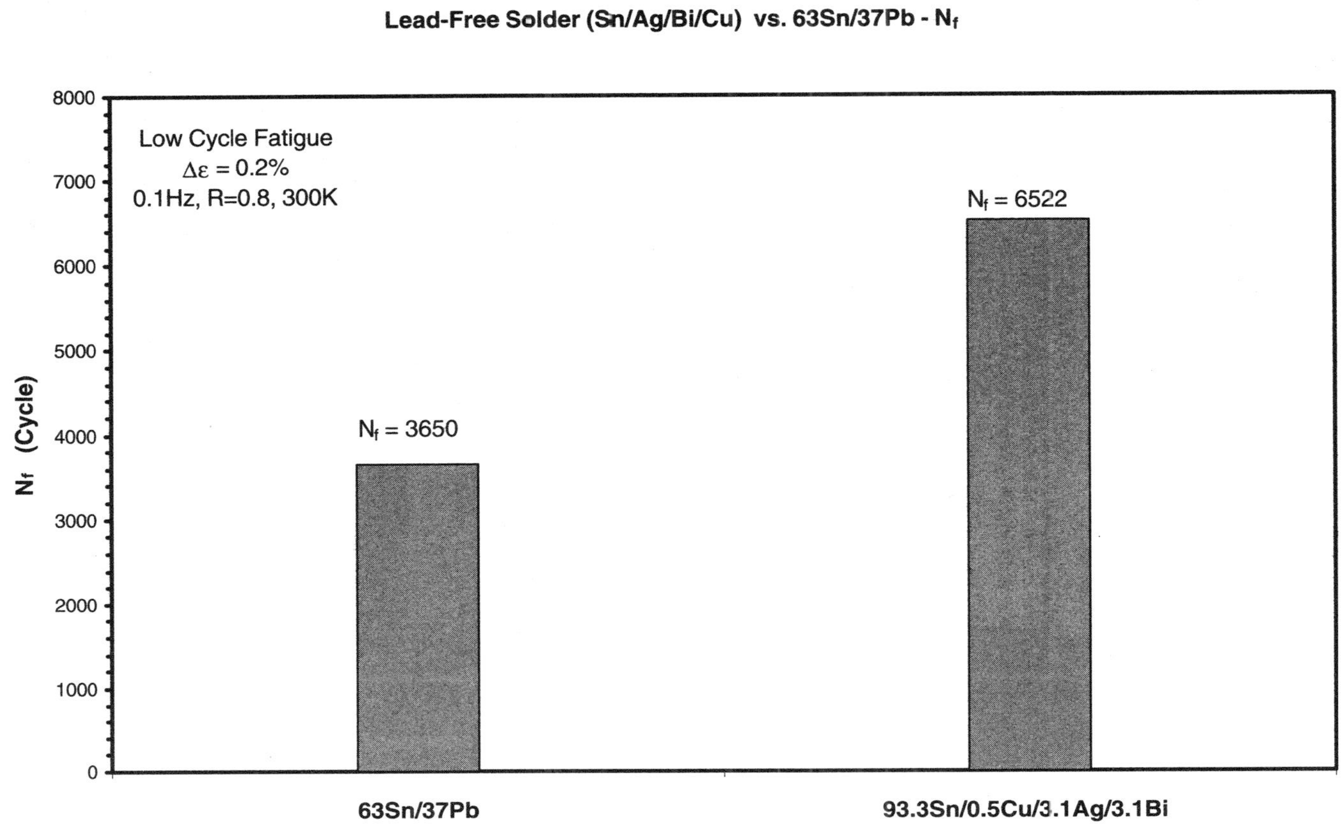

FIGURE 14.19 Comparison of fatigue life of tin-silver-copper-bismuth alloys with 63Sn/37Pb.

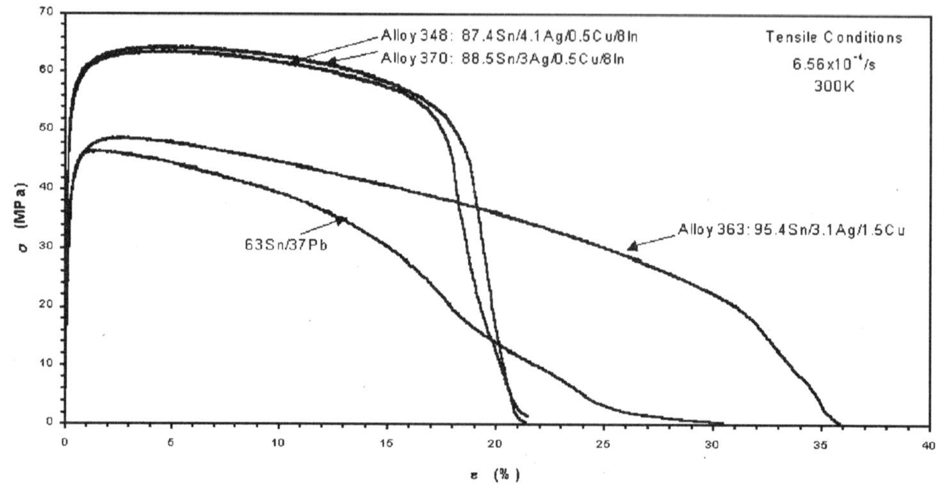

FIGURE 14.20 Tensile stress σ versus strain ε at 300 K and 6.56×10^{-4}/s for tin-silver-copper-indium alloys and 63Sn/37Pb.

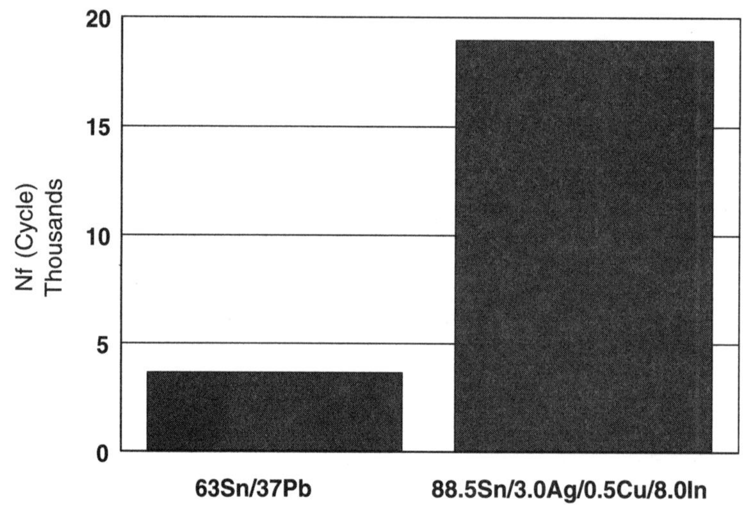

FIGURE 14.21 Comparison of fatigue life of tin-silver-copper-indium alloys with 63Sn/37Pb.

adopted single lead-free solder composition to replace tin-lead eutectic is a natural desire. However, realistically, application-specific and regional adaptation will play a role in selecting the alloy composition on the global landscape.

When selecting materials and processes, simplicity always should be the anchor of principle and practice. To paraphrase Einstein, "Everything should be done as simple as possible, but not simpler." In the end, the selection criteria have to resort to the following priority setting:

1. Performance meeting the manufacturing and application requirements

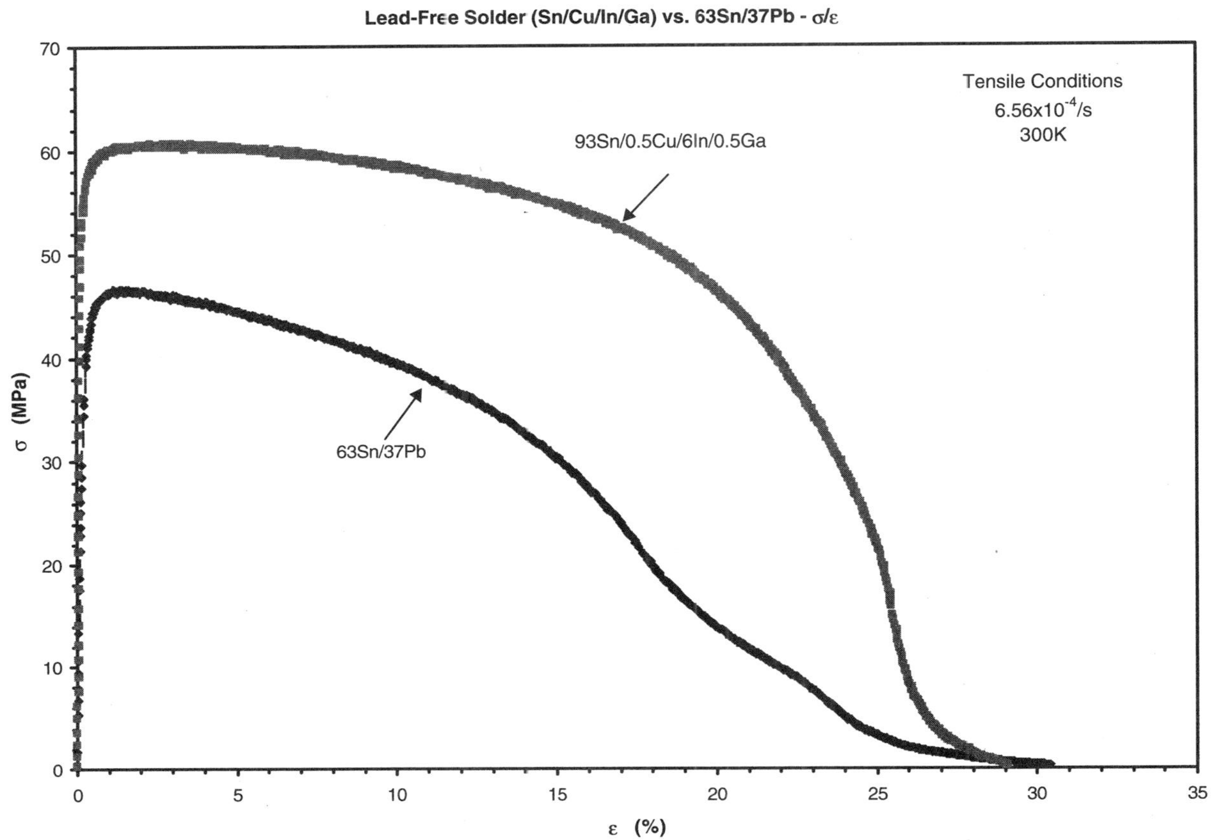

FIGURE 14.22 Tensile stress σ versus strain ε at 300 K and 6.56 $\times$ 10^{-4}/s for tin-copper-indium-gallium alloys and 63Sn/37Pb.

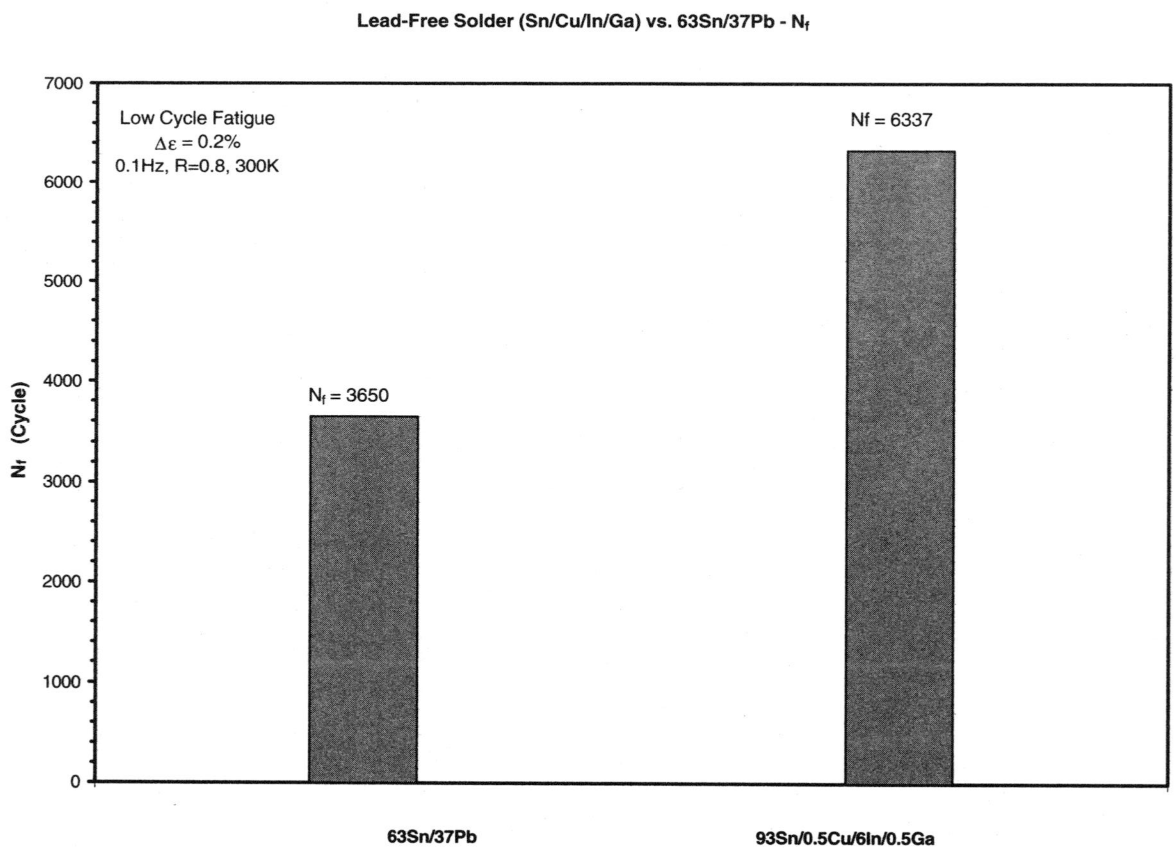

FIGURE 14.23 Comparison of fatigue life of tin-copper-indium-gallium alloys with 63Sn/37Pb.

TABLE 14.9 Ranking of Viable Alloy Compositions by Melting Temperature

Alloy	Melting temperature, °C	N_f
85.2Sn/4.1Ag/2.2Bi/0.5Cu/8.0In	193–199	10,000–12,000
88.5Sn/3.0Ag/0.5Cu/8.0In	195–201	>19,000
93.3Sn/3.1Ag/3.1Bi/0.5Cu	209–212	6000–9000
91.5Sn/3.5Ag/1.0Bi/4.0In	208–213	10,000–12,000
92.8Sn/0.7Cu/0.5Ga/6.0In	210–215	10,000–12,000
95.4Sn/3.1Ag/1.5Cu	216–217	6000–9000
96.2Sn/2.5Ag/0.8Cu/0.5Sb	216–219	6000–9000
96.5Sn/3.5Ag	221	4186
99.3Sn/0.7Cu	227	1125
Reference 63Sn/37Pb	183	3656

TABLE 14.10 Ranking of Viable Alloy Compositions by Fatigue Resistance

Alloy	Melting temperature, °C	N_f
88.5Sn/3.0Ag/0.5Cu/8.0In	195–201	>19,000
91.5Sn/3.5Ag/1.0Bi/4.0In	208–213	
92.8Sn/0.7Cu/0.5Ga/6.0In	210–215	10,000–12,000
85.2Sn/4.1Ag/2.2Bi/0.5Cu/8.0In	193–199	
93.3Sn/3.1Ag/3.1/Bi/0.5Cu	209–212	6000–9000
96.2sn/2.5Ag/0.8Cu/0.5Sb	216–217	
95.4Sn/3.1Ag/1.5Cu	216–217	
96.5Sn/3.5Ag	221	4186
92Sn/3.3Ag/4.7Bi	210–215	3850
99.3Sn/0.7Cu	227	1125
Reference 63Sn/37Pb	183	3650

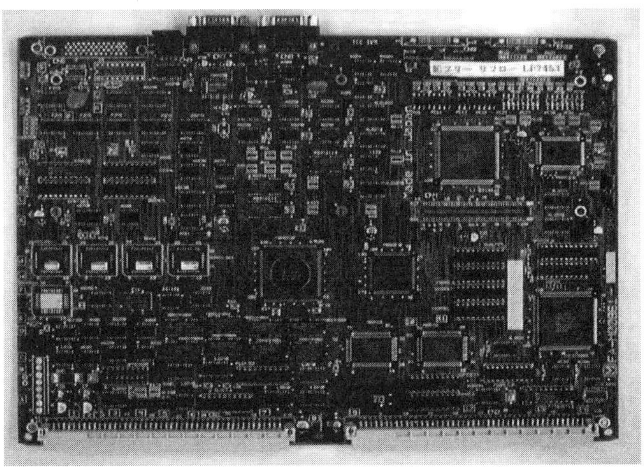

Panasonic Panasert P861

FIGURE 14.24 Manufacturing using a reflow solder of tin-silver-bismuth-indium and a wave solder of tin-copper to produce the Panasonic Panasert image acquisition card.

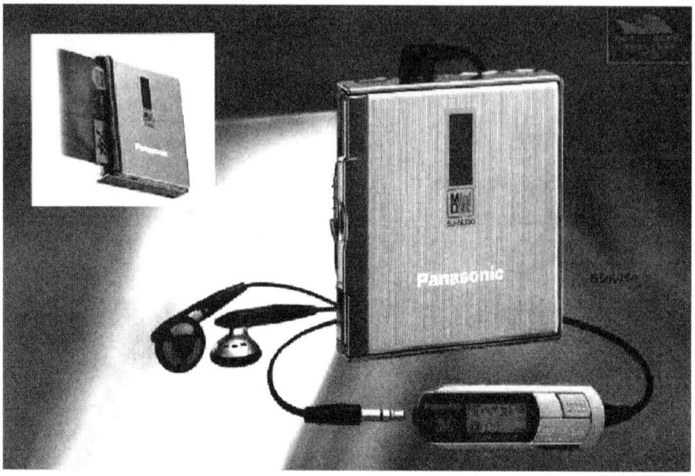

FIGURE 14.25 Panasonic minidisk player MJ30.

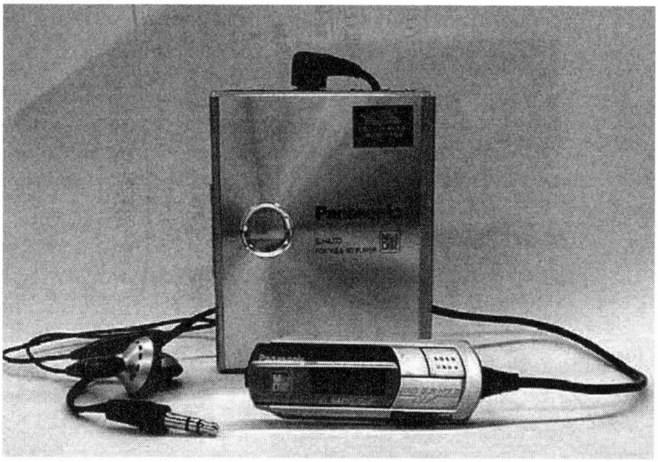

FIGURE 14.26 Panasonic minidisk player MJ70.

FIGURE 14.27 Panasonic minidisk player MR100.

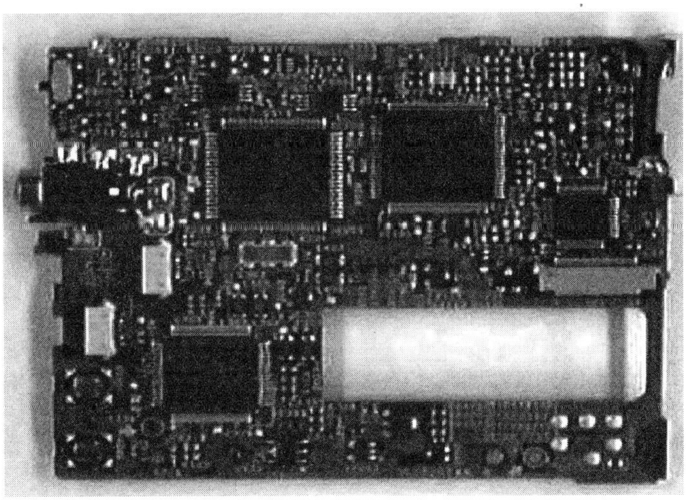

FIGURE 14.28 Main circuit card of the Panasonic minidisk player MJ30.

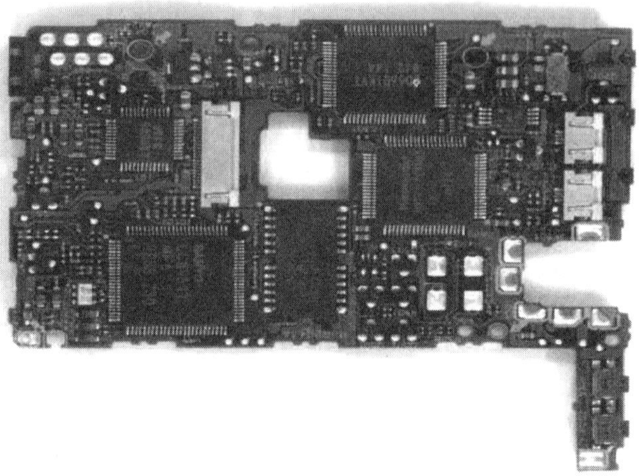

FIGURE 14.29 Main circuit card of the Panasonic minidisk player MJ70.

FIGURE 14.30 Main circuit card of the Panasonic minidisk player MR100.

2. Cost as low as practical
3. Other product virtues

Performance has to come before the cost because a failed product will always incur a higher cost for the manufacturer. With the required performance and desirable cost, the value of the product will be enhanced by additional property virtues.

The expectation of a quick conversion from lead-bearing to lead-free solders across the board is too daunting to fulfill. To implement any new technology, a phase-in period is inevitable. To wait for a global standard alloy is also too high an expectation to meet. After all, the implementation of any new technology or process hardly comes with a standard at the outset. Overall, technological advancement has been made in enhancing creep and fatigue resistance by lead-free research, and the viable alloys are identified for high fatigue-resistant applications.

14.18 REFERENCES

1. Hwang, J. S., *Modern Solder Technology for Competitive Electronics Manufacturing,* McGraw-Hill, New York, 1996, Chap. 3.
2. Hwang, J. S., "Low-Cycle Fatigue versus Thermomechanical Fatigue," *Surface Mount Technology,* January 1995, p. 20.
3. Hwang, J. S., *Environment-Friendly Electronics—Lead-Free Technology,* Electrochemical Publications, Great Britain, 2001.
4. Hwang, J. S., *Solder Paste in Electronics Packaging: Technology and Applications for Surface Mount, Hybrid Circuits and Component Assembly,* Van Nostrand Reinhold, New York, 1989, Chaps. 2 and 3.
5. Backmark, U., Backstrom, N., and Arnberg, L., "Production of Metal Powder by Ultrasonic Gas Atomization," *Powder Metallurgical Institute,* 18 (5), 1986.
6. Hwang, J. S., *Modern Solder Technology for Competitive Electronics Manufacturing,* McGraw-Hill, New York, 1996, Chap. 5.
7. Hwang, J. S., *Modern Solder Technology for Competitive Electronics Manufacturing,* McGraw-Hill, New York, 1996, Chaps. 6 and 12.
8. Hwang, J. S., "Practical Consideration to Minimize BGA Cracks," *Surface Mount Technology,* August 1996, p. 22.
9. Hwang, J. S., "Voids in Solder Joints," *Surface Mount Technology,* September 1996, p. 22.
10. Hwang, J. S., "Effects of Reflow Temperature Profile," *Surface Mount Technology,* June 1996, p. 20.
11. Hwang, J. S., "Optimal Mass Reflow Profile," *Surface Mount Technology,* July 1996, p. 20.
12. Hwang, J. S., "Nitrogen Atmosphere Soldering—Passé or Future," *Surface Mount Technology,* August 1998, p. 18.
13. Hwang, J. S., "Controlled Atmosphere Soldering—Principles and Practice," in *Proceedings NEP-CON West,* Anaheim, CA, 1990, pp. 1539–1546.
14. Cox, N. R., "The Influence of Varying Input Gas Flow on the Performance of a Nitrogen/Convection Oven," in *Proceedings, NEPCON East,* Boston, 1994, p. 323.
15. Hwang, J. S., "Soldering and Solder Paste Prospects," *Surface Mount Technology,* October 1989, p. 45.
16. Ford, J., and Lensch, P. J., "Cover Gas Soldering Leaves Nothing to Clean Off PCB Assembly," *Electronic Packaging and Products,* April 1990, p. 53.
17. Hwang, J. S., "Optimum Oxygen Level for Nitrogen Atmosphere Soldering," *Surface Mount Technology,* March 1995, p. 18.
18. Hwang, J. S., "Reflow Profing—Temperature Measurement," *Surface Mount Technology,* May 1997, p. 22.
19. Hwang, J. S., *Modern Solder Technology for Competitive Electronics Manufacturing,* McGraw-Hill, New York, 1996, Chap. 10.
20. Hwang, J. S., "Design and Use of Solder Paste for System Reliability," *Surface Mount Technology,* March 1997, p. 24
21. Hwang, J. S., *Ball Grid Array and Fine Pitch Peripheral Interconnections,* Electrochemical Publications, Great Britain, 1995, pp. 157–162.
22. Hwang, J. S., *Ball Grid Array and Fine Pitch Peripheral Interconnections,* Electrochemical Publications, Great Britain, 1995, Chap. 7.
23. Hwang, J. S., *Environment-Friendly Electronics—Lead-Free Technology,* Electrochemical Publications, Great Britain, 2001, Chap. 4.
24. Hwang, J. S., and Guo, Z., "Lead-Free Solders for Electronic Packaging and Assembly," in *Proceedings SMI Conference,* San Jose, 1993, p. 732.
25. Hwang, J. S., "Overview of Lead-Free Solders for Electronic Microelectronics," in *Proceedings Surface Mount International,* San Jose, 1994, p. 405.
26. Hwang, J. S., and Koenigsmann, H., "New Developments of Lead-Free Solders," in *Proceedings Surface Mount International,* San Jose, 1997.

27. H-Technologies Group, Internal Reports, Cleveland, 1996, 1997, 1998.
28. Hwang, J. S., *Modern Solder Technology for Competitive Electronics Manufacturing,* McGraw-Hill, New York, 1996, Chap. 15.
29. Lucey, G. K., Wasynczuk, J. A., Clough, R. B., and Hwang, J. S., "Composite Solders," U.S. Patent No. 5,520,752, 1996.
30. Vianco, P. T., and Rejent, J. A., "Tin-Silver-Bismuth Solders for Electronics Assembly," U.S. Patent No. 5439639, August 8, 1995.
31. Seelig, K. F., "Lead-Free and Bismuth-Free Tin Alloy Solder Composition," International Patent No. WO 94/25634, November 10, 1994.
32. Anderson, I. E., Yost, F. G., Smith, J. F., Miller, C. M., and Terpstra, R. L., Iowa State University Research Foundation, Inc., and Sandia Corporation, "Pb-Free Sn-Ag Ternary Eutectic," U.S. Patent No. 5,527,628, June 18, 1996.
33. Hwang, J. S., Guo, Z., "Lead-Free Solders for Electronic Packaging and Assembly," in *Proceedings SMI Conference,* San Jose, 1993, p. 732.
34. Allenby, Ciccarelli, J. P., Artaki, I., et al., "An Assessment of the Use of Lead in Electronic Assembly, in *Proceedings SMI Conference,* San Jose, 1992, p. 1.
35. Hwang, J. S., Lucey, G., Clough, R. B., and Marshall, J., "Futuristic Solders—Utopia or Ultimate Performance," *Surface Mount Technology,* September 1991, p. 40.
36. Hampshire, W. B., "The Search for Lead-Free Solders," in *Proceedings SMI Conference,* 1992, San Jose, p. 729.
37. Greaves, J. B., "Evaluation of Solder Alternatives for Surface Mount Technology," in *Proceedings NEP-CON West,* Anaheim, 1993, p. 1479.
38. Gilleo, K., "The Polymer Electronics Revolution," in *Proceedings NEPCON West,* Anaheim, 1992, p. 1390.
39. Benton, D. H., "Lead Removal from a New Aqueous Cleaning Agent before Discharge," in *Proceedings NEPCON West,* Anaheim, 1994, p. 1346.
40. Dawson, C., and Schultz, C. J., "An Assessment of Lead Exposure to Wave Solder Machine and Solder Pot Preventive Maintenance Workers in an Electronics Assembly Operation," in *Proceedings NEPCON West,* Anaheim, 1994, p. 619.
41. Warwick, J., Vincent, J. H., Harris, P. G., et al., "Screening Studies on Lead-Free Solder Alloys," in *Proceedings NEPCON West,* Anaheim, 1994, p. 874.
42. Napp, D., "NCMS Lead-Free Electronic Interconnect Program," in *Proceedings SMI,* San Jose, 1994.
43. Hwang, J. S., "Innovation, Leadership and Competitives," *Surface Mount Technology,* May 1993, p. 88.
44. Grivas, D., Murty, K. L., and Morris, J. W., "Deformation of Sb-Sn Eutectic Alloys at Relatively High Strain Rates," *Acta Metall Col.,* 27, 1979, p. 731.
45. Tribula, J., Grivas, D., and Frear, R. D., "Microstructure Observation of Thermomechanically Deformed Solder Joints," *Welding Research Supplement,* October 1978, p. 404s.
46. Hwang, J. S., and Vargas, R. M., "Solder Joint Reliability—Can Solder Creep?" In *International Symposium on Microelectronics,* ISHM, 1989.
47. Summece, and Morris, J. W., "Isothermal Fatigue Behavior of Sn-Pb solder Joints," *Transactions of ASME,* 112, June 1990, p. 94.
48. Guo, Z., Sprecher, A. F., and Conrad, H., "Plastic Deformation Kinetics of Eutectic Pb-Sb Solder Joints in Monotonic Loading and Low-Cycle Fatigue," *Journal of Electronic Packaging, Transactions of ASME,* 114, June 1992.
49. Vaynman, Fine, M. F., and Jeannott, D. A., "Low-Cycle Isothermal Fatigue Life of Solder Materials," *Solder Mechanics,* Minerals, Metals and Materials Society, New York, 1991, Chap. 4, p. 155.
50. Frear, R., "Thermomechanical Fatigue in Solder Material," in *Solder Mechanics,* Minerals, Metals and Materials Society, New York, 1991, Chap. 5, p. 192.
51. Marshall, J. L., and Walter, R., "Fatigue of Solders," *International Journal for Hybrid Microelectronics,* 1987, p. 261.
52. Hwang, J. S., Guo, Z., and Lucy, G., "Strengthened Solder Materials for Electronic Packaging," in *Proceedings SMI,* San Jose, 1993, p. 662.
53. Hwang, J. S., *Solder Paste in Electronic Packaging—Technology and Applications in Surface Mount, Hybrid Circuits, and Component Assembly,* Van Nostrand Reinhold, New York, 1989, Chap. 9, p. 282.
54. Nordyne, "Lead Products," *Ceramic Bulletin,* 70 (5), 1991, p. 872.
55. Silverstein, K., "Lead Issues Gain Steam at State and Federal Levels," *Modern Paint and Coatings,* April 1993, p. 30.
56. Hwang, J. S., "A Strong Lead-Free Candidate—SnAgCuBi," *Surface Mount Technology,* August 2000, p. 20.
57. Hwang, J. S., "Another Strong Lead-Free Candidate—SnAgBiIn," *Surface Mount Technology,* September 2000, p. 20.
58. Artaki, I., Jackson, A. M., and Viance, P. T., "Fine Pitch Surface Mount Assembly with Lead-Free, Low Residue Solder Paste," in *Proceedings SMI,* San Jose, 1994, p. 746.

59. U.S. Patent No. 5,229,070.
60. U.S. Patent No. 4,778,733.
61. U.S. Patent No. 5,102,748.
62. U.S. Patent No. 5,538,686.
63. U.S. Patent No. 4,879,096.
64. U.S. Patent No. 5,527,628.
65. U.S. Patent No. WO 94/25634.
66. U.S. Patent No. 4,670,217.
67. U.S. Patent No. 5,520,752.
68. U.S. Patent No. 5,985,211.
69. U.S. Patent No. 6,176,947.
70. Viano, P. T., Artaki, I., and Jones, A. M., "Assembly Feasibility and Reliability Studies of Surface Mount Boards Manufactured with Lead-Free Solders," in *Proceedings SMI Conference,* San Jose, 1994, p. 458.
71. Melton, C., Skipor, A., and Thome, J., "Material and Assembly Issues of Non-Lead Bearing Solder Alloys," in *Proceedings NEPCON West,* Anaheim, 1993, p. 1489.
72. Hosking, M., et al., "Wetting Behavior of Alternative Solder Alloys," in *Proceedings SMI Conference,* Anaheim, 1993, p. 476.
73. Vincent, P., et al., "Alternative Solder for Electronics Assemblies," *Circuit World,* 19 (3m), 1993, p. 32.
74. Harada, J., and Satoh, R., "Mechanical Characteristics of 96.5Sn/3.5Ag Solder in Microbonding," *IEEE Transactions on Component Hybrids Manufacturing Technology,* CHMT-13 (4), 1990, p. 736.
75. Getty, "The Effect of Power and Temperature Cycling on Tin/Bismuth and Indium Solders," *IPC Technical Review,* Northbrook, December 1991, p. 14.
76. Kwoka, M. A., and Foster, D. M., "Lead Finish Comparison of Lead-Free Solders versus Eutectic Solder," in *Proceedings SMI Conference,* San Jose, 1994, p. 485.
77. Abbott, D. C., Brooks, R. M., McLelland, M., and Wiley, J. S., "Palladium as a Lead Finish for Surface Mount Integrated Circuit Packages," *IEEE Transactions on Components, Hybrids and Manufacturing Technology,* 14 (3), September 1991, p. 567.
78. Grudul, H. E., and Carlson, R. R., "Low Temperature Lead-Free Wave Soldering for Complex PCBs," *Electronic Packaging & Production,* September 1992, p. 31.
79. Hwang, J. S., *Environment-Friendly Electronics—Lead Free Technology,* Electrochemical Publications, Great Britain, 2001, Chap. 19.
80. Nikkei Sangyo, Tokyo, Japan, November 1998.
81. Fan, C., Abys, J. A., and Blair, A., "Wirebonding to Palladium Surface Finishes," in *Proceedings NEPCON West,* Anaheim, 1999, p. 1505.
82. Guy, J., Solder Joint Reliability Impact of Using Emersion Metallic Coatings," in *Proceedings NEPCON West,* Anaheim, 1997, p. 1540.
83. Baliga, J., "Tin for No-Lead Solder," *Semiconductor International,* July 1999, p. 74
84. Ormerod, D. H., "The Development and Use of a Modified Immersion Tin as a High Performance Solderable Finish," in *Proceedings NEPCON West,* Anaheim, 1999, p. 1515.
85. Ormerod, D. H., "Production Application for Flat Tin Finishes," in *Proceedings NEPCON West,* Anaheim, 2000, p. 860.
86. Cullen, D. P., "New Generation Metallic Solderability Preservatives: Immersion Silver Performance Results," *Journal of SMT,* Minneapolis, October 1999, p. 17.
87. Parker, L. L., "The Performance and Attributes of the Immersion Silver Solderability Finish," in *Proceedings NEPCON West,* Anaheim, 1999, p. 444.
88. Snowdon, K., "Lead-Free—The Nortel Experience," in *Proceedings IPC Works '99,* Minneapolis, October 1999, p. S-05-1-4.
89. Stafsrom, E., and Wengenroth, K., "OSPs: The Next Generation," in *Proceedings NEPCON West,* Anaheim, 2000, p. 875.
90. Feldmann, K., and Reichenberger, M., "Assessment of Lead-Free Solders for SMT," in *Proceedings APEX,* San Diego, 2000, p. P-MT2/2-3.
91. Courtesy of Dr. Ken Suetsugu and Tom Baggio of Panasonic-Matsushita.

CHAPTER 15
LEAD-FREE SYSTEMS AND PROCESS IMPLICATIONS

Alan Rae
Cookson Electronics

15.1 WHY REMOVE LEAD FROM CIRCUIT BOARDS?

15.1.1 Toxicology of Lead and Lead Removal Legislation

Lead is "widely accepted" as being a "toxic material." What does this mean, and why should we be concerned?

The toxicity of metals is not a simple issue—in fact, any element in the periodic table and all their compounds can be toxic under the wrong set of conditions. Military pilots breathing pure oxygen found this out the hard way in the 1940s. It is a combination of the toxicity of the material, the availability of the material, the metabolic pathways for its incorporation into the body, and the level of exposure that determines whether toxic effects occur.

Large-scale exposure to lead in the population has been through lead-containing paint and lead tetraethyl in gasoline. Other more obscure exposure has been through traditional cosmetics and the addition of lead acetate to wine as a sweetener by the Romans. Over 80 percent of the lead used is now in automobile batteries and presents minimal risk to the public because it is contained and has a well-established recycle route. Lead pipes used for drinking water supplies for many centuries present a small risk, except when the drinking water is "soft," i.e., less alkaline, and no protective coat of calcium minerals builds up.

How does this affect circuit boards? These contain 7 to 8 percent solder, giving a 2 to 3 percent lead concentration. Other lead content in semiconductor packaging or passive components is much less than 1 percent by weight. This lead is quite unlikely to be swallowed (like lead paint or contaminated soil) or inhaled (like lead tetraethyl). The main issue relates to the disposal of circuit boards at the end of their life. *USA Today* published some interesting statistics and projections on June 22, 1999:

- Over 44 percent of U.S. households have personal computers (PCs).
- A PC's life will reduce from 3 years now to 2 years in 2007.
- The number of PCs becoming obsolete each year will grow from the current 20 million to over 60 million by 2003.
- Many obsolete PCs are in storage because there is no easy way to dispose of them.
- Only 11 percent of processors are recycled.

Obviously, in the absence of any other mechanism, PCs and other electronic devices increasingly will find their way into landfills, and therefore, the quantity of lead in landfills will increase.

The next question is, Is the lead mobile, and can it find its way into groundwater? Well, most of the lead already in landfills has come from discarded car batteries (pre-recycling or illegally dumped), TV and computer monitor tubes, and building demolition debris. Yet there is no apparent crisis of lead infiltrating groundwater, and blood lead levels in the United States continue to fall (between the late 1970s and 1980s, the value reduced 78 percent, according to the Centers for Disease Control). Why does the lead not appear to have a significant effect in landfills?

The standard solubility tests used worldwide use deionized water or dilute acid [e.g., the toxicity characteristic leaching procedure (TCLP) test]. Both show lead (and some other metals) to be soluble. But how does this compare with the situation in a landfill?

Lead and its compounds appear to exhibit low mobility. Apart from the many relatively insoluble compounds that can form, such as sulfate and sulfide (the naturally occurring minerals anglesite and galena), ion exchange with, for example, clays is possible. The IPC's lead-free forum (access from http://www.leadfree.org) has had many discussions about this—even to the revelation that a disused lead mine is now used successfully for scuba diving (Bonne Terre MO, http://www.2dive.com/). The relative immobility of lead in river sediments at the centimeter level has even been used to monitor changes in pollution levels.

There is no published environmental impact study that shows that circuit boards present a serious hazard today through leaching into groundwater from landfills.

15.1.2 Current Drivers

There are really three drivers: environmental regulations, takeback regulations, and marketing.

- Environmental regulations in Europe, particularly the draft WEEE (waste electronics) and ELV (end of life vehicle) regulations, have been encouraged by the Scandinavian nations, who wish to eliminate all uses of lead.
- Takeback regulations in Japan do not identify lead as being one of the recyclable materials of choice. Most major electronics companies are setting up their own recycling facilities and would rather not process lead.
- Most of the major electronics manufacturers in Japan have very publicly stated their lead replacement timetable and identified their lead-free products using environmental symbols such as branches and leaves, e.g., Matsushita and NEC, "Eco," and Sony, "Greenplus." These are used to differentiate their products.

15.1.3 The Future

Estimates of the amount of penetration of lead-free solder vary greatly depending on the market, the geographic region, and the opinion of the writer. In Europe, the main push is through direct legislation in member countries and the European Community's (EC) draft directive, which was expected to become law in 2004, but at the time of this writing, the current draft lists 2008 as the target date. Already passed in early 2000, the ELV regulations prevent lead from automobiles entering landfills from 2001 and prevent lead and other materials being used from 2005. There are many exceptions, however. Lead is still usable in copper and aluminum alloys at up to 4 percent and, of course, in batteries, so even at that date European automobiles will not be 100 percent lead-free.

In Japan, the consumer electronics take back regulations become effective in 2001, and Fujitsu, Hitachi, Matsushita, Mitsubishi, NEC, Sony, TDK, Toshiba, NTT, Toyota, and Nissan all have aggressive and public lead reduction or removal programs. Several of these companies also have set up their own recycling facilities. Their suppliers outside Japan are scrambling to find lead-free solutions for subassemblies.

In the United States, there is no current pressure to go lead-free, although the United States started the movement back in 1991 with the Reid bill. However, if you think that this will continue to be the case, I encourage you to look up lead applications on the Environmental Defense Fund Web site (http://www.scorecard.org/). You will see circuit boards mentioned as major users. The Environmental Protection Agency (EPA) has proposed to reduce the reporting limits to 25,000 lb per year to 10 lb. IPC and other organizations are opposing this on the grounds that the paperwork may be excessive for small companies.

15.2 WHERE IS LEAD USED IN CIRCUIT BOARD ASSEMBLIES?

Circuit boards, on average, contain 7 to 8 percent of solder, and the total lead content is typically around 2 to 3 percent.

15.2.1 Board Treatments

Between 50 and 60 percent of boards are surfaced with hot-air-leveled solder, commonly referred to as HAL or HASL. This is the most common board finish, but its use is declining because of competition from less expensive and flatter organic solderability preservative (OSP). The additional thermal load it places on laminates (temperatures of 260°C are not uncommon) are a concern, as are the surface-tension effects, which mean that HSL-treated pads are less planar than those treated with OSP, silver, tin, or nickel-gold.

The HASL surface finish contributes about 20 percent of the lead in the circuit board.

15.2.2 Solder

The most common electronic solders are 63Sn/37Pb and 62Pb/36Pb/2Ag. They contribute about 75 percent of the lead on the circuit board Semiconductor packaging solders typically contain 90Pb/10Sn or 95Sn/5Sn.

15.2.3 Components

Lead in components can be as a key functional chemical constituent (lead magnesium niobate capacitors, lead barium titanate thermistors, or lead zirconate titanate piezos); as a component of a glass frit used as a sintering aid, cover glass, or sealing glass; or in solder bumps and ball grid array (BGA) spheres in packages. Typically, these applications account for less than 5 percent of the contained lead.

Component manufacturers in particular are "between a rock and a hard place." Until the assembly industry standardizes, component manufacturers have no idea whether to design for a sharp peak at high temperature or a long soak at lower temperature that can affect thermal expansion, outgassing, and lead finish dissolution. The assembly industry says it cannot standardize until there is a supply of lead-free components.

15.3 WHAT IS "LEAD-FREE"?

15.3.1 Proposed Limits and Achievable Limits

The question "What is lead-free?" is not trivial. Lead is widely used as an additive in steel and some copper alloys, and the EC WEEE draft regulations recognize this with a series of exemptions. In addition, the long-term reliability of some of these alloys is not proven, and aviation and military applications are exempted.

Denmark and Sweden have proposed eliminating lead completely over the next 20 years, but this will be difficult if not infeasible. In order to maintain an effective recycling circuit for tin, a practicable limit of 0.2 percent is workable. Other proposed limits such as 0.02 percent will tax even the ability of virgin tin producers to reach this level.

15.4 HOW EASY IS IT TO REPLACE LEAD?

Lead is an excellent additive to tin for making solders. It reduces the surface tension of tin to promote wetting and prevents "tin pest" (the transformation from beta to alpha tin at 130°C) even at low concentrations. It lowers the cost of solder and melts at a temperature high enough to allow electronic components to perform reliably but low enough that the fabrication of these components is in a convenient range. There is a clearly defined eutectic composition that melts cleanly and rapidly at 183°C. Tin-lead alloys are malleable and easy to form into wire, powder, and spheres.

15.4.1 Active Components

There are two conflicting requirements on integrated circuits (ICs). First, they are normally soldered into packages at higher temperatures than normal eutectic solder to avoid desoldering of the package during reflow. Second, as circuits move to finer pitch, they are becoming more sensitive to heat, which can encourage diffusion or decomposition of some parts of the structure. The combination of increasing board soldering temperatures, lowering temperature tolerance, and the elimination of the 90Pb/10Sn solder of choice makes life very difficult. The most economic substitute, 95Sn/5Sb, melts at 240°C but antimony (Sb) is under scrutiny because of reliability issues. It looks like tin-silver or tin-silver-copper alloys will be the alloys of choice.

"Popcorning" of packages is caused by rapid heating of water vapor or liquid water within the package. The susceptibility of the package to degradation depends on the thermal profile, the interfacial energy between package and die/substrate, the size of voids or delaminations within the package, and the residual stresses within the package. Surprisingly, water content over a threshold level does not increase susceptibility because the equilibrium vapor pressure is determined only by temperature. These problems are solvable by design, process control, and careful formulation of the transfer-molding epoxy.

Dispensed packaging faces a further challenge. Because these compounds generally are unfilled, their coefficient of thermal expansion (CTE) is around 15 ppm/°C compared with less than 5 ppm/°C for dies. This means that thermal mismatch stresses on subsequent heating are extreme, and a severe thermal cycle can cause delamination.

Active component manufacturers may have to look at a postsoldering attachment using physical attachment methods, conductive adhesives, or spot soldering techniques, where heat is applied using a metal heat source, focused light beam, or laser.

15.4.2 Passive Components

Ceramic capacitors are facing a number of challenges right now, the main one being the switch from precious metal to base metal internal electrodes due to the escalating price of palladium. Lead is mainly used in frits to lower the sintering temperature of ceramics to that of silver-palladium alloys; base metal systems do not require lower temperatures, and so lead can be replaced relatively easily. Large-capacity pieces made of lead-magnesium niobate have a particular niche that either will be exempted or will disappear. Resistors are not normally affected, except that cover glazes can contain some lead.

Substitute ceramic piezo and thermistor compositions are being developed by industry leaders such as TDK; to date, their performance as measured against conventional materials is inferior, but the development of acceptable solutions seems likely.

15.4.3 Interconnects

It is currently very difficult to source lead-free connectors. Resolution of this is just a matter of time as plated leads with alternate finishes become available as demand increases. The most difficult-to-handle option appears to be palladium, which is not easily wetted by lead-free solders. Plated lead finishes under evaluation include tin, tin-bismuth, tin-copper, and tin-silver, as well as nickel-gold.

15.4.4 Board Treatments

There are basically six systems available: HASL, reflowed and hot-air-leveled solder (normally Sn/Cu), OSP, electroless nickel/immersion gold (ENIG), and silver and tin. Each has strengths and weaknesses.

HASL exposes the board to an additional thermal cycle. This can be critical if boards require a high degree of planarity. HASL pads also have a domed profile that is unsatisfactory for fine-pitch features. Against this, however, is the fact that nothing solders like solder! Existing HASL equipment can cope with lead-free solder.

Reflowed finish uses the solder applied as a mask during the final stages of circuit formation. Although this is normally removed, it can be reflowed. The amount of solder applied is less than with HASL, but the extra thermal cycle is again a concern.

OSPs are the most rapidly growing surface finish. They are thin, flat, and inexpensive. They are also the least easily wet by lead-free solders and are prone to oxidation in air reflow. Nitrogen reflow extends the number of cycles tolerated.

Silver coatings applied chemically are prone to tarnish before soldering unless an organic protectant is codeposited. If they are properly applied, solderability and planarity are good. Tin coatings applied chemically with an organic protectant also give excellent protection. Electrolytically deposited tin, in contrast, shows sharp crystal faces that are suspected to be conducive to the formation of "tin whiskers."

ENIG is expensive but solders well. The main concerns are that significant gold dissolution can increase the viscosity of solders, and excessive nickel diffusion can form intermetallics such as Ni_3Sn_4, which, with its shrinkage of more than 10 percent, can increase stresses and be a source of failure.

15.5 THE TECHNOLOGY IMPLICATIONS OF REPLACING LEAD IN SOLDERS

All the studies of lead-free soldering in the United States, Europe, and Japan have shown that, unfortunately, there is no immediate drop-in substitute for 63Sn/37Pb. The alloys showing highest reliability promise processing at higher temperatures and the alloys processing at equivalent temperatures are susceptible to contamination, reliability issues, and cost or supply constraints.

Laminates used in board manufacture can be exposed to thermal cycling during HASL, multiple wave/solder reflows, and rework at temperatures to 260°C and above. Selection of a laminate with a high T_g and other desirable high-temperature properties is vital, especially for large, complex boards that are exposed to multiple cycles.

Higher process temperatures can have a severe impact on components containing volatiles such as electrolytic capacitors and polymer-encapsulated components. Even if the materials of construction are not volatile, they may be at or near their thermal limit, as in the case of wire-wound components or film capacitors. In addition, the change in thermal profile may adversely affect the reliability of thermal shock–sensitive ceramic chip capacitors (MLCCs). Mismatch in CTE with large components (e.g., BGAs) or those with greatly dissimilar CTEs from the circuit board or silicon also will cause problems.

Especially in silicon attachment, tin-lead brings the necessary ductility that allows relaxation of stresses over time. Most lead-free alloys are considerably "stiffer," and tin-bismuth in particular even cannot be drawn into wire.

Self-centering during reflow is a key advantage of eutectic solders. Thermal profiles, solders, fluxes, pick-and-place precision, and pad design will have to be optimized to take this into account. Consumer and telecom equipment has a typical operating environment of −55 to +125°C (0 to 85°C for many consumer goods), but automotive under-the-hood temperatures can approach 200°C. As a result, low-melting solders containing bismuth may be suitable for some applications but not others.

Even within consumer devices there are "hot spots" that even now require the use of high-melting high-lead interconnects. A typical 300-MHz processor can generate up to 40 W that must be channeled and dissipated. At present, it is not clear what can replace the 90Pb/10Sn solder used in wafer bumping, although tin-antimony is a likely candidate.

Many lead-free solders do not have the smooth, shiny finish we associate with tin-lead eutectic because of their larger grain size. They appear dull and whitish or grayish. Although the joints may perform well, the joints look different, and quality assurance procedures will have to be adapted to cope with this.

There are specific defect issues with some lead-free solders, in particular with bismuth-containing alloys. The presence of lead as a contaminant will lead to the formation of a 97°C low-melting phase that will affect reliability. In addition, the bismuth alloys also can show fillet lifting around through-hole connections.

Voids at the interface of copper and the solder are another issue. Although in some consumer applications some void content can be acceptable, this level is totally unacceptable for mission-critical boards. The root cause of voiding has not been identified but seems to be related to surface finish.

In general, lead-free solders do not wet boards or components as well as leaded solders. This means that joints will have a different appearance, with implications for design of inspection systems.

15.5.1 Nonsolder Alternatives

Conductive Adhesives. Most conductive adhesives are based on epoxy filled with conductive silver flake. They have been used very successfully in niche applications such as flexible circuitry but have not replaced mainstream solder because a number of perceived shortcomings:

- Solders under reflow pull components into alignment because of surface tension. Adhesives show no such effect, and so the precision required of pick-and-place equipment is higher than with solders.
- Reliability issues including moisture degradation of the silver flake and heat aging of the epoxy.
- Cure times are relatively long compared with soldering (30 min), although temperatures required are more benign (100 to 150°C).
- Physical strength is generally less than solder.
- Inspection and rework on a large scale are difficult.

This area is a focus for a number of companies who are looking at novel solutions, e.g., acrylic-based pastes or creative filler solutions. Other creative approaches include using anisotropic conductive adhesives employing the (large) single particle bridging concept or dropping conductive spheres on a tacky surface. Conductive polymers without fillers do not as yet have the required conductivity or stability for long-term reliability.

Thermally conductive adhesives have been developed that can replace solder in die-attach and related heat-transfer applications. However, lead-free solder systems are also available for this application, and cost considerations (many thermally conductive materials contain aluminum nitride or boron nitride) may limit the use of the high thermal conductivity adhesives.

Other Techniques. Mechanical attachment systems also have been explored using a vertical compression force to maintain contact, but these have not been used outside specific niches.

15.5.2 Alloy Choice

Some lead-free compositions and their melting ranges are listed in Table 15.1.

Some alloy selection criteria are listed in Table 15.2.

As mentioned earlier, there is no "drop-in" solution. The choice is determined by application type, cost, and the patent position.

Processing Implications. There is currently no drop-in solution, and all commercial materials process at a higher temperature than eutectic solder. This affects laminates, board treatments, fluxes, reflow and wave processing, cleaning, and recycling.

Reliability Implications. There are considerable data in the EC's IDEALS, program and the NCMS lead-free study as well as individual company studies that suggests that lead-free solders can be reliable. IDEALS, or Improved Design Life and Environmentally Aware Manufacturing of Electronics Assemblies by Lead-Free Soldering, was a 3-year EC BRITE-EURAM project anticipating lead reduction and volatile organic compound (VOC) emission legislative control in Europe. Its main objective was to develop soldering processes based on novel lead-free high-strength solders, lead-free board and component metallizations, and VOC-free fluxes. The NCMS (National Center for Manufacturing Sciences) study was a 4-year $10.5 million collaborative R&D program aimed at identifying and evaluating alternatives to eutectic tin-lead solder. The final NCMS report is available at http://lead-free.ncms.org/.

TABLE 15.1 Lead-Free Compositions and Their Melting Ranges.

Alloy	Composition, %	Melting range, °C
Bi/In	33In	109(e)*
Sn/In	50In	118(e)
Sn/Bi	58Bi	138(e)
Sn/Bi/In	20Bi/10In	143–193
Sn/Pb	37Pb	183(e)
Sn/Bi/Zn	8Zn/3Bi	189–199
Sn/Zn	9Zn	199(e)
Sn/Ag/Bi	7.5Bi/2Ag	207–212
Sn/Ag/Cu/Sb	2Ag/0.8Cu/0.5Sb	216–222
Sn/Ag/Cu	3.8Ag/0.7Cu	217(e)
Sn/Ag	3.5Ag	221(e)
Sn/Cu	0.7Cu	227(e)
Sn/Sb	5Sb	232–240
Sn/Au	80Au	280(e)

*(e) is a eutectic.
Source: DTI.

TABLE 15.2 Alloy Selection Criteria

Family	Pluses	Minuses
Sn/Cu	Cost	Melting temperature 227°C,, mechanical properties
Sn/Ag	Reliability, solderability	Cost, melting temperature 221°C
Sn/Ag/Cu (Sb)	Reliability, solderability, mechanical properties	Melting temperature 216°C
Sn/Ag/Bi (Cu) (Ge)	Melting temperature 206°C, solderability, strength, wettability	Fillet lifting, sensitivity to lead contamination
Sn/Zn/Bi	Melting temperature 189°C, close to tin-lead eutectic	Paste shelf life, needs aggressive fluxes, excessive oxidation, corrosion

Source: DTI.

In some cases, the generally higher modulus and lower creep of lead-free solders can give performance advantages, e.g., in high-temperature automotive or lighting applications. However, because of the wetting characteristics of lead-free solder, the joints look different, and care also must be taken to avoid fillet lifting or excessive void formation. Although neither voids nor fillet lifting has been linked conclusively to premature failure, it will be some time before harsh environment or mission-critical systems companies are comfortable with switching from the well-tried tin-lead eutectic.

15.5.3 Processing

Laminates and Inner-Layer Treatment. Most lead-free solders require demanding thermal profiles—a peak temperature of up to 260°C and a hold at up to 180°C. Although simple boards can be constructed of FR-4, complex and/or thick boards will need to be made using higher-T_g laminates to prevent warpage. There may be a cost penalty to this, and until recently, we expected that most consumer manufacturers would try to get by using FR-4. There are, however, a number of high-T_g materials that bring other benefits such as reduced cure time and reduced VOC emissions, and the much improved performance at 260°C and above that will be crucial for complex boards or those requiring multiple reflows. As of this writing, the U.S. industry is switching rapidly to these materials not only to cope with lead-free profiles but also to cope with thermal deformation caused by BGA rework.

Inner-layer black oxide substitutes such as AlphaPrep have been shown under some conditions to increase delamination resistance at temperature by a factor of 2. This is obviously critical to high-temperature performance.

Board Treatments. Approximately 20 percent of the lead in a solder joint is contributed by HASL, the balance coming from the solder applied to the joint (75 percent) and components (5 percent).

Lead-free HASL has been evaluated, and tin-copper shows promise; however, the higher melting temperatures associated with this alloy can mean a dramatically increased thermal load, resulting in potential delamination. Even with tin-lead alloys, the HASL dipping temperature is 235 to 255°C, and there are real concerns over the survival of complex boards. Nevertheless, HASL equipment manufacturers have data that show not only improved flatness but also good compatibility of existing equipment with tin-copper lead-free HASL coatings.

OSP coatings are obviously lead-free but are the most difficult to wet using lead-free solders. Electroless tin coatings can be used, but there are reliability concerns about whisker formation from pure tin.

The silver and tin coatings are not truly electroless processes and in fact are ion exchange/displacement reactions that are driven by the electromotive differential in the metals with which they work. Electroless processes are autocatalytic and deposit on themselves and continue to plate as long as there are metal ions in solution. Immersion processes will only plate until the base metal substrate is fully covered, and no site for nucleation exists. As a result, they will only plate about 10 to 12 μin maximum thickness.

All the alternative chemically prepared coatings provide planarity that is as good as the base electrolytic copper plating, but traditionally, HASL generally does not provide planarity due to the viscosity of molten solder and the physical dynamics of blowing this molten solder off the board.

One of the leading technologies is a coating that couples a codeposited organic inhibitor within a 4- to 6-μin deposit of pure silver (AlphaLevel). This organic inhibitor provides handling and shelf capabilities and stops any tarnishing. The organic layer also encapsulates the silver and deters any opportunity for electromigration.

Tin coatings such as FST also include an organic layer and deposit in sooth textured finishes, avoiding the high-energy sharp crystal edges that can promote tin whisker formation.

Solders. The front-running alloys are tin-copper for wave soldering and tin-silver-copper (plus or minus additives such as antimony) for reflow. Other alloys such as tin-silver or tin-antimony are used in automotive and industrial applications. Bismuth-containing alloys are favored strongly in Japan for consumer applications, where their room-temperature strength is an advantage and their poor high-temperature strength above 100°C is no disadvantage. Tin-bismuth, with a low melting temperature of 139°C, is useful for heat-sensitive systems. Tin-zinc alloys have a very attractive temperature range and cost but may be subject to oxidation and corrosion before and after processing. Many more exotic alloys such as tin-copper-gold or alloys containing indium or germanium are available and will find their own special niches. All these alloys are found in pilot or commercial products now.

There are a number of proprietary alloys covered by patents. The best known of these are patented in the United States and are owned by Ames Laboratories (certain tin-silver-copper compositions), AIM, Inc. ("Castin," certain tin-silver-copper-antimony compositions), and Oatey, Inc. (certain tin-silver-copper-bismuth compositions). There are many other patents in force worldwide, filed by a wide range of companies. Some of the compositional ranges in the patents overlap, and some patents have been granted despite the apparent existence of prior art.

As a result, assembly manufacturers are looking for solder manufacturers who can supply material that is licensed or is demonstrably free of others' patents. Some major electronics companies in Japan have their own patents, and recent information suggests that they may decide to offer free licenses for these proprietary compositions.

The IDEALS program in Europe, Industrial Technology Research Institute (ITRI) Soldertec, and National Electronic Manufacturing Initiative (NEMI) have all come out in favor of tin-silver-copper. Although this will likely be the front runner, the extent of its adoption will not become clear until the patent issues are resolved. At present, it looks like the materials will eventually shake out to

- Tin-silver-copper—general purpose.
- Tin-silver-copper-bismuth—consumer. May be replaced by tin-zinc alloys.
- Tin-silver—automotive.
- Tin-copper—wave-soldered consumer, HASL.

Fluxes. Fluxes must cover the working range of the solder and meet cleanability constraints. Not only must they prepare the way for the joint to form, but they also must prevent oxidation of the solder surface and dross generation, especially in the case of wave soldering. They must not foul or impede probes used in pin testing.

Fluxes typically contain an activator (typically organic acid–based) and a resin (natural or synthetic) in an organic or inorganic vehicle. The activator must react to remove the oxide layer when required, and the resin controls the rheology.

Materials used in pastes should be thixotropic (mobile during printing and then immobile after stencil snap-off, and the paste "bricks" should maintain their shape during heating to avoid "hot slump"). Of course, all these material requirements are contradictory, so development of an effective flux system is at best a compromise.

Printing. There are no special considerations. Concerns about shorter stencil life with harder lead-free solders seem to be unfounded.

Dispensing. There are no special considerations. However, some dispensed systems use unfilled polymers, where thermal mismatch between the resin (15 to 25 ppm °C) and die (3 ppm °C) may become critical.

Wave Soldering. Wave soldering equipment is in intimate contact with the solder for protracted periods. Changes to equipment are usually minor retrofits and can be summarized as

- Nitrogen capability to give improved joint finish, solderability, and wetting
- Flux recovery system to reduce nitrogen use
- Coatings on bath and nozzle to reduce corrosion
- Revised level float design (all lead-free solders are less dense than conventional solder)
- Software revisions for safety interlocks (minimum pump temperatures, etc.)

Care will need to be taken in use and clean-down to avoid contamination of the solder from the previous charge or boards, especially when a composition close to a patented composition is being processed.

Reflow. There are real concerns about the position and size of the process window for lead-free systems, and there is some uncertainty whether older equipment will be able to cope with the increased thermal load and precision required. In particular, it looks like infrared (IR) reflow ovens will in many cases be unable to adapt to the new processing parameters. Japanese manufacturers are using heat shields or protective surface finishes in some cases to protect components.

In most cases, though, recent assembly equipment is "no-lead ready" with little modification. What has to be determined by experimentation is the thermal profile requirements of the individual boards and the controllability and reliability of the equipment at the elevated temperatures needed to process lead-free solders. As with a new board being designed for eutectic alloy, a certain amount of process optimization will be required to make the materials work with acceptable yields.

Nitrogen atmospheres are obviously to be avoided where possible due to cost considerations but will be necessary where flux residues oxidize during processing to yield a finish resistant to cleaning.

Cleaning. Modern high-efficiency cleaning equipment should prove adequate for lead-free systems. Where cleaning is difficult, a switch to a nitrogen atmosphere or a different flux system may be desirable. Some manufacturers are looking at the high ionic content of water-based flux as an advantage—residues under components may be easier to quantify in testing than with no-clean fluxes.

15.5.4 Reclaim

Less than 25 percent of circuit boards make it to the recycling circuit. Of these, some are refurbished, but the majority have precious-metal-rich components such as connectors clipped off and then are sent to copper smelters. This route in itself is challenged because smelters have difficulty handling organic loading, halogens, and some metals found on lead-free boards. In addition, the precious metal content of boards has halved every 5 years over the past 20 years.

A recent United Nations Accord on Emission Levels also will help to add pressure to drive down emission levels from smelters, especially in the case of lead, cadmium, and mercury

Collection rates for appliances in Japan under the 2001 act are expected to be 50 to 90 percent compared with automobiles at nearly 100 percent. Currently, 25 percent of discarded circuit boards are recycled, and the rest go to landfill.

The main source of lead in the current electronic waste stream is actually not solder but is in fact leaded glass used in monitors. This is now regulated in the state of Massachusetts, but the content of leaded glass is expected to decline as unleaded glasses containing zirconia and other substitutes work their way through product lifetimes.

In Japan, major electronic corporations are setting up to recycle electrical goods, with Sony, Mitsubishi, and Matsushita, among others, establishing pilot facilities.

15.5.4 Inspection

Many lead-free solders have a dull appearance on solidification rather than the bright finish we are accustomed to with lead-tin eutectic; in addition, because of the wetting properties of lead-free alloys, the shape of the solder joint may differ. This means that inspection processes and standards will need to be modified in light of performance rather than traditional measures of appearance.

Talking to automatic optical inspection (AOI) manufacturers, this may present a problem, with early AOI equipment using specular reflection, but second-generation equipment will be able to cope.

15.5.5 Rework

Tin-silver-copper-bismuth will form a 96°C low-melting phase if lead-containing solder is used in rework. Manufacturers are looking at marking conventions and color-coded solder mask as ways to visually identify the alloy system on a board to avoid the contamination and reliability issues involved in handling leaded and no-lead lines.

15.6 THE COST IMPLICATIONS OF REPLACING LEAD IN SOLDERS

15.6.1 Background

Although reliability and assembly processing issues have been the main concerns of lead-free assembly, it is also necessary to analyze the relevant cost considerations. The actual incremental increased material costs are not large when represented as a percentage of the value of the electronic materials produced. However, the cost of reduced yield or reduced throughput could be significant. A recent paper by Lasky and Rae,[2] summarized here, analyzed the potential impact on the industry using the cost-estimating Excel program SPACE.

In 1998, Prismark Partners LLC performed a cost "breakdown" of a PC.[3] This personal computer cost breakdown is shown in Figure 15.1. Although it is an approximate estimate and now a little

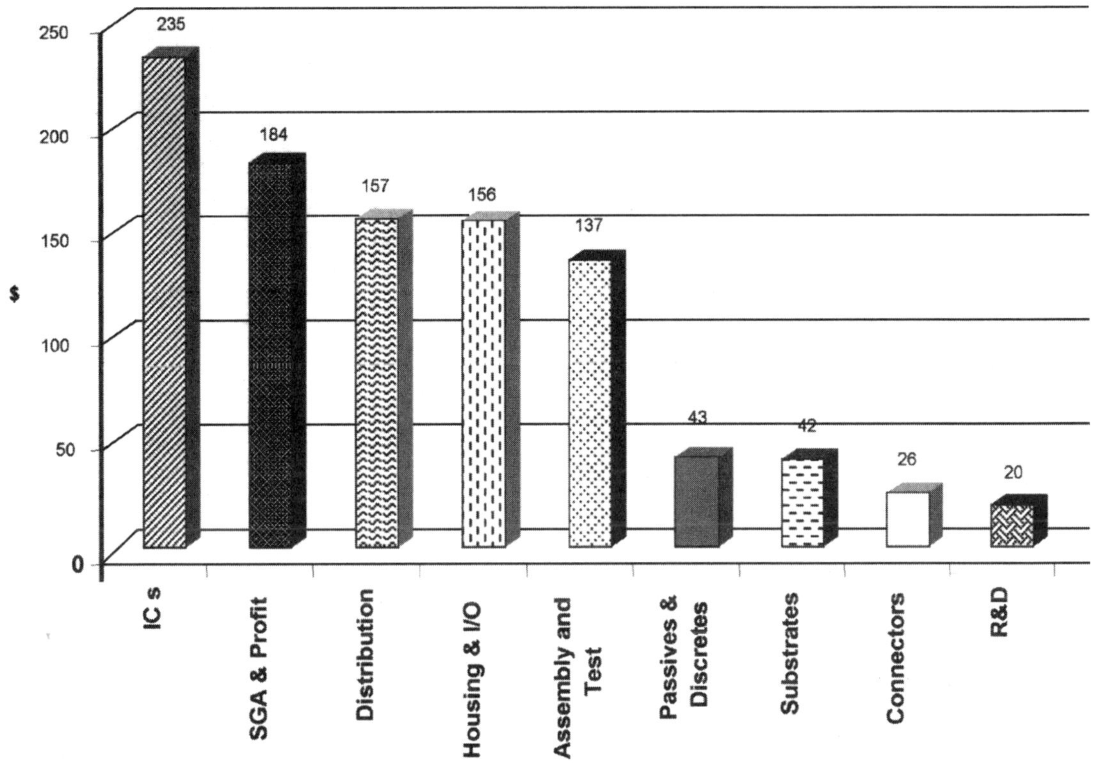

FIGURE 15.1 PC system cost: $1000/PC.

dated, manufacturers we have talked to still consider it fairly representative. If one considers the entire electronics industry to be $1000 billion, these percentages can be used to gauge the value of the services and materials in the electronics industry. This analysis clearly possesses a margin for error, since it would suggest that the semiconductor industry is $235 billion versus the actual $140 billion. However, this estimation provides a reliable order-of-magnitude assessment of the value content of electronics.

15.6.2 Impact of Metal Substitution

Approximately 4000 metric tons (t) of solder paste and 35,000 t of solder bar are consumed annually by the electronics industry. In passing, it is interesting to note that batteries consume 4 million t, bullets 200,000 t, and car balancing weights about 100,000 t of lead each year. Table 15.3 lists some metrics of interest for the candidate metals in lead-free alloys.

From this table and the data mentioned earlier, about $360 million ($90/kg × 4 million kg) of solder paste and $140 million in solder bar (~$4/kg × 35 million kg) are consumed each year. Converting completely to no-lead will increase these numbers very little, the first reason being that the primary factor in solder paste cost is the processing. Using the Sn/Ag4/Cu0.5 alloy, the substituted metal would increase the cost of paste by $8.45/kg ($12.25 − $3.80) to $98.45/kg.

Second, however, since the use of solder is measured by volume and the density of this alloy is less than Sn/Pb37, only 87.9 percent (the ratio of the density of Sn4/Ag0.5/Cu to Sn/Pb37) as much "joint metal" by weight will be needed. Hence the actual cost of paste needed could decrease slightly to $346 million [$360 million × ($98.45/$90) × 0.879]. The cost of solder bar will

TABLE 15.3 Cost and Density of Solder Metals and Alloys

Metal	$/kg (approximate)	Density (g/cc)
Pb	0.55	11.36
Sn	5.70	7.29
Cu	1.50	8.96
Ag	170	10.49
63Sn/37Pb	3.80	8.4
63Sn/37Pb paste	90.00	8.4
95.5Sn/4Ag/0.5Cu	12.25	7.39
99.3Sn/0.7Cu	5.66	7.30

increase moderately to $206 million [$140 million × ($5.66/$3.80) × (7.30/7.39)]. The total cost of all solder increases about 10 percent from $500 million ($360 million + 140 million) to $552 million ($346 million + $206 million). Since this $52 million increase is for the entire $1 trillion electronics industry, the change in materials cost should not be considered great. It is also interesting to note that the cost of soldering materials contributes less than 0.1 percent of the cost of the end electronic products.

15.6.3 Implementation Costs

Implementation costs can vary widely. Some large OEMs may spend tens to hundreds of millions of dollars implementing a lead-free policy. The costs will be in process design and verification, reliability testing, and inventory management (if the operation is part no-lead and part lead, etc.). One of the greatest challenges will be deciding if all assembled products need a complete process redesign and requalification. Because of variability from company to company, this aspect of lead-free conversion is the hardest to estimate. It will be likely that companies will be able to minimize their costs if they collaborate with their materials and assembly equipment suppliers.

15.6.4 Running Costs

One implementation cost that may be insignificant for the board buyer but critical for the EMS provider with paper-thin margins is increased energy use. SPACE shows that facilities and energy costs are typically less than 0.1 percent of finished product cost. Nevertheless, in a rapidly growing industry, space is at a premium—as one manufacturer recently said, "I can't build plants or hire people fast enough."[2]

Another consideration is the lifetime of equipment. Many more nitrogen-enabled ovens are shipped than are in use, ready for the time when nitrogen will be necessary in production.

15.6.5 Yield Loss

To demonstrate the effect of yield loss on cost, we used SPACE to perform a cost analysis on the assembly of a modem over 1 year of production. The modem would sell for $110 and would have an assembled cost in the $104 range. This approximate 6 percent gross margin is typical in the industry, if not slightly high. The assembly parameters for a "baseline" (i.e., before lead-free processing) case are listed in Fig. 15.2. They have been chosen from metrics NEMI has designated as "typical." Yield loss is a legitimate concern in the lead-free processes because the lead-free alloys typically do not wet copper pads as well as Sn/Pb37 solder, and poor wetting is known to increase voiding.

Let us assume that the no-lead process results in lower first-pass yields, 92 percent versus the standard baseline case's 97 percent. The lost yield is completely reworkable in both cases. As another example, let us assume that the first-pass yield is 92 percent, but that only 80 percent of the yield fallout is reworkable. What are the cost implications of both these yield-reduction scenarios? Assuming a year's worth of production, the unit cost increases $0.52 in the 92 percent all-reworkable yield case compared with the baseline case. This scenario results in a decrease in profit of almost $350,000.

Description	Modem		
Assembly Equipment(SP:350K, PP:1500K, RO:180K, T:900K, BH:170K)	$1,700,000.00		
Floor Space and Utilities (sq m) - Cost (per sq m/month)	3682.4	$0.83	
Components (per unit)	$75.00		
Stencil - squeegee - lifetime (cost each, uses)	$600.00	$200.00	10000
Solder Paste ($/g) - number of grams	$0.1000	5	
PCB ($ per unit)	$15.00		
Workers (number, rate per hour)	7	$28.00	
Selling Price ($) - Gross Margin (%) (enter '0' for item not used)	$110.00	0	
Hours per shift - Shifts per day - Days per week	10	2	
Cycle time (seconds)	25		
Downtime (%)-Setup (hours per week)-Maintenance (hours per week)	8	10	1
Workers Supported (number, rate per hour)	26	$42.00	
I/O	950		
Yield first pass (%) - Yield first pass ppm/I/O (N if not used) - Percent Reworkable	97	N	1
Unit RW Materials Cost - Minutes/RW - RW Labor ($/hr)	$0.50	20	$30.
Years Equipment Depreciation - Interest Rate (%)	5	9	
Length of Simulation (weeks) - Number of Boards (enter '0' for item not used)	52	0	

FIGURE 15.2 Manufacturing cost for modem.

However, the worst case turns out to be one in which only 80 percent of the yield loss is reworkable. In this example, the unit cost skyrockets more than $2, and profits plummet by almost $1.4 million. Clearly, it is crucial to have no increase in yield loss when implementing a lead-free policy. If the results of our final case (92 percent first-pass yield, only 80 percent reworkable) were inflicted on the entire $1000 billion electronics industry, it would result in $20 billion of lost profit.

15.6.6 Cycle Time

The $SnAg_4Cu_{0.5}$ alloy melts at 217°C, compared with eutectic tin-lead solder's 183°C. This temperature difference will require the reflow oven to operate at a higher temperature. To minimize the risk of thermal damage to components, some process engineers may modify their processes to run the lead-free reflow oven at a lower temperature by slowing down the reflow oven's belt. Slowing the belt may result in a slower cycle time for the entire line. In our first case we looked at the effect of slowing the cycle time from 25 to 27 s. As a second case, we assumed that in addition to slowing the cycle time down, the higher temperatures in the reflow oven required 1.5 more hours of oven maintenance per week. The results were devastating; with the slower cycle time, almost $1 million less in profit is generated. The modest increase in maintenance time results in an additional $200,000 profit erosion.

15.7 TIME SCALE

Current environmental legislation is aimed at reducing exposure to hazardous materials, in particular lead, mercury, cadmium, hexavalent chromium, and halogen-containing materials. The approaches taken are to limit the material use, assign responsibility to the producer, and encourage reuse and recycling.

15.7.1 Asia

In Japan, the 2001 appliance recycling bill will require the collection and recycling of refrigerators, televisions, air conditioners, and washing machines, with other electronic equipment to follow. The copper, aluminum, iron, tin, zinc, and glass are to be recycled. Note that lead is not mentioned and by implication should not be in the waste stream. The approximate cost to the consumer will be around $30.

In addition, there is a voluntary reduction plan for the lead content of vehicles by Japanese manufacturers. This will eliminate lead in many areas, e.g., heat exchangers, balance weights, and wiring harnesses (but not batteries), and reduce levels to one-half the 1996 level by 2001 and one-third that level by 2003.

Elsewhere in Asia, suppliers to Japanese OEMs will have to comply with the OEMs' published lead-free goals.

15.7.2 Europe

In the EC, the aim of the WEEE (waste electronics) draft directive is to phase out lead and other hazardous materials by January 1, 2008. The range of electronics products covered is comprehensive, with the main exclusions being military and space equipment. The first draft of the WEEE directive was developed in 1998 and as of May 2000 was just "leaked" in draft form and is available from the Institute for Interconnecting Packaging and Circuits (IPC) and International Technology Research Institute (ITRI) Web sites.

Similar to Japan, the ELV draft directive passed in February 2000 deals with vehicle end of life. It also excludes batteries, and there are exemptions for components that are labeled and can be removed separately using a dismantling manual provided by the manufacturer. This directive contains provision for regular amendments. Lead will be prohibited from landfills from 2001 and from many aspects of vehicle manufacture by 2005. This is the only enacted wide-ranging lead-reduction initiative that affects electronics manufacturers.

Individual countries may have even tighter requirements. In Sweden, all lead is to be phased out by 2020. Denmark is proposing to ban the import, sales, and manufacture of materials containing more than 50 ppm lead—a real problem because many substitutes, including tin, normally contain a higher level than this, and removal costs would be prohibitive.

15.7.3 The Americas

Although there is no current legislation planned, it must be borne in mind that the United States actually started the lead-free electronics movement. In the United States, the Reid bill (S. 391, Lead Exposure Reduction Act, 1991) proposed limiting lead content to 0.1 percent but was never passed. Specific states are now taking action to limit electronics in landfills, such as Massachusetts and New Jersey. New reporting requirements from the EPA reduce reportable levels from 25,000 to 10 lb.

Quoted in a 1999 IPC position paper, the CDC reported a 78 percent reduction in the U.S. blood lead levels, from 12.8 to 2.8 μg/dL of blood. The report also quotes a number of studies that failed to show a significant risk from tin-lead solder. Nevertheless, lead is a well-known toxic material, solders contain lead, the number of circuit boards containing lead and entering landfills is increasing, and technology exists to replace lead in this application.

IPC and others have stated that they will not oppose the elimination of lead from electronics—the factors driving this will be more political and marketing-led than science-led.[4] IPC and Electronic Industries Association of Japan (EIAJ), among others, have published lead-free roadmaps to help to guide the industry through the transition.

15.8 COOKSON ELECTRONICS' GREENLINE

Cookson Electronics introduced Greenline, a highly focused, company-wide initiative to provide complete lead-free manufacturing solutions, in 1999. Greenline is the first and most comprehensive technology source in the industry for electronics manufacturers moving to lead-free processing.

15.8.1 Rationale

Through Greenline, Cookson Electronics provides a high level of technical expertise and proven materials and equipment experience to offer front-to-back integration. No other single organization in the electronics industry has the resources to handle all aspects of lead-free, ranging from laminates, process and board treatment chemicals, alloys, fluxes and components to screen printers, ovens, dispensers, cleaning systems, coatings, and recycling systems.

15.8.2 Implementation

Greenline member companies carried out lead-free assembly on a dedicated production line at Productronica '99. Located directly on the show floor, the line manufactured a sophisticated modem assembly with traditional bottom-side components and top-side packages consisting of through-hole devices plus fine-pitch QFPs, BGAs, and micro-BGAs.

Equipment and materials used on Greenline were provided exclusively by Cookson Electronics' companies and their component placement, odd-form placement, and conveyor-equipment partners. The bare boards were provided by Polyclad Laminates and were treated with a lead-free AlphaLevel finish from Enthone Chemicals. Alpha Electronics adhesives were applied by a Camalot Gemini dispenser before wave-solder components were placed. Alpha-Fry Technologies supplied a no-clean, lead-free surface-mount solder paste, applied via Speedline Technologies MPM Rheometric Pump Print Head technology, and reflowed in an Electrovert Bravo oven. No-lead wave-solder bar for wave soldering was applied using Speedline Electrovert's Electra equipment. SCS provided a protective coating system for key components.

Producing a two-board panel every 40 to 45 s during each afternoon, Greenline manufactured approximately 1600 panels (3200 individual boards) during Productronica.

Greenline consolidates Cookson's core electronics manufacturing companies to focus on the lead-free initiative:

Polyclad Laminates: High-temperature resistant laminates

Enthone Technologies: High-temperature HASL oils; lead-free HASL; HASL substitutes based on tin, silver, and organics; solder masks and preparation systems

Alpha-Fry Technologies: Lead-free bars, pastes, wire, and flux; solder and dross reclaim, high-temperature IC packaging materials

Speedline Technologies:

- Camalot: Adhesive dispensing
- MPM: Lead-free paste printing
- Electrovert: High-temperature reflow ovens, wave solder, and compatible cleaning systems
- Accel: Cleaning systems for packaging

SCS: Impervious board coatings

15.9 SUMMARY

In the long term, the switch away from lead-containing materials is probably inevitable. The rate of change will depend on the application and the local conditions, e.g., consumer electronics in Japan first, mission-critical or harsh-environment systems in the United States last. Just about every electronics manufacturer in the world and their suppliers will need to evaluate their lead-free options and develop the capability to process lead-free systems effectively and reliably within the next 5 years.

15.10 REFERENCES

1. Most of the best references, including the DTI reports and WEEE drafts, are available for download or are accessible from the IPC and ITRI Soldertec Web sites: http://www.leadfree.org/ and http://www.lead-free.org/.

2. Lasky, R. C., and Rae, A., "Economic Considerations of Lead-Free Assembly," Second Singapore SMYA Seminar, Singapore, October 1999. SPACE is an internal cost estimating software package developed by Cookson Electronics.
3. Prismark, *The Electronics Industry Report,* Prismark Partners, Cold Spring Harbor, New York, 1997, p. 89.
4. NEMI, 1998 National Electronics Manufacturing Technology Roadmaps, 2214 Rock Hill Road, Suite 110, Herdon, VA 22070, 1996.

ACKNOWLEDGMENTS

I would like to thank numerous colleagues, past and present, for their help in assembling this information.

CHAPTER 16

ELECTRICALLY CONDUCTIVE ADHESIVES FOR SURFACE-MOUNT AND FLIP CHIP PROCESSES: AN ALTERNATIVE TO SOLDER?

Ken Gilleo
Cookson Electronics

David Blumel
McGraw-Hill Companies

16.0 INTRODUCTION

Conductive adhesives represent clean and simple solutions for many and potentially all electronic assembly applications. Decades ago, die-attach adhesives replaced metallurgic connections. Today, conductive adhesive films dominate flat-panel display assembly. Soon, adhesives may provide an alternative to solder, particularly lead-free solder, because they offer practical answers to problems where solder is totally inadequate, such as component assembly, including flip chips (FCs).

Seemingly simple, conductive adhesives contain complicated combinations of customized ingredients. Part chemistry and part material science, the synthetic polymers and inorganic fillers that make up the adhesives work in standard electronic assembly processes and soldering equipment as well.

Many polymers have excellent dielectric properties and find uses as electrical insulators. Conductive adhesives are made by adding conductive fillers to insulating polymers. These polymer-binder and filler composites are very different from other common joining materials.

Compared with inherently conductive solders, adhesives' use of conducting fillers might seem less favorable at first. Still, using conductive fillers with nonconductive binders results in an interesting and valuable feature for adhesives—their electrical properties are independent of their mechanical properties and can be adjusted individually. The polymer binder can be modified for a specific requirement or application characteristic, such as bond strength. The filler also can be selected to provide specific electrical properties, including directional conductivity, a feature not available in solders.

Some interaction occurs between the filler and the binder, but overall, an adhesive's electrical and mechanical properties are independent of each other. Thus the fillers, binders, and additives used in a conductive adhesive can be selected to create a wide range of properties.

16.1 ADHESIVE TYPES

16.1.1 Isotropic Conductive Adhesives

Isotropic conductive adhesives (ICAs) are polymer-based bonding agents that electrically conduct equally in all directions (Fig. 16.1).

Polymers. What are polymers? Polymers are long-chain molecules, like epoxies, acrylics, and urethanes, that are used widely to produce structural products—films, coatings, and adhesives. Although polymers occur naturally, most synthetic ones have properties tailored for specific applications.

Polymers are commonly classified as either thermoset or thermoplastic (Fig. 16.2). Thermosetting polymers are true to their name. Heat "sets" them and permanently changes their properties so that they resist melting and cannot be reshaped. Thermoplastics typically soften and melt when heated, although a few of the very high temperature thermoplastics, such as polyimides, decompose before reaching their melting points. Adhesive binders can be of either type, and each system differs in function, processing, and storage.

Thermoset Polymers and Isotropic Adhesives. Most thermosetting systems used for ICAs contain prepolymers, ingredients that can polymerize. The thermoset prepolymers can be either reactive single-molecular-unit materials called *monomers* or reactive low-molecular-weight polymers called *oligomers.* Since many prepolymers are liquids, no solvents are needed when formulating mixtures of ingredients. These solventless systems are ideal environmentally because there are no emissions during application or curing. Solventless adhesives also minimize the possibility that solvent bubbles will form voids in the cured adhesive.

Epoxy-based adhesives typically contain a liquid epoxy resin and a hardener. When heated, the two ingredients react chemically or cross-link. Cross-links are chemical bonds between adjacent polymer chains that lock the chains in place. The resulting three-dimensional (3D) molecular structure restricts chain movement, produces the characteristic of shape retention, and prevents melting. Still, some degree of softening may occur at high temperatures. For a cross-linked polymer, the softening or deformation point may be the temperature at which thermal degradation occurs.

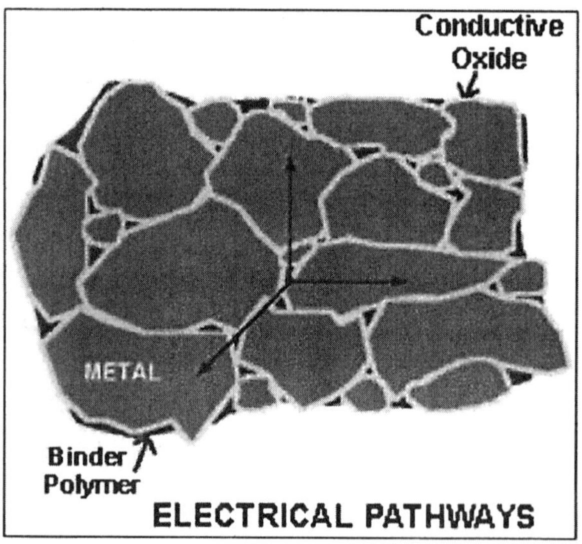

FIGURE 16.1 Isotropic conductive adhesive.

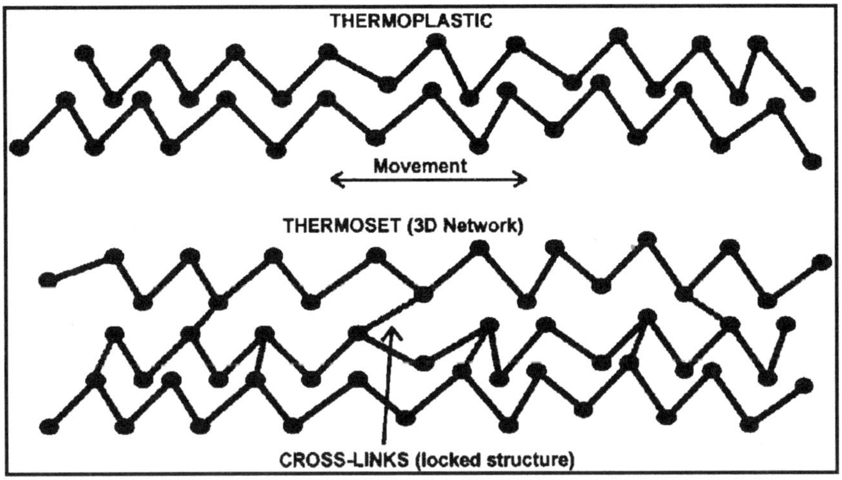

FIGURE 16.2 Thermoset versus thermoplastic polymers.

Thermosetting epoxies have been the workhorse polymers of electronics because of their ease of use, availability, electrical properties, and generally superior bonding properties. And ICAs made with thermosetting epoxies have been used for many years for die attach, a part of the packaging process where bare silicon dies are bonded to lead frames, area array substrates, and chip carriers. Billions of chips are packaged this way.

Die-attach adhesives—typically silver-filled epoxies because of their good electrical conductivity—have been modified for component assembly. The most common systems are thermosetting pastes that can be printed, needle-dispensed, or dip-coated. These materials are highly loaded (about 80 percent by weight) with silver powder and flake. The silver particles make intimate contact with one another so that the conductivity is similar in the *x*, *y*, and *z* axes.

Before use, thermoset adhesives may be stored frozen to prevent them from hardening prematurely, but during use, they must have a reasonable working life, usually 8 hours.

Adhesive curing takes only a few minutes at 130 to 150°C (266 to 302°F). The low-temperature processing and wide compatibility make thermoset ICAs the best assembly choice for flexible circuits. Newer cure systems allow processing as low as 110°C (230°F), making them suitable for heat-sensitive components, such as batteries.

Thermoplastic Polymers and Isotropic Adhesives. Thermoplastic polymers are typically linear long-chain molecules made up of smaller repeating units. Unlike thermosets, they do not have interchain cross-links (see Fig. 16.2). When heated, thermoplastics soften and melt, allowing the individual polymer chains to flow past one another. And when cooled, they "freeze." Thermoplastic polymers can be melted and reshaped without significantly changing their chemical properties unless the temperature is so high that thermal decomposition occurs. In this regard, thermoplastics are a lot like solder.

Isotropic thermoplastic adhesives are available as pastes and films. Thermoplastic pastes have one more processing step than the films—you must evaporate the organic solvent used to dissolve the polymer binder before component assembly.

Thermoplastic-based adhesives offer the advantages of fast processing and easy rework. No chemical reaction occurs during processing. Heating causes a change in physical state—from solid to liquid—in a short time, perhaps less than a second. After processing, the thermoplastic binder will soften or melt if heated to a high enough temperature.

Thermoplastic conductive pastes can be used for component assembly, but their properties are typically not as robust as the thermosets. Thermoplastic adhesives must remain solid at their maximum use temperature, limiting their in-service temperature performance. Thermoplastics also tend to flow or creep—deform over time—under applied force. Cross-linked thermosets resist deformation and have greater mechanical stability.

The thermoplastics form weaker adhesive bonds than the thermosets, which usually start as liquids that can wet out on a surface more thoroughly. Thermoset adhesives also can react with various surfaces to form strong bonds. The superior properties of thermosets compared with thermoplastics offset their handling inconveniences and greater process-control requirements. Still, thermoplastic isotropic adhesives have found use as die-attach adhesives.

Fillers and the Isotropic Conductive Mechanism. Silver is by far the most popular conductive filler, although gold, nickel, copper, and carbon are also used. This might seem a poor choice. Gold, however, a nonoxidizing metal, is too expensive. Nickel oxidizes slowly but is hard, poorly malleable, and difficult to process into flakes of optimal size and shape. Copper, which would appear to be the logical filler, becomes nonconductive as it oxidizes. And carbon's poor conductivity is up to three orders of magnitude lower than silver's.

So why silver? Silver oxide is quite conductive. Almost no change in conductivity occurs as silver particles oxidize. Oxides of most metals are electrical insulators. And oxide on a conductive particle increases junction resistance. Junction resistance is the sum of the electrical resistance due to the bulk of adhesive (volume resistivity) between two conductors, as well as the interface resistance at each of the two interconnects and any connectors or leads making up the circuit.

Junction resistance is the most important electrical criterion of a conductive adhesive because it relates to the actual joints of the assembly. For isotropic adhesives, it varies with the thickness of the junction. For example, a 100-μm-thick (4 mils) junction will have about twice the absolute resistance as one 50 μm thick (2 mils). Surface finishes also influence junction resistance. Gold, a nonoxidizing metal, is often used as a standard because it does not add to the resistance. Oxidizing metals, such as copper, often show higher junction resistance values because their metal oxides are electrical insulators.

Junction stability at elevated temperature and humidity is important. Although the adhesive may be stable, oxide formation on most component surfaces can cause unacceptable increases in junction resistance as an oxide layer "grows" between the adhesive and the component finish (Fig. 16.3). This problem has been solved with special junction-stable adhesives. One successful approach has been to use penetrating filler particles to break through the oxide layer (Fig. 16.4). This type of adhesive has passed 1000 hours of testing at 85 percent relative humidity and 85°C with no increase in junction resistance.[1]

The intrinsic electrical conductivity of an isotropic adhesive is measured by its volume resistivity. This measurement should be independent of junction resistance, which must be measured using the actual assembly's finishes. ICAs can have up to an order of magnitude more volume resistivity than solder, although a few adhesives do have about the same electrical conductivity. The electrical pathway through many discrete conductive particles in close contact accounts for the higher resistivity values. In most cases, volume resistivity values measured after temperature and humidity aging and thermal cycling will improve, presumably because the conductive particles move into more intimate contact.

Filler morphology and metallurgy are critical because the conductive particles form the adhesive's electrical path. The particles must be small enough so that adhesive pastes can be stenciled or otherwise applied. Smaller particles, however, create more electrical junctions per volume of adhesive. And because each junction is a source of electrical resistance, volume resistivity tends to increase. Still, considerable work has gone into creating particles that have low junction resistance.

Silver can be precipitated into a wide range of particle sizes and shapes so that just the right sizes of particles can be produced and used as is, or it can be milled into fine flake, a useful form for boosting conductivity. It is possible to precipitate silver into particles so thin that they are translucent.

A blend of silver flakes and particles improves conductivity in isotropic adhesives. The platelike flakes overlap each other like flagstones, maximizing metal contact. And the small spherical particles fill the spaces between the overlapping flakes.

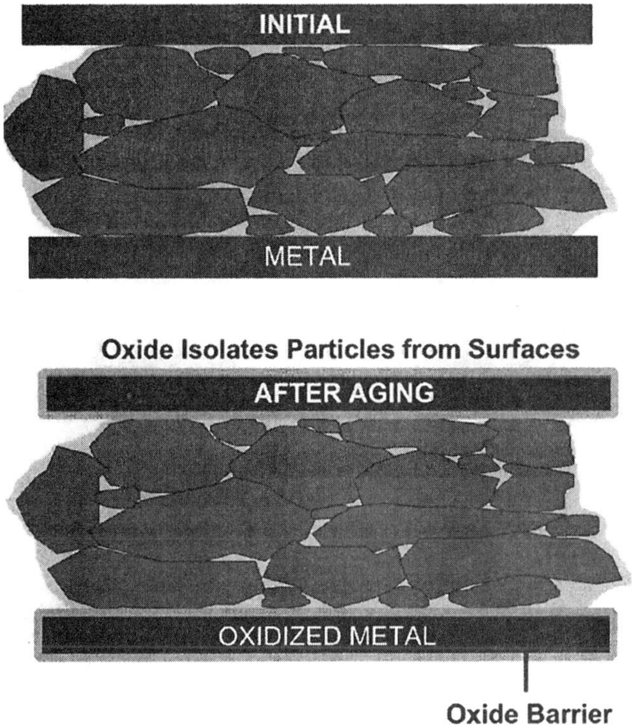

FIGURE 16.3 Isotropic adhesive junction, (*a*) oxide-free and (*b*) after temperature and humidity aging.

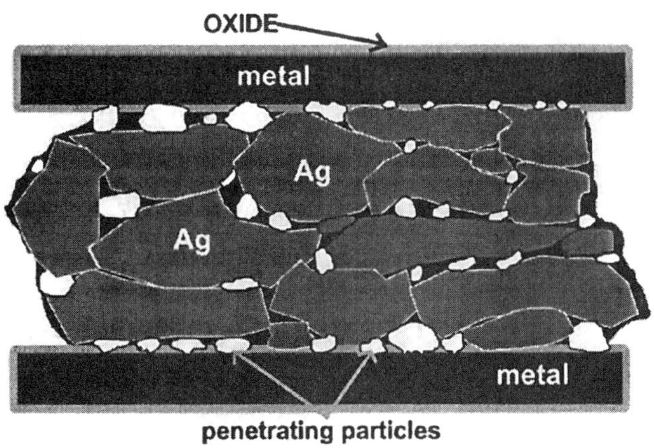

FIGURE 16.4 Proposed model for isotropic adhesive with penetrating particles.

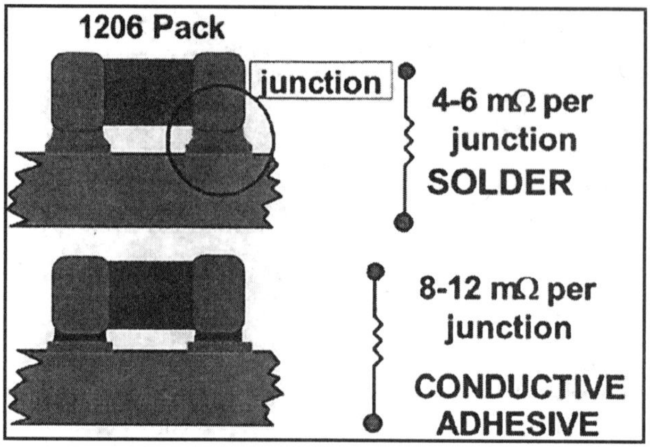

FIGURE 16.5 Typical resistance values for tin-lead solder and silver-filled isotropic conductive adhesive.

Compared with tin-lead solder, ICAs conduct equally but have two to four times higher resistivity (Fig. 16.5). A conductive adhesive junction, with hundreds or thousands of electrical pathways, always has higher resistance than a solder junction, which has a continuous metal path.

Silver-plated copper has found commercial application in conductive inks and should work in adhesives. While compositions made with pure silver particles often show improved conductivity when exposed to heat and humidity, this is not always the case with silver-plated copper flakes. Presumably, the application of heat and mechanical energy forces the particles into more intimate contact when pure silver is used. However, silver-plated copper may have coating discontinuities that allow copper oxidation, which would reduce its conductivity.

16.1.2 Anisotropic Conductive Adhesives

Anisotropic conductive adhesives (ACAs) are polymer-based bonding agents that electrically conduct in one direction, the vertical plane, or *z* plane.

Random Dispersion Type. Most ACAs, also known as *z-axis adhesives,** contain conductive particles dispersed in a polymer binder. Directional conductivity is achieved by using a relatively low loading of conductive filler, ranging from 10 to 40 percent by volume. Low filler loading prevents interparticle contact and conductivity in the adhesive's *xy* plane.

Polymers and Anisotropic Adhesives. Anisotropic adhesives are available as pastes and more commonly as films. An anisotropic film can be applied over the entire circuit because electrical conductivity occurs only between opposing conductors in the *z* plane.

The polymer binder may be thermoset, thermoplastic, or a blend of both. And since selective adhesive application is not required, thermoplastic anisotropic pastes, which require solvent evaporation before heating, can be used.

A thermoplastic binder will soften and bond to the components. Thermoplastics can be remelted because they do not cross-link during the assembly heating process, as do thermosets. They also have excellent storage life and do not require refrigeration, as do the thermosets.

Thermoplastic anisotropic adhesives are noted for fast processing, and the thermosets are noted for higher-temperature performance. Mixed thermoplastic and thermoset anisotropics can have fast-processing and high-temperature performance properties.

---

*Ken Gilleo first described anisotropic conductive adhesives as *z-axis adhesives,* a name that quickly caught on after appearing in the trade literature.

Processing anisotropic adhesives requires specialized bonding equipment that forces the component against adhesive-coated conductors and simultaneously applies heat and pressure. The bonding process squeezes out the nonconductive polymer, allowing the particles to bridge the gap between conductors (Fig. 16.6). There is usually only one layer of conductive particles between the conductors of an anisotropic circuit, and each conductive particle makes electrical contact with the opposing conductors. The polymer binder is hardened by heating, if thermoset, or by cooling, if thermoplastic. The hardened polymer holds the assembly together and helps maintain the pressure contact between conductors and particles.

Fillers and the Anisotropic Conductive Mechanism. Fillers for ACAs are often quite different from those used in isotropic adhesives. A large number of metal-plated conductive particles have been described and produced (Fig. 16.7). The materials can be divided into two broad categories—metal core and nonconductive core. Both types of particles are used in anisotropics.

Nickel has found use in ACAs containing metal-core particles. Solid nickel spheres can be made in practically any size with narrow size distributions. Nickel also can be easily plated with gold to prevent oxidation.

Silver, nickel, and gold plated onto nonmetals are the most common anisotropic fillers. Silver is often used because of its simple plating processes and because it remains conductive in oxidizing environments. Nonconductors, such as glass spheres, can be silver-plated and are available commercially. Recently, plating on plastic spheres has become popular. Plated plastic particles have lower density and are less prone to settling than metal-core particles. Some plastic spheres deform under pressure, improving contact with the bonding surfaces.

While ACAs can be used to join rigid parts, they perform best when one surface is compliant. Anisotropic adhesives have found many applications with flexible circuits because their compliancy allows the bonding pads to be forced into coplanarity with the component leads (Fig. 16.8).

Anisotropic conductive particles are typically less than 25 μm (1 mil) in diameter. The non-coplanarity of the two joined surfaces must be less than the particle diameter when ACAs are used with rigid parts. This makes surface-mount assembly applications very challenging. If a component lead is out of plane by 50 μm (2 mils), the conductive particles cannot make contact unless the lead can be bent closer (Fig. 16.9). However, even if the bond is made successfully, the forced into plane lead will be under tension, and the connection may be unstable.

For anisotropics, the electrical connection is a pressure contact type. The polymer must maintain a certain amount of tensional strain on opposing conductors to keep them in contact. Should the polymer expand, as it would during heating, the contact pressure is reduced. If the polymer expands at a higher rate than the conductor, which is often the case, the pressure contact can open (Fig. 16.10).

Adhesive technologists have attempted to reduce the coplanarity and expansion problems by using metal-coated elastomeric spheres, resilient particles that deform under pressure. Many of the

FIGURE 16.6 Electronic component bonded with *z*-axis adhesive.

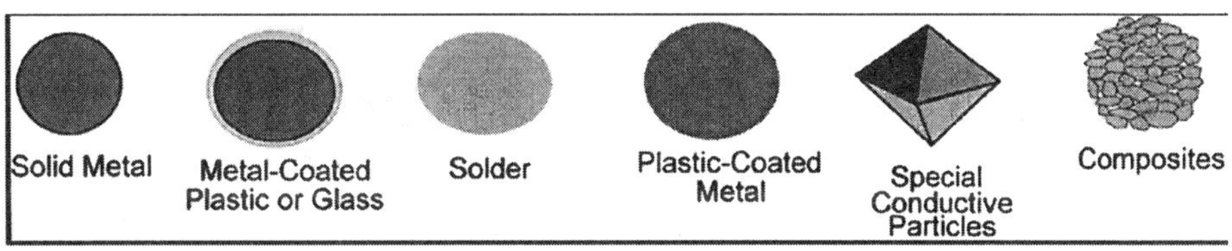

FIGURE 16.7 Anisotropic conductive particles.

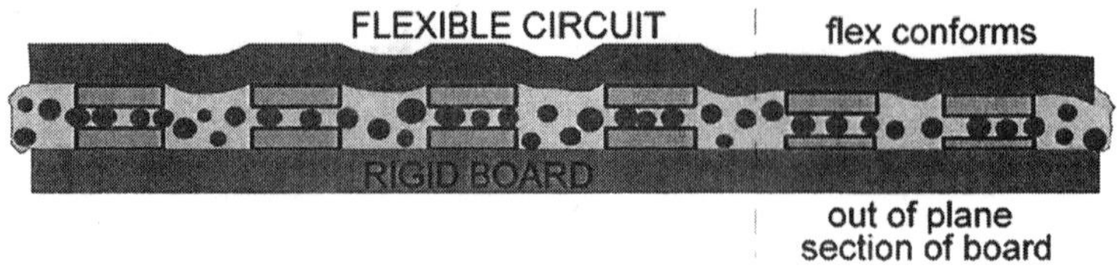

FIGURE 16.8 Flexible z-axis assembly.

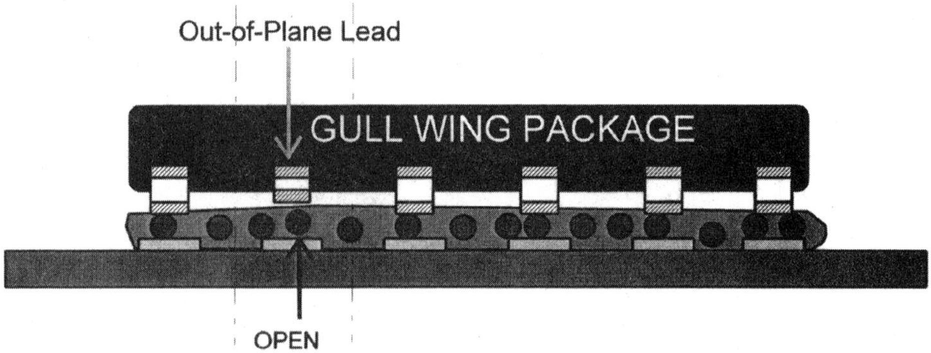

FIGURE 16.9 Noncoplanarity failure in a ridged circuit board.

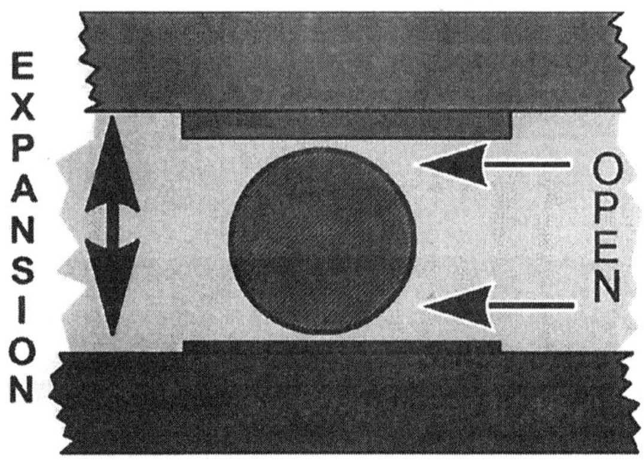

FIGURE 16.10 Effect of heating on conductors and polymers with different coefficients of thermal expansion.

best anisotropic conductors have a plastic core that is coated with either gold or silver. The most popular conductive particle is a 6- to 10-μm (0.24- to 0.4-mil) polymer sphere that has been first plated with nickel and then gold. These conductors can compress under bonding pressure, reducing coplanarity differences somewhat. When compressed, the particles act like small springs, compensating for the thermal expansion of the binder and allowing the adhesive to maintain the force needed for pressure contact during heating.

ACAs are well suited for very fine pitch interconnects in several areas, including flat-panel displays and surface-mount assembly. And z-axis adhesives appear to offer the simplest, most cost-effective means for FC bonding.

Patterned Array Type. Most ACAs use a random dispersion of conductive particles because this is easy to do. Recently, adhesives with patterns of conductors on bondable dielectric (polymeric) films have been introduced.[2] One approach, called *grid array z axis* (GAZA), uses columns of ICA arranged in a grid pattern (Fig. 16.11). The conductive adhesive can be patterned by printing methods, with the dielectric film cast as the last operation, or holes may be formed in a dielectric sheet and filled with conductive adhesive. A limited number of vendors are beginning to supply patterned array adhesives, but the difficulty in manufacturing these materials at low cost has limited the technology. And then there is the issue of producing standardized patterns.

The proliferation of so many different surface-mount packages has made it impractical to provide oriented patterns for all of them, unless custom-made. However, the recent arrival of area array packages, such as the ball grid array (BGA) package, now makes it practical to produce a limited number of standard oriented conductor patterns. Today, there are only a few common and proposed pitches—0.5- (micro-BGA), 1.0-, 1.5-, and 1.75-mm centers. Sheets of anisotropic patterned array films with standard patterns can be cut to match the respective packages. The films may be thermoset, thermoplastic, or mixed systems. For assembly, the films are positioned mechanically and thermocompression bonded.

On the negative side, patterned array films must have greater dimensional stability than the random-type products. And patterned adhesives require alignment with the component, an additional process step.

While it is too early to tell if patterned array anisotropic adhesives will deliver superior performance, they potentially have several benefits over random-type materials. It is almost impossible for a patterned conductive adhesive junction to open, particularly if the dielectric and other materials have similar coefficients of thermal expansion (Fig. 16.12). However, if there is a significant thermal mismatch between the connectors, junction fracture may occur during thermal cycling. Placing all the conductors in an orderly array also maximizes junction conductivity because all the conductive material is connected, compared with a much smaller amount with the random type. The random materials have isolated conductive particles between conductor traces that may degrade high-frequency signals, reduce breakdown voltage, or increase the film's dielectric constant

16.2 SURFACE-MOUNT ASSEMBLY

Using conductive adhesives for surface-mount assembly is quite similar to using solder paste. Overall, the same equipment is used without modification.

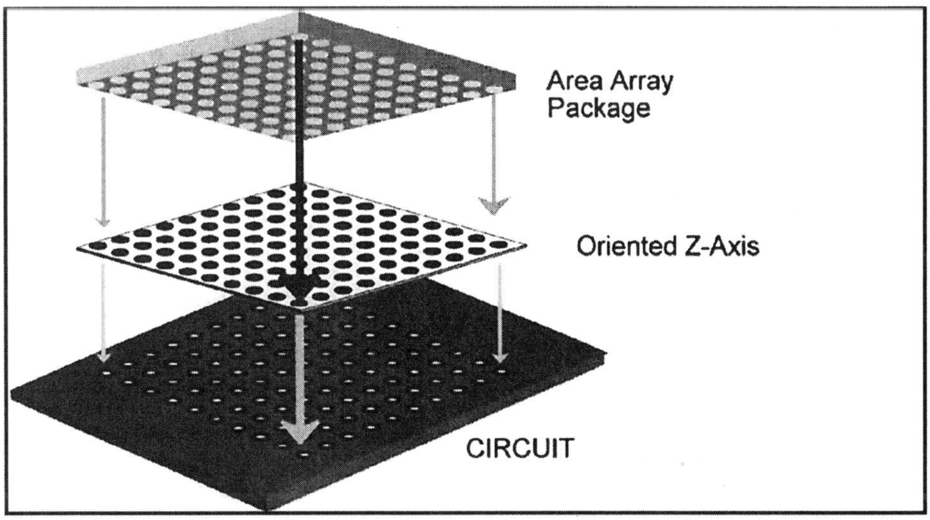

FIGURE 16.11 Schematic structure of a grid array *z*-axis assembly. The arrows indicate the direction of the attachment process.

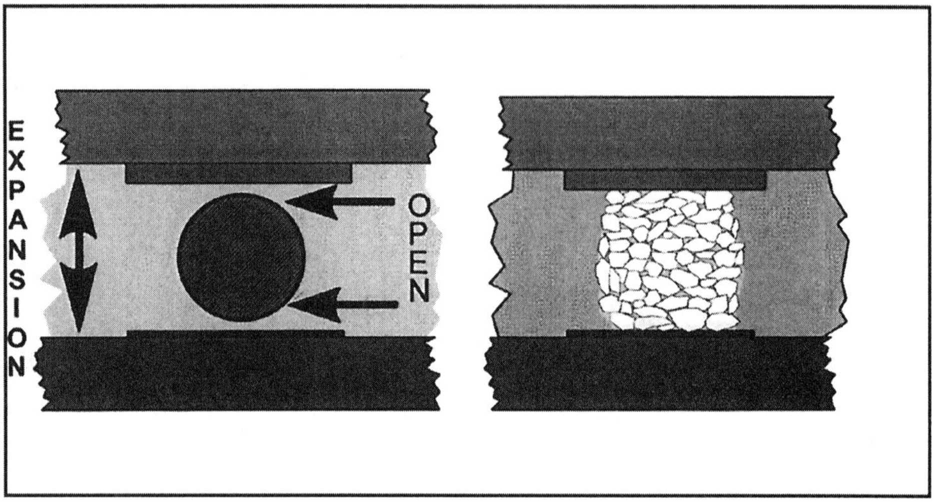

FIGURE 16.12 Thermal expansion failure of an anisotropic adhesive on the left compared with the more heat stable grid array *z*-axis connection on the right.

16.2.1 ICA Assembly Process

Isotropic Paste Application. ICAs require selective application to the circuit pads because they conduct in all directions. All common methods used to deposit solder pastes work equally well for isotropic adhesive pastes, the closest polymer alternatives to solder pastes.

Conductive adhesives contain relatively small [under 10-μm (0.4-mil)] metal powders and flakes. Comparatively, solder pastes contain solder spheres up to 45 μm (1.8 mils) in diameter.

Solder pastes lose about 50 percent of their original volume after processing due to coalescence. Conductive adhesives lose almost no volume. Solder also wicks onto circuit pads and forms fillets with the component leads, whereas adhesives generally stay put. Adhesive junctions require only about half the volume of material compared with junctions made with solder paste.

The density of a silver-epoxy adhesive is about one-third as high as that of solder paste. This one-third density combined with one-half the volume requirement means that only one-sixth as much conductive adhesive, by weight, should be used compared with solder. The rule is to apply a 100- to 150-μm-thick (4 to 6 mils) deposit of ICA for surface-mount junctions. Excessively thick deposits produce poor print definition, add cost, and do not improve performance.

Printing. Screen printing has long been used to selectively apply controlled deposits of pastes. The screen, a woven fabric with mesh openings, is stretched over a frame and coated with an emulsion. Openings in the emulsion are photodefined to form a stencil on the screen that matches the pattern that will be printed on the circuit. A rubber squeegee blade pulls the paste across the emulsion's surface and forces it through the openings onto a substrate.

The thickness of the print is determined by the screen mesh count (threads per centimeter or inch), the emulsion thickness, and the properties of the paste. Decreasing the screen's mesh count and increasing the emulsion's thickness on the screen will increase the thickness of the paste deposit.

In terms of rheology, most ICAs are thixotropic pastes that range in viscosity from about 100,000 to 500,000 cP. At rest, thixotropic materials have high viscosity and resist flow, but when a shear stress is applied over time, they show a reversible lowering in viscosity and become fluid. For screen printing, the squeegee blade must apply sufficient shear force to lower the paste's viscosity in order to push it through the stencil openings onto the circuit pads. Once the paste comes to rest again, it regains its initial viscosity and resists slump or bleed. Typically, the best printing resolution occurs when the lowering of the paste's viscosity on shearing is the greatest and its viscosity recovery at rest is the fastest.

Stencil printing is another suitable method for adhesive application. The stencil is a metal foil with openings made by chemical etching, laser cutting, or electroforming. The stencil thickness determines the print height. Usually, only laser-cut and electroformed stencils are recommended for print thickness requirements of 150 μm (6 mils) or less.

Dispensing. Dispensing machines work well for low-volume production runs, such as prototyping, or when applying pastes to small areas of large circuits. They also work well for applying isotropic pastes to nonplanar substrates, such as molded circuits.

Simple, low-cost, hand-operated dispensing machines deposit adhesives using a syringe with manually applied pressure. Automated dispensing machines use a programmable robotic dispensing head to apply adhesive patterns that normally would be screened or stencil printed. A single-head dispenser may be slower than a screen printer, but it only needs a program to run, no tooling, making dispensing the fastest application process for setup and changeover.

Component Placement. Component placement usually occurs soon after adhesive application. But the long working time of some adhesives permits placement to take place hours later. Placement must be accurate because the surface tension of adhesives is not high enough to orient or move components.

Setup for the placement machine is the same as for solder paste. It is sometimes necessary to slow the downward placement stroke if the adhesive paste gets displaced or if air entrapment occurs. High resistance and low bond strength are symptoms of entrapped air. Slowing the downward motion near the end of the stroke will have little impact on throughput.

Since the adhesive will not wick up and form a fillet on the component leads, the joint must have good integrity in the uncured state. Any severely misaligned components should be repositioned before curing.

Adhesive Curing. The low-temperature processing requirement for isotropic adhesives, typically 130 to 150°C (266 to 302°F), provides a low-stress assembly environment for the circuit and its components. The lower temperature is not that significant for circuit boards that have been designed to withstand the higher soldering temperatures. However, the lower curing temperature may become more attractive if the industry moves to higher-temperature lead-free solders. Also, an increasing number of new component packages may degrade at soldering temperatures. BGAs and thin, low-profile packages may explode if trapped moisture turns to steam during soldering. Packaging experts indicate that assembly below 200°C (392°F) will eliminate this problem. Temperature-sensitive circuits made with polyester film or low-cost molded plastic also benefit from low-temperature processing. One of the major uses of conductive adhesives is in polyester-based flexible circuits, particularly those built using polymer thick-film circuit technology.

16.2.2 ACA Assembly Process

Anisotropic Paste Application. Recall that anisotropic adhesives conduct only in the vertical axis, allowing the complete bonding area or even the entire circuit board to be covered. And the desired adhesive interconnection structure is a single layer of conductive particles.

Anisotropic pastes, which have much lower filler loadings, generally are less viscous and easier to screen print than isotropic pastes. Most anisotropics have a translucent rather than a metallic appearance because of their lower filler loading.

Screen printing is used most commonly to apply anisotropic pastes as long as the conductive particles are 25 μm (1 mil) in diameter or less. Only a thin layer of adhesive—25 to 50 μm (1 to 2 mils)—needs to be printed. Too much adhesive can cause short-out or excessive squeeze-out as the component is pressed against the adhesives.

Stencil printing also can be used to apply anisotropic pastes. However, since there is no need for a fine-pitch, highly accurate print pattern, stencil printing has no advantage over screen printing. It is more important to control the thickness than the x and y dimensions.

Anisotropic Paste Component Placement and Curing. During placement, the component leads are pressed against the adhesive in order to trap the conductive particles between the leads and circuit pads. Pressure must be applied during curing to overcome coplanarity problems with the component leads and the circuit. Curing under force is the standard process.

Typically, connections made with anisotropic adhesives remain reliable only if fine-leaded components are used. The ACA connection is a pressure contact connection, and it does not take much force to open it.

Fine leads act like wires that easily bend and do not fight back too much. However, the 0.625- to 1.25-mm (25- or 50-mil) pitch leads used on many standard components are stiffer and have a greater out-of-plane deviation because of their larger scale. If bent into the plane, the larger leads will have enough spring-back force to fail over time. Also, the out-of-plane leads will continue to exert force after bonding. During accelerated heat aging or humidity exposure, the adhesive's polymer tends to relax and may allow the lead under tension to move enough to cause an electrical failure called an *open.*

Anisotropic Films. Most anisotropic adhesive is supplied as a dry film coated on a release liner with a protective release film on top. The product may be precut to size or supplied in sheet form that is cut by the user. Some automatic bonders have built-in cutters and use rolls of anisotropic adhesive film. A few products require no cutting—heat during the bonding process allows the adhesive to soften or liquefy enough to break cleanly.

To apply the adhesive film, the bonder first removes the top release layer. Next, it presses the exposed adhesive against the bonding site and quickly tacks it in place using minimal heat and pressure.

Just before component placement, the remaining release layer, which protects the adhesive from contamination, is removed. The component is then aligned with the circuit, held against the adhesive film under pressure, and heated for a specified time.

Anisotropic films have been found to be reliable only when the component leads are thin, compliant, and coplanar or the substrate is highly compliant, like a flexible circuit.

Oriented Conductor Films. A limited number of vendors are beginning to supply oriented conductor films that have conductors located in specific patterns and pitches. Sheets of anisotropic oriented conductor films can be die cut to match components having the respective lead configurations.

Compared with placing anisotropic films, the patterns on the oriented films require alignment with the component leads or bumps for BGAs, adding a registration step.

16.3 COMPARISON OF ADHESIVES WITH SOLDERS

Soldering, an ancient metallurgic joining process, is still the most common assembly method used in electronics. And tin-lead solder is the industry standard. A solder junction contains metal phases and intermetallic compounds, all of which conduct electrically (Fig. 16.13).

The solder joining process requires oxide removal on all metal surfaces with a cleaning agent known as a *flux.* During component assembly, the thin surface finishes on the circuit and the components dissolve into the solder and become part of the junction alloy. At the same time, the solder bonds directly to the metal surfaces of the components and the circuit board. After soldering, at least until recently, most fluxes were removed with solvent cleaners. But now, water-cleanable fluxes and no-clean fluxes exist.

While soldering dominates electronic assembly and is expected to in the future,[1] conductive adhesives offer some advantages, including

- Lower-temperature processing
- Finer-pitch printing

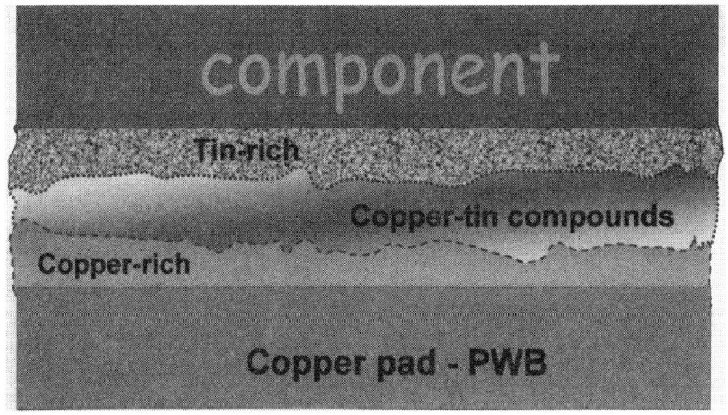

FIGURE 16.13 Solder joint with lead removed for simplicity.

- Less material use
- Thermomechanical fatigue resistance
- No alpha-particle emission
- Less environmental impact

Still, polymer-based adhesives differ significantly in terms of processing and performance, including

- Lower mechanical strength (shock)
- Lower surface tension (not self-aligning)
- Higher contact resistance with oxidized surfaces (some adhesives)

Despite all the advantages and disadvantages associated with using conductive adhesives instead of solder, the assembled product must be electrically and mechanically reliable. To ensure reliability in actual use, conductive adhesives have to maintain bond strength and junction stability after heat aging, thermal cycling, and temperature and humidity testing (Table 16.1).

16.3.1 Reliability

The most important difference between conductive adhesives and solder is that adhesives' mechanical and electrical properties are mostly independent (Table 16.2). For ICAs, a number of reliability issues have been identified, including increased contact resistance values of components with tin-lead finishes after temperature and humidity testing and cracked joints after thermal cycling and drop testing.[4,5]

Mechanical. Unlike solders, ICAs do not form metallurgic interfaces. The electrical pathway is through the adhesive's conductive particles, which contact one another, and the component and circuit surfaces. The overall result is the creation of numerous pathways. And each path has a large number of mechanical contacts.

Although many reinforced polymers are stronger than solder, the high filler loadings are beyond the strength-reinforcing level for isotropic adhesives, which usually have less cohesive strength (within the adhesive) and at times poor adhesive strength (between the adhesive and the substrate).

Acceptable bond and shear strength values are being reported.[5] Still, conductive adhesive bonds do not equal the mechanical values produced by metallurgic junctions. The bond strengths of ICAs range from about 2000 to nearly 5000 psi. And pull-strength, shear, and torque values may be significantly lower than solder.

Another notable difference between adhesives and solders is their surface tensions. Molten solder has very high surface tension. It will wick up and form fillets on bare metal surfaces, causing

TABLE 16.1 Comparison of Solder and Conductive Adhesive Junctions, Typical Values

Characteristic	Tin-lead solder	Isotropic conductive adhesive
Volume resistivity	0.000015 Ω·cm	0.00006 Ω·cm
Junction resistivity	10–15 mΩ	<25 mΩ
Thermal conductivity	30 W/m·K	3–5 W/m·K
Shear strength	100%	40 to 110%
Mechanical shock	Good	Poor
Temperature and humidity (85%/85°C)	No change	Product specific
Finest pitch	0.3 mm (12 mils)	Down to 0.15 mm (6 mils)
Minimum process temperature	210 to 220°C	25 to 150°C
Environmental impact	Negative	Minor
Thermal fatigue	Yes	Can be minimal

TABLE 16.2 Typical Isotropic Conductive Adhesive Performance on Gold-Plated Circuits with Tin-Lead Surface-Mount Components (Results from Alpha Metals, Delta, and DuPont)

Conditions	Electrical	Mechanical
85% RH and 85°C	Poor to good*	Good
Heat aged (150°C)	Good	Good
T-cycle (−25 to 90°C)	Good	Good
T-cycle (−40 to 150°C)	Poor to good	Fair to good†
Volume resistivity	Good	NA

*Largest variable in commercial products.

†Many adhesives outperformed solder.

components to self-align with circuit pads. Adhesives, however, have very low surface tension—they will not self-align or orient components. As a result, adhesives require more accurate component placement than solders. This also means that components that have been designed for solder wicking, particularly those with J leads, are poor choices for adhesives (Fig. 16.14). While J-lead components can be bonded successfully with adhesives, mechanical strength is always lower than with solder bonding, which may be more than twice as strong.

Chip resistors, gullwings, and flatpacks show smaller differences. A typical isotropic adhesive bond strength for a chip resistor will be 80 to 110 percent the strength of a solder joint. Mechanical shock testing, such as dropping a loaded board on edge from a specified height, also shows adhesives to be weaker, especially for larger components.

Thermomechanical Fatigue. One of the major limitations of tin-lead solder is metal fatigue during thermal cycling of assemblies that have a thermal mismatch—a different coefficient of thermal expansion (CTE) between components and the circuit board. Temperature changes cause differential expansion, which stresses the solder joint unless the circuit is very thin and flexible. Over time, solder work hardens and develops a coarse grain structure until it becomes brittle and fractures—a major deficiency.

Polymers do not work harden during temperature cycling, although high-modulus (stiff) materials can fracture under high stress. High metal loadings also make ICAs behave as if they were much stiffer than the polymer's modulus would indicate.

Researchers at DuPont have reported poor temperature cycling results for several conductive adhesives.[7] They also found that high bond strength after thermal cycling did not guarantee good electrical results, suggesting the independence of adhesives' mechanical and electrical properties. A study comparing bond strength and junction resistance (after temperature cycling) showed that there is an optimal modulus range where adhesives perform well mechanically and electrically.[8]

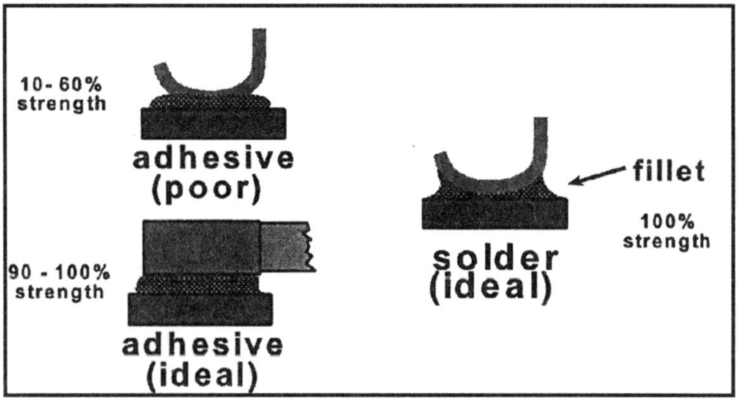

FIGURE 16.14 Component geometry effects.

High-modulus (hard) materials failed bond strength and electrical testing after thermal cycling. Initially, the hard adhesives fractured at the circuit conductor bonds. Further cycling caused components to break off, much the same as happens with solder fatigue.

Low-modulus (soft) materials appeared to absorb stress energy well. Mechanical bonds remained strong. However, the soft materials showed junction instability after cycling. The proposed mechanism for electrical failure in soft adhesives is conductive particle encapsulation by the polymer.[7]

The midrange-modulus adhesives performed well mechanically and electrically. Apparently, the modulus and other physical and chemical characteristics fall within an acceptable range. Good temperature cycle performance is not critical for all applications. Junction stability is.

Electrical. Adhesive junction resistance for common surface-mount components typically ranges from 50 to 100 mΩ, modestly high values that are not a problem for most applications. Still, the adhesive height between the circuit pad and the component must be controlled accurately because greater spacing will increase junction resistance.

Junction instability, caused by component oxidation, is a major problem when using ICAs. Components with solder or reactive metal finishes assembled with conductive adhesives oxidize, partly because polymers tend to be poor oxygen barriers. As the oxide forms, it acts as a barrier between the adhesive particles and the metal of the circuit and component, causing some adhesives to show an increase in junction resistance.

Humidity aging strongly accelerates oxidation because the combination of heat and water promotes the reaction. The most common aging conditions are 85 percent relative humidity and 85°C for up to 1000 h.

Several approaches to solving the component oxidation problem have been tried. The two most common methods are adding oxide-reducing agents to the adhesive and using oxide-penetrating conductive particles.[1]

Epoxy resins cured with acidic hardeners, such as anhydrides, tend to resist oxide formation. The anhydride produces a mild acid in the presence of moisture and acts like a flux. Over time, the acidic materials get depleted, and junction resistance increases.

The most successful approach to junction stability has been the use of electrically conductive oxide-penetrating particles. The process relies on resin shrinkage during curing to force the sharp conductive particles through the metal oxides.[9] Recent studies have confirmed the effects of resin shrinkage on isotropic adhesive conductivity.[10,11]

Adhesives with oxide-penetrating particles showed no increase in junction resistance on solder-coated boards using solder-coated components and only a slight increase on bare copper during temperature and humidity exposure (Fig. 16.15).

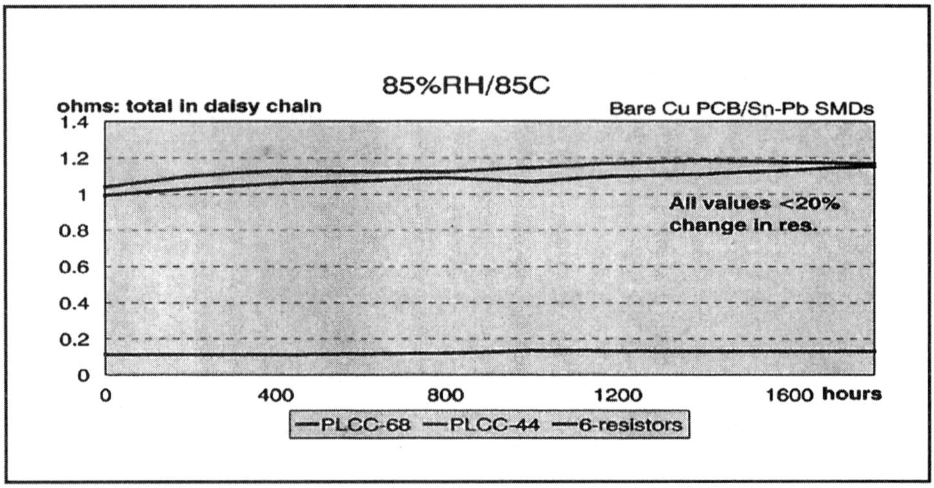

FIGURE 16.15 Change in junction resistance for isotropic particle-penetrating adhesive after temperature and humidity testing. (*Alpha Metals.*)

Oxide formation also can cause junction instability in ACAs. However, since force is used to produce the connection, the conductive particles can penetrate the oxide layers. Some anisotropic adhesives use sharp-edged particles for oxide penetration.

Electromigration. Reactive metals, such as silver or solder, at high humidity and low-voltage dc can undergo an electrochemical reaction called *electromigration* that can cause circuit shorting. Silver migration is a well-recognized problem with silver inks used in the polymer thick-film circuit industry. Silver oxidizes at the surface of the thick film and in the presence of water hydrolyzes to form mobile silver ions. Positive silver ions migrate to a negative electrode under polarized current until a treelike silver growth (dendrite) shorts the circuit (Fig. 16.16).

Electromigration has not been found to occur with silver-filled adhesives at typical digital voltages (5 and 12 V dc). Tests by Poly-Flex (United States), IVF (Sweden), and Delta (Denmark) have not produced dendrite growth even with closely spaced conductive adhesive traces (200-μm space) under dc voltages at various temperature and humidity conditions.[1]

Silver-filled adhesives are less prone to migration than silver inks because of fundamental formulation differences. Silver inks are specifically designed to provide a silver-rich surface. They contain "leafing" silver flakes—treated silver flakes that repel the solvent-binder system. The treated silver floats to the ink's surface on drying. This is a reasonable design for circuit inks that are mainly used for membrane switches. A silver-rich surface provides the maximum surface conductivity required for switch contact. However, a silver-rich surface is also an available source of silver ions.

Conductive adhesives use very different silver particles. Although the particles may be treated to enhance conductivity, their surfaces are not repellent, and silver is not forced to the surface. Besides flakes, silver powders are also used in adhesives, which tend to remain coated with binder, producing a resin-rich surface that encapsulates the silver and makes it less available for migration. Also, cross-linked thermosetting resins greatly reduce silver ion mobility.

Rework. Solder is relatively easy to repair. It melts when heated, allowing for easy component removal and replacement without damage in most cases.

Conductive adhesives can be repaired, but not as easily as solders. Cured thermoset polymers do not melt. Thermosets begin to behave as high-melting thermoplastics at about 50°C (122°F) above their glass transition temperature (T_g). Materials are in the glassy state (ridged) at temperatures below

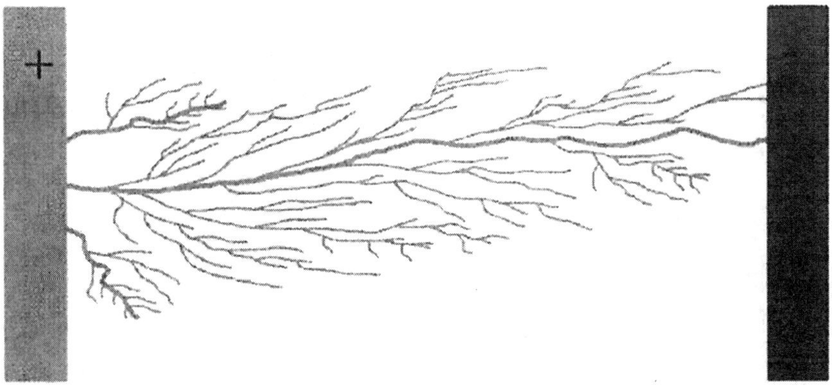

FIGURE 16.16 Dendrite growth.

T_g. Most thermoset adhesives soften at 150 to 180°C (302 to 356°F). Since the adhesive will not melt, some force is needed to remove the component. The heated component may be pulled upward or twisted off. The removal force required for many adhesives is low enough so that no board or component damage occurs.

Cured adhesives cannot be reused for bonding, although uncured adhesive can be placed over cured adhesive. Replacing fine-pitch components requires that the bond sites be planed or flattened mechanically. Wide-pitch, 1.25-mm (50-mil) components typically do not need to have the circuit pads flattened after removal. It is only necessary to apply paste over the cured adhesive before placing the new component.

Assembly defects caused by insufficient material or bent leads can be repaired by leaving the component in place, adding more adhesive, and curing. Cured adhesive can tolerate many baking cycles.

Environmental. Given the absence of lead or any other hazardous metal, adhesives are the most environmentally acceptable joining material available. State-of-the art adhesives are intrinsically clean. When using oxide-penetrating conductive adhesives, no flux or cleaner is needed on any type of board or component. All common electronic finishes have shown good stability, including tarnished copper and copper coated with organic solderability preservatives. Still, components and circuit lands must be free of oil, grease, mold-release compounds, and most organic contaminants.

The Institute for Interconnecting Packaging and Circuits (IPC) group has warned of impending worldwide legislation to replace lead-containing solders and that lead-free solders likely will require assembly temperatures as high as 260°C, a temperature too high for the low-T_g laminates used in consumer electronics.[2]

The low processing temperatures for conductive adhesives make them excellent lead-free alternatives for some applications. An additional benefit to using conductive adhesives in place of lead-containing solders is that conductive adhesive fillers do not emit alpha particles. Lead impurities, such as polonium, emit alpha particles through radioactive decay that can cause functional errors in the nanoscale circuits found in microprocessors and memory chips.

16.4 FLIP CHIP ASSEMBLY

Flip chip (FC) dates back to the early 1960s when the semiconductor industry was figuring out how to package silicon chips, including simple transistors. At that time, two methods emerged, wire bonding and FC. IBM pursued FC on ceramic circuits, called C4 for *controlled collapse chip connection,* for assembling very high density multichip modules (Fig. 16.17). FC, compared with wire bonding, reduced the number of interconnects by half.

Now that FC technology is middle-aged, it is finding wide acceptance on organic substrates (bismaleimide triazine, FR-4, polyimide). The transition from ceramic to organic substrate has required several new developments, including underfill. Underfill is the polymeric answer to the thermal-mismatch problem that occurs when low-expansion chips are joined to relatively high-expansion organic-bound circuitry. The underfill essentially locks the chip to the substrate, restrains *xy* movement, moves the stress to the circuit layer, and protects the chip bumps from thermomechanical strain. Even thin flexible circuits require underfill to obtain high thermal cycle performance.

Underfill also overcomes the low mechanical strength sometimes seen with silver-filled adhesives. The conductive adhesive and underfill combination has high compatibility because both are polymer-based systems.

16.4.1 Wafer Bumping

While solder bumps have been the standard for connecting FC on ceramic, many new interconnect materials are emerging for use with organic substrate. The standard aluminum bonding pad found on most chips is neither solderable nor compatible with conductive adhesives. Electroless nickel plating, mechanical gold studding, and conductive polymer bumping are, however, compatible with conductive adhesives. The joining method often will dictate the type of bumps required on the FC.

Electroless nickel with an immersion gold finish (nickel-gold) is a low-cost bumping option for all types of conductive adhesives. Gold stud bumps are formed by attaching gold wires to the chip's aluminum pads. The wires are then pressed against a flat surface (glass plate) to make the gold studs the same height.

The polymer bumping method requires plating the wafer's aluminum bond pads with a thin layer of nickel-gold or palladium.[13] The bumps are then formed by stencil printing ICA on the plated pads and curing the adhesive at a temperature around 150°C (302°F) (Fig. 16.18). After curing, the wafer is sawed into individual chips ready for FC attachment.

16.4.2 FC Attachment

Both ICAs and ACAs can assemble FCs. Each type of adhesive has advantages and limitations. ICAs produce strong bonds with good electrical performance but must be applied selectively. ACAs can be applied to the entire bond as a film or as a printable paste. Most anisotropics are processed in dry form. The anisotropic process is slower because the FC is bonded in the placement machine.

Isotropic Paste. Isotropic adhesives have a long history and good performance record for surface-mount assembly to flexible circuits. If you imagine an FC as a surface-mount component, assembly with ICAs obviously fits. Silver-filled epoxy-based adhesives are the most popular for FC assembly.

FIGURE 16.17 1960s flip chip. (*IBM.*)

FIGURE 16.18 SEM of a polymer bump over palladium wafer plating. (*KSW-Microtec.*)

However, the adhesive should be made with finer conductive particles—1 to 4 μm (0.04 to 0.16 mil)—than typically required for surface mount because of the much smaller FC junctions.

The substrate's fine FC contact pads require precise adhesive application. Stencil printing has proven adequate for printing adhesive features down to 75 μm (3 mils), provided a high-quality laser-cut or electroformed stencil is used having a thickness of 4 μm or less. Poor printing, due to poor release, will result if the stencil is too thick (high wall-to-open ratio).

After printing, the chip is placed and then cured at the lowest temperature possible. If the chip and substrate are locked together at too high a cure temperature, the cooling stresses can fracture the adhesive joints. Lower-modulus adhesives can greatly reduce cool-down fracturing.

FCs attached with ICAs alone are not robust. The small bond areas do not provide high mechanical strength. And the large thermal mismatch introduced by bonding a low expansion silicon chip to a polymer film will cause joint failure during temperature cycling. The same situation occurs with solder-joined FCs on organic substrates. The answer is to fill the gap between chips with underfill, a slow process and an additional process step. Still, underfill shrinkage during curing tends to improve adhesive conductivity by compressing the joint.

Standard capillary underfill can be used as long as its filler (silica) size is at least one-third the chip's gap height. The CTE of the underfill should match that of the adhesive, which can be twice that of solder.

Isotropic pastes also can be applied by the polymer dip-chip process, an adhesive application method where a bumped chip is dipped into a thin reservoir of adhesive, withdrawn, and then placed on the circuit (Fig. 16.19). As a rule, the adhesive reservoir should be maintained at a thickness less than the bump height. This simple process eliminates one registration step.

The polymer dip-chip process can be run on existing equipment by fitting the FC bonder with a dispensing drum. Electroless nickel bumps are difficult to dip because of their low height. Gold bumps may be ideal because they are nonoxidizing and the bump shape can be controlled accurately.

In Japan, the polymer dip-chip process using gold stud bumped chips, silver isotropic paste, and epoxy underfill is called *bump interconnect technology* (BIT).[14,15] Fujitsu has developed an interesting variation on this technology. Briefly, the chips are dipped into silver paste, the underfill is applied to the substrate, and then the chip is pressed against the underfill-coated substrate while heating, simultaneously connecting the chip bumps to the substrate pads and curing the underfill in one step[16] (Fig. 16.20).

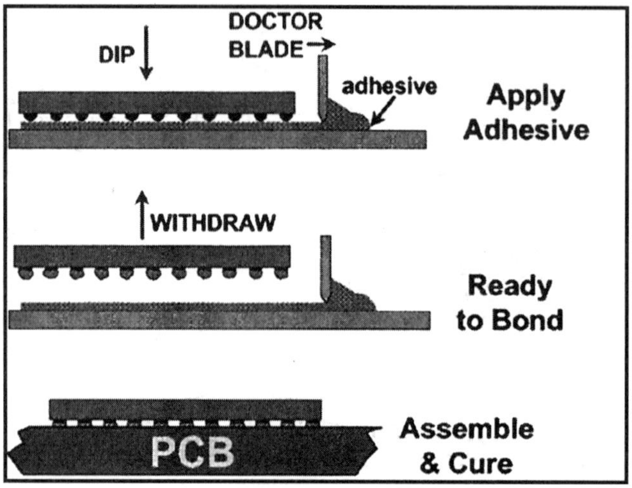

FIGURE 16.19 Polymer dip-chip process.

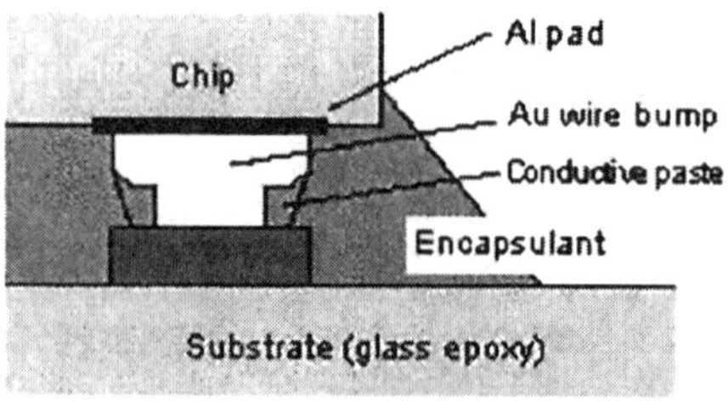

FIGURE 16.20 Cross section of Fujitsu's FC technology. (*Fujitsu.*)

Fujitsu uses this technology in its subnotebook computers to attach the Pentium processor and a three-chip PCI controller to a separate 5 × 5 cm circuit board that plugs into the motherboard. Fujitsu's FC process reduced the typical board area required in notebook computers by at least 75 percent.[17]

Anisotropic Pastes and Films. ACAs have been used in FC assembly for more than a decade. The adhesives have improved steadily as their market, primarily flat-panel displays, has grown. A number of commercial products are available in film form.

For assembly, an anisotropic film is tack-bonded to the circuit, or a paste adhesive is applied to the circuit and dried, or B-staged (partially cured). With the adhesive film or coating in place, the FC is pressure-bonded to the circuit while heating, allowing polymer to flow and conductive particles to connect the chip's bumps and the circuit. Cooling under pressure is often necessary to get a reliable junction.

An example of anisotropic FC assembly is found in the credit card–sized Casio MR-80 radio[18] (Fig. 16.21).

16.5 SUMMARY

ICAs are a real drop-in alternative to solder. They use the same process and equipment. Their low-temperature processing is a major advantage. Still, applications are limited. For wide acceptance, isotropics need improved shock resistance and somewhat better electrical and thermal conductivity.

Good applications for isotropic adhesives include thermally sensitive substrates (Mylar), small components, and FCs. Applications to avoid for now include heavy components on small circuit pads, curved component leads, high mechanical shock, and nonplanar substrates.

ACAs are not well suited for surface-mount applications because they require special bonding equipment. However, they simplify the FC process because no underfill is required.

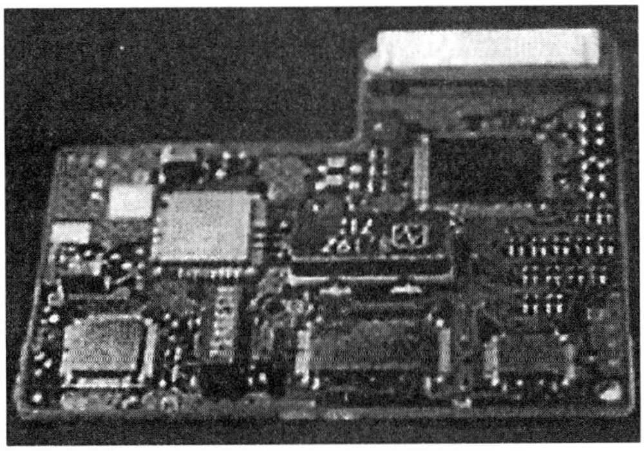

FIGURE 16.21 Multichip module from the Casio MR-80 radio with five FCs assembled with anisotropic adhesive on an organic substrate with gold pads. (*mark Partners LLC.*)

16.6 REFERENCES

1. Gilleo, K., "Poly-Solder: A New Junction-Stable Conductive Adhesive for Rigid Boards," in *Proceedings of Recent Achievements in Conductive Adhesive Joining Technology in Electronics Manufacture,* IVF (Institutet för Verkstadsteknisk Forskning), Gothenburg, Sweden, September 23–24, 1993.
2. Gilleo, K., "Assembly with Conductive Adhesives," in *Proceedings of Surface Mount International,* San Jose, CA, 1994, pp. 279–288.
3. Gilleo, K., "Intrinsically Clean Polymer Bonding: What Are the Trade-Offs," in *Proceedings of Technical Program Surface Mount International,* San Jose, CA, 1993, pp. 655–661.
4. Jagt, R., "Reliability of Electrically Conductive Adhesive Joints for Surface Mount Applications: A Summary of the State of the Art," *IEEE Transactions on Components, Packaging, and Manufacturing Technology, Part A,* 21 (2), June 1998, pp. 215–225.
5. Rubin, H., "Alternate Interconnect Methods Using Conductive Adhesives," in *Surface Mount International Proceedings,* 1993, pp. 748–752.
6. Gilleo, K., "The Polymer Electronics Revolution," in *Proceedings NEPCON West,* February 1992, pp. 1390–1402.
7. Keusseyan, L., and Dilday, J., "Electric Contact Phenomenon in Conductive Adhesive Interconnection," in *International Symposium on Microelectronics (ISHM) Proceedings,* VA, November 1993, pp. 44–49.
8. Gilleo, K., "Evaluating Polymer Solders for Lead-Free Assembly, Circuits Assembly," January 1994, pp. 52–54, 56.
9. Gilleo, K., Corbett, S., Corey, M., "High Stability Solderless Junctions Using Advanced Conductive Adhesives," in *Proceedings NEPCON West,* February 1991, pp. 193–212.
10. Wong, C. P., and Lu, D., "Conductivity Mechanisms of Isotropic Conductive Adhesives (ICAs)," in *1999 International Symposium on Advanced Packaging Materials,* Braselton, CA, March 1999, pp. 2–10.
11. Wong, C. P., and Lu, D., "Effects of Shrinkage on Conductivity of Isotropic Conductive Adhesives," in *1999 International Symposium on Advanced Packaging Materials,* Braselton, CA, March 1999, pp. 295–301.
12. "IPC Roadmap for Lead-Free Electronics Assemblies," 2d draft, November 1999.
13. Seidowski, T., Kriebel, F., and Neumann, N., *Polymer Flip Chip Technology on Flexible Substrates: Development and Applications,* Adhesives in Electronics, Binghamton, NY, 1998.
14. Kira, H., Kobae, K., Kainuma, N., et al., "Method of Producing a Multichip Package Module in Which Rough-Pitch and Fine-Pitch Chips Are Mounted on a Board," U.S. Patent No. 6,006,426, December, 28, 1999.
15. Omoya, K., Oobayashi, T., Sakurai, W. et al., "Semiconductor Unit Package, Semiconductor Unit Packaging Method, and Encapsulant for Use in Semiconductor Unit Packaging," U.S. Patent No. 5,641,996, June 24, 1997.
16. Baba, S., "Low-cost Flip Chip Technology for Organic Substrates," *Fujitsu Science and Technology Journal,* 42(1), September 1998, pp. 78–86.
17. "Subnotebook Computer—Fujitsu Biblo NC," *Prismark Partners LLC Bulletin,* January 1998.
18. "Casio MR-80 Radio—First Use of AFAC Flip Chip," *Prismark Partners LLC Bulletin,* October 1997.

S·E·C·T·I·O·N · 3

EQUIPMENT AND PROCESSES

CHAPTER 17
NEXT-GENERATION FLIP CHIP MATERIALS AND PROCESSING

Daniel F. Baldwin, Ph.D.

17.0 INTRODUCTION

One of the significant developments to improve cost, reliability, and productivity in the electronic packaging industry has been flip chip (FC) technology. This technology has been enhanced further to enable direct assembly to printed wiring boards (PWBs) in high-volume surface-mount assembly called FC-on-board technology. The FC process was first introduced for ceramic substrates as the solid logic technology by IBM in 1962. In 1970, IBM introduced the controlled collapse chip connection (C4 technology) for integrated circuits (ICs) in high-volume FC-in-package (FCIP) production. FC technology is an advanced form of surface mount technology (SMT) in which bare semiconductor chips are turned upside down and hence called flip chips (i.e., active face down) and bonded directly to a PWB or chip carrier substrate. Development of solder-bump FC interconnections was initiated in an attempt to eliminate the expense, low reliability, and low productivity of manual wire bonding of the era. In contrast to wire bonding, which is a peripheral and time-consuming bonding technique in which bonds are formed sequentially, the FC allows all input-outputs (I/Os) to be connected simultaneously. Like its predecessors, FC technology was applied initially to peripheral contacts but quickly progressed to area arrays, which allow for high I/O counts at larger pitches and reduced die size because solder bumps can be put over active device areas on ICs, unlike wire bonds.

17.1 CONCEPT

FC interconnection is the connection of an IC chip to a carrier or substrate with the active face of the chip facing toward the substrate. Interconnection between the chip I/O and substrate is achieved using a bump structure on the chip and a bonding material typically on the substrate forming an electrical interconnection between the chip and the substrate. FC bonding typically involves solder interconnections that make the electrical and mechanical connection between the chip and the carrier although alternate material systems such as conductive adhesives are also used. A schematic representation of an FC interconnection configuration is shown in Fig. 17.1 showing a bumped chip interconnected to a substrate with the active face of the IC toward the substrate surface.

As an example, IBM's C4 FC technology uses high lead solder—usually 93 percent Pb, 7 percent Sn—bumps deposited on solder wettable metal terminals on the active surface of the semiconductor chip that connect to matching wettable substrate pads. During chip assembly, the solder bumps are aligned to the corresponding substrate metal pads and then reflowed at high temperature to simultaneously form electrical and mechanical connections. During reflow, the wetting action of the solder driven by surface tension forces will align the chip's bump pattern to the corresponding substrate pad.

To better understand the FC interconnection system, it is helpful to systematically break down the structural elements of the assembly. One such breakdown is shown in Fig. 17.1. The basic structure of a FC consists of an IC or chip, an interconnection system, and a substrate. The ICs can be made of silicon (the most common), gallium arsenide (GaAs), indium phosphide (IP), silicon germanium (SiGe), etc. The substrate materials could be ceramic (used in IBM's original C4 process), epoxy-glass laminate, polymer thin-film buildup, resin-coated copper (RCC) buildup, glass, silicon, dielectric-coated metal, liquid-crystal polymer, metal-matrix composite, low-temperature cofired ceramic (LTCC), ceramic thick-film, multilayer high-temperature cofired ceramic, etc.

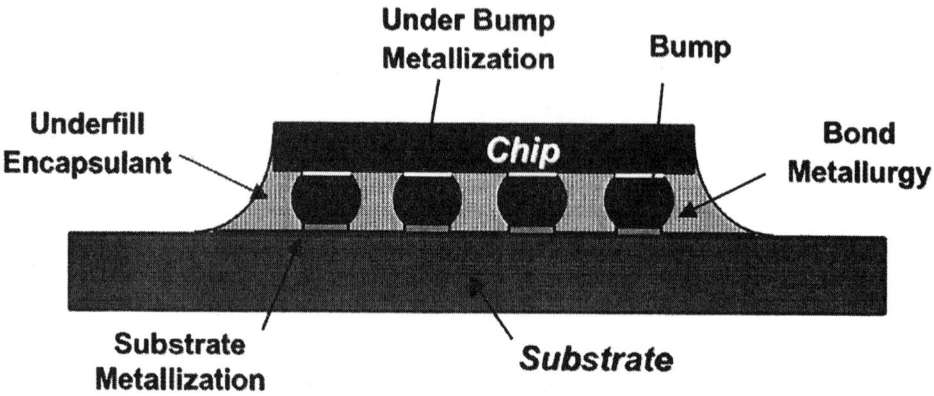

FIGURE 17.1 Schematic of an FC interconnection system.

17.1.1 Limits of Conventional FC Processing

In many cases, the present state of the art FC on-board (FCOB) assembly technology is not capable of achieving the high-throughputs required for integrated high-volume SMT processing and low-cost electronics packaging. Elimination of processing steps enables high-throughput, reduce process complexity, reduce capital equipment requirements, reduce equipment maintenance, and increase process robustness. Processing steps that could be eliminated are flux application, flux residue cleaning, underfill flow, underfill fillet processing, and secondary thermal curing of underfill. Highly populated FC assemblies and large chips compound current FC process limitations. Based on semiconductor and packaging roadmaps, it is clear that trends for larger devices and higher packaging densities will continue, therefore inhibiting cost-effective FC assembly based on conventional FC process techniques. Underfill flow and cure processes dramatically reduce assembly line throughput and therefore exert significant pressure on profit margins. This has been verified by cost modeling and analysis comparing a typical industry FC process, the propposed low-cost, high-throughput process, and surface-mount assembly.[1,16,19,25–30]

The current method for manufacture of FCOB assemblies is time-consuming leading to low throughputs. The processes that increase production time are the flux application process, flux residue cleaning, the underfill dispense process, the underfill fillet dispense process, and underfill cure. The need for underfill derives from the difference in the coefficient of thermal expansion (CTE) between the silicon chip and the organic substrate. The CTE mismatch causes large stresses in the interconnections and eventual failure of the interconnections due to cyclic fatigue induced by thermal cycling. To prevent premature interconnection failure, the space between the chip and the substrate is filled with epoxy, which is then cured, to reduce stress in the interconnection by distributing it over the entire chip area.[5,6,20]

17.1.2 Next-Generation FC Technology Overview

To enable low-cost, high-throughput processing that is compatible with SMT assembly and high-volume electronics packaging, several new processing techniques been developed, including low-cost FC processing based on fast-flow, "snap" cure underfills, high-throughput FC processing using no-flow underfills, and low-cost FC processing based on wafer-applied underfills called wafer-level FC processing.[1,7–11,14,15,19,24–30,36,38–40,42–44]

17.1.2.1 Low-Cost FC Processing Based on Fast-Flow, Snap Cure Underfills. Since underfills are necessary for the reliability of FCOB, key factors to be considered in production are processability (e.g., flow time, cure time, shelf life, flow characteristics) and reliability (e.g., moisture sensitivity, adhesion, thermal mechanical characteristics, corrosion, electromigration). From a processability standpoint, some of the main reasons for incompatibility between FCOB and SMT or FCIP and high-volume electronics packaging are the underfill flow time and the underfill cure time. Conventional underfills can take anywhere from 1 to 10 min to flow under a 5 mm square die while requiring a cure time of 30 to 90 min at temperatures ranging from 125 to 165°C. New commercial fast-flow underfills take between 2 to 30 s to flow under a 5 mm square die. Cure times have also been cut with the introduction of snap-cure underfills. The cure schedules have been reduced to 5 to 10 min at temperatures between 150 and 165°C. A target processing metric for high-throughput FC processing based on fast-flow, snap-cure underfills is a maximum flow time of 5 s on a 5 mm chip and a maximum cure time of 5 min.

The advent of snap-cure underfills has also made possible a new FC assembly process leveraging double-sided board assembly. The new low-cost FC assembly process is shown in Fig. 17.2.[14] The process flow is to print paste for the top-side SMT components, place the SMT components, flux the FC bumps or bond site, place the FCs, reflow the top side forming interconnections, and perform an electrical test. The in-circuit test allows rework of any failed devices prior to dispensing underfill but can be eliminated based on throughput requirements. If there are no defects, the underfill is dispensed on the FC sites, underfill fillets are dispensed if necessary, the board is flipped, the second-side SMT components are processed, and the second-side reflow is used to simultaneously cure the underfill and form the second-side interconnections. The second-side reflow profile is

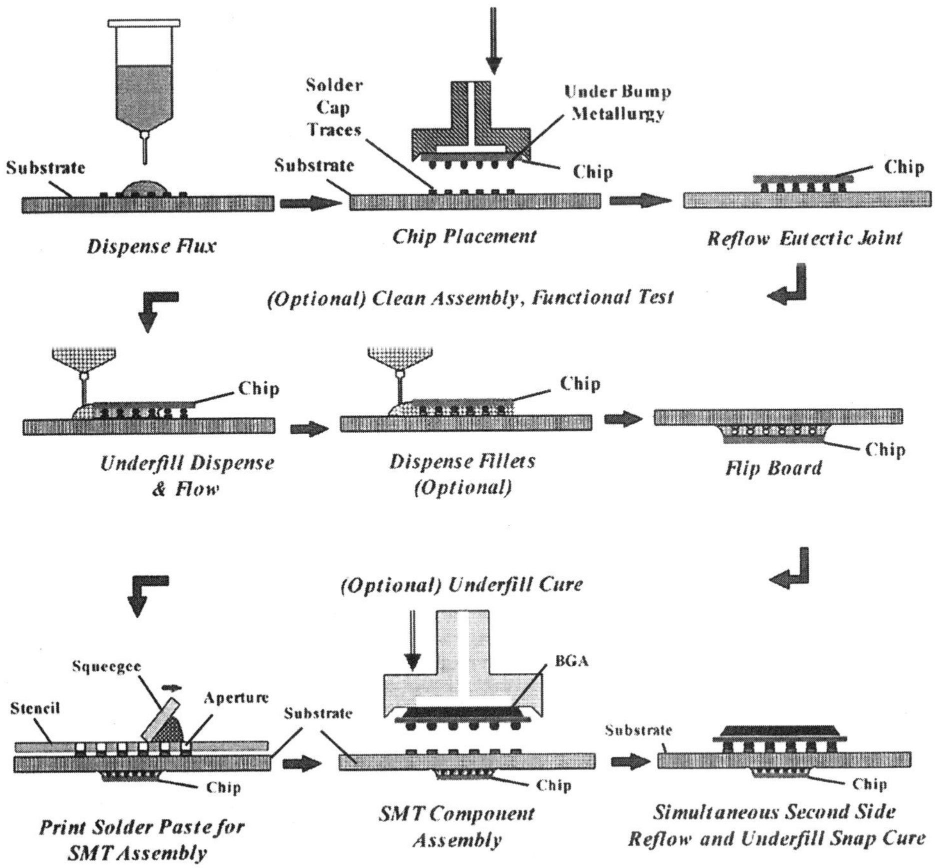

FIGURE 17.2 A fast-flow snap cure underfill process.

slightly modified to accommodate underfill snap-cure in the soak stage of the profile. Typically, this involves increasing the soak time and possibly the soak temperature of a conventional eutectic solder reflow profile. A key element of the process is verifying that the underfills are fully converted during the soak temperature exposure. This is done using differential scanning calorimetry (DSC) analysis to verify underfill conversion. The new FC assembly process is much more conducive to high-throughput, double-sided SMT than conventional FC processes.

17.1.2.2 High-Throughput FC Processing Using No-Flow Underfills. As a concept to achieve high-throughput FCOB and FCIP assembly, a new high-speed FC process has been developed, implementing next-generation FC processing based on large area underfill printing/dispensing, integrated chip-placement and underfill flow, and simultaneous solder interconnection reflow and underfill cure. To enable high-throughput that is compatible with high-speed in-line automated assembly, a new process shown in Fig. 17.3 has been developed.[1,7–10,19,24–30,36,38–40,42–44] Key elements of the

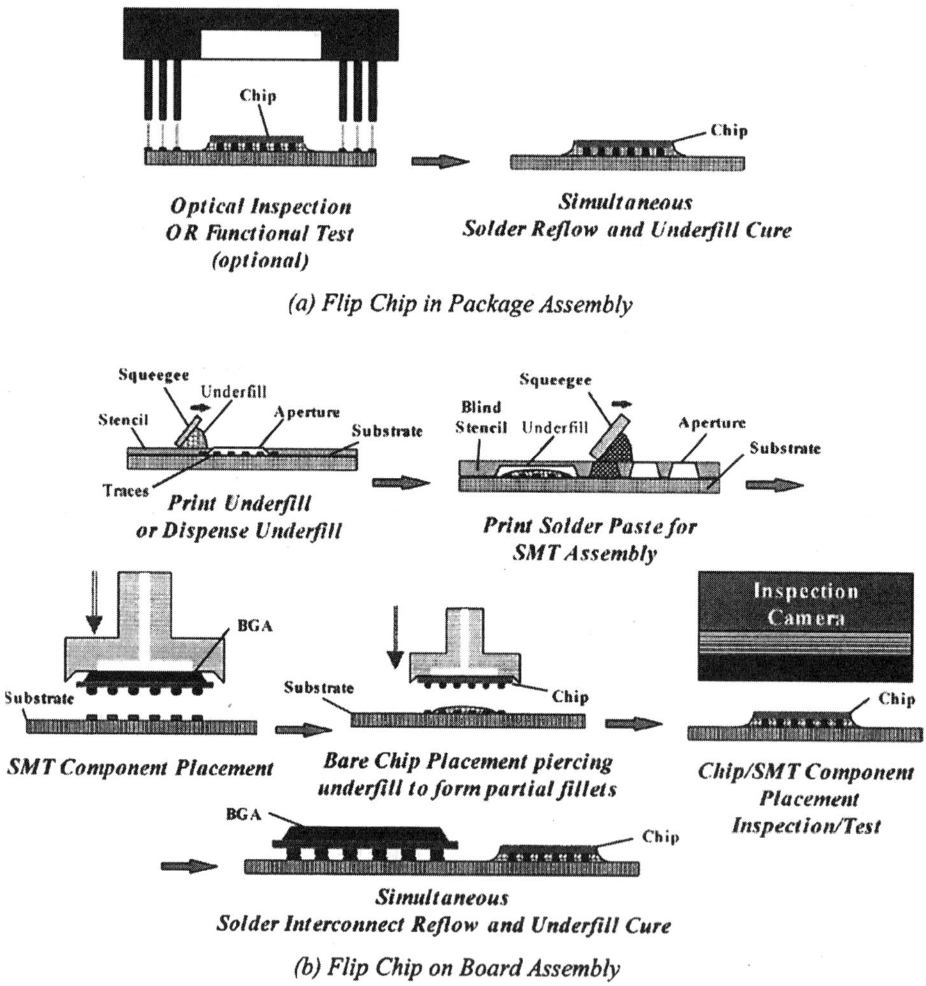

FIGURE 17.3 A high-throughput FC processing using no-flow underfills.

process are elimination of discrete flux application, flux materials management, flux residue cleaning, underfill flow, underfill fillet dispensing, and secondary underfill thermal cure.

The overall process flow begins with a known good substrate and die. A controlled volume of underfill material is stencil printed over the bond pads on the substrate an area-based process compared with conventional point-to-point underfill dispensing. Next, solder paste is printed onto the board using a blind stencil for SMT component assembly. Alternately, the no-flow underfills can be dispensed on the chip site after solder paste printing. Next, the SMT components are placed. The bare chips are then aligned using a vision system to orient the chips relative to bond site fiducials on the substrate. The chips are placed on the substrate, compressing the liquid underfill. Note that the placement forces required are well within those of commercial high-speed FC placement systems and that the process requires no special high-force capability, unlike anisotropic conductive adhesive or compression-bond adhesive processing. A significant advantage here is the elimination of lengthy capillary flow times for underfill processing, particularly with large devices. Finally, the solder interconnections are reflowed simultaneously while curing the polymer encapsulant underfill. Care is taken here to prevent premature gelation of the underfill prior to solder reflow using innovative underfill materials called no-flow underfills that provide latent gelation such that the cure reaction is inhibited until a critical temperature is reached above the solder liquidus temperature. Moreover, it is critical that proper fluxing action be achieved during reflow. The major advantages of the new process are that it increases throughput of FC processing, transforms a chip-based process into an area-based process, reduces the required number of processing steps, and reduces the ratio of wafer to assembly cost by 2 to 5 times over the competing processes.

17.1.2.3 Low-Cost FC Processing Based on Wafer-Applied Underfills: Wafer-Level FC Processing. The key to low-cost and high profit margins in FC assembly is high process throughput. While FC technology has been widely publicized, little attention has been paid to process throughput, whereas the majority of work has concentrated on interconnection technologies. New efforts have focused on the development of process technologies and material systems that could make the FC assembly process transparent to SMT assembly lines and high-volume packaging processes by eliminating FC flux application, underfill application, and underfill cure processes. The innovative process uses reflowable encapsulants similar to no-flow underfills applied to bumped FC wafers forming a wafer scale package and a drop-in solution for low-cost FC assembly.[15]

The new process is unique in that it uses a wafer-level underfill approach, a placement process comparable with chip scale packages (i.e., low placement forces), and underfill systems that enable simultaneous reflow of solder interconnections and cure of the underfill (Fig. 17.4). The basic concept is to apply modified no-flow (reflowable) underfill encapsulants to FC bumped wafers as a form of wafer scale packaging. The wafer-applied underfills dry to a solid film, enabling packing, shipping, handling, and board assembly similar to chip scale packages (CSPs). On reflow, the wafer-applied underfills liquefy, wet the substrate, and provide fluxing action on the solder joints/bond pads, resulting in a high-reliability underfilled FC. The wafer-applied fluxing underfill solid presents several nontrivial challenges, but none of the required properties is mutually exclusive from a materials standpoint.

The new assembly process under development is shown in Fig. 17.5.[11,15] It provides numerous advantages over conventional FC processes because it saves floor space and capital costs by removing underfill dispensers and cure ovens. Moreover, the wafer application of underfill is an enabling technology for fine-pitch FC processing (i.e., ≤ 150 μm). The new process could also increase worldwide FC implementation by reducing the process learning curve and by eliminating special training requirements necessary for current FC dedicated equipment.

17.1.3 Primary Requirements for IC to Package/Substrate Assembly

There are five primary requirements for IC assembly:

1. To provide *acceptable electrical properties,* including capacitance, resistance, and inductance. Each IC assembly has unique electrical characteristics along with mechanical characteristics and manufacturability. For example, wire bonds have long lengths and parallel proximity, resulting in high impedance and longer signal delay times.

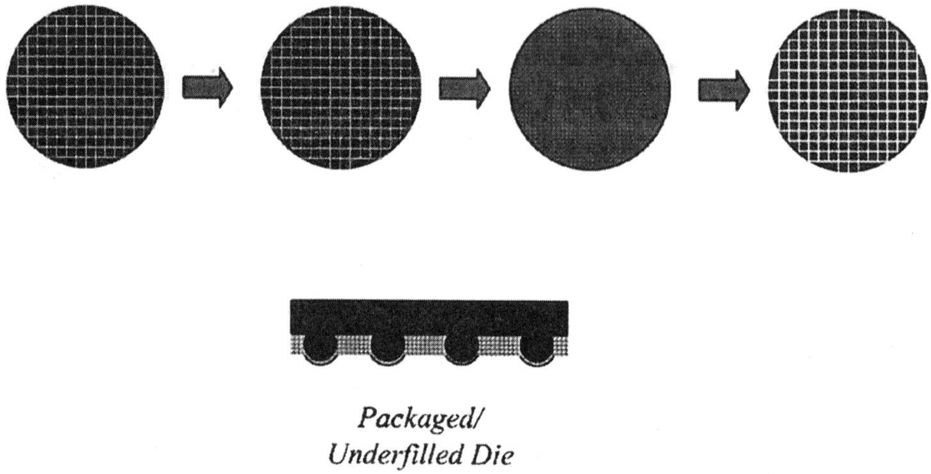

FIGURE 17.4 Wafer-level processing of FC underfills.

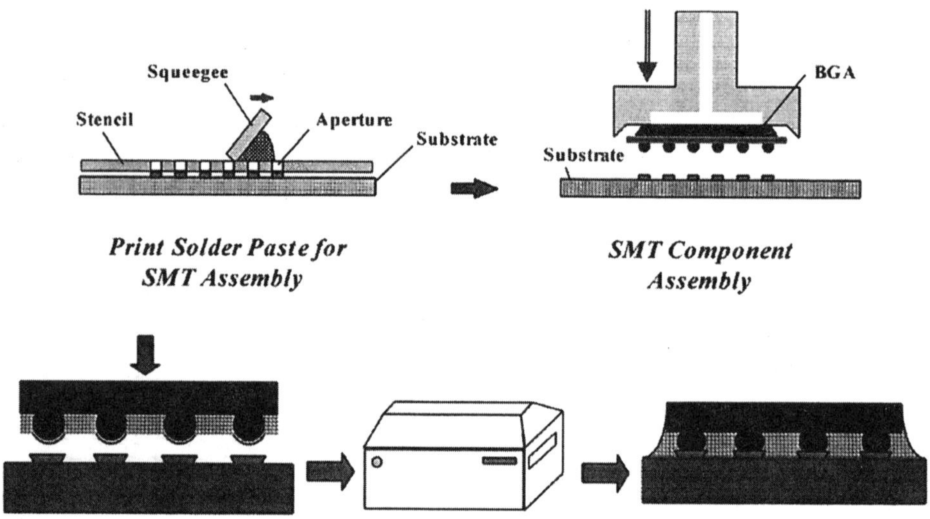

FIGURE 17.5 Wafer-level FC assembly process.

2. IC assembly technologies should provide a *low-cost* solution for providing the electrical interface between the chip and package. Tape automated bonding (TAB) is not typically a low-cost interconnection technique. It requires specialty tooling, equipment, and circuit tapes for production, significantly increasing cost.
3. The third requirement is *high-throughput manufacturing.* For finer pitch packaging applications, TAB provides a low-cycle-time, high-throughput IC assembly approach.
4. The fourth requirement is *high reliability.* FC on-ceramic technology has been a highly reliable interconnection technique. For instance, IBM has used this IC assembly technology for over 40 years with no reported field failures due to thermal cycle fatigue.

5. The final requirement is repairability or replaceability, where the interconnection between the IC and package should provide for removal of a failed IC and replacement with a high-quality new IC or a new part number.

These functional requirements must be met by all IC-to-package/substrate interconnection methods, whether it be wire bonding, TAB, wafer-level packages, or FC interconnection. It is therefore important to consider the degree that each of these functional elements can be met by a particular FC assembly process, particularly in terms of next-generation processes.

17.2 COST ANALYSIS

In order to assess the cost implications of underfill processes and of FC processing in general, activity-based cost modeling has been performed.[1,16,25–27,29,30] Activity-based cost analysis is used to quantify the costs associated with each step of the FC assembly process. Cost model predictions estimate a two to six fold cost reduction over a conventional FC on board assembly process[26,27,29,30] using the new low-cost, high-throughput FC processes of Fig. 17.3. Moreover, the low-cost next-generation FC process also reduces the process cycle time by a factor of two to six. In this case, the conventional process uses capillary flow underfill processing, whereas the low-cost process uses the compression flow underfill process.

17.2.1 Conventional FC Processing Using Capillary Flow Underfills

For the analysis of the conventional FC assembly process, the process flow outlined in Fig. 17.6 is used. A silicon wafer with appropriate solder-wettable metallization is bumped with high-melting-point lead-tin solder, and the bumps are reflowed. The wafer is cleaned and is then diced into individual chips prior to placement. Bond metallurgy made up of a low melting point lead tin solder (typically eutectic) is deposited onto the substrate circuit traces by electroplating or stencil printing. In the case of solder plated substrates, a rosin or no clean flux is deposited onto the chip site, and the chip is placed onto the substrate. The tacky flux holds the aligned chip in position prior to reflow. The assembly is reflowed, and then cleaned to remove any flux residues. Underfill material is dispensed along a single or two adjacent edges of the chip, and is drawn underneath the chip by capillary action. The underfill fillets are applied, and the material is cured in an oven. Modeling of the process was performed for a 5000 panel per day throughput with 30.5 cm (12 in) square substrates populated with 10 1.27 cm (0.5 in) square devices. Cost breakdown results of the baseline process are shown in Fig. 17.7.

Electroplating of the solder (for the interconnection) onto both the wafer (for bumping) and the substrate (for a solder cap) accounts for almost thirty percent of the total assembly cost. These two electroplating steps are expensive due to their limited batch sizes, process time, waste stream environmental clean up, and costly equipment. It was also found that cleaning during FC assembly was expensive due to the numerous cleaning steps involved in the process as well as the limited batch size. The underfilling process accounted for nearly thirty percent of the total assembly cost resulting from the slow processing time, added capital equipment requirements, and material costs.

17.2.2 Low-Cost, High-Throughput FC Processing Using No-Flow Underfills

Cost analysis of the low-cost, high-throughput FC assembly process using the same operating parameters resulted in reduction of the total assembly cost. A breakdown of the costs associated with each process step of Fig. 17.5 is shown in Fig. 17.8. It shows a reduction in process time over previous processes. A cost reduction of approximately 60 percent from the previous benchmark process was achieved as shown in Fig. 17.9. One of the major cost components in the new process was underfill processing, but the overall process cost was reduced. The model showed that the material cost for the underfill step was the major contributor in its expense accounting for approximately ninety percent of the underfill processing costs.

Figure 17.9 shows a comparison of the cost for the conventional FC on board process (specified as the benchmark process), an initial process concept using no-flow underfills, and the low-cost,

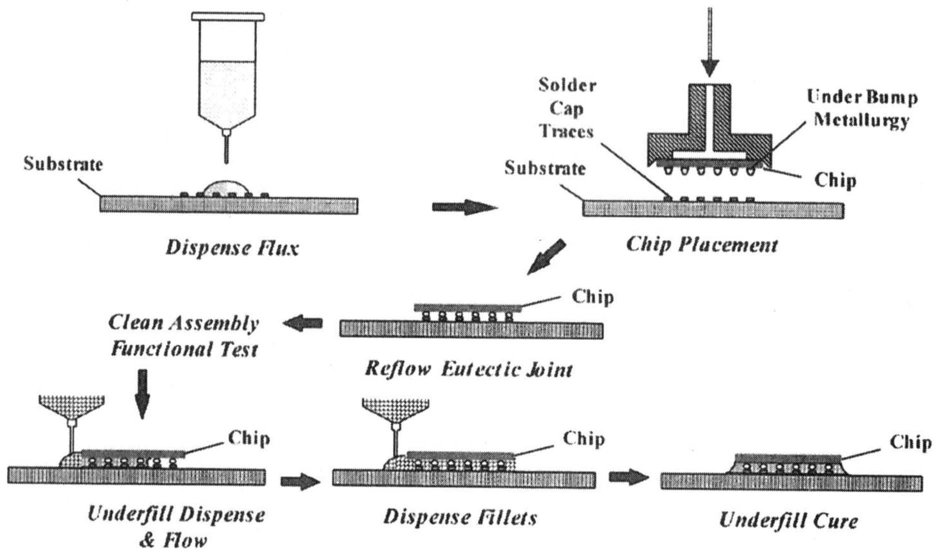

FIGURE 17.6 Conventional FCOB assembly process flow.

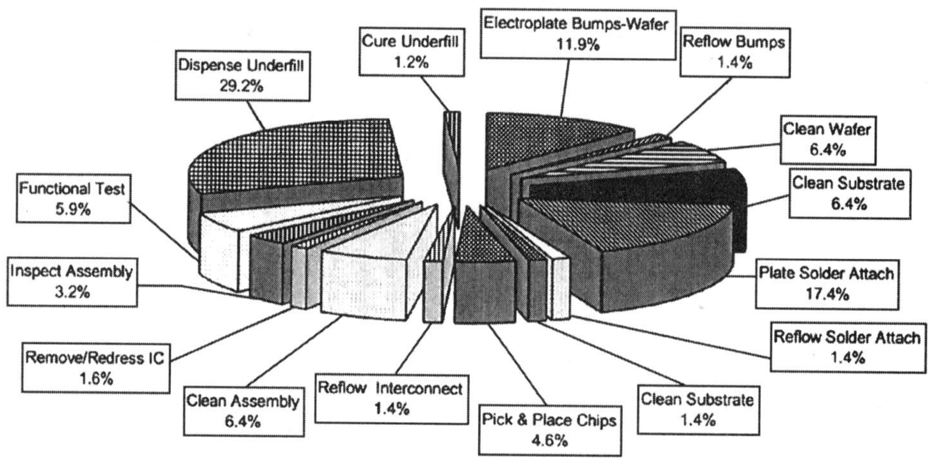

FIGURE 17.7 Breakdown of baseline process costs based on activity based cost model.

high-throughput FC assembly process (specified by the next-generation process). The initial concept process reduced the assembly cost by only twenty percent while the refined process concept reduced cost by approximately sixty percent, representing a significant reduction in assembly process cost. Furthermore, process time has been reduced by over fifty percent as shown in Fig. 17.10. This is mainly due to the reengineering of the process from a point-to-point process to an area-based process. It is important to note that cost and cycle time changed significantly with changes in panel and device sizes.

Cost savings in the low-cost, high-throughput FC assembly process stem from a number of factors. The next-generation process does not require nitrogen inerting during reflow because the encapsulation underfill protects the solder/metallization from oxidation during reflow. Underfill material costs are potentially lower for the high-throughput process in that fillers may not be needed and any

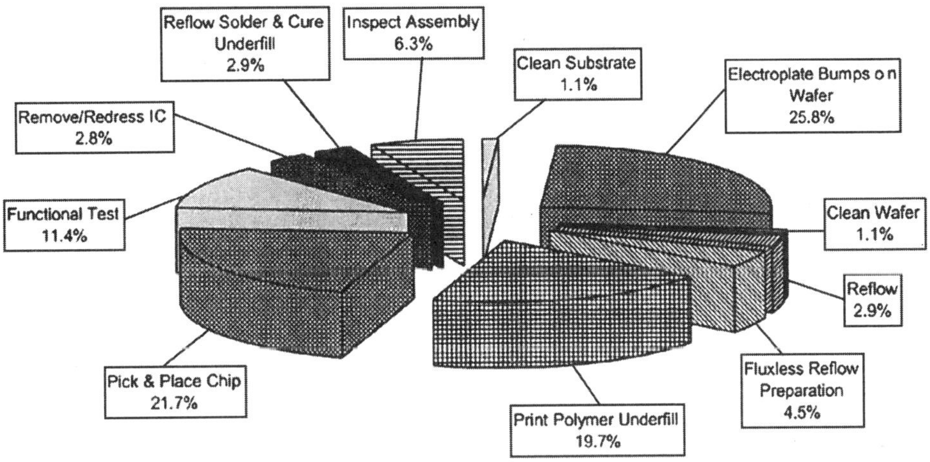

FIGURE 17.8 Breakdown of low-cost high thoughput FC assembly process cost, based on activity-based cost model.

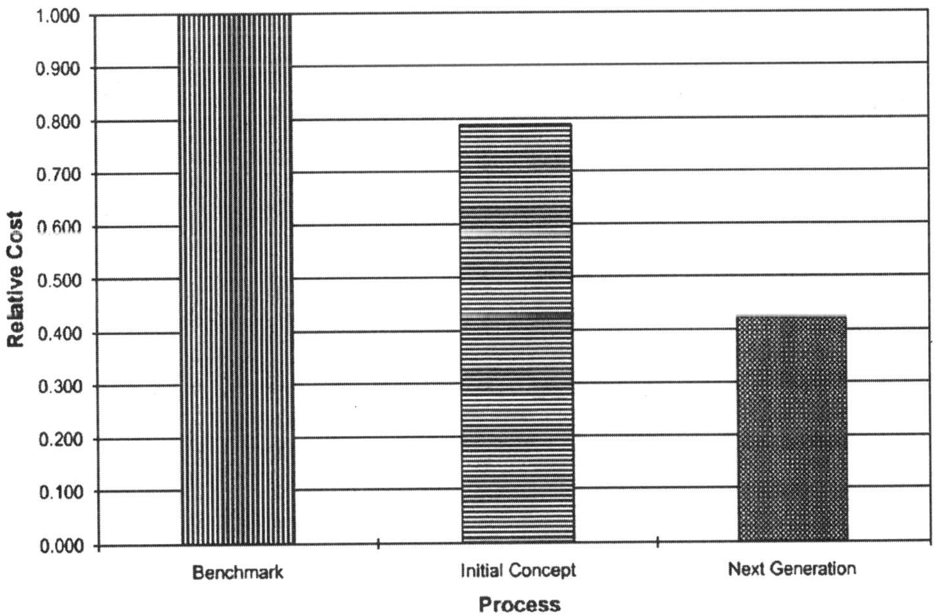

FIGURE 17.9 Comparison of the relative wafer-to-assembly cost for the baseline, initial low-cost concept, and the low-cost next-generation FC assembly processes.

fillers required would not need precision bimodal distributions of spherical particles. Precision filler particles are major cost drivers in conventional capillary flow underfills. Capital equipment and engineering support costs also tend to be lower for the next-generation process due to the reduction of process steps and the ability to use underfill application techniques (stencil printing and dispensing) requiring less precision than for conventional processing. This ultimately reduces the engineering support required for the underfill process.

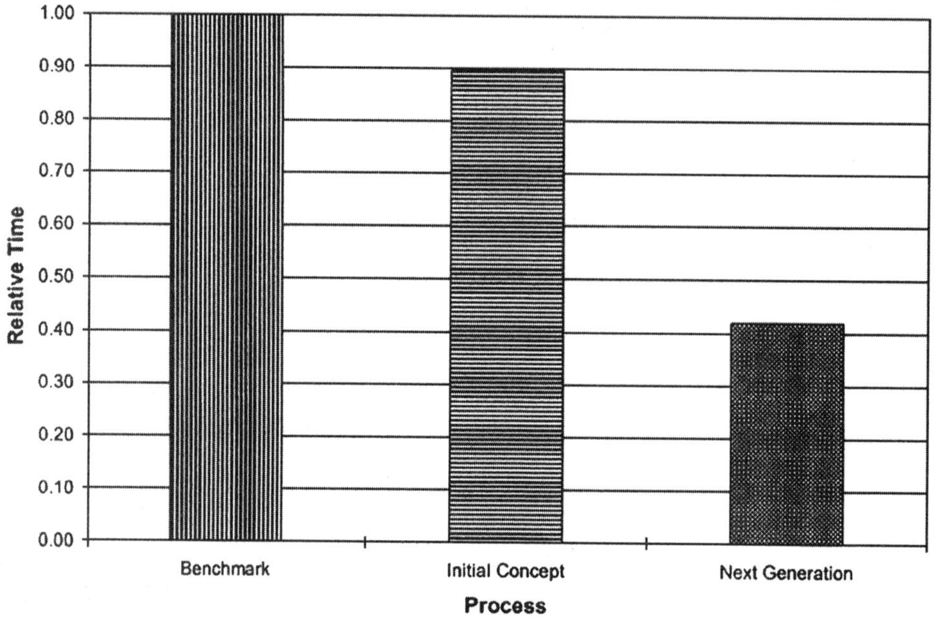

FIGURE 17.10 Comparison of the relative wafer to assembly process cycle times of baseline, initial low-cost concept, and the low-cost, next-generation FC assembly processes.

17.2.3 Cost Implications of Throughput, Chip Size, and Number of Chips per Panel

In order to further understand the cost benefits of the low-cost, high-throughput FC process, cost analysis was used to analyze and compare the conventional and high-throughput processes.[1,25–27,29,30] Parametric studies are presented in Figs. 17.11 and 17.12 illustrating the relative costs of assembly from the bare wafer-level to the final board assembly. The parameters varied included throughput, chip size, and number of chips per panel. The modeling parameters included a 5000-panel-per-day throughput with 30.5 cm (12 in) square substrates populated with ten 1.27 cm (0.5 in) square devices.

The effects of throughput on assembly cost are shown in Fig.17.11. As the throughput increases, the assembly cost decreases. At throughputs below 3000 panels per day, the assembly cost is dramatically decreased with increasing throughput. At throughput levels above 3000 panels per day, relatively little cost benefit is derived with increases in throughput. Moreover, notice that the high-throughput FC process achieves nearly a 60 percent cost reduction over the conventional process over the entire throughput range studied. The substantial reduction in assembly cost for the next-generation FC process is largely due to its characteristically low cycle times. These low cycle times are achieved by the elimination of process steps (relative to the conventional process) and transformation of the FC assembly process into an area array process. While not directly obvious from Fig. 17.11, it is interesting to note that throughput has by far the greatest impact on profitability. Even though the assembly cost decreases relatively little at higher throughputs, relatively small increases in throughput can significantly increase profit margins.

The size of the FC devices used in the large area panel assembly also has a significant effect on assembly cost. Increased device sizes will be characteristic of future device designs and performance demands, as predicted by industry roadmaps. Figure 17.12 shows the results for relative assembly costs with increased device size. The curves for the conventional and high-throughput assembly processes diverge in this plot. Divergence signifies that the next-generation process is very cost effective over the baseline process for increased device size. For smaller chip sizes, a 5 times cost reduction is achieved. As chip size increases, the cost reduction increases to 6 times. Elimination of the time-consuming capillary flow process and transformation of the assembly process from a point-to-point process to an area-based process are the major factors leading to this dramatic cost reduction. In general, printing of the underfill materials enables processing of the entire panel in a single process

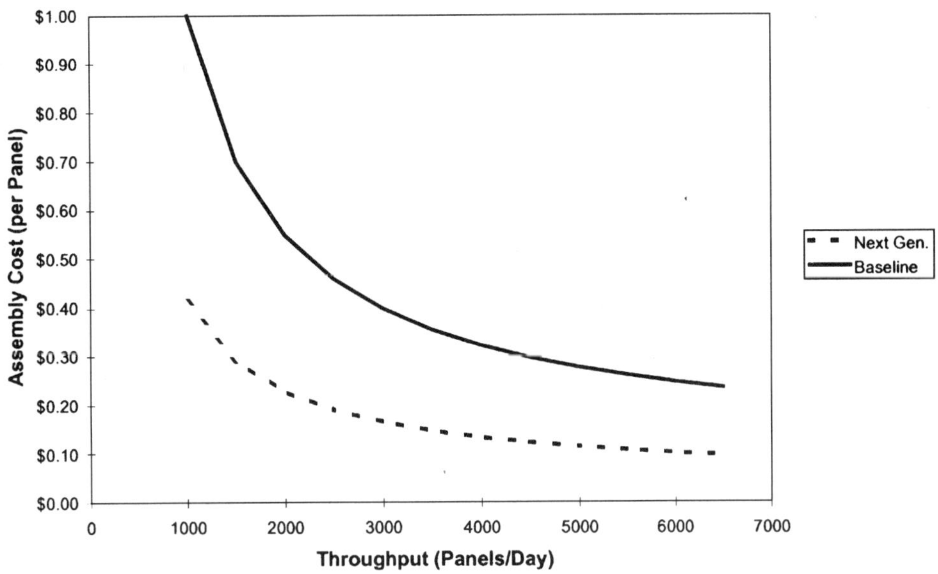

FIGURE 17.11 Effect of throughput on wafer to assembly cost.

step. In contrast, underfill processing based on dispensing processes the individual chip sites on a point-to-point basis, requiring excessive cycle times. In addition, the time-consuming underfill process is essentially eliminated in the next-generation FC process. This follows because compression flow of the underfill occurs simultaneously with chip placement, a required element of the FC assembly process regardless of the underfill process used. The compression flow underfill process also eliminates the chip size effect on underfill processing time. Typically for standard capillary flow underfill processing, underfill flow times increase with the square of the chip size. For compression flow underfill processing, this size effect is eliminated.

Figure 17.13 shows the effect of die per panel on assembly cost. Again, the curves for the two processes diverge, suggesting that the next-generation process will be more cost effective in meeting future miniaturization requirements. In this plot, cost reduction ranged from 2 to 3.3 times, and will continue to increase as more die are added to the panel. Area-oriented processing makes it so that the number of devices per panel has a minimal effect on cost in the next-generation process. The entire panel is processed in one step. This leads to reduced time and lower cost compared with the baseline's point-to-point processing. In general, increased cost due to additional die per panel can be attributed to the additional material required.

17.3 HIGH-THROUGHPUT FC PROCESSING USING NO-FLOW UNDERFILLS

Underfill processing is a time-consuming step in conventional FC assembly processes due to the flow mechanism used to fill the standoff gap between the chip and substrate. The current technique uses a dispensing method where underfill material is dispensed along one edge (or dual adjacent edges) of the chip and allowed to flow under the chip by capillary action. A schematic of the conventional FC assembly process is shown in Fig. 17.14. Based on first-order principles, the fill time can be estimated using Eq. (17.1) assuming the underfill materials are approximately Newtonian, where L is the length of the chip, h is the standoff height, μ is the viscosity, σ is the surface tension, and α is the wetting angle of the underfill, as shown in Fig. 17.15. A comprehensive analysis of underfill processing techniques is presented below.[1] Equation (17.1), illustrates that the underfill

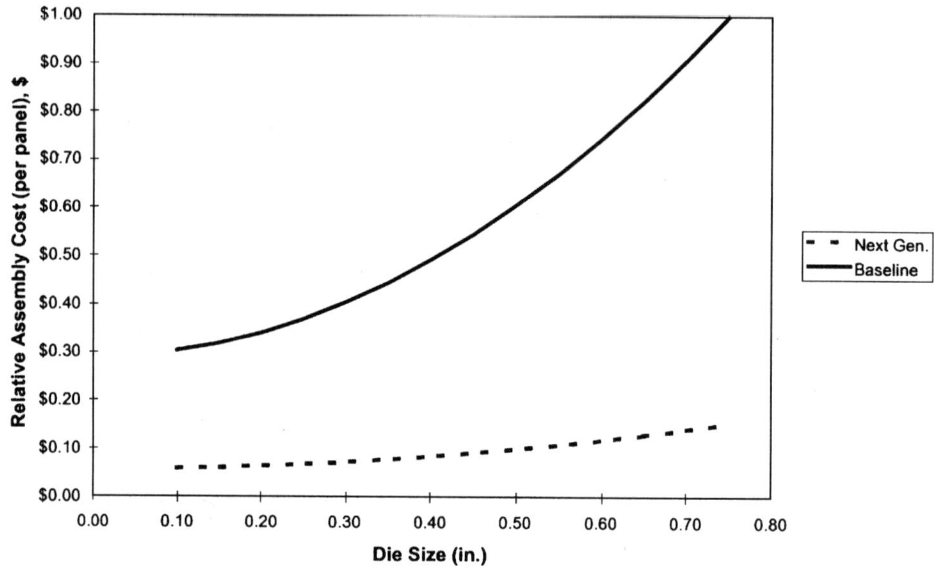

FIGURE 17.12 Effect of die size on relative wafer to assembly cost.

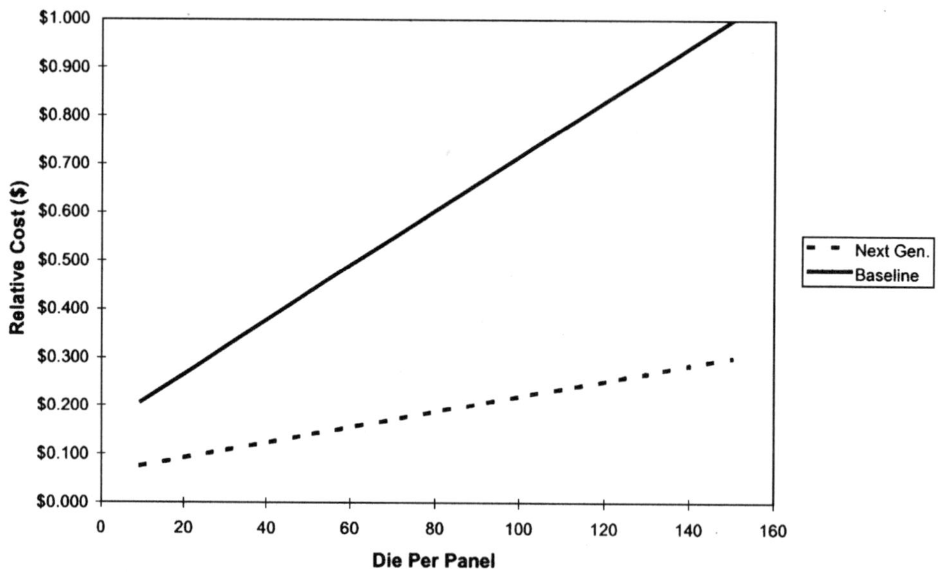

FIGURE 17.13 Effect of number of die per panel on relative wafer to assembly cost.

flow time for a given material increases with the square of the chip size and the inverse of the standoff height. As chip sizes increase and standoff heights decrease (due to the decreasing pitch requirements), indicated by the semiconductor packaging roadmaps, underfill flow times will only become compounded, inhibiting high-throughput production.[1,13,24–30,34]

$$t_{fill} = \frac{3\mu L^2}{\sigma h \cos(\alpha)} \tag{17.1}$$

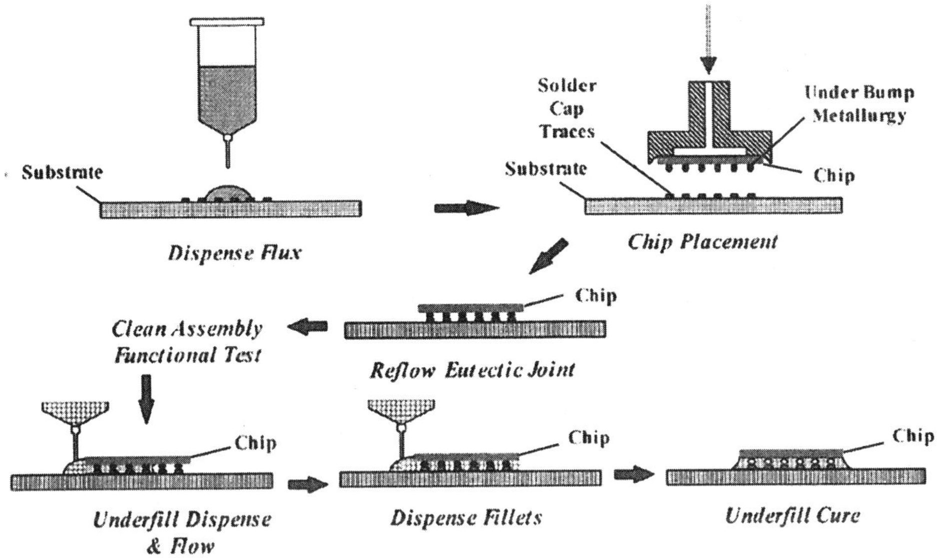

FIGURE 17.14 Conventional FCOB assembly process using capillary flow underfills.

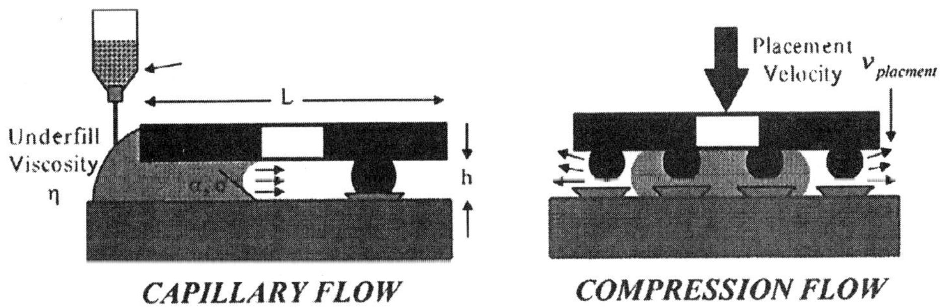

FIGURE 17.15 Underfill flow processing techniques including standard capillary flow and high-throughput compression flow.

With the new low-cost process, the flow time is reduced to the chip placement time, which is given by Eq. (17.2), where h_o is the initial distance of the chip to the substrate, h_f is the final distance of the chip to the substrate once assembled, and $v_{placement}$ is the chip placement velocity[1] (see Fig. 17.15). Equation (17.2) illustrates that in the new process the time required to underfill the chip is no longer dependent on the chip size, chip standoff, and underfill properties. In reality, the underfill flow time is eliminated in that the chip placement time would be incurred regardless of whether the underfill is processed or not.

$$t_{placement} = \frac{(h_o - h_f)}{v_{placement}} \tag{17.2}$$

The new low-cost FC assembly process has several critical requirements. Critical process steps include chip placement and simultaneous solder reflow and underfill cure. With respect to chip placement and reflow, the bumps on the chip must be able to flow through the underfill material, the underfill material must be able to act like a flux when reflowing the solder interconnections, the underfill must maintain a sufficiently low viscosity to enable chip collapse during reflow, and the solder must be able to flow and wet the pad within the same process cycle as the underfill cure. These critical process elements as well as the overall low-cost, high-throughput FC assembly process have been successfully demonstrated. Figure 17.16 shows a cross section of a typical FC test vehicle assembled using the low-cost, high-throughput FC assembly process and no-flow underfill. In addition to demonstrating the feasibility of the process and material, Fig. 17.16 illustrates the ability of the new FC process and no-flow underfill to achieve self-alignment during reflow, indicating the effectiveness of the latent catalyst used in the no-flow underfills.

17.3.1 Process Yield Analysis

The goals of this section are to demonstrate the feasibility of no-flow underfill materials and the low-cost, high-throughput FC process, identify the critical process variables affecting yield, and analyze the effects of the process on different underfill materials.[19,39,40] Reported is the assembly of a series of test vehicles to assess process yield and process defects. The test vehicles were assembled by dispensing a controlled mass of underfill material on the chip site, followed by alignment and placement of the chip onto the substrate pads, inducing a compression underfill flow. Next, the assemblies were reflowed in a commercial reflow oven in an air atmosphere to simultaneously form the solder interconnections and cure the underfill. A series of designed experiments identified the critical process variables to be underfill mass, reflow profile, placement velocity, and underfill type. Of particular interest was the fact that different underfill materials exhibited an affinity for unique reflow profiles to minimize process defects.

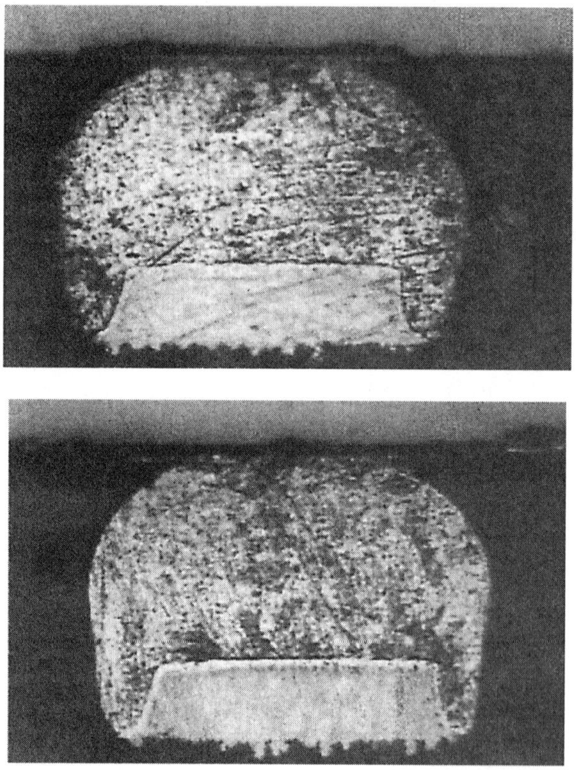

FIGURE 17.16 Cross section of a typical FC assembly using the low-cost high-throughput FC assembly process and no-flow underfill.

17.3.1.1 Test Vehicle Description. Six test vehicles were used in various combinations in the interconnection yield analysis experiments. The test die were supplied by FC Technologies and consisted of eutectic lead-tin (37Pb/63Sn) solder-bump FC devices with daisy-chain structures using a proprietary under-bump metallization (UBM) consisting of aluminum-nickel/vanadium-copper. All substrates were designed at Georgia Tech and consisted of high temperature tetrafunctional FR-4 having a glass transition temperature T_g of 170°C with thicknesses ranging from 0.79 to 1.57 mm. All substrate pads had metallizations consisting of copper, electroplated nickel, and immersion gold. Solder mask thicknesses for each substrate configuration ranged from 20 to 30 μm. Table 17.1 presents the various test vehicles and parameters used in the interconnection yield analysis experiments.

17.3.1.2 Experimental Procedures. This study focuses on evaluating process characteristics over a range of FC test vehicles, assessing the critical process characteristics and design parameters affecting yield, and identifying key factors contributing to process defects. A set of five designed experiments[21] is executed to examine the major process parameters affecting yield. These design experiments seek to determine the effects of key process variables and design parameters on assembly yield. Each experiment is designed to systematically investigate each of the following factors on assembly process yield: FC process variables, chip design variables, substrate design factors, and underfill surface geometry factors. The first experiment evaluates the effect of assembly process variables including chip placement force, chip placement velocity, reflow profile, underfill mass, and no-flow underfill material type on assembly yield. Due to the known interaction between chip placement force and chip placement velocity, the second experiment is intended to evaluate the extent of the interaction between these variables based on the hydrodynamic forces produced during assembly and assess their impact on assembly yield. The third experiment is designed to characterize the effects of FC design parameters including bump geometry layout (area and perimeter array), chip size, and solder bump density (number of bumps per unit area) on assembly yield. The fourth experiment investigates the impact of a key substrate design factor, specifically board thickness, on the assembly yield for FC processing based on no-flow underfills. The final experiment is designed to explore the interaction between no-flow underfill surface geometries and the formation of voids during chip placement. In this way, no-flow underfill surface geometries can be controlled and optimized to ensure minimum voiding during FC assembly.

In whole, these experiments provide a characterization of the low-cost, high-throughput FC assembly process using no-flow underfills, specifically providing knowledge of the feasible process windows and understanding of the key factors that control the process yield.

17.3.1.2.1 Process Variable Main Effects Analysis (Experiment 1). The first design experiment is used to determine the main effects of five process variables on the interconnection yield in test vehicles manufactured using the high-throughput FC assembly process. The experiment examines chip placement force, chip placement velocity, reflow profile, mass of underfill deposited, and type of no-flow underfill material. The intent is to examine the impact of these process variables on the number of solder interconnections that successfully reflow and wet the substrate pads. The PB8-4-0.79 test vehicles are used for this set of experiments (see Table 17.1). Two replicates of the experiment

TABLE 17.1 FC Test Vehicles and Associated Design Parameters

Test vehicle ID	Test chip	Chip size* (mm)	Bump pitch (μm)	I/O count	Bump layout	Substrate thickness (mm)	Nickel thickness (μm)	Copper thickness (μm)	Gold thickness (μm)
PB8-2-1.57	PB8	5.08	203	88	Perimeter	1.57	43	3.8	0.15
PB8-4-0.79	PB8	10.2	203	352	Perimeter	0.79	50	3.8	0.15
PB8-4-1.57	PB8	10.2	203	352	Perimeter	1.57	50	3.8	0.15
FA10-2-1.57	FA10	5.08	254	317	Area	1.57	37	4.4	0.17
FA10-2-0.79	FA10	5.08	254	317	Area	0.79	37	4.4	0.17
FA10-4-0.79	FA10	10.2	254	1268	Area	1.57	34	4.3	0.17

*Chip size is the side measurement of the square test chips.

are performed to allow for calculations of error in the assembly process and to validate the results. The chips are placed at four placement velocities, 0.1, 0.2, 0.3, and 0.4 mm/s and two placement forces, 400 and 900 g. Four underfill masses are investigated 15, 22, 28, and 35 mg. Two sources of no-flow underfill materials are investigated: a commercial no-flow underfill, underfill A, and a Georgia Tech developed no-flow underfill, underfill B,[17,18] along with an adhesion-promoted version of underfill B, underfill C.

Two reflow profiles are studied. One profile consists of a typical eutectic solder profile, termed the *step profile,* having a ramp, soak, peak spike, and cool-down. The second profile is a *rapid ramp profile* consisting of a ramp to the peak temperature (i.e., no soak stage) followed by a slow cool-down and finally a rapid cool down. The standard reflow profile, shown in Fig. 17.17, models a reflow profile used for eutectic solder pastes. It begins with a temperature ramp at 3.0°C/s to a soak temperature between 150 and 160°C, where it dwells for 60 s. The soak activates flux in typical solder pastes. Then a second temperature ramp at, 3.0°C/s, to the peak temperature of 220°C reflows the solder. The spike at the end of the profile keeps the board above 183°C for 80 s while the solder reflows and wets the pads. The rapid ramp profile, shown in Fig. 17.18, increases the temperature of the assembly at a constant rate of 1.5°C/sec up to a peak temperature of 215°C. This keeps the assembly above 183°C for 80 s while the solder reflows and wets the pads.

17.3.1.2.2 Interaction of Placement Force and Placement Velocity (Experiment 2). The hydrodynamic force that the underfill exerts on the chip during placement is proportional to the placement velocity. Therefore, as the placement velocity increases, the force the underfill exerts on the chip, and therefore the placement force, must increase in order to drive the chip through the underfill material to make contact with the substrate. This experiment isolates the placement force and velocity factors to determine if the increased hydrodynamic force of the underfill on the chip has an effect on interconnection yield. The rationale is that if the programmed placement force limit is reached prior to the chip bumps making contact with the substrate pads, the chip will be released some distance above the substrate. Hence contact with the substrate pads is not guaranteed, which could adversely affect interconnection yield. The placement velocities investigated are 0.4, 2.7, and 5.0 mm/s. The placement forces investigated are 400, 1000, and 1600 g. By design, the experiment helps to determine if increasing the programmed placement force allows the chip to penetrate the underfill while traveling at high velocities. The PB8-4-0.79 test vehicles are used for this set of experiments (see Table 17.1). Three replications of the experiments are run to allow for a calculation of error.

17.3.1.2.3 Chip Parameters Main Effects Analysis (Experiment 3). The goal of this experiment is to determine the effect of the chip parameters, such as size and bump pattern, in combination with placement force and velocity on interconnection yield. Four factors are investigated including chip size, bump geometry, placement force, and placement velocity. Two chip sizes, 5.08 by 5.08 mm and 10.16 by 10.16 mm, are used. The bump patterns investigated are perimeter and full area array. The placement velocities studied are 0.1 and 5 mm/s, and the placement forces were 400 and 1600 g. Four types of no-flow underfill materials are investigated: a Georgia Tech developed no-flow underfill, underfill B, an adhesion promoted version of underfill B, underfill C, a 20 percent silica filled version of underfill B, underfill D; and a 40 percent silica-filled version of underfill B, underfill E. The PB8-2-1.57, PB8-4-0.79, FA10-2-1.57, and FA10-4-1.57 test vehicles are used for this set of experiments (see Table 17.1).

17.3.1.2.4 Board Thickness Analysis (Experiment 4). To determine the impact of board thickness on interconnection yield, a simple test is performed. This experiment uses a single FC test vehicle design with two FR-4 substrate thicknesses of 0.79 to 1.57 mm. The primary metrics used for analysis include transmission x-ray microscopy and solder-joint interconnection yield verified by electrical continuity measurements. The FA10-2-0.79 and FA10-2-1.57 test vehicles are used for this set of experiments (see Table 17.1).

17.3.1.2.5 Placement Void Analysis (Experiment 5). Early experiments demonstrating the low-cost, high-throughput FC process[7] exhibited voiding in the underfill attributed to three factors: outgasing of the substrate and chip passivation during reflow, outgasing of the no-flow underfill during reflow, and capture voids formed during placement. It is theorized that as a chip is placed onto a substrate, it is possible for air to become entrapped between the chip and the underfill material on the board. A possible cause of the voids would be the capture of air pockets or bubbles at the chip-underfill interface due to depressions in the underfill surface and/or multiple points of contact between the chip surface and underfill. In support of this theory, acoustic microscopy images of FC test vehicles used to demonstrate process feasibility show large voids in the underfill at the center of die (these initial feasibility studies were performed by McGovern and Baldwin[19]).

To examine whether air is entrapped in the underfill during chip placement, a designed experiment is used to determine if the phenomenon occurs and what factors influence it. Four factors identified for possible interaction are studied to determine their impact on void formation: placement velocity, placement force, underfill viscosity, and bump pattern. Void analysis is performed using a Sonoscan D6000 Scanning Acoustic Microscope after chip placement. Calculation of the percentage voiding is performed using a commercial image analysis software package. The PB8-2-1.57 and FA10-2-1.57 test vehicles are used for this set of experiments (see Table 17.1).

17.3.1.3 Assembly Process. The test vehicle assemblies are produced using a consistent process from experiment to experiment adjusting the process parameters as stated in the experiments section. All substrates are baked for 2 h at 100°C to remove dissolved solvents and moisture. A fixed mass of underfill is dispensed by syringe onto the substrate bond site. Accuracy of the dispensed underfill mass is determined by weighing the samples in a precision balance. The substrate is then placed in a Kulicke and Soffa 6900 FC bonder. The bonder picks the chip, visions the corner bump array, aligns it to the substrate pads based on fiducials, and places it using programmed parameters, including placement velocity, placement force, and bond time. The assembly is then reflowed using a BTU Paragon 98N convection reflow furnace in an air atmosphere. The reflow profiles are optimized for each test vehicle using a KIC Thermal Profiling Prophet system. Acoustic microscopy is performed using a Sonoscan D6000 scanning acoustic microscope to analyze internal defects in the assemblies such as voids, cracks, etc. Interconnection continuity measurements are performed using a Kiethley 210 digital multimeter based on a two-point measurement technique.

A baseline set of process parameters is used for test vehicle assembly, except where the design experiments specify alternate parameters. Baseline placement process parameters include a 1000 g placement force, a 5 mm/s placement velocity, and a 300 ms bond time. Reflow is based on the rapid ramp profile shown in Fig. 17.17.

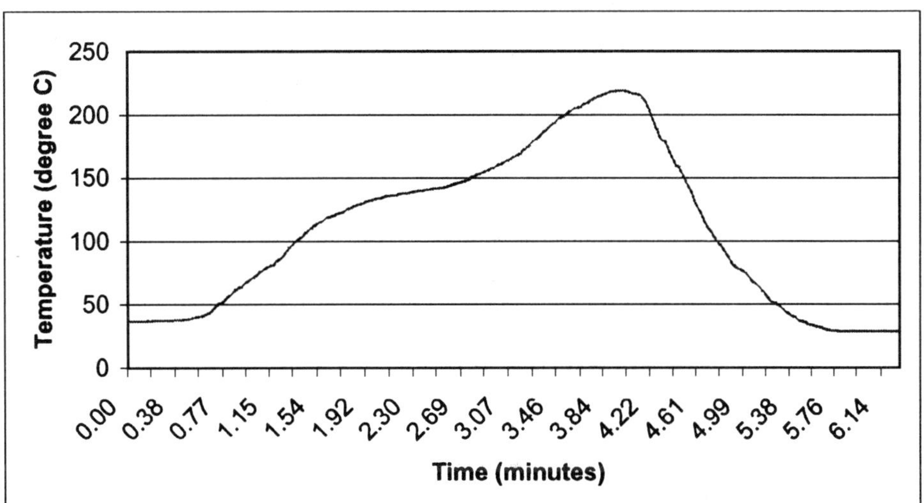

FIGURE 17.17 Conventional step eutectic solder reflow profile.

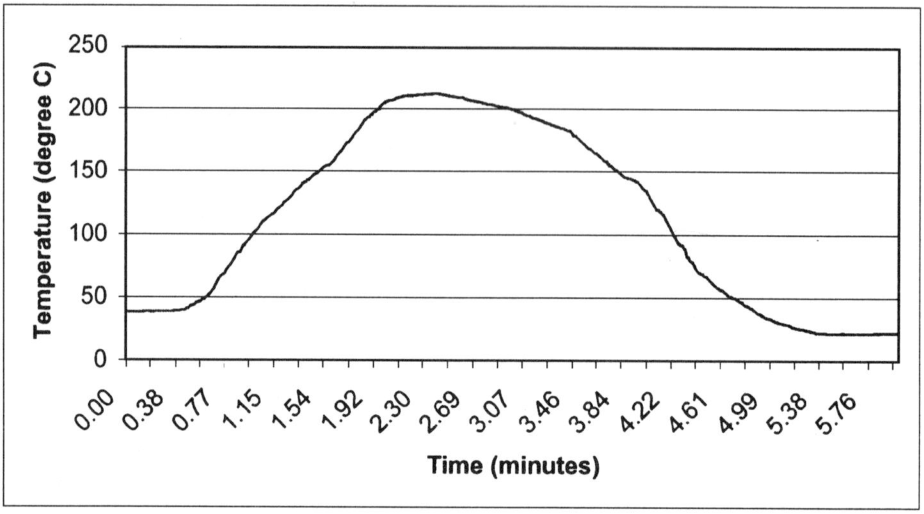

FIGURE 17.18 Rapid ramp reflow profile.

17.3.2 Yield Analysis Results and Discussion

Interconnection yield is determined by electrical resistance continuity measurements of the FC test vehicles via their daisy chain structures. The criterion for interconnection yield failure is a ±100 mΩ continuity loop resistance difference over baseline resistance values previously determined for the test vehicles. Interconnection yield is reported as a percentage of interconnection chains passing the resistance criterion relative to the total number of interconnection chains possible for the particular test vehicle. Depending on the test vehicles used, interconnection failures are based on groups of interconnections or interconnection chains that can be tested. For example, the PB8-2 test vehicle has test points allowing interconnection pairs to be continuity tested. Hence each PB8-2 has 44 interconnection chains from which the yield can be reported.

*17.3.2.1 **Process Variable Main Effects Analysis (Experiment 1).*** Table 17.2 presents the design experiment results of the process variable main effects analysis. An analysis of variance is done on the main effects data. Based on the F-distribution statistic for the five parameters investigated using a 95 percent confidence interval, the reflow profile and the underfill material type have the most significant effect on interconnection yield. Changes in the placement force, the placement velocity, and the underfill mass have relatively little effect on interconnection yield. A typical solder interconnection produced by the low-cost, high-throughput FC process is shown in Fig. 17.19. For the PB8-4-0.79 test vehicles, it is observed that if the underfill mass exceedes 35 mg (i.e., the maximum underfill mass tested), the assemblies have zero interconnection yield.

*17.3.2.2 **Interaction of Placement Force and Placement Velocity (Experiment 2).*** Three PB8-4 chips are assembled for each set of assembly parameters in Table 17.3. The average percentage interconnection yield of the three replicates is listed in Table 17.3. To determine the impact of changes in placement force and placement velocity on interconnection yield, a linear regression is performed on the data using a least squares normal fit. An analysis of variances is used to calculate the contributions of the main effects, the interactions, and the experimental error on interconnection yield. From the analysis of variance, the mean square is 0.0541 for the main effects, 0.1083 for the interactions, and 0.0650 for the error. This yields an F statistic of 0.83 for the main effects and 1.67 for the interactions. Based an F-distribution table, the predicted F statistic for the main effects $F_{0.01,2,10} = 7.56$ is larger than the experimental value of 0.83. Therefore, the main effects of placement force and place-

TABLE 17.2 Yield Results of Assemblies in Experiment 1

Experiment no.	Underfill mass (mg) (mm/s)	Placement velocity profile	Reflow type	Underfill force (g)	Placement replicates	Average percent yield for two
1	15	0.1	Step	B	400	98.9
2	22	0.1	Ramp	B	900	99.0
3	28	0.1	Ramp	A	400	100
4	35	0.1	Step	A	900	98.9
5	15	0.2	Ramp	A	900	100
6	22	0.2	Step	A	400	81.8
7	28	0.2	Step	B	900	96.9
8	35	0.2	Ramp	B	400	100
9	15	0.3	Step	A	900	83.8
10	22	0.3	Ramp	A	400	100
11	28	0.3	Ramp	B	900	99.7
12	35	0.3	Step	B	400	100
13	15	0.4	Ramp	B	400	100
14	22	0.4	Step	B	900	98.9
15	28	0.4	Step	A	400	92.8
16	35	0.4	Ramp	A	900	99.7

FIGURE 17.19 Typical eutectic solder interconnection produced using the low-cost, high-throughput FC process based on no-flow underfills.

TABLE 17.3 Yield Results of Assemblies in Experiment 2

Experiment no.	Velocity (mm/s)	Placement force (g)	Average percent yield for three replicates
1	0.4	400	99.8
2	2.7	400	99.6
3	5.0	400	99.8
4	0.4	1000	100
5	2.7	1000	100
6	5.0	1000	99.8
7	0.4	1600	100
8	2.7	1600	99.8
9	5.0	1600	99.6

ment velocity do not impact yield statistically. Similarly, for the interaction, $F_{0.01,1,10} = 10.04$, which is greater than the experimental value of 1.67. Therefore, the interaction between the placement force and placement velocity does not contribute to the change in yield.[8]

*17.3.2.3 **Chip Parameters Main Effects Analysis (Experiment 3).*** The average interconnection yield measured for the two replicate test vehicle assemblies and the corresponding process variables are presented in Table 17.4. This design experiment shows that only one factor is important relative to interconnection yield under conditions of varying underfill type, chip size, bump layout, placement force, and placement velocity. More specifically, the filler content of the no-flow underfill dominated the interconnection results. The unfilled no-flow underfills yield 100 percent interconnection regardless of process parameters, chip parameters, or underfill chemistry. In contrast, the filled no-flow underfills yield 0 percent interconnection under all conditions. A typical FC solder interconnection cross section using filled no-flow underfill is shown in Fig. 17.20.

*17.3.2.4 **Board Thickness Analysis (Experiment 4).*** All 20 chips assembled had 100 percent interconnection yield indicating that substrate thickness over the range of 0.79 to 1.57 mm had no appreciable effect on interconnection yield based on a 90 percent confidence interval.

*17.3.2.5 **Placement Void Analysis (Experiment 5).*** Test vehicle assemblies are made using the parameters in Table 17.1. Two assemblies are made for each experimental condition. After assembly, each die is inspected using scanning acoustic microscopy with a 180 MHz transducer to inspect underfill voiding due to the placement process. None of the images show evidence of void generation in the no-flow underfill for the assemblies produced. Typical images from the 5.08 by 5.08 mm PB8 and 5.08 by 5.08 mm FA10 test vehicles are shown in Fig. 17.21. In these images, voiding appears as white regions within the gray area of the chip.

FC assembly using the low-cost, high-throughput process based on no-flow underfills is capable of achieving high interconnection yields provided key process design and control measures are integrated into production. The results show that reflow is the most important stage of the process where the reflow profile has the strongest influence on first-pass interconnection yield. Underfill filler content also has a dramatic impact on interconnection yield. The silica filled no-flow underfills studied impede solder interconnection yield. Physically, this result stems from geometric stability. Since the chip is rigid and the surface tension of solder high, the chip will have a tendency not to yield (i.e., chip yield is predicated on all interconnections on the chip yielding) if any three solder bumps, sufficiently separated, capture a filler particle at the interface between a bump and pad. This follows because three points define a plane, and if any three bumps are not able to collapse and contact the substrate pad, the chip will not yield. This phenomenon is reduced if the filler particles are reduced in size. While the probability that a particle is entrapped between a bump and substrate pad is reduced with a reduction is particle size, the overall effect on first-pass chip yield is still likely to exceed acceptable first-pass assembly yields (typically 90 to 99.5 percent at the chip level).

TABLE 17.4 Yield Results of Assemblies in Experiment 3

Experiment no.	Underfill type	Chip size (mm)	Chip type	Placement force (g)	Placement velocity (mm/s)	Average percent yield for two replicates
1	C	5.08	FA10	400	0.1	100
2	B	5.08	FA10	1600	0.1	100
3	D (20 percent filled)	5.08	FA10	1600	5.0	0
4	E (40 percent filled)	5.08	FA10	400	5.0	0
5	C	10.16	FA10	400	5.0	100
6	B	10.16	FA10	1600	5.0	100
7	D (20 percent filled)	10.16	FA10	1600	0.1	0
8	E (40 percent filled)	10.16	FA10	400	0.1	0
9	C	5.08	PB8	400	0.1	100
10	B	5.08	PB8	1600	0.1	100
11	D (20 percent filled)	5.08	PB8	1600	5.0	0
12	E (40 percent filled)	5.08	PB8	400	5.0	0
13	C	10.16	PB8	400	5.0	100
14	B	10.16	PB8	1600	5.0	100
15	D (20 percent filled)	10.16	PB8	1600	0.1	0
16	E (40 percent filled)	10.16	PB8	400	0.1	0

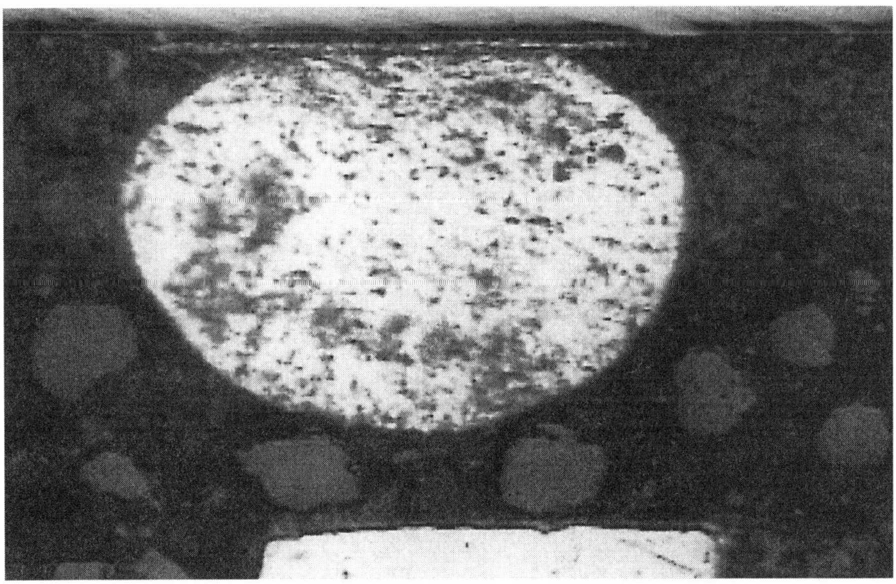

FIGURE 17.20 Solder bump impeded by filler particles from collapsing and forming an interconnection with the substrate pad.

Process parameters such as placement force and placement velocity have relatively little impact on yield over the ranges studied. Underfill mass or volume also has relatively little impact on yield, provided it was keep below a critical value that depends on the chip geometry and bump configuration. Under conditions where the underfill mass/volume is excessive (above a critical value), the assembled FCs are observed to "float" compromising alignment accuracy before and after reflow. The basic physics driving the chip floating phenomenon is discussed below, and a process design model is presented.

FIGURE 17.21 Scanning acoustic microscope image of under fill layer in a perimeter and area array die after chip placement.

17.3.3 Placement Voids

The results of Experiment 5 show no voids in the underfill after the chip placement process. This likely results from the initial underfill geometry being hemispherical in shape. If the underfill forms a hemispherical geometry, as in Fig. 17.22, no air pockets will tend to form during the placement process. As the chip moves into contact with the underfill, it wets at only one point toward the center of the chip. On further motion of the chip, the underfill continues to wet the chip surface from the chip center outward. Locally, the surface tension of the underfill can facilitate the wetting of the chip, driving out the air and eliminating capture voids.

Using the optical microscope, the geometry of the underfill was examined. Visual inspection of the underfill deposits for fluids having viscosities of 0.4, 3.4, 10, and 3400 Pa·s shows no depressions in the surface of the deposited underfills. Without depressions in the underfill deposits, the chips will tend not to have multiple contact points of the underfill with the chip surface. The chip starts with an initial point of contact at the top of the underfill deposit and locally wets the chip surface and bumps as it is moved toward the substrate. Lack of air pocket capture is one of the primary reasons voids are not formed.

The capability to assemble FCs using the low-cost, high-throughput assembly process incorporating no-flow underfill materials was demonstrated. The ability to implement the process on standard SMT assembly equipment, including placement machines and reflow ovens, was shown experimentally. The placement process prefers a low-viscosity, no-flow underfill material. For a given placement force, as underfill viscosity decreases, the allowable placement velocity increases, improving throughput. The low-cost, high-throughput FC assembly process also prefers an underfill with a low surface tension. The yield analysis experiments demonstrated a number of process recipes capable of producing high first-pass yields. It was found that the placement force, the placement velocity, and the underfill mass had relatively little effect on interconnection yield. Excessive underfill volume resulted in chip floating. Chip bump pattern and chip size were also found to have relatively little effect on assembly yield. Capture and assembly-induced void formation during assembly was also found to be minimal when the appropriate underfill geometry is deposited on the substrate.

17.3.4 Summary of Process Yield Analysis

FC technology represents a rapidly advancing area in commercial electronics. To enable low-cost, high-throughput processing compatible with high-speed SMT assembly and high-volume electronics packaging, a new assembly process was developed. This new process eliminates fluxing operations, secondary thermal curing steps, and the need for time-consuming capillary flow underfill processing using a compression flow technique where the underfill is applied prior to chip placement. The innovative process integrates the chip placement and polymer underfill processes using a compression or squeeze flow technique. It results in significantly lower assembly costs and reduced cycle time.

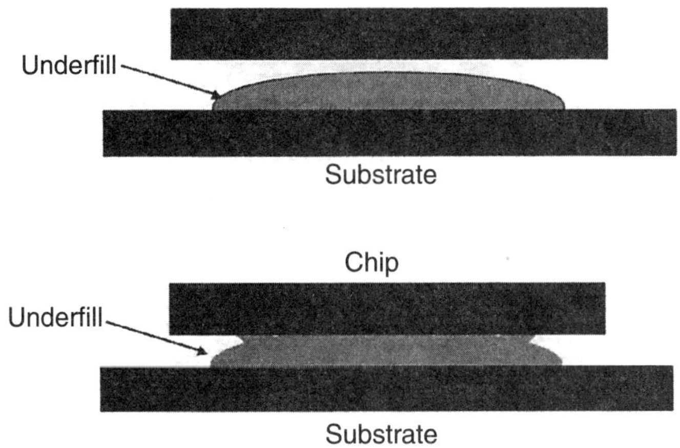

FIGURE 17.22 Chip contacting a spherically shaped underfill.

This work has demonstrated FC assembly using the low-cost, high-throughput process based on no-flow underfills and its ability to achieve high interconnection yields, provided key process design and process control measures were integrated into production. Critical process parameters affecting yield included reflow profile and no-flow underfill filler content. The results also suggest that adding fillers to no-flow underfill systems tended to inhibit robust interconnection formation, significantly lowering chip level yield. Process parameters such as placement force and placement velocity had relatively little impact on yield over the ranges studied.

Two process design models were developed that provide additional insights into designing the process parameters for the low-cost, high-throughput FC process. The compression flow placement model enabled prediction of key placement process parameters such as placement velocity, placement force, and bond time based on the FC design parameters. Moreover, the compression flow placement model was instrumental in analyzing the interaction between placement force and placement velocity. In particular, it was found that bond time was a key process design parameter in order to ensure high yield processing if high placement speeds are required or if placement forces are limited. A chip floating model was developed to determine the key process variables governing floating, including underfill mass, underfill surface tension, and surface wetting characteristics. Chip floating was analyzed to explain chip shifting and motion after placement that significantly affected yield when excess underfill volumes were used. Chip floating was found to depend on the steady-state forces acting on the chip, namely, chip weight, underfill surface tension, hydrostatic pressure on the chip, and pressure differential across the fillet meniscus. In terms of process variables, underfill mass, surface energy, and underfill wetted length had the strongest impact on chip floating. The process design model was developed to understand the basic physics of the floating phenomenon and predict process variables to eliminate process defects. A floating parameter was developed to predict the critical underfill volume resulting in chip floating. The model predictions agreed well with experimental measurements of chip floating with varying underfill masses.

17.3.5 Reliability Analysis

This section will evaluate the reliability of several commercial no-flow underfills and demonstrate the affects of board design and chip design parameters on reliability. Section 17.3.6 will provide a comprehensive failure-mode analysis of the resulting FC assemblies.[36,38–40]

17.3.5.1 Experimental Procedures

17.3.5.1.1 Test Die. Four different test die were used to evaluate the effect of bump pitch, bump layout, and chip size on yield and reliability. All test die were daisy chained devices supplied by FC

Technologies with tin-lead eutectic solder bumps and silicon nitride passivation. The under bump metallization (UBM) consisted of an aluminum, nickel-vanadium, copper system. The daisy-chain structure allowed continuity resistance measurements to be taken. The specifications of the test die are shown in Table 17.5.

17.3.5.1.2 Test Vehicles. Six different test vehicles were used to evaluate the effect of chip size, bond pad metallization, and bond pad definition on yield and reliability. The PB8-4 chip was used on three different test vehicles, and the remaining chip types were used on one test vehicle each. The important variables for each test vehicle are listed in Table 17.6.

All the test vehicles used a high-temperature FR-4 substrate with a glass transition temperature T_g of 180°C. The substrate thickness for the six test vehicles was approximately 0.78 mm. For the organic passivated substrates, the organic solderability preservative (OSP) was 0.05 to 0.13 μm thick. The Cu/Ni/Au boards had an electrolytic nickel layer approximately 3.8 μm thick and an immersion gold layer between 0.05 and 0.30 μm thick. Figure 17.23 shows a schematic of the circular defined bond pad and the rectangular defined bond pad.

17.3.5.1.3 No-Flow Underfill Materials. Seven no-flow underfills were evaluated in this study. All but one were supplied by commercial vendors. All these materials were dispensed during assembly with a controlled geometry and volume. Two of the materials studied required a postcure of 30 min at 160°C (Table 17.7).

17.3.5.2 Assembly Process.

Before assembly, the test vehicles were baked for 1.5 h at 125°C to drive off any entrapped moisture and solvents. Both substrates and die were stored in a desiccant chamber prior to use. A controlled amount of underfill was dispensed on each FC bond site using a CAM/ALOT 3800 automated dispense system with a 22 gauge needle. Two different dispense patterns were used during processing. The underfill amount necessary was determined by the chip size and bump layout, as well as the bump height. Previous work has modeled the relationship between underfill amount and die yield with respect to chip floating and placement parameters.[7,8]

Following underfill dispense, the FCs were placed on the boards using either a K&S 6900 FC bonder or a high-speed Siemens SIPLACE F5 with DCA placement system. Previous work has shown that placement force and dwell (bond) time can have an effect on yield when processing with a no-flow underfill.[6] Table 17.8 lists the placement forces used for the different FC test chips optimized for high yield.

TABLE 17.5 Test Die Used for Reliability Testing

Chip type	Chip size (mm)	Bump count	Bump pitch (μm)	Bump layout
PB8-4	10.2	352	203	Perimeter
PB8-2	5.1	88	203	Perimeter
FA10-4	10.2	1268	254	Area
FA10-2	5.1	317	254	Area

TABLE 17.6 FC Test Vehicles

Test vehicle	Chip type	Metallization	Bond pad definition
TV1	PB8-4	Cu/Ni/Au	Circular Mask
TV2	PB8-4	Cu/Ni/Au	Rectangular Mask
TV3	PB8-4	Cu/OSP	Circular Mask
TV4	PB8-2	Cu/Ni/Au	Rectangular Mask
TV5	FA10-4	Cu/Ni/Au	Rectangular Mask
TV6	FA10-2	Cu/Ni/Au	Rectangular Mask
TV7	PB8-2	Cu/OSP	Rectangular Mask

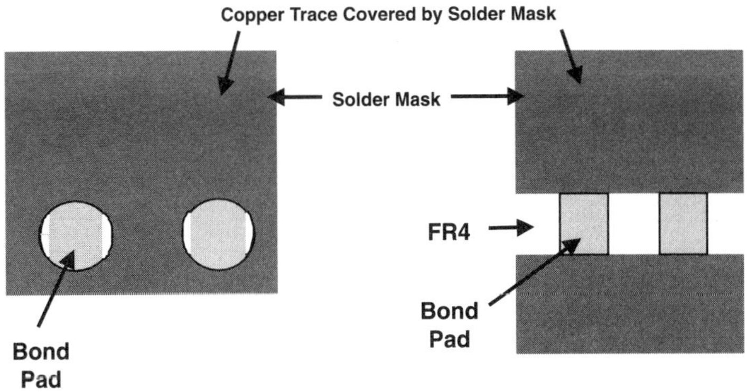

FIGURE 17.23 Circular mask opening (*left*) and rectangular mask opening (*right*).

TABLE 17.7 No-Flow Underfills Analyzed

Underfill	CTE	T_g	Post cure used
A	90ppm/°C	100°C	No
B	70ppm/°C	120°C	Yes
C	75ppm/°C	125°C	No
D	70ppm/°C	90°C	No
E	65ppm/°C	95°C	No
F	70ppm/°C	110°C	No
G	90ppm/°C	95°C	Yes

TABLE 17.8 Placement Forces Used for FC Assembly with No-flow Underfills

Test die	Placement force
PB8-4	800–1600 g
PB8-2	400–1000 g
FA10-4	800–1600 g
FA10-2	400–1000 g

After placement, the substrates were reflowed, simultaneously curing the underfill and forming the solder interconnections. The reflow profile used was determined by the chemistry of the underfill. For this study, three different profiles were used: rapid ramp profile, a conventional step eutectic style profile, and low soak temperature eutectic profile. All reflow profiles used an ambient air environment without nitrogen inerting. A postcure step for 30 min at 160°C was used for materials B and G.

17.3.5.3 Test Methods.

Several different standard accelerated tests were used to evaluate the no-flow underfills. Table 17.9 summarizes the series of tests used.

Liquid-to-liquid thermal shock testing (LLTS) was performed using an ESPEC TSB5 environmental test chamber. Test conditions were from −55 to +125°C with 5 min dwells in each bath. Air to air thermal cycling (AATC) was performed using a Thermotron ATS-320-DD test chamber. Test conditions were from −55 to +125°C with 10 min dwells at each temperature. Transfer times between temperature chambers were less than 10 s in each case. Temperature-humidity (TH) aging was performed in a Thermotron SM-8C test chamber. Test conditions were 85°C and 85 percent relative humidity for a total of 1000 h. Moisture preconditioning tests were done at levels 1 and 3

TABLE 17.9 Reliability Test Methods and Conditions

Reliability test	Industry standard	Test conditions	Cycle time
Air-to-air thermal cycling	JESD22 A104-A	+125°C to −55°C	20 min
Liquid-to-liquid thermal shock	JESD22 A106-A	+125°C to −55°C	10 min
Temperature humidity	JESD22 A101-B	85°C 85 percent RH	100 h
Level 1 preconditioning	J-STD 020-A	85°C 85 percent RH	168 h
Level 3 preconditioning	J-STD 020-A	30°C 60 percent RH	192 h
Autoclave	JESD22-A102	121°C, 100 percent RH, 2.09 atm	192 h

according to J-Standard-020 Revision A. For the moisture preconditioning tests, the assemblies were baked for 24 h at 125 +5/−0°C. They were then aged according to the test level. Level 1 required 168 h at 85°C/85 percent relative humidity and level 3 required 192 h at 30OC/60 percentRH. The test assemblies were then reflowed according to the J-STD and inspected. Autoclave testing was performed in an ESPEC TPC-422M test chamber. Test conditions were 121°C, 100 percent relative humidity, and 2.09 atm absolute pressure for a total of 120 h.

17.3.5.4 C-SAM Analysis, Continuity Testing, and Failure Criteria. During accelerated life testing, test vehicles were tested for electrical continuity and examined using acoustic microscopy. Every 100 h or 100 cycles, electrical continuity was checked. A change in resistance of greater than ±5 percent was classified as a failure. For the autoclaved test vehicles, electrical continuity was measured every 24 h.

Every 200 cycles, a sample of the test vehicles was examined with a Sonoscan D6000 acoustic microscope using a 180- or 230-MHz transducer. Scans were taken at the die-underfill interface to investigate delamination, void growth, die cracking, and underfill cracking. The underfill-to-substrate interface was also scanned for defects. After scanning, test vehicles being subjected to AATC and LLTS were baked out for a minimum of 2 h at 80°C before testing continued. This was done to drive off trapped moisture introduced by the C-SAM, which could induce premature failures.

17.3.6 No-Flow Underfill Reliability Results

All reliability data were analyzed using the Weibull distribution. The Weibull distribution can be used to model early life, intrinsic, and wear-out failures. Weibull distributions return two parameters, the theta parameter (Θ) and the alpha parameter (α) as shown in Eq. (17.3). The theta parameter corresponds to the time to 63.2 percent failure in the sample and is commonly referred to as the mean time to failure (MTTF). The alpha parameter is the shape factor and refers to the slope of the best-fit line. The alpha parameter is associated with the failure rate of the assemblies. An increase in the alpha parameter corresponds to an increase in failure rate. Reliability data are presented in terms of these two values, as well as the number of cycles to first failure.

$$F(t) = 1 - e^{-(t/\Theta)^a} \tag{17.3}$$

17.3.6.1 Liquid-to-Liquid Thermal Shock. Table 17.10 summarizes the reliability data for four no-flow underfills subjected to LLTS using the PB8-4 test die with Cu/Ni/Au metallization and circular solder-mask-defined pads (TV1). Under this set of conditions, underfill E performed the best with the largest MTTF (Fig. 17.24).

Table 17.11 summarizes data for four no-flow underfills on test vehicle 2 (TV2) with a rectangular solder mask opening consisting of a hybrid mask and metal-defined pad (Fig. 17.25). In this case, underfill C performed worse on boards with rectangular openings compared with boards with circular openings (i.e., comparing TV1 and TV2).

Table 17.12 summarizes data for four no-flow underfills tested on test vehicle 3 with a circular solder mask opening and an organic solderability preservative (Fig. 17.26). Comparing the effect of

TABLE 17.10 Weibull Parameters for TV1 (PB84-Au-Circular) Subject to LLTS

Underfill	Θ	α	Cycles to first failure
C	1990	4.4	1000
D	1930	3.9	1000
E	2040	3.5	1000
F	1770	5.5	1100

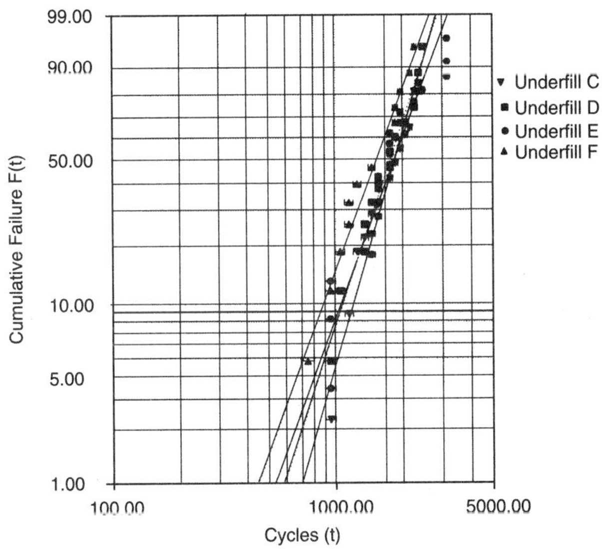

FIGURE 17.24 Weibull plot for TV1 subject to LLTS.

TABLE 17.11 Weibull Parameters for TV2 (PB84-Au-Rectangular) Subject to LLTS

Underfill	Θ	α	Cycles to first failure
A	1550	3.9	600
B	970	3.0	100
C	1180	5.4	600
G	2210	5.1	1200

substrate metallization based on Tables 17.10–17.12 shows that underfill C performed slightly worse on Cu/OSP relative to Cu/Ni/Au, whereas underfills E and F performed slightly better. Again, underfill E was the best performer, followed by underfill C.

Underfills E and F were subject to LLTS testing on TV7 with the results presented in Table 17.13 and Weibull plot in Fig. 17.27. Underfill E performed slightly worse on TV7 compared with TV3 with the first failure occurring 400 cycles earlier and an MTTF approximately 100 cycles lower. Underfill F performed slightly better with first failure occurring 200 cycles later on TV7 and the MTTF was approximately 300 cycles higher than on TV3.

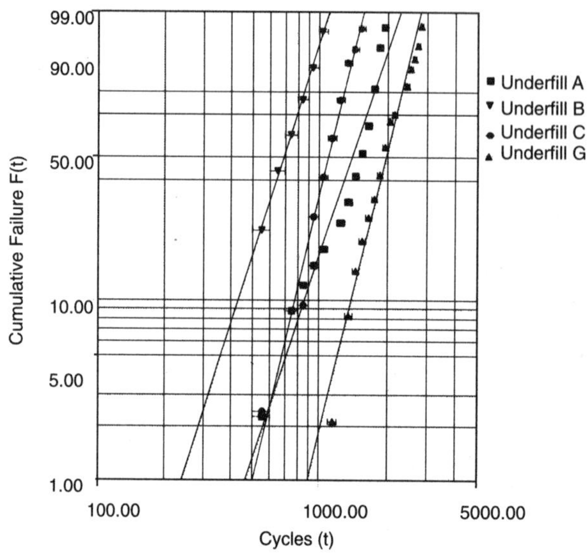

FIGURE 17.25 Weibull plot for TV2 subject to LLTS.

TABLE 17.12 Weibull Parameters for TV3 (PB84-OSP-Circular) Subject to LLTS.

Underfill	Θ	α	Cycles to first failure
A	1380	4.8	600
C	1830	5.5	1000
E	2170	5.2	1100
F	1840	6.0	800

Underfill C was tested on six of the seven test vehicles to examine the effects of bump geometry, chip size, metallization, and pad definition on reliability performance, with the results summarized in Table 17.14 and Fig. 17.28. The MTTF for the FA10-2 was approximately 400 cycles longer than the FA10-4. Similar results hold for the PB8 test die as well, with an approximate difference in MTTF of 300 cycles, with the smaller die having higher reliability. Underfill C was also more reliable on the PB8-4 format with circular-solder-mask-defined pads relative to a hybrid defined pad, with a difference in MTTF of approximately 800 cycles.

17.3.6.2 Air to Air Thermal Cycling. AATC results on TV2 (PB8-4-Au-rectangular) are shown in Table 17.15 and Fig. 17.29. The MTTF in AATC on TV2 was substantially lower for all the underfills tested except for underfill B. Underfills C, A, and G had first failure 300 cycles earlier in AATC, and their MTTF was reduced by 300 to 900 cycles. The more rapid failure rate can be attributed in part to the longer dwell time in AATC, which accelerates solder creep, and solder extrusion and increases solder fatigue. Hybrid pads impart a larger stress concentration, which could also promote more rapid joint fatigue.

Three additional commercial no-flow underfills were tested on TV1 (PB8-4-Au-circular) (Fig. 17.30), and two additional commercial no-flow underfills were tested on TV3 (PB8-4-OSP-circular) (Fig. 17.31). As seen with the other underfill chemistries tested, the time to first failure was lower than it was in LLTS for underfills D, E, and F, as shown in Table 17.16. In addition, the MTTFs for underfills E and F were approximately 400 cycles lower in AATC than LLTS testing. Underfill D had a, MTTF that was approximately 1000 cycles lower in AATC.

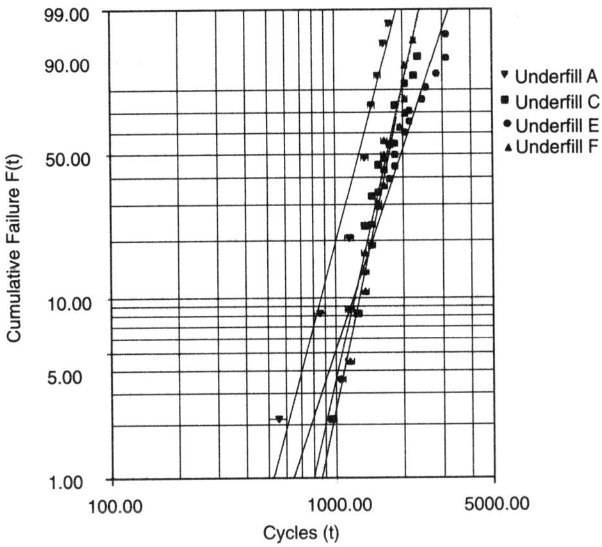

FIGURE 17.26 Weibull plot for TV3 subject to LLTS.

TABLE 17.13 Weibull Parameters for TV7 (PB82-OSP-Rectangular) Subject to LLTS.

Underfill	Θ	α	Cycles to first failure
E	2030	3.1	800
F	2170	4.9	1000

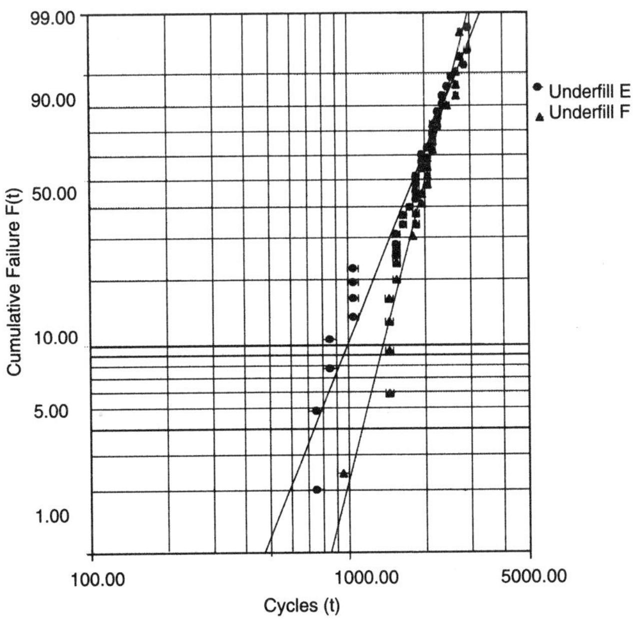

FIGURE 17.27 Weibull plot for TV7 subject to LLTS.

TABLE 17.14 Weibull Parameters for Underfill C on Six Test Vehicles Subject to LLTS.

Chip	Definition	Metallization	Θ	α
PB8-4	Circular	Cu/Ni/Au	1990	4.4
PB8-4	Rectangular	Cu/Ni/Au	1180	5.4
PB8-4	Circular	Cu/OSP	1830	5.5
PB8-2	Rectangular	Cu/Ni/Au	1510	3.3
FA10-4	Rectangular	Cu/Ni/Au	1300	7.8
FA10-2	Rectangular	Cu/Ni/Au	1690	6.0

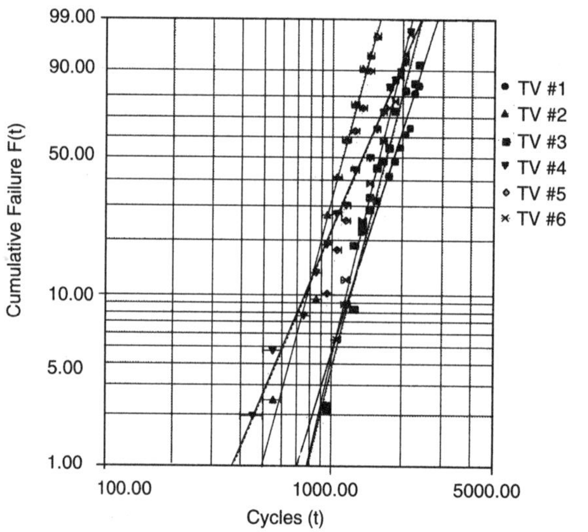

FIGURE 17.28 Weibull plot for underfill C on TV1–TV6 subject to LLTS.

TABLE 17.15 Weibull Parameters for TV2 (PB8-4-Au-Rectangular) Subject to AATC

Underfill	Θ	α	Cycles to first failure
A	800	3.8	300
B	1040	2.8	100
C	820	3.3	300
G	1340	7.8	900

AATC results for underfill E and F are summarized in Table 17.17. Underfill E passed the same number of cycles to first failure in AATC as in LLTS. However, the MTTF was lower by approximately 300 cycles. Underfill F required more cycles to reach first failure in AATC and had an MTTF that was approximately 100 cycles higher than the mean time to failure in LLTS. Additionally, the performance of both underfills E and F was slightly better on TV3 (OSP) than on TV1 (Cu/Ni/Au). This may be due to gold embrittlement in the UBM that increases the failure rate in TV1 during AATC.

On TV7, underfills E and F were tested (Table 17.18 and Fig. 17.32). Underfill F had a much lower reliability in AATC than in LLTS. The MTTF was over 900 cycles lower in AATC, and the first failure occurred 800 cycles earlier. Underfill E had superior performance in AATC, with an increase in MTTF of approximately 300 cycles, and it passed 1500 cycles before reaching first failure.

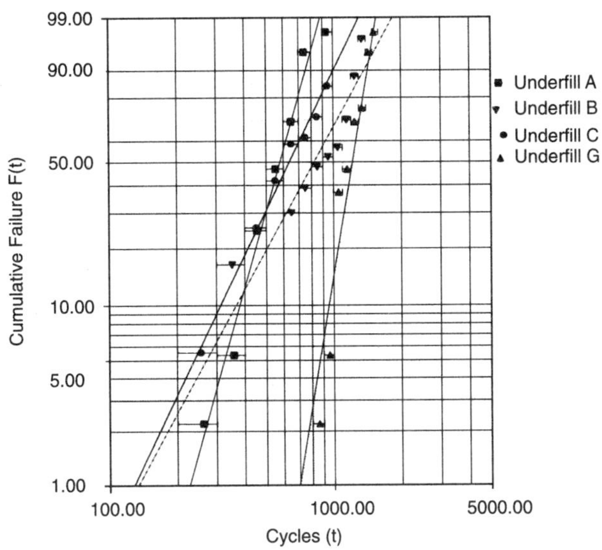

FIGURE 17.29 Weibull plot for TV2 subject to AATC.

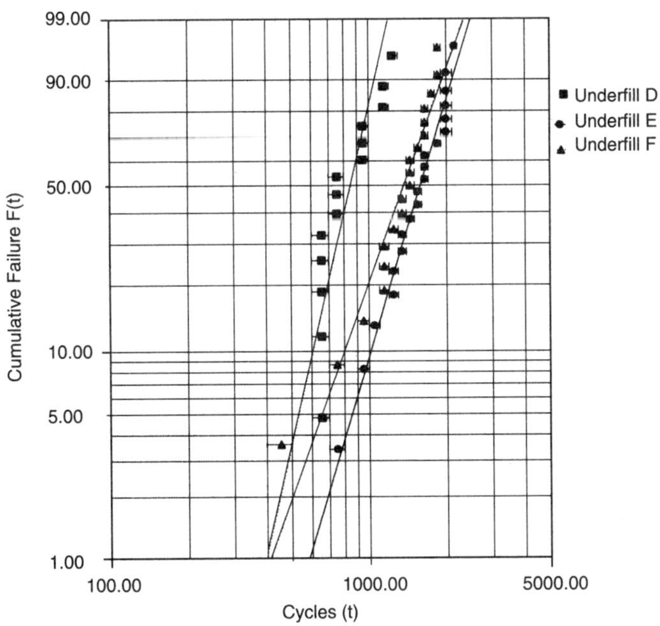

FIGURE 17.30 Weibull plot for TV1 subject to AATC.

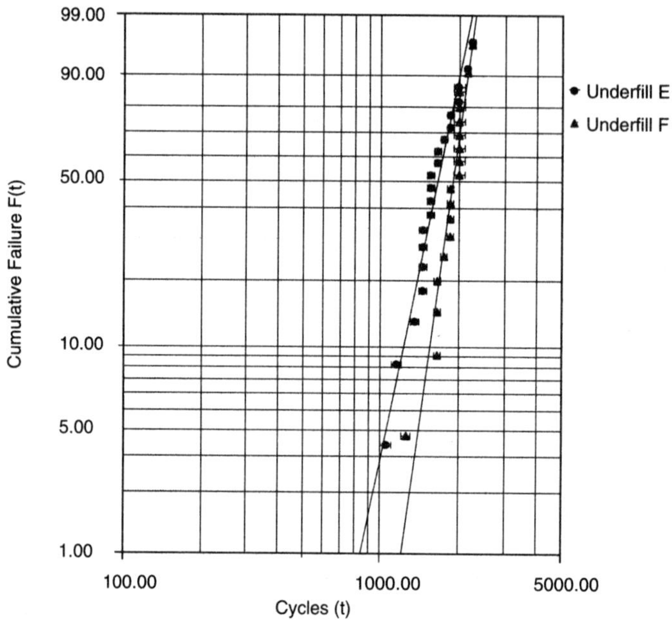

FIGURE 17.31 Weibull plot for TV3 subject to AATC.

TABLE 17.16 Weibull Parameters for TV1 (PB8-4-Au-Circular) Subject to AATC

Underfill	Θ	α	Cycles to first failure
D	910	5.5	600
E	1720	4.2	800
F	1510	3.5	600

TABLE 17.17 Weibull Parameters for TV3 (PB8-4-OSP-Circular) Subject to AATC

Underfill	Θ	α	Cycles to first failure
E	1750	6.3	1100
F	1970	9.4	1300

TABLE 17.18 Weibull Parameters for TV7 (PB82-OSP-Rectangular) Subject to AATC

Underfill	Θ	α	Cycles to first failure
E	2330	6.2	1500
F	1250	1.8	200

17.3.6.3 Temperature Humidity Aging and Autoclave. In addition to LLTS and AATC testing, TH testing was also performed. The commercial no-flow underfills tested did not appear to be susceptible to the 85°C/85 percent relative humidity test condition as shown in Tables 17.19 and 17.20.

Underfills C and A were also subjected to moisture preconditioning based on J-STD-020A standard levels 1 and 3 preconditioning. As the data in Table 17.21 show, both underfills passed level 3

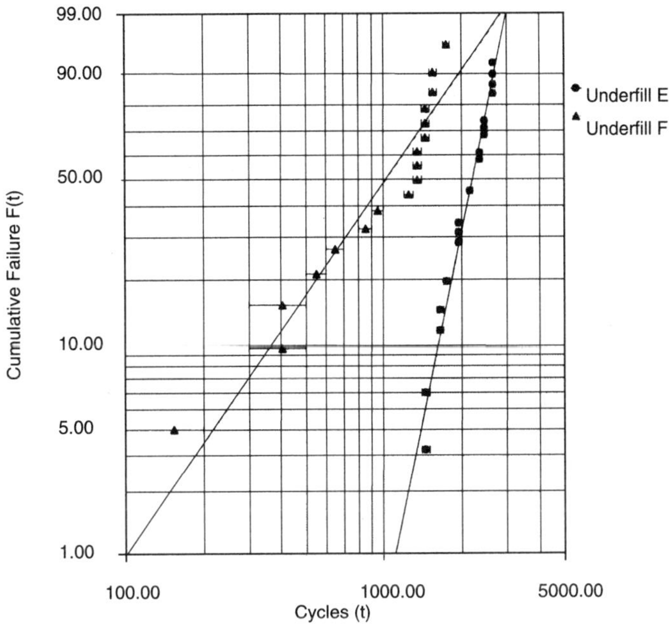

FIGURE 17.32 Weibull plot for TV7 subject to AATC.

TABLE 17.19 TH Reliability Data for TV3 (PB8-4-OSP-Circular)

Underfill	Failures at 1000 h/total sample size
A	0/32
C	0/32
D	0/20
E	0/20
F	0/20

TABLE 17.20 TH Reliability Data for TV1 (PB8-4-Au-Circular)

Underfill	Failures at 1000 h/total sample size
D	0/20
E	0/20
F	0/20

preconditioning. Underfill A came very close to passing level 1 preconditioning, with only one chip failing after reflow. Underfill C had over 50 percent of the assemblies pass level 1 preconditioning.

Underfills A and C were also subject to autoclave testing on TV3. The times to percentage failure are summarized in Table 17.22.

17.3.7 Reliability Data Discussion

In liquid-to-liquid thermal shock (LLTS) testing, using Cu/Ni/Au or Cu/OSP appears to have no significant impact on reliability when all other variables are held constant. Underfills C, E and F were

TABLE 17.21 J-STD 020 Preconditioning Data for TV3 (PB8-4-OSP-Circular)

Underfill	Preconditioning level	Passed chips/total tested
A	1	19/20
A	3	32/32
C	1	11/20
C	3	32/32

TABLE 17.22 Autoclave Test Data for TV3 (PB8-4-OSP-Circular)

Underfill	Percent failure 0 h	Percent failure 24 h	Percent failure 48 h	Percent failure 72 h	Percent failure 96 h	Percent failure 120 h
A	0	0	0	10	45	100
C	0	0	19	93	100	100

tested on PB8-4 test boards with circular mask openings with both Cu/Ni/Au and Cu/OSP metallization (TV1 and TV3). The difference in MTTF was less than 200 cycles for all three underfills. During air to air thermal cycling (AATC), the effect of the metallization on reliability appears to be material-dependent. For underfill E on TV1 and TV3, the difference in MTTF was 25 cycles. For underfill F, the MTTF for the Cu/Ni/Au metallization was 460 cycles lower than the MTTF for Cu/OSP metallization.

For all but two test board configurations, AATC testing resulted in lower MTTF than LLTS testing. This difference ranged from 250 cycles with underfill F on TV1, to 870 cycles on TV2. The two exceptions to this trend were underfill E on TV7 and underfill F on TV3. In these two cases, the MTTF was higher during AATC. It is interesting to note that both these test vehicles used Cu/OSP metallization. For the material systems tested, the Cu/Ni/Au substrates always had lower reliability in AATC testing than in LLTS.

Across all seven underfills tested, there appears to be no direct relationship between the number of cycles to first failure and the MTTF. For example, both underfill A and underfill C experienced their first failure at 600 cycles on TV2 during LLTS testing. However, the difference in MTTF between the two materials was approximately 370 cycles. Another example of this can be seen when comparing underfills D and F on TV1 during AATC testing. Both materials experienced first failure at 600 cycles, but the difference in MTTF was over 600 cycles.

The data obtained using underfill C indicate that chip size and mask design may affect LLTS reliability. When comparing TV2 and TV4, the only difference between the two substrates is the size of the die (10 versus 5 mm), both having the same bump density. The metallization (Cu/Ni/Au) and the mask opening (rectangular) are identical. The MTTF for the small test die was greater by 325 cycles. This difference is also seen when comparing TV5 and TV6. Again, the only difference in board/die configuration is the size of the die (10 versus 5 mm). In this case, the MTTF values differed by 390 cycles in LLTS, with the smaller die being more reliable.

When comparing the reliability data for underfill C on TV1 and TV2, the MTTF was 800 cycles higher for TV1. This suggests that underfill C is more reliable when used on a circular mask opening as opposed to rectangular. However, there are insufficient data to determine whether this large difference is material-dependent.

When comparing the reliability data for underfill C on TV4 (perimeter array) and TV6 (area array) during LLTS, the MTTF was 200 cycles higher for TV6. This difference is not statistically significant enough to indicate that bump density has an effect on reliability.

Of the six no-flow underfills tested, two required a postcure (underfills B and G). Overall, no-flow underfill B was shown to have poor reliability in both LLTS and AATC testing, with first failure occurring after 100 cycles and MTTF values equal to or less than 1000 cycles. No-flow underfill G, conversely, had the highest reliability of all the materials tested on the same test vehicles. These data do not indicate whether a postcure is beneficial to no-flow underfill reliability.

17.3.8 Summary of No-Flow Underfills Reliability

Different test vehicles were used to evaluate the effect of varying chip size, interconnection density, pad surface finish metallization, and solder mask opening design. Accelerated reliability tests performed included liquid/liquid thermal shock (LLTS), air/air thermal cycling (AATC), moisture sensitivity preconditioning, and temperature-humidity aging (TH). This discussion has shown that low-cost FC technology based on select commercial no-flow underfills is reliable, surviving 1000+ LLTS cycles and having MTTFs in excess of 2000 cycles. It has also been demonstrated that bond pad definition can have a large effect on underfill reliability. Small die also appear to be more reliable than large die when all other design factors are equal, while bump density appears to not have a significant affect. No correlation between cycles to first failure and mean time to failure was found. The effect board metallization has on reliability depends on both the test method used and the board/die configuration. There was also no direct correlation between the use of a postcure and overall reliability.

17.3.9 Failure Mode Analysis

This section will describe several failure modes of commercial no-flow underfills and describe the progression of failure.[36,38]

*17.3.9.1 **Solder Fatigue.*** The dominant failure mode for FC assemblies using no-flow underfills is solder fatigue cracking. Ultimately, solder fatigue cracking was responsible for all electrical failures observed in test vehicles subjected to LLTS and AATC. All the electrical failures produced during AATC occurred prior to delamination of the chip-underfill or the board-underfill interfaces. Additionally, all electrical failures observed in assemblies subjected to LLTS occurred either prior to delamination, or, if after delamination showed no significant correlation with the locations of delamination. That is, fatigue cracks causing chip failure were not typically located in regions of the chip experiencing delamination. Based on the AATC and LLTS electrical failure observations, it can be assumed that the mode of interconnection failure was solder fatigue leading to fatigue cracking. Figure 17.33 shows an example of solder fatigue cracking occurring through the UBM side of the bump propagating through the solder and causing an electrical open.

In order to substantiate the experimental fatigue life results, a simple analytical approach was used, based on the Suhir trilayer model.[10] The analytical formulation is based on trilayer structures and can be used to estimate the interfacial shear stresses in a FCOB assembly subjected to a thermal load (such as temperature cycling). Under the assumption of interfacial adhesion, the shear stress can be used to approximate the shear strain in the solder interconnection. The Coffin-Manson relationship can be used to estimate the mean time to solder fatigue failure, based on the plastic strain on the solder interconnection.[11,23,24] Figure 17.34 shows an FCOB trilayer assembly used for the solder fatigue analysis.

The top and bottom materials are represented as materials 1 and 3, which refer to the silicon die and the FR-4 substrate. Suhir showed that the maximum shearing stress (σ_{xz}) occurs at the die-underfill interface and is given by Eq. (17.4).[37]

$$\sigma_{xz} = \frac{\Delta\alpha\Delta T}{\sqrt{\lambda\kappa}} \frac{\sinh kx}{\cosh kl}$$

$$\Delta\alpha = \alpha_3 - \alpha_1$$

$$\Delta T = \text{cycling temperature differential}$$

$$\lambda = \frac{h_1^2}{12D_1} + \frac{h_3^2}{12D_3} + \left(\frac{h_1 + h_3}{2\sqrt{D}}\right)^2$$

$$D_1 = \frac{E_i h_i^3}{12(1 - \nu_i^2)};\ i = 1,3$$

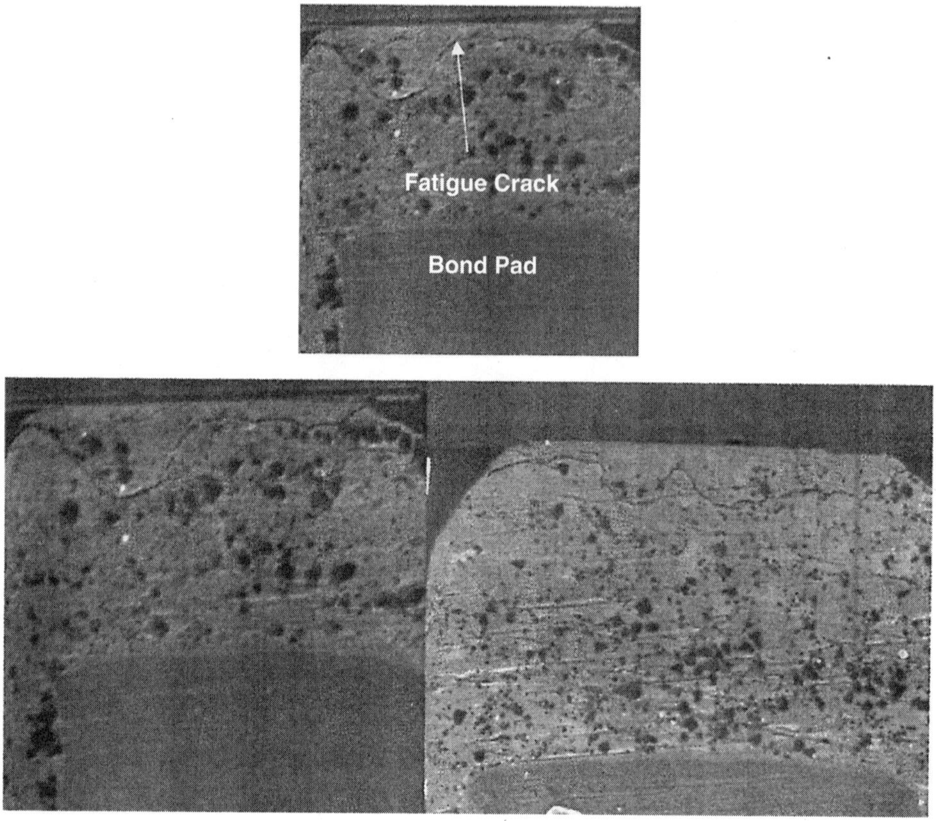

FIGURE 17.33 Scanning electron micrographs of a typical solder fatigue crack.

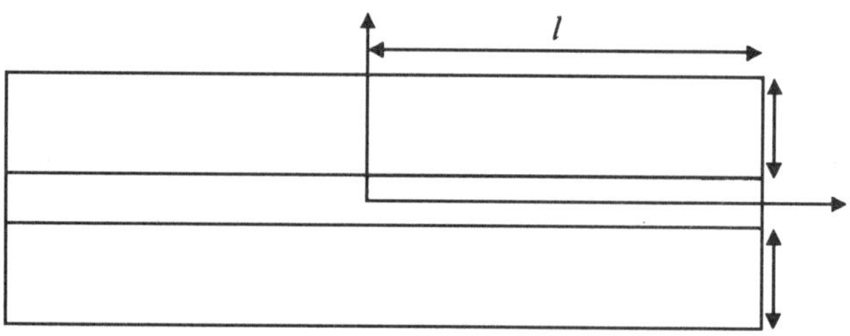

FIGURE 17.34 Structure used for solder fatigue analysis.

$$D = D_1 + D_3$$

$$\kappa = \frac{h_1}{3G_1} + \frac{h_3}{3G_3} + \frac{2h_2}{3G_2}$$

$$\kappa = \sqrt{\frac{\lambda}{\kappa}} \qquad (17.4)$$

where h_i = thickness of the ith material

v_i = Poisson's ratio of the *i*th material
E_i = elastic modulus of the *i*th material
G_i = shear modulus of the *i*th material

A Coffin-Manson relationship (Eq. 17.5) is then used to estimate solder joint fatigue life corresponding to the MTTF.[33,35,37]

$$N_p = 1.2928\ (\epsilon_{solder})^{-1.96} \tag{17.5}$$

Based on the strain values calculated from Eq. (17.4), the Coffin-Manson equation predicts a fatigue life of 2160 cycles. Table 17.23 shows a comparison of the predicted MTTF and the LLTS and AATC experimental results. The analytical prediction of the MTTF provides an order of magnitude approximation of the true fatigue life of an FCOB assembly using no-flow underfills. In the best case for TV1 subjected to LLTS, the predicted life is within 10 percent of the experimental life.

17.3.9.2 Delamination. Delamination refers to a loss of adhesion between surfaces. In the case of an underfilled FC, the adhesion loss is at the chip passivation-underfill interface, the underfill-solder interface, or the underfill-solder mask interface. Delamination in the test vehicles subjected to LLTS testing occurred at the chip-underfill interface. In most cases, the delamination initiated at the corners of the test die. This results from the large stresses generated by the CTE mismatch between the silicon chip, the FR-4 substrate and the no-flow underfill. Underfill delamination typically does not occur until after solder fatigue failure has begun to occur.

Underfill A exhibited chip-underfill delamination in all LLTS assemblies subject to testing and only one assembly failed J-STD-020A level 1 preconditioning due to chip-underfill delamination. Delamination in LLTS testing was very similar between underfill A and underfill C. Typically the delamination was minor and originated predominantly in the chip corners and chip perimeter. Figure 17.35 shows a typical delamination progression found in TV2 assembled with underfill A. In all cases, the first instances of delamination occurred after 1200 cycles. From the time of delamination initiation to chip failure, the regions of delamination remained small and experienced very little growth. Assemblies subjected to AATC did not show any evidence of delamination prior to chip failure.

Underfill B exhibited chip underfill delamination in test vehicles subjected to LLTS and AATC Underfill B delamination occurred the earliest and was the most severe of all underfill materials tested. Figure 17.36 shows typical delamination progression for AATC. The first signs of delamination appeared at 700 cycles, with delamination spreading to all four sides of the chip by 1200 cycles.

Underfill C exhibited chip-underfill delamination in all test vehicles subjected to LLTS. Over half of underfill C assemblies subjected to J-STD-020A level 1 preconditioning failed due to chip-underfill delamination. Delamination occurring in LLTS was minor and originated in the chip corner or along the chip perimeter. Figure 17.37 shows typical delamination progression for TV1 assemblies subjected to LLTS.

Underfill D, when subjected to LLTS for TV2 showed very small amounts of chip-underfill delamination. Delamination began around 1000 cycles and progressed through 1800 cycles. Again, the delamination can be found in the chip corners and perimeter (Fig. 17.38).

TABLE 17.23 Summary of Experimental and Predicted Mean Times to Failure

Reliability Test	Test Vehicle	Experimental MTTF	Predicted MTTF
AATC	2	650	2160
LLTS	2	1150	2160
LLTS	1	2000	2160
LLTS	3	1700	2160

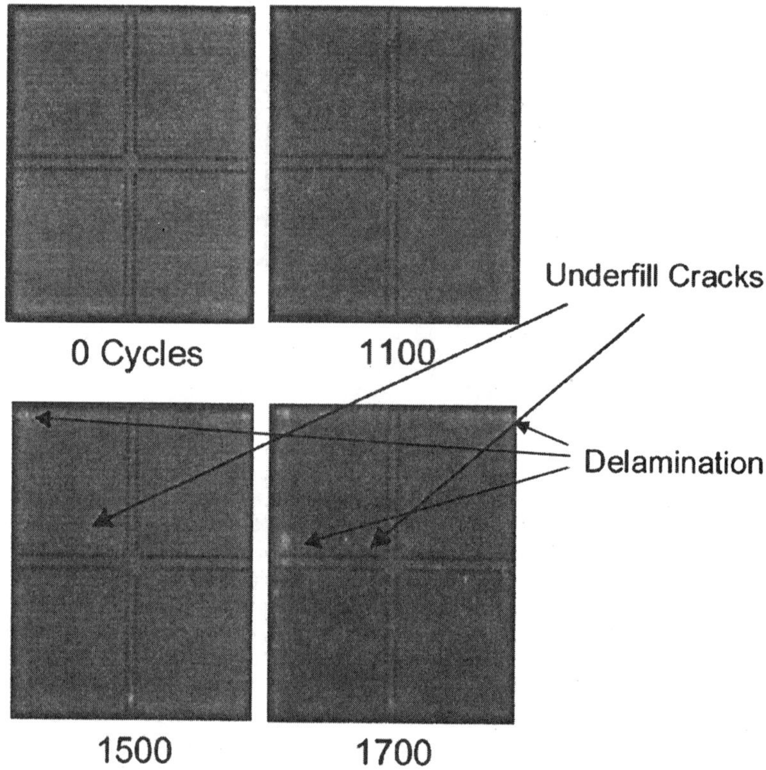

FIGURE 17.35 Delamination progression for underfill A-TV2, subjected to LLTS.

Underfill E showed little to no chip-underfill delamination in LLTS through 1800 cycles. Figure 17.39 shows a progression of acoustic micrographs through 1800 cycles showing no visible delamination. However, underfill cracking was visible at 1400 cycles.

Underfill F showed chip-underfill delamination in assemblies subjected to LLTS. Figure 17.40 shows typical delamination progression occurring along the perimeter of the chip for LLTS assemblies. Delamination initiated around 600 cycles and grew as cycling continued.

Underfill G exhibited chip-underfill delamination only in assemblies subjected to LLTS. Assemblies subjected to AATC did not show any delamination prior to electrical failure. Figure 17.41 shows typical delamination progression for the LLTS assemblies. Delamination first occurred around 1900 cycles and exhibited only minor growth through the next 900 cycles.

17.3.9.3 Underfill Fillet Cracking. Underfill fillet cracking was a common underfill material failure for no-flow underfills, although it can also occur in filled capillary underfills. Fillet cracking occurs mainly in assemblies subjected to LLTS and AATC cycling. Fillet cracks may be classified as one of three types: chip-side cracks, board-side cracks, and complete cracks. Figure 17.42 shows examples of the three crack types.

Chip-side cracks are small cracks that originate at the chip-underfill interface and usually form a V-shaped pattern that extends partially down the fillet. Board-side cracks are short cracks that extend up the fillet or run parallel to the fillet edge. Chip-side cracks can lead to localized areas of delamination between the underfill and the side of the chip, whereas board-side cracks can lead to localized areas of delamination between the underfill and substrate-solder mask. A complete fillet crack extends from the top of the fillet to the substrate and propagates completely through the fillet. In cases of extreme cracking, the underfill fillet separates from the chip, eliminating the stress-relief properties of the underfill.

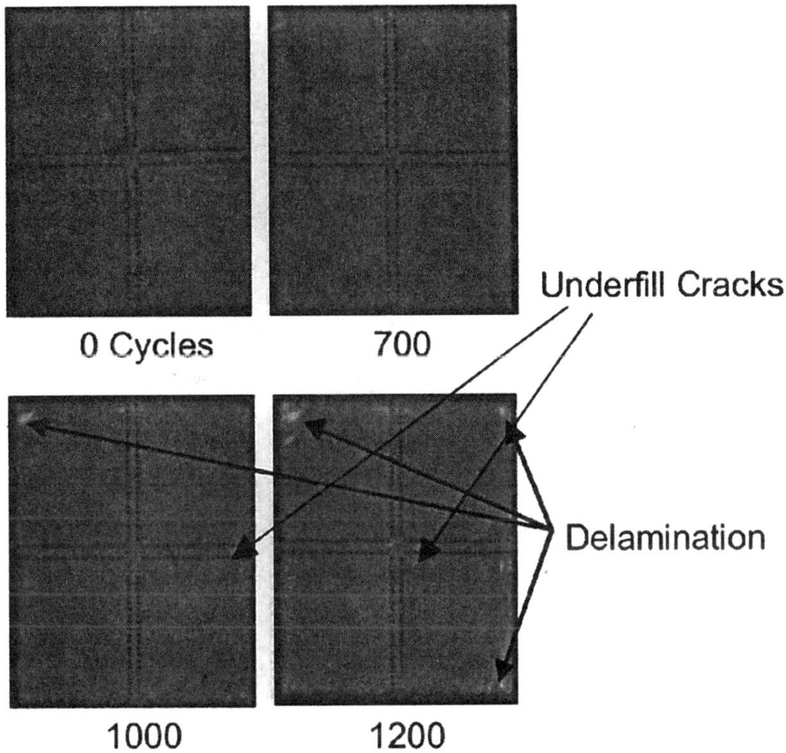

FIGURE 17.36 Delamination progression for underfill B-TV2, subjected to AATC.

Fillet cracking is driven by the large hoop stress produced by the underfill generated during the cooling cycle of LLTS. The net difference between the CTE of the underfill (~70ppm/°C) and the silicon (~3ppm/°C) produces very large hoop tensile and interface shear stress in the fillets.

All underfills tested fillet cracked during cycling. The number of cycles before cracking occurred and the severity of the cracking varied from material to material. For example, underfill B began to fillet crack after 200 cycles LLTS, and underfill E began to fillet crack after 1300 cycles LLTS.

17.3.9.4 Bulk Underfill Cracking. In addition to fillet cracking, cracks may be seen in the bulk underfill underneath the test die occurring during AATC and LLTS testing. Typical bulk underfill cracks in no-flow underfills are shown in Fig. 17.44. It is important to notice that the cracks typically occur adjacent to the solder joints, propagating between the joints. In addition to cracks propagating between solder joints, cracks were also seen to initiate and propagate along the metallization pattern on the test die, as seen in the right-hand second image in Fig. 17.44. Moreover, it has been observed that solder from the joints tends to extrude into underfill cracks during AATC and LLTS cycling, which has the potential to cause electrical shorting, as shown in Fig. 17.45.

Figure 17.46 shows a series of C-SAM images at the chip-underfill interface where fine lines near the solder joints appear and expand as cycling continues. Cross sections show that these fine lines correspond to cracks in the bulk underfill, as illustrated in the figure, indicating that C-SAM analysis can be used to detect bulk underfill cracks.

Bulk underfill cracking has been seen in no-flow underfills as well as in capillary underfills. In addition, for no-flow underfills, bulk cracking occurs before chip failure due to solder fatigue cracking. The chip shown in the acoustic images in Fig. 17.46 failed at 2400 cycles of LLTS, yet underfill cracking is visible in acoustic images at 1400 cycles.

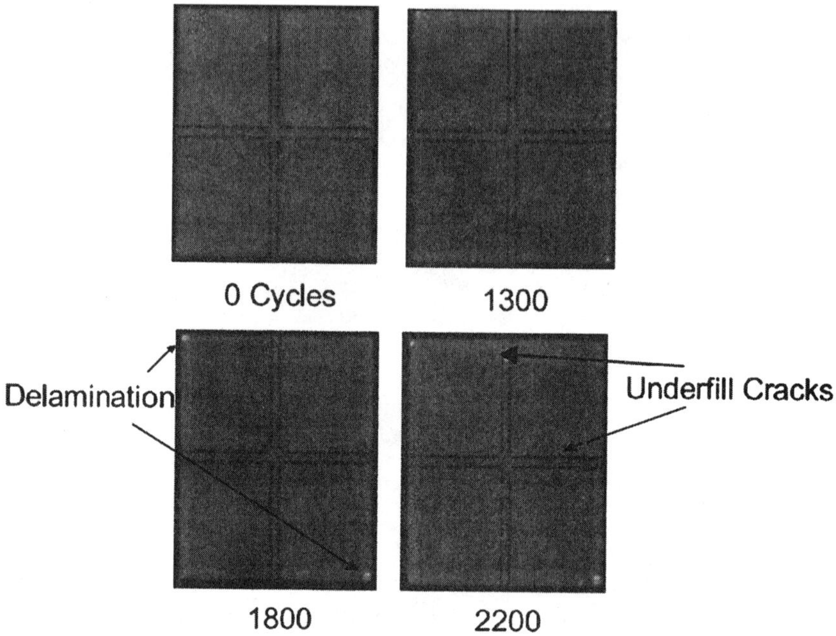

FIGURE 17.37 Delamination progression for underfill C-TV1, subjected to LLTS.

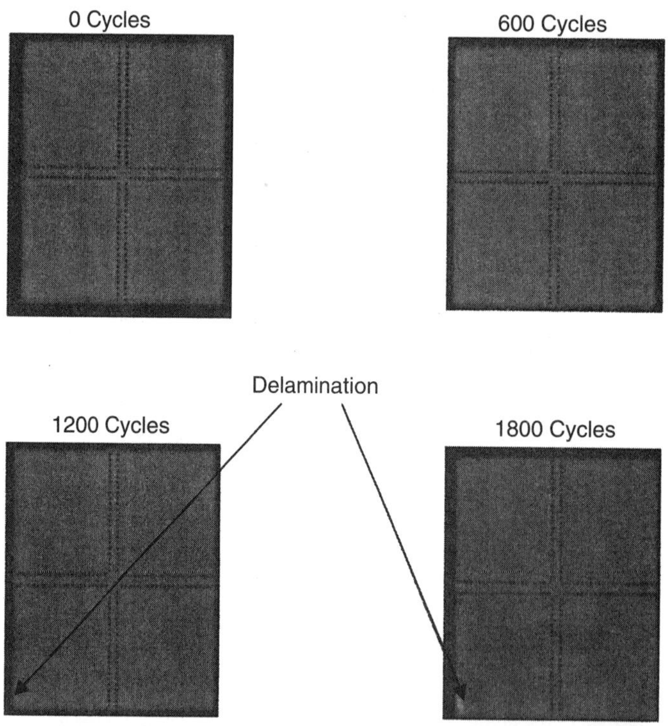

FIGURE 17.38 Delamination progression in underfill D-TV2, subjected to LLTS.

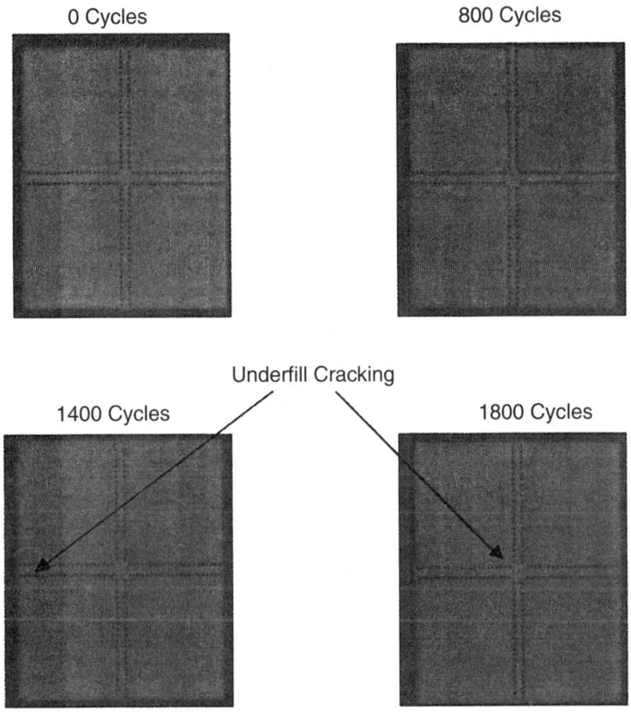

FIGURE 17.39 Cycling progression for underfill E-TV2, subjected to LLTS.

Figure 17.47 shows C-SAM images of a cycled chip showing underfill cracking at the chip-underfill interface and the underfill-board interface.[12]

Bulk underfill cracking was seen in all seven underfills tested to varying degrees. Cracking occurred the earliest and was the most severe in underfill B. Underfills A, C, E, and F all exhibited similar amounts of bulk cracking. Minimal underfill cracking was seen in underfills D and G.

Figure 17.48 shows a schematic of copper circuit trace on a substrate covered by solder mask and underfill. A simple analysis of this microstructure gives one explanation for the underfill and substrate cracks observed. During the reflow process, the underfill material hardens at a temperature of approximately 200°C and locks the assembly in a zero stress state at this temperature. As the assembly cools to room temperature, all four materials attempt to contract different distances based on their respective CTEs. The larger the CTE, the more the material wants to contract. Based on the approximate CTE values presented in Fig. 17.48, both the underfill material and the substrate will contract to a greater extent than the copper trace. The difference in CTE between the substrate and the underfill generates large in-plane stresses in the underfill layer. When the assembly is cycled, it experiences temperatures as low as -55°C, which represents a ΔT of 250°C from the zero stress state. The large temperature change produces large cyclic in-plane stresses that may promote underfill cracking. Sharp corners found on the copper circuitry generate local regions of stress concentration that increase the tendency for crack initiation near the corners of the copper trace. Cracks that form in the solder mask have a tendency to propagate into the bulk underfill layer. Figure 17.48 shows the areas of stress concentration that promote underfill and substrate crack initiation.

Figure 17.49 indicates that underfill cracks can also form between solder bumps. An analysis similar to the one just presented provides one explanation for underfill cracks forming between solder interconnections. Figure 17.49 shows a top-view schematic of solder interconnections surrounded by underfill material. The CTE of the underfill is nearly three times greater than the CTE of the solder. During cycling, the large difference in CTE generates a stress concentration along the circumference of the solder-underfill interface. Any micrcracks in the underfill that exist along the interface can be

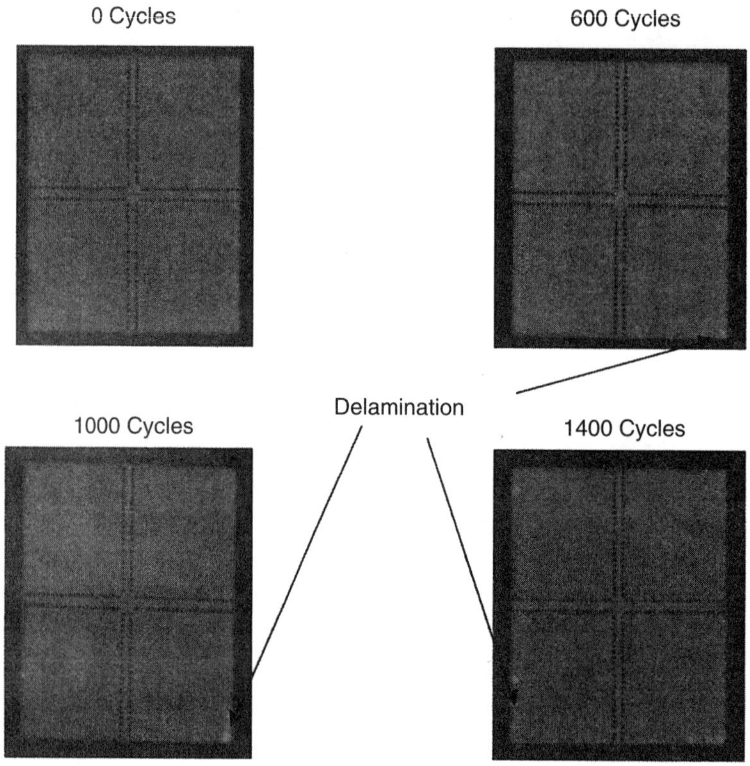

FIGURE 17.40 Delamination progression for underfill F-TV2, subjected to LLTS.

initiation points for underfill cracks. The most common cracks that initiated along the circumference of the solder occurred in the narrowest region between interconnections. This region has the smallest cross-sectional area and therefore is the most likely location for fatigue crack initiation.

17.3.10 Summary of Failure Mode Analysis

The preceding discussion has shown that low-cost FC technology based on select commercial no-flow underfills is reliable, surviving 1000+ LLTS cycles and having MTTF in excess of 2000 cycles. It has also been demonstrated that bond pad definition can have a large effect on underfill reliability. Small die also appear to be more reliable than large die when all other design factors are equal. Solder interconnection density also has a positive effect on reliability, where higher solder interconnection densities tend to be more reliable.

In addition, this discussion has presented a comprehensive failure-mode analysis of FC systems using no-flow underfills. The dominant failure mode was solder fatigue cracking. Due to the high adhesion strength of the no-flow systems, underfill delamination was found to be a secondary failure mode generally occurring after fatigue failure. For the first time, bulk underfill cracking was reported as a dominant failure mode in no-flow materials, driven by local CTE mismatches. Fillet cracks are also reported and discussed.

17.4 WAFER-LEVEL FC PROCESSING

FC technology has been shown to be a reliable method for the direct electrical and mechanical attachment of silicon chips to polymeric substrates. However, the assembly of encapsulated FC circuit

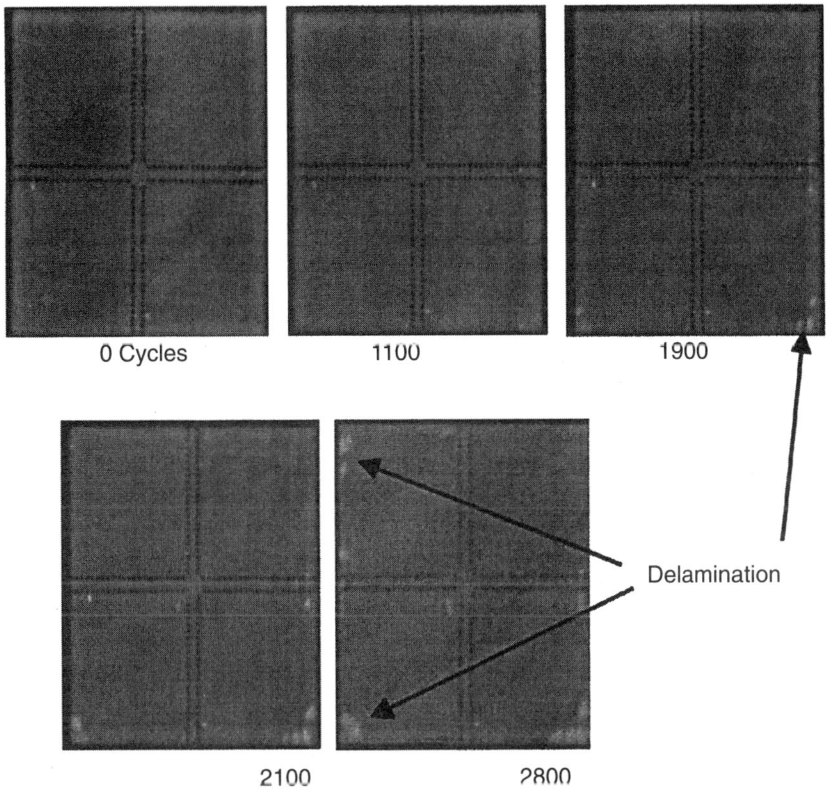

FIGURE 17.41 Delamination progression for TV2 underfill G, subjected to LLTS.

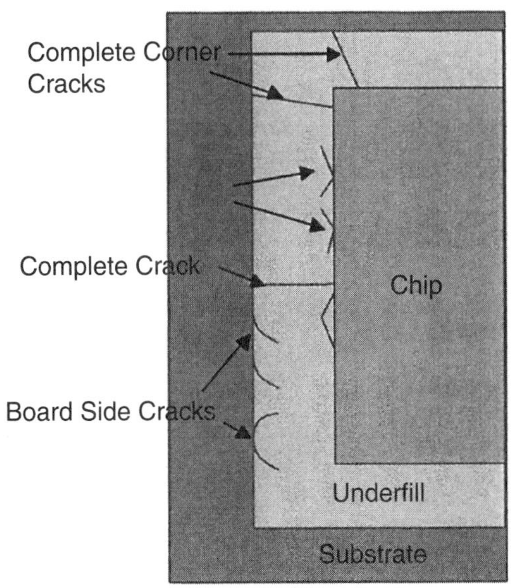

FIGURE 17.42 Underfill fillet crack types.

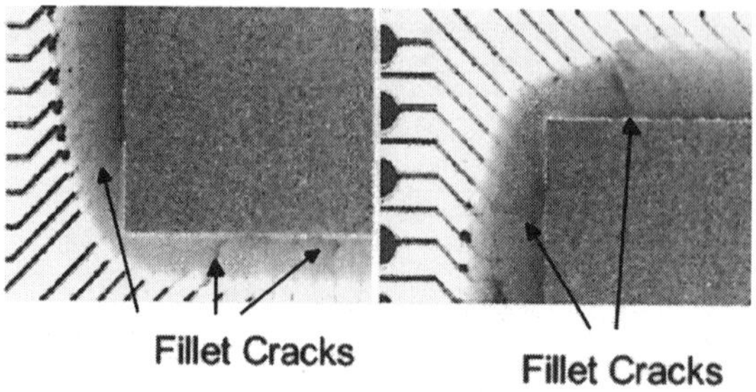

FIGURE 17.43 Typical fillet cracks found in underfill B assemblies subjected to LLTS.

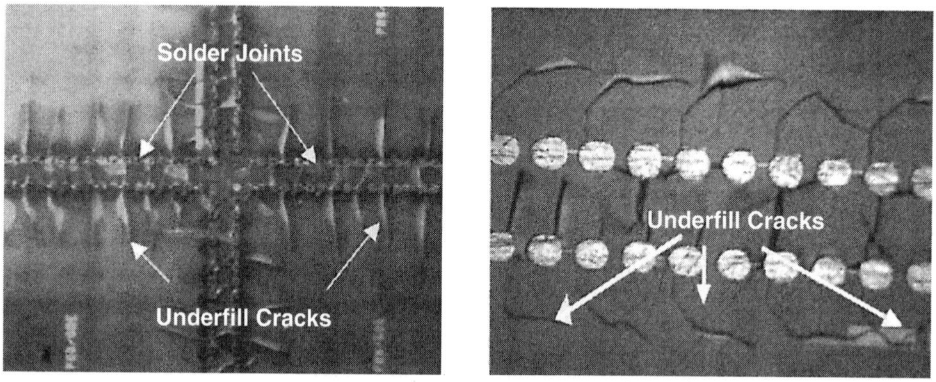

FIGURE 17.44 Examples of bulk underfill cracking.

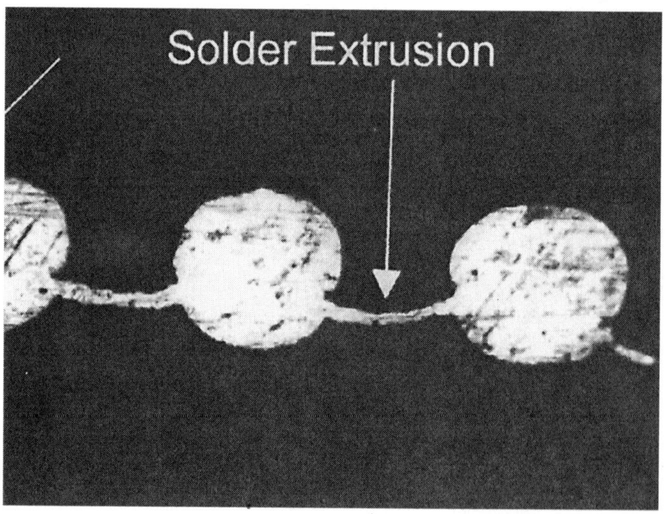

FIGURE 17.45 Example of solder extrusion into underfill cracks between adjacent bumps.

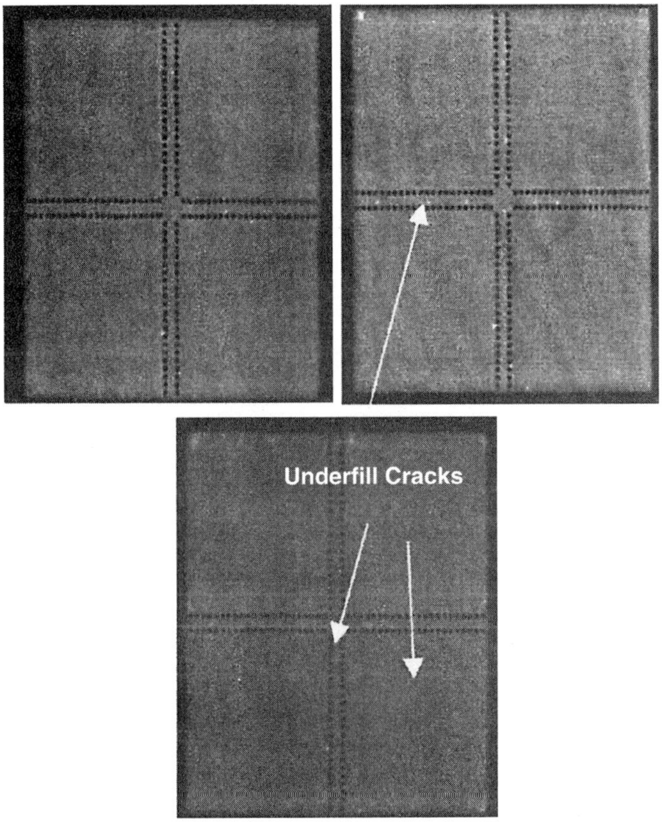

FIGURE 17.46 Underfill E C-SAM images at 0 cycles (*top left*), 1400 cycles (*top right*), and 2200 cycles (*bottom*).

FIGURE 17.47 C-SAM images of TV3 subject to LLTS at the chip-underfill interface (*left*) and the underfill-board interface (*right*).

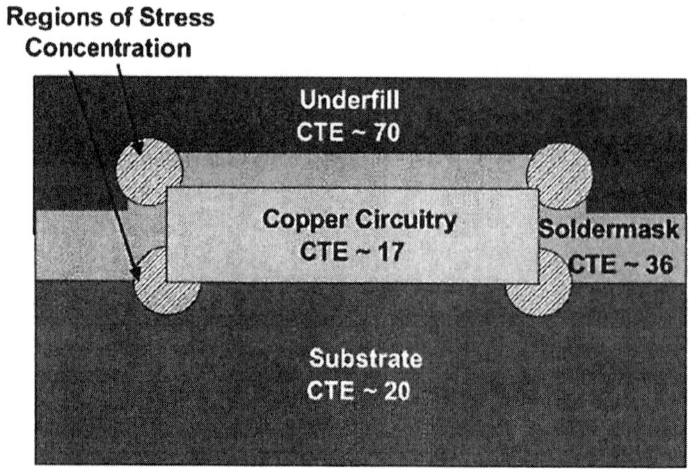

FIGURE 17.48 Stress concentration regions produced by substrate circuitry.

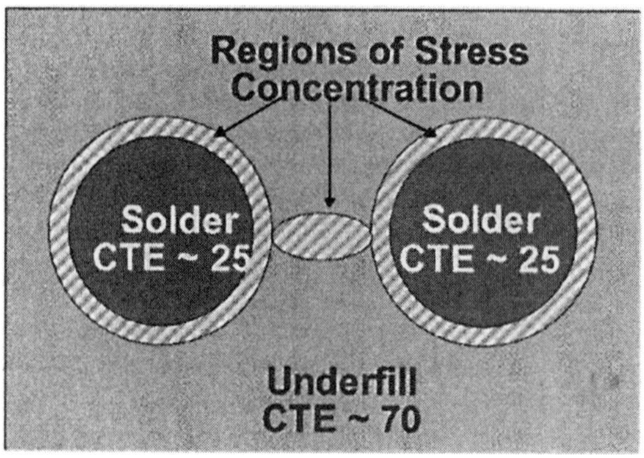

FIGURE 17.49 Stress concentrations caused by solder interconnections.

boards remains a time consuming and expensive process. This discussion focuses on the development of innovative material systems and process technology that can potentially make FC assembly transparent to SMT and high speed packaging lines by eliminating flux application, underfill application, and underfill cure from the assembly process.[15] The process under development is unique in that it uses a wafer-level underfill approach, a placement process comparable with CSPs (i.e., low placement forces), and underfill systems that enable simultaneous reflow of solder interconnections and cure of the underfill. The basic concept is to apply modified no-flow (reflowable) underfill encapsulants to FC bumped wafers as a form of wafer scale packaging. The wafer-applied underfills dry to a solid film, enabling packing, shipping, handling, and board assembly similar to CSPs. On reflow, the wafer-applied underfills liquefy, wet the substrate, and provide fluxing action on the solder-joints bond-pads, resulting in a high reliability underfilled FC. The wafer-applied fluxing underfill solid presents several non-trivial challenges, but none of the required properties are mutually exclusive from a materials standpoint.[11,15] The process is considerably different from other solid underfill film approaches such as substrate-applied dry-film laminated underfills[11] or wafer-applied solid-polymer film interposers and flux films.[11]

The new assembly process under development is shown in Fig. 17.4. It provides numerous advantages over conventional FC processes because it saves floor space and capital costs by removing underfill dispensers and cure ovens. Moreover, the wafer application of underfill is an enabling technology for fine-pitch FC processing (i.e., ≤150 μm). The new process could also increase worldwide FC implementation by reducing the process learning curve and by eliminating special training requirements necessary for current FC dedicated equipment. The program currently underway focuses on the development of a wafer-level application process, determination of the underfill material characteristics for application on bumped wafers, characterization of the FC assembly process, and reliability qualification of FC test vehicle assemblies. This discussion presents an overview of low-cost FC processing based on wafer-applied underfill systems, an analysis of several wafer-level deposition techniques on solder bumped wafers, an examination of critical factors and challenges involved therein, a feasibility analysis of high-speed placement of underfill coated bumped chips, and a characterization of the placement process based on bump and fiducial imaging using the vision system of a commercial placement system. Process challenges arise with the underfill application after bumping. This discussion investigates the challenges associated with polymer film thickness and uniformity, including edge effects, shadowing, and nonconformal coatings over the complex bump topologies of FC wafers. The effects of viscosity, wetting angle, and multiple underfill layers on coating uniformity are also discussed. The process technology, challenges, and preliminary yield analysis associated with the assembly of preunderfilled FCs directly to low-cost substrates is also presented. Of particular interest is how a placement machine handles, images, and places coated chips. Also addressed, but not analyzed, is the reflow of underfill coated wafers. This process step will be analyzed in future publications.

17.4.1 Desired Coating Characteristics

The reflowable underfill may be applied either before or after the prebumped wafer is diced. For the purposes of this discussion, only the case in which the wafer is coated before dicing is examined. In this case, the underfill is applied to a prebumped wafer over the complex surface topology of the bumps and active circuitry.

In order for the coated die to reflow and make a reliable connection, the coating should have certain desired geometric and surface properties. Some application methods produce shadowing effects behind the bumps as seen in Fig. 17.50*a*. This can lead to voiding in the underfill during FC assembly. Other methods promote a conformal coating as shown in Fig. 17.50*b*. This profile creates a thick film coating over the bumps, which can have an impact on the reflow process. A nonconformal geometry, shown in Fig. 17.50*c*, is the preferred coating geometry. During reflow, the underfill liquefies, the flux reduces the surface oxides, the solder wets the substrate pads, and the die collapses, forming the interconnections. As the die collapses, underfill flows under compression forces out past the die edges and wets up the sides of the die, forming fillets.

The coating geometries are determined in part by the zero-shear-rate viscosity of the underfill and by the coating method. If the zero-shear-rate viscosity is low enough, the material will tend to self-level and create the profile shown in Fig. 17.50*c*. The gravity-driven flow enabling self-leveling is governed by wafer-applied underfill rheology under low shear rates. The coating technique also has an impact on the nonconformal coating characteristics of the applied underfill film. Inherently, certain coating techniques promote nonconformal coats based on the physics of wetting. For example, surface tension driven coating techniques tend to promote nonconformal coatings, whereas pressure-driven coating techniques tend to promote conformal coating. Figure 17.51 shows the viscosity of a

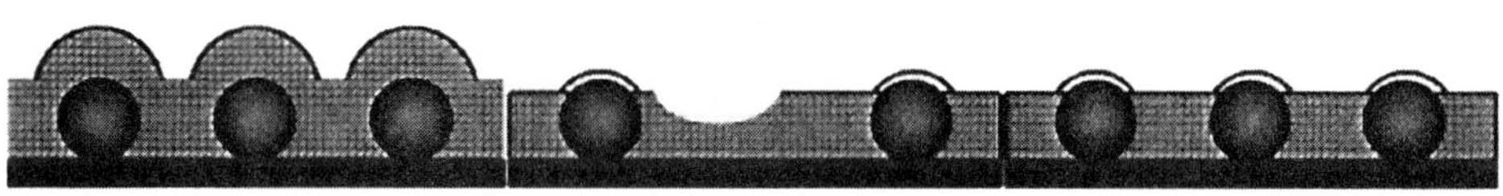

(a) (b) (c)

FIGURE 17.50 Profile of coated die: (*a*) shadowing effect, (*b*) conformal coating, (*c*) desired nonconformal coating.

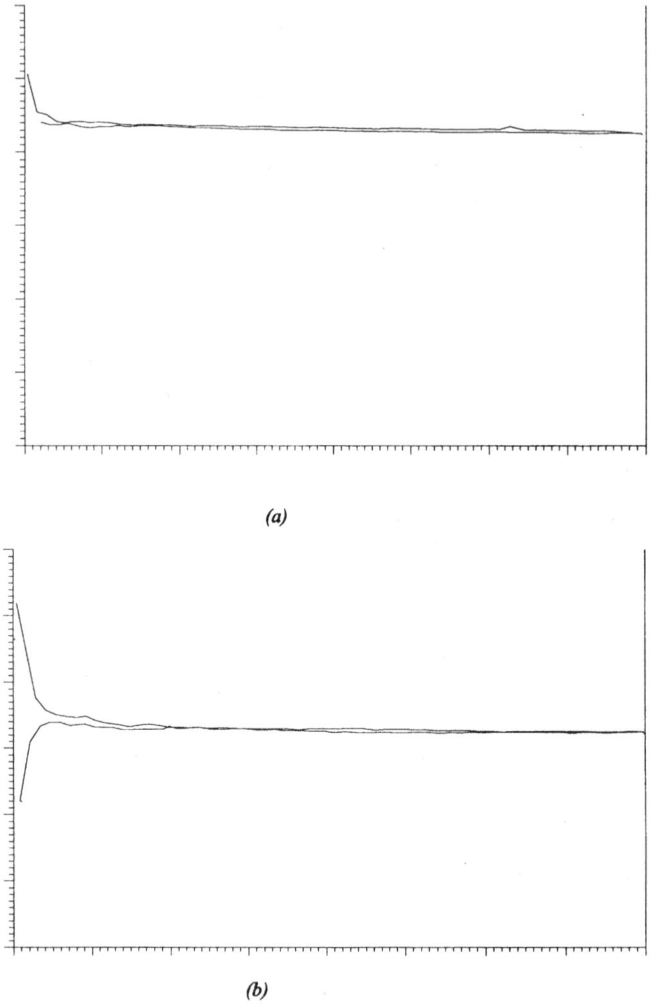

FIGURE 17.51 Apparent viscosity of material A under various shear rates: (*a*) material A and (*b*) material A diluted.

prototype wafer-level underfill as a function of varying shear rate. Both solvent-diluted and undiluted versions of a prototype underfill (material A in Table 17.24) were analyzed as representative materials. The prototype underfill rheology was measured using a TA Instruments rheometer at 25°C. Notice the order-of-magnitude decrease in the high-shear-rate viscosity when the underfill is diluted with 50 percent solvent.

Equally important is the wetting angle of the underfill material. Large wetting angles are especially undesirable. If the material behaves as shown in Fig. 17.52*a* with a wetting angle of larger than 90 degrees, then the underfill has a tendency not to wet the wafer surface or bump surfaces. Moreover, the underfill can be prone to dewetting the surface. Such large wetting angles can result in voiding in the regions where the bumps contact the wafer, undermining the integrity of the underfill to passivation surface. A wetting angle near 90 degrees, shown in Fig. 17.52*b*, improves wafer-level processing but can also be prone to underfill void formation. Wetting angles near 90 degrees also can cause the material to bead at the center of the die, resulting in placement problems during assembly. If the material is thickest at the center of the die, the die could roll or twist out of alignment after placement due to the instability from having a single contact point. Ideally, the wetting angle should be 50 degrees or less, as shown in Fig. 17.52c, for adequate surface quality. Processing coatings with this profile allows for more accurate volume control due to the planarity of the resulting coating geometry. Moreover, this profile exhibits more uniform coatings under conditions of multiple coating passes. Multiple coatings passes tend to exaggerate any previous coating nonuniformities. Numerous material rheologic studies have been performed to assess the required material properties for coating bumped wafers.

Wetting angles measured for prototype wafer-applied underfill materials on eutectic solder, nitride passivation, and solder mask are shown in Table 17.24 based on goniometer measurements taken at room temperature. The dilute versions of materials A and B are 50/50 mixtures by volume of material and solvent. As indicated in the table, the wetting angles remained relatively constant over the different surface materials that were coated. This is interesting in that one would tend to expect different wetting angles between the solder surface and the nitride surface due to their divergent surface energies. However, this does not seem to be the case. Diluting the prototype underfills with solvent resulted in a 50 percent decrease in wetting angle. Material B exhibited the largest wetting angle, approximately 45°. Based on the results to date, it would appear that wetting angles in the range of 35 to 50 degrees should provide adequate performance for coating bumped wafers.

17.4.2 Coating Methods and Materials

Determining the required material properties is the first step in developing the back-end wafer-level packaging process. The next step involves finding a consistent and accurate coating method by which the material may be deposited onto the bumped wafer. The need for a quantitative comparison of different coating methods exists. As a relative comparison of deposited underfill coating, a film quality factor was developed consisting of a *characteristic* film thickness divided by the standard deviation of the surface height variation given by Eq. (17.6). Note that this is a die-level comparison. The characteristic film thickness is taken as a representative film thickness at or near the die center. This was chosen because the maximum film thickness tends to occur at the center of the chip, and

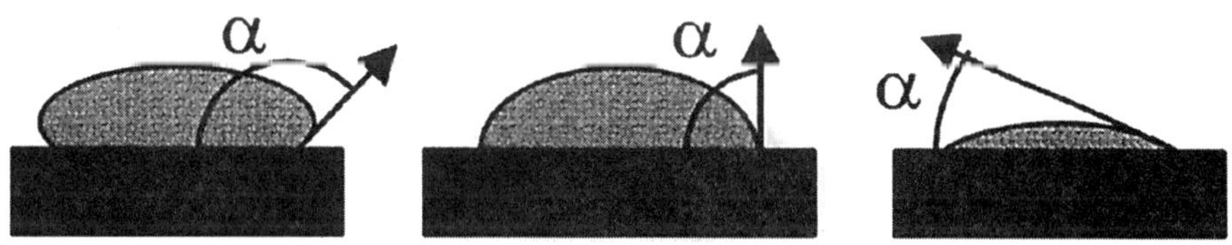

FIGURE 17.52 Profile of three different wetting angles. *a.* $\alpha > 90°$; *b.* $\alpha = 90°$; *c.* $\alpha < 90°$.

TABLE 17.24 Wetting Characteristics of Prototype Wafer-Level Underfills

	Wetting Angle (degrees)								
	Surface and Sample Diameter (cm)								
	Solder mask			Solder-covered substrate			Nitride passivation		
Sample	1	0.5	0.25	1	0.5	0.25	1	0.5	0.25
Material A	39	39	40	36	36	38	39	36	37
Material A Diluted	19	17	18	16	17	16	19	20	15
Material B	47	44	38	42	42	48	45	45	41
Material B Diluted	10	12	12	17	16	16	15	17	18
Material C	37	32	33	32	29	33	30	27	19

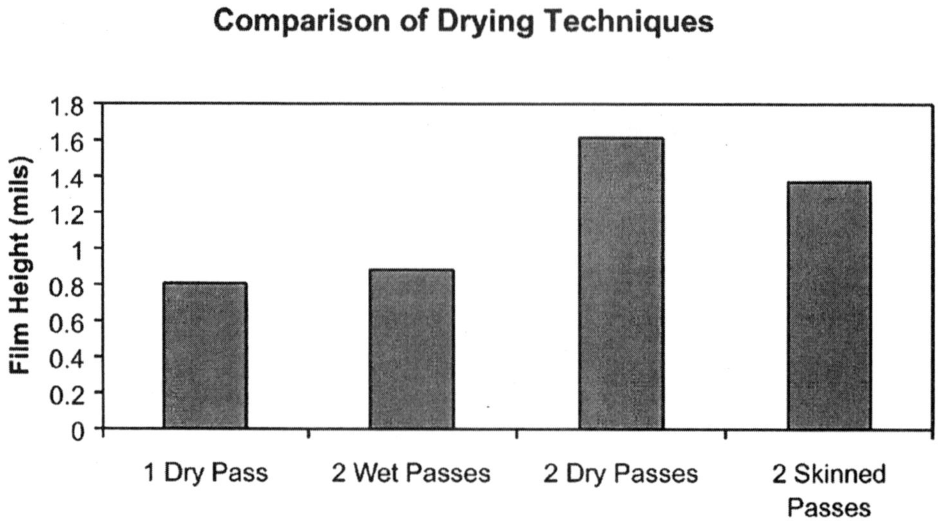

FIGURE 17.53 Film thickness for different drying techniques.

at this location film thickness is easily measured using a stylus contact surface profiling system (for perimeter I/O devices).

$$Q_{film} = \frac{h_{char}}{\sigma_{film}} \tag{17.6}$$

where h_{char} is the characteristic film thickness, and σ_{film} is the standard deviation of the surface height.

Each of the coating methods evaluated has advantages and disadvantages, with some demonstrating clear superiority. The first challenge arises in the thickness of the coating. In order to evaluate the coating, it is useful for the thickness to be comparable with the bump height. Most of the methods under analysis produce coats up to 25 μm in a single pass, whereas the wafer bump height in this study was 150 μm. Therefore, multiple coats are required to achieve the 150 μm value. Since the materials wet onto themselves differently than they wet the passivated wafer, a set of experiments was designed to investigate the differences. The first set compared the additive effects on surface thickness of multiple coatings by varying how dry the first coating was when the second was applied. The results in Fig. 17.53 show that a single coating thickness of 20 μm was achieved for material A after drying. The figure also shows two consecutive coatings with no drying between the first and second coats. In this case, a minimal increase in thickness over a single dried coating was observed. However, if the first coat was completely dry before the second was applied, the resuling coating thickness was essentially twice that of a single coat. If the second coating was applied after the first had dried partially, an intermediate thickness was achieved. This "skinning" process is of little production value because the amount of drying is difficult to control and dramatically affects the coating thickness. Note that all thickness measurements in Fig. 17.53 were taken after the material had been dried according to manufacturer's recommendations.

In the next phase of work, the effect of the different underfill materials and coating methods on surface quality and coating thickness was investigated. Two different materials and three different coating processes were studied. Fig. 17.54 shows the resulting coating thickness and standard deviations at the die level. Notice that material A Diluted yields a coating thickness much less than that of material A for the same coating method. This is largely due to the fact that material A is more viscous than its diluted counterpart. Material A diluted has a viscosity of 0.17 Pa·s, whereas material A has a viscosity of 2.2 Pa·s at room temperature. As expected, the coating thickness is highly sensitive to variations in viscosity. This observation holds true for most coating methods investigated. Also, note that method 2 yields a much thicker coating than method 1 for the same material. This result is not necessarily conclusive because there are unique ways to vary the thickness within each coating method, but it is representative for most experimental conditions studied. It is important to note that the surface quality obtained with method 2 was inferior to that obtained with method 1. The superior coating technique studied in this work was method 3 based on coating quality and thickness. It produces slightly thicker coats than method 1 and generates a surface with significantly less variation.

How the surface roughness and/or film uniformity changes as the number of coats increases is another characteristic of interest. Figure 17.55 shows the maximum film thickness h_{char} and standard deviation of the surface height σ_{film} for eight sites across a solder bumped wafer after a single coating using Method 3. Notice that the film thickness varies by as much as 10 percent and the height standard deviations are around 20 percent of the film thickness. Figure 17.56 shows the results for the film uniformity after five successive coatings completely dried between passes. As evidenced in this figure, the absolute variations in film thickness and standard deviations are the same as those seen with a single coating. This means that relative to the film thickness, the film surface variations have decreased. In other words, the quality factor has increased by a factor of 5. Closer observation of the data actually indicates that the quality factor has increased by a factor of 6, suggesting that this particular underfill material has self-leveling tendencies over the course of multiple coats. A disadvantage of the multiple coats is that they can result in a buildup of material on top of the solder bumps, which has the potential to compromise reflow. Fortunately, each coating pass deposits material on the tip of the solder bump having a thickness one or more orders of magnitude less than that deposited over the unbumped regions of the die.

Figure 17.57 shows a laser profilometer raster scan of a wafer that has been coated with a liquid prototype wafer-level underfill material. The image shows a single die in a wafer and adjacent die edges. The darkest areas are the bumps around the edge of the die adjacent to the saw streets, indicating the highest points in the measurement. Notice that the material is thickest in the central areas of the die, particularly in the upper right corner. The "beading" of this material was common due to

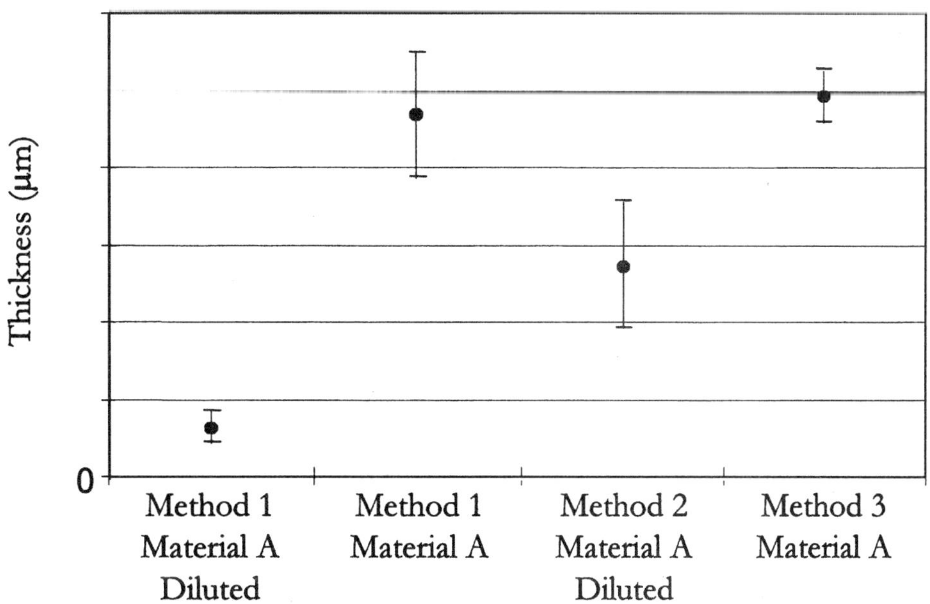

FIGURE 17.54 Comparison of different coating methods and materials.

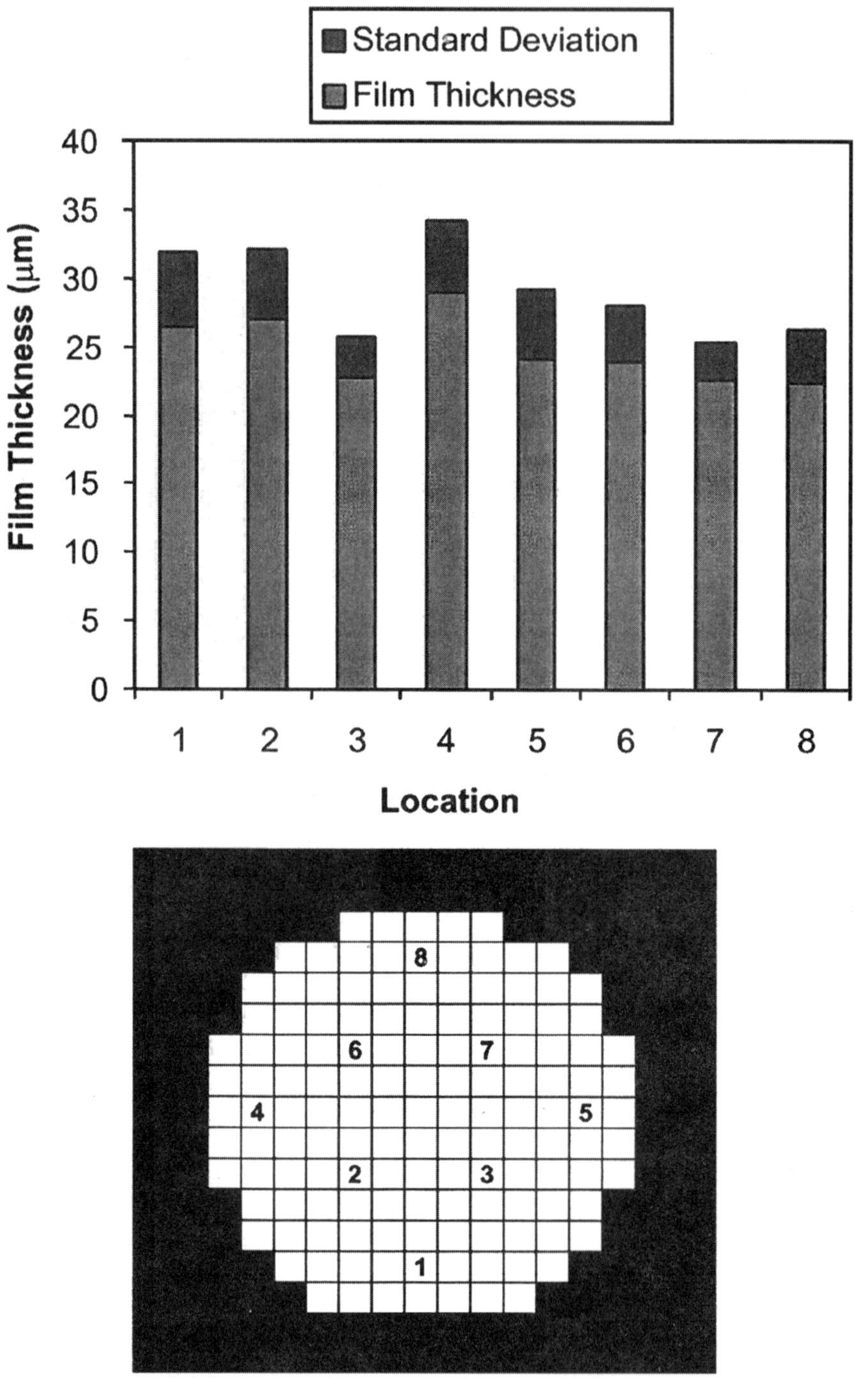

FIGURE 17.55 Film thickness results after one coat.

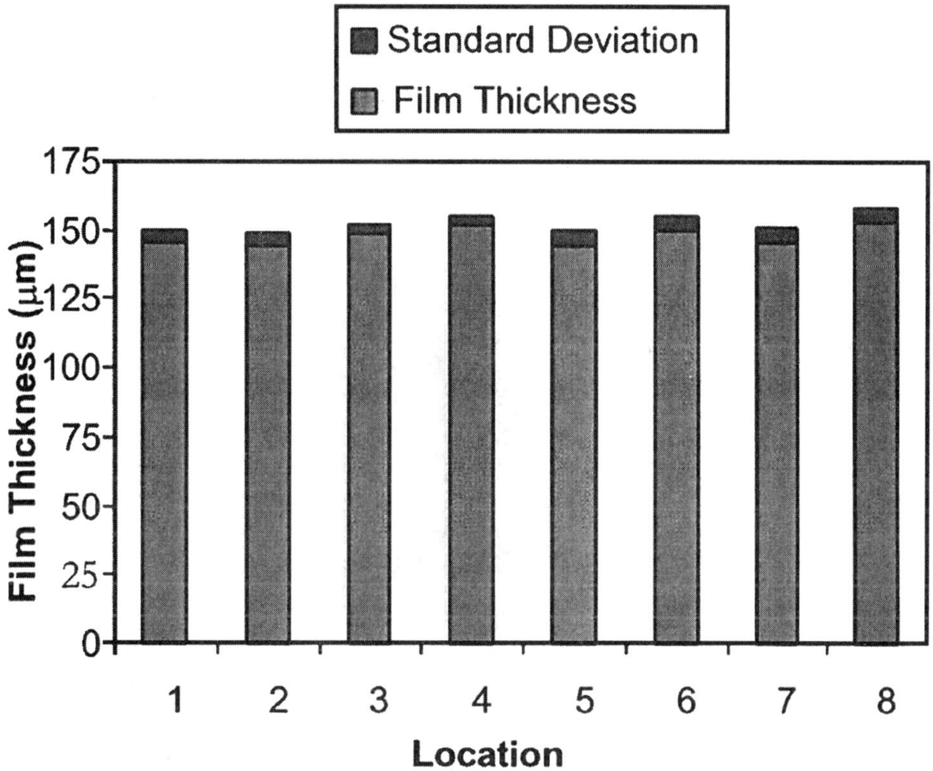

FIGURE 17.56 Film thickness results after five coats.

its rheology and wetting characteristics. This is undesirable because the curved surface caused by the beading reduces the amount of material that may be deposited on the die since the maximum film thickness should be maintained less than the bump height. Fortunately, most other candidate materials studied did not exhibit this beading behavior.

The last issue investigated with respect to coating methods was how thickness may be varied for a particular coating method. For most techniques, the deposition rate or speed was a critical factor for film thickness. Figure 17.58 shows the effect of increasing the application speed for coating method 1. For the range of this experiment, the correlation of increasing speed was nearly linear. As expected, the more viscous material (material A) yields a thicker coat than the less viscous one (material A dilute). The deposition method in Fig. 17.58 corresponds with method 3 in Fig. 17.54. This relationship between speed and thickness is unique in that the slope is positive. Most other methods of application have a negative slope, indicating an inverse relationship between speed and thickness. A number of process cycle time advantages can be derived from this positive coating speed versus thickness relationship.

17.4.3 Chip Imaging and Placement

In order for wafer-level underfill materials and process technology to be successful, placement machines must be capable of accurately aligning and placing the coated chips onto the substrate. Normally, key features on the chips and substrate are located using automated vision systems. These vision points are used as references in the alignment of the chip and the substrate. Key features on the chip are either fiducials or solder bumps. Key features on the substrate are either fiducials or bond

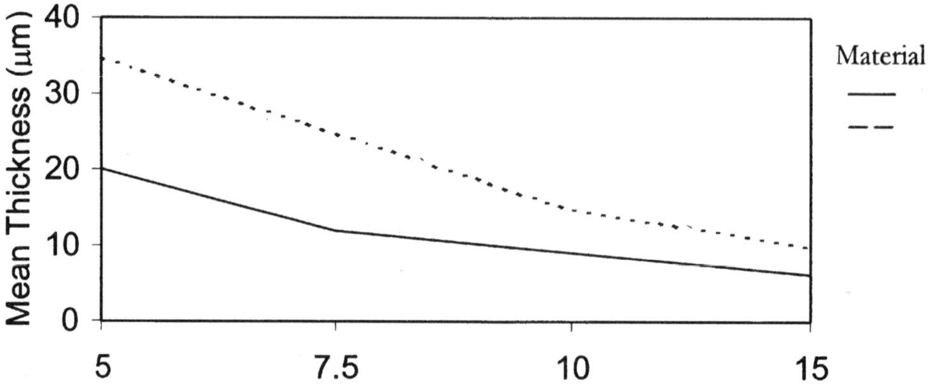

FIGURE 17.57 Raster scan of undiced solder bumped wafer at the die level. Image shows the coating thickness level versus gray scale.

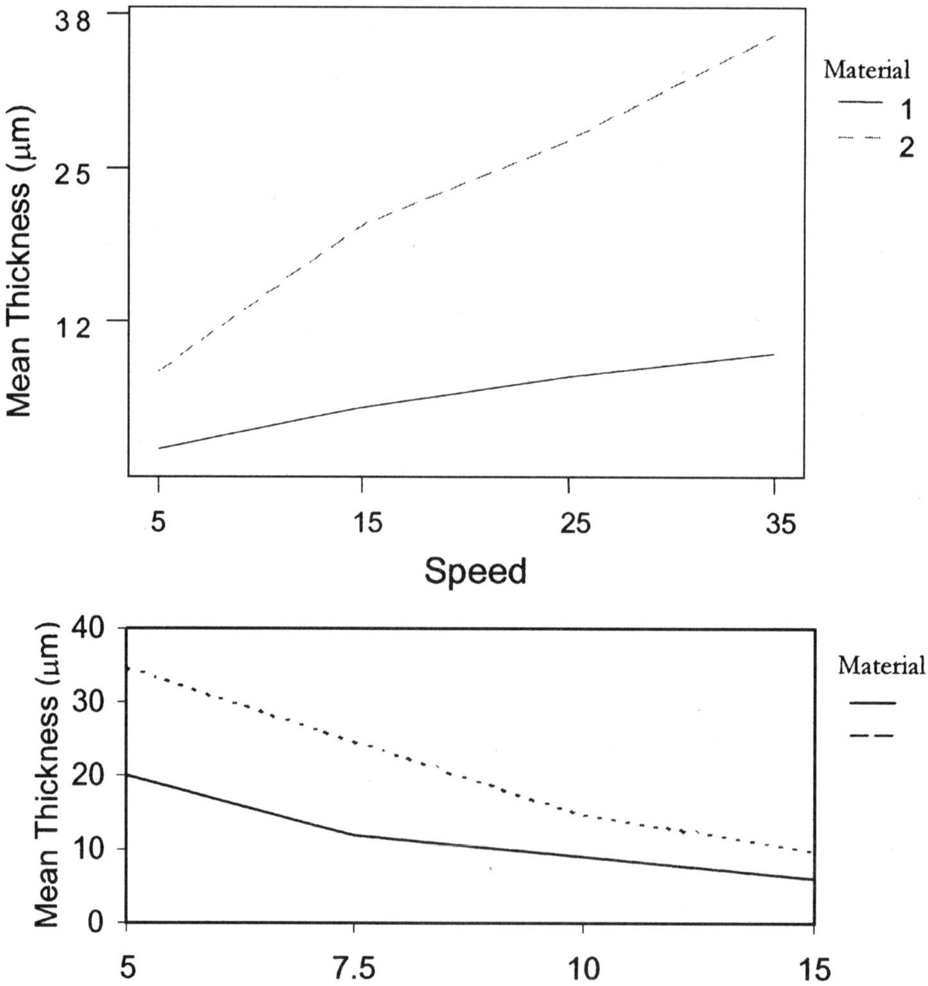

FIGURE 17.58 Film thickness results for different application speeds and materials: (*top*) method 1; and (*bottom*) method 2.

pads. Even if the coating is relatively uniform, the vision system may not be able to locate the alignment points on the chip due to the coating acting as an obstruction. With every coating method under consideration, some material covers the top of the solder balls. Fiducials are covered by the full thickness of the coated underfill material. A key issue to be investigated is whether a typical high-speed FC placement machine can vision and locate these alignment points in the presence of the underfill material coating. Transparent coatings would allow for the best imaging of bumps or fiducials. However, the wafer scale underfills being developed are filled materials. The high solid filler content makes current prototype materials translucent or opaque. Fortunately, many placement machines offer flexibility within their vision system to optimize imaging. Parameters that may be adjusted include illumination angle, light intensity, and image acceptance transform (level transforms).

Figure 17.59 presents an example of how the illumination angles affect feature brightness for coated and uncoated chips. The figure shows four illumination angles x-plane, plane, middle and steep based on the high speed FC placement system used to investigated FC placement of underfill-coated FCs. For each angle, the inside corners of four die are shown. Die 1 is a 200 μm pitch eutectic solder bumped FC device with silicon nitride passivation. Die 3 is a 356 μm pitch high-lead solder bumped FC device with polyimide passivation. For each illumination angle, the coated version of die 1 is shown in the upper left corner, the uncoated version of die 1 is shown in the upper right corner, the coated version of die 3 in the lower left corner, and the uncoated version of die 3 in the lower right corner. The placement machine vision system looks for contrast, which is usually provided by the high reflectivity of the bumps or fiducials compared with the low reflectivity of the die passivation and IC surface. Notice that the bumps on coated die 3 are most visible with steep illumination, whereas the uncoated die 3 bumps are more visible with the other three illumination angles. The coating is most visible in plane illumination for each die. For coated die 1, illumination angles x-plane, middle, and steep provided good contrast, with x-plane providing the highest contrast between the bumps and underfill coating.

17.4.4 Automated Vision Analysis of Underfill Coated Chip

Design experiments were performed to determine the main effects and imaging system configuration that results in the placement machine successfully imaging the chips on a variety of chips and coat-

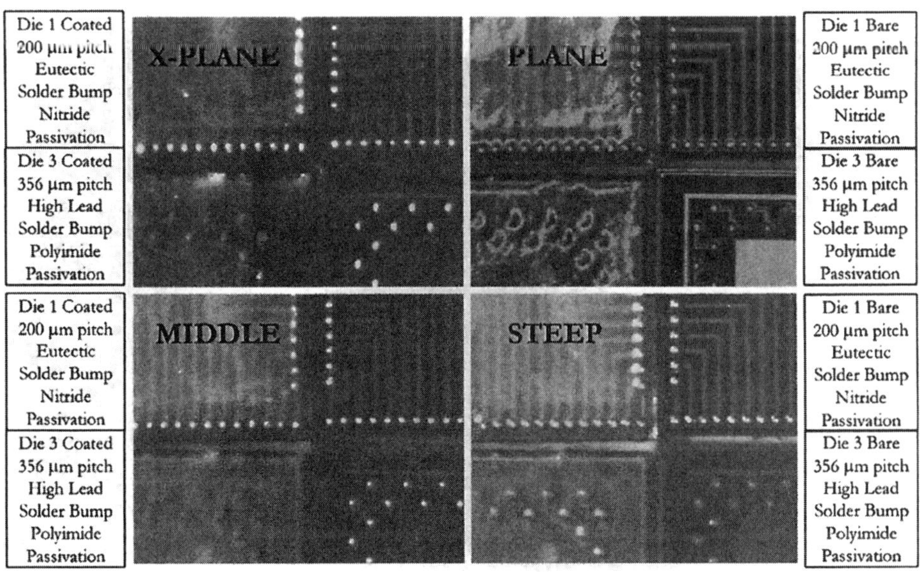

FIGURE 17.59 Comparison of illumination angles of x-plane, plane, middle, and steep for coated and uncoated version of die 1 and 3.

ings. The high speed FC placement system used in this study enabled control of three imaging parameters: illumination angle, light intensity, and image acceptance transform (level transforms). Illumination angle refers to the angle of incidence of light striking the die. The placement system used has preset illumination angles referred to as x-plane, plane, middle, and steep. Each lighting angle used an array of red (680 nm) LEDs to illuminate the die. x-Plane lighting comes from a ring of LEDs located in the same horizontal plane as the die such that the light they emit illuminates the side of the die and the bumps. Plane is the next level of LEDs above x-plane. They are located in a circular array about 80 degrees above the die, but through a system of mirrors, the light strikes the die at an angle of approximately 30 degrees above the horizontal. The middle lighting level is a square array of LEDs. The light they emit strikes the die approximately 50 degrees above the horizontal. The final series of LEDs is steep. This is a vertical light source located much farther from the die than the other light sources and provides light perpendicular to the surface of the die via a mirror system. Light can be delivered from any combination of these angles simultaneously. Figure 17.60 shows a schematic of the lighting system, including the lighting angles, bounce mirrors, and CCD camera.

Another useful tool employed by vision systems is the acceptance transform function, or simple image digital filter. This helps the placement machine vision system recognize the features of the die; it is analogous to a light filter. By allowing certain colors to be emphasized or neglected, image acceptance transforms can create clear, well-defined images in which the important features of the die are recognizable to the vision software. Graphical representations of first-and second-order transforms are shown in Fig. 17.61. In selecting the proper combination of illumination angle, intensity, and transform, a greater percentage of image acceptance is obtained.

The test vehicles studied include five die of each of three types: die 1 with a single row of eutectic solder 200 mm pitch perimeter bumps and silicon nitride passivation, die 2 with two staggered rows of eutectic solder 356 mm pitch perimeter bumps and silicon nitride passivation, and die 3 with two staggered rows of high-lead solder 356 mm pitch perimeter bumps and polyimide passivation. Each of these die was coated with a prototype wafer-applied underfill system. Coating was achieved by stencil printing.

In order to design a relevant experiment, viable values for illumination levels and transform values had to be chosen. This is important because there are thousands of possible combinations, most

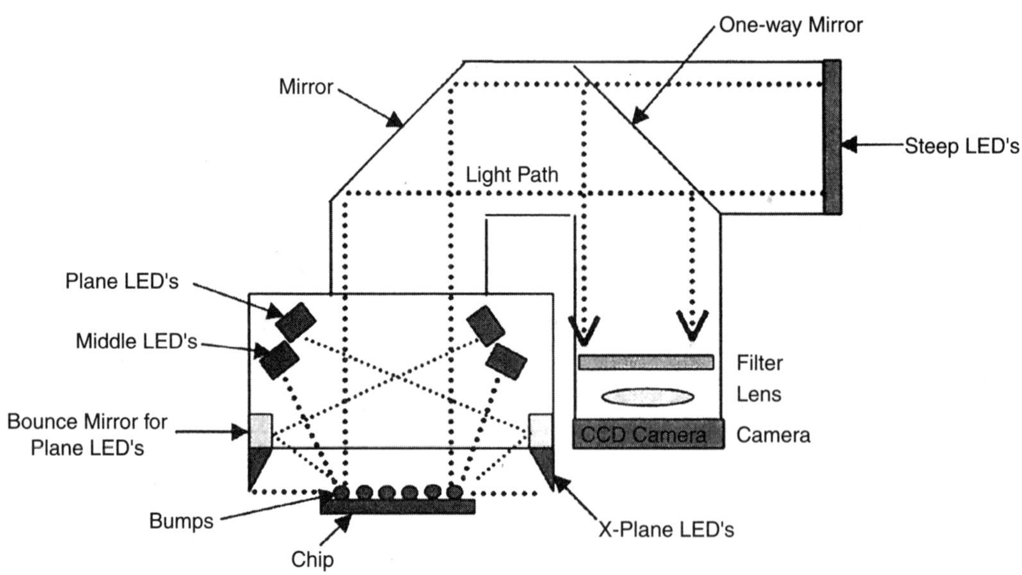

FIGURE 17.60 Commercial high-speed FC placement system vision system schematic.

of which yield a null result. Preliminary tests were conducted to determine the combinations of illumination and transform values that resulted in feature recognition for the largest percentage of die. For each chip feature recognized and accepted, the vision system displays boxes around regions in the image that it considers to be the desired features. If the system targets the correct features for a particular type of die, such as corner bumps in the case of die 2, then that trial is deemed a success. The system either finds all features or none, so there is no ambiguity in declaring a particular test a success or a failure. Since each of the test die has different passivation, bump material, bump location, and bump size, different factor levels were chosen for each.

A five-factor design of experiments was performed to determine which illumination and transform values resulted in the placement machine recognizing bumps on the greatest number of underfill-coated chips. Each of the four illumination angles was a factor, with two levels of illumination each, one level being zero and the other selected such that when all other illumination angles are set to zero and a first-order transform is used, the best possible image of the die is provided. The fifth factor was the acceptance transform, which had four levels. The selected transforms represent both first-order, which is a linear adjustment of color, and three second-order cases, which allow piecewise manipulation of the spectrum (see Fig. 17.61). Each individual combination was tested ten times with each of the 15 test die, with the successful number of recognized images recorded for each combination.

Figures 17.62 through 17.64 show the main effects plots for illumination angles x-plane, plane, middle, and steep for die 1, 2, and 3, respectively. For die 1 shown in Fig. 17.62, plane and x-plane factors are instrumental to successful feature recognition. Middle and steep have little affect. Die 1 had 22 percent of the die recognized, the highest of the chip types tested, indicating that nearly one in four configurations tested resulted in successful bump recognition. The low acceptance percentage results from the fact that most configurations tested resulted in no recognition of the chip bumps or fiducials. A few of the illumination conditions studied yielded complete recognition and acceptance of the chips. For die 2, in Fig. 17.63, Middle was the main lighting angle providing effective illumination. Five percent of these die were recognized, indicating that only 1 in 20 configurations resulting in successful bump recognition. For die 3 shown in Fig. 17.64, plane and middle were the important effects. Eight percent of the high-lead solder, polyimide passivated die were recognized. It is important to note that for Die 3, the chip fiducials (covered with a prototype wafer-applied underfill coating) were imaged instead of the bumps that were used for die 1 and 2. The vision system was unable to image bumps on the bump-die-passivation-underfill combination of die 3.

From these results it can be concluded that certain illuminations made a positive impact on the imaging process, with the specified lighting angles being major factors enabling robust imaging for particular chip, bump, and passivation combinations. In addition, some transforms have proven to be unsuitable for the imaging of underfill-coated die. The overall percentage of successful image recognition configurations was low for this series of experiments. The design of experiments were set up to be primarily a main effects analysis, designed to selectively determine suitable combinations of image parameters and critical illumination parameters. Indeed, so many of the combinations were completely ineffective for imaging the underfill-coated die that their presence in the results lowers the overall success rate substantially. Fortunately, sufficient numbers of the individual combinations resulted in robust image recognition of the underfill-coated chips. Moreover, for the FC systems studied, it was found that bump layout, bump passivation, bump metallurgy, and chip passivation all played influential roles in FC imaging. For die 1 and 2, the eutectic solder bumps were imaged. For die 3 with the high lead-solder bumps and Polyimide passivation, the fiducials on the die were imaged.

Figure 17.65 shows the acceptance transform main effect plots, which was a four-level factor. Die 1 worked best with first-order transforms and second-order transforms with high cutoff values. Visioning of the FCs worked best using second-order transforms at specific illumination angles for die 2 and 3. In general, the vision reproducibility was very repeatable for the die configurations and underfill coatings studied. Consistency and uniformity of the coating were important factors that currently are under further investigation. Based on the current results, perimeter bumps near the edge of the die, such as die 1, were easier to image than perimeter bumps further from the die edge, as with die 2 and 3. High-lead bumps were not as easy to image due to their high lead content and consequent

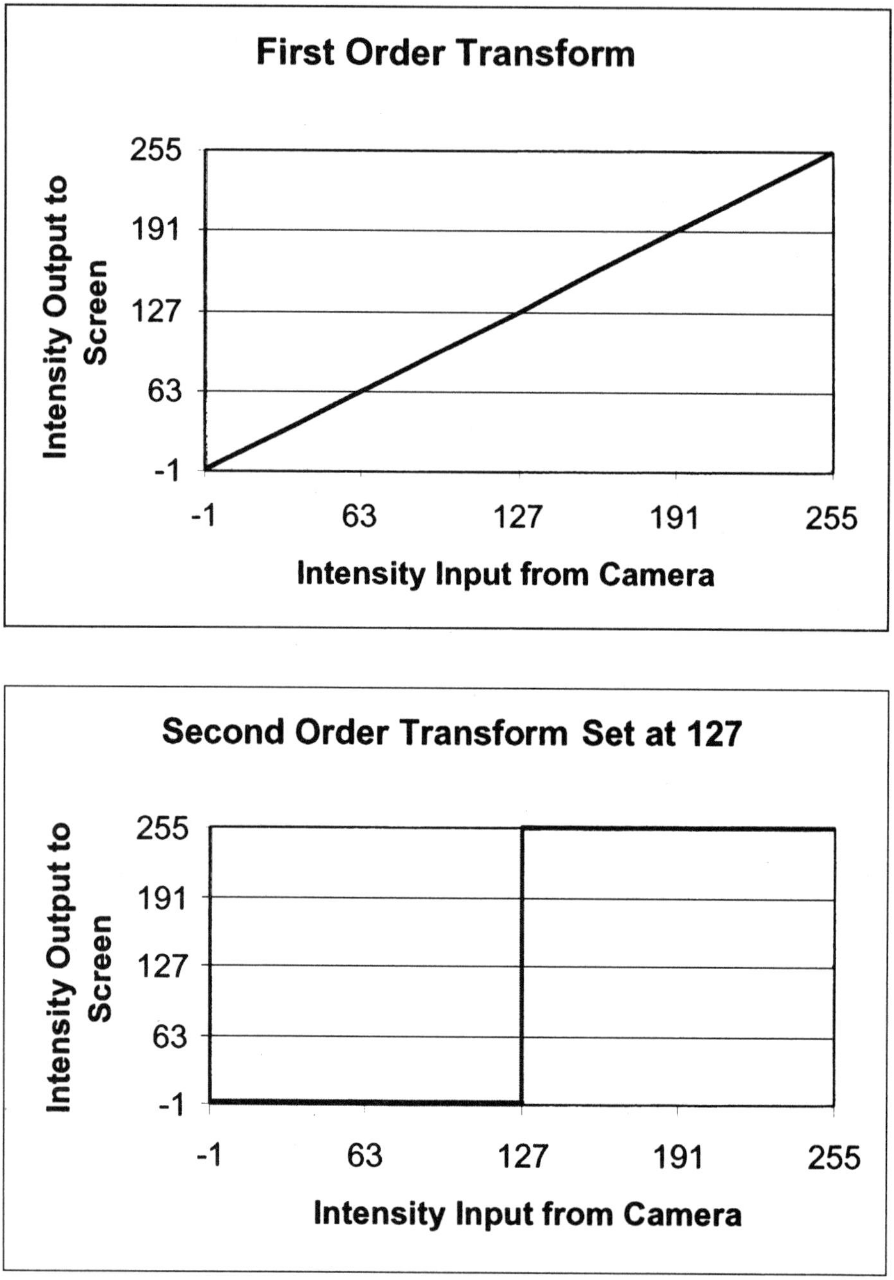

FIGURE 17.61 First- and second-order acceptance transforms for placement system vision.

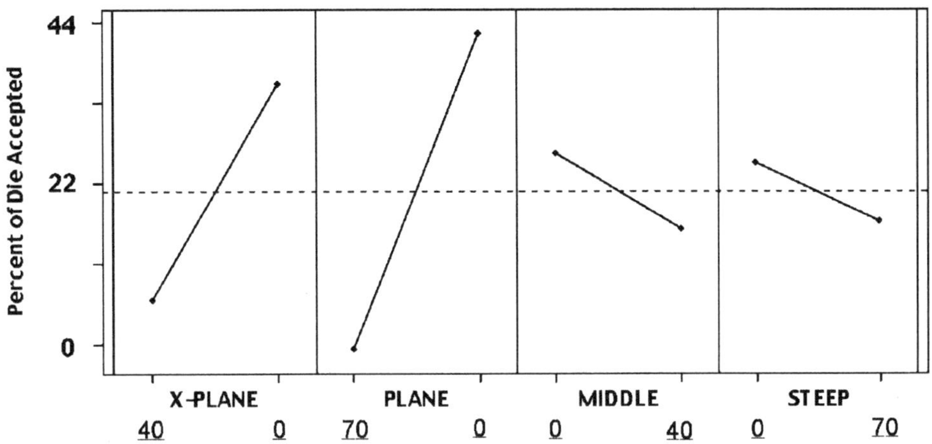

FIGURE 17.62 Main effects plot for die 1, eutectic solder, 200 μm pitch, silicon nitride passivated, underfill-coated die.

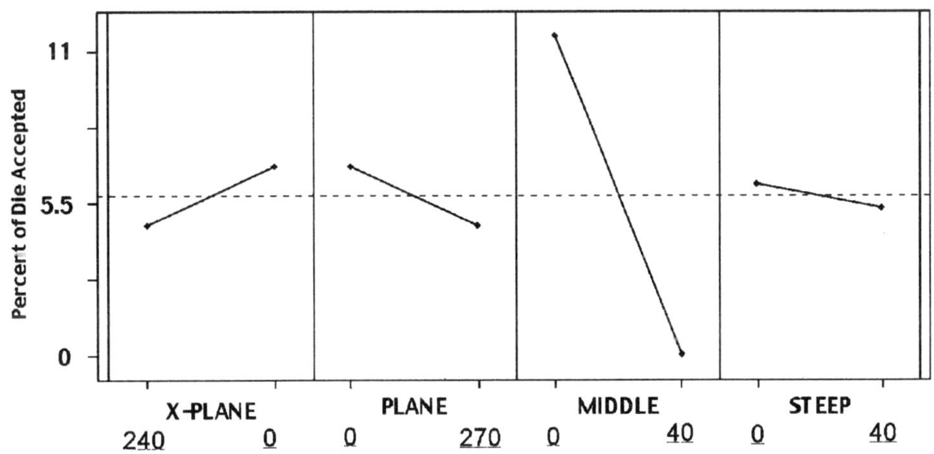

FIGURE 17.63 Main effects plot for die 2, eutectic solder, 356 μm pitch, silicon nitride passivated, underfill-coated die.

lack of reflectivity, evident by the die 3 results. Eutectic bumps provided better reflectivity, giving better contrast and between the passivation and the bumps. It appears that silicon nitride passivation provided high image contrast therefore better imaging of the underfill-coated die compared with polyimide passivation. In general, the silicon nitride was more reflective under the underfill coating, providing higher contrast with the solder bumps. This study indicated that given die with bump topologies and surfaces similar to those used with uniform underfill coatings, imaging of underfill-coated die for high-speed automated FC placement is feasible.

17.4.5 Analysis of Underfill Coated Chip Placement Accuracy

This set of experiments was structured to examine how accurately underfill-coated die can be placed on a substrate. Potentially, the underfill coating can cause misalignment due to refraction through the coating material or shadows formed by the coating topology. Therefore, it is important to determine

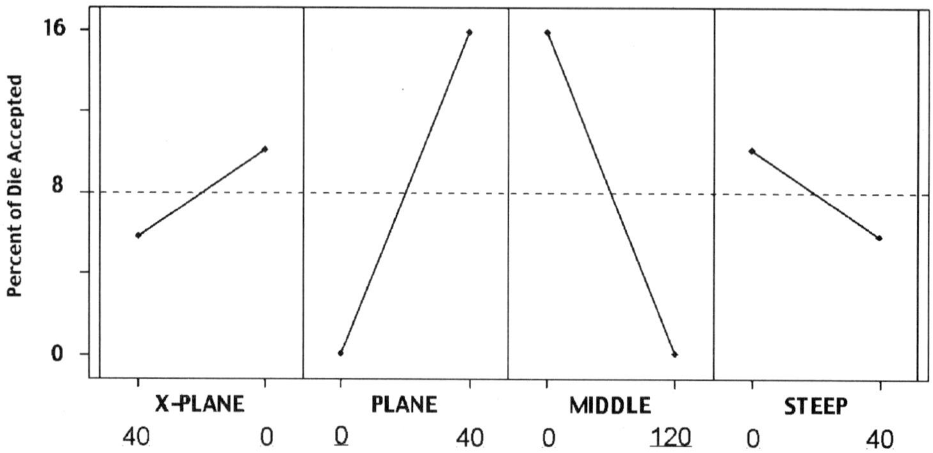

FIGURE 17.64 Main effects plot for die 3, high lead solder, 356 μm pitch, and polyimide passivated, underfill-coated die.

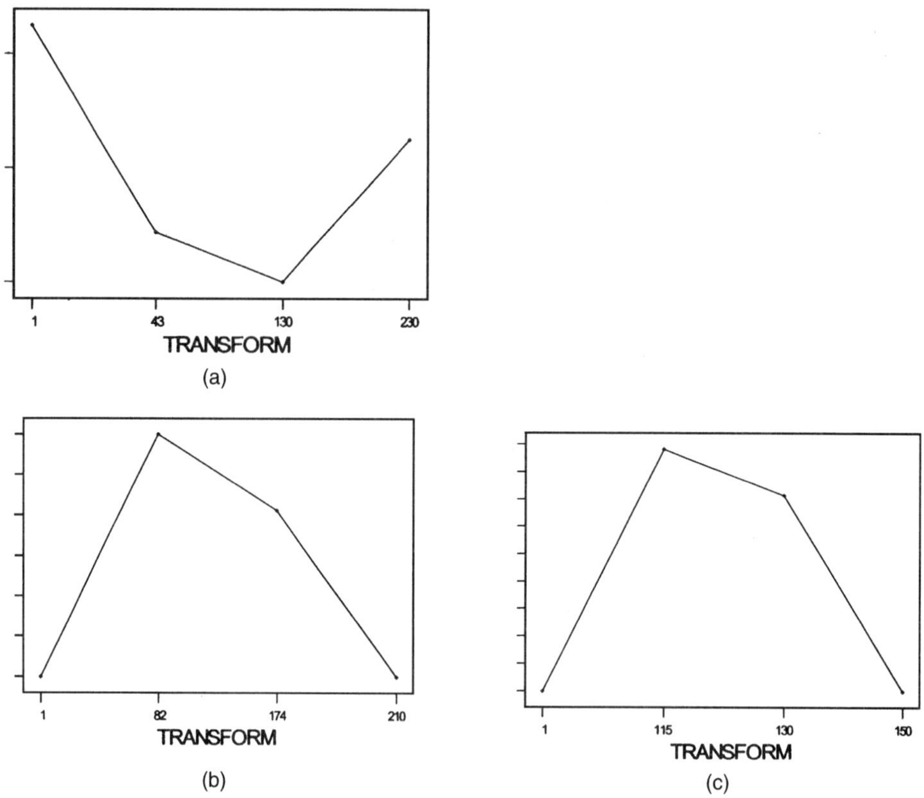

FIGURE 17.65 Main effects plots for (*a*) die 1, (*b*) die 2, and (*c*) die 3 between the vision system acceptance and the acceptance transform factors.

if the machine's image is a correct representation of actual die features because proper placement depends on the ability of the machine's vision system to correctly image bumps or fiducials on the die. Typically, global or local fiducials are used on the substrate. The goal of chip placement is to position the die such that each solder bump is aligned directly above its corresponding pad on the substrate site. This is a necessary but not sufficient condition for all the electrical connections to be formed during reflow. Although perfect alignment during placement is preferred, a die has been placed successfully if its bumps are sufficiently in contact with their respective pads to allow for proper wetting during reflow. This acceptable margin of error has been shown to be less than half the diameter of the minimum substrate pad.[15]

To determine the placement accuracy for die with preapplied underfill, a series of die-to-board placement experiments were performed using a Siemens SIPLACE 80 F5-DCA high-speed FC placement machine. Die 2 and die 3 were placed on matching substrates having six sites, with the die 3 substrates having a eutectic solder cap on the bond sites. Die 1 was placed on matching substrates having eight sites. A tacky adhesion promoter was applied on the substrate at each chip site to secure the chip after placement. X-ray inspection was performed using a Fein Focus Roentgen System providing a transmission x-ray image of the die bumps and substrate pads. The x-ray micrographs were used to determine the relative die-to-board offset. Using the bump's calibrated diameter as a reference distance, the distance between the center of the bump and the center of the pad was measured from the micrograph, providing an x, y, and θ offset measurement based on bump-to-pad distances measured at opposite corners of the die.

The results of the placement experiment indicated that chip placement accuracy and repeatability were equivalent to these of the same chip and substrate combination without preapplied underfill. All the die tested were consistently place within a 50 μm accuracy within the tolerance specifications for the placement machine. Table 17.25 shows the bump-to-pad offsets measured for 10 replicates of five coated FC devices of each die type. Based on these initial FC placement results, there was sufficient alignment of the underfill-coated FCs to allow for reflow. Of the die placed, die 3 showed the greatest offset. It is believed that this consistent placement offset is due to the lower illumination contrast features associated with the high-lead-solder bumps and polyimide passivation.

17.4.6 Wafer-Level FC Processing Summary

The wafer-level underfill FC packaging process, including underfill application, dicing, FC placement, and reflow, is complex, and a number of challenges remain. The investigation presented here has been designed to characterize the material properties necessary to build functional assemblies and demonstrate that the individual steps involved in processing are feasible. Future experiments will be aimed at increasing the surface quality of the underfill coating, evaluating the impact of coating quality on process yield, and increasing the yield of FC assemblies.

TABLE 17.25 Bump-to-Pad Placement Offsets for Six Sites (Measured in Microns)

		Die 1		Die 2		Die 3	
	Die	X Offset	Y Offset	X Offset	Y Offset	X Offset	Y Offset
Lower Right	1	25	25	−25	50	25	50
	2	25	50	−25	50	25	75
	3	0	50	−25	50	25	50
	4	0	50	−25	25	25	50
	5	25	50	−25	25	50	50
Upper Left	1	25	−25	25	25	50	0
	2	25	−25	25	0	50	0
	3	0	−25	0	25	25	25
	4	25	−50	25	25	25	0
	5	25	−25	0	25	50	0

In this study, several wafer-applied underfill coating techniques have been evaluated, and a number have been shown to be capable of depositing uniform nonconformal coatings over bumped wafers. Two preferred coating techniques were identified. A quality factor was defined as a metric to evaluate the various coating and drying techniques. Coating uniformity over 200 mm bumped wafers was also analyzed for select coating techniques exhibiting good uniformity over the wafers as well as locally over bumped die. A process characterization was presented for one of the coating methods studied. Various key material properties including underfill rheology, shear rate effects, and wetting angle were analyzed to determine the necessary material characteristics for uniform nonconformal coatings on bumped wafers.

Vision and illumination of underfill-coated bumped FC test die were also verified. Illumination angle and image acceptance transforms were the most critical factors having an impact on robust imaging of the coated die. Moreover, die configuration and topology were also important, particularly the bump material, the bump reflectivity, the die passivation, and the bump position. In addition to demonstrating the feasibility of imaging and accepting underfill coated FC die, placement accuracy of these innovative FC systems was also verified, exhibiting placement accuracy of underfill-coated die comparable with equivalent uncoated die.

17.5 ANALYSIS OF NEXT-GENERATION FC PROCESSES

Critical to the success of low-cost, high-throughput FC assembly is the ability to design process variables and predict the process window for a given incoming material stream. Several critical process steps have been analyzed and design models developed to enable process compatibility analysis and promote rapid implementation of the technology. The first analysis focuses on underfill encapsulant processing, looking at conventional capillary flow and compression flow underfill processing.

Two critical prereflow process steps are also analyzed. The first is the dynamic chip placement process. Based on computational fluid modeling of underfill compression flow during placement, there exists a theoretical limit to the placement velocity of a chip. The limiting placement speed exists due to the presence of underfill between the chip and the substrate during placement.

The second critical process step occurs after chip placement. Due to steady state forces acting on the chip, it is possible that the chip can "float" above the board if the underfill mass is not properly controlled. A process design model has been developed to understand the basic physics of the floating phenomenon and to predict process variables to eliminate process defects. Using these process design models, a reliable prereflow assembly process can be designed to ensure high first-pass yield.[1,2,4,12,19,22,24,28,31,32,38–40]

17.5.1 Potential Void Formation

In the low-cost FC process, voids in the underfill can form via three mechanisms. The first is due to outgasing during reflow. Common conditions promoting *outgasing voids* are dissolved moisture or solvents in the polymer films covering the chip and substrate surfaces. Outgasing is common from the solder mask, FR-4, polyimide chip passivation, etc. Heat treating (i.e., drying) the boards and chips prior to assembly eliminates this type of voiding. Outgasing can also occur within the underfill itself due to excessive ramp rates or temperature exposure during reflow. Higher ramp rates and elevated temperature exposure can cause the underfill constituents to volatilize during reflow, producing voids. Such voids can be extensive if the reflow profile is not designed correctly. Figure 17.66 shows scanning acoustic micrographs of FC assemblies produced using the new low-cost, high-throughput process at two different reflow profile ramp rates and peak temperatures.[24,25,28]

The second type of potential void formation occurs during compression flow chip placement and is due to *compression voids* forming at the chip surface. Simulation results predict the potential formation of compression voids at the intersection of the active face of the chip and the bumps, as shown in Fig. 17.67. Potential compression voids are at the base of the solder bumps adjacent to the chip. Note that the term potential compression voids is used because the simulations could not be run to the point where the solder bumps and the substrate make contact due to instabilities in the sim-

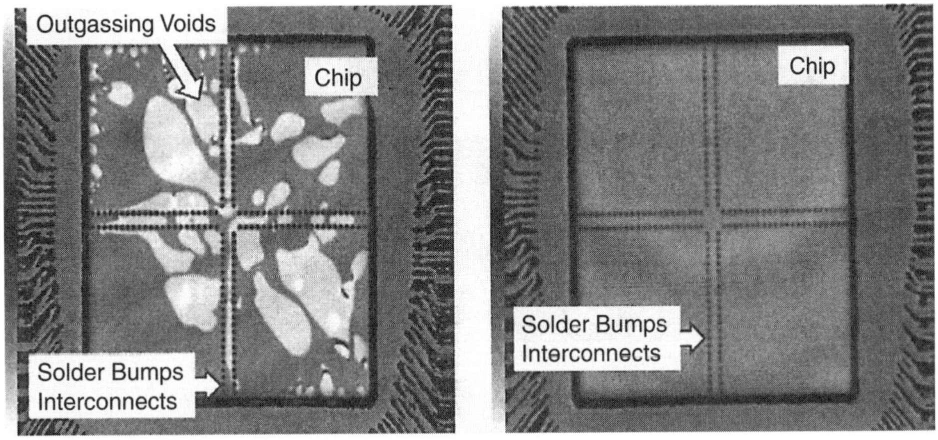

FIGURE 17.66 Scanning acoustic micrographs of low-cost next-generation fc assemblies demonstrating outgasing voids due to excessive reflow ramp rates. (*Left*) high ramp rate profile, (*right*) low ramp rate profile.

ulation software. Simulation results predict that compression voiding is minimal toward the center of the chip and that the number and size of voids increases toward the outer edge of the chip. Potential compression voids tend to be less prevalent at higher underfill compression flow pressures. By increasing the degree of cross-flow or squeeze flow, potential void formation is reduced. In general, void formation decreases with higher temperatures. Underfills with lower viscosity (i.e., lower temperatures in this case) are able to flow into the gaps and cavities more easily, reducing void formation. Exceptions to this trend occur when flow isolation and gravity flow take place. These cases result in "fill starvation" and large capture voids. Larger bumps and larger bump pitch reduce potential compression voiding. Simulation results also predict a strong interaction between the potential formation of compression voids and underfill print/deposition geometry. Lower ratios of the underfill print height to underfill print area result in predominantly compression flow and a higher probability of void formation at the bump-chip interface. Having a larger underfill print height to underfill print area ratio results in predominately squeeze flow and a lower probability of void formation.

Compression voids have been experimentally verified as occurring at the base of the solder bumps adjacent to the chip, as predicted by the flow simulations. Experimental results also show that the base underfill geometry deposited on the board strongly affects void formation. In agreement with the flow simulations, initial experiments indicate that compression and capture voids tend to occur under conditions where the fluid experiences compression flow in the z direction rather than a squeeze flow parallel to the chip and substrate surface. In general, lower print height to print area aspect ratios produce more voids. In contrast, higher print height to print area aspect ratios result in fewer compression and capture voids. The three-dimensional geometry of the deposited underfill is also found to be an important contributor to compression and capture void formation. This effect will be discussed further in future publications.

The third type of potential void formation is due to capture voids formed during placement of the chip where gas pockets (i.e., air in most cases) are captured between the chip and underfill surface. Capture voids, shown in Fig. 17.68 for an actual FC assembly after reflow, are typically larger than compression voids and are not localized to the bumps, like compression voids. The simulation results indicate the conditions under which such voids have the potential of forming.[23] For the lower underfill print height to underfill print area aspect ratio, the mean pressure and velocity profiles exhibit a saddle structure. The pressure profile is low at the center of the chip, increases forming a local maximum, decreases midway out the chip, increases to a second maximum, and decreases at the chip edge. Such a pressure profile implies that the underfill experiences a complex multidirectional flow where the underfill within the first pressure peak flows toward the chip center and the underfill outside the first pressure peak and within the second peak pressure flows toward the chip midpoint. Under these conditions, any capture and/or compression voids formed during chip placement in the

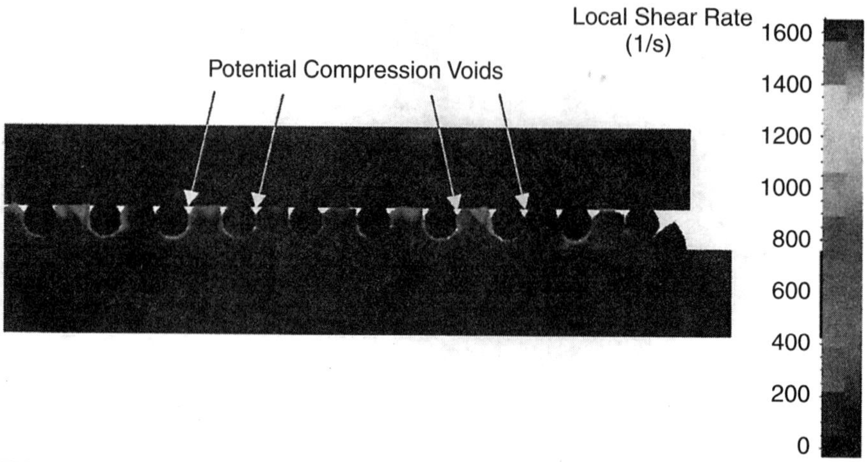

FIGURE 17.67 Compression flow simulation results showing potential void formation.

center or midpoint of the chip are isolated due to the pressure wells. The primary mechanism for eliminating such voids is entraining the interfacial voids in the cross flow of the underfill to expel the voids toward the chip edge. Controlling the underfill print/dispense geometry is also critical to eliminating compression and capture voids. The presence of pressure wells during assembly prevents entrainment of the voids in the cross flow resulting in the presence of capture and compression voids in the final assembly. In contrast, the flow simulation results show that a larger underfill print height to underfill print area aspect ratio results in a monotonically decreasing pressure profile from the chip center to the underfill-free surface at the chip edge. For such a profile, compression and capture void formation is minimized due to the ability of voids to be entrained in the cross-flow of the underfill drawing the void toward the underfill free surface at the chip edge. Preliminary experimental results support the simulation findings and suggest that capture voids are more prevalent in assemblies with a low underfill print height to underfill print area aspect ratio and that void formation decreases with increasing aspect ratio.

17.5.2 Design Guidelines

To reduce placement force requirements, larger bump size and bump pitch are preferred since smaller bump size and bump pitch resulted in higher placement forces. Selection of larger bump size and bump pitch also reduces the potential formation of voids during compression flow. Another design guideline concerns the underfill print height and viscosity as it relates to gravity flow. Results indicated that there is a clear tradeoff between the printed underfill height and the underfill viscosity. Lower viscosity materials in conjunction with higher print heights result in accelerated gravity flow and an underfill starved condition. Higher viscosity materials and lower print heights reduce gravity flow. However, lower underfill print heights reduce the extent of squeeze flow and increase compression flow ultimately leading the potential formation of more voids. Finally, due to its significant impact on pressure and placement force requirements, process temperature should be controlled to prevent gravity flow while providing for acceptable placement forces.[24,28]

The design guidelines derived from the simulation work are summarized below.

- Decreasing bump size and bump pitch results in larger placement forces.
- Potential voiding is reduced as bump size and bump pitch are increased.
- Very low viscosity underfills exhibit gravity flow, which leads to underfill starvation in the gap between the chip and substrate.
- Increased bump size and bump pitch result in smaller pressure gradients, thereby reducing the placement force requirements.

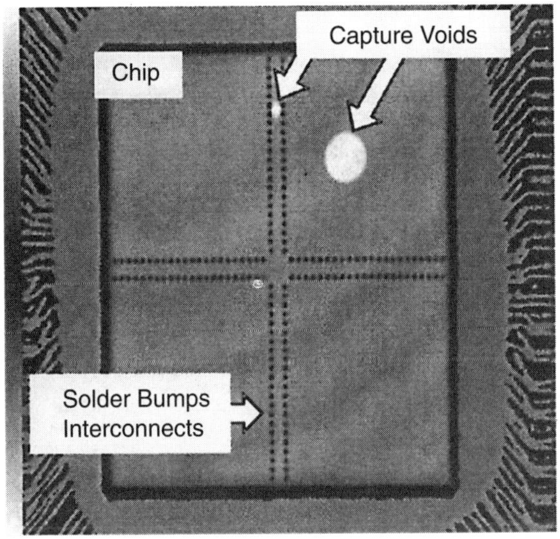

FIGURE 17.68 Scanning acoustic micrograph illustrating capture voids produced during compression flow chip placement.

- Increased temperature reduces underfill pressure and therefore placement forces.
- Placement forces of existing assembly equipment are sufficient for compression flow assembly.
- Higher-temperature processing during compression flow is not feasible done to the lower viscosity of the underfill, resulting in gravity flow and extensive void formation.
- Higher underfill print heights result in more squeezing flow, which provides the maximum pressure at the center of the chip and leads to better gap filling.

17.5.3 Placement Force Characterization of Compression Flow Underfill Processing

It is also important to analyze the flow pressure and placement force requirements for the compression flow underfill process to determine if standard placement systems can achieve the required placement forces. To accomplish this, a squeeze flow analysis between parallel plates can be used. Adopting Leider and Bird's[3,17,18] squeeze flow analysis with appropriate modifications for the flip chip compression flow case, the radial component of the general momentum equation (in cylindrical coordinates) (see Fig. 17.64) yields:

$$\rho\left(\frac{\partial v_r}{\partial t} + v_r\frac{\partial v_r}{\partial r} + \frac{v\theta}{r}\frac{\partial v_r}{\partial \theta} - \frac{v_\theta^2}{r} + v_z\frac{\partial v_r}{\partial z}\right) =$$

$$-\frac{\partial P}{\partial r} - \left(\frac{1}{r}\frac{\partial}{\partial r}(r\tau_{rr}) + \frac{1}{r}\frac{\partial \tau_{r\theta}}{\partial \theta} - \frac{\tau_{\theta\theta}}{r} + \frac{\partial \tau_{rz}}{\partial z}\right) + \rho g \qquad (17.7)$$

For quasi-steady state, symmetric, inertia free flow and neglecting gravity effects, Eq. (17.7) simplifies to Eq. (17.8) where the shear stress gradient is proportional to the radial pressure gradient of the underfill. A schematic of the compression flow system under analysis is shown in Fig. 17.69.

$$\frac{\partial \tau_{rz}}{\partial z} = -\frac{\partial P}{\partial r} \qquad (17.8)$$

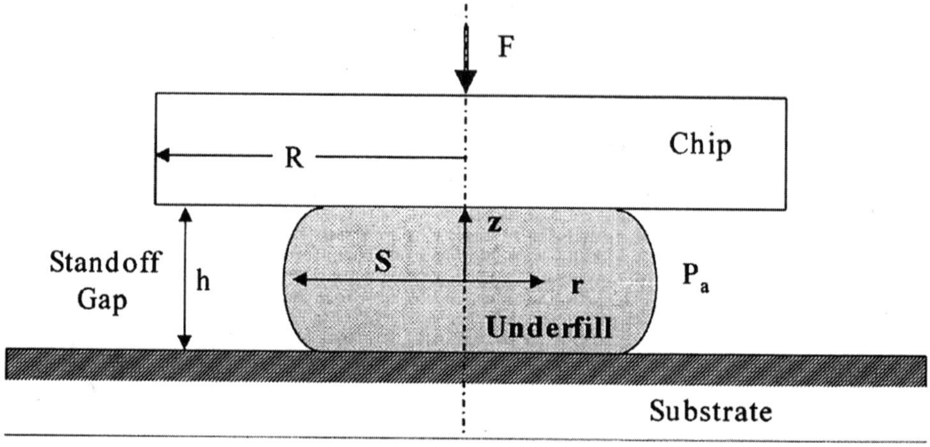

FIGURE 17.69 Compression flow analysis schematic for non-Newtonian squeeze flow between a chip and substrate.

Approximating the underfill flow as a shear thinning (or thickening) non-Newtonian fluid, yields an expression relating the shear stress and velocity gradient over the standoff gap.

$$\eta = m\,\dot{\gamma}_{rz}^{n-1} = m\left(-\frac{dv_r}{dz}\right)^{n-1} \tag{17.9}$$

$$\tau_{rz} = m\left(-\frac{dv_r}{dz}\right)^{n} \tag{17.10}$$

Integrating Eq. (17.8) with respect to z and combining with Eq. (17.10) results in an expression relating the radial pressure gradient and underfill rheology to the velocity gradient over the standoff gap.

$$-\frac{dv_r}{dz} = \left(-\frac{1}{m}\frac{dP}{dr}\right)^{1/n} z^{1/n} \tag{17.11}$$

Integrating Eq. (17.11) with respect to z yields an expression for the radial velocity profile over the standoff gap.

$$v_r = \left(\frac{h}{2}\right)^{1/n+1}\frac{1}{1/n+1}\left(-\frac{1}{m}\frac{dP}{dr}\right)^{1/n}\left[1-\left(\frac{2z}{h}\right)^{1+1/n}\right] \tag{17.12}$$

To relate the radial flow velocity to the chip placement speed, conservation of mass is applied to a cylindrical control volume of radius r encompassing the underfill material. Continuity then yields Eq. (17.13):

$$-\dot{h}\pi r^2 = 2\pi r\int_0^{/2} v_r dz \tag{17.13}$$

Substituting into Eq. (17.13) the radial velocity profile (17.12) and integrating over the standoff gap height yields an expression for the pressure gradient over the chip surface:

$$-\dot{h}\pi r^2 = 2\pi r\left(\frac{h}{2}\right)^{1/n+1}\frac{1}{1/n+1}\left(\frac{1}{m}\frac{dP}{dr}\right)^{1/n}\int_0^{h/2}\left[1-\left(\frac{2}{h}\right)^{1+1/n} z^{1+1/n}\right]dz \tag{17.14}$$

$$\frac{dP}{dr} = -\frac{2^{n+1}m}{h^{2n+1}}\left[\frac{2n+1}{n}\right]^n\left(-\dot{h}\right)^n r^n \tag{17.15}$$

Integrating (17.15) over the wetted radius S gives an expression for the pressure distribution over the chip surface:

$$P = P_a + m\left(\frac{2n+1}{2n}\right)^n \frac{S^{1+n}}{(n+1)} \frac{(-\dot{h})^n}{\left(\frac{h}{2}\right)^{2n+1}}\left[1-\left(\frac{r}{S}\right)^{1+n}\right] \tag{17.16}$$

Next, it is important to estimate the placement forces necessary to achieve standoff heights typical of solder flip chip assembly prior to interconnect reflow (typically 75 to 100 (μm). This follows because for the compression flow underfill processing technique, chip placement and underfill flow occur simultaneously. The placement force can be estimated by integrating the pressure profile (Eq. 17.16) over the chip area. Assuming a constant volume of dispensed underfill V the placement force is given by Eq. (17.17) for non-Newtonian fluids. For Newtonian fluids ($n = 1$ and $m = \mu$), the placement force is given by Eq. (17.18).

$$F = m\pi\left(\frac{2n+1}{2n}\right)^n \frac{(-\dot{h})^n}{(h/2)^{2n+1}} \frac{\left(\frac{V}{\pi h}\right)^{\frac{3+n}{2}}}{3+n} \tag{17.17}$$

$$F = \frac{3\pi\mu\,(-\dot{h})\left(\frac{V}{\pi h}\right)^2}{h^3} \tag{17.18}$$

17.5.4 Simulation Analysis of Compression Flow Underfill Processing

This section focuses on flow simulation studies of the compression flow chip placement process using commercially available fluid flow analysis software.[24,28] It represents a fundamental advancement in compression flow simulation of polymers in its successful application to the complex geometries and surface topologies demanded by miniaturized FC assembly. Here a simulation methodology is developed and simulation studies are conducted to characterize the compression flow of the underfill, estimate required chip placement forces, evaluate the effect of underfill geometry, and assess the potential formation of voids. A major output of this work is process design guidelines for the next-generation FC process. Moreover, these studies give insight into a preliminary process window and outline the range of acceptable process parameters including underfill deposition geometry and underfill viscosity.

17.5.4.1 Finite Element Simulations. Analytical modeling of the compression flow process presented earlier serves as a starting point for the numerical simulations. Placement force estimates based on the analytical model demonstrate the feasibility of the process and give insight into acceptable process parameters (underfill viscosities, placement forces, etc.). Numerical modeling is chosen to address some of the potential yield and reliability issues (i.e., void formation, penetration of underfill by bumped chips, etc.) and gives insight into the flow characteristics of compression flow chip placement. The primary intent of the numerical simulation work is to characterize the compression flow process. The process characterization includes predicting the local shear rates, local velocity profiles, underfill pressure profiles, streamline bending effects, printed underfill shape (aspect ratio) effects, and potential void formation.

To implement numerical flow simulations, first, the parameters associated with compression flow chip placement are identified. The simulation parameters are divided into two categories: assembly and process parameters. *Assembly parameters* are those directly associated with the geometry of the compression flow chip placement process. *Process parameters* are those linked to the underfill material properties and process operating variables, such as material viscosity and process temperature.

The assembly parameters defined for the compression flow chip placement process include the chip geometry, bump geometry, and printed underfill geometry (i.e., the printed aspect ratio). The *aspect ratio* of the printed underfill is defined as the ratio of the print length to printed height (i.e., in two dimensions). These parameters are used to specify the finite-element mesh models for the numerical simulations. In the current work, only two-dimensional compression flow studies are performed.

Process parameters identified for compression flow chip placement include the isothermal process temperature (ultimately influencing the apparent viscosity of the underfill), chip placement speed, and the underfill material rheology. Figure 17.70 illustrates the various process parameters, where T is the temperature of the chip, underfill, and substrate; σ is the underfill surface tension; η is the underfill apparent viscosity; and $\dot{z}$ is the placement speed. Isothermal process temperatures are selected based on typical underfill dispense temperatures ranging from 25 to 80°C. In this preliminary investigation, the compression flow process is modeled as isothermal; however, this may not be the case for the actual assembly process. Future studies will explore the effects of nonisothermal process conditions in the compression flow process. The *chip placement speed* is defined as the z-axis speed of the placement machine's head during the final chip placement, sometimes called the *search speed.* In general, placement systems use high-speed motion to bring the chip close to the final placement position, followed by a slower search speed along the z-axis until a programmed placement force is reached. The placement speeds selected are based on reported specifications for commercial equipment ranging from 0.3 to 1.4 cm/s. Underfill materials are approximated as Newtonian or non-Newtonian fluids depending on their rheologic characteristics. Figure 17.71 presents the experimentally measured rheology of a typical commercial no-flow underfill material. Notice that this no-flow underfill exhibits Newtonian shear thinning and shear thickening characteristics over the shear rates and temperatures studied.

Two additional process parameters, shown in Fig. 17.70, exist for the compression flow process, placement force F_N and underfill surface tension γ. Unfortunately, the software chosen for analysis is unable to allow control of these quantities for a compression flow process. In general, compression flow modeling for polymer flows is extremely difficult due to the moving boundary and the free surface motion. For the software chosen, normal force control for free surfaces is not possible. The same scenario is true for control of the surface tension variable. Surface tension control for free surfaces is not possible for compression flow simulations, but nevertheless surface tension is held constant throughout the simulations. Specifying surface tension is important when comparing the performance of different underfill materials in compression flow processing.

Numerical modeling of the compression flow chip placement process is conducted using commercially available fluid flow software. POLYFLOW version 3.5.3 is chosen because of its advanced compression flow analysis capabilities for polymeric materials. Modeling of this process is essentially the simulation of compression molding with a very complex boundary. Although is another compression molding package is available, POLYFLOW offers the flexibility needed in terms of simulating the complex chip geometry in the compression flow chip placement problem. Compression molding is a relatively new feature in POLYFLOW, and its use in this study of compression flow chip placement is a unique application.

17.5.4.2 Parametric Studies. The response quantities selected for the simulation study include void formation, shear rates, fluid pressure distribution, local shear rates, fluid velocities, and streamlines. POLYFLOW allows for the visualization of both the mold and fluid throughout the simulations. Flow characterization includes color contour maps and cartesian plots of the pressure, velocity, and streamlines of the compression flow along defined sections. In addition, the maximum value of each quantity is recorded and plotted in order to characterize its behavior throughout an entire flow process. Visualization of the flow is achieved by plotting the deformed mesh at various time steps. The location and identification of potential voids are accomplished visually from the simulation output. The finite-element simulations are designed as two-dimensional, axisymmetric, time-dependent flow problems. Modeling is done axisymmetrically to minimize the CPU time required for a simulation.

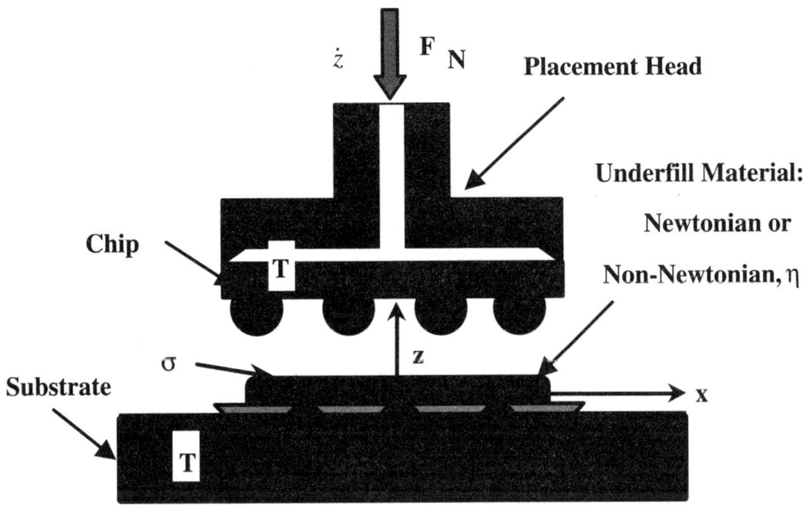

FIGURE 17.70 Process parameters for the simulation of compression flow chip placement.

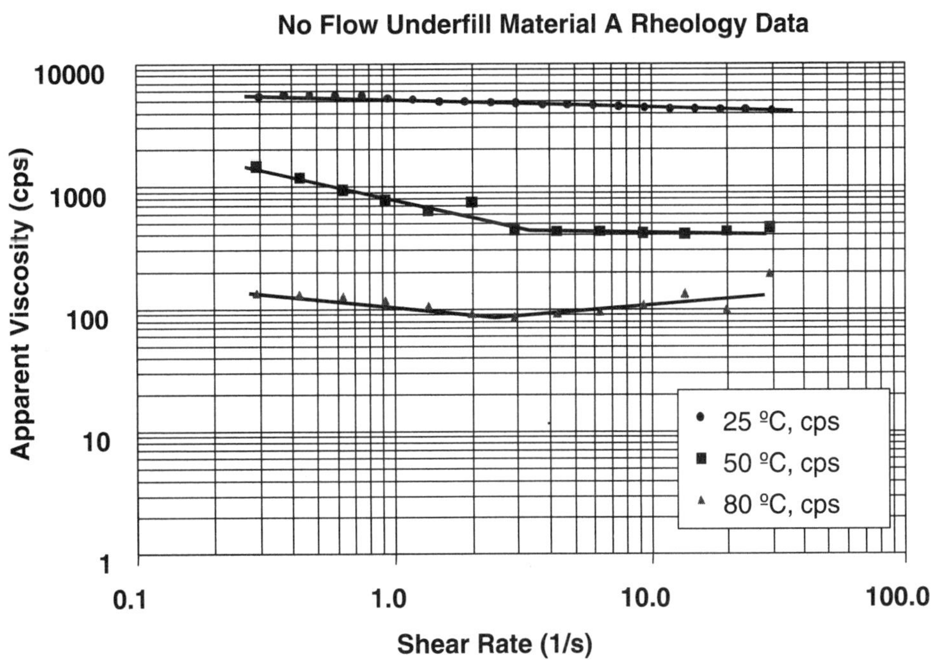

FIGURE 17.71 Rheology of a typical no-flow underfill at various temperatures.

17.5.4.3 Finite Element Mesh and Boundary Conditions. Mesh topology is divided into faces, edges, and vertices for the two-dimensional cases studied. Figure 17.72 is an example of one of the finite-element mesh models generated for the two FC devices studied. A mesh consists of 3 subdomains and 10 boundary sets. Two subdomains represent the chip and substrate, whereas the remaining subdomain (i.e., the fluid subdomain) models the liquid underfill material. The total number of elements for each mesh ranges from 1500 to 2300. Notice in Fig. 17.72 that the mesh of the underfill subdomain is very fine near the bump-fluid interface. A finer mesh is selected to provide more accurate results for the large deformations, especially in the regions of the complex bump geometry.

The fluid subdomain has three different boundary conditions applied to its four boundaries. Boundary 1, defined along the center axis of the chip, is a symmetry condition that allows for the simulation of the axisymmetric geometry. It is defined as having a tangential force and normal velocity equal to zero. Boundaries 2 and 3, the top and outer edges of the underfill-free surfaces, are defined as free surfaces with contact detection. A free surface is essentially a condition where a vanishing surface force is imposed and the fluid cannot cross the free surface. The position of the boundary is not known in advance and is therefore part of the solution. Finally, boundary 4, the interface between the underfill and substrate, is defined as a standard mechanical boundary where either the surface traction or velocity components were specified.

Two mesh models are created for the simulations performed in this work, model I and model II. These models represent two chip geometries with different underfill aspect ratios. Each simulation problem includes the density, gravity, and inertial terms of the underfill material during compression flow. By defining a value for the density, an incompressible fluid is specified. Inclusion of gravity allows for more accurate results and the prediction of gravity flows. Keeping inertial terms in the momentum equations makes the analysis more realistic and comprehensive with respect to fluid motion during rapid compression cycles.

17.5.4.4 Simulation Results. Numerical flow simulations are run for each mesh model. Two underfill materials and six viscosity parameters (i.e., isothermal temperatures) are simulated. Simulation results include the output of the mesh deformation, local shear rate contour, fluid pressure contour, and the fluid velocity profile. The results show the final recorded step of the simulation for the shear rate, pressure, and velocity outputs. The major highlights of these simulations are shown in Figs. 17.73 through 17.81. Figures 17.73 through 17.76 and 17.79 show the simulation results for mesh model I, and Figs. 17.77, 17.78, 17.80, and 17.81 show the results for mesh model II.

Figure 17.73 shows the progression of the fluid flow throughout an entire simulation. The local shear rate is recorded for each simulation, and a typical example is shown in Fig. 17.74 for model I. The shear rate is largest at the fluid-bump interface. This is true for all the compression flow simulations run and is expected due to the bumps piercing the underfill during compression. Looking at Figs. 17.75 and 17.76, the pressure and velocity profiles are plotted along the sections labeled AB. Notice for this model I case with underfill A at 50°C, we have a relatively complex pressure profile.

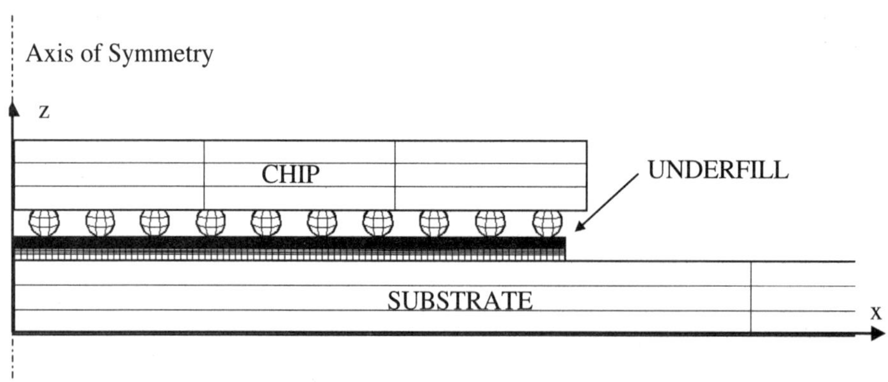

FIGURE 17.72 An example of a compression flow finite element mesh.

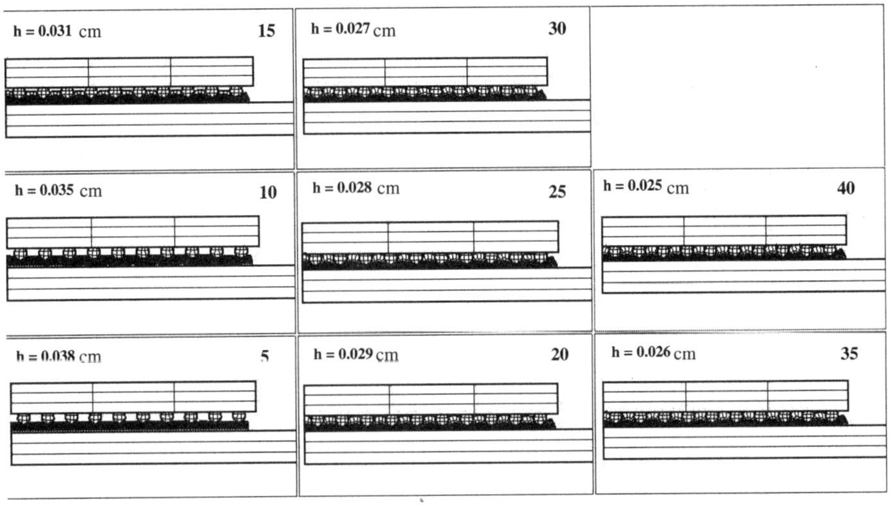

FIGURE 17.73 Progression of the compression flow process—vendor A, model I, T = 50°C.

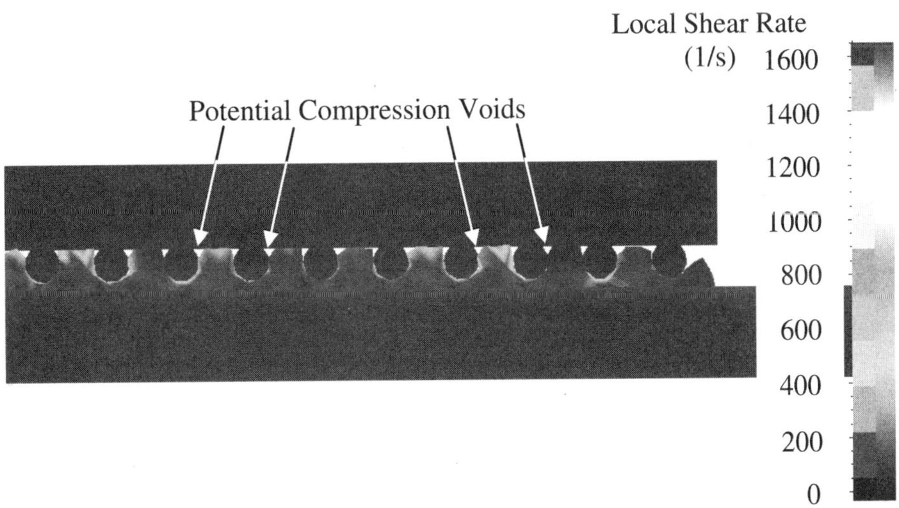

FIGURE 17.74 Local shear rate contour—vendor A, model I, T = 50°C.

Superimposed on the local pressure and velocity spikes under each bump, a saddle profile is seen where the mean pressure is low at the device center and again midway out the half chip length. Pressure peaks are seen just beyond the center and toward the outer edge of the chip. We see a similar trend in the underfill velocity in Fig. 17.76 with localized velocity spikes at the bumps, lower velocity at the center of the chip, lower velocity midway out the half chip length, and velocity peaks between the valleys.

Figure 17.77 shows model II with underfill B processed via compression flow at 80°C. Notice here that the underfill flow proceeds axially along the substrate surface prior to the chip and bumps making contact with the underfill. Under these conditions, the underfill is experiencing gravity-driven flow, where the material flows under its own weight. The velocity profile in Fig. 17.78 shows

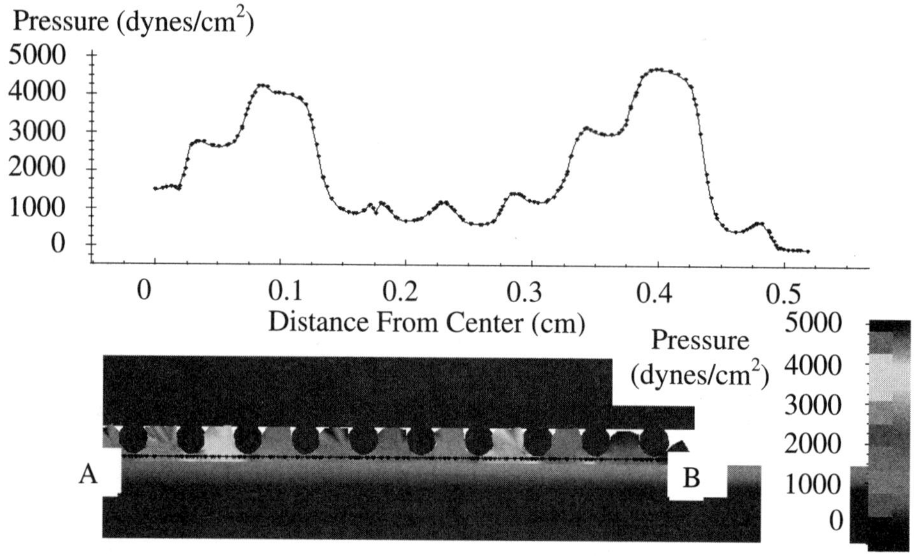

FIGURE 17.75 Fluid pressure contour and section profile (A-B)-vendor A, model I, T = 50°C.

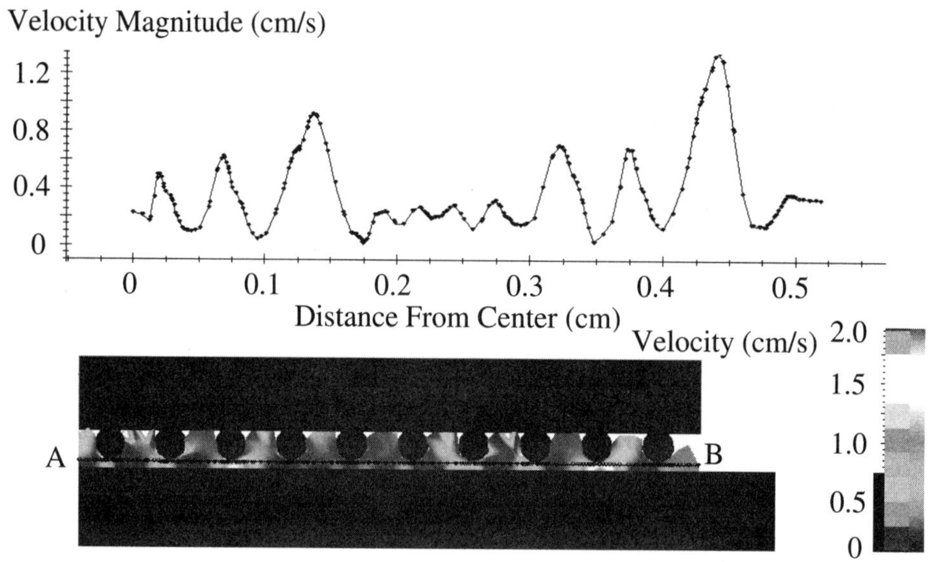

FIGURE 17.76 Velocity contour and section profile (A-B)—vendor A, model I, T = 50°C.

the underfill velocity significantly increasing toward the chip edge. Figure 17.79 also shows gravity flow in this case for model I with underfill A at 80°C. Gravity flow is characteristic of both underfill materials studied at higher temperature.

Figures 17.77, 17.78, 17.80, and 17.81 show the second chip model, model II, simulated. This model has a larger bump size and pitch compared with model I. Based on the simulation results of

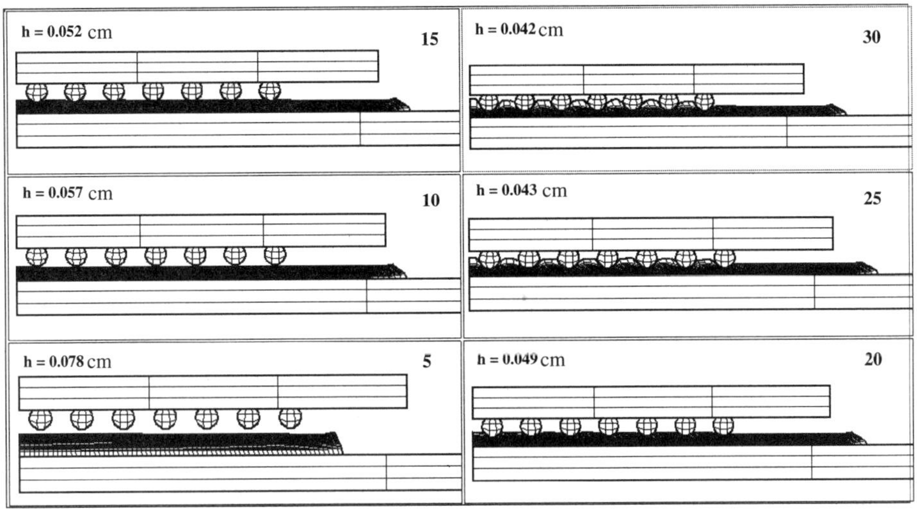

FIGURE 17.77 Gravity flow: progression of the compression flow process—vendor A, model II, T = 80°C.

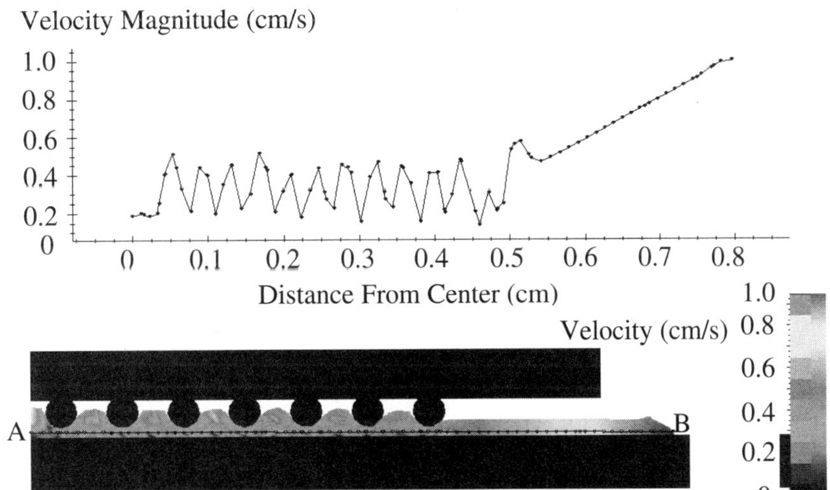

FIGURE 17.78 Velocity contour and section profile (A-B)—vendor A, model II, T = 80°C.

Model II, it is clear that squeeze flow dominated this case. The pressure contour of Fig. 17.80 indicates that the pressure decreases from the chip center to the edge. Similarly, the velocity profile of Fig. 17.81 shows a velocity contour tending to increase to the edge of the chip. It is interesting to compare this with the trend shown in Fig. 17.75 for the same material at the same temperature with a different underfill print height to area aspect ratio.

17.5.4.5 Discussion. Overall, the results of the compression flow simulations provide valuable information on the local shear rates, pressure distributions, and velocity profiles of the various chip configurations and underfill materials studied. The results also provide critical insights into the

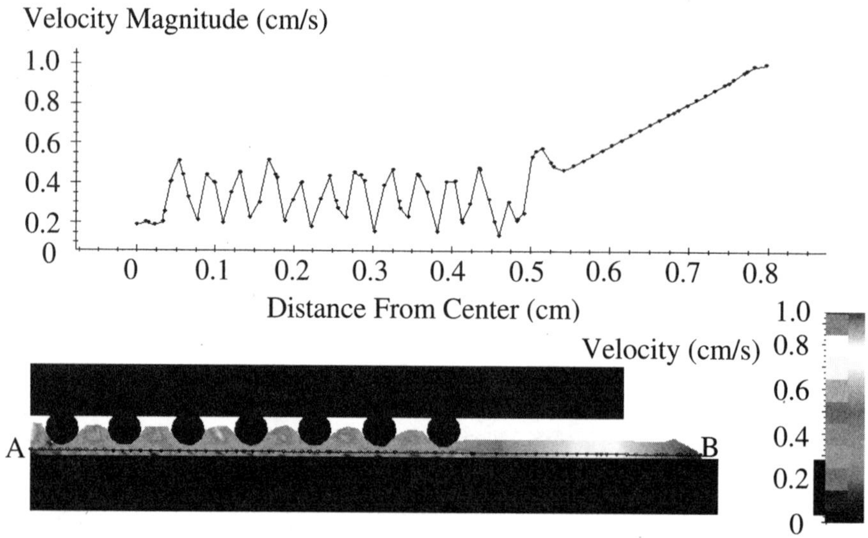

FIGURE 17.79 Velocity contour and section profile (A-B)—vendor A, model I, T = 80°C.

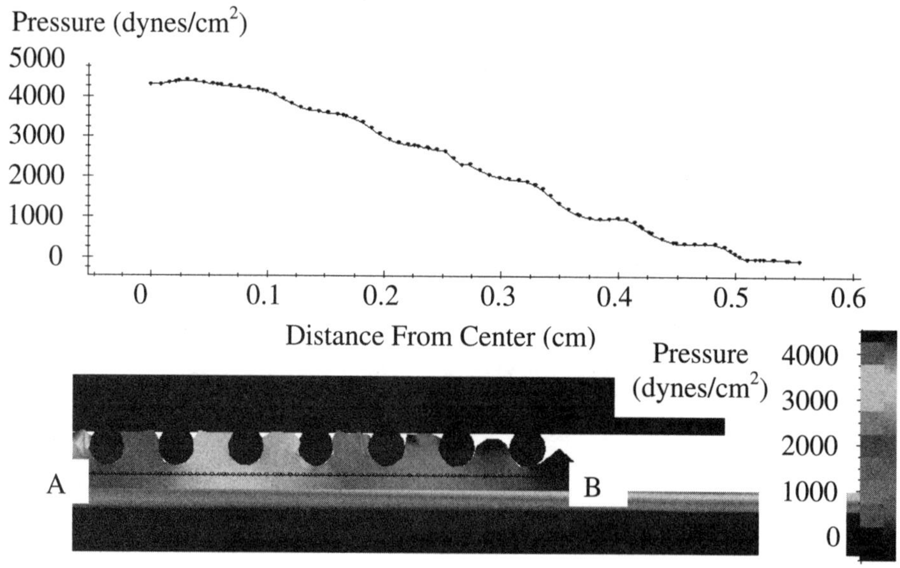

FIGURE 17.80 Fluid pressure contour and section profile (A-B)—vendor A, model II, T = 50°C.

potential formation of voids during next-generation FC processing using compression flow chip placement.

One of the critical elements of the new low-cost next-generation FC process is chip placement. Of particular importance is the ability to predict and understand the forces that must be applied to the chip to compress the underfill to relatively thin gaps (on the order of 25 to 100 μm). In this work, both analytical and numerical simulation results are used to estimate the forces required for the

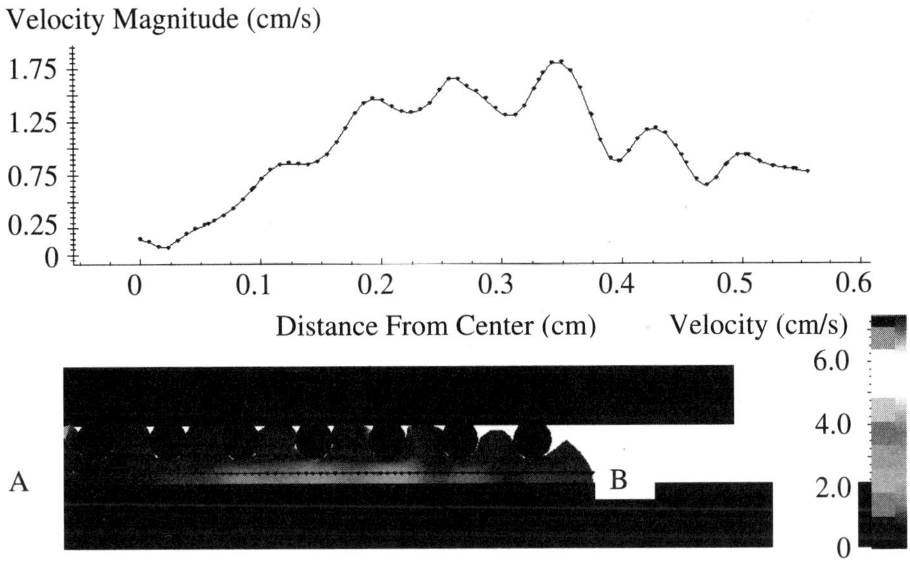

FIGURE 17.81 Velocity contour and section profile (A-B)—vendor A, model II, T = 50°C.

placement machines. While not directly comparable, both the analytical (Fig. 17.69) and numerical simulation (Figs. 17.75, 17.80, and 17.82) results suggest that the placement force capabilities of commercial placement machines are adequate for the next-generation FC process. Thus the analytical and numerical force predictions demonstrate that the process is feasible in terms of the assembly equipment capabilities. All the placement force results are within the capable ranges of existing placement equipment. Figure 17.82 shows the estimated placement forces predicted using the pressure profile simulation results of this study. Estimated placement forces are given for the underfill materials studied (A and B) for both models I and II. Notice in all cases for compression flow processing at 25°C that the required placement forces are less than 100 g well within the placement force capabilities of standard machines.

Another interesting phenomenon predicted by the simulation results is gravity driven flow of the underfill. In some cases, the viscosity of the material is sufficiently low that underfill starvation takes place within the chip-to-substrate gap. This effect is due to gravity flow of the underfill, as illustrated in Fig. 17.77. Gravity flow tends to be governed by the apparent viscosity of the underfill and ultimately temperature due the strong dependence of underfill viscosity on process temperature. Gravity flow is not observed for underfills A and B at 25°C. At 50°C, underfills A and B experience minimal gravity flow, as shown by Figs. 17.80 and 17.81. At 80°C, both underfills A and B experience significant gravity flow, as shown by Figs. 17.78 and 17.79. In particular, notice the relatively sporadic underfill velocities under the chips, as shown in Figs. 17.78 and 17.79. The gravity flow is particularly evident in the underfill segment outside the edge of the chip. In this region, Figs. 17.78 and 17.79 indicate linearly increasing velocity profiles, which are characteristic of uniform radial flow of the underfill due to its own weight. In general, the gravity flow results from the simulations indicate a limit in feasible process temperatures for the next-generation FC process.

Numerical simulations of the compression flow chip placement process are necessary to characterize the very complex flow characteristics of compression flow chip placement in low-cost next-generation FC processing. Moreover, the simulation results aid in identification of potential void formation during processing in order to eliminate this reliability concern. The numerical flow studies predict regions where voiding can occur and gave insight into the flow characteristics that will minimize void formation. Based on the simulation results, a series of design guidelines was formulated.

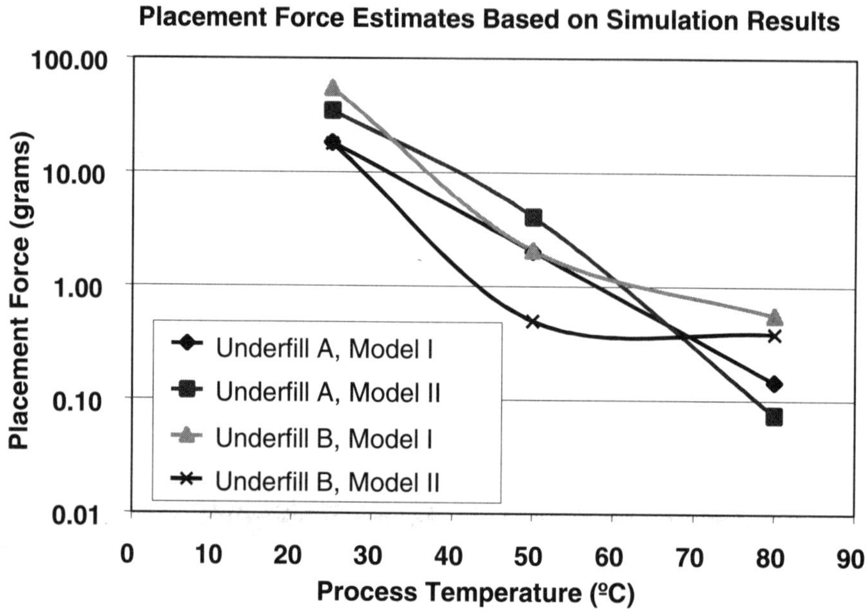

FIGURE 17.82 Placement forces predicted using simulation results for model I, model II, underfill A, and underfill B.

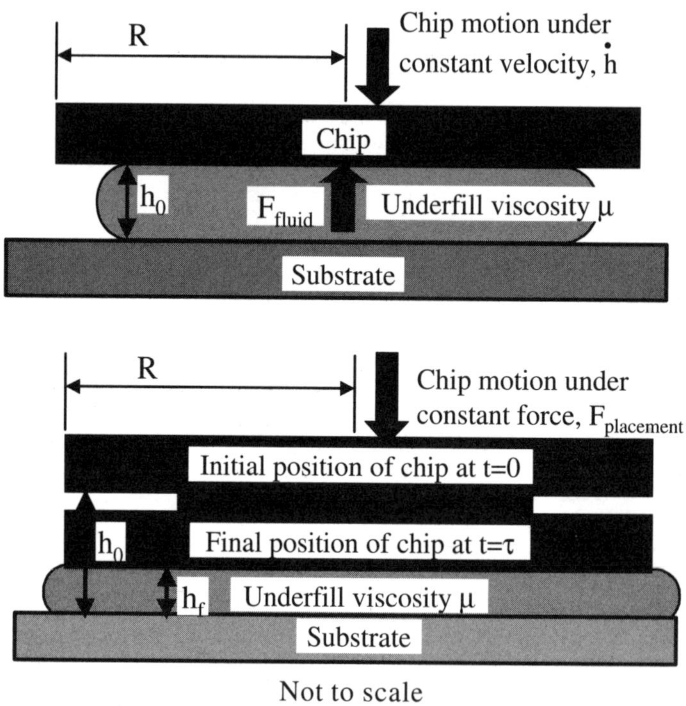

FIGURE 17.83 Schematic of compression flow processing and modeling parameters.

17.5.4.6 Summary of Compression Flow Simulation Analysis. FC technology represents a rapidly advancing area in commercial electronics. In order to ensure adequate reliability, FC assemblies undergo an underfill encapsulation process in which a polymer material is placed between the chip and the substrate. Conventional underfill processing is achieved through chip site to chip site dispensing and underfill flow via capillary action, making it a costly and time-consuming process, particularly as device sizes increase and standoff gaps decrease. To address the limitations associated with conventional FC processing, a low-cost next-generation FC process is proposed. This process eliminates the need for time consuming capillary flow processing using a compression flow technique where the underfill is applied prior to chip placement. The innovative process integrates the chip placement and polymer underfill processes using a compression or squeeze flow technique. It results in significantly lower assembly costs and reduced cycle time.

In general, the compression flow of the underfill material governed assembly yield and reliability for the next-generation process. Both analytical modeling and flow simulation studies were performed based on the compression flow chip placement process. In particular, the flow simulation studies represented a fundamental advancement in compression flow simulation of polymers in its successful application to the complex geometries and surface topologies demanded by miniaturized FC assembly. Simulation studies were conducted to characterize the compression flow of the underfill, estimate required chip placement forces, evaluate the effect of underfill geometry, and assess the potential formation of voids. Three types of voids were identified and experimentally observed that have the potential of forming during compression flow chip placement: compression voids, capture voids, and outgasing voids. With careful control of the underfill geometry, underfill rheology, placement parameters, and reflow process, void formation can be eliminated in the low-cost next-generation FC assembly process. Both analytical modeling and flow simulation results indicated that the placement forces required for compression flow chip placement are well within the ranges capable of standard placement systems. In addition, the results yield design guidelines that gave insight into process parameters such as the limits on underfill deposition geometry, potential of gravity flow, and underfill viscosity. The design guidelines provide a preliminary process window for compression flow chip placement.

17.6 COMPRESSION FLOW PLACEMENT MODEL AND ANALYSIS

The compression flow model is used to design the chip placement process and ensure that the chip makes contact with the substrate during placement.[19,38–40] The model uses a system of two equations that are quasi-steady-state approximations for the squeeze flow of a Newtonian fluid of viscosity μ between two parallel disks of radius R.[3,17,18] The force the underfill exerts on the chip can be approximated with the Stefan equation, Eq. (17.19). This is the quasi-steady-state approximation for the squeeze flow of a Newtonian fluid[3,17,18] between two parallel disks of radius R, a distance h apart, with a fluid of viscosity μ. Equation (17.19) can be used to estimate the force a placement system must exert to place a chip based on the compression flow process. Typically, placement machines move the gantry at high speed to the chip position. The chip is then moved at high speed toward the substrate to a programmed search height. Now the chip is placed onto the substrate at a constant programmed velocity $\dot{h}$ until the force sensor reaches a preprogrammed value $F_{placement}$. The placement machine then maintains the set force for a specified length of time, the dwell time. Using Eq. (17.19), F_{fluid} can be calculated at any height h_1 once the chip is fully wet by the underfill. The effective chip radius, R, can be approximated assuming a circular equivalent area of the rectangular chip given by Eq. (17.20). Equation (17.19) describes the motion of the placement machine until $F_{fluid} = F_{placement}$ at height h_1.

Based on computational fluid dynamics modeling of underfill compression flow during placement, there exists a theoretical limit to the placement velocity of a chip. The reason a limiting speed exists is due to the presence of underfill between the chip and the substrate during placement. There are three forces acting on a chip during the placement process: the force exerted by the placement system on the chip, the weight of the chip, and the hydrodynamic force exerted by the flowing underfill on the chip.

The compression flow model is used to better understand the yield characteristics of the placement process. Specifically, the effects of placement force and placement velocity on interconnection yield were analyzed. In addition, the compression flow model can be used to design the chip placement process and ensure that the chip bumps make contact with the substrate pads during placement. The process parameters are shown in Fig. 17.83. The model uses a system of two equations that are quasi-steady-state approximations for the squeeze flow of a Newtonian fluid between a chip and substrate. The force the underfill exerts on the chip F_{fluid} can be approximated with the Stefan equation, Eq. (17.19) where the chip and substrate surfaces are parallel. In Eq. (17.19), the underfill viscosity is μ, the chip is modeled as having an effective radius of R and the chip and substrate are separated by a distance h_0. The effective chip radius R can be approximated assuming a circular equivalent area of the rectangular chip given by Eq. (17.20).

$$F_{fluid} = \frac{3\pi R^4 \mu (-\dot{h})}{h_0^{\ 3}} \tag{17.19}$$

$$R = \sqrt{\frac{L^2}{\pi}} \tag{17.20}$$

Equation (17.19) can be used to estimate the force a placement system must exert to place a chip based on the compression flow process. Typically, placement machines move the gantry at high speed to the chip position determined by the vision system based on substrate fiducials. The chip is then moved at high speed toward the substrate to a programmed search height. Now the chip is moved toward the substrate at a constant programmed velocity $\dot{h}$ until the placement head force sensor reaches a preprogrammed value $F_{placement}$. The placement machine then maintains the set force for a specified length of time, the dwell time or bond time t. Using Eq. (17.19), F_{fluid} can be calculated at any height h_0 under conditions where the chip is fully wet by the underfill. Similarly, Eq. (17.19) can be used to determine the height h_0 where the underfill hydrodynamic force equals that of the pre-programmed placement force limit, i.e., $F_{fluid} = F_{placement}$.

Once $F_{fluid} = F_{placement}$, the chip moves with a constant force $F_{placement}$ (i.e., rather than a constant speed). The motion of two parallel disks under constant applied force is described by Eq. (17.21). Here τ estimates the time required for the chip to move toward the substrate from h_0 to h_f where h_f is equal to the chip bump height, i.e., $h_f = h_{bumps}$. Equation (17.21) is derived from Eq. (17.19) by solving the differential Eq. in h with an initial condition of $h(t = 0) = h_0$. The parameter τ is the time $F_{placement}$ is applied, h_0 is the initial separation of the chip and substrate at the point, where $F_{fluid} = F_{placement}$, and h_f is the final separation distance between the chip and substrate equal to the bump height.

$$\tau = \frac{3\pi R^4 \mu}{2F_{placement}} \left[\frac{1}{h_f^2} - \frac{1}{h_o^2} \right] \tag{17.21}$$

Equations (17.19) through (17.21) can be used to determine why variations in placement force and placement velocity had no statistical effect on the interconnection yield in experiments 1 through 3. Consider the following case studies based on typical process configurations used in experiments 1 through 3. The viscosity of the underfill is 1.5 Pa·s. The area of the PB8-4 chip is 103.2 mm² having a circular equivalent radius of 5.73 mm. h_0 can be estimated using Eq. 1 given a programmed placement force, $F_{placement} = F_{fluid}$. The largest value for h_0 based on the process parameters in experiments 1–3, occurs at a velocity of 5 mm/s and a placement force of 400 g, (i.e., $F_{placement} = F_{fluid} =$ 3.92 N), yielding $h_0 = 269$ mm. The maximum height of the solder bumps on the PB8-4 chips is $h_{bumps} = h_f = 98$ μm. Based on h_0, h_f and $F_{placement}$, the bond time needed to press the bumps into contact with the substrate can be estimated using $h_0 = 270$ mm, $h = 98$ μm, and $F_{placement} = 3.92$ N in Eq. (17.21). This yields a bond time, τ, of 0.176 s or 176 ms. The bond time for the assemblies in experiments 1 through 3 was 0.3 s. Therefore, there is adequate time to drive the bumps into contact with the substrate pads.

The other extreme case is at the minimum velocity and maximum force used in experiments 1 through 3. In this case, the placement force is 1600 g (i.e., $F_{placement} = F_{fluid} = 15.7$ N), and the placement velocity is 0.1 mm/s. Solving Eq. (17.19) gives the height above the substrate where the placement force is reached h_0 = 46 mm, where the remaining parameters are the same as above. This height is less than the bump height, implying that the chip bumps should make contact with the substrate pads prior to hydrodynamic force of the underfill reaching the placement force limit. In this case, the dwell or bond time is not needed because the placement force is large enough to force contact between the chip bumps and substrate pads during constant velocity chip placement.

17.7 MODELING AND ANALYSIS OF CHIP FLOATING

Under certain process conditions, chips can have a tendency to float during assembly if excess underfill is applied to the substrate.[19,38-40] Such observations are apparent in experiments 1 through 3. In such instances, x-ray inspection shows the chips to be properly aligned with the substrate bond pads before and after reflow, but no interconnections form during reflow. In addition, the chips are observed to easily slide over the substrate when a large amount of underfill is present on the substrate. This yield defect is attributed to a phenomenon called chip *floating*. For a chip to *float*, forces acting on the chip have to be sufficient to lift it off the substrate. Looking at a free body diagram of a chip after placement, Fig. 17.84, there are three forces acting on the chip: the weight of the chip and bumps *W*, the surface tension force F_{st} acting on the perimeter of the chip, and the force due to the pressure difference between the underfill below the chip and bumps P_u and the atmosphere P_a. This force balance forms the basis of the chip floating model and analysis.

Determining the weight of the chip and bumps can be accomplished by weighing the bumped chips prior to assembly. The surface-tension force can be estimated using Eq. (17.22), where *L* is the length of one side of the chip, and γ is the surface tension of the underfill, and α is the fillet contact angle at the top of the chip.

$$F_{st} = 4L\gamma \cos \alpha \tag{17.22}$$

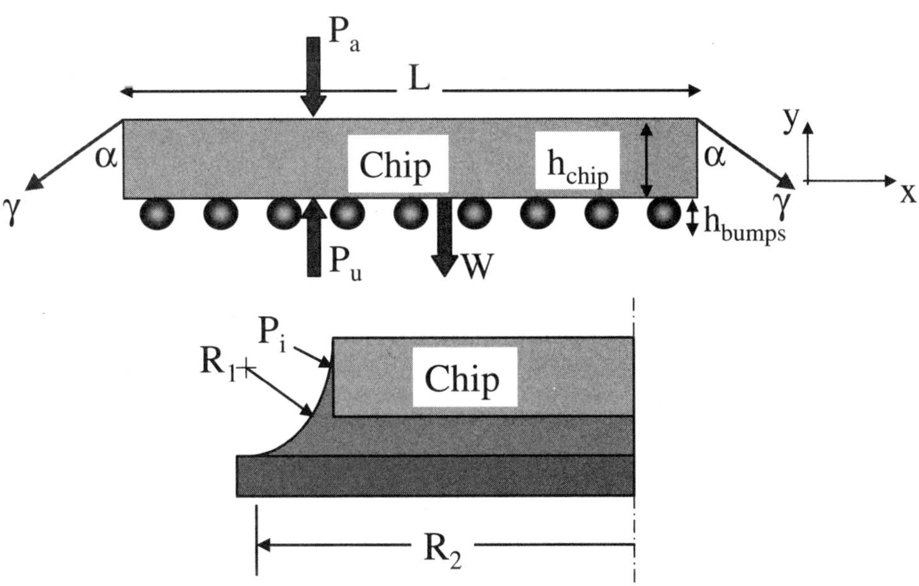

FIGURE 17.84 Free body diagram of chip after placement.

To determine the force caused by the pressure difference $P_a - P_u$, P_u must be broken down into its contributing components. P_u is comprised of two components; the pressure due to the surface energy of the underfill meniscus and the hydrostatic pressure on the active chip surface due to the underfill wetting up the side of the chip. In general, the hydrostatic pressure results from the column of fluid above the active chip surface. The hydrostatic pressure is referenced off the pressure difference due to the surface energy of the underfill meniscus. The Young-Laplace equation (Eq. 17.23) can be used to estimate the pressure difference caused by surface energy and the fillet meniscus. Eq. (17.23) can be used to calculate the magnitude and the sign of the pressure difference at the liquid-air interface based on the principal radii of curvature R_1 and R_2 and the surface energy γ of the underfill. As defined, R_2 is always negative, corresponding to the effective radius of the underfill wetted area on the substrate as shown in Fig. 17.85 (notice the sign convention that the critical radii are positive when measured outside the fluid and negative when measured through the fluid). R_1 can change sign depending on whether the fillet is concave ($R_1 > 0$) or convex ($R_1 < 0$) as shown in Fig. 17.85. P_i is the pressure inside the liquid meniscus at the top of the chip, as shown in the figure.

$$P_a - P_i = \gamma\left(\frac{1}{R_1} + \frac{1}{R_2}\right) \tag{17.23}$$

The hydrostatic pressure contribution can be estimated based on the height of the fillet above the active chip surface and the pressure component exerted by the respective fluid column. Equation (17.24) presents a conservative estimate of the hydrostatic pressure contribution due to the column of underfill. There, a conservative estimate of the pressure acting on the bumps is used, and the underfill fillet is assumed to wet the full height of the chip representing the maximum hydrostatic pressure tending to float the chip. Depending on the chip geometry and bump configuration, the hydrostatic pressure force ranges from 47 to 58 percent of the chip weight based on the test vehicles used in this study. This indicates that the chip geometry and underfill can create a condition where the hydrostatic force is important. In a typical configuration, the top of the chip is 800 μm above the substrate, whereas the gap between the bottom of the chip and substrate is approximately 100 μm. Therefore, a relatively large column of fluid is present where the fillet meets the chip, causing the relatively large hydrostatic pressure. It should also be noted that under process conditions where the chips have partial fillets, not completely wet up the chip side, the contribution of the hydrostatic force is lower, reducing the tendency for chip floating. In this model, a conservative estimate is used to model the fillet as wetting the entire chip thickness.

$$P_{hydrostatic} = \rho_{underfill} g \left[h_{chip}(1 - A_B) + (h_{chip} + h_{bumps}) A_B\right] \tag{17.24}$$

where

$$A_B = \frac{nA_{bump}}{A_{chip}} \tag{17.25}$$

$\rho_{underfill}$ is the density of the underfill, g is acceleration of gravity, h_{chip} is the height of the chip, h_{bumps} is the height of the bumps before reflow, n is the number of bumps, A_{bump} is the cross-sectional area of a bump, and A_{chip} is the area of the active surface of the chip.

The total pressure under the chip P_u can be estimated by adding the fluid pressure at the top surface of the underfill resulting from the meniscus P_i and the hydrostatic pressure acting on the active chip surface and solder bumps, given by Eq. (17.26).

$$P_u = P_i + \rho_{underfill} g \left[h_{chip}\left(1 - A_B\right) + \left(h_{chip}\right) + h_{bumps}) A_B\right] \tag{17.26}$$

The preceding methodology assumes that R_1 is the fillet radius of curvature in the region near the top of the chip. Therefore, the hydrostatic pressure caused by the fluid column along the side of the chip is added to P_i to obtain the pressure of the underfill below the chip and below the individual interconnection bumps. In general, R_1 is not constant along the length of the fillet. The fillet radius of curvature changes to accommodate the increase in pressure due to the hydrostatic pressure at varying depths in the underfill. Summing the forces in the y-direction gives the complete force balance on the chip after placement:

$$F_y = (P_a - P_u)\,A_{chip} + W_{chip} + W_{bumps} + F_{st} = 0 \tag{17.27}$$

Equation (17.27) can be rearranged in the form of a dimensionless parameter Π_f that can be used to predict conditions that promote chip floating. Specifically, Π_f, called for *float parameter,* is the ratio of the "buoyancy forces" acting on the chip to the "adhesion forces" acting on the chip. If Π_f is greater than 1, the chip will have a tendency to float. If Π_f is less than 1, the chip is under adhesion forces tending to hold the chip firmly to the substrate.

$$\Pi_f = \frac{(P_u - P_a)\,A_{chip}}{W_{chip} + W_{bumps} + 4L\gamma\cos\alpha} \tag{17.28}$$

Substituting for P_u and P_i yields:

$$\Pi_f = \frac{\left(\rho_{underfill}g[h_{chip}(1 - A_B) + (h_{chip} + h_{bumps})\,A_B] - \gamma\left(\frac{1}{R_1} + \frac{1}{R_2}\right)\right)A_{chip}}{W_{chip} + W_{bumps} + 4L\gamma cos\alpha} \tag{17.29}$$

where

$$\Pi_f < 1 \quad \text{Adhesion Condition}$$

$$\Pi_f > 1 \quad \text{Floating Condition}$$

The key to predicting floating conditions based on the low-cost, high-throughput FC process and assessing chip floating with respect to the yield experiments presented earlier, centers on developing a relationship between the critical radii R_1 and R_2 and the applied no-flow underfill volume. In general, as the mass of underfill increases, the fillet radius increases until it reaches a point where it will flip from concave to convex, i.e., $R_1 \to \infty$ (see Fig. 17.85) changing sign. At this "critical" underfill mass, $(P_a - P_i)$ changes sign from positive to negative, which can change the sign of the numerator in Eq. (17.29) from negative to positive. This change of sign will cause Π_f to become greater than 0, and for sufficiently large underfill volumes Π_f becomes greater then 1, indicating conditions are favorable for chip floating. Lower underfill volumes ensure concave fillets typically leading to $\Pi_f < 1$ and an adhesion condition between the chip and substrate. It should also be mentioned that the chip floating model assumes the no-flow underfill behaves like a simple fluid, exhibiting negligible viscoelasticity when applied to the substrate prior to reflow curing. The model becomes considerably more complex under conditions where the underfill exhibits appreciable viscoelasticity after deposition on the substrate.

The underfill mass can be estimated using geometric approximations for the underfill fillets. In general, the underfill mass is equal to the mass of underfill under the chip plus the mass of underfill in the fillets less the mass of underfill occupied by the bump interconnections. As a first-order approximation, the fillet is modeled as a circular arc, as shown in Fig. 17.86. The underfill mass associated with the fillet is approximated as symmetric on the four sides of the chip. Based on the fillet geometry of Fig. 17.86 and a chip standoff gap of $(h - h_{chip})$ the underfill mass can be approximated by Eq. (17.30).

$$m_{underfill} = \left(h - h_{chip}\right)A_{chip} + 2L\rho_{underfill}\left(kh^2 \pm 2R_1^{\,2}\sin^{-1}\left(\sqrt{\frac{h^2(k^2+1)}{2R_1}}\right) \pm h^2(k^2+1)\right.$$

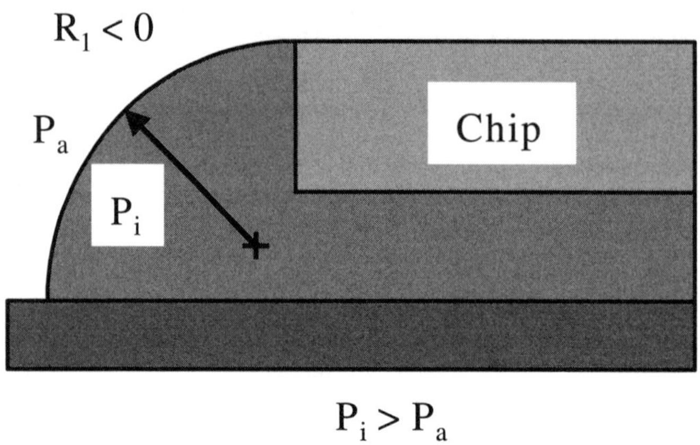

FIGURE 17.85 Influence of fillet shape on pressure difference.

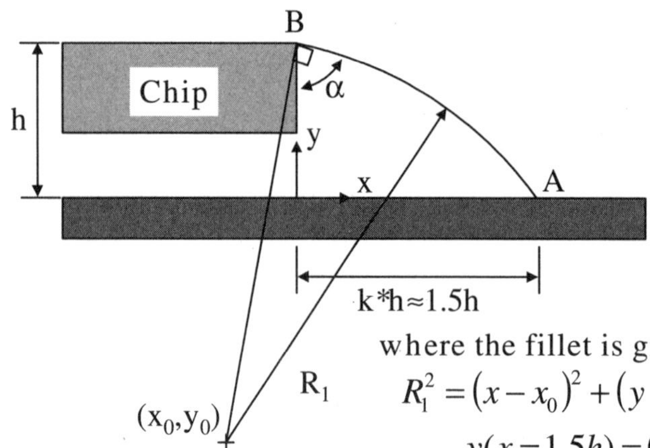

$$R_1^2 = (x - x_0)^2 + (y - y_0)^2$$
$$y(x = 1.5h) = 0$$
$$y(x = 0) = h$$

FIGURE 17.86 Approximation of fillet geometry for no-flow underfill mass estimation.

$$\left(R_1{}^2 - \sqrt{\frac{h^2\,(k^2+1)}{4}}\right)\Bigg) - nV_{bump} \tag{17.30}$$

where k is the underfill wetting factor for the substrate, h is the total standoff gap height from the top of the chip, and V_{bump} is the average volume of a solder bump. Based on experimental observations for the no-flow underfill and substrate-solder mask combinations used, k is typically 1.5.

Assuming a constant fillet radius from the top of the chip to the bottom and using basic geometric relationships, the mass of underfill corresponding to specific values of R_1 and R_2 and hence Π_f can

TABLE 17.26 Parameters for Chip Floating Model

Parameter	Value or range
γ	20 to 50 mN/m
k	1.5
A_{chip}	See Table 17.5
h	800 μm
$W_{chip} + W_{bumps}$	1.33 g (PB8-4)
	0.33 g (PB8-2)
	1.62 (FA10-4)
	0.36 (FA10-2)
$\rho_{runderfill}$	1.1 g/cm^3
h_{bumps}	98 μm (PB8)
	110 μm (FA10)
h_{chip}	700 μm
R_2	−4 mm
n	See Table 17.5

be estimated. This enables prediction of the acceptable underfill volumes or process window for a given chip and underfill system with respect to chip floating. Table 17.26 presents the model parameters used. Figure 17.87 shows a graph of Π_f as a function of underfill mass for various underfill surface energies and FC test chips. The chip parameters (weight, dimensions, etc.) for this graph are based on the commercial FC test chips used in the yield analysis experiments presented earlier, namely, the PB8-2, PB8-4, FA10-2, and FA10-4 test chips. The plot (17.87*d*) shows that the PB8-4 chips should begin to float at an underfill mass of approximately 34.5 mg for an underfill surface energy of 30 mN/m. Experimental observations indicate that the test chip tended to float at an underfill mass of 36 ± 2 mg. Similar predictive capabilities are found for the other FC test vehicles studied that have both area array and perimeter array bump configurations. Specific results are presented in Table 17.27.

While underfill surface energy is a relatively difficult parameter to quantify,[41] notice that the critical underfill mass leading to floating is a relatively weak function of the underfill surface energy. Therefore, even with a poor estimate of the underfill surface energy, the predicted critical underfill amounts do not differ significantly. In contrast, the substrate wetting parameter k has a significant impact on the chip floating model predictions. A sensitivity analysis of k is presented in Fig. 17.88. It is clear from Fig. 17.88 that wetted length of the fillet beyond the edge of the chip can have a significant impact on the critical underfill mass inducing floating, i.e., where $\Pi_f = 1$. As the fillet wets and spreads a greater distance along the substrate, the critical underfill mass inducing floating increases. Therefore, under conditions where the density of components is high, tighter control limits must be held for the deposited underfill volume. However if the keep out distance between components is large, looser underfill volume tolerances can be accommodated without inducing chip floating. A typical k value for the no-flow underfill and substrate-solder mask systems used in this study is 1.5, which was used in the analysis. Careful experiments should be run to characterize k for the particular underfill and substrate-solder mask system used before using the chip floating model to estimate the process window for acceptable no-flow deposition volumes.

The chip floating model serves two purposes. First, it is a process design model enabling rapid estimation of the process window for the no-flow underfill mass to ensure high yield processing. Table 17.27 presents a summary of the results found for the FC test vehicles studied. Second, it helps explain the dramatic yield losses first experienced in experiment 1 prior to the underfill mass range being redesigned. Namely, early trials in experiment 1 showed significant yield loss when the underfill mass exceeded 37 g. The chip floating model predicts a limiting underfill mass of 35 g. In the final experimental matrix, the underfill was varied over a range of 15 to 35 g, minimizing chip floating and the associated yield loss.

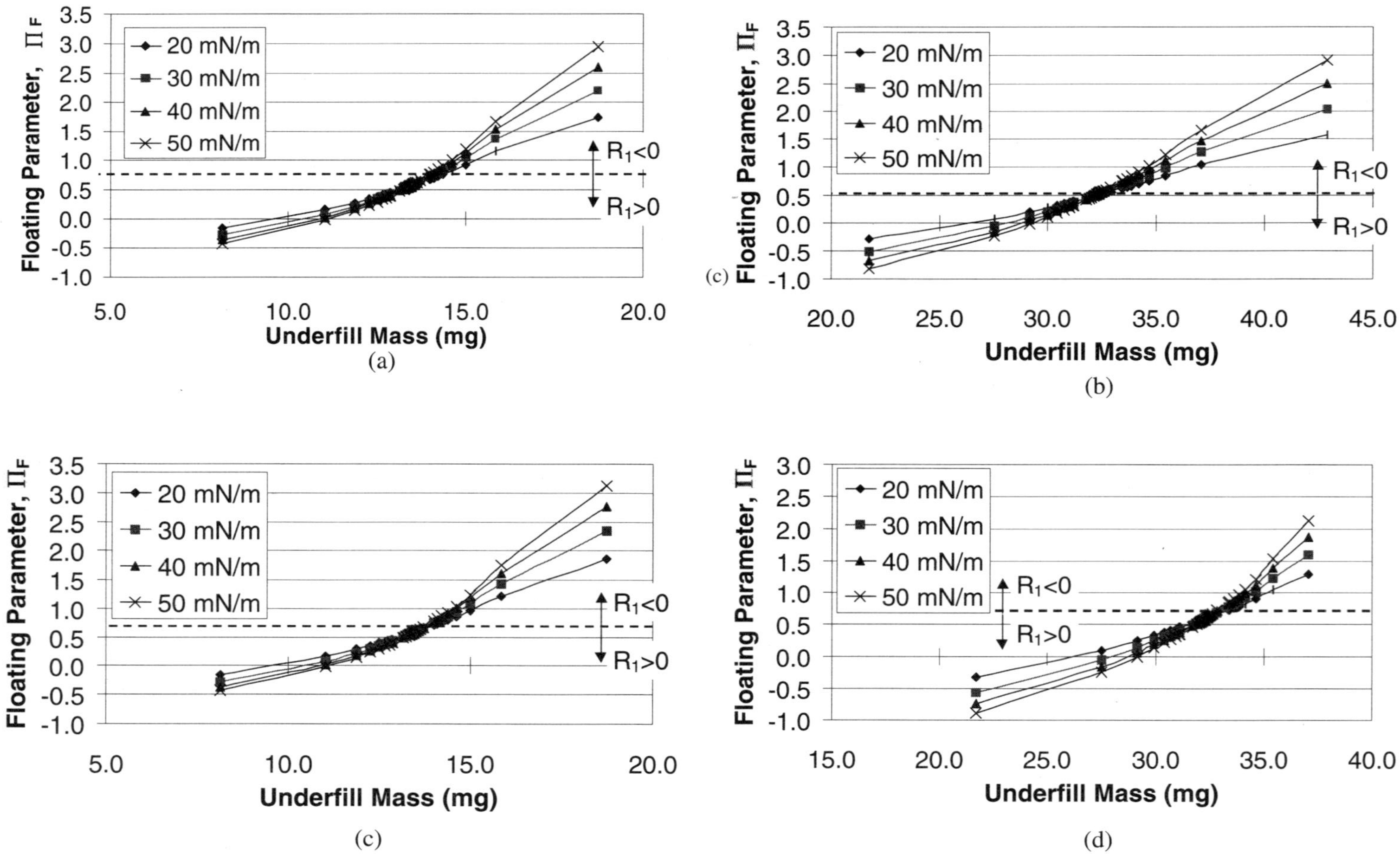

FIGURE 17.87 Floating parameter, Π_f for the FC test chips as a function of underfill mass. (*a*) FA10-2, (*b*) FA10-4, (*c*) PB8-2, and (*d*) PB8-4.

TABLE 17.27 Comparison of Predicted and Measured Underfill Amounts Resulting in Chip Floating

Test Vehicle	Chip size (mm)	I/O count	Predicted underfill mass inducing floating (mg)	Experimentally observed underfill mass inducting floating (mg)
PB8-2	5.08	88	15.0 to 15.8	14 ± 2
PB8-4	10.2	352	34.6 to 35.4	35 ± 2
FA10-2	5.08	317	15.0 to 15.7	16 ± 2
FA10-4	10.2	1268	35.5 to 37.1	37 ± 2

17.8 FC INTERCONNECTION YIELD ANALYSIS DURING THE REFLOW PROCESS

FC solder bumps are produced using a number of different manufacturing processes. One process is based on stencil printing solder paste, after which the solder is reflowed to form a spherical bump. The final heights of the solder bumps are inconsistent (Fig. 17.89) and "normally" distributed. Most of the bumps are in close proximity to the targeted or "mean" size, but depending on the standard, some bumps can be significantly larger or smaller than the targeted size.

Given that the solder bumps on the FC are of varied sizes, the chip must "collapse" during reflow in order for all bumps to make contact with their respective bond pads. In a conventional FC assembly process, the weight of the chip is sufficient to drive the collapse process. In the next-generation

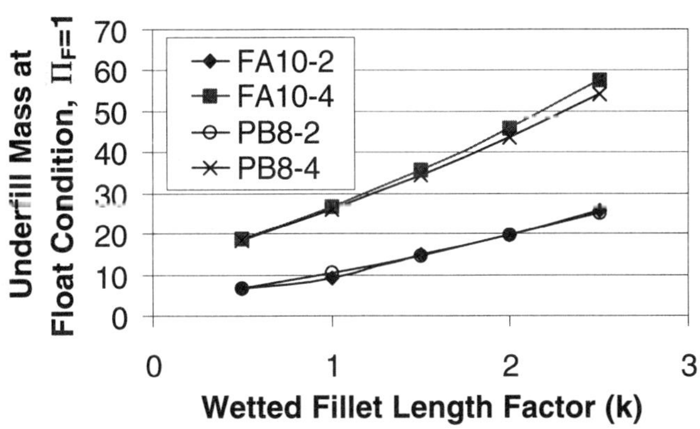

FIGURE 17.88 Underfill mass at the floating condition $\Pi_f = 1$ as a function of wetting length.

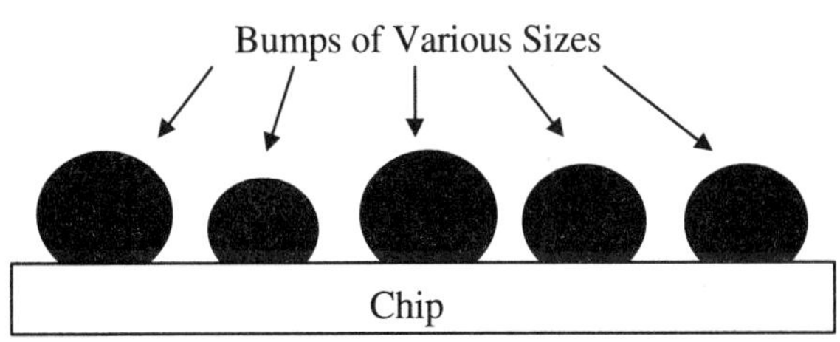

FIGURE 17.89 Chip with varied bump heights.

FC assembly process, a layer of viscous underfill material exerts a force during collapse that resists the collapsing motion.

No-flow underfill materials are formulated such that the underfill does not begin to cure until a latent catalyst has been activated. Ideally, the latent catalyst facilitates underfill curing at a temperature well above the melting point of eutectic solder. This ensures that the chip interconnections reflow, the chip collapses, and all bumps make contact with their respective bond pads prior to underfill hardening. Two potential problems can arise if the underfill begins to harden too early in the reflow process. The underfill viscosity can increase to a point that does not allow the chip to completely collapse. This incomplete collapse results in the shortest bumps not making contact with their respective bond pad. Second, the thin underfill layer between the smallest bumps and their respective bond pads generates a large enough compression flow force to cause the molten solder bumps to deform rather than push through the underfill. To address these issues, a process model was developed that simulates the chip collapse process assuming that the molten solder bumps do not deform as they move through the liquid underfill.

17.8.1 FC Interconnection Yield Analysis

The rigid solder model was developed to determine the time it takes for a complete chip collapse scenario to occur, assuming the mean and standard deviation of the bump height distribution is known. For a successful chip collapse, every bump must make contact with its respective bond pad. Figure 17.90 shows a schematic of the chip collapse process. As mentioned earlier, the heights of the solders bumps are inconsistent. The model assumes that prior to reflow, the chip rests stably on its three tallest bumps.

After chip placement, the assembly is placed in a reflow oven. Figure 6-6 is a schematic of the forces acting on the chip, once the solder has reached its melting point of 183°C.

F_{ss} is the total force due to steady state components: chip weight, underfill surface tension, and adhesion between the chip and board. F_{solder} is the force produced by the surface tension of the wetted solder joints. $F_{underfill}$ is the force exerted by the underfill due to compression flow. The next three sections give detailed explanations of F_{ss}, F_{solder}, and $F_{underfill}$.

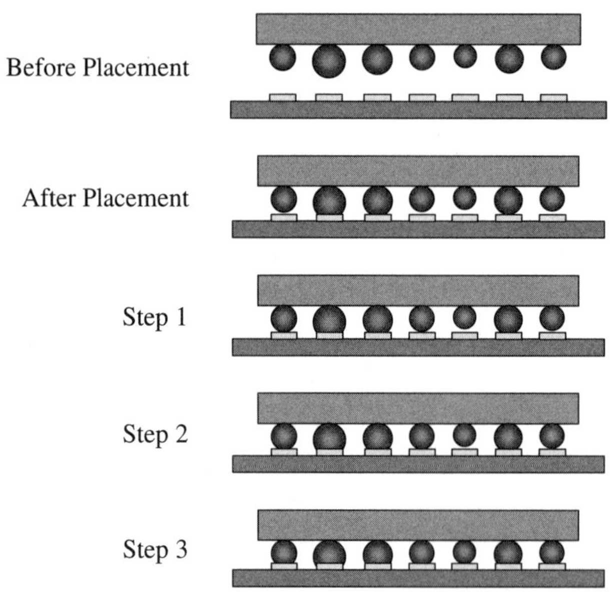

FIGURE 17.90 Chip collapse schematic.

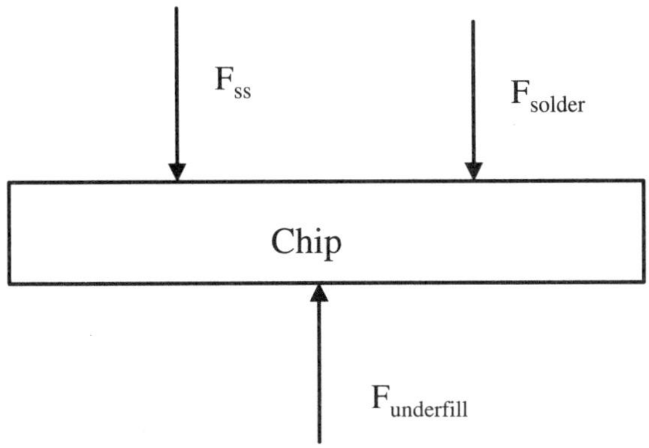

FIGURE 17.91 Free body diagram of collapsing chip.

17.8.2 Steady State Forces

The steady state force acting on a chip after placement is determined using methodology presented by Thorpe et al.[38–40]

Figure 17.92 is a free body diagram of the chip after placement. There are three forces acting on the chip:

1. The weight of the chip
2. Surface tension force F_{st} acting on the perimeter of the die
3. Force due to the pressure difference between the underfill below the die and solder bumps P_u, and the atmosphere P_a.

Determining the weight of the chip is straightforward. The surface tension force can be calculated using Eq. (17.31), where L is the length of one side of the chip, gunderfill is the surface tension of the $\gamma_{underfill}$, and α_c is the fillet contact angle at the top of the chip.

$$F_{st} = 4L\gamma_{underfill}\cos\alpha_c \tag{17.31}$$

To determine the force caused by the pressure difference $P_a - P_u$, P_u must be analyzed. P_u is comprised of two components; pressure due to the surface energy of the underfill, and hydrostatic pressure due to the underfill wetting up the side of the chip. The Young-Laplace equation (Eq. 17.32) can

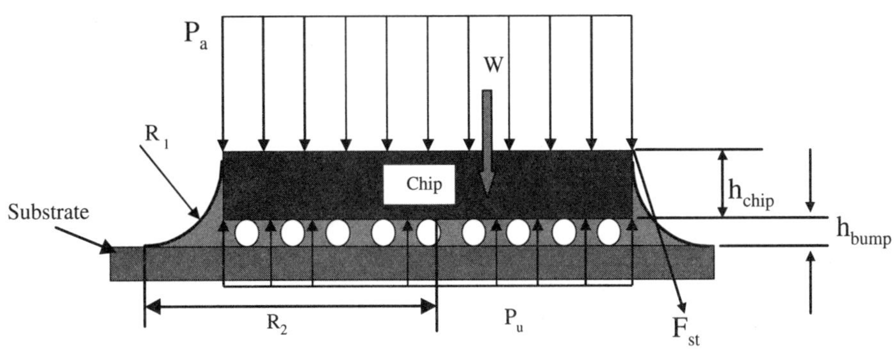

FIGURE 17.92 Free body diagram of a chip after placement.

be used to determine the pressure difference caused by surface energy and the fillet meniscus. This equation and Fig. 17.93 are used to calculate the magnitude and the sign of the pressure difference at the liquid-air interface based on the principal radii of curvature R_1 and R_2 and the surface energy γ of the underfill. P_i is the pressure inside the liquid at the fillet liquid-air interface at the top of the chip and R_1 and R_2 are shown on Fig. 17.93.

$$P_a - P_i = \gamma\left(\frac{1}{R_1} + \frac{1}{R_2}\right) \tag{17.32}$$

The hydrostatic pressure can then be added to P_i to obtain the total P_u, given by Eqs. (17.33) and (17.34).

$$P_u = P_i + \rho_{underfill}g\,[h_{chip}\,(1 - A_F) + (h_{chip} + h_{bumps})\,A_F] \tag{17.33}$$

$$A_F = \frac{\text{(Cross Sectional Area of Solder Interconnect)} \times \text{(\# of interconnects)}}{\text{Area of Die}} \tag{17.34}$$

The preceding methodology assumes that R_1 is the radius of curvature of the fillet in the region near the top of the chip. Therefore, the hydrostatic pressure caused by the column of fluid along the side of the chip is added to P_i to obtain the pressure of the underfill below the chip and below the individual interconnection bumps. R_1 is not constant along the length of the fillet. It changes radius of curvature to accommodate the increase in pressure due to the hydrostatic pressure at varying depths in the underfill. Summing the forces in the y-direction gives the complete force balance:

$$\Sigma F_y = (P_a - P_u)\,A_{die} + W + F_{st} \tag{17.35}$$

For the rigid solder model, the steady state force F_{ss} is equal to ΣF_y.

17.8.3 Solder Force

When a solder ball contacts and wets a bond pad, the equilibrium shape of the interconnection is dictated by surface energy forces. The interconnection attempts to minimize its total surface energy. In doing so, the solder joint exerts either a pushing or pulling force on the chip. Figure 17.94 shows examples of both scenarios. If the interconnection is shorter than the equilibrium height, then it will produce a force that pushes up on the chip. If the interconnection is taller than the equilibrium height, then it will produce a force that pulls down on the chip.

Extensive research has been done in the area of solder self-alignment, including the normal reaction forces just shown.[12] They developed an explicit regression model that can calculate a normalized normal reaction force for one interconnection as a function of normalized chip misalignment, joint height, and the aspect ratio between the joint height and pad radius. The regression model has

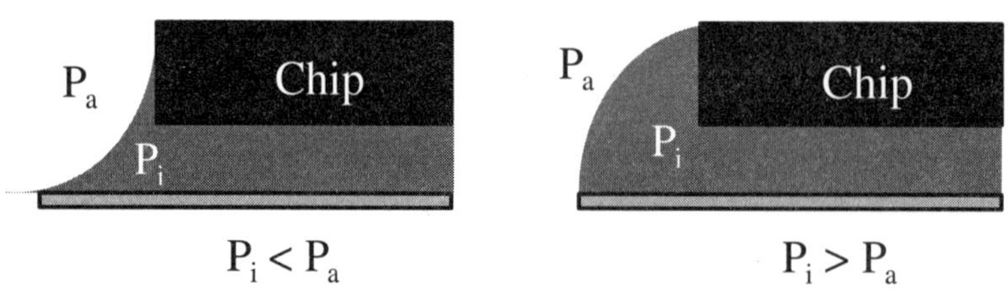

FIGURE 17.93 Effect of fillet curvature of underfill pressure.

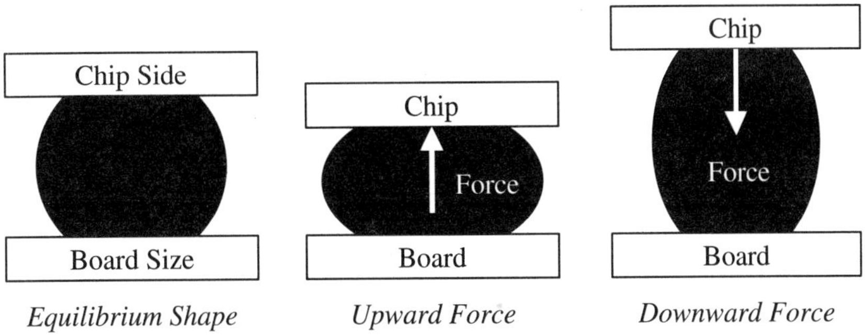

FIGURE 17.94 Force exerted by molten solder interconnections.

three nondimensional input parameters, N, P^*, H^* and one nondimensional output, F_N^*. Physically, N is a measure of the aspect ratio of the joint, P^* is the misalignment between the solder bump and the substrate pad, and H^* is the instantaneous joint height.

N is calculated using Eq. (17.36), P^* is assumed to be zero for simplicity, and H^* is calculated using Eqs 17.37 and 17.38.

$$N = \frac{V}{\pi R_p^3} \tag{17.36}$$

V is the volume of the solder bump, and R_p is the chip pad and substrate bond pad radius. These two pad radii must be equal to use the regression model.

$$H^* = \frac{H}{H_c} \tag{17.37}$$

$$H_c = \frac{V}{\pi R_p^2} \tag{17.38}$$

The coefficients for the regression model are a function of N and are presented in Eqs. 17.39 through 17.42.

$$a_0 = -907.38 + 2441.73N - 2823.98N^2 + 1439.13N^3 - 266.43N^4 \tag{17.39}$$

$$a_1 = 1881.44 - 5357.26N + 6525.01N^2 - 3431.96N^3 + 648.69N^4 \tag{17.40}$$

$$a_2 = -1192.24 + 3685.92N - 4799.78N^2 + 2622.64N^3 - 507.91N^4 \tag{17.41}$$

$$a_3 = 224.22 - 786.56N + 1122.26N^2 - 642.96N^3 + 128.19N^4 \tag{17.42}$$

F_N^* is calculated using Eq. (17.43), and the dimensioned F_N (N) is calculated using Eq. (17.44).

$$F_N^* = a_0 + a_1H^* + a_2H^{*2} + a_3H^{*3} \tag{17.43}$$

$$F_N = F_N^* = F_N^*\gamma R_P \tag{17.44}$$

F_N is the force that one interconnection exerts on the chip. F_{solder} is calculated by summing the F_N contribution from all bumps that have contacted and wetted a substrate bond pad (Eq. 17.45).

$$F_{solder} = \sum_i^k F_{Ni} \tag{17.45}$$

where F_{Ni} = the normal force contribution from interconnection 'i'
k = the number of bumps in contact with a bond pad

17.8.4 Underfill Force

Hydrostatic squeeze film theory provides an approximation for the force that the underfill exerts on a collapsing chip. Basic viscous squeeze film theory assumes that a circular disk of radius R is approaching another circular disk of equal radius (Fig. 17.95).

Because the chip is not circular, an equivalent radius approximation must be made. Eq. (17.46) is used to calculate the equivalent circular radius of a square chip. L is the length of one side of the chip.

$$R = \sqrt{\frac{L^2}{\pi}} \tag{17.46}$$

The force that the underfill exerts on the upper disk (chip) is given by Eq. (17.47).

$$F_{underfill} = \frac{3\pi\mu\, hR^4}{h^3} \tag{17.47}$$

where μ = the viscosity of the underfill
h = the instantaneous separation distance between the disks
h^* = the velocity of disk 2 (chip)
R = the equivalent chip radius

To use Eq. (17.47), the collapse velocity of the chip is needed. This value is difficult to obtain. Solving the differential Eq. (17.47) for h results in Eq. (17.48), which gives the time of collapse from height h_2 to height h_1.

$$\tau = \frac{3\pi\mu R^4}{2F_{total}} \left[\frac{1}{h_{i+1}^{\;2}} - \frac{1}{h_i^2} \right] \tag{17.48}$$

where

$F_{total} = F_{ss} + F_{solder}$
h_i and h_{i+1} are subsequent bumps in the collapse process

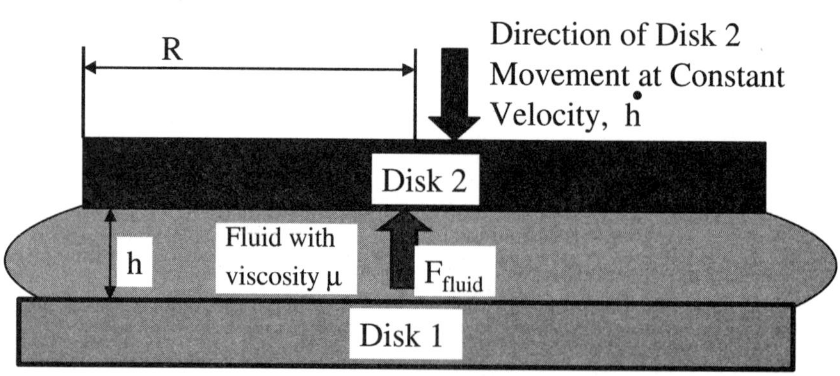

FIGURE 17.95 Compression flow with constant velocity.

17.8.5 Chip Collapse Time

The collapse time results are based on a numerical simulation of the collapse process. The simulation begins by sampling N bump heights from a "normal" distribution with mean μ and standard deviation σ. The bumps are numbered 1 to N, where h_i is the height of bump i, and $h_1 > h_2 > h_3 > \ldots > h_N$.

The collapse process is then simulated in an iterative stepwise manner. The distance steps are not constant. Each step is equal to the difference in height between bump i and bump $i+1$. For each step, F_{solder} is a unique value, based on the number of bumps in contact with bond pads, and the current shape of those interconnections. F_{solder} is added to the constant F_{ss} to find F_{total} for each step. Equation (17.48) is then used to determine the time of collapse from bump i to bump $i+1$. An example of steps 1 and 2 of the simulation is given below.

After placement, the chip is assumed to be resting on its three tallest bumps. The first distance step is the difference in height between bump 3 and bump 4 ($h_3 - h_4$) (Fig. 17.96). F_N for bumps 1, 2, and 3 is calculated based on the current height of h_3. F_{solder} is the summation of the F_N contributions from bumps 1 through 3. The collapse time from h_3 to h_4 is calculated using Eq. (17.48) where h_i is equal to h_3 and h_{i+1} is equal to h_4. At the conclusion of the first distance step, bumps 1 through 4 are wetted to their respective bond pads. The second distance step is the difference in height between bumps 4 and 5. F_{solder} for step two is the summation of the F_N contributions from bumps 1 through 4. The collapse time for step two is again given by Eq. (17.48), where h_i is equal to h_4 and h_{i+1} is equal to h_5.

This iterative step process continues until the Nth bump has made contact with a bond pad. A summation of the collapse times for each step gives the total collapse time.

17.8.6 Underfill Viscosity Effect

The most important results from the model examine the effect of underfill viscosity on collapse time. These results determine an allowable viscosity such that a complete chip collapse can occur. Figure 17.97 is a plot of collapse time versus underfill viscosity. Each line in the plot represents a standard deviation (in microns) of the initial bump height distribution. The results are based on the following assumptions: a PB8-4 chip with total steady state forces of 3 mN, solder surface tension of 300 mN/m, an initial "normal" bump height distribution with a mean of 98 μm, and a constant underfill viscosity. A steady-state force of 3 mN was chosen based on earlier modeling done by McGovern and Baldwin.[19] McGovern and Baldwin determined that using an optimal volume of underfill for the PB8-4 chip produced a steady-state force of 3 mN.

Figure 17.97 indicates that for a given bump height standard deviation, a linear relationship exists between viscosity and collapse time. As expected, the lower the underfill viscosity, the faster the chip collapse will occur. The figure also shows that the slope of collapse time versus viscosity decreases

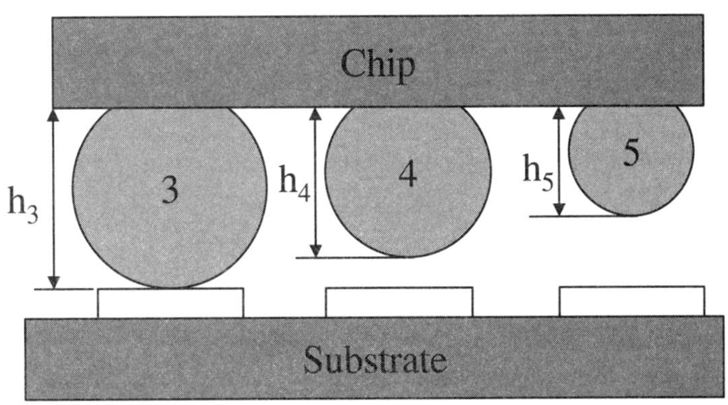

FIGURE 17.96 Bump height notation.

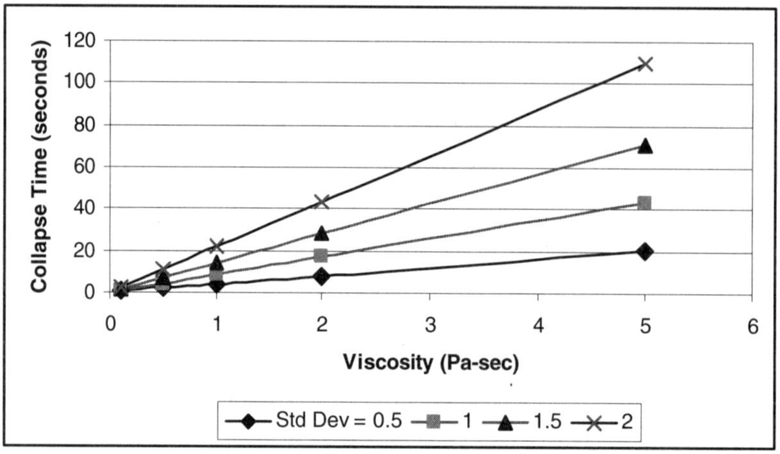

FIGURE 17.97 Collapse time versus underfill viscosity.

as standard deviation decreases. Therefore, a height distribution with a small standard deviation is less affected by increases in viscosity than a distribution with a larger standard deviation.

17.8.7 Processing Windows for Complete Chip Collapse

The results from the rigid solder model can be used to generate process windows that predict the conditions for which a complete chip collapse is likely. A process window graph allows a potential user of the high-throughput FC assembly process to understand the interactions of three important variables in the collapse process. The process windows presented in Fig. 17.98, relate underfill viscosity, maximum allowable collapse time, and initial bump height standard deviation. The shaded areas in the graph define combinations of underfill viscosity, maximum allowable collapse time, and bump height standard deviation that result in a successfully collapsed chip. Figure 17.98 shows that the maximum allowable collapse time parameter corresponds to the *x* axis, and the underfill viscosity corresponds to the *y* axis. The four shaded regions (AD) in the graph represent the bump height standard deviation. Regions A, B, C, and D define the processing window for a bump height standard deviation of 0.5 mm. Regions B, C, and D define the processing window for a standard deviation of 1.0 μm, regions C and D define the window for a standard deviation of 1.5 μm, and region D defines the process window for a standard deviation of 2.0 μm.

The solder bump manufacturer controls the bump height standard deviation, and in most cases, the user of the bumped FCs has little control over this parameter. The underfill viscosity and maximum allowable collapse time are the two parameters that users of the high-throughput process may have control over.

One approach to determining the maximum allowable collapse time is to find the length of time above 183°C that the underfill remains a constant viscosity. Typically, no-flow underfill materials remain at a relatively constant viscosity for x amount of time above 183°C, after which the viscosity quickly increases. If the chip collapse process takes longer than this time of constant viscosity, the likelihood of complete collapse is reduced. The time of constant viscosity depends on the underfill and the reflow profile. The underfill supplier may be able to adjust the chemistry to produce a desired time of constant viscosity, or the reflow profile can be adjusted to produce the desired time.

An alternative approach for estimating the maximum allowable collapse time is based strictly on the reflow profile. Reflow profiles are characterized by a number of different parameters, one of which is the time above 183°C. Knowing that the chip can only collapse while the solder intercon-

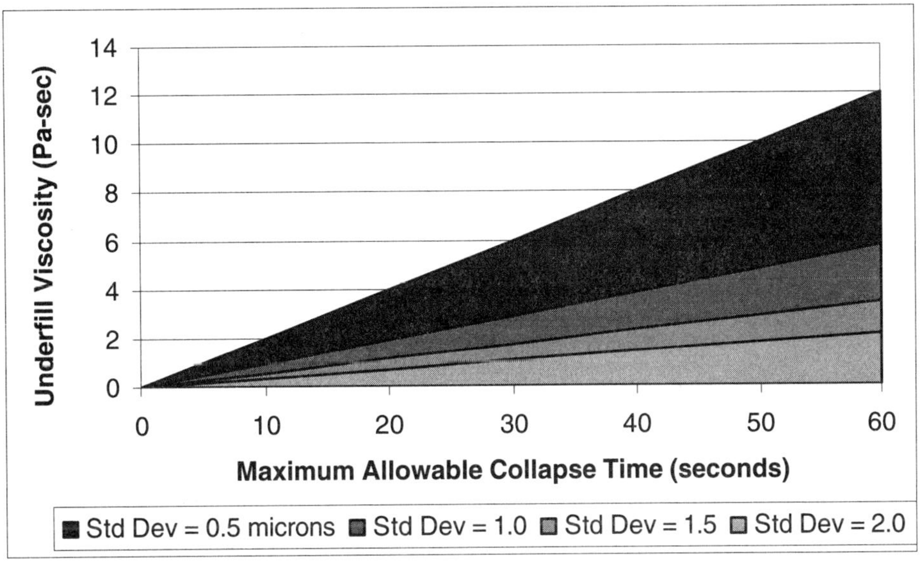

FIGURE 17.98 Process windows for successful chip collapse.

nections are molten, the time above 183°C reflow parameter can be used as the maximum allowable collapse time. It should be noted that in most cases, the underfill material will harden while the assembly is still experiencing reflow temperatures in excess of 183°C.

The process windows shown in Fig. 17.98 allow a user to specify a range of one parameter based on known ranges of the other two parameters. For example, if the bump height standard deviation is 0.5 μm, then any combination of allowable collapse time and underfill viscosity that lies in shaded regions AD will result in a completely collapsed chip. Alternatively, if known values of underfill viscosity and maximum allowable collapse time correspond to a point in region B, then the bump height standard deviation can be anywhere between 0 and 1.5 μm.

In general, low underfill viscosity, long allowable collapse times, and low bump height standard deviation increase the likelihood of a complete chip collapse.

17.8.8 Interconnection Density Effect

The rigid solder model can also be used to relate interconnection density and collapse time. Figure 17.99 shows collapse time versus number of bumps on a chip. The assumptions for this plot are a 10.16 × 10.16 mm chip, steady state force of 3 mN, bump height distribution with a mean of 98 μm, and a standard deviation of 2 μm. The data show that as the number of bumps per chip area increases, the collapse time decreases. Adding bumps increases the total amount of downward force produced by the solder interconnections, thus increased producing a faster collapse.

17.8.9 Bumps in Contact Due to Placement

The model results to this point have assumed that only the three tallest bumps are in contact with the substrate prior to reflow. During chip placement, it is possible to increase the placement force such that more than three solder balls come in contact with the substrate. Figure 17.100 shows the benefit of increasing placement force. If the tallest 10 percent of bumps are in contact with a bond pad prior to reflow, the collapse time is reduced by as much as 35 percent. If 50 percent of the bumps are in contact with bond pads, the collapse time is reduced by more than 50 percent.

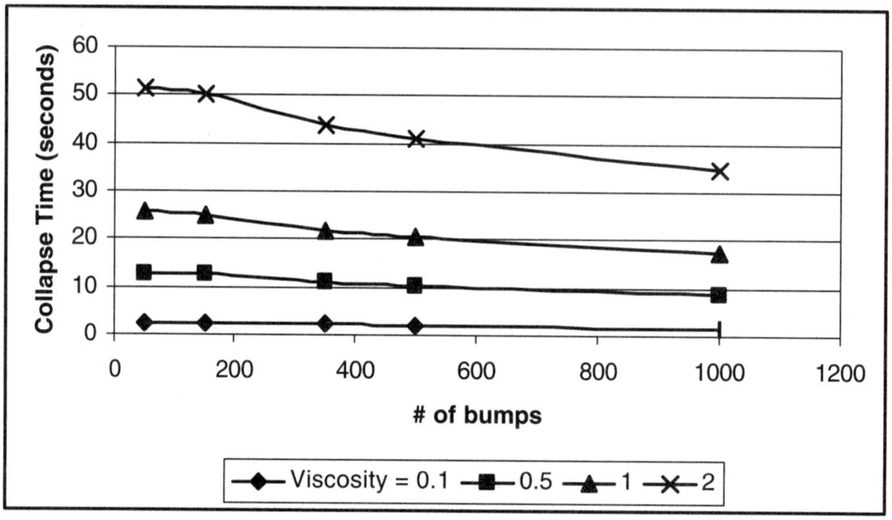

FIGURE 17.99 Collapse time versus number of bumps.

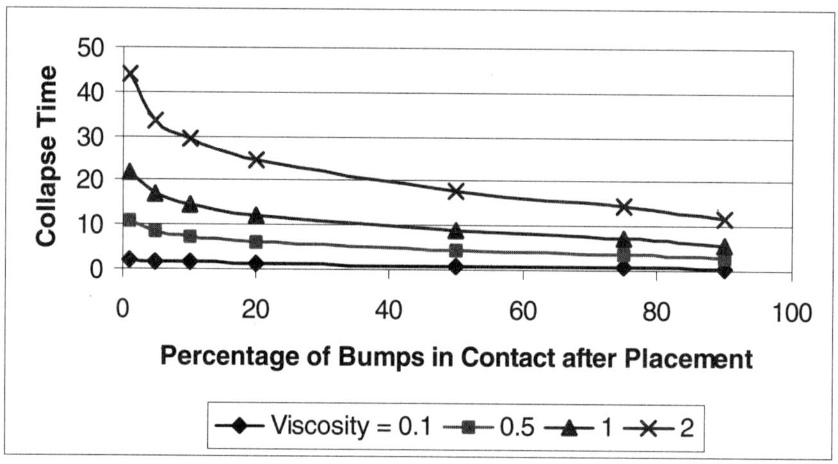

FIGURE 17.100 Collapse time versus percent of bumps in contact.

17.9 SUMMARY

FC technology represents a rapidly advancing area in commercial electronics. In order to ensure adequate reliability, FC assemblies undergo an underfill encapsulation process in which a polymer material is placed between the chip and the substrate. Conventional underfill processing is achieved through chip site to chip site dispensing and underfill flow via capillary action, making it a costly and time-consuming process, particularly as device sizes increase and standoff gaps decrease. To address the limitations associated with conventional FC processing, a low-cost, high-throughput FC process was developed. This process eliminates the need for time consuming capillary flow processing using a compression flow technique where the underfill is applied prior to chip placement. The innovative process integrates the chip placement and polymer underfill processes using a compression or squeeze flow technique. It results in significantly lower assembly costs and reduced cycle time.

The low-cost, high-throughput FC assembly process has converted the assembly process from a point-to-point process to an area-oriented process. This has resulted in decreased assembly time, allowing for increased throughput at lower cost. Parametric studies of FC assembly processes using the activity-based cost model highlighted the benefits of the low-cost, high-throughput assembly process. It has been shown that the assembly cost is reduced with increases in throughput. Furthermore, increased chip size showed minimal increases in cost due to the elimination of the time consuming capillary flow process, whereas the benchmark FC assembly process exhibits substantial cost increases. Cost reductions of 5 to 6 times were demonstrated. In the high-throughput process, underfill flow is no longer dependent on the chip size due to the compression flow chip placement process. Increased chips per panel showed little increase in cost for the next-generation process as a result of area-oriented processing. Cost reductions of 2 to 3 times were demonstrated.

In general, the compression flow of the underfill material governed assembly yield and reliability for the low-cost, high-throughput process using no-flow underfills. Both analytical modeling and flow simulation studies were performed based on the compression flow chip placement process. Simulation studies were conducted to characterize the compression flow of the underfill, estimate required chip placement forces, evaluate the effect of underfill geometry, and assess the potential formation of voids. Three types of voids were identified and experimentally observed that have the potential of forming during compression flow chip placement: compression voids, capture voids, and outgasing voids. With careful control of the underfill geometry, underfill rheology, placement parameters, and reflow process, void formation can be eliminated in the low-cost, high-throughput FC assembly process. Both analytical modeling and flow simulation results indicated that the placement forces required for compression flow chip placement are well within the ranges capable of standard placement systems. In addition, the results yield design guidelines that gave insight into process parameters such as the limits on underfill deposition geometry, potential of gravity flow, and underfill viscosity. The design guidelines provided an preliminary process window for compression flow chip placement.

The capability to assemble FCs using the low-cost, high-throughput assembly process incorporating no-flow underfill materials was demonstrated. The ability to implement the process on standard SMT assembly equipment, including placement machines and reflow ovens, was shown experimentally. The placement process was modeled based on parallel plate squeeze flow analysis both for constant placement speed and constant placement force conditions. Since placement machines have maximum placement forces, these models can be used to estimate whether a placement machine is compatible with the high-throughput assembly process for a given chip size and no-flow underfill material.

The placement process prefers a low to moderate viscosity, no-flow underfill material. Based on the compression flow model, for a given placement force, as underfill viscosity decreases, the allowable placement velocity increases, improving throughput. The die "floating" model was based on a steady-state force balance on the die after placement. The model determines an acceptable range of underfill mass such that the die remains in contact with the substrate after placement. Theoretical and actual data show that there exists a wide processing window for dispensed/printed underfill mass. Acceptable values for no-flow underfill viscosity and surface tension can be determined using the placement process models and the die floating analysis. For a given placement force, the allowable placement velocity increases as underfill viscosity decreases improving throughput. The low-cost, high-throughput assembly process also prefers an underfill with a low surface tension. From the die floating model, the higher the underfill surface tension, the more likely a die will float. It was also determined experimentally that the placement process does not cause voids in the underfill provided that appropriate deposition geometry is used.

Excellent reliability can be achieved using the low-cost, high-throughput FC process using no-flow underfill materials for standard FC on laminate applications. Liquid-to-liquid thermal shock reliability testing results indicate that no-flow materials have the ability to survive in excess of 1000 air-to-air and liquid-to-liquid cycles (−55 to 125°C). In general, the reliability data indicate that the high-throughput assembly process using no-flow underfills can satisfy the reliability requirements for many electronics applications.

17.10 REFERENCES

1. Baldwin, D., and Pascarella, N., "Manufacturability of Underfill Processing for Low-cost Flip Chip," *Structural Analysis in Microelectronics and Fiber Optics,* EEP-Vol. 21, ASME, New York, 1997, pp. 21–31.
2. Baldwin, D. F., and Beerensson, J. T., "Thermal Management in Direct Chip Attach and Chip on Board Assemblies," *Journal of Electronic Packaging,* Vol. 121 (4), ASME, New York, 1999, pp. 222–230.
3. Bird, R. B., Armstrong, R. C., and Hassager, O., *Dynamics of Polymeric Liquids, Vol. 1: Fluid Mechanics,* John Wiley & Sons, New York, 1977.
4. Chen, R., and Baldwin, D. F., "Smart Tooling for Assembly of Thin Flexible Systems," Proceedings of the 5th International Symposium and Exhibition on Advanced Packaging Materials, Processes, Properties, and Interfaces, Braselton, GA, IMAPS, New York, March, 1999, pp. 268–274.
5. Doot, R. K., "Motorola's First DCA Product: The Gold Line Pin Pager," Proceedings of the Electronic Components and Technology Conference, Orlando, FL, IEEE, New York, 1996, pp. 535–539.
6. Engelmaier, W., "Fatigue Life of Leadless Chip Carrier Solder Joints During Power Cycling," *IEEE Transactions on Components, Hybrids, and Manufacturing Technology,* Vol. CHMT-6 (3), 1983, pp. 232–237
7. Gamota, D., and Melton, C., "Materials to Integrate the Solder Reflow and Underfill Encapsulation Processes for Flip Chip on Board Assembly," *IEEE Transactions on Components, Packaging, and Manufacturing Technology,* Part C, Vol. 21 (1), January, 1998.
8. Gamota, D., and Melton, C., "Reflowable Material Systems to Integrate the Reflow and Encapsulant Dispensing Process for Flip Chip on Board Assemblies," International Conference Electronics Assembly Materials Process Challenges, Atlanta, GA, IPC-TP-1098, IPC, Chicago, IL, 1996.
9. Gamota, D., and Melton, C., "The development of reflowable materials systems to integrate the reflow and underfill dispensing processes for DCA/FCOB assembly," *IEEE Transactions on Components, Packaging & Manufacturing Technology,* Part C Manufacturing, Vol. 20 (3), 1997, pp. 183–187.
10. Gamota, D., and Melton, C., "Materials to Integrate the Solder Reflow and Underfill Encapsulation Processes for Flip Chip on Board Assembly," *IEEE Transactions on Components, Packaging, and Manufacturing Technology,* Part C, Vol. 21 (1), January 1998.
11. Gilleo, K., and Blumel, D., "The Great Underfill Race," in 1998 International Symposium on Microelectronics, Philadelphia, PA, IMAPS, New York, 1996, pp. 701–706.
12. Goldman, L. S., "Self-Alignment Capability of Controlled-Collapse Chip Bonding," in Proceedings of the Electronic Components Conference, Orlando, FL, IEEE, New York, 1972, pp. 332–339.
13. Han, S., Wang, K., and Cho, S., "Experimental and Analytical Study on the Flow of Encapsulant During Underfill Encapsulation of Flip-Chips," Proceedings of the 46th Electronic Components and Technology Conference, Orlando, FL, May, IEEE, New York, 1996, pp. 327–334.
14. Houston, P. N., Baldwin, D. F., Crane, L. N., and Konarski, M., "Low Cost Flip Chip Processing and Reliability of Fast Flow, Snap Cure Underfills," Proceedings of the Electronic Components and Technology Conference, San Diego, CA, IEEE, New York, May 1999, p. 61.
15. Johnson C. D., and Baldwin, D. F., "Wafer Scale Packaging Based On Underfill Applied At The Wafer Level For Low-Cost Flip Chip Processing," Proceedings of the Electronic Components and Technology Conference, San Diego, CA, IEEE, New York, May 1999, p. 950.
16. Lasky, R. C., and Baldwin, D. F., "Throughput: The Critical Cost Variable in DCA Assembly," Proceedings of the SMTA 3rd Annual National Symposium, Research Triangle Park, NC, 1996.
17. Leider, P. J., and Bird, R. D., "Squeezing Flow Between Parallel Disks. I. Theoretical Analysis," *Industrial Enginnering Chemical Fundamentals,* Vol. 13, 1974, pp. 336–341.
18. Leider, P. J., "Squeezing Flow Between Parallel Disks. II. Experimental Results," *Industrial Engineering Chemical Fundamentals,* Vol. 14, 1974, pp. 342–346.
19. McGovern, L. P., and Baldwin, D. F., "High-throughput Low-cost Flip Chip on Board Assembly Processing," *Electronics Packaging and Production,* Vol. 38 (2), February 1998, pp. 68–76.
20. Mizutani, M., "A Study of a New Flip Chip Packaging Process for Diversified Bump and Land Combination," Proceedings of the Electronic Components and Technology Conference, Seattle WA, IEEE, New York, 1998, pp. 316–319.

21. Montgomery, D. C., *Design and Analysis of Experiments,* John Wiley & Sons, New York, 1997.
22. Palaniappan, P., and Baldwin, D. F., "In Process Stress Analysis of Flip Chip Assemblies During Underfill Cure," *Applications of Experimental Mechanics to Electronic Packaging,* EEP-Vol. 22/AMD-Vol. 226, ASME, New York, 1997, pp. 7–14.
23. Palaniappan, P., Selman, P., Baldwin, D. F., Wu, J., and Wong, C. P., "Correlation Of Flip Chip Underfill Process Parameters And Material Properties With In-Process Stress Generation," *IEEE Transactions on Electronics Packaging Manufacturing,* Vol. 22 (1), IEEE, New York, January 1999, pp. 53–62.
24. Pascarella, N., and Baldwin, D. F., "Compression Flow Modeling of Underfill Encapsulants for Low-cost Flip Chip Assembly," Proceedings of the 48th Electronic Components and Technology Conference, Seattle, WA, May, IEEE, New York, 1998, pp. 463–470.
25. Pascarella, N. W., and Baldwin, D. F. , "Manufacturing Analysis of Underfill Processing for Low-cost Flip Chip," *Journal of Electronics Manufacturing,* Vol. 8 (1), ASME, New York, October 1998, pp. 39–50.
26. Pascarella, N. W., and Baldwin, D. F., "Advanced Encapsulation Processing for Low-cost Flip Chip Assembly—A Cost Analysis," 3rd International Symposium on Advanced Packaging Materials Processing Properties and Interconnection, Braselton, GA, IMAPS, New York, March 1997.
27. Pascarella, N. W., and Baldwin, D. F., "Advanced Encapsulation Processing for Low Cost Electronics Assembly—A Cost Analysis," *Advances in Electronic Packaging 1997,* EEP-Vol. 19–1, ASME, New York, 1997, pp. 359–363.
28. Pascarella, N. W., and Baldwin, D. F., "Compression Flow Modeling of Underfill Encapsulants for Low-cost Flip Chip Assembly," *IEEE Transactions on Components, Packaging, and Manufacturing Technology—Part C,* Vol. 21 (4), IEEE, New York, October 1998, pp. 325–335.
29. Pascarella, N. W., and Baldwin, D. F., "Cost Analysis of Low Cost High Throughput Next Generation Flip Chip Assembly," *International Journal of Microcircuits and Electronic Packaging,* Vol. 20, IMAPS, New York, 1997, pp. 571–577.
30. Pascarella, N. W., and Baldwin, D. F., "Cost Analysis of Low-cost High-throughput Next-generation Flip Chip Assembly," *International Journal of Microcircuits and Electronic Packaging,* Vol. 20, 1997, pp. 571–577.
31. Rodriguez, G., and Baldwin, D. F., "Analysis of Solder Paste Release in Fine Pitch Stencil Printing Processes," *Journal of Electronic Packaging,* Vol. 121, September, ASME, New York, 1999, pp. 1–11.
32. Rodriguez, G., and Baldwin, D. F., "Analysis of Solder Paste Release in Fine Pitch Stencil Printing Processes," *Advances in Electronic Packaging 1999,* EEP-Vol. 26-1 ASME, New York, 1999, pp. 649–660.
33. Schubert, A., Dudek, R., Leutenbauer, R., Doring, R., Kloeser, J., Oppermann, H., Mechel, B., Reichl, H., Baldwin, D. F., Qu, J., Sitaraman, S., Swaminathan, M., Wong, C. P., and Tummala, R., "Do Chip Size Limits Exist for DCA?," Proceedings of the 5th International Symposium and Exhibition on Advanced Packaging Materials, Processes, Properties, and Interfaces, Braselton, GA, IMAPS, New York, March 1999, pp. 150–162.
34. Schwiebert, M., and Leong, W., "Underfill Flow as Viscous Flow Between Parallel Plates Driven by Capillary Action," *IEEE Transactions on Components Packaging and Manufacturing Technology—Part C,* Vol. 19, 1996, pp. 133–137.
35. Sitaraman, S., Hanna, C., Mechaelides, S., Palaniappan, P., and Baldwin, D. F., "Numerical and Experimental Study of the Evolution of Stresses in Flip Chip Assemblies During Assembly and Thermal Cycling," Proceedings of the Electronic Components and Technology Conference, May 1999.
36. Smith, B. A., Houston, P. N., Baldwin, D. F., and Thorpe, R., "A Reliability Analysis of No Flow Underfill Materials," Proceedings of the Electronic Components and Technology Conference, Las Vegas, NV, IEEE, New York, May 2000, pp. 1719–1730.
37. Suhir, E., "Calculated Thermal Induced Stresses in Adhesively Bonded and Soldered Assemblies," Proceedings of the 1986 International Symposium on Microelectronics, Chicago, IL, IEEE, New York, pp. 383–392.
38. Thorpe, R., Baldwin, D. F., and McGovern, L. P., "High Throughput Flip Chip Processing and Reliability Analysis Using No-Flow Underfills," Proceedings of the Electronic Components and Technology Conference, San Diego, CA, IEEE, New York, May 1999, p. 419.
39. Thorpe, R., Baldwin, D. F., and McGovern, L. P., "Yield Analysis and Process Modeling of Low Cost, High Throughput Flip Chip Assembly Based on No-Flow Underfills," IEEE Transactions on Electronics Packaging Manufacturing, Vol. 24 (2), April 2001, pp. 123–135.
40. Thorpe, R., McGovern, L. P., and Baldwin, D. F., "Analysis Of Process Yield In Low-cost Flip Chip On Board Assembly Processes," Thermo-Mechanical Characterization of Evolving Packaging Materials and Structures, ASME, New York, EEP—Vol. 24, 1998, pp. 27–33.
41. Van Krevelen, D. W., *Properties of Polymers,* Elsevier, New York, 1972.
42. Wong, C. P., and Baldwin, D. F., "Novel No-flow Underfills for Low-Cost Flip-Chip Applications: Materials and Processes," *Future Circuits International,* Issue 3, London, 1998, pp. 67–70.

43. Wong, C. P., and Shi, S., "Study of the Fluxing Effects on the Properties of No-Flow Underfill Materials for Flip Chip Applications," Proceedings of the 48th Electronic Components and Technology Conference, Seattle, WA, IEEE, New York, 1998.
44. Wong, C. P., Shi, S., and Jefferson, G., "High Performance Low-cost Underfills for Flip-Chip Applications," Proceedings of the 47th Electronic Components and Technology Conference, San Diego, CA, IEEE, New York, 1997.

CHAPTER 18
FLIP CHIP ASSEMBLY AND UNDERFILLING

R. Wayne Johnson
Auburn University

18.0 INTRODUCTION

IBM first introduced flip chip (FC) technology in 1964 in the solid logic technology (SLT) hybrid modules in the System 360 mainframe.[1] The technology was developed by IBM to replace wire bonding as a means of interconnecting semiconductor dies to thick-film metallization on alumina. The dies were three-terminal transistors with Au/Ni-plated Cu balls embedded in a Pb/Sn solder bump on the three input-output (I/O) pads of the transistor. A Cr/Cu/Au interface layer was deposited between the Al transistor bond pads and the solder bump. The device was assembled to the hybrid substrate by inverting (flipping) the dies and reflow soldering the copper balls to corresponding metal pads on the substrate. The copper balls maintained a constant standoff between the dies and the substrate after reflow.

As the I/O count of the dies increased, the copper spheres were replaced with high-lead tin-lead alloy solder bumps.[2] The solder balls collapse somewhat during the reflow soldering process, balancing the weight of the chip and the surface tension forces of the molten solder. This phenomenon gave rise to the IBM terminology *controlled collapse chip connection* (C4). The flow of the solder (collapse of the chip-substrate spacing) was controlled during reflow by controlling the solder volume and the wettable metal exposed on the dies and the substrate.

The advantages of the FC assembly process for high-volume manufacturing were recognized by the automotive industry. FC-on-ceramic (FCOC) was adopted by the automotive industry in the late 1970s as a high-volume surface-mount assembly technology for applications such as ignition modules. The dies typically were small (<3 mm) with a low I/O count (<20). The solder bumps primarily were high-lead alloys, although some eutectic bumps were used. The large pitch of the bumps allowed solder paste to be printed on the thick-film metal pads. The FC die was then placed and reflowed along with other surface-mount components. With the high-lead bumps, a high-lead solder paste was used. The thick-film ceramic substrate allowed reflow at temperatures in excess of 325°C. The FCOC assembly process paralleled the development of surface-mount technology (SMT) on laminate with plastic leaded packages such as plastic leaded chip carriers (PLCCs), quad flatpacks (QFPs), and small outline transistors (SOTs).

FC technology offers other advantages as semiconductor device speed, size, and I/O count increase. For high-speed applications, the small solder balls add minimal parasitics (resistance, capacitance, and inductance) and propagation delays to the electrical signal path. In addition, the elimination of the die package permits the dies to be placed close together on the substrate, further reducing interconnection parasitics and propagation delay. Elimination of the die package and close spacing of dies are also an advantage in portable electronics, where size and weight are critical. Finally, as the die I/O count continues to increase, perimeter electrical interconnection by wire bonding becomes limited. In FC technology, I/Os can be arrayed over the entire area of the die, dramatically increasing the number of I/Os.

While there is a coefficient of thermal expansion (CTE) mismatch between alumina (6.5 ppm/°C) and silicon (3 ppm/°C), millions of reliable, under-the-hood modules were built using small dies on ceramic without the need for underfill. Millions of FC solder joints also were used in mainframe computers without failure. Increasing either the size of the die or the CTE of the substrate (switching from ceramic to laminate at approximately 16 ppm/°C) increases the strain on the solder joints during thermal cycling or thermal shock (Fig. 18.1). To improve reliability, underfill was added to the FC assembly process.[3–5] Underfill is a polymer material (typically a filled epoxy) that fills the space between the die and the substrate. The underfill bonds the die and the substrate together, forming a trilayer structure. The net effect is to reduce the strain on the individual solder joints by creating a structure that warps as the temperature changes (Fig. 18.2).

The objective of this chapter is to discuss the materials and processes for FC assembly with underfill.

18.1. PROCESS OVERVIEW

The application and cure of the underfill material are the primary assembly steps that have an impact on FC assembly process flow. The underfill can be applied after the FC is placed and reflowed (capillary flow) or prior to die placement. Fluxing underfills are applied to the substrate just prior to die placement, whereas wafer-applied underfills are coated onto the wafer prior to singulating the wafer into individual dies. These three options are introduced below and will be detailed in later sections of this chapter.

18.1.1 Capillary Flow Underfills

Liquid capillary flow underfill materials are dispensed along the edge of the die after the die has been reflow soldered to the substrate. Capillary action pulls the liquid under the die, as illustrated in Fig. 18.3. After the liquid underfills the die, a second dispense may be required to form a fillet around the edge of the die. Self-filleting is observed with some materials, eliminating this step. Following the dispense step, the underfill must be cured. Cure times and temperatures vary from 5 to 90 min at 150 to 165°C depending on the underfill chemistry. Capillary flow underfills are available commercially from a number of suppliers and are used most commonly today.

18.1.2 Fluxing Underfills

Fluxing underfills include the fluxing chemistry for the soldering step in the underfill. The liquid is applied to the substrate, typically by dispensing. The die is then placed and reflowed. The underfill may cure during the reflow cycle, or a postreflow cure may be required depending on the material.

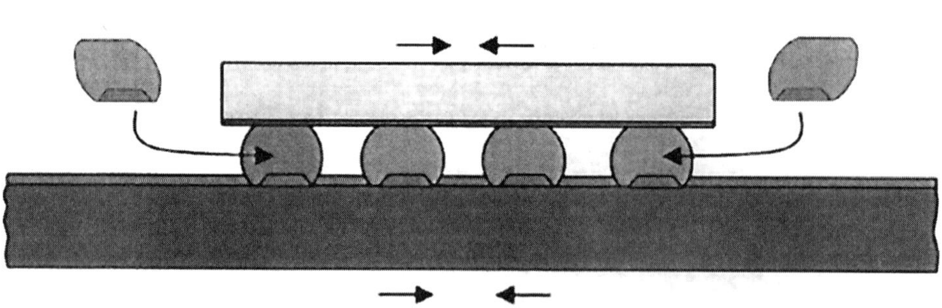

FIGURE 18.1 Stress on solder joints due to CTE mismatch when assembly is cooled from solder solidification temperature to room temperature. The substrate contracts more than the silicon die during cool-down.

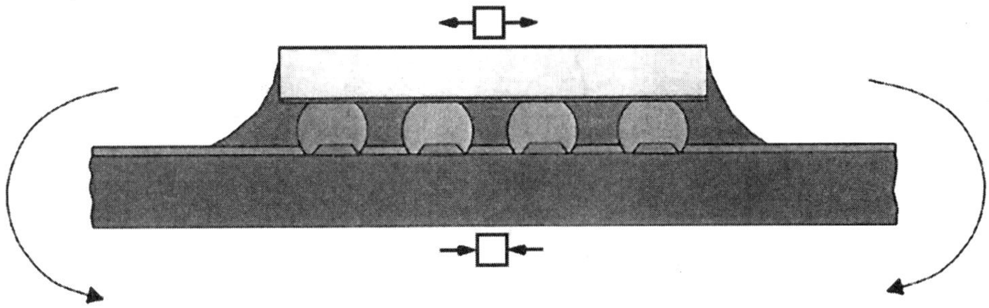

FIGURE 18.2 Warpage when underfilled structure is cooled from underfill cure temperature to room temperature.

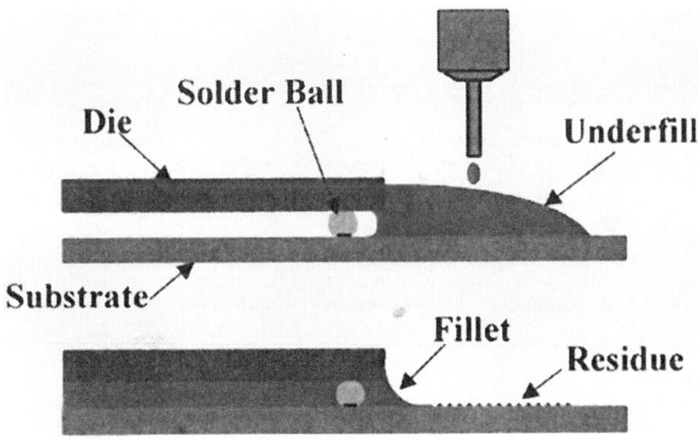

FIGURE 18.3 Capillary underfill process.

The process is illustrated in Fig. 18.4. Fluxing underfills are available commercially from a number of suppliers.

18.1.3 Wafer-Applied Underfills

Rather than applying the underfill one die at a time during the assembly operation, wafer-applied underfills are coated onto all dies on a wafer at one time. After coating, the underfill must be a dry (nontacky) film for handling. During the reflow cycle, the underfill liquefies, providing fluxing action. The underfill may cure during the reflow cycle or may require a postreflow cure step. The process is illustrated in Fig. 18.5. Wafer-applied underfills are in the developmental stage and not available commercially.

18.2 SUBSTRATE DESIGN

With packaged SMT components, the design process is simplified through industry standardized I/O configurations and recommended substrate pad designs. However, FC I/O configurations are custom to each die; therefore, substrate pad designs have not been standardized. The design will depend on the I/O pitch, the number of I/Os, the arrangement of the I/Os, and the substrate technology. Within this parameter space, design options exist and are discussed in the following section.

18.2.1 Layout

While FC bump patterns can be irregular, single-row perimeter, two-row perimeter, and full area array are most common. Each presents its own design and substrate fabrication challenges.

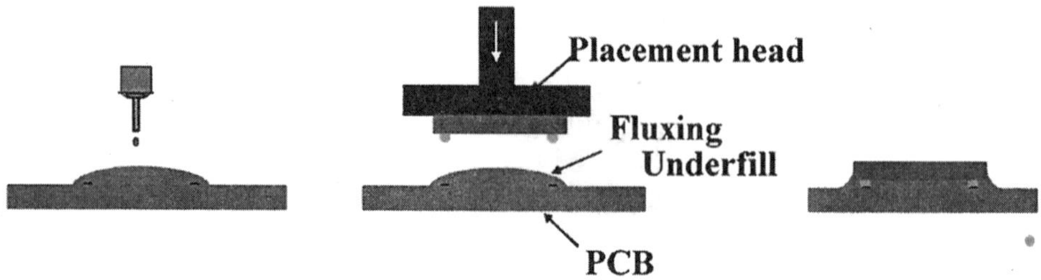

FIGURE 18.4 Assembly with fluxing underfill.

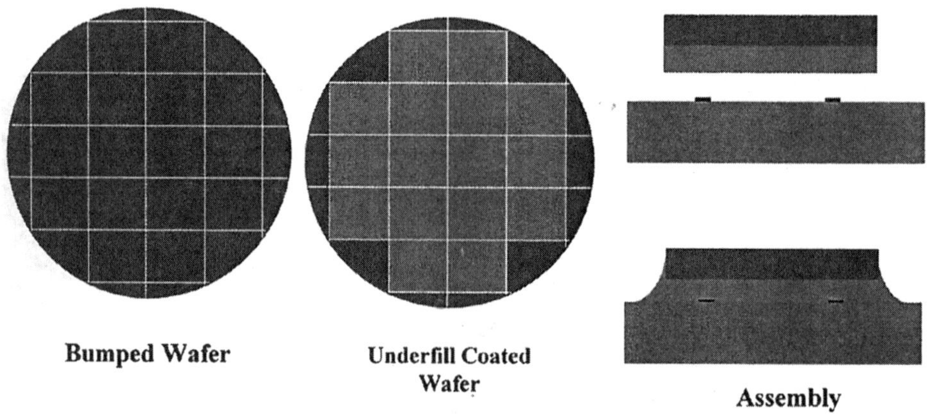

FIGURE 18.5 Wafer-applied underfill assembly process.

Single-Row Perimeter. FC dies with a single row of perimeter solder bumps are the simplest to route. The printed wiring board (PWB) technology must provide line and space pitch equal to the pitch of the solder bumps, as illustrated in Fig. 18.6. In this design, the copper trace width and the solder mask opening define the solderable copper area exposed to the solder ball. Increasing the solderable area decreases the gap height between the die face and the solder mask. In turn, decreasing this gap height increases the time required for capillary underfills to flow, adversely affecting the underfill process. If the gap becomes sufficiently small, capillary underfill may not flow into the gap.

For fine-pitched solder bumps, decreasing the solderable area increases the risk of shorting between bumps. If the solder mask opening (trench) becomes too narrow, the solder ball will rest on the edges of the solder mask and not make contact with the copper trace. An open solder joint after reflow can be the result.

PWB manufacturing variations must be considered. Variations in copper etching (trace width) and solder mask developing (trench width) from board to board will lead to variations in gap height. In turn, this will change the volume of material required to underfill the die. In production, the underfill dispense volume is not changed from board to board. Therefore, the underfill fillet must provide the reservoir to ensure that the die is completely underfilled. With gap variations there will be a corresponding variation in the amount of fillet formed. Consistent PWB fabrication is important.

A second issue in PWB fabrication is the solder mask registration. As illustrated in Fig. 18.7, if the solder mask is shifted to the right, the solder balls on the two sides can be aligned, but the solder balls on the top and bottom are misaligned. The placement in Fig. 18.7*a* assumes that copper-defined fiducials are used. If solder mask–defined fiducials are used, the placement result is shown in Fig. 18.7*b*. In both cases, the solder balls do not line up with the pads. At the top and bottom of the die, the solderable pads are defined by the copper trace in the x direction and the solder mask in

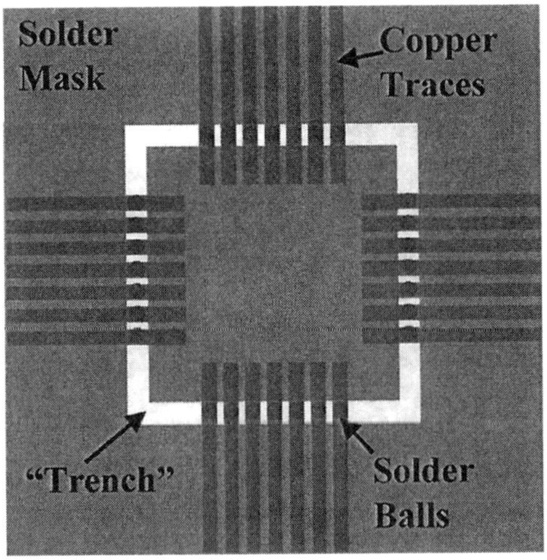

FIGURE 18.6 "Trench" design for single-row perimeter FC dies.

the y direction. On the other two sides of the die, the solderable pads are defined by the solder mask in the x direction and the copper trace in the y direction. Thus no placement shift can compensate for solder mask misregistration. The surface tension of the molten solder during the reflow process will tend to center the chip, creating approximately even solder joint distortion on all sides, as shown in Fig. 18.8. The impact of this distortion on reliability has not been documented. With fine-pitch solder bumps, solder mask-to-copper misregistration increases the likelihood of solder bridging.

An alternate design can be used to address misregistration issues (Fig. 18.9). In this design, the die placement is based on copper fiducials. The maximum solder mask misregistration must be considered in the determination of the length of the exposed copper fingers. Overetching of the copper is more critical in this design because overetching reduces not only the trace width but also the trace length. Overetching combined with a solder mask shift to the left can result in very small pads on the right-hand side. If the solder mask is misregistered as shown in Fig. 18.9*b,* the gap height will be less on the left side (more solderable copper area). This design approach decreases the potential for bridging of fine-pitch FC bumps. The die-to-PWB gap height is increased in this design by eliminating the solder mask under the die. Other designs for single-row perimeter dies are possible, but these two are the most common.

Two-Row Perimeter. The first determining factor for a two-row perimeter design is the FC I/O pitch relative to the pitch of the PWB. Can a PWB trace be routed between adjacent solder ball pads (Fig. 18.10)? The design in Fig. 18.10 is similar to Fig. 18.6. Isolated solder mask openings (as shown) or two trenches can be used depending on the I/O pitch and the solder mask resolution. The two-trench approach would impede underfill flow. Consistent control of copper etching, solder mask resolution, and solder mask-to-copper registration are again critical.

Figure 18.11 is a two-row perimeter design with characteristics similar to Fig. 18.9. The solder mask misregistration should be less than 50 percent of the line-to-line spacing on the PWB. Depending on pitch, this may be challenging.

If the PWB technology will not support one trace between solder bump pads, the second row of bumps must be routed toward the center of the die and down through plated through-holes or vias. If conventional PWB technology with plated through-holes for z-axis interconnections is to be used, the plated through-holes under the dies must be tented or plugged. This is necessary to prevent the underfill from flowing out through the hole during the underfill process. Plated through-holes typically are tented with dry-film solder mask. However, FC PWBs are fabricated almost exclusively with liquid photoimagable (LPI) solder mask to achieve thinner coatings and better resolution. LPI

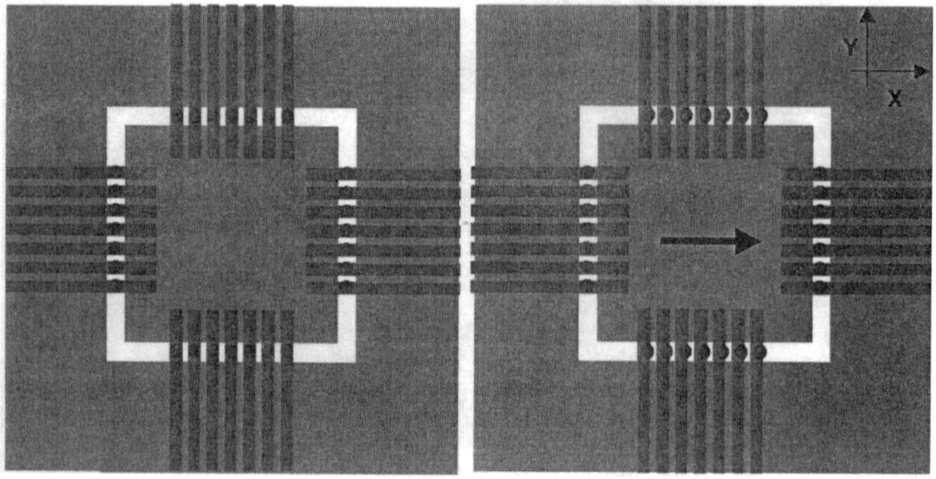

a *b*

FIGURE 18.7 Solder mask misregistration to the right. (*a*) Placement based on copper-defined fiducials; (*b*) Placement based on solder mask–defined fiducials.

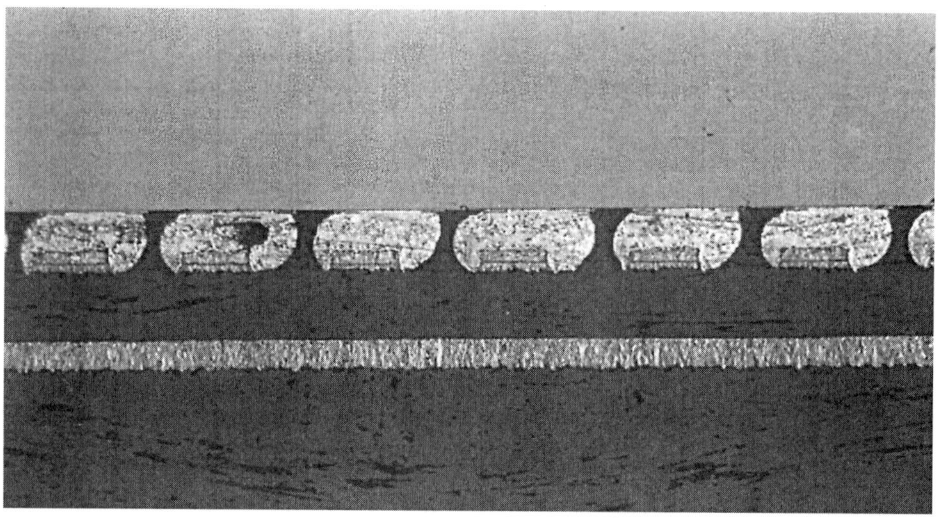

FIGURE 18.8 Solder joint distortion due to solder mask misregistration.

solder mask cannot tent plated through-holes. Thus plugged holes typically are used for FC PWBs. After plating, the holes are filled (plugged) with epoxy and cured. The LPI solder mask can then coat over the plugged hole. The design is illustrated in Fig. 18.12. Copper etching and solder mask-to-copper registration are critical to this design.

Area Array. With smaller dies or full area arrays, the pad diameters for the plated through-holes are too large to allow routing. High-density interconnect (HDI) substrates are required. The substrates are fabricated by buildup of sequential dielectric and copper layers processed on a laminate core. The HDI structure is illustrated in Fig. 18.13. This figure is of the IBM surface laminar cir-

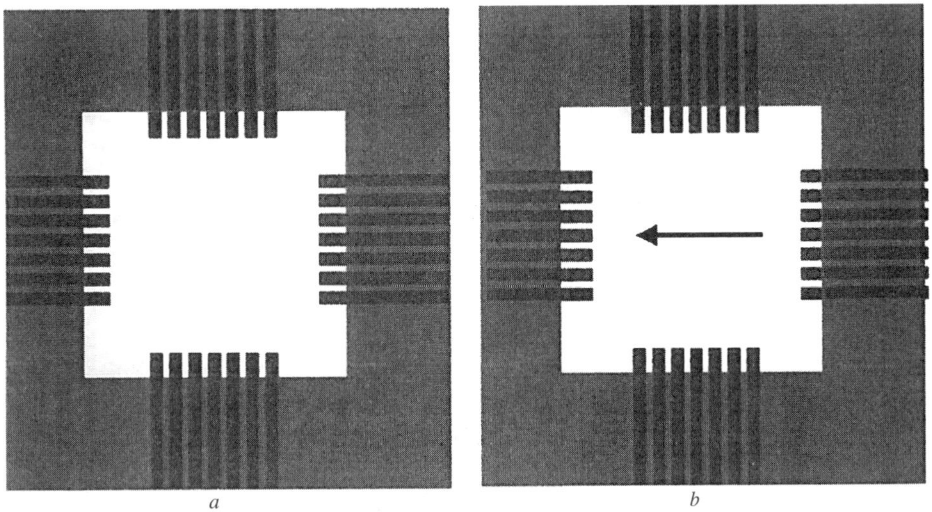

FIGURE 18.9 Alternate design for single-row perimeter FC: (*a*) perfect copper-to-solder mask registration; (*b*) solder mask is misregistered to the left.

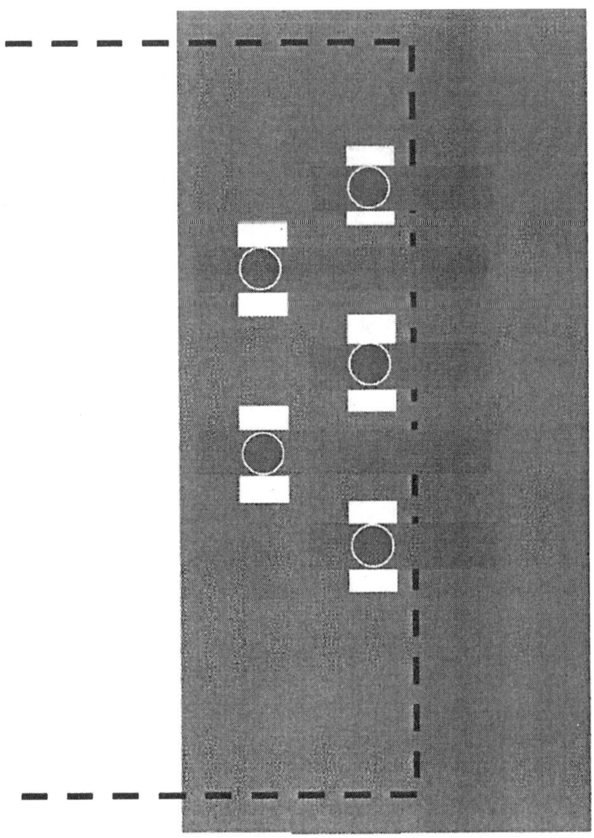

FIGURE 18.10 Two-row perimeter design.

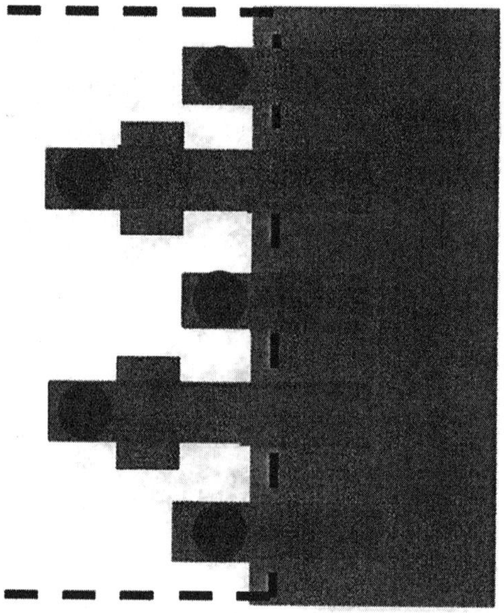

FIGURE 18.11 Alternate two-row perimeter design.

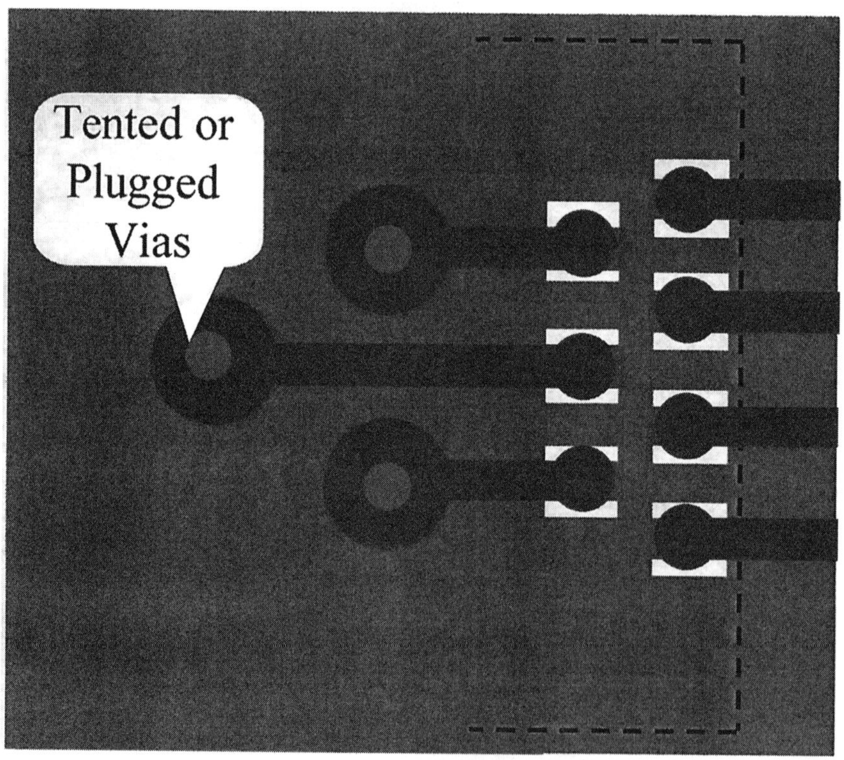

FIGURE 18.12 Two-row perimeter design with plugged plated through-holes.

cuitry (SLC) technology, one of the first HDI approaches. The SLC technology used a liquid, photoimageable dielectric layer with additive copper to fabricate the HDI layers.

A number of HDI approaches subsequently have been developed. The starting dielectric material may be either liquid (nonreinforced) or a dry film (reinforced or nonreinforced). The formation of vias is by photoimaging, laser, or plasma, as shown in Table 18.1. The vias are through only one layer of dielectric and are less than 125 μm in diameter (Fig. 18.14). Additive processing is used commonly for copper deposition/patterning. In some approaches, the vias are filled with a conductor-filled epoxy. This is an advantage in "pad in via" designs to be discussed below.

With the smaller pads and vias achievable with HDI technology, area array FC dies can be routed. The pitch of the vias must equal that of the solder bumps on the dies. If the via pad can fit in the interstitial space between solder bump pads, a "dog bone" design similar to ball grid array (BGA) and chip scale packaging (CSP) patterns can be used (Fig. 18.15).

The ultimate in density is achieved with a "pad in via" design (Fig. 18.16). One concern with pad in via design is the potential for voids trapped in the solder joint (Fig. 18.17). The impact of these voids on reliability has not been determined. The second concern is the increased solder wettable area, which decreases the standoff height. The vias may be either filled or plated solid to avoid entrapped voids during solder reflow. In addition to density, the absence of a solder mask and potential for misregistration is a benefit of this design.

18.2.2 Copper and Solder Mask Requirements

Voids (air pockets) can be trapped during underfill flow at steep topology features such as the edge of a thick copper trace. Thinner copper reduces the probability of voids in the underfill at the base of the copper trace. As the pitch of the FC device decreases, thinner copper is also required to fabricate finer lines and spaces. The thickness of the surface copper layer is the combination of starting foil thickness and the plating thickness required for reliable plated through-holes in conventional PWB fabrication. Foil thicknesses less than ¼ oz/sq. ft. (8.3 μm) are available but more expensive and difficult to process. Typical plated through-hole copper thicknesses range from ½ oz/sq. ft. (16.6 μm) for consumer/portable electronics applications to 1 oz/sq. ft. (33.2 μm) for automotive applications. The thickness of the plated through-hole copper required is a function of the total board thickness and the reliability (thermal cycle) requirements. In most HDI structures, no beginning foil is used, and the surface copper thickness is determined by the plating thickness.

Electroless nickel/immersion gold is the most commonly used surface finish for FC assembly. For long-term reliability, the gold volume within the solder joint should be less than 3 percent and preferably less than 1 percent to avoid embrittlement of the joint and Au/Sn intermetallic formation. As the solder ball volume decreases, the maximum gold thickness decreases. Other finish alternatives include silver and tin. Organic solderability preservative (OSP) coatings also are used commonly for FC boards. The number of reflow cycles and high-temperature process steps and the processing sequence should be reviewed if OSPs are to be used. The effectiveness of OSPs decreases with

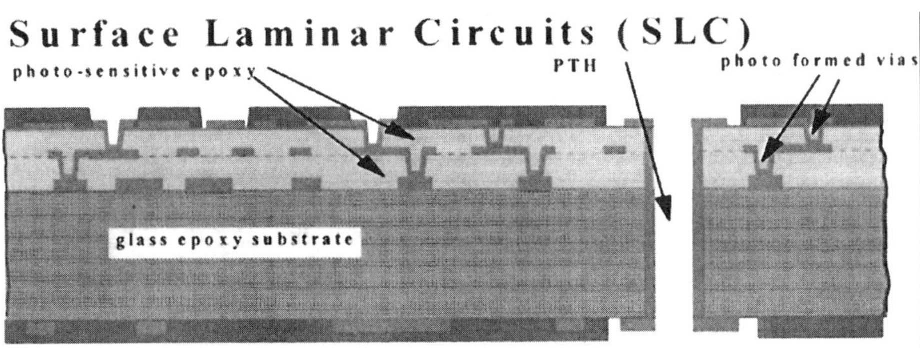

FIGURE 18.13 Example of high-density interconnect (HDI) substrate.

TABLE 18.1 HDI Dielectric Types and Via Processing

Dielectric type	Form	Via processing
Non reinforced	Liquid	Photoimage, laser
	Dry film	Photoimage, laser, plasma
Reinforced	Dry film	Laser

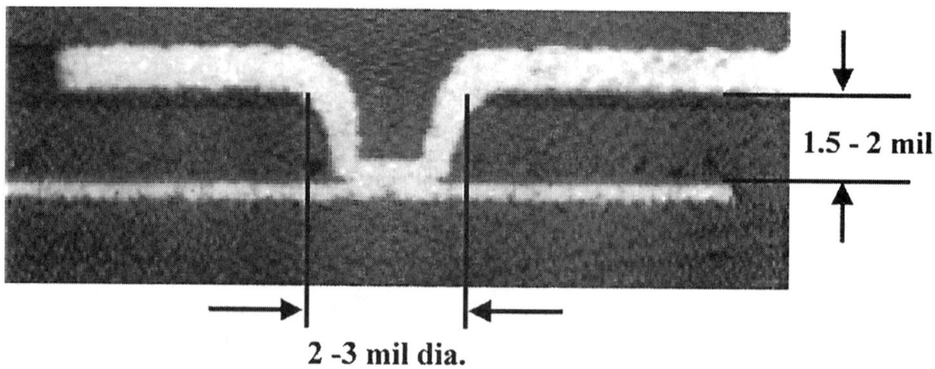

FIGURE 18.14 Example of HDI via.

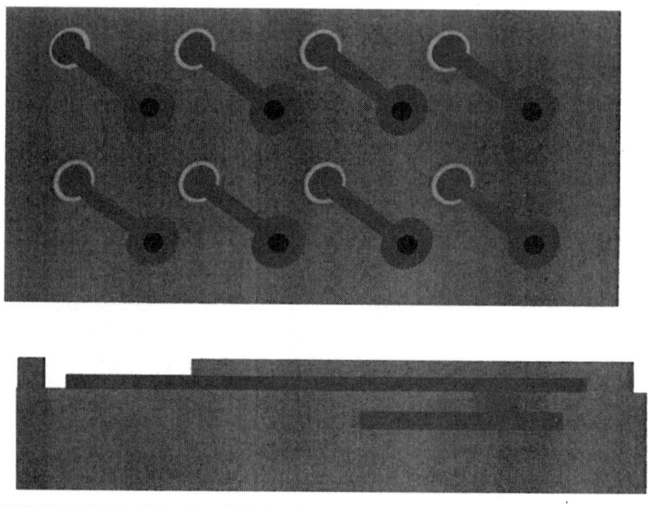

FIGURE 18.15 "Dog bone" design for area array FC using an HDI substrate.

multiple high-temperature exposures. Some OSP coatings are formulated to withstand more high-temperature exposures.

The solder mask has an impact on the assembly process and yield. As discussed in the preceding section, solder mask registration is critical. While the industry standard registration tolerance is 3 mils (75 μm), fine-pitch FCs may require a tolerance of less than 2 mils (50 μm). This is a challenge for the industry, particularly if processing 18 × 24-in panels for manufacturing efficiency. Solder mask resolution is also important. In many designs, the solderable area is partially defined by the solder mask opening. Increasing the solderable area decreases the gap height, making the underfill process more difficult. If the solder mask opening is too small, solder bridging can result with fine-pitch assemblies. In the extreme case, the solder mask opening may be too small for the solder ball

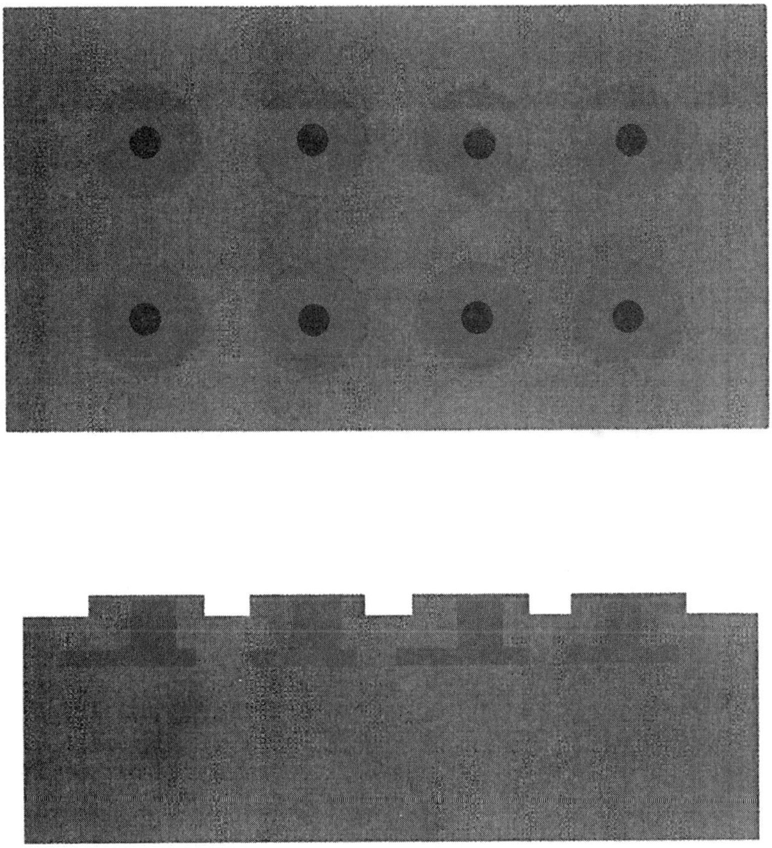

FIGURE 18.16 Pad-in-via design with HDI substrate.

to touch the copper pad, resulting in an open connection. The solder mask thickness has an impact on the gap height. The solder mask thickness over the copper traces should be less than 12 μm. It is a challenge to apply thin solder mask without having skips or exposed copper, particularly at the edge of traces.

Proper board design, materials selection, and fabrication are important to FC assembly yield and reliability. Each design is unique, and compromises must be considered.

18.3 ASSEMBLY WITH CAPILLARY FLOW UNDERFILL

The process flow for FC assembly with capillary flow underfill is illustrated in Fig. 18.18. This process assumes that the flux residue is not cleaned after reflow, which is typical of most FC assemblies. Each of the process steps and the associated materials will be discussed in the following subsections.

18.3.1 Die Presentation

Dies can be presented for pick and place either as sawn wafers, in waffle packs, or in tape and reel. By placing dies directly from the wafer, intermediate handing steps are eliminated. Since wafers are sawn bump side up, the dies must be picked from the bump side and inverted for placement. This adds an extra step to the pick-and-place operation. Care must be taken that ejector pins used to assist

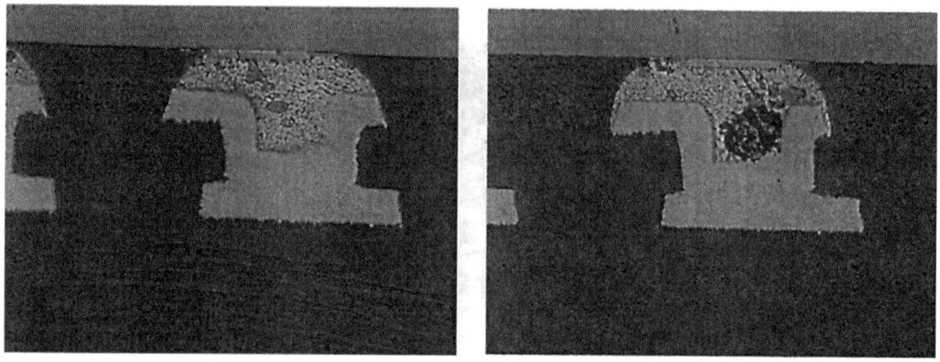

FIGURE 18.17 Example of pad-in-via design with and without entrapped void. (*Jabil Circuits.*)

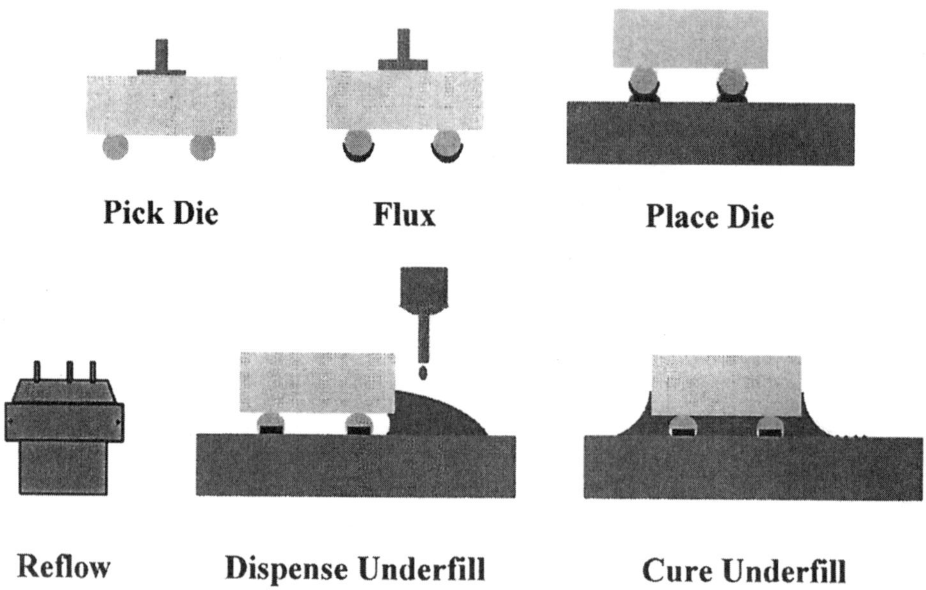

FIGURE 18.18 FC assembly with capillary flow underfill.

in the removal of dies from the expanded tape do not cause microcracks or flaws in the backside of the dies. Motorola has shown that ejector pin damage of the die backside can lead to die fracture during assembly or in the field.[6] This is an issue both for pick and place from wafers and for placing of dies into waffle packs or tape from the wafer.

Waffle packs are used commonly for FC dies, but are limited in the number of dies they contain. For high-volume assembly, tape and reel is preferred, since a single reel will hold hundreds of dies. In both cases, the dies are inverted, solder balls down for direct pick and place. The individual die cavities in the waffle pack or tape should be only slightly larger than the dies to avoid rotation and translation of the dies in the cavity.

18.3.2 Flux and Flux Application

In standard SMT assembly, solder paste is stencil printed, and the components are placed into the paste. Flux, not solder paste, generally is used in the assembly of FC dies. Flux application is required prior to placement of the FC. Fluxing options are listed in Table 18.2. Dip fluxing is illustrated in Fig. 18.19. The dipping operation is performed on the pick-and-place system and adds to the FC placement cycle time. The variables to control are flux depth, dip force, and hold time. Increasing these variables increases the amount of flux transferred to the solder balls. This, in turn, increases solder wetting and self-centering of dies placed off center. However, increasing the volume of flux increases the amount of residue, which may interfere with underfill flow and decrease reliability if a no-clean process is used. The viscosity and surface tension of the flux also affect flux transfer. Since the flux is exposed to the atmosphere as a thin layer, the evaporation rate for volatile components must be extremely low at room temperature.

Stencil or screen printing also has been used for flux application.[7] If both FC and SMT components are to be assembled to the board, the flux is printed first. A step stencil (Fig. 18.20) is then used to print the solder paste. Flux printing can apply flux to multiple sites on a board in one pass and eliminates the flux dip step from the pick-and-place operation, increasing throughput at pick and place. The cost analysis of print versus dip depends on the product and the assembly line.

Stencil/screen design, flux printing characteristics, flux slumping, and flux life on the stencil are important processing considerations when developing a flux printing process. The design of the stencil or screen and printing characteristics determine the flux volume. Slumping of the flux can cause solder bridging of fine-pitch FCs. Similar to dip fluxing, the evaporation rate for volatile components must be extremely low at room temperature. Otherwise, the print characteristics will change during the production run.

Spray, jetting, and flood dispense are used with no-residue (low) fluxes. These are low-viscosity, low-solids-content fluxes that are applied over the entire area of the chip site. The coverage for flood dispense depends on substrate design/topology and the wetting characteristics of the substrate surfaces. Spray dispense and jetting are more uniform and less dependent on topology. Some pick-and-place systems are equipped with flood dispense capability. Separate fluid dispense systems can be used for flood, jet, and spray dispense.

The wetting characteristics of the flux can be evaluated by cross-sectioning solder joints and examining the solder-pad interface (Fig. 18.21). The die shear strength and the failure mode also can be used to evaluate wetting. With good solder wetting, the failure point for die shear of FC-on-laminate (FCOL) should be either at the die solder ball interface (Fig. 18.22) or at the copper pad–laminate interface for small pad (fine-pitch) designs. In Fig. 18.22, the wetting at some pad sites was minimal, and the failure was at the solder ball–substrate pad interface. Since some wetting did occur, electrical continuity was measured after assembly. Thus electrical testing alone is not sufficient for process development. With fine-pitch FC dies, the copper pad is small, and pad adhesion to the laminate is the limiting factor in shear strength.

With all fluxes and application methods, the flux must provide sufficient tack to hold the FC dies in place prior to melting of the solder in the reflow process. Mechanical shock and vibration in the conveyor systems, as well as airflow in the reflow oven, can cause chip movement.

No-clean fluxes are used most often in FC assembly because cleaning in the small FC gap is difficult. No-clean fluxes leave residue after the reflow process. Figure 18.23 is a thermal gravimetric analysis (TGA) plot of a typical no-clean flux used for FC assembly.[8] The TGA thermal profile approximates a reflow cycle. After the thermal cycle, approximately 35 percent by weight of the flux remains as residue. As mentioned previously, flux residue can impede the flow of underfill material. Chemical interaction between the underfill and the flux residue during underfill curing also can degrade adhesion and reliability. Fluxes and underfills must be evaluated as a materials system.

TABLE 18.2 Flux Application Options

Option	Flux	Comment
Dip	Medium viscosity, no-clean or solvent clean	Increases placement cycle time
Print	Medium viscosity, no-clean or solvent clean	High throughput for boards/panels with many FC dies
Spray or jetting	Low viscosity, no (low) residue	Flux applied over entire chip site including nonsoldering areas
Flood dispense	Low viscosity, no (low) residue	Flow of flux substrate topology-dependent

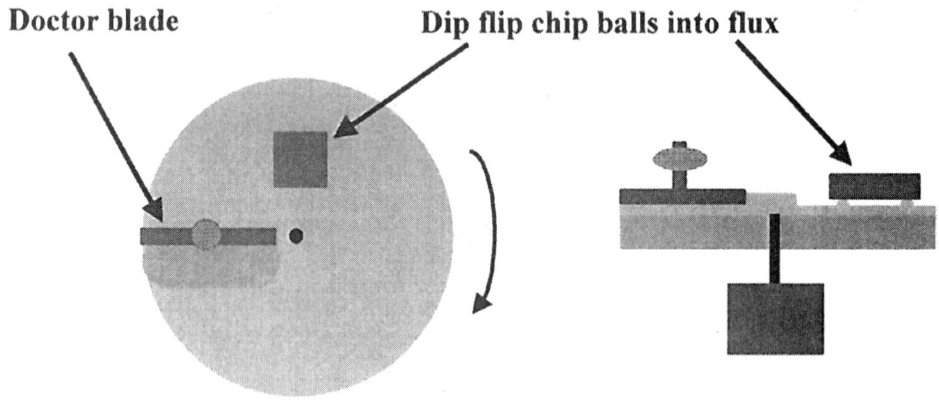

FIGURE 18.19 Dip fluxing.

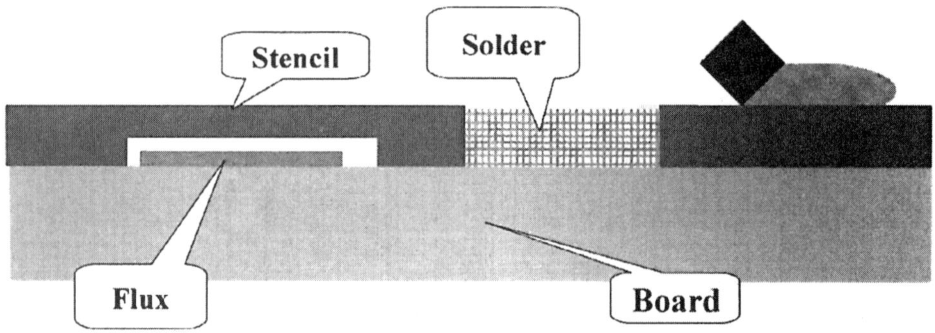

FIGURE 18.20 Example of flux and solder paste printing for mixed SMT assembly.

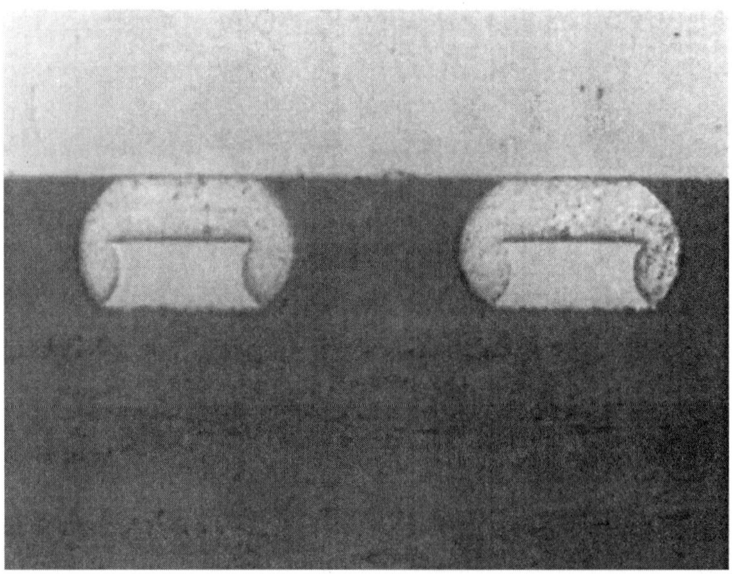

FIGURE 18.21 Cross section showing excellent solder wetting of copper pad.

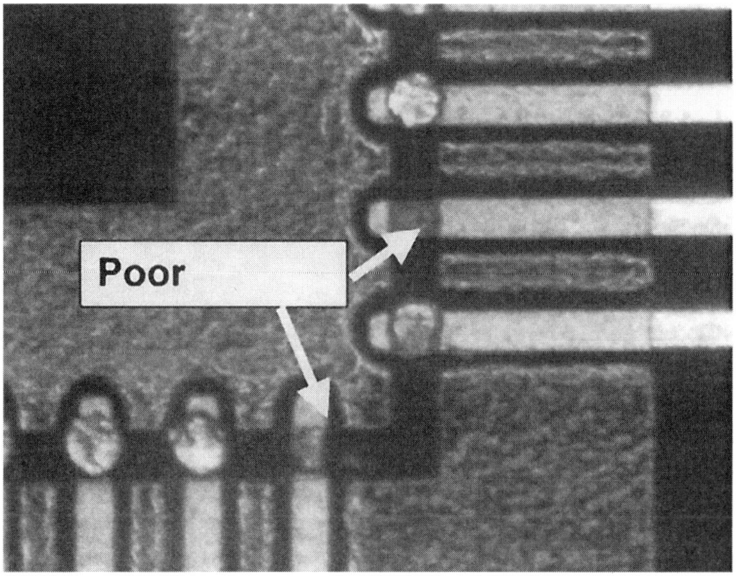

FIGURE 18.22 Example of poor solder wetting at two sites. The assembled FC die was sheared from the PWB. With good solder wetting, the solder ball remains on the PWB pad. The failure is at the die-to-solder ball inferface.

No-residue (low) fluxes are being evaluated to minimize these issues. These materials leave less than 2 percent by weight residue when tested using the same TGA temperature profile used in Fig. 18.23. Another no-clean alternative is epoxy flux. The chemistry of these fluxes incorporates epoxy groups. After the reflow cycle is completed, the flux "residue" is an adherent epoxy. The challenge in the formulation of epoxy fluxes is that the acids common to fluxes react with the epoxy groups to cross-link the material. Thus the fluxing activity and shelf-life must be balanced.

If cleaning is required, solvent cleaning typically is used. The surface tension of water is too high for cleaning in the small gap under the FC. Centrifuge-based cleaning systems have been shown to be effective at cleaning in small gaps. For under-the-hood automotive applications, Delphi-Delco has described the evaluation of cleaning processes and materials for FC-on-laminate using quartz dies and dyed fluxes.[9] If flux cleaning is used, a process-control method must be developed because inspection is not a production option.

18.3.3 Pick and Place

Pick and place for FC dies is similar to that for other fine-pitch SMT devices, with added emphasis on the vision system and placement accuracy. Standard wafer sawing tolerances do not allow FC placement based on visioning of the die outline. Exceptions are very course pitch FC dies or very tight sawing specifications. The CCD camera in the vision system must have sufficient resolution to recognize individual solder balls. Sufficient contrast between the solder balls and the die surface is also required. Vertical lighting (Fig. 18.24) typically leads to reflection from both the solder balls and the die surface and often cannot be used. Because of the curvature of the solder balls and the resulting angles of reflection, angled lighting can provide the required contrast. Transforms are used to further increase the contrast and image recognition. If dip fluxing is used, fluxing before or after visioning must be considered, as well as the impact of the flux on the image.

The required placement accuracy (maximum misalignmnet) is often specified at 20 percent, as defined in Fig. 18.25. Lewis has shown that high assembly yields can be achieved with higher levels of misplacement due to the self-centering nature of molten solder.[10] As noted previously, increasing the volume of flux increases self-centering and improves assembly yield. X-ray can be used to

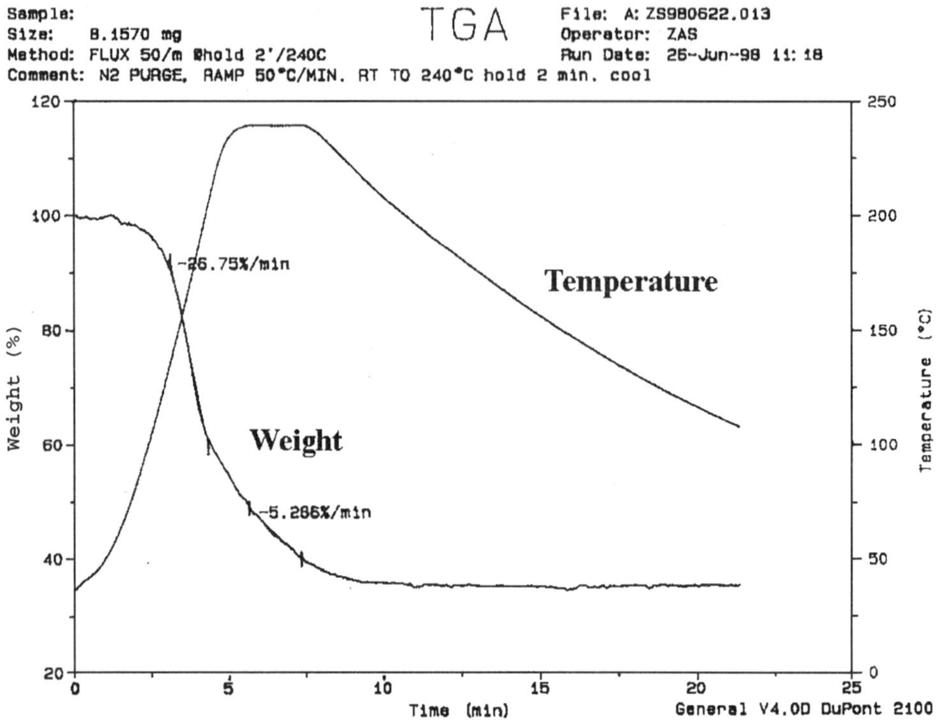

FIGURE 18.23 TGA plot of a no-clean flux.

measure the initial placement accuracy (to characterize the placement system capability) as well as to measure the alignment after reflow (to measure the limits of self-centering).

18.3.4 Reflow

The reflow temperature profile for FC assembly should follow the manufacturer's recommended profile for the flux used and is similar to standard SMT reflow profiles. Depending on the flux and board surface finish, air or nitrogen reflow can be used. In forced convection reflow ovens, the force of the gas flow on the dies must not be sufficient to move the dies. This will be a function of die size, the number of solder balls, the tackiness of the flux, and the change in tackiness during the reflow until the solder melts and wets the PWB metallization. As mentioned previously, vibration in conveyors or reflow ovens also can cause die movement and low assembly yields.

X-ray is often used to characterize the solder joints after reflow. Figure 18.26 is a typical x-ray image after reflow. Bridging between adjacent solder balls is easily detected by x-ray. X-ray (Fig. 18.27) also can detect voids in the solder joint. The voids may have existed in the solder balls as originally formed on the die or result from trapped flux vapors during the reflow profile. Void formation during reflow can be reduced or eliminated by flux selection and reflow profile modifications to minimize vapor generation once the solder ball liquefies. The impact of voids on FC reliability has not been quantified. Voids in solder balls are also common to BGAs and CSPs. An industry accepted limit of 20 percent or less void content generally is accepted for BGAs.

X-ray also can be used to measure self-alignment during the soldering process and to detect shorts between solder balls. Solder bridging can result from die misplacement or movement after placement prior to reflow. Insufficient wettable substrate pad area also can lead to bridging of fine-pitch solder balls. This is aggravated by solder mask misalignment, as has been mentioned previously.

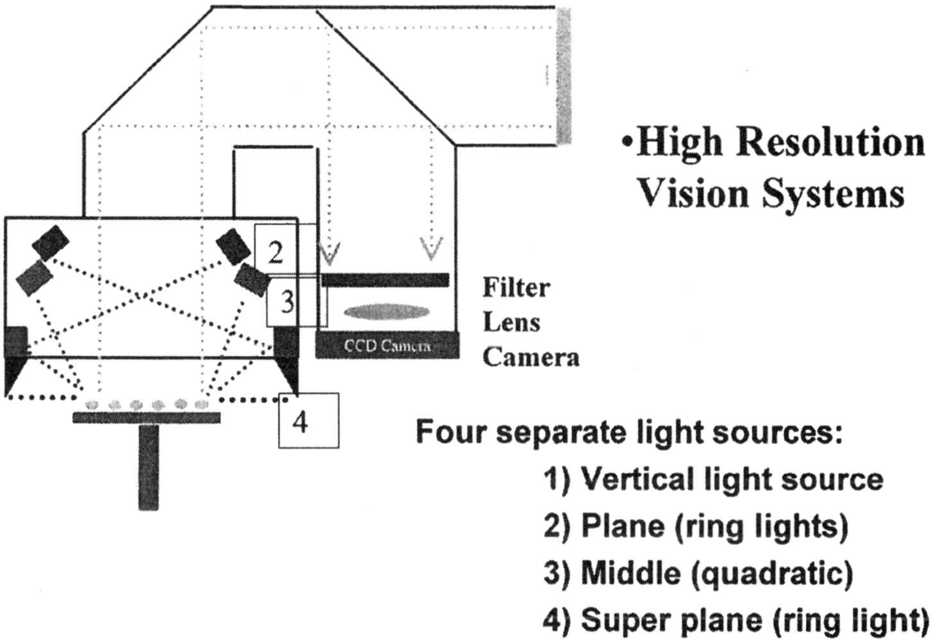

FIGURE 18.24 Example of vision system lighting options for FC dies. (*Siemens Energy and Automation.*)

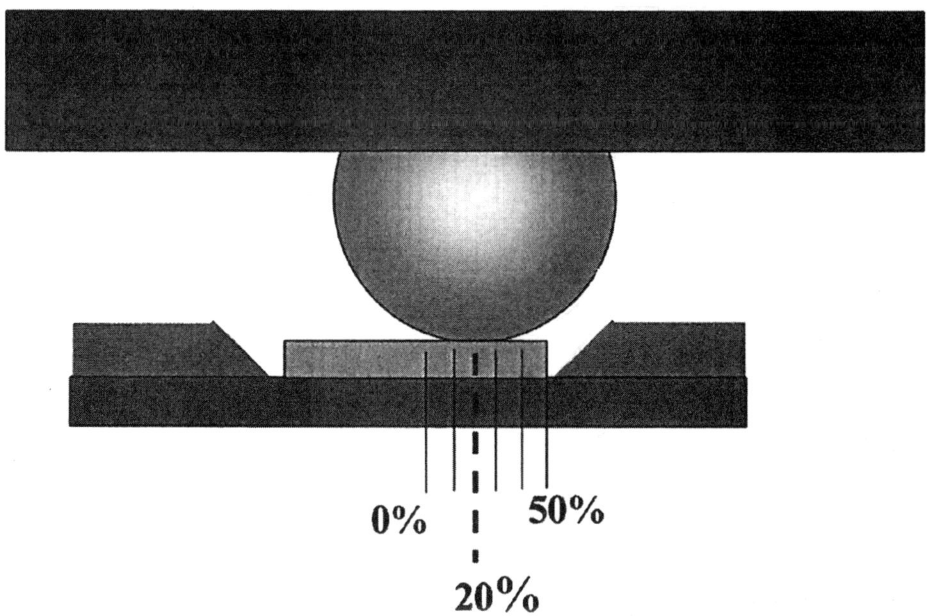

FIGURE 18.25 Definition of percent misalignment. (*Siemens Energy and Automation.*)

18.3.5 Substrate Dehydration

If the PWB has absorbed moisture from the environment, this moisture may be liberated and create moisture bubbles (voids) in the underfill during the underfill curing process. Moisture from the atmosphere is absorbed by both the PWB laminate and the solder mask. In an in-line SMT process, the underfill dispense follows the reflow cycle. This reflow cycle is effective in removing a significant amount of moisture absorbed by the PWB (Fig. 18.28). In this test, the PWBs (bare and with several different solder masks) were exposed to 85 percent relative humidity at 85°C for 168 h to fully saturate the boards. The weight gain was measured after humidity exposure, after one pass through the reflow cycle, and after one pass through the reflow cycle followed by humidity exposure at 60 percent relative humidity at 30°C for 6 h, all relative to the initial "dry" weight of the test coupon. The reexposure to 60 percent relative humidity at 30°C for 6 h after reflow was to simulate a delay between reflow and the underfill dispense related to possibly board testing prior to underfill or some production delay. As can be seen, little moisture is absorbed during this 6-h period. The reflow cycle drives off the "easy to remove" moisture. The lower-temperature underfill cure cycle is not likely to cause subsequent moisture release. Depending on the underfill and the solder mask selected, a dehydration bake may not be necessary with in-line FC assembly. However, individual material combinations should be tested.

18.3.6 Underfill Dispense and Cure

The manufacturer provides the underfill in syringes of various sizes. The underfill is loaded into the syringe under vacuum and may be centrifuged to ensure that there are no bubbles in the syringe. The underfill is shipped frozen and should be stored at −40°C until ready to use. The underfill must be thawed completely prior to use. Thawing should be accomplished by placing the syringe at room temperature. Heating the syringe for speed thawing is not recommended.

Automated dispense systems are used to dispense a controlled volume of underfill in a precise pattern. The volume of underfill dispensed is critical to ensure complete underfilling of the die and

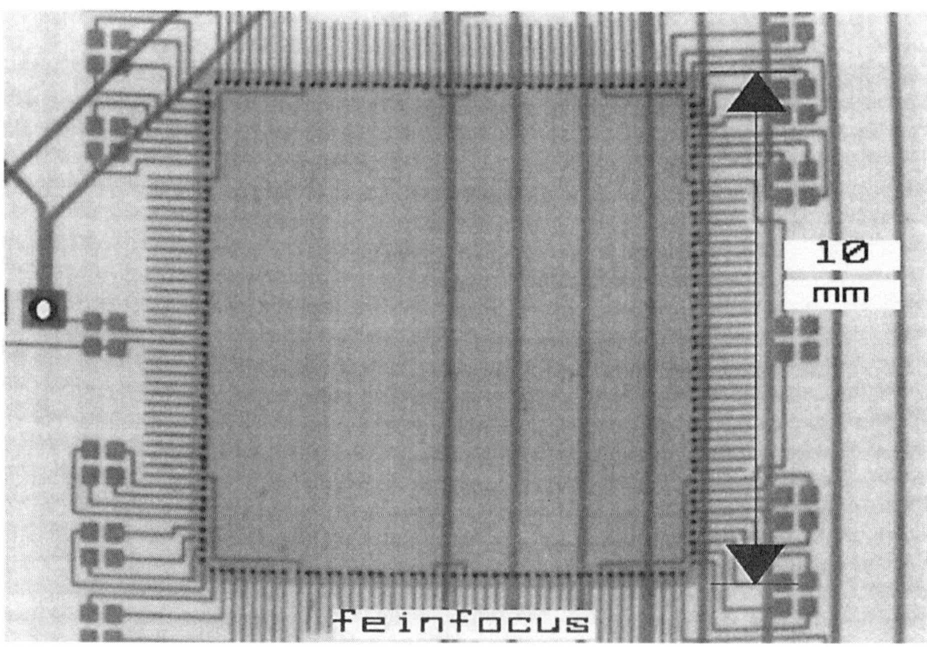

FIGURE 18.26 X-ray image of an FC assembly. (*FeinFocus.*)

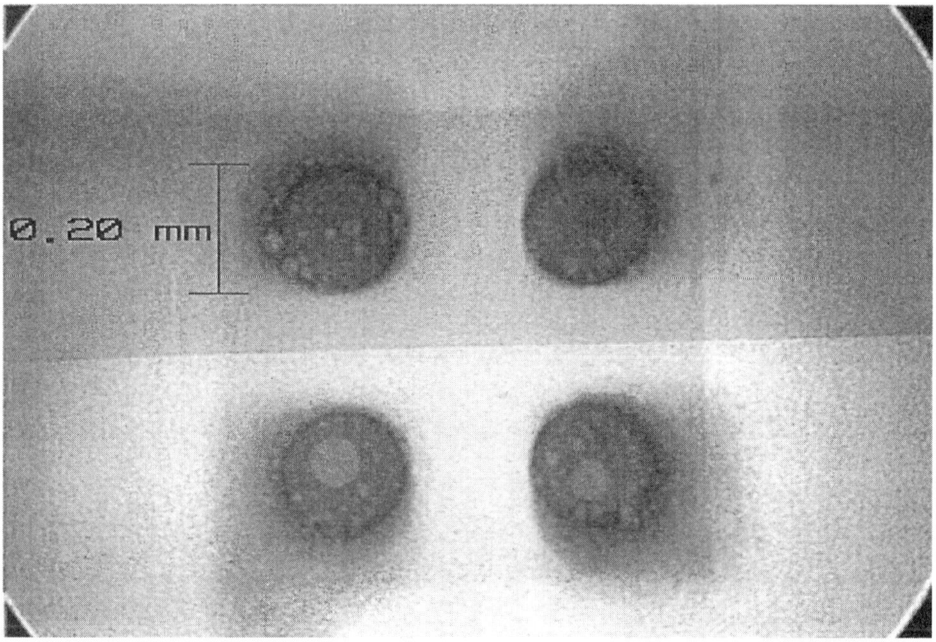

FIGURE 18.27 Example of voids in FC solder joints as detected by x-ray. (*FeinFocus.*)

fillet formation. In production, the gap height and hence the volume of underfill required to fill the space between the die and the board will vary somewhat from board to board. The fillet provides the reservoir to accommodate this variation. Volumetric control is achieved using a rotary valve or positive-displacement dispense valve. Conventional time-pressure dispense valves do not provide the required volume control and repeatability. Since the viscosity of the underfill may increase with time at room temperature, the dispense parameters may require minor adjustments over a period of time to maintain a constant volume. With current production dispense systems, the parameters can be adjusted automatically based on calibration using either volumetric measurement or weight-based systems. The density of the underfill is required in the weight-based system to calculate actual dispense volume. A baseline volume or weight of underfill is dispensed with a known set of dispense parameters. For a rotary-valve system, these parameters would be dispense time and valve rpms. At a later calibration, the same parameters would be used, and the corresponding volume or weight would be measured. If the amount of material dispensed was less than during the baseline case, the rpms would be increased to compensate.

The dispense pattern is critical to achieving complete underfilling without entrapping voids under the dies. The dispense pattern is a function of the underfill, the die size, the die bump pattern, and the PWB topology and is determined experimentally. Common dispense patterns are shown in Fig. 18.29.

The dispense needle must be placed close to the die so that material contacts the gap between the die and the PWB. Capillary forces then pull the material under the die. The needle travel speed, the valve rpms for a rotary-valve system, the needle gauge, and the travel length determine the volume of underfill dispensed. From a production point of view, needle travel speed should be maximized. However, if the underfill does not flow rapidly under the die, a large volume of material may accumulate beside the die. This material may get on top of the die or flow away from the die before it has an opportunity to flow under the die. Board heating typically is used to increase the flow rate under the die. PWB temperatures range from 60 to 100°C depending on the underfill material and the die size. Heating the underfill initially lowers its viscosity but also causes curing of the underfill, which increases its viscosity. For large dies, the underfill time is longer, and

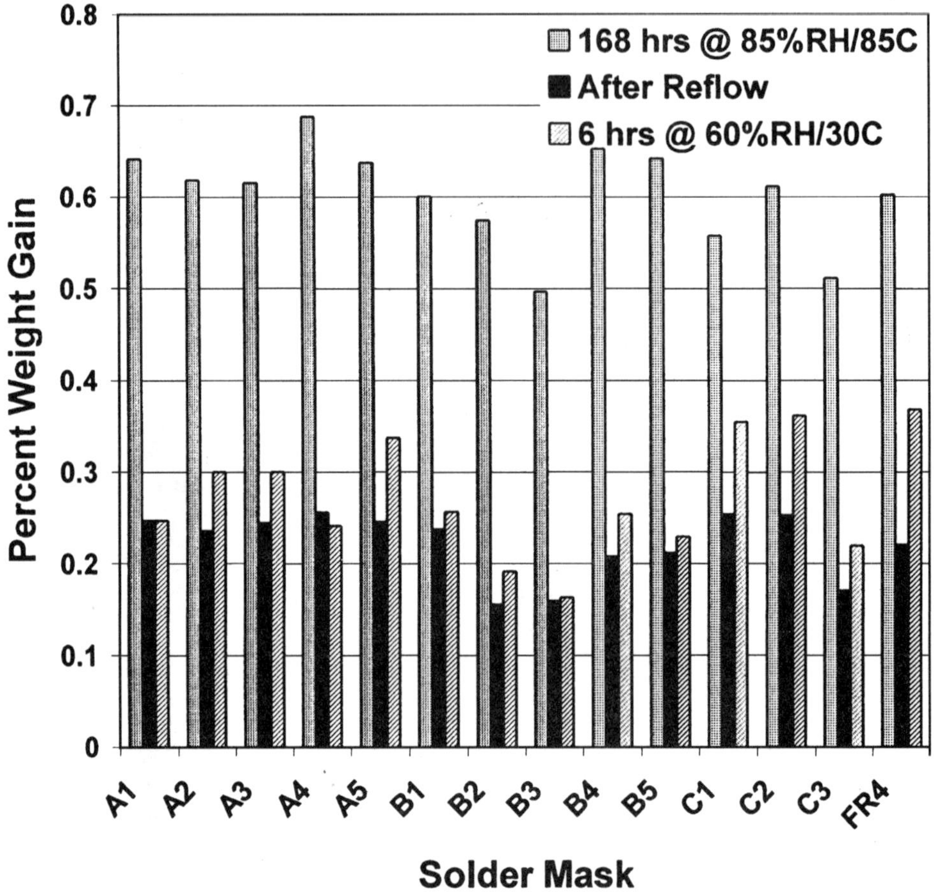

FIGURE 18.28 Percent weight gain after 168 h of exposure at 85 percent relative humidity at 85°C, after one reflow cycle, and after 6 h of exposure at 60 percent relative humidity at 30°C.

partial underfill curing can be a concern at the higher PWB temperatures. A design keep-out region must be defined around the FC die for underfill dispense. If the dispensed underfill contacts adjacent components, some of the underfill may be "pulled" to that component. The variation in the amount pulled would lead to a lack of underfill volume control for the FC die and possibly incomplete underfilling.

A particular challenge with perimeter bumped dies is that the solder bumps tend to accelerate the flow of the underfill around the perimeter of the die. If the flow around the perimeter exceeds the flow under the area of the die, a void will be trapped under the die as the two perimeter flow fronts meet. An example of this is shown in Fig. 18.30, which is a scanning acoustic microscopic image through the FC die. The light area is due to reflection of the ultrasonic wave at the discontinuity caused by a large void in the underfill. The gray area indicates complete underfill. The length of the line or the line segments of the L pattern must be adjusted to limit this "racing" of material around the die perimeter.

Following the initial dispense of the underfill, a fillet dispense may be required to form a fillet around the die on the sides of the die where underfill was not dispensed originally (Fig. 18.31). Time

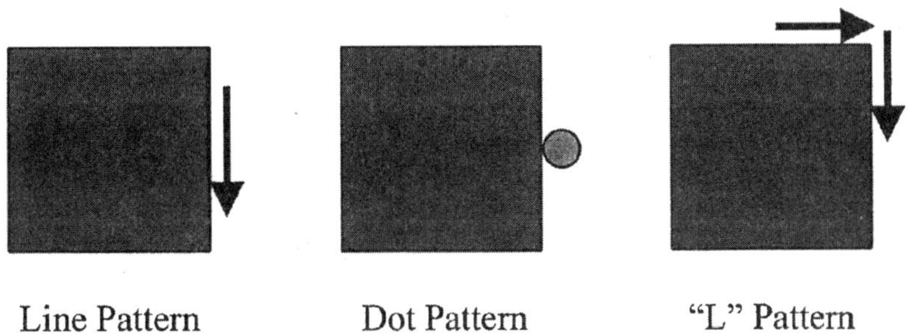

FIGURE 18.29 Line, dot, and L dispense patterns.

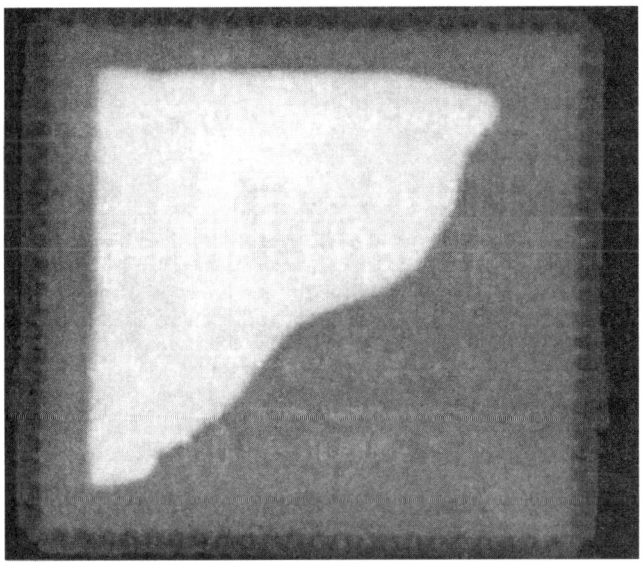

FIGURE 18.30 Scanning acoustic microscopic image of an underfilled FC die. The underfill "raced" around the perimeter solder bumps, trapping a void under the die.

must be allowed for the underfill initially dispensed to flow completely under the die to prevent trapping a void under the die with the fillet dispense. Many fast-flow underfills are self-filleting, eliminating the need for this step. The surface tension and wetting characteristics of the underfill continue to pull material under the die to form a fillet on the nondispensed sides of the die.

Underfill time is a critical issue in high-volume production. Newer-generation "fast flow" underfills have been introduced to increase production throughput. This is a particular issue as the die size increases. Equation (18.1) relates the flow time t for a viscous fluid between two parallel plates with a spacing of h. The flow time increases as the square of the distance traveled and linearly with decreasing gap height. Thus doubling the die size increases the flow time by a factor of 4. The trend to finer-pitch solder balls on the die (higher I/O counts) leads to a reduction in solder ball volume, resulting in lower gap heights, further increasing the flow time. The equation simplifies the actual FC structure by ignoring the presence of solder balls and their impact on increasing flow speed due to additional surface and capillary forces.

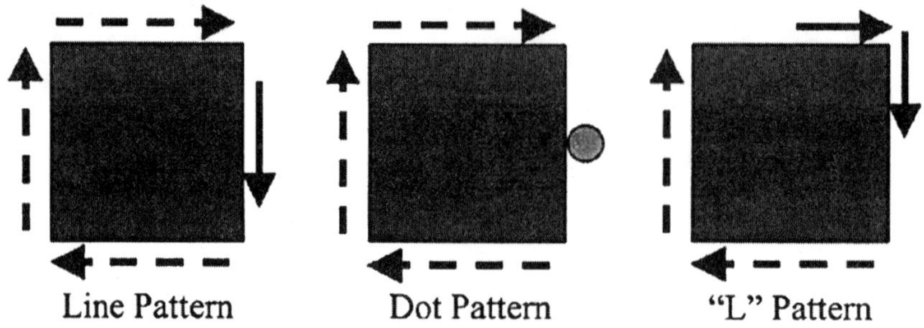

FIGURE 18.31 Underfill dispensing patterns. Dotted lines indicate fillet dispense pattern.

$$t = \frac{3\mu L^2}{h\gamma \cos \theta} \tag{18.1}$$

where t = flow time
μ = absolute viscosity
L = distance traveled
h = gap height
γ = surface tension
θ = wetting angle

Figure 18.32 illustrates a common method to measure underfill flow speeds. The underfill is dispensed between two glass plates with a known separation, and the time and distance traveled are measured. This test method is equivalent to the conditions (parallel plates, no solder balls) described by Eq. (18.1), and the data can be plotted as t versus L^2. The wetting angle for the surfaces used in the test setup may differ from the wetting angle of the surfaces in an actual FC assembly.

Sandia National Laboratories developed an alternate approach to measuring flow speed.[12] The test uses a V groove machined into an aluminum plate (Fig. 18.33). Markers are also machined into the aluminum to facilitate measurement of underfill flow distance. An equation was developed to relate flow distance to the flow time (Eq. 18.2). The underfill is dispensed into a reservoir, and the flow distance versus time is measured. The aluminum piece can be heated to investigate the effects of temperature on flow speed. An example is shown in Fig. 18.34.

$$X^2 = (\gamma/\mu)\, df(\alpha, \theta)\, t \tag{18.2}$$

where X = distance traveled
t = time
γ = surface tension of fluid
μ = viscosity of fluid
d = characteristic depth of groove
$f(\alpha, \theta)$ = function of groove and contact angle

After underfill dispense, the material must be cured. Cure conditions vary from supplier to supplier and range from 150 to 165°C for 5 to 90 min. "Snap" cure underfills with cure times of less than 10 min can be cured using in-line conveyor ovens, whereas the 90-min cures usually are performed in batch convection ovens.

Scanning acoustic microscopy (SAM) is used commonly to develop underfill processes and dispense patterns as well as in-process sampling and process requalification. High-frequency transducers (>100 MHz) are used to provide the necessary resolution to detect small defects. Figure 18.30 is a C-mode SAM image of a good underfill dispense with no voids, whereas Fig. 18.33 shows small voids in the underfill. Voids can lead to early delamination, which in turn leads

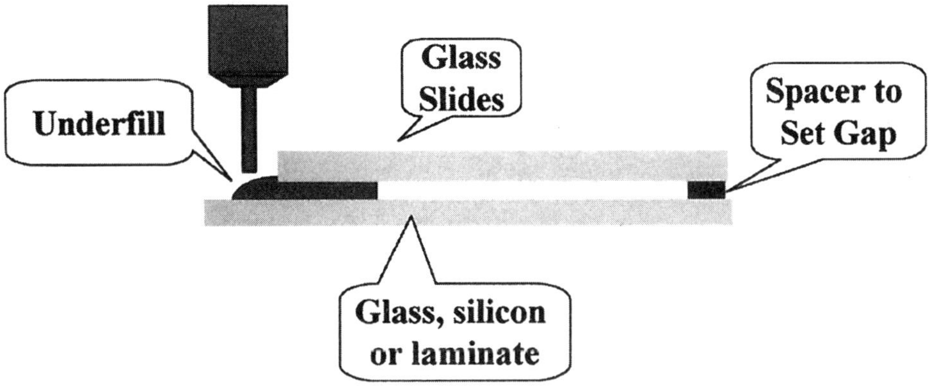

FIGURE 18.32 Test setup for measuring underfill flow rate.

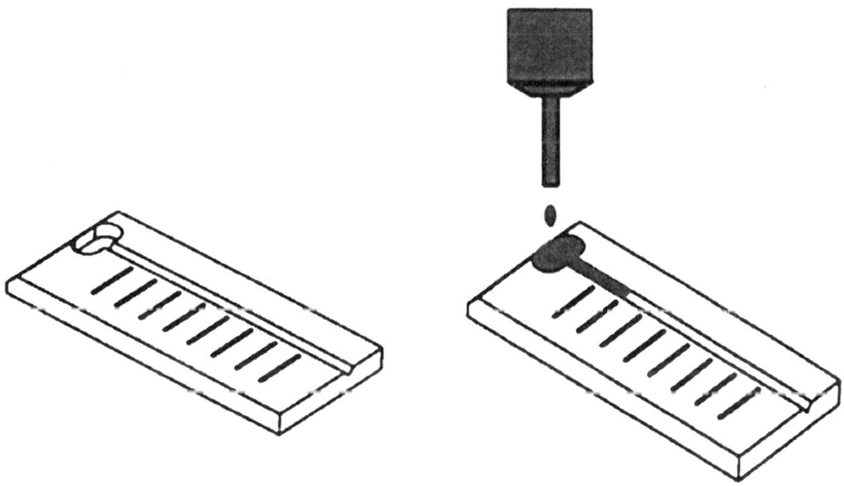

FIGURE 18.33 Sandia National Laboratories underfill flow test method.

to electrical failures. Solder extrusion from the solder balls into adjacent voids during thermal cycling also leads to early electrical failures. SAM also can be used to detect filler separation during the underfill flow process (see Fig. 18.33). Filler separation results in a nonuniform CTE within the underfill. In particular, the resin-rich region in contact with the silicon dies will have a higher CTE than the filler-rich area. Cross sections and flat sections (polishing either the PWB or the die away) are used to complement and verify SAM results. Figure 18.30 is a cross section illustrating filler separation.

18.3.7 Rework

The removal and replacement of a defective FC is relatively straightforward prior to underfill. The die can be heated using a hot-air rework station until the solder balls melt and the die can be removed. With continued heating, the excess liquid solder remaining on the pads after die removal can be vacuumed or wicked away. To attach a replacement die, flux is applied to the solder balls on the die by dipping, the die is aligned using split optics, and the solder is reflowed by hot air. The process is similar to BGA or CSP removal and replacement. However, the optics and placement resolution must be greater for FC rework. If boards are to be tested prior to underfill, care must be used in handling and probing. Flexing of the PWB, particularly the thin PWBs used in portable products, can fracture the delicate FC solder joints.

In many cases, testing is not done prior to underfill dispense and cure. In addition, rework of field-return units is often a consideration. The purpose of underfill is to strongly bond the FC die to the PWB. However, if rework is required after underfill, some mechanism must exist for decreasing the underfill structural strength for die removal and site preparation. Two approaches have been taken to address rework. The first is to chemically attach the underfill material. This process requires application of the appropriate chemical to the die site and penetration of the chemical into the underfill under the die. The chemical must not attach the PWB or surrounding components.

The second approach is thermally reworkable underfills.[13,14] The underfill is designed to break down when heated to solder reflow (rework) temperature. This breakdown of the thermoset network is a result of the incorporation of a monomer that has a special linkage designed to break apart on heating. The breakdown of the network reduces the adhesion to the die, allowing easy removal of the die. The underfill is also now ready for easy removal from the PWB for site preparation. Using a conventional hot-air rework station, the die is heated to typical reflow temperatures of 220°C, and the die is removed with a slight twisting motion. After die removal, significant underfill material remains on the PWB and must be removed prior to placement of the new die (Fig. 18.35).

Mechanical brushing is used to clean the die site. A custom cleaning machine has been designed and fabricated (Fig. 18.36). The machine provides *x, y,* and *z* control. Several different brush styles and materials were tested. The flat-end horsehair brush provided the best cleaning with the least amount of damage to the board and the brush. A flat-end nylon brush wore away before one die site was cleaned. Disk-style brushes are not recommended because the disk is more difficult to control in

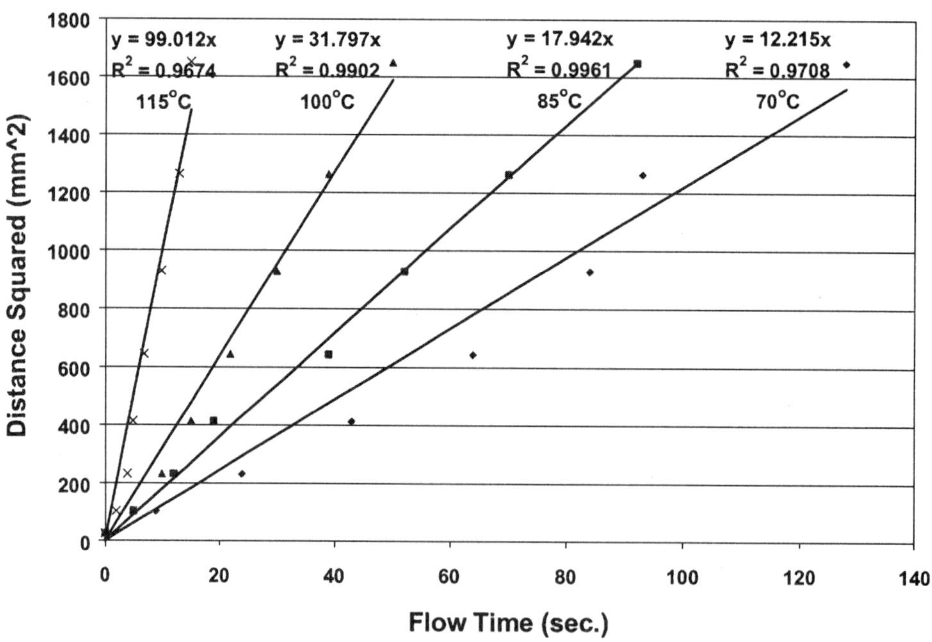

FIGURE 18.34 Distance squared versus flow time as a function of temperature for a commercial "fast flow" underfill.

the small work area and because its hard core can cause damage to the pads. Figure 18.37 shows the site after preparation by brushing. The die site is then ready for die placement, reflow, and underfill.

Since the thermally reworkable underfill is triggered at the rework (reflow) temperature, dies that have been underfilled with this material cannot be subjected to subsequent reflow cycles in the manufacturing process. Reworkable underfills with higher trigger temperatures are being developed to allow reflow after underfill. These materials will require a higher rework temperature.

18.4 ASSEMBLY WITH FLUXING UNDERFILLS

To reduce the number of process steps associated with FC assembly, Pennisi[15] patented an adhesive and encapsulant material with fluxing properties. The idea was to incorporate fluxing activity into the underfill material and then dispense sufficient quantity of the fluxing underfill on the die site prior to die placement to serve as both the flux for soldering and, when cured, the underfill. The assembly concept is illustrated in Fig. 18.38. The flux application step is replaced by the fluxing underfill dispense. The postreflow underfill dispense and associated capillary flow time are eliminated. The fluxing underfill may cure during the reflow cycle, eliminating another process step, or it may require a postreflow cure depending on the material.

The formulation of a fluxing underfill is challenging. Acids typically are found in solder fluxes to provide the fluxing action of removing metal oxides to facilitate metallurgical wetting. However, acids initiate the cross-linking of epoxies. By increasing the acid content, the fluxing activity increases, but the shelf life decreases. Shelf life, fluxing activity level, and lot-to-lot consistency are issues being addressed by the materials suppliers, and newer material systems are being introduced.[16]

Fluxing underfills have no or low filler content. If filler particles are trapped between the solder ball and the PWB pad during placement, they may prevent the solder ball from touching the PWB pad during reflow, and wetting will not occur. Addition of filler to the fluxing underfill can decrease assembly yield. Without filler, the CTE of the fluxing underfill is significantly higher than that of traditional capillary underfills. Reliability studies have shown that even with the higher CTE, fluxing underfills meet the thermal cycling requirements for many applications.[17]

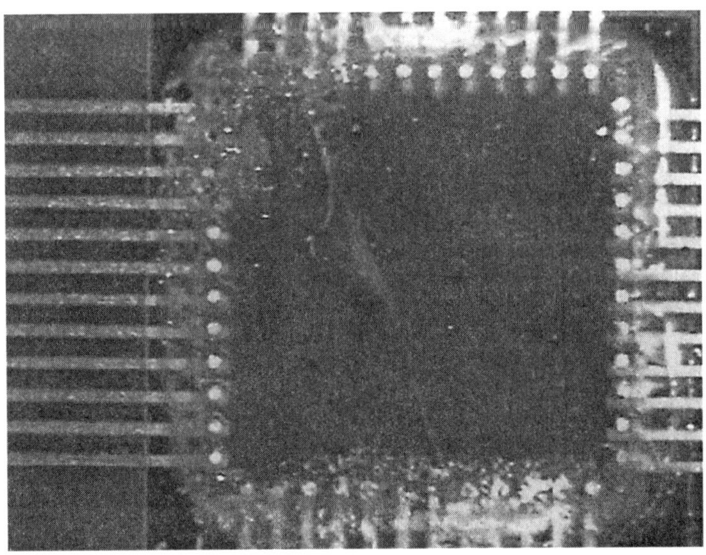

FIGURE 18.35 Die site after die removal.

FIGURE 18.36 Automated brushing machine for site preparation.

FIGURE 18.37 Die site after brush cleaning.

18.4.1 Substrate Dehydration

As discussed previously, PWB laminate and solder mask absorb moisture from the atmosphere. In the capillary underfill approach, the PWB has passed through a reflow cycle just prior to underfill dispense, and much of the absorbed moisture has been removed, potentially eliminating the need for a dehydration bake. With fluxing underfills, the reflow cycle follows the dispense of the fluxing underfill, and the potential exists for voids to form due to moisture release from the PWB or solder mask during the reflow cycle. A second source of gas-generated voids in the fluxing underfill is out-

FIGURE 18.38 Fluxing underfill assembly process.

gassing from incompletely cured solder mask. A dehydration bake is recommended prior to fluxing underfill dispense to remove moisture and to ensure that the solder mask is cured completely. In a two-sided SMT assembly process with the FC assembly on side 2, dehydration may not be necessary if the board goes immediately from side 1 reflow to side 2 assembly. If a dehydration step is required, this may be an added process step compared with the capillary flow underfill process and should be included in cost models.

18.4.2 Application of Fluxing Underfill

The fluxing underfill is applied to the substrate by dispensing. Stencil printing has been proposed as a higher-speed application method, but this approach does not produce the smooth, dome-shaped deposit that is required for void-free die placement.[18] During placement, the fluxing underfill should contact the center of the die first and then spread radially as the die continues to move toward the PWB. Other dispense patterns have been used.[16] Volume control is important in dispensing fluxing underfill. Insufficient volume will result in nonwetting by some solder balls and incomplete underfilling, whereas excess fluxing underfill can cause the die to "float." If the die floats and the solder balls do not contact the pads on the PWB during reflow, opens will occur. Shifting of the die with respect to the PWB pads also can result from die floating.

18.4.3 Die Placement

During placement of an FC die in fluxing underfill, the underfill must flow outward as the die nears the PWB in a "squeeze flow" process. As the underfill is squeezed between the die and the PWB, it generates a force opposing the placement force. To understand the interaction of variables in the squeeze-flow process, the placement of a flat, circular disk on a substrate with a newtonian fluid in between can be considered analogous to the placement of an FC die in fluxing underfill. Assuming that the disk and substrate are parallel and that the volume between the disk and substrate is filled with underfill, the Stefan equation (Eq. 18.3) can be used to approximate the force.[19,20]

$$F_{\text{underfill}} = \frac{3\pi R^4 \mu (-v)}{h_0^3} \tag{18.3}$$

where $F_{\text{underfill}}$ = force exerted by underfill
R = radius of the disk
μ = viscosity of underfill
v = placement velocity
h_0 = height between disk and substrate

From this equation, the force increases with increasing placement velocity, underfill viscosity, and disk (die) size. The force also increases as the disk approaches the substrate (h_0 decreases). If $F_{\text{underfill}}$ equals the placement force before the solder balls contact the PWB pads (h_0 > solder ball height), then sufficient hold time must be programmed into the placement program to allow the die under constant force to traverse the remaining distance. For practical underfill viscosities, die sizes, and placement velocities, excessive placement force or hold times are not required to place FC dies into fluxing underfills with high yields.

As mentioned previously, the fluxing underfill should have a smooth, domed shape, and the center of the die should first make contact with the underfill. As the die proceeds downward, the underfill is squeezed outward, and the underfill contact circle with the die surface will expand outward, dispelling air as it advances. If the outward movement is too fast, there is a potential to form voids behind the solder balls as the underfill flows outward. Placement experiments should be conducted to optimize the placement process for high electrical interconnection yield and no voiding.

18.4.4 Reflow

The reflow profile for a fluxing underfill is critical to high assembly yields. Figure 18.39 is an example of a good solder joint using fluxing underfill. The viscosity of the fluxing underfill must remain low until the solder balls melt to allow collapse of the die as the solder wets the PWB pads. While the viscosity of the underfill decreases with temperature, the thermal energy input to the fluxing underfill during the early stages of the reflow profile accelerates cross-linking of the polymer, increasing the viscosity. Figure 18.40 is a cross section illustrating the effect if the underfill viscosity increases before the solder melts.

Three approaches can be taken to address the profile issue. The first approach is to limit the cross-linking rate of the underfill by selection of the catalyst and the amount used. These underfills can be reflowed in a conventional SMT reflow profile (Fig. 18.41). With a slower cross-linking rate, complete curing of the underfill is not achieved during the reflow cycle, and a postreflow cure is required. This reduces the sensitivity to the reflow profile but adds a cure step similar to the capillary underfills. In fact, due to the slow cross-linking rate, the postreflow cure time may be longer than the cure time for "snap" cure capillary flow underfills, reducing the process advantage of fluxing underfills.

The second approach is the use of a "volcano" reflow profile, as shown in Fig. 18.42. This profile rapidly heats the assembly to the soldering temperature, minimizing the thermal exposure of the underfill prior to melting of the solder. In a mixed assembly with SMT components and solder paste to be reflowed simultaneously with the fluxing underfill, the compatibility of the SMT components and solder paste with this fast-temperature-rise profile must be established.

The third approach is to use a catalyst that does not initiate cross-linking until temperatures rise slightly above the melting point of the solder. Work is ongoing to develop and refine these materi-

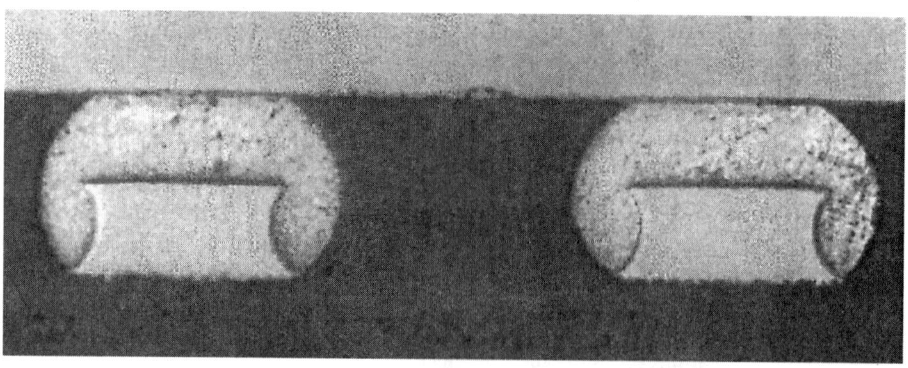

FIGURE 18.39 Cross section illustrating good solder joint formation when the fluxing underfill.

FIGURE 18.40 Cross section illustrating poor solder joint formation when the fluxing underfill viscosity increases too much in the reflow profile prior to melting of the solder.

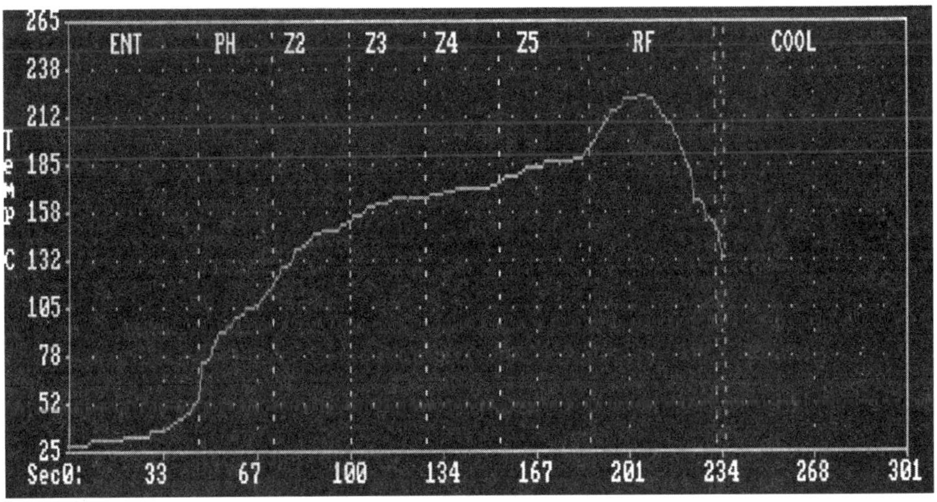

FIGURE 18.41 Typical SMT reflow profile.

als. New materials are being introduced into the market. With this approach, the underfill is much less sensitive to the temperature and duration of the soak in a conventional SMT reflow profile. These underfills typically cure during the reflow cycle. With mixed SMT assembly, the challenge will be getting large SMT components to temperature without the FC dies and underfill soaking at too high a temperature prior to solder melting. Advanced profiling tools are available to assist in profile development.

18.4.5 Rework

Reworkable, fluxing underfills have not been commercialized but are under development. It has been noted that the fluxing underfills that do not fully cure during reflow can be "reworked" prior to the postreflow cure. Since the underfill has not cured fully, it has not achieved its full adhesion and mechanical strength, and the die may be removed. Further work is required to develop truly reworkable fluxing underfills.

18.5 WAFER-APPLIED UNDERFILLS

To achieve an FC manufacturing process that is transparent to a standard SMT assembly line, the fluxing underfill should be applied at the wafer level and should cure during the reflow cycle.[21] The capillary, fluxing, and wafer-applied assembly approaches are compared in Fig. 18.43. Applying the fluxing underfill to the wafer simultaneously coats hundreds of dies, eliminating the need for underfill dispense for each die individually at the assembly stage. The materials and coating and assembly processes are being developed currently in the industry. While significant progress has been made, wafer-applied underfills are not available commercially at the time of this writing.

18.5.1 Application to Wafer

Finding a consistent and accurate coating method by which the material may be deposited onto the wafer is crucial to the successful implementation of wafer-applied underfill. Several techniques potentially could be used for coating a bumped wafer: (1) screen/stencil printing, (2) pad printing, and (3) extrusion coating.

Screen/Stencil Printing. The stencil and screen-printing processes are used widely in the electronics industry for the fabrication of PWBs, flexible circuits, and other related items. The challenge in this process is the irregular surface topology presented by the solder bumps. A uniform coating thickness that does not coat the tops of the solder balls is required.

Pad Printing. Pad printing is an offset gravure printing method, where the gravure or cliché is patterned with the artwork to be printed, and ink is transferred from the cliché to the part using a silicon pad. First, artwork is transferred to a stainless steel cliché plate using a photoetch process. Ink is spread onto the cliché, and a silicone pad is pressed against the cliché and then lifted, with the ink

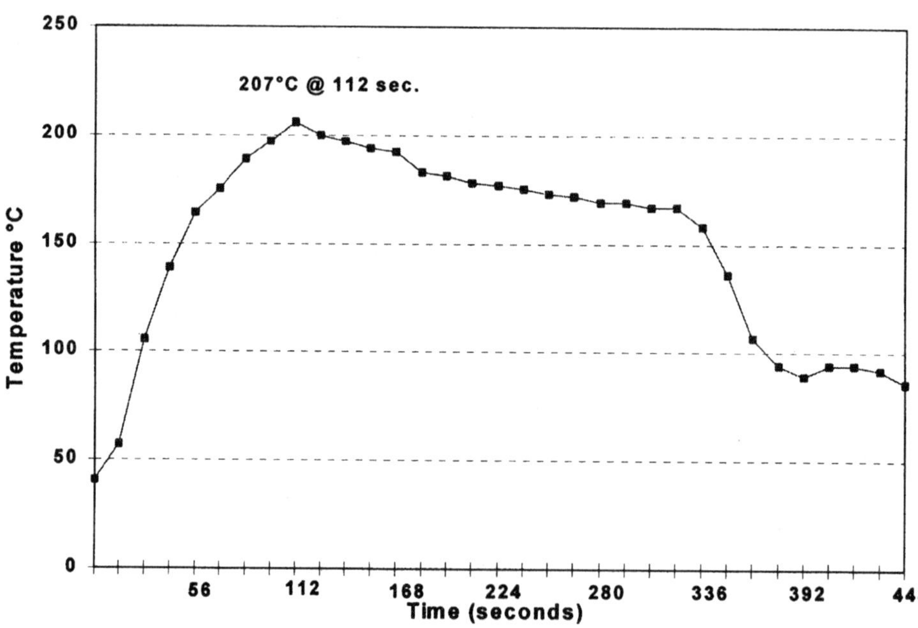

FIGURE 18.42 Example of "volcano" reflow profile. (*Emerson & Cumming.*)

adhering to the silicone pad. The part to be printed is then aligned under the pad, and the pad is pressed against the part and lifted off, with the ink transferring to the part. A thin film of ink is deposited on the part, with a fine resolution and a locational accuracy of ±25.4 μm.

Pad printing has the ability to print over irregular surfaces, with the silicon pad conforming to the shape of the substrate. It is this characteristic that gives it great potential for coating a bumped wafer.

Extrusion Coating. Extrusion coating is the direct application of process fluids to a substrate via a patented, specially designed fluid pumping and delivery system. This creates a highly uniform coating over the entire substrate with virtually no waste. The extrusion coating process has been used extensively for thick-film application on semiconductors wafers up to 300 mm in diameter. This unique technique holds great potential for coating bumped wafers.

Experimental Wafer Coating. In recent work, the wafer-applied underfill was applied successfully on a 101.6-mm wafer. A thin uniform coating layer with clean solder balls and clean saw streets was produced (Fig. 18.44).

After coating, the underfill is B-staged to produce a tack-free surface compatible with die placement in tape and reel for shipping, storage, and automated pick and place. The shelf life and required

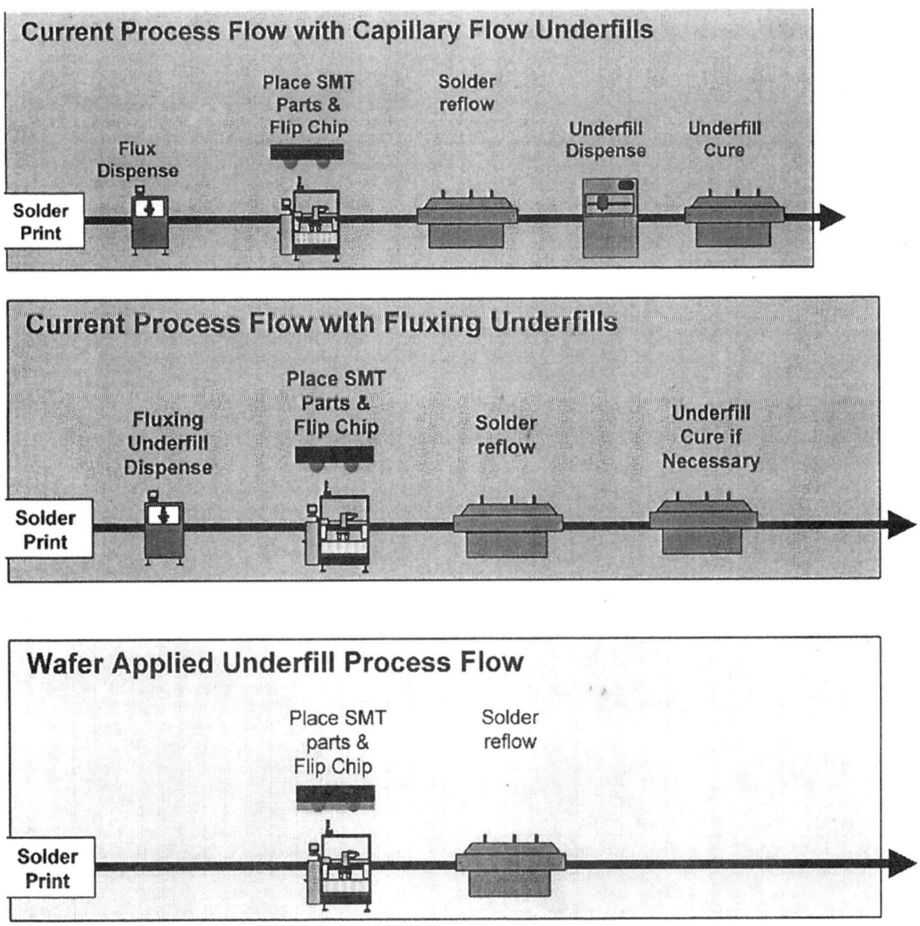

FIGURE 18.43 Comparison of capillary, fluxing, and wafer-applied underfill assembly processes.

storage conditions for the B-staged underfill are still being determined but must be compatible with transportation and inventory management.

One additional issue that is being addressed is the sawing of the wafer. In a typical wafer-sawing operation, water is used to remove sawing debris and cool the saw blade. The compatibility of the B-staged underfill with water must be understood. If the coating process coats underfill into the saw streets, the saw characteristics of the underfill must be examined, particularly any tendency to "gum" the saw blade.

18.5.2 Placement

The placement issues with a wafer-applied underfill are vision recognition and holding the die in place after placement. Vision studies have been performed using standard pick-and-place equipment used for traditional FC placement. In evaluating underfill color, black was found to be the best (Fig. 18.45). As the materials and coating process mature, additional vision studies will be required, but the results to date are encouraging.

Tack is required to hold the die in place during the time between placement and melting of the solder. In FC assembly with capillary underfill, a tacky flux is used, whereas the fluxing underfill provides the tack in that assembly process. The wafer-applied underfill is tack-free after B-staging. However, it can be softened by heating to produce a tacky surface. It is proposed that a focused heat source or soft-beam laser be used to slightly heat the die just prior to placement. The necessary pick-and-place equipment modifications are being investigated.

18.5.3 Reflow

The reflow cycle for wafer-applied underfill is a standard SMT profile. The underfill cures during the reflow cycle, eliminating the need for a postreflow cure step. Figure 18.46 shows a cross section of a solder joint reflowed with wafer-applied underfill. Good solder wetting is observed.

18.5.4 Rework

The ultimate wafer-applied underfill will incorporate reworkability, but it is too early in the development cycle to demonstrate this capability. If the wafer-applied underfill must be exposed to multiple reflow cycles in the assembly process flow, a higher rework temperature will be required if a thermally activated rework chemistry/process is used.

18.6 RELIABILITY

The reliability requirements of FC assemblies fall into two broad categories: (1) component level and (2) board or product level. If FC technology is used to assemble a die into a package that is to be

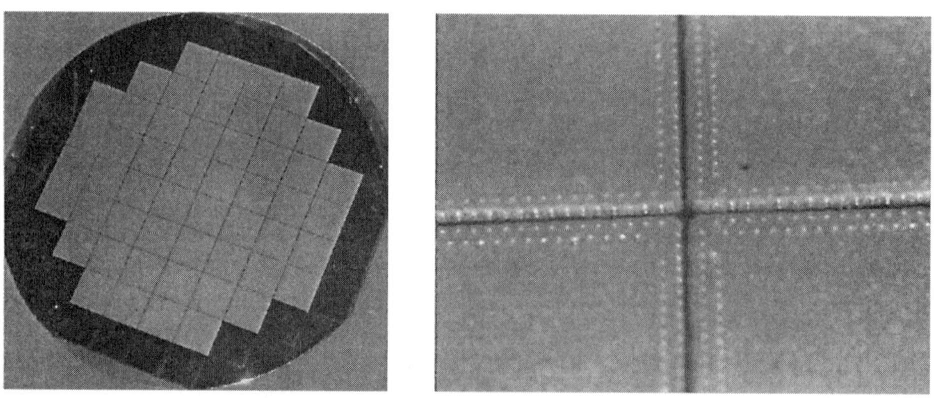

FIGURE 18.44 Photographs of wafer-applied underfill.

assembled subsequently onto a board, component-level testing is required. An example would be an FC reflowed onto a laminate carrier to create an FC BGA package (Fig. 18.47). As a component, the FC must first survive the board-level assembly process and then still meet the reliability requirements of the final product. The primary concerns are the effects of moisture and reflow cycles on reliability. This will be discussed in the next subsection.

In a board-level assembly, the FC must meet the reliability requirements of the final product. The environmental exposure and expected product life vary significantly for mainframe computers, pagers, and automotive engine controllers. The potential field failure modes for the product must be determined and appropriate laboratory test requirements defined. Since the actual field-use conditions cannot be employed for reliability testing due to the test time required, accelerated laboratory conditions are used. It is important in defining laboratory testing that new failure modes that do not occur in the field are not introduced while accelerating the test. Potential FC failure modes and test methods are shown in Table 18.3.

In field use, electronic products are not exposed to a single test condition, such as high temperature, but see combined environmental stresses. Laboratory tests are typically a single test condition (thermal cycling, high temperature, humidity, etc.). There has been some effort to define tests that expose the assembly to multiple tests in a prescribed sequence. However, with ever-decreasing time-to-market pressures, sequential testing often is not practical. Qualification testing of new materials and processes is time-consuming and expensive. Sometimes, potentially superior new materials are not used because of the time and cost associated with their qualification.

Effort has been devoted to developing "physics of failure" models to evaluate new product designs in an attempt to reduce test time and cost. This requires complete, accurate material properties, which often are not available. A combination of modeling and stress testing is used commonly in the design for reliability and qualification processes.

18.6.1 Component-Level Testing

As mentioned previously, component-level testing adds assembly-related exposure of the component to the product-level reliability testing requirements. The primary issue is the exposure to multiple

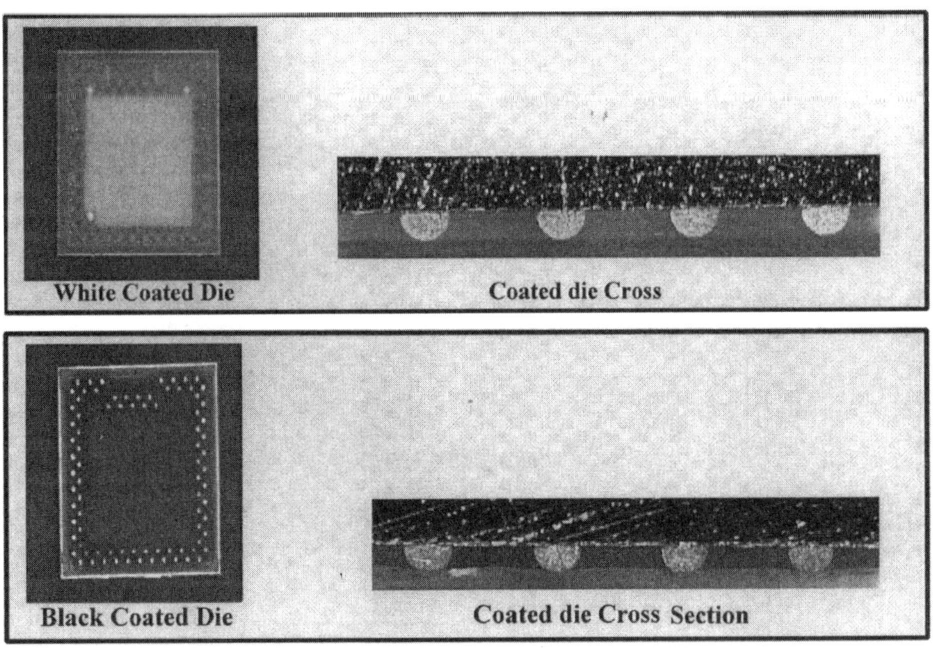

FIGURE 18.45 Comparison of visual contrast with white and black underfill.

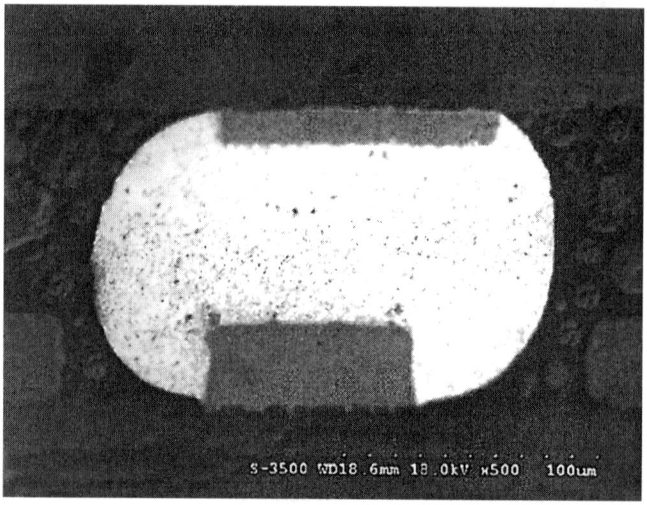

FIGURE 18.46 Cross section of a solder joint assembled with wafer-applied underfill.

reflow cycles. If the component package absorbs moisture from the environment, the rapid heating during the reflow cycle can liberate this absorbed moisture in the form of pressurized steam that can delaminate or crack the package. This assembly failure mode is known as *popcorning* because the mechanism is the same for popping popcorn. In an FC package, the underfill can absorb moisture. Moisture exposure can reduce the adhesion of the underfill. If a laminate substrate is used in the package, the laminate and the solder mask also may absorb moisture. During reflow, delamination at the die-underfill and the underfill-laminate interfaces may occur.

Moisture Sensitivity. The JEDEC Solid State Technology Association (*www.jedec.org*) has developed test methods and an industry-accepted definition of moisture sensitivity (J-STD-020A). The moisture exposure test conditions and factory storage conditions are listed in Table 18.4. If a component passes level 1 test conditions, it is deemed moisture-insensitive and can be stored indefinitely at 30°C or less at 90 percent relative humidity. At level 3, the component may be exposed to 30°C or less at 60 percent relative humidity for only 72 h. Moisture-sensitive components are dehydration baked to remove moisture and sealed in dry bags by the manufacturer for shipment and storage. Once the dry bag is opened, the component must be used in assembly within the time limit indicated by its moisture sensitivity level. The moisture sensitivity level is marked on the dry bag. If the component is exposed to the atmosphere longer than the time specified, it must be dehydrated prior to use to avoid popcorning during reflow.

To determine the sensitivity level of a package, the components are first electrically tested and inspected. The inspection includes a visual inspection for cracks and scanning acoustic imaging for any delamination. Then the components are exposed to the time, temperature, and humidity conditions specified for the sensitivity level of interest in the right columns (soak requirements) of Table 18.4. If the sensitivity level is not known, tests must be preformed at multiple levels to determine the lowest level the package can pass. After the temperature and humidity exposure, the components are subjected to three reflow cycles defined in the J-STD-020A. The reflow cycles are to follow no sooner than 15 min after removal from the humidity chamber and not longer than 4 h after removal. Following the three reflows, the components are electrically tested and inspected. Electrical failure, cracking, and delamination are criteria for failure at a given sensitivity level.

Preconditioning. For environmental testing of components, moisture preconditioning is often specified. The objective of moisture preconditioning is to induce any potential degradation of the

TABLE 18.3 Potential FC Failure Mechanisms and Test Methods

Failure mode	Test methods
Underfill delamination	Thermal cycling, thermal shock, humidity, highly accelerated stress testing (HAST)
Solder fatigue	Thermal cycling, thermal shock, power cycling, flexing
Creep	High-temperature storage, thermal cycling, thermal shock
Corrosion	Humidity, HAST
Electromigration	High-temperature storage with current flow
Excess intermetallic formation	High-temperature storage
Die fracture	Thermal cycling, thermal shock, flexing
Silicon cratering	Thermal cycling, thermal shock, flexing
Alpha particles	Electrical testing

TABLE 18.4 JEDEC J-STD-020A Definition of Moisture Sensitivity Level

	Assembly floor life		Soak requirements	
Level	Time	Conditions	Time (h)	Conditions
1	Unlimited	≤30°C/85%RH	168	85°C/85%RH
2	1 y	≤30°C/60%RH	168	85°C/60%RH
2a	4 wks	≤30°C/60%RH	696*	30°C/60%RH
3	168 h	≤30°C/60%RH	192*	30°C/60%RH
4	72 h	≤30°C/60%RH	96*	30°C/60%RH
5	48 h	≤30°C/60%RH	72*	30°C/60%RH
5a	24 h	≤30°C/60%RH	48*	30°C/60%RH
6	Time on label	≤30°C/60%RH	Time on label	30°C/60%RH

*Standard soak time includes a default value for semiconductor manufacturer's exposure time between bake and bag plus the maximum time allowed out of the bag at the distributor's facility of 24 h.

component due to assembly to simulate the condition of the component on an actual assembly prior to environmental testing. To moisture precondition the component, it is exposed to the time, temperature, and humidity test level (right columns of Table 18.4) consistent with the moisture sensitivity level rating of the component. Following exposure, the component is exposed to three reflow cycles (240°C peak) within 0.5 to 4 h. The component is then tested (electrical, visual, and acoustic microscopy) and ready for subsequent environmental testing (thermal shock, thermal cycling, etc.).

Multiple melting of the solder ball during reflow cycles can lead to excess intermetallic formation. Depending on the FC underbump metallurgy, the intermetallic will be either Sn/Cu or Sn/Ni. The growth rate for Sn/Cu intermetallics is greater than for Sn/Ni intermetallics. Excess intermetallic formation can lead to failure by delamination, dewetting, or embrittlement. To evaluate intermetallic formation, solder ball shear strength and cross sections after multiple reflow cycles can be examined. Figure 18.48 plots the solder ball shear strength as a function of reflow cycles. In this case, no degradation in strength was observed.

18.6.2 Environmental and Board-Level Testing

Thermal Cycling and Thermal Shock. Electronic products normally are exposed to variations in environmental temperatures in actual use. The mismatch in CTEs of the materials used to construct the product results in mechanical stresses as the temperature varies. Solder is susceptible to failure under stress by two modes. Fatigue failure results from cyclic stress, whereas creep is a failure under constant stress or load. The creep failure rate increases with increasing temperature. Figure 18.49 shows creep failure of FC solder joints held at constant force at 165°C.

FIGURE 18.47 Example of an FC BGA.

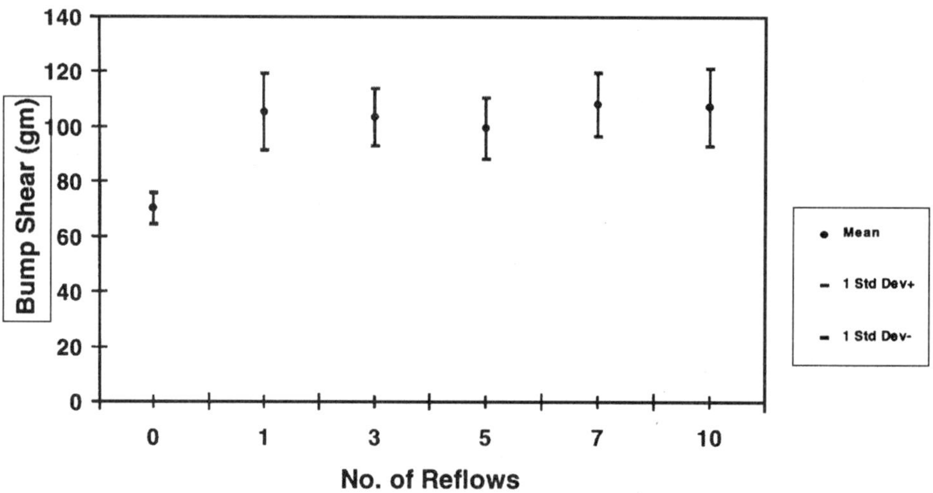

FIGURE 18.48 Solder ball shear strength as a function of reflow cycles. (*Flip Chip Technologies.*)

Failure of FC solder joints in thermal cycling and thermal shock testing results from a combination of fatigue and creep. The role of each failure mechanism depends on the materials, the temperature range, the temperature ramp rate, and the hold times (particularly the high-temperature hold time). In laboratory testing, the ramp rate and the temperature range are increased, whereas the hold times are decreased to accelerate the test and reduce test time. Direct correlation of laboratory test results with actual field life is complex.

The difference between thermal cycling and the thermal shock is the *ramp rate*. In thermal shock, the transition time from hot to cold and back is very rapid. In liquid-to-liquid thermal shock, the test boards are moved from one liquid held at high temperature to a second liquid held at low temperature. The transfer of heat by liquid contact is very efficient. For air-to-air thermal shock, the test boards are automatically shuttled between two chambers. The air in the two chambers is maintained at the two test temperatures. While the rapid transfer between chambers is comparable between liquid-to-liquid and air-to-air testing, the heat transfer rate is less for the air-to-air system. Since the thermal mass of the test boards and any fixtures or test cables has an impact on the amount of heat transferred, thermocouples are required to measure the actual temperature ramp rate and to ensure that all parts reach the temperature extremes.

Thermal cycling often uses a single chamber that is alternately heated and cooled. The temperature ramp rate is significantly slower. While this may better represent the actual temperature cycle rate in the field, the cycle time is longer, extending the test time. There is increasing use of thermal shock testing in the industry to accelerate testing to meet product development cycle requirements.

As mentioned in Sec. 18.1, underfill is used to increase the thermal cycle/thermal shock life of FC solder joints. During thermal cycling/thermal shock, there is also stress on the underfill globally and locally at the underfill-die and underfill-laminate interfaces. A typical failure mode is delamination of the underfill (most commonly at the die-underfill interface) and then rapid failure of the nearby solder joint. An example is shown in Fig. 18.50. With high adhesion underfills that do not match the CTE of the solder, solder joint fatigue failures ultimately can occur without delamination (Fig. 18.51).

As discussed in the underfill dispense section, voids in the underfill are a processing defect. During thermal cycling, extrusion of solder into voids between adjacent solder balls can lead to shorts, as shown in Fig. 18.52.

If the CTE of the underfill is higher than that of the solder, the solder is compressed by the underfill on the cold side of the cycle and is in tension in the z direction on the hot side of the cycle. Either the solder joint ultimately will fail open or the underfill may crack and solder will extrude into the crack, forming a short (Fig. 18.53).

Cracking of the underfill fillet and cracking of the PWB also are observed in thermal cycle and thermal shock testing. Propagation of these cracks can lead to either delamination or cracking of large silicon dies. Solder extrusion has been observed in PWB cracks at the base of solder balls. This extruded solder could lead to shorts within the PWB, particularly in HDI boards.

In situ resistance monitoring is required for accurate thermal shock or thermal cycle testing of underfilled FC dies. Failures (high resistance/open circuits) have been observed with daisy-chain test dies in the hot bath hundreds of cycles before failure is observed at room temperature or at the cold extreme. This can be explained using the illustration in Fig. 18.54. From finite-element modeling, the maximum stress on the solder joint occurs at the low-temperature extreme of the cycle because the "stress-free" temperature should be near the cure temperature of the underfill. However, at the cold temperature extreme, the contraction of the underfill (the CTE of the solder is lower than that

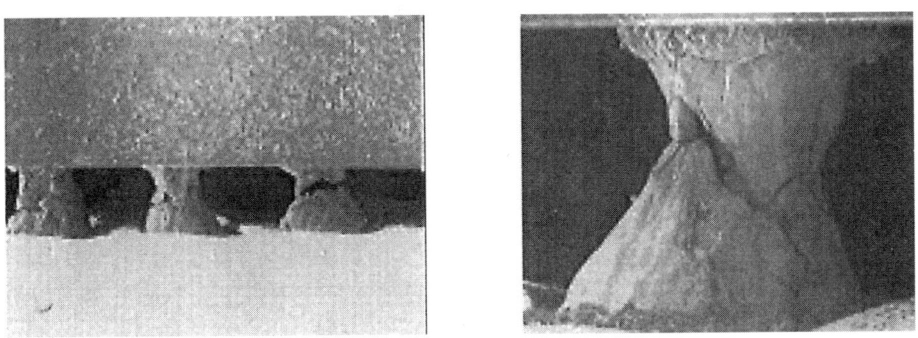

FIGURE 18.49 Creep failure of FC solder joints with a constant force at 165°C. (*Flip Chip Technologies.*)

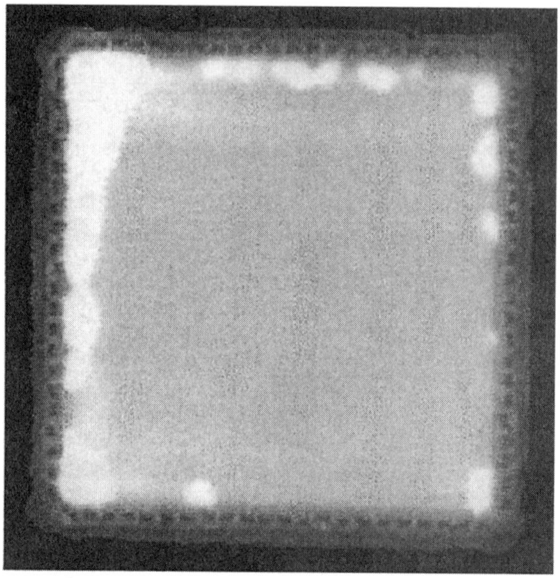

FIGURE 18.50 C-SAM image of underfill delamination due to thermal shock cycling.

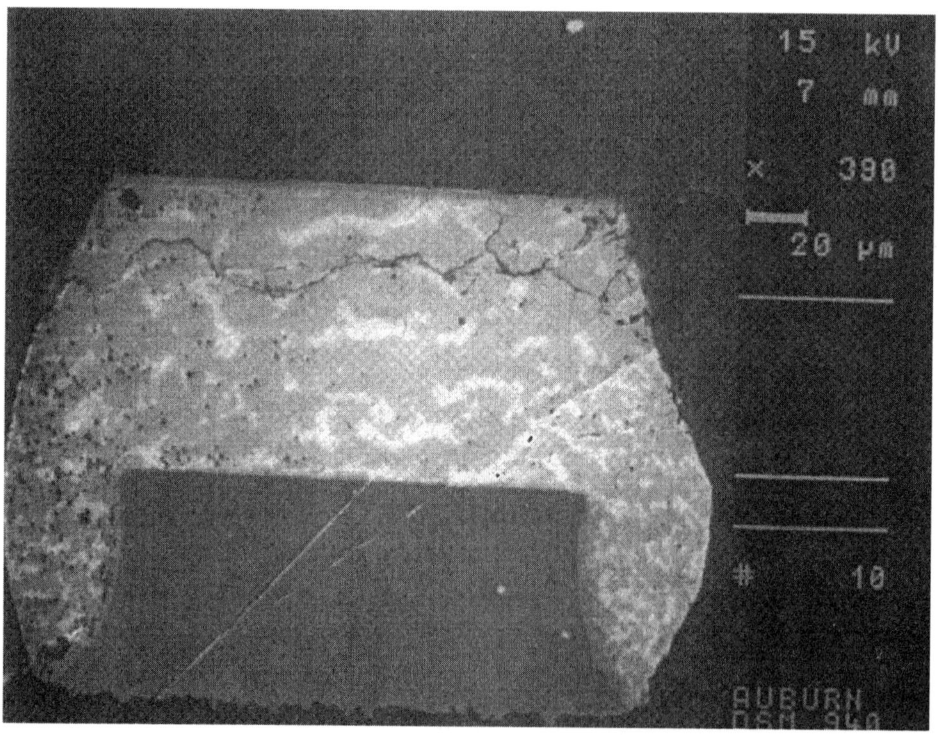

FIGURE 18.51 Solder joint fatigue failure. There was no delamination of the underfill.

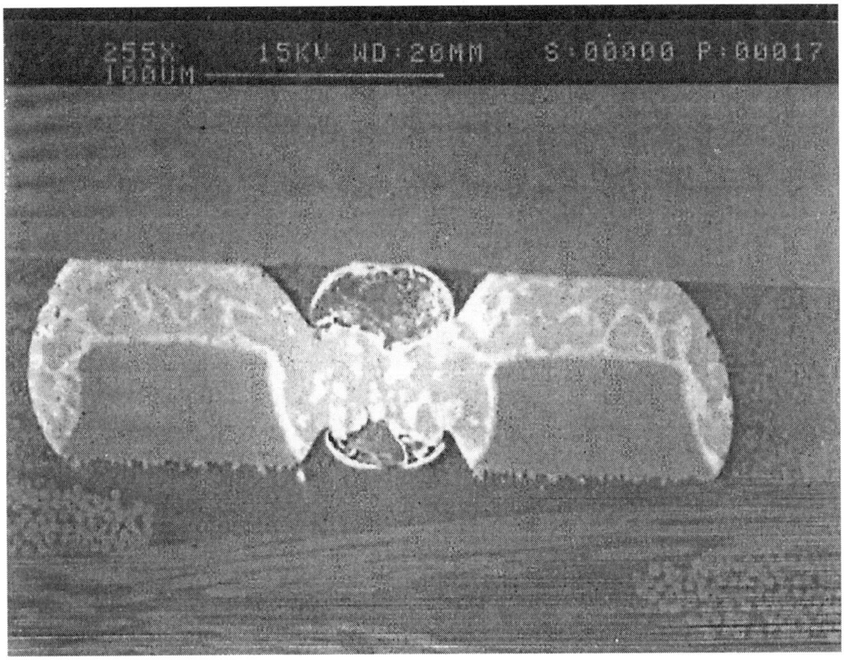

FIGURE 18.52 Cross section of a short due to extrusion of solder into the void between adjacent solder balls.

of the underfill) will hold the cracked solder joint together, maintaining electrical continuity. On heating to the high-temperature extreme, the expansion of the underfill will open the crack, and a failure can be measured. Thus periodic measurements for continuity at room temperature will not discover high-temperature intermittent opens and will overestimate the reliability of the connections.

Thermal cycling and thermal shock also can lead to die fracture. The underfill couples the silicon die to the PWB. As the structure is cycled, it bends or warps. Since the PWB contracts most, the backside of the silicon die is placed in tension during the cold side of the cycle. Any defects in the backside of the die such as microcracks from ejector pins can serve as stress concentrators, and a crack can propagate through the silicon. An example of die cracking is shown in Fig. 18.55.

Power Cycling. When an electronic product is turned on, the semiconductor devices dissipate power and get hot first. Thus the silicon die starts to expand before the underfill, solder, and PWB do. Power cycling can result in stress on the assembly and failure. Power cycling tests should be conducted for dies that dissipate significant power and that are subject to power cycling in the product application.

Flex Testing. Many portable products have integrated keypads. Pressing of the keypad flexes the thin PWB. FCs on the opposite side of the PWB would experience cyclic mechanical stresses. Underfill significantly increases the reliability of these applications.

Temperature-Humidity-Bias Aging. Electronic products are exposed to atmospheric moisture. Temperature-humidity-bias aging is used to evaluate the product reliability. The potential failure mode is corrosion in the presence of humidity and electrical bias. Temperature increases the moisture level and is an accelerating factor in corrosion. The epoxy underfills are formulated with very low extractable chlorine levels to minimize the potential for corrosion. No-clean flux residues are

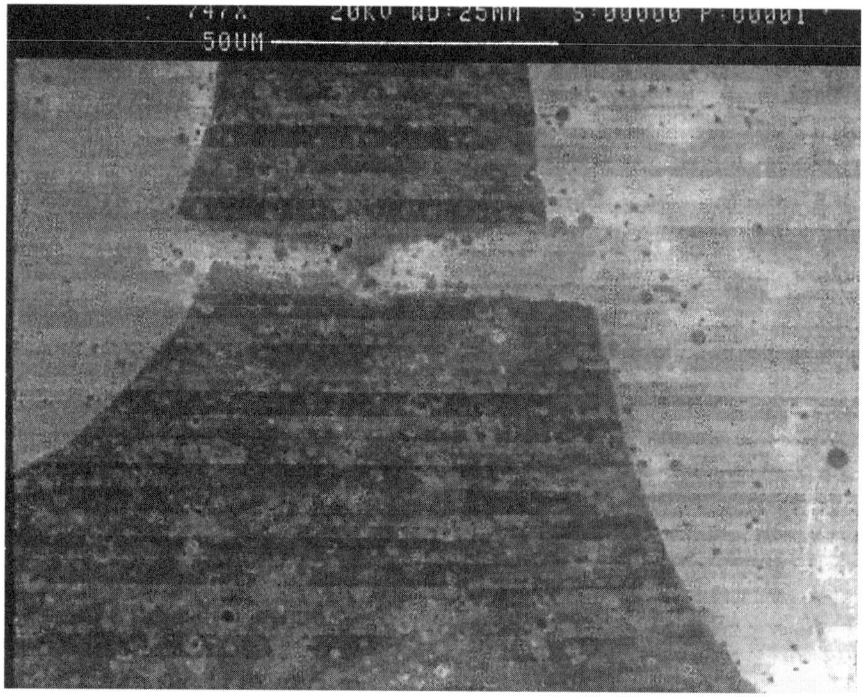

FIGURE 18.53 Flat section (die polished away) showing crack and solder extrusion in underfill between adjacent solder balls after 3200 thermal shock cycles.

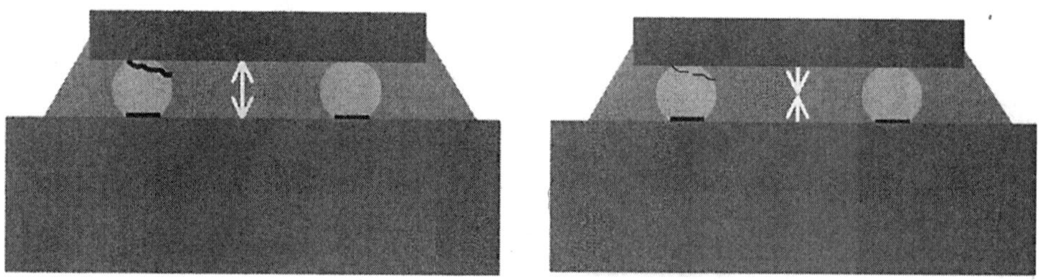

FIGURE 18.54 Illustration of the need for in situ monitoring.

also formulated to be chemically inert after the reflow cycle. A common test is 85 percent relative humidity at 85°C and 5 V for 1000 h. Failure is detected by electrical failure, shorts, and opens. Humidity and temperature also can result in degradation of underfill adhesion. Delamination can occur.

Autoclave and HAST. An *autoclave* is a pressure vessel used to accelerate humidity-temperature testing. A humidity level of 85 percent relative humidity can be achieved at 121°C under 2 atm of pressure. Other humidity, temperature, and pressure combinations also can be used. Adhesion loss and delamination are the common failure modes. Autoclave testing sometimes is used to rapidly screen underfill materials for moisture sensitivity. A highly accelerated stress test (HAST) is the addition of bias to the autoclave test. HAST accelerates humidity-temperature-bias aging tests.

High-Temperature Storage and Electromigration. High-temperature storage examines potential decomposition of the underfill and intermetallic growth in the solder joint. Figure 18.56 is a cross section of an FC solder joint after 3000 h of storage at 150°C. The intermetallic formation at the solder-to-PWB pad and the solder-to-underbump metallurgy (UBM) is evident. As mentioned previously, intermetallic formation can lead to dewetting, delamination, and embrittlement. Factors affecting reliability are UBM, UBM thickness, bump alloy, assembly processes, and service temperature. The UBM, bump alloy, and service temperature determine the intermetallic growth rate. The assembly process (number of reflow cycles, reflow profile, etc.) has an impact on the original intermetallic formation during assembly. Finally, the UBM thickness determines the amount of metal available for consumption. Since copper forms intermetallics with Sn/Pb eutectic solder at a faster rate than nickel, copper UBMs must be thicker than nickel UBMs.

If current flows through the solder bumps at elevated temperature, electromigration may occur. *Electromigration* is the movement of metal by momentum transfer from electrons. Increasing the current density (number of electrons per unit area) and temperature increases electromigration. In Sn/Pb eutectic solder, the Sn and Pb grains are separated by electromigration due to their different masses, as evident from Fig. 18.57. The concentration of Pb moves from the bottom of the solder ball to the top when the direction of current flow is reversed. From the plot, the maximum current a solder ball can carry decreases with increasing temperature and decreasing UBM/via size (increased current density).

Alpha Particles. An *alpha particle* is the nucleus of a helium atom (He^{4+}) that is emitted through radioactive decay. Trace amounts of thorium (Th), uranium (U), polonium (Po), and lead isotopes (^{214}Pb and ^{210}Pb) are the primary sources of alpha particles in tin-lead solder. A high-energy alpha particle (8 MeV) could generate up to 2.5×10^6 electron-hole pairs and cause soft errors in a sensitive semiconductor device, particularly memory. The *soft error* is not permanent damage to the device, but the data stored in the device are corrupted. To reduce soft errors, low-alpha-emission lead or lead- and bismuth-free solders can be used.

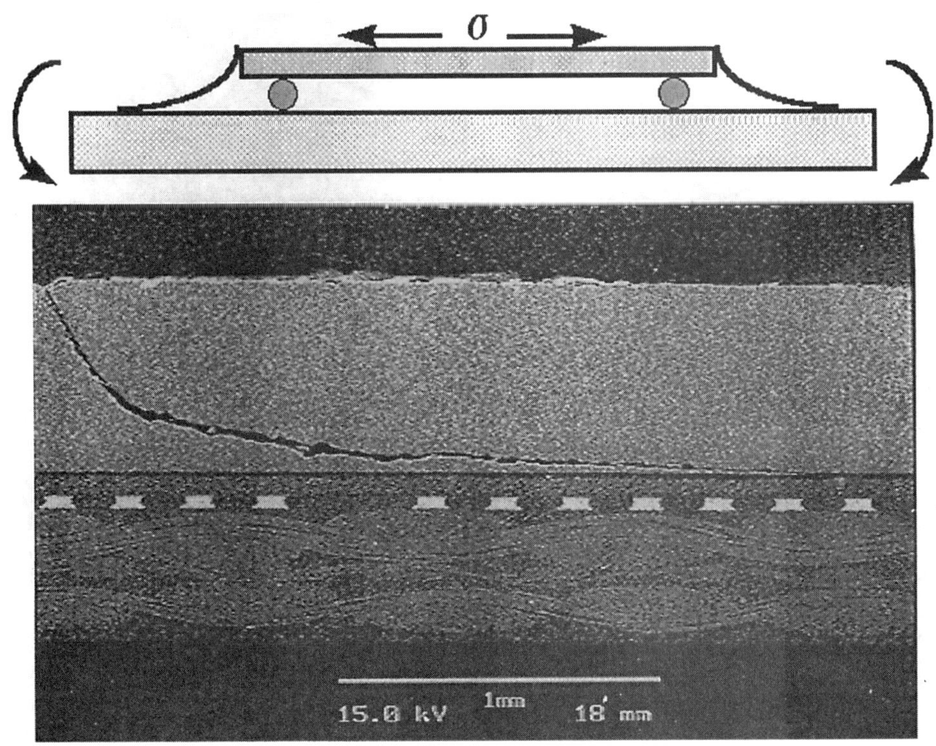

FIGURE 18.55 Example of FC die fracture. (*Flip Chip Technologies.*)

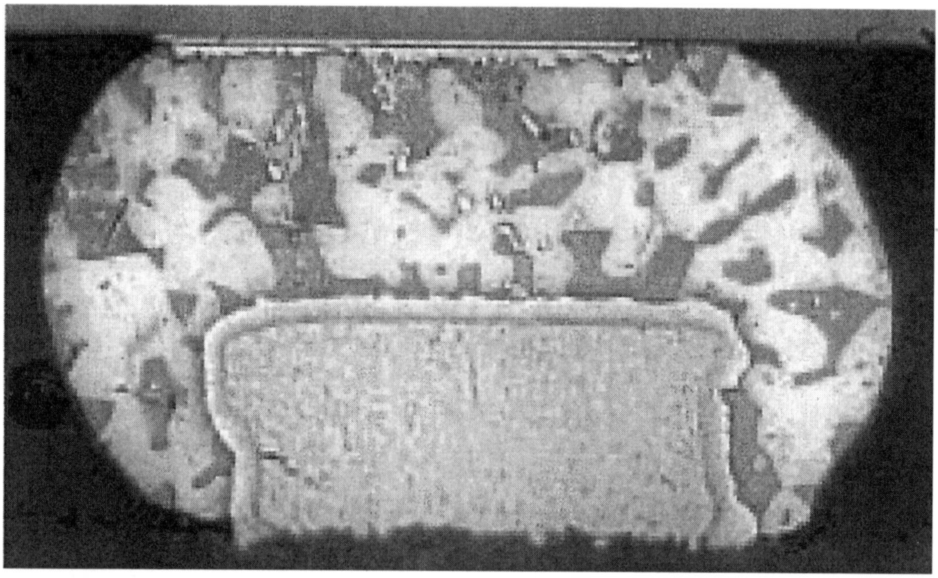

FIGURE 18.56 Cross section of solder ball aged at 150°C for 3000 h. (*Flip Chip Technologies.*)

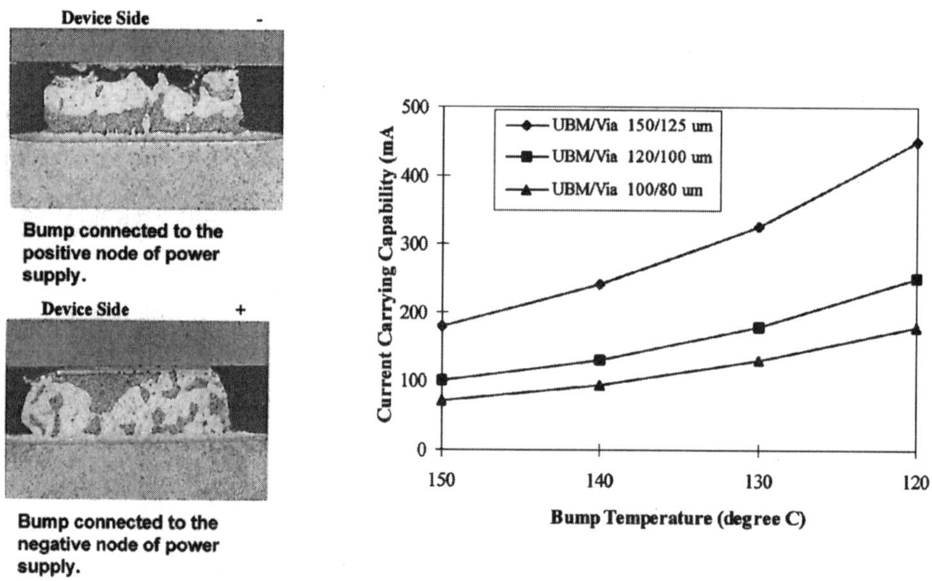

FIGURE 18.57 Electromigration test results and cross sections. (*Flip Chip Technologies.*)

18.7 SUMMARY

FC technology has been in use for nearly four decades. The continuing increases in semiconductor speed, die size, and I/O counts coupled with the explosion of the portable electronics market make FC an attractive assembly technology. While the current assembly process with capillary underfills is more complex than standard SMT assembly with packaged components, advances in materials and processes such as wafer-applied underfills may soon address this. Another limiting factor, the availability of bumped dies, is being addressed by the third-party bumping companies. Redistribution of perimeter I/O dies designed for wire bonding significantly adds to the bumping cost. In the future, more dies will be designed for FC assembly instead of wire bonding. This has already happened with microprocessor dies to meet the I/O requirements. Testing and burn-in of bumped dies still remain a hurdle to widespread use of FCs, but this too is being addressed by the semiconductor manufacturers.

The use of FC assembly will continue to grow as other approaches reach their limits and the issues mentioned herein are resolved.

18.8 REFERENCES

1. Totta, P. A., and Sopher, R. P., "SLT Device Metallurgy and Its Monolithic Extension," *IBM Journal of Research and Development,* May 1969, pp. 226–238.
2. Tummala, R. R., and Ramaaszewski, E. J. (eds.), *Microelectronics Packaging Handbook,* Van Nostrand Reinhold, New York, 1989, pp. 376–380.
3. Suryyanarayana, D., Hsiai, R., Gall, T. P., and McCreart, J. M., "Enhancement of Flip Chip Fatigue Life by Encapsulation," *IEEE Transactions on Components, Hybrids, and Manufacturing Technology,* 14 (1), March 1991, pp. 218–223.
4. Suryyanarayana, D., Wu, T. Y., and Varcoe, J. A., "Encapsulants Used in Flip Chip Packages," *IEEE Transactions on Components, Hybrids, and Manufacturing Technology,* 16 (8), December 1993, pp. 858–862.
5. Clementi, J., McCreary, J., Niu, T. M., et al., "Flip Chip Encapsulation on Ceramic Substrates," in *Proceedings of the 43rd Electronics Components Conference,* Orlando, FL, June 1–4, 1993, pp. 175–181.
6. Hu, K. X., Yeh, C. P., Doot, R., Skipor, A. F., and Wyatt, K. W., "Die Cracking in Flip-Chip-on-Board Assembly," in *Proceedings of the 45rd Electronics Components Conference,* Las Vegas, NV, May 21–25, 1995, pp. 293–304.
7. Novak, P., "Process Limit Testing on Fluxes Used for Flip Chip Soldering," in *Proceedings of the 1998 Surface Mount International,* San Jose, CA, August 23–27, 1998, pp. 275–280.
8. Qi, J., Johnson, R. W., Yaeger, E., et al., "Manufacturability Issues in Flip Chip on Laminate Assembly," *International Journal of Microcircuits and Electronic Packaging,* 22 (3), 1999, pp. 270–279.
9. Sellers, R., Gibson, M., and Gullion, D., "Cleaning for High Reliability Flip-Chip-on-Laminate Assembly," in *Proceedings of the 1998 Surface Mount International,* San Jose, CA, August 23–27, 1998, pp. 325–336.
10. Lewis, B., "Process Characterization and the Effect of Process Defects on Flip Chip Reliability," in *Proceedings of the Electronics Assembly Process (APEX) Conference,* Long Beach, CA, March 12–16, 2000, pp. AP2/3-1–AP2/3-6.
11. Schwiebert, M. K., and Leong, W. H., "Underfill Flow as Viscous Flow Between Parallel Plates Driven by Capillary Action," *IEEE Transactions on Components, Packaging and Manufacturing Technology, Part C,* 19 (2), April 1996, pp. 133–137.
12. Emerson, J. A., and Jones Adkins, C. L., "The Tools You Need for Development and Production When Using Underfill Materials," in *Proceedings of the 1998 Surface Mount International Conference,* San Jose, CA, August 23–27, 1998, pp. 270–274.
13. Crane, L., Torres-Filho, A., Ober, C., and Johnson, W., "Development of Reworkable Underfills, Materials, Reliability and Processing," *IEEE Transactions on Advanced Packaging,* 22 (2), 1999, pp. 111–115.
14. Kulkarni, P., Johnson, R. W., Yaeger, E., et al., "Manufacturing of Flip Chip-on-Laminate," in *Proceedings of the APEX Technical Program,* Long Beach, CA, March 12–16, 2000, pp. AP3/2/1–AP3/2/5.
15. Pennisi, R., "Adhesive and Encapsulant Material with Fluxing Properties," U.S. Patent No. 5,128,746, July 7, 1992.
16. Katze, D., "No-Flow Fluxing Underfill Process Characterization," in *Proceedings of the APEX Technical Program,* San Diego, CA, January 16–18, 2001, pp. MT2-21–MT2-29.
17. Previti, M. A., "No-Flow Underfill: A Reliability and Failure Mode Analysis," in *Proceedings of the APEX Technical Program,* San Diego, CA, January 16–18, 2001, pp. MT2-11–MT2-15.

18. Pascarella, N. W., and Baldwin, D. F., "Compression Flow Modeling of Underfill Encapsulants for Low Cost Flip Chip Assembly," *IEEE Transactions on Components, Packaging, and Manufacturing Technology, Part C,* 21 (4), 1998, pp. 325–335.
19. Leider, P. J., and Bird, R. D., "Squeezing Flow Between Parallel Disks. I. Theoretical Analysis," *Industrial Engineering Chemistry Fundamentals*, 13, 1974, pp. 336–341.
20. Leider, P. J., and Bird, R. D., "Squeezing Flow Between Parallel Disks: II. Experimental Results," *Ind. Eng. Chem. Fundam.,* 13, 1974, pp. 342–346.
21. Chason, M., Danvir, J., Yala, N., et al., "Development of Wafer Scale Applied Reworkable Fluxing Underfill for Direct Chip Attach," in *Proceedings of the APEX Technical Program,* San Diego, CA, January 16–18, 2001, pp. AT6-11–AT6-15.

CHAPTER 19
BGA AND CSP REWORK: WHAT IS INVOLVED?

Paul Wood and Howard Rupprecht
Metcal, Inc.

19.0 INTRODUCTION

As array package components become increasingly accepted in mainstream production, one of the most commonly asked questions is, "How can I rework these new components?" Many existing installed QFP rework machines may not have the required features for successful ball grid array (BGA) or chip scale package (CSP) rework. While rework of these components is not difficult, there are certain pitfalls that can be avoided if the engineer understands the method of interconnection, the different rework options, and the application of the product being reworked.

19.1 REWORK PROCESS CONSIDERATIONS

19.1.1 The Need for Process Control

If we compare a peripheral lead package such as a QFP with an array package such as a BGA or CSP, the fundamental difference is that one has leads on the perimeter and the other has an array of solder balls on its underside—functionally the packages are the same. If we are reworking a QFP and a defect occurs, it has always been possible to "touch up" that defect with a conventional soldering iron, repairing the fault. Engineers traditionally have had this safety net that often has meant that less process control was applied to the rework operation.

With any array package, when it is assembled, there is no possible way to "touch up" a defective solder joint—making complete package removal mandatory. Since none of us want to be perpetually removing and replacing devices, a greater level of process control is required to ensure that we achieve the perfect result on the first attempt.

Packaging trends such as increased input-output (I/O) count and reduced pitch have heightened the need for this level of process control, with components on the market with 1600 solder balls and pitches as low as 0.3 mm. On a package such as this, low DPM levels are absolutely essential for successful rework as well as production.

Another thing to consider is that rework normally is carried out at the end of the production process, possibly at final test stage. At this point any printed wiring board (PWB) assembly has reached its highest built-in value, making it a significant financial risk if a board is scrapped through uncontrolled rework.

19.1.2 What Is the Interconnection Method?

The primary consideration for any engineer who is reworking an array package is to know how that package is interconnected to the PWB or other substrate. There are many different package types from various manufacturers on the market, and understanding these is the key.

We have all heard the acronyms PBGA, CBGA, CCGA, CSP, and TBGA, but what do they mean in terms of interconnection? Components generally fall into one of two categories: those with a collapsible (melting) solder ball and those with a nonmelting ball or column.

Collapsible Solder Balls. The most common type of interconnection currently in use is the collapsible solder ball (Figs. 19.1 and 19.2). This ball generally is made from eutectic tin-lead solder (63/37 typically) and is designed to melt during the assembly reflow process. The actual component will "float" on the array of molten solder balls during reflow but will solidify to support the component. This is currently the accepted method on most plastic BGAs, most CSPs, and some tape BGAs, but there are exceptions to the rule.

Nonmelt Solder Balls and Columns. These normally are made from 90Sn/10Pb solder and generally are used in applications where the component is physically heavy, is in a high-reliability environment, or has extraordinary heat-dissipation requirements. In production reflow, these solder balls remain solid (they melt at approximately 300°C) and serve to support the component at the standoff height intended (Fig. 19.3). The solder paste printed on the PWB will melt and form the necessary intermetallic connection.

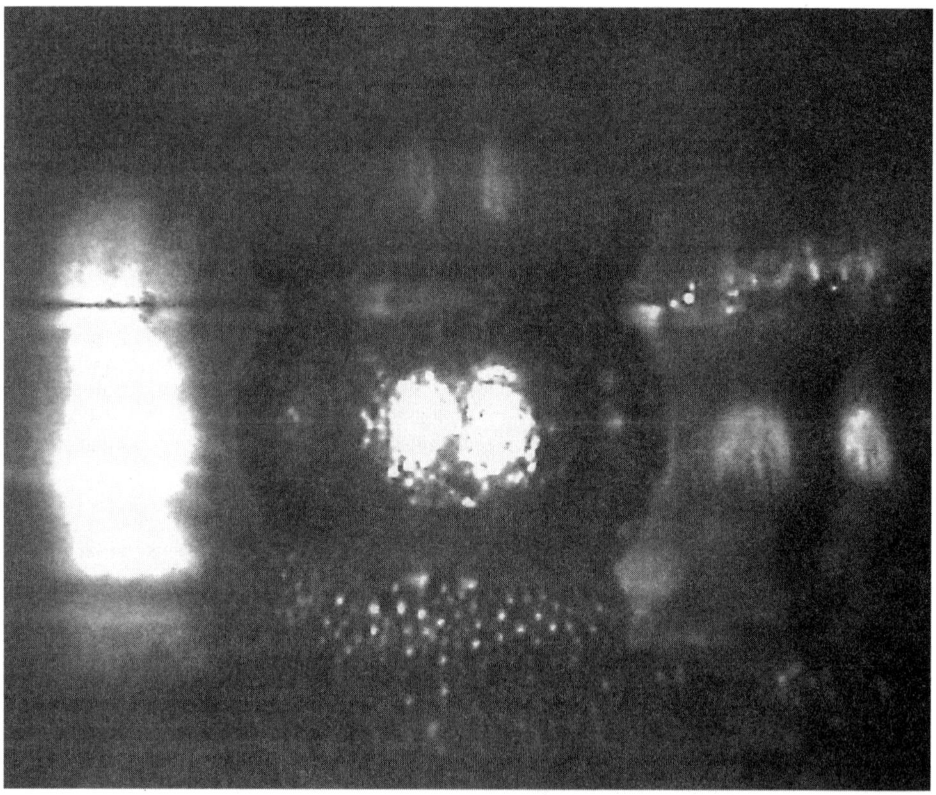

FIGURE 19.1 Eutectic solder ball and paste before reflow.

FIGURE 19.2 Eutectic solder ball and paste after reflow.

FIGURE 19.3 Nonmelt solder ball interconnection after reflow.

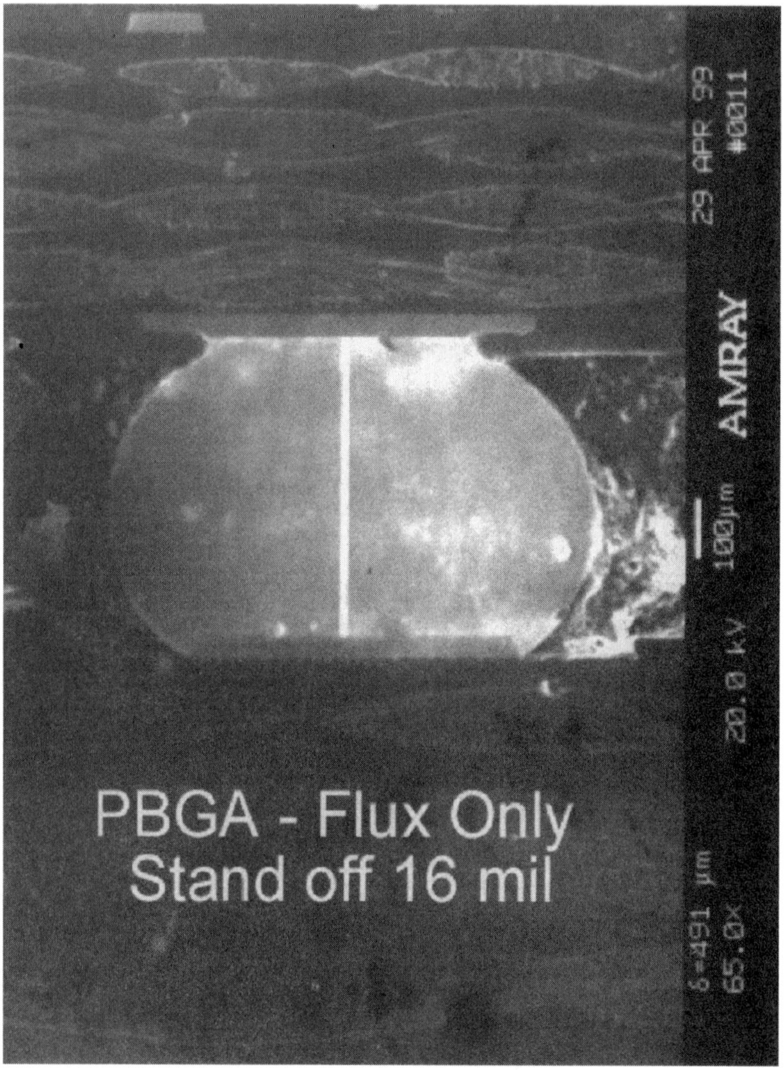

FIGURE 19.4 Reflowed solder joint with flux only.

This type of connection generally is found on ceramic BGAs, ceramic column grid arrays, and some tape BGAs and CSPs. Again, there are exceptions to the rule, and the component specification sheet should always be checked if one is unsure.

19.1.3 Do I Use Solder Paste or Flux?

The primary choice an engineer has when reworking is whether or not he or she uses solder paste or flux for component reconnection. In all cases, the main manufacturing process would be to use solder paste, since array package assembly is essentially the same process as surface mount technology (SMT).

If the solder ball/column of the component is nonmelt solder, then solder paste must be applied to make the solder joint (Fig. 19.4). The engineer can either print or dispense solder paste onto the pad area prior to component replacement. This is directly copying the production process.

If the component has a collapsible eutectic solder ball, the engineer has a choice of using either paste or flux (Fig. 19.5), depending on his or her own considerations. Things to consider:

- *Standoff height.* The height that the component stands off the board will depend mainly on the volume of solder. Normal standoff height of a 30-mil solder ball attached with paste would be about 20 mils after reflow. If only flux was used, this may reduce to 16 mils or less. If you have any concerns about meeting JEDEC standards, cleaning under the package, or heat dissipation, then this may not be acceptable.
- *Reliability.* Various independent tests have shown that there will be a difference in long-term reliability depending on standoff height. As a rule, it is generally considered that the higher the solder joint, the more compliant or flexible it will be.

FIGURE 19.5 Reflowed solder joint with solder paste.

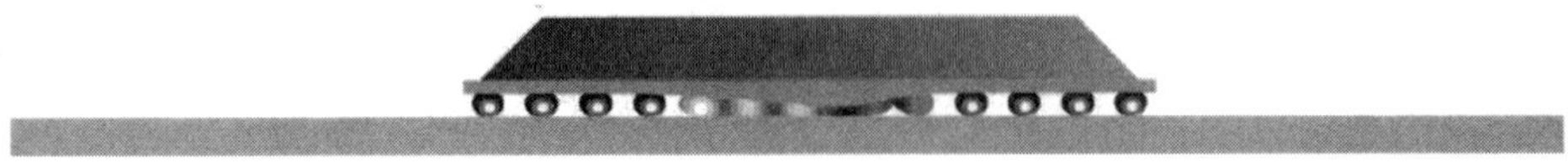

FIGURE 19.6 Popcorn effect, delamination of component.

In practical terms, if you are working with a product that is destined for domestic use for a limited time, then flux probably will prove the easiest to apply and will offer more than adequate long-term reliability. Conversely, if you manufacture long-service, harsh-environment products, such as products for use in the aerospace industry, then flux probably would not be an acceptable method.

19.2 BASIC REWORK STEPS

19.2.1 Prebaking Components and PWBs

This step is often overlooked due to its time-consuming nature, but it can be extremely important in successful rework. Most plastic packages and PWBs are hydroscopic and will absorb moisture from the atmosphere during their life cycle. If the assembly is heated rapidly, as it would be during a rework operation, there may be a possibility of the moisture expanding to delaminate the component internally. This effect is known as *popcorning* and will lead to component failure.

If a board or component is suspect, it should be baked in an oven at low heat ($\approx$100°C) for a number of hours to allow the parts to dry slowly prior to any rework operation. A number of large computer manufacturers rigorously enforce this process in their repair centers.

Prevention is always better than cure, and the key with components is to ensure that they are stored correctly. Most components are supplied in sealed packages, and once the packages are opened, the components should be used quickly or resealed for long-term storage.

Popcorning is characterized by a blister on the underside of the component as a result of the internal expansion (Fig. 19.6). Under x-ray inspection, you would normally see solder shorts toward the center of the device.

19.2.2 Component Removal

The primary heating method used by most rework systems is forced air/gas convection, although there are still some infrared systems available. Rework has followed reflow oven technology, where convection is preferred due to better temperature uniformity across dissimilar surface textures and colors. Convection is also considered to be easier to control for profiling and offering compatible ramp rates and dwell times.

To remove a component, use of a reflow nozzle that surrounds the component is preferred (Fig. 19.7). Generally, adjacent components around the nozzle should not reflow, although this will be impractical on very densely packed boards such as mobile phones and pagers. Using a low-velocity airflow system can minimize the problem of adjacent reflow (Fig. 19.8).

The removal process should be designed carefully so that the PWB does not become thermally stressed or warped. Generally, a removal profile can be shorter than a replacement profile because we are not concerned about solder paste characteristics, and in some cases, we will be disposing of the removed component. Care should be taken to allow even preheating on thermally demanding PWBs. This ensures good temperature uniformity, especially on the inner layers of a multilayer PWB (Fig. 19.9).

Most rework systems use a vacuum pickup to physically lift the component at the end of the removal cycle. It is important to ensure minimal vacuum pressure because this may cause solder collapse and extra cleaning operations.

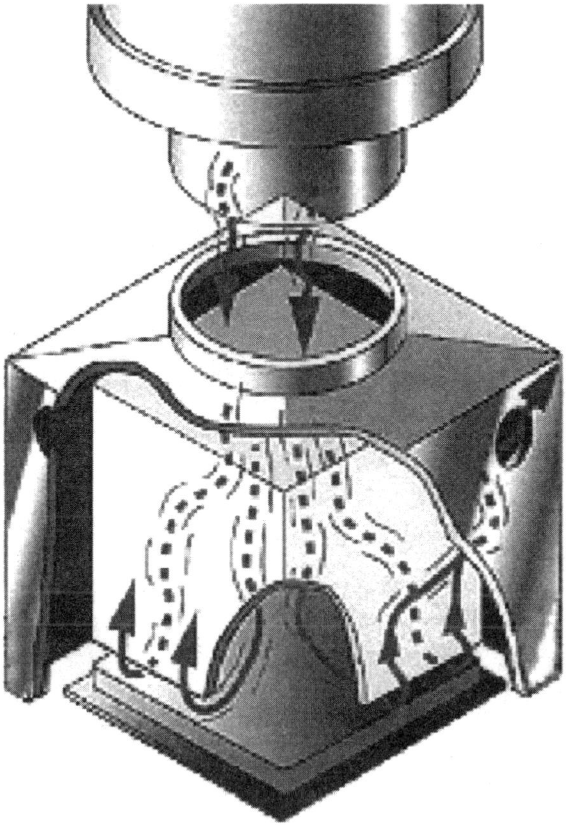

FIGURE 19.7 Principle of microoven reflow head.

19.2.3 Pad Cleaning and Preparation Methods

Once a component has been removed, generally, a certain amount of cleanup will be required to prepare the area for reapplication of solder paste or flux. Ironically, this is the process that presents the greatest risk to the PWB, normally as a result of pad, track, or solder resist damage (Fig. 19.10).

There are four recognized methods of solder removal:

1. *Desolder gun.* This method is time-consuming, and direct contact with the conduction tip may cause solder resist damage.
2. *Solder wick with blade tip.* This method is used commonly and can be very successful with a skilled operator using a controlled soldering iron (Fig. 19.11). Pad finish generally is flat, making the surface ideal for stencil printing.
3. *Solder reapplication with hoof tip.* This method allows the operator to reapply solder to the pads using a concave soldering tip (Fig. 19.12). This method is suitable for flux-only applications and on small pads; a high degree of uniformity can be achieved.
4. *Solder "scavengers."* These systems effectively hot-air-level the pads while sucking off any excess solder. Results are very consistent, although the cost of equipment may be an issue. This type of system is ideal for removal of nommelt solder balls.

Once excess solder removal is complete, a cleaning operation will be required to remove flux residues prior to paste printing or new flux application.

FIGURE 19.8 Microoven on densely packed PWB.

19.2.4 Material Deposition Techniques

Once the decision to apply either solder paste or flux has been made, the engineer has to decide the most appropriate method of material deposition. There currently are four recognized methods for doing this.

1. *Brush application.* This method is the quickest option but is only suitable for applying fluxes. In most cases, a gel flux should be used because this will provide some adhesion to the component during reflow and will not evaporate too quickly. Brush application has the advantage of being quick and requires no component-specific tooling. Unfortunately, this process will leave large amounts of residue, probably necessitating a subsequent cleaning operation. The amount of flux applied is also imprecise and operator-dependent—too much flux may cause the component to float, and too little will result in unsoldered joints.
2. *Dispensing.* This process can be done either manually or automatically and can be used for both solder paste and flux. Manual dispensing is considered impractical due to its time-consuming nature and operator dependency. Automatic dispensing with an *xy* table will yield a consistently good result, but equipment costs may make this economically unfeasible.
3. *Printing.* This method can be done either manually or using vision assistance and generally is used for solder paste deposition. The process requires component/board-specific stencils that match the land pattern and any surrounding components. Advances have been made in single-component printing, making it relatively easy to print solder paste even on fine-pitch CSPs. A system that uses vision-mounted stencils will be much easier to align under high magnification. Coplanarity should be able to be adjusted to ensure absolute flatness of the stencil to the PWB. The actual print cycle is done using a metal squeegee, manually drawn across the stencil surface (Figs. 19.13 and 19.14). With minimal operator training, a high degree of consistency can be achieved.

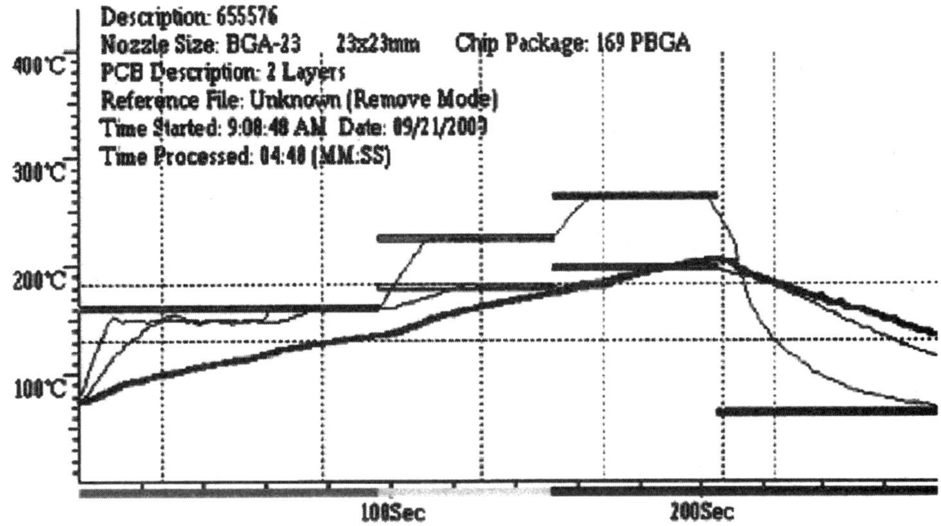

Time	Air	T.C.	PreH
00:27	150°C	100°C	152°C
01:18	161°C	129°C	162°C
02:09	225°C	162°C	181°C
02:48	264°C	181°C	186°C
03:27	241°C	205°C	195°C
03:44	130°C	184°C	180°C

FIGURE 19.9 Graph of typical reflow profile.

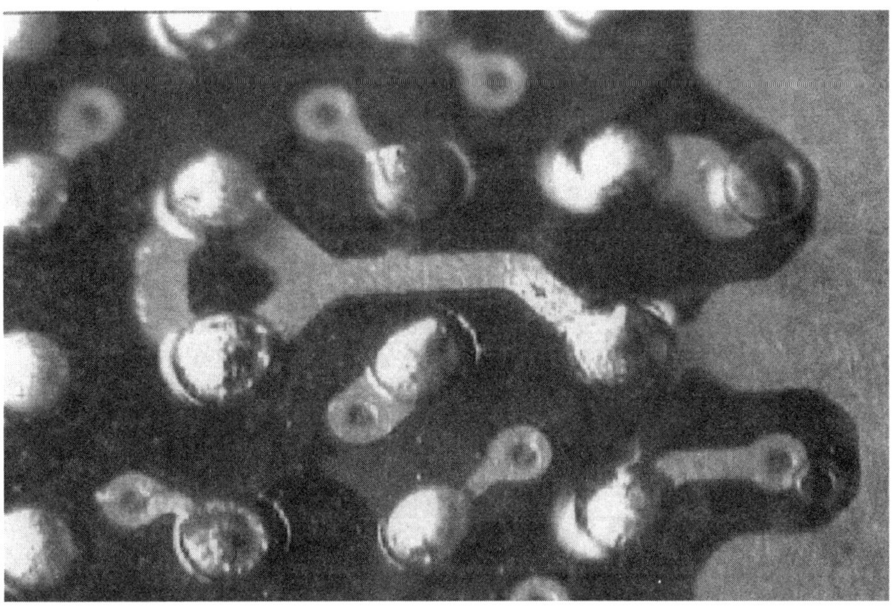

FIGURE 19.10 Solder resist damage.

FIGURE 19.11 Removal of excess solder using blade tip.

FIGURE 19.12 Using a concave soldering tip to clean pads.

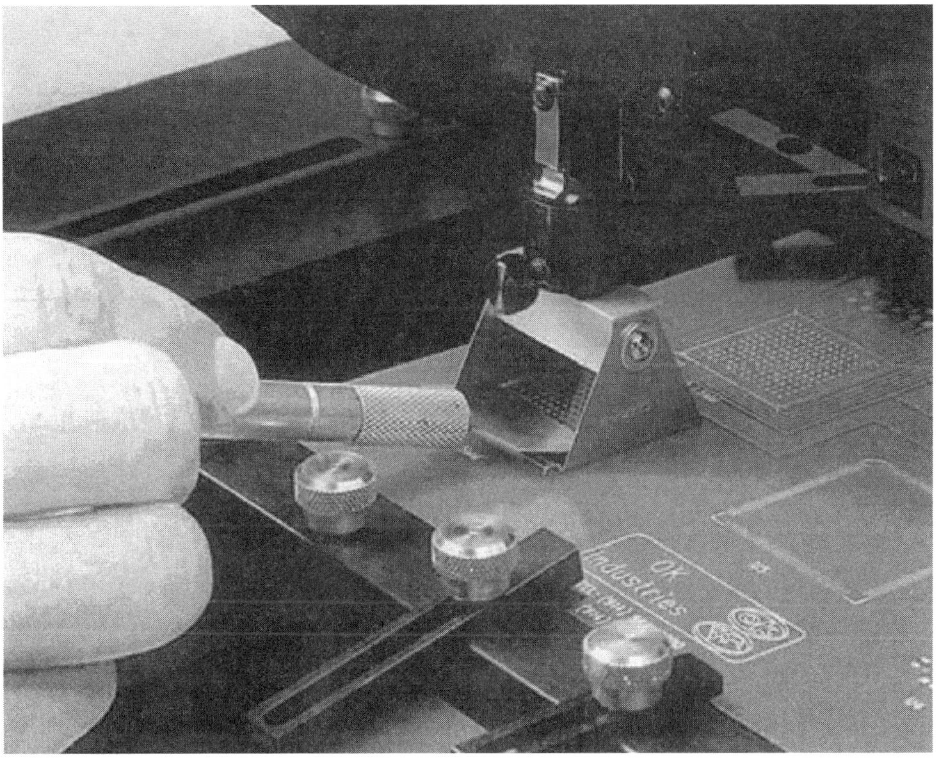

FIGURE 19.13 Manual solder paste printing.

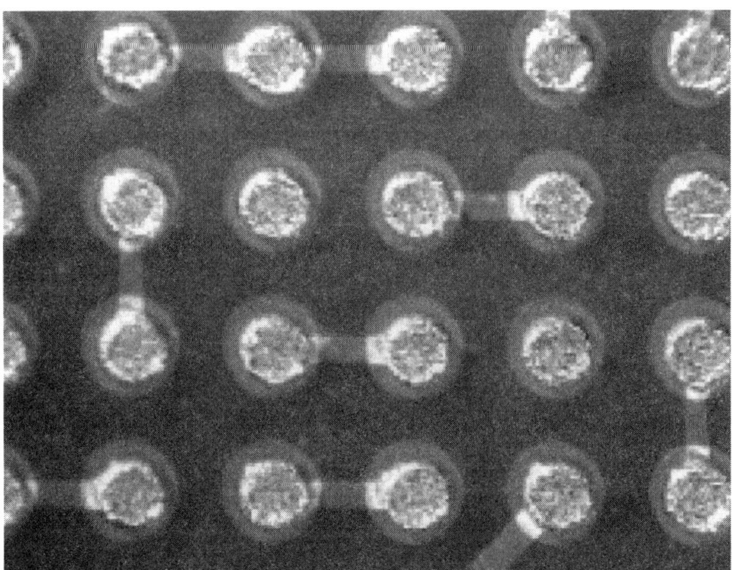

FIGURE 19.14 Photograph of successful manual print.

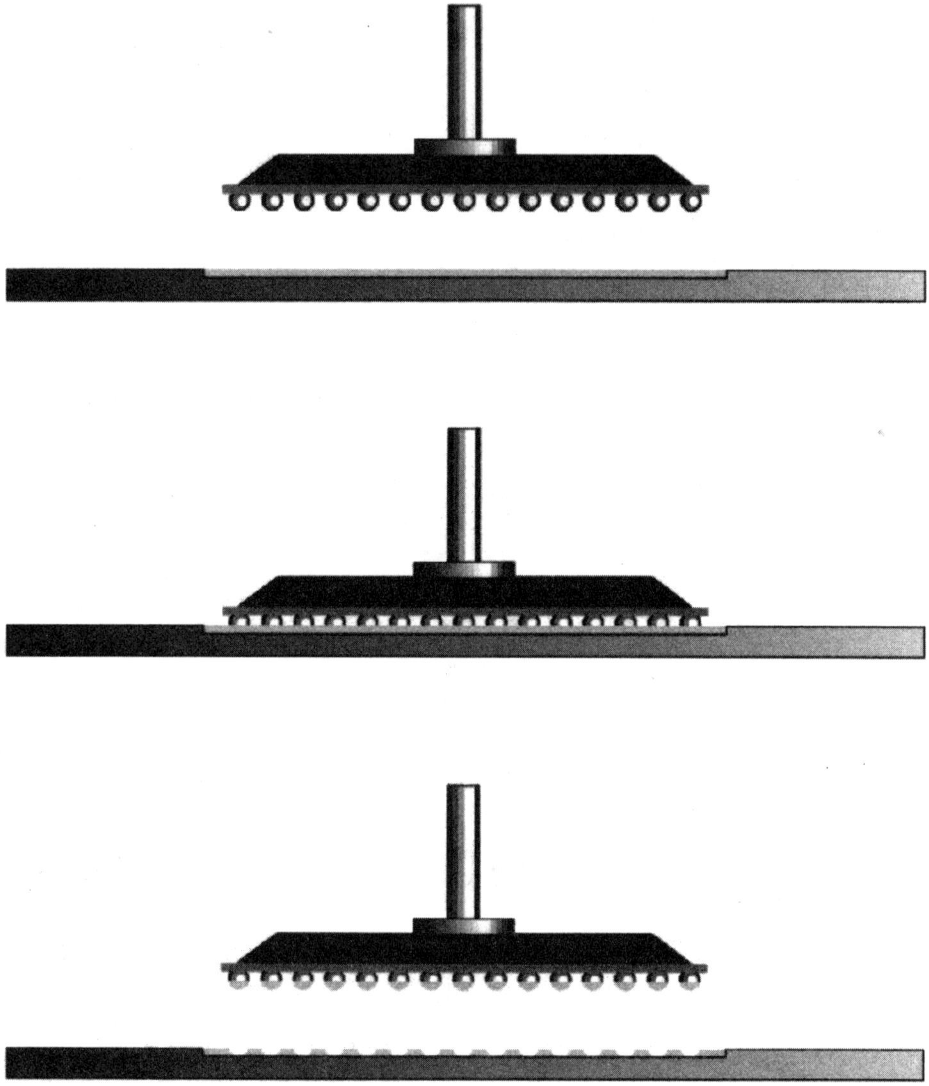

FIGURE 19.15 The steps of the dip transfer process.

4. *Dip transfer techniques.* This process involves dipping the component to be placed into a known depth of gel flux (results with solder paste are not practical) (Fig. 19.15). The technique is relatively new, being pioneered for use in this application by Salford University and OK International/Metcal, Inc. The advantages are that a very consistent flux deposit can be achieved very quickly. The process is becoming widely accepted as an efficient means of applying flux.

19.2.5 Component Placement

Array packages initially had relatively large ball pitches (typically 60 mils); this, combined with the strong self-centralization characteristics of the BGA, made hand placement a workable option for simple components. As component ball pitches continue to get smaller, vision-assisted placement is becoming essential in reliable rework.

The most common placement method in rework is to use a split-field prism system (Fig. 19.16). This allows the user to look simultaneously at the underside of the component and the topside of the PWB. The two images can be superimposed and accurately aligned on a monitor before placement. Care should be taken to ensure that the component is placed at the correct pressure.

19.2.6 Component Reflow Profiling

During component replacement, we try to emulate the original production reflow profile to match the required profile characteristics of the solder paste or flux being used. This normally would involve designing a four-stage profile as follows:

1. *Preheat stage.* The purpose of this stage is to get heat into the assembly and bring it up to flux activation temperature. A typical target would be to reach 120°C within 30 to 90 s, depending on the type of board.
2. *Soak stage.* The purpose of the soak stage is to allow the fluxes to activate at an elevated temperature, removing oxides and contaminants prior to soldering. Time is important because an overly extended soak time may "dry" out the solder paste, causing reoxidation. A typical soak would range from 120 to 170°C and would last between 30 and 90 s.
3. *Reflow stage.* In the reflow stage, the requirement is to bring the component to a temperature of between 200 and 210°C to allow soldering (63Sn/37Pb melts at 183°C). This time generally would be between 30 and 60 s.
4. *Cooling stage.* Having a cooling stage is important because overly extended reflow times will cause changes in the solder joint structure that could affect reliability. There is also a practical

FIGURE 19.16 A split-field imaging system.

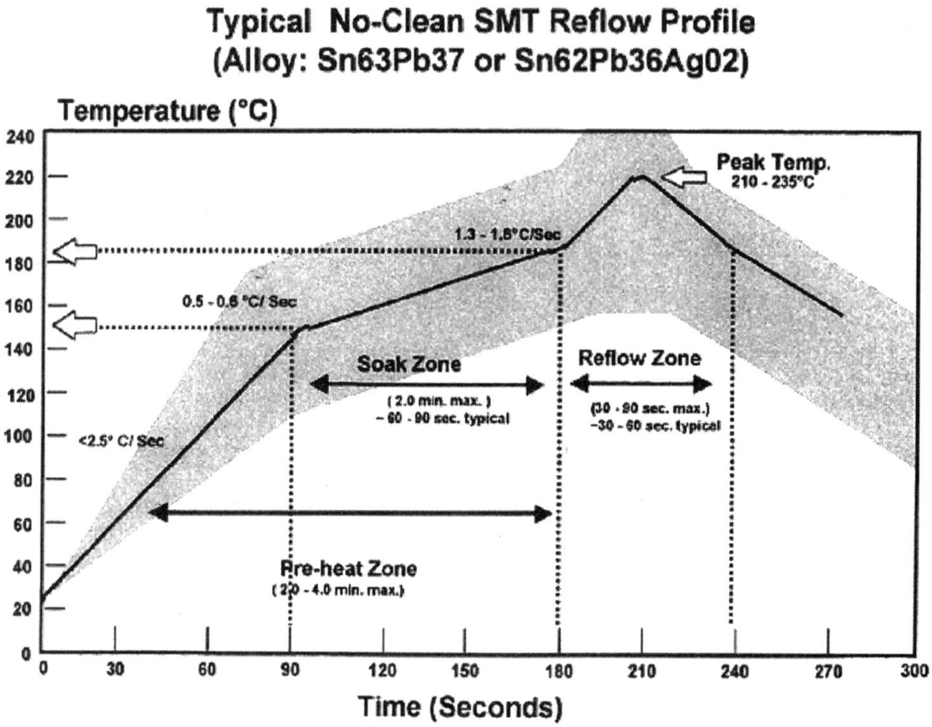

FIGURE 19.17 Typical solder paste process window. (*Kester solders.*)

side to ensuring that the board is back below reflow temperature before operator handling. The rework system should direct cold air through a separate air chamber because continuation through the heaters, even after they have been switched off, will promote extended reflow (Fig. 19.17).

19.3 IDENTIFICATION OF PROCESS YIELD PROBLEMS

If we consider profiling a PWB for reflow in production, we generally are concerned with achieving an even ΔI across all the joints being soldered. This is accomplished by evenly heating both the top and bottom sides of the whole board, generally through a multizoned convection reflow oven. The capacity of the oven would be appropriate to the size or thermal mass of the PWB and the production throughput requirements.

When we consider reworking an array package such as a BGA or CSP, it is recognized that process control is of paramount importance to achieve a successful result. As discussed earlier in this chapter, the normal way to approach the rework is to try to emulate the production reflow profile for the individual component being removed or replaced. The nature of the operation dictates that we only wish to reflow the component being reworked, and therefore, we have to heat a specific area of the circuit board. Selective heating of an area of board can inadvertently cause process failure even when all the parameters appear to be correct.

19.3.1 Balancing Profile Settings

On any reflow soldering system that heats both sides of the PWB assembly, the required solder-joint temperature is a function of how much heat is applied to both the top and bottom sides. It is quite

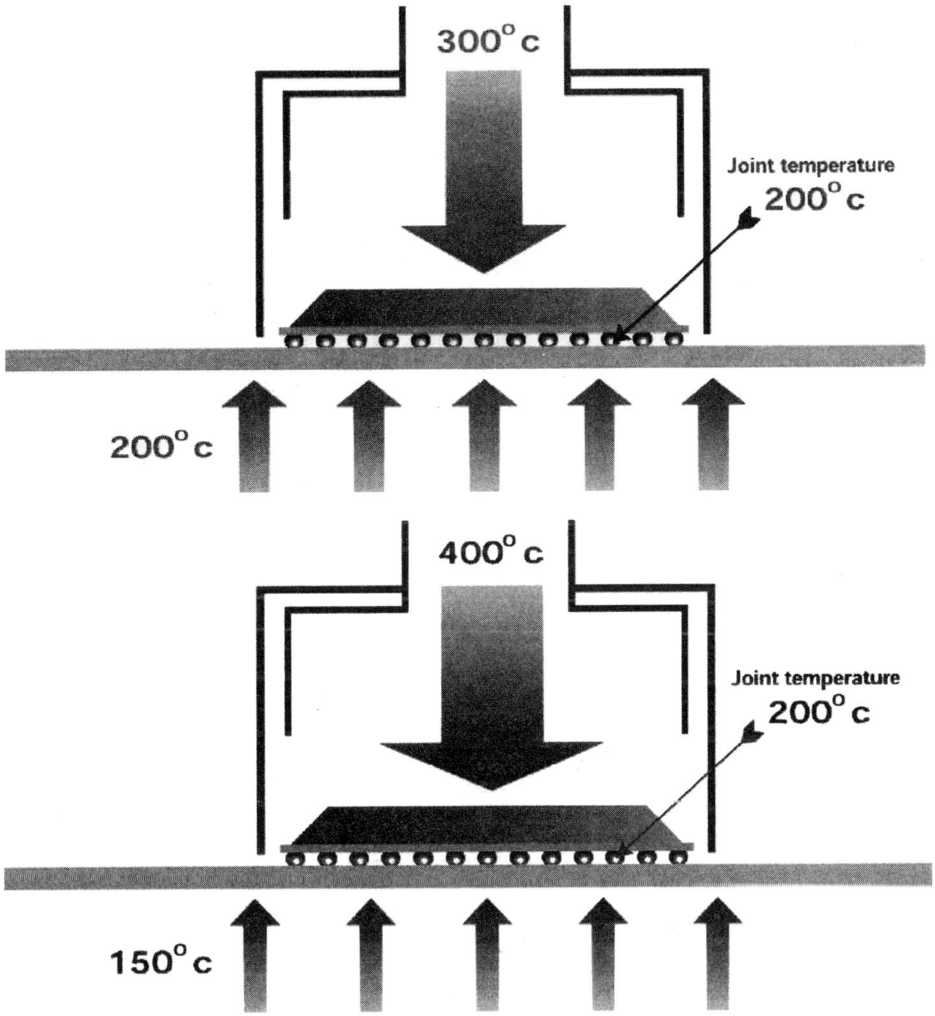

FIGURE 19.18 Same solder-joint temperature, different settings.

possible to achieve the same solder-joint temperature with a number of heat settings (Fig. 19.18). The following two examples illustrate this.

Practical Case Study. The graph in Fig. 19.19 shows a trace of a profile taken from under a 492 PBGA on a PC motherboard where the thermocouple trace appears to be normal—hitting the desired preheat, soak, and reflow temperatures. In this case, with either solder paste or flux, a very low pass rate was achieved.

The graph in Fig. 19.20 is the final profile that was used and has the original trace overlaid. As you can see, the two traces are very similar, but the heating has been done in a totally different way. In this case, we have reduced the top air temperature significantly while increasing the under-board heat to the whole board.

The root of the yield problem was that the PBGA warped during reflow. An excessive source temperature on the topside of the PWB caused this.

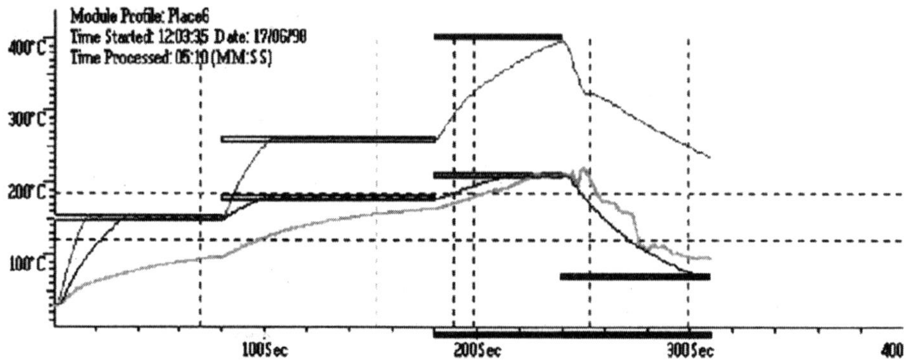

Time	Air	T.C.	PreH
01:10	150°C	093°C	151°C
02:32	259°C	155°C	180°C
03:09	295°C	171°C	185°C
03:18	324°C	179°C	195°C
04:13	322°C	206°C	170°C
04:59	249°C	098°C	075°C

FIGURE 19.19 Unsuccessful profile.

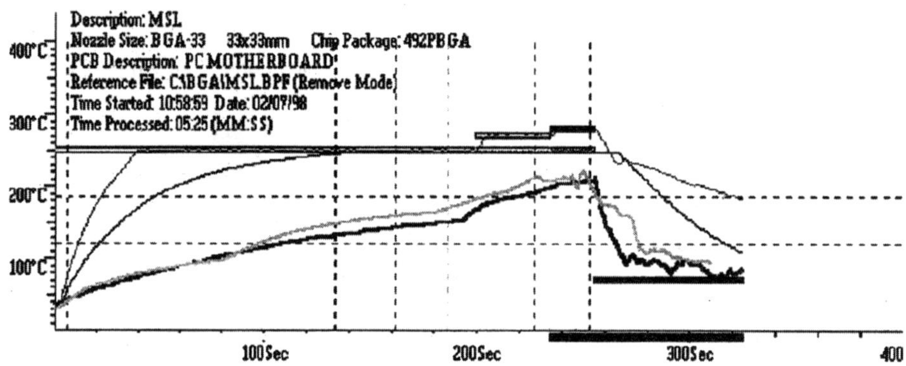

Time	Air	PreH	New	Old
00:06	078°C	057°C	042°C	040°C
02:14	249°C	245°C	131°C	147°C
02:42	249°C	246°C	142°C	157°C
03:06	249°C	246°C	149°C	168°C
03:47	269°C	246°C	190°C	210°C
04:13	278°C	246°C	206°C	206°C

FIGURE 19.20 Successful profile.

The following subsection offers an explanation of the kind of faults that can occur when everything appears to be set correctly (i.e., perfect screen print, no board damage, and accurate placement) but the yields are low.

19.3.2 Problems Caused by Warping

A major cause of defects is warpage of either the PC board or the component during reflow. Even minor deformation can lead to defects such as unsoldered joints and solder shorts. Component lifting by even 0.005 in across the device is enough to cause an open circuit. Another factor is that the larger the device, the more prone the device will be to these problems. Even if the board/component survives the process with no apparent defect, then the joint will be constantly under strain as the board returns to its normal shape—causing long-term reliability problems.

Figure 19.21 shows unsoldered joints caused by PWB warpage, the defect being caused by component warpage. The joints in the center of the component have lifted above the PWB and were not able to make the solder joint. This is more typical when only flux is used because there is no paste to make up the void area. Another characteristic is that the opens generally will be located toward the center of the device.

Figure 19.22 shows solder shorts. In this case, the deformation has caused the solder paste and the solder balls to combine into solder shorts. This is more typical with solder paste and normally is characterized by shorts on the device periphery, especially in the corners when the device becomes "dog-eared" during reflow.

The solution to the problem of warping generally is related to the profile settings, particularly insufficient under-board preheat temperature combined with too high a top temperature. Some boards, however, are badly balanced thermally, and this also can lead to problems.

On large PWBs, under-board support also can be a factor, with the PWB sagging under its own weight. Note that downward pressure should not be applied by eiher the reflow head or vacuum because this will exacerbate any problems.

19.3.3 Simultaneous Reflow Issues

The reason that manufacturers strive to attain simultaneous reflow is that having a large ΔT across a component can lead to rework and production defects. Some factors can be influenced directly by the profile settings, whereas others are determined by the physical makeup of the PWB assembly. The first factor that will influence how "simultaneous" the reflow is will be profile temperatures at the point of reflow—more specifically, how quickly we pass from solid to liquid state of the solder (Fig. 19.23).

Figure 19.24 shows a graph section of a component's solder joints going from solid to liquid at a melting temperature of 183°C. We are assuming that the ΔT achieved across the component is 10°C.

The rate of temperature rise (ramp rate) will determine the duration of reflow from start to finish. This is directly proportional to the input temperatures set on the profile. In this case, the ramp rate is 0.5°C/s.

The solution would involve changing the profile input temperatures to achieve a higher ramp rate. This, in turn, would produce a more simultaneous reflow, as illustrated in Fig. 19.25.

There are other factors that will influence ΔT on any PWB assembly, and these are otten out of the control of the rework machine manufacturer. Position of the component on the PWB is a major

FIGURE 19.21 Unsoldered joints caused by PWB warpage.

FIGURE 19.22 Solder shorts caused by PWB warpage.

FIGURE 19.23 Uneven reflow caused by poor temperature uniformity.

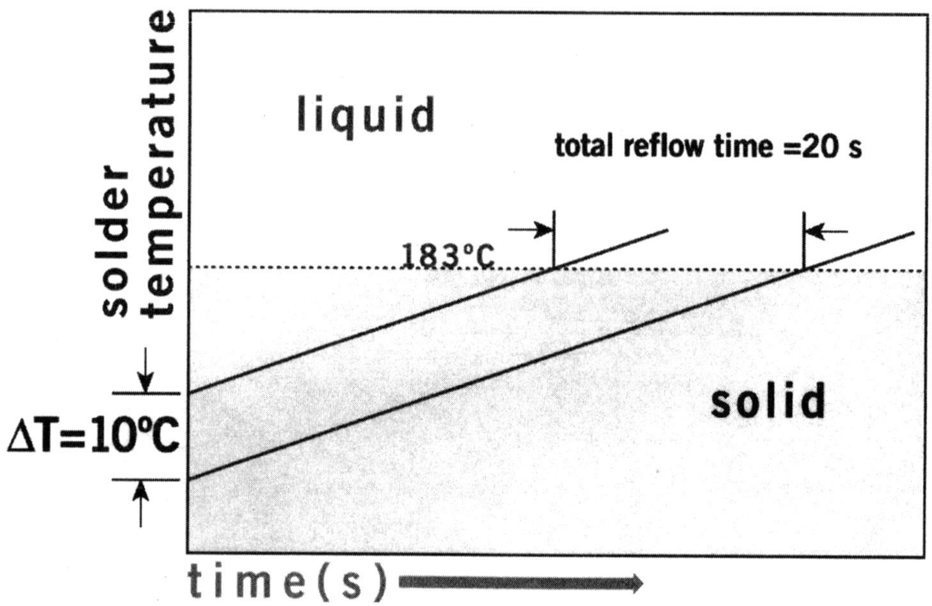

FIGURE 19.24 Slow transition from solid to liquid.

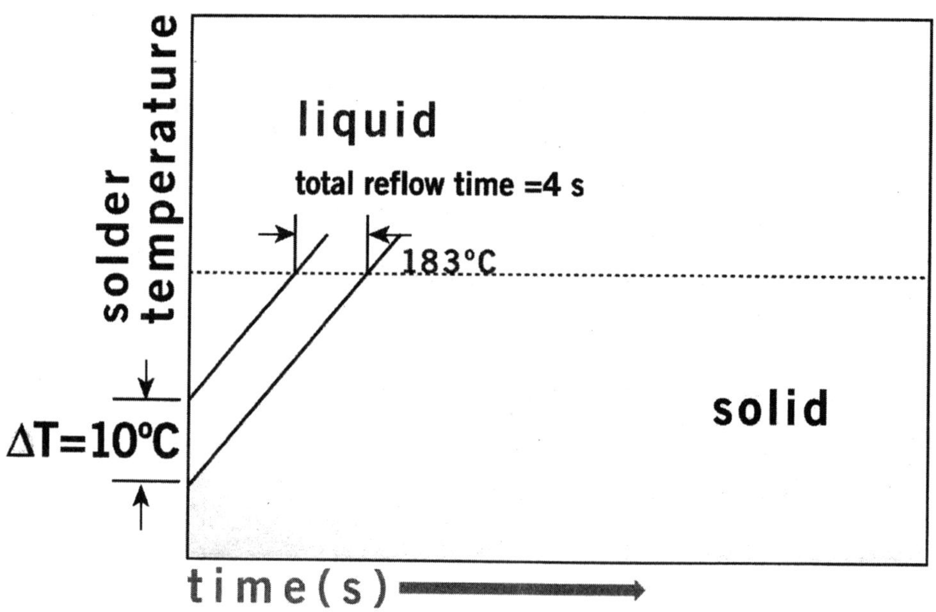

FIGURE 19.25 Faster transition from solid to liquid, more simultaneous.

influence; if a component is placed at the edge of a board, the temperature of the outer two sides generally will be greater than the temperature of the inner sides. This is due to greater heat dissipation toward the center of the PWB. Another factor is the area to which some solder balls are connected. It is quite common on a BGA for a center array of solder balls to be connected to a ground plane specifically to dissipate heat from the component die. These balls may require extra time to reflow.

19.4 COMPONENT REBALLING CONSIDERATIONS

One of the most commonly asked questions is, "Can I reuse components, and how do I reball the components?" While this is obviously relevant, there are very few companies who are currently reballing, and the following things should be considered before starting.

Component reusability questions need to be answered by the component supplier. The supplier will know what amount of thermal cycling its components will stand. If we consider a BGA assembled on a double-sided SMT PWB that was removed, reballed, and replaced, it would have had to undergo potentially six reflow cycles during this process, in addition to any ball reflow during the component manufacture:

1. Reflow assembly, topside
2. Reflow assembly, bottom side
3. Removal of component
4. Removal of excess solder from component
5. Reflow of new solder balls
6. Replacement of component

Another consideration would be the amount of handling and potential electrostatic discharge (ESD) damage that could occur during the reballing process.

Considering the preceding, in many cases, the reballing process is not viable commercially due to the extended labor times. We do, however, come across cases in which, because component value is very high or because of component unavailability, it is necessary to reball.

There are two basic methods most commonly used for reballing array packages: preforms and reballing fixtures. These methods can be used in conjunction with the features of the chosen rework machine.

19.4.1 Reballing Using Solder Preforms

One method for reballing involves using solder preforms (normally an array of solder balls to match the component, either held in a paper sheet or joined together). These are available from some of the solder manufacturers; there is also a product available called SolderQuik (Fig. 19.26). These preforms, in the case of eutectic balls, need to be soldered to the device in a controlled environment (reflow oven or head) using flux. When using high-temperature nonmelt solder balls, the preform needs to be attached to the device using screen-printed or dispensed solder paste. The typical process steps for attaching an array of solder spheres using the SolderQuik BGA preform are listed below:

1. *Removal of residual solder.* The pads on the part need to be prepared for the replacement solder balls using the methods described earlier in this chapter.
2. *Flux application.* Prior to placement, the correct flux should be applied to the pad areas by dispensing, printing, or more commonly, brush application.
3. *Preform placement.* Placement of the new preform is generally a manual operation where alignment is controlled by either edge registration or a fixture to align the component pad with the preform.
4. *Reflow.* After the part is fluxed and the preform placed, the next stage is to reflow the part. Again, reflow should be multizone controlled to match the profile of the flux or solder paste being used. When the reflow cycle is complete, the new array of solder balls should have been attached successfully.

FIGURE 19.26 SolderQuik and other preforms. (*Winslow Automation and the Indium Corporation.*)

5. *Carrier removal and component cleaning.* At this point, the carrier material needs to be removed, which can be accomplished by letting the carrier soak in some deionized water for a few seconds and then peeling away. The part then needs to be cleaned of flux residue, which can be done by almost any standard aqueous method. If you remove the paper with a batch cleaner, this step probably will not be necessary. The paper also can be removed easily when hot if a no-clean flux gel is used. This prevents the need to soak the paper off and is of great benefit because no prebaking of the component will be needed.

19.4.2 Reballing Using Fixtures

This method normally involves component-specific stencils and tooling to allow paste or flux to be printed and loose solder balls to be located in the correct position before being reflowed—again in a controlled environment. Solder balls of different sizes and compositions are available from most of the solder paste manufacturers.

Figures 19.27 through 19.29 show the process being used with printed solder paste and nonmelt, high-temperature solder balls on a CBGA component.

1. *Stencil printing of solder paste.* Figure 19.27 shows solder paste being printed onto the component. The part is held in a fixture, and using a chemically etched stencil, the solder paste is applied manually.
2. *Application of solder balls.* In the second stage, Fig. 19.28 shows that the paste-printing stencil has been removed and has been replaced with a larger stencil that will dictate the position of each solder ball. Loose solder balls are then poured onto this stencil and fall into their intended positions above each printed pad. When all the apertures have been filled, any excess solder balls simply can be brushed off or poured away.
3. *Reflow.* After all the balls have been placed, the component is ready for reflow. The method shown in Fig. 19.29 is to completely remove the stencil and use forced convection from above to reflow the solder paste attaching the nonmelt solder balls.

Some methods advocate leaving the ball-location stencil in position during the reflow operation because a ball's natural reaction is to self-centralize; this will make little or no difference. The stencil also may act as a heat sink, which will increase the required reflow temperature.

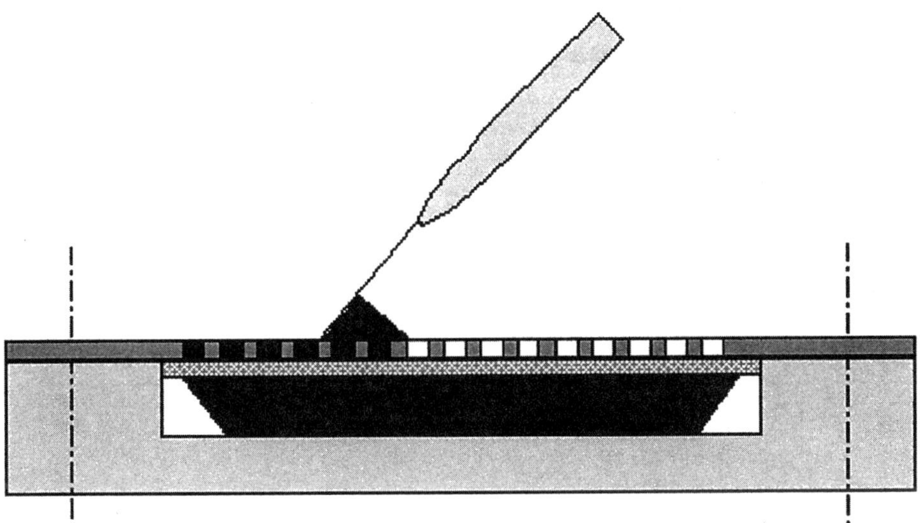

FIGURE 19.27 Printing solder paste onto the component.

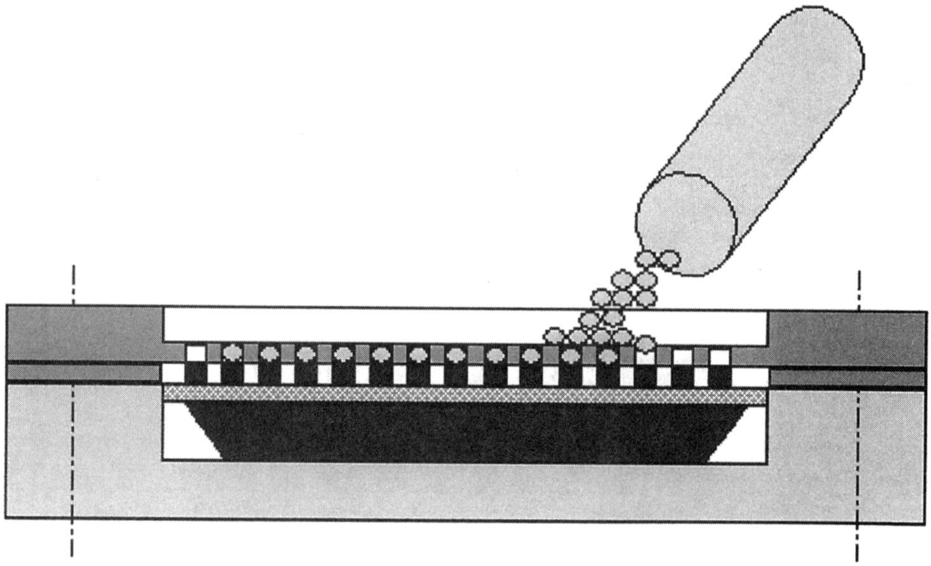

FIGURE 19.28 Application of loose solder balls.

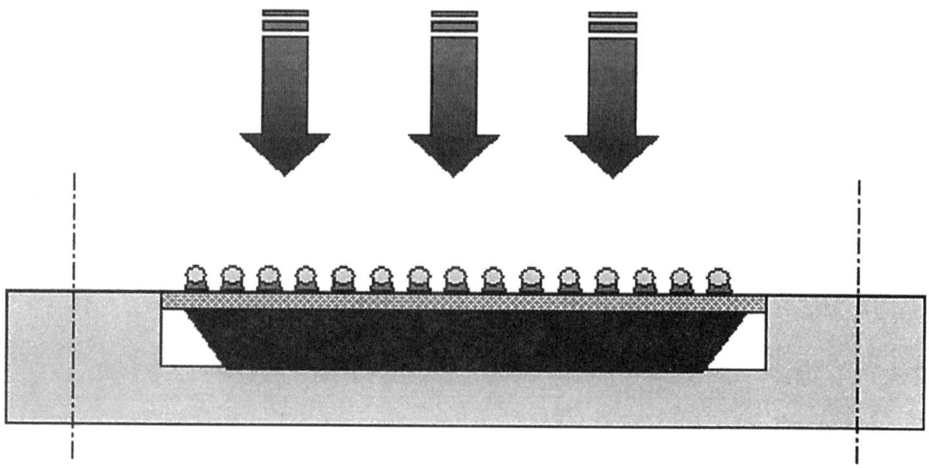

FIGURE 19.29 Solder ball reflow.

The reballed part is removed from the fixture and is ready to be attached to the PWB as a normal component.

19.5 ANALYSIS USING X-RAY INSPECTION

On an array package, because most of the solder joints are hidden from view, the only comprehensive way to evaluate results is to use an x-ray inspection system. The following subsections illustrate some of the common visible faults and their respective causes.

19.5.1 Solder-Joint Size Variability

When looking down on a BGA assembly, the uniformity of the solder joints can be a good indicator of process problems (Fig. 19.30). In a perfect result, we would expect all solder joints to be approximately the same size and shape. Variability could indicate the following defects:

- Component or PWB warpage
- Popcorning
- Insufficient solder paste deposits
- Excessive solder paste deposits

19.5.2 Solder-Joint Shape

Solder-joint shape can be an indicator of process or design problems but also can act as a tool to identify wetting. On an assembly with round pads that are "solder mask–defined," we would expect all joints to be round in shape. If the pads were teardrop-shaped, it could be indicative of the following defects:

- PWB solder resist misalignment (Fig. 19.31)
- Damage to solder resist causing solder migration along tracks

On an assembly with pads that are not solder mask–defined, we would expect to see a less uniform joint shape, where solder has wetted adjoining tracks. This would not be considered a defect but would indicate different PWB characteristics. This effect can be used to the engineer's advantage if pads are deliberately designed to be teardrop-shaped. In this case, a teardrop-shaped joint would indicate successful wetting of the solder joint.

19.5.3 Voids

The effect of voids in a solder joint depends very much on the size and quantity present. The main factor that contributes to generating voids is an incompatibility between the reflow profile and the materials used (Fig. 19.32).

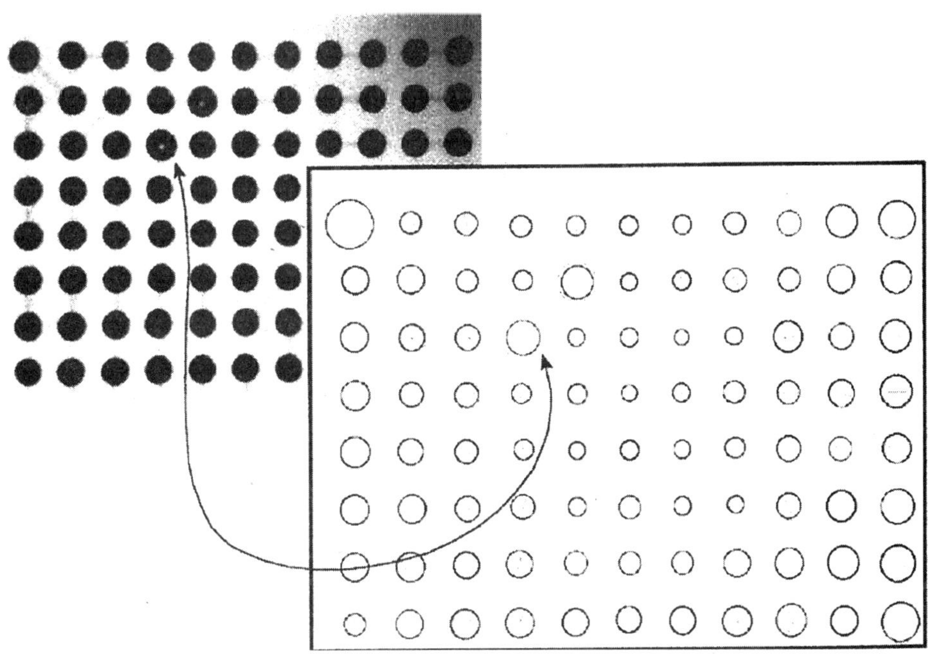

FIGURE 19.30 Variable solder-joint sizes. (*Glenbrook Technologies.*)

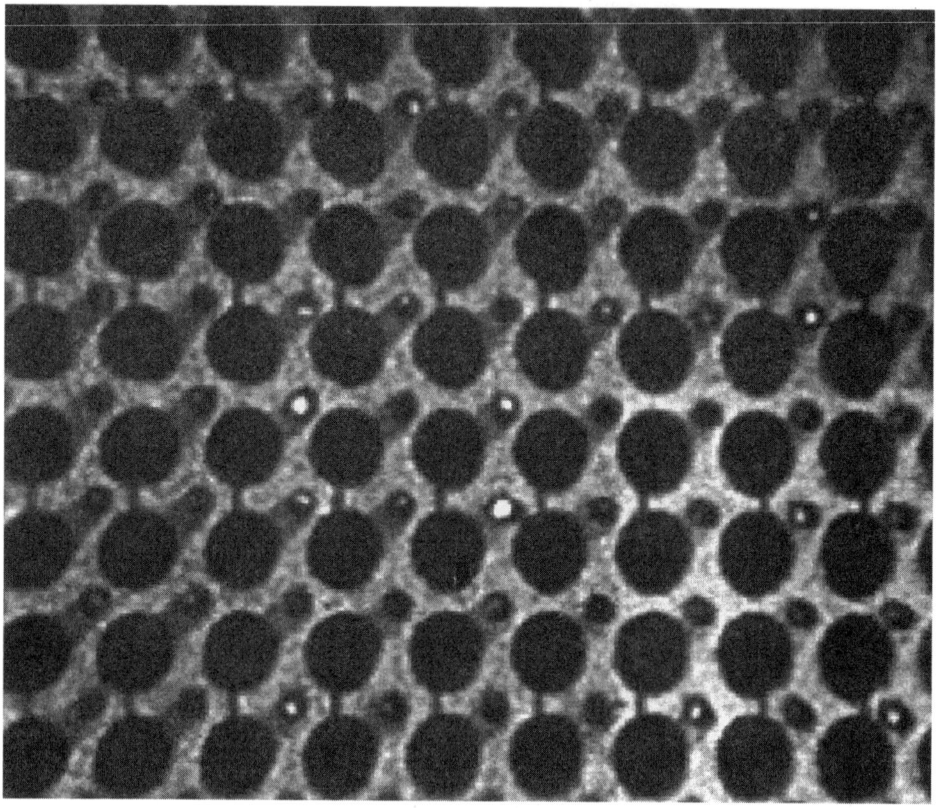

FIGURE 19.31 Misshaped joints due to misalignment. (*Glenbrook Technologies.*)

If a profile is too quick, you may see a number of small, consistent voids, although these are considered relatively harmless to long-term reliability. If large and inconsistent voids are observed, this would be detrimental to the long-term reliability of the solder joint, with an increased risk of cracks propagating throughout the joint. Incorrect profile settings would cause this.

19.5.4 Solder Shorts

Solder shorts (Fig. 19.33) represent a complete assembly failure and can be caused by the following factors:

- PWB or component warpage
- Excessive or inconsistent solder paste deposits
- Damage to solder resist on adjoining tracks
- Incorrectly placed components

19.5.5 Open Circuits

This defect can be particularly difficult to identify because it may not be apparent from a two-dimensional view (Fig. 19.34). As stated previously, inconsistent joint size may be an indicator of this. The primary causes of opens are as follows:

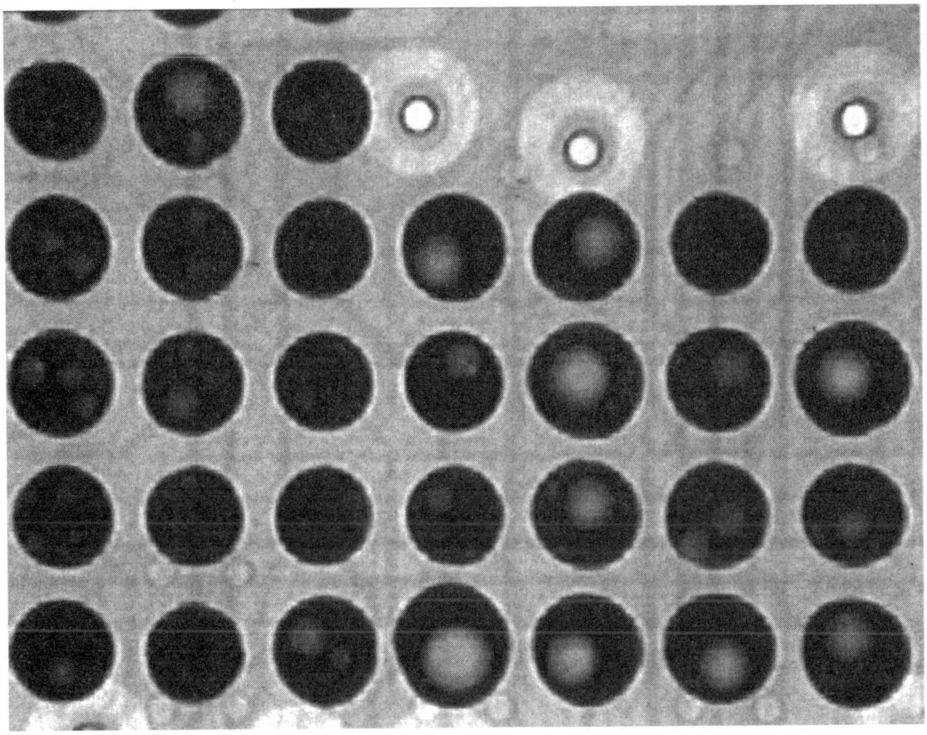

FIGURE 19.32 Voids in solder balls. (*Glenbrook Technologies.*)

- PWB or component warpage
- Insufficient solder paste or flux deposit
- Missing solder balls on the component

19.6 PWB DESIGN GUIDELINES FOR REWORK

As manufacturers of rework machines, we find it unfortunate that in most cases design to facilitate easy rework is a rare occurrence. If the following rules are observed, then rework problems often can be avoided:

1. *Clearance for reflow nozzles.* If some clearance is left around the perimeter of the device, it will be much easier to locate the nozzle and shroud the component completely. This will result in better ΔT characteristics and minimal adjacent reflow. A suggested keep-out zone would be 80 mils. Where a product demands a tightly packed board, this may not be possible. In this case, the engineer will have to accept adjacent reflow as a consequence.
2. *Clearances for stencils.* If you intend to print solder paste in rework, it is imperative that adequate clearance is allowed for the stencil. A suggested keep-out zone of 100 mils normally would be acceptable. If this is not provided, it may mean that the operator has to remove surrounding components to facilitate printing.

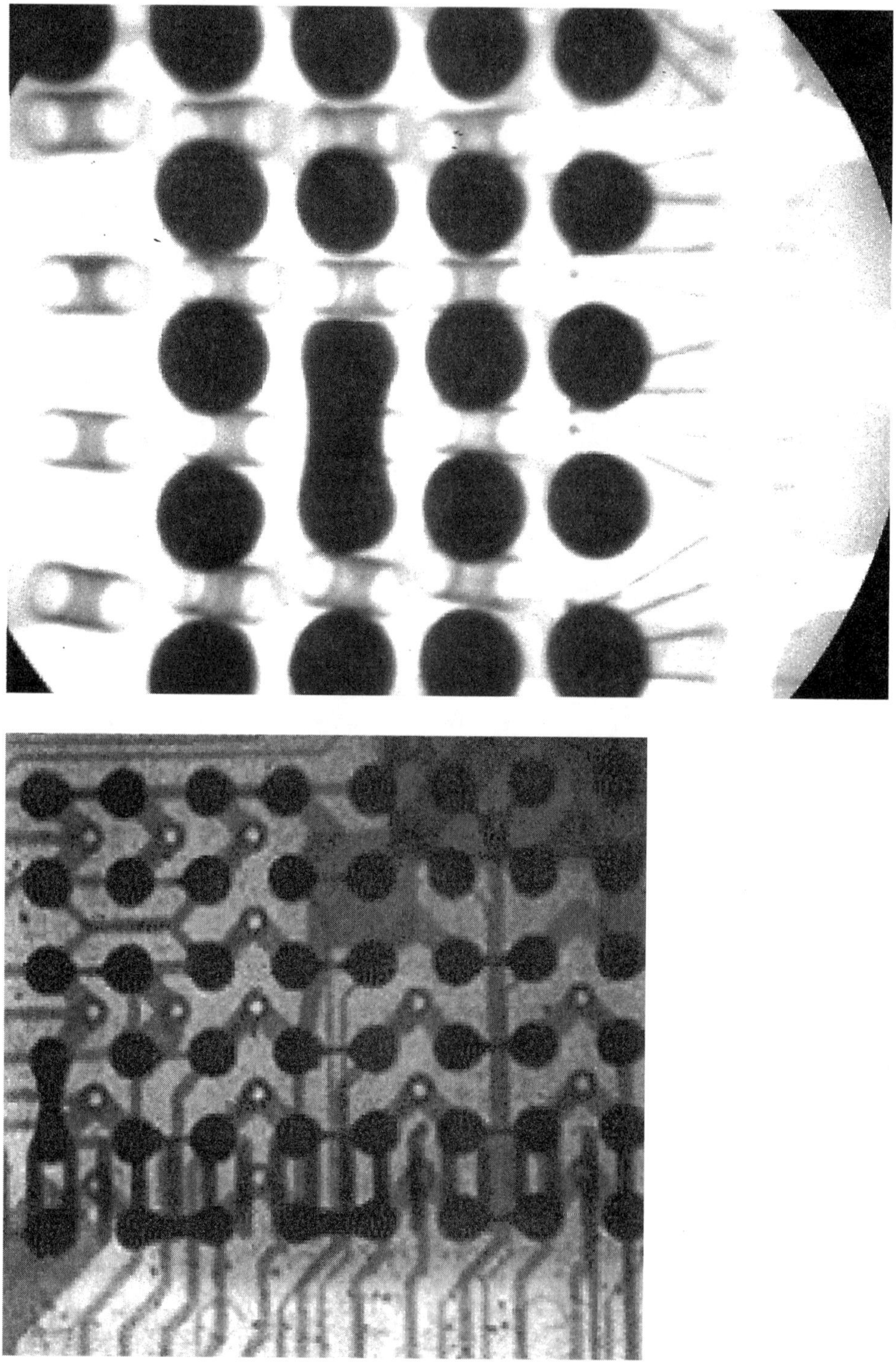

FIGURE 19.33 Solder shorts. (*Glenbrook Technologies.*)

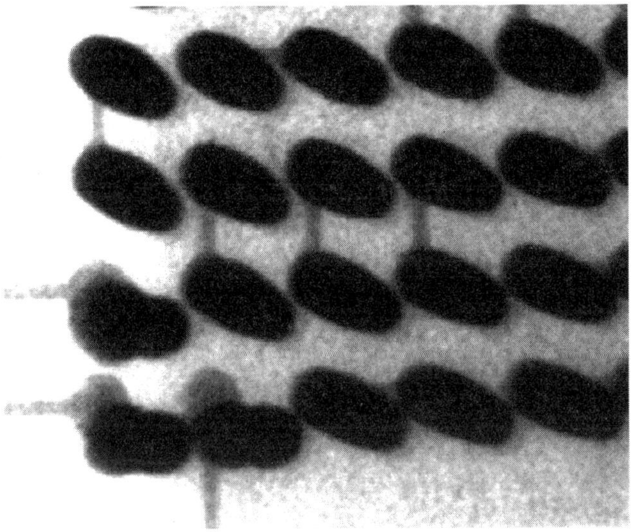

FIGURE 19.34 Formation of an open circuit (3D view).

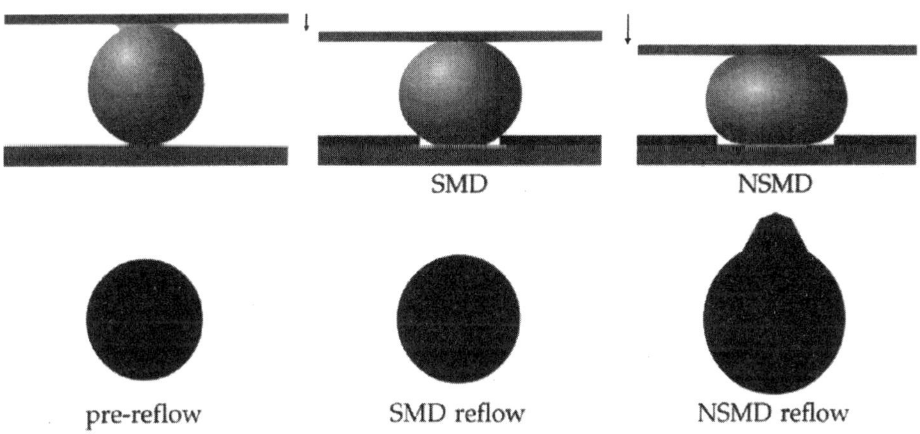

FIGURE 19.35 Pad solder mask definition can facilitate inspection. (*Glenbrook Technologies.*)

3. *Pad design.* As stated previously, pad design can affect wetting and inspection characteristics. It also can be of benefit when removing nonmelt solder ball components. If the pad on the board is smaller than the pad on the component, then the solder balls will tend to adhere to the component during removal, making the cleaning operation easier (Fig. 19.35).
4. *Via holes.* Open via holes in pads will cause migration of solder and low standoff heights after reflow.
5. *Positions of components.* The position of an array component may influence the ΔT characteristics during reflow. It may be a problem to position a large array at the edge of a PWB because simultaneous reflow will be impractical.

CHAPTER 20
BGA ASSEMBLY RELIABILITY

Raza Ghaffarian
Jet Propulsion Laboratory

20.0 INTRODUCTION

The dominant trend within electronics today is the drive toward increasingly smaller feature sizes on integrated circuit (IC) chips. Ten years ago, feature sizes of 1 μm were the norm; by the mid-1990s, 0.35 μm was the norm, and research and development ICs are at 0.25 μm and under. This has led to a drastic increase in the number of input-outputs (I/Os) on the chip, particularly for application-specific ICs (ASICs). Indeed, 300 to 500 I/Os per chip are no longer uncommon, and now ASICs having up to and higher than 1000 I/Os are under development. Figure 20.1 shows different surface-mount components from plated through-hole (PHT) to chip scale packages (CSPs) in the trend toward miniaturization.

This great increase in I/Os pushes on the packaging engineer to arrive at a suitable way of interconnecting the IC to a printed wiring board (PWB) or to a ceramic substrate. Mounting on an organic substrate, typically referred to as a PWB, is much less expensive than mounting on a ceramic substrate. Concomitant with the increase in I/Os, or pins, is the drive to achieve higher component densities, thereby increasing board functionality. The current trend in electronic packaging is

- Increasing pin count
- Increasing component density (more functionality per unit area)
- Lighter weight

Several different approaches are being employed to meet the challenge of mounting high-pin-count ICs to substrates and packages on PWBs. These approaches can be considered a form of direct chip attach (DCA), flip chip (FC), and use of package ICs rather the use of bare IC dies. Each of these approaches has its own merits and drawbacks.

- Attach the IC die to the bare PWB and wire bond from the die bonding pads directly to bonding pads of the PWB. This is commonly referred to as *chip-on-board* (COB) or *chip-and-wire DCA*. ICs can be mounted first on polyimide tape, and then the ICs can be gang bonded to the PWB using the hot-bar soldering technique. This is commonly referred to as *tape automated bonding* (TAB) or *tape carrier package* (TCP) *DCA*.
- Permanently attach small solder bumps to the bottom of the IC die, flip it over, and then mount it onto suitable mounting pads on the PWB. This is commonly referred to as *direct flip chip* (FC).
- Mount the IC internally, wire bond or FC, on a flexible/rigid organic or ceramic substrate and package the chip into a suitable package material. The package then has small solder bumps applied to the bottom, is flipped over, and mounted onto suitable mounting pads on the PWB. This is commonly referred to as *ball grid array* (BGA) technology when the ball pitches (distance

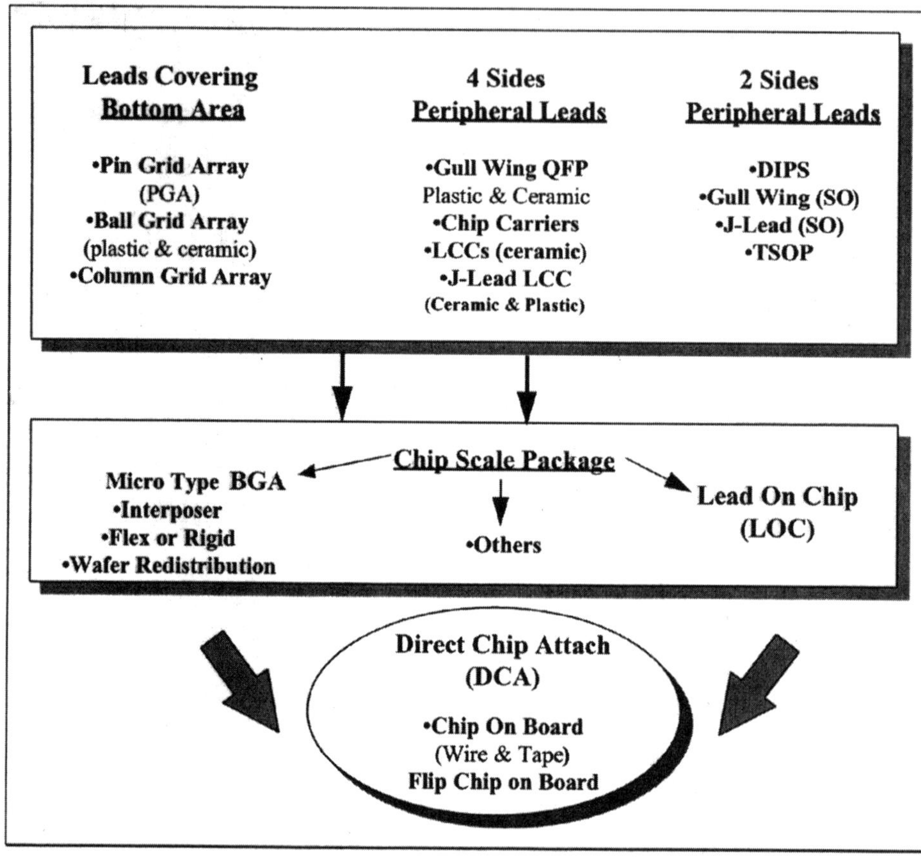

FIGURE 20.1 Electronics package miniaturization trends.

between two adjacent ball centers) equal 1 mm and higher, normally 1.27 mm. If the package dimensions are very nearly those of the IC itself, this technology is being called *chip scale packaging* (CSP) with pitches equal to or less than 0.8 mm. The principal advantage of BGA and CSP is their combination of a package to protect the IC and their close similarity to FC. Figure 20.2 shows a comparison of a COB component with a CSP.

Each approach has, of course, its own unique set of advantages and drawbacks. Many technical advances in materials now allow the use of these technologies in more demanding environments requiring higher reliability.

20.1 BALL GRID ARRAY (BGA)

BGA combines the best features of FC but ensures that the die is fully enclosed by a suitable packaging materiel to protect it from hostile environments. BGAs are the package analogue of FC technology (controlled collapse chip connection, or C4) developed by IBM for denser interconnection at the IC packaging level. The typical pitch for these is 1.27 mm (0.050 in).

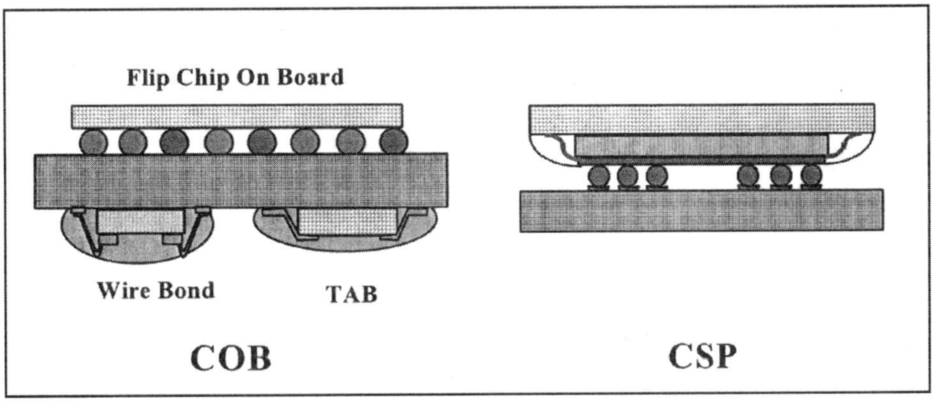

FIGURE 20.2 Comparison of a COB and a CSP.

Because the IC is mounted within a suitable packaging material, the chip can be burned in as any conventional package. In addition, the package provides direct protection of the IC, and it can be made to provide hermeticity if necessary. Because the package is an area array, it uses less PWB real estate than peripheral array-mounted packages such as the quad flatpak (QFP). Generally, an underfill is not used. On the other hand, conventional QFPs, with their gullwing leads, consume considerable board real estate.

BGAs offer several distinct advantages over FP (fine pitch) and UFP (ultrafine pitch) surface-mount components (SMCs) having gullwing leads, including

- BGAs are capable of high pin counts, generally more than 200.
- BGAs have larger lead pitches, which significantly reduces the manufacturing complexities for high-I/O parts.
- Higher packaging densities are achievable because the lead envelope for the gullwing leads is not applicable in the case of BGAs; thus it is possible to mount more packages per board.
- BGAs have faster circuitry speed than gullwing SMCs because the terminations are much shorter.
- BGAs have better heat dissipation than gullwing SMCs.
- Conventional surface-mount technology (SMT) production techniques such as stencil printing and component mounting can be employed.

BGAs are also robust in processing. This stems from their higher pitch (1.27 mm, 0.050 in, typical), better lead rigidity, and self-alignment characteristics during reflow processing. This latter feature, self-alignment during reflow, is very beneficial and opens the process window considerably.

BGAs, however, are not compatible with multiple solder processing methods, and individual solder joints cannot be inspected and reworked using conventional methods. In high-reliability applications, inspection of solder joints visually has been a standard practice and is a key factor for providing confidence in solder-joint reliability. Advanced inspection techniques including x-rays need to be further developed to provide such confidence for BGAs and grid CSPs.

To summarize, the three main drawbacks of both FCs and BGAs are similar:

- Lack of direct visual inspectability
- Lack of individual solder-joint reworkability
- Interconnect routing between the chip and the PWB necessitating a multilayer PWB

20.2 BGA AND QFP COMPARISON

Figure 20.3 shows a size comparison of a 256-pin gullwing QFP and a 256-pin BGA. In this figure, the top of the 256-pin QFP and the bottom of the 256-pin BGA are shown. A reduction by 2.5 times can be achieved when comparing BGA parts with QFP parts. When considering PWB area requirements, the area required on the PWB for solder lands is included. For a QFP, this area would be calculated by measuring from lead tip to lead tip of opposing sides.

The 256-pin BGA shown is a plastic package with a peripheral solder-ball array. When BGA packages first entered the market, they were configured in a full solder-ball array layout. Thermal cycling tests of various package sizes showed that the full solder-ball array layout was susceptible to high stresses under the area of the die periphery resulting in premature failure of the solder joints in this area. Also, the fatigue life of the solder spheres for a BGA part is directly related to the die size used. These high stresses are caused by the constraining effect of the semiconductor die (typically silicon) on the laminate part package. The highest stresses are found in the solder spheres located under the corners of the silicon device.

A silicon die has an approximate coefficient of thermal expansion (CTE) of 2 to 3 ppm/°C, whereas the plastic part package has a CTE of approximately 17 to 20 ppm/°C. The silicon die strains the part laminate material, which has a close CTE match with the board material. To aid in reducing the stresses induced under the semiconductor die, peripheral solder-ball array BGA parts were developed. In the peripheral solder-ball array parts, the solder spheres are placed around the periphery of the package, away from the die edges. This reportedly results in better reliability performance during thermal cycling.

Figure 20.4 shows the difference in pitch for the 256-pin QFP and the 256-pin BGA. In the figure, the BGA has been placed directly on top of the QFP. The QFP requires a pitch of 0.020 in to achieve a lead count of 256, whereas the BGA part requires 0.050 in to achieve this same lead count. The greater pitch of the BGA relative to the equivalent QFP package allows for a far more robust

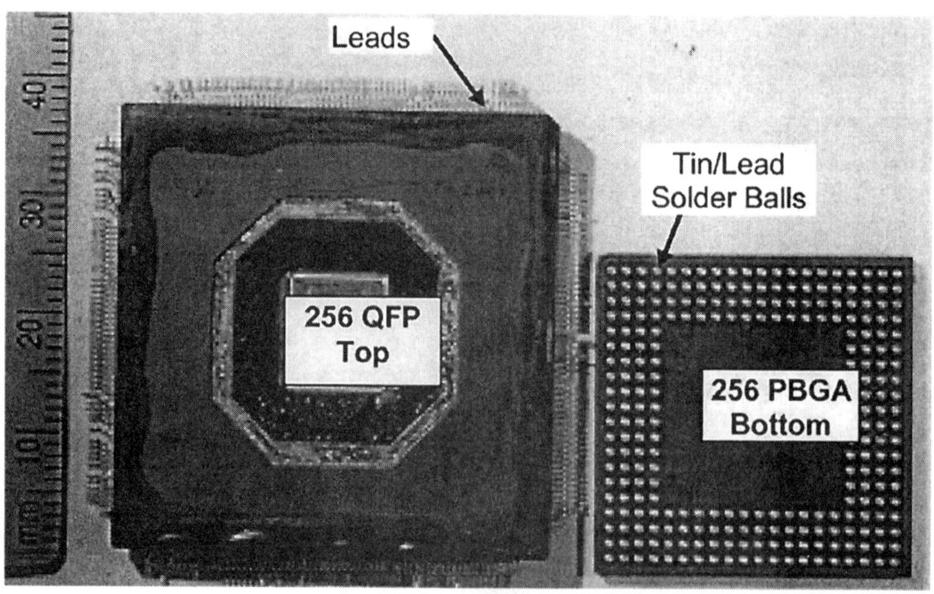

FIGURE 20.3 Comparison of a 256-I/O plastic BGA and 256-I/O ceramic QFP.

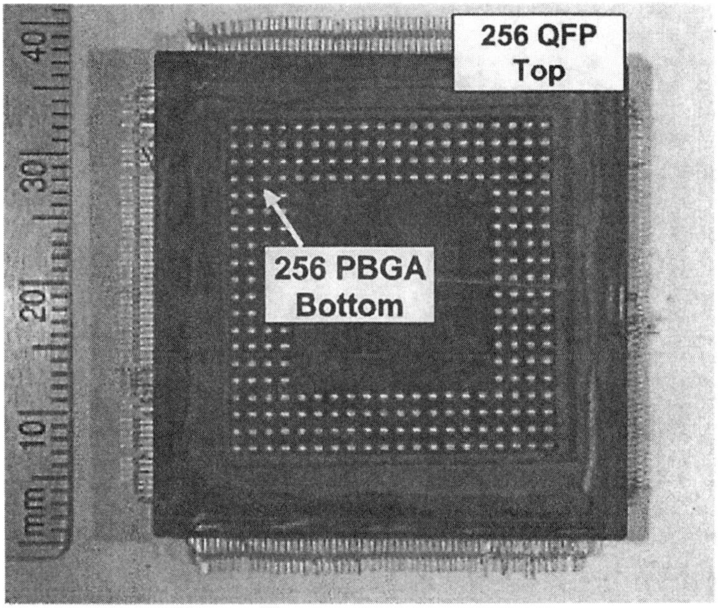

FIGURE 20.4 Pitch difference for BGAs (1.27 mm) and QFPs (0.508 mm), same I/O count.

manufacturing cycle. Solder shorting and assembly part placement issues are substantially reduced with BGA packages.

Significant differences also exist between fine-pitch (0.020 in) QFPs and BGA parts in manufacturing processing. QFPs require lead forming, lead degolding, lead tinning, coplanarity inspection, coplanarity adjustment, specialized packaging and handling procedures, and specialized PWB placement processes. BGAs require package flatness inspection and solder sphere planarity inspections (both part types require PWB flatness verification). The handling and transport of QFPs are extremely critical because the leads are damaged very easily, with skew and nonplanarity resulting. BGA leads are very robust when handled with reasonable care.

The downside to BGAs for high-reliability applications is the inability to inspect all solder joints. Since conventional visual inspection techniques are not adequate for BGA solder joints, process and material controls combined with simple verification techniques following assembly are required to be performed to ensure the quality of solder-joint interconnections.

20.3 RELIABILITY OF BGA ASSEMBLY BY JPL-LED BGA CONSORTIUM

Reliability of BGAs was assessed as part of the Jet Propulsion Laboratory (JPL) BGA consortium effort. Nearly 200 test vehicles, each with four packages, were assembled and tested using an experimental design (Figs. 20.5 and 20.6). The most critical variables incorporated in the experiment were package types (ceramic and plastic), board materials (FR-4 and polyimide), surface finishes (OSP, HASL, and Ni/Au), solder volumes (low, standard, and high), and environmental conditions.[1–3]

The two test vehicle assembly types were plastic (PBGAs) and ceramic (CBGAs) packages. Both FR-4 and polyimide PWBs with six layers 1.57 mm thick (0.062 in) were used.

Plastic packages covered the range from overmolded pad array carrier (OMPAC) to superBGAs (SBGAs). These were (Table 20.1):

(a) (b)

FIGURE 20.5 Test vehicles: (*a*) type 1 (300 I/Os) and (*b*) type 2 (600 I/Os).

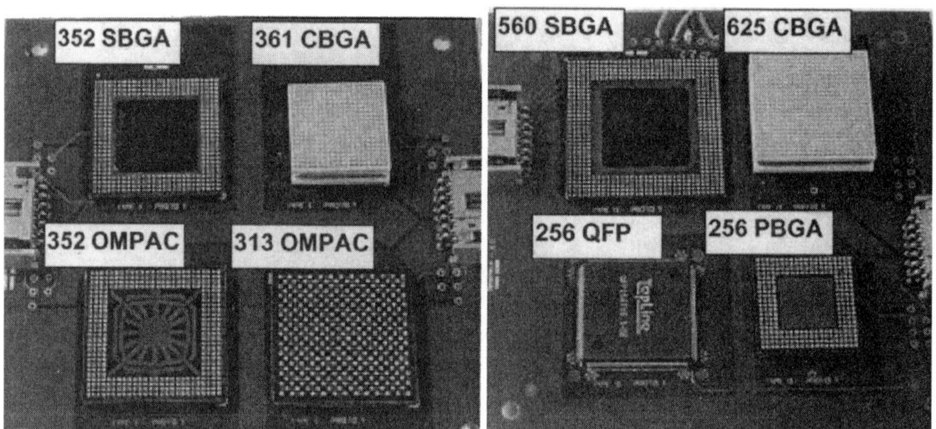

a) Type 1 BGA configuration and I/Os **b) Type 2 BGA configuration and I/Os**

FIGURE 20.6 Test vehicles with unassembled parts topping assembled ones: (*a*) type 1 BGA configuration and I/Os; (*b*) type 2 BGA configuration and I/Os.

TABLE 20.1 Packaging Characteristics

Part type	I/O	Materials configurations	Size (mm^2)	Wiring	Pitch	Selected die size
PBGA 300	352	Plastic/peripheral/superBGA	35	Daisy	1.27	13.3
PBGA 300	352	Plastic/peripheral/OMPAC	35	Daisy	1.27	13.3
CBGA 300	361	Ceramic/full array	25	Daisy	1.27	
PBGA 300	313	Plastic/full array/OMPAC	35	Daisy	1.27	13.3
PBGA 600	560	Plastic/peripheral/superBGA	42	Daisy	1.27	15.25
CBGA 600	625	Ceramic/full array	32.5	Daisy	1.27	
PBGA 300	256	Plastic/peripheral	27	Daisy	1.27	10.8
GW	256	Plastic/gull wing	30.6	Daisy	0.4	

- Two peripheral SBGAs, 352 and 560 I/Os
- Peripheral OMPAC, 352 I/Os, PBGA, 352 and 256 I/Os
- Depopulated full-array PBGA, 313 I/Os
- 256-I/O QFPs, 0.4 mm pitch

In SBGAs, the IC die is attached directly to an oversized copper plate, providing better heat dissipation efficiency than standard PBGAs. The solder balls for plastic packages were eutectic (63Sn/37Pb).

Ceramic packages with 625 and 361 I/Os also were included in my evaluation. Ceramic packages had high-melting solder balls with a composition of 90Pb/10Sn and 0.035-in diameters. These balls were attached to the ceramic substrate with eutectic solder (63Sn/37Pb). During BGA assembly, package-side eutectic solder and the PWB-side eutectic paste are reflowed to provide the electromechanical interconnects.

Plastic packages had dummy dies and daisy chains, with the daisy chains on the PWB designed to complete the package chain loop to be able to monitor critical solder-joint regions. Most packages had four daisy-chain patterns, the 560 I/O package had five, and the QFP had one.

Full assembly was implemented after process optimization from the trial test. The following procedures were followed:

- PWBs were baked at 125°C for 4 h prior to screen printing.
- Two types of solder pastes were used, a rosin mild activation (RMA) and a water-soluble one.
- Pastes were screen printed, and the heights were measured by laser profilometer. Three levels of paste were included in the evaluation: standard, high, and low. Stencils were stepped to 50 percent to accommodate assembling ceramic, plastic, and fine-pitch QFP packages in the type 2 test vehicle.
- A 10-zone convection oven was used for reflowing.
- The first assembled test vehicle (TV) using an RMA reflow process was inspected visually and x-rayed to check solder-joint quality.
- All assemblies were x-rayed.

Two test vehicles were assembled.

- Type 1, ceramic and plastic BGA packages with nearly 300 I/Os.
- Type 2, ceramic and plastic BGA packages with nearly 600 I/Os. Also used were a 256-I/O leaded ceramic and a 256-I/O plastic BGA package for evaluating and directly comparing manufacturing robustness and reliability.
- Assemblies with water-soluble flux were cleaned using an Electrovert H500. Those with RMAs were cleaned used isopropyl alcohol (IPA) and a 5% saponifier.
- All fine-pitch QFPs had to be reworked for bridges.

20.4 BGA THERMAL CYCLING

For cost-effectiveness and engineering efficiency considerations, it was desirable to have the shortest possible test period. However, adequate high-temperature dwells are necessary for solder creep and relaxation to occur. On the other hand, excessively high-temperature dwells may cause excessive aging, resulting in brittle fracture. The ramping rates of heating and cooling usually are limited by the capacity of the environmental test chamber.

Four different thermal cycle profiles were used (Figs. 20.7 through 20.10):

- *Cycle A.* The cycle A condition ranged from −30 to 100°C and had an increase/decrease heating rate of 2 to 5°C/min and dwell of about 20 min at the high temperature to ensure near-complete creeping. The duration of each cycle was 82 min.

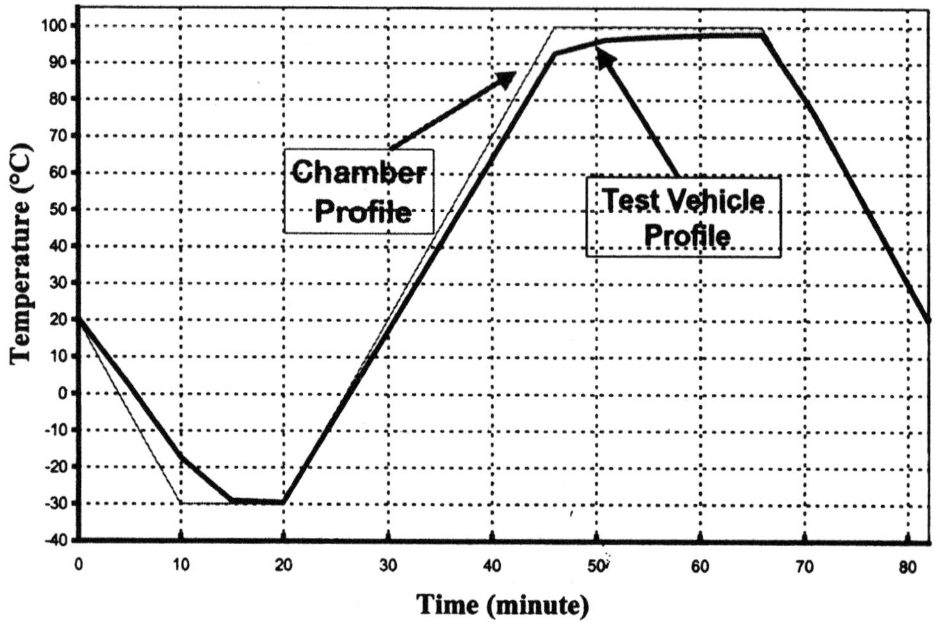

FIGURE 20.7 JPL thermal cycle profile (−30–100°C)—cycle A.

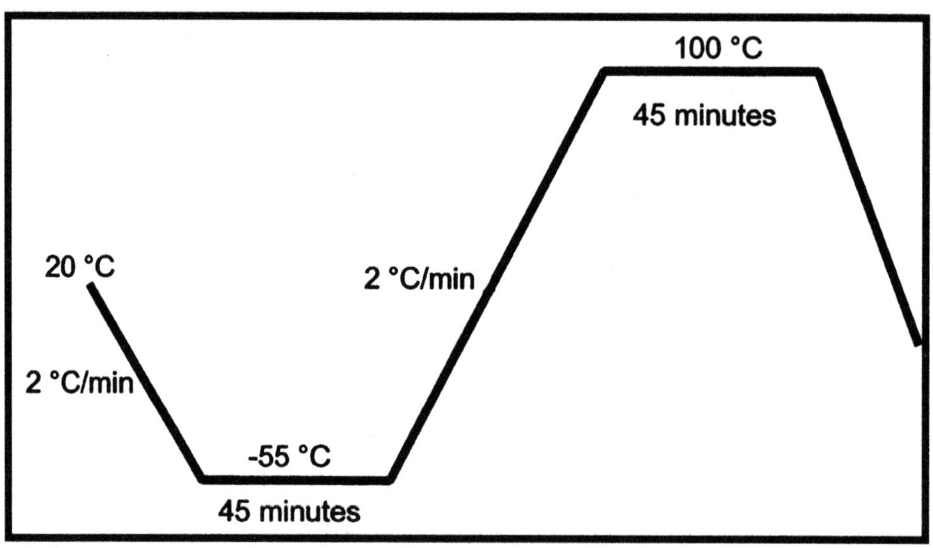

FIGURE 20.8 JPL thermal cycle profile (−55–100°C)—cycle B.

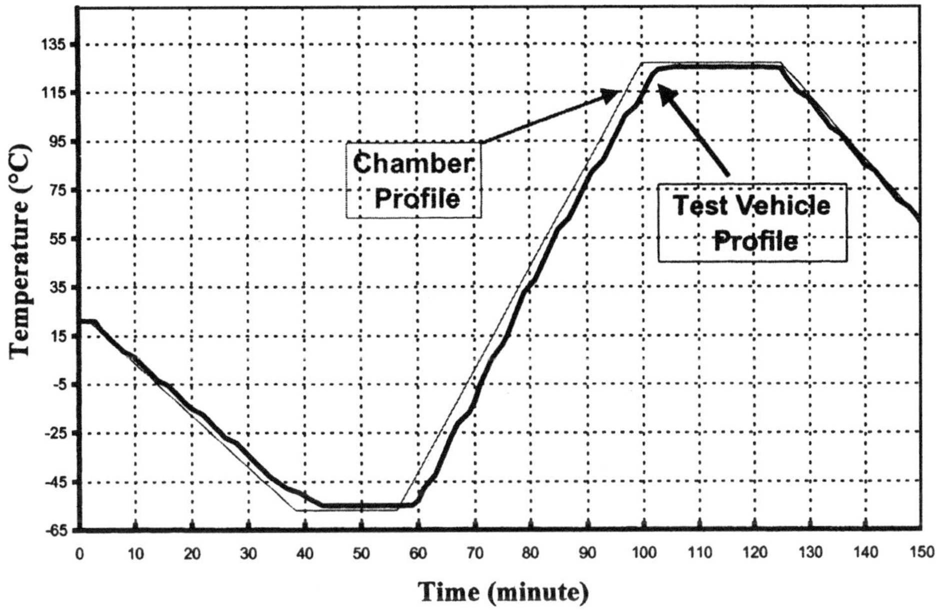

FIGURE 20.9 JPL thermal cycle profile (−55 to 125°C)—cycle C.

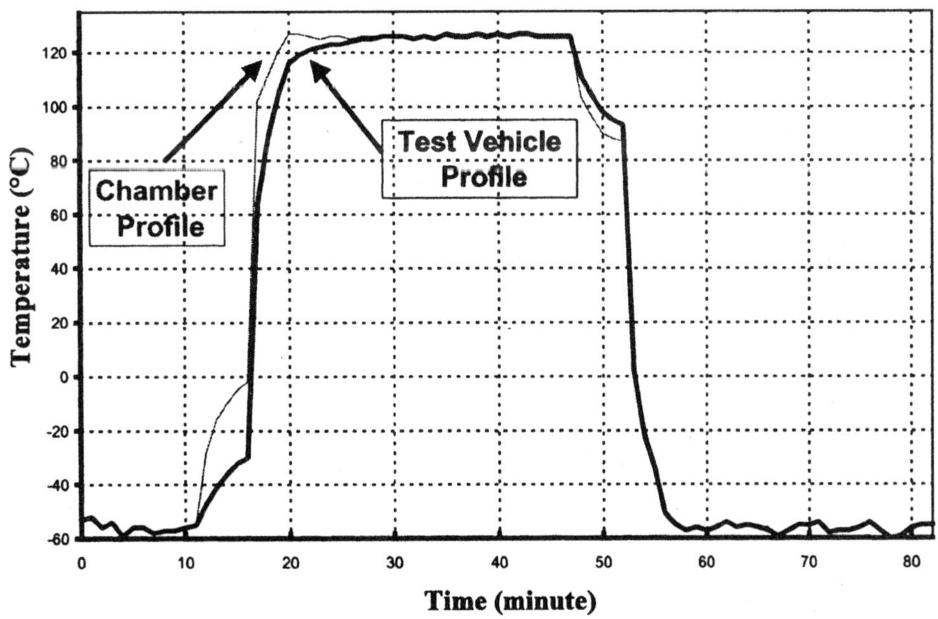

FIGURE 20.10 Boeing thermal cycle profile (−55 to 125°C)—cycle D.

- *Cycle B.* The cycle B condition ranged from −55 to 100°C with long duration. The heating and cooling rates were 2 to 5°C/min with an oven dwell setting of 45 min at the two extreme temperatures. The duration of each cycle was 246 min.
- *Cycle C.* The cycle C condition ranged from −55 to 125°C with 2 to 5°C/min heating/cooling rate. Dwell at extreme temperatures was at least 10 min, with a duration of 159 min for each cycle.
- *Cycle D.* The cycle D condition ranged from −55 to 125°C, the same as cycle C but with very high heating/cooling rates. It also could be considered a thermal shock because it used a three-region chamber: hot, ambient, and cold. Heating and cooling rates were nonlinear and varied between 10 and 15°C/min with dwells at extreme temperatures of about 20 min. The total cycle lasted approximately 68 min.

20.5 FAILURE CRITERIA

The criteria for an open solder joint specified in IPC-SM-785, Sec. 7.8, were used as guidelines to interpret electrical interruptions. Generally, once the first interruption was observed, there were many additional interruptions within 10 percent of the cycle life. In several instances, a few nonconsecutive early interruptions were not followed by additional interruptions until much later stages of cycling. This was found more with plastic packages.

Figure 20.11 plots the total number of electrical interruptions (daisy-chain opens) versus the number of cycles for the cycle D condition. For ceramic BGAs, the rate of interruptions (slope) was

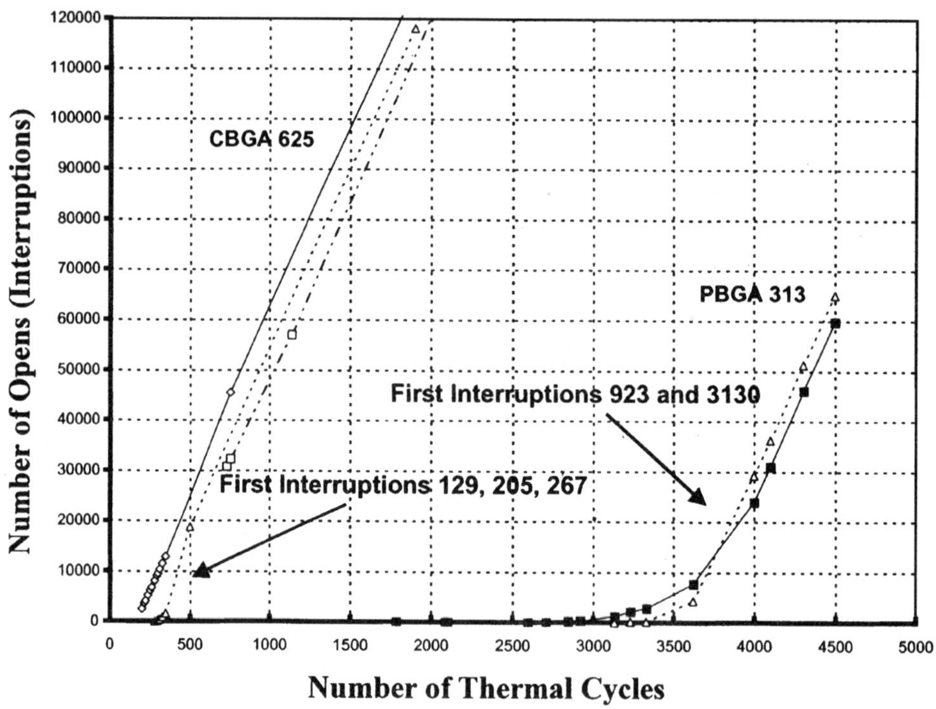

FIGURE 20.11 Number of electrical interruptions detected with thermal cycles (−55 to 125°C) for ceramic and plastic BGAs.

approximately constant. This was not the case for the plastic BGAs. For a PBGA with 313 I/Os, the first interruption was at 923 cycles, with no additional interruptions until about 3000 cycles. Further erratic interruptions were observed between 3000 and 3700 cycles, followed by linear interruptions having approximately the same slope as the CBGAs. Another PBGA showed the first interruption at 3130 cycles and become open at about 3300 cycles.

20.6 CBGA FAILURE MECHANISMS

20.6.1 Damage Monitoring

For conventional SMT with peripheral interconnections, the pass/fail criteria for high-reliability applications rely on visual inspection of solder joints at 10× to 50× magnifications. For the BGAs on the test vehicles, only edge balls, those not blocked by other components, were inspected visually. A series of single assemblies that were cut from the test vehicles also were used for both visual and scanning electron microscope (SEM) inspection to better define visual criteria for acceptance of solder joints as well as to monitor damage progress under different cycling environments.

20.6.2 Failure Mechanism for Ceramic Assemblies

Figure 20.12 shows representative SEM and cross-sectional micrographs for a CBGA 625-I/O package after 348 cycles of cycle A conditions. Cross-sectional photos are for the perimeter balls, two corners, and a center ball. SEM photos for package and board sides of a corner (maximum distance to neutral point, or DNP) and the edge center ball are also included. Note the large and small pin holes in the eutectic solder of the corner pin near the board site (lower right SEM photo). Generally, pin holes formed prior to crack initiation.

Figure 20.13 compares the corner solder joint of Fig. 20.6 and a typical failure photo for the B cycling conditions for a CBGA 361-I/O package after 350 cycles. Note that the failures for the cycle A

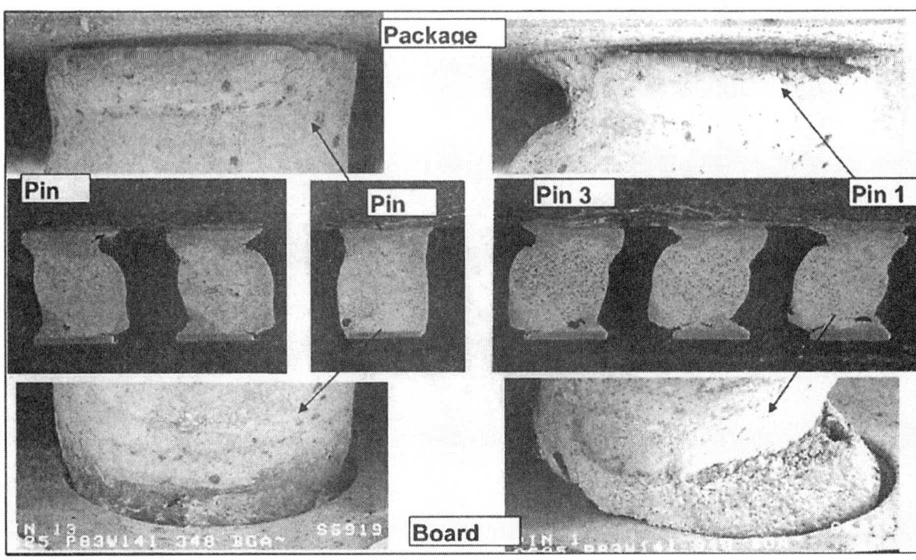

FIGURE 20.12 SEM and cross-sectional SEM photographs of CBGA 625 after 348 cycles (−30 to 100°C). Balls (pins) 1 and 25 are the corner balls with the maximum DNP, and ball 13 is the center of the edge balls.

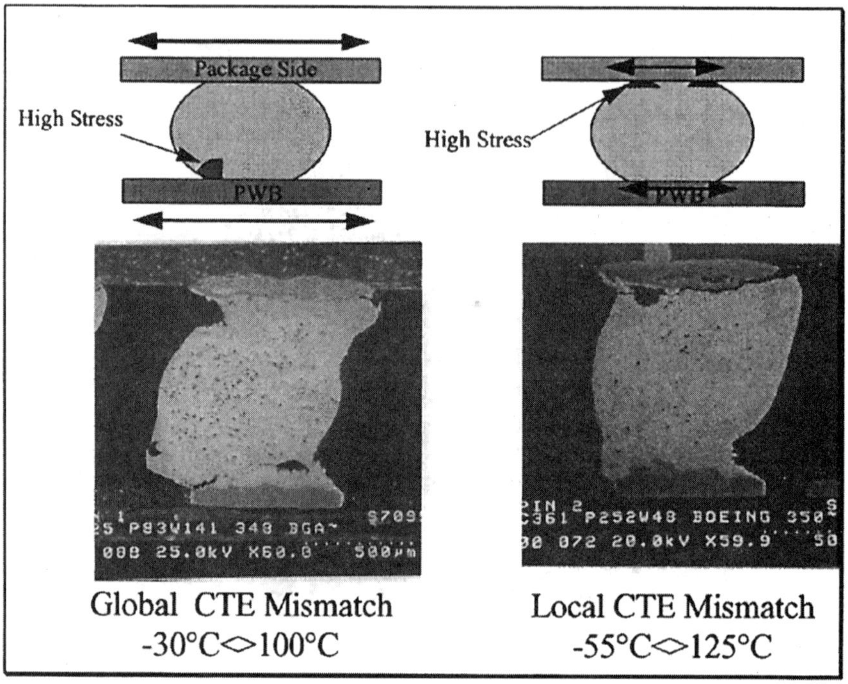

FIGURE 20.13 Cross sections of failure sites for CBGA 625 after 350 cycles under cycle A and D conditions.

conditions were from the PWB and for the cycle B conditions were from the package sites. These were the typical failure mechanisms for the two conditions and are similar to those reported in literature.[4]

Failure mechanism differences could be explained by global or local stress conditions. Modeling indicates that the high-stress regions shifted from the board to the packages themselves when stress conditions changed from global to local. For A cycling with slow heating/cooling ramping, which allowed the system to reach uniform temperature, damages could indicate a global stress condition. For D cycling with rapid heating/cooling, damage could indicate a local stress condition.

20.7 PBGA FAILURE MECHANISMS

Among the plastic packages, the PBGA 313-I/O package with depopulated full-array balls was first to fail under both D and A thermal cycling conditions. Figure 20.14 shows various cross sections of this package's balls from corner to center at 4682 A cycles. These photographs are of the balls under the die, where most damage occurs due to local CTE mismatch. Photographs with and without voids also were included for comparison. Voids appear to have concentrated at the package interface under the die. Cracking propagation occurred at package or board interfaces for sections with or without voids.

The sections with voids were opens, indicated by seepage of mounting materials into the voids. Except for the interface-connecting cracks, there appeared to be no crack propagation among the voids. The solder joints for this package also were inspected visually to identify indicators of defects. None were apparent. Even when these were inspected at 100× by SEM, no gross damage was observed, as evidenced by the top left SEM photograph of Fig. 20.14.

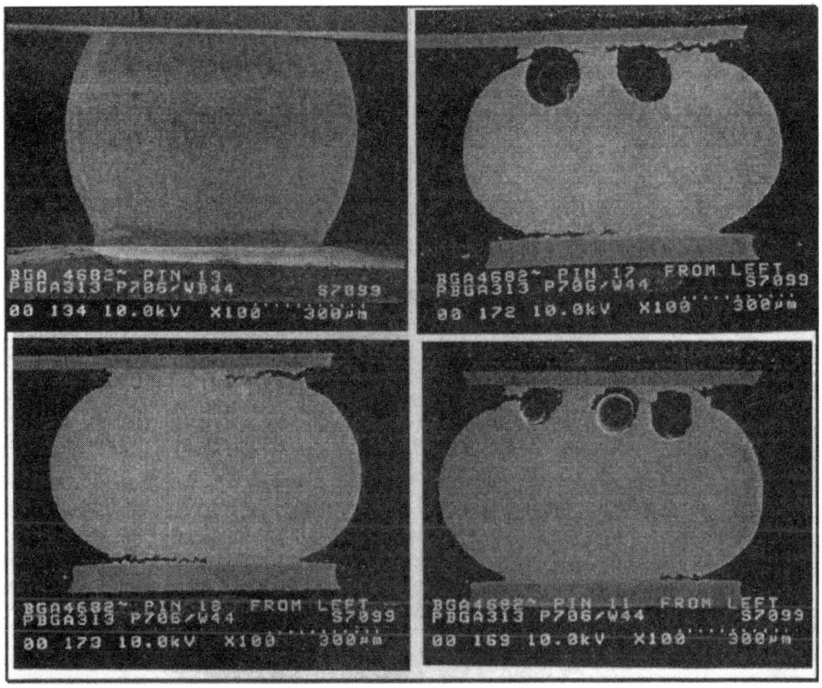

FIGURE 20.14 SEM and cross sections of PBGA 313-I/O package after 4682 cycles under cycle A conditions (−30 to 100°C).

Package attachment become very weak after 4500 D cycles, and the package could be detached easily from the board. Packages were detached, and images of balls that remained on package were mapped. The ball map distributions for several detached PBGA 313-I/O packages are shown in Fig. 20.15. All joint failures were not from the board interface, and there appeared to be a trend in their distribution.

For this depopulated full-array package, most balls were separated from the package interface, as evidenced by the low number of dark spots that represents the balls on the package. This is even more noticeable for the balls under the die. More joint failures at the package under the die portion could have occurred from the extreme local CTE mismatch of these joints. Similarly, mapping for other plastic packages was performed, and trends are being studied. For ceramic packages, mapping was not possible because most of the packages were already separated from the board at 4500 cycles. Failures were mostly at the package interface, since detached packages generally had no balls on them. In addition, several test vehicle boards had a large number of balls remaining at the ceramic package locations.

Figure 20.15 shows cycles to first failure for four daisy chains of PBGA 313. The chains were for balls under the die, the edge die row, outer/inner rows, and outer rows. As expected, the majority of the cycles to first failure were from the inner row, those under the die. Those which lasted longest were from the outer rows. There were overlaps, however, in distribution, and distinctions between sites of failure cannot be clearly defined. Similar plots also were generated for other packages.

For SBGA 352, package daisy chains were divided equally into four regions, each covering a corner of the package. For this package, cycles to first failure were random from any of the four daisy chains. For PBGA 256 and SBGA 560, both with peripheral-array configuration, it appeared that cycles to first failure were not from the inner to outer rows.

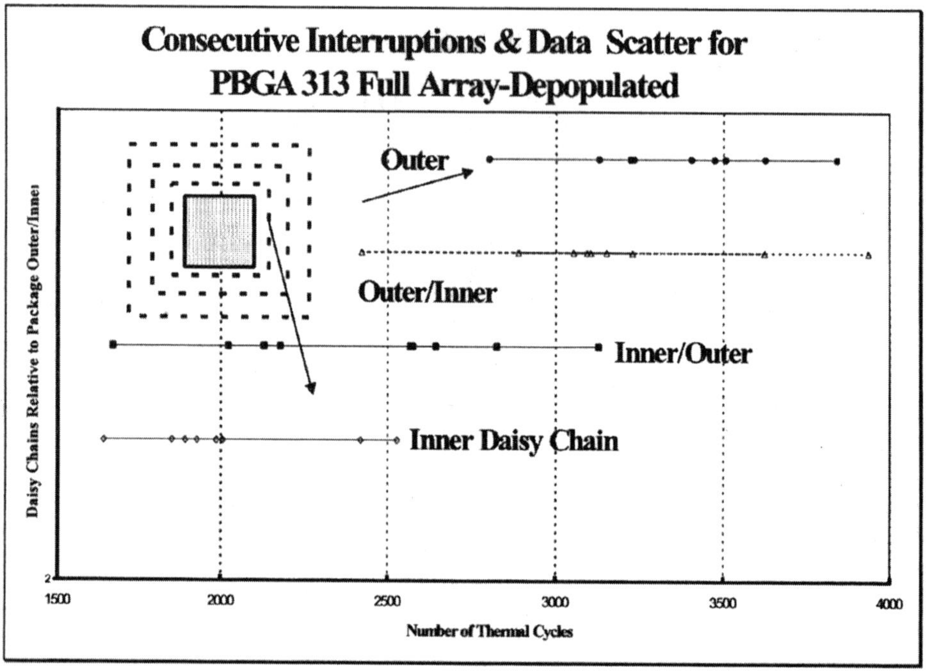

FIGURE 20.15 Cycles to first failure for inner to outer daisy chains of PBGA 313.

20.8 PBGA THERMAL CYCLING RESULTS

Figure 20.16 shows cycles to first failure for all plastic packages: PBGA 313 and 256, OMPAC 352, and SBGA 352 and 560. The results are for assemblies on FR-4 and polyimide PWBs. Preliminary results for two assemblies showed generally higher cycles to first failure for polyimide under cycle B conditions. The most recent PBGA 313 assemblies that failed under cycle A conditions are also included in the plots for comparison.

These plots were generated by ranking cycles to first failures from low to high and then approximating the failure-distribution percentiles using a median plotting position, $F_i = (i - 0.3)/(n + 0.4)$,[5] where i is the ranking number and N is the total of test vehicles. No attempts were made to generate Weibull parameters because of sample size and wide distribution.

Several trends were identified. PBGA 313, with a depopulated full array with solder balls under the die, was the first plastic package to show signs of failure at both cycling conditions. The scatter for both was narrow, indicating a large m Weibull parameter.

The next package assembly failure was SBGA 560. This package and the PBGA 256 with failure data at the extreme cycles showed very large scatter, indicating a low m value. Scatter in PBGAs, in general, could be attributed to slow failure progress. Scatter in failure data for SBGA 560 could be due to the immaturity of the package at the time of use. These packages were taken from those assembled in the early stage of production. The reasons for a large scatter for the PBGA 256 cycles to first failure data are not known at this time. Both OMPAC and SBGA 352 failed at later cycles than SBGA 560, and SBGA 560 was between a low tail for PBGA 313 and a high tail for PBGA 256. The 352-I/O package showed low scatter in cycles to first failure data.

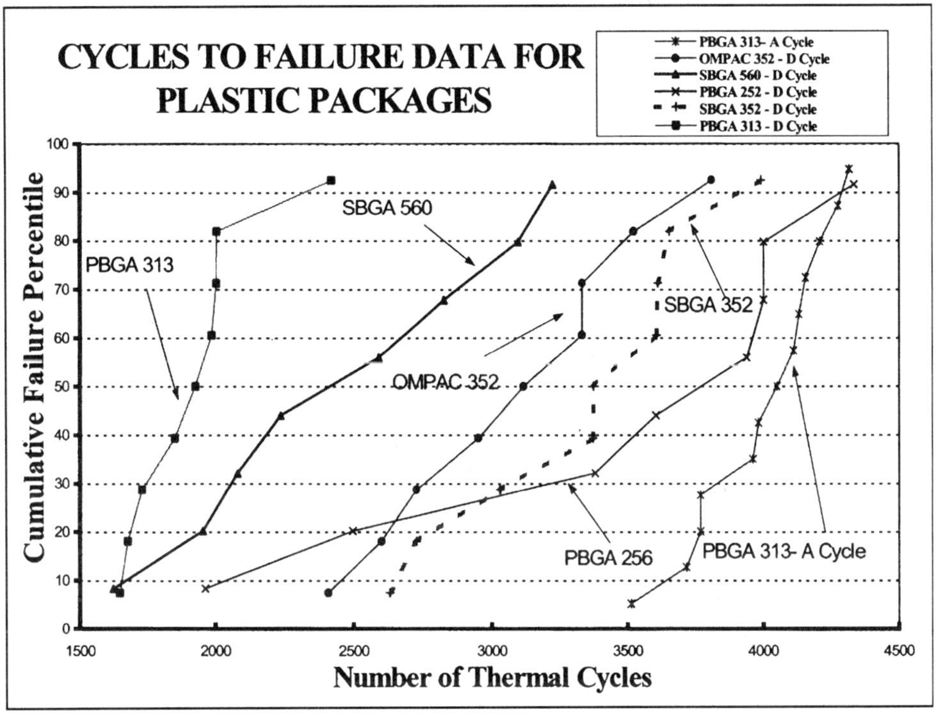

FIGURE 20.16 Cumulative cycling to first failure for PBGA assemblies (cycle B, −55 to 125°C) and PBGA 313 (cycle A, −30 to 100°C).

20.9 CBGA THERMAL CYCLING RESULTS

Figure 20.17 shows cycles to first failure for CBGA 625 under four different thermal cycling conditions. Again, to generate plots, the cycles to first failure were ranked from low to high, and failure distribution percentiles were approximated using a median plotting position, $F_i = (i - 0.3)/(n + 0.4)$.

Weibull parameters for cycles to first failure also were generated and plotted in continuous graphs. Weibull parameters for the four conditions are listed in Table 20.2. In addition to cycling conditions, it includes manufacturing and assembly variables that possibly affected the m values.

The two-parameter Weibull distribution was used to characterize failure distribution. The Weibull cumulative failure distribution was used to fit cycles to failure data. The equation is

$$F(N) = 1 - \exp\left[-(N/No)^m\right]$$

where $F(N)$ = cumulative failure distribution function

N = number of thermal cycles

No = scale parameter that commonly is referred to as characteristic life and is the number of thermal cycles with 63.2 percent failure occurrence

m = shape parameter—a large m is approximately inversely proportional to the coefficient of variation (CV) by $1.2/CV$; that is, as m increases, spread in cycles to failure decreases

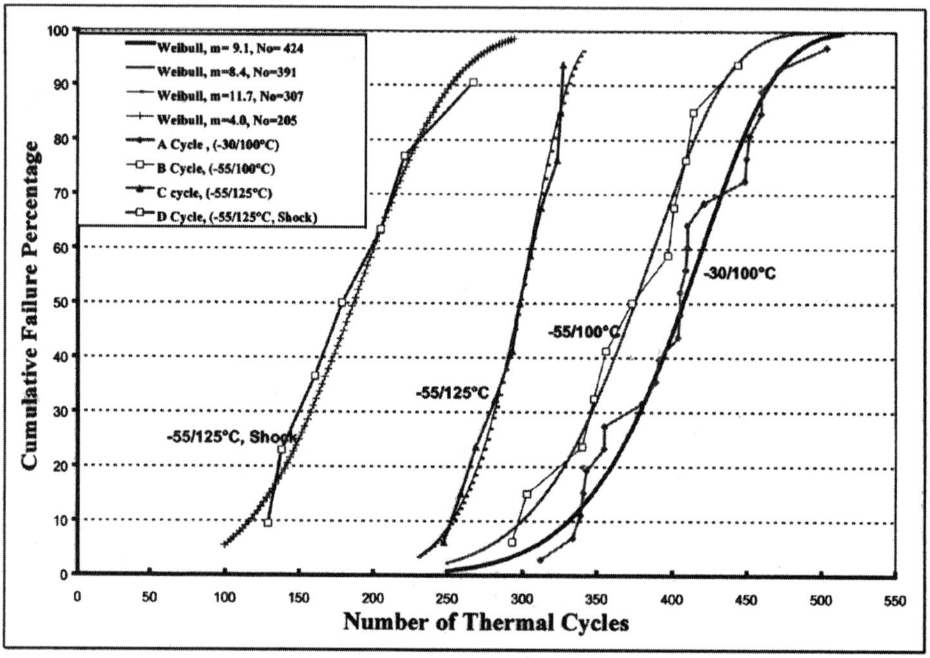

FIGURE 20.17 Cycles to first failure for CBGA 625 assemblies under different cycling conditions.

TABLE 20.2 Weibull Parameters for CBGA 625 Under Different Thermal Cycling Conditions

Cycling condition	Weibull *No* (N50%)	Weibull *m*	Comments
A (−30–100°C)	424 (407)	9.1	23 test vehicles (TVs), includes FR-4, polyimide PWB, and three solder volumes
B (−55–100°C)	391 (373)	8.4	11 TVs, FR-4 and polyimide
C (−55–125°C)	307 (298)	11.7	11 TVs, different assembler
D (−55–125°C, near thermal shock)	205 (189)	4.0	7 TVs, 3 on FR-4, 3 on polyimide, 1 with high solder volume

This equation, in double logarithm format, results in a straight line. The slope of the line will define the Weibull shape parameter.

The widest spread, i.e., $m = 4$, was observed for those assemblies cycled under condition D. The other three conditions had m values ranging from 8.4 to 11.7, with the narrowest spread, $m = 11.7$, for condition C. Near-thermal shock with two hot and cold extreme temperatures and the use of a limited number of test samples are possible causes of low values for m under condition D. Only seven test vehicles were tested: three on polyimide, three on FR-4, and one with solder volume above the norm. Test vehicles tested under condition C were manufactured at another facility under different manufacturing conditions.

20.9.1 Fatigue Equation

The Coffin-Manson relationship is a simple relationship in which cycles to first failure are inversely proportional to creep strain. Its modified version includes the effects of frequency as well as the maximum temperature in the form of

$$(N_1/N_2)\ \alpha\ (\Delta\gamma_2/\Delta\gamma_1)^{\beta}\ (f_1/f_2)^{\kappa}\exp\ [1414\ (1/T_1\ -\ 1/T_2)]$$

- N_1 and N_2 are cycles to first failure under two plastic strain conditions. β is the fatigue exponential and generally is assumed to be equal to 1.9.
- $\Delta\gamma$ is proportional to $(DNP/h)\ \Delta\alpha\Delta T$, where DNP is the distance from neutral point, h is equal to the solder-joint height, $\Delta\alpha$ is the difference in the CTE of package and PWB, and ΔT is the cycling temperature range.
- f_1 and f_2 are frequencies. κ is the frequency exponential varying from 0 to 1, with value 0 for no frequency effect and 1 for the maximum effect, which depends on materials and testing conditions. A value equal to 1/3 is commonly used to extrapolate the laboratory accelerated thermal cycles to first failure data with short duration (high frequency) to on/off field operating cycle with long duration (low frequency), i.e., a shorter field cycles-to-first-failure projection.
- T_1 and T_2 are absolute maximum temperatures under the two cycling conditions.

20.9.2 Projections

The preceding relationship was used to correlate the test results for CBGA 625 generated under the preceding four thermal cycling conditions. Cycles to 50 percent failure data are shown in Fig. 20.18. Projections to lower temperature ranges from two data points and a single data point using the preceding relationship also were plotted. Projections were compared with a test result for the same package thermal cycled in the range of 0 to 100°C (Martin et al., *SMTA International,* 1997). The following key points were considered:

- *Effect of frequency.* The effect of frequency was considered to be minor because frequency ranges in our study were very similar and their dwells at maximum temperatures were at least 10 min.

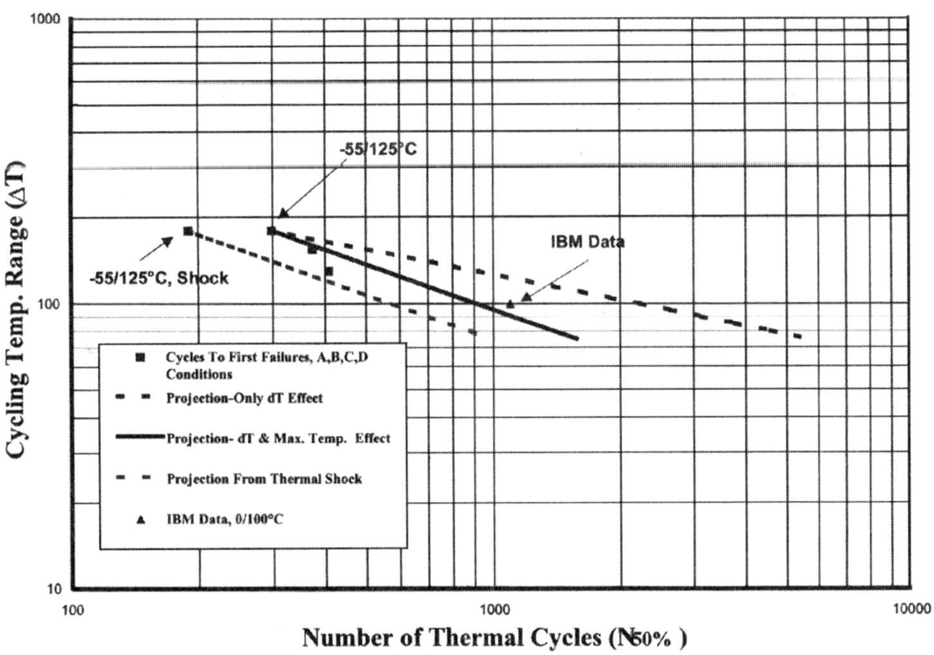

FIGURE 20.18 Effect of temperature range on 50 percentile failure cycles for CBGA 625 and projection using a modified Coffin-Manson relationship.

Creep was assumed to be almost complete for the two maximum cycling temperatures of 100 and 125°C. These temperatures are well above the creep temperature of the eutectic Pb/Sn solder.

- *Projection from cycles C and B (−55 to 125°C and −55 to 100°C) to lower cycling ranges.* The projection lines use the Coffin-Manson relationship. It includes plots from projection with and without considering the effect of maximum temperature. Projections with correction for maximum temperature better approximate the test results under conditions C and A as well as an IBM data point.
- *Projection from near-thermal shock data.* Cycles to first failure under thermal shock conditions were much lower than those under low-heating/cooling-rate cycling; therefore, projection to other temperature ranges provides conservative failure cycles. Characteristic changes in solder behavior under rapid heating and cooling as well as failure mechanism changes for thermal shock conditions may not realistically represent most field applications.
- *Projection from −55 to 100°C and −30 to 100°C to lower temperatures.* This projection is not shown in the figure, but it is conservative because the slope between the two data points is much deeper than the slope connecting data sets for the −55 to 125°C and −55 to 100°C cycles. This means that damage induced by fatigue due to a decrease in cold temperature is not as severe as creep induced by an increase in hot temperature.

20.10 SHOCK AND THERMAL CYCLING SYNERGISM EFFECTS ON RELIABILITY OF CBGAs

For electronics in commercial applications, commonly, thermal cycling tests are performed to simulate on/off conditions. However, most electronic systems are exposed to other environments, including mechanical fatigue and random vibration. For example, vibration occurs during transportation, and mechanical fatigue occurs by repeated use of key punching for portable electronics. Occasional high shocks could occur due to accidental drops. Drop test and mechanical fatigue are beginning to be considered for qualification of electronic assemblies, especially for newer CSP assemblies.

In addition to much harsher thermal requirements for aerospace and military applications, generally assemblies are required to meet severe dynamic loads and vibration fatigue cycling. Therefore, there is a strong need to understand assembly behavior under such stress conditions. Very limited data are available for vibration and especially shock behavior of BGA assemblies. To understand behavior for space environments, several test vehicles were subjected to dynamic testing representative of a launch, 3 min at 3 axis vibration/shock.

Synergistic stresses including thermal cycles representative of the following environmental conditions were considered:

- Prelaunch thermal cycling due to manufacturing, repair, screening, storage, and transportation
- Dynamic testing representative of the launch environment, including sinusoidal vibration, transient vibration, pyroshock, and acoustic vibration
- Thermal cycling representative of internal and external temperature changes of a spacecraft

Effects of several test variables were considered:

1. Thermal cycling before vibration/shock and after
2. Increase in rigidity of board by addition of thin rigid strips in one case and bonding to a rigid plate in another case
3. Change in thermal history condition by exposure to thermal cycles before vibration/shock

This section presents experimental results as well as analyses for two CBGA assemblies (361 and 625 I/Os) subjected to either or both thermal cycling and vibration.

20.10.1 Literature Review of Dynamic Behavior of BGAs

Cycles to first failure after vibration of CBGAs and column ceramic grid arrays (CCGAs) were presented by M. Cole of IBM.[6] Balls and columns were 90Pb/10Sn and were 0.89 mm (0.035 in) in diameter for CBGAs and 0.5 mm (0.02 in) in diameter for CCGAs and 2.2 mm (0.050 or 0.087 in) in height. Packages were assembled on FR-4 boards with single test specimens mounted on 110 × 90-mm boards, clamped by screws at 101 × 75-mm locations. Mil-STD 810E was used to generate impact and vibration data for the test vehicles.

No failures of CBGAs and CCGAs were observed after vibration with a heat sink of 73 g. CCGAs with heat-sink weights of 100 and 150 g failed, whereas CBGAs did not. Cracks were induced in CBGAs in the eutectic solder either in package or board sites when subjected in 20 to 2000 Hz with 7.73 g. It was reported that for CBGAs, crack initiations were similar to those of accelerated thermal cycling (ATC) but with no deformation typically present in ATC. Also, thermal mismatch induced both shear and tensile stress, but vibration induced primarily tensile stress and did not cause deformation.

It was observed that CBGA assemblies with heat sinks of lower weight than 150 g that were subjected to shock and vibration did not show any degradation in thermal fatigue life. No statistical differences between those with initial shock and vibration and those without any a priori test were observed.

In another study,[7] CBGAs with 256, 625, and 1089 I/Os were subjected to thermal cycling and vibration evaluation. The 1089-I/O package was fabricated with balls rather commonly supplied in columns because they were used on metal matrix-restraining core rather than polymeric board such as FR-4. For restrained board with a CTE (6 ppm/°C) close to the ceramic package (7.2 ppm/°C), local mismatch between solder and package/board was considered to play a primary role rather than global mismatch. Assemblies on restrained board were subjected to thermal cycling in the range of −55 to 125°C (a 5°C/min ramp and a 30-min dwell at each extreme). No failures were reported to 1500 cycles.

Vibration tests also were performed on three CBGAs at three levels:

- Level I was the military avionics equipment workmanship vibration level of environmental stress screening (ESS).
- Level II was an enveloped subassembly based on several avionics programs.
- Level III was the same as level II with power spectral density (PSD) increased by 3 dB.

Test results are listed in Table 20.3. All assemblies passed the level I, whereas only CBGAs with 254 I/Os passed the level II vibration test. Finite-element prediction for level I agreed with the experimental results but for other levels had mixed agreement.

20.10.2 Vibration

Three test vehicles with three levels of rigidity representing three loading conditions were stacked with spacers and subjected to random vibration. The bottom TV was bonded to an aluminum plate, the middle had center and edge stiffener strips, and the top had no stiffeners. The test vehicles were clamped from the two sides of the PWB, the sides with no connectors, as shown in Fig. 20.19. The block of three stacked test vehicles was mounted on a very stiff, thick aluminum plate on the shaker table with their natural frequencies well above the test vehicle assembly range.

TABLE 20.3 Vibration and Thermal Cycling for CBGAs

Vibration level	Duration (min)	Input level (g)	Response (g)	CBGA 256 I/O	CBGA 625 I/O	CBGA 1089 I/O
I	10	6.06	21	Pass (4/4)	Pass (2/2)	Pass (3/3)
II	60	12.8	37	Pass (4/4)	Pass (0/2)	Pass (0/3)
III	120	18.07	53	Pass (1/4)	N/A	N/A

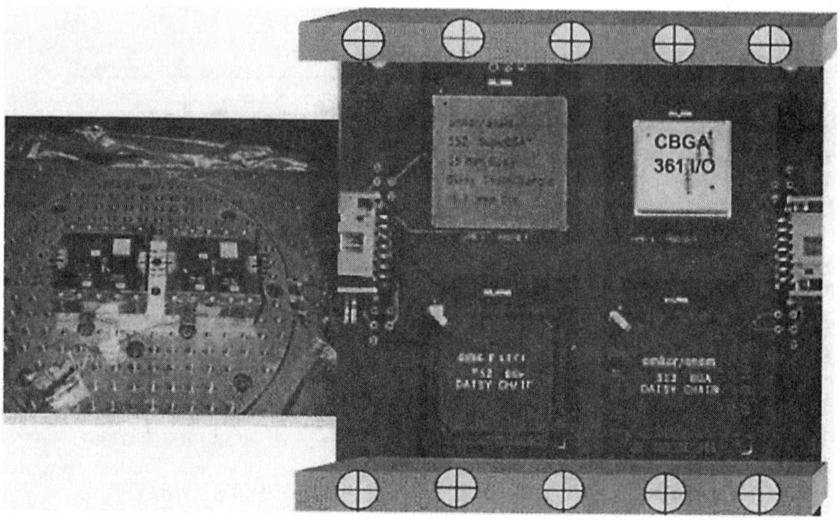

FIGURE 20.19 Random vibration test setup, (*Left*) Two three-stack of test vehicle on vibration table. (*Right*) Enlarged test vehicle with the CBGA 361 at the top right corner.

Initially, a very low vibration spectrum was sweeped to determine natural frequencies for the test vehicle. The board was subjected to a vibration spectrum in the range of 200 to 2000 Hz.

20.10.3 Natural Frequency Measurement/Projection Test Results

Plots for natural frequencies were generated. A simplified finite-element analysis was performed to predict natural frequencies and stress/strain conditions at the solder joints.[8] Ten percent model damping was used for the random response calculation. Projections for the first, second, and third natural frequencies are compared with measured values, as listed in Table 20.4.

20.10.4 Damage Induced by Thermal Cycling Alone

Both board and package interface cracking was observed with increasing number of thermal cycles. Figure 20.13 shows typical failures for the two cycling conditions, A and D. Failure under cycle A conditions generally was from the PWB and for cycle D conditions from the package sites. Failure mechanism differences could be explained either by global or local stress conditions.

Modeling indicates that the high-stress regions shifted from the boards to the packages themselves when stress conditions changed from global to local. For cycle A conditions, with slow heating/cooling ramping, which allowed the system to reach a uniform temperature, damage could indicate a global stress condition. For cycle D conditions, with rapid heating/cooling, damage could indicate a local stress condition.

TABLE 20.4 Measured and Finite-Element Projection of Natural Frequencies and Stress/Strain for CBGA 361

TV condition	Frequency mode 1 (modes 2,3)	Amplitude (g^2/Hz)	Maximum stress (psi)	Maximum strain (μstrain)
Measured values for no stiffener	356 (688,1700)	190	N/A	N/A
FEA for TV with no stiffener	381 (594,795)	170	945	315
FEA for TV with stiffeners	593 (820,1625)	43	558	185
FEA for TV bonded to Al plate	3155 (3520,4700)	N/A	25	8

20.10.5 Damage Induced by a Priori Vibration

Damage induced by vibration was different in some cases, as shown in Figs. 20.20 and 20.21. Appearance of tensile deformation from the center of high balls as shown was significantly different from that observed for thermal cycling conditions. Similar to thermal cycle conditions, damage was more dominant for the balls with higher distance to neutral point (DNP), especially for the corner balls. However, additional microcracks in eutectic solder joints, different from the norm for thermal cycling, were induced by tensile and shear load during random vibration.

20.10.6 Synergism of Vibration and Thermal Cycling

Cycles to first failures for three assemblies with levels of rigidity and a priori vibration conditions are shown in Table 20.5. Test data for cycles to first failure after vibration for the most severe conditions,

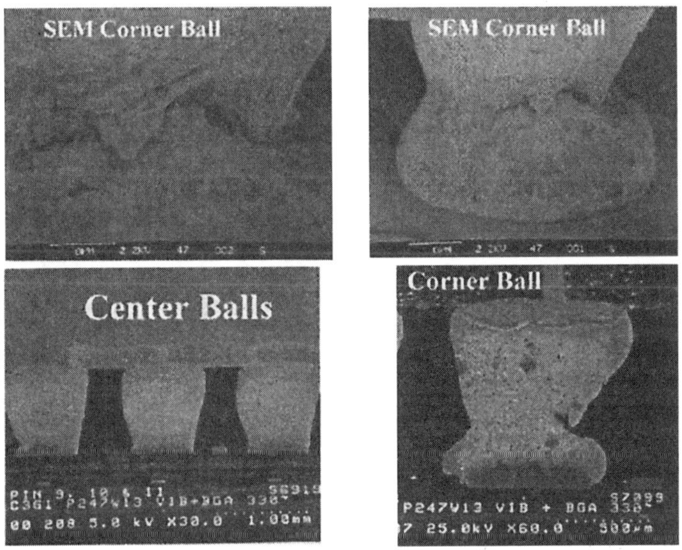

FIGURE 20.20 Thermally cycled samples after vibration. Note tensile failure for the corner ball and minimum damage for the center balls.

FIGURE 20.21 Thermal cycled samples after vibration. Note mechanical crack propagation at two locations and minimum damage at a center ball.

TABLE 20.5 Cycles to First Failure after Random Vibration

TV condition	ID, vibration (3 axes), and thermal cycle type	Thermal cycles to failure
No stiffener under vibration	4, vib. + cycle B	4, failed between 292 and 326 cycles
	50, vib. + cycle B	50, failed at 400 cycles
	34, vib. + cycle A	34, no failure to 434
With stiffeners	33, vib. + cycle A	33, failed between 292 and 326 cycles
	13, vib. + cycle A	13, failed at 330 cycles
Bonded to Al plate	31, vib. + cycle A	31, No apparent change to virgin test vehicles

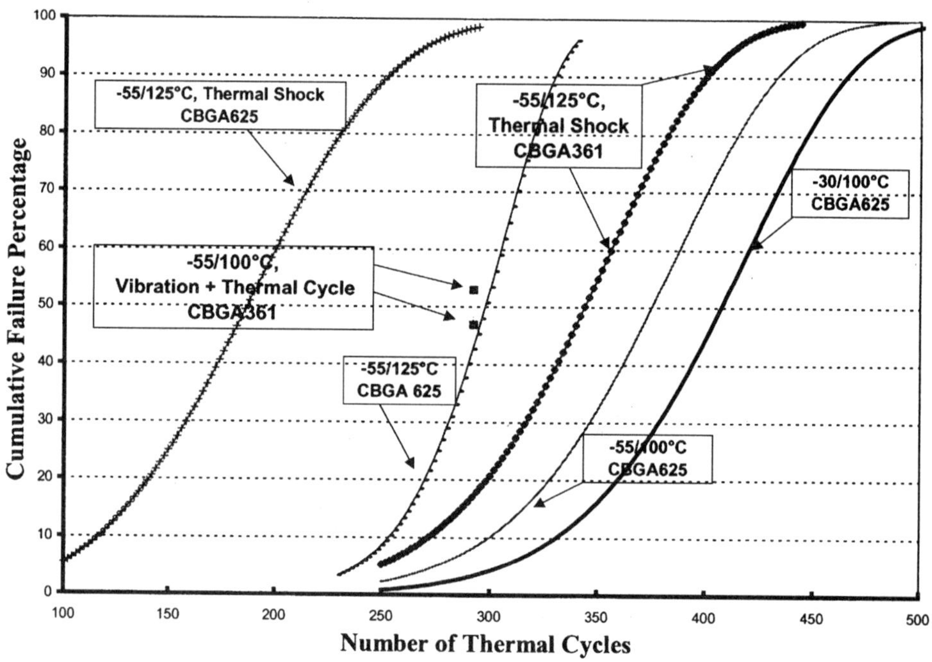

FIGURE 20.22 Cycles to first failure data for two CBGAs under four thermal conditions with two assemblies with a priori vibration conditions.

i.e., the test vehicles with no rigidity enhancement, were marked in Fig. 20.22 for comparison only. It is difficult to draw a statistically meaningful conclusion because of insufficient sample size.

20.11 BGA CONCLUSIONS FOR QUALITY AND RELIABILITY ASSURANCE PROGRAMS

- Plastic assemblies with I/O counts from 254 to 560 on FR-4/polyimide PWBs showed an order of magnitude larger number of cycles to first failure than ceramic assemblies with I/O counts from 361 to 625. They were in thousands of cycles for plastics, whereas in hundreds for ceramics, in cycling temperature ranges of −55 to 125°C, −55 to 100°C, and −30 to 100°C. Not reported here, vibration and mechanical shock exposures prior to thermal cycling were shown to have varying effects on the subsequent cycling to failure.

- Near-thermal shock conditions induced the most damage on CBGA assemblies compared with thermal cycling. Up to 50 percent reductions in cycles to first failure were observed when heating/cooling rates significantly increased for the thermal cycling in the range of −55 to 125°C.
- Failures for thermal cycled CBGAs were either from board or package sites in the eutectic solder joints. Those subjected to vibration showed tensile deformation in high-melt balls and tensile and shear failures in eutectic solder.
- An increase in the maximum cycling temperature from 100 to 125°C had a more damaging effect than a decrease in the minimum temperature by the same amount (−30 to −55°C), keeping the ΔT constant. Fatigue, rather than creep, is considered to be the main damage mechanism below one-half the absolute melting temperatures.
- Cycles to first failure can be projected reasonably using a simple modified Coffin-Manson relationship and accelerated thermal cycle test results.
- Assemblies with three-level rigidity passed a launch random vibration/shock condition. However, cycles to first failures after vibration were affected by the rigidity of the board, a significant reduction (>50 percent) for less rigid and minimum reduction for highly rigid assemblies after vibration.

20.12 REFERENCES

1. Ghaffarian, R., "Ball Grid Array Packaging Guidelines," distributed by Interconnect Technology Research Institute (ITRI), August 1998; available at *http://www.ITRI.org.*
2. Ghaffarian, R., and Kim, N., "Ball Grid Reliability Assessment for Aerospace Applications," in *Proceedings of the 30th International Symposium on Microelectronics,* Philadelphia, PA, October 1997, pp. 396–400.
3. Ghaffarian, R., "CBGA/PBGA Package Planarity and Assembly Reliability," in *The 1997 International Flip Chip, Ball Grid Array, TAB and Advanced Packaging Symposium Proceedings,* Semiconductor Technology Center, Neffs, PA, February 17–19, 1997.
4. Lau, J. (ed.), *Ball Grid Array Technology,* McGraw-Hill, New York, 1995.
5. Kapur, K. C., and Lamberson, L. R., *Reliability in Engineering Design,* John Wiley & Sons, New York, 1977.
6. Cole, M., Kastberg, E., and Matin, G., "Shock and Vibration Limits for CBGA and CCGA," in *Surface Mount International Proceedings,* Surface Mount Technology Assoc., Edina, MN, 1996, pp. 89–94.
7. Marsico, J. W., AIL Systems, private communications, 1999.
8. Pitarresi, J., University of New York at Binghamton, private communications.

ACKNOWLEDGMENTS

Part of the research described in this chapter was carried out by the Jet Propulsion Laboratory, California Institute of Technology, under a contract with the National Aeronautics and Space Administration.

Special thanks to Drs. D. Mulville, NASA HQ Code AE, M. Greenfield, NASA HQ Code Q, H. K. Detweiler, Director of Safety and Mission Assurance Directorate (SMAD), T. Gindorf, Manager of Assurance Technology Office, P. Barela, Manager of Quality Assurance Office, and P. Zulueta, Group Supervisor of Applications Engineering, for their vision, invaluable support, and continuous encouragement. I wish to acknowledge in-kind contributions and cooperative efforts of JPL-led BGA consortium team members with my deepest appreciation to others who contributed to meet the objectives of the programs.

CHAPTER 21
DIE ATTACH AND REWORK

Ken Gilleo
Cookson Electronics

21.0 INTRODUCTION

The semiconductor die or chip typically is bonded to a substrate for several critical reasons. First, an unattached die is subject to much greater mechanical abuse during handling of the package and assembly during manufacturing. The final end product can expect considerable abuse during its lifetime, mostly as mechanical shock. Portable products must be especially robust because users have come to expect a system that can be mistreated, including dropping. Attaching the die to a lead frame, chip carrier, or circuit board makes the system very much more robust and immune to most mechanical breakage of both the die and the interconnection structure, which can be very fragile. The attached die is also better able to dissipate heat to the substrate, particularly when a metal lead frame or thermally conductive metal or ceramic package is used. The die-attach material also can be designed with electrical conductivity in order to ground the back of the die.

There are three specific categories of die-attach materials: metallurgic, inorganic (ceramic), and polymeric. The earlier materials were either pure metals or alloys that provided an excellent thermal path. Metallic die-attach systems are still used but represent only a small niche today. Ceramic adhesives, most notably silver-glass, also have been used for some time. However, polymer adhesives are the dominant materials for today and are used widely for area array packaging. More than three decades ago, polymer-based adhesives began to displace metals as the preferred attachment materials. The polymer adhesives provided a simple, reliable, and automatable means of die attachment.

Polymers can be divided into two major classes, thermosets and thermoplastics. Both are used for die attach, and each has a unique set of useful attributes. Thermosets, the de facto standard, are easier to use but form permanent bonds that preclude repair. Thermoplastics are easily reworkable but may require modified die bonders or special equipment. Before moving into the various die-attach processes, a brief discussion of polymers will be provided.

21.1 POLYMER BASICS

Polymers are molecules made of long chemical chains with high molecular weights. A single polymer chain can be made up of hundreds or thousands of atoms, making polymers some of the largest molecules found in nature or synthesized. However, since the individual polymer chains in the thermoset class are interlinked, these polymers can contain millions of atoms. The long, chainlike structures are the key to the unusual strength, elasticity, and general resilience so common to many polymers. Many polymers, both natural and synthetic, are stronger than metals on a weight basis. High strength, low density, strong adhesion, and excellent electrical properties make them the best choice for die attach in the vast majority of applications.

21.1.1 Thermoplastic versus Thermoset Polymers

Polymers are easily classified into two major categories: thermoplastics and thermosets. Thermoplastics are capable of softening or even melting when heated but will return to a solid state when cooled. The process is repeatable and does not alter the polymer characteristics to any extent. However, heating can "anneal" films to reduce stress, but this is a physical rather than a chemical change. Heating simply imparts enough thermal energy (molecular vibration) to allow the long polymer chains to move freely past one another and take on new shapes. Cooling reduces molecular motion to a level where chains no longer move freely.

Thermosets do not display this reversible phase-change behavior. During the polymerization, or chain-building, step, thermoset polymers form links, or chemical bonds, between adjacent chains. The result is a three-dimensional network that is much more rigid than the linear thermoplastic structure. The interlinked chains are not free to move when heat is applied, and the thermoset, as the name implies, is "set" into a permanent shape after polymerization. The level of cross-linking can be varied. Materials with high cross-link densities are hard, rigid, and somewhat brittle substances. Heating to high temperatures can soften thermosets with low cross-link densities, but they do not melt, and their original shape is retained.

Both types of polymers are used in electronics, but the thermosets, like phenolic and epoxy resins, have a long history as wiring board and packaging materials. Thermoplastics, with the exception of polyester and polyimide film, popular flexible circuitry dielectrics, have only recently gained popularity in electronics. Advanced thermoplastics now can provide the high melting points and impressive dimensional stability that earlier polymers could not deliver. State-of-the-art thermoplastics now perform in the same temperature ranges as the thermosets. Low moisture absorption and low thermal coefficients of expansion are now possible, making thermoplastics the preferred type of polymer for such applications as molded circuits, encapsulation, and die attach. It is the *remeltable* quality that sets thermoplastics apart and provides special processing advantages. Table 21.1 compares thermoset and thermoplastic characteristics.

TABLE 21.1 Thermoplastic versus Thermoset Properties

Property	Thermoplastic	Thermoset
Initial physical state	Solid	Liquid or low-melting solid
Polymerization	Prepolymerized by supplier	Polymerized by user
Bonding temperatures	Low to high; near softening point	Can be at ambient
Cure time	0; already cured	Variable; minutes to hours
Bonding time	Can be < 1 s	> 1 minute
Ionic content	Near 0	Can be > 50 ppm
Storage	Room temperature	Usually freezer for one-part
Available forms	Solid film, solvent paste	Solventless paste, solid preform
Reworkability	Good to excellent	Generally poor
Property variability	Narrow range	Moderate, depends on cure
Types	One-part only	One- or two-part
Outgassing	Low	Low to high
Solvent resistance	Low to moderate	Typically high
High-temperature performance	Moderate	Moderate to high
Dimensional stability	Moderate	Can be high

21.2 DIE-ATTACH MATERIALS

21.2.1 Metallurgic; Solders

Gold has been used for several decades to bond silicon chips to metal lead frames. A common "solder" consists of 94 percent gold and 6 percent silicon. The common form is ribbon or preforms cut to the die size. Heat is applied by the die pickup collet that also vibrates in the *xy* plane to "scrub" the chip and metal together. This allows a hard bond to form. The bonding temperature is over 400°C, and the process takes 5 to 10 s. While the resulting bond has excellent thermal and electrical properties, the high processing temperature and relatively long processing time are considered to be deficiencies for metallurgic die attachment. However, performance is unmatched by other types of materials.

21.2.2 Glass Die-Attach Materials

Metal-glass materials emerged later as a specialty bonding system. Silver-glass blends have been used for some time and are still used today, although volume continues to diminish. The silver-glass adhesives have good thermal conductivity and high stability, thus providing outstanding reliability. However, the glass bonding agents also require high bonding temperatures, even though the temperature has been reduced over the years. The bonding temperatures are too high for use with organic chip carriers. Like metal, the glass adhesives are considered to be premier high-reliability systems and are used with some ceramic packages.

21.2.3 Polymer Die-Attach Adhesives

Metal Fillers. Silver-filled epoxy thermosets are the largest class of die-attach adhesives in use today. They provide efficient processing, low-temperature bonding, electrical conductivity, reasonably good thermal conductivity, and generally high reliability. Epoxies are the most commonly used binder or matrix material because of a good balance of properties. The most common form is a one-part paste that is thermally cured but stored in a freezer to prevent premature polymerization that can occur at ambient temperatures. Newer polymers are also used, including polyimides, cyanoacrylates, cyanate esters, and silicones. Some of the resins, such as the cyanate esters, can be coreacted with epoxies However, the epoxies continue to dominate even though the alternates may have modestly better properties such as very low moisture absorption. However, the high demands required for the increasingly thin plastic ball grid arrays (BGAs) may open up a market for these more advanced die-attach systems. Lower moisture absorption, which would reduce explosive delamination (called *popcorning*) on soldering, may be more important if higher-temperature lead-free solders are implemented.

Silver-filled reworkable thermoplastics are also used and are supplied in paste, ribbon, and die-sized preforms. The ribbons and films, unlike the epoxy thermosets, can be stored at room temperature because they are already polymerized. The thermoplastic resins are often totally free of undesirable ionic contaminants that are nearly always present in epoxies due to their formation during synthesis. Gold is also used as a filler for very high reliability applications where silver's high chemical activity could be an issue.

Inorganic Fillers. Inorganic compounds such as aluminum nitride (AlN) are also used as fillers for die-attach adhesives where electrical insulation is required but thermal conductivity is important. Fillers also reduce the coefficient of thermal expansion (CTE), thus providing a closer thermomechanical match to the die. Silicon has a CTE of about 2.5 ppm/°C, whereas polymers can approach more than 100 ppm/°C. Epoxies fall within 70 to 90 ppm/°C range, and thermoplastics can be twice as high. The addition of fillers with small CTE values lowers the CTE value of the adhesive to a more reasonable number ranging from 10 to 20 ppm/°C.

Unfilled. Pure polymer die-attach materials are used in some cases, but these are mostly thermoplastics where temporary attach is the goal. The short duration required for testing may not require fillers. Absence of filler allows faster and lower-temperature bonding of thermoplastic films. Die removal of the tested and burned-in dies is often simpler without filler. However, filled die-attach adhesives are preferred for permanent bonds.

21.3 DIE-ATTACH APPLICATION METHODS

21.3.1 Needle Dispensing

Needle dispensing is by far the most popular process for applying polymeric die-attach adhesives. The dispenser can be a stand-alone automatic system or one that is integrated into the bonder. Force is applied to the adhesive paste by means of air pressure or a mechanical pump. Precision positive-displacement pumps offer the best control and repeatability. Modern equipment will deposit the desired amount of adhesive and also can be programmed to apply a specific pattern that can be useful for larger dies. Today, some area array packages contain multiple chips so that more sophisticated dispensing robots are required to place the correct amounts of paste in the desired locations. Figure 21.1 shows a needle-dispensing machine intended for high volume that can be used for die-attach adhesives (see Chap. 22 for more details).

21.3.2 Stenciling

Stenciling, also called *stencil printing,* can be used to mass-apply accurate amounts of adhesive paste for high-throughput applications such as area array packaging. A metal stencil is produced in a thickness equal to the desired deposition height. Openings are laser cut or chemically etched into the stencil metal that are equal to the die size and in the locations that match the array. Chip carrier substrate is often produced in sheets with hundreds of die bond sites. This is the ideal format for stenciling because all the bond sites can be coated accurately with die-attach adhesive in a single fast step. Area array substrate is also used in strip form, where a dozen or more die bond sites are presented. A

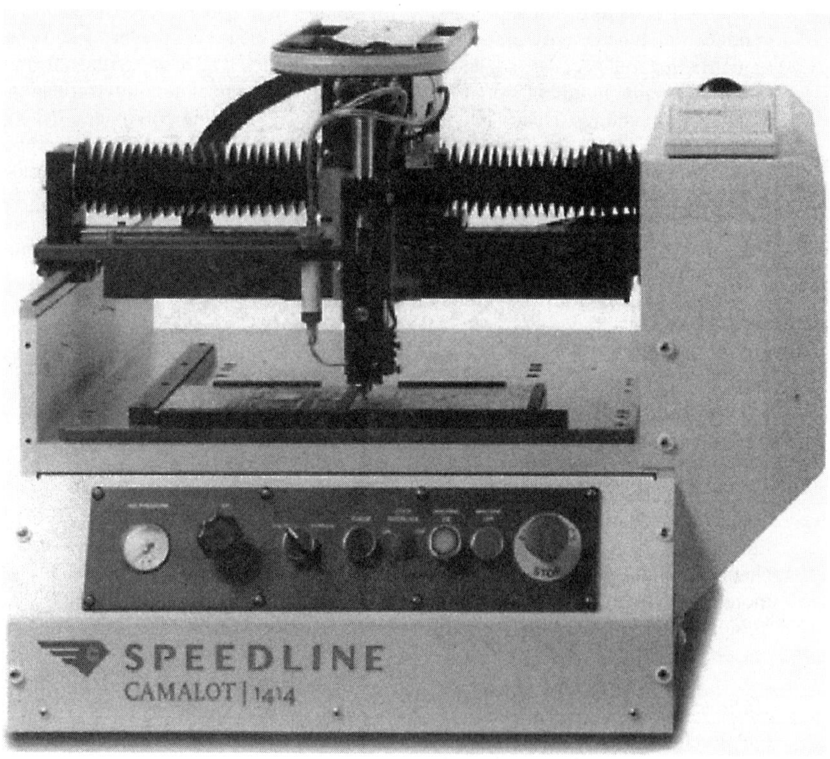

FIGURE 21.1 Needle-dispensing machine.

smaller stencil printer can be used that has automatic feed to accommodate the strip format (see Chap. 23 for more details).

21.3.3 Screen Printing

Screen printing is similar to the stenciling process and may use the same equipment. However, the printing tool has a continuous screen mesh across the holding frame. Openings are produced with a polymer emulsion applied over the entire screen and photoimaged into a pattern of openings. Printing occurs where emulsion has been removed to leave an opening in the screen mesh. Screen printing is more limiting than stenciling because the adhesive must be able to pass through the mesh and transfer cleanly to the substrate. Thermoplastic adhesives tend to be stringy and do not screen print very well, especially since dissolving the polymer in solvent makes the paste. Solvent evaporation of paste on the screen can cause blocking. Thermoset adhesives generally do not contain solvents but rather liquid reactants that do not tend to evaporate. This makes thermoplastics a better candidate.

Screen printing is more suitable for thinner deposits than stenciling and may be a better choice where less than 2 mils (50 μm) of deposit is needed. Stencils with a thickness of even less than 3 mils (75 μm) can be difficult to produce and handle. A 50-μm-thick stencil can be a hazard because such thinness effectively produces a knifelike edge. Screen printing is quite precise even down to 10 or 12 μm, which is not practical with stenciling.

21.3.4 Fluid Jetting

Fluid jetting is an off-contact dispensing process where a short-duration pressure wave or impulse is applied to fluid in a dispenser. A small needlelike orifice serves as the directing barrel. A piezoelectric "piston" can provide the right impulse to cause a droplet of adhesive to be ejected at the substrate. Pulses can be rapid, ranging up to many hundreds or thousands per second. The high rate in effect produces a seemingly steady stream of material that can be applied in the desired thickness and shape by moving either the head or the substrate. Presently, jetting is considered too slow for general die-attach adhesive use, but increasing rates may make it a better option.

21.3.5 Spin Coating on Wafer

Wafer coating can only be used with adhesives that can be hardened but remain bondable. While thermosets can be partially cured, or B-staged, to a solid that can be bonded later, the thermoplastics are the preferred class for this process. In fact, some thermoplastic die-attach adhesives are specifically designed for wafer use. Thermoplastic paste is deposited onto the back of the spinning wafer near the center. The spinning action produces flow out by means of centrifugal force to yield a thin, precision coating. The coating is then dried by solvent evaporation, typically by baking in an oven. The dried wafer now contains a solid die-attach material on the backside. The wafer can be singulated into individual dies by sawing. Many thermoplastic adhesives can be sawn without problems, provided that the saw blade does not overheat to the extent that melting occurs. The die is now ready to bond in an efficient process that only requires placing the chip with a heated pick tool, or thermode. Precoated dies can be bonded in less than 1 s, making this the fastest die bond method. The chip carrier may be preheated to increase the adhesion level and reduce bonding time. Stenciling and screen printing also can be used for wafer coating.

21.3.6 Film Application

Both thermosets (B-staged) and thermoplastics are available in films. The thermosets have a limited useful life unless kept in cold storage. The thermoplastic films, since they are already polymerized, have a nearly infinite life but should be used within 6 to 12 months because they can absorb contaminants and fillers can age. Room temperature storage in sealed plastic bags or in a dry box is preferred. The film typically is cut to match the die size before proceeding. The cut adhesive can be applied to the die or substrate using a process called *tack bonding*. Here, a lower temperature is used with just enough dwell time to cause the film to adhere. The cut film also can be placed on the substrate, the die placed on the film, and the stack can be bonded with heat and pressure. Films also can be applied at wafer level, but this is considered more difficult than liquid coating. The dies are singulated by sawing, as described previously.

Sizing, Cutting, Singulating. Film is available in several forms that can require cutting, blanking, shearing, and other mechanical processes. Rolls of film can be cut to reels of the desired width with automatic equipment. Some bonders can cut film to the right length to produce die-sized preforms, assuming that the film width is correct. Film punching equipment also can be used, but tooling is required for every size. Preforms are often placed in waffle packs for presentation to the bonder, and this can be done automatically by the film blanker. Vendors offer all forms that have been described.

Laser cutting has become increasingly popular because no tooling is required. Many die-attach films, especially the thermoplastics, cut very well. There can be problems with B-staged thermosets because the laser heating can polymerize the cut zone.

Film Attaching Process. The film is usually prebonded to the die or substrate before the die-attach bonding process, but this is not essential because the film also can be sandwiched between the substrate and die just prior to bonding. However, bonding the film accurately to the die is preferred because this will produce high placement precision and preclude the possibility of the loose film moving. The film tacking process requires just enough heat, pressure, and dwell to get some level of bonding. Since the thermoplastics are already polymerized, the process window can be quite wide.

Final Die-Attach Bonding. The final bond with film adhesives typically uses higher temperature and pressure to get complete die-to-substrate contact and to force any air pockets out to the die edge. The key parameters for dry-film bonding are temperature, pressure, and time. All three are interactive, and the user may trade one variable against another. For example, bond time can be reduced if the temperature or pressure is increased. Time can be reduced to a minimum, even down to milliseconds, if both pressure and temperature are increased. Conversely, increasing one or both of the other factors can reduce either pressure or temperature. This interactivity allows the process to be optimized over a wide range of products and applications. Some minimum pressure must be applied during the bonding process even though the value may be small; the weight of the die alone unfortunately is insufficient. Since common epoxy thermoset die-attach materials are fluids when the die is placed, no added pressure is required. Standard die-attach bonders therefore have been designed for minimal force application. Thermoplastic die-attach processing often requires equipment that can apply significant force. Although some existing die-attach equipment may apply adequate die bond pressure for thermoplastics, many are not suitable. Modified or totally new equipment may be needed. However, the need for equipment investment can be justified because the resulting benefits of reworkability, better precision, and faster bonding are realized.

Thermoplastic adhesives are suitable for all types of area array packages, especially multiple-chip types, where removing a die may be critical. The bonding surface mechanism is primarily a mechanical interlocking process and therefore is more dependent on morphology than on material composition. The triad of bonding parameters creates a wide processing window that enables thermoplastics to fit into various assembly manufacturing systems, particularly those with processing restrictions such as temperature sensitivity and time constraints. Thermoplastics thus can span the full range of packages: ceramic, metal, rigid organic, and flexible.

Adequate bonding is achieved by bringing the thermoplastic adhesive to the right viscosity range. Figure 21.2 relates viscosity to temperature and shows the generic bonding process window. Thermoplastic adhesives can bond over a broad process window with more than a 100°C span. Bonding is performed above the glass transition temperature T_g of the polymer but below the decomposition point. These two points determine the adhesive limits, but the substrate and die may set the upper bond limits if either is temperature-sensitive. Bonding is accomplished by applying pressure to the softened or liquefied polymer. It is important that adhesive be forced into the surface microstructure of the substrate and die for maximum bond strength. Bonding results from both atomic forces and mechanical linkage. The greater the penetration of the adhesive into the microstructure, the higher is the bond strength. Shortening the dwell time or reducing temperature and pressure can

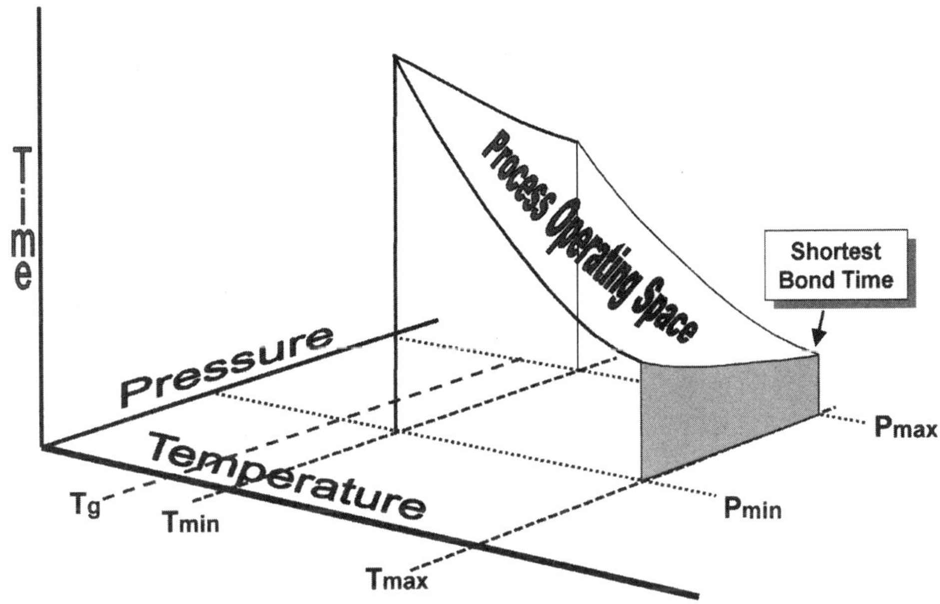

FIGURE 21.2 Thermoplastic bonding window.

reduce penetration and bond strength. The time factor, or dwell, allows heat to distribute at the interface and the fluid polymer to penetrate the surface microstructures. The process is fast, clean, and efficient, and the bond is stable and reliable.

Above T_g, the polymer changes from a harder, glassy state to a softer, rubbery form. Increasing the temperature continues to soften the plastic, which is seen as a drop in viscosity. Viscosity continues to drop as temperature is increased, but the effect will level off. Some thermoplastics become high-viscosity, deformable masses, whereas others change to thin fluids. However, it is not necessary to reduce the polymer to a fluid for bonding. A viscosity level must only be reached where the polymer can be forced into the substrate microstructure under pressure to wet these surfaces. As the plastic cools to ambient temperature, the viscosity increases, and the mechanical bond structure is locked in place. Under typical bonding conditions, very little squeeze-out occurs because the viscosity is still relatively high at suitable bonding temperatures. Filler also will tend to hold the adhesive in place and prevent undesirable squeeze-out. Thermoset adhesives, on the other hand, experience a much greater drop in viscosity on heating, and spreading can occur.

The triad of key bonding parameters—temperature, pressure, and time—is interactive, as stated earlier, but the parameters are also nonlinear, as can be seen from Fig. 21.2. Temperature has the most pronounced effect. The interaction between process parameters is shown in Fig. 21.3. The minimum operating temperature must be somewhere above the glass transition temperature in a region where viscosity begins to drop off (T_{min}). The upper limit (T_{max}) must lie some safe level below the polymer decomposition temperature. Pressure is determined by the minimum that will cause polymer flow, wetting, and subsequent bonding. The high-pressure limit is less well defined and may be limited by equipment, substrate, and components. At very high pressure, some squeeze-out can occur, however.

Thermoplastic paste adhesive bonding normally does not require pressure because the adherends are wet by the adhesive during component placement into the wet paste. When the thermoplastic adhesive is used in dry form, pressure is required to force the softened polymer into the surface microstructure. However, thermoplastic pastes are limited to small dies of about 3 to 4 mm (250 mils) because solvent must escape during the drying process. Larger dies can trap solvent or voids can form because the solvent is unable to vent through the longer path to the edge of the die. Figure 21.4 diagrams the various die bonding processes for thermosets and thermoplastics.

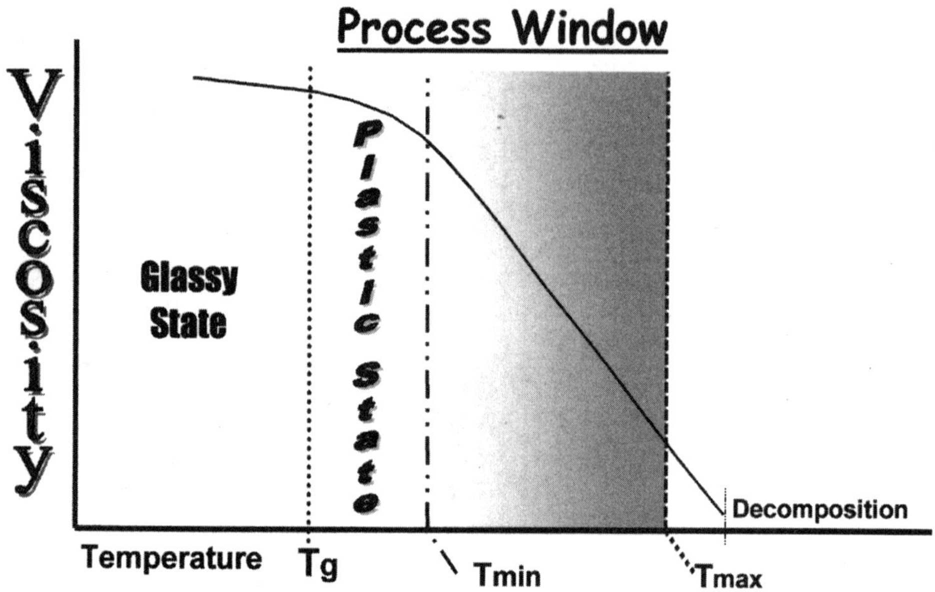

FIGURE 21.3 Graph of bonding parameters.

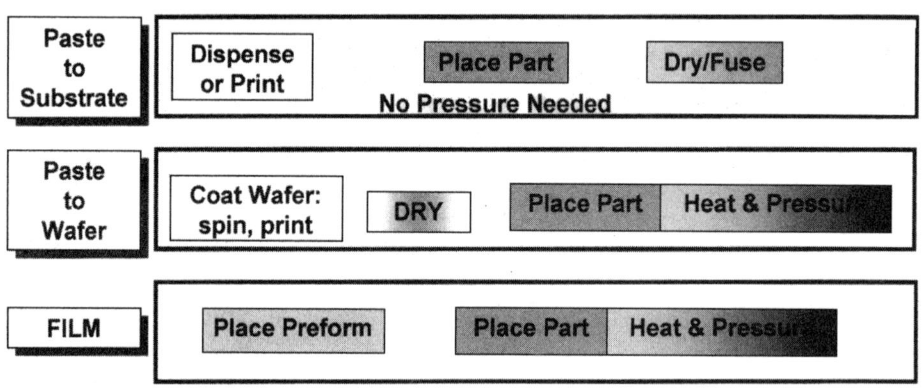

FIGURE 21.4 Block diagram of die-attach processes.

21.4 APPLICATIONS (USES)

21.4.1 BGAs

BGAs typically use dies that are adhesively bonded to the substrate, which can be organic, ceramic, or metal. Both thermoset and thermoplastic die-attach adhesives are used. Reworkable thermoplastic adhesives are especially useful for multiple-chip packages.

21.4.2 Lid Seal

While not a true die-attach process, the same adhesives are being used for lid sealing of hermetic-type packages instead of welding or soldering. However, the resulting package will not be a full hermetic because gases can move slowly through the polymer adhesive. Sealed packages using polymer adhesives are referred to as *near-hermetic* or *quasi-hermetic*. The use of die-attach-type adhesives simplifies the sealing process and reduces temperature. Many systems show high reliability, even with the slow gas leak resulting from the polymer's moderate barrier characteristics. Molecular scavengers, called *getters,* can be placed inside the package to absorb contaminants, including moisture, during the life of the package. The adhesives can be coated onto the lid and dried or provided as preforms that look like tiny gaskets. The precoated lids, as would be expected, offer a very manufacturing-friendly format, and the method has gained popularity.

21.4.3 Known Good Dies

Die testing before packaging remains a significant challenge as chips become more complex and speeds increase.[1] Testing can reveal bad dies but may not indicate the maximum performance speed. Chips often are rated by maximum clock speed and priced accordingly. A very high speed chip may be housed in a top-of-the-line package. It is therefore important to have a method that can ensure known good dies (KGDs) with a high level of confidence. Burned-in dies also may be required and generally command a premium. So-called infant mortality occurs in the early life of the semiconductor, and burn-in often can sort out this type of defective product.

Thermoplastic die-attach adhesives can be used to temporarily attach a die to a test platform for burn-in. The attached die is electrically connected by probes or by weak, temporary wire bonds. The chip is powered up and exposed to accelerated environmental aging conditions. This can be high-temperature storage, thermal cycling, or a combination. The die is given a final test and removed from the test platform. The die-attach adhesive is debonding by heating to its rework temperature as force is applied. Alternatively, a solvent-sensitive adhesive can be used that debonds with a specific solvent. More information is given in Sec. 21.7.

21.5 PROPERTIES

21.5.1 Bond Strength

Bond strength is a common metric for good adhesive performance. An attached die can be sheared or pushed off using a commercial die shear tester that slowly applies an increasing level of force to the side of the die until the bond is broken. Bond strength can range from an initial value of 2500 psi to slightly more than 4000 psi for thermoplastics. Thermoset materials fall into a similar range. However, heat aging often will boost thermoplastic strength, presumably by annealing the adhesive and/or allowing more microstructure penetration and bonding. Table 21.2 shows bond strengths for common thermoplastics.

TABLE 21.2 Die-Attach Adhesive Bond Strength*

Substrate	Conditioning	Initial bond strength, psi	Final bond strength, psi
Ceramic	150°C, 1000 h	2500	4800
Gold-plated ceramic	150°C, 1000 h	3100	4900
Aluminum	150°C, 1000 h	4100	4500
Ceramic	85%RH, 85°C, 1000 h	2500	2800
Gold-plated ceramic	85%RH, 85°C, 1000 h	3100	2600
Aluminum	85%RH, 85°C, 1000 h	4100	3600
Ceramic	−65 to +150°C, 500 cycles	2500	4000
Gold-plated ceramic	−65 to +150°C, 500 cycles	3100	3700
Aluminum	−65 to +150°C, 500 cycles	4100	4800

*Bonding conditions were 180°C for 10 s at 135 g on 80 × 80-mil Ni/Au-plated Kovar tabs.

21.5.2 Bonding Temperature/Curing

Thermoset die-attach adhesives cure at 150 to 175°C, and the bake time can be from 30 min to 2 h. "Snap" cure products can cure is as little as 5 min, however. Thermoplastics have bond times of only a few seconds or less because they area already polymerized. Table 21.3 shows bonding temperatures for a range of thermoplastics compared with T_g.

21.5.3 Outgassing

Outgassing is the release of vapors or gases from hardened die-attach adhesives that is mostly detrimental, especially if the agents can degrade the die or the electrical interconnect structure. Epoxies, because of their complex curing chemistry, can release a variety of materials. Unreacted hardener and resin also can be released, especially as the package is heated. It is nearly impossible for all the polymer reactants to be consumed completely during polymerization, so a trace amount may be released as volatiles. Anhydrides, common epoxy hardeners, can be a problem because they react with moisture to form organic acids that can corrode structures.

Thermoplastics, however, show very low outgassing. This is so because they have been polymerized in highly controlled reactors that generally preclude the presence of unreacted starting materials. The thermoplastics subsequently are manufactured into films or pellets, and this heating process further removes any volatiles that potentially could outgas.

21.5.4 Electrical Conductivity

Electrical conductivity is often desired to ground the die to a conductive platform. Silver is the most common filler used to achieve electrical conductivity, although gold powder may be used for very high reliability situations such as military packaging. Die-attach adhesives optimized for electrical conductivity have volume resistivity values in the range of 0.00001 to about 0.0005 Ω·cm.

21.5.5 Coefficient of Thermal Expansion

While it is theoretically desirable to match the adhesive's coefficient of thermal expansion (CTE) with that of the low-expansion die (~2.5 ppm/°C), this is not really necessary because of the elastic nature of many die-attach adhesives that can be expressed as a modulus. The actual CTE values for commercial products can range from a low of about 15 ppm/°C to a high of more than 200 ppm/°C. More common values are around 30 to 80 ppm/°C. Filler significantly reduces the CTE because the percentage of polymer is reduced and both metal and ceramic fillers have relatively low CTE values.

21.5.6 Glass Transition Temperature T_g

This value is often engineered to be above 125°C so that no phase changes occur during thermal cycling, which typically is run from −55 to 125°C. However, this is not an absolute rule, and good adhesives can have lower values. It was shown in Table 21.3 that T_g values could be well below ambient provided the modulus is also relatively low.

TABLE 21.3 Thermoplastic Bonding Temperatures

Fillers	T_g, °C	Bond temperature, °C
None	−40	100–150
Ag, AlN, none	25	150–200
Ag, AlN, none	45	160–220
Ag, AlN, none	85	160–250
None	145	200–230
Ag, Au, AlN, none	180	325–400
None	280	350–450
Thermoset (ref.)	100–150	150–175

21.5.7 Thermal Conductivity Properties

A good thermal pathway is often desired to allow die-generated heat to dissipate to the package. While solders can have values of 30 W/m·K or higher, polymer adhesives are an order of magnitude lower, ranging from 2 to 4 W/m·K. However, the large die area and the small distance between the die and substrate offset the concern over lower thermal conductivity. Thermal resistance is directly proportional to thickness, so thin bond lines make good pathways. Typical thermal fillers are silver and aluminum nitride (not electrically conductive). Beryllium compounds have conductivity values higher than many metals but are not common because of toxicity during manufacturing and other problems. Carbon nanostructures could find use as die-attach fillers in the future once better manufacturing methods are developed.

21.5.8 Thermal Stability

Both thermosets and thermoplastics can handle a minimum of 300°C before any decomposition begins. High-performance thermoplastic and thermoset systems may not decompose below 400°C. Special polyimides may exceed 500°C. However, modern area array packages typically are made from organic substrate that should not exceed about 260°C during processing. The die-attach adhesive will not be the weakest thermal link. The service temperature rating of die-attach adhesives is usually given as 150°C based on a standard test even though these materials may be able to handle higher values.

21.5.9 Modulus

Modulus, a measure of material stiffness, can be varied over a wide range for die-attach adhesives. Table 21.4 shows the range of values found in commercial thermoplastic products. Thermosets tend to be stiffer materials with values as high as 500,000 psi or even more. Adhesives with a modulus value of less than 100,000 psi are considered low-stress stain-relief materials.

21.5.10 Solubility/Chemical Resistance

Cured and hardened adhesives should not dissolve or even swell in common cleaning solvents. Today, solvent cleaning is much less common, especially for removing solder flux. Aqueous systems are now used where cleaning is deemed necessary, but no-clean solder pastes and fluxes are used commonly for environmental and productivity benefits. Thermosets are very resistant to even strong polar solvents. Thermoplastics typically will dissolve only in very polar solvents that are not used in packaging plants. Since most packages provide a barrier to the environment, die-attach adhesives would not be expected to encounter solvents once the area array package is completed. It may be noted that solvent-reworkable thermoplastic die-attach adhesives are designed to be sensitive to alcohols and are not necessarily sensitive to other solvents.

21.5.11 Debonding Characteristics

See Sec. 21.7.

TABLE 21.4 Thermoplastic Die-Attach Adhesive Modulus

Fillers	T_g, °C	Bond temperature, °C	Modulus, psi
None	−40	100–150	60,000
Ag, AlN, none	25	150–200	60,000
Ag, AlN, none	45	160–220	500,000
Ag, AlN, none	85	160–250	400,000
None	145	200–230	150,000
Ag, Au, AlN, none	180	325–400	360,000
None	280	350–450	370,000

21.5.12 Test Methods

MIL-STD-883D has been used commonly to test die-attach adhesives and is considered a more severe set of conditions than should be expected in commercial use. Most die-attach adhesive suppliers test and qualify to this specification.

21.6 RELIABILITY

The performance of thermoplastic adhesives is much more consistent and reproducible than that of thermosets for fundamental reasons. First, a thermoplastic has been fully polymerized by the manufacturer under highly controlled conditions of temperature, pressure, reaction rate, and time. Thermoset adhesives are polymerized by the end user, and final properties vary with conditions that cannot always be controlled. Thermoplastic adhesives can be bonded in a dry state that significantly reduces the possibility of voids so often experienced with epoxies and polyimide thermosets. Voids and adhesive stress are the major causes of failures related to die attach.[2] However, another increasingly important attribute of thermoplastic adhesives is their low modulus and corresponding elastic characteristics.[3] The result is very low stress on the die. The advent of larger and larger dies makes low stress imperative, especially when dies are mounted directly to organic circuit substrates. Table 21.5 gives the properties of a range of thermoplastics now in commercial use.[4]

21.6.1 Heat Aging

Heat aging is not a problem for most die-attach adhesives. The thermoplastics are stable, and the thermosets may cross-link slightly and increase the modulus, but this is acceptable. Thermoplastics, as was seen earlier, often will develop stronger bonds on heat aging due to annealing or increased microstructure penetration.

21.6.2 Thermocycling

Thermocycling should not have an adverse effect on die-attach bonds unless the material has a high modulus, the die is large, and the CTE is high.

21.6.3 Temperature and Humidity (T&H)

Humidity exposure is the real "litmus test" of polymer systems. Most die-attach adhesives will absorb some moisture, and this may lower the bond strength slightly. The effect is generally reversible. However, the T&H exposure also can boost strength for some systems, as seen in Table 21.2.

TABLE 21.5 Thermoplastic Die Attach Adhesive Properties

Fillers	T_g, °C	Bond temperature, °C	Rework, min·°C	Die shear, psi @25°C	Thermal conductivity, W/m · K	Modulus psi
None	−40	100–150	110	1700	0.20	60,000
Ag, AlN, none	25	150–200	160	2200	0.3–3.0	60,000
Ag, AlN, none	45	160–220	170	2600	0.3–3.0	500,000
Ag, AlN, none	85	160–250	170	3000	0.3–3.0	400,000
None	145	200–230	210	3800	0.22	150,000
Ag, Au, AlN, none	180	325–400	350	4000	0.3–3.0	360,000
None	280	350–450	400	4900	0.25	370,000

Note: Typical electrical conductivity is 1×10^{-4} Ω·cm for metal-filled and 1×10^{12} Ω·cm for nonmetal. Typical thermal conductivity is 3.0 W/m · K for silver and 1.0 W/m · K for AlN.

21.6.4 Failure Modes and Mechanisms

Bond failure of attached dies is uncommon and should not be expected in normal situations. Die cracking or bond delamination is an indication that the wrong adhesive has been used or that the process is out of control. Cracking suggests that a lower-modulus adhesive should be used. Low bond strength indicates that contamination may be present or that voiding is reducing bond area. Voiding can have many sources, and help from the adhesive supplier usually can solve the problem. Plastic BGA packages, however, can experience a number of moisture-related problems that relate to the die-attach adhesive. Popcorning, the explosive evolution of steam from trapped moisture during solder heating, can be related to the adhesive. However, this is a complex issue where substrate, encapsulant, adhesive, and design must be examined as a system.

21.7 REWORK

21.7.1 Thermosets

Thermosets are not considered to be reworkable in the common sense of the term. However, thermosets have been designed that can undergo depolymerization reactions. Some are made from epoxy resin with a thermally unstable link that will break down at 225 to 250°C. Other polymer systems have been designed that are attacked by specific agents such as acids. However, in most regards, only the thermoplastics are reworkable in the practical sense.

21.7.2 Thermoplastics

Removal. Rework can be essential in many multichip package (MCP) applications. Large, complex dies can have low wafer yields, perhaps under 20 percent. Test-at-speed is required but difficult in the bare die state, although several known good die (KGD) schemes are in use and there is continual progress. The fixed softening point and predictable temperature-viscosity relationship of thermoplastic adhesives ensure easy rework. The adhesive is brought to a temperature where the reduced viscosity allows the die to be removed by low-force shearing or torquing with a vacuum collet or simply by lifting a corner. The same time-temperature-pressure relationship used for bonding can be applied to rework because no property change occurred during the attachment process. The temperature-viscosity curve previously shown in Fig. 21.1 also can be used to estimate removal force because it is related directly to viscosity.

Although the thermoplastic adhesive is chemically unchanged after die repair, shape change may have occurred, and new material is recommended for rebonding unless reuse of the same adhesive pad is verified by testing. After die removal, the adhesive pad may remain on the die or substrate. The old adhesive pad can be removed by peeling it away with tweezers during heating. Not only does a clean surface result, the first adhesive actually may remove contamination so that the reworked bond is stronger.

Special temporary thermoplastic adhesives can be used when the intent is to bond, test, and remove the die. These materials can have the same electrical and thermal performance as the products intended for permanent die attach. Specific solvents affect the temporary adhesives. The bonding process is the same as that described earlier. After test and burn-in, the die is readily removed by adding solvent to the bond site. The adhesive swells and loses strength so that the die is removed with very little force. The die can now be bonded to the final assembly with standard thermoplastic die-attach adhesive. The temporary adhesives are being used in certain KGD applications. Note that the intent is not to dissolve the adhesive but to cause debonding by swelling. This results in a clean, noncontaminating process.

Rebonding. Since a thermoplastic die-attach film can be removed cleanly, no special cleaning step or preparation is needed. Rebonding simply involves starting over with a new die and fresh adhesive.

21.8 SUMMARY AND CONCLUSIONS

Die attachment is an important step in manufacturing area array packages. Nearly all use adhesives. The exception is the flip chip (Chap. 5, 17, 18), where the die is connected from the face. While thermoset die-attach adhesives are the common standard, thermoplastics offer two advantages for area array packaging. The first is higher performance because of lower ionics and lower modulus where beneficial. The second is easy die removability and the ability to rework. This feature is especially valuable where an expensive package can be salvaged or multiple dies are used where the odds of a bad die upsetting the economics are significant.

21.9 REFERENCES

1. Martin, S., "A Practical Approach to Producing Known-Good Die," in *ICEMM Proceedings, '93*, IMAPS, Washington, DC, April 1993, pp. 139–151.
2. Trigwell, S., "Failure Mechanisms of Wire and Die Bonding," *Solid State Technology,* May 1993, pp. 45–49.
3. Corbett, S., and Ongley, P. E., "New Innovations in Thermoplastic Die Attach Adhesives for Microelectronic Packaging," *Proceedings—ISHM-UK European Multichip Module Conference,* London, England, 1996.
4. Gilleo, K., Cinque, T., Corbett, S., et al., "Die Attachment for Today's Packaging Challenges," *Electronic Packaging and Production,* Feb. 1994, pp. 109, 110, 112.

CHAPTER 22
LIQUID ENCAPSULATION EQUIPMENT AND PROCESSES

C. Lazinsky
Speedline Technologies

K. Gilleo
Cookson Electronics

22.0 INTRODUCTION

The area array revolution has brought on a plethora of new packaging designs, new and modified materials, innovative processes, and advanced encapsulation equipment. This chapter briefly describes the package types, provides the basics of the materials, and highlights the most common processes before going into the details of equipment for liquid encapsulation. Major advances in materials, processes, and equipment have occurred during the 1990s, and the pace of change may even be accelerating today. Dispensers are now available that apply material with vacuum injection to eliminate air. Very high capacity machines are also available that employ up to four dispensing heads that can run the same or different liquid encapsulants. All these developments have occurred while the equipment has become more robust, increasingly precise, and easier to set up.

22.1 MODERN TRENDS IN PACKAGING

The advent of area array packaging has greatly increased the use of liquid encapsulants and has resulted in new processes. Although transfer molding using solid epoxy molding compound (EMC) is still dominant, liquid encapsulation has become popular for many chip scale packages (CSPs) and larger plastic ball grid arrays (PBGAs). Liquids offer certain advantages that offset their slightly higher cost. Direct chip attach (DCA) assembly, better known as *flip chip* (FC), uses liquid dispensing almost exclusively for underfill. The next subsection will describe the basic packaging concepts and processes.

22.1.1 COB

The chip-on-board (COB) process involves first attaching the bare die or semiconductor chip to a circuit board or to a chip carrier packaging substrate. A polymeric die-attach adhesive is used commonly, as described in Chap. 21. Next, first-level electrical connections are made between the chip pads and substrate using wire bonding. The system is now ready to function, but it is unprotected from the environment and could not survive very long in service. A polymeric material must now be added around and over the die and all the wire bond structure. This material is called *encapsulant,* and the process is referred to as *encapsulation.* The encapsulant must be applied in liquid form to surround and cover the chip and wire structure. Both solid and liquid encapsulants can be used here, but this chapter is focused on liquid. (See Chap. 11 for details on materials and Chap. 23 for encapsulation using solid molding.)

COB has been used for many decades for ceramic hybrid and later for building modules on organic circuit boards. However, these products were circuits, not packages, and they typically contained an assortment of packaged components along with the one or more bare dies that were wire bonded. However, when IBM and Motorola commercialized the ball grid array (BGA) package, the COB concept was applied to packaging. Motorola's Ompack used COB as the first-level connection method to the organic chip carrier. The COB structure was then overmolded using the common transfer molding process. However, transfer molding requires significant and costly tooling. The molds can take months to build, costing tens of thousands of dollars. Since the new BGA packaging was undergoing continual design changes in the early 1990s, an encapsulation process with low tooling, simple setup, and good economics for short runs was sought. The answer was to dispense liquid encapsulant onto the COB area.

22.1.2 BGAs

There are two basic BGA designs: *die-up* and *die-down* (also called *cavity-up* and *cavity-down,* but this terminology is less accurate). Both constructions can use liquid encapsulant. In fact, the die-down design cannot be encapsulated easily without using liquid dispensing because the package actually has a recessed cavity. BGAs made from organic, ceramic, and metal-ceramic platforms can all be encapsulated with liquids using needle dispensing. BGAs with the die-down configuration are simpler to fill because the BGA substrate often can contain the liquid if there is a deep enough cavity. However, the die-up BGA built on a planar carrier requires an additional step of first adding a retaining structure to hold the liquid encapsulant. The retaining methods will be covered in Sec. 22.3. Flux also may be needle dispensed onto BGA substrate for the ball attachment step, although it is more common to apply flux to the solder balls while they are held in a vacuum pickup tool. Camalot Matrixx applies flux with pin transfer and then uses positive pin placement to place solder spheres accurately. This avoids contaminating the spheres and eliminates potential clogging of any vacuum mechanism.

22.1.3 Flip Chips (FCs)

Nearly all modern FC packages require underfill, and most underfill is applied by liquid dispensing, although molded underfill methods have been commercialized in the past several years. Underfill materials and processes have been covered in several other chapters.

22.1.4 Chip Scale Packages (CSPs)

CSPs come in a wide variety of designs and therefore can require different encapsulation methods. The simplest constructions use the die-up design, similar to the larger BGAs. The only difference is one of dimensions, where the CSP must be kept within a limit where the package area does not exceed chip area by more than 10 percent. This type of CSP can be encapsulated by the same methods used for the larger BGA kin. A retaining structure is added, and the area around and over the COB structure is filled with liquid encapsulant.

More complex structures such as the Tessera μBGA require special processes. As shown in Chap. 6, the encapsulant is applied to the inside of the package through an opening. Special dispensing machines with vacuum assistance have been designed and are in use for this type of package.

22.2 ENCAPSULATION MATERIALS

Although a large number of liquid encapsulants are on the market, most are based on epoxies, which have been the workhorse of electronics for many decades. However, the encapsulants contain enough ingredients to make the number of combinations endless. Furthermore, different classes of liquids are available for specific purposes, such as forming a retaining dam. Other classes of polymers are being offered as liquid encapsulants with very good properties. If brominated epoxy resin is banned, nonepoxy systems with inherent flame retardancy could come into use. Brominated compounds impart flame retardancy and are considered safe. However, there seems to be an initiative to ban bromine compounds without any scientific support or even a rationale. Bromine and other halogens are "mined" from the ocean and are absolutely required for life.

22.2.1 Basic Chemistry

The chemistry of encapsulants is described in detail in Chap. 11. Briefly, the liquid encapsulants are a solventless blend of polymer precursors and filler. In the case of the common epoxy system, hardener and epoxy resin react together to produce long polymer chains that are solids that cannot melt. The polymerized structure is a very good barrier to most but not all external agents. A small amount of moisture and gases penetrate the polymer enclosure. The barrier properties are good enough to protect all but a few types of semiconductors. The liquid starting point of these encapsulants means that they can be pumped and dispensed by machines and then hardened to a permanent solid at a later stage. The liquid epoxies have a composition very similar to the EMCs that are so common. The filler in both is simple silica (SiO_2) that is used primarily to reduce the coefficient of thermal expansion. Wetting and flow agents also may be added.

22.2.2 Functionality

There are several types of encapsulants, and each has a different function. Retaining wall or damming compounds are used to form the wall around the die that is needed to hold back the liquid encapsulant that is often called the *fill.* The precured properties are very important because the damming compound must stay in place. The viscosity is high and must remain so even during the curing stage, which can be as high as 175°C. The uncured retaining dam also must hold the low-viscosity liquid fill encapsulant so that both materials can be cured together in a single step. The cured properties of the retaining dam usually are similar to those of the fill, although this is not a requirement because the dam does not contact the die. The ideal system of dam and fill will fuse together during the thermal cure to provide a strong, resistant, and aesthetically appealing structure. The appearance is important because the package will be seen and handled by assemblers and others. Also, a smooth finish is need so that nomenclature can be printed on the package and an automatic assembly machine can handle the package.

Packages that do not need an added retaining wall can be encapsulated with fill material directly. Low viscosity allows the material to flow around the fine-wire structure and also helps reduce air entrapment. However, too low a viscosity can cause the filler to settle before the cure step is complete. Heating nearly always causes the prepolymer system to drop in viscosity and thus accelerates settling.

22.2.3 Performance Characteristics

First, all polymer-enclosed packages are considered nonhermetic. The die is exposed to small amounts of gases that diffuse through the polymer structure. The most significant gases in any measurable quantity are oxygen and water vapor. Oxygen generally is not a problem for common silicon chips that are already protected from further oxidation by silicon oxide. The aluminum (or copper) pads on the chip readily oxidize, but the resulting oxide is self-limiting. Moisture is a larger problem, however, because it serves as a medium for many chemical reactions, especially corrosion. Ions such as chloride (Cl^-) are especially damaging in the presence of water and can degrade chip pads and the wire bond. Chloride is present as a by-product of nearly all epoxies, and too high a level will cause reliability problems. Liquid encapsulants are made with epoxies that are highly purified to remove chloride and other ions so that the finished product will have an ion level of 10 ppm or less. The moisture-barrier properties of a good encapsulant will only allow a fraction of a percent of water absorption, even in a high-humidity environment. Therefore, the epoxies provide a highly reliable enclosure, provided that chloride is kept low and moisture resistance is high. Encapsulants have continued to improve in both regards, and these products are considered to be highly reliable.

There is another reliability concern for BGAs made with organic substrates that involves performance during the heating in the solder assembly process. The small amount of moisture within the encapsulant and the chip carrier can outgas at an explosive rate during the soldering step. Enough pressure can be generated to cause delamination or even cracking of the encapsulant in a phenomenon called *popcorning.* While the source of moisture entry can be the package, any reduction in absorption by the encapsulant should improve the situation. The industry has continued to improve moisture resistance of epoxies, often by adding water repellants. However, some epoxy encapsulants contain copolymers such as cyanate esters that improve moisture resistance and other properties. The package may require predrying prior to soldering depending on moisture-absorption properties and storage conditions.

22.2.4 Handling and Storage

Liquid encapsulants are commonly supplied as a premix of hardener and resin that is reactive even at room temperature. The reaction rate depends directly on temperature, however. Raising the temperature to around 150°C will cause the polymerization to complete in 30 min or even less. Conversely, cooling the encapsulant to −40 to −50°C will retard the reaction and thus provide a storage lifetime of up to 12 months. The vast majority of liquid encapsulants are shipped with ice packs or dry ice and stored in freezers by users. The material must be brought to room temperature before dispensing, and the time required depends on the size and construction of the container. Many products are sealed in plastic bags that should not be opened until the product has warmed up. This keeps any water that condenses out of the atmosphere away from the primary container, which typically is a syringe designed to fit a needle dispenser.

Epoxies are not very hazardous, but they should be handled with the respect afforded any chemical. The liquids can irritate skin and cause dermatitis in some people, and gloves are recommended if there is any chance of skin contact. The materials are not very volatile but should still be handled with good ventilation. Any pretesting should be done in a fume hood, especially if the material will be heated. Many liquid epoxy encapsulants use acid anhydride hardener that can be released during open heating. The anhydride reacts with moisture in the air or in the nasal passages to form weak but irritating organic acids. The uncured epoxies are soluble in alcohols, such as isopropyl alcohol (IPA), and most organic solvents. They are not soluble in water but can still be washed off with detergent, which will emulsify them. As with all chemicals, they should be disposed of in keeping with state and federal laws. Since solid plastics are safer and easier to deal with than reactive liquid precursors, it is a common practice to harden any of the waste liquid.

22.3 ENCAPSULATING METHODS

22.3.1 Overmolding

Overmolding, or *transfer molding,* is the more common method for encapsulating standard packages. This is covered in Chap. 23 and is discussed briefly here for reference as the alternative to liquid dispensing. The method is capital-intensive, and a good machine can cost over $1 million. However, the method is very well suited for high volume because it is a parallel process that can produce hundreds of parts per cycle.

22.3.2 Overmolding Equipment

The modern molding machine is large and costly but has very high output that justifies this method for high-volume production. Figure 22.1 presents a simple diagram of a molding press. Solid epoxy molding compound preform is preheated and then physically transferred to the mold, where it is liquefied and injected under pressure. The liquid epoxy enters the mold cavities that contain the electronic packages to be encapsulated. The heated mold is held at a temperature that causes the liquid epoxy resin-hardener compound to polymerize into a permanent solid. The mold is opened, and the parts are ejected. The parts may be exposed to more heat in a postcuring operation. The process is efficient for high-volume production but not well suited to short runs.

22.3.3 Needle Dispensing

Needle dispensing must deal with liquid materials, but the viscosity can range from water-thin (alcohol and flux) to high-viscosity pastes. The process can be as simple as using a syringe that relies on human force or as complex as an automatic dispensing machine that even calibrates the volume or weight of material and adjusts for changes in its properties. The key feature of dispensing is that no tooling is required. The programming is the virtual tooling.

Modern equipment can check fiducials on the substrate, calibrate material, position the needle, set the height, and dispense the correct volume of material. Needle dispensing is the most versatile of all the encapsulation methods, but its serial nature means that it is not inherently as fast as mass methods such as pin transfer. Viscosities can be very high as seen with solder paste, conductive adhesives, and damming compounds or very low, as found in fluxes and underfills. Another valuable feature is that material can be applied in the presence of components or even on three-dimensional (3D) structures. In fact, many applications involve placing material next to, into, or onto components already in place. While this data-driven "no tooling" method has superb versatility and precision, the most significant limitation has been modest throughput. Fortunately, the productivity issue has been solved, and this will be the theme in the remaining sections.

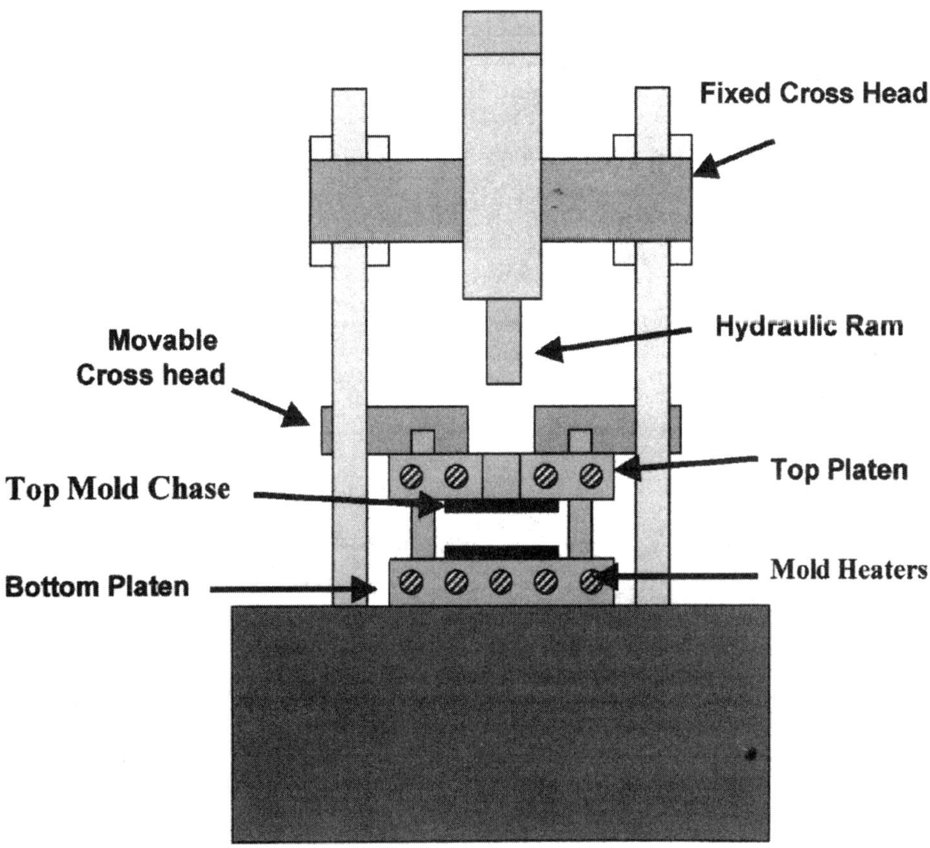

FIGURE 22.1 Transfer molding press.

22.3.4 Needle Dispensing Equipment

Basic requirements are accurate positioning of the needle over the target zone, automatic height adjustment, and precise control of the dispensing process. The amount of material deposited and its shape are determined by a number of factors, including needle height, pump rate, pressure, temperature, and rate of needle withdrawal (*z* height). In many ways, needle dispensing is a complex process requiring more machine control than other methods.

The dispensing machine consists of a frame that holds the subsystems, a platform that produces *xy* positioning movement, a moving head that contains the needle and dispensing control, a pump with fluid delivery, and the master control system. Some machines move the base that holds the substrate instead of the needle. Fully automated equipment has substrate handling to move parts in and out of the dispensing zone. The dispenser may have a vision system, heaters, and even an atmosphere-control enclosure.

Components of the subsystems include

1. Machine platform
2. *xyz* positioner
3. Control head
4. Fluid handling system including the pump
5. Total system control unit

Each section is critical, and their design must be balanced and well integrated. Figure 22.2 shows the basic machine elements.

22.4 PLATFORM DESIGN AND CONSTRUCTION

A sturdy and robust platform is at the top of the list. Just as a reliable building or vehicle must begin with a sound foundation or frame, the same holds true for durable and reliable assembly-line equipment. Any frame movement can translate to motion errors as well as maintenance problems associated with vibration, distortion, and fatigue. At the same time, the footprint must kept at a minimum. The platform, often taken for granted, is more than just the foundation; it is the frame that must be well thought out to support and contain all the subsystems that make up the total machine. Some frames are cast from metal and "aged" so that there will be no dimensional changes once the machine has been constructed. While the frame only has to support modest weight, the ability to resist movement during machine operation is important. Thus, while some dispensing machines may seem to be "overbuilt," this is a prudent design feature and a sign of true quality.

22.5 XY POSITIONING MECHANISMS

The dispensing needle must move in an *xy* direction in relationship to the substrate, and this requires that either the needle assembly or the piece parts be moved. The most popular design concept is a stationary piece under a moving head. The design challenge is to provide fast movement with high precision and repeatability. However, the system also must be capable of very slow movement required for dispensing damming compound in a rectangular pattern or underfill along the edge of a tiny chip.

22.5.1 Conventional Positioning Assemblies

Standard dispensing machines typically are comprised of a single gantry, an *xyz* motion table, that accommodates one or two valves and a single- or dual-lane conveyor system to shuttle parts in and out of the equipment. Accuracy down to 2 mils (50 μm) is required today. There are many architectures, but most involve separate *x* and *y* drives that work in unison to move the dispensing head to a precise location. Each directional unit has a motor and a mechanical movement with a controller. The platform also includes a z-axis movement to lift and drop the dispensing head (Fig. 22.2).

A common motion system may use ball screws and servo motors with closed loop encoders to provide smooth point-to-point movement with input-output zones to reduce substrate transport time. Lower-cost systems still use pulleys and cables, but the penalty is reduced accuracy and repeatability. More than one needle dispenser may be configured onto the gantry.

22.5.2 *z* Positioning (Vertical Control)

The tip of the dispensing needle must be positioned at the right height above the substrate, and the distance may be different for various applications and materials. Height control is also important if large boards or long strips of packaging substrate are to be managed and the deposit is to have optimal accuracy. The z height, the space between the needle and the substrate, is more important than many people realize. Too great a distance and a small drop may not even deposit, whereas too small a distance will deform the shape and even interfere with flow. Improper z spacing also can cause tailing as the needle moves away. Effective designs would employ an active touch probe with a transducer to accurately access pressure and thereby control height to the desired preset values. Alternatively, a noncontact laser sensor may be employed to measure z height on densely populated boards or on delicate substrates. Since the substrate can have warp, it is desirable to have a means of adjusting for the height variation. Height-sensing technology can be used, but a mechanical "foot" has proven to solve the problem much more simply.

22.5.3 Multigantry or Modular-Style Systems

Since needle dispensing is a serial process applying only one "dot" at a time, efforts have been made to boost the rate to better compete with screen printing, which is a parallel deposition method. Note

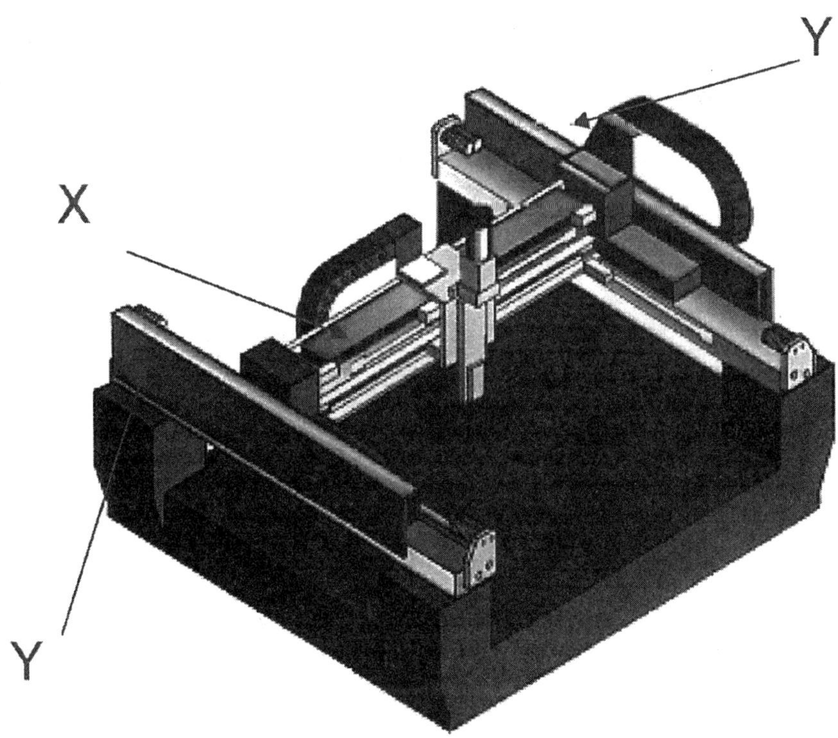

FIGURE 22.2 Typical single-gantry platform of a dispensing system.

that screen printing can be used for surface mount device (SMD) attachment adhesive dots, but the method is not practical for underfilling and the encapsulation process in most situations. One alternative to increase the overall production capacity is to link multiple machines. Each one of them requires a handling system and an area to create a work cell. Clean-room floor space is at a premium, and any solution to avoid redundancy in the line (such as reducing the handling system and conveyor) can save real estate. Camalot XyflexPro is a compact, modular dispenser that can be linked to one, two, three, or four units in-line to provide required throughput. All machines are controlled from a single host computer and user interface. They can be converted easily from SMD assembly to semiconductor packaging applications in the field. Dual-lane single-head dispensers are another possibility, but they are used very rarely because of additional cost for the handling system for a minimal gain in units per hour. The last alternative is to design a multiple-gantry machine that minimizes the footprint but still achieves greater units per hour (uph). Traditional gantry designs where the *y* axis carries the *x* and *z* axes, determine the overall machine footprint. Innovative designs using two *y*-axis gantries allow a four-head system to occupy the floor space of only one and a half times the size of a standard dispenser yet deliver three to four times more units per hour (Fig. 22.3).

Since the gantry occupies the greatest area in the system, it dictates the amount of floor space required by the dispenser. Incorporating traditional *xy* gantries into a multihead system leads to an extremely large platform, which would occupy the same floor space as linking multiple dispensers side by side. Using this innovative gantry design, the Camalot Xyflex is a four-head dispensing system that occupies about $1\frac{1}{2}$ times the floor space of a single-head dispenser and dispenses at rates up to three to four times faster (application-dependent). This is accomplished with a gantry design that is driven by a single computer that helps the system to achieve its speed with four independently operating heads, each with a dispense unit installed on a *y*-axis gantry only (Fig. 22.4).

FIGURE 22.3 Multiple gantries (Speedline Camalot Xyflex).

High throughput is attained without the higher purchase cost and system complexity of linking two or more dispensers in series. Gantry robustness and accuracy are also critical to achieving high-volume, high-yield production. Lead-screw-driven gantry systems historically have proven more accurate at high speed and more robust than alternative designs. With the demands for higher units per hour, the gantry must now keep up with the pump dispense rate. Most dispensing valves are capable of dispense rates greater than the ones use in production, but dispensers with superior gantry systems are going to achieve higher yields and greater units per hour. Dispensing patterns can be run at greater speed without overshoot or misalignment and without causing mechanical defects (i.e., wire-bond sweep). Innovative dispensing patterns are also an important part of void-free encapsulation. Advanced software allows the end user to optimize and balance the dispenser's multiple heads. Two modes are used frequently: parallel mode, in which every gantry executes the same pattern, or serial mode, in which each head executes a section of the program. Typically, for a BGA encapsulation, one or two heads dispense the dam, and two to three heads deposit the fill pattern.

22.6 DISPENSING HEAD

The head contains the vision camera, the z probe, and the all-important dispensing unit (DU). This is where the action is—where the material meets the product. The dispensing unit must respond quickly and accurately regardless of the rheology of the material. A needle calibration station can be used where the highest accuracy and consistency are important. Quick change is also required for high throughput. The lowest mass would facilitate fast, accurate movement. At the same time, a rugged structure is a fundamental requisite.

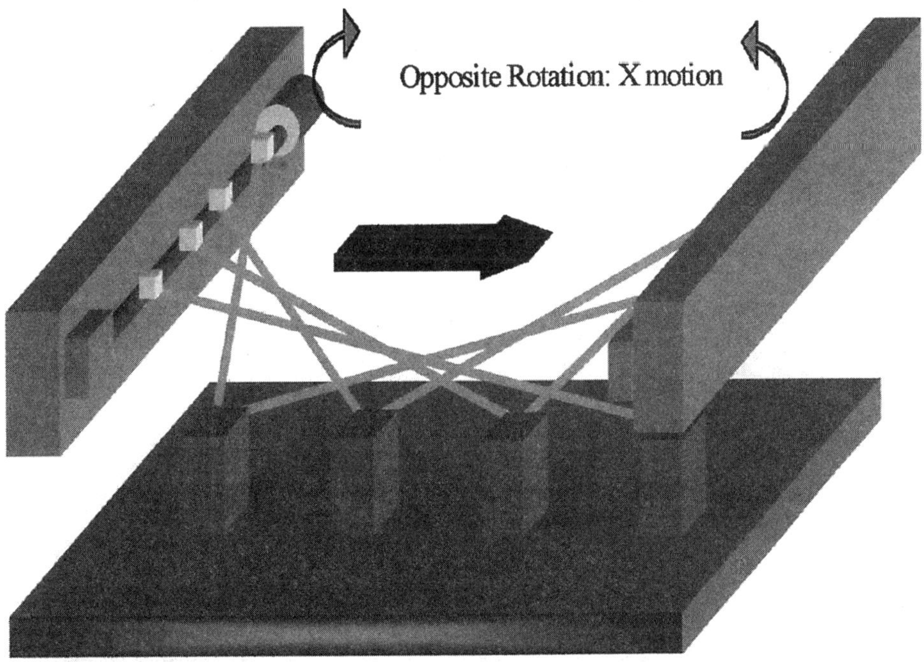

FIGURE 22.4 Modular dispensing system (Camalot XyflexPro).

22.7 NEEDLES

The industry is well supplied with manufacturers of precise needles covering a wide range of diameters.

22.8 PUMPS

The pump is the heart of the dispensing machine, and much engineering has gone into improving designs and constructions. The simplest mechanism is to apply manual pressure to a plunger, and this can be automated with a piston pump. Air pressure also can be used, and such systems are known as *pneumatic* pumps. Auger pumps add more precision and control and are popular in automatic dispensing machines. Desire for maximum control, however, has led to several pump concepts that fall under the heading of *positive-displacement.* Each pump style will be covered in the following subsections.

22.8.1 Pneumatic Pumps

The simplest method for rudimentary automation is the pneumatic approach, where air pressure pushes liquid out of a syringe. Precision is only moderate, but the mechanism is quite simple and is used as an adjunct to manual assembly. The manual plunger is removed from a disposable plastic syringe, and the end is fitted with an adapter that connects to an air controller. A simple pressure controller, often with a settable timer, is used to regulate air pressure and duration.

22.8.2 Piston-Type Pumps

Piston pumps apply force directly to the liquid material and are considered to be positive-displacement pumps. They can be single- or multiple-piston types and work well with thin and high-viscosity materials. This type of pump, however, has found favor for lower-viscosity underfills and fill-type encapsulants. Volume repeatability is better than ± 1 percent, and recharge time is about 1 s. Figure 22.5 shows a cross section of a single-piston pump. These pumps can be used with disposable plastic syringes for lower cost of use.

Multipiston pumps have been developed more recently to provide a faster recharging time and more accuracy at higher dispensing rates. In fact, the flow rate is continuous, so recharge time essentially drops to zero. These pumps are used for high-rate dispensers of encapsulants and underfills. Advanced designs have heated pump chambers so that the viscosity of the encapsulant or underfill can be reduced and kept constant. Figure 22.6 shows the multipiston design and how it works.

Why have dispensing machines moved to new pumps such as the multipiston type? A single-piston pump is capable of achieving less than 1 percent volume repeatability, and the smaller micropackages can require increased tolerances. Different designs of single-piston pumps are available, but all use the same principle. A chamber is filled with material that is expelled with a piston. Even though the repeatability and accuracy are good, a recharge step is required when the chamber is empty. It typically takes 1 to 10 s to recharge a 2-cc chamber, which results in productivity loss. Using an Auer boat (parts carrier) containing five BGA packages requiring 900 mg of material, a recharge step is required to finish the entire boat, thus interrupting production for a minimum of 10 s.

Technology has evolved to the multiple-piston pump design to eliminate the recharge step inherent in the single-piston pump design. While one chamber is being emptied, one or two others are being recharged simultaneously. In a dual-chamber piston pump, the recharge rate is equal to or faster than the dispense rate. One commercial multipiston pump incorporates three pistons driven by a cam–gate valve assembly. This continuous-flow design allows the pump to dispense material through one chamber onto the parts and simultaneously recharge the other two chambers. The slow recharge rate allows the material to feed the chamber uniformly without creating a vacuum. Its patented design prevents dripping from occurring. On the other hand, commercially available standard rotary-gate-valve designs usually allow air and material to migrate into the assembly when

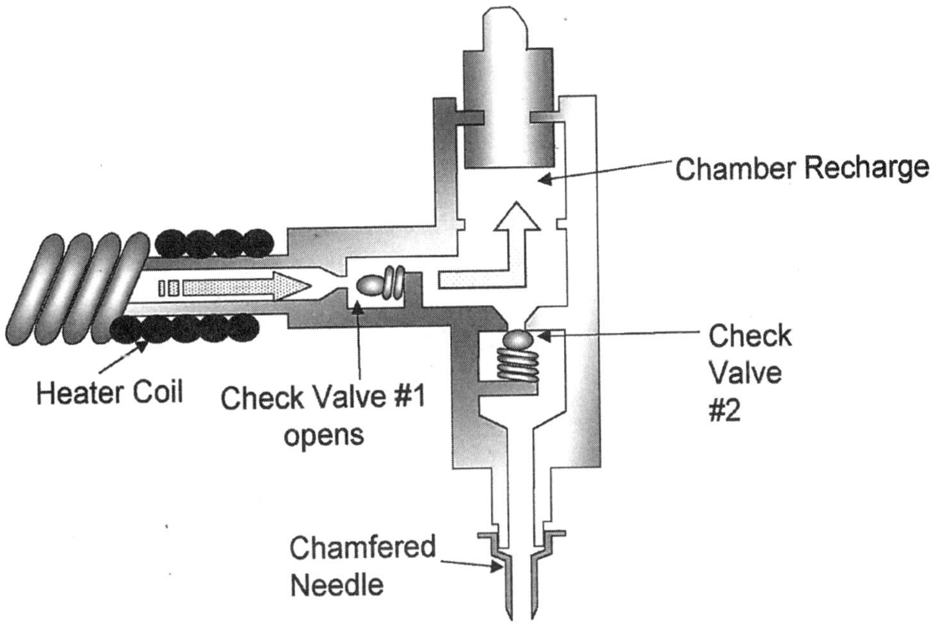

FIGURE 22.5 Single-piston pump.

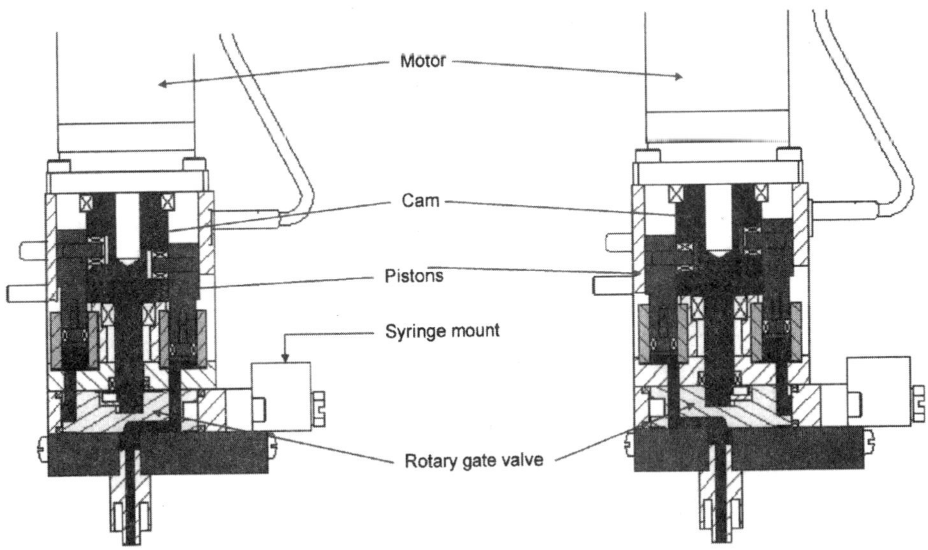

FIGURE 22.6 Multipiston pump.

moving from chamber to chamber. The entrapped air causes dripping or inconsistent dispensing, especially after the chamber is refilled. The multipiston pump design maximizes units per hour (since no recharge is needed) and provides less than ±1 percent weight repeatability. Combined with a weight-calibration system, customers are able to achieve process capability index (Cpk) numbers greater than 2. Figure 22.7 shows the high level of repeatability.

22.8.3 Rotary Pumps

Rotary pumps are considered to be positive-displacement pumps and are able to meet high precision demands at high speeds. Repeatability is excellent even with a very fast rate. Cleaning is simple, and changeover is fast. Multiple nozzles can be configured. Figure 22.8 shows a rotary pump assembly with two syringes.

22.8.4 Auger Pumps

The traditional auger design for semiconductor applications is perceived to be outdated technology, but it is still the preferred technology for dispensing the retaining dam. Very high viscosity materials perform marginally with piston pump designs, whereas augers are capable of transferring viscous material with ease. Typically, dual-helix augers and components exposed to the material are made of carbide to enhance their strength and longevity. Again, these pumps must be fast, accurate, and repeatable to produce a dam of consistent height, width, and cosmetic perfection (i.e., no bulge at the start or end of the dam).

The auger design provides steady pressure and excellent repeatability, and the mechanism does not produce mechanical friction heating that will change viscosity. However, the lower cartridge assembly can be heated for materials that should be applied at elevated temperatures. Pressure is very constant and can be held at a low value for very fluid materials. One design keeps the motor at a constant speed, and a clutch is used to transmit motion to the pump when required. The dripping tendency is very low with augers. Figure 22.9 shows an auger pump and head configuration.

Modern auger pumps have a double-helix thread for higher precision, and the lead screw and insert are made of carbide to reduce wear caused by mineral fillers in most encapsulants. Advanced dispensing machines use a closed-loop rmp controller.

22.8.5 Fluid Jet Pumps

The fluid jet approach to deposition involves a noncontact method where liquid is "shot" from the head at the substrate. Strictly speaking, this is not a needle dispenser but is included under pumps

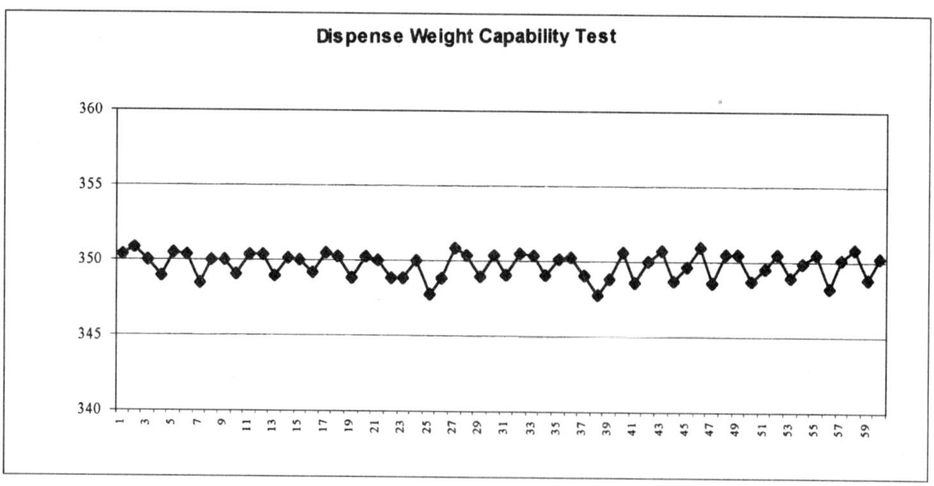

FIGURE 22.7 Repeatability.

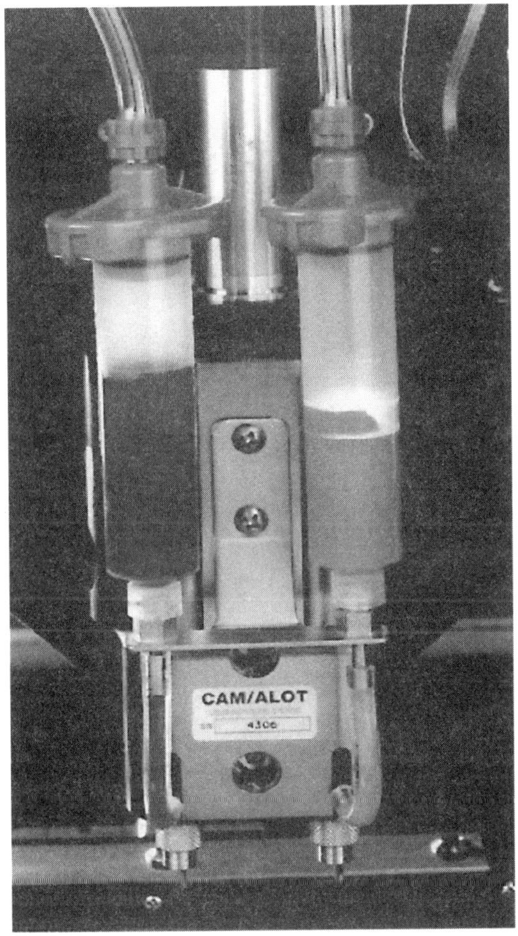

FIGURE 22.8 Rotary pump assembly.

for thoroughness. Fluid jet dispensing is much less versatile than needle dispensing and is used primarily to apply dots of adhesive for SMD assembly. The method has not found significant use in the encapsulation area.

22.8.6 Overall Fluid-Handling System

Earlier dispensing designs consisted of a fluid reservoir, a pump connected to the reservoir, valving, and a needle. Today's systems are much more integrated into a dispensing head. The basic fluid-handling system consists of a reservoir or container in a holder, at least one pump, and a valving mechanism. Advanced equipment should detect low material levels and even report changes in material properties, especially a viscosity drift. While a number of pump types are used, rotary-auger or positive-displacement pump designs are best. Air pressure mechanisms are considered inferior. Multipiston and rotary pump designs were considered, and both versions were implemented. The pumping system must respond quickly and repeatably, dealing with very small or large volumes. A constant-pressure design was considered ideal. The wide variety of materials, from peanut butter-like pastes to watery fluids, requires an extremely versatile pump system.

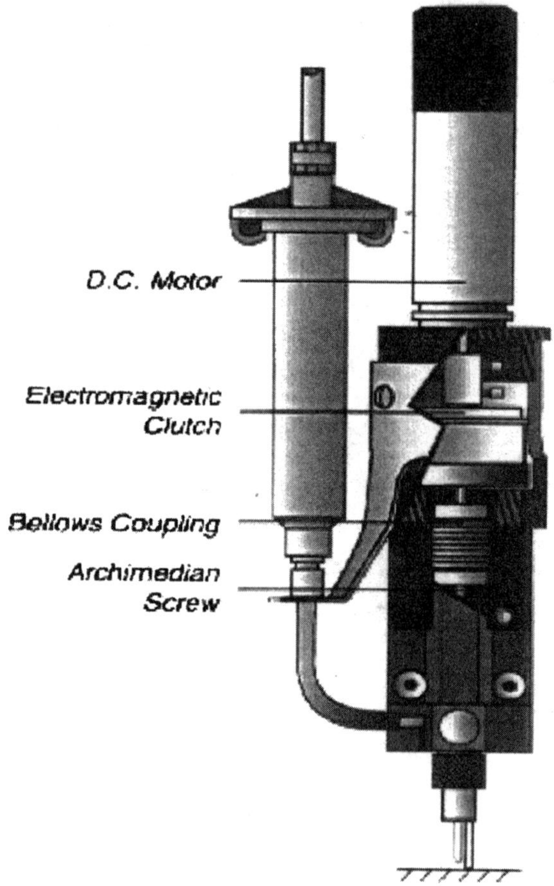

FIGURE 22.9 Auger pump assembly.

22.9 SYSTEM CONTROL

22.9.1 Advanced Process/Software Controls

The alignment system controls the *xy* positioner motion so that material will be deposited precisely where desired even if the substrate is dimensionally unstable. The best system uses a digital camera, advanced illumination, and sophisticated vision-recognition software along with the appropriate computer hardware. Such an intelligent vision system is able to detect inconsistencies in the substrate and "no component" areas, as well as do a quality check on the deposited material.

All the subsystems must work in concert with good balance. The central system is the nerve center and mission control for the entire machine. High-speed data access and analysis are essential. The system must deal with simultaneous input, make a sound decision, and initiate the correct action. The most modern, state-of-the-art technology is employed. Fiberoptics, found in advanced aircraft such as the Boeing 777, is used for maximum rate, extreme reliability, and immunity from interference.

The technical advancements in materials are forcing the dispensing equipment industry not only to be faster and more accurate but also to provide new software advancement to help the end user improve productivity and reduce down time. The new high-flow or single-pass materials require equipment that can handle higher dispensing rates and can still maintain superior positioning accuracy. Most underfill materials have a pot life of 8 h, which basically means that after 8 h of exposure at room temperature, the material doubles in viscosity. The viscosity being a function of time and temperature, it is critical to maintain the temperature of the material in the syringe constant at room temperature. The material can then be heated just prior dispensing to improve its flow characteristics. Several software features are available to help optimize pot life. Timers can be set to warn the operator that the material is reaching its limits. Alarms are also available to ensure that the substrate has reached the optimal flow temperature of the material. The fast cure materials have a tendency to thicken in the needle tip if the flow is interrupted for too long, causing poor underfill during the next dispense cycle. An autopurge timer ensures that the material is moved for a settable amount of time at regular intervals.

Also, when the size of the die requires several successive passes, it becomes critical to provide material in a timely manner so that there is no material flow interruption and limited air entrapment risk. Timer commands can be inserted into the pattern to ensure that the proper interval between successive passes is respected. Additionally, advanced routines allow the dispenser to process as many parts as possible in a given period of time before returning to the first part to complete the process. Similarly, when dealing with multiple array panels or partially filled trays, new software routines are now capable of optimizing the process regardless of the number of dies or parts. Most of the cycle time is spent dispensing material and not waiting for it to flow. The new features offer the end user an automatic optimization and process-balancing system that maximizes the material characteristics but minimizes cycle time and down time.

Along with the linear positive-displacement pump, volumetric and weight-measurement systems were developed to target and monitor the process by either weight or volume. Weight-calibration tools integrated into the dispenser are more common. They monitor either the flow rate or the weight of the pattern to be dispensed. Typically, when the integration of the scale is well designed, settling time of the tool is fast. The finest resolution can then be used (to a tenth of a milligram), increasing the accuracy of the system. Vibrations, air drafts, and static can generate false readings. Even a half-milligram error caused by an air draft or vibrations can offset the setup of the machine and result in scrap parts. On the more advanced systems, a calculator is integrated to help the user set up the pattern for the right volume. Target weight, tolerance, and weight-calibration frequency are also programmable parameters. The weight adjustment is achieved by automatically adjusting either the travel speed of the dispense head or the speed (rpms) of the pump. On the more advanced systems, averaging can be used when the amount dispensed is low, thus minimizing the effect of weight-scale limitation (to a tenth of milligram). For example, at 10 mg, a variation of 0.1 mg is 10 percent.

22.9.2 Human Interface

While dispensing machines have become more complex, the human interface has been designed to be friendlier. Multiple screens are used with high-resolution graphics that show just what is happening and what function is being queried or altered. Today's machines are easy to program and have a graphic interface with real-time process monitoring.

22.9.3 Computers

Today's fast and sophisticated dispensing machines use a relatively powerful computer to control functions, handle the operator interface, monitor the process data, and store both programs and data. These computers are industrial hardened and can have uninterrupted power supplies so that no data are lost in the event of factory power disruptions. Redundancy of critical systems is found in the best machines. Quick change-out and "hot swapping" of computer boards and controllers are valuable, especially if there is easy access.

22.9.4 Software

Today's software has a high-level graphic interface to make the system friendlier and less prone to operator errors. The entire process can be monitored in real time. The setup screen takes the operator through the process step by step and "flags" unlikely parameter inputs. Figure 22.10 shows a setup screen. Powerful software allows the end user to create complex patterns with multiple passes without any idle time. Self-diagnosis and preemptive troubleshooting are valuable in the software.

22.9.5 System Calibration

Advanced dispensing machines have calibration-integrated hardware and software that can be set for periodic calibration of material weight, volume, or both (Fig. 22.10). When used in a multiple-gantry system, greater units per hour are achieved with minimal machine footprint. Along with the linear positive-displacement pump, volumetric and weight-measurement systems were developed to target and monitor the process either by weight or by volume.

22.10 THE DISPENSING PROCESS

22.10.1 Material Constraints and Considerations

A large number of materials are used, and dispensers have been designed to handle products ranging from viscous damming compounds to very fluid fast-flow underfills. The industry also has specialized in and built machines for BGA encapsulation, others for underfill, and still others with vacuum assist for filling internal regions of CSPs.

22.10.2 Substrates

Substrates vary in both material composition and shape. The most common packaging substrates are ceramic, rigid organic, and flexible circuitry. While single packages are used, productivity demands have led to the use of strips and even reels of substrate in the case of flexible carriers.

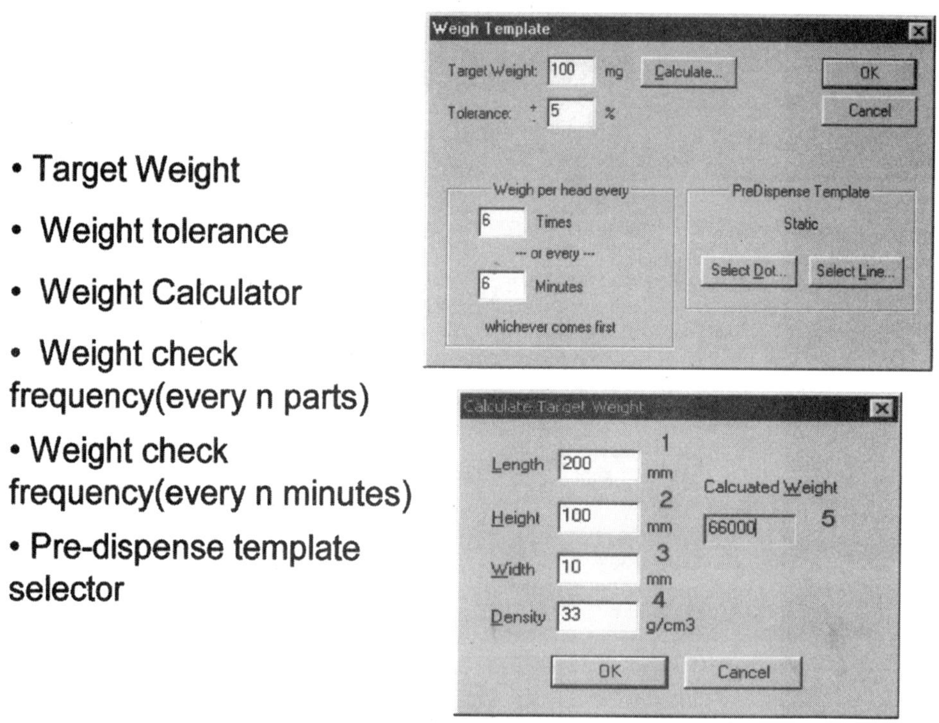

FIGURE 22.10 Calibration screen.

22.10.3 Applications and Processes

The two main applications areas for liquid dispensing in packaging are underfill for the flip-chip-in-package (FCIP) and encapsulation of the many types of BGAs. While many of the CSPs are little more than small BGAs, some have unique structures that require different dispensing methods. The μBGA, for example, requires a vacuum-assisted filling process because an internal cavity is involved, and air must be displaced. Table 22.1 lists a few of the products where liquid dispensing is employed. However, the use of area array packaging has become so pervasive that it includes nearly all electronics except very low end products that still use low-density feed-through devices.

22.10.4 Chip-on-Board (COB)

From a packaging perspective, COB is a common method for building most BGAs and some CSPs. The chip can be electrically interconnected to the package by three basic methods, and each represents a different challenge. The most common, wire bond connections, can present the problem of covering a large number of very fine and closely spaced wires. Air can be trapped, and filler in the encapsulant can aggravate the problem by blocking space between closely spaced wires. The dispensing head can be programmed to move around the wire structure in a spiral pattern or one that has been found to reduce any propensity for air entrapment. In addition, a dam may be formed first around the perimeter of the package if there is no recess to hold the filling encapsulant.

Tape carriers, really flexible circuitry, are becoming more common as area array packages. In a sense, the wires for bonding are built into the circuit or carrier. An array of fine metal beam leads is formed to correspond to the chip pad layout. The beam leads are bonded to the chip pads by a tape automated bonding (TAB) bonder that applies heat and force. All the bonds may be made simultaneously (gang bonding) or only one side at a time. For very fine pitch, the wires can be bonded one at a time like conventional wire bonding. The resulting chip and wire type of connection is easier to encapsulate that the traditional wire-bonded chip. The array is much more orderly and planar, making it unlikely that any two wires will be too close to permit encapsulant flow. Since there is only a single set of bonds, those at the chip, there is less impedance to flow.

The densest connection means is direct chip attach (DCA) or FC, where the chip is flipped over and interconnected directly to the packaging substrate. The bumped chip usually is soldered, but some processes use conductive adhesives. The assembly will now have a gap between the chip and substrate that must be filled with encapsulant called *underfill*. This is covered in several other chapters, and the reader should refer to the index for the particular underfill process of interest. The underfilled package may now be ready to ship. However, the back of the die is exposed, and some packages are further encapsulated to hide the chip. However, this seals in heat and is not recommended for any integrated circuits (ICs) that generate substantial heat. Since the back of the die is not very sensitive to handling, leaving it exposed allows heat to be removed. For example, a heat spreader or heat sink can be connected to the die either as part of the package or after assembly.

TABLE 22.1 Typical Large-Scale Electronics Assemblers' Applications

Industry	Typical end product
Communications	TV cable boxes Telephone modems Satellite applications
Automotive	Air bag sensors Car radios Brake systems
Consumer electronics	Personal digital assistants (PDAs) Mobile phones Pagers Personal computers
Medical	Analyzers/diagnosis equipment Hearing aids

22.10.5 Dam/Fill

The retaining dam application step involves first applying a viscous paste called *damming compound.* A square or rectangular pattern is applied close to the edge of the package to retain the more fluid fill encapsulant in the second step. The dispensing machine is programmed to automatically form the dam and then fill the area. Two separate machines can be used, but a multihead type is ideal. Large BGAs with high-density wire bonds require several dispensing steps. A first layer of material is dispensed on top of the wires of every package, typically at a moderate speed, to avoid any sweep. This allows the epoxy to adequately flow around the wires and prevent any air entrapment. A second pass dispensed at a higher rate to finish the encapsulation then follows. Figure 22.11 shows the dam and fill steps.

(a)

(b)

FIGURE 22.11 (*a*) Applying the dam. (*b*) Fill step.

20.10.6 Underfill

The aim of the preceding steps is to be able to bring the needle of the valve as close to the die edge as possible in order to avoid contamination around the die perimeter. It is also necessary to position the needle tip just below the lower surface of the die to maximize the flow underneath the die. As long as the position of the needle is close enough to the edge of the die and at the correct height, the material dispensed along the edge of the die will flow underneath rapidly and uniformly.

Figure 22.12 shows the results of insufficient, excessive, and correct dispensing. However, it is better to err on the side of excessive rather than insufficient dispense, but with the equipment available today, it is possible to control dispense volume to within ± 2 percent.

Another issue that has been debated recently is the need to carry out a fillet dispense around the other three sides of the die. Until recently, this was thought to be purely cosmetic and a waste of valuable dispensing time and material. However, recent studies have shown that if the three-side fillet dispense is not carried out, the dispense side has a much larger fillet. During curing of the underfill material, the die may be subjected to uneven stresses that in some cases have lifted the die up and fractured the interconnections on one side. The needle height and position are important, and Fig. 22.13 shows the correct relationship to the chip.

It may be necessary with larger dies to make more than one pass along the edge of the die, and in other cases, the pattern of the pass also may vary. The way in which the needle is moved along the edge of the die and in which sequence can prevent air entrapment. For dies smaller than 3 mm^2, one pass may be sufficient. For dies larger than 3 mm^2 and smaller than 6 mm^2, a two-sided pass may be preferred in an L shape. These special L-shaped passes may be done in different ways, as shown in Fig. 22.14. Figure 22.15 shows the flow-out process. Also see the various FC and underfill chapters.

Temperature Control. When dispensing underfill materials onto any substrate, the temperature of the surface and the die must be 70 to 90°C. One reason for this is that it will create an upward-moving

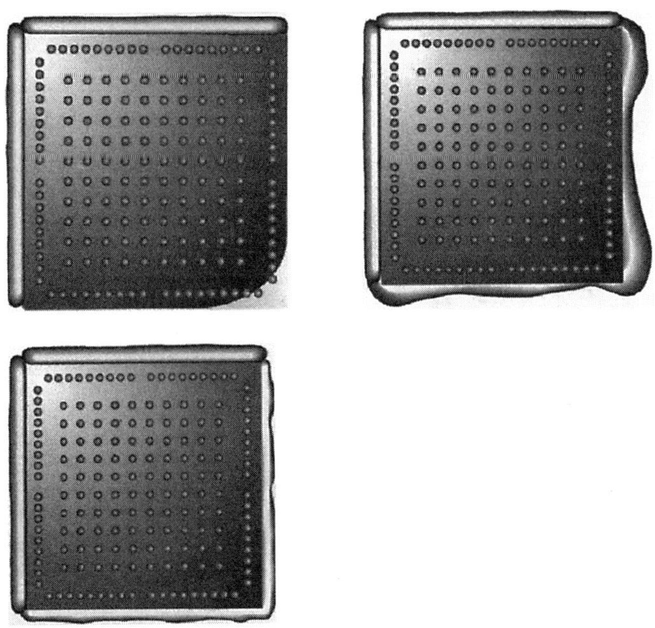

FIGURE 22.12 (*a*) Insufficient underfill. (*b*) Excessive underfill. (*c*) Correct amount of underfill.

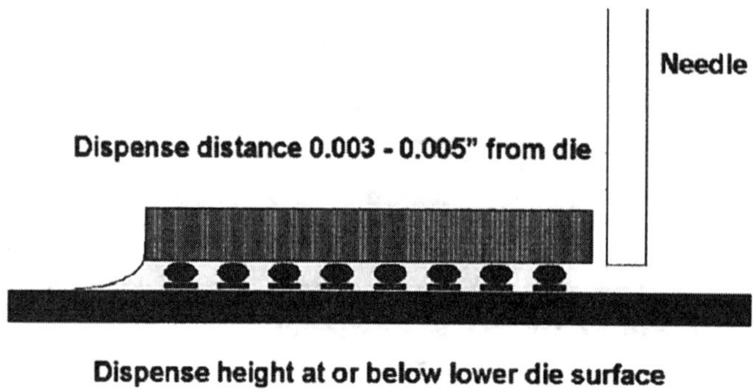

FIGURE 22.13 Needle position.

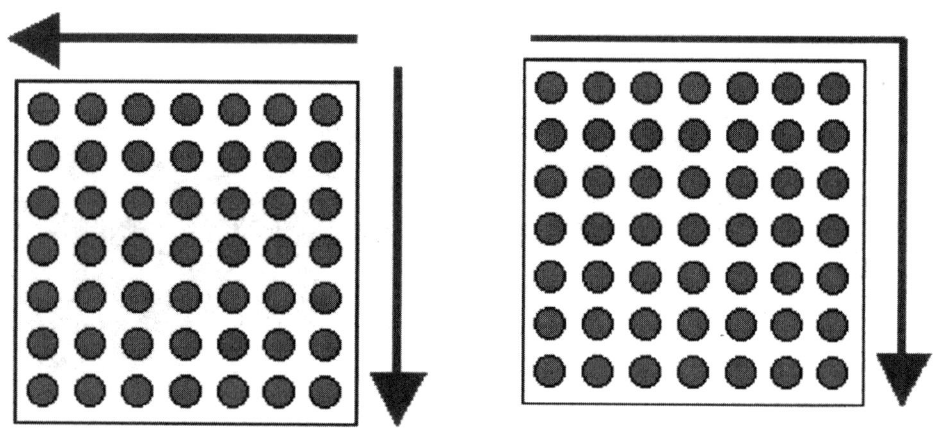

FIGURE 22.14 Dispensing patterns.

airflow from underneath the die and the surface of the substrate, degassing the surface and helping to eliminate air bubbles trapped underneath the die, which will result in long-term failure. The second reason for heating the surface of the substrate and the die is to ensure accurate flow of material underneath the die. Varying or nonuniform temperatures of either the substrate or the die can result in a nonlinear flow and thus air entrapment. The temperature of the substrate must be controlled using a closed-loop system to within ±5°C for noncontact systems or better than ±2°C for contact systems using heated vacuum chucks underneath the substrate. The substrate must be reheated to the ideal temperature before it arrives in the dispensing zone; a preheat station therefore is necessary. At the dispensing station, a second heating system is used to maintain the temperature of the substrate and die during dispensing. A third station, which is optional depending on the application, may be used to help the material flow correctly or gel slightly before being transported to a batch oven for curing.

The choice between noncontact and contact heating systems is a fairly simple one between flexibility or dedicated tooling. For all film-based applications, the only choice is contact heating using vacuum hot chucks under the film at the preheat, maintain heat, and flow stations. For all those applications using Auer boats or any dedicated transport tooling, as well as lead frame applications, the

FIGURE 22.15 Flow-out process.

contact heating system is the preferred method. Contact heating has the advantages of a more unified temperature distribution and the ability to heat only part of the substrate. In FC applications, for example, there may exist more than one size of die, perhaps ranging from 3 up to 12 or even 20 mm^2. By manufacturing dedicated tooling, it is possible for one set of tooling to handle the whole range of components. The only drawback to using contact heating systems is their inherent inflexibility, but often the advantages, such as speed of temperature ramp-up and temperature uniformity, outweigh the disadvantage of a 10-min tooling changeover.

In applications using standard FR-4 printed wiring boards (PWBs) as the substrate, such as FC underfill for contract houses where board sizes may vary considerably, noncontact heating can be used successfully. Although there is a tradeoff between the speed of temperature ramp-up and thermal uniformity, the possibility of changing the dimension of the substrate as required, without stopping the production, is an advantage for subcontractors. With the use of FC technologies becoming more and more widespread in mainstream electronics, manufacturers will be required to offer flexible noncontact heating systems with the advantages of contact heating where possible.

22.11 ENCAPSULATING WITH VACUUM ASSIST FOR CSPs AND OTHER PACKAGES

As semiconductor packaging continues to shrink in size and grow in number of inputs-outputs (I/Os) and complexity, more demands are placed on the equipment supplier to develop process solutions to aid the back-end semiconductor packaging industry to increase encapsulation yields and throughput. As the package sizes of CSPs get smaller, they become more difficult to encapsulate with standard dispensing processes. MCPs also are improved by vacuum encapsulation because their complexity tends to trap air. Two methods of increasing encapsulation yields and throughput have emerged recently. Vacuum-assisted dispensing and vacuum-assisted injection are now used as a result of new down-sizing assembly techniques. Encapsulating materials use highly refined mixing and packaging methods to avoid even the slightest introduction of air into encapsulation materials loaded into the final shipping media (e.g., syringes and cartridges). Thus the need for vacuum is not a result of problems with materials but rather is due to the ever-increasing complexity of the microelectronic assemblies they are used in or on.

One of the other problems with trapped air can show up just after the encapsulation process while curing the dispensed encapsulation. After encapsulant is dispensed into or onto the package, it must be placed in a cure oven to raise the temperature in the curing or hardening process. During this curing process, most encapsulation materials tend to decrease slightly in viscosity with the initial onset of the heated cure cycle. Then during the cure cycle, the encapsulation material's viscosity begins to increase rapidly to the hard or cured state. During this process, as the initial drop in viscosity occurs, trapped air can rise slowly to the surface of the component, leaving a pinhole that extends from point of origin to the surface. This pinhole will allow moisture and air to reach the components and cause premature failure of the device.

Vacuum applied during the encapsulation process can reduce void-type defects significantly. The encapsulation process for some CSPs was unsuccessful until vacuum was used. The extra care taken to slowly cover the wire bonds and capacitor was, however, time-consuming. The vacuum-modified dispensing process was an effort to reduce dispense time (and increase throughput). One process involves simple layering—initial layer, degas, and then the finish layer. Throughput is increased significantly, and the quality of the finished surface remains unblemished.

The Tessera μBGA package offers the size, weight, and performance advantages of an FC in a mechanically robust package that is compatible with existing SMD placement equipment. The μBGA is a popular CSP licensed by Tessera to over two dozen companies worldwide. This package uses a special low-modulus encapsulant (silicon-based). This material flows around the nubbin spacer columns that separate the silicon die and the flexible circuit tape. This material has a high viscosity, and although it is easy to dispense, the flow of this material is hampered by entrapped air underneath the die. Tessera and Speedline Technologies engineers have developed a method of dispensing material under vacuum that aids in the flow and allows complete coverage underneath the die.

The main advantage of a CSP, such as the Tessera μBGA package, is its inherent flexibility or compliance between the interconnect solder bumps/spheres on the bottom of the package and the die that overcomes issues related to CTE mismatch. This compliant interlayer or buffer (low-modulus silicone encapsulant) gives this package its compliant properties and makes underfill used in FC manufacturing redundant, which allows faster throughput times in circuit board assembly. Other than the fact that there is no longer a requirement for underfilling, the main advantage is that the packages are smaller than standard BGA packages but only 25 percent larger than an FC or bare die. Add to this the fact that the packages can be tested more easily than bare dies and can be placed with conventional pick-and-place production equipment, and we have a winning combination.

The way in which the compliant interlayer is achieved differs from package to package; however, one of the most popular at this moment is the μBGA package. This package uses a special low-modulus encapsulant, developed by Dow Corning, around the nubbin spacers that separates the silicon die and the flexible circuit tape. Although this material is easy to dispense, its flow underneath the die is hampered due to the viscosity of the material (50,000 to 60,000 cP) and to entrapped air between the silicon die and the flexible interposer. We cannot rely on capillary action as we do with FC underfill encapsulation, which has a viscosity typically in the less than 10,000-cP range.

Because of the high viscosity of this low-modulus encapsulant, approximately 50,000 to 60,000 cP, the material tends to flow very slowly. Unlike FC underfill, heating the part or the material inhibits (rather than enhances) the flow because of the curing characteristics of the encapsulant. The process thus required the dispensing platform to dispense a bead of encapsulant along three sides of the die, which are arranged on the circuit tape in a matrix format.

Vacuum dispensing equipment was designed by Speedline Technologies, with support from CSP maker Tessera, for dispensing in a vacuum environment and the associated manufacturing process. This system is not a single-chamber system, which has proven to be very slow. The system discussed in this section is equipped with three independent vacuum chambers, with the dispensing head actually working inside the main chamber at levels of vacuum equaling or exceeding 28 in Hg (710 torr).

The Speedline System 3900 has a small predispense vacuum chamber at the entrance end of the machine in which the boat, strip, or individual part is brought to vacuum at high speed. Once the required level of vacuum is obtained in the predispense chamber, an airtight interlock door is opened so that the part can move into the dispensing chamber, which is kept under constant vacuum while the machine is in production. The door separating the predispense chamber from the main dispense chamber is then closed, and the predispense chamber is brought back to atmospheric pressure so that it can receive another strip or boat. Inside the dispense chamber, an *xyz* movement system and dispensing pump with specifically engineered vacuum-capable components and lubricants dispenses the required pattern. Air is absent from the chamber, so no air can be trapped under the die. However, this alone does not ensure complete flow of the material under the die. Once the dispensing cycle is completed, the airtight interlock door between the main dispense chamber and the postdispense chamber opens, and the boat, strip, or individual part moves into the postdispense vacuum chamber. The door between the main dispensing chamber and the postdispense chamber then closes. In this third chamber, the environment is programmed to ramp slowly back to normal atmospheric pressure. As the pressure changes from vacuum back to normal atmosphere, the differential in chamber pressure forces the dispensed encapsulation material under the die into the void created while the device was under vacuum.

22.12 PERFORMANCE

The subcontracting business has increased dramatically as original equipment manufacturers (OEMs) continue to outsource more work. Subcontractors are now required to demonstrate their capability to produce better-quality components because they now compete against each other. Statistical tools such as CpK are now being used to qualify processes. In a BGA encapsulation, the critical dimensions are height and flatness of the package after curing. Because the wire bonds need to be covered and the surface of the encapsulation must be flat, the amount dispensed throughout the shift needs to be tightly controlled and monitored. In the past, 1 to 2 percent weight repeatability was required without CpK or accuracy requirements. Today, statistical process control and parameters are being used to qualify a process. CpK is a statistical number used to qualify a process. It indicates the number of defects that may be expected from a process. In the past, a CpK of 1.33 (or 31.5 defects per million) was the standard. Today, a CpK of 1.66 to 2 (100 defects per billion) is more commonly demanded from leading subcontractors. CpK relates the measurement of equipment capability as a ratio to the tolerance range established by the customer for the particular process. A higher CpK ratio is better because fewer devices need to be scrapped. To minimize the number of defects per million given a set of tolerances (USL, LSL), established by the customer, the process variations must be minimal. For example, to get a CpK of 2 given a USL and LSL of 2 percent, the 3SD (standard deviation) variation of the process has to be less than or equal to 1 percent (when mean is equal to the target).

22.12.1 Positioning Accuracy

An *xy* position accuracy of 2 mils (50 μm) is required today, and this can be met readily with the present equipment. Several gantry systems are even more precise. Vision systems make sure that positioning is accurate even when substrate is not precisely dimensioned. High precision is most critical for FC underfilling, where the needle must be located at a specific distance from the edge.

The *z*-height accuracy is also very important. Excessive distance from the substrate can produce too small a drop and may not even permit deposition. Insufficient clearance can deform the shape and even interfere with flow. Improper *z* spacing also can cause tailing as the needle moves away. Effective designs employ an active touch probe with a transducer to accurately access pressure and thereby control height to the desired preset values.

22.12.2 Shot Accuracy

Advanced dispensing machines have calibration-integrated hardware and software to ensure a high level of absolute shot-size accuracy. Weight-calibration tools integrated into the dispenser are becoming more common. Resolution of 0.1 mg can be achieved with an advanced system, and this level is becoming essential. Even a half-milligram error caused by an air draft or vibrations can offset the setup of the machine and result in scrap parts.

22.12.3 Application Rate

Multigantry systems with up to four dispensers and high-volume pumps now provide high application rates with a reasonable footprint.

22.12.4 Selecting the Right Equipment

The large variety of basic dispensing machine types and models within the categories makes the selection process more complicated than it was even a few years ago. The following subsections represent important areas for consideration.

Flexibility. How many types of materials will be run? Must a single dispenser run everything from flux to underfill? High machine flexibility is required when one dispenser must do many tasks. Although flexible equipment is available, there is always a compromise. As has been shown in earlier sections, some pumps are better for low-viscosity materials, whereas others are more suitable for thick pastes. One must balance flexibility against quality and productivity. Production may be served best by setting up dedicated lines.

Footprint. Factory or even laboratory floor space has considerable cost associated with it. Clean rooms add even more cost from both an initial construction figure and the annual upkeep. Fortunately, the industry has worked hard to decrease footprint while adding capability. Multigantry machines appear to provide the most efficient use of floor space.

Equipment Cost. While most salespeople will suggest that capital cost is the least important factor in determining annual running costs, productivity analysis, found in other chapters (see Chap. 31), bears this out. Productivity differences soon overwhelm initial investment dollars. Throughput is a major multiplier, and even a 10 percent increase offsets many thousands of dollars of investment. Another value to factor in is uptime. A more costly but more robust production machine quickly gives a return on investment if there is low downtime. Downtime should include actual time for repairs as well as maintenance. Modern features, such as self-diagnosis and hot swapping of control cards, reduce both factors.

22.13 SUMMARY AND CONCLUSIONS

Needle dispensing has become a key process to produce several types of area array products. Many BGAs use the dam and fill method of encapsulation that can only be done with needle dispensing. The die-down (cavity-down) packages also require this type of dispensing method. FCIP designs also use needle dispensing extensively.

Modern needle dispensers are highly automated and are self-calibrating. Software provides an easy interface for fast setup. Programs are stored for recall by part number, and many systems come with a library of common dispensing patterns. The latest equipment can provide very fast rates. The highest-productivity machines have up to four gantries and heads that can supply different materials that may be required in one package.

22.14 REFERENCES

1. Lee, C., "Ultra-High Speed Dispensing of Surface Mount Materials to Enhance Production Throughput," paper presented at APEX 2000, Long Beach, CA, 2000.
2. Miquel, B., and Lee, C., "Advanced Process Control for BGA Encapsulation and Flip Chip Underfill," paper presented at Semicon Taiwan, September 2000.
3. Lawing, C., "Preventing Voids in BGA Packages," *Chip Scale Review,* March 1998.
4. Lawing, C., "Vacuum Dispensing of Encapsulants for μBGA Manufacturing," *Electronics Packaging and Production,* February 1998.

CHAPTER 23
MOLDING FOR AREA ARRAY PACKAGES

K. Gilleo
Cookson Electronics

B. Cotterman
Cookson Semiconductor Packaging Materials

23.0 INTRODUCTION

Electronic packaging must provide physical protection during component storage, handling, and assembly to the circuit board and long-term operation within the finished product. Environmental protection is also required, especially resistance to moisture and ionic materials. The end user typically will expose the package to extremes of temperature, mechanical shock, and harsh environments. The finished package must have proper dimensions for machine handling while also allowing identification nomenclature to be applied. All these requirements can be achieved using polymers in the transfer-molding process. Plastic packages make up a large segment of area array packages. Both larger plastic ball grid arrays (PBGAs) and smaller chip scale packages (CSPs) use molding processes. Transfer molding is by far the predominant process employed because it has a long and successful history in the production of all types of packages. Transfer-molding processes and materials will be described in detail with this chapter. For now, it will suffice to describe the method that uses solid preforms of polymer precursors that are melted, injected into molds, and permanently hardened by thermally induced polymerization.

Encapsulants were described in Chap. 11, so the discussion here will only be a summary. The most common materials are epoxy mold compounds (EMCs). The mixture contains reactive epoxy resins, a hardener, at least one reaction accelerator, flame retardants, a filler, a lubricant, a coupling agent, a flow promoter, a stress modifier, an ion getter (chemical scavenger), and a colorant. The basic epoxy-hardener prepolymer mixture consists of low-melting solids that will liquefy, flow, and polymerize to a permanent solid in the heated mold. Table 23.1 lists the effects produced by the components.

The solid materials are mixed, melted, and formed into preforms. A large EMC plant may make a powder or blocks of material that are stored and later formed into the desired preforms, often at other facilities. The in-process material must be kept at low temperature to avoid slowly polymerizing.

23.1 MOLDING MATERIALS

Molding materials and other encapsulants were described in Chap. 11, so a brief discussion is provided here. Epoxies are the predominate polymers used for transfer molding for a number of reasons.

TABLE 23.1 Molding Compound Composition

Component	Percent	Function(s)
Polymer bulk		
Coupling agent	<1	Bond between hardener and filler
Resin	5–20	Provide cross-linking reaction
Hardener (curing agent)	3–10	Provide cross-linking reaction
Accelerators		
Catalyst	<2	Accelerate cross-linking reaction
Fillers		
Silica	70–92	Improve physical properties
Additives		
Flame retardant	<3	To meet UL-94 flammability requirement
Pigment	<1	Color for marking purposes
Lubricant	<1	Improve mold release
Stress modifier	<3	Reduce internal package stress
Flow promoter	<3	Improve compound flow length
Adhesion promoter	<2	Improve compound adhesion to different surfaces
Ion getter	<1	Improve corrosion resistances

First, the cured epoxy provides very good mechanical and electrical characteristics. Perhaps the most important feature of epoxies is that starting materials can be solids. In fact, the range of resins and hardeners is so broad that the melting point can be formulated precisely. This is important for EMCs because the materials must be melted and transferred to the mold. The cure rate also can be tailored by the judicious choice of resins, hardeners, catalysts, and accelerators. This is important because material must remain in a liquid state long enough to be transferred to the molds. Also, the rheology of the molding material under heating must change in a way that meets the requirements of the molding equipment. Heating reduces viscosity but also triggers the polymerization reactions that increase viscosity. These two viscosity change forces must be balanced by skillful formulation.

The EMC contains significant mineral filler, typically silica, to reduce thermomechanical expansion. While standard EMCs have a filler level in the 70 to 75 percent range, advanced BGA materials have levels as high as 85 percent by weight. The filler size, shape, and quantity are critical for good performance during the molding process and for finished properties of the package. Once again, there must be skillful balancing of the formulation. Increasing the filler level decreases the coefficient of thermal expansion (CTE) but increases the viscosity. If the viscosity of the melted EMC is too high, flow will be impeded, and the molds will not fill properly. However, smoother and more spherical filler will reduce viscosity at the same loading. This allows molding compounds to be produce with good flow properties but low CTE values. High filler loadings are especially beneficial for area array packaging because stress is reduced and a more planar package is produced. Higher filler means that the total polymer content is lower, and this reduces shrinkage. Epoxies and most other polymers shrink on curing. Shrinking, as would be expected, tends to warp the packaging substrate, especially when it is a thin organic material. High filler loadings using smooth spherical filler offset shrinking to give the flattest package. Although some warpage is acceptable, a warped area array package is more difficult to assemble because not all the solder balls may make contact with the circuit board. Low flammability that is required for some packages is achieved by adding brominated hardeners and resins at levels up to 5 percent by weight of total composition.

Molding compounds are provided in a variety of solid forms designed to fit available equipment and mold sizes. Products for area array packaging, while similar to standard molding compounds, tend to be optimized for this type of package. One of the most important optimizations is low stress to minimize warpage that can be improved with biphenyl epoxy resins that have less shrinkage and the addition of elastomerics, such as silicone rubbers. The particles are extremely small and may be less than 0.25 μm in diameter. Specialty EMCs are also offered with such properties as low popcorning and higher thermal conductivity, which is achieved with different fillers with greater thermal conductivity that include aluminum nitride, boron nitride, silicon carbide, and even diamond dust.

The preforms are made by mixing the solid ingredients (dry mix), melting them, further mixing (melt mix) in the liquid phase, and then cooling back to a solid before pulverizing or granulizing into a powder. The powder may be stored at low temperatures as an intermediate raw material. The mix must be very thorough for best results. Powder is shaped into preforms by applying heat and pressure before shipping to customers. Preforms typically are stored at 5°C. Figure 23.1 shows typical EMC preforms.

23.2 TRANSFER-MOLDING EQUIPMENT

Molding machines vary from manual types to fully automatic. Manual molding has obvious limitations, such as high labor requirements, longer cycle times, and lower yield. Automatic gang pot molding is 100 percent automatic, requiring very little attention. Yield is virtual 100 percent, cycle time is short, and waste is very small. This process is used for high-value-added packages and high volumes. Figure 23.2 shows a simple molding machine diagram, whereas Figure 23.3 represents a production transfer-molding machine with a loader/unloader for automatic handling. Figure 23.4 is a diagram of a mold.

23.3 MOLDING PROCESSES

The chip carrier is first prepared, and this involves attaching a die and making the electrical connections, usually by wire bonding. Although single-package structures can be molded, it is more

FIGURE 23.1 EMC preforms.

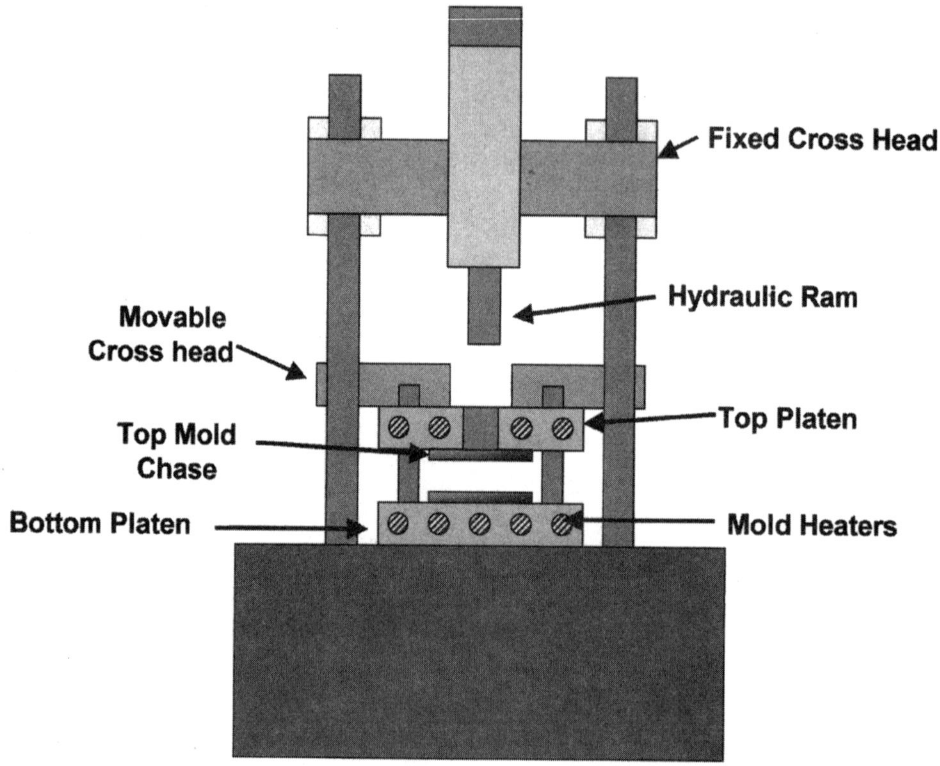

FIGURE 23.2 Transfer molding press diagram.

common to use strips or even full sheets of product for manufacturing efficiency. Strips loaded with dies are placed in the open mold, which is preheated. Solid preforms of EMC are placed into the mold. The mold is closed, and the cycle begins. The EMC melts, and a plunger pushes or transfers the liquid to the mold cavities. The material must be fluid enough so that the dies and fine-connection structure are surrounded by the encapsulant. Pressure and flow rate are regulated so that no damage occurs to the thin wire bonds (wire sweep damage) but so that the material fills the mold before thickening and hardening occur. The heat from the mold causes the polymerization reaction to accelerate to polymerize the EMC into a permanent solid that cannot be melted again. Cycle times range from 60 to 120 s for conventional molds and from 60 to 90 s for smaller multiplunger gang pot systems for high-value parts. The conventional molding process yields from 50 to 1000 packages per cycle, whereas the gang pot multiplunger method produces only 4 to 40 parts.

Equipment used in the molding process is continually changing, so when a problem is encountered, the solution is not always readily apparent. Any of the following variables can affect the encapsulation process:

- Mold temperature profile
- Transfer pressure
- Transfer ram speed
- Clamp pressure

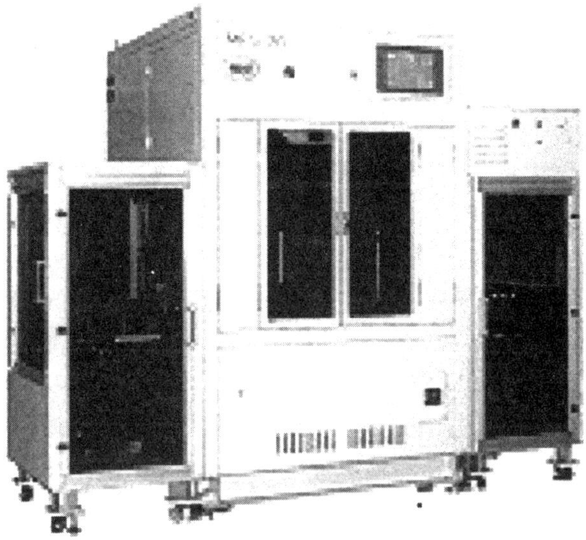

FIGURE 23.3 Transfer molding machine.

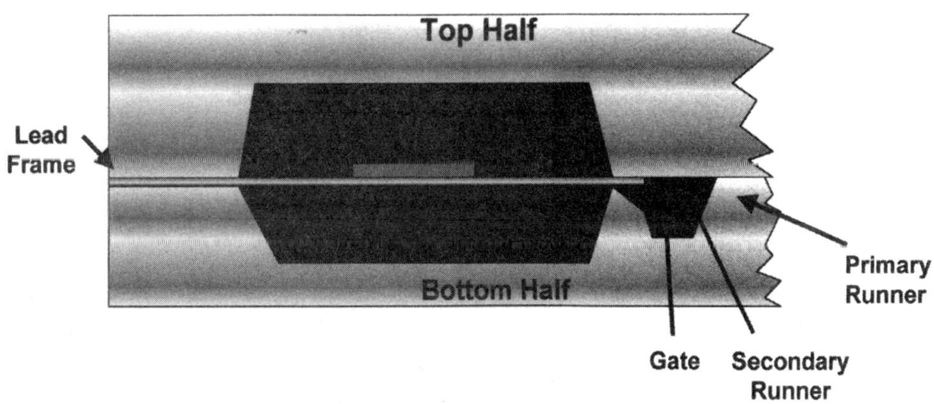

FIGURE 23.4 Mold.

- Preheat time and temperature
- Cure time and temperature
- Operator technique
- Material variations
- Material handling variations
- Tooling designs
- Preform quality

The next subsection will go into detail describing the parameters and the rationale for the changes.

23.3.1 Recommended Transfer-Molding Parameters

A daily log is invaluable in setting a standard for optimal molding parameters and will assist in problem solving. EMCs can be molded over a range of temperatures from 165 to 188°C (300 to 370°F). Molding parameters should be optimized for each process depending on mold design, device type, package type, and molding compound characteristics. Generally, low transfer pressures and slow transfer times are recommended for most encapsulation materials. Molding pressures for encapsulation materials typically are within the range of 500 to 1500 psia (pounds per square inch, actual).

The use of dielectically preheated preforms is important for minimizing cure cycles and improving moldability. Preheat temperatures of 80 to 100°C (175 to 210°F) are recommended. Uniform preheat temperature can best be achieved by using roller preheaters. Uniform-quality preforms, exhibiting a density in the range of 80 to 90 percent of molded density (0.8 to 0.9 preform bulk density), also improve preheat consistency.

Mold Temperature. It is very important that the temperature profile, across both the top and bottom of the mold, be within ±3°C (5°F). To obtain a temperature profile, measure the temperature of at least six points across the mold surface. This tight temperature differential is essential to maintain uniform molding conditions for all cavities of the mold.

For example, if one cavity block is 5.5°C (10°F) hotter than another, significantly different molding conditions will exist within the same cycle. This difference might result in insufficient fill in one cavity block and excessive flash in another. Similarly, good parts may be obtained from one side of the mold and parts exhibiting cosmetic defects may be produced on the other.

The temperature differential between top and bottom dies should not be more than 5.5°C (10°F). This will ensure uniform flow and cure and will reduce bowing of large parts.

The temperature profile should be measured periodically throughout the day and recorded on a process sheet. Adjustment of temperature should be made as needed.

Transfer Pressure. The actual transfer pressure, measured at the bottom of the transfer ram, should be checked and recorded routinely. A commonly used technique is to place a force gauge under the transfer ram to measure the total applied force directly. This will serve to verify the hydraulic pressure gauges on the press and will indicate that the pot and plunger frictions are constant. If low or erratic readings are obtained, this could indicate friction or improper fit between the transfer ram and the pot.

The molding press transfer pressure gauge should be read periodically throughout the day and recorded on a process sheet. These readings should be validated routinely (e.g., weekly) per the force gauge method described above.

Transfer Ram Speed. In setting the transfer ram speed, it is essential that the ram speed control, and not some other factor, control the maximum rate of flow of the material. If the ram is set too fast, slight variations in preheat or other parameters can cause the mold to fill more rapidly and result in voids in the molded parts. The transfer ram speed should be fast enough to ensure that all cavities fill before gelling takes place but not so fast as to cause air entrapment.

Transfer ram speed should be checked periodically throughout the day and recorded on the process sheet.

Preheat Time and Temperature. To obtain maximum preheat efficiency, the preforms should be heated to 80 to 100°C (175 to 210°F). Standard high-frequency preheating equipment is recommended. Preheaters operating at low frequencies require longer heating cycles and are not recommended for use with Plaskon epoxy molding compounds. When the preheater's electrodes are adjusted properly, Plaskon epoxy molding compound preforms should heat in approximately 20 to 30 s.

If heating is too rapid, a runaway exotherm could cause premature curing of the preforms. Conversely, if the preheat cycle is too long, the preforms may begin to cure slightly while still in the preheater, resulting in marginal cavity filling and/or damage to the device being encapsulated. The preform should be soft and pliable but not so soft as to stick to the operator's glove. The transfer time between completion of preheating and activation of the transfer ram press should be minimal.

The preheat time and temperature should be checked periodically throughout the day and recorded on the process sheet. If any changes in these parameters are made during operation, the changes also should be noted on the process sheet.

Operator Technique. The press operator can be a major source of variations in the molding process. The operator's performance should be monitored from time to time. The supervisor should check preheat time and temperature, transfer time, molding pressure on the hydraulic gauge, cure time, and other variables periodically to see that the proper molding conditions are being used.

Material Variation. Variations in the molding compound that would result in molding difficulties are rare. If a problem develops, all molding parameters should be checked. In most cases, simple adjustments in the transfer pressure or speed and/or cure time will solve the problem. If the difficulty persists, contact your local Plaskon sales representative.

Once a systematic pattern of in-process checks has been established, these historical data are invaluable in tracing molding problems. After adequate information is established, the frequency of calibration can be reduced for those parameters which are shown to remain constant. For additional information on solving molding problems, refer to the troubleshooting guide (Sec. 23.5).

Postcure of Molded Parts. Postcure is recommended to obtain optimal properties of molded parts. A typical postcure schedule of 4 h at 175°C (347°F) usually will maximize the physical properties of the encapsulant. However, each semiconductor device–encapsulant assembly should be evaluated at several different postcure schedules to determine the best postcure time and temperature for individual applications.

Plasticity or Flow. Materials for encapsulation are supplied with various melt viscosities, as indicated by the flow range listed on the individual data sheets. The most common method of measuring plasticity or flow characteristics of Plaskon epoxy molding compounds is ASTM-D3123-72, "Spiral Flow of Low Pressure Thermosetting Molding Compounds." For most semiconductor encapsulants, spiral flow is run at 177°C (350°F) with 1000 psi transfer pressure.

Calculation of Molding Pressure. Use the following formula to calculate actual hydraulic pressure (psia) being exerted on the molding compound:

Hydraulic pressure ÷ transfer cylinder diameter^2 × pressure = molding compound transfer plunger diameter^2 (psig) in the pot (psia)

Example

Diameter of transfer cylinder = 4 in

Diameter of transfer plunger = 1.75 in

Gauge pressure of hydraulic press = 150 psig

Therefore:

$$\frac{4^2}{1.75^2} \times 150 = 784 \text{ psia}$$

Testing Methods. The following test methods are used to determine the properties and specifications of epoxy molding compound encapsulants.

Property	Procedure
Physical	
Specific gravity	ASTM-D792*
Spiral flow and ram follower gel time	S-724-TO†
Flammability rating per UL-94-VO	PD-1007†
Ash content	U-815-TO†
Hydrolyzable chloride	U-814-TO
Water extract conductance and pH	PD-1026
Nitrogen content	U-819-TO
Thermal	
Expansion coefficients and glass transition temperatures	U-858-TO
Mechanical	
Flexural strength and modulus	PD-1025
Tensile strength	PD-1027
Hot hardness and mold appearance	PD-1008
Electrical	
Dielectric strength	ASTM-D149
Dielectric constant	ASTM-D150
Dissipation factor	ASTM-D150

*ASTM = American Society for Testing Materials.

†Cookson Semiconductor Packaging Materials Test Methods. Copies available on request.

Shipping, Handling, and Storage. To maximize retention of the spiral flow of the material, Plaskon epoxy molding compounds should be shipped by refrigerated truck or packed in dry ice if shipped by air. Temperature recorders, using 30-, 60-, or 90-day charts, are available, and it is suggested that one be included in each shipment to monitor temperature during transit. Plaskon can furnish disposable temperature recorders on request.

Shipping and storage should be at or below 5°C (40°F). A cold storage room can be designed so that new material enters one door and material to be used is withdrawn from a door at the opposite end. This ensures use of the oldest material first. Allow 2 ft^3 for every 50-lb pail of Plaskon epoxy molding compound.

Plaskon Electronic Materials' semiconductor-grade epoxy molding compounds are supplied in moisture-resistant plastic pails designed for overseas shipment. These heavy-duty, self-locking polyethylene pails provide excellent resistance to moisture ingress and physical damage during shipment or storage.

After removal from cold storage, the pails should be kept closed and allowed to equilibrate to ambient conditions before opening. Recommended time is 24 h. This ensures that the material will preheat uniformly and prevents moisture from condensing on the molding compound. Care should be exercised in opening pails to see that no moisture drops onto the molding compound.

Shelf Life. If the molding compound is to be kept for extended periods, it should be stored at 5°C (40°F) or below. Materials stored at room temperature (21°C, 70°F) will retain at least 70 percent of their original flow characteristics for no more than 2 weeks. Material always should be stored in tightly closed containers.

Mold Design. The mold design for epoxy molding compounds, with few exceptions, is similar to the mold design used for other thermosetting resins. Epoxy molding compounds have low-melt viscosities, which means that molds must be designed with close tolerances to minimize flashing around knockout pins, core pins, and the pot and transfer ram. Shallow, wide gates and small runner systems are recommended to minimize spurting and "jetting."

Adequate venting will ensure uniform density and minimize molding problems. Venting should be provided for each cavity.

When "breaking in" new molds, carnauba wax or some other release agent should be applied to the mold surface. If molds previously have used materials containing a different lubricating system (or other molding compounds, e.g., phenolic), general-purpose melamine cleaning compounds should be used to purge the molds before using epoxy molding systems.

Mold Cavities. Relatively low shrinkage is one of the characteristics of Plaskon epoxy molding compounds, and it must be taken into account when designing a mold. Mold cavities should be designed with adequate draft around all areas perpendicular to the parting line. This is especially true when the part to be ejected is very delicate, as are most semiconductor devices. Adequate draft minimizes the ejection pressure required to remove parts from molding cavities.

Gating. Gates approximately 1/8 in wide can be as thin as 0.010 to 0.015 in. Gate width is limited by lead-frame and package designs. Gates should be located in the thickest sections, if possible.

Venting. To ensure adequate venting of entrapped air, vents should be added to each cavity of the mold. For epoxy molding compounds, vents about 0.0005 to 0.001 in deep and about 1/4 in wide should prove satisfactory.

Molding Surface. The mold should be made of high-quality, heat-treated tool steel having a Rockwell hardness of at least 60C. The molding surface should be highly polished, with a finish of about 10 to 15 μin on all molding surfaces, including areas of the parting line that may be exposed to flash. All the molding surfaces should then be plated with a hard, highly polished chrome deposit.

Lead Slots. Lead slots should conform as closely as possible to the lead dimensions. Tolerances in this area should be approximately 0.0005 in or less. Minimizing flashing around the leads facilitates molding and ejection and alleviates lead deflashing after molding. Where practical, an interference fit should be employed.

Knockout System. All production molds should have some type of mechanical knockout system. If possible, the knockouts should be located in the cavities with the maximum surface area exposed for ejection from the mold. If more than one knockout pin is used per cavity, it is essential that the pins remove the part in a level manner. Knockout pins should be placed against solid, thick, or inflexible sections. Whenever possible, knockout pins also should be placed in runner and cull areas.

Mold Cleaning. To produce well-molded parts, molds must be maintained in top condition at all times. Specific mold cleaning compounds are described in separate brochures (26AI05 and 26AI06).

Preforming. EMCs can be formed into various size and weight preforms. This provides several advantages in handling the molding compound. Preforms

- Minimize entrapped air in transfer molding
- Reduce dusting
- Improve material handling
- Optimize dielectric preheating

If preforms are made in the molder's plant, the pressures listed in Table 23.2 are recommended to achieve the noted hardness.

Mold Maintenance. To produce well-molded parts, molds must be maintained in top condition at all times. Regularly scheduled in-process mold cleaning is the best way to maintain productivity. Molds should be cleaned periodically with either compression or transfer molding of a mold-cleaning compound. Mold-cleaning compounds designed specifically for this job are available.

In-Process Cleaning. EMCs are formulated to release easily from all areas of the mold. Before starting a new molding cycle, the upper and lower mold halves should be inspected for any loose

TABLE 23.2 Preform Pressures and Hardness

Degree of hardness	Pressure, t/in^2 (approx.)	Shore D hardness
Minimum	2.5	62–75
Average	5	69–80
Maximum	7.5	74–82

plastic. Brushing the surfaces of the mold with a brass or high-temperature-nylon brush should remove excess plastic. Stubborn pieces can be removed with a brass pick. After brushing, any remaining loose plastic can be blown off the mold surface with clean, dry compressed air. It is very important that no excess plastic be left on the mold from the previous shot, particularly in the vent area. Such residue in the vents can result in incomplete cavity fill on successive shots.

Caution: Never use any material other than brass, beryllium, copper, or high-temperature nylon (melting point $>$ 400°F) for mold cleaning. Harder materials can scratch the mold surface, and the scratches will appear on the molded epoxy surface.

Note: Do not remove any plastic that accumulates in the grooves of the transfer plunger. These grooves are designed to fill with plastic and act as a seal to prevent plastic from flowing upward past the plunger.

Periodic Cleaning. The most effective cleaning method for removing mold stains is with the compression molding of Plaskon Melamine Mold Cleaning Compound. This melamine molding compound does not harm mold surfaces but will attract deposits causing surface staining. For those operations in which transfer-mold-cleaning compounds are preferred, materials such as Plaskon's Nikalet ECR are recommended.

As an alternative to compression cleaning, the industry offers a transfer-grade melamine cleaning compound for cleaning. Typical transfer-molding pressures for 14-lead DIP molds range from 3000 to 4500 psi. This procedure may require more shots to obtain as clean a mold as when a compression mold cleaning compound is used.

Whenever possible, it is better not to use lead frames in cleaning. When lead frames are not used, the compound will flow into vents and land areas, cleaning these parts of the mold as well as the cavities and runners. It should be noted, however, that some molds can be damaged if lead frames are not in place during molding. Whether using preforms or powder, the cleaning compound should be preheated. The transfer time should be approximately 15 s.

23.3.2 Area or Flood Molding

Flood molding is a newer process that provides several advantages over the conventional transfer-molding approach. A general-purpose single-cavity mold is used that can hold the entire sheet of substrate that is loaded with dies. This means that different-sized packages can be molded with one general-purpose tool. The EMC is melted as with the conventional process and injected into the single large cavity. The hot mold cures the epoxy. The mold opens, and the molded sheet is ejected. The dies must now be singulated into separate packages. The sheet containing hundreds of dies is sawn in a process similar to wafer sawing. The PBGAs appear similar to those molded in separate cavity tools except that the edges are flat. Although flood molding adds the extra step of sawing, the economics are very good because molding is faster and tooling cost and time are reduced to near zero. The flood-molding process has become increasingly popular because of faster molding and then elimination of custom tooling.

23.3.3 Injection Molding

While thermoset polymers such as epoxies have dominated molding, high-temperature thermoplastics can and are being used. Thermosets dominate because they can be formulated as low-melting solids that harden to nonmelting solids. However, since heating can always remelt thermoplastics, they are a greater challenge. The main problem is selecting a plastic with a high enough melting point for package performance but not so high that it cannot be melted for the injection-molding process. Several thermoplastics have been found acceptable and are used to mold some packages today.

Injection molding is a very common process that is used to make the thousands of plastic parts used everywhere from toys to automobile parts. The process has some similarities to transfer molding but is really quite different. Granular plastic pellets are fed into the molding machine through a hopper or even by means of a piping system. The pellets are melted and then moved toward the mold by a screw that mixes the hot mass while generating force. The mold is closed, and a nozzle is inserted into the mold that will inject hot molten plastic. The plastic is injected under high pressure while the two halves of the mold are clamped together. The liquid flows through runners into the cavity. Unlike the transfer mold, the injection mold is usually not heated, although it may be kept at some optimal temperature that allows the liquefied plastic to flow completely before it solidifies. Fluid typically is circulated through the mold or through heat exchangers in contact with the mold to remove heat from the hot plastic. The mold is opened, and the finished parts are ejected.

Injection molding is a simpler process in that the thermoplastic will not cure like the thermoset. This means that the plastic can remain hot for long periods of time. In fact, the barrel that contains the screw serves as a reservoir for material. Injection molding is a fast and efficient process that is well supported by a large machine and tool industry. However, because the finished product can still soften and distort at higher temperatures, the method has not gained widespread use. One concern is that poorly controlled solder ovens could damage the package.

23.3.4 Molding Underfill (MUF)

Materials. Some suppliers have started to offer molded underfill (MUF) materials that are based on special EMCs used in molds designed for this special process. The EMC must convert to a low-viscosity fluid on heating that will flow between the small gap between a flip chip (FC) and a packaging substrate. The final properties must meet those achieved by the common liquid underfills. In some cases, the mold is designed to allow product to flow under and over the chip so that it is an encapsulant as well as an underfill. The MUF material is similar to EMCs but must have filler particles that are small enough that they will fit in the small gap. The material must mold into a spacing down to a minimum of 2.5 mils (0.06 mm) and fill completely without voiding or other defects.

Tooling. The mold is designed to hold the FC assembled to the packaging substrate, but with gating that allows input to the edge of the chip. A vent may be required to allow air to escape from the chip gap. Some processes require a small vent hole in the substrate underneath the chip. One alternative is to use a vacuum.

The MUF Process. This is a one-step process that simultaneously underfills and overmolds to provide a robust, one-step total encapsulation solution, as shown in Fig. 23.5.

Optimal molding conditions for the MUF process have been identified as

Preform temperature: 75–80°C

Transfer time: 15 s

Material pressure: 350–400 psi

Mold temperature: 165°C

In-mold cure time: 105 s

MUF Performance. While transfer-molded underfill is relatively new, the process appears to run well, and good performance has been measured. Tests by Cookson Semiconductor Packaging Materials with large FCs did not show any problems with voiding or incomplete filling. Figures 23.6 and 23.7 show results after molding.

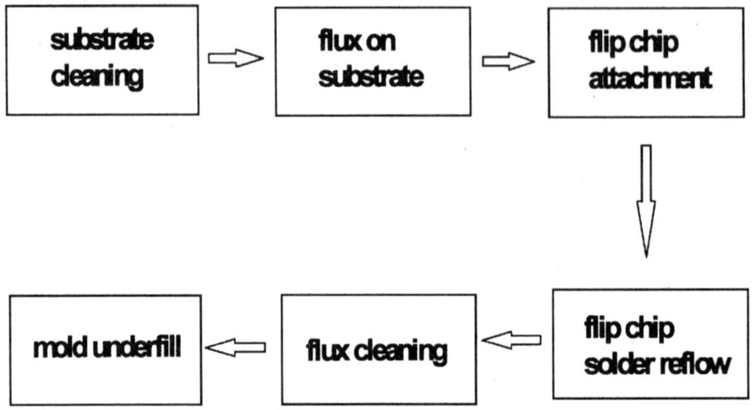

FIGURE 23.5 Molded underfill process steps.

Chips with 2.5-mil bumped spacing (400 × 600-mil bumped die, bump pitch 8 mils, bump diameter 5 mils, gap between bumps 3 mils) are used with bumps intact, and all the gaps and spacing were filled completely with no voids (microscope). Further tests showed that the gap filling limitation for the current formulation (153.405) was 2.1 mils.

The MUF approach is a fast process with no liquid-related issues (particle settling, nonhomogeneous). MUF is a traditional process (both material and equipment) that is a more automated, simple, robust process and less expensive relative to present underfill and overmold processes and materials. The method is ideal for FC BGAs, where strips of loaded substrate can be fed into a transfer-molding machine. Higher filler levels can be used than possible with liquids because force is used to push the material under the chip.

23.3.5 Preventing Molding Process Problems

The first step in preventing any molding problem is the use of proper molding techniques and maintenance. Sometimes, however, even with recommended molding techniques and mold maintenance, problems still arise. This section presents recommendations for molding techniques and mold maintenance, as well as troubleshooting for the prevention and solution of these molding problems.

Important Transfer-Molding Parameters

Effect of molding parameters

- Mold temperature
- Transfer pressure
- Transfer ram speed
- Preheat time and temperature
- Operator technique
- Material variation

Postcure of molded parts

Plasticity or flow

Pressure

Testing methods to ensure quality

- Physical
- Thermal

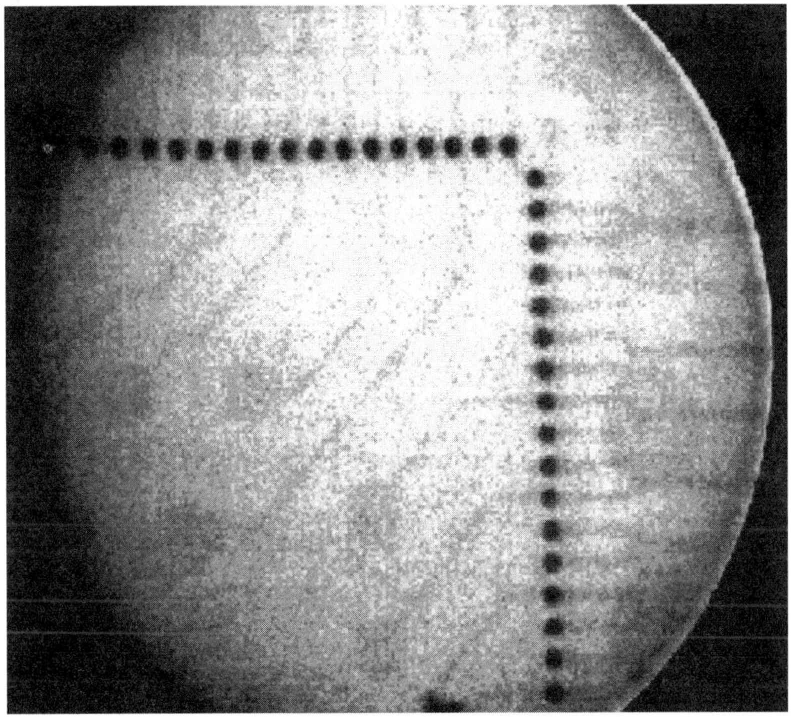

FIGURE 23.6 X-ray of molded underfill part.

- Mechanical
- Electrical

Shipping and handling of material

Storage of material

Shelf life, age of material

Mold design

- Mold cavities
- Grating
- Venting
- Molding surface
- Lead slots
- Knockout system
- Mold cleaning
- Preforming

23.4 TROUBLESHOOTING GUIDE

The following troubleshooting section is presented to address commonly encountered molding problems with a **cause** → remedy format.

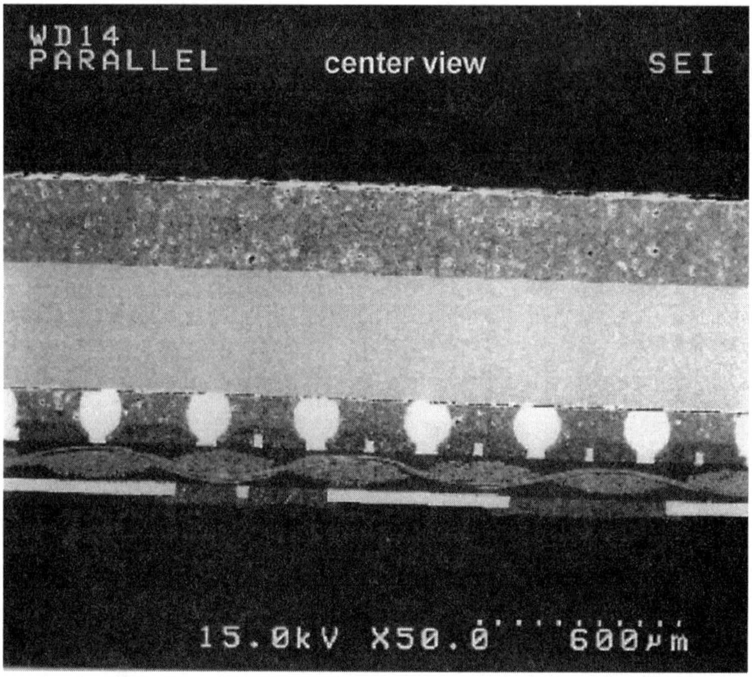

FIGURE 23.7 SEM of MUF CSP.

23.4.1 Incomplete Fill and Knit Lines

Short or Overcharge. Check cull thickness to ensure proper charge weight. Check preform weights. An excessive charge weight can cause filling problems when the material does not attain optimal viscosities during transfer.

Incorrect Transfer Pressure. Check pressure with force gauge to verify pressure gauge, and check wear on plunger and pot. May need to adjust pressure or align plunger.

Improper Mold Temperature. Verify that heating elements are operating effectively and that mold is heating evenly by obtaining accurate pyrometer or temperature salt readings over the mold surface. Low temperatures can cause higher viscosities, whereas high temperatures can cause premature gelling. Temperatures that are either too low or too high can result in incomplete filling.

Incorrect Transfer Speed. Adjust the transfer speed. A fast transfer speed can cause turbulent flow, resulting in air entrapment. A transfer speed that is too low can permit material to gel before the mold is filled completely.

Improper Preheat Temperature. Closely control preheat temperature to avoid filling problems associated with high melt viscosities (low temperature) or premature gelling (high temperature).

Poor Venting. Check to verify that the vents are clean. Adequate venting is required to ensure that air in the runners and cavities is not entrapped. Vents normally are located opposite the gate area, and dimensions should be optimized for each mold. Some molds may require multiple vents.

Short Flow Material. Optimize preheat, mold temperature, and/or transfer pressure. Consult vendor to optimize spiral flow, cure rate, and melt viscosity. Check in-house storage conditions.

23.4.2 Voids and Blisters

Fast Transfer Time. Reduce the transfer time to allow the trapped air to escape from the runners and cavities through the vents.

Incorrect Preheat Temperature. Ensure that preheat temperature is optimized and that inserts, if used, are not cold.

Low Melt Viscosity. Consult vendor to obtain higher melt viscosity or longer gel time materials. Check transfer speed control of press. Material with a very low melt viscosity or a high plunger speed setting can cause turbulent filling and entrapped air in the runners and cavities.

Preform Density. Ensure that preforms are sufficiently dense (at 0.83 of molded density) by measuring dimensions and weight and calculating actual preform density. Consult vendor and/or preformer if required.

Molding Temperature. You may need to adjust molding temperature if blisters appear in molded parts. Excessive temperature and/or insufficient cure can cause blisters.

Poor Venting. Adequate venting is required to prevent air entrapment inside runners and cavities. Clean vents if needed.

High Volatiles. Proper handling procedures must be used to ensure that material is not contaminated with high moisture levels. To prevent condensation, material coming from cold storage should remain at room temperature for 24 h prior to opening pails.

23.4.3 Flashing and Resin Bleed

Transfer Pressure and Time. Reduce the time in which the transfer pressure is acting on the cavities. Lower the transfer pressure.

Clamp Pressure. Check the clamp pressure, and increase it if needed.

Worn Molds. Replace worn parts.

Out-of-Spec Inserts. Ensure that insert tolerances are maintained properly.

External Release Agent. Minimize the use of external release agents that can act as a plasticizer for the molding compound.

Low-Viscosity Molding Compound. Materials with very low viscosities can generate more flash and bleed. Higher-viscosity products may be necessary. Reducing the preheat temperature and reducing the transfer pressure can help to reduce flash.

23.4.4 Mold Staining and High Molding Temperature

Lower molding temperature but do *not* sacrifice cure. Use increased curing time if needed. Implement periodic thorough cleaning.

External Release Agent. Minimize or avoid using an external release agent. Use molding compound and longer cure time (5 min) for conditioning of the mold after cleaning.

Molding Compound. Consult vendor for products with better staining resistance.

23.4.5 Sticking

Low Mold Temperature. Check temperature profile of the mold.

High Mold Temperature. Reduce mold temperature. Check for hot spots. Excessive mold temperature accelerates degradation of mold-release agents that can form a buildup on the mold surface and interfere with release.

Short Cycle Time. Increase cure time to ensure effective cure.

Changing Materials. When changing molding compounds, thorough cleaning and conditioning of the mold are required.

Mold Design. Materials vary in shrinkage, and the draft angles may need to be modified.

Molding Compound. Inadequate release agent levels in the molding compound can cause sticking.

Moisture Contamination. Anhydride materials will stick if contaminated with moisture. Check handling procedures and aging time.

23.5 TOOLING

The tooling consists of a precision metal mold that can have many hundreds of mold cavities. Tooling can be expensive, especially for high-volume molds with several hundred cavities. The lead time can extend for several months. However, newer methods, such as flood or area molding, have reduced tooling substantially. Flood molding involves using a generic mold that is loaded with package carriers. Molding compound is injected into the mold to encapsulate all the packages as if they were a single giant package. The entire array is removed, and sawing or singulating forms the individual packages. This adds a new process, but savings in mold costs and a fast process offset the cost of mechanical singulation.

23.6 PRODUCTIVITY AND COST

23.6.1 Volume Dependency

Transfer molding is intended to be a relatively high-volume process. While capital equipment cost and tooling cost are considered high, transfer molding provides substantial savings over liquid dispensing once the volumes reach several million units per year. When volumes are low and many different package sizes are being made, liquid dispensing is the better process.

23.6.2 Advantages of Area Molding

As mentioned earlier, area (flood) molding can offer very good economics for reducing tooling costs. A few generic molds can be used for a variety of different-sized packages. This also reduces time to market because there is no wait for tooling that can take several months for large molds with individual cavities. The disadvantage of area molding is that mechanical singulation is required. The array, which is made up of hundreds of packages joined by the hardened molding compound, must be singulated mechanically. Sawing, the same process used to singulate semiconductor chips, is used here. Laser, water jet cutting, and other processes have not been able to displace the mechanical saw method. The sawing process leaves a slightly rougher edge than molding, thus reducing the cosmetic appearance slightly. Overall, the area molding process has been successful and probably will become the standard package method for BGAs and CSPs.

23.6.3 Liquid versus Solid Encapsulants

Transfer molding provides the best cosmetics and the most mechanical precision for packaging. While liquid dispensing is more convenient for small runs, cosmetics are not nearly as good for die-up packaging, where a retaining dam must be applied. EMCs also provide slightly higher performance than liquids that must flow by gravity. The simple rule for choice is to use liquid encapsulants for shorter runs or where molding is not suitable, such as with die-down cavity packages. Transfer molding is the de facto standard for high volume, and its use will increase because of the ability to achieve good, low-volume economics with area molding.

23.7 RELIABILITY

Transfer molding is a tried and true method for producing packages. The method has been used successfully for many decades and should be considered the de facto standard for producing plastic packages. Transfer molding was used on the earliest PBGA packages starting with Motorola's OMPACK. In fact, the name stands for *o*ver*m*olded *pack*age. The molding materials, equipment, tools, and processes have been optimized, and substantial test data support the reliability of this method. While reliability of the molded BGA is high, the very nature of the package design creates some minor problems that are addressed by proper handling and storage. Since the bottom of the substrate is often not encapsulated, the packaging barrier properties are influenced by the substrate properties. The most significant problem has been moisture entry through the substrate. Absorbed moisture can produce the undesirable phenomenon known as *popcorning*. During heating from the solder assembly process, instantaneous conversion of this moisture to steam can delaminate the molding compound from the substrate or even cause encapsulant fracture.

While the development of very low moisture absorption materials eventually may solve this problem, the process solution is to keep the package dry or to dry it just prior to assembly. Not all packages are affected by popcorning, but those which are known to be sensitive can be stored in a dry box. The alternative is to predry in an oven at 125°C for 2 h.

23.8 SUMMARY AND CONCLUSIONS

Molding is the most common method for encapsulating area array packages such as the PBGA and some of the CSPs. Transfer molding, the same process used for nearly all packages, is the dominant one here, although needle dispensing of liquid encapsulants is also used. The large, thin area array package has offered more challenges than the common perimeter-lead packages, however. The very concept of area array requires a relatively flat package for successful board assembly. The construction of the PBGA package, where molding compound typically is applied to just the top, sets up the mechanism for warpage if materials are not balanced. Mold compound shrinkage, while less important for fully encapsulated designs, will promote warpage with top or overmolded BGAs.

The encapsulant producers have responded to area array packages with low-stress, reduced-warpage EMCs. Today's EMCs designed for area array produce flat packages. The warpage has been so reduced that full sheets of die-loaded substrate can be overmolded using the newer flood-molding process. The ability to mold hundreds of packages in a single cycle with one mold has substantially boosted productivity. Area or flood molding is now a popular process for high-volume BGAs and μBGAs.

This segment of the packaging industry is one of the most innovative, since area array packages are leading-edge enabling technology. The previously mentioned flood-molding process is one example of innovation, but there are others. Molded underfill for FC packages is the newest and will grow in importance as increasing lead counts move us from wire bonding to FC in-package (FCIP). The response of the materials, equipment, and tooling companies will continue as part of the activities that make molding and area array packaging a rapidly growing business.

23.9 FUTURE EXPECTATIONS

We should expect molding compounds to continue to improve, especially as regards a reduction in moisture absorption. Improvements in low-warp materials also will continue because this feature will be more in demand as packages grow larger because of larger dies and multiple components. This feature is also of value for the flood-molding process, although it is less critical because the sheets of packages are sawn.

Depending on legislative activities, bromine may be eliminated as a flame retardant, although no negative environmental case has been made for the elimination. However, should the use of bromine be curtailed, the industry could move to nonepoxy chemistry.

Improvements in molding processes will continue. The most notable new processes are flood molding and molded underfill. We can expect larger arrays to be molded in a cycle with parts per shot exceeding 1000. Singulating will improve, and perhaps methods other than sawing will be adopted. Work will continue aimed at simplifying the MUF process and incorporating flood molding to boost productivity.

CHAPTER 24
SCREEN PRINTING AND STENCILING

Joe Belmonte
Cookson Performance Solutions

24.0 INTRODUCTION

Surface mount technology (SMT) is at the leading edge of assembling high-pin-count components. However, small as they are when compared with their through-hole counterparts, SMT components can be reduced in size considerably by reducing their lead pitch. Today, the lead pitch of the most common SMT packages is at 0.050-in (0.65-mm) centers. The availability of some fine-pitch 0.020-in (0.5-mm) lead spacing and below enables designers to make more efficient use of printed wiring board (PWB) real estate. The trend is toward SMT packages with lead spacing as small as 0.012 in (0.3 mm). The majority of the fine-pitch SMT components are called *quad flat packs* (QFPs). The QFP has input and output pins on all four sides of the component. This type of component is called a *perimeter array package* because all the input and output points are on the perimeter of the component. The QFP packages are used for active components. Several new active component packages have been introduced over the last few years, such as ball grid arrays (BGAs) and micro ball grid arrays. The BGAs and micro-BGAs are called *area array packages* because they use the entire bottom of the device for the input and output points. The requirement for smaller and lighter portable products such as cell phones and laptop computers has introduced the use of nonpackaged components. There are several types of nonpackaged components such as the *flip chip* (FC) and *chip-on-board* (COB).

At the same time, smaller and smaller discrete components (resistors and capacitors) are being introduced. The new discrete components are 0.040 in (1 mm) × 0.020 in (0.5 mm) (called 0402s) and a much smaller component that is 0.020 in (0.5 mm) × .010 in (0.25 mm) (called 0201s).

The introduction of these smaller and smaller components has increased the difficulty of the solder paste printing process. In order to realize all the benefits that can come from reductions in the size of SMT packages, advances in component packaging must be accompanied by advances in the materials, equipment, and process controls required for assembly on a production-line basis.

While most modern SMT facilities are capable of processing fine-feature components in small lots, the development of a production-volume process for fine-pitch components is a much more challenging objective. The full range of variables in materials, equipment, process controls, and process parameters must be studied, properly specified, and implemented to produce an acceptable process. The time required for process development will be reduced greatly if available materials and equipment satisfy the application requirements.

In printing solder pastes, printing defects are caused by errors in one or a combination of the x, y, and Θ positions of the solder paste on the substrate, the amount of solder paste deposited, or the makeup of the solder paste itself. These errors, in turn, are due to one or a combination of deficiencies in screen printer capability and errors in stencil selection or design, solder paste selection, or process parameter settings.

The solder paste printing process is the most critical process in the entire SMT manufacturing operation. Most assembly operations calculate the quality of the manufacturing process by counting the defects produced versus the total number of opportunities for defects. For example, a 132-pin QFP will have 133 opportunities for a defect: 132 for the 132 solder joints for each component, one for each lead, and one for the component itself (positioning, missing, functionality, etc.). Controlling the soldering processes and especially the solder paste printing process will produce the highest possible process quality.

There is no one "recipe" (equipment operating parameters, solder paste type, stencil design, etc.) that can define an assembly process that is optimal for a wide range of fine-pitch applications. The purpose of this chapter is to identify the factors that determine a fine-pitch process and provide some guidance to the manufacturing engineer who must develop an assembly process that will produce high yields and bring new products to market more quickly.

24.1 SCREEN PRINTERS

The base element for any solder paste printing process is the printer itself. There are numerous suppliers of screen printers, each having an array of models and features to choose from (Fig. 24.1). The best choice for fine-pitch processing is a printer that satisfies four key criteria:

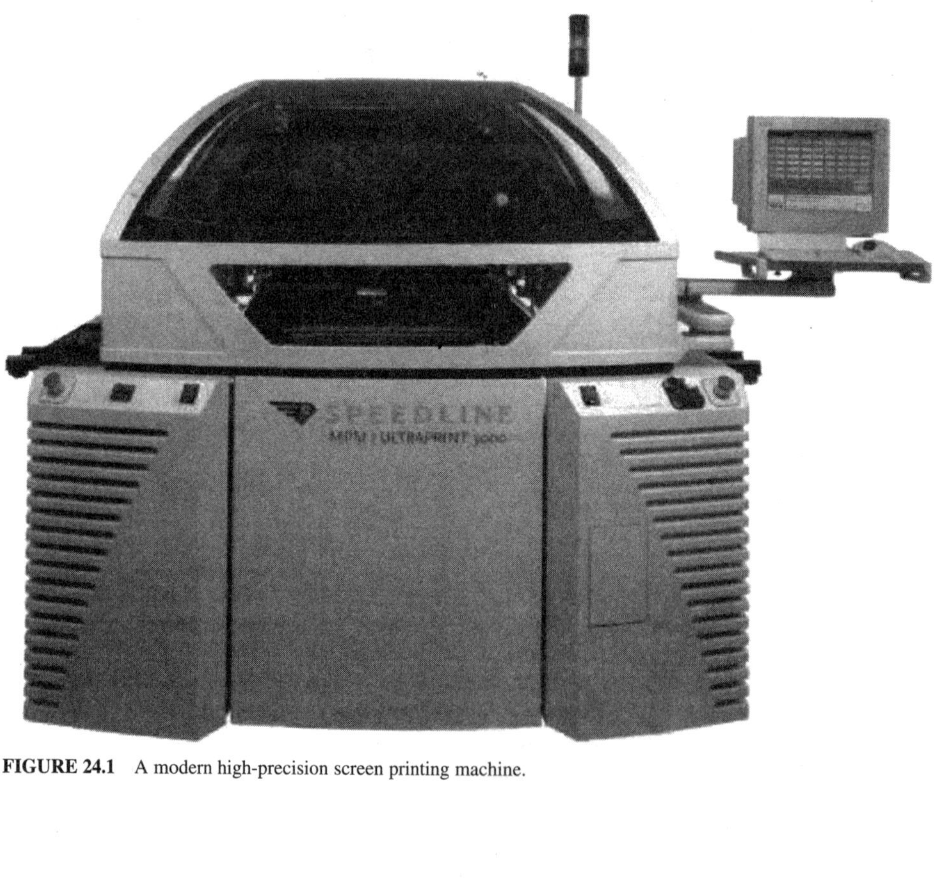

FIGURE 24.1 A modern high-precision screen printing machine.

- Adequate positioning capability
- Mechanical stability
- Consistency and accuracy of setup
- Availability of options to minimize operator intervention

24.1.1 Adequate Positioning Capability

The printer must be capable of aligning the substrate to the stencil with precision and repeatability that are consistent with the application. Because of the dimensions inherent in fine-pitch processing, positioning capability is the most important factor in printer selection.

For a 0.020-in (0.5-mm) lead-pitch component, the typical pad width is 0.012 in (0.4 mm), and to provide a reasonable aspect ratio between the stencil openings and the stencil thickness, the opening width might be 0.010 in (0.25 mm). These dimensions allow only 0.001 in (0.025 mm) between either side of the opening and the edge of the pad. If the printer alignment system is not capable of maintaining this relationship within the 0.001-in tolerance and solder paste is deposited closer to the edge of a pad, there is a possibility of bridging to an adjacent pad and producing a defective assembly.

In volume production, the printer must be able to reproduce this precise alignment, at production speeds, each time the process is repeated. To ensure that it can meet the requirements of the application, the printer alignment process must be capable of greater precision than is required to meet the physical alignment specifications. This more stringent requirement on the process recognizes that no process is exactly centered on the theoretical mean and that the process mean may shift to the left or right of center.

A 6σ process, which can ensure low defect rates on fine-pitch processes, requires that the process itself have twice the capability required to meet the application specifications. The standard index used to denote process capability is *Cp,* where *Cp* is calculated as the quotient of the allowable specification divided by 6 times the process variation expressed by its standard deviation (hence the term 6σ). For a 6σ process, this quotient must be at least $Cp = 2$:

$$Cp = \frac{\text{process specification}}{6 \times \text{standard deviation of process}} = 2 \tag{24.1}$$

A detailed explanation of this well-known relationship can be found in any standard text on statistical process control. The point is that to produce repeatable alignment at assembly, the printer itself must be designed, built, and verified with even greater precision.

24.1.2 Mechanical Design

The ability of the printer to provide the necessary placement accuracy and repeatability requires a robust printer design. Structurally, the machine must be rigid enough to prevent relative motion between any axis of motion and the rest of the machine. Otherwise, there is a potential for variations in the system response that can result in assembly defects. The placement repeatability built into high-end printers ranges from ±0.0005 in (0.0127 mm) to a low of ±0.0003 in (0.00762 mm). Rigidity is also required to ensure that all movement of the stencil in relation to the board is accomplished in planes that are parallel within narrow limits.

The most efficient and reliable alignment system will incorporate discrete motion control for each axis. Ideally, the alignment process is controlled by a closed-loop vision system that has the capability to view multiple fiducial points on the stencil or substrate. This combination of alignment and vision, which can provide the highest level of alignment available, is a prerequisite for low-defect, high-volume fine-pitch processing. (See Table 24.1.)

Computer-based design systems and analytical techniques, coupled with experience in the industry, can make it possible to achieve the mechanical design objectives while maintaining efficiency in machine size and weight. Machine size is often a critical issue, because floor space can be an expensive part of the cost equation. The best equipment manufacturers also perform rigorous verification tests on their stencil printers before shipment and installation on a customer's floor. For example, Speedline/MPM Corporation performs stringent acceptance procedures, including a unique glass-plate test that is accurate to one-millionth of an inch, to ensure repeatability of each system prior to shipping.

TABLE 24.1 Fiducial Mark Recommendations

Quantity:	3 (minimum)
Location:	Corners of the board
Size:	.040" (1 mm) dia. pad
Clearance:	.120" (3 mm) diameter (no soldermask, traces, via holes and legends/silk screening)
Edge Distance:	.250" (6.35 mm) min

TABLE 24.2 Squeegee Force, Downstop, and Speed Recommendations

	English Units	
	Trailing Edge Metal Blades	Trailing Edge Poly Blades
Force	1–1.5 lb. per linear inch of squeegee	1–2 lb. per linear inch of squeegee
Downstop	0.070"–0.085"	0.055"–0.065"
Speed	0.5 in/sec to 6.0 in/sec	1 in/sec to 6.0 in/sec
	Metric Units	
	Trailing Edge Metal Blades	Trailing Edge Poly Blades
Force	.45 to .68 kg per 25.4 mm of squeegee	.45 to .91 kg per 25.4 mm of squeegee
Downstop	1.7 mm–2.16 mm	1.4 mm–1.65 mm
Speed	12.7 mm/sec to 152 mm/sec	25.4 mm/sec to 152 mm/sec

24.1.3 Consistency and Accuracy of Setup

The ability to achieve an accurate and repeatable setup is essential to ensure consistency and high yields on a volume-production basis. Board location mechanism and supports must be easily located and consistent in their operation. Stencil loading and clamping should be quick, easy, and repeatable. Finally, process parameters such as squeegee pressure, speed, and downstop must be maintained easily and not be operator-dependent. (See Table 24.2.)

24.1.4 Automation and Minimizing Operator Intervention

Maximizing the use of automatic operations and minimizing the need for operator intervention not only make setup easier but also ensure that the operations will be consistent and repeatable. Some advanced stencil printers have features such as a programmable print head that allows the operator to program squeegee downstop and pressure and automatically levels the blade before printing. There is no need for the operator to constantly monitor squeegee pressure and downstop, and parameters automatically become part of an individual board file, available to be recalled at any time. Automatic solder paste dispensing systems dispense solder paste onto the stencil in preprogrammed amounts and at preselected intervals. This feature reduces the potential for solder paste to dry out on the stencil and, in so doing, promotes more efficient use of the paste. In addition, dried paste can result in clogged stencils, defective printing, and lower process yields. Automatic stencil wiping and vacuuming of the apertures provide unassisted cleaning of the stencil underside and apertures at user-programmable intervals. This feature can prevent the transfer of built-up solder paste to nonpad areas and the resulting exposure to bridging and lower process yields without operator intervention.

24.2 WIRING BOARD DESIGN

PWB design and the many combinations of materials and fabrication methods used in PWB construction are beyond the scope of this chapter. This section briefly discusses three elements of PWB design that directly affect the printing process:

- Pad size
- Pad plating
- Solder mask type

The size of the solderable pad on the PWB directly determines the required thickness of the solder paste deposition. To ensure that the solder paste will release from the stencil, the surface tension between the paste and the pad must be greater than the surface tension between the paste and the stencil. The pad size must provide the required surface area while staying as small as possible to closely match the lead size. Some generally accepted lead-pitch and pad-size relationships are shown in the stencil design section (Sec. 24.3).

The copper surface and the material plated on copper pads directly affect the printability and solderability of the pad. Copper, tin, tin-lead, and gold will each perform differently when printed with various solder paste formulations in terms of adhesion and wetting during reflow. In general, determination of performance levels for specific combinations of solder paste and pad metalliazation requires experimentation.

The type of mask used on the PWB can affect its overall flatness and therefore the deposition quality. To the extent possible, the pad and the mask should be at the same height. Differential heights can prevent a good "gasket" between the stencil and the board during printing, resulting in paste/flux bleed around stencil openings and the need for more frequent stencil underwipings.

24.3 STENCIL DESIGN

Five main elements define a typical stencil design:

- Material
- Thickness
- Image pattern
- Aperture size
- Fabrication method

There is no single combination of these elements that can be recommended as the best choice for fine-pitch printing. Instead, the various options available must be considered in the context of the overall assembly process.

Stainless steel is the most popular stencil material in use, due to its ability to resist the harsh chemicals used to replace fluorocarbons for cleaning. When printing 0.025-in (0.65-mm) pitch and above, standard technology chemically etched stainless steel can provide satisfactory results. However, when the progression is made down to 0.020- (0.5-mm), 0.015- (0.4-mm), and 0.012-in (0.3-mm) pitches, this combination of material and process is not acceptable for a high-yield assembly process. The fundamental problem with chemically etched stainless steel is that the resulting chamfered and rough aperture walls hamper the release of solder paste. Without efficient solder paste release, clogging of stencil apertures will occur, and consistent, defect-free printing below 0.020-in pitch will not be possible. (See Fig. 24.2.)

Laser-cut stainless steel stencils can produce surfaces that are tapered and have the consistently smooth walls required for 0.012- or 015-in pitch processing. The resulting apertures have a trapezoidal

FIGURE 24.2 Magnified view of chem-etched stencil aperture.

shape that facilitates the release of solder paste. Although laser-cut stainless steel stencils can cost slightly more than those which are chemically etched, their solder paste release is generally better. (See Fig. 24.3.)

Another stencil fabrication technology is called *E-fab.* This technology "grows" the stencil. The advantage of an E-fab stencil is that it can be produced in a variety of thicknesses and has a unique gasketing feature. E-fab stencils also release the highest percentage of solder paste from their apertures when compared with laser-cut and chemically etched stencils. (See Fig. 24.4.)

The key element in all stencil design is to ensure the aperture size is reduced from the pad size for all fine-pitch components. The purpose of reducing the aperture size is to seal or gasket the stencil aperture to the component pad on the PWB. Each method of stencil fabrication creates its own aperture wall shape (see Fig. 24.5).

Experimentation has shown that the best technique for achieving optimal print quality and solder-joint formation is to reduce the width of a stencil aperture, relative to pad size, by approximately 20 percent. The resulting reduction in width of the printed paste compensates for spreading that occurs when the component lead is placed into it. The lead will displace a volume of paste that is approximately equal to the lead area. If too much paste is on the pad and the displaced volume spreads to the interpad area, it can form solder balls and cause shorting between pads. Excess paste also can adhere to the bottom of a stencil, where it can build up and be printed on interpad or other undesirable areas.

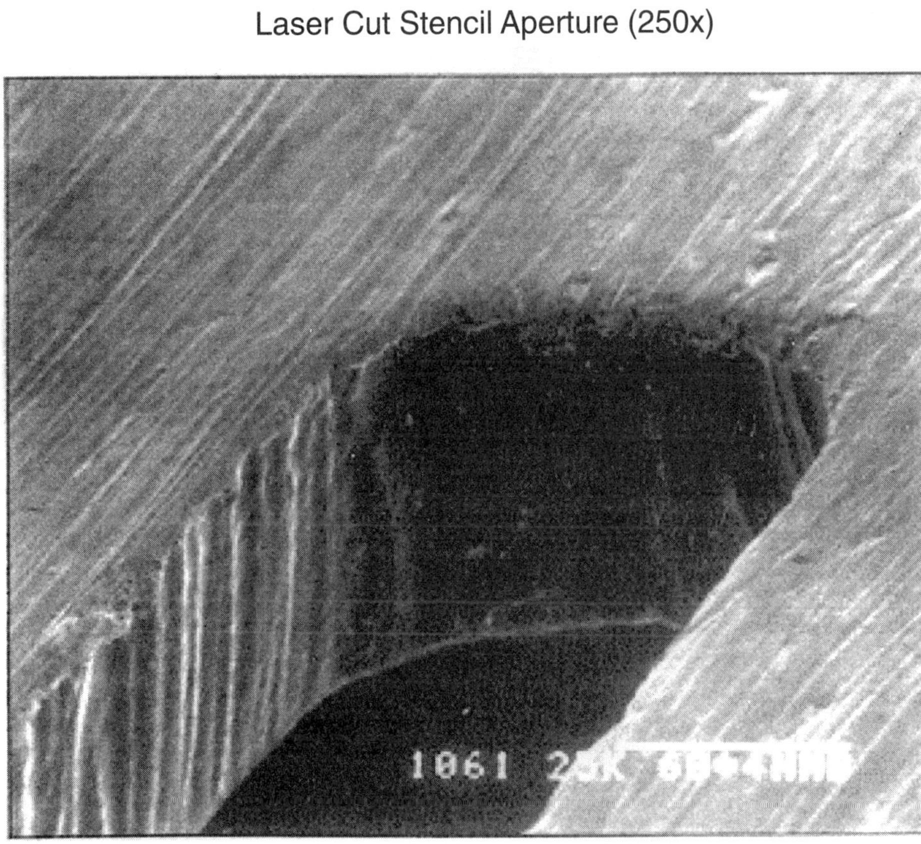

FIGURE 24.3 Magnified view of laser cut stencil aperture.

Again, this condition can result in solder balls and shorting problems unless the buildup is removed by cleaning. The correct width reduction in the stencil opening can prevent paste from bleeding into gaps between the pad and the solder mask and reduce the frequency at which the underside of the stencil must be cleaned.

The guidelines in Table 24.3 are offered for the pitch, pad, and aperture size relationship (dimensions are in inches).

Aspect ratio is the ratio between stencil thickness and stencil width (Fig. 24.6). When using chemically etched stainless steel, the stencil width should be a minimum of 1.5 times the thickness of the stencil. For laser-cut stainless steel or E-fab stencils, this ratio can be reduced to a minimum of 1.2 times the stencil thickness because of their more efficient release of solder paste.

24.4 SOLDER PASTE SELECTION

Selection of a solder paste to optimize fine printing and performance depends on several key factors, including

- Rheology

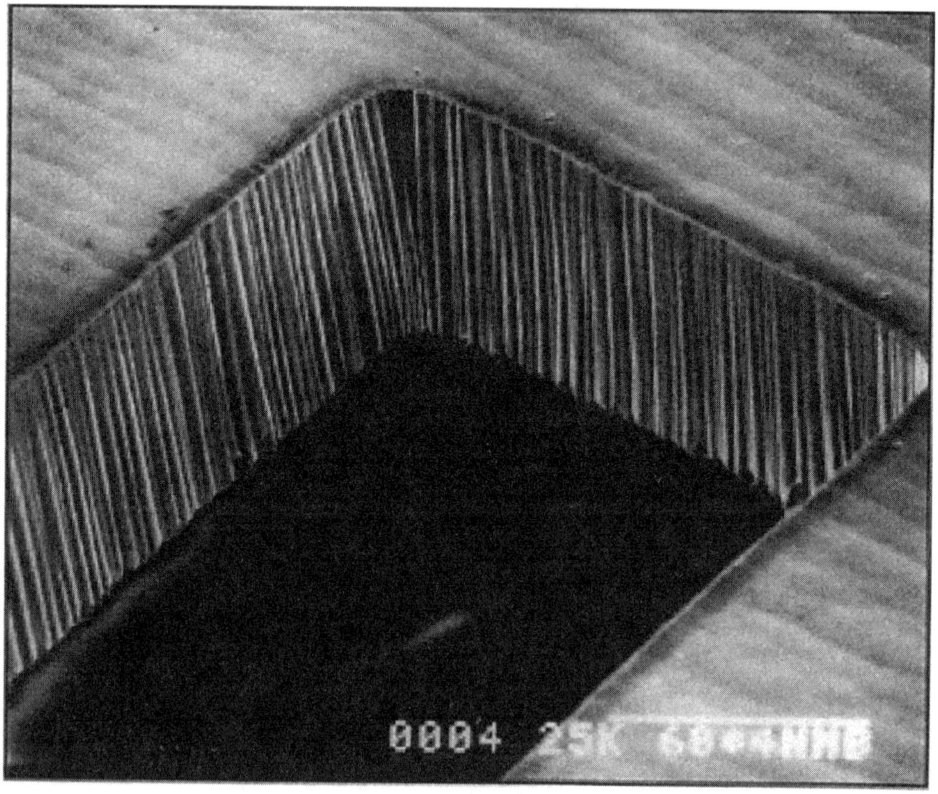

E-FAB XL Stencil Aperture (250x)

FIGURE 24.4 Magnified view of an E-Fab stencil aperture.

TABLE 24.3 Recommended Component Pad Aperture Size.

		English Units		
Pitch	Pad Width	Aperture	Stencil thick	A.R.
.025"	.015"	.012"	.006"	2.0
.020"	.012"	.009–.010"	.005–.006"	1.7
.016"	.010"	.007–.008"	.005"	1.4
.012"	.008"	.005–.006"	.004–.005"	1.2
		Metric Units		
Pitch	Pad Width	Aperture	Stencil thick	A.R.
.64 mm	.38 mm	.30 mm	.15 mm	2.0
.51 mm	.30 mm	.23–.25 mm	.127–.15 mm	1.7
.41 mm	.25 mm	.18–.20 mm	.127 mm	1.4
.30 mm	.20 mm	.127–.15 mm	.1–.127 mm	1.2

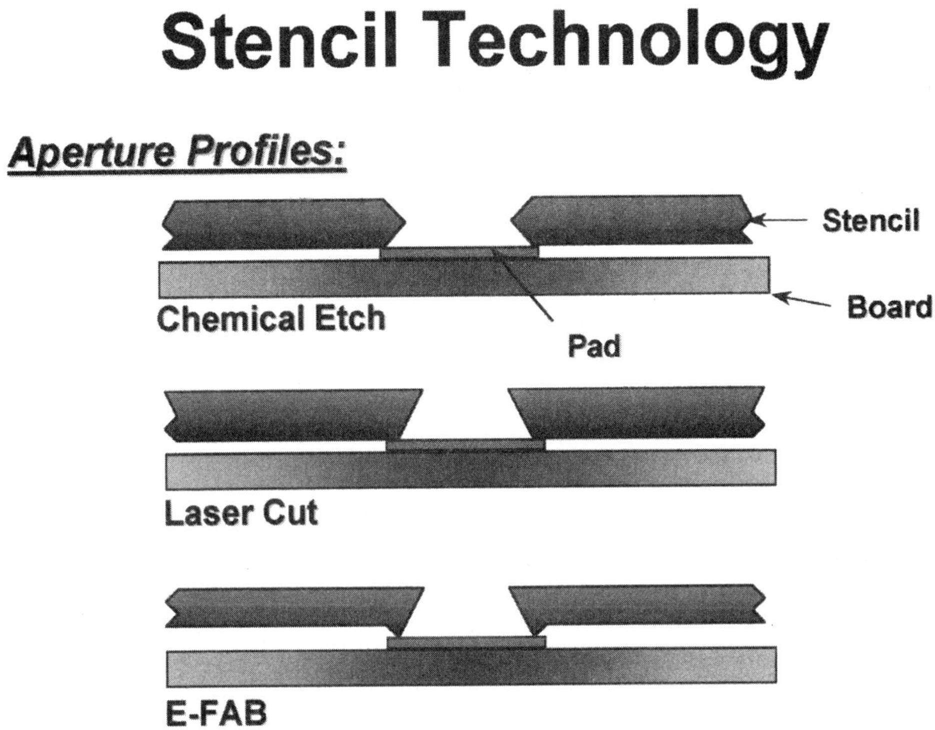

FIGURE 24.5 Illustrated view of stencil aperture cross sections.

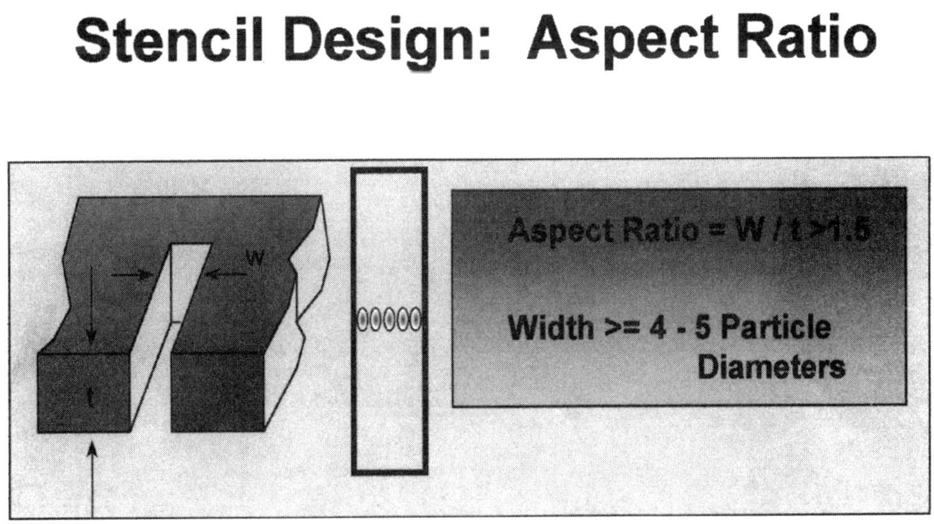

FIGURE 24.6 Stencil aperture aspect ratio.

- Viscosity
- Solder powder particle size
- Jar-to-jar or cartridge-to-cartridge consistency

The rheologic properties that are desired in a solder paste for stenciling are determined by the shear history that the solder paste experiences. Because solder pastes are thixotropic, they tend to decrease in viscosity during the printing process. When formulating solder pastes for fine-pitch, high-speed printing processes, it is especially important to consider the rate of this thinning, as determined by the solder paste's shear sensitivity. In addition to having to thin and recover in viscosity after being passed through the stencil aperture, the solder paste also must be able to withstand the rolling motion of being pushed back and forth at high speeds on the stencil. If too much shear thinning occurs during this process, the result is a low viscosity that recovers too slowly, causing slump in fine-pitch applications.

The solder powder influences the print quality, slumping, and wetting properties. Specifically, the shape and size of solder powder particles directly affect printability and tack retention. These properties are enhanced by formulating with a solder powder that consists of a narrow range of particle sizes but with a sufficient number of small particles to fill spaces that otherwise might attract flux and promote separation (Fig 24.7).

Parameters to consider during flux selection include

- Stencil life of the solder paste
- Solder paste temperature sensitivity
- Overall process robustness

Reduced stencil life as a result of solvent evaporation is an issue because of the repeated exposure of large solder paste surface areas during high-speed printing. Temperature sensitivity can be an issue because the solder paste is occasionally exposed to temperature fluctuations during its working life on the stencil. And finally, overall process robustness to ensure proper printability, reflow, and solderability is an important issue in high-throughput operations in order to reduce variability and achieve hands-free operation.

Solder Paste

ASTM Mesh Designation	Opening Size (um)	(in)
200	74	0.0027
250	58	0.0023
325	44	0.0017
400	37	0.0015
500	30	0.0012
625	20	0.00078

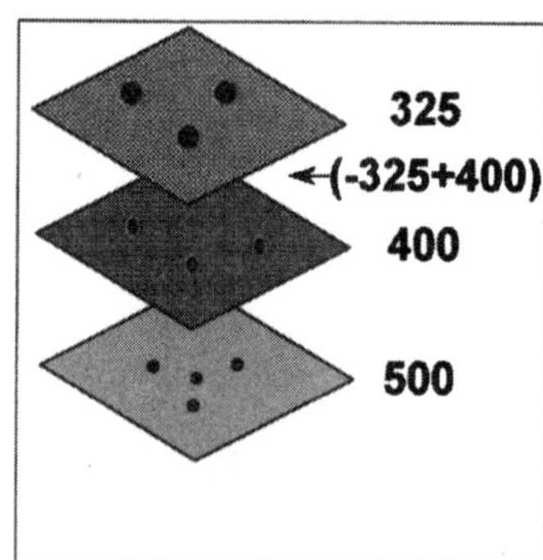

Any particles with a nominal diameter less than 1.7 mils will pass through the 325 mesh size (-325) and will the get caught in the finer mesh size (e.g. +400 or +500).

FIGURE 24.7 Solder paste particle size.

It is clear that the selection of solder paste for these operations goes well beyond meeting the requirements outlined in the IPCJ standard. Solder paste designers must consider all the process demands outlined earlier as well as ensure that the solder paste can be manufactured with demonstrated repeatability and consistency.

24.5 PROCESS DEVELOPMENT

It is no longer reasonable to define high-yield production-volume processes by trial and error. Many aspects of designing and optimizing a high-yield printing process require efficient, accurate experimentation. Statistically designed experiments are used primarily to obtain the maximum amount of information for the minimum cost in time and resources. Conclusions drawn from experiments are used to determine the best course of action in establishing the process. In this way, the controllable variables of the process can be set at optimal levels in an objective manner supported by data and can produce the desired outcome. Two commonly used experimental design methods for process development are Taguchi and factorial design of experiment (DOE).

In the past, the same set of printing process parameters might be used regardless of board size or type. It is no longer possible to make recommendations for specific parameter settings that can be applied without question. For fine-pitch printing, parameters must be tailored individually to each application. However, it is possible to offer some general guidance as a starting point for experimentation and process development.

Downstop level depends on the machine type, with the major consideration being the design of the machine's print head. In general, the downstop should be set to a level that enables the squeegee to wipe the stencil clean while allowing the squeegee head to be compliant as it moves across the board. Downstop level becomes critical when metal blades are used and when, due to head design, downstop affects squeegee pressure. When using flexible metal blades, increases in squeegee pressure may require adjustment in the squeegee attack angle to arrive at the optimal setting.

The use of metal squeegee blades enables a more controlled and consistent print height across the entire board area. However, good results can be obtained with urethane squeegee blades if there are no relatively large pads on the board. Using a harder urethane material minimizes the potential for scooping on all but very large pads. A durometer of 90 to 95 on the Shore D scale provides the best performance. No appreciable difference in print definition or paste column height has been found with 0.031 in (0.79-mm) pitch components and below, and components down to 0.012-in pitch have been printed successfully. Metal squeegees, however, perform much better in relation to height variations on parts such as large capacitor pads and printed holes.

Squeegee pressure is based on size and material. For urethane squeegees, the pressure should range between 15 (6.8 kg) and 30 (13.6 kg) lb or approximately 1.6 lb/in (0.73 kg) of squeegee. When using a metal squeegee, the range of allowable pressures is wider, and the pressure can be set to a lower value. As a general rule, the pressure should be set at the minimum value that ensures that the stencil is wiped clean. It may be necessary to perform some experimentation to arrive at the optimal pressure setting.

Squeegee speeds are determined primarily by the solder paste that is being printed. A range of 0.5 to 2.0 in/s (12.7 to 51 mm) is typical, with higher viscosity levels requiring speed at the upper end to force paste in front of the stencil to roll and flow into the stencil openings. When printing ultrafine-pitch, a range of 0.5 to 1.0 in/s (12.7 to 25.4 mm) has been found to provide optimal results. New solder pastes that have been developed over the last few years can be printed at a faster speed, some as high as 8 in/s (203 mm). However, several factors, including stencil design, pad design, support tooling, etc., must be considered to implement a high-speed solder paste printing process.

24.6 REFLOW OVEN CONSIDERATIONS

Fine-pitch processing has also had an impact on the reflow soldering equipment. The uniformity of the heating process is critical to ensuring minimal temperature differentials among the many different

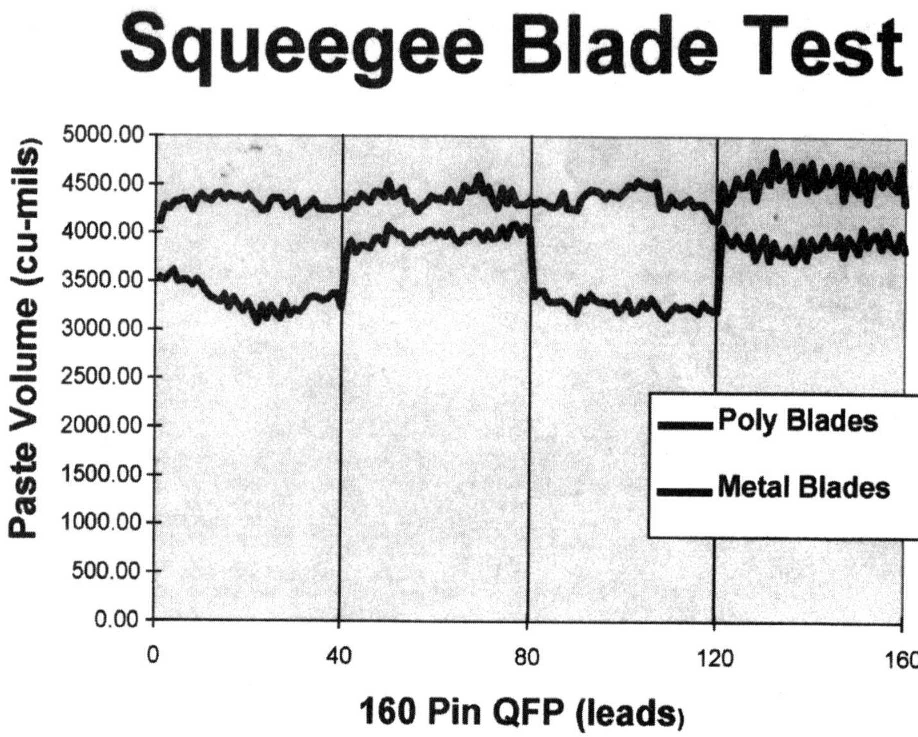

FIGURE 24.8 Squeegee blade solder volume test.

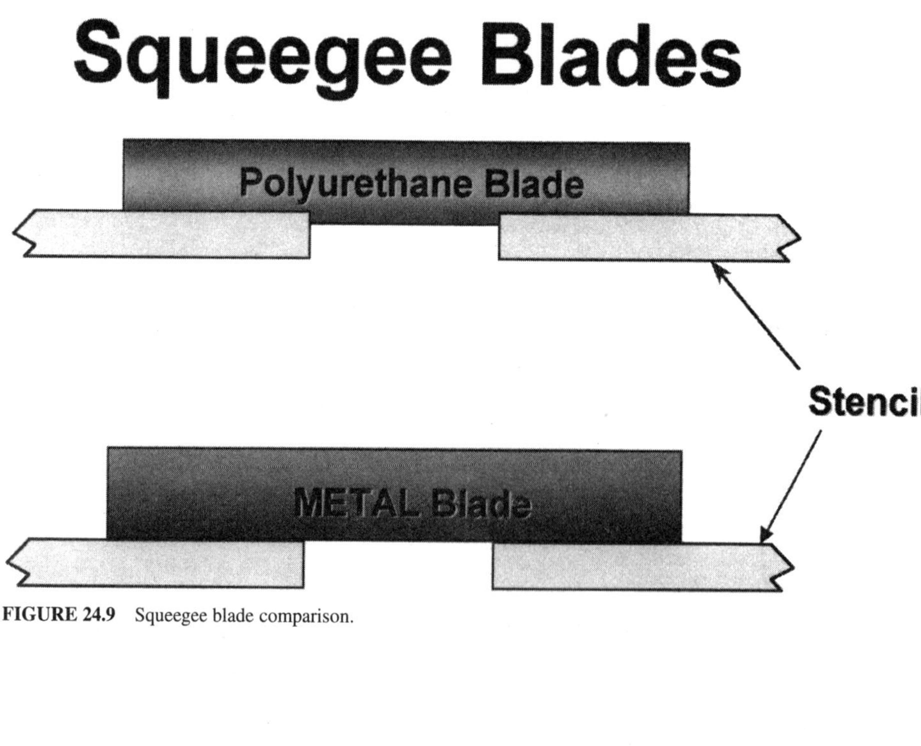

FIGURE 24.9 Squeegee blade comparison.

component masses that constitute modern SMT assemblies. Properly applied, high-volume forced-convection heating has the potential for achieving very small temperature deltas, even on complex, massive assemblies. In fact, only high-volume forced-convection ovens have the capability to achieve a one-to-one relationship between the heater temperatures and the product temperature in the soak zone. This capability eliminates the possibility of overheating temperature-sensitive components because the heat transfer is accomplished by convection, and nothing can reach a higher temperature than the air-gas heating medium.

Fine-pitch technology requires tight process control, particularly the time above liquidus (TAL) temperature during reflow soldering. Intermetallic formation is accelerated while the metals being joined are above the solder liquidus temperature. Because of the relatively small volume of solder in fine-pitch joints, it is especially important to minimize the growth of the brittle intermetallic compound layer by controlling the time above liquidus. Just a few years ago, the accepted criterion for TAL was in the 60- to 120-s range, but today, with fine-pitch technology, reflow users expect to see TAL capability of 30 to 60 s.

To be viable in today's market, reflow ovens also must have nitrogen inerting capability. Although the actual percentage of nitrogen users is small, there are some real benefits for the use of nitrogen in fine-pitch soldering. It has been well documented that solder wetting time is decreased and wetting forces are increased in an inert atmosphere. The increased wetting force and higher surface tension under nitrogen have proven beneficial in reducing bridging on fine-pitch joints. Solder paste formulated for fine-pitch use has smaller solder particle sizes that, because of the greater exposed surface area, are more prone to oxidation during the reflow process. An inert environment will prevent oxidation and ensure optimal wetting.

24.7 CONCLUSION

The development of a high-yield production process for fine-pitch components must be approached as a complete process, where the combination of equipment, materials, operator proficiency, and processes produces the desired results. Each new variation of devices, lead spacing, substrate, and schedule concerns requires a thoughtful determination of specific requirements and how best to satisfy them. By using proven equipment, materials, processes, and the scientific method, process development time is reduced, and products are brought to market more quickly. At the same time, the results of designed experiments identify limits and controls needed to ensure high process yields.

CHAPTER 25
CRITERIA FOR PLACEMENT AND PROCESSING OF ADVANCED PACKAGES

Günter Schiebel
Siemens Production and Logistics Systems AG

25.0 INTRODUCTION

Advanced packages such as the ball grid array (BGA), chip scale package (CSP), and flip chip (FC) to a continuously increasing degree will be incorporated into future surface mount tehcnology (SMT) board designs. Currently micro-BGAs/CSPs with a minimum ball pitch of 0.5 mm and a typical ball diameter of 0.25 to 0.3 mm are attracting more and more interest. National Semiconductor's approach to a CSP, the micro-surface mount device (micro-SMD), already has ball/bump diameters in the range of 0.14 to 0.18 mm. This can be interpreted as an indicator that CSP and FC technologies are converging.

Due to the strongly growing consumption figures of all these packages, particularly in the automotive, telecommunications, computer, watches and smart card/RFID applications, *throughput is becoming more and more an issue.* This chapter gives a brief overview of some applications where high-speed mounting or even processing of advanced packages is required. The chapter addresses as well the various criteria for achieving high throughputs with area array packages and bare dies for chip-on-board (COB) using advanced chipshooters. The key criteria are the machine concept (placement principle), component feeding, type of flux and adhesive application, component vision technique, and printed wiring board (PWB) vision technique.

The 0201 package with an outline dimension of 0.6 × 0.3 mm is about to find increasing applications in portable devices. It also has to be considered as a package that requires advanced process capabilities combined with high speed. The placement and overall process challenges are really tough when component spacings are tight. The key placement machine requirements for 0201 packages are covered within this chapter.

One of the key issues that has to be considered in context with the placement rate is most definitely placement accuracy. Physical laws do not allow achieving extreme values in terms of speed and accuracy within a certain technical approach. Nevertheless, different placement principles are showing some significantly different capabilities regarding the *combination of speed and accuracy.*

25.1 AREAS FOR HIGH-SPEED ADVANCED PACKAGE MOUNTING WITH SOME KEY CRITERIA

25.1.1 Portable Communications Products (Handies, Pagers), Base Stations, and Cameras

In manufacturing such products, a clear package-form trend toward 0201 passives, CSPs, and FCs is visible. The placement accuracy is determined by the FCs and the 0201 packages (substrate type: organic, rigid and flex).

A tendency within high-end consumer products such as handies is to remove fine-pitch/fine-line technology from the base board using the submodule approach. A very popular submodule in the future will be the *Bluetooth* module for short-distance wireless interconnection of mainly portable devices (substrate type: organic and ceramic).

25.1.2 Automotive Industry

Most characteristic for many under-the-hood applications is the high number of bare chips (10 to 15 percent of the component total). Typically, all bare dies have different part numbers. The European automotive industry still applies COB technology. The trend there is pointing toward FC technology, which is already widely used in the American automotive industry. Delphi, for example, already had adopted this most innovative technology in the 1970s. High-speed FC mounting is a big issue in this arena (substrate type: organic and ceramic).

25.1.3 Computers, Memory Modules, Disk Drive Read-Write Heads

Motherboards are showing the greatest variety of package forms. The traditional integrated circuit (IC) packages, quad flat packs (QFPs), and all the SO variations are step-by-step substituted by standard BGAs and CSPs. This application requires revolver-head-based high-speed placement equipment, which can do a reliable BGA/CSP placement using ball centering.

To achieve significantly shorter memory access times, memory module designs must more and more include CSPs. Low-cost, high-volume assembly in conjunction with extreme yields is required in this area.

Since disk drive read-write head flex circuits typically are populated with an FC plus a couple of SMD passives, the use of FC-capable shooter systems seems to be ideal (substrate type: FR-4 for motherboards and memory modules, rigid-flexible for disk drive read-write heads).

25.1.4 Hearing Aids

Today, for these types of products, COB technology is still dominating. However, FC technology is being adopted more and more. A hearing aid circuit typically contains up to three bare dies and approximately 10 passive components, which will have, to an increasing degree, 0201 dimensions (substrate type: organic, rigid or flexible).

25.1.5 Smart Cards/Smart Labels

Smart cards and smart labels (*r*adio*f*requency *id*entification, or RFID) offer exciting application possibilities in the future. The worldwide production volume, particularly of RFIDs, in the near future will be in the range of several billion units per year. Undoubtedly, the ideal manufacturing technique is FC. A smart label exists of a flexible substrate, which serves as an antenna and a carrier for one FC. In case of a contactless smart card, the flexible substrate with the FC is embedded in the card-sized piece of plastic. The FC has three or four bumps. The active two bumps are interconnected with the antenna. The FC placement accuracy requirement is relatively low (range of 90 to 100 μm). The challenge here is to achieve the highest possible throughput because the target costs for smart labels are extremely low. Another typical requirement in this arena is the FC presentation in wafer format (additional die-sorting costs are not acceptable) (substrate type: flexible only).

25.1.6 Low-Cost Watches

Similar to hearing aids, where space is very limited, in low-cost watch manufacture FC technology instead of COB will be applied to an increasing degree. The key driving factor for using FCs is the cost-saving potential when production is based on an FC shooter capable of wafer feeding. Low-cost watches, like smart cards and RFIDs, can be pure FC applications (no passives) (substrate type: organic, rigid or flex).

25.2 POSITIONING AND PLACEMENT PRINCIPLES

25.2.1 Basics of Positioning

The following design and positioning criteria are regarded as having a major influence on positioning accuracy of the key axes *x* and *y* and rotation of the vacuum nozzle:

- Measuring principle
- Resolution of the linear scales or rotary encoders
- Distance between the guide plane and the PWB plane
- Drive concept
- Mechanical stiffness of the *xy* gantry system
- Control of mechanical oscillations

Positioning and Path-Measuring Principles. Figure 25.1 shows some typical approaches for positioning and path measuring in SMD placement machines. The drive elements in the main axes *x* and *y* are dc, ac, or linear motors. The movements within the placement head typically are carried out with dc or ac motors. The advantage of an ac motor over a dc motor is the fact that it is maintenance-free (brushless). Consistently high positioning accuracy of the *x* and *y* axes can only be achieved via *direct path measurement* using linear glass or steel scale path-measurement systems.

The positioning electronics can determine the current position of the placement head at any time during the positioning sequence and thus respond correspondingly precisely. To put it drastically, using rotatory measurement systems coupled to the *xy* drives results in measurement at the wrong place. All errors such as mechanical play or wear in the drive system exert an adverse effect on placement accuracy. Direct path measurement is also preferable over indirect path measurement at the drive motor for the high-accuracy rotational axes for vacuum nozzles. A resolution of 1 to 2.5 μm for the *xy* axes for an exact placement of fine-pitch packages is today's standard. In the case of highly accurate placement systems, the distance between the guide plane and the PWB plane usually exerts a direct effect on placement accuracy. In view of trigonometric laws, this distance should be as small as possible.

Drive Concept. In the early days of surface mount technology (SMT),the main axis drives were mostly based on ac or dc motors in conjunction with a spindle gear. Other systems were based on belts as a transmission element (Fig. 25.2). Belt drives gained in importance because they allowed for higher accelerations and decelerations due to lower a mechanical mass (inertia) involved. The most advanced drive technique is doubtless the linear motor, which is basically a brushless synchronous three-phase ac motor. The great advantage of the linear motor is the fact that it includes the transmission element (gear) and as a result of this it does not show any mechanical wear and loosening. The use of linear motors is most beneficial when very high masses have to be positioned with high dynamics.

Compensation of Systematic Mechanical Deviations by Means of Mapping. Eliminating the remaining systematic error caused by the very slightest nonlinearities within the guides would

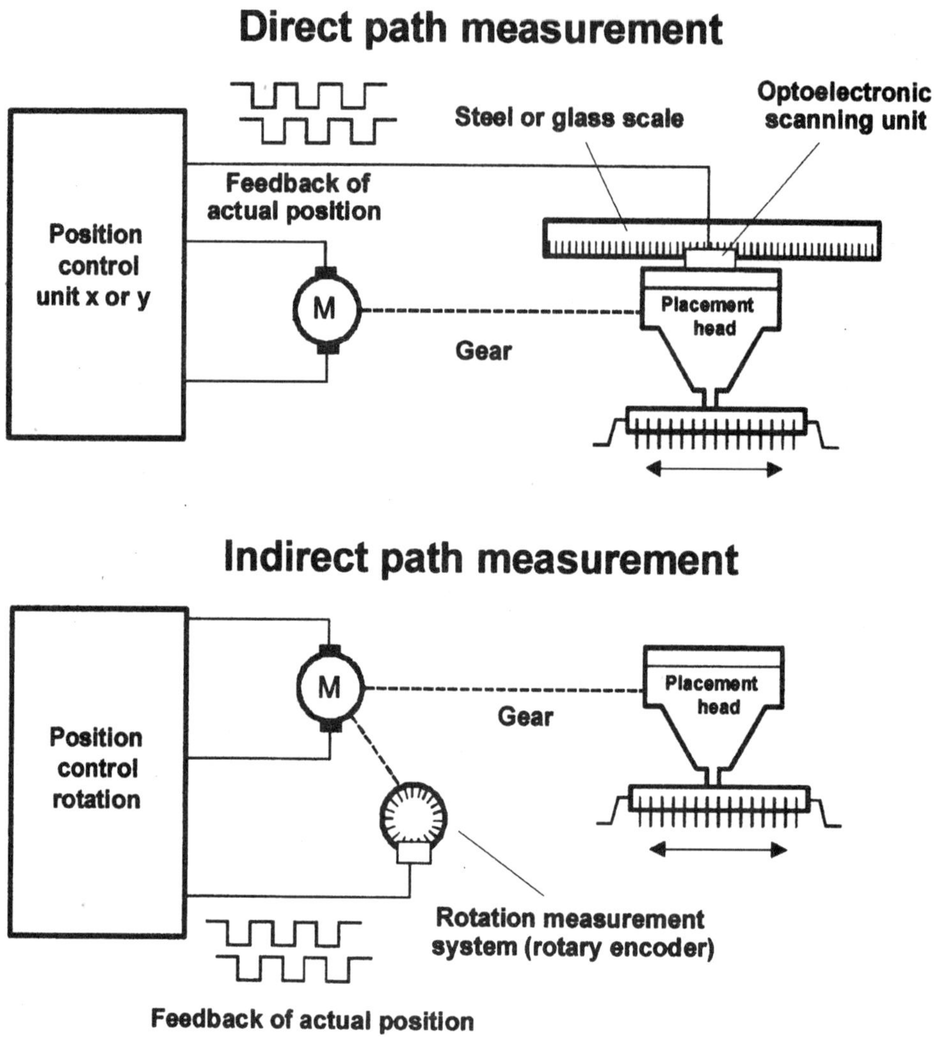

FIGURE 25.1 Path-measurement principle in SMD placement systems.

involve much more work that can be justified. It is definitely better to integrate a self-calibration system into the SMD placement system to compensate for these errors. During the calibration run, the system uses a high-precision reference glass mapping plate with a grid of evaporated reference structures that are imaged by the PWB camera, which is mounted near the placement head(s) on the *xy* positioning system. A number of position-dependent *xy* correction values are obtained by comparing the captured actual positions with the nominal values. By extrapolating correction values stored in a table, it is possible to conduct a precision correction of every placement position.

25.2.2 Pick and Place

This placement principle clearly provides the best placement accuracy as long as the machine is equipped with a minimum number of placement heads (Fig. 25.3). The smaller the number of moving elements (bearings), the better are the preconditions for achieving high accuracy. The quality of

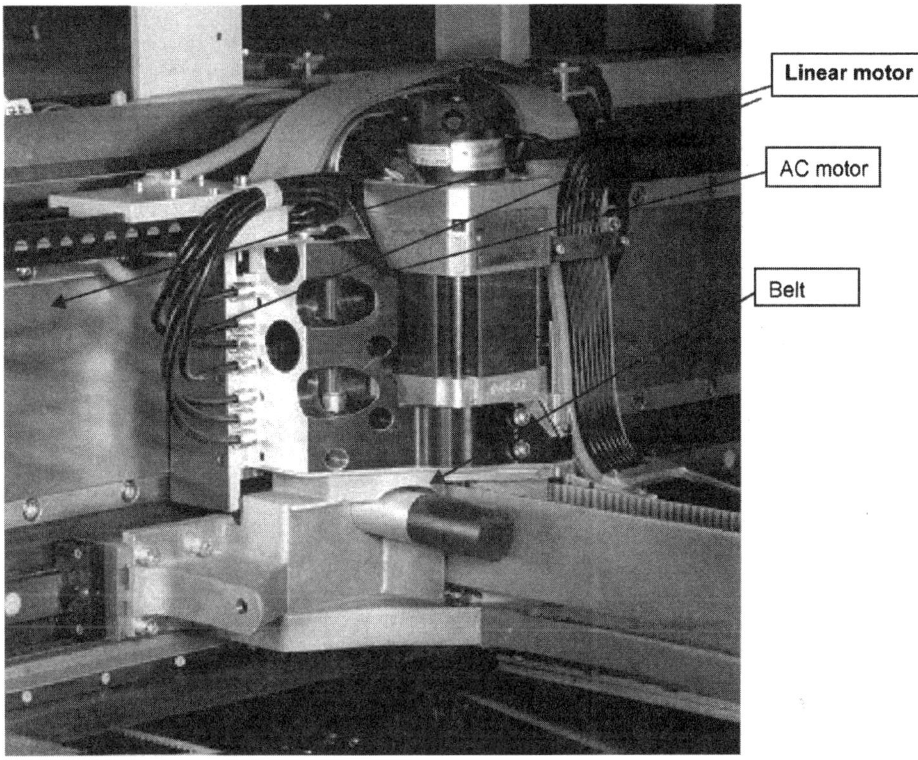

FIGURE 25.2 Combination of a linear motor and ac motor with belt drive.

the key positioning axes x, y, and Θ is very important for overall placement accuracy. In pick-and-place machines, the placement head(s) is(are) typically carried by an xy gantry system. Within the placement head, the most important axis is the rotational axis, but the precision of z-axis movement should not be neglected. In high-performance pick-and-place systems, z-axis movement control typically is achieved with a microprocessor using sensors for both the length of the vertical stroke and the required placement force. In conjunction with advanced processes such as FC placement into no-flow underfill (see Sec. 25.9), z-axis dynamics are moving more and more into the foreground.

A key advantage of the pick-and-place principle is that the precision placement head can be freely positioned in x and y, which allows for all specific lateral movements, e.g., for waffle pack pickup or multiple measurements on a component over a stationary upward-looking camera. This stationary camera for optical component centering is more or less standard in pick-and-place systems.

The most advanced pick-and-place systems today are specified with an xy accuracy in the range of ± 15 μm based on 4σ. The key disadvantage of high-accuracy pick-and-place systems, which generally provide only one high-precision placement head, is the very limited placement rate. The placement rate of such system is usually below 2000 components per hour (cph), excluding any additional process activities such as FC fluxing.

25.2.3 Multiple Pick-and-Place Chipshooter

Within such shooter systems, typically 16 independent pick-and-place heads are arranged in a serial configuration (Fig. 25.4). Since 16 pick-and-place heads can work in parallel, very high placement rates in the range of 100,000 cph and more can be achieved. Optical component centering normally is carried out with an integrated laser-based system. This so-called laser-align system, based on a

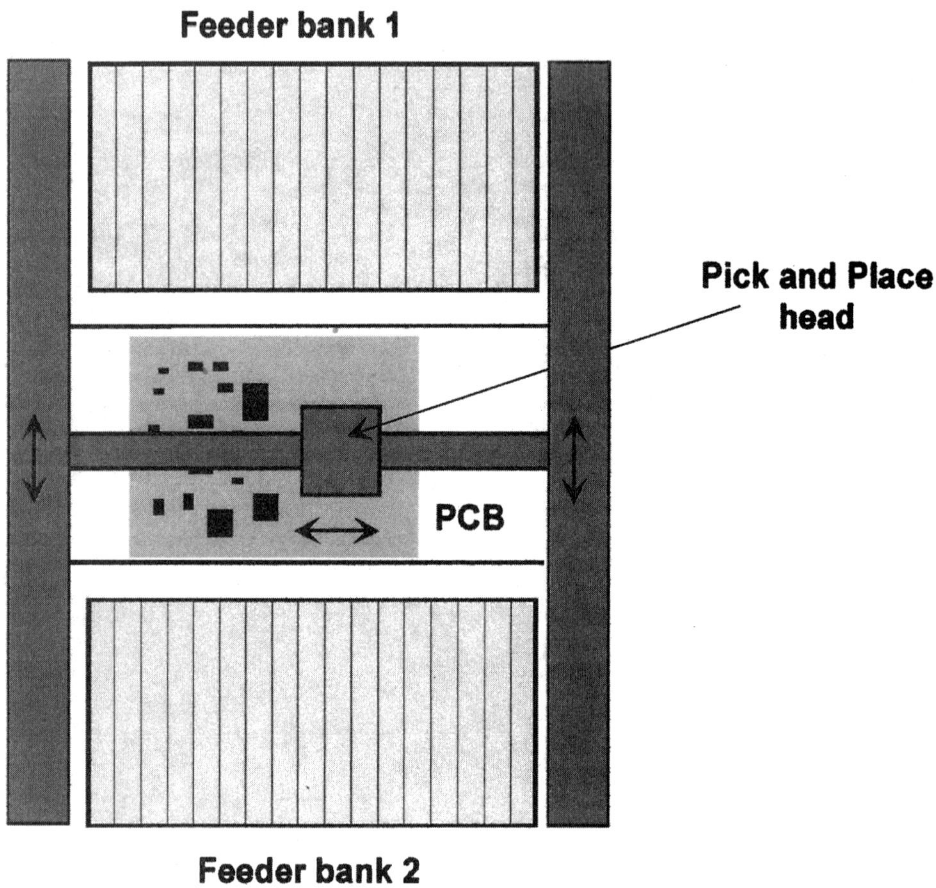

FIGURE 25.3 Pick-and-place principle.

laser-light curtain, provides outline-measuring capability. The placement accuracy of such chip-shooter systems, however, is limited. Only values in the range of more than 100 μm/4σ are achievable currently.

25.2.4 Collect and Pick and Place

This system provides (on a single gantry) a high-precision pick-and-place head in combination with a multinozzle revolver head (Fig. 25.5). While the pick-and-place head is responsible for the placement of large BGAs/QFPs, odd-shaped components, and very challenging fine-pitch FCs the high-speed job with smaller and less accuracy demanding components such as μBGAs and CSPs down to a ball pitch of 0.5 mm (20 mils) is done by the revolver (shooter) head.

This placement principle is called *collect and pick and place.* The component camera for the pick-and-place head is stationary, whereas the revolver-head camera is mounted on the gantry. The benchmark throughput of such a collect and pick-and-place system is currently 12,500 cph.

This high-performance placement system has a so-called head modularity for different revolver heads. It can be equipped with a 12-nozzle head for packages in a dimensional range between 0.25 $\times$ 0.5 mm and 18.7 $\times$ 18.7 mm or a 6-nozzle head. This 6-nozzle head is available in two versions. One version provides a large field of view for high-speed mounting (benchmark 8500 cph) of mainly larger packages including QFPs and BGAs up to 32 $\times$ 32 mm.

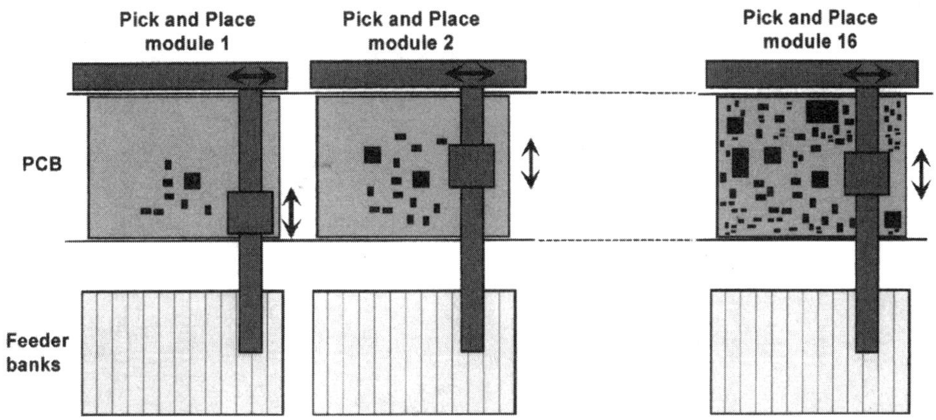

FIGURE 25.4 Theoretical configuration of a multiple pick-and-place head shooter system.

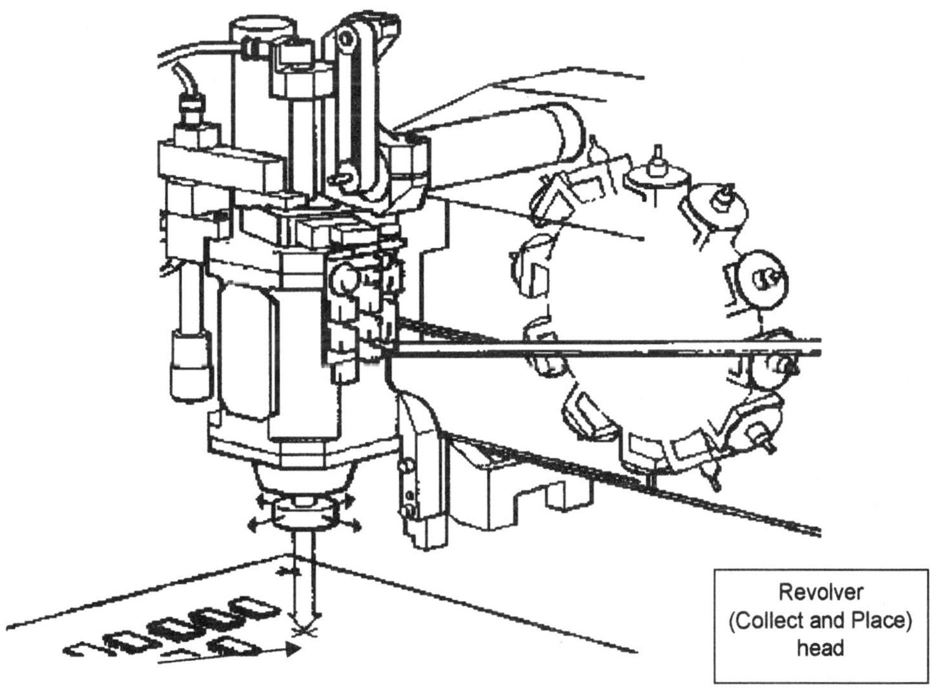

FIGURE 25.5 Collect and pick and place.

The other version is a 6-nozzle direct chip attach (DCA) revolver head for high-speed mounting of bare dies for COB, FCs, and most challenging micro-BGAs and CSPs based on ball/bump centering with ball/bump diameters down to 110 μm on bump pitches in the range of 200 μm (8 mils). The maximum placement rate for such applications with a collect and pick-and-place machine is currently 8500 cph. When dipping is a part of the revolver-head FC process, the throughput will be in the range of 3600 cph (throughput depends strongly on the number of components on the PWB).

25.2.5 Collect and Place

In a *collect-and-place* shooter system, each of the two or four revolver heads is carried by an *xy* gantry (Fig. 25.6). *Thus the revolver heads, currently equipped with 12 nozzles, have random access to the feeding and placement area.* Today, such systems are specified in a dual-beam version for a throughput of 25,000 cph.

The head modularity mentioned earlier is also available for collect-and-place shooter systems. Thus high-speed mounting of the complete SMD package-form spectrum including advanced packages such as the 0201s and FCs is possible.

A four-gantry collect-and-place chipshooter system currently provides a benchmark throughput of 50,000 cph (Fig. 25.7). In the future, such high-performance four-gantry chipshooter systems also will have a so-called head modularity. Then such collect-and-place chipshooters can be configured with different types (combinations) of revolver heads. In the case of DCA revolver heads mounting CSPs, bare dies for COB and most challenging FCs with extreme speeds will be possible.

A key advantage of the collect-and-place principle is the stationary PWB (substrate), which guarantees that already placed components are not shifted as a result of inertia. The stationary placement area is also mandatory for substrates in tape format (reel-to-reel technique).

Other advantages of collect-and-place chipshooters include the following:

- Since the revolver heads can be positioned freely over the feeding area, a component pickup from matrix trays is possible. This chipshooter approach also allows for a die feeding directly from the wafer (see Sec. 25.5).
- Microprocessor-controlled *z*-axis movements in revolver heads allow for process-specific adaptations for 0201 and FC placement into no-flow underfill (see corresponding sections).
- Future multigantry collect-and-place chipshooters also will provide dipping as a material-transfer method for fluxes and conductive adhesives.

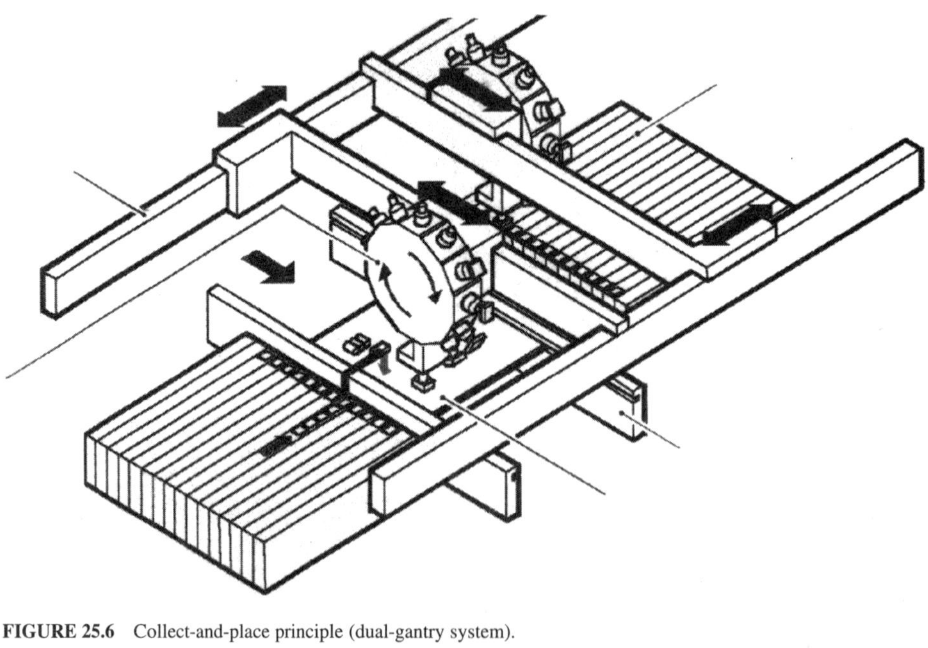

FIGURE 25.6 Collect-and-place principle (dual-gantry system).

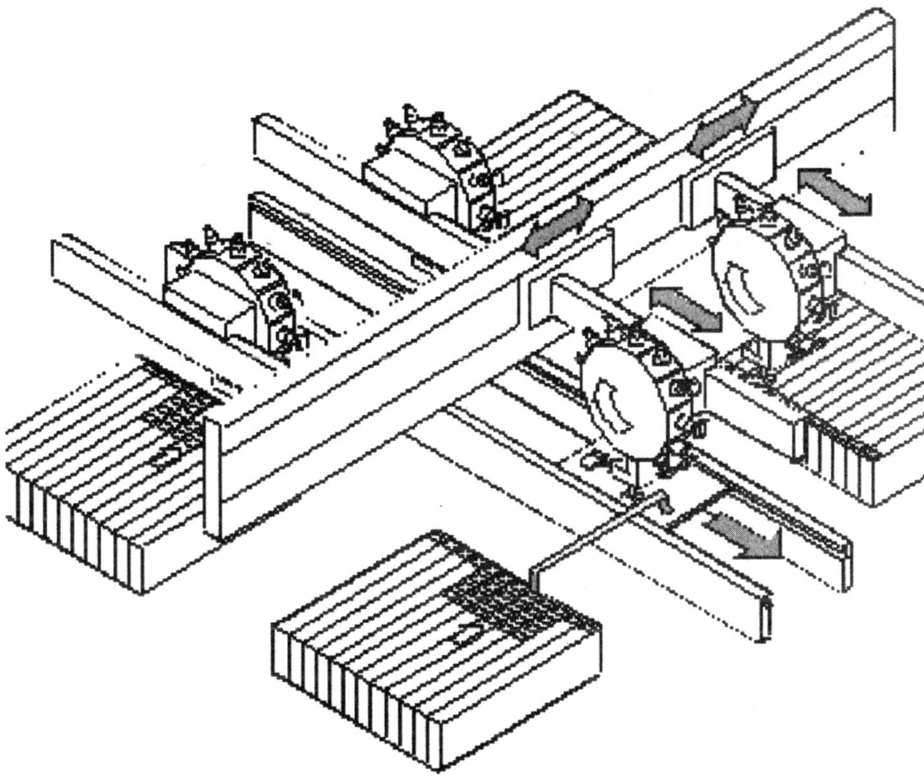

FIGURE 25.7 Collect-and-place principle (four-gantry system).

25.2.6 Other Multinozzle Systems

Dual-gantry systems, depending on the supplier, are available with up to 16 nozzles (Fig. 25.8). The optical component centering is carried out in three ways:

- Components are moved individually over a stationary camera system (one per gantry).
- In the placement head, an integrated camera is moved under all components on the nozzles (scanning approach).
- Laser-side illumination and imaging per nozzle and component allows only for outline centering.

25.2.7 Traditional Chipshooters

Such a system is determined by a horizontally rotating a turret head, which simultaneously picks the components from a moving feeder bank and places them onto a moving PWB (Fig. 25.9). Thus, very high theoretical placement rates in the range of 50.000 cph are enabled. For high-speed mounting of area array packages, traditional chipshooters can only be used to a very limited degree.

The limitations are as follows:

- No component pickup is possible out of the matrix tray (waffle pack) and wafer.
- Spring-loaded z-axis movement of vacuum nozzle does not allow a time-optimized and reliable bare die pickup from surf tape (tacky tape).

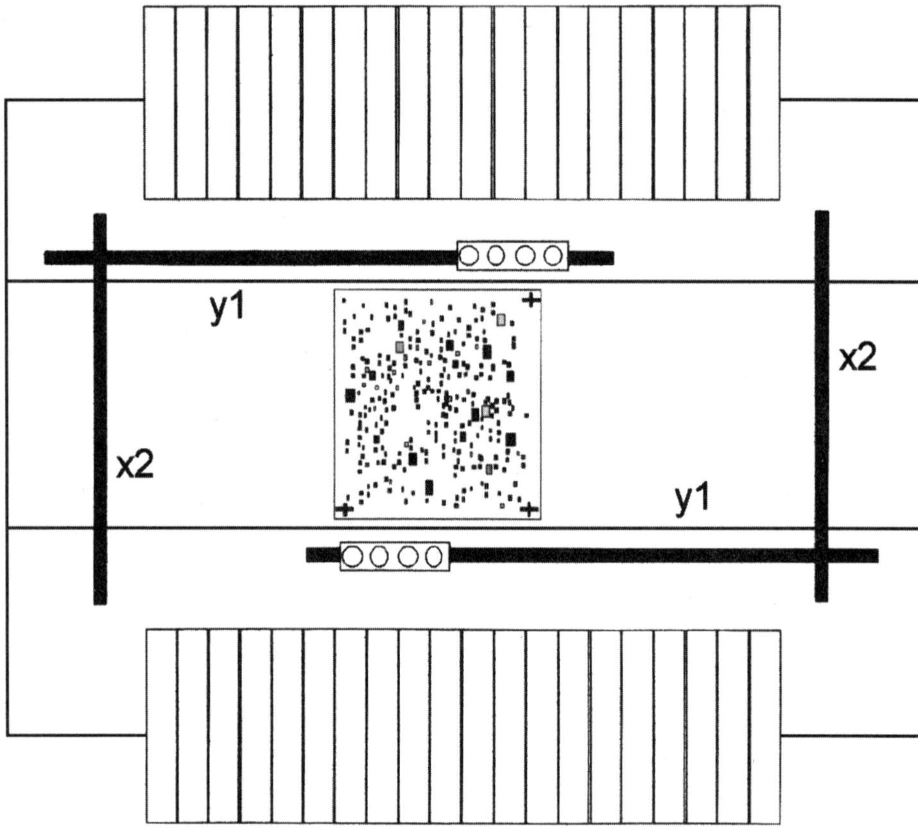

FIGURE 25.8 Non-revolver-head multinozzle placement system.

- Placement accuracy is not sufficient for most area array packages, typical values being clearly higher than 100 μm at 4σ.
- FC dipping for fluxes and conductive adhesives is not possible.

It seems that traditional chipshooters can only be used for high-speed mounting of area array packages with pitches greater than 0.8 mm in conjunction with package outline centering and standard embossed tape as a feeding format.

25.3 PLACEMENT ACCURACY

25.3.1 Statistical Basics and Definitions

In the early days of SMT, it was not a problem when an SMD placement machine supplier specified the placement accuracy of a certain machine, e.g., at ±60 μm, without any additional information. In this case, the question remains, How many of the placements may be inside or outside of the specified tolerance band of ±60 μm?

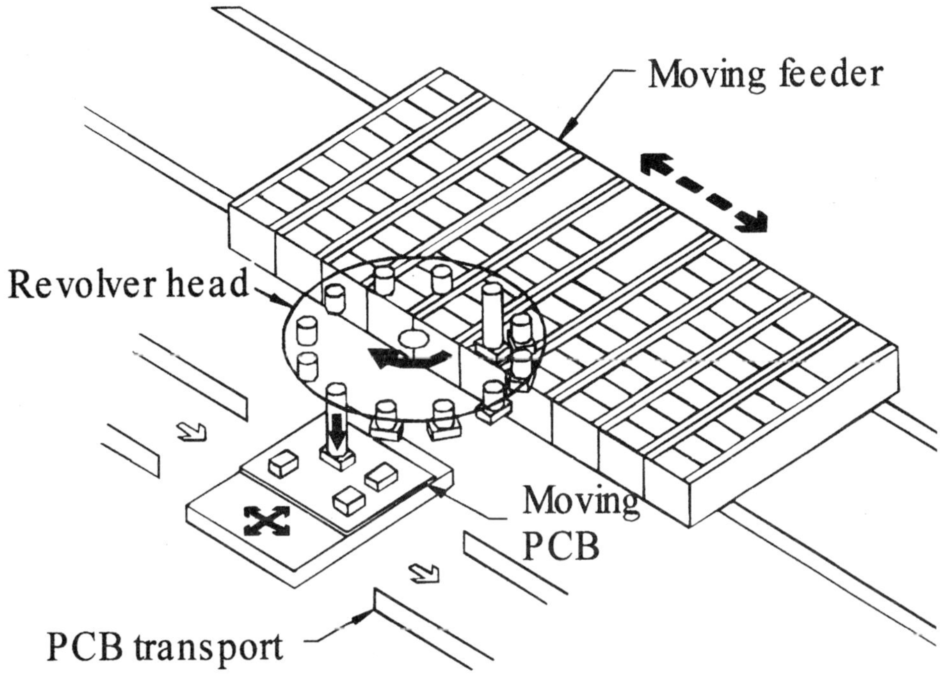

FIGURE 25.9 Traditional chipshooter principle.

When the positional deviations are distributed normally, which is nearly always the case, the rules of the gaussian bell curve apply (Fig. 25.10). Two criteria are important with this curve: the shape, which results in a certain *standard deviation sigma* (σ), and the location of the *mean value* (*xm,* respectively μ) situated on the x axis. The mean value x_m is determined mathematically as follows:

$$x_m = \frac{1}{n} \sum_{i=1}^{n} x \tag{25.1}$$

The standard deviation σ, which is determined graphically by the distance between the point(s) of inflection and the centerline, can be determined mathematically as follows:

$$\sigma = \sqrt{\frac{1}{n-1} \sum_{n-1}^{1} (x - x_m)^2} \tag{25.2}$$

Important note: When comparing statistically correctly defined placement accuracy data, it is essential to check that the statistical basis is the same. If this is not the case, one of the two statements has to be converted. The easiest procedure is to calculate back to the 1σ value and multiply the result by the target sigma number. For example, a value of 90 μm/3σ definitely corresponds with 120 μm/4σ because the standard deviation is 30 mm.

Example:

System 1: 50 μm/4σ. This corresponds to 12.5 μm/1σ.

System 2: 40 μm/3σ. This corresponds to 13.3 μm/1σ.

System 1 therefore is somewhat more accurate because it has a 3σ accuracy of 37.5 μm.

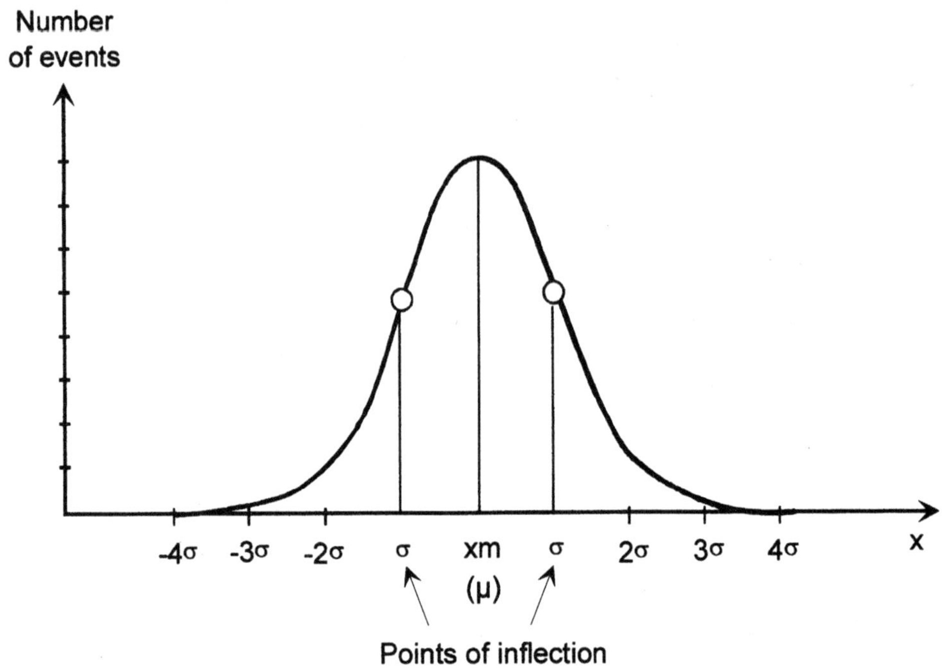

FIGURE 25.10 Gaussian bell curve.

25.3.2 Machine Capability Test (MCT)

The two core variables for placement accuracy, standard deviation and deviation of the mean value from the nominal value of the placement system, are measured as part of a *machine capability test* (MCT). The MCT is performed in the following sequence: First, a certain minimum number of precision glass components are placed on a precision glass plate. Then the placement deviations of all placed glass components are determined in x, y, and Θ using a high-precision measurement machine. The measurement machine then calculates the standard deviations σ and placement offsets μ (deviation of average value from nominal value) in the relevant positioning axes x, y, and Θ.

The MCT result represented graphically in Fig. 25.11 shows the following core placement accuracy values:

Standard deviation $\sigma = 8\ \mu m$
Placement offset $\mu = 6\ \mu m$

We can expect the placement deviations to correspond to the normal gaussian distribution, allowing conversion to a wider statistical basis of, say, 3σ or 4σ. With reference to the statistical basis 4σ so often used, the placement system specified above would have a precision of 32 μm.

In a second step, the suitability of the machine for a specific requirement is then assessed by placing the derived accuracy in relation to a required tolerance limit. The *machine capability index cmk* has proven most suitable for this. It is generally used to assess the process capability of machines. The machine capability index *cmk* can be used to calculate the placement accuracy once an upper and lower specification limit (USL and LSL) has been defined.

Since specification limit values usually are symmetric, we can perform the calculation with a simplified specification limit SL = USL = −LSL, as shown in Fig. 25.11.

The following machine capability index results for the conditions presented in Fig. 25.11 and a specification limit of 50 μm defined by the user of the equipment:

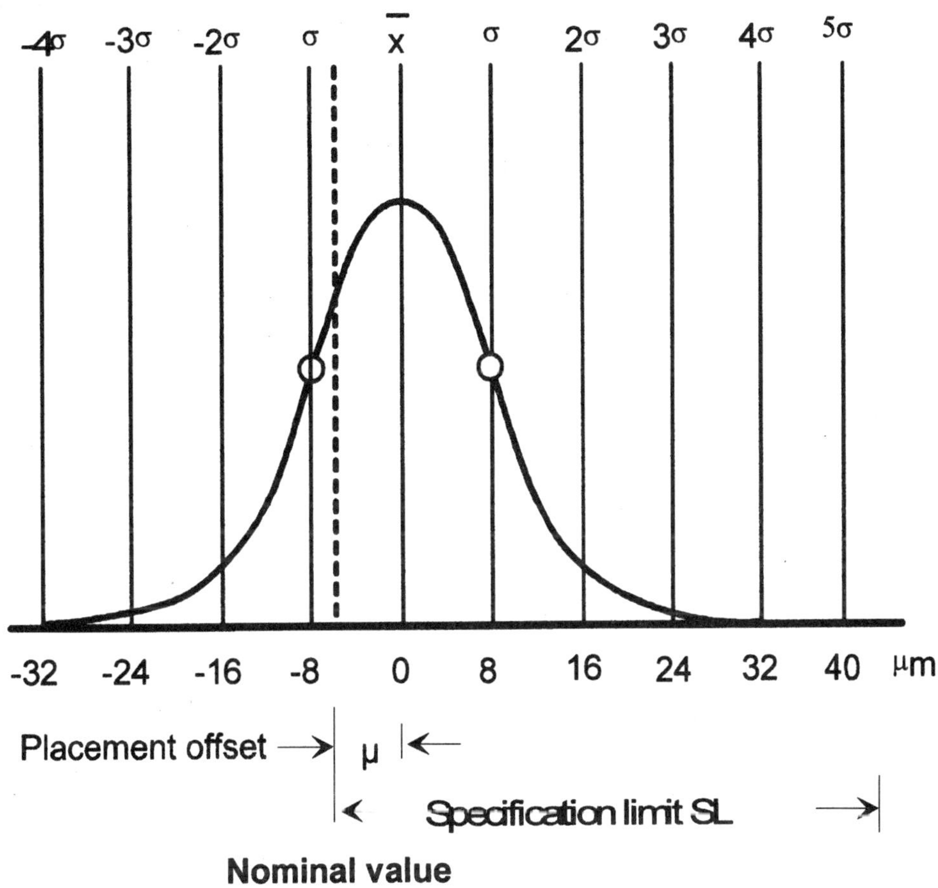

FIGURE 25.11 Graphic representation of an MCT result.

$$cmk = \frac{\textit{specification limit} - \textit{placement offset}}{3 \times \textit{standard deviation}} = \frac{3\ SL - \mu}{3\sigma} \tag{25.3}$$

The machine capability index *cmk* therefore evaluates the spread and the average deviation (placement offset μ) of the placement position in relation to three times the value of the standard deviation σ.

Important note: A comparison of capability indices is only permissible when based on the same specification limit.

25.3.3 Placement Accuracy Requirements for Advanced Packages

PWBs and component tolerances naturally have a very strong influence on product quality.

Leaded Packages (QFPs, SOPs, etc.). The following example of a tolerance analysis for the placement of a fine-pitch QFP with a pitch of 20 mils (0.508 mm) demonstrates that placement accuracy is just one link in the long tolerance chain (Fig. 25.12). Together with the existing geometric conditions and company-specific rules, the tolerances play a major role in determining the production yield that can be achieved.

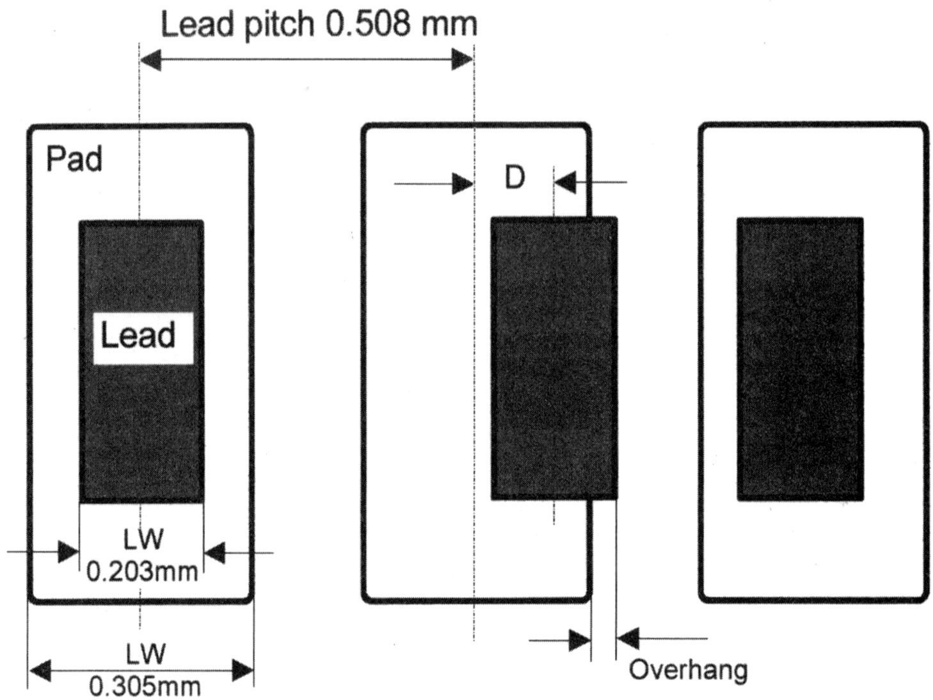

FIGURE 25.12 Example with a lead pitch of 20 mils (0.508 mm).

The values and dimensions used in the tolerance analysis are assumptions and are by no means to be regarded as dogma.

Lead pitch (component):	TLP approximately ±0.070 mm
Lead width (component):	TLW approximately ±0.050 mm
Pad pitch (PWB):	TPP approximately ±0.020 mm
Pad width (PWB):	TPW approximately ±0.020 mm
Local fiducials (PWB):	TLW approximately ±0.040 mm

By quadratically superimposing these tolerances, we obtain the total tolerance:

$$Ttot = \sqrt{TLP^2 + TPP^2 + TLF^2\left(\frac{TLW}{2}\right)^2 1 \left(\frac{TPS}{2}\right)^2} = \pm 0.0873\ mm \qquad (25.4)$$

For the company-specific rule that the maximum lead overhang at the pad amounts to one-third the lead width, the permissible deviation D of lead on the pad obtained is

$$D = PW/2 - LW/2 + \text{overhang} = (0.305/2 - 0.203/2 + 0.203/3)\ mm = 0.119\ mm \qquad (25.5)$$

The placement accuracy required for the placement system therefore is as follows:

$$PACCreq = \pm\sqrt{D^2 - T_{tot}^{\ 2}} = \sqrt{(0.119\ mm)^2 - (0.873\ mm)^2} = \pm 0.0805\ mm \qquad (25.6)$$

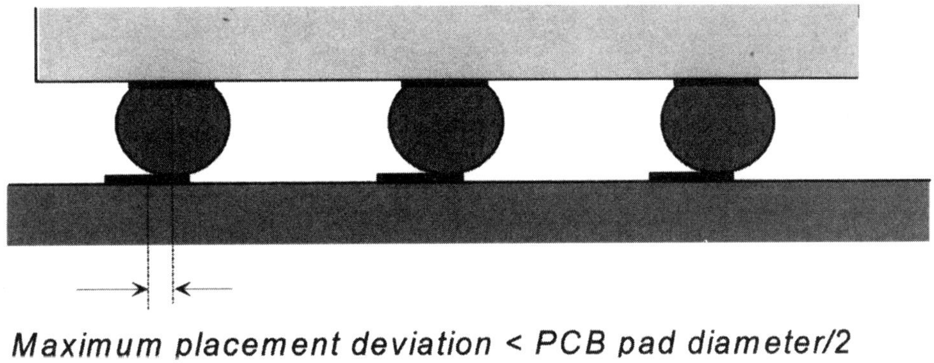

FIGURE 25.13 Placement accuracy requirement for CSPs and solder-bumped FCs.

Area Array Packages (BGAs, CSPs, and FCs). For most area array packages, the placement accuracy requirement is strongly relaxed compared with leaded packages (QFPs/SOPs) with the same pitch (Fig. 25.13). This is one of their key advantages.

The maximum acceptable placement error theoretically equals half the PWB (substrate) pad diameter in cases of (mostly used) circular pads without solder mask. A misplacement of over half the PWB pad diameter still ensures a mechanical contact of the ball/bump and PWB pad, respectively, with the solder paste deposit, and thus a perfect self-alignment should happen. The minimum placement accuracy required is

$$PACC \leq \frac{PCBpad\phi}{2} \tag{25.7}$$

In the case of a CSP pad diameter of 250 μm, the placement accuracy requirement is

$$PACC \leq \frac{PCBpad\phi}{2} = \frac{250}{2}\ \mu m = 125\ \mu m \tag{25.8}$$

However, the placement accuracy expectation is in reality higher. To achieve a good process capability index (cpk), the users are demanding 4σ placement accuracies of clearly better than 100 μm (100 μm/4σ = 75 μm/3σ).

FCs with a bump diameter of 100 μm (4 mils): ≤50 μm (50 μm/4σ = 37.5 μm/3σ).

Within Siemens' Production and Logistics Systems Division, experiments have shown that FCs, which are (by intention) placed *40 percent off the pad, give an extremely good yield.* In the case of FC bumps, which have a placement deviation of 50 percent off the pad diameter, the yield is noticeable lower.

0201 packages. Overall 0201 process quality is determined by the layout design, PWB tolerances, solder paste printing, placement, and the reflow process. All these factors have a mutual impact on each other.

The 0201 placement accuracy requirements for the following example are calculated based on the IPC (Institute of Interconnecting and Packaging Electronic Circuits; formerly Institute of Printed Circuits) standard for chip components, substrate pad, and component tolerances and the assumption that no self-alignment during reflow soldering occurs.

Boundary Conditions, Definitions, and Explanations in Terms of Placement Accuracy Requirements. Permitted deviations according to IPC J-STD-001B and IPC A-610B:

Maximum side overhang: 50 percent of component width for class 1 (general electronic products) and class 2 (dedicated service electronic products, including telecommunications equipment)

End overhang: Not permitted for classes 1, 2, and 3 (high-performance electronics, including military and harsh-environment applications)

Product tolerances:

- Pad: ±25 μm
- Component length: ±30 μm (according to Murata)

The layout design in Fig. 25.14 is based on a 150-μm pad spacing in the lateral (y) and longitudinal (x) directions. This layout design should allow for a solder resist between the pads in case of a high-quality PWB manufacturing. The resulting component spacings (gaps) are 310 and 200 μm in x and y.

Maximum deviations under ideal conditions (without product tolerances):

- *DEVYtheor:* ±100 μm
- *DEVXtheor:* ±80 μm

Considering the preceding defined product tolerances by applying the standard statistical laws (quadratic combination), the required placement accuracy in the more critical longitudinal direction x is

$$PACCrequ = \pm\sqrt{DEVXtheor^2 - \left(\frac{PADtol}{2}\right)^2 - \left(\frac{COMPtol}{2}\right)^2} = \pm\sqrt{80^2 - 12.5^2 - 15^2} = \pm 77.58\ \mu m \tag{25.9}$$

One has to be aware that for a placement accuracy statement based on 4σ the dpm rate is already 60. The preceding considerations show that it is very tough to achieve overall defects per minute (dpm) rates of clearly below 100 for tight layout designs.

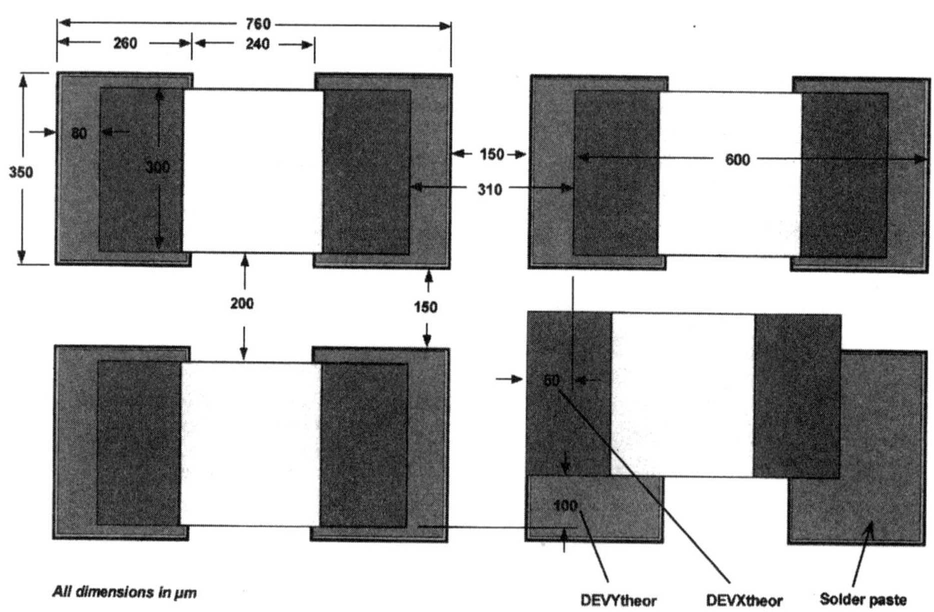

FIGURE 25.14 Layout example.

25.4 PLACEMENT MACHINE VISION TECHNIQUE

A central aspect of modern SMD placement systems is doubtless the vision technique. Such a placement machine vision system generally consists of a PWB vision system and a component vision system.

25.4.1 PWB Vision System

In combination with the sometimes extreme placement accuracy requirements (in particular with FCs), PWB fiducial recognition must not be underestimated. With large, low-cost PWBs, local fiducials should be used. In the hybrid area, where the ceramic boards have an extremely low nonlinear distortion, local fiducials normally are not required. However, global fiducial recognition can, in part, be very difficult due to color conflicts.

A major factor in determining the reliability of vision systems is the vision processing method used. The competing types are the contour sequence, center-of-gravity, and correlation methods. To achieve maximum reliability, the correlation method is always advisable. The point of maximum correlation or coincidence in terms of the nominal and actual image is decisive for the position of the structure (Fig. 25.15). Disruptive factors, which partially alter the image of the object, do not negatively affect the accuracy of position recognition. Using this method, it is also possible to use nonhomogeneous and asymmetric fiducial markings. The measuring accuracy of PWB vision systems must be at least in the range of 1 to 2 μm.

Figure 25.16 shows a concrete application with a vision system for PWB and component recognition integrated into a collect and pick-and-place system. After the PWB is moved into the placement area, a CCD camera mounted on the placement head sequentially detects the position of the two or three fiducial markings on the PWB. Afterwards, all placement positions are corrected in accordance with the incorrectness of the *xy* position rotation and the distortion of the PWB. Local fiducials near the placement position are recommended in order to enhance the positioning accuracy where nonlinear distortion of the PWB is involved.

Naturally, this camera also can recognize ink spots during panel production. Individual circuits marked no-go are excluded subsequently from the placement process.

25.4.2 Component Vision. After a component is picked up from the feeder unit by the placement head and briefly presented to the component vision camera, the camera recognizes the lead/ball/bump structure from below, and the analysis unit applies a suitable algorithm to determine the offset of the component in *x*, *y*, and *U* based on the center of the vacuum nozzle. In order to meet user demands in terms of strongly varying materials and surface properties with FCs and other area array packages, extremely powerful component vision systems are required. Component vision

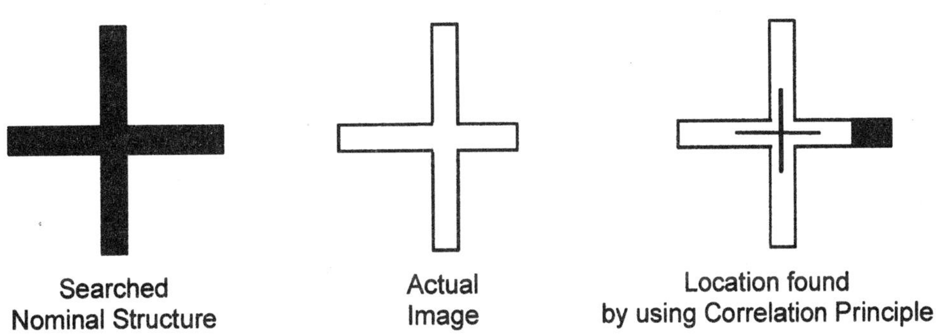

FIGURE 25.15 Correlation principle for PWB position recognition.

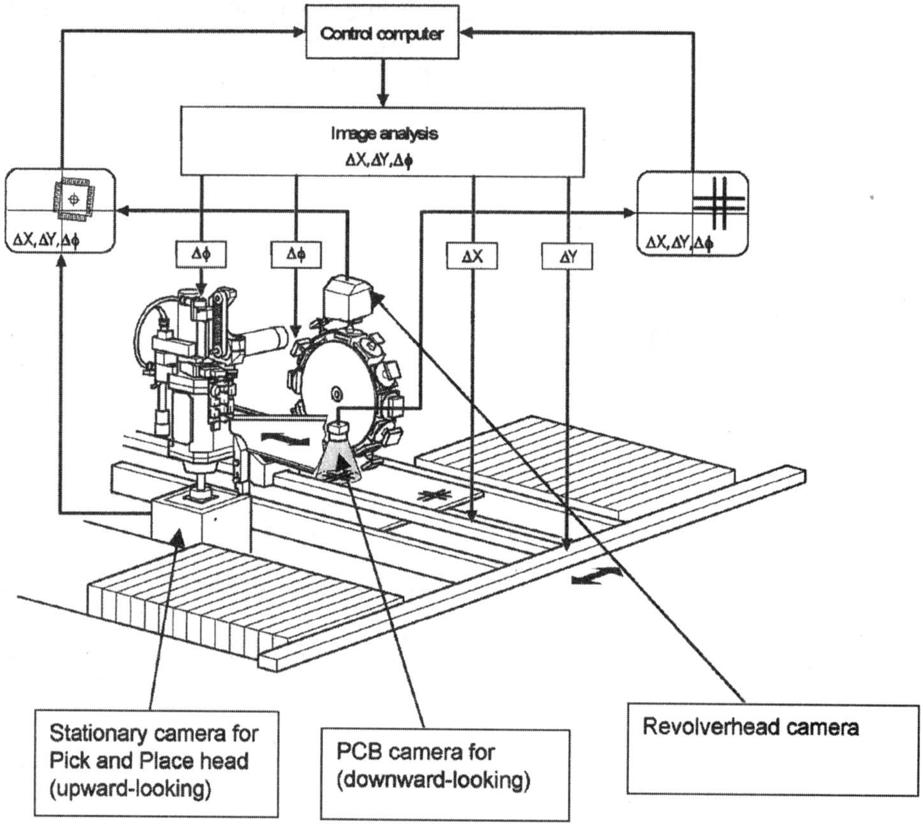

FIGURE 25.16 Complete SMD placement machine vision system.

system capabilities (and vision capabilities in general) depend on both lighting (camera) technique and algorithms applied by the evaluation unit.

Outline centering using backlighting or laser-side illumination (laser align) should be suitable for area array packages with ball/bump diameters of 0.3 mm or more. A negative impact on placement quality, however, is presented by the wafer sawing tolerances (average approximately 25 μm) in the case of FCs and substrate tolerances in the case of BGAs, micro-BGAs, and CSPs.

Lighting Technique for Ball/Bump Centering (Frontlighting). In order to eliminate FC sawing and BGA/μBGA/CSP substrate tolerances, ball/bump centering is mandatory. Optical ball/bump centering is only possible with a *frontlight system.* A maximum of recognition reliability can be achieved only by means of a sophisticated flexible lighting technique, since even the best evaluation algorithms are worthless if the captured image is of too poor quality.

The primary objective in the component camera therefore must be first to illuminate the object with well-proportioned quantities of light from different directions (angles) with the result that an adequate contrast between the relevant structures (balls/bumps) and the background environment can be achieved. The required high variance in lighting can be provided by using at least three different light sources. Each light source is determined by a specific light emission angle.

The collect-and-place revolver-head camera system equipped with four LED light sources shown in Fig. 25.17 is capable of optically centering the complete SMD package-form range, including FCs (bump centering). A maximum of flexibility exists when the intensity of each light source can even be programmed in fine increments. The package-form library contains the individual, ideal lighting for each package form.

The quality of the FC image in Fig. 25.18 captured by the preceding revolver-head camera is nearly perfect. The bump-find algorithms easily can suppress the still slightly visible routing structures.

Since high-resolution cameras for 0201 packages and FCs have a small field of view (FOV), they can only handle smaller packages up to dimensions of approximately 13 × 13 mm. When a wider package dimensional range is a required, cameras with a larger FOV must be available. As a result of this, powerful SMD placement machines including chipshooters must have different component cameras integrated or available in certain combinations as an option. Within traditional chipshooter systems, this requirement is covered by one camera with two or three FOVs. Collect-and-place chipshooters, where the component camera is integrated in a highly compact and small revolver head, are meeting this requirement with the earlier addressed head modularity. Then, using a dedicated head configuration, the throughput of a chipshooter machine or line can be optimized for a certain package-form range.

High-performance SMD placement systems, which have to deal with the complete package-form range, including fine-pitch FCs, must have at least two component cameras. The FC camera must have a different lighting approach and a much higher camera resolution (magnification) than a standard camera.

Orientation Check. A further strong argument for ball/bump centering is the orientation check (commonly called *pin 1 recognition*) for area array packages. This is the only SMD placement

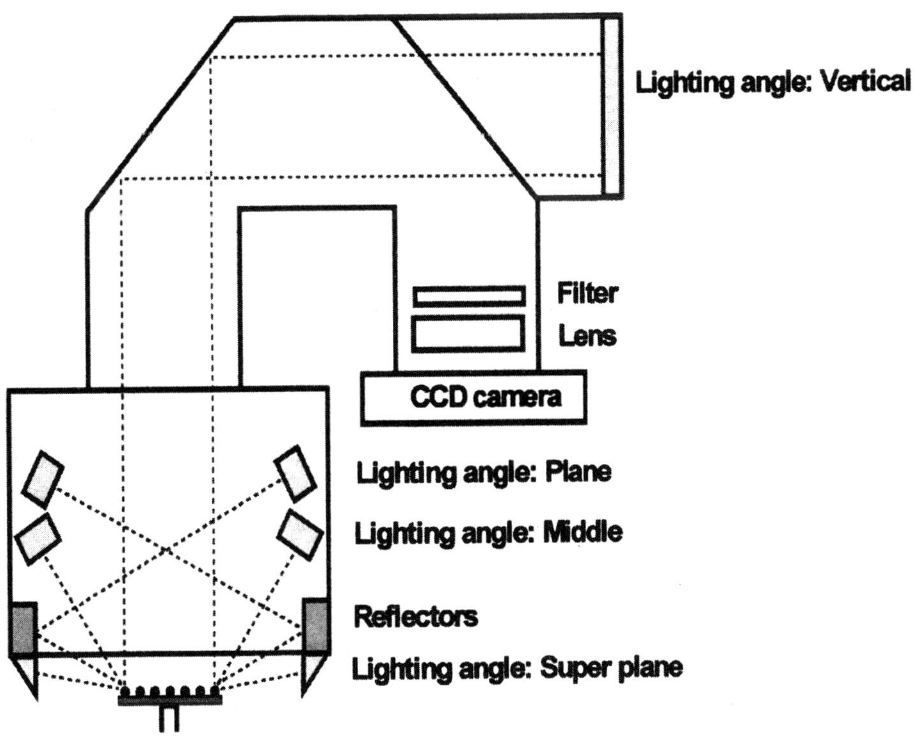

FIGURE 25.17 Revolver-head camera system with flexible lighting.

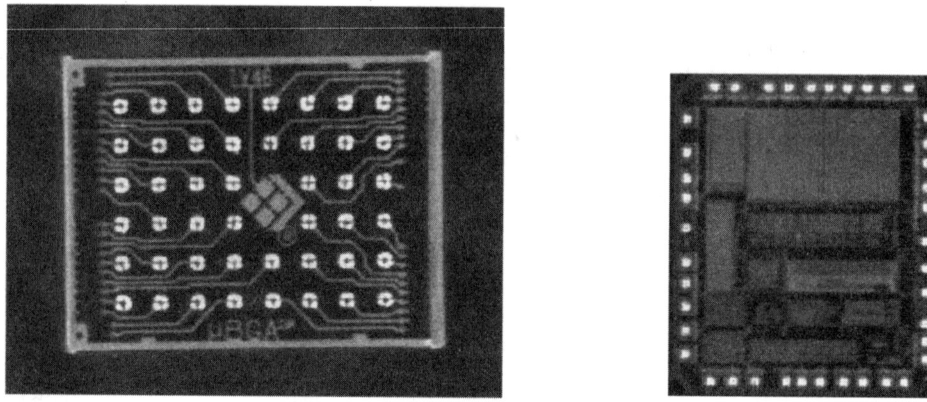

FIGURE 25.18 Revolver-head camera image of a Tessera μBGA and a functioning Siemens/Infineon FC.

machine vision feature that can reliably prevent placing these packages with the wrong orientation. The orientation check is automatically included in the placement process when ball/bump centering is carried out with nonregular (nonsymmetric) ball/bump arrays. While in FC technology the bump arrays mostly are already nonsymmetric (see Fig. 25.18), this important precondition for a high-yield, fully automatic SMT production mostly with BGAs/micro-BGAs and CSPs is not yet in place. The Tessera μBGA in Fig. 25.18 is a positive exception.

Algorithms. For ball/bump centering of the sometimes very challenging area array packages, algorithms suitable for standard SMDs can scarcely be applied. Therefore, very complex, time-consuming, but very interference-tolerant contour search methods are more advantageous.

CSP/BGA Ball and FC Bump Inspection. Despite the fact that the robustness of the balls and bumps is one of the greatest advantages of area array packages, component vision ball/bump inspection capability is sometimes requested. With a powerful, flexible lighting approach and specific inspection algorithms, a ball/bump inspection with respect to deformation and presence or absence is possible to a limited degree. One has to be aware that the primary job of a component vision system is the precise and fast centering of the various packages. Since FC bump recognition (centering) today is clearly more difficult than with BGAs and CSPs, FC bump inspection capabilities are even more limited.

Furthermore, there is a strong contradiction between full array ball/bump inspection and a good placement rate. In order to achieve acceptable bump find calculation times and thus high placement rates in most applications, only a few (e.g., five) balls/bumps in each corner of the package should be programmed. While the optical centering of an area array package over a total of 20 corner balls might take approximately 0.15 s, a full array inspection of a couple of hundred balls or bumps might take a second.

The best place to carry out a real ball/bump inspection should be the package manufacturer's site, and the only equipment must be dedicated inspection systems, which typically cost several hundred thousand U.S. dollars.

25.4.3 Lead Coplanarity Measurement

A particularly important aspect in fine-pitch processing is QFP lead coplanarity. Using a coplanarity module integrated into SMD placement systems, it is possible to recognize lead coplanarity errors with a high degree of resolution (Fig. 25.19).

The leads are scanned sequentially, with the component being held on the vacuum nozzle while it is moved through a stationary laser beam (Fig. 25.20). Measurement is based on the triangulation principle, with a part of the diffuse reflected laser light being used to generate a height-dependent signal in a position-sensitive device (PSD) or a CCD sensor. Special analysis algorithms are used to

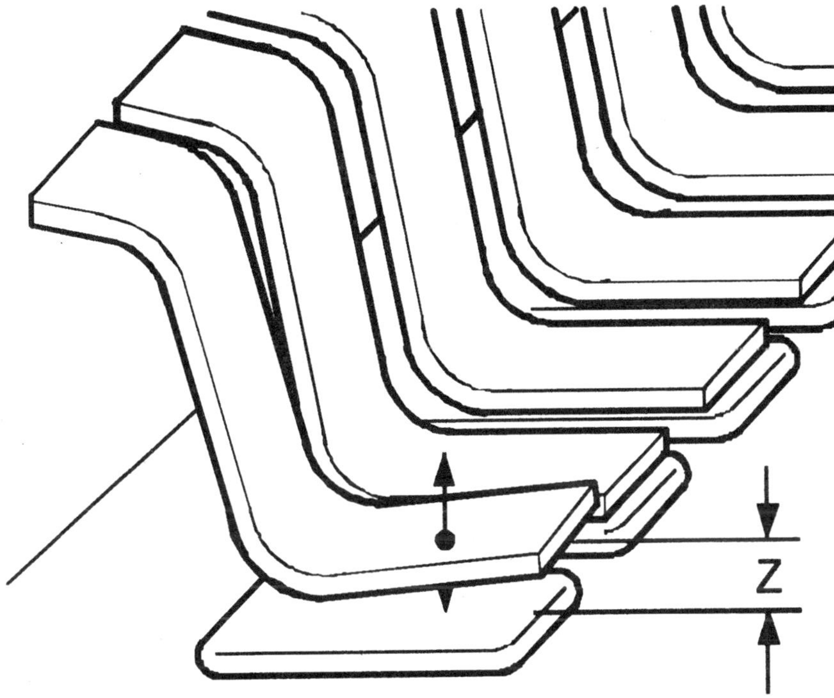

FIGURE 25.19 Lead coplanarity inspection.

determine the correct placement plane, which represents the reference plane for the programmable tolerance band of the vertical lead skew.

The level of the acceptable vertical lead screw is proportional to the thickness of the applied solder paste layer. It is advisable to take possible PWB warpage into account in the tolerance band when processing large, high-quality ICs. The following typical coplanarity values can be applied: 100 μm for a lead pitch of 0.65 mm and 80 μm for a lead pitch of 0.5 mm.

Compared with various shadow-based camera methods, coplanarity measurement using lasers has the decisive advantage of a clearly higher accuracy. Values in the range of 20 μm/3σ can be achieved. A disadvantage, however, is the relatively high time consumption for the scanning process.

25.5 ADVANCED PACKAGE FEEDING

A very important criterion in bare die and area array package processing is component feed on the SMD placement machine. Feeding of BGAs and CSPs is not critical and typically is done with JEDEC and or 4 × 4 in matrix trays and standard embossed tapes. In high-volume applications, the trays should be handled with waffle pack changers. In the case of bare chips, the following four feeding methods can be used.

25.5.1 Embossed Tape

This is ideal for mass production. There is no negative impact on placement rate. The standard tape does not find full acceptance in the FC area so far. This is surprising because it appears to have no real technical drawbacks. One reason might be the fact that FC dimensions are not standardized,

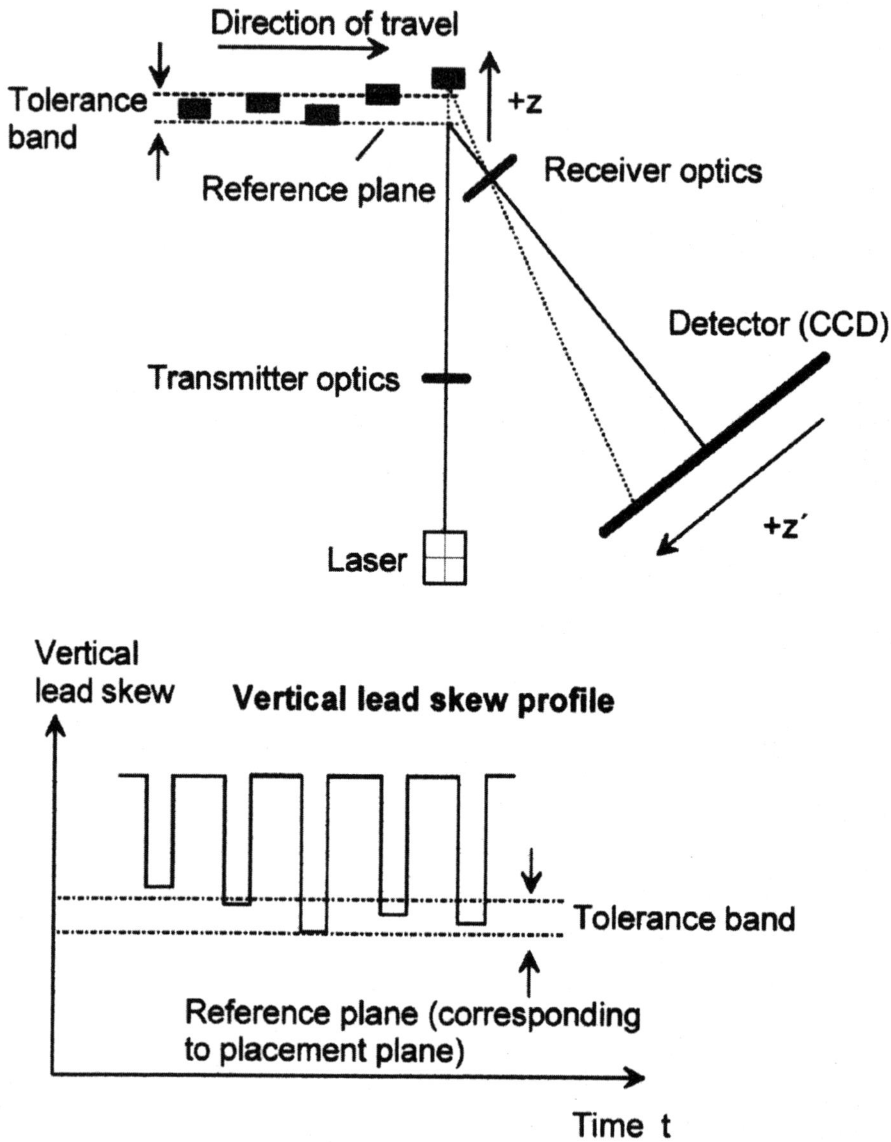

FIGURE 25.20 Principle of lead coplanarity inspection.

which causes slightly higher tape material costs. A concern in the COB arena is sometimes the fact that the continuous contact of the die with the tape cavity can produce silicon dust or even splinters. This runs contrary to the cleanliness requirements of the COB process. To reduce or even eliminate this problem, the tape industry is working on improved tape designs.

25.5.2 Surftape

In a surftape setup, the bare chips are attached face up on a tacky area (Fig. 25.21). Surftape requires a more complicated and relatively expensive feeder. If bare chip pickup out of the same feeder is required, a pickup cycle time of at least 0.5 s will reduce the placement rate of high-speed bare chip placement systems. Today, an additional disadvantage seems to be the more costly tape material compared with standard tapes. Surftape can be a good solution for high-volume, flexible production with bare dies in the COB arena.

In the case of FCs, the dies must be flipped (inverted) before they can be placed face down onto the substrate. Chip turnover is accomplished by a feeder internal pickup-and-flip head. An FC surftape feeder, of course, is more complicated and thus more expensive than a surftape feeder for bare dies for COB. The flip job prolongs the pickup cycle time. Particularly in high-volume production, surftape feeders are not the best solution for presenting FCs to an SMD placement machine.

25.5.3 Matrix Tray

The 2 × 2 in matrix tray is currently very common in the bare die arena. It is ideal for low- and medium-volume production and/or high component variance. A waffle pack changer can greatly enhance the flexibility but is considerably more expensive.

25.5.4 Wafer

Bare die feeding directly from the wafer definitely results in lower component purchasing prices due to the eliminated die-sorting job. Wafer feeding within next-generation chipshooters is a tremendous challenge. Since the number of passives on a substrate will decrease and the number of bare dies or wafer-level CSPs on a substrate or panel definitely will increase, this challenge must be overcome. Several different ways are possible.

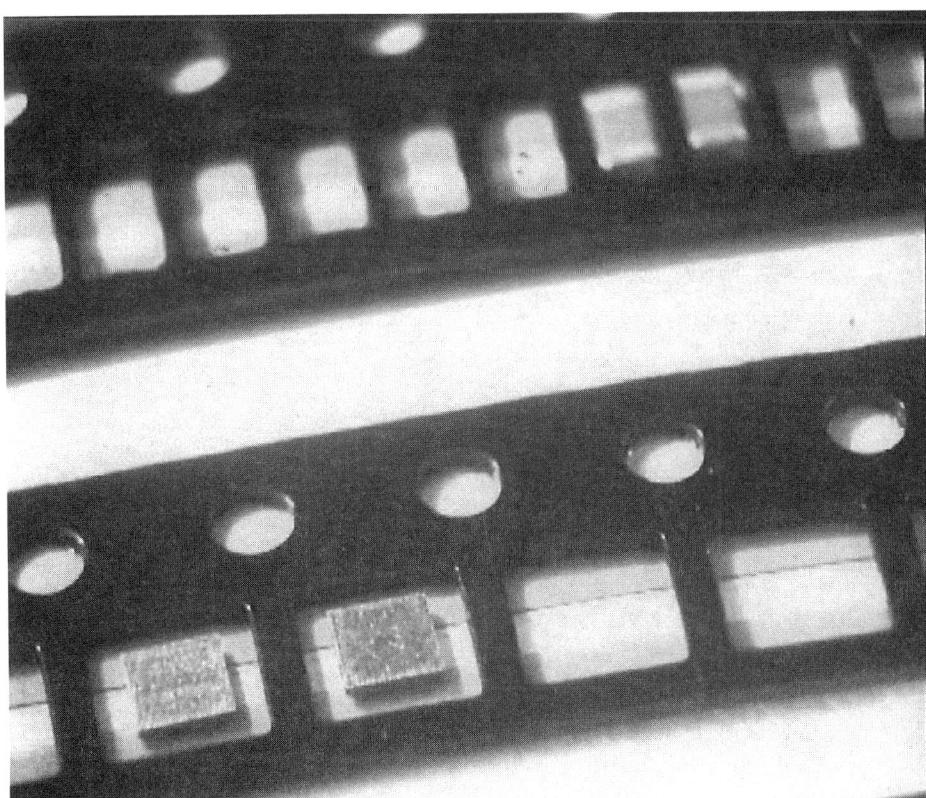

FIGURE 25.21 Surftape.

Wafer Feeder Module. Such a system can be considered to be a complex SMD placement machine feeder for one type of die (Fig. 25.22). The wafer is loaded manually from the front side. Loading and processing are carried out in a vertical orientation. The dies are ejected, picked up, and placed onto a conveyor belt. Depending on the bare die technology applied within the placement machine, the dies can be placed onto the conveyor either face up (for COB) or face down (for FC technology). The pickup rate of the placement machine's revolver head, however, is higher than the die-placing rate of the wafer feeder. To enable high-volume bare die applications such as RFID manufacturing, two or more wafer feeders can be installed on the feeder banks of an advanced chipshooter (see practical examples in Sec. 25.10).

This wafer feeder approach can be applied for single-chip as well as multichip applications, which exist in the automotive electronics industry.

Stationary System Integrated in the Feeder Bank as a Future Solution? Within such a solution, the wafer is situated within the access area of the pick-and-place head or revolver head. The dies could be handed over to the placement head by means of an intermediate head or even be picked up directly by a revolver head. With time-consuming large-die pickup, the use of an intermediate, independent pickup head is advantageous to avoid a significant throughput derating. With small dies in high-speed applications, direct pickup of the dies from the wafer may be the most economic method. The required flip job in the case of bumped dies could be carried out within the revolver head using a flip station (patent pending).

25.6 FLUX APPLICATION FOR FCS

The standard mass reflow process for solder-bumped FCs using ovens requires flux. Today, advanced SMD placement machines must have a built-in flux application. The two common built-in methods are as follows.

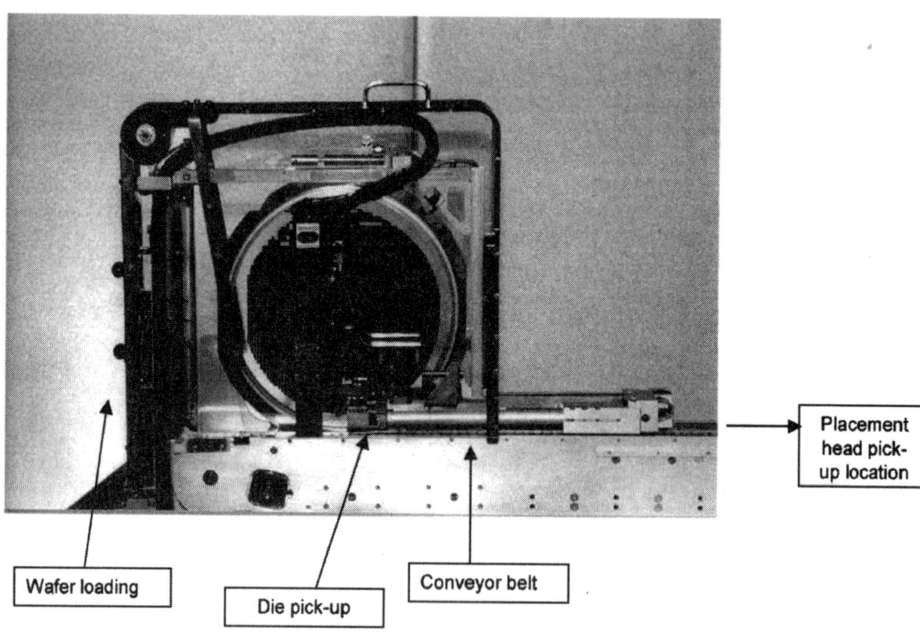

FIGURE 25.22 Wafer feeder module.

25.6.1 Fluxing of the Substrate at the FC Site

The dispensing unit is mounted in the vicinity of the placement head (Fig. 25.23). Immediately prior to FC placement, the placement site is wetted with flux. The required amount of flux, which is dispensed onto the center of the placement site, is determined by the dimensions of the FC and the wetting properties of flux and substrate material.

To avoid nonwetted pads as a result of wetting behavior tolerances, the flux film diameter (proportional to programmed volume) must be chosen high enough (Fig. 25.24). This reliable and proven

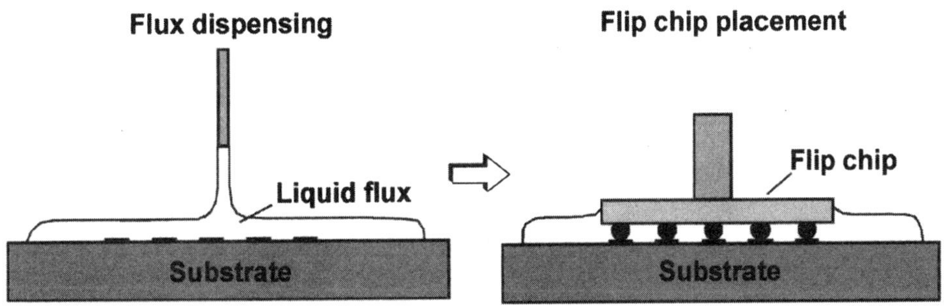

FIGURE 25.23 Flux dispensing for low-viscosity fluxes.

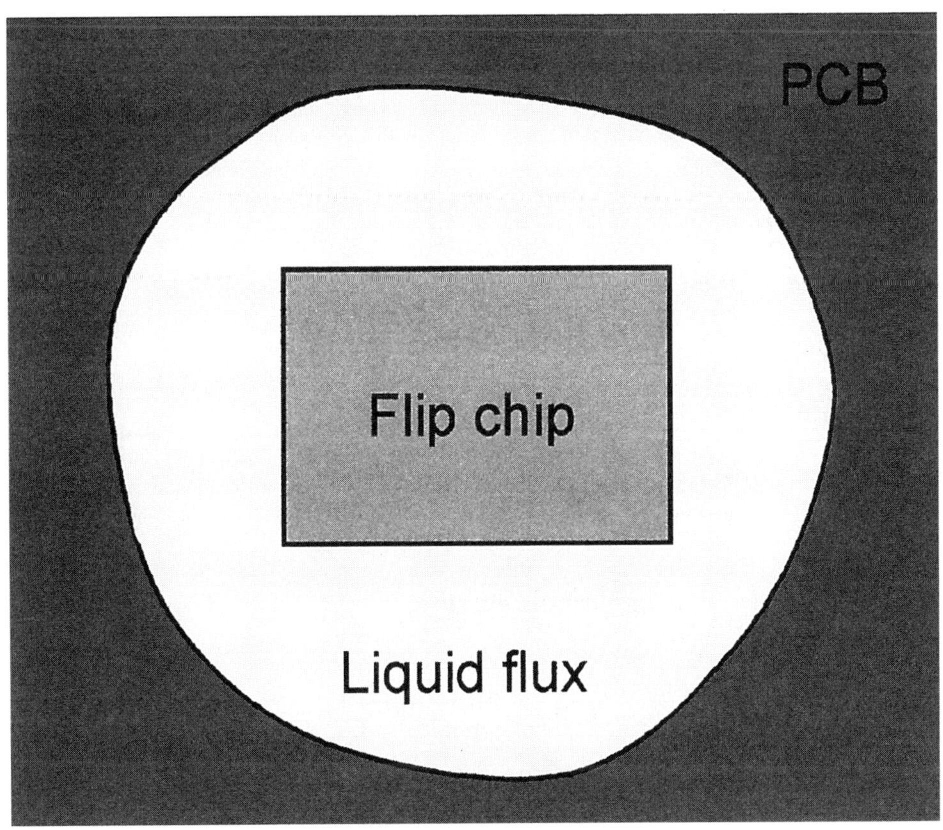

FIGURE 25.24 Wetting of the liquid flux.

method, which at present is used widely, is only suitable for low-viscosity fluxes. In order to enable an effective underfilling in a no-clean process, the flux must be a no-clean (no-residue) material. Liquid fluxes, which typically contain little solids, are most suitable for a no-clean process.

Important note: Liquid fluxes naturally cannot have a good fixing property. The conveyor movement of the placement system immediately after the last FC placement cycle can cause a shifting of the FCs away from the original position (as a result of inertia). There are two ways to overcome this problem:

- Waiting time of a couple of seconds prior to PWB transport (negative impact on throughput). During this waiting time, the quick evaporation of the flux on the FC perimeter increases adhesion strongly.
- Acceleration and deceleration of conveyor movements must be adaptable to the flux adhesion. Sufficiently smooth conveyor movements do not cause FC shifts.

The key disadvantage of flux dispensing is the relatively high cycle-time penalty. For each component to be fluxed, the placement cycle time has to be increased by approximately 1.5 s.

25.6.2 Dip Fluxing of the FC

Prior to placement at the final location, the FC is positioned onto a flux carrier plate. The flux carrier typically is a rotating drum on which a thin film (e.g., 55 μm) of flux is adjusted by a doctor blade (Fig. 25.25). This principle is most suitable for high-viscosity fluxes. The amount of flux involved in the process is brought to a minimum by only fluxing the bump undersides. Dip fluxing is not well suited for fluxes with a high evaporation rate of the volatile ingredients.

The following two process sequences are possible:

- *Optical bump centering, bump dip fluxing, placement.* This has the disadvantage that the mechanical contact of the FC bumps with the flux carrier can have a negative impact on placement accuracy.
- *Bump dip fluxing, optical bump centering, placement.* There is a very slight risk that the optical bump image can be influenced adversely by the flux material.

It seems that dipping prior to optical centering is the preferable sequence (Fig. 25.26).

25.7 CSP DIP FLUXING

Currently, there are some concerns about processing CSPs with ball diameters of 0.3 or even 0.25 mm using standard SMT. The critical process step obviously is the solder paste printing for products with mixed technology (standard SMDs plus CSPs), where the stencil thickness commonly is 150 μm. In order to release the solder paste reliably out of the cylindrical stencil openings, however, a

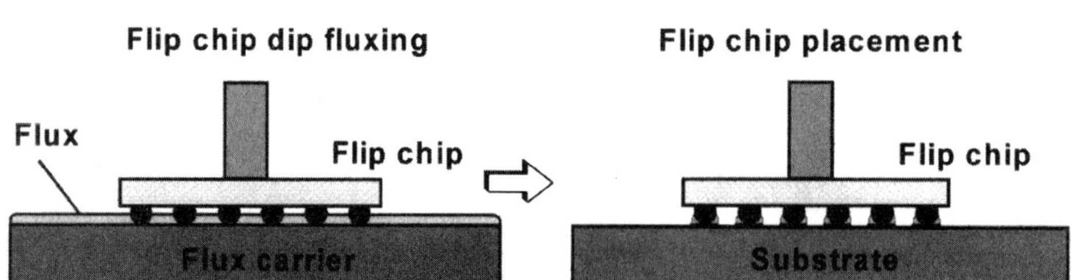

FIGURE 25.25 Dip fluxing principle.

FIGURE 25.26 Installation of a dip module on a advanced chipshooter feeder bank.

stencil thickness in the range of 120 μm is required. This problem can be bypassed by applying the classic FC process *without solder paste printing at the CSP placement site.* A very economical and reliable process sequence within an advanced shooter system can be CSP pickup (plus time-neutral optical ball centering), dip fluxing of the CSP ball array (see preceding figures), and placement.

25.8 ADHESIVE APPLICATION FOR FCS

Isotropic conductive adhesive (ICA) FC technology will gain considerably in importance because this seems to be the ideal interconnection technique in the smart card/RFID business. The adhesive dipping process is considered to be a flexible, low-cost alternative to the stencil printing process.

Since the bump height with adhesive-joined FCs can be rather low compared with standard solder bumps, the application of a very small conductive adhesive deposit at the bumps is required. The use of special unfilled underfillers is also necessary.

The test FCs from Fraunhofer IZM with electroless nickel bumps on a pitch of 200 μm were processed with a flex substrate. Prior to placement, the FC bumps were dipped into a thin film of a silver-filled conductive adhesive.

The optimal thickness of the adhesive film used was determined to be 25 μm for nickel bumps with a bump height of 45 μm. Figure 25.27 shows a thin layer of adhesive between the FC bump and substrate. The figure also gives an impression of the well-shaped adhesive reservoirs around the bumps.

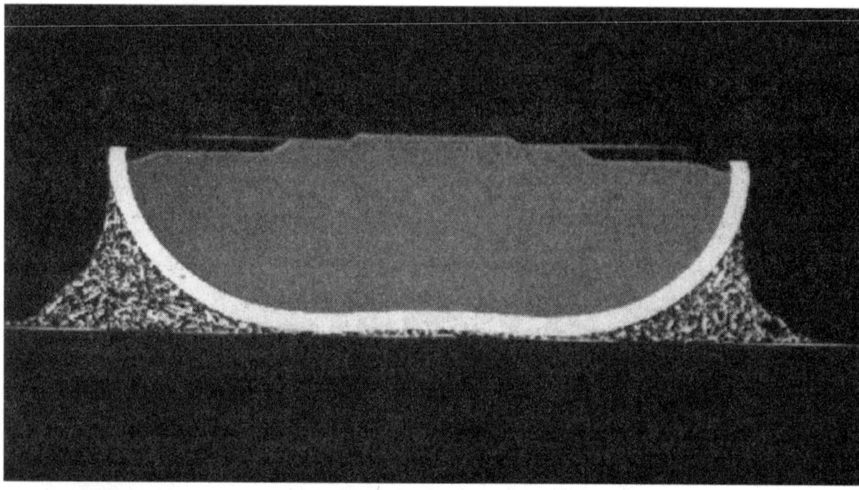

FIGURE 25.27 FC interconnection using ICA. (*Fraunhofer IZM.*)

25.9 FC PLACEMENT INTO NO-FLOW UNDERFILL (FLUX/UNDERFILL)

Lots of companies from all around the world are currently working on no-flow underfill materials. Such an underfill, which additionally operates as a flux, can be stencil printed or dispensed to the substrate. During the reflow step, the epoxy is at least partially cured. The use of a no-flow underfill can speed up the whole FC process dramatically.

To ensure a void-free wetting of the active FC area, the downstroke velocity must come to a certain material-dependant minimum when the FC touches down on top of the flux/underfill deposit (Fig. 25.28). This requirement can only be met with an intelligent, microprocessor-controlled *z*-axis positioning machine, which is provided by advanced collect-and-place chipshooters. Other chipshooters, where the *z*-axis movement is spring-driven, are not suitable for these innovative future processes.

25.10 TYPICAL APPLICATIONS OF AN ADVANCED CHIPSHOOTER IN SMART CARD OR RFID MANUFACTURING

Figure 25.29 shows a flexible approach that does not require an adhesive printing machine. The machine internal process sequence is:

1. The revolver head collects 12 FCs from two wafer feeders.
2. The revolver head moves over the dip module and carries out an adhesive application with all FCs. Each FC bump structure is dipped into the thin film of conductive adhesive at different positions on the drum (disk). During the sequential adhesive dipping, the drum stands still.
3. The revolver head moves over to the flex panel for placing the 12 FCs. At the same time, at least one rotation of the drum is activated to planarize the conductive adhesive film.
4. The revolver head moves back to the wafer feeders for picking the next 12 FCs.

Steps 1 through 4 will be completed alternatingly with both revolver heads. The practical throughput of the configuration in Fig. 25.29 will be in the range of 6000 units/h (including all

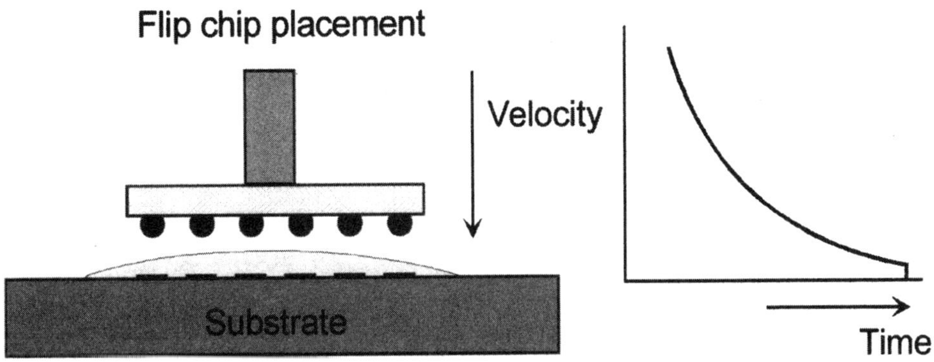

FIGURE 25.28 Positioning of an FC into no-flow underfill.

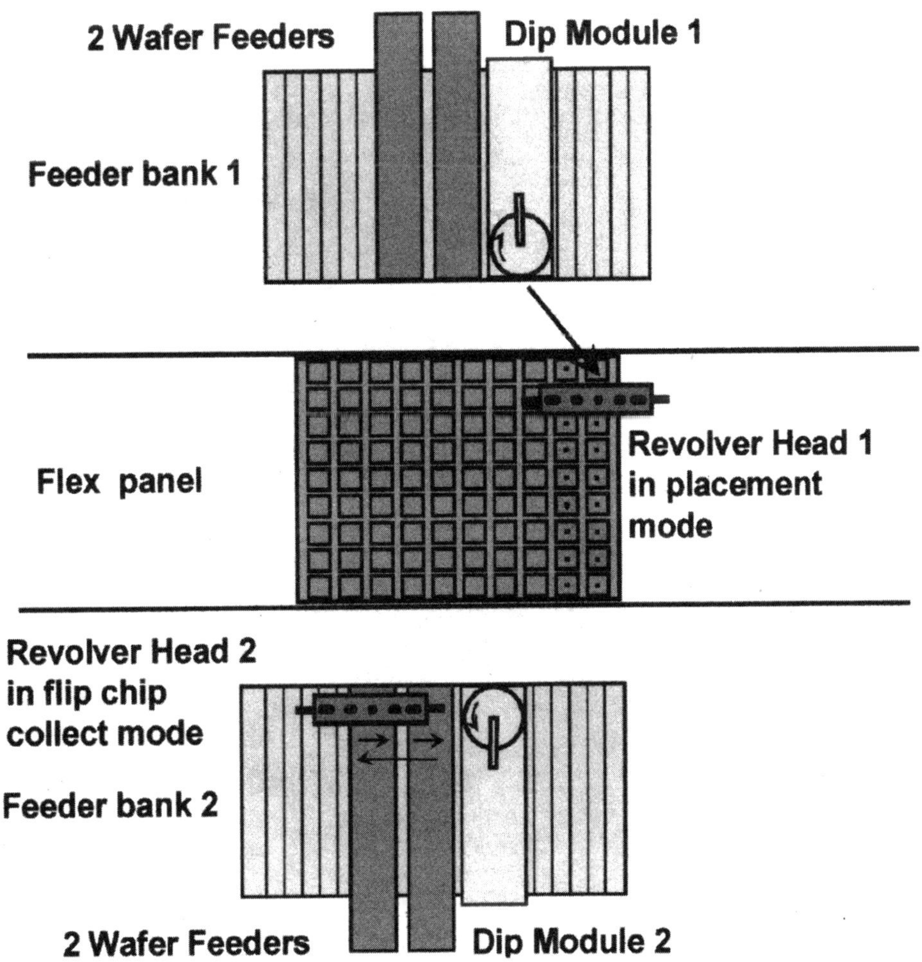

FIGURE 25.29 FC shooter for RFID manufacturing including adhesive dipping.

secondary times and adhesive applications). In the case of a 200-unit panel with smaller substrates, the practical throughput will rise into the range of 7500 units/h.

The approach illustrated in Fig. 25.30 is a solution for highest throughput, lower flexibility, and higher equipment investment. The adhesive application is realized with a screen printer or a pin-transfer machine that is situated in front of the FC shooter.

The practical throughput of the configuration in Fig. 25.30 will be in the range of approximately 11,000 units/h (including all secondary times). To achieve the maximal throughput, the FC chip-shooter will require six wafer feeders in total. In the case of a 200-unit panel with smaller substrates, the practical throughput will rise into the range of 12,500 units/h.

25.11 0201 CHIPSHOOTER PLACEMENT

Today's standard passive component package form is the 0603 (inch basis). This package form is in widespread use, and the related SMT process is well understood. Resistors and capacitors are the majority of components placed on SMT PWBs (50 to 90 percent of all components).

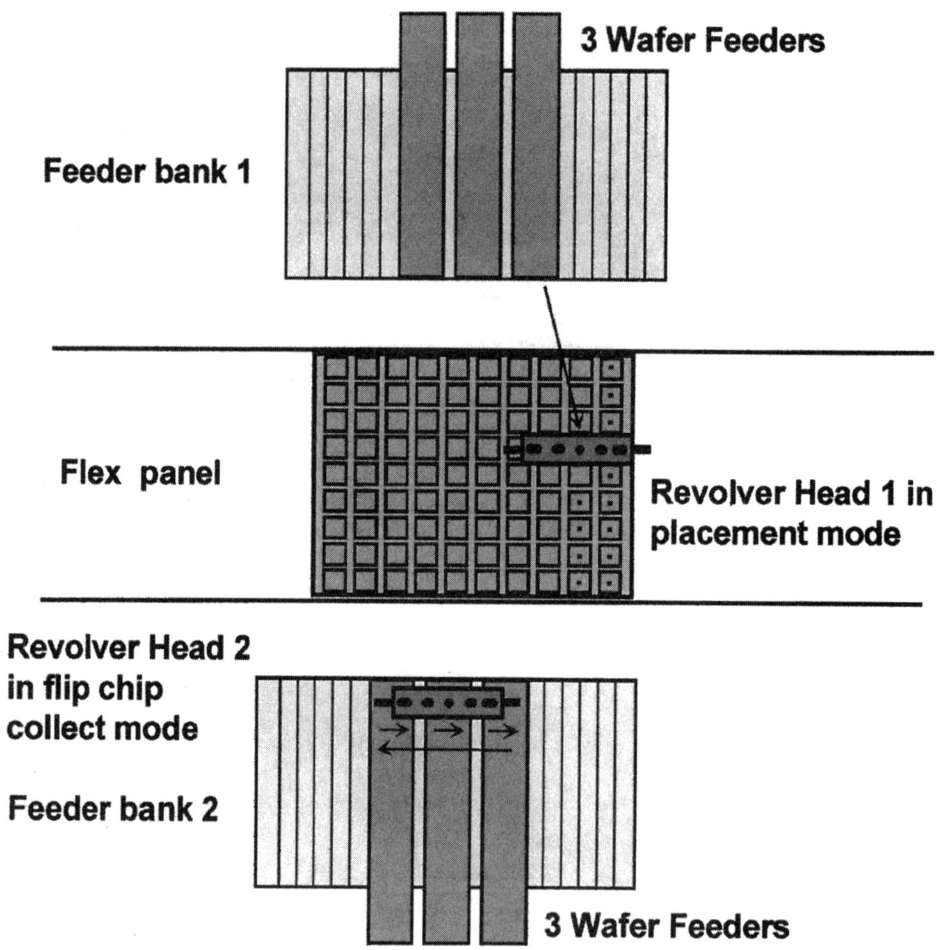

FIGURE 25.30 FC shooter for RFID manufacturing without adhesive dipping.

The market requires smaller and smaller products, especially in areas where weight and size matter. These are products such as handheld digital cameras, mobile phones, palmtops, medical products such as hearing aids, etc. To achieve the required size reduction and performance improvement of the products, it is extremely effective to shrink the passive components (resistors and capacitors).

Publications forecast that within the next 2 years, package form 0402 (1 × 0.5 mm) will be the standard. The next step is to move toward 0201 (0.6 × 0.3 mm) packages (Fig. 25.31).

In fact, 0402 is already a standard in some of the driving industry sectors—especially in the mobile phone sector. This is precisely the area where the pressure to introduce 0201 technology is extremely high (Fig. 25.32). While during the year 2000 only a few Japanese mobile phone

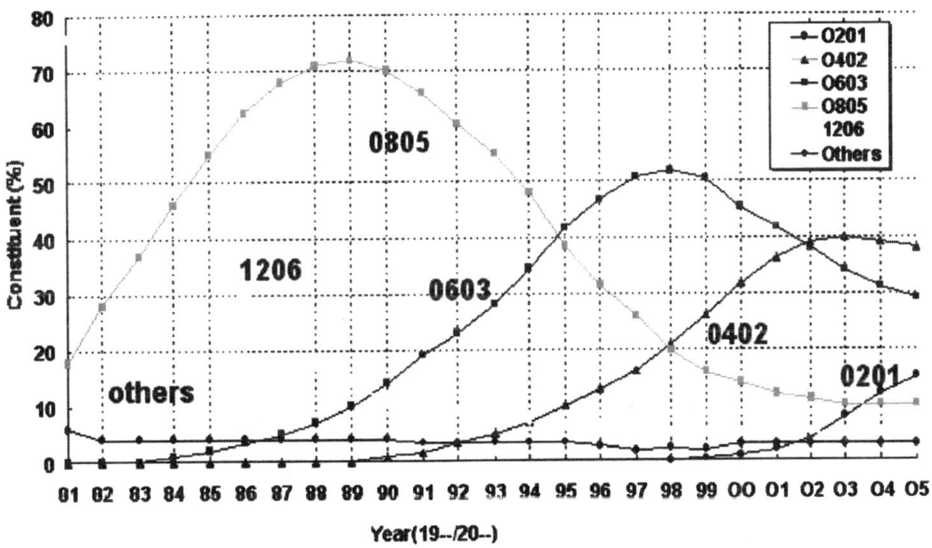

FIGURE 25.31 Passive component package-form trends (dimensions are inch-based).

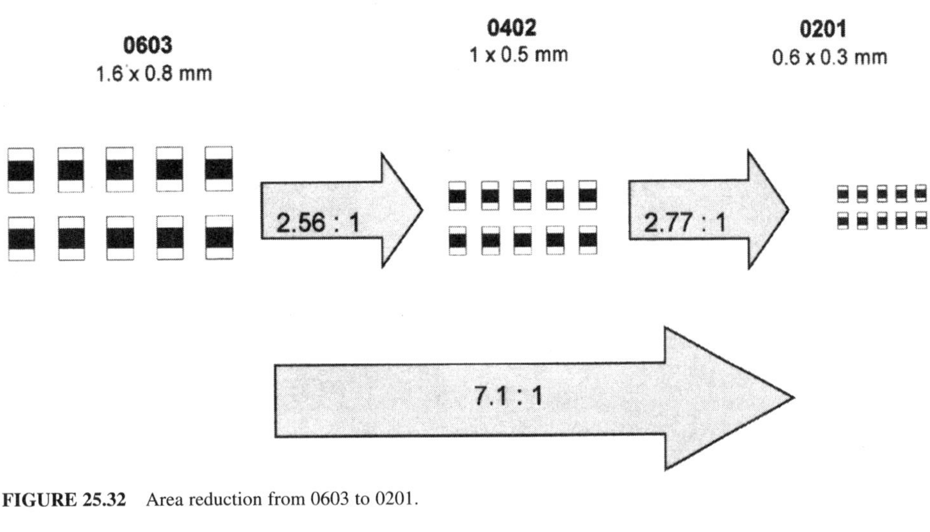

FIGURE 25.32 Area reduction from 0603 to 0201.

manufacturers started to apply 0201 packages in mass production, their Western competitors used this year mainly for process optimization and prototyping.

By year-end 2001, all big players in this market segment will have to use 0201 passives in their high-end products. Within each of the three elementary links of the SMT process chain—solder paste printing, SMD placement, and reflow soldering—the 0201 package causes in part a dramatic increase in challenges.

The most critical areas within the most challenging placement process step are as follows:

Feeding/Pickup. The first step to ensure accurate placement is to pick up the components reliably from the feeder module. This is a complex operation that can greatly affect the quality, robustness, and cost of the 0201 placement process. Components must be fed accurately to the placement system before they can be processed. Standard feeder modules, which work reliably for larger components, often have positioning tolerances on the order of magnitude of ± 300 μm, which already equals to the width of a 0201 component. Thus the first task is to determine the precise position of the component in the tape pocket. This is normally done in one of two ways: either by determining the precise mechanical position of the feeder module over a feeder fiducial mark or by measuring the position of the tape pocket using a PWB vision camera (Fig. 25.33). To achieve very high reliability during the 0201 first pickups, a tape pocket recognition by the PWB camera system is mandatory.

To maintain a constantly high pickup reliability, so-called online learning of the component pickup position is indispensable. Herewith, the pickup positions are continuously optimized in x and y directions when the component vision system recognizes a tendency toward a lateral shift of the component on the nozzle.

Vacuum Check and Optical Centering. The vacuum system has to be sensitive enough to recognize missing components on the nozzle. The vision camera resolution in a chipshooter system must be high enough to allow for a reliable optical centering, including verification of the dimensional tolerance, which is in the range of ± 10 percent.

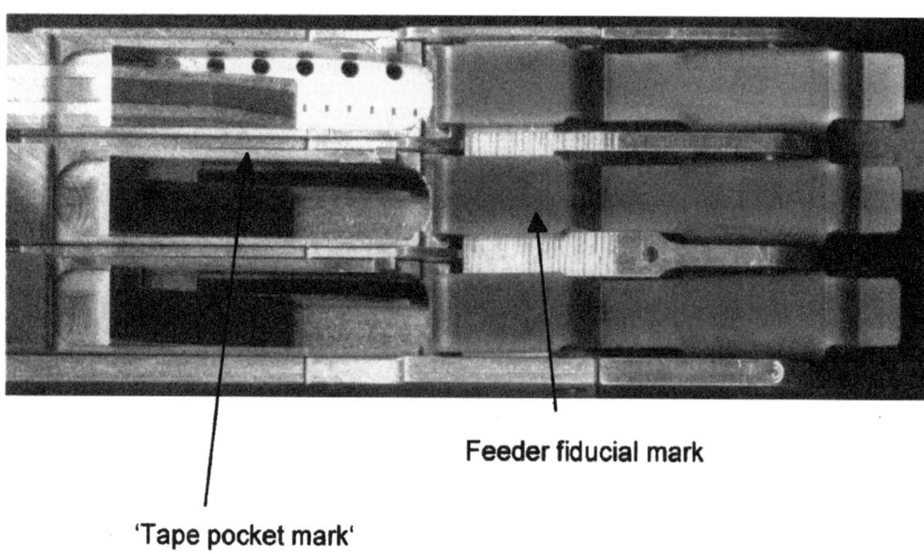

FIGURE 25.33 Feeder reference marks.

Component Sensor. When pickup nozzles reach certain minimum dimensions, the vacuum sensing system may not be reliable enough anymore. Then the use of an additional optical sensor (light curtain) can help to improve pickup and placement reliability.

Smooth Placement. Since the 0201 pads and, as a result of this, the solder paste deposits are extremely small, it is advantageous to place the 0201 packages smoothly (Fig. 25.34). Chipshooters in which the *z*-axis movement is microprocessor-controlled can reduce the touchdown velocity to a defined low value. Thus solder balling can be reduced to a minimum or even eliminated. Of course, the consequence will be a slight throughput derating.

Nozzles. The component should protrude as little as possible beyond the outline of the component. This will prevent contamination and minimize placement shadow. The nozzles should be made from a highly stable material, such as ceramic, in order to guarantee reliable pickup and placement over long periods.

25.12 REEL-TO-REEL VERSUS PANEL TECHNIQUE

In electronics production including flexible substrates, a reel-to-reel conveyor system for SMD placement systems is demanded to an increasing degree. This makes more sense the smaller the width of the substrate. The major difficulty is to adjust processes and coordinate the conveyor movement in the SMD placement machine and reflow/curing oven in such a manner that already placed components are not misplaced before they are bonded to the flex circuit.

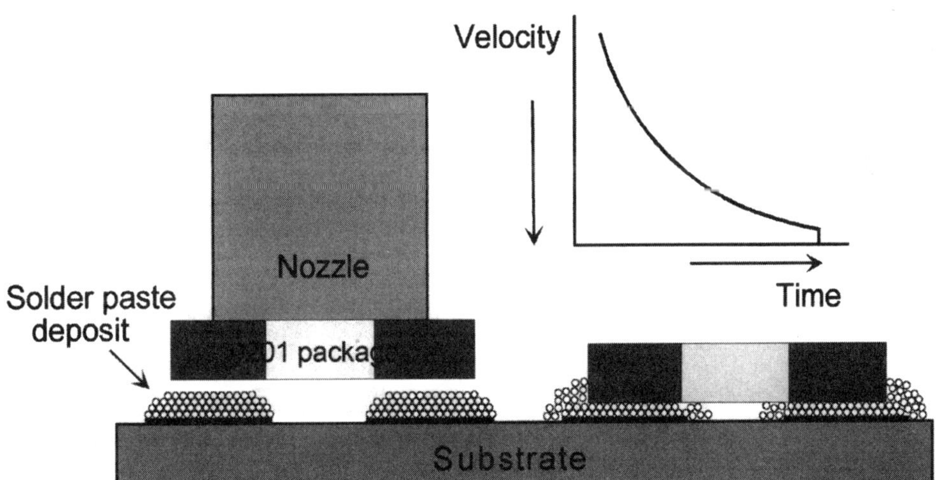

FIGURE 25.34 Soft 0201 placement into solder paste.

CHAPTER 26
OVENS IN ELECTRONICS

Marc Apell
Speedline Electrovert

26.0 INTRODUCTION

Ovens have been used in manufacturing processes for thousands of years, primarily for ceramics and metallurgy, and generally were referred to as *furnaces*. Early practitioners developed furnaces that could melt most metals, including gold, but these were simple systems with almost no control. Today's industrial ovens use high technology for precise monitoring and control. Electronic package manufacturing and component assembly are heavily dependent on ovens that can control both temperature and the atmosphere. Automatic monitoring and control are essential for high-yield manufacturing and efficiency.

Electronics may be viewed as the conversion of all types of materials into useful devices and systems by the application of chemistry, physics, and good engineering principles. Chemical reactions depend on energy to drive and regulate them. Electronics relies on many polymerization reactions to make and harden such things as laminates, solder masks, plating resists, encapsulants, and underfills. Assembly processes use ovens to melt solder and harden adhesives. Area array packaging manufacturing is even more dependent on ovens because solder balls are attached to ball grid arrays (BGAs) and chip scale packages (CSPs) by running the package through a reflow oven to melt the solder and thereby attach the array of balls or microspheres. This chapter briefly discusses industrial ovens used in electronics.

26.1 TYPES OF OVENS

There are many ways of heating up electronic materials and components. Radiation (infrared), convection, and conduction can transfer heat. All three methods are used, and they are often combined. Ovens are designed to provide a specific heating profile where the temperature rises and then drops according to a programmed thermal process. Solder assemblies use a profile because the process is rather complex, with several stages. Polymerization reactions are more tolerant, and constant-temperature ovens can be used. One way of approaching the discussion of ovens is to categorize them by the materials to be processed. The two primary areas of interest for electronics are soldering and encapsulation curing. We will begin with solder reflow, the most critical heating process for area array package production and assembly.

26.1.1 Reflow Ovens

The term *reflow* is derived from the fact that solder material was first formed into a specific shape or coated onto a component or printed wiring board (PWB) by melting in most cases. Solder balls were first formed in a materials plant and then shipped to a packaging foundry. The arrays of solder balls, also called *spheres,* are metallurgically bonded to the bottom of the package by reflowing, or remelting, the solder in a reflow oven. The finished area array package is then attached to the PWB by once again reflowing the solder in a reflow oven designed specifically for component assembly. Several oven designs can be used.

Heated Conveyor (Conduction Oven). Different methods have been developed and tried over the years to reflow the solder and create a metallurgical bond. One of the first reflow methods was direct heating of the base or conveyor, sometimes called a *hot-belt machine.* This method is much like it sounds. Electronic assemblies with surface-mount components are conveyed over a hot surface in order to warm the board and components and melt the solder paste. Although very simple, the process has major limitations. Hot-belt methods transfer heat energy to the electronics package via conductive heat. This method requires the package to contact the energy source during the process. This effectively eliminates the capability of having double-side-populated packages due to components that can be misplaced on the bottom side during the process. Another disadvantage is the temperature differentials of the package. Due to differences in the thermal conductivity of the components, board, mass, and materials do not all see the same temperatures at the same time, otherwise stated as a *temperature delta.* Heating and cooling slopes may be dramatically different, resulting in poor solder attachments.

Vapor Phase. Vapor-phase reflow has been an effective method of reflowing electronic assemblies. A vapor-phase reflow oven uses heated chemical vapors to condense on a package conveying through the oven. The process of the vapor condensing on the package allows the heat energy stored in the vapor to transfer to the product. This process is very controllable and allows for an efficient transfer of heat energy to the product. In the past, military applications specifically called for packages to be produced in vapor-phase reflow ovens due to the controlled process.

Although very precise and controllable, vapor-phase reflow is used rarely in today's electronics assembly. Costs of operating this type of equipment and the chemicals required during operation are very high. The consumable chemicals and subsequent waste removal are expensive and requires heavy maintenance.

Three general types of heat-transfer methods are used commonly in reflow ovens today. They are infrared (IR) only, IR with convection support, and forced convection dominant. These methods will be described in the following sections.

IR Only. In IR-specific systems, lamps or heated elements are used to generate energy in the IR wavelengths. This energy produces heat when absorbed by the product.

Although IR is an efficient means of heat transfer, problems arise when using only IR for reflow soldering. IR heat transfer depends on wavelength energy and product color, and the heat is absorbed selectively by the product. Color determines a material's heat absorption characteristic. Dark objects, especially dull black bodies, strongly absorb heat, whereas light or shiny objects reflect much of this radiant energy. The result is different heating rates for various components on a circuit board that is being reflow soldered.

The problem with this method is that objects with the darkest color on a PWB are not necessarily the objects one wants to heat most. This is especially true of component packages that are typically black with shiny solder joints. As a result, black-bodied components are heated more quickly than solder joints or other components of different color. This effect gives rise to large temperature differentials across the product in IR heating.

Control problems are also inherent in IR-only systems. Emitter temperatures must be set in excess of the desired product temperature, resulting in difficulty determining the required temperature setpoints to achieve the desired profile. Additionally, large temperature fluctuations occur based on product loading.

The combination of large temperature differentials and control problems makes process control in IR-only systems very difficult.

IR with Support. Convection support was added to IR heat transfer to reduce the aforementioned problems. Convection heat transfer is accomplished through molecular motion. In the case of a reflow oven, this means that gas molecules heated by the oven transfer heat when they come in contact with the product. Unlike IR heat transfer, convection heat transfer is not color-selective. This is beneficial in terms of the temperature differential problem encountered with IR-only systems.

Reflow ovens heating through IR with convection support use a small volume of gas to deliver convection heat transfer to the product. However, because the total impinged gas volume is much too low to convey the required heat-transfer, IR heating is the dominant heat-transfer mechanism in this system.

Because IR is the dominant source of heat transfer, the problems mentioned earlier with IR, while reduced, are still present. The problem of large temperature differentials is reduced but not eliminated. Although process consistency is improved by more stable temperature control, it is still not as stable as required.

Convection-Dominant Systems. Because of the preceding problems with IR and IR with convection support heat transfer, convection-dominant systems are the most popular choice in reflow ovens today. Convection-dominant systems use forced convective heat transfer with a very minor amount of IR heat transfer to heat the product. These systems move a much larger gas volume than IR with convection support systems. Additionally, the heaters employed are designed to heat gas rather than act solely as IR emitters. Forced-convection heat transfer is not color-sensitive, as is IR. This means that the large product temperature differentials seen with IR do not occur in convection-dominant systems.

Control of convection dominant systems is much more precise. Gas stream temperatures impinging on the product are close to the desired product temperature. Control of the heater system is achieved readily by monitoring this gas stream temperature (typically with a thermocouple). Process consistency is much greater as a result. Increased process consistency means fewer defects as the result of an "in control" process.

26.1.2 Nitrogen versus Air Atmosphere

Most applications of reflow use the air environment to process product. This method uses air from within the factory and allows it to be heated for transfer to the product. Using the air environment in the process of reflow allows less maintenance on the machine and lower operating costs. Greater exhaust is allowed in the process that effectively draws out the flux volatiles created within the oven during the reflow process. The exhaust keeps the flux volatiles from collecting within the reflow oven and thus greatly reduces the requirements for maintenance on the oven.

The use of nitrogen, a gas that reduces oxygen, within the reflow oven has some distinct advantages and disadvantages. Oxidation that occurs naturally on electronic assemblies and components can make soldering very difficult. This fine film of oxidation effectively can "block" the formation of solder joints. Elevated temperatures during the reflow process act as a catalyst for oxidation in a regular air process. This usually does not affect the typical process and products if components and boards are stored and handled properly. The flux in the paste normally is enough to break down the oxidation that has formed. The introduction of nitrogen in the reflow process reduces the oxygen levels in the process and decreases the chance of oxidation. Nitrogen is very beneficial for some coatings used to protect the board. Many of these assemblies can become discolored or damaged in multiple air reflow processes. Nitrogen acts as a protection and does not allow oxygen to break down the surface finish. Nitrogen is also very beneficial for fine-pitch components. The introduction of nitrogen in a reflow process reduces surface tension on the solder during reflow. This allows better wetting properties and is more effective with finer-pitch products.

26.1.3 Newer Oven Technologies

Most soldering methods require heating the entire assembly, as was noted in the preceding discussion. However, many applications would be better served by localized heating. Although localized heating equipment has long been available for repair and rework, this has not been the general case for initial soldering. Recently, laser soldering has gained attention, and customers have asked the oven industry to provide such equipment. A new laser soldering oven is currently being introduced and is shown in Fig. 26.1.

26.2 GOALS AND ELEMENTS OF A PROFILE

26.2.1 Chemical Reactions during Heating

Flux Reactions. Numerous chemical reactions occur during assembly. Flux is activated so that it can reduce any metal oxides found on the PWB and components. The oxide is reduced back to the original metallic state that is required for soldering. Water can be a by-product of the flux reduction reaction but is eliminated quickly as a vapor effluent. Some fluxes, especially no-clean types, go through a deactivation stage after solder joints have formed.

No-clean epoxy flux, one of the newest classes, is a good example. This class of flux contains liquid epoxy resins and hardener. The hardener is also a fluxing agent such as an organic acid. During the early heating stage, the acid reduces the metal oxides and serves only as a flux. The heating process continues, typically by moving the assembly through different oven zones. A solder reflow temperature is reached, and solder joints form. The organic acid that is no longer required as a flux begins to polymerize by reacting with the epoxy resins. The acid is totally consumed by the polymerization reaction, leaving a neutral, noncorrosive residue.

Alloys and Liquidus Points. Although eutectic tin-lead is the present standard of electronics, many other alloys are used. The increasing interest in lead-free alloys is also having an impact (see Chap. 15). This means that a modern oven must be capable of accommodating the heating profiles of many dif-

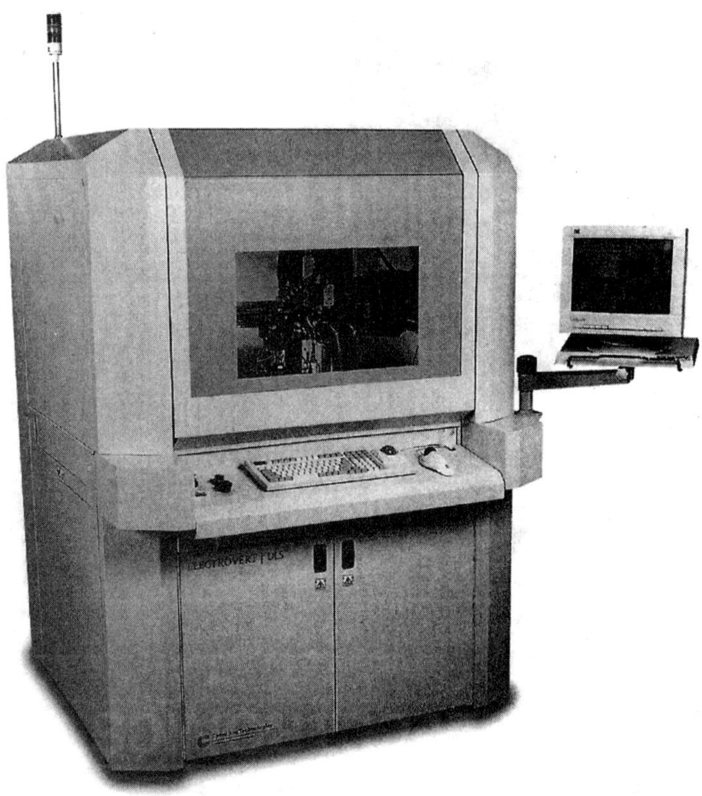

FIGURE 26.1 Laser soldering machine. (*Speedline/Electrovert.*)

ferent alloys. This is accomplished by using multiple heating zones with computer control to allow heating-profile programs to be generated and stored.

26.2.2 Polymerization

Polymerization reactions are used in electronics to convert liquids to solids and to modify material properties. Many area array packages use encapsulants that must be polymerized to nonmelting plastics. Liquid encapsulants, including flip chip (FC) underfills, are conveniently applied to packages as free-flowing liquids made up of monomers and fillers. Once in place, the liquids are polymerized into hard plastics by thermally induced polymerization. Encapsulants typically will polymerize at temperatures ranging from 130 to about 165°C. The polymerization rate increases with temperature, but degradation can occur if the rate is too high. The curing, or polymerization, oven therefore must have good control, although the heating profile may not be as demanding as required for soldering.

26.2.3 Impact of Time and Temperature on Chemistry

Chemical reactions are both time- and temperature-dependent. The general rule for many reactions is a doubling of reaction rate for every 10°C increase in temperature. Reactions follow precise mathematical relationships that are well understood by chemists. Analytical equipment is used routinely in laboratories to follow reactions and determine time-temperature relationships and their effects on properties. Once the optimal conditions are determined in the laboratory, they can be duplicated in production ovens, provided that there is adequate control. Better oven control translates to adequate process control and optimal final properties. This is especially true in the manufacture of plastic BGAs. Several other chapters deal with encapsulation methods and chemistry.

26.2.4 Joint Formation

Solder joint formation is highly dependent on temperature control once the correct materials have been identified. The heating profile must be correct to allow the flux to activate and the solder to flow. Every stage in the profile, from heating to cool down, will affect the solder joint. This area is covered in more detail in Chap. 14.

Mechanical Bond. Various bond interfaces are formed during component manufacturing and assembly. They can be divided into two classes: polymer bonds and metallurgic joints. Both are highly dependent on temperature control and heating profiles. Polymerization of underfills and encapsulants requires good temperature control. The bond strength of underfills can change dramatically over a relatively small temperature range. Polymer bond strength often increases with elevating temperatures, and it can be advantageous to push the temperature to a maximum limit. However, since degradation will occur above a specific temperature, good oven control is essential for optimal results. See Chap. 11 for more on the chemistry of polymers.

Likewise, temperature and profile, often in a more complex fashion than polymers, influence solder bond strength. Flux must activate properly, solder paste must coalesce and fuse, flow must occur, and flux, in many cases, must deactivate without interfering with solder-joint formation. All factors require good oven control.

Wetting. Solder wetting is influenced by temperature. Too low of a temperature can retard wetting, resulting in poor joint formation. Too high of a temperature can cause excessive flow. This topic is covered in Chap. 14.

Oven Influence on Appearance. The oven influences all solder joint, component, and PWB appearances. Inadequate control of temperature and the oven atmosphere can strongly affect the appearance of solder joints and coatings. This goes beyond cosmetics, since joint appearance has long been a criterion of reliability. However, good control also can improve solder appearance so that the finished product has a better perception by the customer.

Components and PWBs can change color if exposed to excessive heating. Even if the color-degraded assembly is fully functional, a poor appearance typically will convey the notion of an inferior product. One should not underestimate the influence of perception.

26.3 SETTING UP A PROFILE

During the reflow process, heat is applied within the oven to bring the assembly to the proper temperature to solder, but without damaging the product. In order to verify the process of reflow soldering, a profile is used to determine process setup. A *profile* is a visual data log of the time and temperature of each sensor as it is run through the heating process. By reviewing this profile, one is able to visually see exactly what energy is being applied to the product.

A typical reflow profile is made up of several different sections. This includes the initial ramp, soak, spike to reflow, reflow, and cool-down of the product. As a general rule of thumb, it is desired to have a ramp-up and cool-down in temperature for the product no greater than 3°C/s to prevent damage to the board and/or components.

During the heating of the product within the oven, many things occur that will affect the outcome of the product. The initial ramp is a quick rise in temperature as the product enters the oven to bring the paste to the desired soak temperatures required to begin activation of the paste and evaporation of the solvents within the paste material. This soak temperature is most desired just under the melting point of the paste material—183°C for eutectic solders. The soak zone has two purposes. The first is to bring the board, components, and material to an even temperature close to the melting point of the solder paste. This will allow an easier transition into the reflow zone so that the assembly will see the same temperature at the same time. Another purpose is for the flux to become active on the assembly. The flux activates at this temperature and begins the process of cleaning the pads through oxidation in order to have clean surfaces for the solder to adhere to. Spike to reflow is another transition point of temperature where the assembly temperature is raised above the melting point of the solder and a liquidus state of the paste occurs. Once the solder paste is above the liquidus state, the components enter the reflow zone, commonly referred to as the *time above liquidus* (TAL). This is a critical state within the oven because the temperature gradients on the assembly must be minimized and the TAL must be kept within the parameters specified by the paste manufacturer. The peak temperature of the product also is reached during this period. This is the point at which the assembly reaches the maximum temperature within the oven. It is desired to have all points on the assembly reach the same peak temperature at the same time and rate to ensure that all parts are subjected to the same environment within the oven. After the reflow zone, the product is quickly cooled to solidify the solder joints and cool the assembly. This is also a critical state that requires control of the cooling slope. Cooling too fast may damage the assembly, whereas cooling too slow will increase the TAL and can cause brittle solder joints.

Two common types of profiles are used when reflow soldering assemblies (Fig. 26.2). They are commonly referred to as the *soak* and *tent profiles.* The soak profile was described earlier. There is a period within the oven in which the assembly is subjected to the same temperature for a period of time in order to soak the assembly to the same temperature. The tent profile is a ramp of temperature from the time the assembly enters the oven until the assembly reaches the desired peak temperature, and then the cool-down occurs.

Different profiles are based on the type of solder paste used within the assembly. Depending on the chemical makeup of the solder pastes, the manufacturer will suggest the best profile to achieve maximum performance. This information may be found by contacting the paste manufacturer. The most common solder paste formulations include water-soluble (OA), RMA, and no-clean solder pastes.

26.4 PROCESS IMPROVEMENT AND TROUBLESHOOTING

26.4.1 Developing a Profile

The development of a profile is not very difficult, but it does take an understanding of the reflow process and the steps to control it. The first step in developing a profile for a particular product is to review the paste specification. This is provided by the paste manufacturer. The paste specification will list the heating and cooling requirements and the desired profile to correctly form the electromechanical bond desired in the reflow process.

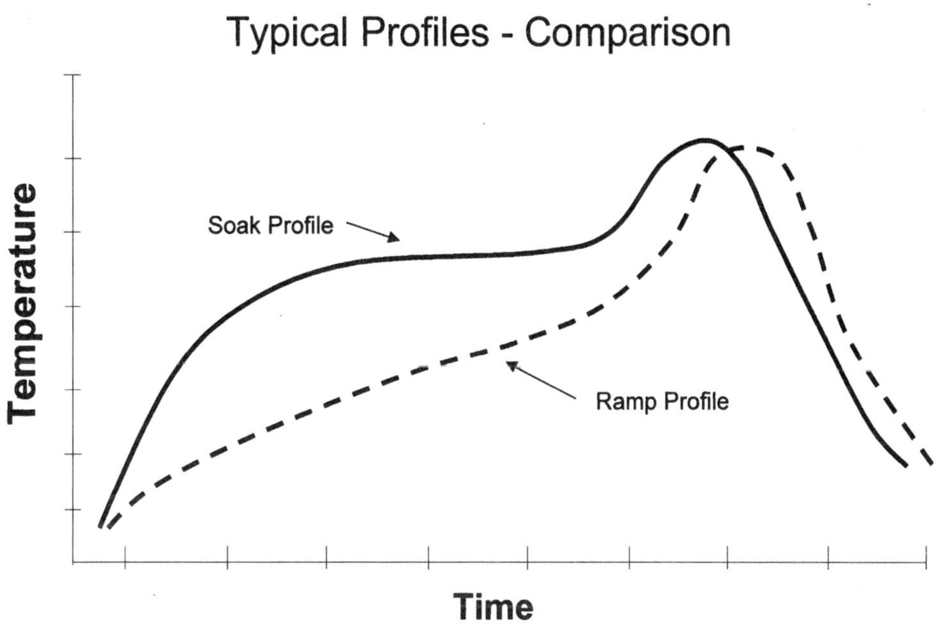

FIGURE 26.2 Comparison of typical profiles. (*Speedline/Electrovert.*)

Once the paste specification is reviewed, the product being soldered also should be reviewed. Special attention should be paid to the size, weight, and mass of the product. It is strongly suggested that thermocouples be permanently attached to the product so that a profiling device can capture the results during the reflow process. When determining the location to attach the thermocouples, it is best to select areas with a variety of masses. Selecting large-, medium-, and small-mass components will best represent the heating within the product.

Once the paste specification is determined and the product mass is reviewed, one would begin by estimating the temperatures required to meet the paste specification. At this point, it is the process of loading the temperatures into the oven and running profiles. After each run, the profile can be reviewed to the desired specification, and then the settings of the oven can be altered to get to the desired profile. The process of setting up the desired profile can be somewhat time-consuming, and because of this, new software programs in conjunction with data-logging devices are available today to assist with the development of profiles with less effort and time. These devices use prediction programs in the software to determine the product temperatures at different oven temperatures.

Working with the oven supplier and a data-logging supplier will assist in the development of a robust product profile.

26.4.2 Factory Profile Management

To ensure product quality and throughput, it is important to verify that the reflow oven is performing consistently. This general process is commonly referred to as *statistical process control* (SPC) and can be applied very easily to the reflow process. This is the common practice of running profiles for oven performance and ensuring that the results fall within the set control limits. This process also will alert the operator if the process falls outside the process window. The most common tracked results are soak time, peak temperature, and TAL (time above liquidus).

The first step is creating a profiling device. This can be a product or a device specifically created for profile management. The device should be reviewed periodically for quality because repeated processing can degrade the performance of the measuring device. Once the device is run through the oven and the profile is determined, the run should be repeated periodically to ensure that the oven is performing the same from day to day. Many users run a profile at the start of each shift to ensure that the oven is within the process requirements. If it is found that the process parameters are trending outside the requirements, a problem may be occurring within the oven or profiling device. Before looking into the oven, the profiling device should be checked to ensure that false readings are not occurring. If there are no problems with the profiling device, the next step is to investigate the oven for any possible issues.

A well-developed SPC program will highlight any process problems immediately and before large numbers of product are subjected to a poor process. Readers are encouraged to further investigate SPC programs in conjunction with their companies' quality department.

26.4.3 Oven Maintenance and Heating Verification

Through a developed SPC program, as referenced earlier, users of ovens can be aware of potential problems prior to defective product. As effective as SPC programs are for detecting problems in a process, the best measure is a preventive maintenance program to ensure that the machine is operating as required. Each oven manufacturer should supply the user with a periodic maintenance program to ensure that the machine operates effectively. Following the oven maintenance routine will allow any potential issues with the machine to be found prior to their affecting machine performance. Another inspection of the machine is the process verification, described earlier. Setting up an SPC program that verifies the process of the machine will assist in uncovering any machine performance issues prior to affecting product.

26.4.4 Bridging and Opens

After the reflow process, many times this is the first product inspection or test during the assembly process. During the inspection and testing, it may be found that there are defects such as bridging and open joints. Many people quickly blame the oven. However, although the oven may have an affect, it is best to begin at the start of the process with components and materials.

Bridging is the result of leads or solder joints soldering together. At times, this is the result of too much solder paste or improper placement of the component. It is relatively rare to have the reflow oven create bridging unless unwanted component movement occurs within the oven. This may be the result of too much turbulence within the oven.

Open solder joints are commonly traced back to the materials used or the solder paste printing process. In the case of component contamination, the material may not effectively "wet" to the surface of the board or component. It may be as simple as a poor stencil design, a worn or clogged stencil, or a spoiled paste material that has dried out.

When defects occur within the assembly process, it important to begin to review process at the beginning of the line and eliminate each potential cause.

26.5 CURING OVENS

Curing ovens also play an important role in the production of electronic packages. These ovens come in a variety of sizes and offer capabilities tailored to meet the process requirements of the material. These systems can be batch or in-line and should be selected based on the process they will perform. Curing ovens can perform such applications as underfill, prebake, adhesive curing, and encapsulation.

As with reflow ovens, curing ovens are equipment that combines time and temperature to a product to fulfill a process need. The first step in selecting a cure oven is determining the product to be cured. The materials manufacturer supplies this information and lays out the profile required to properly process the material. These can vary from a 3- to 5-min "snap" cure material to encapsulations that take 3 to 4 h to allow any air pockets to gas out prior to hardening.

As with reflow ovens, a process verification SPC program can be set up to ensure proper machine performance. By data logging the process and reviewing and charting the results, machine performance can be determined and corrected prior to affecting end assemblies.

26.6 SUMMARY AND CONCLUSIONS

It is clear that reflow and curing ovens play a vital role in the assembly of electronic packages. The proper use and application of an oven allow for a controllable process and consistent quality of products. When first selecting an oven, it is best to determine and review the process being applied. Ensure that the machine is capable and built with components and features that will allow a long, trouble-free life. Selecting the oven is not the only step. Careful thought must be given to the product and materials being processed within the oven. The oven is only a piece of the puzzle that is successful reflow processing. Figure 26.3 shows a modern reflow oven.

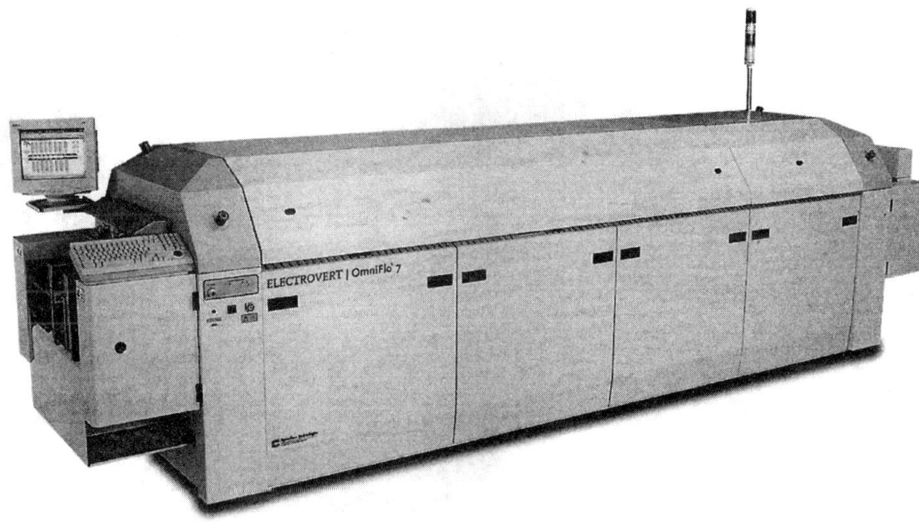

FIGURE 26.3 Convection reflow oven. (*Speedline/Electrovert.*)

CHAPTER 27
PROCESS DEVELOPMENT, CONTROL, AND OPTIMIZATION

A. James McLenaghan
CPS Cookson

27.0 INTRODUCTION

Historically, process improvements have focused on measuring and improving process quality by reacting to defects at the product level. By contrast, today's quality techniques have shifted to the use of problem solving and statistical thinking to become proactive in preventing defects.[1] This way of thinking establishes strategies to achieve defect reduction and improve process stability and leads to the implementation of corrective actions to eliminate the root cause of any problem. These techniques highlight the importance of expanding the historical definition of continuous improvement to include all the major functions of the manufacturing process.

Continuous improvement and a method of quantifying it, i.e., the ability to measure a definite improvement in quality, are very important to customer satisfaction. Documentation of the corrective actions taken and improvements made can reassure customers that steps are being taken to correct problems and maintain control of the process. Such documentation allows the customer to feel that he or she is receiving a product with a high degree of quality that will continue to improve.

Focusing on improving process quality through the use of problem solving and statistical thinking will result in reduced rework/scrap costs, increased productivity, increased market share, open lines of communication, and a higher degree of customer satisfaction. It is essential to remember that this program supports the end customer as well as the internal needs of the corporation in achieving its quality and productivity goals.

27.1 COMPONENTS OF PROBLEM SOLVING WITH STATISTICAL THINKING

Thinking statistically uses a broad range of quality tools and techniques that

- Support a preventive and continuous improvement program
- Establish process-control strategies
- Improve process stability
- Prevent defects from occurring

The major elements of the program are

1. Problem recognition
2. Project selection
3. Utilization of the problem-solving techniques and statistical methods displayed in Table 27.1

Recognizing a problem is the first element of the program. Within this element we need to first understand the process by developing a macro map of the major steps of the process flow of an operation (such as a manufacturing line). For each major process step, develop a micro map that breaks down the process into several consecutive substeps (identifying all input parameters, process parameters, and output parameters).

The term *process* means different things to different people. Thus it is important to clearly define what we mean by process. A process can be defined as the logical organization of people, materials, environment, equipment, and procedures into work activities designed to produce a specified end result (Fig. 27.1).

TABLE 27.1 Problem Solving with Statistical Thinking

Step	Process	What to do?
1	Understand the process and baseline	Understand the process: Process mapping: define input parameters, process parameters and output parameters Recognizing a problem In-process defect data Customer return data Pareto charts of defects by process step Control charts to understand process variation Customer Satisfaction Surveys Baseline the process defect rates Defect rate by process step Customer return rate In-process and customer return scrap/rework costs
2	Project selection	Project selection Brainstorming Prioritization matrix
3	Problem solving and statistical methods	Utilize problem-solving techniques Problem-solving techniques such as Ford 8D Process Cause and effect diagrams to identify potential causes Process FMEAs to identify potential causes Using data to define and describe the problem What is the problem? What is the impact on the process? Is it continuous or random? Is it a new problem or recurring one? Describe in quantifiable terms, who, what, when, where, why, and how many? Identify root causes, potential solutions, and implement corrective action Test potential solution(s) and verification through experimentation and statistical analysis How to choose the "right statistical tools" Repeatability and reproducibility study SPC and capability study Comparative methods Design of experiment Verify corrective action: Analyze data and compare to goal (meets goal?) Document solution and quantify improvements in quality levels and costs savings: Institutionalize corrective action Monitor process performance Select next problem

Input parameters are those parameters or elements which are introduced into a process. Output parameter(s) are the resulting output as product or services created by the activities or steps of the process.[2]

Once we understand the process flow, we need to quantify and baseline the process defect rate for each process step, the customer return rate, and the scrap/rework costs. This can be achieved by identifying the parameters that will be the leading indicators to the quality level being produced by the process and that have the most significant impact on the next process step, client or customer. Each defect will be attributed to a process step, and then Pareto charts are developed to identify the problem that offers the greatest potential for improvement. Other techniques that can be used are control charts to understand the variation within each process step by monitoring the output parameter or customer-satisfaction surveys to capture and integrate the voice of the customer.

27.2 SELECTING PROCESS PERFORMANCE MEASURES

It is a fact that process performance measures are useful in managing process activities. If one is to affect change, one has to measure process progress. The present status must be known and compared with the goal.

Measuring process performance is important because

1. If you can measure it, you can understand it.
2. If you can understand it, you can control it.
3. If you can control it, you can improve it.
4. Processes are critical to executing strategies and plans aimed at organizations goals.
5. Processes must be controlled, predictable, and improved to achieve organizational goals.

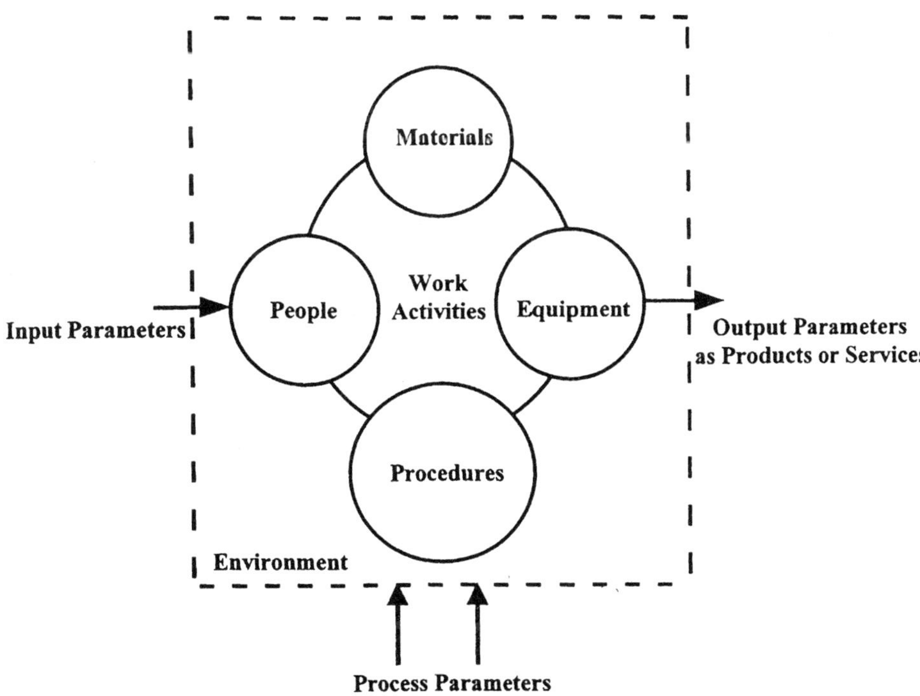

FIGURE 27.1 Elements of a process.

Process performance measures must be selected that will adequately describe this status and which can be used to express progress in moving from the present status to the desired goal. Process performance measures help to develop realistic plans, communicate effectively, identify and correct problems, track objectives, and justify decisions. Selecting the correct process performance measures must begin with a need to know something about a process. Common questions that help in selecting the process measures and how to obtain the data are as follows:

1. What do we need to know?
2. What format do we need the data in?
3. What information is needed to generate the data we need?
4. What types of analytical tools are required to generate the information we need?
5. Where can we get this information?

Simply being measurable is not always sufficient. Criteria such as percent defective or yield are usually easy to quantify and measure. On the other hand, measurement of defects per unit or parts per million (ppm) is considered more complex and may involve several departments of a manufacturing organization to implement this reporting system.

Some typical examples of performance measurements for manufacturing operations are listed in Table 27.2. The following provides only as a starting point to identify potential performance measurements that can used in manufacturing operations:

Productivity = percentage indicator that measures the output (earned hours) versus input (available hours)

$$= \frac{\text{earned hours}}{\text{available hours}} \times 100$$

Utilization = percentage indicator measuring the amount of time an operator was used to produce the product versus the total hours worked

$$= \frac{\text{actual hours}}{\text{available hours}} \times 100$$

Efficiency = percentage indicator measuring an operator's performance on an operation

TABLE 27.2 Performance Measurements

Schedule	Productivity	Costs	Quality
Yield	Percent efficiency	Cost savings and reduction	Percent defective
Variance	Percent utilization	Profit improvement	Rework
Pieces per hour	Percent	Costs avoidance	Scrap
Units per day	Productivity	Inventory	DPU
Shortages		Rework (material and labor)	EPU
Backlog		Scrap	PPM
Lots processed		Labor efficiency	Inspector and appraiser effectiveness
Lots defective		Materials	Costs of quality
Forecast			Reduction in SPC out of Control Conditions
Build plan			Reduction in variation (Cp/Cpk)
Percent lots accepted			
Errors per opportunity			

$$= \frac{\text{earned hours}}{\text{actual hours}} \times 100$$

Defects per unit (dpu) = the total number of all defects within the process, divided by the number of units produced by the process during the same time period

$$= \frac{\text{total defects}}{\text{total number of units}}$$

Defect rate (ppm) = a measurement that enables us to normalize the defect rate of product of different complexity; the total number of defects divided by the total number of opportunities for error times 1,000,000

$$= \frac{\text{total defects}}{\text{total opportunities}} \times 1{,}000{,}000$$

First pass yield (FPY) = the number of units that pass the test without failure

$$= \frac{\text{number of units passed first time}}{\text{number of units tested first time}}$$

27.3 QUALITY PROCESS DEFINITIONS

The following are typical quality process definitions:[3]

Critical measurement system: Defined as any measurement system that is used on a daily basis to determine if the product/technology/service meets the customer requirements as defined in the respective test specifications.

Process: A process consists of a machine, an operation, or a combination of machines and operations that produces an output impacting product quality. A key/critical process is basically a valued-added process in the macro map that inhibits achieving the defined quality requirements This does not include processes that verify the actions of previous processes. A process may include one specific type of product, families of products, or all products produced by a machine or during an operation. A process has one or more defined inputs and consists of a series of operations and/or machinery that produce one or more outputs.

Process output: The end result of a process. It may include only one specific output on a single product, or it may include all products and all outputs produced by the machine(s) and/or operation(s) included in the process.

Input variables: A measurement system or a parameter in a process that is used in the creation of process output. Input variables include materials, methods, measurements, machines, people, and environment.

Macro map: The process of charting the coarse flow of an operation (such as a manufacturing line).

Micro map: Breaks down a process step in the macro map into several consecutive substeps (identifying all inputs and outputs).

Variation control: A technique to minimize and control process variation, where continuous data are used on a real-time basis.

Critical output: Any process output impacting the quality of the next process or of the final product.

Stabilized process: A process meeting the stability requirement.

Instability index (St): A metric that provides a means of estimating whether assignable causes are present in the process. The instability index does this by measuring the fraction of points that are not within control for a given process.

Optimized process requirements: An optimized process produces products at very high quality levels. To be considered optimized, a process must meet stability requirements on all controlling critical inputs and monitored critical outputs. Cp and Cpk values must be 2.0 or greater for all critical variable outputs. Critical attribute outputs must be 3.4 ppm or less. Automatic feedback loops and specialized control charting methods are acceptable for use with an optimized process.

Measurement systems analysis: A measurement system analysis (MSA) is typically called a gauge repeatability and reproducibility (GR&R) study. A gauge R&R study is a mathematical procedure to quantify the amount of variability that is introduced to a process by the act of measuring a particular characteristic. MSA allows us to account for the variability introduced by the measurement system. This variability is known as measurement error.

Repeatability: Gauge repeatability is the variation in measurements obtained with one gauge when used several times by one operator while measuring the identical characteristic on the same part.

Reproducibility: Gauge reproducibility is the variation in the average of the measurements made by different operators using the same gauge when measuring identical characteristic on the same part.

Bias: The difference between the observed averages of measurements and the reference value.

Stability: The total variation in the measurements obtained with a gauge when measuring an identical characteristic on the same part over an extended time period.

Precision-tolerance ratio: The P/T ratio is the ratio between the measurement precision estimate and the difference between the lower specification limit and the upper specification limit of the characteristic being measured.

27.4 PROJECT SELECTION

When most organizations start the quality improvement movement, all managers wrestle with the natural urge to start several projects right away without understanding the impact on quality of each project. An organization that takes the time to set the stage properly will reap the benefits of improved process quality, increased productivity, higher employee morale, and better customer satisfaction. The first step is to select which projects to work on by analyzing the defect rate by process step to determine where to focus engineering resources to obtain the maximum return on investment. When initiating quality improvement programs, it is important to work on the problems that have the most impact, yielding "a success story" in defect reduction, rework/scrap reduction, and increased productivity. Techniques that can be used are brainstorming sessions with the managers and process engineers. When conducting a brainstorming session, the following guidelines are useful to determine which project to work on to effectively use your engineering resources:

1. Identify customer issues and process defect rates, and develop defect Pareto charts by process step.
2. Identify problems that have been around for a long time and are persistent.
3. Identify problems that significantly impact quality, cost, and/or cycle time.
4. Identify problems that are causing increased factory costs due to rework/reprocessing of defective products or services.
5. State the problem, quantify the statement with data, and determine the nature of the project by forecasting outcomes, expectations, and the impact on your customers.

A prioritization matrix is used to set priorities on a list of choices by discussing and weighting each problem. This method is used frequently after a brainstorming session to help the team to focus on the most important project.

The most common errors in selecting a project are:

- Selecting a process that has a small impact on quality, productivity, and costs.
- Selecting a desired solution instead of a process.
- Selecting a process in transition.
- Selecting a system to study, not a process to improve.

Once the projects have been identified and prioritized, identifying whom to involve in each project is just as important as solving the problem. The problem solving team plays an important role in finding the root cause of the problem. When selecting members for the team, it is important that team members have knowledge of the product and process involved with the problem. The team should be cross-functional, with representatives from different departments who can provide knowledge to help solve the problem. The team members also should have been trained in problem-solving techniques such as the Ford 8D problem-solving process shown in Table 27.3.

Using the Ford 8D problem-solving techniques, the team describes the problem in quantifiable terms by analyzing existing data and implementing a containment plan to isolate the problem from the customer. Cause-and-effect diagrams or failure mode and effect analysis (FMEA) can be used to identify potential causes of the problem. The team should isolate and verify the root cause by testing each potential cause against the problem description. This process will result in identifying the potential causes of the problem, which should be verified through experimentation and statistical analysis. It is important to select the correct statistical method(s) to verify the potential solution that will eliminate the root cause of the problem.

Figure 27.2 displays the types of statistical tools normally used to evaluate measurement systems, monitor process variation, determine whether samples are drawn from the same population, and identify the significant process parameters and optimization.

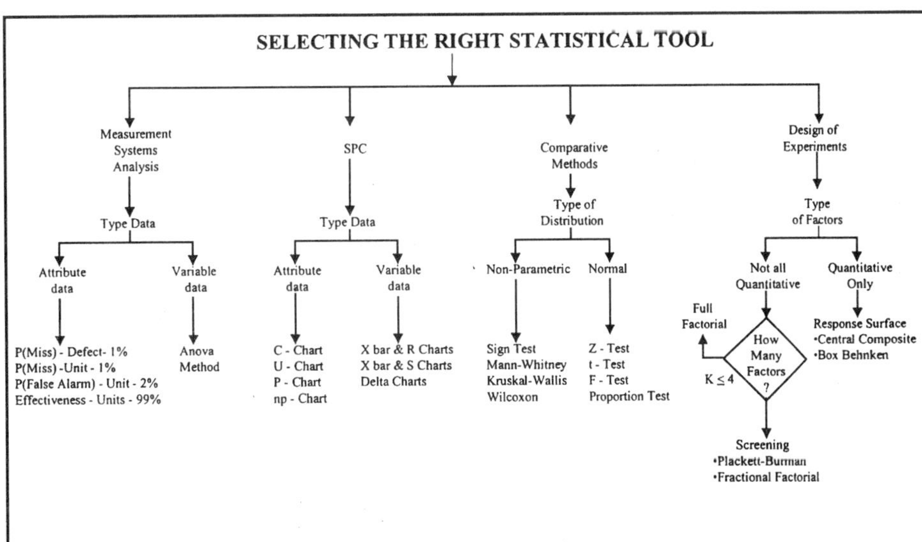

FIGURE 27.2 Selecting the right statistical tool.

TABLE 27.3 Ford 8D Team Problem-Solving Techniques

Step	Process	What to do?
1	Use teaming techniques.	Establish a team of individuals with the process and product knowledge, allocated time, authority, and skill in the discipline to solve problems and implement corrective actions. The team must have a designated leader.
2	Describe the problem.	Specify the problem by identifying in quantifiable terms who, what, when, where, why, and how many defects by analyzing existing data.
3	Implement and verify interim (containment) actions.	Define and implement containment actions to isolate the problem from the customer until permanent corrective action is implemented. Then verify the effectiveness of the corrective action.
4	Define and verify root cause.	Identify all potential causes that could explain why the problem occurred. Isolate and verify the root cause by testing each potential cause against the problem description and test data. Test each potential cause through experimentation and statistical analysis.
5	Verify corrective action.	Through experimentation and statistical analysis, verify corrective action that will not cause undesirable side effects. Evaluate the degree of problem reduction or elimination using preproduction tests.
6	Implement permanent corrective action.	Define and implement the best permanent corrective action. Choose ongoing controls to ensure that the root cause is eliminated. Monitor long-term effects and implement contingency actions.
7	Prevent recurrence.	Revise the process systems, operating procedures, and practices to prevent recurrence of the problem.
8	Congratulate the team.	Recognize the collective efforts of the team members.

27.5 PROCESS CONTROL AND ANALYSIS METHODS

Since the early 1980s, the use of statistical methods in process work is common in electronic manufacturing organizations.[4] Using statistical methods has been a means for obtaining process improvements and gaining and ensuring competitive advantage. Statistical methods range from informal investigations to well-planned studies involving a series of specific experiments. There are certain properties all statistical studies must have to provide an effective basis for action:

- A well-defined objective
- Description of the statistical technique that will be used
- Sampling methodology
- Significance level
- List of fixed factors
- List of factors under study
- Measurement equipment that will be used
- Gauge repeatability and reproducibility study
- Strategy for how the study will be performed

27.5.1 Measurement Systems Analysis

Prior to performing data collection for any statistical study, a measurement systems analysis should be performed on the measurement equipment or measurement method. A measurement system analysis or what is typically called a *gauge repeatability and reproducibility* (GR&R) *study* is a mathematical procedure to quantify the amount of variability that is introduced to a process by the act of measuring a particular characteristic. Measurement variation that occurs during data collection can mislead management, problem-solving teams, designers, operators, engineers, suppliers, and all others who rely on data for decision making.

What Is Measurement System Variation?

- Variation that is introduced into the total process by the act of collecting data.
- Fluctuations in data due to the process chosen for taking measurements.
- Inability of a measurement process to repeatably measure a single data point.
- Inability of a measurement process to reproduce the measurement of a single data point when used by multiple operators.

This variation can be analyzed to determine the amount that is present and the sources of the variation within the measurement process. Once the measurement process is recognized as a process similar to a manufacturing process or a service process, many of the same techniques, plus several specific analysis methods, can be used to better understand the variation present. The following are methods used to evaluate measurement systems.

Accuracy of the measurement process is the difference between the observed average of the measurement and the true value. Gauge repeatability, illustrated in Fig 27.3, is the variation in measurements obtained by using one gauge multiple times one operator measuring the identical characteristics on one part. That is, measuring the same exact item, the same way, multiple times by the same operator. Gauge reproducibility, shown in Fig. 27.4, is the variation in the average of the measurements made by different operators using the same gauge when measuring identical characteristics on the same part.

Measurement system analysis allows us to account for the variability introduced by the measurement system. This variability is known as *measurement error.* The precision-tolerance (P/T) ratio is the ratio between the measurement precision estimate and the difference between the lower specification limit and the upper specification limit of the characteristic being measured.

The measurement precision estimate includes the repeatability and reproducibility of the measurement system being evaluated. The acceptability of a measurement system is based on the following criteria:

Acceptability Criteria

- 10 percent or less is considered preferred.
- 11 to 20 percent is considered acceptable.

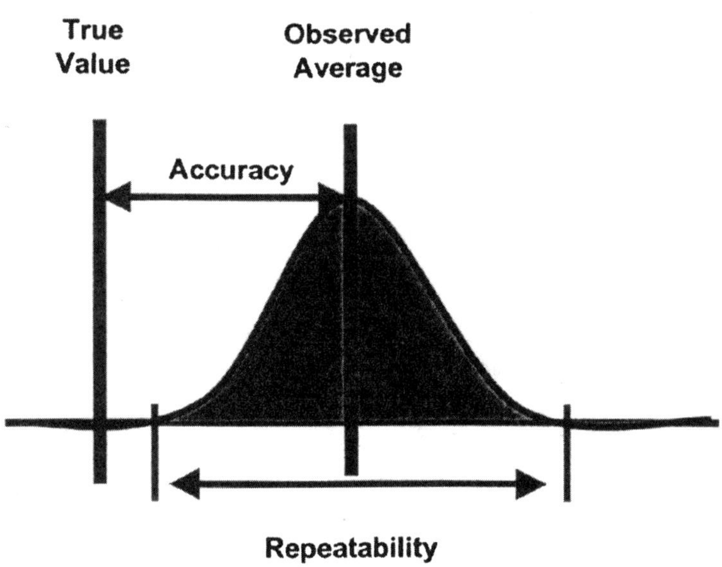

FIGURE 27.3 Repeatability.

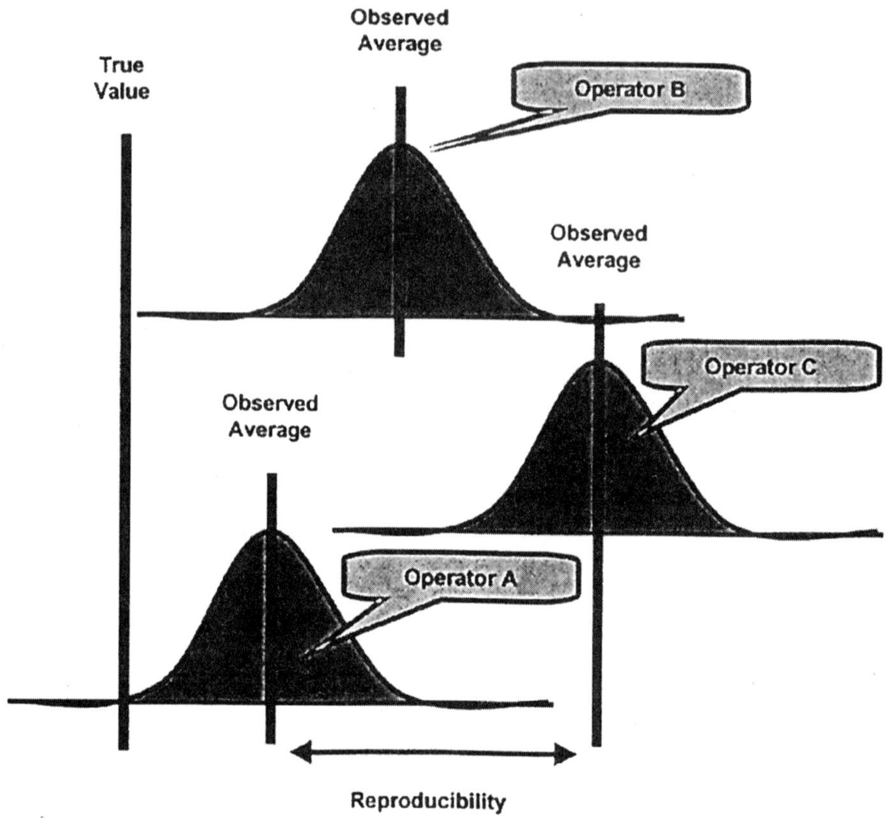

FIGURE 27.4 Reproducibility.

- 21 to 30 percent is considered marginal and may not be acceptable depending on the importance of the application.
- Over 30 percent is considered not acceptable.

Analysis-of-Variance Method. The analysis of variance (ANOVA) method is recommended for variable data. ANOVA is a standard statistical technique used to analyze the measurement error and other sources of variability of data in a measurement system. In the analysis of variance, the variation can be decomposed into four categories: parts, operators, interaction between parts and operators, and the replication error due to the measurement system itself. The statistical expressions used in the ANOVA method are summarized in Table 27.4.

$$SS_o = \sum_{i=1}^{o} \frac{y^2 i}{pn} - \frac{y^2}{opn}$$

$$SS_{TOT} = SS_{OP} + SS_{PART} + SS_{OxP} + SS_{ERR}$$

$$SS_{TOT} = \sum_{i=1}^{o} \sum_{j=1}^{p} \sum_{k=1}^{n} y_{ijk}{}^2 - \frac{y^2}{opn}$$

TABLE 27.4 Analysis of Variance

Source	*DF*	*SS*	*MS*	*F*
Operator	$k - 1$	SS_o	$SS_o/K - 1 = MS_o$	MS_o/MS_e
Part	$n - 1$	SS_p	$SS_p/n - 1 = MS_p$	MS_p/MS_e
Optr*part	$(n - 1)(k - 1)$	SS_{op}	$SS_{op}/(n - 1)(k - 1) = MS_{op}$	MS_{op}/MS_e
Gauge (error)	$nk(r - 1)$	SS_{ERR}	$SS_{ERR}/nk(r - 1) = MS_e$	
TOTAL	$nkr - 1$	SS_{TOT}		

$$SS_P = \sum_{j=1}^{P} \frac{y_j^2}{on} - \frac{y^2}{opn}$$

$$SS_{OxP} = \sum_{i=1}^{o} \sum_{j=1}^{P} \frac{y_{ij}^2}{n} - \frac{y^2}{opn} - SS_O - SS_P$$

$$SS_{ERR} = SS_{TOT} - SS_O - SS_P - SS_{OxP}$$

where o = number of operators
p = number of parts
n = number of trials

Variance Component Estimation

- Calculates accurate variance estimates using mean-squares computations.
- The ANOVA technique partitions the total observed variance into fundamental components.
- The variance of the operator-part interaction is obtained.
- Determines the significance of ANOVA terms (operator, part, interaction).
- Can accommodate variable sample size.

$$\hat{\sigma}^2_{ERR} = MS_{ERR}$$

$$\hat{\sigma}^2_{OxP} = \frac{MS_{OxP} - MS_{ERR}}{n}$$

$$\hat{\sigma}^2_P = \frac{MS_P - MS_{OxP}}{on}$$

$$\hat{\sigma}^2_O = \frac{MS_O - MS_{OxP}}{pn}$$

Attribute Data Method. Measurement systems analysis also can be performed using attribute data. The attribute method is used to evaluate if an inspector or test station has the ability to detect conforming or nonconforming parts repeatedly. The emphasis is on the effectiveness of the inspector or tester in detecting conforming or nonconforming parts repeatedly and the degree of bias toward rejecting conforming parts or accepting nonconforming parts. The attribute method evaluates the following:

Effectiveness. The ability of an appraiser to accurately detect conforming and nonconforming parts.

Probability of a miss [P(miss)]. The chance of not rejecting a nonconforming part. This is a serious error because a nonconforming part that is accepted causes high customer impact and dissatisfaction.

Probability of a false alarm [P(FA)]. The chance of rejecting a conforming part. This type of error is not as serious as a miss; however, it does cause rework and reinspection when they are unnecessary. This does increase rework costs.

Bias (B). The measure of a person's tendency to classify an item as conforming or nonconforming and is a function of *P*(miss) and *P*(*FA*).

What Are the Possible Causes of Gauge Repeatability and Reproducibility Errors? The most likely causes for lack of repeatability are

1. The measurement equipment needs maintenance.
2. The measurement equipment has insufficient resolution for the measurement requirements.
3. The way the part is positioned for measuring needs to be improved.
4. The technique for measuring needs to be improved.

The most likely causes for lack of reproducibility are

1. The operators are not trained how to use the measurement equipment.
2. There is insufficient resolution.
3. The equipment has not been calibrated.
4. Measurements were taken in two different environments.

27.5.2 Statistical Process Control (SPC) and Capability

SPC analyzes process variability and determines when special cause variation is present. A key aspect of the improvement process is to identify critical characteristics to monitor the performance of the process. SPC is an engineering tool that optimizes engineering problem-solving time by monitoring the performance of a process and alerting the engineer to a real process change. At the same time, an SPC chart reveals the nature of the problem and the type of solution required to solve the problem. Control charts are used to identify sources of variability, either common cause or special cause, and to isolate the source of defects and determine appropriate preventative solutions (Fig. 27.5).

Special Cause Variation. Those causes which affect the process sporadically:

- Special events that cause a specific change in process performance
- More easily identified and controlled
- Can be eliminated

Common Causes. Those causes which are consistently present in the process:

- Inherent variation caused by the regular and consistent operation of the process steps, procedures, skill sets, etc.
- More difficult to identify
- Requires a significant change in the process to reduce
- Cannot be eliminated but can be reduced by design of experiments

The primary advantage of SPC is real-time feedback to the process. SPC is used widely for maintaining a process in a state of statistical control, monitoring process parameters, and establishing process capability. Statistical Stability is consistent, predictable performance of a process operating within the common variation produced by the chosen design of the process and its respective inputs, process steps, and environment.

- *Stable process.* The process is stable when the mean and the variance calculated for any span of time are statistically equal to the calculated mean and variance for any other span of time, as shown in Fig. 27.6.

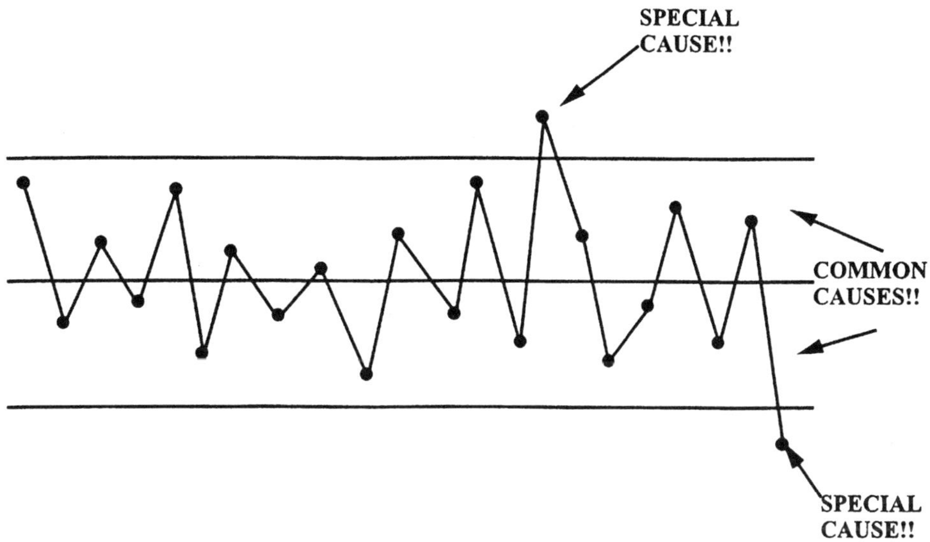

FIGURE 27.5 Identification of special and common cause variation.

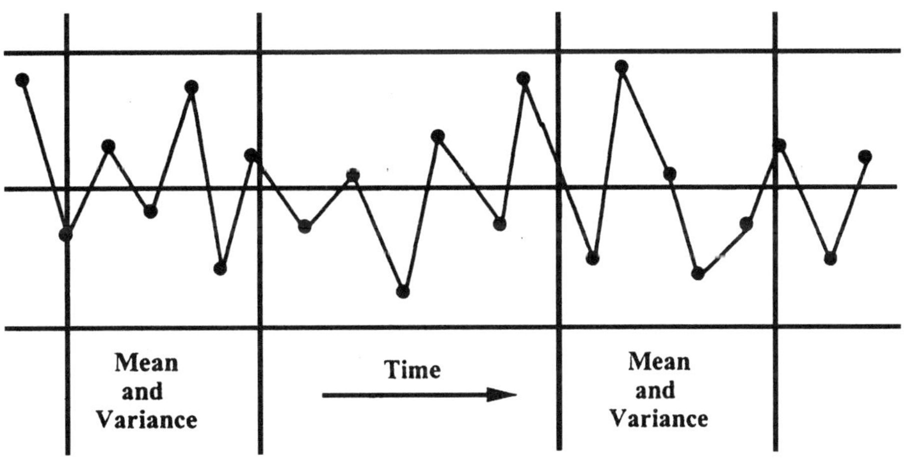

FIGURE 27.6 In control, stable, and predictable process.

- *Variance.* An estimate of the variability, which is equal to the standard deviation squared.

To determine if a process parameter is statistically stable, there is a metric called an *instability index* that can be used. The instability index (*St*) provides a means of estimating whether assignable causes are present in the process parameter being monitored. The instability index does this by measuring the fraction of points that are not within control for a given process.

$$\text{St} = \frac{\text{number of out-of-control data points}}{\text{total number of data points}} \times 100$$

The minimum requirement for a stable process is that the instability index must be 2 percent or less.

One way to demonstrate this is with two consecutive control charts having a total of 50 subgroups, 25 subgroups per chart, and showing no points out of control on the most current chart and no more than one point out of control on the previous chart when a process is under variation control and data are being monitored statistically on a real-time basis. Figure 27.7 displays the effect on a process when special cause variation is present and, after the implementation of corrective actions, the process is statistically in control.

Prior to calculating the process capability indices (*Cp*/*Cpk*), the process should demonstrate statistical control. The following capability index formulas assume a normal distribution.

Cp. The *Cp* is a ratio between the width of the tolerance (USL − LSL) and the width of the normal distribution (6 standard deviations).

$$Cp = \frac{\text{USL} - \text{LSL}}{6\text{s}}$$

The *Cp* only measures whether or not the distribution will fit within the specification limits. It does not account for where the distribution's center is located in relation to the tolerance or specifications.

Cpk. The *Cpk* takes into consideration both the width of the distribution and the location of the distribution by evaluating the distance from the upper and lower specs, respectively, to the mean. The *Cpk* is the "worst case" or the calculation that has a mean closest to the specification limit. To arrive at a *Cpk*, first a *Cpu* and a *Cpl* must be calculated. The *Cpu* uses the upper specification limit, and the *Cpl* uses the lower specification limit. The lowest or minimum value between the *Cpu* and *Cpl* becomes the *Cpk*.

$$Cpu = \frac{\text{USL} - \overline{X}}{3\sigma}$$

$$Cpl = \frac{\text{LSL} - \overline{X}}{3\sigma}$$

$$Cpk = \min(Cpu, Cpl)$$

What are the ideal values for *Cp* and *Cpk?* To achieve 6σ capability, the following *Cp* and *Cpk* values must be met: $Cp \geq 2$, $Cpk \geq 2.0$. Examples of *Cp* and *Cpk* values and the position of the distribution in relation to the specification limits are shown in Fig. 27.8.

There are several different types of SPC charts designed to monitor various types of data from different distributions.[5] These are divided into two groups:

- *Attribute charts.* Designed to monitor categorical data that describe failures of parts or services to meet specifications.
- *Variable charts.* Designed to monitor continuous data that are measurable and representative of the quality and performance of the process being monitored.

The control chart selection map shown in Fig. 27.9 is used to identify which type of control chart to use based on the type of data available.

By establishing control charts throughout the manufacturing process as well as at the end of the process, the manufacturing operators can be alerted to process changes that have an impact on product quality. Table 27.5 is a list of recommended control charts to monitor product quality performance throughout the surface mount technology (SMT) manufacturing process.

When using the control charts in Table 27.5, there is an opportunity of uncovering undesirable process changes prior to an entire lot of product being manufactured, thereby realizing a reduction in manufacturing repair costs. When out-of-control conditions are identified on the control charts as a violation of the Shewhart rules, feedback to the process must be immediate, with follow-up corrective actions implemented to realize defect reductions. This is the critical area of real-time feedback, where there is a difference between success and failure. Minimizing the time lapse between the occurrence of an out-of-control condition and the identification and correction of the root cause is the key to a successful SPC program.

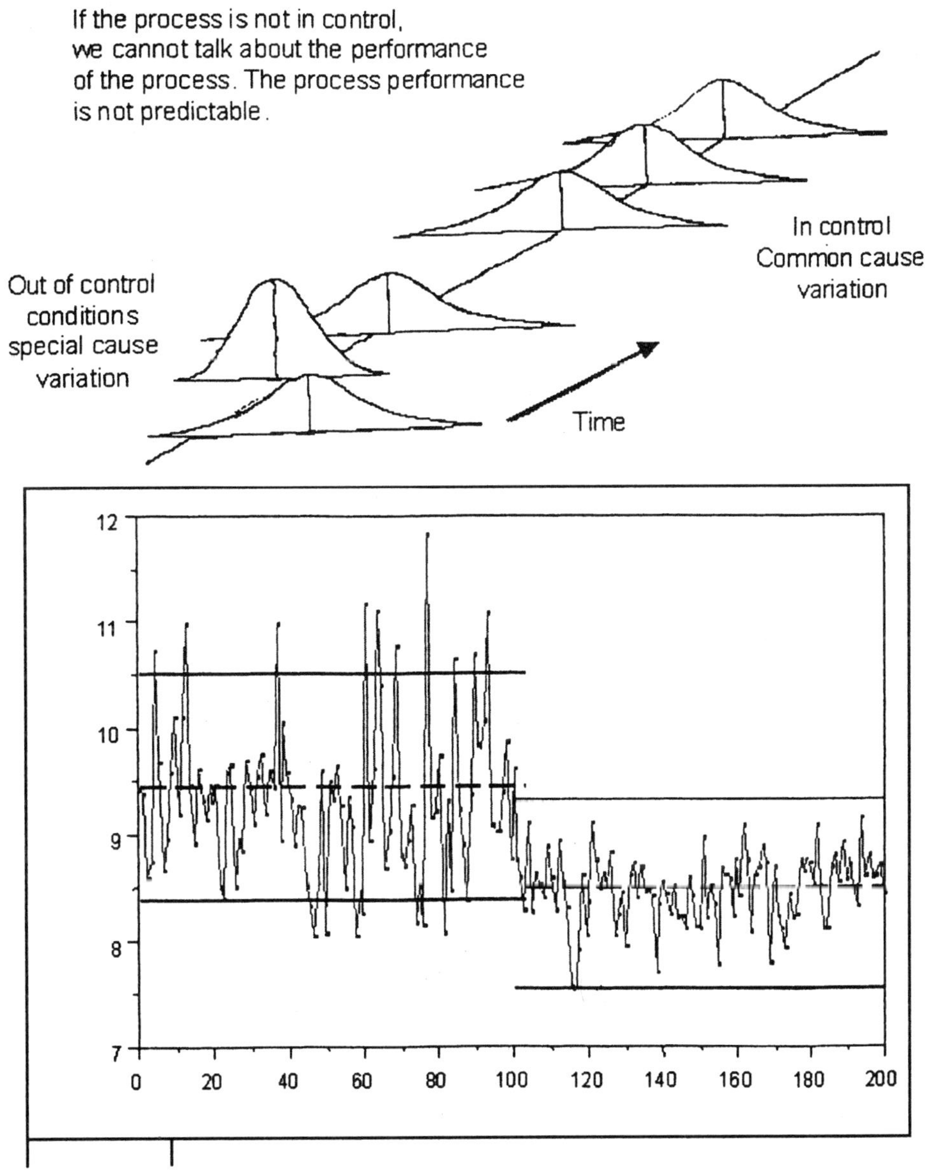

$\bar{X}$ Control Chart

FIGURE 27.7 Understanding and controlling process variation.

Process control charts tell those in direct control of the operation when an abnormal event has occurred, such as a process shift or a sudden change in performance. It is recommended that the process be stopped immediately for all out-of-control situations. The out-of-control action plan can be used to tell the operator what specific action to take to bring the process back to a state of statistical control. The out-of-control action plan also helps operators to solve problems on their own.

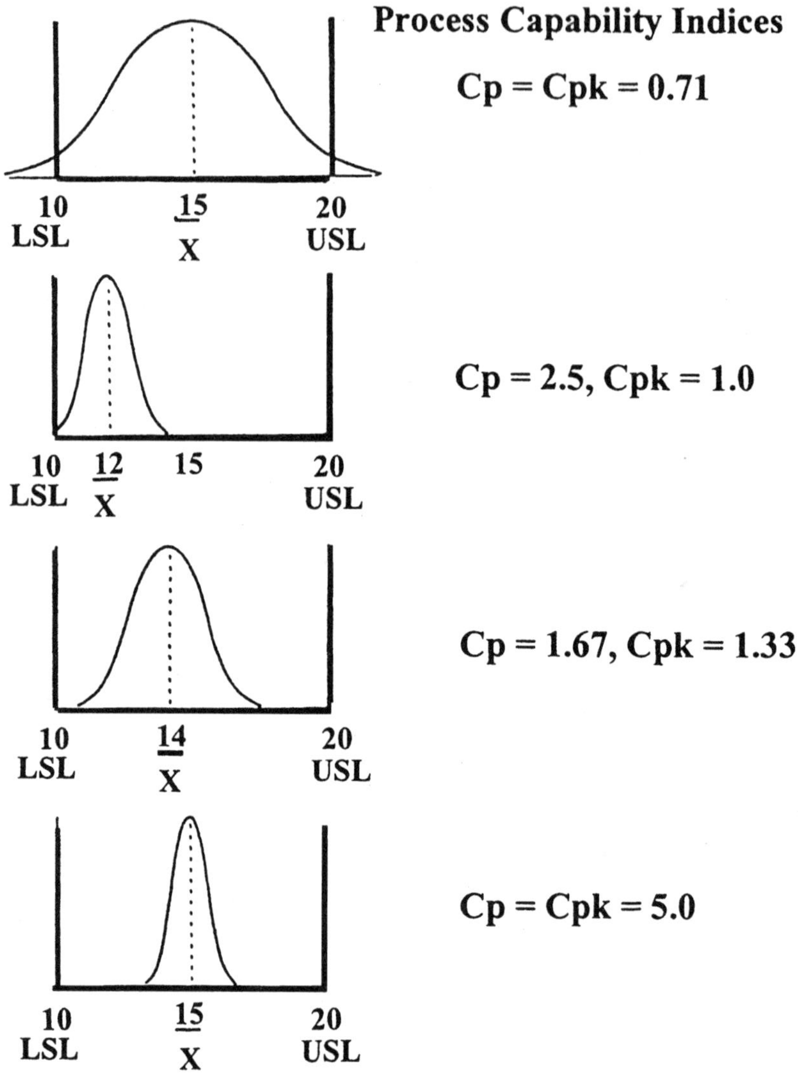

FIGURE 27.8 Process capability indices.

When a process goes out of control, the operator identifies the defect or defects and initiates the action indicated on the out-of-control action plan. The documented defects and their respective corrective actions offer the operator a troubleshooting guide. One technique that can be used to develop the out-of-control action plan is a process failure mode and effects analysis (FMEA). Generally, the FMEA assesses known or potential process defects (failure modes), the effects of the defect, potential cause(s) of the defect, and recommended corrective actions to take to eliminate the defect. Only after corrective action is implemented and recorded should the process be allowed to restart.

Comparative Methods. To carry out a statistical test, we need a theory to test against another theory. Most hypothesis tests concern the parameters of a distribution but sometimes will be interested in comparing one distribution against another. Any theory about a distribution for a measurement is known as

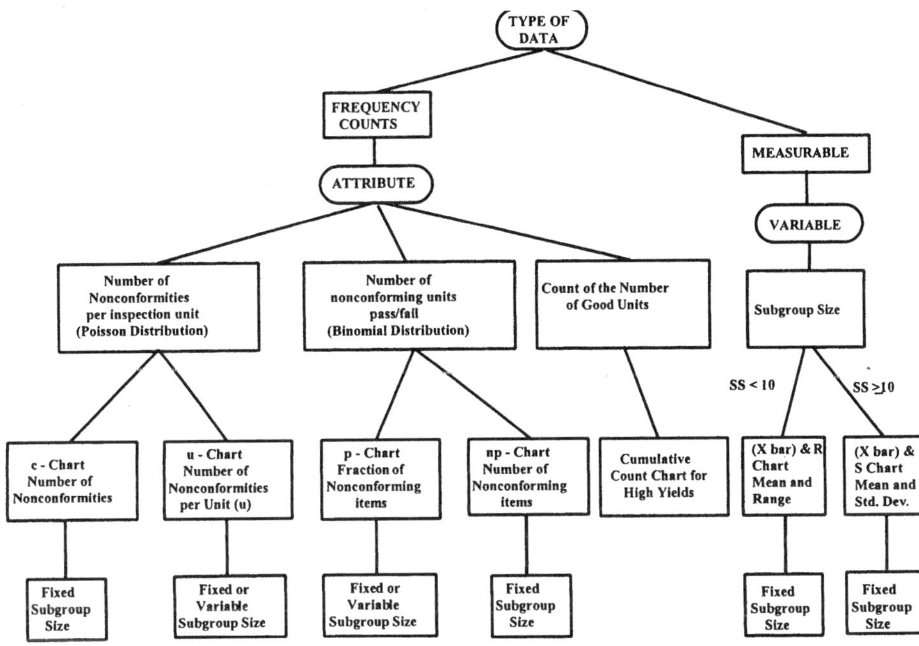

FIGURE 27.9

TABLE 27.5 Suggested SMT Control Charts

SMT process	Characteristic	Suggested type of control chart
Stencil printer	Solder paste volume	$\bar{X}$, R control chart
Placement	Placement accuracy x and y axis	$\bar{X}$, R control chart
	Nozzle pressure	$\bar{X}$, R control chart
Reflow	Peak temperature	$\bar{X}$, R control chart
	Time over liquidus	$\bar{X}$, R control chart
Postreflow	Quantity of defects	C control chart
Board-level test-repair	Quantity of defects	C control chart
Unit assembly	Quantity of defects	C control chart
Unit test-repair	Quantity of defects	C control chart

a *hypothesis.* If the information provided completely defines a distribution (states the form of the distribution and gives a value to all the parameters of the distribution), it is known as a *simple hypothesis.*

In general, sample averages, counts, variances, or proportions will have different values even if they are drawn from a single population. Hypothesis testing allows us to determine whether the differences displayed in the samples are due to sampling variations or to real differences in the population averages, counts, variances, or proportions.

27.5.3 Design of Experiments

A *design of experiment* (DOE) is a test in which we make changes to input parameters of a process to observe and measure changes in the output parameter. Multiple input parameters are systematically changed to observe changes in the output parameter of a process, system, or product. DOEs make efficient use of time and resources and maximize the usefulness of the results. We can study simultaneously the effect of many factors that may influence product or process performance through the use of DOEs. The objective is to determine the true cause-and-effect relationship between the input parameters and the output parameter. DOE is an iterative process. Table 27.6 shows the iterative nature of experimentation.

Careful planning and identification of which type of DOE to use are important. The approach taken in experimentation depends on the type of experiment chosen. To understand the variation in a process, we need to identify which factors have the most influence on the output of the process. Typically, there are several factors in a process that can be investigated. To identify which factors are important, screening designs can be performed. A screening design is aimed at reducing the number of potential influential factors to the vital few. Screening designs allow us to focus process improvement activities on the most important process factors. Screening designs do not allow all interactions to be estimated. Screening designs are a good application for investigating how input machine parameters affect the product being produced. Based on the results of the experimentation, you can understand the relationship of each input parameter to the output response. Typical designs for screening are fractional factorials and Plackett-Burman designs.

Response surface designs are used to optimize a process. The goal is to determine the best factor values in order to optimize process settings to minimize, maximize, or hit a target. The mathematical equation to describe this relationship is

$$Y = f(X_1, X_2, \ldots, X_k) + e$$

Typically, this is represented by a response surface or by a contour plot, as shown in Fig. 27.10.

Optimization can be used to improve process capability indices (*Cp*/*Cpk*), increase productivity, improve quality, and reduce rework/scrap costs. The purpose of optimization is to identify the "best" input parameter settings to achieve the optimal value for the mean and variance of the process output responses. Typical response surface designs are central composite and Box-Behnken.

The central composite design is the most efficient experimental design for sequential experimentation. You can create blocked or unblocked central composite designs. Central composite designs consist of $2k$ factorial or "cube" points, where K is the number of factors, axial points (also called *star points*), and center points. A central composite design with two factors is shown in Fig. 27.11. Points on the diagram represent the experimental runs that would be performed:

TABLE 27.6 Nature of Experimentation

Step	Process	Knowledge
1	Screening designs	Identify important independent parameters
2	Full factorial	To find important interactions
3	Response surface	To determine the best factor settings in order to optimize process settings to minimize, maximize, or hit a target

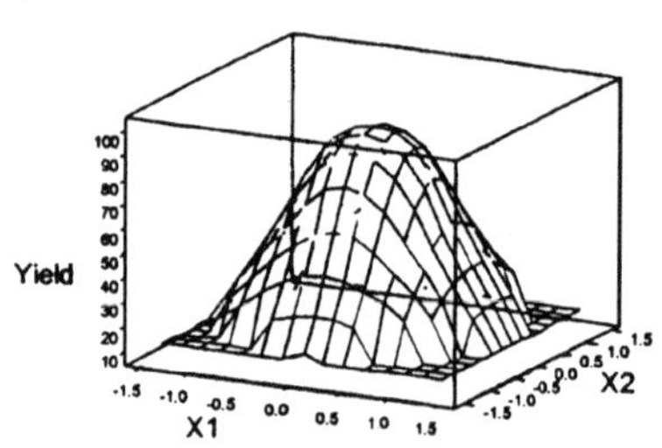

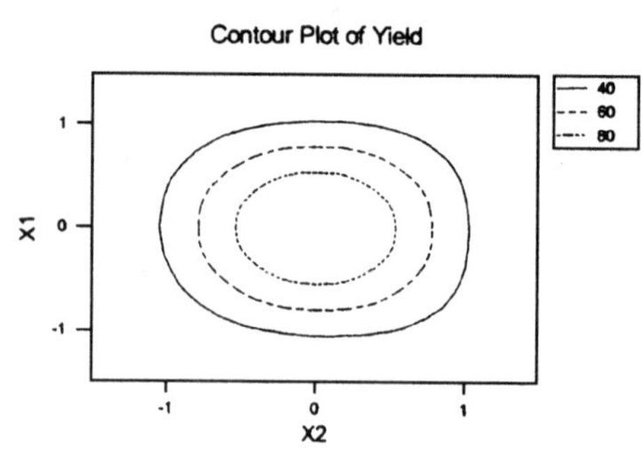

FIGURE 27.10 Response surface plots.

The points in the "cube" portion of the design are coded -1 and $+1$. The points in the axial or star portion of the design are $(0, -\alpha)$ $(0, \alpha)$ $(\alpha, 0)$. The design center is at (0, 0). Central composite design includes points at the corners of the cube, center points, and star points that include additional treatment conditions than the Box-Behnken designs.

The Box-Behnken designs in Fig. 27.12 are used when performing nonsequential experiments. That is, you are only planning to perform the experiment once. Box-Behnken designs have fewer design points that are less expensive to run than central composite designs with the same number of factors. Box-Behnken designs do not have axial points and do not cover all the corners of the cube. They are used when there is no interest in predicting responses at the extremes, i.e., at the corners of the cube.

Prior to performing any experimentation, the process should demonstrate stability and statistical control.[6] A measurement systems analysis also should be performed on the measurement equipment for the response parameter. Careful planning and execution of the experimentation are critical to the success of the experimental study. The following is an eight-step process in conducting DOEs:

1. *State objective.* A concise description of why the experiment is being run. What are the critical characteristics (measurable attributes that have a significant impact on the function, performance of a product, or process) and the desired outcome (eliminate root cause, reduce variation, hit a target, maximize/minimize response, or identify critical parameters).

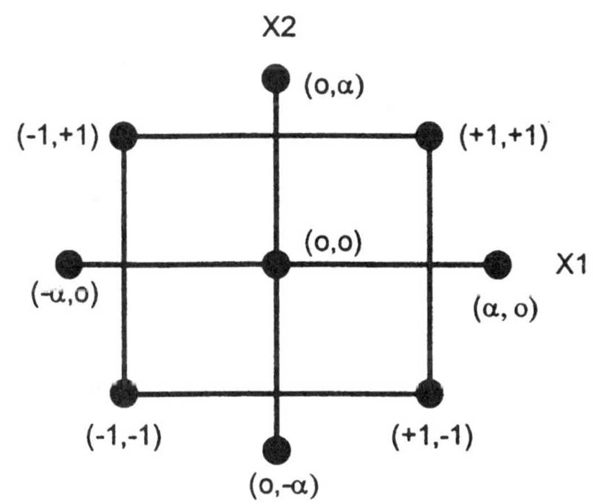

FIGURE 27.11 Central composite design.

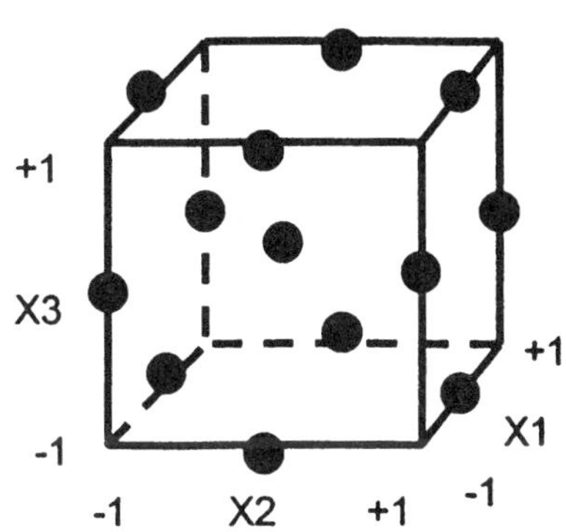

FIGURE 27.12 Box-Behnken design.

2. *Select output responses.* Quantitative response measures continuous data, whereas a qualitative response measures categorical data. Prior to data collection, perform a measurement system analysis on the measurement equipment or measurement method.
3. *Select input variables (factors).* Input variables are changeable process settings or parameters that are varied to determine the effect on the output response. Input parameters can be determined by developing a modified cause-and-effect diagram.
4. *Select type of experiment.* Select the best experiment to meet the overall objective, number of replicates, and if need to, block and define the alias structure.
5. *Develop experiment strategy.* This is a statistical strategy document that is an interim report including the objectives, fixed factors, type of experiment, experiment matrix, response parameters, experiment process, experiment process times, experiment materials required, and schedule of the activities.
6. *Run the experiment.* The experiment should be run in randomized order.
7. *Data analysis and final report.* Quantify the effects that the input factors have on the output response(s). The final report should include the contents of the statistical strategy document, experimental conclusions, and future actions.
8. *Run confirmation.* Confirmation runs are to validate the predication model throughout the design space. Confirmation runs should be determined for at least three locations within the design space.

27.6 REFERENCES

1. Chaudhry, A. M., *To Be a Problem Solver, Be a Classicist,* Quality Progress, American Society for Quality, Milwaukee, WI, 1999.
2. Hradesky, J. L., *Productivity and Quality Improvement,* McGraw-Hill, New York, 1988.
3. Scholtes, P. R., *The Team Handbook,* Joiner Associates, Oriel, Inc., Madison, WI, 1988.
4. Statistical Quality Solutions, *Statistical Process Control for SMT Process,* Massina, WI, 2000.
5. American Industry Action Group, *Measurement Systems Analysis,* 2d ed., Southfield, MI, February 1995.
6. Barrentine, L., *Concepts for R&R Studies,* American Society for Quality Press, Surface Mount Technology Associates, Edina, MN, 1991.

S•E•C•T•I•O•N • 4

ECONOMICS AND PRODUCTIVITY

CHAPTER 28
METRICS: THE KEY TO PRODUCTIVITY

Ronald Lasky
Cookson Performance Solution

Daniel Baldwin
Georgia Institute of Technology

Brian Lewis
Siemens USA

Joe Belmonte
Cookson Performance Solutions

28.0 INTRODUCTION

Electronics has become the largest industry in the world at $1 trillion per year. It has grown at four times the rate of the gross domestic product of the countries in which it exists. Hence it is the industry of choice for developing nations to accelerate the establishment of strong economies.

As electronic assembly matures, there is no more critical issue than productivity. If productivity worldwide were increased by only 5 percent, the $1 trillion dollar electronics business would produce an extra $50 billion worth of products. One of the keys to improving productivity is to use the most effective metrics to track a facility's productivity. This chapter will propose some of these metrics that resulted from a productivity investigation performed by Cookson Electronics, the Georgia Institute of Technology, and Siemens Electronic Assembly Equipment.

28.1 THE CASE FOR PRODUCTIVITY

In order to better understand the importance of productivity, a case study will be performed using a typical product. In this case, a communication modem will be analyzed with respect to its productivity during assembly.

The recent development of real-time cost-estimating software such as SPACE™ has enabled a simple calculation of the effects of productivity on profit.[1] To demonstrate this effect, consider a modem assembled with the parameters in Table 28.1.

TABLE 28.1 Input Parameters for the Modem Case Study Assembly Cost Analysis

Description	Baseline case for comparison		
Assembly equipment (SP:200K, PP:900K, RO:100K, T:400K, BH:100K)	$1,700,000.00		
Floor space and utilities (sq m)—cost (per sq m/month)	3682.4	$0.83	
Components (per unit)	$75.00		
Stencil, squeegee, lifetime (cost each, uses)	$600.00	$200.00	1000000
Solder paste ($/g), number of grams	$0.1000	5	
PWB ($ per unit)	$15.00		
Workers (number, rate per hour)	7	$28.00	
Selling price ($), gross margin (%) (enter 0 for item not used)	$110.00	0	
Hours per shift, shifts per day, days per week	10	2	6
Cycle time (seconds)	25		
Down time (%), setup (hours per week), maintenance (hours per week)	8	10	12.5
Workers supported (number, rate per hour)	26	$42.00	
I/O	950		
Yield first pass (%), yield first pass ppm/I/O (*N* if not used), percent reworkable	97	N	100
Unit RW materials cost, minutes/RW, RW labor ($/h)	$0.50	20	$30.00
Years equipment depreciation, interest rate (%)	5	9	
Length of simulation (weeks), number of boards (enter 0 for item not used)	52	0	

By entering these and other detailed parameters into the SPACE™ cost-estimating software, the cost analysis results for 1 year's worth of production are given in Tables 28.2 and 28.3.

This case study analysis indicates that the particular surface-mount assembly line can realize more that $4 million in gross profit in 1 year based on a cycle time of the line of 25 s (0.42 min). Using the SPACE™ software, one can vary the cycle time to determine its effects on gross profit. The result is shown in Figure 28.1.

The effect of cycle time on gross profit is stunning. A reduction in cycle time from 0.5 to 0.25 min (30 to 15 s) results in an increase in gross profit from $2 million to $12 million, a sixfold increase. The reason for this nonlinear growth with cycle time reduction is that while there are linearly more units to sell (i.e., in this case, two times more units), each unit also has a lower cost. In low-margin businesses, this effect can be strikingly strong.

TABLE 28.2 Assembly Cost Analysis Results for the Modem Case Study

Cost	$ per board	%	Total $
Components	$ 75.000	72.194	$49,362,750.00
Labor	$ 12.211	11.755	$ 8,037,120.00
PCB	$ 15.000	14.439	$ 9,872,550.00
Consumables	$ 0.501	0.482	$ 329,611.54
Machine cost	$ 0.798	0.769	$ 525,475.61
Rework	$ 0.315	0.303	$ 207,323.55
Floorspace, utilities	$ 0.061	0.058	$ 39,842.25
Total	$103.886	100.000	$68,334,830.70

TABLE 28.3 Profit Results for the Modem Case Study

Unit cost	Unit profit	Units produced	No. weeks	Total profit
$103.89	$6.11	658170	52	$4,024,027.05

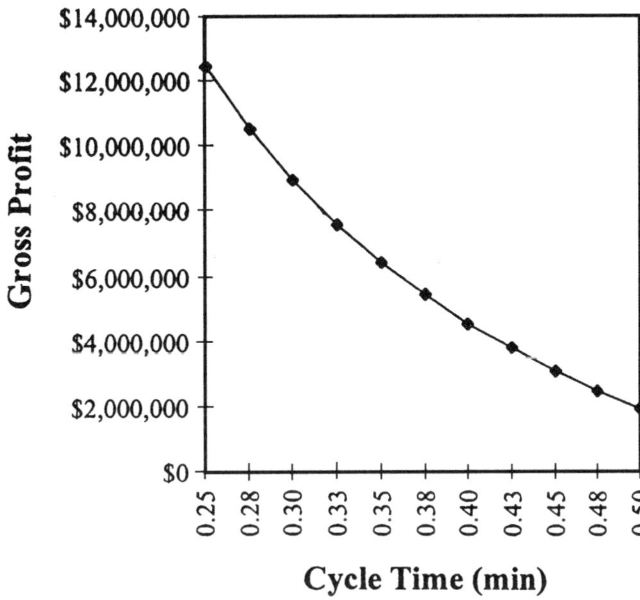

FIGURE 28.1 Impact of cycle time on gross profit for the modem case study.

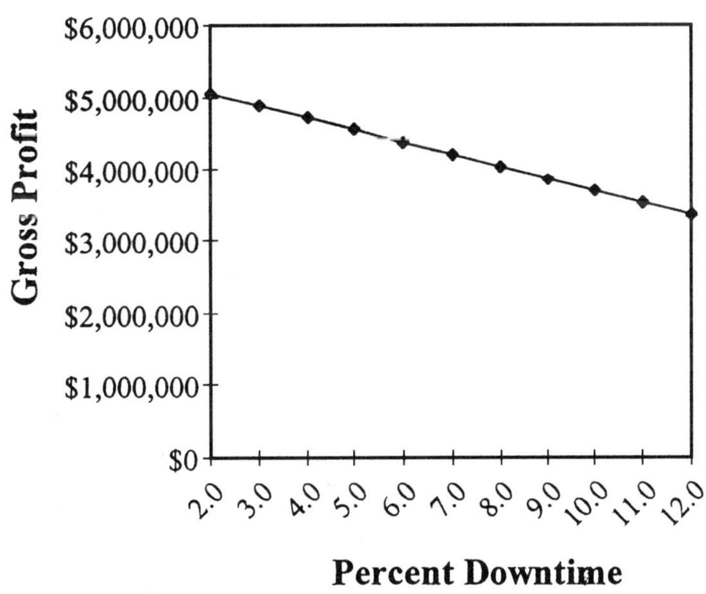

FIGURE 28.2 Impact of line down time on gross profit for the modem case study.

The reduction in down time of the line also can have a dramatic effect on profits. In our modem example from Table 28.1, the total line down time was 8 percent. SPACE™ will allow us to vary the down time to determine its effect on gross profit. The results of this analysis are shown in Fig. 28.2. From this graph, we see that the difference in gross profit generated by this line would be about $1.7 million if the down time were improved from 12 to 2 percent.

Having made this case for productivity, consider now some suggested metrics to track and improve productivity.

28.2 PRODUCTIVITY METRICS

Considering the wide variety of manufacturing lines that exist, a recommended productivity plan that is universally applicable is difficult to define. However, measuring and tracking of the right productivity metrics are most of the battle and are the first steps in developing a productivity improvement plan. After these metrics are settled on, measured, and tracked, the right productivity improvement plan for any specific situation is usually obvious. The minimum metrics we recommend measuring and tracking are

1. Nonmaterial assembly cost per I/O or nonmaterial assembly cost per component
2. Inventory turns per year
3. Average setup time
4. Unscheduled line down time
5. Work in process time
6. Average cycle time
7. Average defect level

Each of these metrics is discussed briefly below.

28.2.1 Nonmaterial Assembly Cost per Input-Output (I/O)

The first metric is also called the *board assembly conversion cost* and is given by

$$\text{Conversion cost} = \frac{\text{cost to run facility per year (nonmaterial costs)}}{\text{total number of I/Os assembled per year}}$$

A best-in-class value for this metric is approximately $0.013 per I/O. Basically, this metric represents the total annual nonmaterial costs required to run the facility normalized with respect to the total number of I/Os assembled per year. This normalized cost provides an effective means of measuring cost competitiveness, particularly for organizations assembling a wide mix of component I/Os and/or predominately medium- to high-I/O components.

28.2.2 Nonmaterial Assembly Cost per Component

The second metric is the nonmaterial assembly cost per component, also called the *cost per placement,* given by

$$\text{Cost per placement} = \frac{\text{cost to run factory per year (nonmaterial costs)}}{\text{number of components per year}}$$

A best-in-class value for this metric is approximately $0.10 per component. Basically, this metric represents the total annual nonmaterial costs required to run the facility normalized with respect to the total number of components placed per year. This normalized cost provides an effective means of measuring cost competitiveness, particularly for organizations assembling low- to moderate-I/O components.

28.2.3 Inventory Turns per Year

This metric tracks the number of times per year that the materials (components, paste, etc.) turn over. It is particularly important in tracking the availability of capital and the liability associated with material costs billed FOB at the manufacturing site. Typically, this metric is targeted at greater than 10 inventory turns per year in order to minimize capital outlay for component inventory.

28.2.4 Average Setup Time

This metric tracks the average time it takes to set up the line between the end of one job and the beginning of the next scheduled job, assuming that the board has been run previously on the line (i.e., it is not the first time setup, and no process development is needed). This would include the time to change all setups of each active cell in the line, particularly the placement machines and stencil printer. The setup time also would include the time spent to verify the yield of a first pass of product through the line. This metric is particularly relevant for medium- to high-mix production strategies, where non-value-added setup time should be minimized.

28.2.5 Unscheduled Line Down Time

This metric measures the average time the line is down due to reasons other than scheduled preventive maintenance (PM). This includes any time the cell is not performing its set function and the time the line is not producing product during scheduled production times. Any time a machine is down due to operator-related problems, machine-related problems, software-related problems, material stocking, etc. would fall into this area. Minimizing this metric is warranted. Excessive unscheduled line down time is an indicator that a task force should be assembled to break down the factors contributing to this non-value-added processing time in order to identify the major contributing factors that need to be addressed in a productivity improvement plan.

28.2.6 Work-in-Process Time

This represents the time required for production of a product lot. It accounts for the time from the point the product hits the manufacturing floor to the point it is shipped or warehoused. In certain, medium-volume, high-mix manufacturing strategies, work-in-process time can be an important metric where multiple lines share common process equipment such as panel routing or hand stuff (e.g., two SMT lines feeding into one router). In these cases, floor space can be a premium, leading to problems where long work-in-process times result in large queues and congested floor space significantly limiting line capacity.

28.2.7 Average Cycle Time

This metric measures the longest average cycle time or the beat rate of the line averaged over the products produced. It is measured on a machine-specific basis and typically represents a specific machine or process step for a given product having the longest cycle time. Cycle time is measured as the in-to-out time of a particular machine or process step to produce one panel. For example, cycle time for a placement system would be measured as the cumulative time from the point where the panel leaves the staging area upstream of the placement process (often placement systems stage boards on a conveyor within the placement system) to the point the board leaves the placement system conveyor. Cycle time is particularly relevant for identifying and/or tracking the process step or machine, which is the rate-limiting step in the line. Under most circumstances, it is desired to have the machine(s) representing the largest capital expenditure defining cycle time for the process. Typically, this corresponds to the placement machines.

28.2.8 Average Defect Level

This metric tracks the number of defects resulting from the assembly process and incoming materials. It can be tracked on a number of levels and at various points in the assembly process. Often defects are measured in the following locations:

1. Postprinting inspection of the solder paste position and volume
2. Postplacement inspection of component placement location
3. In-circuit testing after the reflow process
4. Visual inspection and functional testing after any hand-stuff processes and wave solder, which includes any board-to-board assembly required for functional testing.

The defects identified will be process-related; e.g., postplacement defects might include excessive x, y, and Θ offsets, missing components, etc. It is important to choose defect criteria that track the total process defects rather then just those which are easily measured or addressed in terms of process improvements. The average defect level is a direct indicator of the assembly process and materials quality. The most common points within the assembly line where defects are tracked are in circuit test and/or functional test. Moreover, every defect represents a factor that increases the work-in-process time, resulting in lost production time and/or repair/rework processing time. In all cases, higher average defect levels lead to lower productivity and lower profit margins.

At a minimum, the preceding seven parameters should be tracked, and an improvement plan to address these metrics should be formulated. A costing-estimating software tool such as SPACE™ should be used to estimate costs and track them accurately.

Other parameters that may be valuable to track depending on the particular production strategy implemented are

1. Placement capacity
2. Cost per unit of placement capacity
3. Placement machine utilization
4. Cost of ownership
5. First-pass yield
6. Response time
7. Changeover time
8. Mean time between *x*
9. Product mix
10. Work in process (WIP)
11. Placements/year/square foot factory floor space

1. *Placement capacity* is defined by the following expression.

$$\text{Placement capacity/m}^2 = \frac{\text{hourly machine rate as published in manufacturers brochure}}{\text{published floor space requirement}}$$

Placement capacity is an important metric to track in cases were strategies for capital equipment expenditures, production expansion, and/or expanded work-order acceptance are being considered. Placement capacity assumes that the placement systems define the cycle time of the production line. It is particularly useful in evaluating placement system and driving maximization of the line capacity with minimum floor space. This metric has been proposed by NEMI for evaluating assembly process productivity.[2]

2. *Cost per unit of placement capacity* is defined by the following expression:

$$\text{Cost/unit of placement capacity} = \frac{\text{Machine capital cost including feeders}}{\text{hourly machine rate as published}}$$

This metric provides a convenient means of comparing placement systems during capital equipment and production expansion evaluations. This metric has been proposed by NEMI for evaluating assembly process productivity.[5]

3. *Placement machine utilization,* as defined by the NEMI roadmap,[5] is given by the following expression:

$$\text{Utilization \%} = \frac{\text{actual weekly shipments}}{\text{vender-published machine rate} \times 168\ \text{h}}$$

This particular metric tracks the placement machine utilization relative to the published vender machine placement rates. Typical utilization rates are currently 10 to 35 percent. Targeted machine utilization rates over the next several years are 65 to 85 percent.

4. *Cost of ownership* is a system-level cost measured at the machine level or at the line level. It can be calculated as the sum of the amortized machine cost plus the annual utility costs plus the annual maintenance cost plus the annual disposable components cost (nonmaterial disposables such as nozzles, squeegees, nitrogen, etc.) plus loss production time costs due to unscheduled down time. Additional cost components that are less frequently captured within the cost of ownership are the quality-lost costs due to estimated defect rates, change over time, programming time, debugging time, etc.

5. *First-pass yield* can be measured in terms of three factors: first-pass board yield, component yield, and solder joint yield.

First-pass board yield is by far the most common yield metric tracked in production. It basically measures the number of fully functional boards produced on a given line prior to any rework or repair normalized with respect to the total number of boards produced. This metric is a direct indicator of process quality and is extremely useful in driving total quality improvement in the production facility.

$$\text{First-pass board yield} = 1 - \frac{\text{number of boards requiring repair per month}}{\text{number of boards assembled per month}}$$

First-pass component yield, while less common, is a secondary metric that can be useful in tracking down cells and/or processes within a production facility, process design issues, and board-level design issues that are negatively impacting process quality.

$$\text{First-pass component yield} = 1 - \frac{\text{number of components requiring repair per month}}{\text{number of components assembled per month}}$$

First-pass solder-joint yield is the least often used of the three metrics and basically tracks the number of solder-joint defects with respect to the number of opportunities for forming interconnects. In many cases, this metric is a "feel good" metric in that the yield numbers are almost always high, particularly for production facilities assembling dense-I/O components. It can be useful in tracking process quality and interconnect formation quality in cases where very high I/O components are assembled such as large I/O ball grid arrays (BGAs), quad flat packs (QFPs), and area array flip chips (FCs). This metric can be used to drive process improvement strategies in cases where interconnect formation is experiencing probabilistic failures.

$$\text{First-pass solder-joint yield} = 1 - \frac{\text{number of I/Os requiring repair per month}}{\text{number of I/Os assembled per month}}$$

6. *Response time* is a metric used to measure the turn times on new orders for both new and previously manufactured products. It is typically measured as the time from which an order is received for a new product to the time production is initiated. For contract manufactures, this metric can be particularly important to track and to develop a strategy for minimizing.

7. *Changeover time* is a metric used to track the time required to initiated production of a new product design that has not been manufactured previously. It is measured as the time from which a new product is first set up for production on a line to the time where first-pass board yield reaches a stable value for some predetermined minimum number of boards or panels. Inherently, this metric includes the typical process learning curves associated with implementing and debugging a new product on a line. Moreover, it is an indicator of the time spent on the production line performing process development rather than producing product.

8. *Mean time between x:*

(a) *Failure.* Known as MTBF, this metric tracks the average time between failures of a specific machine performing its service and requiring service or maintenance. Any circumstance where the equipment does not respond to a change in its status would be flagged as a failure.

(b) *Repair.* Known as MTBR, this metric tracks the average time between repairs that need to be performed on a particular tool to bring it from a down state to an up state where it can perform its function. Many times MTBR can be measured by mean time to repair (MTTR), which tracks the time to bring a tool from a down state to a performing state.

(c) *Interrupts.* Known as MTBI, this metric tracks the average time that elapses between machine interrupts. This time is measured when a tool delays its normal operation due to machine-related problems, i.e., software crashes, machine accessory defects, etc., and requires operator assistance rather than service or maintenance.

(d) *Assist.* Known as MTBA, this metric tracks the time between equipment delays that require outside assistance to regain performing status. This assist time also would include the time that elapses before an operator determines that an assist is required.

9. *Product mix* tracks the number of independent panel designs processed on a particular line or within a particular production facility on an annual basis. This metric is particularly important to track when trying to benchmark one production facility relative to another. It is also very helpful to track to allow better.

10. *Work in process* (WIP) tracks the number of panels in production at any given instant in time. In order to maximize inventory turns, minimize inventory, and reduce production costs, it is desirable to minimize WIP. WIP is particularly important to track in production facilities where multiple process lines share common process equipment. In these cases, large WIP leads to large queues and larger inventories then desired.

11. *Placement/year/square foot of factory floor space* is a metric used by some companies that accounts for the overhead costs within a particular production facility based on the floor space used for production. This metric is commonly cited in cases where companies want to shrink their production lines without sacrificing throughput or capacity.

28.3 SUMMARY

One of the keys to improving productivity is to use key metrics to track a facility's productivity. This chapter has discussed some recommended metrics for tracking productivity based on a series of productivity audits performed by the authors. The proposed metrics can form the cornerstone for developing and implementing a productivity improvement strategy.

28.4 REFERENCES

1. SPACE™, developed by Ron Lasky of Cookson Electronics.
2. NEMI, *National Electronics Manufacturing Technology Roadmap,* National Electronics Manufacturing Initiative, Herndon, VA, 1996.

CHAPTER 29
COST ESTIMATING FOR ELECTRONIC ASSEMBLY

Ronald C. Lasky
Cookson Performance Solutions

Nan Jiang
Cookson Electronics

Joseph Travers
Speedline Technologies

29.0 INTRODUCTION

In 1997, electronics passed the automotive industry as the largest industry in the world. Figure 29.1 is a graph detailing the value of electronics production throughout the world. These data include telecommunications, consumer electronics, information processing, automotive, military, and manufacturing electronics. The current annual growth rate of electronics is in the 8 to 10 percent range. This rate is about four times higher than that considered to be a robust growth rate for the overall gross domestic product. To capitalize on this much greater growth rate, developing nations have sought to master electronics manufacturing. This emphasis has resulted in the "miracle" economies of such countries as Singapore, Hong Kong, and South Korea.

In addition to its rapid growth rate, electronics is also the greatest value-added manufacturing industry. Starting with only a $3 billion investment of raw materials (oil, sand, water, lead, copper, etc.), goods valued at $1000 billion are produced. This progression of added value is shown in Figure 29.2.

The Food Chain

- Electronic equipment: computer, consumer, communications, industrial/instrument, military
- Electronic components: semiconductors, PWBs, connectors, passives, etc.
- Electronic materials: silicon, laminates, ceramic parts, photomasks, lead frames, resins, paste, process chemicals
- Industrial materials: ferrosilica, metal strip, pert/inorganic chemicals, powders
- Raw materials: oil, gases, water, silica, bauxite, etc.

In addition to being the largest industry in the world, electronics has spawned the greatest value-added industry in history, the software industry. This vital industry is currently valued at $300 billion.

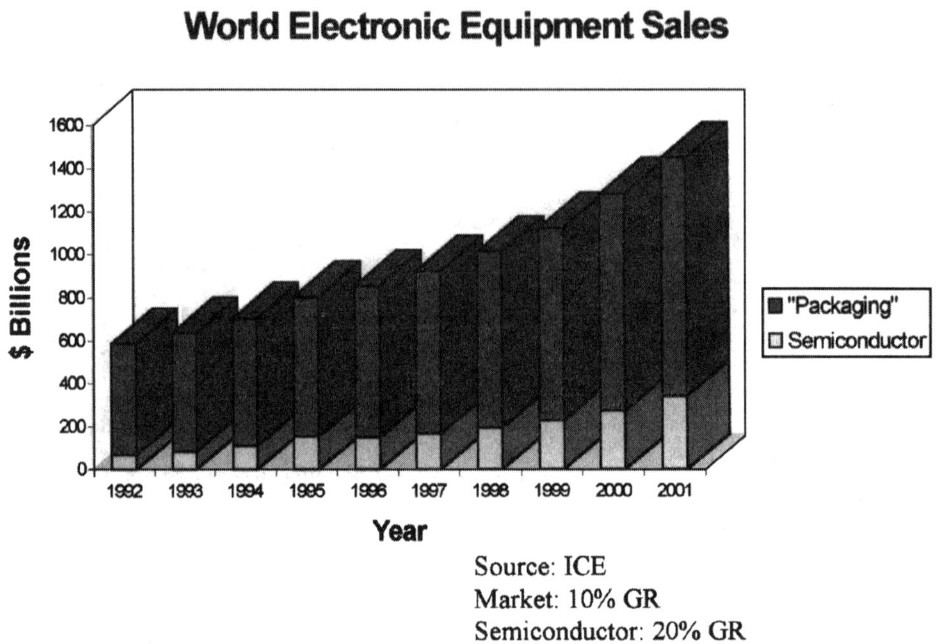

FIGURE 29.1 World electronic equipment sales.

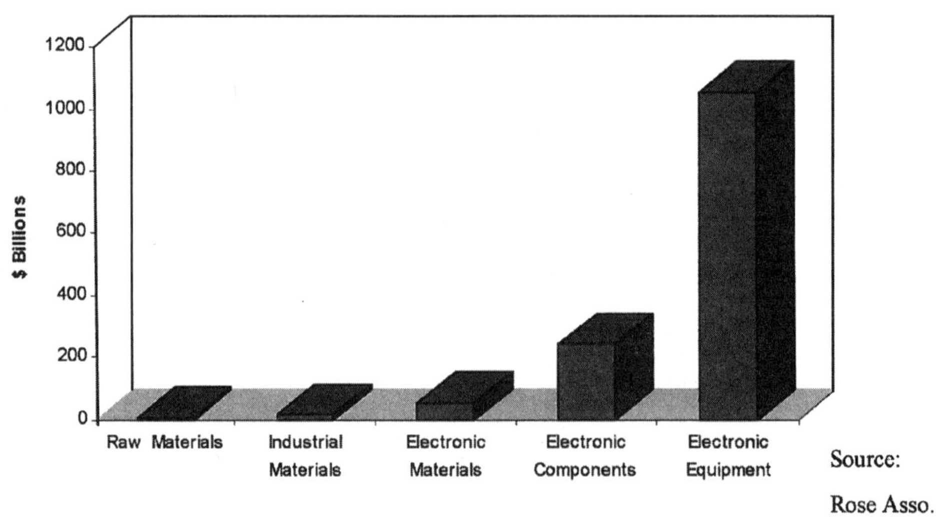

FIGURE 29.2 Electronics: The greatest value-added industry in the world.

With all these impressive accolades, it is most interesting to consider a quote by an" industry consultant. "The electronics industry is today where the automobile industry was in 1915."

29.1 COST: THE ISSUES

Reducing cost and increasing profit are often part of the same agenda, but not always. If the market for a product is strong, and the selling price is over $95.00, it is better to produce 1 million units per year at a $92.50 cost than 0.5 million units at cost of $92.00. Hence cost and profit are related but can vary widely depending on external conditions. While great demand accompanied by a high selling price makes cost less critical to profits, low demand and its associated low selling price make cost critical not only to profits but also to survival. This simple argument demonstrates the amazingly complex world of maximizing profit in electronics assembly.

A real-world example of this analogy can be seen in IBM prior to 1991. The gross margin on IBM's mainframe computers was between 6:1 and 8:1 (i.e., a computer costing $1 million dollars to produce would sell for between $6 and $8 million). Demand was so high that IBM could sell as many computers as it could produce. It was logical, then, that IBM could afford to pay a premium to ensure that its integrated circuit (IC) facility in East Fishkill, New York, produced enough ICs for the mainframes. Hence IBM invested heavily in this facility for years to support Fishkill's mission to produce "dies" at any cost. Fishkill became replete with much capital equipment and staff. In the early 1990s, it was the largest, most well equipped semiconductor facility in the world. By 1993, when IBM's margins and demand shrank, this modus operandi could not continue to be supported. Today, few, if any, ICs are manufactured at Fishkill.

The direct variables affecting assembly cost/profit are

1. The cost of the bill of material (BOM)
2. The cost of labor
3. The depreciation or lease cost of the assembly equipment
4. Cost of utilities and floor space
5. Cost of consumables
6. Cost of money (i.e., interest on loans to buy goods or services used)

These variables are referred to as *direct variables* because their cost can be reduced directly by negotiating a lower price or by using less of the specific variable. If all else stays the same, it always increases profit to reduce direct-cost variables.

It is more difficult to elucidate the effect of *indirect variables* on profit. These variables are

1. Throughput
2. Yield
3. Product sales price
4. Product demand
5. Inventory turns

In a brisk market—where demand and selling price are high—it almost always pays to increase productivity, even at the expense of considerable capital equipment. This situation was the case at IBM for years. It does not always pay to improve yields,[2] especially if the yield loss is reworkable. Improving yields may improve unit cost and profit, but not if productivity is affected. On the other hand, it always pays to have fast inventory turns. We will examine cost simulations in future sections that support these points.

29.2 DETERMINING COST AND POTENTIAL PROFIT

One of the most popular current techniques of determining cost is *activity-based costing* (ABC), which, once implemented, is simple to use. ABC is performed by determining the cost of the activities of assembling a product and adding to that the cost of the materials. Hence, for printed wiring board (PWB) assembly, one adds the cost of stencil printing, the cost of pick and place of the components, the cost of reflow, etc. and adds these costs to the cost of the PWB, the components, and the consumables. With this approach it is easy to estimate the total cost once the BOM is determined. This approach has weaknesses in that the true cost of an activity can vary depending on whatever step in the process is rate-limiting. In other words, it assumes that the activities of assembling can be costed separately, which they cannot. Because ABC does not view the assembly process as a system, it has a weakness: It helps determine approximate cost but not necessarily throughput or profit. The advantage of ABC, however, is that it can "tax" a component that takes more time to place and therefore increases cost. ABC is explained in detail in section 29.8.

Recently, assembly cost-estimating software has been developed.[2] This software, called SPACE™, enables rapid costing of any assembly process. It even has a feature to balance the line between chipshooters and flexible placement machines. It can perform minimum-cycle-time calculations but not complete line balancing. It does not afford ABC's ease of "taxing" a process step component. However, SPACE™ allows analysts to easily perform "what if" scenarios that more closely approximate the assembly line as a system. Throughout this chapter we will use SPACE™ to help us understand the critical issues associated with cost estimating, controlling cost, and maximizing profit.

29.3 LINE-LEVEL AND FACTORY-LEVEL COST

The National Electronics Manufacturing Initiative (NEMI) has delineated the costs of assembly steps in a chapter of its recently released *Technology Roadmap.*[3] These cost numbers were higher than those we obtained by using SPACE™. In analyzing the differences, it became clear that the NEMI team assumed staffing at the factory level. In factory-level staffing, the number of employees supported by each line is determined by dividing the total number of employees in the factory by the number of lines at the factory. Hence, if there are eight assembly lines at the factory and 240 employees, 30 employees are supported by each line. We refer to this approach as *factory-level cost estimating.* In our initial cost estimates, we only included the people working on and supporting the line. Typical numbers in this scenario are from 4 to 7 people. We call this *line-level cost estimating.*

When analyzing how to reduce costs from an assembly perspective, factory-level cost estimating is at a disadvantage. Adding the overhead of the entire factory (i.e., personnel, food service, product design) creates a situation that suggests that the greatest cost reduction would involve factors unrelated to the assembly process. For example, streamlining personnel, vending out the cafeteria, and reducing product design costs may reduce costs in this approach more than improving assembly-line productivity. Therefore, we believe that line-level cost estimating indicates the effects of cost actions related to assembly more accurately than factory-level cost estimating. We will, however, use both approaches in the following sections.

29.4 EXAMPLE: A MODEM

To give us our initial cost-estimating experience, we will start with the assembly of modems. The following table lists the cost parameters:

Item	Cost
Assembly equipment	$1.7 M
Floor space + utilities	$40,000/yr
Components	$75/unit
Stencil, squeegee	$600, $200, life = 1,000,000
Solder paste, consumables	$0.7/unit
PWB	$15/unit
7 workers	$28/h (includes benefits)
Selling price	$110
Shifts	12 10-h shifts per week
Cycle time	25 s
Down time, setup, maintenance	8% (10 h/wk–12.5 h/wk)
Workers supported	26 @ $42/h each
I/O	950
Yield	97% first pass, reworkable

29.4.1 SPACE™ Costing

We will assume that the assembly line produces these products for 1 year. We will first perform a cost calculation at the line level using the SPACE™ software and ABC (Table 29.1).

This table is from SPACE™. It lists the unit cost of the specific category, say, components, the percent of the unit cost, and the total cost for this category in the simulation. Hence, of the total unit cost of $93.63, the PWB and components are $90.00, or more than 95 percent of the cost of the product. It is easy to understand, then, how important it is to have effective purchasing agents negotiating the cost of the BOM. In this example, with a selling price of $110.00, even a 3 percent reduction in these costs ($2.70) increases profits by 15 percent or more. Labor is a relatively small cost at $2.65, but we must remember that this analysis has been performed at the line level (i.e., labor in the remainder of the company is not included). It surprises many people that the assembly equipment (machine cost) is such a low percentage of the cost. In this case, machine coast accounts for only $0.70 per unit produced. The consumables, rework, utilities, and space costs are also listed.

The number of boards produced in 1 year on this line would be 658,170, resulting in approximately $10.77 M in gross profit at the $110.00 selling price. The assembly cost per I/O is about $0.0038. NEMI states[3] that a reasonable assembly cost per I/O figure at the factory level should be $0.015; hence we also choose to examine the assembly cost per I/O at the factory level. By adding the 26 workers supported by the line at $42/h, we obtain the results in Table 29.2.

The factory-level approach dramatically alters costing results. The labor cost now skyrockets from $1.86 to $12.21, gross profit plummets to $3.96 million, and assembly cost per I/O is now

TABLE 29.1 SPACE ™ Output for Line-Level Analysis

A	B	C	D	E	F	G
20	Cost element	Unit cost	Percent	Total cost	Metrics	
21	Components	$ 75.00	80.100	$ 49,362,750.00	Total boards	658170
22	Labor	$ 1.86	1.985	$ 1,223,040.00	Unit cost	$ 93.63
23	PWB	$ 15.00	16.020	$ 9,872,550.00	Sell price	$ 110.00
24	Consumables	$ 0.70	0.748	$ 460,719.00	Gross profit	$ 10,772,307.30
25	Machine cost	$ 0.70	0.746	$ 460,010.15	NM cost/I/O	$ 0.0038
26	Rework	$ 0.32	0.336	$ 207,323.55	NM cost/plcmt	$ 0.0378
27	Floorspace, utilities	$ 0.06	0.065	$ 40,000.00	Line efficiency	73.250%
28	Total	$ 93.63	100.000	$ 61,586,392.70	Cycle time (s)	25

TABLE 29.2 SPACE™ Output for Factory-Level Analysis

A	B	C	D	E	F	G
20	Cost element	Unit cost	Percent	Total cost	Metrics	
21	Components	$ 75.00	72.125	$ 49,362,750.00	Total boards	658170
22	Labor	$ 12.21	11.743	$ 8,037,120.00	Unit cost	$ 103.99
23	PWB	$ 15.00	14.425	$ 9,872,550.00	Sell price	$ 110.00
24	Consumables	$ 0.70	0.673	$ 460,719.00	Gross profit	$ 3,958,227.30
25	Machine cost	$ 0.70	0.672	$ 460,010.15	NM cost/I/O	$ 0.0147
26	Rework	$ 0.32	0.303	$ 207,323.55	NM cost/plcmt	$ 0.1457
27	Floorspace, utilities	$ 0.06	0.058	$ 40,000.00	Line efficiency	73.250%
28	Total	$ 103.99	100.000	$ 68,400,472.70	Cycle time (s)	25

$0.0147. Since this value is in the range of NEMI's "best of breed" figure, our assumption of 26 workers supported by the line appears to be reasonable.

This analysis reflects, however, the previously mentioned error inherent in the factory-level costing approach. As anticipated, factoring in the people supported by the line shifts the potential cost savings from the line to other parts of the operation.

It is interesting to determine the gross profit as a function of cycle time. Using SPACE™, we can perform this analysis rapidly because it has a "scenario" feature that allows one to generate a graph of numerous dependent variables (such as net and gross profit) versus several independent variables (such as yield and cycle time). SPACE™ performs this analysis by varying the one independent variable (in our case, cycle time) while keeping the other variables constant. The results are shown in Fig. 29.3.

From this figure we see that the gross profit at a cycle time of 25 s (0.417 min) is about $4 million, as calculated earlier. This graph indicates the striking effect of cycle time on gross profit. By reducing cycle time to 20 s (0.33 min), gross profit increases to about $7.1 million, an increase of over 50 percent. Why does this striking effect occur? The first reason is that there are 25 percent more units produced to sell. Second, the unit cost of the products produced is reduced. Hence gross profit grows exponentially as cycle time is reduced.

29.4.2 Modem Cost with ABC

Activity-based costing can result in simple cost estimates. One only needs to add the cost of the materials to the cost of the activities. In our example we will assume the total cost of activities is rolled into the pick-and-place process step. The cost of the activities is only determined at the factory level; hence, to determine the cost of placing a component, we simply add the nonmaterial costs to run the factory over a given period of time and divide this cost by the number of components placed during this time. Since some components require more time to place than others (e.g., a 200-I/O quad flat pack versus a passive discrete), they would share a larger portion of the placement cost. This aspect of ABC is a strong point because it allows more difficult packages to be "taxed" accordingly.

Returning to our example from the last subsection, we remember that the total material cost was $90.70. To perform ABC, we need to know the number of components of each type and the cost to place them. These data are given in Table 29.3.

By adding the material cost and the component-related cost, we arrive at the same cost, $103.99. In this simple example we assume that rework is built into the ABC structure. The extreme cost of placing the highly leaded PQFP (>100 I/Os) is due to the fact that a lead coplanarity test must be performed before the part is placed. This additional step slows down the process down, thereby increasing cost. A detailed discussion of how ABC costs are calculated is contained in Section 29.8.

29.4.3 SPACE™ and ABC Compared

Considering the ease of ABC, one would wonder why anyone would want to use SPACE™. ABC is simple to use and has the advantage that one can readily "tax" a hard-to-place component. However,

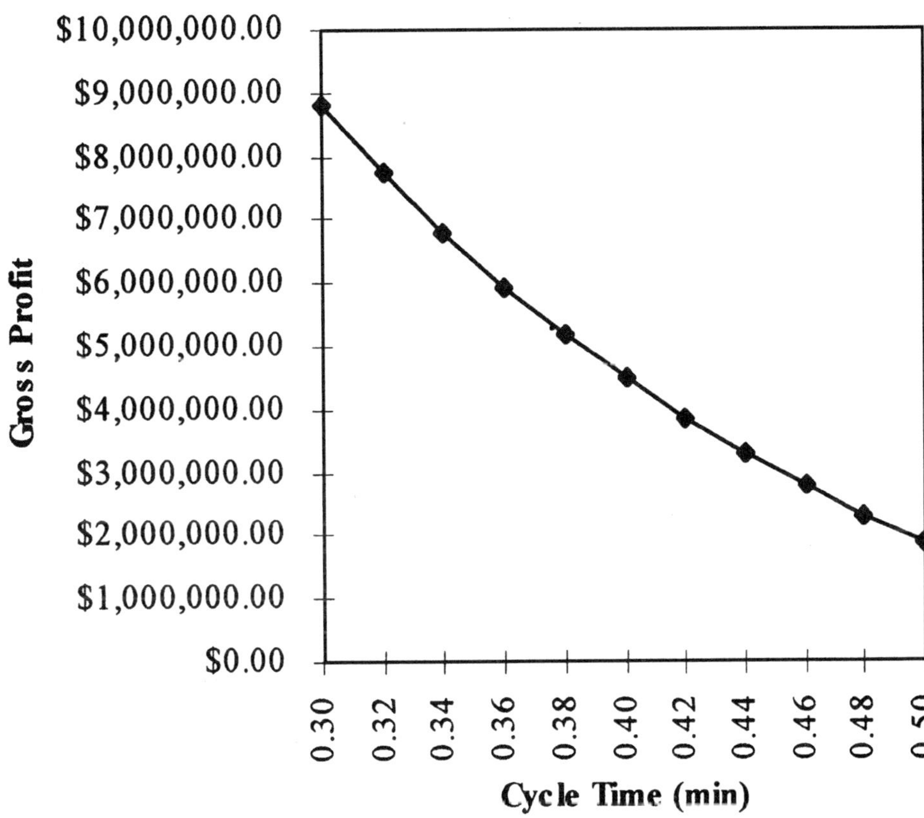

FIGURE 29.3 Gross profit versus cycle time at the factory level.

TABLE 29.3 Component and Placement Costs

Part type	Number	Unit placement cost ($)	Subtotal ($)
Passive	488	0.022	10.73
SOICs	40	0.03	1.20
PQFP < 100 I/O	22	0.04	0.88
PQFP > 100 I/O	4	0.12	0.48
Grand total	—	—	13.29

one must bear in mind that a team of cost analysts was required to develop the placement costs. Also, activity costs represent an average of the entire manufacturing site's activity costs, making it difficult to determine cost-reduction methods for a specific process. In addition, ABC cannot readily perform the cost scenarios that SPACE™ can.

Not only can SPACE™ be used to perform scenarios, but doing so requires only the first-order information related to the specific process at hand (i.e., BOM, specific machines related to the process, the precise number of workers needed for the assembly, etc.). This need, however, does require one to know the cycle times of the individual machines. Currently, the cycle times can only be determined accurately by physically setting up the line. Added to this disadvantage is the fact that SPACE™ does not provide a "tax" for parts that slow the process down. Again, one has to determine the process cycle time. Since SPACE™ is a new software product, we expect that these minor shortcomings will be improved in future versions.

All things considered, SPACE™ is a much more powerful tool than ABC because it allows one to perform the "what if" scenarios to reduce cost and increase profit. In the following sections we will demonstrate the use of SPACE™ to improve profit in numerous types of assembly.

29.5 MORE ILLUSTRATIVE EXAMPLES

To learn more about cost estimating for assembly and the drivers of cost, we will use SPACE™ to explore modified versions of our example in the last section.

29.5.1 Case 1: Add Capital Equipment to Reduce Cycle Time

Let us return to the example first introduced in Sec. 29.4. For convenience, we reproduce the pertinent parameters below. We will be looking at factory-level costs, as in Table 29.2.

Item	Cost
Assembly equipment	$1.7 M
Floor space + utilities	$40,000/yr
Components	$75/unit
Stencil, squeegee	$ 600, $200, life = 1,000,000
Solder paste, consumables	$0.7/unit
PWB	$15/unit
7 workers	$28/h (includes benefits)
Selling price	$110
Shifts	12 10-h shifts per week
Cycle time	25 s
Down time, setup, maintenance	8% (10 h/wk–12.5 h/wk)
Workers supported	26 @ $42/h each
I/O	950
Yield	97% first pass, reworkable

Let us assume that one wants to determine the effect (regarding cost and profit) of reducing cycle time by 2 s and adding a $1 million chipshooter into the line. These changes increase the assembly equipment cost to $2.7M and decrease the cycle time to 23 s. SPACE™ now gives the output in Table 29.4

Gross profit is now $4.76 million versus $3.96 million, and unit cost has dropped almost $0.65. Hence, despite the cost of the chipshooter, profits increased by 20 percent. The reason for this favorable situation is that since the cycle time was reduced by about 8 percent, 8 percent more product is produced. In other words, the increased number of units produced not only reduces unit cost but also provides more product to sell. Therefore, *profit increases exponentially as cycle time decreases.* This statement is perhaps the most important statement in this chapter.

29.5.2 Case 2: Increase Assembly-Line Workers' Pay $5/h to Decrease Cycle Time by 1 s and Improve Yield by 1 Percent

In this example, management meets with the assembly-line workers and offers an incentive. If the workers can increase first-pass yield by 1 percent and reduce cycle time by 1 s, they will get a $5/h

TABLE 29.4 Cost Output from SPACE ™ after Increasing Assembly Equipment by $1 Million and Reducing Cycle Time by 2 s

A	B	C	D	E	F		G
20	Cost element	Unit cost	Percent	Total cost	Metrics		
21	Components	$ 75.00	72.576	$ 53,655,150.00	Total boards		715402
22	Labor	$ 11.23	10.871	$ 8,037,120.00	Unit cost	$	103.34
23	PWB	$ 15.00	14.515	$ 10,731,030.00	Sell price	$	110.00
24	Consumables	$ 0.70	0.677	$ 500,781.40	Gross profit	$	4,764,776.82
25	Machine cost	$ 1.03	1.001	$ 740,010.15	NM cost/I/O	$	0.0140
26	Rework	$ 0.32	0.305	$ 225,351.63	NM cost/plcmt	$	0.1390
27	Floorspace, utilities	$ 0.06	0.054	$ 40,000.00	Line efficiency		73.250%
28	Total	$ 103.34	100.000	$ 73,889,443.18	Cycle time (s)		23

TABLE 29.5 Cost Output from SPACE™ after Raising Assembly-Line Workers' Wages $5/h and Improving Cycle Time to 24 s and First-Pass Yield to 98 Percent

A	B	C	D	E	F		G
20	Cost element	Unit cost	Percent	Total cost	Metrics		
21	Components	$ 75.00	72.426	$ 51,419,550.00	Total boards		685594
22	Labor	$ 11.91	11.505	$ 8,168,160.00	Unit cost	$	103.55
23	PWB	$ 15.00	14.485	$ 10,283,910.00	Sell price	$	110.00
24	Consumables	$ 0.70	0.676	$ 479,915.80	Gross profit	$	4,419,819.31
25	Machine cost	$ 0.67	0.648	$ 460,010.15	NM Cost/I/O	$	0.0143
26	Rework	$ 0.21	0.203	$ 143,974.74	NM cost/plcmt	$	0.1412
27	Floorspace, utilities	$ 0.06	0.056	$ 40,000.00	Line efficiency		73.250%
28	Total	$ 103.55	100.000	$ 70,955,520.69	Cycle time (s)		24

raise. Hence, in this scenario, labor is $33/h (the workers do not get this amount, since benefits are in this figure), cycle time is 24 s, and first-pass yield is 98 percent. The results are shown in Table 29.5.

In this scenario, unit cost is reduced $0.44, and profit is increased from $3.96 million to $4.42 million. The first reason that unit cost and profit improve is the same as in Sec. 29.5.1: More product is produced by reducing cycle time. The yield improvement also has the same effect. *The workers on the assembly line support the entire factory; hence their productivity is of prime importance. The quality and treatment of these workers should be outstanding. Hence, in this case, a 10 percent increase in the workers' salary (which is about 1% of the final cost) results in 11 percent increase in gross profits.*

29.5.3 Case 3: Improving Yield at the Cost of Throughput

Senior management is quite enthusiastic to improve first-pass yields. A task force is formed to study the problem, and after 3 months, the group concludes that the solution is to perform a lead coplanarity test on one of the 208-I/O quad flat packs (QFPs). This test causes the first-pass yield to skyrocket to 99.5 percent, but the pick-and-place process must be slowed by 2 s, resulting in a new cycle time of 27 s. Keep in mind that in our example all yield loss is reworkable. The cost results are shown in Table 29.6.

The results to cost are disastrous. Unit cost increases more than $0.75, and gross profit plummets to less than $3.2 million. The reason for this result is that increasing yield does not increase the number of units made (all products lost in first-pass yield loss are reworkable), and increasing cycle time reduces the amount of product assembled. Hence both the unit cost goes up and the profit goes down exponentially. *Improving yield is a noble goal, but it must never be accomplished at the expense of throughput.*

TABLE 29.6 Cost Output from SPACE™ after Rising First-Pass Yield to 99.5 Percent While Adding 2 s to the Cycle Time

A	B	C	D	E	F	G
20	Cost element	Unit cost	Percent	Total cost	Metrics	
21	Components	$ 75.00	71.591	$ 45,706,275.00	Total boards	609417
22	Labor	$ 13.19	12.589	$ 8,037,120.00	Unit cost	$ 104.76
23	PWB	$ 15.00	14.318	$ 9,141,255.00	Sell price	$ 110.00
24	Consumables	$ 0.70	0.668	$ 426,591.90	Gross profit	$ 3,192,623.55
25	Machine cost	$ 0.75	0.721	$ 460,010.15	NM cost/I/O	$ 0.0155
26	Rework	$ 0.05	0.050	$ 31,994.39	NM cost/plcmt	$ 0.1538
27	Floorspace, utilities	$ 0.07	0.063	$ 40,000.00	Line efficiency	73.250%
28	Total	$ 104.76	100.000	$ 63,803,246.44	Cycle time (s)	27

29.6 A REAL-WORLD EXAMPLE

In the real world, one usually does not assemble one product for a year. More typically, one will assemble a certain number of a specific product. When this product run is finished, the line will be modified, and a new product will be assembled. To address this type of situation, we will use SPACE™ to analyze three product runs. All the costs will be at the factory level. We will start by assembling 12,680 of the modems discussed earlier. For convenience, we repeat the table of data relevant to this product.

Item	Cost
Assembly equipment	$1.7 M
Floor space + utilities	$40,000/yr
Components	$75/unit
Stencil, squeegee	$ 600, $200, life = 1,000,000
Solder paste, consumables	$0.7/unit
PWB	$15/unit
7 workers	$28/h (includes benefits)
Selling price	$110
Shifts	12 10-h shifts per week
Cycle time	25 s
Down time, setup, maintenance	8% (10 h/wk–12.5 h/wk)
Workers supported	26 @ $42/h each
I/O	950
Yield	97% first pass, reworkable

After assembling these units, we then assemble 1210 low-end computer motherboards that require a 50-s cycle time and have 90 percent first-pass reworkable yield. The selling price for the motherboard is $230.00. The PWB cost is $30.00, and the component BOM cost is $150.00. On finishing this product, we assemble 49,073 controller units that have a 20-s cycle time and 98 percent reworkable first-pass yield. The PWB cost for this unit is $10, and the BOM cost is $60. The controller unit's selling price is $90.00.

By analyzing this scenario with SPACE™, we find that the modems take 7 days to assemble, the motherboards take 1.4 days, and the controller units take 21.7 days. Hence production of the three product runs consumes 1 month. If desired, extra setup time could have been allowed for the analysis. The cost analysis is listed in Tables 29.7 through 29.9.

TABLE 29.7 Modem Cost and Profit Figures

A	B	C	D	E	F	G
20	Cost element	Unit cost	Percent	Total cost	Metrics	
21	Components	$ 75.00	72.107	$ 951,000.00	Total boards	12680
22	Labor	$ 12.24	11.763	$ 155,139.91	Unit cost	$ 104.01
23	PWB	$ 15.00	14.421	$ 190,200.00	Sell price	$ 110.00
24	Consumables	$ 0.70	0.673	$ 8,876.00	Gross profit	$ 75,928.28
25	Machine cost	$ 0.70	0.674	$ 8,889.49	NM cost/I/O	$ 0.0147
26	Rework	$ 0.32	0.303	$ 3,994.20	NM cost/plcmt	$ 0.1460
27	Floorspace, utilities	$ 0.06	0.059	$ 772.12	Line efficiency	73.250%
28	Total	$ 104.01	100.000	$ 1,318,099.61	Cycle time (s)	25

TABLE 29.8 Motherboard Cost Figures

A	B	C	D	E	F	G
20	Cost element	Unit cost	Percent	Total cost	Metrics	
21	Components	$ 150.00	72.130	$ 181,500.00	Total boards	1210
22	Labor	$ 24.66	11.860	$ 29,844.42	Unit cost	$ 207.96
23	PWB	$ 30.00	14.426	$ 36,300.00	Sell price	$ 230.00
24	Consumables	$ 0.70	0.337	$ 847.00	Gross profit	$ 26,671.27
25	Machine cost	$ 1.42	0.683	$ 1,718.28	NM cost/I/O	$ 0.0294
26	Rework	$ 1.05	0.505	$ 1,270.50	NM cost/plcmt	$ 0.2912
27	Floorspace, utilities	$ 0.12	0.059	$ 148.53	Line efficiency	73.250%
28	Total	$ 207.96	100.000	$ 251,480.20	Cycle time (s)	50

TABLE 29.9 Controller Unit Cost Figures

A	B	C	D	E	F	G
20	Cost Element	Unit cost	Percent	Total cost	Metrics	
21	Components	$ 60.00	73.807	$ 2,944,380.00	Total boards	49073
22	Labor	$ 9.77	12.024	$ 479,687.09	Unit cost	$ 81.29
23	PWB	$ 10.00	12.301	$ 490,730.00	Sell price	$ 90.00
24	Consumables	$ 0.70	0.861	$ 34,351.10	Gross profit	$ 427,264.35
25	Machine cost	$ 0.56	0.688	$ 27,464.77	NM cost/I/O	$ 0.0119
26	Rework	$ 0.21	0.258	$ 10,305.33	NM cost/plcmt	$ 0.1176
27	Floorspace, utilities	$ 0.05	0.060	$ 2,387.36	Line efficiency	73.250%
28	Total	$ 81.29	100.000	$ 3,986,918.29	Cycle time (s)	20

We see that in this scenario total gross profit was about $530K on sales of $6.09 million, or about 8.7 percent gross margin.

29.6.1 References

1. Lassen, Charles, Prismark Partners LLC.
2. SPACE™, Speedline Technologies, 16 Forge Park, Franklin, MA 02038.
3. NEMI, 2214 Rock Hill Road, Suite 110, Herndon, VA, *1996 Technology Roadmap*, pp. 41–49.

29.7 INTRODUCTION TO SURFACE MOUNT TECHNOLOGY AS A REFERENCE TO COST ESTIMATING

Surface mount technology (SMT) has been the electronics industry's critical manufacturing technology for the past 15 years. It offers many advantages over the traditional through-hole (TH) technology, such as increased component density, improved electrical performance, reduced cost, and higher quality.

In order for a manufacturer to adapt the SMT technology successfully, various aspects of the technology must be clearly understood. This section is written with the hope of providing a fundamental, concise introduction to SMT. It is separated into three subsections: an introduction of the various package types, descriptions of typical SMT line configurations, and a presentation of the major pieces of equipment used in SMT production.

29.7.1 Package Types

An *electronic package* is defined as a housing that interconnects electronic devices and isolates them from the environment. Traditionally, the packaging scheme has been divided into three levels (Fig. 29.4). The first level houses the devices and acts as a space transformer from the small, closely spaced contact pads on the chip to the large wiring network on the PWB. The second level of packaging is the PWB card or package. The third level of the package consists of the PWB motherboard.

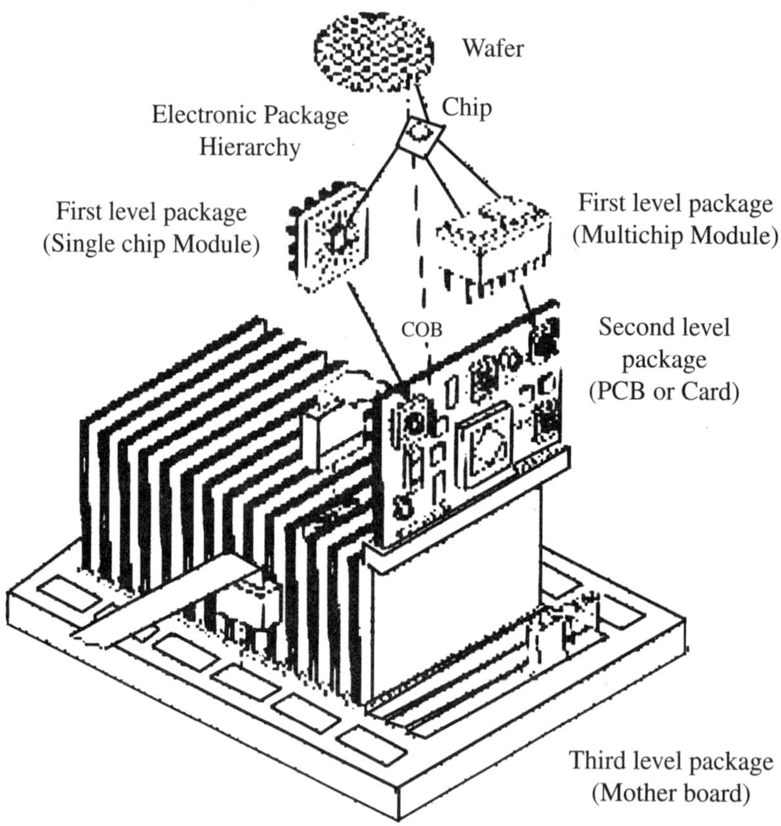

FIGURE 29.4 The first three levels of packaging.

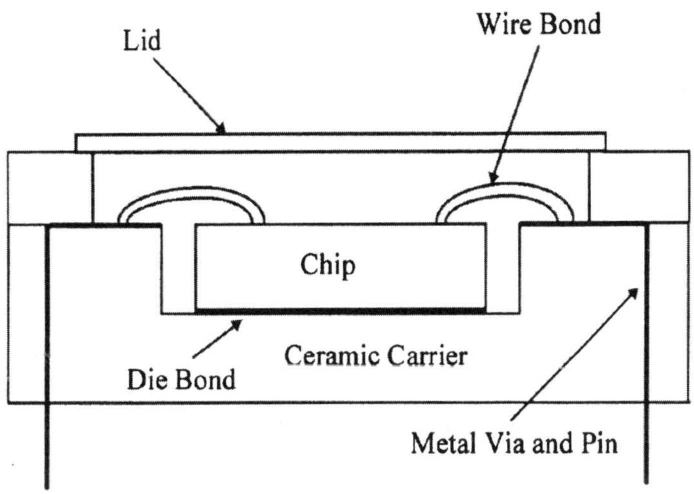

a) Wire bonded First Level Package

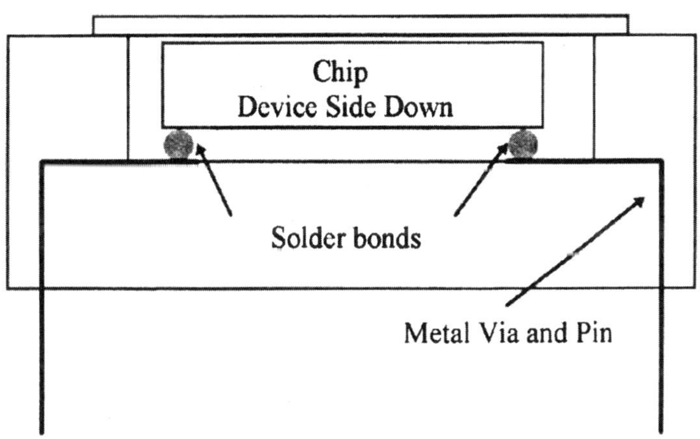

b) "Flip Chip" First Level Package

FIGURE 29.5 Two types of first-level packages: wire bonded and flip chip (the second-level package is a ceramic DIP).

First-Level Package. The major concern at this level is the electrical connection between the IC and the second-level package. There are two methods to perform this interconnection, and they are shown in Fig. 29.5. The first method, wire bonding, ultrasonically bonds 0.001-in-thick gold wires from the chip to the carrier. The problem with wire bonding is that very closely spaced interconnections are difficult to wire reliably. Also, parasitic inductance (crosstalk) between adjacent wires can occur and scramble the signals. The second method, flip chip (FC) technology, directly solders the chip, device side down, to a chip carrier. The surface tension of the solder balls holds the chip up.

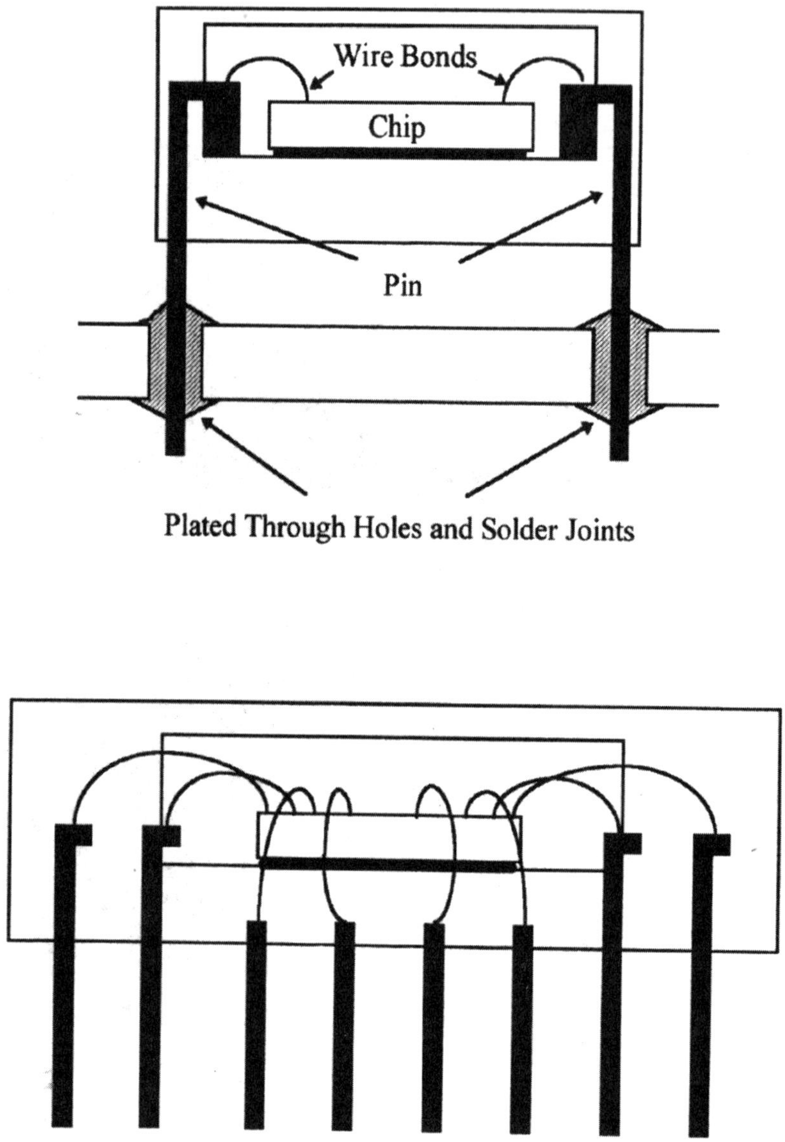

FIGURE 29.6 Two side views of a dual in-line package (DIP).

Because this design is wireless, it eliminates the problem of wire crosstalk. The chip self-aligns on reflow of solder; therefore, the complexity of wiring closely spaced contact pads is eliminated in FC technology. These solder joints, however, are highly susceptible to thermal fatigue failures.

Attaching to the Second-Level Package. One of the first packaging designs used to attach to the second level was the dual in-line package (DIP). The DIP consists of a rectangular enclosure made of plastic or ceramic with two rows of leads on opposite sides of the rectangle, as shown in Fig. 29.6. This package is inserted and soldered into plated through-holes (PTH) on the PWB. A PTH is a small hole drilled through the PWB that has been plated with a thin, metallized layer. The problem with

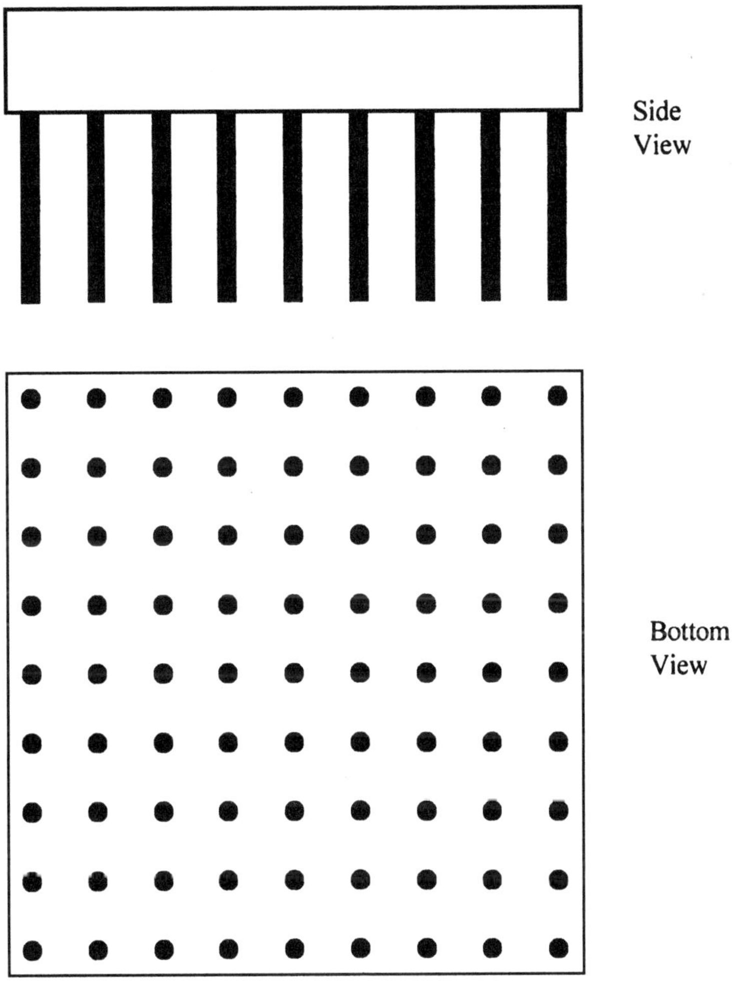

FIGURE 29.7 Two views of pin grid arrays (PGAs).

the DIP is that it takes up a lot of PWB real estate. As the number of I/O leads increases, the DIP also must increase in length. There is also a signal-propagation delay associated with traveling over long distances and the possibility of parasitic inductance between pads. There is a desire, therefore, to design a compact-sized package with a large number of leads.

The second design possibility is a compact carrier called the *pin grid array* (PGA). In the PGA (Fig. 29.7), I/O leads are pins on the bottom of the package rather than on the edge. This results in a package of greatly reduced size. One problem with the PGA is that parasitic inductance is still possible across pin leads. In addition, this package design is much more expensive than the DIP and is difficult to replace. To replace a PGA, all the solder joints on the pins must be remelted. If the PGA is pulled before full melting has occurred, a PTH can be lost, resulting in the loss of the entire board.

The most promising design is the leaded, or leadless, chip carrier mounted using SMT. This square-shaped chip carrier is made of either plastic or ceramic and has I/O connections on all four sides, as shown in Fig. 29.8. It is surface mounted with solder bonding the pads on the PWB and the metallized connection of the carrier. As compared with a DIP with the same number of leads, a chip carrier has a 60 percent decrease in area. Because no expensive PTHs are required on the PWB, the

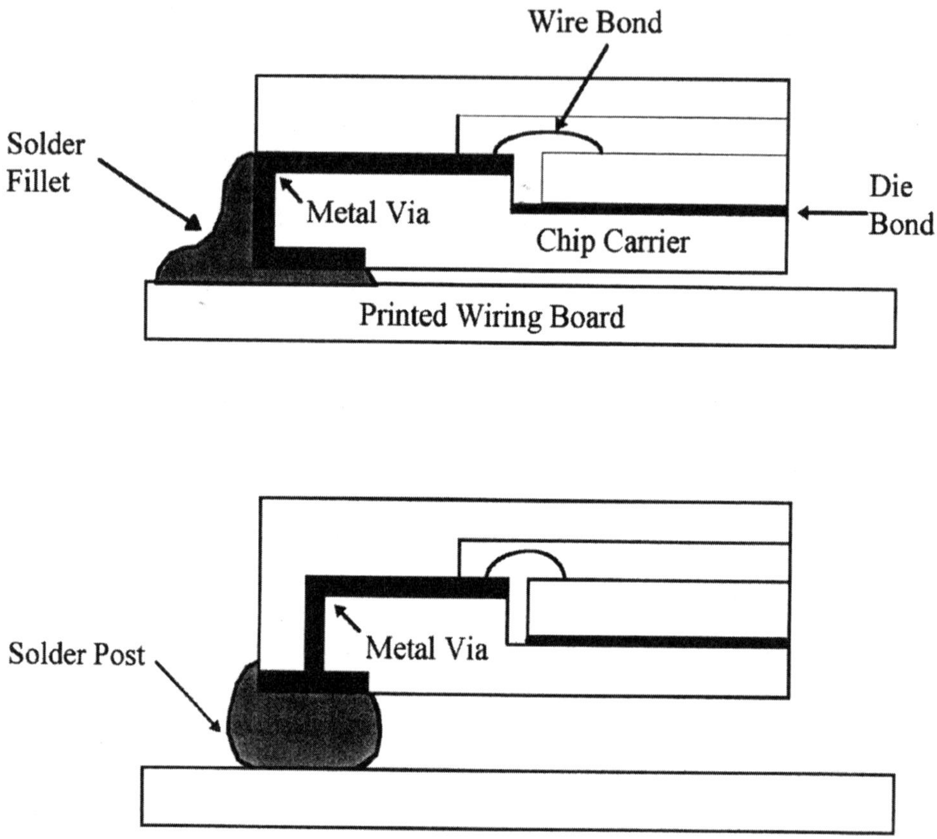

FIGURE 29.8 Two examples of leadless chip carriers (LCCs).

chip carrier is less expensive than the DIP as well as the PGA. It is also possible to solder carriers to both sides of the PWB. The short line lengths in the package result in low parasitic impedance in surface mount. In addition to all these advantages, the chip carrier's rugged, pinless (there are no pins that can be bent) construction makes it easy to handle.

Leaded versus Leadless Chip Carriers in SMT. Two types of leaded carriers are shown in Fig. 29.9. A leadless chip carrier (LCC) is shown in Fig. 29.8. The leaded carrier has the advantages that flux removal from under the carrier is easy and the lead takes up some of the strain in thermal cycling. A leaded carrier, however, does increase lead lengths and therefore is slower at signal transport.

Plastic versus Ceramic Chip Carriers in SMT. Plastic packages are inexpensive and facilitate size reduction of consumer products. In the United States, however, the primary use of SMT chip carriers is by the military, and the packages are made of alumina. These ceramic packages, although expensive, have very high reliability. The small size, light weight, and rugged build of the ceramic chip carrier are ideal for high-speed operations in missiles and spacecraft when extremely high reliability is required. The reliability of the ceramic package hinges on the hermiticity of the package. Properly sealed, a ceramic package is impermeable to the environment. A plastic carrier is much more susceptible to the environment, and errors can occur in device function from impurities polluting the high-grade Si or GaAs chips.

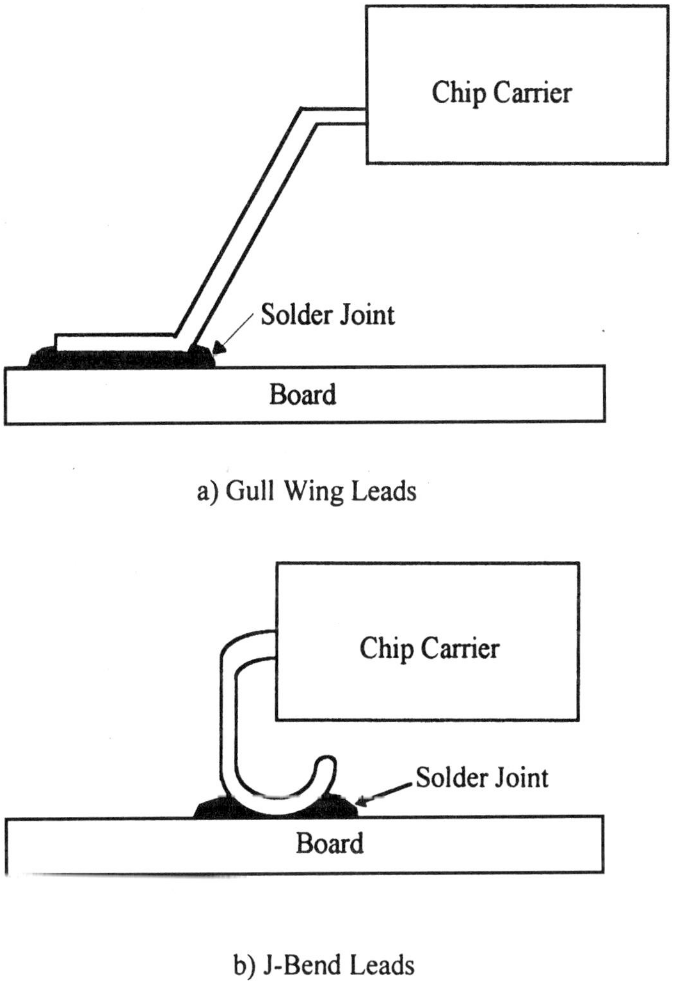

FIGURE 29.9 Two examples of leaded chip carriers.

29.7.2 Typical SMT Line Configurations Used to Assemble First-Level Packages to PWB

There are three distinctive types of SMT assembly currently being used. They are commonly referred to as types I, II, and III. *Type I SMT assembly* involves packages with only surface-mounted components on the top side of the board. *Type II SMT assembly* includes packages with both SMT and through-hole components. *Type III SMT assembly* includes packages with SMT components on both sides of the PWB. Each of these assemblies requires a unique set of process steps.

The flowchart of a type I assembly can be seen in Fig. 29.10. The solder paste (normally an Sn/Pb alloy) is first screened or printed with a stencil printer on top of the lands and pads where the components will be placed. The components are then placed on top of the solder paste with a placement machine. This is followed by a reflow step, during which solder melts and forms a metallurgic bond between the leads of the component and the substrate. The last step of the process is cleaning off the board, which can be skipped if a no-clean paste is used. For double-sided SMT

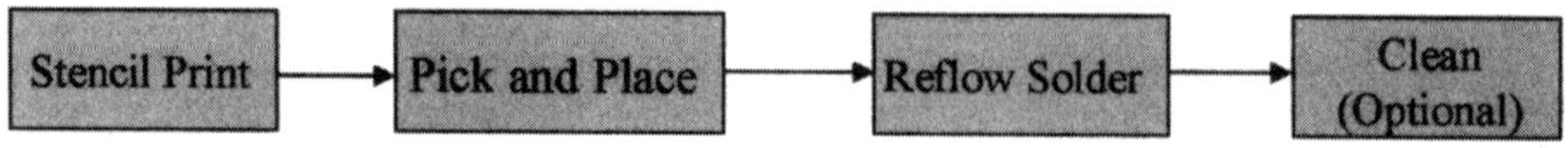

FIGURE 29.10 Type I SMT assembly.

FIGURE 29.11 Type III SMT assembly.

assembly, the board is turned over after the type I assembly process, and the entire sequence is repeated. The joints on the top side are reflowed again, and the surface tension of the solder holds the components in place.

The process sequence for type III assembly can be seen in Fig. 29.11. First of all, the through-hole components are inserted and their leads clinched. The board is turned over, and adhesive is placed on the board, followed by placement of the components with a placement machine. The adhesive is then cured in a oven, and both leaded and surface-mount components are reflowed in a single wave solder step.

Type II assembly is essentially a combination of the type I and III assembly, so it goes through the process sequence of type I followed by the sequence of type III. The flowchart of type II assembly can be see in Fig. 29.12.

29.7.3 SMT Equipment

Four major processes are used in a typical SMT line. They are stencil printing, placement of components, adhesive dispensing, and thermal processing. The function and critical operating parameter of each one will be discussed.

Stencil printing is the SMT process of applying solder paste to the circuit board using a stencil and squeegee. The squeegee pushes solder paste across the top of the stencil positioned above the circuit board, forcing the solder paste into apertures in the stencil. When the stencil is lifted from the board, solder bricks are left on the individual SMT land pads. Stainless steel is the most common material used to make stencils, but beryllium copper, Alloy 42, and molybdenum are also used. There are many different methods of making stencils, but the most common are chemical etching, laser cutting, and plating. Stencils are mounted to a frame using a stretched mesh border. This provides the tension to maintain a taught stencil yet allow deflection of the stencil under the pressure applied with the squeegee. Some of the other important parameters in stencil printing are squeegee pressure, squeegee speed, stencil snap-off, and squeegee angle. An example of a stencil printing machine is shown in Fig. 29.13.

After solder paste has been applied on the bare circuit board, the components are positioned with a placement machine. Two types of placement machines currently are used: the turret-style chipshooter and the flexible/fine-pitch placer. Instead of picking the first part from a component feeder, centering the nozzle, placing the board, and processing the second component in a sequential fashion, the turret-style chipshooter performs the steps in parallel with each other. This greatly increases machine throughput, since the fastest chipshooter today can place parts onto a board at a rate of more than 40,000 components per hour. An alternative design to the turret-style chip shooter is the flexible/fine-pitch placer. Most flexible placement machines rely on an overhead gantry positioning system on which the placement head is mounted, possibly with multiple spindles. Components are picked from stationary feeders and delivered to the board for placement. Component registration prior to placement can be accomplished with either mechanical centering jaws or a machine vision system. The main advantage of the flexible placer is that it can cover the

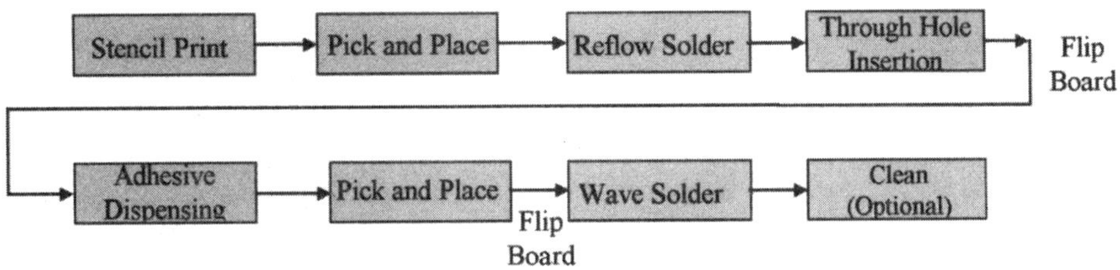

FIGURE 29.12 Type II SMT assembly.

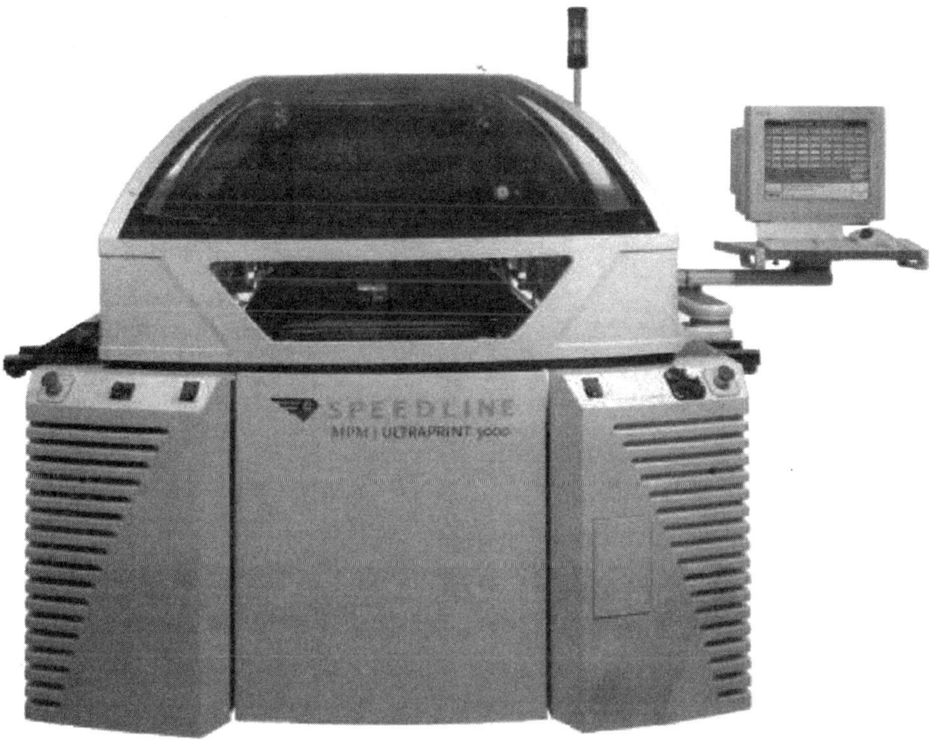

FIGURE 29.13 Automatic stencil printer. (*Speedline/MPM.*)

complete range of packages, but the main disadvantages of the system is that it can only operate at about 30 percent of the maximum speed of the chipshooter. An example of a flexible/fine-pitch placer is shown in Fig. 29.14.

Adhesive dispensers are used for bottom-side application of adhesive dots. The board must be flipped over and passed through a wave solder machine. Three basic dispensing technologies are being used today in the electronics assembly industry, each of which excels when used in specific applications. They are time/pressure dispensers, rotary-screw-valve dispensers, and positive-displacement dispensers. Of these dispensing technologies, the time/pressure dispenser method is the oldest. In this method, the syringe holds the material, and a plunger applies pressure to the material to push it through the tip. The higher the pressure and the longer it is applied, the greater amount

FIGURE 29.14 Automatic placement machine. (*Siemens.*)

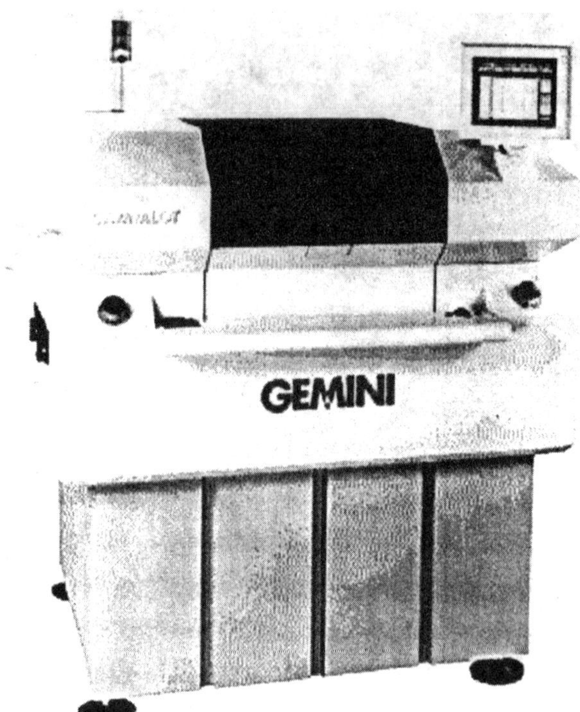

FIGURE 29.15 Adhesive dispenser. (*Speedline/Camalot.*)

is dispensed. The second method, the rotary-screw-valve dispenser, also known as the auger-screw dispenser, is an extremely popular technique. With this dispenser, an auger screw controls the material flow through the dispensing tip. Material gets fed through to a motorized auger screw that rotates and forces the material through the tip. The third method, a positive-displacement pump, uses a piston moving through a chamber filled with adhesive to push controlled volumes of the adhesive out through the nozzle. Low-pressure air is applied continuously to keep the pump chambers filled. An example of an adhesive dispenser is shown in Fig. 29.15.

After the components have been placed onto the board with either solder paste or adhesive, the boards are fed through a thermal processor, and metallurgic bonds are formed. Two distinctive types of processes typically are used: reflow soldering and wave soldering. In reflow soldering, deposited solder paste is reflowed (melted) to form an electrical and mechanical connection between the PWB and the component leads. Four types of reflow techniques are used: vapor phase, infrared, natural convection, and forced-air convection. In comparison with reflow soldering, which typically is used for SMT components, wave soldering was designed for mass soldering of through-hole components on a PWB. In the wave soldering process, components on the board with leads accessible from the bottom of the board are passed over a flux dispenser and then heat, which activates the cleaning agents in the flux. The boards are then passed over waves of molten solder to create a connection between the leads and the pad. An example of a reflow oven is shown in Fig. 29.16.

This section is only intended to give you a brief introduction to the field of SMT. It is not by any means comprehensive in terms of scope or depth. Interesting readers may want to refer to works by Lau,[1] Prasad,[2] or Lasky et al.[3] for more details on the subject.

29.7.4 References

1. Lau, J. H., *Handbook of Fine Pitch Surface Mount Technology,* Van Nostrand Reinhold, New York, 1994.
2. Prasad, R., *Surface Mount Technology,* Chapman & Hall, New York, 1997.
3. Lasky, R. C., Seraphim, D. P., and Li, C. Y., *Principles of Electronic Packaging,* McGraw-Hill, New York, 1989.

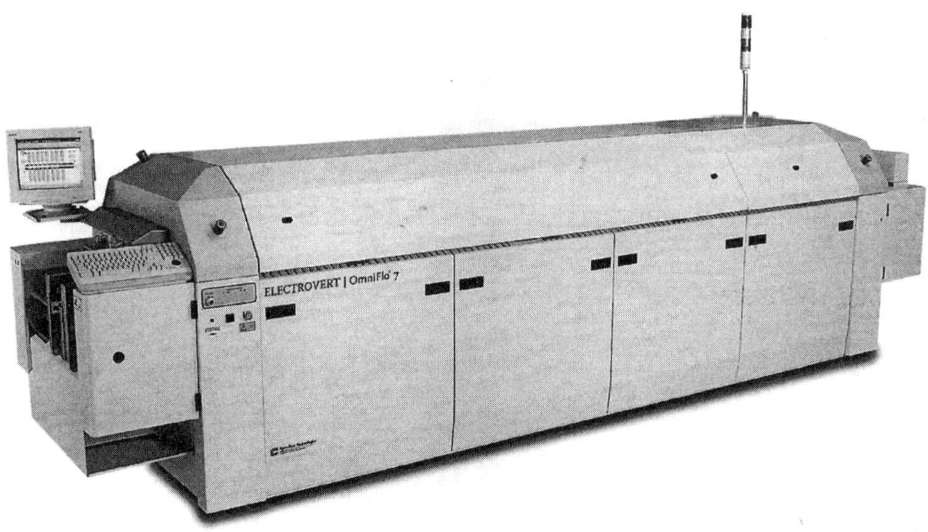

FIGURE 29.16 A reflow oven. (*Speedline/Electrovert.*)

29.8 APPENDIX: ACTIVITY-BASED COSTING

Activity-based costing (ABC) in its most basic form is a tool that allocates overhead in a more precise manner to products and processes. The ABC concept is used to create a relationship between overhead costs and production activities to accurately assign these costs to a given product. The objectives of ABC are (1) inventory valuation and the associated value to cost of sales, (2) cost control and performance measurement, and (3) decision costing and the design for manufacturability. The primary emphasis is that activities to build product consume resources and that products consume those same activities.

Properly implemented ABC systems build complete confidence in all direct organizations such as finance, manufacturing, and engineering. The goal for finance is to integrate this system into its current cost-management system to cost products. Manufacturing will use this system to drive process/build improvements, and engineering will use it to design products using ABC-approved components by defining a rating system that fairly allocates overhead.

In finance, the best ABC systems integrate the application to its information systems. Finance values inventory using driver rates that most likely reflect the cost of the activity to produce them. The rates usually are established annually and are driven by the comprehensive master budget. Systems actually carry the proper ABC overhead based on activity/driver rates related to them. They are not done on the side by some cost accountant and then reallocated to products. Dual-cost systems are not cost-effective and do not gain the confidence of all organizations involved. Manufacturing provides the activities used to cost the products and updates accordingly based on process improvement. Engineering provides the proper design by using process-approved and cost-effective components.

In order to set up an ABC model, we need to (1) define activities, (2) determine manufacturing budgets, (3) determine manufacturing management-allocated costs, (4) choose cost drivers, (5) determine driver volume based on production plan, (6) calculate driver rates, and (7) determine driver usage for each product.

Let's look at a manufacturing process at a motherboard factory. The activities will be split between production line board insertions and a board test activity at the end of the line. The supply management activity will be based on a component rating system that allocates overhead. The budgets will be based on a $10 million revenue plan, and we will assume that the manufacturing management team allocates 50 percent of its time on this product line. The production cost drivers will be component insertions on the raw PWB. Specifically, we will use two types of automatic insertions: complex placement and simple automatic insertion. Operators will place the third type of insertion manually. We will have an in-circuit test that will be either complex or simple. The supply management cost drivers will rate components as a preferred part, adequate part, restricted part, or unqualified part. The annual volume of motherboards will be set at 100,000 units, and we will manufacture 60,000 of motherboard A and 40,000 of motherboard B.

Supply Management Department Budget for Year 1997

Personnel	
8 planner/buyers	$600,000
4 component engineers	$350,000
25% of senior management support	$250,000
Facility/IS allocation	$650,000
Miscellaneous expenses	$200,000
TOTAL BUDGET	$2,050,000

Supply Management Resource

Allocation matrix	Percent of time dedicated	Budget dollars
Preferred parts	20%	$410,000
Adequate parts	35%	$718,000
Restricted parts	25%	$512,000
Unqualified parts	20%	$410,000
TOTAL	100%	$2,050,000

Preferred Parts (P). These parts have been subject to the highest vendor quality survey. Rating categories include quality, on-time delivery, and engineering component rating.

Adequate Parts (A). These parts have been subject to the second highest vendor quality survey. Rating categories include quality, on-time delivery, and engineering component rating.

Restricted Parts (R). These parts have been subject to the third highest vendor quality survey. Rating categories include quality, on-time delivery, and engineering component rating. These parts have been known to fail, and reliability and delivery have been suspect.

Unqualified Parts (U). These parts have been subject to the fourth highest vendor quality survey. Rating categories include quality, on-time delivery, and engineering component rating. These parts have been known to fail, and reliability and delivery have been suspect. In many cases these parts have not been subject to quality ratings. They are normally very new and have not had enough time in the field to evaluate. They also include very poor components that should replaced, but they have been designed into the product, and changing vendors would be somewhat costly.

The next task will be to analyze for each product the number of each type of receipt and determine the driver rates for each supply management receipt.

ABC Supply Management Overhead Cost Analysis

Bill of materials	Motherboard A		Motherboard B
Raw PWB	0.3825		3.4595
Complex components			
350 (P) @ 0.0077	2.6950	100 (P)	0.7700
50 (A) @ 0.0246	1.2300	190 (A)	4.6740
0 (R) @ 0.0731	0.0000	100 (R)	7.3100
0 (U) @ 0.2278	0.0000	10 (U)	2.2780
Simple components			
400 (P) @ 0.0077	3.0800	100 (P)	0.7700
110 (A) @ 0.0246	2.7060	300 (A)	7.3800
0 (R) @ 0.0731	0.0000	75 (R)	5.4825
0 (U) @ 0.2278	0.0000	35 (R)	7.9730
Total ABC supply management overhead	10.0935		40.0970

By using a component rating system that allocates overhead to each part, one can fairly allocate overhead to products. In the preceding example, both products are motherboards, but motherboard B has chosen components that have not been highly rated by supply management. In many cases we spend significant time on components that are of less quality and more time scheduling than others do. In traditional cost systems that allocate overhead with straight percentages, there is no differentiation in allocation between good and bad components. Using ABC, one can reward and drive improvement by rating components. The design engineer will need to work with supply management to improve the specific components or to change to preferred components.

Listed below are the resources required to run a three-shift production line to produce 100,000 motherboards. For simplicity, we are going to use percent of time to allocate costs, but normally you would set up a work center per activity and define costs more precisely with specific depreciation.

Production Department Budget for Year 1997

Personnel	
24 operators/test techs	$624,000
2 process engineers	$150,000
25% of senior management support	$250,000
Facility/IS allocation/depreciation	$1,150,000
Miscellaneous expenses/expendables	$600,000
TOTAL BUDGET	$2,774,000

Production Resource

Allocation matrix	Percent of time dedicated	Budget dollars
Complex	40%	$1,109,600
Simplex	30%	$832,200
Manual	15%	$416,100
In-circuit test	15%	$416,100
TOTAL	100%	$2,774,000

ABC Production Overhead Cost Analysis

Bill of Materials	Motherboard A		Motherboard B
Complex components			
400 @ 0.0277	11.0800	400	11.0800
Simple components			
490 @ 0.0173	8.4770	470	8.1310
Manual components			
20 @ 0.1486	2.9720	40	5.9440
In-circuit test			
1 low	1.7338	1 high	7.8019
Total ABC production overhead	24.2630		32.9570

Summary Cost Analysis Using ABC to Allocate Overhead for the Two Motherboards

	Motherboard A	Motherboard B
Supply management activities	10.0935	40.0970
Production activities	24.2630	32.9570
Total overhead per board	34.3570	73.0540

Using ABC to cost overhead differs from using traditional accounting methods that base allocation on percentages of materials or production hours. ABC rewards process improvement by defining activities that can be measured and improved. This company should try to move customers from motherboard B or should try to match as many processes or components as motherboard A.

In summary, the keys to activity-based costing are integration into mainstream information systems, allowing production, purchasing, and engineering to define activities that should be used as drivers to allocate costs; and revisiting drivers frequently to ensure that they are still valid. ABC is the tool to (1) establish a common bond between the operational and financial groups, (2) address total product costs in common, easily understood, and recognizable terms as an activity in the production cell, and (3) monitor and drive continuous improvements.

ABC will allow for the following performance measurements:

- Providing increased understanding of activity cell costs
- Providing decision costing for product design engineers
- Providing decision costing for process design engineers
- Providing activity center measurements actual versus budget
- Measuring manageable segments of the process
- Providing incentive to engineering to eliminate costly activities
- Providing incentive to use preferred components in design

The overall goal of any business should be to integrate finance into the business and not vice versa. ABC allows business to focus attention on real opportunities and problems.

S • E • C • T • I • O • N • 5

THE FUTURE

CHAPTER 30
THE FUTURE OF ELECTRONIC PACKAGING

Steve Anderson
Silicon Bandwidth, Inc.

30.1 THE CONVERGENCE OF MARKETS: COMPUTING, COMMUNICATIONS, CONSUMER

It is clear that today's world is focused on miniaturization, with the result that the convergence phenomenon is increasingly evident in markets, technologies, and interconnect. As the markets are converging among computing, communications, and consumer devices and applications, the convergence is creating an entirely new category of lower-cost, high-volume Internet or portable information appliances. The convergence likewise drives an unprecedented increase in both the volume of new devices and the actual rate of introduction of new devices. The results are major changes in the way that semiconductor packages are developed, introduced, and customized.

With the advent of voice-over-IP, or digital voice telephone traffic over the Internet, it has become increasingly difficult to separate the "standard" telecommunications markets and companies from the emerging data communications markets and their associated companies. The startling increase in numbers of startup firms and home-based businesses with Internet capability has leveled long-standing barriers between large and small businesses, furthering the impetus for these smaller enterprises to purchase and effectively implement new, attractively cost-effective technologies. The quick price decline in these technology products has moved the decision point for many electronics goods out of the corporate environment and into the hands of consumers, whether the new products are destined for personal or business use.

30.2 THE CONVERGENCE OF TECHNOLOGIES: OPTICS, ELECTRONICS, MECHANICS

Semiconductor technology has provided the "engine" of the microelectronic age for the past three decades but is rapidly converging with the allied technologies of optics and micromachines. Semiconductor technology, according to the much-quoted Moore's law, has been doubling the performance of devices every 18 months or so for decades and has produced countless new developments in computing and communications, a fact confirmed by a visit to any electronics mass-market retailer.

Optics, however, is increasingly becoming the technology of choice to carry communications and data because it permits doubling of performance roughly every 9 months, half that of semiconductors. The worldwide demand for increased bandwidth and connectivity has further intensified efforts to develop vastly improved optical and optoelectronic packaging solutions. Likewise, it appears that micromachines are the next step in creating replacements for relays, switches, and filter networks typically constructed of passive components (Fig. 30.1).

The new world of micromachines promises to revolutionize nearly every industry, from medicine, power generation, robotics, automotive, and aerospace to entertainment, biotechnology, and electronics, among dozens of others. Micromachine technology (Fig. 30.2) has been developing for the past 30 years and at present is used heavily in automotive sensor applications. Along with their large-scale application in automotive air bag sensors, they are increasingly a key technology in ink-jet printers, a high-volume marketplace. The future will see low-cost microelectromechanical circuits that can be integrated with high-speed electronics to deliver a broad range of new applications for the handheld market, ranging from tunable radiofrequency (rf) antennas to crystal replacements and from tunable filter networks to moving mirrors for all-optical networks.

30.3 THE CONVERGENCE OF INTERCONNECT: PACKAGE, BOARD, CONNECTOR

System-level interconnect traditionally has consisted of three separate industries and processes: packaging, board technology, and connector technology. When combined, the disparate technologies

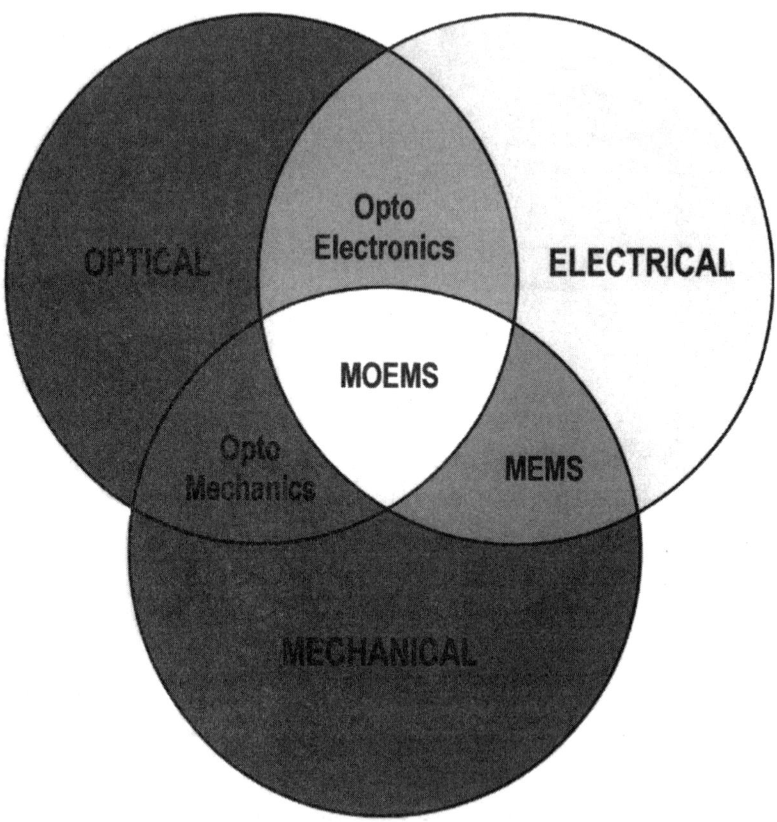

FIGURE 30.1 The convergence of technologies.

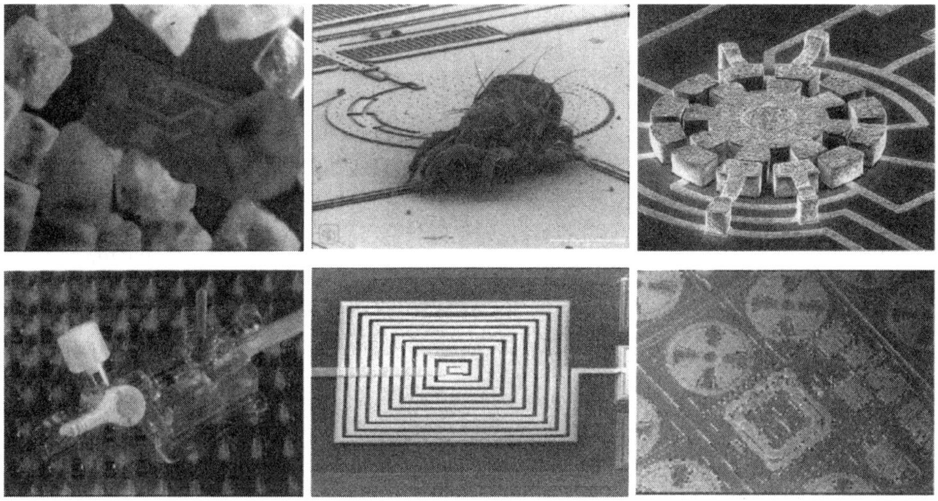

Courtesy:
Dr. Victor M. Bright, Colorado University
Sandia Nat'l Laboratories, Intelligent Micromachine Initiative

FIGURE 30.2 Micromachine technology.

constitute a system. The relentless trend toward miniaturization mentioned earlier is forcing major changes in the manner in which these three industries variously interrelate, compete, and cooperate. This convergence has the dual effect of creating opportunities for more highly customized packaging solutions as well as upping the requirement for further integration in systems.

One of the most common complaints of today's electronic system designer is the annoying delay in high-speed systems caused by excessive delays in system wiring due to the many layers of electrical interconnections necessary to route electrical signals between and among components and board and system interfaces. This delay can be markedly reduced with even closer integration of package, board, and connector.

The packaging of silicon circuits has undergone major development and advance at approximate 10-year intervals. The dual-in-line plastic (DIP) and ceramic package styles evolved during the early days of the integrated circuit (IC) in the 1960s and 1970s to accommodate increased levels of integration. The replacement of these packages in the 1980s with surface mount technology (SMT) packages such as the small outline integrated circuit package (SOIC), the plastic leaded chip carrier package (PLCC), and the quad flat pack package (QFP) allowed higher lead counts and higher circuit integration. Last decade's replacement by the ball grid array package (BGA) and its derivatives pushed lead counts to over 1000 leads and permitted the very dense circuit structures in use today. However, while admittedly increasing the density of silicon in each package, these new package technologies did not significantly reduce the overall number of interconnects in the system, a challenge that until now has remained largely unmet.

The demand for miniaturized handheld systems, on the other hand, has begun to reverse this trend by skillfully combining various package, board, and connector elements. The next generation of packaging should provide for vertical stacking of silicon, the integration of mixed functions such as passive and active components, and the integration of CMOS ICs with optics and MEMS device technologies. New package technologies such as system-in-package, or SiP (Fig. 30.3), are prime examples of a number of passive components either embedded in or surface mounted to the packaging substrate, while multiple ICs are placed within the package.

What has been primarily a flat, two-dimensional packaging world is now opening up to become a three-dimensional (3D) world. The introduction of plastic premolded cavity packages aims to pro-

Courtesy: Amkor, Inc.

FIGURE 30.3 MEMs fingerprint sensor system-in-package (SiP).

vide a new level of integration and electrical performance. Increasingly, the manufacturers of cellular phones and similar handheld appliances are stacking SRAM and Flash memory ICs on top of one another to provide the same functionality in one board footprint area as the two typical in traditional packaging. In fact, some manufacturers are stacking up to four ICs in the same package.

The addition of a small flex cable connector (Fig. 30.4) within a package, for example, eliminates the requirement for a separate connector on the system board or, worse yet, soldering of the flex cable to the package, which may give rise to unreliable interconnections. Merging a high-density substrate with embedded passive components inside one of the 3D plastic packages should provide the industry with the next level of integration and, in some cases, move closer to the goal of a whole system or subsystem in one package.

30.4 THE MARCH OF IC MINIATURIZATION

Relentless shrinking of IC feature sizes has been a central fact in the electronics industry since Mr. Moore's sixties-era observation that transistor density was doubling at 18- to 24-month intervals. Today, the advent of the independent foundry has accelerated the availability of leading-edge silicon technology without the staggering capital costs necessary to add multi-billion-dollar wafer fabrication facilities. Additionally, these independent foundries can offer a wide variety of special processes and product capabilities through intellectual property (IP) libraries developed internally or licensed from third-party IP providers.

This availability of "outside" IP has the effect of encouraging the rapid development of many innovative new designs. Ready IP availability also permits designers to design into systems the latest low-voltage silicon technology, requiring in turn significant advances in noise immunity and signal integrity, gains that can only be made with highly sophisticated solutions.

As evidenced in earlier chapters, the number of IC package varieties has likewise increased quite dramatically during the past decade with the advent of BGA technology. The BGA design incorporates a multilayer substrate within the package, in contrast to the traditional single-layer metal leadframe substrate. These new BGA technologies, coupled with continued advances in smaller-sized chip scale package (CSP) technologies, have combined to offer unprecedented new opportunities for package and device customization.

These imaginative packaging technologies also have had the effect of dramatically increasing package density, which at present can exceed 300 I/Os per square centimeter, matching the ever-smaller die sizes that result from reduced lithography. The net result has been a remarkable series of

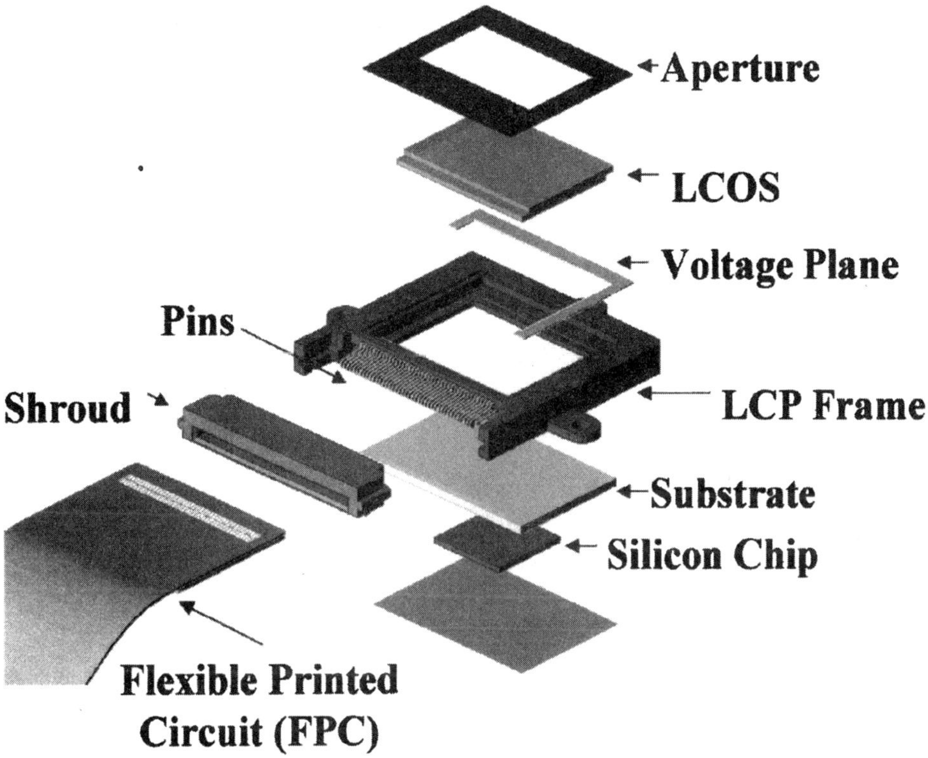

FIGURE 30.4 Liquid-crystal-on-silicon MCM with flex cable. (*Silicon Bandwidth, Inc.*)

advances in design, materials, and manufacturing techniques that moves closer to the goal of smaller, cheaper, better.

30.5 PACKAGING EVOLUTION AND FUTURE DIRECTION

Interestingly, a look back through the evolution of semiconductor packaging reveals that major changes have occurred at intervals of approximately 10 years. At the zenith of each of these intervals, there have been dramatic increases in the capability of the marketplace to use newly available silicon in a more cost-effective manner and at volumes substantially higher than those of the previous decade.

Beginning in 1960 with the advent of the IC, we can identify a move away from single-transistor packaging to relatively large yet simplistic dual-sided ceramic and metal packages, most of which did an excellent job of protecting the emerging IC technologies from the ambient environment.

Unfortunately, these early packages were not only highly labor-intensive, but they used expensive materials and required huge manufacturing overhead structures. The 1970s saw the emergence of plastic dual-sided metal lead-frame packages that enabled manufacturers to dramatically scale up their volume production capability, at the same time affording adequate environmental protection to the silicon technology. Since these successive evolutions in packaging adhered to existing standards of through-hole circuit board component mounting and wave-soldering technologies, no major system-level assembly changes were required, and assembly remained essentially unchanged.

Beginning in the decade of the 1980s, a new packaging methodology arose, termed *surface mount technology* (SMT). SMT offered much improved silicon wiring density as well as changes in mounting on the surface of the circuit board. While the technology was a welcome step forward, it forced a major retooling of the industry's circuit board assembly techniques. Although component drilling was not required, a new manufacturing infrastructure was necessary to accurately place the components on the circuit board and to then solder them to the circuit board using advanced reflow technology. It was an inevitable change, given that the increases in silicon technology interconnection density continued to adhere to Moore's law.

Each time these technology advances occurred, the number and sheer variety of packaging choices increased dramatically, further complicating inventory management but at the same time making possible entirely new classes of electronic appliances. SMT technology ushered in a more scalable packaging technology in which the size of the package is readily scaled to an increase or decrease in silicon size, at the same time maintaining industry standardization. The industry overall benefited from the vastly increased range of interconnect density delivered in a much broader selection of form factors.

During the era of the 1990s, the introduction of "array" technologies, most commonly seen in the form of the BGA, gained wide acceptance. BGA package technology improved on the older SMT by dramatically increasing the packaging density once again, with the addition of BT- or FR-4-type circuit boards within the package. BGA technology places arrays of wiring connections not only on the perimeter of the package, as was the industry custom during the prior decade, but also across the bottom of the package.

The addition of a "true" circuit board to the package versus simply a single-layer metal lead frame in the SMT designs meant that, for the first time, major reductions in system-level circuit boards were possible, and the industry now had cost-effective package solutions for scaling beyond 240 leads. The high-density patches usually required on these system-level circuit boards in order to handle their few but highly dense IC components could for the first time be placed within an IC package, thereby opening up the possibility of a reduction in system-level board layer count or size.

Most significant was the fact that a new level of electrical optimization and silicon integration could occur outside the silicon itself, a major leap forward in design and manufacturing (Fig. 30.5).

The outlook for the present decade portends yet another quantum evolution in the package, foreseeing a transition to 3D packaging and true system-in-a-package technologies, sparking sophisticated new levels of miniaturization. Most significant, however, is the integration of new optics and micromachine technologies that in themselves will radically change the world forever. Packaging plays an even more significant role in this latest transition, since the world cannot enjoy the benefits of new low-cost consumer appliances without the cost-effective packaging of mixed-signal, mixed-technologies, and high-speed devices scalable at very high manufacturing levels.

We are moving very quickly to the industry-wide adoption of 3D packaging, system-in-a package, designs that integrate connectors within the package, or even programmable packages.

30.6 PERFORMANCE REQUIREMENTS DRIVE PACKAGING EVOLUTION

The industry finds itself guided by each successive generation of the SIA *Semiconductor Roadmap of Technology.* The latest version incorporates many elements of interconnection technology, due to the proliferation of outsourcing from silicon foundry service companies. The larger implication of this more rapidly advancing roadmap is that individual companies cannot afford the time (or the financial resources) to individually solve current technology development issues. More than 50 percent of all the available wafer capacity worldwide is now positioned to support a 0.35-μm and smaller transistor feature size, meaning that a huge amount of new packaging technology, particularly CSP and BGA, is required.

One of the most significant changes to occur in the area of packaging in the past few years has been the rapid reduction in package size, coupled with a dramatic increase in the wiring density of

Courtesy: Amkor, Inc., Rangestar, Inc. (right)

FIGURE 30.5 System-in-package designs (integral antenna, right). (*Amkor, Inc., Rangestar, Inc.*)

the package. Much of the industry's attention remains focused on the reduction in transistor spacing, which boosts the performance and functionality of today's devices. Consistent with these transistor spacing reductions is a parallel increase in interconnection density, power-dissipation capability, and electrical-matching capability in this latest generation of semiconductor packaging. In addition, we are seeing a growing demand for even newer generations of packaging technology to meet the needs of the emerging optics, micromachines, and nanotechnology industry sectors.

30.7 THE ROLE OF PACKAGING

Described throughout the preceding chapters, packaging (Fig. 30.6) serves an important yet often-underrated role in making possible many of today's new miniaturized devices. The goal of packaging is threefold: to sell silicon by protecting the silicon from the environment, to connect the signal and power paths to the system-level components, and perhaps most important, to serve as a translator between the very small wiring pitches of the silicon and the very large wiring pitches of the system-level circuit boards.

Increasingly, packaging is being customized to fit into a much wider range of handheld equipment and appliances, many of which derive their form factors and even their performance from the rapid advances in low-cost application-specific package technologies. These advances underlie such innovations as multiple dies within a package, package stacking, improved thermal management, and even newer features such as sensing light, sensing environment, and of greatest importance, scaling to the volume levels required for consumer-level applications.

The recent leaps forward in optic technologies, along with the apparently insatiable demand for the increased bandwidth that derives from integrated optic and electronic technologies, will together further demonstrate the importance of packaging.

30.8 MEETING THE CHALLENGE: PACKAGING AS AN ENABLING TECHNOLOGY

A number of cautions are in order. The sale of silicon actually may be inhibited by the shortcomings of the package for any of the following reasons:

1. *Scalability.* If the packaging technology cannot be scaled into large-volume production using best-known methods.

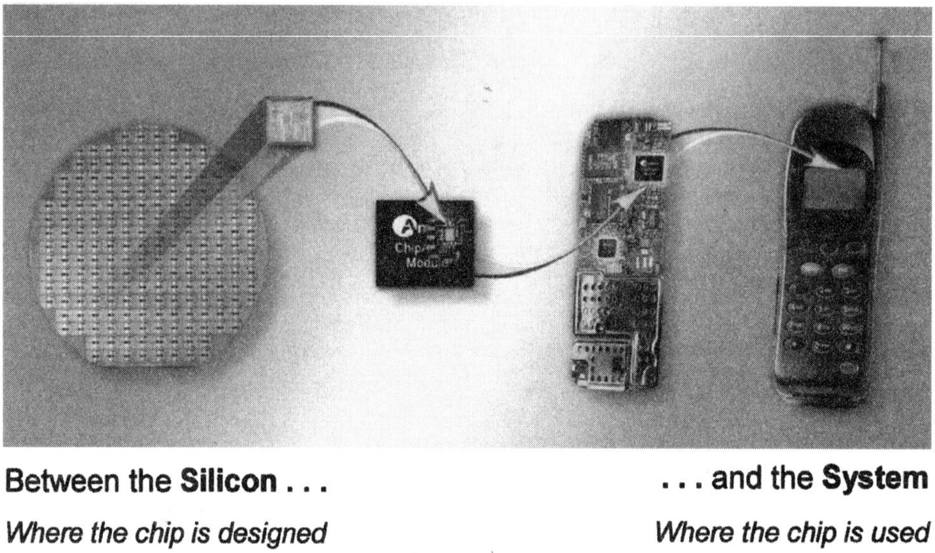

Courtesy: Amkor, Inc.

FIGURE 30.6 The role of packaging.

2. *Reliability.* If the silicon is not well protected from the environment.
3. *Testing.* If the packaging fails to allow adequate contacts for proper device-level testing or if, in the case of newer device technologies, there are inadequacies in the power or frequency simulation necessary to ensure that the device will meet operating specifications when installed in a system under load.
4. *Electrical and thermal.* If the selection of packaging in any way degrades the electrical performance of the silicon device being protected or fails to quickly move a significant amount of thermal energy away from the silicon in order to ensure reliable operation.
5. *Interconnection technology.* If the selection of interconnection technology such as gold or aluminum wire bonding or flip chip solder bonding fails to ensure that proper electrical performance is achieved or if environmental stresses go uncontrolled. For example, a package that continues to function dependably after an automobile crash or in ambient temperatures from arctic to desert is quite different from a package designed for the benign environment of a desktop or living room.
6. *Cost-effectiveness.* If packaging costs are too high with respect to the costs of the silicon or if thermal, electrical, or environmental factors cannot be controlled without expensive fixturing.
7. *Handling.* If the device cannot be handled properly at both the assembly location and the final system assembly location. The packaging selected should enable current system manufacturers to leverage all their existing infrastructures without massive investments in new assembly technology and equipment.

The "best" package choice, therefore, is one that optimizes all these elements for the application in question yet allows for scaling to high-volume manufacturing and second-source arrangements. The nature of today's diverse range of large and small appliances dictates an even more highly focused role for packaging as an application-specific silicon-to-system solution.

30.9 NEXT-GENERATION PACKAGING INTERCONNECTION TECHNOLOGIES

30.9.1 Environmentally Friendly Package Technologies

Solder plating, BGA and CSP solder balls, flip chip solder balls, and SMT solder paste have all been formulated with high percentages of lead. On the positive side, the lead has been instrumental in raising assembly yields to remarkable levels, in turn aiding the overall growth in the market for electronics.

On the other hand, an increasingly vocal movement to recycle systems, particularly in Japan and Europe, is prompting all manufacturers to investigate better, more environmentally friendly methods of assembly, along with satisfactory replacement materials for the lead, bromine, and antimony that are considered environmentally unsafe today. While it is still not clear what the legislative pressures and the ultimate legal requirements will be in each world market, the movement toward improved materials that are nonetheless cost-effective is a certainty.

30.9.2 Flip Chip Interconnection Technology

After a protracted period of slow adoption, flip chip (FC) technology (Fig. 30.7) is finally beginning to replace wire bonding as the preferred silicon-to-package interconnection method in high-performance devices. While this technology was developed over 30 years ago, its use until recently has been limited to those few large vertical system manufacturers whose concentration has been in extremely high-speed processors and associated ASIC devices. Today's high-speed workstations and high-speed Internet routing and switching equipment require FC technology in their packaging to ensure proper operation.

The number of devices that used FC technology in 2000 represented only about 2 percent of the total packages built. It should be pointed out, however, that these devices represent very advanced high-lead-count (800 to 3000 leads) ASIC, system-on-a-chip (SOC), and very high-speed processors. The advent of rf devices operating above 2.4 GHz is also adding increased momentum to the replacement of wire bonding with FC technology for low-lead-count devices. FC technology makes

FIGURE 30.7 BGA module with FC.

possible the same degree of improvement in die-to-package interconnect density that the adoption of BGA packages did 10 years ago.

The use of FC technology certainly will grow dramatically in the future, since continued die lithography reductions will favor an array pattern of solder bumps rather than the perimeter wire bonding or perimeter FC designs that demand associated tight bond pad pitches. Reduced die lithography also brings with it lower operating voltages, thereby reducing the noise margins of previous device generations. FC interconnect can provide a physically shorter and lower inductance signal path to meet these requirements.

30.9.3 Wafer-Level Package Technologies

A number of new wafer-level technologies (Fig. 30.8) have appeared from such companies as Form Factor, Motorola, Shellcase, and Tessera, along with most of the mainstream packaging companies such as Amkor and ASE, as well as other packaging suppliers. The FC bumping companies, such as K&S, FlipChip Technologies, Unitive Electronics, and others, have been quick to adopt these designs as well. Whether these technologies use a wire-bond spring, a thin-film elastomer, or solder-bump redistribution technology, they all promise to reduce the costs of packaging further by combining the packaging and test operations immediately after wafer fabrication, with all operations performed on the wafer level.

30.9.4 3D Package Technologies

The increase in handheld devices of all types is driving the demand for even more integration and even higher-density packages, particularly CSPs. To address this requirement without increasing the number of packages, the cellular phone market and other memory-intensive market segments have seen increased stacking of the dies within a single package (Fig. 30.9). While this solution meets the

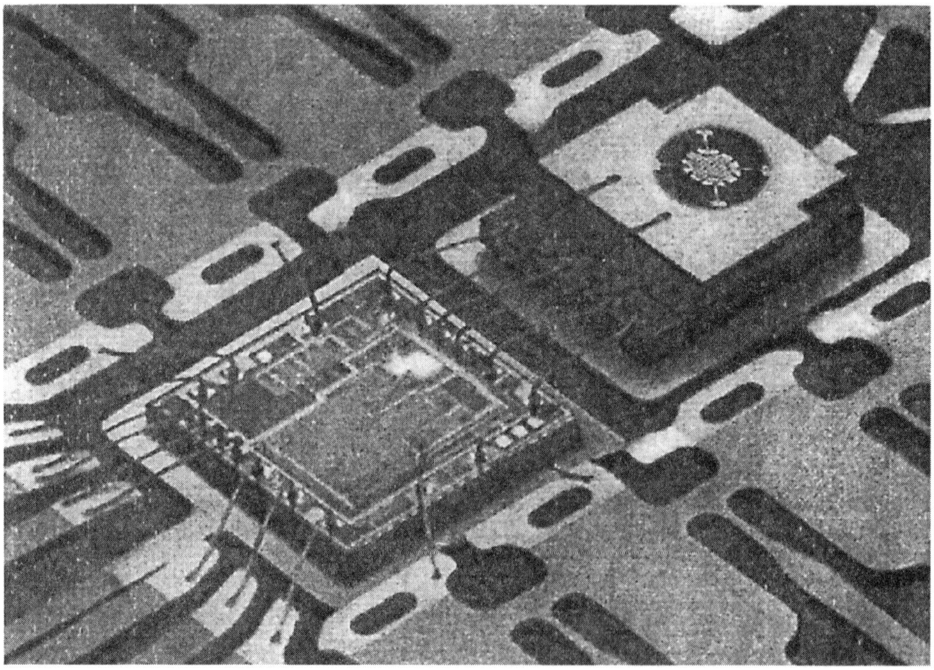

FIGURE 30.8 Wafer-level package. (*Motorola, Inc.*)

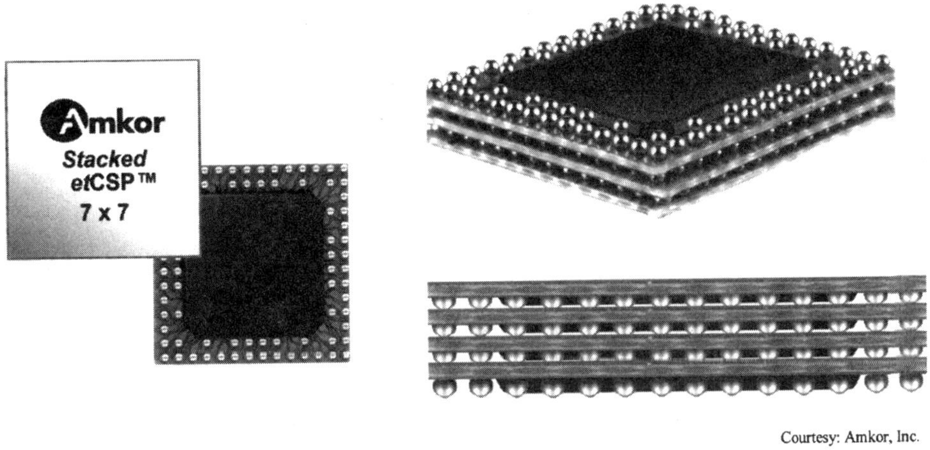

FIGURE 30.9 Stacked package. (*Amkor, Inc.*)

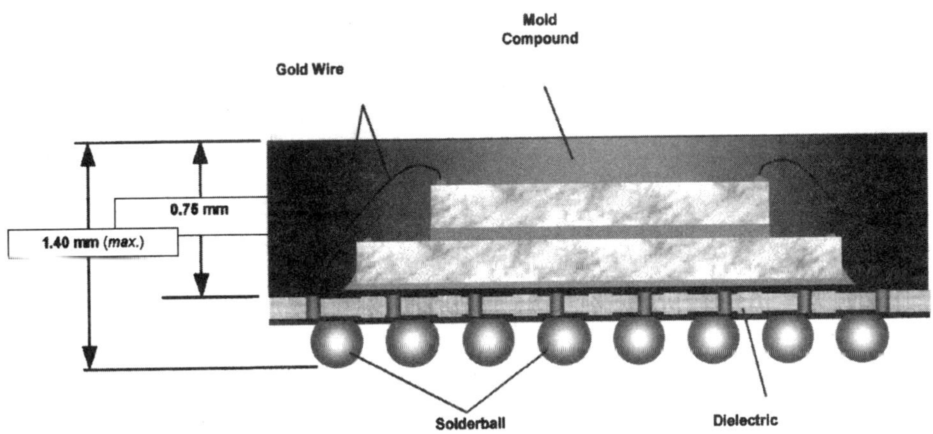

FIGURE 30.10 Stacked CSP.

challenge of size reduction, it nonetheless creates new challenges in mounting two dies, one on top of the other, in thin packages.

While many companies are investigating methods to stack multiple memory and even mixed technologies such as microprocessors and memory in a single package, a number of companies are designing "stackable" packages that will radically increase the density of single-die packaging (Fig. 30.10). Evolutionary improvements in material design steadily afford package designers with new opportunities to radically change the direction and shape of packaging.

Although ceramic preformed packages have been available for many years, they are being replaced by new generations of high-performance, low-cost plastic packages that substantially advance management of electrical signals, improve wiring transformation from die to circuit board, and perhaps most exciting, increase the capability of mounting multiple dies within one package.

While packaging efforts to date have focused on increasing stacking density through the use of multilayer laminates or ceramic substrates, alternative low-cost packages are emerging that stack wire-bond surfaces using premolded thermoplastic package configurations. These new configurations offer a variety of attractive options for the silicon designer, including

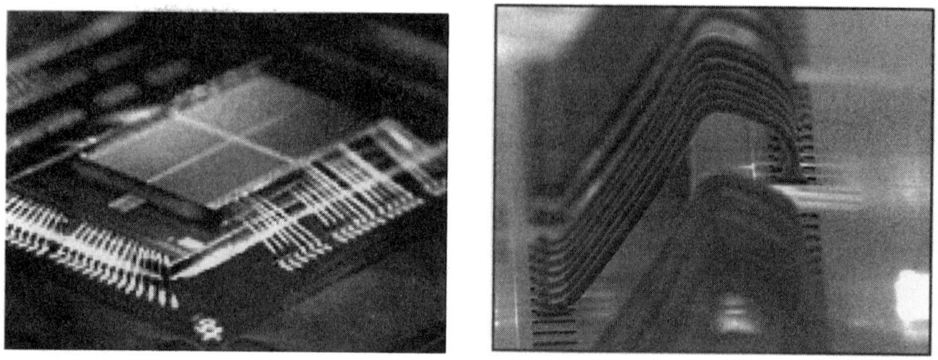

Courtesy: Amkor, Inc.

FIGURE 30.11 Wire-bonded CSP.

- Wire bonding (Fig. 30.11) that for the first time can have differential lines for managing low-voltage signals
- Low-temperature processing that is ideal for new display, imaging, and low-temperature materials
- A combination of very low-inductance and high-current capability, owing to short yet large leads
- Surface mount or "socket-able" packages
- Up to four die stacks with wire-bonding layers in the plane of the die layer

If you consider that the packaging world has been primarily two-dimensional for the past few decades, the advent of these high-volume 3D packaging technologies represents a major shift in focus and a quantum improvement for high-density or truly volumetric packaging.

30.9.5 Optical Package Technologies

Among the most mature of semiconductor packages, and the first employed in the early days of the IC, was the metal flat pack package, a design that, despite its age, continues to be used widely for optical receivers, transmitters, and transceivers. The stepped-up industry pressure for more automated optic packaging, coupled with steadily increasing volumes to support a growing optical market for metro-area connectivity (and its demand for increases in bandwidth), will together intensify the focus on packaging that meets demanding optics and optoelectronic device requirements.

Present-day state-of-the-art in optical packaging (Fig. 30.12) is roughly where IC packaging technology found itself more than 25 years ago. Given, on the one hand, the present low volume of device requirements and, on the other, the extremely rigorous telecommunication environments in which optical components must operate reliably, the reasons for the slower technology development path begin to emerge.

Now that the Internet is in common use throughout the world, the requirement for advanced optical packaging is increasing at a dramatic rate. It is now a universal industry expectation that nearly every section of the optical backbone can use advanced packaging and semiconductor technology development.

Optical packaging challenges the most fundamental materials and methods used over the past decades, since not only electricity but also light itself must be managed by the package vehicle. Materials well suited for two-dimensional thin semiconductor packages are now expected to handle 3D devices as well and to permit full interaction of the light with the outside environment.

30.9.6 System-in-Package Technology

The next logical progression in packaging is to combine a number of ICs with passive components mounted on a single substrate (Fig. 30.13). While similar in some ways to the now-mature "hybrid"

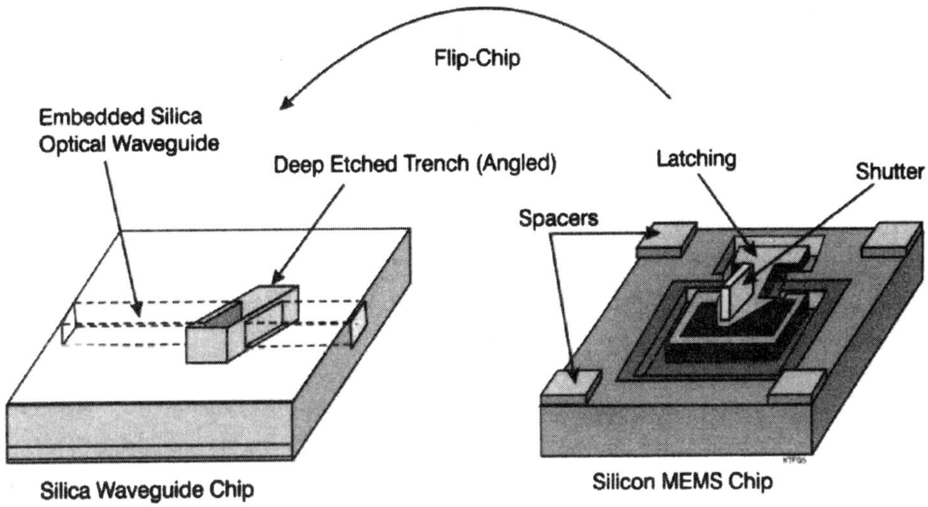

Courtesy: Cronos, Inc.

FIGURE 30.12 Optical MEMS.

or multichip module packaging, this approach differs somewhat in that extremely high-volume "commodity" silicon devices typically are used and are mixed with low-cost passive components.

The advantage gained with these separate high-volume components is, of course, that they are readily available, and in combination with other components, a complex system can be assembled without the time and cost associated with an integrated system-on-chip solution. System-in-package (SiP) technology also can reduce the difficulty of surface mounting large numbers of passives on circuit boards by placing a portion of the passives directly inside the package itself, thereby improving the electrical performance, the reliability, and the cost structure of the system overall.

Each new generation of IC lithography drives faster transistors and passives closer to the actual silicon, thereby improving signal integrity and performance. Device voltages have seen steady decreases over the past decade, falling below 5.0 V and projected to soon drop below 1.0 V, very close to the noise levels themselves. These new optic components will have to convert electricity to light—and back again—much the way that familiar analog-to-digital and digital-to-analog devices operate. Some will have to amplify light before sending it onto its next destination, whereas others will have to redirect and switch large arrays of light. The future of optic packaging is very bright (no pun intended) and will serve to energize and accelerate the packaging developments of the past decade.

30.9.7 Connector-in-Package Technology

A new concept is now emerging that combines two traditional interconnection methods into a single design: the concept of a connector built into the package. This convergence of packaging and connectors, and even multilayer circuit boards embedded in the package, is evidence of an increasing trend toward a mounting convergence of technologies. Fueled once again by intensified miniaturization and improved performance, processor modules camera modules and memory cards are among the first examples of package modules that combine connectors within the larger packaging structure. What is most new and noteworthy is the integration of very-fine-pitch connectors within a semiconductor package that makes possible much smaller and much more reliable system components. 2000 has witnessed the introduction of a new style of microdisplay package with an embedded connector that eliminates the traditional soldering of the flex tape to the package yet allows easy insertion during manufacturing, thereby saving time and money and markedly improving reliability.

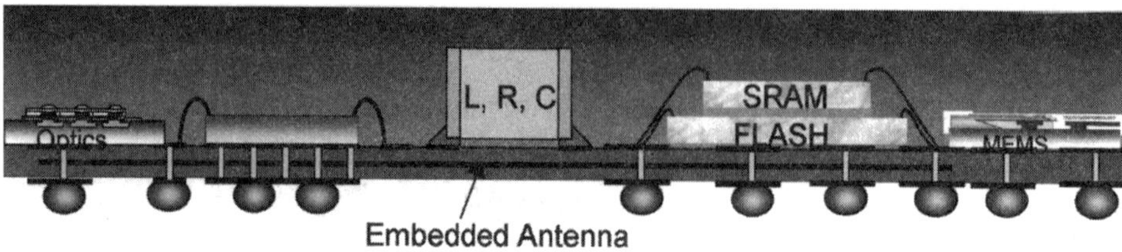

FIGURE 30.13 System-in-package.

30.10 MEMS PACKAGE TECHNOLOGY

Another of the exciting new technologies to emerge from more than three decades of basic research is the entirely new class of microelectronics and mechanical devices termed *microelectromechanical systems* (MEMS) (Fig. 30.14).

These MEMS devices are leading the way to a completely new level of miniaturization, similar in some ways to the effect of the transistor in the early days of the electronics industry. Leveraging off mature silicon processing technology and employing the scalability and volume manufacturing—and even the older facilities—of the semiconductor industry, these new technologies are set to revolutionize the entire world of electronics and to propel us into a true mechanical/electrical convergence.

MEMS devices can be manufactured using older semiconductor fabrication facilities, since their geometries at this time do not require submicron lithography and the associated critical processing requirements of leading-edge processors and memory devices. Another attractive advantage of the MEMS technology is that it can allow the integration of previously separate mechanical functions within the same substrate as the active silicon devices or on a separate silicon substrate that can be readily combined into a single package. The massive advantages gained by leveraging the existing 6σ or 7σ semiconductor-processing factories around the world cannot be overestimated.

Design tools and the methodology for integrating MEMS functions within standard silicon structures are still in their infancy. The rapid growth of the optical and wireless market segments is expected to fuel increased development for improved MEMS design tools, design libraries, and packaging solutions.

While MEMS devices can be fabricated in great numbers, the primary challenge to cost-effective and reliable delivery for final applications is the persistent lack of cost-effective packaging solutions. These devices are literally tiny machines with moving parts. Current high-volume semiconductor and electronics packaging options are neither designed for nor capable of low-cost implementation or high-volume scaling. As is the case with optics technology, most companies designing MEMS today are forced to resort to 1960s-style ceramic preformed packaging.

Nevertheless, a few high-volume options are being introduced to protect the fragile MEMS device topside with a cap constructed of ceramic, silicon, plastic, or metal. This covered MEMS device can then be assembled in standard semiconductor high-volume plastic packages. One drawback to this approach is the increased cost to mount the protective cap. Other difficulties in assembling MEMS devices entail such issues as higher cleanliness levels in the manufacturing environment and more demanding test procedures.

Soon we may see the emergence of systems that are *true* hybrids of optical, mechanical, and electrical systems, and we may soon see versions that combine all three elements within one package, amounting to a true and literal convergence. Laboratories around the world are working to integrate these mechanical and electrical and even optical functions to create innovative, wholly new systems. This capability could offer unprecedented levels of integration, miniaturization, cost-effectiveness, and reliability.

Today we are noting the rapid adoption of MEMS devices in automobiles for controlling everything from air bag collision detectors (Fig. 30.15) and vehicle antisway and rollover protection sen-

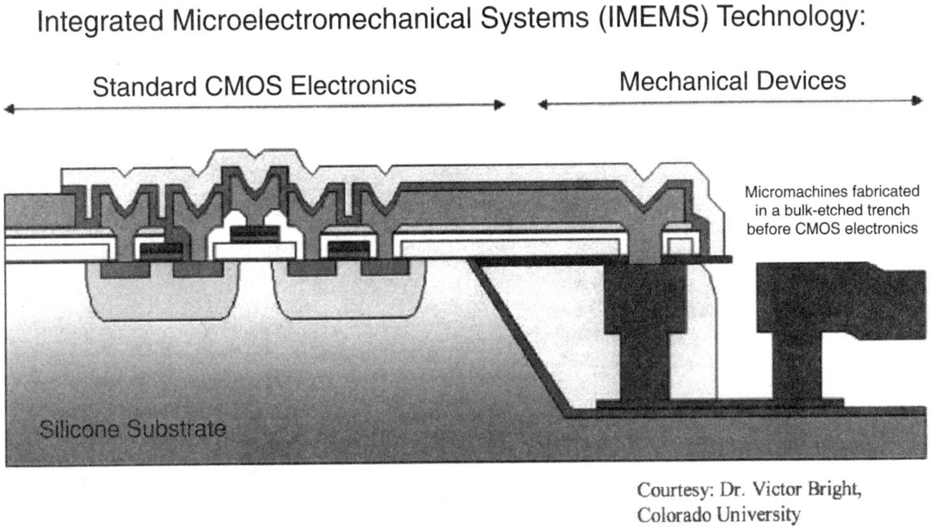

FIGURE 30.14 MEMS technology.

sors to the sensing and detection of noxious gases and adjusting the fuel-air mixtures necessary for combustion system optimization. MEMS likewise are fast finding their way into many sectors of the toy industry, as evidenced by game controllers that can sense motion via human tilt and rotate inputs and return vibratory feedback to the operator. Camcorders and digital cameras continue to use MEMS technology to offer amazing new autostabilization capabilities.

Behind these recent developments is the introduction of higher-volume, cost-effective packaging as described earlier. Recent advances in alternative plastic packaging materials such as liquid-crystal polymers (LCP) open up new avenues to create packaging that has preformed shapes that allow MEMS devices to interoperate with other silicon devices, such as optics and high-speed electronic ICs, while providing a versatile and dependable interface to the outside world. Connectors integrated into these types of packages may redefine the way in which system design is typically accomplished, allowing even more flexible and reusable configurations in package design.

One MEMS design in development at a number of laboratories is a so-called lab-on-a-chip device (Fig. 30.16) that could, in a matter of minutes, analyze blood, DNA samples, or even a pathogen such as a virus or bacterium, providing instant feedback and more accurate results than has been possible up to now. One need only look to our aging population to see how quickly this concept will meet with immediate acceptance.

The packaging for this new class of device must be extremely low in cost, even disposable, and incorporate materials that are not only recyclable but also compatible with moderate-temperature assembly methods, given the sensitive technology inside. This type of package may well be emblematic of the next generation of packaging that must provide visual feedback and reliable environmental protection of the device while at the same time affording access to the biologic sample being tested.

Those companies involved in exploiting MEMS designs have noted widened interest in their technology, as the need for vastly improved optics and rf switching technologies has become clear. The ability to fabricate tiny mirrors (Fig. 30.17) and tiny rf switches means that the two technologies can together offer radically improved performance levels. The proven feasibility of an array of multiple tiny mirrors is radically changing the display market by enabling a very precise array of pixels to be fabricated. The tiny image is then "blown up" to display the very large high-quality image required for next-generation HDTV systems. These mirror arrays also can help enable a true optical system without electrical conversion. The packaging for the mirror switches must permit clear opti-

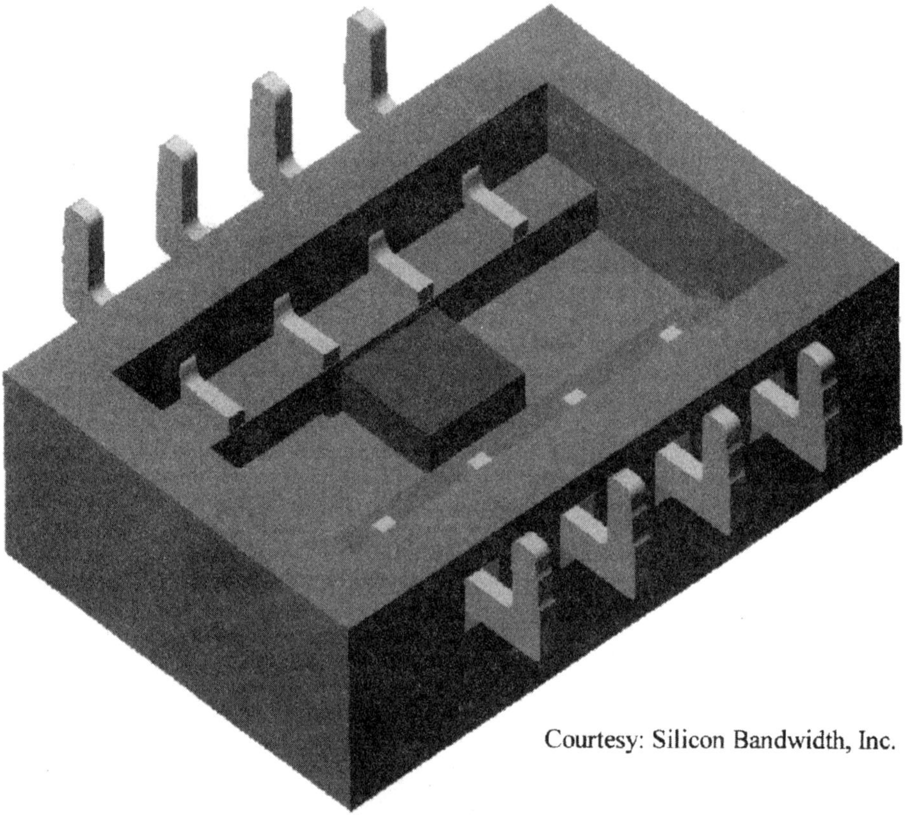

FIGURE 30.15 MEMS air bag sensor.

cal transmission to the mirror yet still protect the silicon mirrors from the environment. Today's ceramic packaging solutions are too expensive to allow for low-cost implementation, but developments are aimed at designing new packages that will have desirable cost parameters and reliability characteristics.

30.10.1 MEMS and in Vitro Applications: Bioelectrical Packages

Related applications actually may see the connection of a MEMS package with biologic cells or even DNA strands, thereby connecting a microprocessor to nerve junctions or other biologic structures, in order to repair or manage diseases such as Alzheimer's or Parkinson's. These bioelectrical packages will have vastly more stringent design requirements than any package that now exists. The concept of miniaturizing mechanical functions also should allow for a wide range of entirely new appliances, such as GPS on your person, personal real-time water and air quality monitors, superdense gigabyte storage systems on your wrist, etc. Tiny jets or gas-burning generators are being fabricated at labs at MIT for applications such as home power generation or for miniature MEMS-powered aircraft for use in monitoring enemy movements or weather patterns.

MEMS technology developments and applications are expected to underlie dramatically increased functionality for future systems, for second-generation silicon integration, and for entirely new levels of hyperminiaturization. The packaging challenges are admittedly formidable, but given the relentless demand for improved packaging in optics and submicron technologies, the field should experience heightened focus and increased development.

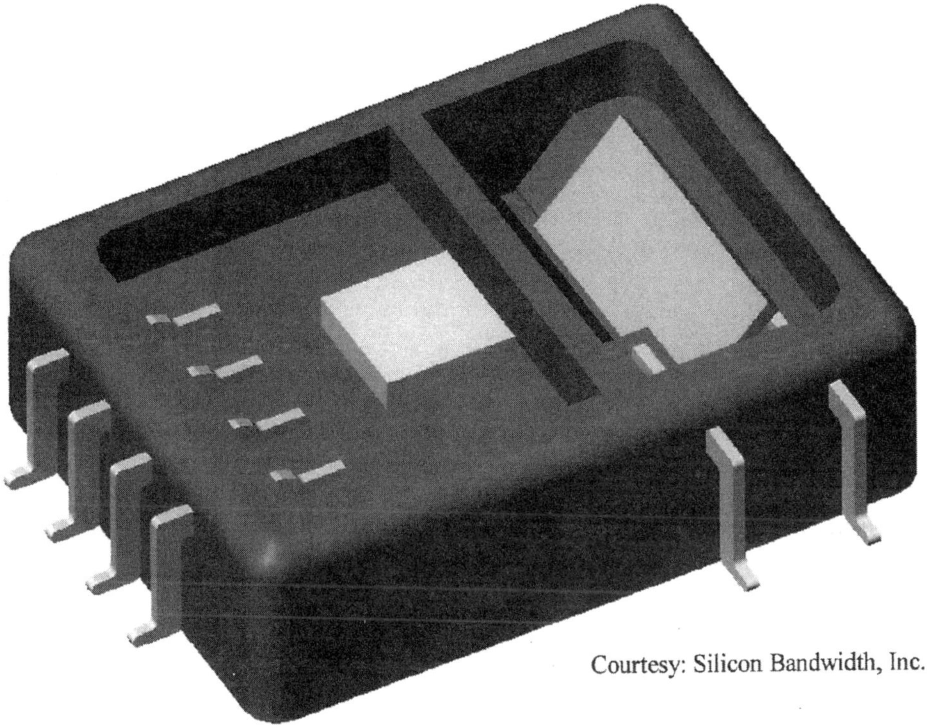

FIGURE 30.16 Lab-on-a-chip.

FIGURE 30.17 Micromirrors.

30.11 SUMMARY

The relentless shrinking of IC feature sizes continues to necessitate packaging size reductions and performance improvements, both made possible through ever more rapid advancements in increasing density wiring on the package substrates. These high-density package trends likewise call for correspondingly fine-line-width lithography. Further, it appears likely that the increased stacking of silicon to achieve vastly greater silicon densities will be a development that continues to gain momentum.

Package design trends initiated years ago with the integration of multiple transistors have further accelerated, surpassing even the integration of PC chipset functions, and have clearly advanced toward the total integration of a system-in-a-package. Beyond even this advanced technology plateau, a plethora of customized formats and application-specific packages beckons, all of them specially tailored to meet specific form-factor requirements.

The inexorable march to integrate multiple technologies such as optics, micromachines, and high-speed, high-bandwidth electronics will continue unabated, bringing with it a new set of challenges in creating the next generation of packaging platforms. The mixture of alternative silicon technology, such as gallium arsenide, indium phosphide, silicon-on-sapphire (SOS), and scores of others, coupled with new developments such as plastic transistors or amorphous memories, points to the need for continued development of packaging solutions that can deliver the resulting devices to the marketplace at the right cost, timing, and reliability levels.

The rapid advancement of technology along the electrical, optical, and mechanical fronts, coupled with the miniaturization of systems, points to a new era of exciting system-level packaging developments that will solidify many corporate joint efforts, along with cooperation among many of the world's foremost educational institutions.

The convergences described throughout this book highlight the ever-increasing need for continued work along all technology fronts, augmented by close cooperation through industry consortiums and professional groups. The rapid acceptance of new digital appliances throughout the world's consumer base is being endlessly accelerated with the maturing of new generations of children growing into adults who desire and seek out new technologies that can enhance their daily lives.

And while economic cycles will always be a fact of life that affects development cycles, there can remain no doubt that we are entering yet another golden era of electronics, this one brought about by the merging of optics and micro- and nanomachines and the continued reduction of transistor scaling.

New high-volume packaging opportunities continue to emerge in industry sectors as unlikely as solar roof tiles or as obvious as personal video communicators—Dick Tracy's wrist TV. Plastic transistor technologies could replace today's newspaper with a reusable plastic sheet material posing unheard-of challenges to the packaging designer. High-volume lab-on-a-chip could revolutionize the detection and treatment of the world's medical ailments as our populations age and our health care costs surge. Personal wired and wireless device technologies inevitably will explode, with corresponding advances in packaging that make new devices cost-effective and available to much of the world's population.

The technologies discussed throughout this chapter could well revolutionize the concept of system design, since it may someday be possible to "snap together" a system configured to an individual's precise desires and preferences. An early glimpse of this approach became evident with the recent introduction of the Lego Mindstorms electronic system building block toy set. Designed in collaboration with some MIT professors and their students, this flexible concept of system design offers a tantalizing hint of what might be possible in the years ahead.

Market and social forces driving such causes as lead-free solder formulations and environmental recycling likely will add impetus for the electronics and packaging industry to rethink the manner in which systems and packages are now built and to look to emerging technologies for solutions. The fact that we now are dealing simultaneously with three "laws" of technology advancement both complicates and challenges the role of new technologists in optimizing systems based on a variety of factors.

- *Moore's law* of transistor density—doubling approximately every 14 to 18 months
- *Optics Law* of performance improvement—doubling every 9 months
- *Storage law* of density—doubling every 12 months

These challenges, along with the emergence of the many diverse new technologies now available to the electronics industry, offer nearly unlimited opportunities to packaging engineers and researchers to make significant contributions to enabling a truly wonderful new world of applications and systems.

CHAPTER 31
THE FUTURE OF SMT PROCESS EQUIPMENT

Phil Zarrow
ITM, Inc.

31.0 INTRODUCTION

The perception of what the future holds for the surface mount technology (SMT) assembly process and, in particular, the equipment used has always seemed to be guided by what the industry perceived as the most pressing problems at the time. In the late 1980s and early 1990s, there was a push for "lights-out manufacturing," where the equipment would virtually run itself, 7 days a week and 24 hours a day, without assistance or supervision by humans. While technically such a scheme is possible, the feasibility remained questionable, and we rarely hear discussion of such a quest anymore.

Thus, based on the current directions and trends, as well as perceived needs of the industry, at the dawn of the twenty-first century, what does it look like the future of SMT assembly equipment might be?

31.1 SOLDER PASTE AND ADHESIVE PRINTING

At this juncture, printing remains the most human-interaction-dependent process of the key assembly areas. A number of operator "judgments" preclude proper setup and ongoing operation to attain high printing yields. Accordingly, the solder paste deposition process still usually accounts for over 50 percent of assembly process errors.

While many printed wiring board (PWB) assemblies will benefit from advances in solid solder deposit technologies (discussed elsewhere in this book), deposition of solder paste and adhesives, both conductive and nonconductive, will likely remain a staple of the SMT PWB assembly process in most instances.

One very late twentieth-century innovation that leapfrogged the solder paste printing state of the art was the direct-pressure printing head. Rather than relying on a squeegee, with all its respective variables (pressure, speed, wear, angle, etc.), pushing a roll of solder paste via vector force and using the pressure of that roll to force the solder paste through the stencil apertures, a more direct pressure approach is used. The head contains a reservoir of solder paste, and an internal platen pushes on the solder paste, forcing it out of a slot as the head transverses the image portion of the stencil. In addition to further controlling solder paste dispensing, the solder paste, being contained in the head (as opposed to residing on the stencil), is not exposed to the ambient air and its respective temperature and relative humidity, thus extending the usability of the solder paste—both before and after deposition. Also, savings are realized with virtually no solder paste left on the stencil after a pass, and there is no longer dependence on the operator to maintain the proper roll volume. Figure 31.1 shows a state-of-the-art stencil printer with pump head in place of a squeegee.

Another key advantage, besides partial fillet of the fishbone of stencil printing variables, is the resulting increase in actual printing speed. Accelerating the flood speed will no longer jeopardize the pressure constituents on the head (since pressure is now independent of the head's horizontal movement, which is not the case with a conventional squeegee).

Direct-pressure print heads have been capable of print speeds of over 7 in/s (170 mm/s) with moderate-density PWBs, with some fine-pitch. This is in comparison with an "industry average" print speed of 1 to 2 in/s (25 to 50 mm/s) with the traditional squeegee methodology.

The direct-pressure printing methodology is having the most profound effect on the shape of the printer of the future. Throughput speed would now be limited by the mechanical action and duration of stencil positioning on the board. The next-generation high-speed printer will see the stencil mounted on a drum rather than a frame. Contained within that drum will be a derivation of the direct-pressure print head. Boards will be processed through the printer on a continuous basis, and solder paste (or adhesive) will be deposited on the board as it passes under the rotating drum (much like the old mimeograph machines of a bygone era). On the on-load portion of the machine, prior to reaching the print drum/stencil, a vision system will acquire the fiducials on the board to ensure proper registration with the drum/stencil.

Courtesy of MPM/Speedline

FIGURE 31.1 Screen printer with Rheopump head.

31.2 COMPONENT PLACEMENT

The late 1990s saw component technology both go from the sublime to the ridiculous and offer up some relief. Of the latter, area array packages such as ball grid arrays (BGAs), chip scale packages (CSPs), and flip chips (FCs) offered up a very reasonable and manufacturing process friendly solution to high-I/O integrated circuit (IC) packaging. These devices have more robust interconnects as compared with fine-pitch-leaded devices and, for the most part, use concurrent process engineering in that they are adaptable to the print-place-reflow assembly process.

On the other hand, passive component packaging became far more challenging. In the early 1990s, 0402 components made their debut. At the time, they presented a major design-for-manufacturability dilemma in that none of the pick-and-place machines on the market could handle and place them. Nevertheless, circuit designers loved them, and 0402s rapidly found their place in board designs, particularly in radiofrequency (RF) applications. The component placement machine manufacturers rose to the occasion, and soon, accurate and repeatable pick-and-place capability could be either retrofitted or upgraded to. Just when assemblers thought it was "safe to go back in the water," 0201s made their debut. Again, designers came up with board layouts with these little wonders only to confound the assemblers who lacked machinery to place them. By the end of the century, a number of machine manufacturers claimed bragging rights on their ability to handle 0201s, but they remain an enigma to high placement yields.

Embedded component technology has begun to emerge on a broader scale in the new century. In addition, we will see other technologies, likely older ones that have been resurrected for modern applications, such as printed resistors and the like, as we have been using for decades in thick-film applications. Technologies such as system-on-a-chip also will reduce the requirement for small, discreet components and thus lessen the dependence on and requirement for dedicated passive placement machines (chipshooters). However, the requirement for accurate high-speed passive component placement will not go away.

Expect to see less of a reliance on the horizontal turret concept that symbolizes the traditional chipshooter. This concept, a hallmark of Japanese-developed systems that specialize in high-speed placement of passives delivered on 8- and 12-mm tape, including Fuji, Panasonic, Sanyo, and TDK, uses a multispindled head arranged in a carousel configuration. While rotating horizontally and with pick-and-place motion in the z axis, the head basically remains stationary with the feeders moving to supply the head and the board moving in the x and y axes for placement location. While this methodology has been able to attain placement rates exceeding, in some applications, 50,000 placements per hour, there are some imposing limitations. Constant feeder-bank motion means that the machine has to be brought to a halt to replenish depleted feeders. The machine also has a very large factory floor footprint, and assemblers have become very conscious of maximizing productivity from factory real estate.

A more enduring placement machine type is the parallel-sequential system. Here the PWB assembly travels down a conveyor past a series of heads, each responsible for placing components from certain specific feeders. Hence boards are populated in an assembly-line fashion. This methodology has been applied to both passive dedicated machines (Philips FCM and Fuji QP-1) and, more recently, placement of high-I/O components (Fuji QP-3). Since global vision recognition takes place at the entrance of the machine, through-put rate is greater than when a placement head has to check each assembly prior to population. Since the feeders remain stationary throughout the placement operation, they can be replenished "on the fly," thus keeping the system in a constant, efficient running state. With potential throughput rates exceeding, in some applications, 80,000 placements per hour, the parallel-sequential configuration also takes up considerably less floor space than its rival horizontal turret system.

Greater demands have been put on flexible placement systems as well. These are the systems that, rather than being dedicated to strictly high-speed placement of passives, also must handle high-I/O, multipin devices such as QFPs, PLCCs, TSOPs, BGAs, CSPs, and even FCs. CSP and FC, with tighter array pitch, require a greater degree of placement accuracy, as do very-fine-pitch leaded devices, all this while requiring faster placement rates, smaller machine footprint, and easier setup and programming.

With the widerspread adaptation of intrusive reflow soldering (simultaneous reflow soldering of through-hole components on a PCBA alongside SMCs), some placement systems have been called on to place (insert) certain through-hole components, such as connectors, sockets, and LEDs, along with a variety of surface-mount components. While a number of level II and level III machines have a limited capability for some through-hole components, the flex cell is such a machine specifically configured to do a range of SMCs and either through-hole components or FCs. Universal Instruments GSM, in certain configurations, pioneered the modern flex-cell concept, with other machine manufacturers following suit with their own flex-cell offerings (such as Philips' ACM and Celtronix).

The mantra of component placement has always been "speed, accuracy, flexibility." With this in mind, it becomes clear that the pick-and-place machine of the future will incorporate the most optimal features of all classes of machines. When the "key ingredients" of high speed, accurate, repeatable, flexible component placement are combined, what emerges is a multihead sequential platform configuration. Here, a machine module may contain as many as four individual heads, each on its own *xy* gantry. These heads are multitooled (pickups) in a vertically arranged rotary configuration. Moving to pick from stationary feeders and placing on stationary PCBAs, the boards progress from one head to the other for optimal component mix/balance on parallel dual conveyors. The epitome of this configuration of the future is Siemens HS-50, which is already setting a trend, with similarly configured offerings being presented by other equipment manufacturers, most notably those which formerly specialized in horizontal turret systems. The multimodule platform can handle passives at speeds rivaling horizontal turret systems but have the accuracy and repeatability to handle fine-pitch leaded and arrayed packages as well.

This represents and evolution away from dedicated passive placement and flexible placement machine centers to single (or multiple) multimodule systems that can handle the entire range of components in different modules. Encompassing the aforementioned mantra, this is most likely the placement platform of the future. Figure 31.2 shows a state-of-the-art SMT printing line that is intended specifically for process and equipment improvement and development. The line can handle the smallest components, including FCs, and dispense underfill.

31.3 REFLOW SOLDERING

During the mid-1990s, reflow soldering technology reached a technological plateau with regard to heat-transfer capability. Almost all reflow ovens currently use convection heat transfer as their pri-

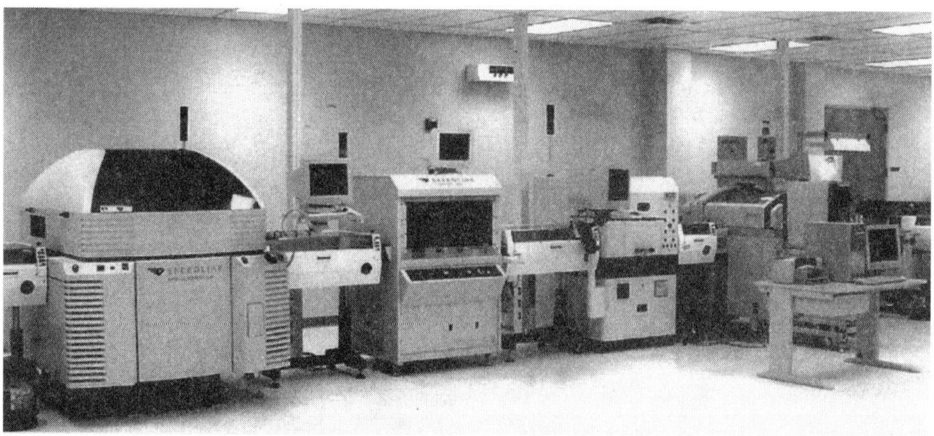

FIGURE 31.2 State-of-the-art SMT/FC line. Courtesy of Cookson Performance Solutions

mary PCBA heating methodology. Essentially, the leading machines on the market have attained the maximum amount of heated gas velocity and volume and thus have effectively "topped out" heat-transfer efficiency with this methodology. Despite this, convection-dominant (also known as forced-convection) ovens seem quite capable for present and near-future PCBA reflow and curing applications. While far from near asymptotic, there is ample energy to achieve acceptable across-the-PCBA gradients to accommodate even the most complex and massive applications.

If a number of manufacturers adapt lead-free solder alloys, there will be a requirement for higher operating temperature capabilities. Many of the most promising alloys have melting points of around 217°C (as opposed to 183 and 179°C of 63Sn/37Pb and 62Sn/36Pb/2Ag, respectively, which are currently the most popular alloys in use. This means a reflow (full liquidus) temperature of around 235°C (as opposed to the current 205 to 210°C). Again, the thermal capabilities and construction of most of the leading reflow ovens on the market are ready for this partial eventuality. Even if soldering in an inert (nitrogen) atmosphere is required for such processing, the capability for tight atmosphere containment has been in place for several years now.

Reflow soldering of no-clean solder paste flux formulations has required consideration of keeping residue from condensing on cooler parts of the oven system and impeding performance. While some work remains with the solder paste manufactures to improve their respective chemistries, we can expect to see more efficient flux-management systems on future reflow ovens. Expect to see improvements in other peripheral issues such as center board support systems, software, and profile tracking and prediction.

A number of oven manufacturers have developed vertical reflow ovens. Keeping with the theme of a desire for a smaller factory footprint, most of these ovens have been applied for adhesive and underfill/globtop curing that normally requires longer thermal durations. However, some vertical ovens have been applied for reflow applications. While this is not anticipated to be a major trend, a number of facilities may well adapt this methodology.

In the near future we can see more assemblers adapt the intrusive reflow (reflow of through-hole component) process. The present generation of ovens has this capacity, and most through-hole components are constructed of plastics compatible with the associated thermal excursions (even the elevated ones associated with no-lead reflow applications). This further diminish the demand for and application of wave-soldering, a process that was never intended nor completely adapted to SMT soldering.

Selective soldering, particularly using laser technology, also will increase in use. Besides the obvious applications of soldering the few component that might be on an assembly that cannot withstand reflow oven thermal excursions, this technology will find a number of other applications, including PCBA rework and repair (especially when combined with a nominal pick-and-place capability), part shortage makeup soldering, and prototyping. At this juncture, diode laser systems appear to offer the most potential for powering these systems. Figure 31.3 shows an automatic laser diode–powered soldering machine that is just being introduced.

Ultimately, an alternative and more optimal mass reflow technology will emerge. Microwave is a very like candidate. Having been applied to curing applications, it is only a matter of time before it is adapted successfully to reflow. Such a process would allow for near instantaneous soldering of the interconnections without exposing the components and PWB substrate to excessive thermal excursions. Figure 31.4 shows a reflow oven that is fitted with instrumentation to allow careful measurements that will be used to allow the design of future equipment.

31.4 AUTOMATED OPTICAL INSPECTION

In 1992, less than 20 percent of the electronic assembly industry was incorporating inspection in their assembly process. Those which did mostly relied on visual inspection performed by humans—very subjective and error-prone. At that time, less than 5 percent of the industry performed postprint inspection—automatic or manual. By the end of the decade, at least 50 percent of the industry was employing in-process inspection, and postprint inspection was up to 20 percent.

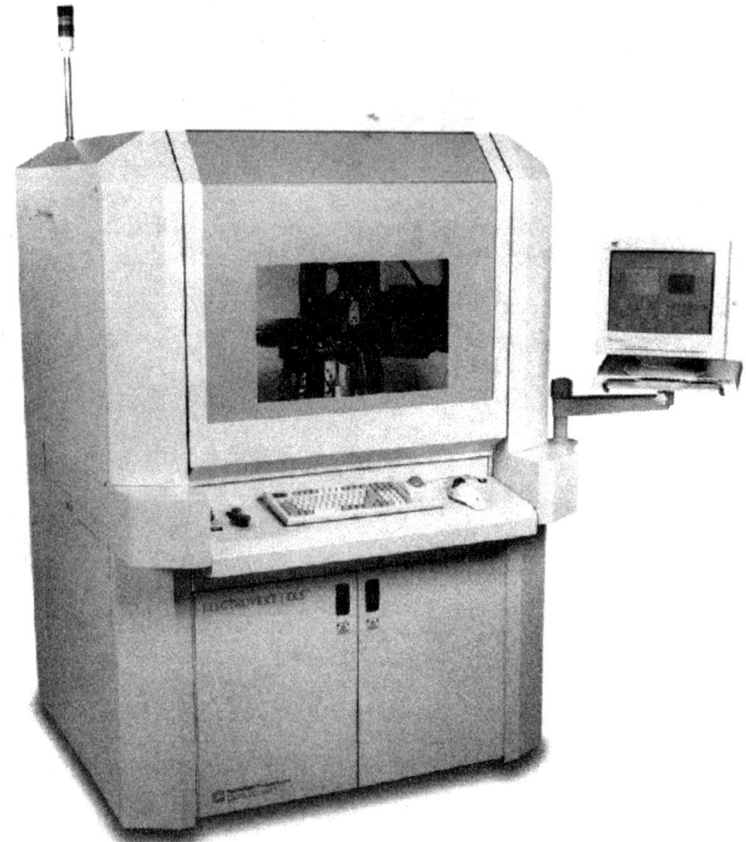

Courtesy of Electrovert/Speedline

FIGURE 31.3 Laser diode soldering machine.

Much of this increased acceptance and adaptation of the concept is due to realization of the sheer logic of it. Catch defect-generation problems where and when they occur in the process—print, place, or reflow—and correct the problem and the source of the problem. Of course, the equipment itself improved and continues to advance as enabling technologies evolve. Higher-resolution cameras backed by faster microprocessors performing more complex algorithms have gone a long way toward making the "known good board" technique feasible and effective. This methodology has been used for three-dimensional postprint, postplacement, and postsoldering (reflow and wave soldering) inspection with ever-increasing ease and accuracy.

X-ray inspection technology also has improved. Higher resolution and x-ray laminography technique have allowed for not just detection but actual measurement and quantification of voids and area array solder defects. Expect to see faster, less expensive, and easier to program inspection systems of all types. Figure 31.5 shows a real-time x-ray machine with magnification to revolve micro-BGA package and FC structures. Improved x-ray equipment has been essential for the development of area array processes, packages, and design rules.

All of this will make in-circuit testing (ICT) obsolete. A standard method of detecting process-related defects for years, ICT is running out of steam. Coverage never reaches 90 percent, even when the program is developed alongside the design. Most users attain 60 percent coverage, and even this can be hindered by no-clean residues and other sources of false readings.

Courtesy of Electrovert/Speedline

FIGURE 31.4 Reflow oven with instrumentation.

SMT has made boards far more complex and dense, with test node requirements in some applications far exceeding the approximately 5000 limit of ICT. As mentioned previously, while ICT can only catch defects after the board has been soldered (if it catches the defects at all), when it is most expensive to rework, in-process automated optical inspection (AOI) catches defects at the most logical phases of the process. ICT is very slow and is truly a process bottleneck, whereas AOI capabilities are quite compatible with most line throughput speeds—and will continue to get faster.

The most difficult challenge remains in postsoldering inspection. While the current systems can quickly analyze the presence and volume of components and solder, they still cannot analyze the shape of the solder joint. This is a tall order because a good solder joint is shiny, and quantifying all the variables of an "acceptable" solder joint is difficult. Nevertheless, we can expect this obstacle to be overcome in the very near future.

31.5 ENTERPRISE SYSTEMS

Equipment software also will expand in capabilities to a great extent. Beyond easier programming of the machine center, interprocess communication will be a part of the new order. Besides the ability to monitor machine status and thus PCBA build status from remote locations as well as defect monitoring in real-time, system integration will be even further. Material handling will no longer be a distant relative of assembly but rather integrated with it. "Dock-to-stock" will evolve into "dock-

FIGURE 31.5 X-ray inspection machine.

to-machine." All aspects of material handling, including ordering, directing, kitting, and bill-of-material generation, will be done automatically and for the most part paperlessly. Tracking of components throughout the assembly process, including moisture-sensitive devices, will all be done electronically. Of the latter, machine centers will go further to control the manufacturing environment (such as temperature and humidity).

31.6 CONCLUSION

Our brave new world of SMT equipment will not be without humans. While operator dependency and error generation will be reduced, the trained and process-savvy operator will still be an important presence on the line. Fewer operators and technicians, as well as better and cleaner material handling and flow, will result in SMT assembly being cheaper, faster, and better.

INDEX

ABOUT THE EDITOR

KEN GILLEO, PH.D., General Technologist for Cookson Electronics, is a widely respected expert on microelectronics packaging. The author of *Handbook of Flexible Circuitry* and *Polymer Thick Film*, as well as over 300 professional journal articles on packaging, circuitry, and materials, he holds 40 U.S. patents and currently serves as VP of Technology for the SMTA (Surface Mount Technology Association).